tangents and secants

29. $\int \tan x \, dx = \ln |\sec x| + C$

30. $\int \sec x \, dx = \ln |\sec x + \tan x| + C$

31. $\int \tan^2 x \, dx = \tan x - x + C$

32. $\int \sec^2 x \, dx = \tan x + C$

33. $\int \tan^3 x \, dx = \frac{1}{2} \tan^2 x + \ln |\cos x| + C$

34. $\int \sec^3 x \, dx = \frac{1}{2} \sec x \tan x + \frac{1}{2} \ln |\sec x + \tan x| + C$

35. $\int \tan^n x \, dx = \frac{\tan^{n-1} x}{n-1} - \int \tan^{n-2} x \, dx$

36. $\int \sec^n x \, dx = \frac{\sec^{n-1} x \sin x}{n-1} + \frac{n-2}{n-1} \int \sec^{n-2} x \, dx$

37. $\int \sec x \tan x \, dx = \sec x + C$

cotangents and cosecants

38. $\int \cot x \, dx = \ln |\sin x| + C$

39. $\int \csc x \, dx = \ln |\csc x - \cot x| + C$

40. $\int \cot^2 x \, dx = -\cot x - x + C$

41. $\int \csc^2 x \, dx = -\cot x + C$

42. $\int \cot^3 x \, dx = -\frac{1}{2} \cot^2 x - \ln |\sin x| + C$

43. $\int \csc^3 x \, dx = -\frac{1}{2} \csc x \cot x + \frac{1}{2} \ln |\csc x - \cot x| + C$

44. $\int \cot^n x \, dx = -\frac{\cot^{n-1} x}{n-1} - \int \cot^{n-2} x \, dx$

45. $\int \csc^n x \, dx = -\frac{\csc^{n-1} x \cos x}{n-1} + \frac{n-2}{n-1} \int \csc^{n-2} x \, dx$

46. $\int \csc x \cot x \, dx = -\csc x + C$

(table continued at the back)

C A L C U L U S

One and Several Variables

Sir Isaac Newton

S. L. SALAS   EINAR HILLE

CALCULUS

One and Several Variables
6th Edition

WILEY

John Wiley and Sons/New York/Chichester/Brisbane/Toronto/Singapore

In fond remembrance of
Einar Hille

Text and cover design: Hudson River Studio
Production: Hudson River Studio
Editorial supervision: Deborah Herbert
Copyediting: Lilian Brady

ISBN 0-471-61195-6

10   9   8   7   6   5   4   3   2   1

PREFACE

Over the years we have heard a continuing murmur of criticism: SALAS/HILLE does not have enough contact with science and engineering, not enough physical applications. We have finally addressed the problem.

In this edition you will find simple physical applications scattered throughout the text and here and there, listed as optional, some applications that are not so simple. Perhaps some of these may pique the interest of the more serious student.

Notwithstanding the increased presence of applications, the book remains a text on mathematics, not science or engineering. The subject is calculus and the emphasis is on three basic ideas: limit, derivative, integral. All else is secondary; all else can be omitted.

S. L. SALAS

ACKNOWLEDGMENTS

Many have contributed to this edition. First of all I want to thank my collaborator (and now good friend) John D. Dollard of the University of Texas. Many of the improvements in the last six chapters (13–18) originated with him. My guru for physical applications was Richard W. Lindquist of Wesleyan University. Our head-to-head meetings were always productive and always fun. The computer programs that appear in the text are the generous contribution of Colin C. Graham of Northwestern University.

Good ideas came from old friends. Edwin Hewitt suggested an elegant little argument relating to simple harmonic motion. [Proof of (7.8.2)] W. W. Comfort called to my attention a remarkable property of the sphere of which I had been totally unaware. (Exercise 25, p. 559) James D. Reid simplified a proof that had stood for several editions. (Theorem 7.2.2)

We learn from our critics. The following reviewers (their criticisms were not always gentle) guided me through this edition.

David Ellis	San Francisco State University
William P. Francis	Michigan Technological University
Lew Friedland	SUNY at Geneseo
Colin C. Graham	Northwestern University
Ed Huffman	Southwest Missouri State University
John Klippert	James Madison University
Michael McAsey	Bradley University
Hiram Paley	University of Illinois, Urbana
Dennis Roseman	University of Iowa
Stephen J. Willson	Iowa State University

Each edition has developed from the previous one. The present book owes much to those who reviewed previous editions and encouraged me to further effort. I find a certain nostalgic pleasure in recalling their names and affiliation.

John T. Anderson	Hamilton College
Elizabeth Appelbaum	University of Missouri, Kansas City
Victor A. Belfi	Texas Christian University
W. W. Comfort	Wesleyan University
Louis J. Deluca	University of Connecticut
Garret J. Etgen	University of Houston
Eugene B. Fabes	University of Minnesota
Fred Gass	Miami University (Ohio)
Adam J. Hulin	Louisiana State University, New Orleans
Max A. Jodeit, Jr.	University of Minnesota
Donald R. Kerr	University of Indiana
Harvey B. Keynes	University of Minnesota
Jerry P. King	Lehigh University
Clifford Kottman	Simpson College
Hudson Kronk	SUNY, Binghamton
John W. Lee	Oregon State University
David Lovelock	University of Arizona
Stanley M. Lukawecki	Clemson University
Giles Maloof	Boise State University
Robert J. Mergener	Moraine Valley Community College
Carl David Minda	University of Cincinnati
C. Stanley Ogilvy	Hamilton College
Bruce P. Palka	University of Texas, Austin
Mark A. Pinsky	Northwestern University
Gordon D. Prichett	Babson College
Jean E. Rubin	Purdue University
John Saber	Babson College
B. L. Sanders	Texas Christian University
Ted Scheick	Ohio State University
Thomas Schwartzbauer	Ohio State University
Donald R. Sherbert	University of Illinois, Urbana
George Springer	University of Indiana
Robert D. Stalley	Oregon State University
Norton Starr	Amherst College
James White	University of California, Los Angeles
Donald R. Wilken	SUNY, Albany

In particular I want to acknowledge my debt to John T. Anderson (who collaborated with me on the fifth edition), to Harvey B. Keynes (who made major contributions to the fourth edition), to Donald R. Sherbert (whose reviews sustained me over many years), and to George Springer (who gave me the encouragement I needed at the very beginning some twenty years ago).

My thanks to Edward A. Burke. He designed the book, he designed the cover, and he supervised production. Valerie Hunter, Mathematics editor at Wiley, kept the process going. It was a pleasure to work with her and I look forward to working with her again.

SPECIAL ACKNOWLEDGMENT

For years now my son Charles G. Salas has collaborated with me on all my projects. Everything I write passes through his hands. He criticizes, objects, and suggests improvements. Usually I agree immediately. Sometimes I don't and we fight it out. More often than not I find that he is right and the book is improved. Once again, thank you, Charlie.

A Note from the Publisher: Answers/solutions to all the odd-numbered exercises are available in a *Student's Solution Manual.*

THE CHANGES

CHAPTERS 1–12: ONE VARIABLE

• Greater variety of exercises on limits, including calculator exercises. • More attention to numerical approximations. Calculator exercises and, thanks to Professor Colin Graham of Northwestern University, some computer programs written in BASIC. (These programs serve to illustrate certain procedures explained in the text and may prove helpful to some students. All the computer programs are *optional*. Access to a computer is not necessary for an understanding of the text.) • Early consideration of infinite limits and limits as $x \to \pm \infty$. • Fine-tuning of sections on maxima-minima and simpler prescriptions for graphing. • More emphasis on motion with constant acceleration. Conservation of energy during free-fall. • Introduction to angular velocity and uniform circular motion. • Early explanation of the role of symmetry in integration (odd functions, even functions). • Weighted averages; the mass of a rod, center of mass. • A new section on the centroid of a region culminates in Pappus's theorem on volumes. • New material on gravitational attraction. *(optional)* • Strong emphasis on exponential growth and decline. • The section on simple harmonic motion has been revised to give the student more insight into oscillatory phenomena. • The chapter on differential equations has been discontinued. Those parts most relevant to elementary calculus have been rewritten and distributed in the appropriate chapters. • Detailed examination of the conic sections in polar coordinates. *(optional; necessary only for a later section on planetary motion, which too is *optional*)* • The centroid of a curve and Pappus's theorem on surface area. (The centroid of a solid of revolution is introduced in the exercises.) • A brief explanation of the gravitational force exerted by a spherical shell. *(optional)* • The inverted cycloid as the tautochrone and the brachystochrone. *(optional)*

CHAPTERS 13–18: SEVERAL VARIABLES

• Although vectors are still introduced as ordered triples of real numbers, we have placed increased emphasis on working with vectors and vector functions in a component-free manner. • Earlier introduction to the cross product. The cross product is now defined geometrically, and its components are derived from that definition. • The chapter on vectors contains a brief introduction to matrices and determinants (2 by 2 and 3 by 3 only). • An elementary discussion of curvilinear motion from a vector viewpoint (followed by some rudimentary vector mechanics) culminates in an *optional* section on Kepler's three laws of planetary motion. • Two intermediate-value theorems that prove useful in later chapters. • Students are invited to derive Snell's law of refraction from Fermat's principle of least time. • Early exploitation of symmetry in multiple integration. • Moments of inertia are introduced by examining the rotation of a rigid body. Frequent use of the parallel axis theorem. • All the material on changing variables in multiple integration has been totally rewritten. A more unified treatment ends with Jacobians and the general theorem. • A revised introduction to line integrals precedes an *optional* section on the work–energy formula and the conservation of mechanical energy. • More attention is given to line integrals with respect to arc length (mass of a wire, center of mass, moments of inertia). • Green's theorem for Jordan regions is followed by Green's theorem for regions bounded by two or more Jordan curves. • Surfaces are introduced in parametrized form. Care is taken to help the student understand various ways of parametrizing the more common surfaces. • Paramount in our discussion of surfaces is the fundamental vector product. • Surface area and surface integrals are defined initially for parametrized surfaces. [Surfaces of the form $z = f(x, y)$ then appear as a special case.] • Surface integrals are used first to calculate the mass, center of mass, and moments of inertia of a material surface. Later we focus on two-sided surfaces and the notion of flux. • The basic differential operators are presented in terms of the operator ∇. The divergence and curl of a vector field $\mathbf{v}$ are introduced as $\nabla \cdot \mathbf{v}$ and $\nabla \times \mathbf{v}$. The Laplacian is written ∇^2. • The divergence theorem first stated for a solid bounded by a single surface is extended to solids bounded by two or more surfaces. (The idea is then applied to find the flux of the electric field that surrounds a point charge.) • Stokes's theorem is made more intelligible by working first with polyhedral surfaces.

THE GREEK ALPHABET

A	α	alpha
B	β	beta
Γ	γ	gamma
Δ	δ	delta
E	ϵ	epsilon
Z	ζ	zeta
H	η	eta
Θ	θ	theta
I	ι	iota
K	κ	kappa
Λ	λ	lambda
M	μ	mu
N	ν	nu
Ξ	ξ	xi
O	o	omicron
Π	π	pi
P	ρ	rho
Σ	σ	sigma
T	τ	tau
Υ	υ	upsilon
Φ	ϕ	phi
X	χ	chi
Ψ	ψ	psi
Ω	ω	omega

CONTENTS

Gottfried Leibniz

C H A P T E R

INTRODUCTION

1.1 WHAT IS CALCULUS?

To a Roman in the days of the empire a "calculus" was a pebble used in counting and in gambling. Centuries later "calculare" came to mean "to compute," "to reckon," "to figure out." To the mathematician, physical scientist, and social scientist of today calculus is elementary mathematics (algebra, geometry, trigonometry) enhanced by *the limit process*.

Calculus takes ideas from elementary mathematics and extends them to a more general situation. Here are some examples. On the left-hand side you will find an idea from elementary mathematics; on the right, this same idea as extended by calculus.

Elementary Mathematics	*Calculus*

---

slope of a line
$y = mx + b$

slope of a curve
$y = f(x)$

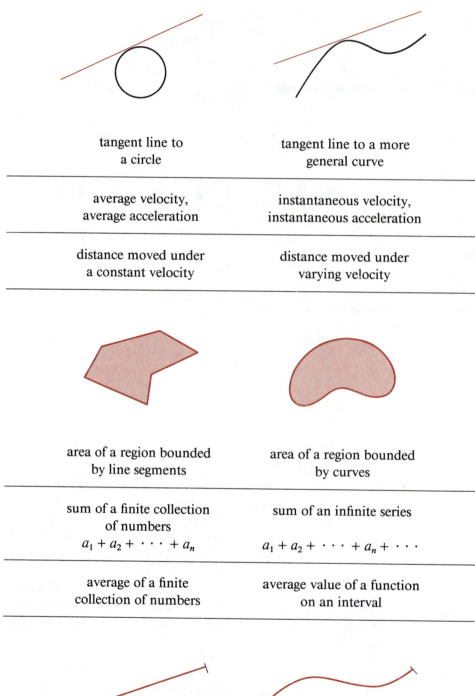

| tangent line to a circle | tangent line to a more general curve |

| average velocity, average acceleration | instantaneous velocity, instantaneous acceleration |

| distance moved under a constant velocity | distance moved under varying velocity |

| area of a region bounded by line segments | area of a region bounded by curves |

| sum of a finite collection of numbers $a_1 + a_2 + \cdots + a_n$ | sum of an infinite series $a_1 + a_2 + \cdots + a_n + \cdots$ |

| average of a finite collection of numbers | average value of a function on an interval |

| length of a line segment | length of a curve |

center of a circle

centroid of a region

volume of
a rectangular solid

volume of a solid
with a curved boundary

surface area of
a cylinder

surface area of
a more general solid

tangent plane to
a sphere

tangent plane to
a more general surface

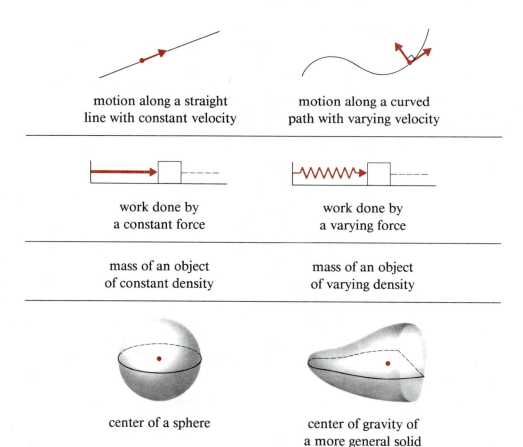

motion along a straight
line with constant velocity

motion along a curved
path with varying velocity

work done by
a constant force

work done by
a varying force

mass of an object
of constant density

mass of an object
of varying density

center of a sphere

center of gravity of
a more general solid

It is fitting to say something about the history of calculus. The origins can be traced back to ancient Greece. The ancient Greeks raised many questions (often paradoxical) about tangents, motion, area, the infinitely small, the infinitely large — questions that today are clarified and answered by calculus. Here and there the Greeks themselves provided answers (some very elegant), but mostly they provided only questions.

After the Greeks, progress was slow. Communication was limited, and each scholar was obliged to start almost from scratch. Over the centuries some ingenious solutions to calculus-type problems were devised, but no general techniques were put forth. Progress was impeded by the lack of a convenient notation. Algebra, founded in the ninth century by Arab scholars, was not fully systemized until the sixteenth century. Then, in the seventeenth century, Descartes established analytic geometry, and the stage was set.

The actual invention of calculus is credited to Sir Isaac Newton (1642–1727) and to Gottfried Wilhelm Leibniz (1646–1716), an Englishman and a German. Newton's invention is one of the few good turns that the great plague did mankind. The plague forced the closing of Cambridge University in 1665 and young Isaac Newton of Trinity College returned to his home in Lincolnshire for eighteen months of meditation, out of which grew his *method of fluxions*, his *theory of gravitation*, and his *theory of light*. The method of fluxions is what concerns us here. A treatise with this title was written by Newton in 1672, but it remained unpublished until 1736, nine years after his death. The new method (calculus to us) was first announced in 1687, but in vague general terms without symbolism, formulas, or applications. Newton himself seemed reluctant to publish anything tangible about his new method, and it is not surprising that the development on the Continent, in spite of a late start, soon overtook Newton and went beyond him.

Leibniz started his work in 1673, eight years after Newton. In 1675 he initiated the basic modern notation: dx and $\int$. His first publications appeared in 1684 and 1686. These made little stir in Germany, but the two brothers Bernoulli of Basel (Switzerland) took up the ideas and added profusely to them. From 1690 onward calculus grew rapidly and reached roughly its present state in about a hundred years. Certain theoretical subtleties were not fully resolved until the twentieth century.

1.2   NOTIONS AND FORMULAS FROM ELEMENTARY MATHEMATICS

The following outline is presented for review and easy reference.

I. SETS

the object x is in the set A (*x* is an element of *A*)	$x \in A$
the object x is not in the set A	$x \notin A$
subset, containment	$A \subseteq B$
union	$A \cup B$
intersection	$A \cap B$
Cartesian products	$A \times B,\ A \times B \times C$
empty set	$\varnothing$
the set of all x for which property P holds	$\{x : P\}$

(These are the only notions from set theory that you will need for this book. If you are not familiar with them, see Appendix A.1 at the back of the book.)

II. REAL NUMBERS

Classification

integers	$0, 1, -1, 2, -2, 3, -3$, etc.
rational numbers	p/q with p and q integers, $q \neq 0$; for example, $\frac{2}{5}, -\frac{9}{2}, \frac{4}{1} = 4$.
irrational numbers	real numbers that are not rational; for example, $\sqrt{2}, \pi$.

Order Properties

(i) Either $a < b$, $b < a$, or $a = b$.   (trichotomy)

(ii) If $a < b$ and $b < c$, then $a < c$.

(iii) If $a < b$, then $a + c < b + c$ for all real numbers c.

(iv) If $a < b$ and $c > 0$, then $ac < bc$.
   If $a < b$ and $c < 0$, then $ac > bc$.

(Techniques for solving inequalities are reviewed in Section 1.4.)

Density

Between any two real numbers there are infinitely many rational numbers and infinitely many irrational numbers. In particular, *there is no smallest positive real number.*

Absolute Value

$$\textit{definition} \qquad |a| = \left\{ \begin{array}{ll} a, & \text{if } a \geq 0 \\ -a, & \text{if } a \leq 0 \end{array} \right].$$

$$\textit{other characterizations} \qquad |a| = \max\{a, -a\}; \quad |a| = \sqrt{a^2}.$$

$$\textit{geometric interpretation} \qquad |a| = \text{distance between } a \text{ and } 0.$$
$$|a - c| = \text{distance between } a \text{ and } c.$$

properties   (i) $|a| = 0$ iff $a = 0$.†

(ii) $|-a| = |a|$.

(iii) $|ab| = |a||b|$.

(iv) $|a + b| \leq |a| + |b|$.   (the triangle inequality)††

(v) $\big||a| - |b|\big| \leq |a - b|$.   (a variant of the triangle inequality)

(The solution of inequalities involving absolute values is reviewed in Section 1.5.)

Intervals

Suppose that $a < b$. The *open interval* (a, b) is the set of all numbers between a and b:

$$(a, b) = \{x : a < x < b\}.$$

---

† By "iff" we mean "if and only if." This expression is used so often in mathematics that it's convenient to have an abbreviation for it.

†† The absolute value of the sum of two numbers cannot exceed the sum of their absolute values, just as in a triangle the length of a side cannot exceed the sum of the lengths of the other two sides.

The *closed interval* $[a, b]$ is the open interval (a, b) together with the endpoints a and b:

$$[a, b] = \{x : a \leq x \leq b\}.$$

There are seven other types of intervals:

$$(a, b] = \{x : a < x \leq b\}.$$
$$[a, b) = \{x : a \leq x < b\}.$$
$$(a, \infty) = \{x : a < x\}.$$
$$[a, \infty) = \{x : a \leq x\}.$$
$$(-\infty, b) = \{x : x < b\}.$$
$$(-\infty, b] = \{x : x \leq b\}.$$
$$(-\infty, \infty) = \text{set of real numbers.}$$

This interval notation is easy to remember: we use a bracket to include an endpoint; otherwise a parenthesis. The symbol ∞, read "infinity," has no meaning by itself. Just as in ordinary language we use syllables to construct words and need assign no meaning to the syllables themselves, so in mathematics we can use symbols to construct mathematical expressions, assigning no separate meaning to the individual symbols. While ∞ has no meaning, (a, ∞), $[a, \infty)$, $(-\infty, b)$, $(-\infty, b]$, $(-\infty, \infty)$ do have meaning, and that meaning is given above.

Boundedness

A set S of real numbers is said to be

   (i) *bounded above* iff there exists a real number M such that

$$x \leq M \qquad \text{for all } x \in S.$$

   M is called an *upper bound* for S.

   (ii) *bounded below* iff there exists a real number m such that

$$m \leq x \qquad \text{for all } x \in S.$$

   m is called a *lower bound* for S.

   (iii) *bounded* iff it is bounded above and below.

For example, the intervals $(-\infty, 2]$ and $(-\infty, 2)$ are bounded above but not below. The set of positive integers is bounded below but not above. The intervals $[0, 1]$, $(0, 1)$, and $(0, 1]$ are bounded (both above and below).

III. ALGEBRA AND GEOMETRY

General Quadratic Formula

The roots of a quadratic equation

$$ax^2 + bx + c = 0, \qquad a > 0$$

are given by

$$x = \frac{-b \pm \sqrt{b^2 - 4ac}}{2a}.$$

Factoring Formulas

$$a^2 - b^2 = (a + b)(a - b),$$
$$a^3 - b^3 = (a - b)(a^2 + ab + b^2),$$
$$a^3 + b^3 = (a + b)(a^2 - ab + b^2).$$

Polynomials

A *polynomial* is a function of the form

$$P(x) = a_n x^n + a_{n-1} x^{n-1} + \cdots + a_1 x + a_0$$

where n is a positive integer. If $a_n \neq 0$, the polynomial is said to have *degree n*. Recall the *factor theorem*: if P is a polynomial and c is a real number, then

$$x - c \text{ is a factor of } P(x) \quad \text{iff} \quad P(c) = 0.$$

Factorials

$$0! = 1, \quad 1! = 1, \quad 2! = 2 \cdot 1, \quad 3! = 3 \cdot 2 \cdot 1, \cdots.$$

In general, for each positive integer n,

$$n! = n(n - 1) \cdots (2)(1).$$

Elementary Figures

triangle

area $= \frac{1}{2}bh$

equilateral triangle

area $= \frac{1}{4}\sqrt{3}\, s^2$

rectangle

area = lw
perimeter = $2l + 2w$
diagonal = $\sqrt{l^2 + w^2}$

rectangular solid

volume = lwh
surface area = $2lw + 2lh + 2wh$

square

area = x^2
perimeter = $4x$
diagonal = $x\sqrt{2}$

cube

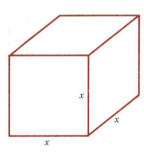

volume = x^3
surface area = $6x^2$

circle

area = πr^2
circumference = $2\pi r$

sphere

volume = $\frac{4}{3}\pi r^3$
surface area = $4\pi r^2$

right circular cylinder

volume = $\pi r^2 h$
lateral area = $2\pi rh$
total surface area = $2\pi r^2 + 2\pi rh$

cone

volume = $\frac{1}{3}\pi r^2 h$
slant height = $\sqrt{r^2 + h^2}$
lateral area = $\pi r\sqrt{r^2 + h^2}$
total surface area = $\pi r^2 + \pi r\sqrt{r^2 + h^2}$

IV. ANALYTIC GEOMETRY

Rectangular Coordinates: The Distance and Midpoint Formulas

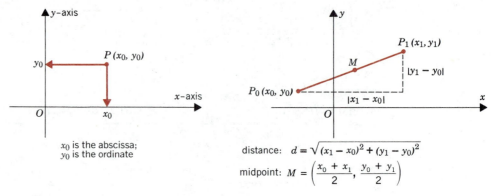

x_0 is the abscissa;
y_0 is the ordinate

distance: $d = \sqrt{(x_1 - x_0)^2 + (y_1 - y_0)^2}$

midpoint: $M = \left(\dfrac{x_0 + x_1}{2},\ \dfrac{y_0 + y_1}{2} \right)$

Figure 1.2.1   **Figure 1.2.2**

Lines

(i) Slope of a nonvertical line:

$$m = \frac{y_1 - y_0}{x_1 - x_0}$$

where $P_0(x_0, y_0)$ and $P_1(x_1, y_1)$ are any two points on the line;

$$m = \tan \theta$$

where θ, called the *inclination* of the line, is the angle between the line and the x-axis measured counterclockwise from the x-axis. (Figure 1.2.3)

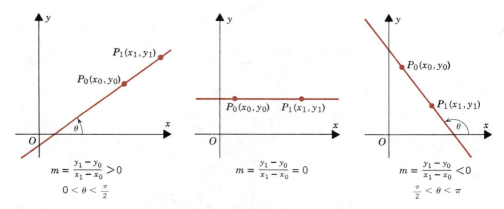

$m = \dfrac{y_1 - y_0}{x_1 - x_0} > 0$

$0 < \theta < \dfrac{\pi}{2}$

$m = \dfrac{y_1 - y_0}{x_1 - x_0} = 0$

$m = \dfrac{y_1 - y_0}{x_1 - x_0} < 0$

$\dfrac{\pi}{2} < \theta < \pi$

Figure 1.2.3

(ii) Equations for lines:

general form $Ax + By + C = 0$, with A and B not both 0.

$$\text{vertical line} \quad x = a.$$

$$\text{slope-intercept form} \quad y = mx + b.$$

$$\text{point-slope form} \quad y - y_0 = m(x - x_0).$$

$$\text{two-point form} \quad y - y_0 = \frac{y_1 - y_0}{x_1 - x_0}(x - x_0).$$

$$\text{two-intercept form} \quad \frac{x}{a} + \frac{y}{b} = 1.$$

(iii) Two nonvertical lines $l_1: y = m_1 x + b_1$ and $l_2: y = m_2 x + b_2$ are

$$\text{parallel} \quad \text{iff} \quad m_1 = m_2;$$
$$\text{perpendicular} \quad \text{iff} \quad m_1 m_2 = -1.$$

The angle α (taken between 0° and 90°) formed by two intersecting nonvertical lines can be obtained from the relation

$$\tan \alpha = \left| \frac{m_2 - m_1}{1 + m_1 m_2} \right|.$$

(Some review problems on lines are given in Section 1.3.)

Equations of Conic Sections

$$\text{circle of radius } r \text{ centered at the origin} \quad x^2 + y^2 = r^2.$$

$$\text{circle of radius } r \text{ centered at } P(x_0, y_0) \quad (x - x_0)^2 + (y - y_0)^2 = r^2.$$

$$\text{parabolas} \quad x^2 = 4cy, \quad y^2 = 4cx.$$

$$\text{ellipse} \quad \frac{x^2}{a^2} + \frac{y^2}{b^2} = 1.$$

$$\text{hyperbolas} \quad \frac{x^2}{a^2} - \frac{y^2}{b^2} = 1, \quad xy = c.$$

(The conic sections are discussed extensively in Chapter 9.)

V. TRIGONOMETRY

Angle Measurement

$$\pi \cong 3.14159.\dagger$$

$$2\pi \text{ radians} = 360 \text{ degrees} = \text{one complete revolution.}$$

$$\text{One radian} = \frac{180}{\pi} \text{ degrees} \cong 57.296 \text{ degrees.}$$

$$\text{One degree} = \frac{\pi}{180} \text{ radians} \cong 0.0175 \text{ radians.}$$

(For a discussion of radian measure, see Appendix A.2.)

---

† We use $\cong$ to indicate approximate equality.

Sectors

Figure 1.2.4 displays a sector in a circle of radius r.   If θ is the central angle measured in radians, then

the length of the subtended arc $= r\theta$,   the area of the sector $= \frac{1}{2}r^2\theta$.

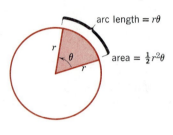

Figure 1.2.4   **Figure 1.2.5**

Sine and Cosine

Take θ in the interval $[0, 2\pi]$.   In the unit circle (Figure 1.2.5) mark off the sector OAP with a central angle of θ radians:

$\cos\theta = x$-coordinate of P, $\sin\theta = y$-coordinate of P.

These definitions are extended periodically:

$$\cos\theta = \cos(\theta + 2n\pi),   \sin\theta = \sin(\theta + 2n\pi).$$

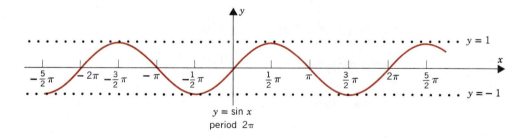

$y = \sin x$
period 2π

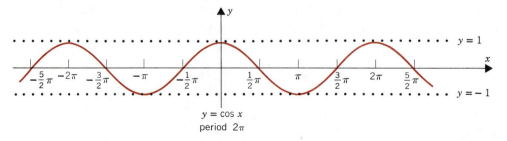

$y = \cos x$
period 2π

Figure 1.2.6

Other Trigonometric Functions

$$\tan \theta = \frac{\sin \theta}{\cos \theta}, \qquad \cot \theta = \frac{\cos \theta}{\sin \theta},$$

$$\sec \theta = \frac{1}{\cos \theta}, \qquad \csc \theta = \frac{1}{\sin \theta}.$$

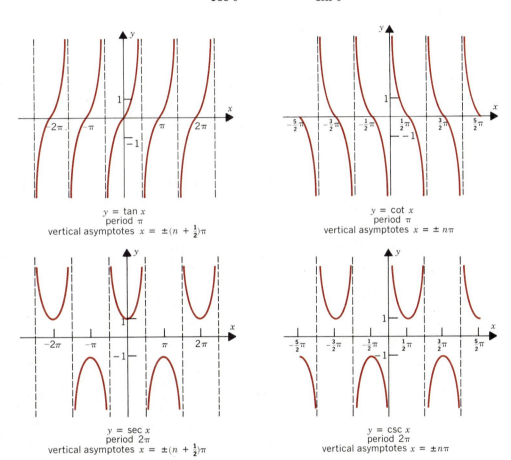

$y = \tan x$
period π
vertical asymptotes $x = \pm(n + \frac{1}{2})\pi$

$y = \cot x$
period π
vertical asymptotes $x = \pm n\pi$

$y = \sec x$
period 2π
vertical asymptotes $x = \pm(n + \frac{1}{2})\pi$

$y = \csc x$
period 2π
vertical asymptotes $x = \pm n\pi$

Figure 1.2.7

In Terms of a Right Triangle

For θ between 0 and $\frac{1}{2}\pi$,

$$\sin \theta = \frac{\text{opposite side}}{\text{hypotenuse}}, \qquad \cos \theta = \frac{\text{adjacent side}}{\text{hypotenuse}},$$

$$\tan \theta = \frac{\text{opposite side}}{\text{adjacent side}}, \qquad \cot \theta = \frac{\text{adjacent side}}{\text{opposite side}},$$

$$\sec \theta = \frac{\text{hypotenuse}}{\text{adjacent side}}, \qquad \csc \theta = \frac{\text{hypotenuse}}{\text{opposite side}}.$$

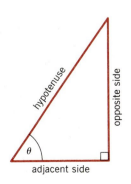

Figure 1.2.8

Some Important Identities

(i) *unit circle*

$$\cos^2 \theta + \sin^2 \theta = 1, \qquad \tan^2 \theta + 1 = \sec^2 \theta, \qquad \cot^2 \theta + 1 = \csc^2 \theta.$$

(ii) *addition formulas*

$$\sin(\theta_1 + \theta_2) = \sin \theta_1 \cos \theta_2 + \cos \theta_1 \sin \theta_2,$$
$$\sin(\theta_1 - \theta_2) = \sin \theta_1 \cos \theta_2 - \cos \theta_1 \sin \theta_2,$$
$$\cos(\theta_1 + \theta_2) = \cos \theta_1 \cos \theta_2 - \sin \theta_1 \sin \theta_2,$$
$$\cos(\theta_1 - \theta_2) = \cos \theta_1 \cos \theta_2 + \sin \theta_1 \sin \theta_2;$$
$$\tan(\theta_1 + \theta_2) = \frac{\tan \theta_1 + \tan \theta_2}{1 - \tan \theta_1 \tan \theta_2}, \qquad \tan(\theta_1 - \theta_2) = \frac{\tan \theta_1 - \tan \theta_2}{1 + \tan \theta_1 \tan \theta_2}.$$

(iii) *double-angle formulas*

$$\sin 2\theta = 2 \sin \theta \cos \theta, \qquad \cos 2\theta = \cos^2 \theta - \sin^2 \theta.$$

(iv) *half-angle formulas*

$$\sin^2 \theta = \tfrac{1}{2}(1 - \cos 2\theta), \qquad \cos^2 \theta = \tfrac{1}{2}(1 + \cos 2\theta).$$

For an Arbitrary Triangle

area $\tfrac{1}{2}ab \sin C.$

law of sines $\dfrac{\sin A}{a} = \dfrac{\sin B}{b} = \dfrac{\sin C}{c}.$

law of cosines $a^2 = b^2 + c^2 - 2bc \cos A.$

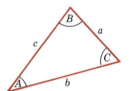

Figure 1.2.9

VI. INDUCTION

AXIOM

Let S be a set of integers. If 1 is in S and if k is in S implies that $k + 1$ is in S, then all the positive integers are in S.

In general, we use induction when we want to prove that a certain proposition is true for all positive integers n. To do this, first we prove that the proposition is true for $n = 1$. Next we prove that, if it is true for $n = k$, then it is true for $n = k + 1$. It follows then from the axiom that the proposition is true for all positive integers n.

(We will use induction only sparingly. For a discussion of it see Appendix A.3.)

1.3   SOME PROBLEMS ON LINES

Problem 1. Find the slope and y-intercept of each of the following lines:
$$l_1: 20x - 24y - 30 = 0, \qquad l_2: 2x - 3 = 0, \qquad l_3: 4y + 5 = 0.$$

Solution. The equation of l_1 can be written
$$y = \tfrac{5}{6}x - \tfrac{5}{4}.$$

This is in the form $y = mx + b$. The slope is $\tfrac{5}{6}$, and the y-intercept is $-\tfrac{5}{4}$.

The equation of l_2 can be written
$$x = \tfrac{3}{2}.$$

The line is vertical and the slope is not defined. Since the line does not cross the y-axis, it has no y-intercept.

The third equation can be written
$$y = -\tfrac{5}{4}.$$

The line is horizontal. Its slope is 0 and the y-intercept is $-\tfrac{5}{4}$. The three lines are drawn in Figure 1.3.1. ■

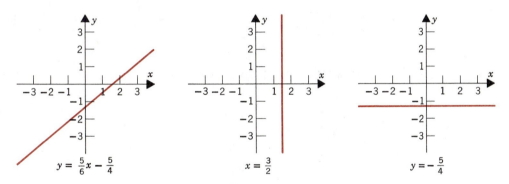

Figure 1.3.1

Problem 2. Write an equation for the line l_2 that is parallel to
$$l_1: 3x - 5y + 8 = 0$$

and passes through the point $P(-3, 2)$.

Solution. We can rewrite the equation of l_1 as
$$y = \tfrac{3}{5}x + \tfrac{8}{5}.$$

The slope of l_1 is $\tfrac{3}{5}$. The slope of l_2 must also be $\tfrac{3}{5}$. (*Remember:* For nonvertical parallel lines $m_1 = m_2$.)

Since l_2 passes through $(-3, 2)$ with slope $\tfrac{3}{5}$, we can use the point-slope formula and write the equation as
$$y - 2 = \tfrac{3}{5}(x + 3). \quad ■$$

Problem 3. Write an equation for the line that is perpendicular to

$$l_1 : x - 4y + 8 = 0$$

and passes through the point $P(2, -4)$.

Solution. The equation for l_1 can be written

$$y = \tfrac{1}{4}x + 2.$$

The slope of l_1 is $\tfrac{1}{4}$. The slope of l_2 is therefore -4. (*Remember:* For nonvertical perpendicular lines $m_1 m_2 = -1$.)

Since l_2 passes through $(2, -4)$ with slope -4, we can use the point-slope formula and write the equation as

$$y + 4 = -4(x - 2). \quad \square$$

Problem 4. Find the inclination θ of the line

$$3x - 4y + 2 = 0.$$

Solution. Our first step is to find the slope. Writing the equation as

$$y = \tfrac{3}{4}x + \tfrac{1}{2},$$

we see that $m = \tfrac{3}{4}$. Thus we have $\tan \theta = \tfrac{3}{4} = 0.7500$. By Table 5 at the back of the book, $\theta \cong 37°$. $\quad \square$

The Angle Between Two Lines

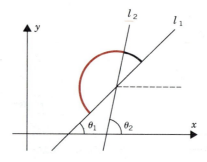

Figure 1.3.2

Figure 1.3.2 shows two intersecting lines l_1, l_2 with inclinations θ_1, θ_2. In the figure $\theta_1 < \theta_2$. The two lines form two angles, $\theta_2 - \theta_1$ and $180° - (\theta_2 - \theta_1)$. If these two angles are equal, the two lines meet at $90°$. If the two angles are not equal, then the smaller of the two (the one less than 90°) is called the *angle between l_1 and l_2*.

Problem 5. Find the point where the lines

$$x - 4y + 7 = 0 \quad \text{and} \quad 2x - 5y + 31 = 0$$

intersect, and then determine the angle between the two lines.

Solution. To find the point of intersection we solve the two equations simultaneously. As you can check, this gives $x = \frac{13}{3}$, $y = \frac{17}{3}$. The point of intersection is $P(\frac{13}{3}, \frac{17}{3})$.

To find the angle of intersection, we find the inclination of each line:

$$x - 4y + 7 = 0 \qquad\qquad 2x - 5y + 31 = 0$$
$$4y = x + 7 \qquad\qquad\qquad 5y = 2x + 31$$
$$y = \tfrac{1}{4}x + \tfrac{7}{4} \qquad\qquad\qquad y = \tfrac{2}{5}x + \tfrac{31}{5}$$
$$\tan\theta = \tfrac{1}{4} = 0.25 \qquad\qquad \tan\theta = \tfrac{2}{5} = 0.40$$
$$\theta \cong 14°. \qquad\qquad\qquad\qquad \theta \cong 22°.$$

The angle between the lines is about $22° - 14° = 8°$. □

EXERCISES 1.3

(The odd-numbered exercises have answers at the back of the book.)

Find the slope of the line through the given points.

1. $P_0(-2, 5)$, $P_1(4, 1)$.   **2.** $P_0(4, -3)$, $P_1(-2, -7)$.   **3.** $P_0(4, 2)$, $P_1(-3, 2)$.
4. $P_0(5, 0)$, $P_1(0, 5)$.   **5.** $P(a, b)$, $Q(b, a)$.   **6.** $P(4, -1)$, $Q(-3, -1)$.
7. $P(x_0, 0)$, $Q(0, y_0)$.   **8.** $O(0, 0)$, $P(x_0, y_0)$.

Find the slope and y-intercept.

9. $y = 2x - 4$.   **10.** $6 - 5x = 0$.   **11.** $3y = x + 6$.   **12.** $y = 4x - 2$.
13. $4x = 1$.   **14.** $6y - 3x + 8 = 0$.   **15.** $7x - 3y + 4 = 0$.   **16.** $4y = 3$.
17. $7y - 5 = 0$.   **18.** $4y = x - 5$.

Write an equation for the line with

19. slope 5 and y-intercept 2.   **20.** slope 5 and y-intercept -2.
21. slope -5 and y-intercept 2.   **22.** slope -5 and y-intercept -2.

Write an equation for the horizontal line 3 units

23. above the x-axis.   **24.** below the x-axis.

Write an equation for the vertical line 3 units

25. to the left of the y-axis.   **26.** to the right of the y-axis.

Find an equation for the line that passes through the point $P(2, 7)$ and is

27. parallel to the x-axis.   **28.** parallel to the y-axis.
29. parallel to the line $3y - 2x + 6 = 0$.   **30.** perpendicular to the line $y - 2x + 5 = 0$.
31. perpendicular to the line $3y - 2x + 6 = 0$.
32. parallel to the line $y - 2x + 5 = 0$.

Find the inclination of the line.

33. $x - y + 2 = 0$.   **34.** $6y + 5 = 0$.   **35.** $2x - 3 = 0$.
36. $x + y - 3 = 0$.   **37.** $3x + 4y - 12 = 0$.   **38.** $9x + 10y + 4 = 0$.

Write an equation for the line with

39. inclination 30°, y-intercept 2.   **40.** inclination 60°, y-intercept 2.
41. inclination 120°, y-intercept 3.   **42.** inclination 135°, y-intercept 3.

Determine the point(s) where the line intersects the circle.

43. $y = x$, $x^2 + y^2 = 1$.   **44.** $y = mx$, $x^2 + y^2 = 4$.
45. $4x + 3y = 24$, $x^2 + y^2 = 25$.   **46.** $y = mx + b$, $x^2 + y^2 = b^2$.

Find the point where the lines intersect and determine the angle of intersection.

47. $l_1: 4x - y - 3 = 0$, $l_2: 3x - 4y + 1 = 0$.
48. $l_1: 3x + y - 5 = 0$, $l_2: 7x - 10y + 27 = 0$.
49. $l_1: 4x - y + 2 = 0$, $l_2: 19x + y = 0$.
50. $l_1: 5x - 6y + 1 = 0$, $l_2: 8x + 5y + 2 = 0$.
51. Determine the slope of the line that intersects the circle $x^2 + y^2 = 169$ only at the point (5, 12).
52. Suppose that l_1 and l_2 are two nonvertical lines. If $m_1 m_2 = -1$, then l_1 and l_2 intersect at right angles. Show that, if l_1 and l_2 do not intersect at right angles, then the angle α between l_1 and l_2 is given by the formula

(1.3.1)
$$\tan \alpha = \left| \frac{m_2 - m_1}{1 + m_1 m_2} \right|.$$

HINT: Use the identity

$$\tan (A - B) = \frac{\tan A - \tan B}{1 + \tan A \tan B}.$$

1.4   INEQUALITIES

(All our work with inequalities is based on the order properties of real numbers as given on page 6.) In this section and the next, we work with the sort of inequalities that abound in calculus, inequalities that involve a variable.

To solve an equation in x is to find the set of numbers x for which the equation holds. To solve an inequality in x is to find the set of numbers x for which the inequality holds.

The way we solve an inequality is very similar to the way we solve an equation, but there is one important difference. We can maintain an inequality by adding the same number to both sides, or by subtracting the same number from both sides, or by multiplying or dividing both sides by the same *positive* number. But if we multiply or divide by a *negative* number, then the inequality is *reversed*:

$x - 2 < 4$ gives $x < 6$, $x + 2 < 4$ gives $x < 2$, $\frac{1}{2}x < 4$ gives $x < 8$,

but

$$-\tfrac{1}{2}x < 4 \quad \text{gives} \quad x > -8.$$
$$\uparrow$$
$$\rule{1cm}{0.4pt}\text{note}$$

Problem 1. Solve the inequality

$$\tfrac{1}{2}(1 + x) \leq 6.$$

Solution. The idea is to "unwrap" the x. We can eliminate the $\tfrac{1}{2}$ by multiplying through by 2. This gives

$$1 + x \leq 12.$$

Subtracting 1 from both sides, we get

$$x \leq 11.$$

The solution is thus the interval $(-\infty, 11]$. ☐
$$11$$

Problem 2. Solve the inequality

$$-3(4 - x) < 12.$$

Solution. Multiplying both sides of the inequality by $-\tfrac{1}{3}$, we have

$$4 - x > -4. \qquad \text{(the inequality has been reversed)}$$

Subtracting 4, we get

$$-x > -8.$$

To isolate x, we multiply by -1. This gives

$$x < 8. \qquad \text{(the inequality has been reversed again)}$$

The solution is the interval $(-\infty, 8)$. ☐
$$8$$

There are generally several ways to solve a given inequality. For example, the last inequality could have been solved this way:

$$-3(4 - x) < 12,$$
$$-12 + 3x < 12,$$
$$3x < 24, \qquad \text{(we added 12)}$$
$$x < 8. \qquad \text{(divided by 3)}$$

The solution, of course, is once again $(-\infty, 8)$.

The usual way to solve a quadratic inequality is to factor the quadratic.

Problem 3. Solve the inequality

$$\tfrac{1}{5}(x^2 - 4x + 3) < 0.$$

Solution.   First we eliminate the outside factor $\frac{1}{5}$ by multiplying through by 5: this gives

$$x^2 - 4x + 3 < 0.$$

Factoring the quadratic, we have

$$(x - 1)(x - 3) < 0.$$

The product $(x - 1)(x - 3)$ is zero at 1 and 3. We mark these points on a number line (Figure 1.4.1). The points 1 and 3 separate three intervals:

$$(-\infty, 1), \qquad (1, 3), \qquad (3, \infty).$$

On each of these intervals the product $(x - 1)(x - 3)$ keeps a constant sign:

$$\text{on } (-\infty, 1) \quad \text{[to the left of 1]} \quad \text{sign of } (x - 1)(x - 3) = (-)(-) = +;$$
$$\text{on } (1, 3) \quad \text{[between 1 and 3]} \quad \text{sign of } (x - 1)(x - 3) = (+)(-) = -;$$
$$\text{on } (3, \infty) \quad \text{[to the right of 3]} \quad \text{sign of } (x - 1)(x - 3) = (+)(+) = +.$$

```
+ + + + + + + + + + + + +  0  – – – – – – – – – –  0  + + + + + + + + + + +
———————————————————┼————————————————————┼———————————————————▶
                   1                    3
```

Figure 1.4.1

The product $(x - 1)(x - 3)$ is negative only on $(1, 3)$. The solution is the open interval $(1, 3)$. ☐

More generally, consider an expression of the form

$$(x - a_1)(x - a_2) \cdots (x - a_n) \qquad \text{with } a_1 < a_2 < \cdots < a_n.$$

Such an expression is zero at $a_1, a_2, \ldots, a_n$. It is positive on those intervals where an even number of terms are negative, and it is negative on those intervals where an odd number of terms are negative.

As an example, take the expression

$$(x + 2)(x - 1)(x - 3).$$

This product is zero at $-2, 1, 3$. It is

negative on $(-\infty, -2)$,	(3 negative terms)
positive on $(-2, 1)$,	(2 negative terms)
negative on $(1, 3)$,	(1 negative term)
positive on $(3, \infty)$.	(0 negative terms)

See Figure 1.4.2.

```
– – – –  0  + + + + + + + + + + +  0  – – – – – – –  0  + + + +
———————┼———————————————————————┼—————————————————┼—————————▶
      −2                       1                 3
```

Figure 1.4.2

Problem 4. Solve the inequality

$$(x + 3)^5(x - 1)(x - 4)^2 < 0.$$

Solution. We view $(x + 3)^5(x - 1)(x - 4)^2$ as the product of three factors: $(x + 3)^5$, $(x - 1)$, $(x - 4)^2$. The product is zero at -3, 1, and 4. These points separate the intervals

$$(-\infty, -3), \qquad (-3, 1), \qquad (1, 4), \qquad (4, \infty).$$

On each of these intervals the product keeps a constant sign. It is

positive on $(-\infty, -3)$,	(2 negative factors)
negative on $(-3, 1)$,	(1 negative factor)
positive on $(1, 4)$,	(0 negative factors)
positive on $(4, \infty)$.	(0 negative factors)

See Figure 1.4.3.

Figure 1.4.3

The solution is the open interval $(-3, 1)$. ▢

Problem 5. Solve the inequality

$$x^2 + 4x - 2 \leq 0.$$

Solution. To factor the quadratic we first complete the square by adding and subtracting 4:

$$x^2 + 4x + 4 - 4 - 2 \leq 0$$
$$(x^2 + 4x + 4) - 6 \leq 0$$
$$(x + 2)^2 - 6 \leq 0.$$

We can factor this quadratic as the difference of two squares:

$$(x + 2 + \sqrt{6})(x + 2 - \sqrt{6}) \leq 0.$$

The product is 0 at $-2 - \sqrt{6}$ and $-2 + \sqrt{6}$. It is negative in between. The solution is the closed interval $[-2 - \sqrt{6}, -2 + \sqrt{6}]$. ▢

Remark. The roots of the quadratic $x^2 + 4x - 2 = 0$ could have been obtained from the general quadratic formula. ▢

In solving inequalities that involve quotients we use the fact that

(1.4.1) $\dfrac{a}{b} > 0$ iff $ab > 0$ and $\dfrac{a}{b} < 0$ iff $ab < 0.$

Problem 6.   Solve the inequality

$$\frac{x+2}{1-x} > 1.$$

Solution

$$\frac{x+2}{1-x} - 1 > 0,$$

$$\frac{x+2-(1-x)}{1-x} > 0,$$

$$\frac{2x+1}{1-x} > 0,$$

$$(2x+1)(1-x) > 0, \qquad\qquad \text{[by (1.4.1)]}$$

$$(2x+1)(x-1) < 0, \qquad\qquad \text{(we multiplied by } -1\text{)}$$

$$(x+\tfrac{1}{2})(x-1) < 0. \qquad\qquad \text{(divided by 2)}$$

The product $(x+\tfrac{1}{2})(x-1)$ is zero at $-\tfrac{1}{2}$ and 1. As you can check, it is negative only on the open interval $(-\tfrac{1}{2}, 1)$. This open interval is the solution to our inequality.   ▨

Problem 7.   Solve the inequality

$$\frac{(x+3)^5(x-1)}{(x-4)^2} < 0.$$

Solution.   By (1.4.1) this inequality has the same solution as the inequality

$$(x+3)^5(x-1)(x-4)^2 < 0,$$

which we solved in Problem 4.   ▨

EXERCISES 1.4

Solve the following inequalities.

1. $2 + 3x < 5.$

2. $\tfrac{1}{2}(2x + 3) < 6.$

3. $16x + 64 \leq 16.$

4. $3x + 5 > \tfrac{1}{4}(x - 2).$

5. $\tfrac{1}{2}(1 + x) < \tfrac{1}{3}(1 - x).$

6. $3x - 2 \leq 1 + 6x.$

7. $x^2 - 1 < 0.$

8. $x^2 + x - 2 \leq 0.$

9. $4(x^2 - 3x + 2) > 0.$

10. $x^2 + 9x + 20 < 0.$

11. $x(x - 1)(x - 2) > 0.$

12. $x(2x - 1)(3x - 5) \leq 0.$

13. $x^3 - 2x^2 + x \geq 0.$

14. $x^2 - 4x + 4 \leq 0.$

15. $x^2 + 1 > 4x.$

16. $2 - x^2 \geq -4x.$

17. $\tfrac{1}{2}(1 + x)^2 < \tfrac{1}{3}(1 - x)^2.$

18. $2x^2 + 9x + 6 \geq x + 2.$

19. $1 - 3x^2 < \frac{1}{2}(2 - x^2).$ **20.** $6x^2 + 2x \le (x - 1)^2.$ **21.** $\frac{1}{x} < x.$

22. $x + \frac{1}{x} \ge 1.$ **23.** $\frac{x}{x - 5} \ge 0.$ **24.** $\frac{x}{x + 5} < 0.$

25. $\frac{x}{x - 5} > \frac{1}{4}.$ **26.** $\frac{1}{3x - 5} < 2.$ **27.** $\frac{x^2 - 9}{x + 1} > 0.$

28. $\frac{x^2}{x^2 - 4} < 0.$ **29.** $x^3(x - 2)(x + 3)^2 < 0.$ **30.** $x^2(x - 3)(x + 4)^2 > 0.$

31. $x^2(x - 2)(x + 6) > 0.$ **32.** $x^3(x + 3)(x - 5) > 0.$ **33.** $5x(x - 3)^2 < 0.$

34. $7x(x - 4)^2 < 0.$ **35.** $\frac{2x}{x^2 - 4} > 0.$ **36.** $\frac{x^2 - 9}{3x} > 0.$

37. $\frac{x - 1}{9 - x^2} < 0.$ **38.** $\frac{x^2 - 4}{(x + 4)^2} < 0.$ **39.** $\frac{1}{x - 1} + \frac{4}{x - 6} > 0.$

40. $\frac{3}{x - 2} - \frac{5}{x - 6} < 0.$ **41.** $\frac{2x - 6}{x^2 - 6x + 5} < 0.$ **42.** $\frac{2x + 8}{x^2 + 8x + 7} > 0.$

43. $\frac{x + 3}{x^2(x - 5)} < 0.$ **44.** $\frac{x^2 - 4x}{x + 2} > 0.$ **45.** $\frac{x^2 - 4x + 3}{x^2} > 0.$

46. Arrange the following in order: $1, \ x, \ \sqrt{x}, \ 1/x, \ 1/\sqrt{x}$ given that $1 < x.$

47. Exercise 46 given that $0 < x < 1.$

48. Compare $\sqrt{\dfrac{x}{x + 1}}$ and $\sqrt{\dfrac{x + 1}{x + 2}}$ given that $x > 0.$

49. Show that $2ab \le a^2 + b^2.$

50. Let a and b be nonnegative numbers. Show that
$$\text{if} \quad a^2 \le b^2 \quad \text{then} \quad a \le b.$$

51. Let a and b be nonnegative numbers. Show that
$$\text{if} \quad a \le b \quad \text{then} \quad \sqrt{a} \le \sqrt{b}.$$

52. Let a and b be nonnegative numbers. Show that their *geometric mean* $\sqrt{ab}$ cannot exceed their *arithmetic mean* $\frac{1}{2}(a + b).$

53. Show that
$$\text{if} \quad 0 \le a \le b \quad \text{then} \quad \frac{a}{1 + a} \le \frac{b}{1 + b}.$$

54. Let a, b, c be nonnegative numbers. Show that
$$\text{if} \quad a \le b + c \quad \text{then} \quad \frac{a}{1 + a} \le \frac{b}{1 + b} + \frac{c}{1 + c}.$$

1.5 INEQUALITIES AND ABSOLUTE VALUE

Here we take up some inequalities that involve absolute values. With an eye toward Chapter 2 we introduce two Greek letters: δ (delta) and ϵ (epsilon). You can find the complete Greek alphabet in the front matter.

Recall that for each real number a

(1.5.1)
$$|a| = \begin{cases} a, & \text{if } a \geq 0 \\ -a, & \text{if } a \leq 0 \end{cases}, \qquad |a| = \max\{a, -a\}, \qquad |a| = \sqrt{a^2}.$$

We begin with the inequality

$$|x| < \delta$$

where δ is some positive number. To say that $|x| < \delta$ is to say that x lies within δ units of 0, or, equivalently, that x lies between $-\delta$ and δ. Thus

(1.5.2) $\quad |x| < \delta \quad \text{iff} \quad -\delta < x < \delta.$

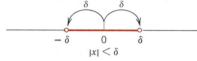

To say that $|x - c| < \delta$ is to say that x lies within δ units of c, or, equivalently, that x lies between $c - \delta$ and $c + \delta$. Thus

(1.5.3) $\quad |x - c| < \delta \quad \text{iff} \quad c - \delta < x < c + \delta.$

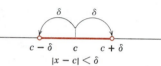

Somewhat more delicate is the inequality

$$0 < |x - c| < \delta.$$

Here we have $|x - c| < \delta$ with the additional requirement that $x \neq c$. Consequently

(1.5.4) $\quad 0 < |x - c| < \delta \quad \text{iff} \quad c - \delta < x < c \quad \text{or} \quad c < x < c + \delta.$

Thus, for example,

$$|x| < \tfrac{1}{2} \quad \text{iff} \quad -\tfrac{1}{2} < x < \tfrac{1}{2}; \qquad\qquad\qquad \textit{Solution: } (-\tfrac{1}{2}, \tfrac{1}{2})$$

$$|x - 5| < 1 \quad \text{iff} \quad 4 < x < 6; \qquad\qquad\qquad \textit{Solution: } (4, 6)$$

$$0 < |x - 5| < 1 \quad \text{iff} \quad 4 < x < 5 \quad \text{or} \quad 5 < x < 6. \quad \textit{Solution: } (4, 5) \cup (5, 6). \ \blacksquare$$

Problem 1. Solve the inequality

$$|x + 2| < 3.$$

Solution. The inequality $|x + 2| < 3$ holds iff

$$|x - (-2)| < 3 \quad \text{iff} \quad -2 - 3 < x < -2 + 3 \quad \text{iff} \quad -5 < x < 1.$$

The solution is the open interval $(-5, 1)$. $\blacksquare$

Problem 2. Solve the inequality

$$|3x - 4| < 2.$$

Solution. Since

$$|3x - 4| = |3(x - \tfrac{4}{3})| = |3||x - \tfrac{4}{3}| = 3|x - \tfrac{4}{3}|,$$

the inequality can be rewritten

$$3|x - \tfrac{4}{3}| < 2.$$

This gives $|x - \tfrac{4}{3}| < \tfrac{2}{3}$. Therefore

$$\tfrac{4}{3} - \tfrac{2}{3} < x < \tfrac{4}{3} + \tfrac{2}{3},$$
$$\tfrac{2}{3} < x < 2.$$

The solution is the open interval $(\tfrac{2}{3}, 2)$. □

Let $\epsilon > 0$. If you think of $|a|$ as the distance between a and 0, then obviously

(1.5.5) $|a| > \epsilon$ iff $a > \epsilon$ or $a < -\epsilon.$

Problem 3. Solve the inequality

$$|2x + 3| > 5.$$

Solution. As you saw, in general

$$|a| > \epsilon \quad \text{iff} \quad a > \epsilon \quad \text{or} \quad a < -\epsilon.$$

So here

$$2x + 3 > 5 \quad \text{or} \quad 2x + 3 < -5.$$

The first possibility gives $2x > 2$ and thus

$$x > 1.$$

The second possibility gives $2x < -8$ and thus

$$x < -4.$$

The total solution is therefore the union

$$(-\infty, -4) \cup (1, \infty).$$ □

We come now to one of the fundamental inequalities of calculus: for all real numbers a and b,

(1.5.6) $$|a + b| \le |a| + |b|.$$

This is called the *triangle inequality* in analogy with the geometric maxim "in a triangle the length of each side is less than or equal to the sum of the lengths of the other two sides".

Proof of the Triangle Inequality. The key here is to think of $|x|$ as $\sqrt{x^2}$. Note first that

$$(a + b)^2 = a^2 + 2ab + b^2 \le |a|^2 + 2|a||b| + |b|^2 = (|a| + |b|)^2.$$

Comparing the extremes of the inequality and taking square roots, we have

$$\sqrt{(a+b)^2} \le |a| + |b|. \qquad \text{(Exercise 50, Section 1.4)}$$

The result follows from observing that

$$\sqrt{(a+b)^2} = |a+b|. \quad \square$$

Here is a variant of the triangle inequality that also comes up in calculus: for all real numbers a and b,

(1.5.7)
$$\big| |a| - |b| \big| \le |a - b|.$$

The proof is left to you in the exercises.

EXERCISES 1.5

Solve the following inequalities.

1. $|x| < 2$.   **2.** $|x| \ge 1$.   **3.** $|x| > 3$.

4. $|x - 1| < 1$.   **5.** $|x - 2| < \frac{1}{2}$.   **6.** $|x - \frac{1}{2}| < 2$.

7. $|x + 2| < \frac{1}{4}$.   **8.** $|x - 3| < \frac{1}{4}$.   **9.** $0 < |x| < 1$.

10. $0 < |x| < \frac{1}{2}$.   **11.** $0 < |x - 2| < \frac{1}{2}$.   **12.** $0 < |x - \frac{1}{2}| < 2$.

13. $0 < |x - 3| < 8$.   **14.** $0 < |x - 2| < 8$.   **15.** $|2x - 3| < 1$.

16. $|3x - 5| < 3$.   **17.** $|2x + 1| < \frac{1}{4}$.   **18.** $|5x - 3| < \frac{1}{2}$.

19. $|2x + 5| > 3$.   **20.** $|3x + 1| > 5$.   **21.** $|5x - 1| > 9$.

Find the inequality of the form $|x - c| < \delta$ the solution of which is the given open interval.

22. $(-2, 2)$.   **23.** $(-3, 3)$.   **24.** $(0, 4)$.

25. $(-3, 7)$.   **26.** $(-4, 0)$.   **27.** $(-7, 3)$.

Determine all values of $A > 0$ for which the statement is true.

28. If $|x - 2| < 1$, then $|2x - 4| < A$.   **29.** If $|x - 2| < A$, then $|2x - 4| < 3$.

30. If $|x + 1| < A$, then $|3x + 3| < 4$.   **31.** If $|x + 1| < 2$, then $|3x + 3| < A$.

32. Prove that for all real numbers a and b

$$|a - b| \le |a| + |b|.$$

33. Prove (1.5.7).   HINT: Calculate $\big| |a| - |b| \big|^2$ and use the fact that $ab \le |a| \, |b|$.

34. Show that equality holds in (1.5.6) iff $ab \ge 0$.

1.6 FUNCTIONS

The fundamental processes of calculus (called *differentiation* and *integration*) are processes applied to functions. To understand these processes and to be able to carry them out, you have to be thoroughly familiar with functions. Here we review some of the basic notions.

Domain, Range, Graph

First we need a working definition for the word *function*. Let D be a set of real numbers. By a *function* on D we mean a rule that assigns a unique number to each number in D. The set D is called the *domain* of the function. The set of assignments that the function makes (the set of values that the function takes on) is called the *range* of the function.

Here are some examples. We begin with the squaring function

$$f(x) = x^2, \qquad x \text{ real.}$$

The domain of f is explicitly given as the set of all real numbers. As x runs through all the real numbers, x^2 runs through all the nonnegative numbers. The range is therefore $[0, \infty)$. In abbreviated form we write

$$\text{dom } (f) = (-\infty, \infty) \quad \text{and} \quad \text{ran } (f) = [0, \infty)$$

and say that f *maps* $(-\infty, \infty)$ *onto* $[0, \infty)$.

Now take the function

$$g(x) = \sqrt{x + 4}, \qquad x \in [0, 5].$$

The domain of g is given as the closed interval $[0, 5]$. At 0, g takes on the value 2:

$$g(0) = \sqrt{0 + 4} = \sqrt{4} = 2.$$

At 5, g takes on the value 3:

$$g(5) = \sqrt{5 + 4} = \sqrt{9} = 3.$$

As x runs through all the numbers from 0 to 5, $g(x)$ runs through all the numbers from 2 to 3. The range of g is therefore the closed interval $[2, 3]$. In abbreviated form

$$\text{dom } (g) = [0, 5], \qquad \text{ran } (g) = [2, 3].$$

The function g maps $[0, 5]$ onto $[2, 3]$.

If f is a function with domain D, the *graph* of f is by definition the set of all points

$$P(x, f(x)) \qquad \text{with } x \in D.$$

The graph of the squaring function

$$f(x) = x^2, \qquad x \text{ real}$$

is the parabola shown in Figure 1.6.1. The graph of the function

$$g(x) = \sqrt{x + 4}, \qquad x \in [0, 5]$$

is the arc drawn in Figure 1.6.2.

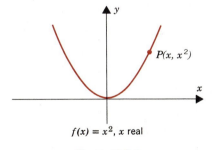

$f(x) = x^2$, x real

Figure 1.6.1

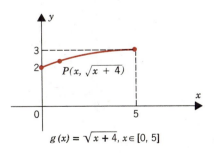

$g(x) = \sqrt{x + 4}$, $x \in [0, 5]$

Figure 1.6.2

Sometimes the domain of a function is not explicitly given. Thus, for example, we might write

$$f(x) = x^3 \quad \text{or} \quad f(x) = \sqrt{x} \qquad \text{(Figure 1.6.3)}$$

without further explanation. In such cases, take as the domain the maximal set of real numbers x for which $f(x)$ is itself a real number. For the cubing function, take dom $(f) = (-\infty, \infty)$, and for the square root function, take dom $(f) = [0, \infty)$.

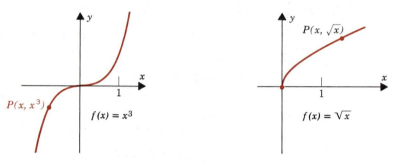

Figure 1.6.3

Problem 1. Find the domain and range of

$$f(x) = \frac{1}{\sqrt{2-x}} + 5.$$

Solution. First we look for the domain. To be able to form $\sqrt{2-x}$ we need $2 - x \geq 0$ and therefore $x \leq 2$. But at $x = 2$, $\sqrt{2-x} = 0$ and its reciprocal is not defined. We must therefore restrict x to $x < 2$. The domain is $(-\infty, 2)$.

Now we look for the range. As x runs through $(-\infty, 2)$, $\sqrt{2-x}$ takes on all positive values and so does its reciprocal. The range of f is therefore $(5, \infty)$. ∎

Algebraic Combinations of Functions

Functions with a common domain can be added and subtracted:

$$(f+g)(x) = f(x) + g(x), \qquad (f-g)(x) = f(x) - g(x).$$

They can be multiplied:

$$(fg)(x) = f(x)g(x).$$

If $g(x) \neq 0$, we can form the reciprocal:

$$\left(\frac{1}{g}\right)(x) = \frac{1}{g(x)},$$

and also the quotient:

$$\left(\frac{f}{g}\right)(x) = \frac{f(x)}{g(x)}.$$

With α and β real, we can form *scalar multiples*:

$$(\alpha f)(x) = \alpha f(x)$$

and *linear combinations*:

$$(\alpha f + \beta g)(x) = \alpha f(x) + \beta g(x).$$

Example 2. If

$$f(x) = x^3 + x - 1 \quad \text{and} \quad g(x) = 3x^2 + 4,$$

then

$$(f + g)(x) = (x^3 + x - 1) + (3x^2 + 4) = x^3 + 3x^2 + x + 3,$$
$$(f - g)(x) = (x^3 + x - 1) - (3x^2 + 4) = x^3 - 3x^2 + x - 5,$$
$$\left(\frac{1}{g}\right)(x) = \frac{1}{3x^2 + 4}, \quad \left(\frac{f}{g}\right)(x) = \frac{x^3 + x - 1}{3x^2 + 4}, \quad (6f)(x) = 6(x^3 + x - 1),$$
$$(6f - 5g)(x) = 6(x^3 + x - 1) - 5(3x^2 + 4) = 6x^3 - 15x^2 + 6x - 26. \quad \square$$

Example 3. Now consider two functions defined piecewise:

$$f(x) = \begin{cases} 1 - x^2, & x \le 0 \\ x, & x > 0 \end{cases}, \quad g(x) = \begin{cases} -2x, & x < 1 \\ 1 - x, & x \ge 1 \end{cases}.$$

To find $f + g, f - g, fg$, we must fit the pieces together; namely, we must break up the domain of both functions in the same manner:

$$f(x) = \begin{cases} 1 - x^2, & x \le 0 \\ x, & 0 < x < 1 \\ x, & 1 \le x \end{cases}, \quad g(x) = \begin{cases} -2x, & x \le 0 \\ -2x, & 0 < x < 1 \\ 1 - x, & 1 \le x \end{cases}.$$

The rest is straightforward:

$$(f + g)(x) = f(x) + g(x) = \begin{cases} 1 - 2x - x^2, & x \le 0 \\ -x, & 0 < x < 1 \\ 1, & 1 \le x \end{cases},$$

$$(f - g)(x) = f(x) - g(x) = \begin{cases} 1 + 2x - x^2, & x \le 0 \\ 3x, & 0 < x < 1 \\ -1 + 2x, & 1 \le x \end{cases},$$

$$(fg)(x) = f(x)g(x) = \begin{cases} -2x + 2x^3, & x \le 0 \\ -2x^2, & 0 < x < 1 \\ x - x^2, & 1 \le x \end{cases}. \quad \square$$

EXERCISES 1.6

Find the number(s), if any, where f takes on the value 1.

1. $f(x) = |2 - x|$.   **2.** $f(x) = \sqrt{1 + x}$.   **3.** $f(x) = x^2 + 4x + 5$.

4. $f(x) = 4 + 10x - x^2$.   **5.** $f(x) = 1 + \sin x$.   **6.** $f(x) = \tan x$.

7. $f(x) = \cos 2x$.   **8.** $f(x) = -1 + \cos x$.

Find the domain and the range.

9. $f(x) = |x|$.   **10.** $g(x) = x^2 - 1$.   **11.** $f(x) = 2x - 1$.

12. $g(x) = \sqrt{x} - 1$.   **13.** $f(x) = \dfrac{1}{x^2}$.   **14.** $g(x) = \dfrac{1}{x}$.

15. $f(x) = \sqrt{1 - x}$.   **16.** $g(x) = \sqrt{x - 1}$.   **17.** $f(x) = \sqrt{1 - x} - 1$.

18. $g(x) = \sqrt{x - 1} - 1$.   **19.** $f(x) = \dfrac{1}{\sqrt{1 - x}}$.   **20.** $g(x) = \dfrac{1}{\sqrt{4 - x^2}}$.

21. $f(x) = |\sin x|$.   **22.** $f(x) = \sin^2 x + \cos^2 x$.   **23.** $f(x) = 2 \cos 3x$.

24. $f(x) = \sqrt{\cos^2 x}$.   **25.** $f(x) = 1 + \tan^2 x$.   **26.** $f(x) = 1 + \sin x$.

Sketch the graph.

27. $f(x) = 1$.   **28.** $f(x) = -1$.   **29.** $f(x) = 2x$.

30. $f(x) = 2x + 1$.   **31.** $f(x) = \frac{1}{2}x$.   **32.** $f(x) = -\frac{1}{2}x$.

33. $f(x) = \frac{1}{2}x + 2$.   **34.** $f(x) = -\frac{1}{2}x - 3$.   **35.** $f(x) = \sqrt{4 - x^2}$.

36. $f(x) = \sqrt{9 - x^2}$.   **37.** $f(x) = 3 \sin 2x$.   **38.** $f(x) = 1 + \sin x$.

39. $f(x) = 1 - \cos x$.   **40.** $f(x) = \tan \frac{1}{2}x$.   **41.** $f(x) = \sqrt{\sin^2 x}$.

42. $f(x) = -2 \cos x$.

Sketch the graph and specify the domain and range.

43. $f(x) = \begin{cases} -1, & x < 0 \\ 1, & x > 0 \end{cases}$.   **44.** $f(x) = \begin{cases} x^2, & x \leq 0 \\ 1 - x, & x > 0 \end{cases}$.

45. $f(x) = \begin{cases} 1 + x, & 0 \leq x \leq 1 \\ x, & 1 < x < 2 \\ \frac{1}{2}x + 1, & 2 \leq x \end{cases}$.   **46.** $f(x) = \begin{cases} x^2, & x < 0 \\ -1, & 0 < x < 2 \\ x, & 2 < x \end{cases}$.

47. Given that

$$f(x) = \sqrt{x} + \frac{1}{\sqrt{x}} \quad \text{and} \quad g(x) = \sqrt{x} - \frac{2}{\sqrt{x}},$$

find   (a) $6f + 3g$,   (b) fg,   (c) f/g.

48. Given that

$$f(x) = \begin{cases} x, & x \leq 0 \\ -1, & x > 0 \end{cases} \quad \text{and} \quad g(x) = \begin{cases} -x, & x < 1 \\ x^2, & x \geq 1 \end{cases},$$

find   (a) $f + g$,   (b) $f - g$,   (c) fg.

49. Given that

$$f(x) = \begin{cases} 1 - x, & x \leq 1 \\ 2x - 1, & x > 1 \end{cases} \quad \text{and} \quad g(x) = \begin{cases} 0, & x < 2 \\ -1, & x \geq 2 \end{cases},$$

find   (a) $f + g$,   (b) $f - g$,   (c) fg.

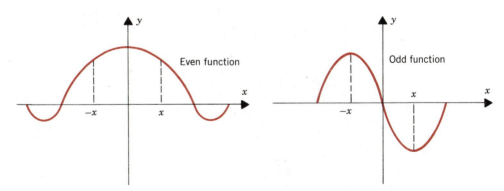

Figure 1.6.4

Sketch the graphs of the following functions with f and g as in Figure 1.6.4.

50. $2f$.   **51.** $\frac{1}{2}f$.   **52.** $-f$.   **53.** $2g$.   **54.** $\frac{1}{2}g$.   **55.** $-2g$.

56. $f + g$.   **57.** $f - g$.   **58.** $f + 2g$.

Even Functions, Odd Functions

(1.6.1)

A function f is said to be *even* iff $f(-x) = f(x)$ for all $x \in \operatorname{dom}(f)$.

A function f is said to be *odd* iff $f(-x) = -f(x)$ for all $x \in \operatorname{dom}(f)$.

The graph of an even function is symmetric about the y-axis. (Figure 1.6.5) The graph of an odd function is symmetric about the origin. (Figure 1.6.6)

Figure 1.6.5   **Figure 1.6.6**

Determine whether the function is odd, even, or neither.

59. $f(x) = x^3$.   **60.** $f(x) = x^2$.   **61.** $g(x) = x(x - 1)$.

62. $g(x) = x(x^2 + 1)$.   **63.** $f(x) = \dfrac{x^2}{1 - |x|}$.   **64.** $f(x) = x + \dfrac{1}{x}$.

65. $f(x) = \sin x$.   **66.** $f(x) = \tan x$.   **67.** $f(x) = \cos x$.

68. What can you conclude about the product of two odd functions?

69. What can you conclude about the product of two even functions?

70. What can you conclude about the product of an odd function and an even function?

71. A function f defined for all real numbers is defined as follows for $x \geq 0$:

$$f(x) = \begin{cases} x, & 0 \leq x \leq 1 \\ 1, & 1 < x \end{cases}.$$

How is f defined for $x < 0$ (a) if f is even? (b) if f is odd?

72. Exercise 71 with $f(x) = x^2 - x$, $x \geq 0$.

73. Given that the function f is defined for all real numbers, show that the function $g(x) = f(x) + f(-x)$ is an even function.

74. Show that every function defined for all real numbers can be written as the sum of an even function and an odd function.

1.7   THE COMPOSITION OF FUNCTIONS

In the last section, you saw how to combine functions algebraically.   There is another way to combine functions, called *composition*.   To describe it, we begin with two functions f and g and a number x in the domain of g.   By applying g to x, we get the number $g(x)$.   If $g(x)$ is in the domain of f, then we can apply f to $g(x)$ and thereby obtain the number $f(g(x))$.

What is $f(g(x))$?   It is the result of first applying g to x and then applying f to $g(x)$. The idea is illustrated in Figure 1.7.1.

If the range of g is completely contained in the domain of f (namely, if each value that g takes on is in the domain of f), then we can form $f(g(x))$ for *each x* in the domain of g and in this manner create a new function.   This new function — it takes each x in the domain of g and assigns to it the value $f(g(x))$ — is called the *composition* of f and g and is denoted by the symbol $f \circ g$.   What is this function $f \circ g$?   It is the function that results from first applying g and then f.   (Figure 1.7.2)

Figure 1.7.1

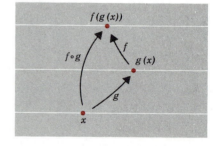

Figure 1.7.2

DEFINITION 1.7.1   COMPOSITION

If the range of the function g is contained in the domain of the function f, then the *composition* $f \circ g$ (read "f circle g") is the function defined on the domain of g by setting

$$(f \circ g)(x) = f(g(x)).$$

Example 1. Suppose that

$$g(x) = x^2 \qquad \text{(the squaring function)}$$

and

$$f(x) = x + 1. \qquad \text{(the function that adds 1)}$$

Then

$$(f \circ g)(x) = f(g(x)) = g(x) + 1 = x^2 + 1.$$

In other words, $f \circ g$ is the function that *first* squares and *then* adds 1.
   On the other hand,

$$(g \circ f)(x) = g(f(x)) = [f(x)]^2 = (x + 1)^2.$$

Thus $g \circ f$ is the function that *first* adds 1 and *then* squares. Notice that $g \circ f$ is *not* the same as $f \circ g$. ☐

Example 2. If

$$g(x) = \frac{1}{x} + 1 \qquad \text{(g adds 1 to the reciprocal)}$$

and

$$f(x) = x^2 - 5, \qquad \text{(f subtracts 5 from the square)}$$

then

$$(f \circ g)(x) = f(g(x)) = [g(x)]^2 - 5 = \left(\frac{1}{x} + 1\right)^2 - 5.$$

Here $f \circ g$ *first* adds 1 to the reciprocal and *then* subtracts 5 from the square. ☐

Example 3. If

$$f(x) = \frac{1}{x} + x^2 \quad \text{and} \quad g(x) = \frac{x^2 + 1}{x^4 + 1},$$

then

$$(f \circ g)(x) = f(g(x)) = \frac{1}{g(x)} + [g(x)]^2 = \frac{x^4 + 1}{x^2 + 1} + \left(\frac{x^2 + 1}{x^4 + 1}\right)^2. ☐$$

   It is possible to form the composition of more than two functions. For example, the triple composition $f \circ g \circ h$ consists of first h, then g, and then f:

$$(f \circ g \circ h)(x) = f(g(h(x))).$$

We can go on in this manner with as many functions as we like.

Example 4. If

$$f(x) = \frac{1}{x}, \qquad g(x) = x + 2, \qquad h(x) = (x^2 + 1)^2,$$

then

$$(f \circ g \circ h)(x) = f(g(h(x))) = \frac{1}{g(h(x))} = \frac{1}{h(x) + 2} = \frac{1}{(x^2 + 1)^2 + 2}. \quad \square$$

To apply some of the techniques of calculus you will need to be able to recognize composites. You will need to be able to start with a function, say $F(x) = (x + 1)^5$, and recognize how it is a composition.

Problem 5. Find functions f and g such that $f \circ g = F$ if

$$F(x) = (x + 1)^5.$$

A Solution. The function consists of first adding 1 and then taking the fifth power. We can therefore set

$$g(x) = x + 1 \hspace{4cm} \text{(adding 1)}$$

and

$$f(x) = x^5. \hspace{4cm} \text{(taking the fifth power)}$$

As you can see,

$$(f \circ g)(x) = f(g(x)) = [g(x)]^5 = (x + 1)^5. \quad \square$$

Problem 6. Find functions f and g such that $f \circ g = F$ if

$$F(x) = \frac{1}{x} - 6.$$

A Solution. F takes the reciprocal and then subtracts 6. We can therefore set

$$g(x) = \frac{1}{x} \hspace{4cm} \text{(taking the reciprocal)}$$

and

$$f(x) = x - 6. \hspace{4cm} \text{(subtracting 6)}$$

As you can check,

$$(f \circ g)(x) = f(g(x)) = g(x) - 6 = \frac{1}{x} - 6. \quad \square$$

Problem 7. Find three functions f, g, h such that $f \circ g \circ h = F$ if

$$F(x) = \frac{1}{|x| + 3}.$$

A Solution. F takes the absolute value, adds 3, and then inverts. Let h take the absolute value:

$$\text{set} \quad h(x) = |x|.$$

Let g add 3:

$$\text{set} \quad g(x) = x + 3.$$

Let f do the inverting:

$$\text{set} \quad f(x) = \frac{1}{x}.$$

With this choice of f, g, h, we have

$$(f \circ g \circ h)(x) = f(g(h(x))) = \frac{1}{g(h(x))} = \frac{1}{h(x) + 3} = \frac{1}{|x| + 3}. \quad \square$$

In the next problem we start with the same function

$$F(x) = \frac{1}{|x| + 3},$$

but this time we are asked to break it up as the composition of only two functions.

Problem 8. Find two functions f and g such that $f \circ g = F$ if

$$F(x) = \frac{1}{|x| + 3}.$$

Some Solutions. F takes the absolute value, adds 3, and then inverts. We can let g do the first two things by setting

$$g(x) = |x| + 3$$

and then let f do the inverting by setting

$$f(x) = \frac{1}{x}.$$

Or, we could let g just take the absolute value,

$$g(x) = |x|,$$

and then have f add 3 and invert:

$$f(x) = \frac{1}{x + 3}. \quad \square$$

EXERCISES 1.7

Form the composition $f \circ g$.

1. $f(x) = 2x + 5$, $g(x) = x^2$.

2. $f(x) = x^2$, $g(x) = 2x + 5$.

3. $f(x) = \sqrt{x}$, $g(x) = x^2 + 5$.

4. $f(x) = x^2 + x$, $g(x) = \sqrt{x}$.

5. $f(x) = \dfrac{1}{x}$, $g(x) = \dfrac{1}{x}$.

6. $f(x) = x^2$, $g(x) = \dfrac{1}{x - 1}$.

7. $f(x) = x - 1$, $g(x) = \dfrac{1}{x}$.

8. $f(x) = x^2 - 1$, $g(x) = x(x - 1)$.

9. $f(x) = \dfrac{1}{\sqrt{x} - 1}$, $g(x) = (x^2 + 2)^2$.

10. $f(x) = \dfrac{1}{x} - \dfrac{1}{x + 1}$, $g(x) = \dfrac{1}{x^2}$.

Form the composition $f \circ g \circ h$:

11. $f(x) = 4x$, $g(x) = x - 1$, $h(x) = x^2$.

12. $f(x) = x - 1$, $g(x) = 4x$, $h(x) = x^2$.

13. $f(x) = x^2$, $g(x) = x - 1$, $h(x) = x^4$.

14. $f(x) = x - 1$, $g(x) = x^2$, $h(x) = 4x$.

15. $f(x) = \dfrac{1}{x}$, $g(x) = \dfrac{1}{2x + 1}$, $h(x) = x^2$.

16. $f(x) = \dfrac{x + 1}{x}$, $g(x) = \dfrac{1}{2x + 1}$, $h(x) = x^2$.

Find f such that $f \circ g = F$ given that

17. $g(x) = \dfrac{1 + x^2}{1 + x^4}$, $F(x) = \dfrac{1 + x^4}{1 + x^2}$.

18. $g(x) = x^2$, $F(x) = ax^2 + b$.

19. $g(x) = 3x$, $F(x) = 2 \sin 3x$.

20. $g(x) = -x^2$, $F(x) = \sqrt{a^2 + x^2}$.

Find g such that $f \circ g = F$ given that

21. $f(x) = x^3$, $F(x) = \left(1 - \dfrac{1}{x^4}\right)^2$.

22. $f(x) = x + \dfrac{1}{x}$, $F(x) = a^2 x^2 + \dfrac{1}{a^2 x^2}$.

23. $f(x) = x^2 + 1$, $F(x) = (2x^3 - 1)^2 + 1$.

24. $f(x) = \sin x$, $F(x) = \sin \dfrac{1}{x}$.

Form $f \circ g$ and $g \circ f$.

25. $f(x) = \sqrt{x}$, $g(x) = x^2$.

26. $f(x) = 3x + 1$, $g(x) = x^2$.

27. $f(x) = 2x$, $g(x) = \sin x$.

28. $f(x) = 2x$, $g(x) = \frac{1}{2}$.

29. $f(x) = \begin{bmatrix} 1 - x, & x \le 0 \\ x^2, & x > 0 \end{bmatrix}$, $g(x) = \begin{bmatrix} -x, & x < 1 \\ 1 + x, & x \ge 1 \end{bmatrix}$.

30. $f(x) = \begin{bmatrix} 1 - 2x, & x < 1 \\ 1 + x, & x \ge 1 \end{bmatrix}$, $g(x) = \begin{bmatrix} x^2, & x < 0 \\ 1 - x, & x \ge 0 \end{bmatrix}$.

31. For $x \ne 0, 1$ define

$$f_1(x) = x, \qquad f_2(x) = \dfrac{1}{x}, \qquad f_3(x) = 1 - x,$$

$$f_4(x) = \dfrac{1}{1 - x}, \qquad f_5(x) = \dfrac{x - 1}{x}, \qquad f_6(x) = \dfrac{x}{x - 1}.$$

This family of functions is *closed* under composition; that is, the composition of any two of these functions is again one of these functions. Tabulate the results of composing these functions one with the other by filling in the table shown in Figure 1.7.3. To indicate that $f_i \circ f_j = f_k$ write "f_k" in the ith row, jth column. We have already made two entries in the table. Check out these two entries and then fill in the rest of the table.

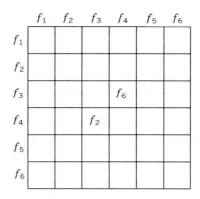

Figure 1.7.3

32. Find the minimal family of functions that contains the given function and is closed under composition.
(a) $f(x) = 1 - x.$ (b) $f(x) = 1 + x.$ (c) $f(x) = -x^3.$

1.8   ONE-TO-ONE FUNCTIONS; INVERSES

One-to-One Functions

A function can take on the same value at different points of the domain. Constant functions, for example, take on the same value at all points of the domain. The squaring function takes on the same value at $-c$ as it does at c; so does the absolute-value function. The function

$$f(x) = 1 + (x - 3)(x - 5)$$

takes on the same value at $x = 5$ as it does at $x = 3$:

$$f(3) = 1, \qquad f(5) = 1.$$

Functions for which this kind of repetition does *not* occur are called *one-to-one*.

DEFINITION 1.8.1

A function f is said to be *one-to-one* if and only if there are no two points of the domain at which it takes on the same value:

$$f(x_1) = f(x_2) \quad \text{implies} \quad x_1 = x_2.$$

Thus, if f is one-to-one and x_1, x_2 are different points of the domain, then

$$f(x_1) \neq f(x_2).$$

The functions

$$f(x) = x^3 \quad \text{and} \quad g(x) = \sqrt{x}$$

are both one-to-one. The cubing function is one-to-one because no two numbers have the same cube. The square-root function is one-to-one because no two numbers have the same square root.

There is a simple geometric test for one-to-oneness. Look at the graph of the function. If some horizontal line intersects the graph of the function more than once, then the function is not one-to-one (Figure 1.8.1). If, on the other hand, no horizontal line intersects the graph more than once, then the function is one-to-one (Figure 1.8.2).

f is not one-to-one: $f(x_1) = f(x_2)$

Figure 1.8.1

f is one-to-one

Figure 1.8.2

Inverses

I am thinking of a real number and its cube is 8. Can you figure out the number? Obviously, yes. It must be 2.

Now I am thinking of another number and its square is 9. Can you figure out the number? No, not unless you are clairvoyant. It could be 3, but it could also be -3.

The difference between the two situations can be phrased this way: In the first case we are dealing with a one-to-one function — no two numbers have the same cube. So, knowing the cube, we can work back to the original number. In the second case we are dealing with a function that is not one-to-one — different numbers can have the same square. Knowing the square, we still can't figure out what the original number was.

Throughout the rest of this section we will be looking at functional values and tracing them back to where they came from. So as to be able to trace these values back, we will restrict ourselves entirely to functions that are one-to-one.

We begin with a theorem about one-to-one functions.

THEOREM 1.8.2

If f is a one-to-one function, then there is one and only one function g that is defined on the range of f and satisfies the equation

$$f(g(x)) = x \qquad \text{for all } x \text{ in the range of } f.$$

Proof. The proof is easy. If x is in the range of f, then f must take on the value x at some number. Since f is one-to-one, there can be only one such number. We have called that number $g(x)$. ☐

The function that we have named g in the theorem is called the *inverse* of f and is usually denoted by f^{-1}.

DEFINITION 1.8.3 INVERSE FUNCTION

Let f be a one-to-one function. The *inverse* of f, denoted by f^{-1}, is the unique function that is defined on the range of f and satisfies the equation

$$f(f^{-1}(x)) = x \qquad \text{for all } x \text{ in the range of } f.$$

Problem 1. You have seen that the cubing function

$$f(x) = x^3$$

is one-to-one. Find the inverse.

Solution. We set $t = f^{-1}(x)$ and solve the equation $f(t) = x$ for t:

$$f(t) = x \qquad\qquad (1.8.3)$$
$$t^3 = x \qquad\qquad (f \text{ is the cubing function})$$
$$t = x^{1/3}.$$

Substituting $f^{-1}(x)$ back in for t, we have

$$f^{-1}(x) = x^{1/3}.$$

The inverse of the cubing function is the cube-root function. ☐

Remark. We substitute t for $f^{-1}(x)$ merely to simplify the calculations. It is easier to work with the symbol t than with the string of symbols $f^{-1}(x)$. ☐

Problem 2. Show that the linear function

$$f(x) = 3x - 5$$

is one-to-one. Then find the inverse.

Solution. To show that f is one-to-one, let's suppose that

$$f(x_1) = f(x_2).$$

Then

$$3x_1 - 5 = 3x_2 - 5$$
$$3x_1 = 3x_2$$
$$x_1 = x_2.$$

The function is one-to-one since

$$f(x_1) = f(x_2) \quad \text{implies} \quad x_1 = x_2.$$

(Another way to see that this function is one-to-one is to note that the graph is a line of slope 3 and as such cannot be intersected by a horizontal line more than once.)

Now let's find the inverse. To do this, we set $t = f^{-1}(x)$ and solve the equation $f(t) = x$ for t:

$$f(t) = x$$
$$3t - 5 = x$$
$$3t = x + 5$$
$$t = \tfrac{1}{3}x + \tfrac{5}{3}.$$

Substituting $f^{-1}(x)$ back in for t, we have

$$f^{-1}(x) = \tfrac{1}{3}x + \tfrac{5}{3}. \quad \square$$

Problem 3. Show that the function

$$f(x) = (1 - x^3)^{1/5} + 2$$

is one-to-one. What is the inverse?

Solution. First we show that f is one-to-one. Suppose that

$$f(x_1) = f(x_2).$$

Then

$$(1 - x_1^3)^{1/5} + 2 = (1 - x_2^3)^{1/5} + 2$$
$$(1 - x_1^3)^{1/5} = (1 - x_2^3)^{1/5}$$
$$1 - x_1^3 = 1 - x_2^3$$
$$x_1^3 = x_2^3$$
$$x_1 = x_2.$$

The function is one-to-one since

$$f(x_1) = f(x_2) \quad \text{implies} \quad x_1 = x_2.$$

To find the inverse we set $t = f^{-1}(x)$ and solve the equation $f(t) = x$ for t:

$$f(t) = x$$
$$(1 - t^3)^{1/5} + 2 = x$$
$$(1 - t^3)^{1/5} = x - 2$$
$$1 - t^3 = (x - 2)^5$$
$$t^3 = 1 - (x - 2)^5$$
$$t = [1 - (x - 2)^5]^{1/3}.$$

Substituting $f^{-1}(x)$ back in for t, we have

$$f^{-1}(x) = [1 - (x - 2)^5]^{1/3}. \quad \square$$

By definition f^{-1} satisfies the equation

(1.8.4) $\boxed{f(f^{-1}(x)) = x \quad \text{for all } x \text{ in the range of } f.}$

It is also true that

(1.8.5) $\boxed{f^{-1}(f(x)) = x \quad \text{for all } x \text{ in the domain of } f.}$

Proof. Take x in the domain of f and set $y = f(x)$. Since y is in the range of f,

$$f(f^{-1}(y)) = y. \qquad\qquad (1.8.4)$$

This means that

$$f(f^{-1}(f(x))) = f(x)$$

and tells us that f takes on the same value at $f^{-1}(f(x))$ as it does at x. With f one-to-one, this can only happen if

$$f^{-1}(f(x)) = x. \quad \square$$

Equation (1.8.5) tells us that f^{-1} "undoes" the work of f:

if f takes x to $f(x)$,   then f^{-1} takes $f(x)$ back to x. (Figure 1.8.3)

Figure 1.8.3   **Figure 1.8.4**

Equation (1.8.4) tells us that f "undoes" the work of f^{-1}:

if f^{-1} takes x to $f^{-1}(x)$,   then f takes $f^{-1}(x)$ back to x. (Figure 1.8.4)

The Graphs of f and f^{-1}

There is an important relation between the graph of a one-to-one function f and the graph of f^{-1}. (See Figure 1.8.5.) The graph of f consists of points of the form $P(x, f(x))$. Since f^{-1} takes on the value x at $f(x)$, the graph of f^{-1} consists of points of the form $Q(f(x), x)$. Since, for each x, P and Q are symmetric with respect to the line $y = x$, the graph of f^{-1} is the graph of f reflected in the line $y = x$.

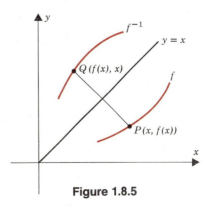

Figure 1.8.5

Problem 4.   Given the graph of f in Figure 1.8.6, sketch the graph of f^{-1}.

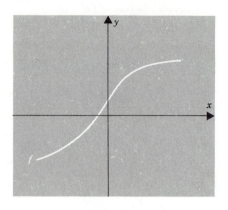

Figure 1.8.6

Figure 1.8.7

Solution.   First draw in the line $y = x$.   Then reflect the graph of f in that line.   See Figure 1.8.7.   □

EXERCISES 1.8

Determine whether or not the given function is one-to-one and, if so, find its inverse.

1. $f(x) = 5x + 3$.

2. $f(x) = 3x + 5$.

3. $f(x) = 4x - 7$.

4. $f(x) = 7x - 4$.

5. $f(x) = 1 - x^2$.

6. $f(x) = x^5$.

7. $f(x) = x^5 + 1$.

8. $f(x) = x^2 - 3x + 2$.

9. $f(x) = 1 + 3x^3$.

10. $f(x) = x^3 - 1$.

11. $f(x) = (1 - x)^3$.

12. $f(x) = (1 - x)^4$.

13. $f(x) = (x + 1)^3 + 2$.

14. $f(x) = (4x - 1)^3$.

15. $f(x) = x^{3/5}$.

16. $f(x) = 1 - (x - 2)^{1/3}$.

17. $f(x) = (2 - 3x)^3$.

18. $f(x) = (2 - 3x^2)^3$.

19. $f(x) = \dfrac{1}{x}$.

20. $f(x) = \dfrac{1}{1 - x}$.

21. $f(x) = x + \dfrac{1}{x}$.

22. $f(x) = \dfrac{x}{|x|}$.

23. $f(x) = \dfrac{1}{x^3 + 1}$.

24. $f(x) = \dfrac{1}{1 - x} - 1$.

25. $f(x) = \dfrac{x + 2}{x + 1}$.

26. $f(x) = \dfrac{1}{(x + 1)^{2/3}}$.

27.   What is the relation between f and $(f^{-1})^{-1}$?

Sketch the graph of f^{-1} given the graph of f.

28.

29.

30.

31.

32.

33.

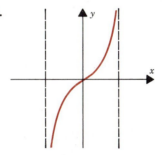

34. (a) Show that the composition of two one-to-one functions is one-to-one.
(b) Express $(f \circ g)^{-1}$ in terms of f^{-1} and g^{-1}.

1.9 A NOTE ON MATHEMATICAL PROOF

The notion of proof goes back to Euclid's *Elements*, and the rules of proof have changed little since they were formulated by Aristotle. We work in a deductive system where truth is argued on the basis of assumptions, definitions, and previously proved results. We cannot claim that such and such is true without clearly stating the basis on which we make that claim.

A theorem is an implication; it consists of a hypothesis and a conclusion:

$$\text{if (hypothesis)} \ldots, \text{then (conclusion)} \ldots.$$

Here is an example:

$$\text{if } a \text{ and } b \text{ are positive numbers, then } ab \text{ is positive.}$$

A common mistake is to ignore the hypothesis and persist with the conclusion: to insist, for example, that $ab > 0$ just because a and b are numbers.

Another common mistake is to confuse a theorem

$$\text{if } A, \text{ then } B$$

with its converse

$$\text{if } B, \text{ then } A.$$

The fact that a theorem is true does not mean that its converse is true: While it is true that

> if a and b are positive numbers, then ab is positive

it is *not* true that

> if ab is positive, then a and b are positive numbers.

$[(-2)(-3)$ is positive, but -2 and -3 are not positive.$]$

A third, more subtle mistake is to presume that the hypothesis of a theorem represents the only condition under which the conclusion is true. There may well be other conditions under which the conclusion is true. Thus, for example, not only is it true that

> if a and b are positive numbers, then ab is positive

but it is also true that

> if a and b are negative numbers, then ab is positive.

In the event that a theorem

> if A, then B

and its converse

> if B, then A

are both true, then we can write

> A if and only if B or more briefly A iff B.

We know, for example, that

> if $x \geq 0$,   then $|x| = x$;

we also know that

> if $|x| = x$,   then $x \geq 0$.

We can summarize by writing

> $x \geq 0$ iff $|x| = x$.

A final point. One way of proving

> if A, then B

is to assume that

(1) A holds and B does not hold

and then arrive at a contradiction. The contradiction is taken to indicate that (1) is false and therefore

> if A holds, then B must hold.

Some of the theorems of calculus are proved by this method.

Calculus provides procedures for solving a wide range of problems in the physical and social sciences. The fact that these procedures give us answers that seem to make sense is comforting, but it is only because we can prove our theorems that we can be confident in the results. Accordingly, your study of calculus should include the study of some proofs.

CHAPTER HIGHLIGHTS

1.1 What Is Calculus?

calculus as elementary mathematics (algebra, geometry, trigonometry) enhanced by the limit process

Sir Isaac Newton (p. 4) Gottfried Wilhelm Leibniz (p. 4)

1.2 Notions and Formulas from Elementary Mathematics

sets (p. 5) real numbers (p. 5) algebra and geometry (p. 8)
analytic geometry (p. 10) trigonometry (p. 11) induction (p. 14)

1.3 Some Problems on Lines

writing an equation in x and y for a line specified geometrically; finding the intercepts, slope, and inclination of a line given by an equation in x and y; calculating the angle between two lines

1.4 Inequalities

To solve an inequality is to find the set of numbers for which the inequality holds.

We can maintain an inequality by adding the same number to both sides, or by subtracting the same number from both sides, or by multiplying or dividing both sides by the same *positive* number. But if we multiply or divide by a negative number, then the inequality is *reversed*.

1.5 Inequalities and Absolute Value

$$|a| = \begin{cases} a, & \text{if } a \geq 0 \\ -a, & \text{if } a \leq 0 \end{cases} = \max\{a, -a\} = \sqrt{a^2}$$
$$|x - c| < \delta \quad \text{iff} \quad c - \delta < x < c + \delta$$
$$0 < |x - c| < \delta \quad \text{iff} \quad c - \delta < x < c \quad \text{or} \quad c < x < c + \delta$$
$$|a + b| \leq |a| + |b|, \qquad \big||a| - |b|\big| \leq |a - b|$$

1.6 Functions

domain, range, graph (p. 26)

algebraic combination of functions (p. 28)

even function, odd function (p. 31)

1.7 The Composition of Functions

If the range of a function g is contained in the domain of a function f, then the composition $f \circ g$ is the function defined on the domain of g by setting

$$f \circ g = f(g(x)).$$

1.8 One-to-One Functions; Inverses

1.9 A Note on Mathematical Proof

C H A P T E R

LIMITS AND CONTINUITY

2.1   THE IDEA OF LIMIT

Introduction

We could begin by saying that limits are important in calculus, but that would be a pathetic understatement. *Without limits calculus simply does not exist. Every single notion of calculus is a limit in one sense or another.*

What is instantaneous velocity?   It is the limit of average velocities.

What is the slope of a curve?   It is the limit of slopes of secant lines.

What is the length of a curve?   It is the limit of the lengths of polygonal paths.

What is the sum of an infinite series?   It is the limit of finite sums.

What is the area of a region bounded by curves?   It is the limit of areas of regions bounded by line segments.   See Figure 2.1.1 below.

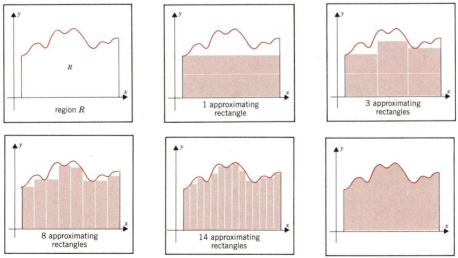

Area as a limit of sum of areas of rectangles by taking more and more rectangles

Figure 2.1.1

The Idea of Limit

We begin with a number *l* and a function *f* defined *near the number c but not necessarily defined at c itself.* A rough translation of

$$\lim_{x \to c} f(x) = l \qquad \text{(the limit of } f(x) \text{ as } x \text{ tends to } c \text{ is } l\text{)}$$

might read

as x approaches c, f(x) approaches l

or, equivalently,

for x close to c but different from c, f(x) is close to l.

We illustrate the idea in Figure 2.1.2. The curve represents the graph of a function *f*. The number *c* appears on the *x*-axis, the limit *l* on the *y*-axis. As *x* approaches *c* along the *x*-axis, *f(x)* approaches *l* along the *y*-axis.

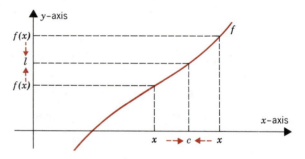

Figure 2.1.2

In taking the limit as *x* approaches *c*, it does not matter whether or not *f* is defined at *c* and, if so, how it is defined there. The only thing that matters is how *f* is defined *near c*. For example, in Figure 2.1.3 the graph of *f* is a broken curve defined peculiarly at *c*, and yet

$$\lim_{x \to c} f(x) = l$$

because, as suggested in Figure 2.1.4,

as x approaches c,   f(x) approaches l.

Figure 2.1.3

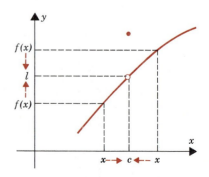

Figure 2.1.4

Numbers x near c fall into two natural categories: those that lie to the left of c and those that lie to the right of c. We write

$$\lim_{x \to c^-} f(x) = l \qquad \text{(the left-hand limit of } f(x) \text{ as } x \text{ tends to } c \text{ is } l)$$

to indicate that

as x approaches c from the left, $f(x)$ approaches l.

Similarly,

$$\lim_{x \to c^+} f(x) = l \qquad \text{(the right-hand limit of } f(x) \text{ as } x \text{ tends to } c \text{ is } l)$$

indicates that

as x approaches c from the right, $f(x)$ approaches l.†

For an example, see Figure 2.1.5. Note that in this case the left-hand limit and the right-hand limit are not the same.

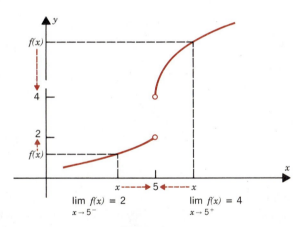

$$\lim_{x \to 5^-} f(x) = 2 \qquad \lim_{x \to 5^+} f(x) = 4$$

Figure 2.1.5

The left- and right-hand limits are called *one-sided limits.* As must be intuitively obvious (and everything we say in this section is only intuitive)

$$\lim_{x \to c} f(x) = l$$

if and only if both

$$\lim_{x \to c^-} f(x) = l \qquad and \qquad \lim_{x \to c^+} f(x) = l.$$

---

† The left-hand limit is sometimes written $\lim_{x \uparrow c} f(x)$ and the right-hand limit, $\lim_{x \downarrow c} f(x)$.

Example 1. For the function f graphed in Figure 2.1.6

$$\lim_{x \to (-2)^-} f(x) = 5 \quad \text{and} \quad \lim_{x \to (-2)^+} f(x) = 5$$

and therefore

$$\lim_{x \to -2} f(x) = 5.$$

It does not matter that $f(-2) = 3$.

Examining the graph of f near $x = 4$, we find that

$$\lim_{x \to 4^-} f(x) = 7 \quad \text{whereas} \quad \lim_{x \to 4^+} f(x) = 2.$$

Since these one-sided limits are different,

$$\lim_{x \to 4} f(x) \quad \text{does not exist.} \quad \square$$

Figure 2.1.6

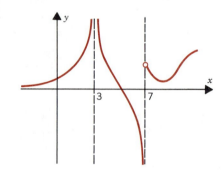

Figure 2.1.7

Example 2. Consider the function f graphed in Figure 2.1.7. As x approaches 3 from either side, $f(x)$ becomes arbitrarily large. Becoming arbitrarily large, $f(x)$ cannot stay close to any fixed number l. Therefore

$$\lim_{x \to 3} f(x) \quad \text{does not exist.}$$

Now let's focus on what happens near 7. As x approaches 7 from the left, $f(x)$ becomes arbitrarily large negative (i.e., less than any preassigned number). Under these circumstances $f(x)$ can't possibly approach a fixed number. Thus

$$\lim_{x \to 7^-} f(x) \quad \text{does not exist.}$$

Since the left-hand limit at $x = 7$ does not exist,

$$\lim_{x \to 7} f(x) \quad \text{does not exist.} \quad \square$$

Our next example is somewhat exotic.

Example 3. Let f be the function defined as follows:

$$f(x) = \begin{cases} 1, & x \text{ rational} \\ 0, & x \text{ irrational} \end{cases}.$$

(Figure 2.1.8)

Now let c be any real number. As x approaches c, x passes through both rational and irrational numbers. As this happens, $f(x)$ jumps wildly back and forth between 0 and 1 and thus cannot stay close to any fixed number l. Therefore $\lim_{x \to c} f(x)$ does not exist. ☐

Figure 2.1.8

We go on with examples. Where a graph has not been provided, you may find it useful to sketch one.

Example 4

$$\lim_{x \to 3} (2x + 5) = 11.$$

As x approaches 3, $2x$ approaches 6, and $2x + 5$ approaches 11. ☐

Example 5

$$\lim_{x \to 2} (x^2 - 1) = 3.$$

As x approaches 2, x^2 approaches 4, and $x^2 - 1$ approaches 3. ☐

Example 6

$$\lim_{x \to 3} \frac{1}{x - 1} = \frac{1}{2}.$$

As x approaches 3, $x - 1$ approaches 2, and $1/(x - 1)$ approaches $\frac{1}{2}$. ☐

Example 7

$$\lim_{x \to 2} \frac{1}{x - 2} \quad \text{does not exist.} \quad \square \qquad \text{(Figure 2.1.9)}$$

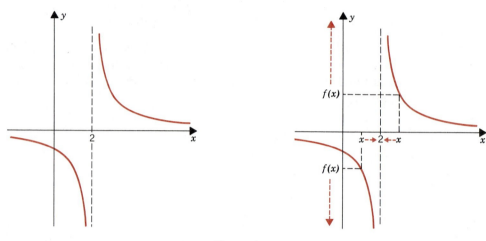

Figure 2.1.9

Before going on to the next examples, remember that, in taking the limit as x approaches a given number c, it does not matter whether or not the function is defined at the number c and, if so, how it is defined there. The only thing that matters is how the function is defined *near* the number c.

Example 8

$$\lim_{x \to 2} \frac{x^2 - 4x + 4}{x - 2} = 0.$$

At $x = 2$, the function is undefined: both numerator and denominator are 0. But that doesn't matter. For all $x \neq 2$, and therefore *for all x near 2*,

$$\frac{x^2 - 4x + 4}{x - 2} = x - 2.$$

It follows that

$$\lim_{x \to 2} \frac{x^2 - 4x + 4}{x - 2} = \lim_{x \to 2} (x - 2) = 0. \quad \square$$

Example 9

$$\lim_{x \to 2} \frac{x - 2}{x - 2} = 1.$$

At $x = 2$, the function is undefined: both numerator and denominator are 0. But, as we said before, that doesn't matter. For all $x \neq 2$, and therefore for all x near 2,

$$\frac{x-2}{x-2} = 1 \qquad \text{so that} \qquad \lim_{x \to 2} \frac{x-2}{x-2} = \lim_{x \to 2} 1 = 1. \quad \square$$

Example 10

$$\lim_{x \to 2} \frac{x-2}{x^2 - 4x + 4} \quad \text{does not exist.}$$

This result does not follow from the fact that the function is not defined at 2, but rather from the fact that for all $x \neq 2$, and therefore for all x near 2,

$$\frac{x-2}{x^2 - 4x + 4} = \frac{1}{x-2}$$

and, as you saw before,

$$\lim_{x \to 2} \frac{1}{x-2} \quad \text{does not exist.} \quad \square$$

Example 11

$$\text{If} \quad f(x) = \begin{cases} 3x - 1, & x \neq 2 \\ 45, & x = 2 \end{cases}, \qquad \text{then} \quad \lim_{x \to 2} f(x) = 5.$$

It doesn't matter to us that $f(2) = 45$. For $x \neq 2$, and thus for all x near 2,

$$f(x) = 3x - 1 \qquad \text{so that} \qquad \lim_{x \to 2} f(x) = \lim_{x \to 2} (3x - 1) = 5. \quad \square$$

Example 12

$$\text{If} \quad f(x) = \begin{cases} -2x, & x < 1 \\ 2x, & x > 1 \end{cases}, \qquad \text{then} \quad \lim_{x \to 1} f(x) \quad \text{does not exist.}$$

The limit does not exist since the one-sided limits are different:

$$\lim_{x \to 1^-} f(x) = \lim_{x \to 1^-} (-2x) = -2, \qquad \lim_{x \to 1^+} f(x) = \lim_{x \to 1^+} 2x = 2. \quad \square$$

Example 13

$$\text{If} \quad f(x) = \begin{cases} -2x, & x < 1 \\ 2x, & x > 1 \end{cases}, \qquad \text{then} \quad \lim_{x \to 1.03} f(x) = 2.06.$$

Notice that for values of x sufficiently near 1.03, the values of the function are computed using the rule $2x$ whether x is to the left or the right of 1.03. $\quad \square$

If you have found all this very imprecise, you are absolutely right. It is imprecise, but it need not remain so. One of the great triumphs of calculus has been its capacity to formulate limit statements with precision, but for this precision you will have to wait until Section 2.2.

EXERCISES 2.1

In Exercises 1 – 12 you are given a number c and the graph of a function f.   Use the graph of
f to find

$$\text{(a) } \lim_{x \to c^-} f(x) \qquad \text{(b) } \lim_{x \to c^+} f(x) \qquad \text{(c) } \lim_{x \to c} f(x) \qquad \text{(d) } f(c).$$

1. $c = 2$.

2. $c = 3$.

3. $c = 3$.

4. $c = 4$.

5. $c = -2$.

6. $c = 1$.

7. $c = 1$.

8. $c = -1$.

9. $c = 3$.

10. $c = 3$.

11. $c = 2$.

12. $c = -2$.

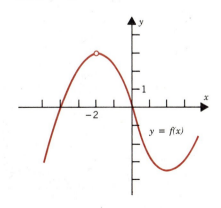

Exercises 13 and 14: state the values of c for which $\lim\limits_{x \to c} f(x)$ does not exist.

13.

14.

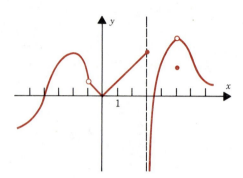

Exercises 15–50: decide on intuitive grounds whether or not the indicated limit exists, and evaluate the limit if it does exist.

15. $\lim\limits_{x \to 0} (2x - 1)$.

16. $\lim\limits_{x \to 1} (2 - 5x)$.

17. $\lim\limits_{x \to -2} x^2$.

18. $\lim\limits_{x \to 4} \sqrt{x}$.

19. $\lim\limits_{x \to -3} (|x| - 2)$.

20. $\lim\limits_{x \to 0} \dfrac{1}{|x|}$.

21. $\lim\limits_{x \to 1} \dfrac{3}{x + 1}$.

22. $\lim\limits_{x \to 0} \dfrac{4}{x + 1}$.

23. $\lim\limits_{x \to -1} \dfrac{2}{x + 1}$.

24. $\lim\limits_{x \to 2} \dfrac{1}{3x - 6}$.

25. $\lim\limits_{x \to 3} \dfrac{2x - 6}{x - 3}$.

26. $\lim\limits_{x \to 3} \dfrac{x^2 - 6x + 9}{x - 3}$.

27. $\lim\limits_{x \to 3} \dfrac{x - 3}{x^2 - 6x + 9}$.

28. $\lim\limits_{x \to 2} \dfrac{x^2 - 3x + 2}{x - 2}$.

29. $\lim\limits_{x \to 2} \dfrac{x - 2}{x^2 - 3x + 2}$.

30. $\lim\limits_{x \to 1} \dfrac{x - 2}{x^2 - 3x + 2}$.

31. $\lim\limits_{x \to 0} \left(x + \dfrac{1}{x} \right)$.

32. $\lim\limits_{x \to 1} \left(x + \dfrac{1}{x} \right)$.

33. $\lim\limits_{x \to 0} \dfrac{2x - 5x^2}{x}$.

34. $\lim\limits_{x \to 3} \dfrac{x - 3}{6 - 2x}$.

35. $\lim\limits_{x \to 1} \dfrac{x^2 - 1}{x - 1}$.

36. $\lim\limits_{x \to 1} \dfrac{x^3 - 1}{x - 1}$.

37. $\lim\limits_{x \to 1} \dfrac{x^3 - 1}{x + 1}$.

38. $\lim\limits_{x \to 1} \dfrac{x^2 + 1}{x^2 - 1}$.

39. $\lim\limits_{x \to 0} f(x)$; $f(x) = \begin{cases} 1, & x \neq 0 \\ 3, & x = 0 \end{cases}$.

40. $\lim\limits_{x \to 1} f(x)$; $f(x) = \begin{cases} 3x, & x < 1 \\ 3, & x > 1 \end{cases}$.

41. $\lim\limits_{x \to 4} f(x)$; $f(x) = \begin{cases} x^2, & x \neq 4 \\ 0, & x = 4 \end{cases}$.

42. $\lim\limits_{x \to 0} f(x)$; $f(x) = \begin{cases} -x^2, & x < 0 \\ x^2, & x > 0 \end{cases}$.

43. $\lim\limits_{x \to 0} f(x)$; $f(x) = \begin{cases} x^2, & x < 0 \\ 1 + x, & x > 0 \end{cases}$.

44. $\lim\limits_{x \to 1} f(x)$; $f(x) = \begin{cases} 2x, & x < 1 \\ x^2 + 1, & x > 1 \end{cases}$.

45. $\lim\limits_{x \to 0} f(x)$; $f(x) = \begin{cases} 2, & x \text{ rational} \\ -2, & x \text{ irrational} \end{cases}$.

46. $\lim\limits_{x \to 1} f(x)$; $f(x) = \begin{cases} 2x, & x \text{ rational} \\ 2, & x \text{ irrational} \end{cases}$.

47. $\lim\limits_{x \to 2} f(x)$; $f(x) = \begin{cases} 3x, & x < 1 \\ x + 2, & x \geq 1 \end{cases}$.

48. $\lim\limits_{x \to 0} f(x)$; $f(x) = \begin{cases} 2x, & x \leq 1 \\ x + 1, & x > 1 \end{cases}$.

49. $\lim\limits_{x \to 1} \dfrac{\sqrt{x^2 + 1} - \sqrt{2}}{x - 1}$.

50. $\lim\limits_{x \to 5} \dfrac{\sqrt{x^2 + 5} - \sqrt{30}}{x - 5}$.

51. (Calculator) Guess

$$\lim_{x \to 0} \frac{\sin x}{x} \qquad \text{(radian measure)}$$

after evaluating the quotient at $x = \pm 1, \pm 0.1, \pm 0.01, \pm 0.001$.

52. (Calculator) Guess

$$\lim_{x \to 0} \frac{\cos x - 1}{x} \qquad \text{(radian measure)}$$

after evaluating the quotient at $x = \pm 1, \pm 0.1, \pm 0.01, \pm 0.001$.

53. (Calculator) Guess

$$\lim_{x \to 1} \frac{x^{3/2} - 1}{x - 1}$$

after evaluating the quotient at $x = 0.9, 0.99, 0.999, 0.9999$ and $x = 1.1, 1.01, 1.001, 1.0001$.

2.2 DEFINITION OF LIMIT

Our work with limits in Section 2.1 was very informal.   It is time to be more precise.

Let f be a function and let c be a real number. We do not require that f be defined at c, but we do require that f be defined at least on a set of the form $(c - p, c) \cup (c, c + p)$ with $p > 0$. (This guarantees that we can form $f(x)$ for all $x \ne c$ that are "sufficiently close" to c.)   To say that

$$\lim_{x \to c} f(x) = l$$

is to say that

$\lvert f(x) - l \rvert$ can be made arbitrarily small	you pick $\epsilon > 0$; $\lvert f(x) - l \rvert$ can be made less than ϵ
simply by requiring that	simply by requiring that
$\lvert x - c \rvert$ be sufficiently small but different from zero.	$\lvert x - c \rvert$ satisfy an inequality of the form $0 < \lvert x - c \rvert < \delta$ for δ sufficiently small.

Putting the various pieces together in compact form we have the following *fundamental definition*.

DEFINITION 2.2.1   THE LIMIT OF A FUNCTION

Let f be a function defined at least on some set of the form $(c - p, c) \cup (c, c + p)$.

$$\lim_{x \to c} f(x) = l \quad \text{iff} \quad \begin{cases} \text{for each } \epsilon > 0 \text{ there exists } \delta > 0 \text{ such that} \\ \text{if } \ 0 < \lvert x - c \rvert < \delta, \qquad \text{then } \ \lvert f(x) - l \rvert < \epsilon. \end{cases}$$

Figures 2.2.1 and 2.2.2 illustrate this definition.

Figure 2.2.1

Figure 2.2.2

In general, the choice of δ depends upon the previous choice of ϵ. We do not require that there exist a number δ which "works" for *all* ϵ but, rather, that for *each* ϵ there exist a δ which "works" for it.

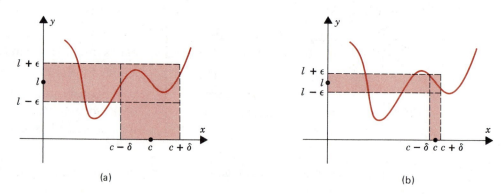

Figure 2.2.3

In Figure 2.2.3, we give two choices of ϵ and for each we display a suitable δ. For a δ to be suitable, all the points within δ of c (with the possible exception of c itself) must be taken by the function to within ϵ of l. In part (b) of the figure we began with a smaller ϵ and had to use a smaller δ.

The δ of Figure 2.2.4 is too large for the given ϵ. In particular the points marked x_1 and x_2 in the figure are not taken by the function to within ϵ of l.

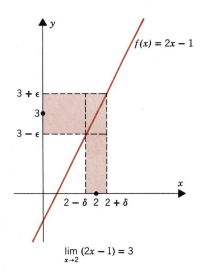

$$\lim_{x \to 2} (2x - 1) = 3$$

Figure 2.2.5

Figure 2.2.4

Next we apply the ϵ, δ definition of limit to a variety of functions. If you have never run across ϵ, δ arguments before, you may find them confusing at first. It usually takes a little while for the ϵ, δ idea to take hold.

Example 1. Show that

$$\lim_{x \to 2} (2x - 1) = 3.$$ (Figure 2.2.5)

Finding a δ. Let $\epsilon > 0$. We seek a number $\delta > 0$ such that,

$$\text{if} \quad 0 < |x - 2| < \delta, \quad \text{then} \quad |(2x - 1) - 3| < \epsilon.$$

What we have to do first is establish a connection between

$$|(2x - 1) - 3| \quad \text{and} \quad |x - 2|.$$

The connection is simple:

$$|(2x - 1) - 3| = |2x - 4|$$

so that

(*) $$|(2x - 1) - 3| = 2|x - 2|.$$

To make $|(2x - 1) - 3|$ less than ϵ, we need only make $|x - 2|$ twice as small. This suggests that we choose $\delta = \frac{1}{2}\epsilon$.

Showing that the δ "works." If $0 < |x - 2| < \frac{1}{2}\epsilon$, then $2|x - 2| < \epsilon$ and, by (*), $|(2x - 1) - 3| < \epsilon$. ☐

Remark. In Example 1 we chose $\delta = \frac{1}{2}\epsilon$, but we could have chosen any positive $\delta \leq \frac{1}{2}\epsilon$. In general, if a certain δ "works," then any smaller $\delta > 0$ will also work. $\square$

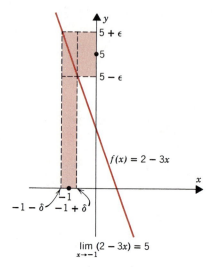

$$\lim_{x \to -1} (2 - 3x) = 5$$

Figure 2.2.6

Example 2. Show that

$$\lim_{x \to -1} (2 - 3x) = 5. \qquad \text{(Figure 2.2.6)}$$

Finding a δ. Let $\epsilon > 0$. We seek a number $\delta > 0$ such that

$$\text{if} \quad 0 < |x - (-1)| < \delta, \qquad \text{then} \quad |(2 - 3x) - 5| < \epsilon.$$

To find a connection between

$$|x - (-1)| \quad \text{and} \quad |(2 - 3x) - 5|,$$

we simplify both expressions:

$$|x - (-1)| = |x + 1|$$

and

$$|(2 - 3x) - 5| = |-3x - 3| = 3|-x - 1| = 3|x + 1|.$$

We can conclude that

$$(**) \qquad\qquad |(2 - 3x) - 5| = 3|x - (-1)|.$$

We can make the expression on the left less than ϵ by making $|x - (-1)|$ three times as small. This suggests that we set $\delta = \frac{1}{3}\epsilon$.

Showing that the δ "works." If $0 < |x - (-1)| < \frac{1}{3}\epsilon$, then $3|x - (-1)| < \epsilon$ and, by (**), $|(2 - 3x) - 5| < \epsilon$. $\square$

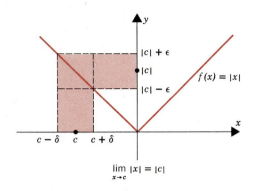

$$\lim_{x \to c} |x| = |c|$$

Figure 2.2.7

Example 3

(2.2.2)

$$\lim_{x \to c} |x| = |c|.$$

(Figure 2.2.7)

Proof. Let $\epsilon > 0$. We seek $\delta > 0$ such that

$$\text{if} \quad 0 < |x - c| < \delta, \qquad \text{then} \quad \big||x| - |c|\big| < \epsilon.$$

Since

$$\big||x| - |c|\big| \le |x - c|, \tag{1.5.7}$$

we can choose $\delta = \epsilon$; for

$$\text{if} \quad 0 < |x - c| < \epsilon, \qquad \text{then} \quad \big||x| - |c|\big| < \epsilon. \quad \square$$

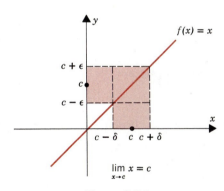

$$\lim_{x \to c} x = c$$

Figure 2.2.8

Example 4

(2.2.3)

$$\lim_{x \to c} x = c.$$

(Figure 2.2.8)

Proof. Let $\epsilon > 0$. Here we must find $\delta > 0$ such that

$$\text{if} \quad 0 < |x - c| < \delta, \qquad \text{then} \quad |x - c| < \epsilon.$$

Obviously we can choose $\delta = \epsilon$. $\quad \square$

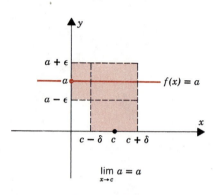

Figure 2.2.9

Example 5

(2.2.4)

$$\lim_{x \to c} a = a.$$

(Figure 2.2.9)

Proof. Here we are dealing with the constant function

$$f(x) = a.$$

Let $\epsilon > 0$. We must find $\delta > 0$ such that

$$\text{if}\quad 0 < |x - c| < \delta, \qquad \text{then}\quad |a - a| < \epsilon.$$

Since $|a - a| = 0$, we always have

$$|a - a| < \epsilon$$

no matter how δ is chosen; in short, any positive number will do for δ. ☐

Usually ϵ, δ arguments are carried out in two stages. First we do a little algebraic scratchwork, labeled "finding a δ" in some of the examples above. This scratchwork involves working backward from $|f(x) - l| < \epsilon$ to find a $\delta > 0$ sufficiently small so that we can begin with the inequality $0 < |x - c| < \delta$ to arrive at $|f(x) - l| < \epsilon$. This first stage is just preliminary, but it shows us how to proceed in the second stage. The second stage consists of showing that the δ "works" by verifying that for our choice of δ the implication

$$\text{if}\quad 0 < |x - c| < \delta, \qquad \text{then}\quad |f(x) - l| < \epsilon$$

is true. The next two examples are more complicated and therefore may give you a better feeling for this idea of working backward to find a δ.

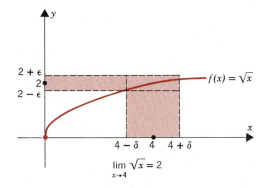

Figure 2.2.10

Example 6

$$\lim_{x \to 4} \sqrt{x} = 2.$$ (Figure 2.2.10)

Finding a δ. Let $\epsilon > 0$. We seek $\delta > 0$ such that

$$\text{if} \quad 0 < |x - 4| < \delta, \quad \text{then} \quad |\sqrt{x} - 2| < \epsilon.$$

First we want a relation between $|x - 4|$ and $|\sqrt{x} - 2|$. To be able to form $\sqrt{x}$ at all we need $x \geq 0$. To ensure this we must have $\delta \leq 4$. (Why?)

Remembering that we must have $\delta \leq 4$, let's go on. If $x \geq 0$, then we can form $\sqrt{x}$ and write

$$x - 4 = (\sqrt{x})^2 - 2^2 = (\sqrt{x} + 2)(\sqrt{x} - 2).$$

Taking absolute values, we have

$$|x - 4| = |\sqrt{x} + 2||\sqrt{x} - 2|.$$

Since $|\sqrt{x} + 2| > 1$, we have

$$|\sqrt{x} - 2| < |x - 4|.$$

This last inequality suggests that we simply set $\delta = \epsilon$. But remember now the previous requirement $\delta \leq 4$. We can meet all requirements by setting $\delta = $ minimum of 4 and ϵ.

Showing that the δ "works." Let $\epsilon > 0$. Choose $\delta = \min\{4, \epsilon\}$ and assume that

$$0 < |x - 4| < \delta.$$

Since $\delta \leq 4$, we have $x \geq 0$ and can write

$$|x - 4| = |\sqrt{x} + 2||\sqrt{x} - 2|.$$

Since $|\sqrt{x} + 2| > 1$, we can conclude that

$$|\sqrt{x} - 2| < |x - 4|.$$

Since $|x - 4| < \delta$ and $\delta \leq \epsilon$, it does follow that

$$|\sqrt{x} - 2| < \epsilon. \quad \blacksquare$$

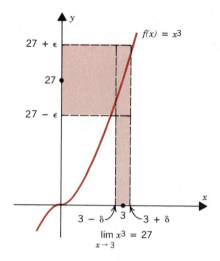

Figure 2.2.11

Example 7

$$\lim_{x \to 3} x^3 = 27. \qquad \text{(Figure 2.2.11)}$$

Finding a δ. Let $\epsilon > 0$. We seek $\delta > 0$ such that

$$\text{if} \quad 0 < |x - 3| < \delta, \qquad \text{then} \quad |x^3 - 27| < \epsilon.$$

The needed connection between $|x - 3|$ and $|x^3 - 27|$ is found by factoring:

$$x^3 - 27 = (x - 3)(x^2 + 3x + 9)$$

and thus

(∗) $$|x^3 - 27| = |x - 3||x^2 + 3x + 9|.$$

At this juncture we need to get a handle on the size of $|x^3 + 3x + 9|$ for x close to 3. For convenience let's take x within one unit of 3.

If $|x - 3| < 1$, then $2 < x < 4$ and

$$|x^2 + 3x + 9| \leq |x^2| + |3x| + |9|$$
$$= |x|^2 + 3|x| + 9$$
$$< 16 + 12 + 9 = 37.$$

It follows by (∗) that

(∗∗) $$\text{if} \quad |x - 3| < 1, \qquad \text{then} \quad |x^3 - 27| < 37|x - 3|.$$

If, in addition, $|x - 3| < \epsilon/37$, then it will follow that

$$|x^3 - 27| < 37(\epsilon/37) = \epsilon.$$

This means that we can take $\delta = $ minimum of 1 and $\epsilon/37$.

Showing that the δ "works." Let $\epsilon > 0$. Choose $\delta = \min\{1, \epsilon/37\}$ and assume that

$$0 < |x - 3| < \delta.$$

Then

$$|x - 3| < 1 \quad \text{and} \quad |x - 3| < \epsilon/37.$$

By (∗∗)

$$|x^3 - 27| < 37|x - 3|,$$

which, since $|x - 3| < \epsilon/37$, gives

$$|x^3 - 27| < 37(\epsilon/37) = \epsilon. \quad \square$$

There are many different ways of formulating the same limit statement. Sometimes one formulation is more convenient, sometimes another. In any case, it is useful to recognize that the following are equivalent:

(2.2.5)

(i) $\lim\limits_{x \to c} f(x) = l.$ (ii) $\lim\limits_{x \to c} (f(x) - l) = 0.$

(iii) $\lim\limits_{x \to c} |f(x) - l| = 0.$ (iv) $\lim\limits_{h \to 0} f(c + h) = l.$

The equivalence of these four statements is obvious. It is, however, a good exercise in ϵ, δ technique to show that (i) is equivalent to (iv).

Example 8. For $f(x) = x^3$, we have

$$\lim_{x \to 2} x^3 = 8, \qquad \lim_{x \to 2} (x^3 - 8) = 0,$$

$$\lim_{x \to 2} |x^3 - 8| = 0, \qquad \lim_{h \to 0} (2 + h)^3 = 8. \quad \square$$

The limit statements

$$\lim_{x \to c} f(x) = l \quad \text{and} \quad \lim_{x \to c} |f(x)| = |l|$$

are *not* equivalent. While the first statement implies the second (Exercise 34a), the second statement does not imply the first (Exercise 34b).

We come now to the ϵ, δ definitions of one-sided limits. These are just the usual ϵ, δ statement, except that, for a left-hand limit, the δ has to "work" only to the left of c and, for a right-hand limit, the δ has to "work" only to the right of c.

DEFINITION 2.2.6   LEFT-HAND LIMIT

Let f be a function defined at least on an interval of the form $(c - p, c)$.

$$\lim_{x \to c^-} f(x) = l \quad \text{iff} \quad \begin{cases} \text{for each } \epsilon > 0 \text{ there exists } \delta > 0 \text{ such that} \\ \text{if } \ c - \delta < x < c, \quad \text{then } \ |f(x) - l| < \epsilon. \end{cases}$$

DEFINITION 2.2.7   RIGHT-HAND LIMIT

Let f be a function defined at least on an interval of the form $(c, c + p)$.

$$\lim_{x \to c^+} f(x) = l \quad \text{iff} \quad \begin{cases} \text{for each } \epsilon > 0 \text{ there exists } \delta > 0 \text{ such that} \\ \text{if } \ c < x < c + \delta, \quad \text{then } \ |f(x) - l| < \epsilon. \end{cases}$$

As indicated in Section 2.1, one-sided limits give us a simple way of determining whether or not a (two-sided) limit exists:

(2.2.8) $$\lim_{x \to c} f(x) = l \quad \text{iff} \quad \lim_{x \to c^-} f(x) = l \quad \text{and} \quad \lim_{x \to c^+} f(x) = l.$$

The proof is left as an exercise. The proof is easy because any δ that "works" for the limit will work for both one-sided limits and any δ that "works" for both one-sided limits will work for the limit.

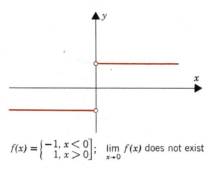

$$f(x) = \begin{cases} -1, & x < 0 \\ 1, & x > 0 \end{cases}; \quad \lim_{x \to 0} f(x) \text{ does not exist}$$

Figure 2.2.12

Example 9

If $f(x) = \begin{cases} -1, & x < 0 \\ 1, & x > 0 \end{cases}$,   then $\lim_{x \to 0} f(x)$ does not exist.   (Figure 2.2.12)

Proof. It is clear that

$$\lim_{x \to 0^-} f(x) = -1 \quad \text{and} \quad \lim_{x \to 0^+} f(x) = 1.$$

Since these one-sided limits are different, $\lim_{x \to 0} f(x)$ does not exist.   □

EXERCISES 2.2

Decide in the manner of Section 2.1 whether or not the indicated limit exists. Evaluate the limits which do exist.

1. $\lim_{x \to 1} \dfrac{x}{x + 1}$.

2. $\lim_{x \to 0} \dfrac{x^2(1 + x)}{2x}$.

3. $\lim_{x \to 0} \dfrac{x(1 + x)}{2x^2}$.

4. $\lim_{x \to 4} \dfrac{x}{\sqrt{x + 1}}$.

5. $\lim_{x \to 4} \dfrac{x}{\sqrt{x + 1}}$.

6. $\lim_{x \to -1} \dfrac{1 - x}{x + 1}$.

7. $\lim_{x \to 1} \dfrac{x^4 - 1}{x - 1}$.

8. $\lim_{x \to 2} \dfrac{x}{|x|}$.

9. $\lim_{x \to 0} \dfrac{x}{|x|}$.

10. $\lim_{x \to 1} \dfrac{x^2 - 1}{x^2 - 2x + 1}$.

11. $\lim_{x \to -2} \dfrac{|x|}{x}$.

12. $\lim_{x \to 9} \dfrac{x - 3}{\sqrt{x} - 3}$.

13. $\lim_{x \to 2} f(x)$ if $f(x) = \begin{bmatrix} 3, & x \text{ an integer} \\ 1, & \text{otherwise} \end{bmatrix}$.   **14.** $\lim_{x \to 3} f(x)$ if $f(x) = \begin{bmatrix} x^2, & x < 3 \\ 7, & x = 3 \\ 2x + 3, & x > 3 \end{bmatrix}$.

15. $\lim_{x \to \sqrt{2}} f(x)$ if $f(x) = \begin{bmatrix} 3, & x \text{ an integer} \\ 1, & \text{otherwise} \end{bmatrix}$.   **16.** $\lim_{x \to 2} f(x)$ if $f(x) = \begin{bmatrix} x^2, & x \le 1 \\ 5x, & x > 1 \end{bmatrix}$.

17. Which of the δ's displayed in Figure 2.2.13 "works" for the given ϵ?

Figure 2.2.13

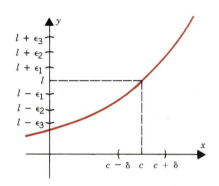

Figure 2.2.14

18. For which of the ϵ's given in Figure 2.2.14 does the specified δ work?

For the following limits, find the largest δ that "works" for a given arbitrary ϵ.

19. $\lim_{x \to 1} 2x = 2$.   **20.** $\lim_{x \to 4} 5x = 20$.   **21.** $\lim_{x \to 2} \frac{1}{2}x = 1$.   **22.** $\lim_{x \to 2} \frac{1}{5}x = \frac{2}{5}$.

Give an ϵ, δ proof for the following limits.

23. $\lim_{x \to 4} (2x - 5) = 3$.   **24.** $\lim_{x \to 2} (3x - 1) = 5$.   **25.** $\lim_{x \to 3} (6x - 7) = 11$.

26. $\lim_{x \to 0} (2 - 5x) = 2$.   **27.** $\lim_{x \to 2} |1 - 3x| = 5$.   **28.** $\lim_{x \to 2} |x - 2| = 0$.

29. Let f be some function of which you know only that

$$\text{if} \quad 0 < |x - 3| < 1, \quad \text{then} \quad |f(x) - 5| < 0.1.$$

Which of the following statements are necessarily true?

(a) If $|x - 3| < 1$, then $|f(x) - 5| < 0.1$.

(b) If $|x - 2.5| < 0.3$, then $|f(x) - 5| < 0.1$.

(c) $\lim_{x \to 3} f(x) = 5$.

(d) If $0 < |x - 3| < 2$, then $|f(x) - 5| < 0.1$.

(e) If $0 < |x - 3| < 0.5$, then $|f(x) - 5| < 0.1$.

(f) If $0 < |x - 3| < \frac{1}{4}$, then $|f(x) - 5| < \frac{1}{4}(0.1)$.

(g) If $0 < |x - 3| < 1$, then $|f(x) - 5| < 0.2$.

(h) If $0 < |x - 3| < 1$, then $|f(x) - 4.95| < 0.05$.

(i) If $\lim_{x \to 3} f(x) = l$, then $4.9 \le l \le 5.1$.

30. Suppose that $|A - B| < \epsilon$ for each $\epsilon > 0$. Prove that $A = B$.
HINT: Consider what happens if $A \neq B$ and $\epsilon = \frac{1}{2}|A - B|$.

Give the four limit statements displayed in 2.2.5, taking

31. $f(x) = \dfrac{1}{x - 1}$, $c = 3$.

32. $f(x) = \dfrac{x}{x^2 + 2}$, $c = 1$.

33. Prove that

(2.2.9)
$$\lim_{x \to c} f(x) = 0 \qquad \text{iff} \qquad \lim_{x \to c} |f(x)| = 0.$$

34. (a) Prove that
$$\text{if} \quad \lim_{x \to c} f(x) = l \qquad \text{then} \quad \lim_{x \to c} |f(x)| = |l|.$$

(b) Show that the converse is false. Give an example where
$$\lim_{x \to c} |f(x)| = |l| \qquad \text{and} \qquad \lim_{x \to c} f(x) = m \neq l$$

and then give an example where
$$\lim_{x \to c} |f(x)| \quad \text{exists} \qquad \text{but} \qquad \lim_{x \to c} f(x) \quad \text{does not exist.}$$

35. Give an ϵ, δ proof that statement (i) of 2.2.5 is equivalent to (iv).

36. Give an ϵ, δ proof of (2.2.8).

37. (a) Show that $\lim_{x \to c} \sqrt{x} = \sqrt{c}$ for $c > 0$.   HINT: If x and c are positive, then
$$0 \leq |\sqrt{x} - \sqrt{c}| = \frac{|x - c|}{\sqrt{x} + \sqrt{c}} < \frac{1}{\sqrt{c}}|x - c|.$$

(b) Show that $\lim_{x \to 0^+} \sqrt{x} = 0$.

Give an ϵ, δ proof for the following limits.

38. $\lim_{x \to 2} x^2 = 4$.   **39.** $\lim_{x \to 1} x^3 = 1$.   **40.** $\lim_{x \to 3} \sqrt{x + 1} = 2$.   **41.** $\lim_{x \to 3^-} \sqrt{3 - x} = 0$.

Prove the following statements.

42. If $g(x) = \begin{cases} x, & x \text{ rational} \\ 0, & x \text{ irrational} \end{cases}$,   then $\lim_{x \to 0} g(x) = 0$.

43. If $f(x) = \begin{cases} 1, & x \text{ rational} \\ 0, & x \text{ irrational} \end{cases}$,   then for no number c does $\lim_{x \to c} f(x)$ exist.

44. $\lim_{x \to c^-} f(x) = l$ iff $\lim_{h \to 0} f(c - |h|) = l$.   **45.** $\lim_{x \to c^+} f(x) = l$ iff $\lim_{h \to 0} f(c + |h|) = l$.

2.3   SOME LIMIT THEOREMS

As you saw in the last section, it can be rather tedious to apply the ϵ, δ limit test to individual functions. By proving some general theorems about limits we can avoid some of this repetitive work. The theorems themselves, of course (at least the first ones), will have to be proved by ϵ, δ methods.

We begin by showing that, if a limit exists, it is unique.

THEOREM 2.3.1   THE UNIQUENESS OF A LIMIT

If $\lim_{x \to c} f(x) = l$ and $\lim_{x \to c} f(x) = m$,   then $l = m$.

Proof. We will show that $l = m$ by proving that the assumption that $l \neq m$ leads to the contradiction

$$|l - m| < |l - m|.$$

Let's assume that $l \neq m$. Then $|l - m|/2 > 0$. Since

$$\lim_{x \to c} f(x) = l,$$

we know that there exists $\delta_1 > 0$ such that

$$(1) \qquad \text{if} \quad 0 < |x - c| < \delta_1, \quad \text{then} \quad |f(x) - l| < \frac{|l - m|}{2}.$$

(Here we are using $\dfrac{|l - m|}{2}$ as ϵ.) Since

$$\lim_{x \to c} f(x) = m,$$

we know that there exists $\delta_2 > 0$ such that

$$(2) \qquad \text{if} \quad 0 < |x - c| < \delta_2, \quad \text{then} \quad |f(x) - m| < \frac{|l - m|}{2}.$$

Now take x_1 as a number that satisfies the inequality

$$0 < |x_1 - c| < \text{minimum of } \delta_1 \text{ and } \delta_2.$$

Then by (1) and (2)

$$|f(x_1) - l| < \frac{|l - m|}{2} \qquad \text{and} \qquad |f(x_1) - m| < \frac{|l - m|}{2}.$$

It follows that

$$|l - m| = |[l - f(x_1)] + [f(x_1) - m]|$$

$$\leq |l - f(x_1)| + |f(x_1) - m|$$

by the triangle —↑
inequality

$$= |f(x_1) - l| + |f(x_1) - m| < \frac{|l - m|}{2} + \frac{|l - m|}{2} = |l - m|. \quad \square$$

$|a| = |-a|$ —↑

THEOREM 2.3.2

If $\lim_{x \to c} f(x) = l$ and $\lim_{x \to c} g(x) = m$,   then

(i) $\lim_{x \to c} [f(x) + g(x)] = l + m$,

(ii) $\lim_{x \to c} [\alpha f(x)] = \alpha l$ for each real α,

(iii) $\lim_{x \to c} [f(x)g(x)] = lm$.

Proof. Let $\epsilon > 0$. To prove (i) we must show that there exists $\delta > 0$ such that

$$\text{if} \quad 0 < |x - c| < \delta, \qquad \text{then} \quad |[f(x) + g(x)] - [l + m]| < \epsilon.$$

Note that

(*)
$$|[f(x) + g(x)] - [l + m]| = |[f(x) - l] + [g(x) - m]|$$
$$\leq |f(x) - l| + |g(x) - m|.$$

We will make $|[f(x) + g(x)] - [l + m]|$ less than ϵ by making $|f(x) - l|$ and $|g(x) - m|$ each less than $\frac{1}{2}\epsilon$. Since $\epsilon > 0$, we know that $\frac{1}{2}\epsilon > 0$. Since

$$\lim_{x \to c} f(x) = l \quad \text{and} \quad \lim_{x \to c} g(x) = m,$$

we know that there exist positive numbers δ_1 and δ_2 such that

$$\text{if} \quad 0 < |x - c| < \delta_1, \qquad \text{then} \quad |f(x) - l| < \tfrac{1}{2}\epsilon$$

and

$$\text{if} \quad 0 < |x - c| < \delta_2, \qquad \text{then} \quad |g(x) - m| < \tfrac{1}{2}\epsilon.$$

Now we set $\delta = $ minimum of δ_1 and δ_2 and note that, if $0 < |x - c| < \delta$, then

$$|f(x) - l| < \tfrac{1}{2}\epsilon \quad \text{and} \quad |g(x) - m| < \tfrac{1}{2}\epsilon$$

and thus by (*)

$$|[f(x) + g(x)] - [l + m]| < \epsilon.$$

In summary, by setting $\delta = \min \{\delta_1, \delta_2\}$, we found that,

$$\text{if} \quad 0 < |x - c| < \delta, \qquad \text{then} \quad |[f(x) + g(x)] - [l + m]| < \epsilon.$$

Thus (i) is proved. For proofs of (ii) and (iii), see the supplement at the end of this section. ☐

The results of Theorem 2.3.2 are easily extended (by mathematical induction) to any finite collection of functions; namely, if

$$\lim_{x \to c} f_1(x) = l_1, \quad \lim_{x \to c} f_2(x) = l_2, \ldots, \quad \lim_{x \to c} f_n(x) = l_n$$

then

(2.3.3) $\lim\limits_{x \to c} [\alpha_1 f_1(x) + \alpha_2 f_2(x) + \cdots + \alpha_n f_n(x)] = \alpha_1 l_1 + \alpha_2 l_2 + \cdots + \alpha_n l_n$

and

(2.3.4) $\lim\limits_{x \to c} [f_1(x) f_2(x) \cdots f_n(x)] = l_1 l_2 \cdots l_n.$

It follows that for every polynomial $P(x) = a_n x^n + \cdots + a_1 x + a_0$

(2.3.5) $$\boxed{\lim_{x \to c} P(x) = P(c).}$$

Proof.   We already know that

$$\lim_{x \to c} x = c.$$

Applying (2.3.4) to $f(x) = x$ multiplied k times by itself, we have

$$\lim_{x \to c} x^k = c^k \qquad \text{for each positive integer } k.$$

We also know that $\lim\limits_{x \to c} a_0 = a_0$. If follows now from (2.3.3) that

$$\lim_{x \to c} (a_n x^n + \cdots + a_1 x + a_0) = a_n c^n + \cdots + a_1 c + a_0;$$

that is,

$$\lim_{x \to c} P(x) = P(c). \quad \square$$

Examples

$$\lim_{x \to 1} (5x^2 - 12x + 2) = 5(1)^2 - 12(1) + 2 = -5,$$

$$\lim_{x \to 0} (14x^5 - 7x^2 + 2x + 8) = 14(0)^5 - 7(0)^2 + 2(0) + 8 = 8,$$

$$\lim_{x \to -1} (2x^3 + x^2 - 2x - 3) = 2(-1)^3 + (-1)^2 - 2(-1) - 3 = -2. \quad \square$$

We come now to reciprocals and quotients.

THEOREM 2.3.6

If $\lim\limits_{x \to c} g(x) = m$ with $m \neq 0$,   then $\lim\limits_{x \to c} \dfrac{1}{g(x)} = \dfrac{1}{m}.$

Proof.   See the supplement at the end of this section. $\square$

Examples

$$\lim_{x \to 4} \frac{1}{x^2} = \frac{1}{16}, \qquad \lim_{x \to 2} \frac{1}{x^3 - 1} = \frac{1}{7}, \qquad \lim_{x \to -3} \frac{1}{|x|} = \frac{1}{|-3|} = \frac{1}{3}. \quad \square$$

Once you know that reciprocals present no trouble, quotients become easy to handle.

THEOREM 2.3.7

If $\lim\limits_{x \to c} f(x) = l$ and $\lim\limits_{x \to c} g(x) = m$ with $m \neq 0$, then $\lim\limits_{x \to c} \dfrac{f(x)}{g(x)} = \dfrac{l}{m}$.

Proof. The key here is to observe that the quotient can be written as a product:

$$\frac{f(x)}{g(x)} = f(x) \, \frac{1}{g(x)}.$$

With

$$\lim_{x \to c} f(x) = l \quad \text{and} \quad \lim_{x \to c} \frac{1}{g(x)} = \frac{1}{m},$$

the product rule [part (iii) of Theorem 2.3.2] gives

$$\lim_{x \to c} \frac{f(x)}{g(x)} = l \, \frac{1}{m} = \frac{l}{m}. \quad \square$$

As an immediate consequence of this theorem on quotients you can see that, if P and Q are polynomials and $Q(c) \neq 0$, then

(2.3.8)
$$\lim_{x \to c} \frac{P(x)}{Q(x)} = \frac{P(c)}{Q(c)}.$$

Examples

$$\lim_{x \to 2} \frac{3x - 5}{x^2 + 1} = \frac{6 - 5}{4 + 1} = \frac{1}{5}, \qquad \lim_{x \to 3} \frac{x^3 - 3x^2}{1 - x^2} = \frac{27 - 27}{1 - 9} = 0. \quad \square$$

There is no point looking for a limit that does not exist. The next theorem gives a condition under which a quotient does not have a limit.

THEOREM 2.3.9

If $\lim_{x \to c} f(x) = l$ with $l \neq 0$ and $\lim_{x \to c} g(x) = 0$, then $\lim_{x \to c} \dfrac{f(x)}{g(x)}$ does not exist.

Proof. Suppose, on the contrary, that there exists a real number L such that

$$\lim_{x \to c} \frac{f(x)}{g(x)} = L.$$

Then

$$l = \lim_{x \to c} f(x) = \lim_{x \to c} \left[g(x) \cdot \frac{f(x)}{g(x)} \right] = \lim_{x \to c} g(x) \cdot \lim_{x \to c} \frac{f(x)}{g(x)} = 0 \cdot L = 0,$$

which contradicts our assumption that $l \neq 0$. ☐

Examples. From Theorem 2.3.9 you can see that

$$\lim_{x \to 1} \frac{x^2}{x - 1}, \qquad \lim_{x \to 2} \frac{3x - 7}{x^2 - 4}, \qquad \text{and} \qquad \lim_{x \to 0} \frac{5}{x}$$

all fail to exist. ☐

Problem 1. Evaluate the limits that exist:

(a) $\lim_{x \to 3} \dfrac{x^2 - x - 6}{x - 3}$,   (b) $\lim_{x \to 4} \dfrac{(x^2 - 3x - 4)^2}{x - 4}$,   (c) $\lim_{x \to -1} \dfrac{x + 1}{(2x^2 + 7x + 5)^2}$.

Solution. In each case both numerator and denominator tend to zero, and so we have to be careful.

(a) First we factor the numerator:

$$\frac{x^2 - x - 6}{x - 3} = \frac{(x + 2)(x - 3)}{x - 3}.$$

For $x \neq 3$,

$$\frac{x^2 - x - 6}{x - 3} = x + 2.$$

Thus

$$\lim_{x \to 3} \frac{x^2 - x - 6}{x - 3} = \lim_{x \to 3} (x + 2) = 5.$$

(b) Note that

$$\frac{(x^2 - 3x - 4)^2}{x - 4} = \frac{[(x + 1)(x - 4)]^2}{x - 4} = \frac{(x + 1)^2 (x - 4)^2}{x - 4}$$

so that, for $x \neq 4$,

$$\frac{(x^2 - 3x - 4)^2}{x - 4} = (x + 1)^2 (x - 4).$$

It follows then that

$$\lim_{x \to 4} \frac{(x^2 - 3x - 4)^2}{x - 4} = \lim_{x \to 4} (x + 1)^2(x - 4) = 0.$$

(c) Since

$$\frac{x + 1}{(2x^2 + 7x + 5)^2} = \frac{x + 1}{[(2x + 5)(x + 1)]^2} = \frac{x + 1}{(2x + 5)^2(x + 1)^2},$$

you can see that, for $x \neq -1$,

$$\frac{x + 1}{(2x^2 + 7x + 5)^2} = \frac{1}{(2x + 5)^2(x + 1)}.$$

By Theorem 2.3.9,

$$\lim_{x \to -1} \frac{1}{(2x + 5)^2(x + 1)} \quad \text{does not exist,}$$

and therefore

$$\lim_{x \to -1} \frac{x + 1}{(2x^2 + 7x + 5)^2} \quad \text{does not exist either.} \quad \square$$

Remark. In this section we have phrased everything in terms of two-sided limits. Although we won't stop to prove it, *all these results carry over to one-sided limits.* $\square$

EXERCISES 2.3

1. Given that

$$\lim_{x \to c} f(x) = 2, \qquad \lim_{x \to c} g(x) = -1, \qquad \lim_{x \to c} h(x) = 0,$$

evaluate the limits that exist.

(a) $\lim_{x \to c} [f(x) - g(x)]$.

(b) $\lim_{x \to c} [f(x)]^2$.

(c) $\lim_{x \to c} \dfrac{f(x)}{g(x)}$.

(d) $\lim_{x \to c} \dfrac{h(x)}{f(x)}$.

(e) $\lim_{x \to c} \dfrac{f(x)}{h(x)}$.

(f) $\lim_{x \to c} \dfrac{1}{f(x) - g(x)}$.

2. Given that

$$\lim_{x \to c} f(x) = 3, \qquad \lim_{x \to c} g(x) = 0, \qquad \lim_{x \to c} h(x) = -2,$$

evaluate the limits that exist.

(a) $\lim_{x \to c} [3f(x) + 2h(x)]$.

(b) $\lim_{x \to c} [h(x)]^3$.

(c) $\lim_{x \to c} \dfrac{h(x)}{x - c}$.

(d) $\lim_{x \to c} \dfrac{g(x)}{h(x)}$.

(e) $\lim_{x \to c} \dfrac{4}{f(x) - h(x)}$.

(f) $\lim_{x \to c} [3 + g(x)]^2$.

3. When asked to evaluate

$$\lim_{x \to 4} \left(\frac{1}{x} - \frac{1}{4} \right)\left(\frac{1}{x - 4} \right),$$

I reply that the limit is zero since $\lim\limits_{x \to 4} [1/x - \frac{1}{4}] = 0$ and cite Theorem 2.3.2 (limit of a product) as justification.   Verify that the limit is actually $-\frac{1}{16}$ and identify my error.

4. When asked to evaluate

$$\lim_{x \to 3} \frac{x^2 + x - 12}{x - 3},$$

I reply that the limit does not exist since $\lim\limits_{x \to 3} (x - 3) = 0$ and cite Theorem 2.3.9 (limit of a quotient) as justification.   Verify that the limit is actually 7 and identify my error.

Evaluate the limits that exist.

5. $\lim\limits_{x \to 2} 3$.

6. $\lim\limits_{x \to 3} (5 - 4x)^2$.

7. $\lim\limits_{x \to -4} (x^2 + 3x - 7)$.

8. $\lim\limits_{x \to -2} 3|x - 1|$.

9. $\lim\limits_{x \to \sqrt{3}} |x^2 - 8|$.

10. $\lim\limits_{x \to 3} 2$.

11. $\lim\limits_{x \to -1} \dfrac{x^2 + 1}{3x^5 + 4}$.

12. $\lim\limits_{x \to 2} \dfrac{3x}{x + 4}$.

13. $\lim\limits_{x \to 0} \left(x - \dfrac{4}{x} \right)$.

14. $\lim\limits_{x \to 2} \dfrac{x^2 + x + 1}{x^2 + 2x}$.

15. $\lim\limits_{x \to 5} \dfrac{2 - x^2}{4x}$.

16. $\lim\limits_{x \to 0} \dfrac{x^2 + 1}{x}$.

17. $\lim\limits_{x \to 0} \dfrac{x^2}{x^2 + 1}$.

18. $\lim\limits_{x \to 2} \dfrac{x}{x^2 - 4}$.

19. $\lim\limits_{h \to 0} h \left(1 - \dfrac{1}{h} \right)$.

20. $\lim\limits_{h \to 0} h \left(1 + \dfrac{1}{h} \right)$.

21. $\lim\limits_{x \to 2} \dfrac{x - 2}{x^2 - 4}$.

22. $\lim\limits_{x \to 2} \dfrac{x^2 - 4}{x - 2}$.

23. $\lim\limits_{x \to -2} \dfrac{(x^2 - x - 6)^2}{x + 2}$.

24. $\lim\limits_{x \to 1} \dfrac{x - x^2}{1 - x}$.

25. $\lim\limits_{x \to 1} \dfrac{x - x^2}{(3x - 1)(x^4 - 2)}$.

26. $\lim\limits_{x \to 1} \dfrac{x^2 - x - 6}{(x + 2)^2}$.

27. $\lim\limits_{x \to -2} \dfrac{x^2 - x - 6}{(x + 2)^2}$.

28. $\lim\limits_{x \to 0} \dfrac{x^4}{3x^3 + 2x^2 + x}$.

29. $\lim\limits_{h \to 0} \dfrac{1 - 1/h^2}{1 - 1/h}$.

30. $\lim\limits_{h \to 0} \dfrac{1 - 1/h^2}{1 + 1/h^2}$.

31. $\lim\limits_{h \to 0} \dfrac{1 - 1/h}{1 + 1/h}$.

32. $\lim\limits_{h \to 0} \dfrac{1 + 1/h}{1 + 1/h^2}$.

33. $\lim\limits_{t \to -1} \dfrac{t^2 + 6t + 5}{t^2 + 3t + 2}$.

34. $\lim\limits_{x \to -3} \dfrac{5x + 15}{x^3 + 3x^2 - x - 3}$.

35. $\lim\limits_{t \to 0} \dfrac{t + a/t}{t + b/t}$.

36. $\lim\limits_{x \to 1} \dfrac{x^2 - 1}{x^3 - 1}$.

37. $\lim\limits_{x \to 1} \dfrac{x^5 - 1}{x^4 - 1}$.

38. $\lim\limits_{t \to 0} \dfrac{t + 1/t}{t - 1/t}$.

39. $\lim\limits_{h \to 0} h \left(1 + \dfrac{1}{h^2} \right)$.

40. $\lim\limits_{h \to 0} h^2 \left(1 + \dfrac{1}{h} \right)$.

41. $\lim\limits_{x \to -2} \dfrac{(x^2 - x - 6)^2}{(x + 2)^2}$.

42. $\lim\limits_{x \to -2} \dfrac{x^2 - x - 6}{(x + 2)^2}$.

43. $\lim\limits_{t \to 1} \dfrac{t^2 - 2t + 1}{t^3 - 3t^2 + 3t - 1}$.

44. $\lim\limits_{x \to -4} \left(\dfrac{3x}{x + 4} + \dfrac{8}{x + 4} \right)$.

45. $\lim\limits_{x \to -4} \left(\dfrac{2x}{x + 4} + \dfrac{8}{x + 4} \right)$.

46. $\lim\limits_{x \to -4} \left(\dfrac{2x}{x + 4} - \dfrac{8}{x + 4} \right)$.

47. Evaluate the limits that exist.

(a) $\lim\limits_{x \to 4} \left(\dfrac{1}{x} - \dfrac{1}{4} \right)$.

(b) $\lim\limits_{x \to 4} \left[\left(\dfrac{1}{x} - \dfrac{1}{4} \right) \left(\dfrac{1}{x - 4} \right) \right]$.

(c) $\lim\limits_{x \to 4} \left[\left(\dfrac{1}{x} - \dfrac{1}{4} \right) (x - 2) \right]$.

(d) $\lim\limits_{x \to 4} \left[\left(\dfrac{1}{x} - \dfrac{1}{4} \right) \left(\dfrac{1}{x - 4} \right)^2 \right]$.

48. Evaluate the limits that exist.

(a) $\lim_{x \to 3} \dfrac{x^2 + x + 12}{x - 3}$.

(b) $\lim_{x \to 3} \dfrac{x^2 + x - 12}{x - 3}$.

(c) $\lim_{x \to 3} \dfrac{(x^2 + x - 12)^2}{x - 3}$.

(d) $\lim_{x \to 3} \dfrac{x^2 + x - 12}{(x - 3)^2}$.

49. Given that $f(x) = x^2 - 4x$, evaluate the limits that exist.

(a) $\lim_{x \to 4} \dfrac{f(x) - f(4)}{x - 4}$.

(b) $\lim_{x \to 1} \dfrac{f(x) - f(1)}{x - 1}$.

(c) $\lim_{x \to 3} \dfrac{f(x) - f(1)}{x - 3}$.

(d) $\lim_{x \to 3} \dfrac{f(x) - f(2)}{x - 3}$.

50. Given that $f(x) = x^3$, evaluate the limits that exist.

(a) $\lim_{x \to 3} \dfrac{f(x) - f(3)}{x - 3}$.

(b) $\lim_{x \to 3} \dfrac{f(x) - f(2)}{x - 3}$.

(c) $\lim_{x \to 3} \dfrac{f(x) - f(3)}{x - 2}$.

(d) $\lim_{x \to 1} \dfrac{f(x) - f(1)}{x - 1}$.

51. Show by example that $\lim_{x \to c} [f(x) + g(x)]$ can exist even if $\lim_{x \to c} f(x)$ and $\lim_{x \to c} g(x)$ do not exist.

52. Show by example that $\lim_{x \to c} [f(x)g(x)]$ can exist even if $\lim_{x \to c} f(x)$ and $\lim_{x \to c} g(x)$ do not exist.

Exercises 53–59: true or false?

53. If $\lim_{x \to c} [f(x) + g(x)]$ exists but $\lim_{x \to c} f(x)$ does not exist, then $\lim_{x \to c} g(x)$ does not exist.

54. If $\lim_{x \to c} [f(x) + g(x)]$ and $\lim_{x \to c} f(x)$ exist, then it can happen that $\lim_{x \to c} g(x)$ does not exist.

55. If $\lim_{x \to c} \sqrt{f(x)}$ exists, then $\lim_{x \to c} f(x)$ exists.

56. If $\lim_{x \to c} f(x)$ exists, then $\lim_{x \to c} \sqrt{f(x)}$ exists.

57. If $\lim_{x \to c} f(x)$ exists, then $\lim_{x \to c} \dfrac{1}{f(x)}$ exists.

58. If $f(x) \le g(x)$ for all $x \neq c$, then $\lim_{x \to c} f(x) \le \lim_{x \to c} g(x)$.

59. If $f(x) < g(x)$ for all $x \neq c$, then $\lim_{x \to c} f(x) < \lim_{x \to c} g(x)$.

60. (a) Verify that $\max \{f(x), g(x)\} = \frac{1}{2}\{[f(x) + g(x)] + |f(x) - g(x)|\}$.
(b) Find a similar expression for $\min \{f(x), g(x)\}$.

61. Let $h(x) = \min \{f(x), g(x)\}$ and $H(x) = \max \{f(x), g(x)\}$. Show that

$$\text{if } \lim_{x \to c} f(x) = l \quad \text{and} \quad \lim_{x \to c} g(x) = l, \quad \text{then} \quad \lim_{x \to c} h(x) = l \quad \text{and} \quad \lim_{x \to c} H(x) = l.$$

HINT: Use Exercise 60.

62. Suppose that $\lim_{x \to c} f(x) = l$.

(a) Show that, if l is positive, then $f(x)$ is positive for all $x \neq c$ in an interval of the form $(c - \delta, c + \delta)$.
(b) Show that, if l is negative, then $f(x)$ is negative for all $x \neq c$ in an interval of the form $(c - \delta, c + \delta)$.
HINT: Use an ϵ, δ argument setting $\epsilon = l$.

63. Given that $\lim_{x \to c} g(x) = 0$ and $f(x)g(x) = 1$ for all real x, prove that $\lim_{x \to c} f(x)$ does not exist.

HINT: Suppose that $\lim_{x \to c} f(x)$ exists (say it is equal to l) and deduce a contradiction.

64. (*The stability of limit*) Start with a function f. Now change the value of f at a billion points or, if you prefer, at ten billion points. Call the new function g.
   (a) Show that, if $\lim_{x \to c} f(x) = l$, then $\lim_{x \to c} g(x) = l$.
   (b) Show that, if $\lim_{x \to c} f(x)$ does not exist, then $\lim_{x \to c} g(x)$ does not exist.

*SUPPLEMENT TO SECTION 2.3

Proof of Theorem 2.3.2 (ii). We consider two cases: $\alpha \neq 0$ and $\alpha = 0$. If $\alpha \neq 0$, then $\epsilon/|\alpha| > 0$ and, since

$$\lim_{x \to c} f(x) = l,$$

we know that there exists $\delta > 0$ such that,

$$\text{if}\quad 0 < |x - c| < \delta, \qquad \text{then}\quad |f(x) - l| < \frac{\epsilon}{|\alpha|}.$$

From the last inequality we obtain

$$|\alpha||f(x) - l| < \epsilon \quad \text{and thus}\quad |\alpha f(x) - \alpha l| < \epsilon.$$

The case $\alpha = 0$ was treated before in (2.2.4). $\square$

Proof of Theorem 2.3.2 (iii). We begin with a little algebra:

$$\begin{aligned}
|f(x)g(x) - lm| &= |[f(x)g(x) - f(x)m] + [f(x)m - lm]| \\
&\leq |f(x)g(x) - f(x)m| + |f(x)m - lm| \\
&= |f(x)||g(x) - m| + |m||f(x) - l| \\
&\leq |f(x)||g(x) - m| + (1 + |m|)|f(x) - l|.
\end{aligned}$$

Now let $\epsilon > 0$. Since $\lim_{x \to c} f(x) = l$ and $\lim_{x \to c} g(x) = m$, we know

(1) that there exists $\delta_1 > 0$ such that, if $0 < |x - c| < \delta_1$, then

$$|f(x) - l| < 1 \quad \text{and thus}\quad |f(x)| < 1 + |l|;$$

(2) that there exists $\delta_2 > 0$ such that

$$\text{if}\quad 0 < |x - c| < \delta_2, \qquad \text{then}\quad |g(x) - m| < \left(\frac{\frac{1}{2}\epsilon}{1 + |l|}\right);$$

(3) that there exists $\delta_3 > 0$ such that

$$\text{if}\quad 0 < |x - c| < \delta_3, \qquad \text{then}\quad |f(x) - l| < \left(\frac{\frac{1}{2}\epsilon}{1 + |m|}\right).$$

We now set $\delta = \min \{\delta_1, \delta_2, \delta_3\}$ and observe that, if $0 < |x - c| < \delta$, then

$$|f(x)g(x) - lm| < |f(x)||g(x) - m| + (1 + |m|)|f(x) - l|$$

$$< (1 + |l|)\left(\frac{\frac{1}{2}\epsilon}{1 + |l|}\right) + (1 + |m|)\left(\frac{\frac{1}{2}\epsilon}{1 + |m|}\right) = \epsilon. \quad \square$$

by (1) —   by (2)   by (3)

Proof of Theorem 2.3.6. For $g(x) \neq 0$,

$$\left| \frac{1}{g(x)} - \frac{1}{m} \right| = \frac{|g(x) - m|}{|g(x)||m|}.$$

Choose $\delta_1 > 0$ such that

$$\text{if } \quad 0 < |x - c| < \delta_1, \qquad \text{then} \quad |g(x) - m| < \frac{|m|}{2}.$$

For such x,

$$|g(x)| > \frac{|m|}{2} \quad \text{so that} \quad \frac{1}{|g(x)|} < \frac{2}{|m|}$$

and thus

$$\left| \frac{1}{g(x)} - \frac{1}{m} \right| = \frac{|g(x) - m|}{|g(x)||m|} \leq \frac{2}{|m|^2} |g(x) - m|.$$

Now let $\epsilon > 0$ and choose $\delta_2 > 0$ such that

$$\text{if } \quad 0 < |x - c| < \delta_2, \qquad \text{then} \quad |g(x) - m| < \frac{|m|^2}{2} \epsilon.$$

Setting $\delta = \min \{\delta_1, \delta_2\}$, we find that

$$\text{if } \quad 0 < |x - c| < \delta, \qquad \text{then} \quad \left| \frac{1}{g(x)} - \frac{1}{m} \right| < \epsilon. \quad \square$$

2.4   CONTINUITY

In ordinary language, to say that a certain process is "continuous" is to say that it goes on without interruption and without abrupt changes. In mathematics the word "continuous" has much the same meaning.

The idea of continuity is so important to calculus that we will discuss it with some care. First we introduce the idea of continuity at a point (or number) c, and then we talk about continuity on an interval.

Continuity at a Point

DEFINITION 2.4.1

Let f be a function defined at least on an open interval $(c - p, c + p)$. We say that f is *continuous at* c iff

$$\lim_{x \to c} f(x) = f(c).$$

If the domain of f contains an interval $(c - p, c + p)$, then f can fail to be continuous at c only for one of two reasons: either $f(x)$ does not have a limit as x tends to c (in which case c is called an *essential discontinuity*), or it does have a limit, but the limit is not $f(c)$. In this case c is called a *removable discontinuity*; the discontinuity can be removed by redefining f at c. If the limit is l, define f at c to be l.

The function depicted in Figure 2.4.1 is discontinuous at c because it does not have a limit at c. The discontinuity is essential.

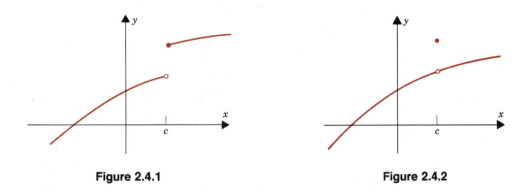

Figure 2.4.1   **Figure 2.4.2**

The function depicted in Figure 2.4.2 does have a limit at c. It is discontinuous at c only because its limit at c is not its value at c. The discontinuity is removable; it can be removed by lowering the dot into place (by redefining f at c).

Figure 2.4.3

In Figure 2.4.3, we have tried to suggest the Dirichlet function

$$f(x) = \begin{cases} 1, & x \text{ rational} \\ 0, & x \text{ irrational} \end{cases}.$$

At no point c does f have a limit. It is therefore everywhere discontinuous and every point is an essential discontinuity.

Most of the functions that you have encountered so far are continuous at each point of their domains. In particular, this is true for polynomials,

$$\lim_{x \to c} P(x) = P(c), \tag{2.3.5}$$

for rational functions (quotients of polynomials),

$$\lim_{x \to c} \frac{P(x)}{Q(x)} = \frac{P(c)}{Q(c)} \quad \text{if} \quad Q(c) \neq 0, \tag{2.3.8}$$

and for the absolute-value function,

$$\lim_{x \to c} |x| = |c|. \tag{2.2.2}$$

In one of the exercises you showed (or at least were asked to show) that

$$\lim_{x \to c} \sqrt{x} = \sqrt{c} \quad \text{for each } c > 0.$$

This makes the square-root function continuous at each positive number. What happens at 0, we discuss later.

With f and g continuous at c, we have

$$\lim_{x \to c} f(x) = f(c), \qquad \lim_{x \to c} g(x) = g(c)$$

and thus, by the limit theorems,

$$\lim_{x \to c} [f(x) + g(x)] = f(c) + g(c), \qquad \lim_{x \to c} [\alpha f(x)] = \alpha f(c) \quad \text{for each real } \alpha$$

$$\lim_{x \to c} [f(x)g(x)] = f(c)g(c), \qquad \text{and, if } g(c) \neq 0, \quad \lim_{x \to c} \frac{f(x)}{g(x)} = \frac{f(c)}{g(c)}.$$

In summary, we have the following theorem:

THEOREM 2.4.2

If f and g are continuous at c, then

    (i) $f + g$ is continuous at c,   (ii) αf is continuous at c for each real α,

(iii) fg is continuous at c,   (iv) f/g is continuous at c provided $g(c) \neq 0$.

Parts (i) and (iii) can be extended to any finite number of functions.

Example 1. The function

$$F(x) = 3|x| + \frac{x^3 - x}{x^2 - 5x + 6} + 4$$

is continuous at all real numbers other than 2 and 3. You can see this by noting that

$F = 3f + g/h + k$ where

$$f(x) = |x|, \quad g(x) = x^3 - x, \quad h(x) = x^2 - 5x + 6, \quad \text{and} \quad k(x) = 4.$$

Since f, g, h, and k are everywhere continuous, F is continuous except at 2 and 3, the numbers where h takes on the value 0. □

Our next topic is the continuity of composite functions. Before getting into this, however, let's take a look at continuity in terms of ϵ, δ. A direct translation of

$$\lim_{x \to c} f(x) = f(c)$$

into ϵ, δ terms reads like this: for each $\epsilon > 0$ there exists $\delta > 0$ such that

$$\text{if} \quad 0 < |x - c| < \delta \quad \text{then} \quad |f(x) - f(c)| < \epsilon.$$

Here the restriction $0 < |x - c|$ is unnecessary. We can allow $|x - c| = 0$ because then $x = c$, $f(x) = f(c)$, and thus $|f(x) - f(c)| = 0$. Being 0, $|f(x) - f(c)|$ is certainly less than ϵ.

In summary, an ϵ, δ characterization of continuity at c reads as follows:

(2.4.3) $\quad f$ is continuous at c iff $\begin{cases} \text{for each } \epsilon > 0 \text{ there exists } \delta > 0 \text{ such that} \\ \text{if} \quad |x - c| < \delta \quad \text{then} \quad |f(x) - f(c)| < \epsilon. \end{cases}$

In simple intuitive language

$\quad\quad f$ is continuous at c iff   for x close to c, $f(x)$ is close to $f(c)$.

We are now ready to take up the continuity of composites. Remember the defining formula: $(f \circ g)(x) = f(g(x))$.

THEOREM 2.4.4

If g is continuous at c and f is continuous at $g(c)$, then the composition $f \circ g$ is continuous at c.

The idea here is simple: with g continuous at c, we know that

$$\text{for } x \text{ close to } c, \quad g(x) \text{ is close to } g(c);$$

from the continuity of f at $g(c)$, we know that

$$\text{with } g(x) \text{ close to } g(c), \quad f(g(x)) \text{ is close to } f(g(c)).$$

In summary,

$$\text{with } x \text{ close to } c, \quad f(g(x)) \text{ is close to } f(g(c)). \quad \square$$

The argument we just gave is too vague to be a proof. Here in contrast is a proof. We begin with $\epsilon > 0$. We must show that there exists some number $\delta > 0$ such that

$$\text{if} \quad |x - c| < \delta, \quad \text{then} \quad |f(g(x)) - f(g(c))| < \epsilon.$$

In the first place, we observe that, since f is continuous at $g(c)$, there does exist a number $\delta_1 > 0$ such that

(1) $\qquad\qquad$ if $\quad |t - g(c)| < \delta_1, \qquad$ then $\quad |f(t) - f(g(c))| < \epsilon.$

With $\delta_1 > 0$, we know from the continuity of g at c that there exists a number $\delta > 0$ such that

(2) $\qquad\qquad$ if $\quad |x - c| < \delta, \qquad$ then $\quad |g(x) - g(c)| < \delta_1.$

Combining (2) and (1) we have what we want: by (2)

$$\text{if} \quad |x - c| < \delta, \qquad \text{then} \quad |g(x) - g(c)| < \delta_1$$

so that by (1)

$$|f(g(x)) - f(g(c))| < \epsilon. \quad \square$$

This proof is illustrated in Figure 2.4.4. The numbers within δ of c are taken by g to within δ_1 of $g(c)$ and then by f to within ϵ of $f(g(c))$.

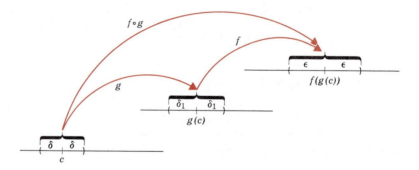

Figure 2.4.4

It is time to look at some examples.

Example 2. The function

$$F(x) = \sqrt{\frac{x^2 + 1}{(x - 8)^3}}$$

is continuous at all numbers greater than 8. To see this, note that $F = f \circ g$, where

$$g(x) = \frac{x^2 + 1}{(x - 8)^3} \quad \text{and} \quad f(x) = \sqrt{x}.$$

Now take $c > 8$. The function g is continuous at c, and since $g(c)$ is positive, f is continuous at $g(c)$. By the theorem F is continuous at c. $\quad \square$

The continuity of composites holds for any finite number of functions. The only requirement is that each function be continuous *where it is applied*.

Example 3.   The function

$$F(x) = \frac{1}{5 - \sqrt{x^2 + 16}}$$

is continuous everywhere except at $x = \pm 3$, where it is not defined.   To see this, note that $F = f \circ g \circ k \circ h$, where

$$f(x) = \frac{1}{x}, \quad g(x) = 5 - x, \quad k(x) = \sqrt{x}, \quad h(x) = x^2 + 16$$

and observe that each of these functions is being applied only where it is continuous. In particular, f is being applied only to nonzero numbers and k is being applied only to positive numbers.   □

Just as we considered one-sided limits, we can consider one-sided continuity.

DEFINITION 2.4.5   ONE-SIDED CONTINUITY

A function f is called

> *continuous from the left at c*   iff $\lim\limits_{x \to c^-} f(x) = f(c)$.

It is called

> *continuous from the right at c*   iff $\lim\limits_{x \to c^+} f(x) = f(c)$.

In Figure 2.4.5 we have an example of continuity from the right at 0; in Figure 2.4.6 we have an example of continuity from the left at 0.

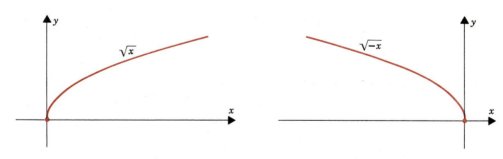

Figure 2.4.5   **Figure 2.4.6**

It is clear from (2.2.8) that a function f is continuous at c iff it is continuous there from both sides.   Thus

(2.4.6)

> f is continuous at c iff $f(c)$, $\lim\limits_{x \to c^-} f(x)$, $\lim\limits_{x \to c^+} f(x)$ all exist and are equal.

This is sometimes referred to as the *slot-machine test* for continuity.   We apply it below to a piecewise-defined function.

Problem 4. Determine the discontinuities of the following function:

$$f(x) = \begin{cases} x^3, & x \le -1 \\ x^2 - 2, & -1 < x < 0 \\ 3 - x, & 0 \le x < 2 \\ \dfrac{4x - 1}{x - 1}, & 2 \le x < 4 \\ \dfrac{15}{7 - x}, & 4 < x < 7 \\ 5x + 2, & 7 \le x \end{cases}.$$

Solution. It should be clear that f is continuous on the open intervals $(-\infty, -1)$, $(-1, 0)$, $(0, 2)$, $(2, 4)$, $(4, 7)$, and $(7, \infty)$. All we have to check is the behavior of f at $x = -1, 0, 2, 4$, and 7. To do so, we apply (2.4.6).

At $x = -1$, f is continuous since $f(-1) = (-1)^3 = -1$,

$$\lim_{x \to -1^-} f(x) = \lim_{x \to -1^-} x^3 = -1, \quad \text{and} \quad \lim_{x \to -1^+} f(x) = \lim_{x \to -1^+} (x^2 - 2) = -1.$$

Our findings at the other four points are displayed in the chart below. Try to verify each entry.

c	$f(c)$	$\lim\limits_{x \to c^-} f(x)$	$\lim\limits_{x \to c^+} f(x)$	conclusion
0	3	-2	3	discontinuous
2	7	1	7	discontinuous
4	not defined	5	5	discontinuous
7	37	does not exist	37	discontinuous

Note that the discontinuity at $x = 4$ is removable; if we define $f(4) = 5$, f is now continuous at $x = 4$. The discontinuities at $x = 0, 2$, and 7 are essential since the limit does not exist at these points. ☐

Continuity on [a, b]

From a function defined on a closed interval $[a, b]$ the most continuity that we can possibly expect is

1. continuity at each point c of the open interval (a, b),
2. continuity from the right at a, and
3. continuity from the left at b.

Any function f that fulfills these requirements is called *continuous on $[a, b]$*.

As an example take the function

$$f(x) = \sqrt{1 - x^2}.$$

Its graph is the semicircle displayed in Figure 2.4.7. The function is continuous on $[-1, 1]$ because it is continuous at each number c in $(-1, 1)$, continuous from the right at -1, and continuous from the left at 1.

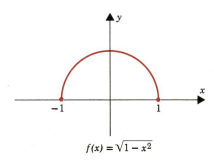

$$f(x) = \sqrt{1 - x^2}$$

Figure 2.4.7

Functions that are continuous on a closed interval $[a, b]$ have some important special properties not shared by functions in general. We will focus on two such properties in Section 2.6.

EXERCISES 2.4

Determine whether or not the function is continuous at the indicated point. If not, determine whether the discontinuity is essential or removable.

1. $f(x) = x^3 - 5x + 1$; $x = 2$.

2. $g(x) = \sqrt{(x-1)^2 + 5}$; $x = 1$.

3. $f(x) = \sqrt{x^2 + 9}$; $x = 3$.

4. $f(x) = |4 - x^2|$; $x = 2$.

5. $f(x) = \begin{cases} x^2 + 4, & x < 2 \\ x^3, & x \geq 2 \end{cases}$; $x = 2$.

6. $h(x) = \begin{cases} x^2 + 5, & x < 2 \\ x^3, & x \geq 2 \end{cases}$; $x = 2$.

7. $g(x) = \begin{cases} x^2 + 4, & x < 2 \\ 5, & x = 2 \\ x^3, & x > 2 \end{cases}$; $x = 2$.

8. $g(x) = \begin{cases} x^2 + 5, & x < 2 \\ 10, & x = 2 \\ 1 + x^3, & x > 2 \end{cases}$; $x = 2$.

9. $f(x) = \begin{cases} -1, & x < 0 \\ 0, & x = 0 \\ 1, & x > 0 \end{cases}$; $x = 0$.

10. $f(x) = \begin{cases} 1 - x, & x < 1 \\ 1, & x = 1 \\ x^2 - 1, & x > 1 \end{cases}$; $x = 1$.

11. $h(x) = \begin{cases} \dfrac{x^2 - 1}{x + 1}, & x \neq -1 \\ -2, & x = -1 \end{cases}$; $x = -1$.

12. $g(x) = \begin{cases} \dfrac{1}{x + 1}, & x \neq -1 \\ 0, & x = -1 \end{cases}$; $x = -1$.

13. $f(x) = \begin{cases} -x^2, & x < 0 \\ 1 - \sqrt{x}, & x \geq 0 \end{cases}$; $x = 0$.

14. $g(x) = \dfrac{x(x + 1)(x + 2)}{\sqrt{(x - 1)(x - 2)}}$; $x = -2$.

Sketch the graph and classify the discontinuities (if any).

15. $f(x) = \dfrac{x^2 - 4}{x - 2}$.

16. $f(x) = \dfrac{x - 3}{x^2 - 9}$.

17. $f(x) = |x - 1|$.

18. $h(x) = |x^2 - 1|$.

19. $g(x) = \begin{bmatrix} x - 1, & x < 1 \\ 0, & x = 1 \\ x^2, & x > 1 \end{bmatrix}$.

20. $g(x) = \begin{bmatrix} 2x - 1, & x < 1 \\ 0, & x = 1 \\ x^2, & x > 1 \end{bmatrix}$.

21. $f(x) = \max\{x, x^2\}$.

22. $f(x) = \min\{x, x^2\}$.

23. $f(x) = \begin{bmatrix} -1, & x < -1 \\ x^3, & -1 \le x \le 1 \\ 1, & 1 < x \end{bmatrix}$.

24. $g(x) = \begin{cases} 1, & x \le -2 \\ \frac{1}{2}x, & -2 < x < 4 \\ \sqrt{x}, & 4 \le x \end{cases}$.

25. $h(x) = \begin{bmatrix} 1, & x \le 0 \\ x^2, & 0 < x < 1 \\ 1, & 1 \le x < 2 \\ x, & 2 \le x \end{bmatrix}$.

26. $g(x) = \begin{cases} -x^2, & x < -1 \\ 3, & x = -1 \\ 2 - x, & -1 < x \le 1 \\ x^2, & 1 < x \end{cases}$.

27. $f(x) = \begin{cases} 2x + 9, & x < -2 \\ x^2 + 1, & -2 < x \le 1 \\ 3x - 1, & 1 < x < 3 \\ x + 6, & 3 < x \end{cases}$.

28. $g(x) = \begin{cases} x + 7, & x < -3 \\ |x - 2|, & -3 < x < -1 \\ x^2 - 2x, & -1 < x < 3 \\ 2x - 3, & 3 \le x \end{cases}$.

Each of the functions f below is defined everywhere except at $x = 1$. Where possible, define f at 1 so that it becomes continuous at 1.

29. $f(x) = \dfrac{x^2 - 1}{x - 1}$.   **30.** $f(x) = \dfrac{1}{x - 1}$.   **31.** $f(x) = \dfrac{x - 1}{|x - 1|}$.   **32.** $f(x) = \dfrac{(x - 1)^2}{|x - 1|}$.

33. Let $f(x) = \begin{cases} x^2, & x < 1 \\ Ax - 3, & x \ge 1 \end{cases}$. Find A given that f is continuous at 1.

34. Let $f(x) = \begin{cases} A^2x^2, & x \le 2 \\ (1 - A)x, & x > 2 \end{cases}$. For what values of A is f continuous at 2?

35. Give necessary and sufficient conditions on A and B for the function

$$f(x) = \begin{cases} Ax - B, & x \le 1 \\ 3x, & 1 < x < 2 \\ Bx^2 - A, & 2 \le x \end{cases}$$

to be continuous at $x = 1$ but discontinuous at $x = 2$.

36. Give necessary and sufficient conditions on A and B for the function in Exercise 35 to be continuous at $x = 2$ but discontinuous at $x = 1$.

Define a function that is everywhere continuous from

37. the left but discontinuous from the right at $x = \frac{1}{2}$.

38. the right but discontinuous from the left at $x = \frac{1}{2}$.

Define the function at 5 so that it becomes continuous at 5.

39. $f(x) = \dfrac{\sqrt{x + 4} - 3}{x - 5}$.

40. $f(x) = \dfrac{\sqrt{x + 4} - 3}{\sqrt{x} - 5}$.

41. $f(x) = \dfrac{\sqrt{2x - 1} - 3}{x - 5}$.

42. $f(x) = \dfrac{\sqrt{x^2 - 7x + 16} - \sqrt{6}}{(x - 5)\sqrt{x + 1}}$.

At what points (if any) are the following functions continuous?

43. $f(x) = \begin{cases} 1, & x \text{ rational} \\ 0, & x \text{ irrational} \end{cases}$.

44. $g(x) = \begin{cases} x, & x \text{ rational} \\ 0, & x \text{ irrational} \end{cases}$.

45. $g(x) = \begin{cases} 2x, & x \text{ an integer} \\ x^2, & \text{otherwise} \end{cases}$.

46. $f(x) = \begin{cases} 4, & x \text{ an integer} \\ x^2, & \text{otherwise} \end{cases}$.

47. (*Important*) Prove that

$$f \text{ is continuous at } c \quad \text{iff} \quad \lim_{h \to 0} f(c + h) = f(c).$$

48. (*Important*) Let f and g be continuous at c. Prove that if

(a) $f(c) > 0$, then there exists $\delta > 0$ such that $f(x) > 0$ for all $x \in (c - \delta, c + \delta)$.

(b) $f(c) < 0$, then there exists $\delta > 0$ such that $f(x) < 0$ for all $x \in (c - \delta, c + \delta)$.

(c) $f(c) < g(c)$, then there exists $\delta > 0$ such that $f(x) < g(x)$ for all $x \in (c - \delta, c + \delta)$.

49. (*Essential discontinuities are not easy to erase.*) Start with any function f that has an essential discontinuity at c. Now change the value of f at a billion points in any manner you want. Call the new function g. Show that g suffers from the same discontinuity.

2.5 THE PINCHING THEOREM; TRIGONOMETRIC LIMITS

Figure 2.5.1

Figure 2.5.1 shows the graphs of three functions f, g, h. Suppose that, as suggested by the figure, for x close to c, f is trapped between g and h. (The values of these functions at c itself are irrelevant.) If, as x tends to c, both $g(x)$ and $h(x)$ tend to the same limit l, then $f(x)$ also tends to l. This idea is made precise in what we call *the pinching theorem.*

THEOREM 2.5.1 THE PINCHING THEOREM

Let $p > 0$. Suppose that, for all x such that $0 < |x - c| < p$,

$$h(x) \leq f(x) \leq g(x).$$

If

$$\lim_{x \to c} h(x) = l \quad \text{and} \quad \lim_{x \to c} g(x) = l,$$

then

$$\lim_{x \to c} f(x) = l.$$

Proof. Let $\epsilon > 0$. Let $p > 0$ be such that

$$\text{if} \quad 0 < |x - c| < p, \qquad \text{then} \quad h(x) \le f(x) \le g(x).$$

Choose $\delta_1 > 0$ such that

$$\text{if} \quad 0 < |x - c| < \delta_1, \qquad \text{then} \quad l - \epsilon < h(x) < l + \epsilon.$$

Choose $\delta_2 > 0$ such that

$$\text{if} \quad 0 < |x - c| < \delta_2, \qquad \text{then} \quad l - \epsilon < g(x) < l + \epsilon.$$

Let $\delta = \min\{p, \delta_1, \delta_2\}$. For x satisfying $0 < |x - c| < \delta$, we have

$$l - \epsilon < h(x) \le f(x) \le g(x) < l + \epsilon$$

and thus

$$|f(x) - l| < \epsilon. \quad \square$$

Remark. Obvious modifications of the pinching theorem hold for one-sided limits. There is little reason to spell out the details. In any event we will be working here with two-sided limits. $\square$

We come now to some trigonometric limits. All calculations are based on radian measure.

As our first application of the pinching theorem we will prove that

(2.5.2)

$$\lim_{x \to 0} \sin x = 0.$$

Proof. To follow the argument, see Figure 2.5.2.†

Figure 2.5.2

For small $x \ne 0$

$$|\sin x| = \text{length of } \overline{BP} < \text{length of } \overline{AP} < \text{length of } \overparen{AP} = |x|.$$

---

† Recall that in a circle of radius r a central angle of θ radians subtends an arc of length $r\theta$. In Figure 2.5.2 we have $r = 1$.

For such x

$$0 < |\sin x| < |x|.$$

Since

$$\lim_{x \to 0} 0 = 0 \quad \text{and} \quad \lim_{x \to 0} |x| = 0,$$

we know from the pinching theorem that

$$\lim_{x \to 0} |\sin x| = 0 \quad \text{and therefore} \quad \lim_{x \to 0} \sin x = 0. \quad \square$$

From this it follows readily that

(2.5.3)

$$\lim_{x \to 0} \cos x = 1.$$

Proof. In general, $\cos^2 x + \sin^2 x = 1$. For x close to 0, the cosine is positive and we have

$$\cos x = \sqrt{1 - \sin^2 x}.$$

As x tends to 0, $\sin x$ tends to 0, $\sin^2 x$ tends to 0, and therefore $\cos x$ tends to 1. $\quad \square$

Next we show that the sine and cosine are everywhere continuous; which is to say, for all real numbers c,

(2.5.4)

$$\lim_{x \to c} \sin x = \sin c \quad \text{and} \quad \lim_{x \to c} \cos x = \cos c.$$

Proof. By (2.2.5) we can write

$$\lim_{x \to c} \sin x \quad \text{as} \quad \lim_{h \to 0} \sin (c + h).$$

This form of the limit suggests that we use the addition formula

$$\sin (c + h) = \sin c \cos h + \cos c \sin h.$$

Since $\sin c$ and $\cos c$ are constants, we have

$$\lim_{h \to 0} \sin (c + h) = (\sin c)(\lim_{h \to 0} \cos h) + (\cos c)(\lim_{h \to 0} \sin h)$$
$$= (\sin c)(1) + (\cos c)(0) = \sin c.$$

The proof that $\lim_{x \to c} \cos x = \cos c$ is left to you. $\quad \square$

The remaining trigonometric functions

$$\tan x = \frac{\sin x}{\cos x}, \qquad \cot x = \frac{\cos x}{\sin x}, \qquad \sec x = \frac{1}{\cos x}, \qquad \csc x = \frac{1}{\sin x}$$

are all continuous where defined. Justification? They are all quotients of continuous functions.

We turn now to two limits, the importance of which will become clear in Chapter 3:

(2.5.5)

$$\lim_{x \to 0} \frac{\sin x}{x} = 1 \quad \text{and} \quad \lim_{x \to 0} \frac{1 - \cos x}{x} = 0.$$

Proof.   We will show that

$$\lim_{x \to 0} \frac{\sin x}{x} = 1$$

by some simple geometry and the pinching theorem.

For small $x > 0$ (see Figure 2.5.3)

$$\text{area of triangle } OAP = \frac{1}{2} \sin x,$$

$$\text{area of sector} = \frac{1}{2} x,$$

$$\text{area of triangle } OAQ = \frac{1}{2} \tan x = \frac{1}{2} \frac{\sin x}{\cos x}.$$

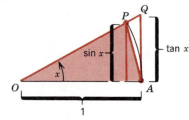

Figure 2.5.3

Since triangle $OAP \subseteq$ sector $\subseteq$ triangle OAQ (and these are all proper containments), we have

$$\frac{1}{2} \sin x < \frac{1}{2} x < \frac{1}{2} \frac{\sin x}{\cos x},$$

$$1 < \frac{x}{\sin x} < \frac{1}{\cos x},$$

$$\cos x < \frac{\sin x}{x} < 1. \qquad \text{(took reciprocals)}$$

This inequality was derived for small $x > 0$, but, since

$$\cos (-x) = \cos x \quad \text{and} \quad \frac{\sin (-x)}{-x} = \frac{-\sin x}{-x} = \frac{\sin x}{x},$$

this inequality also holds for small $x < 0$.

We can now apply the pinching theorem.   Since

$$\lim_{x \to 0} \cos x = 1 \quad \text{and} \quad \lim_{x \to 0} 1 = 1,$$

we can conclude that

$$\lim_{x \to 0} \frac{\sin x}{x} = 1.$$

We can now show that

$$\lim_{x \to 0} \frac{1 - \cos x}{x} = 0.$$

For small $x \neq 0$, $\cos x \neq 1$. Therefore we can write

$$\frac{1 - \cos x}{x} = \left(\frac{1 - \cos x}{x}\right)\left(\frac{1 + \cos x}{1 + \cos x}\right)†$$

$$= \frac{1 - \cos^2 x}{x(1 + \cos x)}$$

$$= \frac{\sin^2 x}{x(1 + \cos x)}$$

$$= \left(\frac{\sin x}{x}\right)\left(\frac{\sin x}{1 + \cos x}\right).$$

That

$$\lim_{x \to 0} \frac{1 - \cos x}{x} = 0$$

follows from noting that

$$\lim_{x \to 0} \frac{\sin x}{x} = 1 \quad \text{and} \quad \lim_{x \to 0} \frac{\sin x}{1 + \cos x} = 0. \quad \square$$

We are now in a position to evaluate a variety of trigonometric limits.

Problem 1. Find

$$\lim_{x \to 0} \frac{\sin 4x}{x}.$$

Solution. You know that

$$\lim_{x \to 0} \frac{\sin x}{x} = 1.$$

From this it follows that

$$\lim_{x \to 0} \frac{\sin 4x}{4x} = 1 \qquad \text{(see Exercise 32)}$$

and

$$\lim_{x \to 0} \frac{\sin 4x}{x} = \lim_{x \to 0} 4\left(\frac{\sin 4x}{4x}\right) = 4. \quad \square$$

---

† This "trick" is a fairly common procedure with trigonometric expressions. It is much like using "conjugates" to revise algebraic expressions:

$$\frac{3}{4 + \sqrt{2}} = \frac{3}{4 + \sqrt{2}} \cdot \frac{4 - \sqrt{2}}{4 - \sqrt{2}} = \frac{3(4 - \sqrt{2})}{14}.$$

Problem 2. Find

$$\lim_{x \to 0} x \cot 3x.$$

Solution. We begin by writing

$$x \cot 3x = x \, \frac{\cos 3x}{\sin 3x}.$$

Since

$$\lim_{x \to 0} \frac{\sin x}{x} = 1 \quad \text{and equivalently} \quad \lim_{x \to 0} \frac{x}{\sin x} = 1,$$

we would like to "pair off" $\sin 3x$ with $3x$. Thus, we write

$$x \, \frac{\cos 3x}{\sin 3x} = \frac{1}{3} \left(\frac{3x}{\sin 3x} \right) (\cos 3x).$$

Now we take the limit:

$$\lim_{x \to 0} x \cot 3x = \frac{1}{3} \left(\lim_{x \to 0} \frac{3x}{\sin 3x} \right) \left(\lim_{x \to 0} \cos 3x \right) = \frac{1}{3} (1)(1) = \frac{1}{3}. \quad \square$$

Problem 3. Find

$$\lim_{x \to \frac{1}{4}\pi} \frac{\sin (x - \frac{1}{4}\pi)}{(x - \frac{1}{4}\pi)^2}.$$

Solution

$$\frac{\sin (x - \frac{1}{4}\pi)}{(x - \frac{1}{4}\pi)^2} = \left[\frac{\sin (x - \frac{1}{4}\pi)}{(x - \frac{1}{4}\pi)} \right] \cdot \frac{1}{x - \frac{1}{4}\pi}.$$

We know that

$$\lim_{x \to \frac{1}{4}\pi} \frac{\sin (x - \frac{1}{4}\pi)}{x - \frac{1}{4}\pi} = 1.$$

Since $\lim\limits_{x \to \frac{1}{4}\pi} (x - \frac{1}{4}\pi) = 0$, you can see by Theorem 2.3.9 that

$$\lim_{x \to \frac{1}{4}\pi} \frac{\sin (x - \frac{1}{4}\pi)}{(x - \frac{1}{4}\pi)^2} \quad \text{does not exist.} \quad \square$$

Problem 4. Find

$$\lim_{x \to 0} \frac{x^2}{\sec x - 1}.$$

Solution. The evaluation of this limit requires a little imagination. We cannot deal with the fraction as it stands because both numerator and denominator tend to zero with x and it is not clear what happens to the fraction as x tends to 0. We can, however, rewrite the fraction in more amenable form:

$$\frac{x^2}{\sec x - 1} = \frac{x^2}{\sec x - 1}\left(\frac{\sec x + 1}{\sec x + 1}\right)$$

$$= \frac{x^2(\sec x + 1)}{\sec^2 x - 1} = \frac{x^2(\sec x + 1)}{\tan^2 x}$$

$$= \frac{x^2 \cos^2 x \,(\sec x + 1)}{\sin^2 x}$$

$$= \left(\frac{x}{\sin x}\right)^2 (\cos^2 x)(\sec x + 1).$$

Since each of these expressions has a limit as x tends to 0, the fraction we began with has a limit:

$$\lim_{x \to 0} \frac{x^2}{\sec x - 1} = \lim_{x \to 0}\left(\frac{x}{\sin x}\right)^2 \cdot \lim_{x \to 0} \cos^2 x \cdot \lim_{x \to 0}(\sec x + 1) = (1)(1)(2) = 2. \quad \square$$

EXERCISES 2.5

Evaluate the limits that exist.

1. $\lim\limits_{x \to 0} \dfrac{\sin 3x}{x}$.

2. $\lim\limits_{x \to 0} \dfrac{2x}{\sin x}$.

3. $\lim\limits_{x \to 0} \dfrac{3x}{\sin 5x}$.

4. $\lim\limits_{x \to 0} \dfrac{\sin 3x}{2x}$.

5. $\lim\limits_{x \to 0} \dfrac{\sin x^2}{x}$.

6. $\lim\limits_{x \to 0} \dfrac{\sin x^2}{x^2}$.

7. $\lim\limits_{x \to 0} \dfrac{\sin x}{x^2}$.

8. $\lim\limits_{x \to 0} \dfrac{\sin^2 x^2}{x^2}$.

9. $\lim\limits_{x \to 0} \dfrac{\sin^2 3x}{5x^2}$.

10. $\lim\limits_{x \to 0} \dfrac{\tan^2 3x}{4x^2}$.

11. $\lim\limits_{x \to 0} \dfrac{2x}{\tan 3x}$.

12. $\lim\limits_{x \to 0} \dfrac{4x}{\cot 3x}$.

13. $\lim\limits_{x \to 0} x \csc x$.

14. $\lim\limits_{x \to 0} \dfrac{\cos x - 1}{2x}$.

15. $\lim\limits_{x \to 0} \dfrac{x^2}{1 - \cos 2x}$.

16. $\lim\limits_{x \to 0} \dfrac{x^2 - 2x}{\sin 3x}$.

17. $\lim\limits_{x \to 0} \dfrac{1 - \sec^2 2x}{x^2}$.

18. $\lim\limits_{x \to 0} \dfrac{1}{2x \csc x}$.

19. $\lim\limits_{x \to 0} \dfrac{2x^2 + x}{\sin x}$.

20. $\lim\limits_{x \to 0} \dfrac{1 - \cos 4x}{9x^2}$.

21. $\lim\limits_{x \to 0} \dfrac{\tan 3x}{2x^2 + 5x}$.

22. $\lim\limits_{x \to 0} x^2(1 + \cot^2 3x)$.

23. $\lim\limits_{x \to 0} \dfrac{\sec x - 1}{x \sec x}$.

24. $\lim\limits_{x \to \pi/4} \dfrac{1 - \cos x}{x}$.

25. $\lim\limits_{x \to \pi/4} \dfrac{\sin x}{x}$.

26. $\lim\limits_{x \to 0} \dfrac{\sin^2 x}{x(1 - \cos x)}$.

27. $\lim\limits_{x \to \pi/2} \dfrac{\cos x}{x - \frac{1}{2}\pi}$.

28. $\lim\limits_{x \to \pi} \dfrac{\sin x}{x - \pi}$.

29. $\lim\limits_{x \to \pi/4} \dfrac{\sin\left(x + \frac{1}{4}\pi\right) - 1}{x - \frac{1}{4}\pi}$.

30. $\lim\limits_{x \to \pi/6} \dfrac{\sin\left(x + \frac{1}{3}\pi\right) - 1}{x - \frac{1}{6}\pi}$.

31. Show that $\lim_{x \to c} \cos x = \cos c$ for all real numbers c.

32. Show that

$$\text{if} \quad \lim_{x \to 0} f(x) = l, \quad \text{then} \quad \lim_{x \to 0} f(ax) = l \quad \text{for all } a \neq 0.$$

HINT: Let $\epsilon > 0$. If $\delta_1 > 0$ "works" for the first limit, then $\delta = \delta_1/|a|$ "works" for the second limit.

Give an inequality to which we can apply the pinching theorem to obtain the indicated limit.

33. $\lim_{x \to 0} x f(x) = 0$ given that $|f(x)| \leq M$ for all $x \neq 0$.

34. $\lim_{x \to 0} f(x) = 0$ given that $\left| \dfrac{f(x)}{x} \right| \leq M$ for all $x \neq 0$.

35. $\lim_{x \to c} f(x) = l$ given that $\left| \dfrac{f(x) - l}{x - c} \right| \leq M$ for all $x \neq c$.

36. $\lim_{x \to 0} f(x) = 1$ for $f(x) = \begin{bmatrix} 1 + x^2, & x \text{ rational} \\ 1 + x^4, & x \text{ irrational} \end{bmatrix}$.

37. $\lim_{x \to 0} x \sin \dfrac{1}{x} = 0$. **38.** $\lim_{x \to \pi} (x - \pi) \cos^2 \dfrac{1}{x - \pi} = 0$.

39. $\lim_{x \to c} \sqrt{x} = \sqrt{c}$ for each positive number c.

Program for trigonometric limits (BASIC)

This program features the limits

$$\lim_{x \to 0} \frac{\sin x}{x} \quad \text{and} \quad \lim_{x \to 0} \frac{1 - \cos x}{x}.$$

The program requests you to enter a starting value for x. It evaluates the expressions above at the value x, then divides x by 2 and evaluates the expressions at the resulting value, etc. This procedure is iterated as many times as is indicated by your input n to the question "How many times?".

Remark: This program can be modified to find the limit of any function f at any point a, provided that the limit exists. Simply replace the expressions above by the expression $f(a + x)$.

```
10 REM Program estimates limits of sin(x)/x and (1 − cos(x))/x as x → 0
20 REM copyright © Colin C. Graham 1988-1989

100 INPUT "Enter starting value:"; x
120 INPUT "How many times?"; n
130 REM 13 gives a nice display on many computers

200 PRINT "x", "sin(x)/x", "(1 − cos(x))/x"

300 FOR j = 1 TO n
310     PRINT x, SIN(x)/x, (1 − COS(x))/x
320     x = x/2
330 NEXT j

500 INPUT "Do another? (Y/N):"; a$
510 IF a$ = "Y" OR a$ = "y" THEN   100
520 END
```

2.6   TWO BASIC THEOREMS

A function that is continuous on an interval does not "skip" any values, and thus its graph is an "unbroken curve." There are no "holes" in it and no "gaps." This is the idea behind the *intermediate-value theorem*.

THEOREM 2.6.1   THE INTERMEDIATE-VALUE THEOREM

If f is continuous on $[a, b]$ and C is a number between $f(a)$ and $f(b)$, then there is at least one number c between a and b for which $f(c) = C$.

We illustrate the theorem in Figure 2.6.1. What can happen in the discontinuous case is illustrated in Figure 2.6.2. There the number c has been "skipped." You can find a proof of the intermediate-value theorem in Appendix B. We will assume the result and use it.

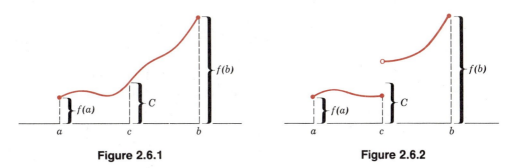

Figure 2.6.1   **Figure 2.6.2**

Suppose that f is continuous on $[a, b]$ and

$$f(a) < 0 < f(b) \qquad \text{or} \qquad f(b) < 0 < f(a).$$

Then, by the intermediate-value theorem, we know that the equation $f(x) = 0$ has at least one root in $[a, b]$. For simplicity let's assume that there is only one such root and call it c. How can we estimate the location of c? The intermediate-value theorem itself gives us no clue. The simplest method of finding the approximate location of c is called the *bisection method*.

The Bisection Method

The method is an iterative process. A basic step is iterated (carried out repeatedly) until the goal is reached, in this case an approximation for c that is as accurate as we wish.

It is standard practice to label the elements of successive iterations by using as subscripts $n = 1, 2, 3$, etc. We begin by setting $x_1 = a$ and $y_1 = b$. We now bisect $[x_1, y_1]$. If c is the midpoint of $[x_1, y_1]$, our search is over. If not, then c lies in one of the halves of $[x_1, y_1]$. We call that half $[x_2, y_2]$. If c is the midpoint of $[x_2, y_2]$, then our search is over. If not, then c lies in one of the halves of $[x_2, y_2]$. We call that half

$[x_3, y_3]$ and continue.   The first three iterations for a particular function are depicted in Figure 2.6.3.

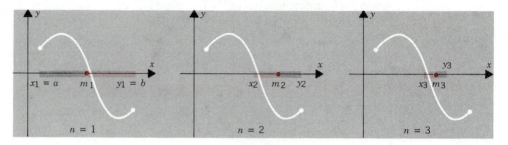

Figure 2.6.3

After n bisections we are examining the midpoint m_n of the interval $[x_n, y_n]$.   We can thus be certain that

$$|c - m_n| \leq \frac{1}{2}(y_n - x_n) = \frac{1}{2}\left(\frac{y_{n-1} - x_{n-1}}{2}\right) = \cdots = \frac{b - a}{2^n}.$$

Thus m_n approximates c within $(b - a)/2^n$.   If we want m_n to approximate c within a given ϵ, then we must carry out the iteration to the point where

$$\frac{b - a}{2^n} < \epsilon.$$

Example 1.   The function $f(x) = x^2 - 2$ is zero at $x = \sqrt{2}$.   Since $f(1) < 0$ and $f(2) > 0$, we can estimate $\sqrt{2}$ by applying the bisection method to f on the interval $[1, 2]$.   Let's say we want a numerical estimate accurate to within 0.001.   To achieve this degree of accuracy we must carry out the iteration to the point where

$$\frac{2 - 1}{2^n} < 0.001,$$

or, equivalently, to the point where $2^n > 1000$.   As you can check, this condition is met by taking $n \geq 10$.   Thus 10 bisections will suffice.

$x_1 = 1, \ y_1 = 2, \ m_1 = 1.5$:
$$f(x_1) < 0, \quad f(y_1) > 0, \quad f(m_1) = (1.5)^2 - 2 = 0.25 > 0$$
$x_2 = 1, \ y_2 = 1.5, \ m_2 = 1.25$:
$$f(x_2) < 0, \quad f(y_2) > 0, \quad f(m_2) = (1.25)^2 - 2 = -0.4375 < 0$$
$x_3 = 1.25, \ y_3 = 1.5, \ m_3 = 1.375$:
$$f(x_3) > 0, \quad f(y_3) < 0, \quad f(m_3) = (1.375)^2 - 2 = -0.109375 < 0$$
$x_4 = 1.375, \ y_4 = 1.5, \ m_4 = 1.4375$:
$$\text{etc.}$$

The complete computations are summarized in Table 2.6.1.   To increase readability, all the entries in the table have been rounded off to 5 decimal places.

TABLE 2.6.1

n	x_n	y_n	m_n	$f(m_n)$
1	1.00000	2.00000	1.50000	0.25000
2	1.00000	1.50000	1.25000	-0.43750
3	1.25000	1.50000	1.37500	-0.10938
4	1.37500	1.50000	1.43750	0.06641
5	1.37500	1.43750	1.40625	-0.02246
6	1.40625	1.43750	1.42187	0.02171
7	1.40625	1.42187	1.41406	-0.00043
8	1.41406	1.42187	1.41797	0.01064
9	1.41406	1.41797	1.41601	0.00508
10	1.41406	1.41601	1.41503	0.00231

We can conclude that $m_{10} = 1.41503$ approximates $\sqrt{2}$ within 0.001. ☐

We come now to the maximum-minimum property of continuous functions.

THEOREM 2.6.2   THE MAXIMUM-MINIMUM THEOREM

If f is continuous on $[a, b]$, then f takes on both a maximum value M and a minimum value m on $[a, b]$.

The situation is illustrated in Figure 2.6.4. The maximum value M is taken on at the point marked x_2, and the minimum value m is taken on at the point marked x_1.

Figure 2.6.4

It is interesting to note that in the maximum-minimum theorem the full hypothesis is needed. If we drop the continuity requirement, the conclusion does not follow. As an example, we can take the function

$$f(x) = \begin{cases} 3, & x = 1 \\ x, & 1 < x < 5 \\ 3, & x = 5 \end{cases}.$$

The graph of this function is pictured in Figure 2.6.5. The function is defined on $[1, 5]$, but it takes on neither a maximum nor a minimum value. If, instead of dropping the continuity requirement, we drop the requirement that the interval be of the form $[a, b]$, then again the result fails. For example, in the case of

$$g(x) = x \qquad \text{for } x \in (1, 5),$$

g is continuous at each point of $(1, 5)$ but again there is no maximum value and no minimum value. See Figure 2.6.6.

Figure 2.6.5

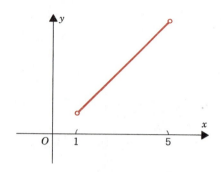

Figure 2.6.6

A proof of the maximum-minimum theorem appears in Appendix B. Techniques for determining the maximum and minimum values of functions are described in Chapter 4. These techniques require an understanding of "differentiation," the subject to which we devote Chapter 3.

EXERCISES 2.6

Sketch the graph of a function f that is defined on $[0, 1]$ and meets the following conditions. (if possible)

1. f is continuous on $[0, 1]$, minimum value 0, maximum value $\frac{1}{2}$.

2. f is continuous on $[0, 1)$, minimum value 0, no maximum value.

3. f is continuous on $[0, 1)$, no minimum value, maximum value 1.

4. f is continuous on $[0, 1]$, the maximum value at $x = 0, \frac{1}{2}, 1$, the minimum value at $x = \frac{1}{4}, \frac{3}{4}$.

5. f is continuous on $(0, 1)$, takes on the values 0 and 1 but does not take on the value $\frac{1}{2}$.

6. f is continuous on $[0, 1]$, takes on the values -1 and 1 but does not take on the value 0.

7. f is continuous on $[0, 1]$, maximum value 1, minimum value 1.

8. f is continuous on $[0, 1]$ and nonconstant, takes on no integer values.

9. f is continuous on $[0, 1]$, takes on no rational values.

10. f takes on only two distinct values.

11. f is continuous on $(0, 1)$, takes on only two distinct values.

12. f is not continuous on $[0, 1]$, takes on both a maximum value and a minimum value and every value in between.

13. f is continuous on $[0, 1]$, takes on only two distinct values.

14. f is continuous on $(0, 1)$, takes on only three distinct values.

15. f is discontinuous at $x = \frac{1}{2}$ but takes on both a minimum value and a maximum value.

16. f is continuous except at $x = \frac{1}{2}$ but has no minimum value and no maximum value.

17. (*Fixed-point property*) Show that, if f is continuous and $0 \leq f(x) \leq 1$ for all $x \in [0, 1]$, then there exists at least one point c in $[0, 1]$ for which $f(c) = c$.   HINT: Apply the intermediate-value theorem to the function $g(x) = x - f(x)$.

18. Let n be a positive integer.
   (a) Prove that, if $0 \leq a < b$, then $a^n < b^n$.
   (b) Prove that every nonnegative real number x has a unique nonnegative nth root $x^{1/n}$.   HINT: The existence of $x^{1/n}$ can be seen by applying the intermediate-value theorem to the function $f(t) = t^n$ for $t \geq 0$. The uniqueness follows from part (a).

19. Given that f and g are continuous and $f(0) < g(0) < g(1) < f(1)$, prove that, for some point c between 0 and 1, $f(c) = g(c)$.

20. The intermediate-value theorem can be used to prove that each polynomial equation of odd degree

$$x^n + a_{n-1}x^{n-1} + \cdots + a_1 x + a_0 = 0 \qquad \text{with } n \text{ odd}$$

has at least one real root. Prove that the cubic equation

$$x^3 + ax^2 + bx + c = 0$$

has at least one real root.

(Calculator) Use the bisection method to approximate the root of $f(x) = 0$ in the given interval to within 0.001.

21. $f(x) = x^2 - 3$; [1, 2].

22. $f(x) = 5 - x^2$; [−3, −2].

23. $f(x) = x^2 + x - 1$; [0, 1].

24. $f(x) = x^2 - x - 1$; [1, 2].

25. $f(x) = x^3 + x - 10$; [1, 4].

26. $f(x) = 2x^3 + 5x - 7$; [−1, 2].

27. $f(x) = \cos x - x$; [0, 1].

28. $f(x) = \cos x - 2x$; [0, 1].

Program for finding roots by bisection   (BASIC)

This program is meant to "mechanize" the procedure given in the text for finding the root of an equation of the type FNf(x) = 0, where FNf is the computer's name for a function. The procedure uses repeated bisection of smaller and smaller intervals on which the function is defined. The program is reasonably self-explanatory, and probably the best way to understand it is to study the instructions it gives. Note that you are meant to replace the given sample function FNf by the function of your choice, and that you must specify the initial interval in question by giving its endpoints. You should convince yourself that the statements 320–360 do result in the selection of a subinterval containing a root of the equation.

```
10 REM Finding roots by bisection
20 REM copyright © Colin C. Graham 1988-1989

30 REM change the line "def FNf = ..."
40 REM to fit your function
50 DEF FNf(x) = x^2 − 1

100 INPUT "Enter left-hand endpoint:"; left
110 INPUT "Enter right-hand endpoint:"; right
120 INPUT "Enter the number of times to iterate:"; ntimes
130 REM    13 gives nice display on many computers

200 middle = (left + right)/2
210 PRINT "left"; "       "; "right"; "       "; "FNf(middle)"

300 FOR j = 0 TO ntimes
310     middle = (left + right)/2
320     PRINT left; " "; right; " "; FNf(middle)
330     IF (FNf(left) > 0 AND FNf(middle) > 0) THEN left = middle
340     IF (FNf(left) > 0 AND FNf(middle) <= 0) THEN right = middle
350     IF (FNf(left) <= 0 AND FNf(middle) > 0) THEN right = middle
360     IF (FNf(left) <= 0 AND FNf(middle) <= 0) THEN left = middle
370 NEXT j
380 middle = (left + right)/2

400 PRINT left; " "; right; " "; FNf(middle)

500 INPUT "Do it again? (Y/N)"; a$
510 IF a$ = "Y" OR a$ = "y" THEN 100
520 END
```

2.7 ADDITIONAL EXERCISES

Evaluate the limits that exist.

1. $\lim\limits_{x \to 3} \dfrac{x^2 - 3}{x - 3}.$

2. $\lim\limits_{x \to 2} \dfrac{x^2 + 4}{x^2 + 2x + 1}.$

3. $\lim\limits_{x \to 3} \dfrac{(x - 3)^2}{x + 3}.$

4. $\lim\limits_{x \to 0} \left(\dfrac{1}{x} - \dfrac{x}{1 - x} \right).$

5. $\lim\limits_{x \to 3} \dfrac{x^2 - 9}{x^2 - 5x + 6}.$

6. $\lim\limits_{x \to 2^+} \dfrac{x - 2}{|x - 2|}.$

7. $\lim\limits_{x \to 3^-} \dfrac{x - 3}{|x - 3|}.$

8. $\lim\limits_{x \to -2} \dfrac{|x|}{x - 2}.$

9. $\lim\limits_{x \to 0} \left(\dfrac{1}{x} - \dfrac{1 - x}{x} \right).$

10. $\lim\limits_{x \to -1} \dfrac{x^2 - x - 2}{x + 1}.$

11. $\lim\limits_{x \to 1^-} \sqrt{|x| - x}.$

12. $\lim\limits_{x \to 3^+} \dfrac{\sqrt{x} - 3}{|x - 3|}.$

13. $\lim\limits_{x \to 2^+} \dfrac{x^2}{x + 2}.$

14. $\lim\limits_{x \to 1} \dfrac{|x - 1|}{x}.$

15. $\lim\limits_{x \to 1} \dfrac{x^3 - 1}{|x^3 - 1|}.$

16. $\lim\limits_{x \to 1^+} \sqrt{|x| - x}.$

17. $\lim\limits_{x \to -1} \dfrac{1 + |x|}{1 - |x|}.$

18. $\lim\limits_{x \to 1} \dfrac{1 - 1/x}{1 - 1/x^2}.$

19. $\lim\limits_{x \to -1} \dfrac{1 + 1/x}{1 + 1/x^2}.$

20. $\lim\limits_{x \to 1} \dfrac{1 + 1/x}{1 - x^2}.$

21. $\lim\limits_{x \to 1} \dfrac{1 - 1/x}{1 - x^2}.$

22. $\lim\limits_{x \to 1} \dfrac{1 - 1/x^2}{1 - x}.$

23. $\lim\limits_{x \to 0^+} \dfrac{1 + \sqrt{x}}{1 - \sqrt{x}}.$

24. $\lim\limits_{x \to -1} \dfrac{1 + 1/x^2}{1 - x}.$

25. $\lim\limits_{x \to 4} \left[\left(\dfrac{1}{x} - \dfrac{1}{4} \right) \left(\dfrac{1}{x - 4} \right) \right].$

26. $\lim\limits_{x \to 1^-} \dfrac{\sqrt{x} - 1}{x - 1}.$

27. $\lim\limits_{x \to 0^-} \dfrac{\sqrt{x} - 1}{x - 1}.$

28. $\lim\limits_{x \to 0^+} \dfrac{\sqrt{x} - 1}{x - 1}.$

29. $\lim\limits_{x \to 3^+} \dfrac{x^2 - 2x - 3}{\sqrt{x - 3}}.$

30. $\lim\limits_{x \to 3^+} \dfrac{\sqrt{x^2 - 2x - 3}}{x - 3}.$

31. $\lim\limits_{x \to 2^+} \left(\dfrac{1}{x - 2} - \dfrac{1}{|x - 2|} \right).$

32. $\lim\limits_{x \to 2^-} \left(\dfrac{1}{x - 2} - \dfrac{1}{|x - 2|} \right).$

33. $\lim\limits_{x \to 4} \dfrac{\sqrt{x + 5} - 3}{x - 4}.$

34. $\lim\limits_{x \to 2} \dfrac{x^3 - 8}{x^4 - 3x^2 - 4}.$

35. $\lim\limits_{x \to 0} \dfrac{5x}{\sin 2x}.$

36. $\lim\limits_{x \to 0} \dfrac{\tan^2 2x}{3x^2}.$

37. $\lim\limits_{x \to 0} 4x^2 \cot^2 3x.$

38. $\lim\limits_{x \to 0} x \csc 4x.$

39. $\lim\limits_{x \to 0} \dfrac{x^2 - 3x}{\tan x}.$

40. $\lim\limits_{x \to 0} \dfrac{x}{1 - \cos x}.$

41. $\lim\limits_{x \to \pi/2} \dfrac{\cos x}{2x - \pi}.$

42. $\lim\limits_{x \to 0} \dfrac{\sin 3x}{5x^2 - 4x}.$

43. $\lim\limits_{x \to 0} \dfrac{5x^2}{1 - \cos 2x}.$

44. $\lim\limits_{x \to -\pi} \dfrac{x + \pi}{\sin x}.$

45. $\lim\limits_{x \to 2} \dfrac{x^2 - 4}{x^3 - 8}.$

46. $\lim\limits_{x \to 3} \left(3 - \dfrac{1}{x} \right) \left(\dfrac{2x}{1 - 9x^2} \right).$

47. $\lim\limits_{x \to 2^+} \dfrac{\sqrt{x - 2}}{|x - 2|}.$

48. $\lim\limits_{x \to 2^-} \dfrac{x - 2}{|x^2 - 4|}.$

49. $\lim\limits_{x \to 2} \left(\dfrac{6}{x - 2} - \dfrac{3x}{x - 2} \right).$

50. $\lim\limits_{x \to 2} \left(\dfrac{2x}{x^2 - 4} - \dfrac{x^2}{x^2 - 4} \right).$

51. $\lim\limits_{x \to -1^+} \dfrac{|x^2 - 1|}{x + 1}.$

52. $\lim\limits_{x \to 3} \dfrac{\sqrt{x + 1} - 2}{x - 3}.$

53. $\lim\limits_{x \to 2} \dfrac{1 - 2/x}{1 - 4/x^2}.$

54. $\lim\limits_{x \to -4^+} \dfrac{|x + 4|}{\sqrt{x + 4}}.$

55. $\lim\limits_{x \to 2^+} \dfrac{\sqrt{x^2 - 3x + 2}}{x - 2}.$

56. $\lim\limits_{x \to 1} \dfrac{x^2 - 1}{x^5 - 1}.$

57. $\lim\limits_{x \to 2} \left(1 - \dfrac{2}{x} \right) \left(\dfrac{3}{4 - x^2} \right).$

58. $\lim\limits_{x\to 1^+}\dfrac{x^2-3x+2}{\sqrt{x-1}}$.

59. $\lim\limits_{x\to 7}\dfrac{x-7}{\sqrt{x+2}-3}$.

60. $\lim\limits_{x\to 3}\dfrac{1-9/x^2}{1+3/x}$.

61. $\lim\limits_{x\to 2} f(x)$ if $f(x) = \begin{cases} x^2, & x \text{ an integer} \\ 3x, & \text{otherwise} \end{cases}$.

62. $\lim\limits_{x\to 3} f(x)$ if $f(x) = \begin{cases} x^2, & x \text{ rational} \\ 3x, & x \text{ irrational} \end{cases}$.

63. $\lim\limits_{x\to -2} f(x)$ if $f(x) = \begin{cases} 3+x, & x<-2 \\ 5, & x=-2 \\ x^2-3, & x>-2 \end{cases}$.

64. $\lim\limits_{x\to 2} f(x)$ if $f(x) = \begin{cases} x+1, & x<1 \\ 3x-x^2, & x>1 \end{cases}$.

65. (a) Sketch the graph of

$$f(x) = \begin{cases} 3x+4, & x<-1 \\ 2x+3, & -1<x<1 \\ 4x+1, & 1<x<2 \\ 3x-2, & 2<x \\ x^2, & x=-1,1,2 \end{cases}.$$

(b) Evaluate the limits that exist.

(i) $\lim\limits_{x\to -1^-} f(x)$. (ii) $\lim\limits_{x\to -1^+} f(x)$. (iii) $\lim\limits_{x\to -1} f(x)$.

(iv) $\lim\limits_{x\to 1^-} f(x)$. (v) $\lim\limits_{x\to 1^+} f(x)$. (vi) $\lim\limits_{x\to 1} f(x)$.

(vii) $\lim\limits_{x\to 2^-} f(x)$. (viii) $\lim\limits_{x\to 2^+} f(x)$. (ix) $\lim\limits_{x\to 2} f(x)$.

(c) (i) At what points is f discontinuous from the left?
(ii) At what points is f discontinuous from the right?
(iii) Where are the removable discontinuities of f?
(iv) Where are the essential discontinuities of f?

66. Exercise 65 with

$$f(x) = \begin{cases} -2x, & x<-1 \\ x^2+1, & -1<x<1 \\ \sqrt{2x}, & 1<x<2 \\ x, & 2<x \\ |x^2-3|, & x=-1,1,2 \end{cases}.$$

67. Find $\lim\limits_{x\to c} f(x)$ given that $|f(x)| \le M|x-c|$ for all $x \ne c$.

68. Given that $\lim\limits_{x\to 0} (f(x)/x)$ exists, prove that $\lim\limits_{x\to 0} f(x) = 0$.

Give an ϵ, δ proof for the following limits.

69. $\lim\limits_{x\to 2} (5x-4) = 6$. **70.** $\lim\limits_{x\to 3} (1-2x) = -5$. **71.** $\lim\limits_{x\to -4} |2x+5| = 3$.

72. $\lim\limits_{x\to 3} |1-2x| = 5$. **73.** $\lim\limits_{x\to 9} \sqrt{x-5} = 2$. **74.** $\lim\limits_{x\to 2^+} \sqrt{x^2-4} = 0$.

75. Prove that there exists a real number x_0 such that $x_0^5 - 4x_0 + 1 = 7.21$.

76. Let f be a continuous function and n a positive integer. Show that, if $0 \le f(x) \le 1$ for all $x \in [0, 1]$, then there exists at least one point c in $[0, 1]$ for which $f(c) = c^n$.

CHAPTER HIGHLIGHTS

2.1 The Idea of Limit

Intuitive interpretation of $\lim\limits_{x\to c} f(x) = l$:

as x approaches c, f(x) approaches l

or, equivalently,

for x close to c but different from c, f(x) is close to l.

Restricting our attention to values of x to one side of c, we have *one-sided limits*:

$$\lim_{x \to c^-} f(x) \quad \text{and} \quad \lim_{x \to c^+} f(x).$$

2.2 Definition of Limit

limit of a function (p. 57)   left-hand limit (p. 65)   right-hand limit (p. 65)
four formulations of the limit statement (p. 65)

2.3 Some Limit Theorems

uniqueness of limit (p. 69)   stability of limit (p. 77)

The "arithmetic of limits": if $\lim_{x \to c} f(x) = l$ and $\lim_{x \to c} g(x) = m$, then

$$\lim_{x \to c} [f(x) + g(x)] = l + m, \quad \lim_{x \to c} [\alpha f(x)] = \alpha l \quad \text{for each real } \alpha,$$

$$\lim_{x \to c} [f(x)g(x)] = lm, \quad \lim_{x \to c} \frac{f(x)}{g(x)} = \frac{l}{m} \quad \text{provided } m \neq 0.$$

If $m = 0$ and $l \neq 0$, then $\lim_{x \to c} \frac{f(x)}{g(x)}$ does not exist.   Similar results hold for one-sided limits.

2.4 Continuity

continuity at c (p. 78)   essential discontinuity, removable discontinuity
one-sided continuity (p. 83)   (p. 79)
continuity on $[a, b]$ (p. 84)   slot-machine test (p. 84)

The sum, difference, product, quotient, and composition of continuous functions are continuous where defined.   Polynomials and the absolute-value function are everywhere continuous.   Rational functions are continuous on their domains.   The square-root function is continuous at all positive numbers and continuous from the right at 0.

2.5 The Pinching Theorem; Trigonometric Limits

pinching theorem (p. 87) $\lim_{x \to 0} \dfrac{\sin x}{x} = 1$ $\lim_{x \to 0} \dfrac{1 - \cos x}{x} = 0$

Each trigonometric function is continuous on its domain.

2.6 Two Basic Theorems

intermediate-value theorem (p. 95)
bisection method (p. 95)
maximum-minimum theorem (p. 97)

2.7 Additional Exercises

DIFFERENTIATION

3.1   THE DERIVATIVE

Introduction: The Tangent to a Curve

We begin with a function f, and on the graph we choose a point $(x, f(x))$. See Figure 3.1.1. What line, if any, should be called tangent to the graph at that point?

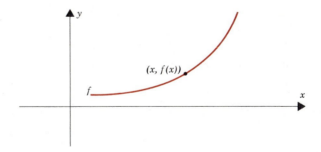

Figure 3.1.1

To answer this question, we choose a small number $h \neq 0$ and on the graph mark the point $(x + h, f(x + h))$. Now we draw the secant line that passes through these two points. The situation is pictured in Figure 3.1.2, first with $h > 0$ and then with $h < 0$.

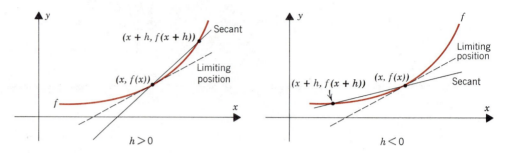

Figure 3.1.2

As h tends to zero from the right (see the figure), the secant line tends to a limiting position, and it tends to the same limiting position as h tends to zero from the left. The line at this limiting position is what we call "the tangent to the graph at the point $(x, f(x))$."

Since the approximating secants have slope of the form

(*) $$\frac{f(x + h) - f(x)}{h},$$ (check this out)

you can expect the tangent, the limiting position of these secants, to have slope

(**) $$\lim_{h \to 0} \frac{f(x + h) - f(x)}{h}.$$

While (*) measures the steepness of the line that passes through $(x, f(x))$ and $(x + h, f(x + h))$, (**) measures the steepness of the graph at $(x, f(x))$ and is called the "slope of the graph" at that point.

Derivatives and Differentiation

In the introduction we spoke informally about tangent lines and gave a geometric interpretation to limits of the form

$$\lim_{h \to 0} \frac{f(x + h) - f(x)}{h}.$$

Here we begin the systematic study of such limits, what mathematicians call the theory of *differentiation*.

DEFINITION 3.1.1

A function f is said to be *differentiable at x* iff

$$\lim_{h \to 0} \frac{f(x + h) - f(x)}{h} \quad \text{exists.}$$

If this limit exists, it is called the *derivative of f at x* and is denoted by $f'(x)$.†

---

† This prime notation goes back to the French mathematician Joseph Louis Lagrange (1736 – 1813). Another notation will be introduced later.

We interpret

$$f'(x) = \lim_{h \to 0} \frac{f(x+h) - f(x)}{h}$$

as the *slope of the graph at the point* $(x, f(x))$.   The line that passes through the point $(x, f(x))$ with slope $f'(x)$ is called the *tangent line at* $(x, f(x))$.   This is the line that best approximates the graph of f near the point $(x, f(x))$.

It is time we looked at some examples.

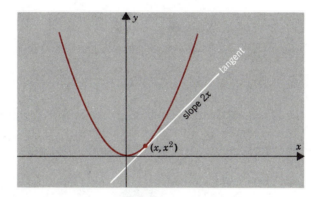

Figure 3.1.3

Example 1. We begin with the squaring function

$$f(x) = x^2.$$ (Figure 3.1.3)

To find $f'(x)$ we form the *difference quotient*

$$\frac{f(x+h) - f(x)}{h} = \frac{(x+h)^2 - x^2}{h}.$$

Since

$$\frac{(x+h)^2 - x^2}{h} = \frac{x^2 + 2xh + h^2 - x^2}{h} = \frac{2xh + h^2}{h} = 2x + h,$$

we have

$$\frac{f(x+h) - f(x)}{h} = 2x + h,$$

and therefore

$$f'(x) = \lim_{h \to 0} \frac{f(x+h) - f(x)}{h} = \lim_{h \to 0} (2x + h) = 2x. \quad \square$$

Example 2. In the case of a linear function

$$f(x) = mx + b$$

we have

$$f'(x) = m.$$

In other words, the slope is constantly m.   To verify this, note that

$$\frac{f(x+h) - f(x)}{h} = \frac{[m(x+h)+b] - [mx+b]}{h} = \frac{mh}{h} = m.$$

It follows that

$$f'(x) = \lim_{h \to 0} \frac{f(x+h) - f(x)}{h} = m. \quad \blacksquare$$

The derivative

$$f'(x) = \lim_{h \to 0} \frac{f(x+h) - f(x)}{h}$$

is a two-sided limit.   Therefore it cannot be taken at an endpoint of the domain.   In our next example we will be dealing with the square-root function.   Although this function is defined for all $x \geq 0$, you can expect a derivative only for $x > 0$.

square root function

Figure 3.1.4

Example 3.   The square-root function

$$f(x) = \sqrt{x}, \qquad x \geq 0 \qquad \text{(Figure 3.1.4)}$$

has derivative

$$f'(x) = \frac{1}{2\sqrt{x}}, \quad \text{for } x > 0.$$

To verify this, we begin with $x > 0$ and form the difference quotient

$$\frac{f(x+h) - f(x)}{h} = \frac{\sqrt{x+h} - \sqrt{x}}{h}.$$

To remove the radicals from the numerator we multiply both the numerator and the denominator by $\sqrt{x+h} + \sqrt{x}$.   This gives

$$\frac{f(x+h)-f(x)}{h} = \left(\frac{\sqrt{x+h}-\sqrt{x}}{h}\right)\left(\frac{\sqrt{x+h}+\sqrt{x}}{\sqrt{x+h}+\sqrt{x}}\right)$$

$$= \frac{(x+h)-x}{h(\sqrt{x+h}+\sqrt{x})} = \frac{1}{\sqrt{x+h}+\sqrt{x}}$$

and thus

$$f'(x) = \lim_{h \to 0} \frac{f(x+h)-f(x)}{h} = \frac{1}{2\sqrt{x}}. \quad \square$$

To *differentiate* a function is to find its derivative.

Example 4. Here we differentiate the function

$$f(x) = \frac{1}{x}, \qquad x \neq 0.$$

(See Figure 3.1.5.) We form the difference quotient

$$\frac{f(x+h)-f(x)}{h} = \frac{\dfrac{1}{x+h} - \dfrac{1}{x}}{h}$$

and simplify:

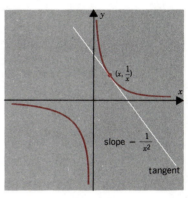

a hyperbola

Figure 3.1.5

$$\frac{\dfrac{1}{x+h}-\dfrac{1}{x}}{h} = \frac{\dfrac{x}{x(x+h)} - \dfrac{x+h}{x(x+h)}}{h} = \frac{\dfrac{-h}{x(x+h)}}{h} = \frac{-1}{x(x+h)}.$$

Thus

$$f'(x) = \lim_{h \to 0} \frac{f(x+h)-f(x)}{h} = \lim_{h \to 0} \frac{-1}{x(x+h)} = -\frac{1}{x^2}. \quad \square$$

Evaluating Derivatives

Problem 5. Find $f'(-2)$ given that

$$f(x) = 1 - x^2.$$

Solution. We can first find $f'(x)$:

$$f'(x) = \lim_{h \to 0} \frac{f(x+h)-f(x)}{h}$$

$$= \lim_{h \to 0} \frac{[1-(x+h)^2]-[1-x^2]}{h} = \lim_{h \to 0} \frac{-2xh-h^2}{h} = \lim_{h \to 0} (-2x-h) = -2x$$

and then substitute -2 for x:

$$f'(-2) = -2(-2) = 4.$$

We can also evaluate $f'(-2)$ directly:

$$f'(-2) = \lim_{h \to 0} \frac{f(-2 + h) - f(-2)}{h}$$

$$= \lim_{h \to 0} \frac{[1 - (-2 + h)^2] - [1 - (-2)^2]}{h} = \lim_{h \to 0} \frac{4h - h^2}{h} = \lim_{h \to 0} (4 - h) = 4. \quad \square$$

Problem 6. Find $f'(-3)$ and $f'(1)$ given that

$$f(x) = \begin{cases} x^2, & x \le 1 \\ 2x - 1, & x > 1 \end{cases}.$$

Solution. By definition

$$f'(-3) = \lim_{h \to 0} \frac{f(-3 + h) - f(-3)}{h}.$$

For all x sufficiently close to -3, $f(x) = x^2$. Thus

$$f'(-3) = \lim_{h \to 0} \frac{(-3 + h)^2 - (-3)^2}{h} = \lim_{h \to 0} \frac{(9 - 6h + h^2) - 9}{h} = \lim_{h \to 0} (-6 + h) = -6.$$

Now let's find

$$f'(1) = \lim_{h \to 0} \frac{f(1 + h) - f(1)}{h}.$$

Since f is not defined by the same formula on both sides of 1, we will evaluate this limit by taking one-sided limits. Note that $f(1) = 1^2 = 1$.

To the left of 1, $f(x) = x^2$. Thus

$$\lim_{h \to 0^-} \frac{f(1 + h) - f(1)}{h} = \lim_{h \to 0^-} \frac{(1 + h)^2 - 1}{h}$$

$$= \lim_{h \to 0^-} \frac{(1 + 2h + h^2) - 1}{h} = \lim_{h \to 0^-} (2 + h) = 2.$$

To the right of 1, $f(x) = 2x - 1$. Thus

$$\lim_{h \to 0^+} \frac{f(1 + h) - f(1)}{h} = \lim_{h \to 0^+} \frac{[2(1 + h) - 1] - 1}{h} = \lim_{h \to 0^+} 2 = 2.$$

The limit of the difference quotient exists and is 2:

$$f'(1) = \lim_{h \to 0} \frac{f(1 + h) - f(1)}{h} = 2. \quad \square$$

Tangent Lines and Normal Lines

We begin with a differentiable function f and choose a point (x_0, y_0) on the graph. As you know, the tangent line through this point has slope $f'(x_0)$. To get an equation for this tangent line, we use the point-slope formula

$$y - y_0 = m(x - x_0).$$

In this case $m = f'(x_0)$ and the equation becomes

(3.1.2)
$$\boxed{y - y_0 = f'(x_0)(x - x_0).}$$

The line through (x_0, y_0) that is perpendicular to the tangent is called the *normal line*. Since the slope of the tangent is $f'(x_0)$, the slope of the normal is $-1/f'(x_0)$, provided of course that $f'(x_0) \neq 0$. (*Remember:* for perpendicular lines neither of which is vertical, $m_1 m_2 = -1$.) As an equation for this normal line we have

(3.1.3)
$$\boxed{y - y_0 = -\frac{1}{f'(x_0)}(x - x_0).}$$

Some tangents and normals are pictured in Figure 3.1.6.

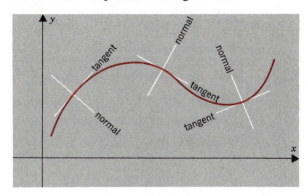

Figure 3.1.6

Problem 7. Find equations for the tangent and normal to the graph of

$$f(x) = x^2$$

at the point $(-3, f(-3)) = (-3, 9)$.

Solution. Earlier we saw that

$$f'(x) = 2x.$$

At the point $(-3, 9)$ the slope is $f'(-3) = -6$. As an equation for the tangent we have

$$y - 9 = -6(x - (-3)), \quad \text{which simplifies to} \quad y - 9 = -6(x + 3).$$

The equation for the normal can be written

$$y - 9 = \tfrac{1}{6}(x + 3). \quad \square$$

If $f'(x_0) = 0$, Equation (3.1.3) does not apply. If $f'(x_0) = 0$, the tangent line at (x_0, y_0) is horizontal (equation $y = y_0$) and the normal line is vertical (equation $x = x_0$). In Figure 3.1.7 you can see several instances of a horizontal tangent. In each case, the point of tangency is the origin, the tangent line is the x-axis ($y = 0$), and the normal line is the y-axis ($x = 0$).

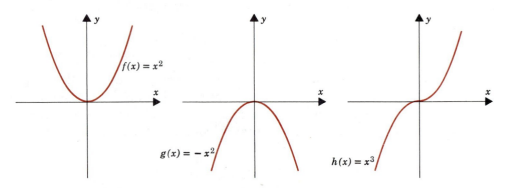

Figure 3.1.7

A Note on Vertical Tangents

It is also possible for the graph of a function to have a vertical tangent. Figure 3.1.8 shows the graph of the cube-root function $f(x) = x^{1/3}$. The difference quotient at $x = 0$,

$$\frac{f(0 + h) - f(0)}{h} = \frac{h^{1/3} - 0}{h} = \frac{1}{h^{2/3}},$$

increases without bound as h tends to zero. The graph has a vertical tangent at the origin (the line $x = 0$) and it has a horizontal normal (the line $y = 0$).

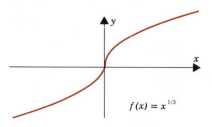

Figure 3.1.8

Vertical tangents are mentioned here only in passing. They are discussed in detail in Section 4.8.

Differentiability and Continuity

A function can be continuous at some number x without being differentiable there. For example, the absolute-value function $f(x) = |x|$ is continuous at 0 (it is everywhere continuous), but it is not differentiable at 0:

$$\frac{f(0 + h) - f(0)}{h} = \frac{|0 + h| - |0|}{h} = \frac{|h|}{h} = \begin{cases} -1, & h < 0 \\ 1, & h > 0 \end{cases},$$

so that

$$\lim_{h \to 0^-} \frac{f(0 + h) - f(0)}{h} = -1, \qquad \lim_{h \to 0^+} \frac{f(0 + h) - f(0)}{h} = 1$$

and thus

$$\lim_{h \to 0} \frac{f(0 + h) - f(0)}{h} \quad \text{does not exist.} \quad \square$$

In Figure 3.1.9 we have drawn the graph of the absolute-value function. The lack of differentiability at 0 is evident geometrically. At the point $(0, 0) = (0, f(0))$ the graph changes direction abruptly and there is no tangent line.

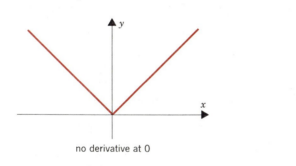

no derivative at 0

Figure 3.1.9

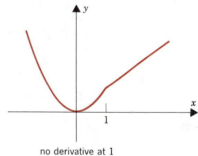

no derivative at 1

Figure 3.1.10

You can see a similar abrupt change of direction in the graph of

$$f(x) = \begin{cases} x^2, & x \le 1 \\ x, & x > 1 \end{cases} \qquad \text{(Figure 3.1.10)}$$

at the point $(1, 1)$. Once again f is everywhere continuous (you can verify that), but it is not differentiable at 1:

$$\lim_{h \to 0^-} \frac{f(1 + h) - f(1)}{h} = \lim_{h \to 0^-} \frac{(1 + h)^2 - 1}{h} = \lim_{h \to 0^-} \frac{h^2 + 2h}{h} = \lim_{h \to 0^-} (h + 2) = 2,$$

$$\lim_{h \to 0^+} \frac{f(1 + h) - f(1)}{h} = \lim_{h \to 0^+} \frac{(1 + h) - 1}{h} = \lim_{h \to 0^+} 1 = 1.$$

Since these one-sided limits are different, the two-sided limit

$$f'(1) = \lim_{h \to 0} \frac{f(1 + h) - f(1)}{h} \quad \text{does not exist.} \quad \square$$

Although not every continuous function is differentiable, every differentiable function is continuous.

THEOREM 3.1.4

If f is differentiable at x, then f is continuous at x.

Proof. For $h \neq 0$ and $x + h$ in the domain of f,

$$f(x + h) - f(x) = \frac{f(x + h) - f(x)}{h} \cdot h.$$

With f differentiable at x,

$$\lim_{h \to 0} \frac{f(x + h) - f(x)}{h} = f'(x).$$

Since $\lim\limits_{h \to 0} h = 0$, we have

$$\lim_{h \to 0} [f(x + h) - f(x)] = \left[\lim_{h \to 0} \frac{f(x + h) - f(x)}{h} \right] \cdot \left[\lim_{h \to 0} h \right] = f'(x) \cdot 0 = 0.$$

It follows that

$$\lim_{h \to 0} f(x + h) = f(x) \qquad \text{(explain)}$$

and thus (by Exercise 47, Section 2.4) that f is continuous at x. ☐

EXERCISES 3.1

Differentiate each of the following functions by forming a difference quotient

$$\frac{f(x + h) - f(x)}{h}$$

and taking the limit as h tends to 0.

1. $f(x) = 4$.	**2.** $f(x) = c$.	**3.** $f(x) = 2 - 3x$.
4. $f(x) = 4x + 1$.	**5.** $f(x) = 5x - x^2$.	**6.** $f(x) = 2x^3 + 1$.
7. $f(x) = x^4$.	**8.** $f(x) = 1/(x + 3)$.	**9.** $f(x) = \sqrt{x - 1}$.
10. $f(x) = x^3 - 4x$.	**11.** $f(x) = 1/x^2$.	**12.** $f(x) = 1/\sqrt{x}$.

For each of the functions below, find $f'(2)$ by forming the difference quotient

$$\frac{f(2 + h) - f(2)}{h}$$

and taking the limit as $h \to 0$.

13. $f(x) = (3x - 7)^2$.	**14.** $f(x) = 7x - x^2$.	**15.** $f(x) = 9/(x + 4)$.
16. $f(x) = 5 - x^4$.	**17.** $f(x) = x + \sqrt{2x}$.	**18.** $f(x) = \sqrt{6 - x}$.

Find equations for the tangent and normal to the graph of f at the point $(a, f(a))$.

19. $f(x) = x^2$; $a = 2$.   **20.** $f(x) = \sqrt{x}$; $a = 4$.

21. $f(x) = 5x - x^2$; $a = 4$.   **22.** $f(x) = 5 - x^3$; $a = 2$.

23. $f(x) = |x|$; $a = -4$.   **24.** $f(x) = 7x + x^2$; $a = -1$.

25. $f(x) = 1/(x + 2)$; $a = -3$.   **26.** $f(x) = 1/x^2$; $a = -2$.

Draw the graph of each of the following functions and indicate where it is not differentiable.

27. $f(x) = |x + 1|$.   **28.** $f(x) = |2x - 5|$.

29. $f(x) = \sqrt{|x|}$.   **30.** $f(x) = |x^2 - 4|$.

31. $f(x) = \begin{cases} x^2, & x \le 1 \\ 2 - x, & x > 1 \end{cases}$.

32. $f(x) = \begin{cases} x^2 - 1, & x \le 2 \\ 3, & x > 2 \end{cases}$.

Find $f'(c)$ if it exists.

33. $f(x) = \begin{cases} 4x, & x < 1 \\ 2x^2 + 2, & x \ge 1 \end{cases}$; $c = 1$.

34. $f(x) = \begin{cases} 3x^2, & x \le 1 \\ 2x^3 + 1, & x > 1 \end{cases}$; $c = 1$.

35. $f(x) = \begin{cases} x + 1, & x \le -1 \\ (x + 1)^2, & x > -1 \end{cases}$; $c = -1$.

36. $f(x) = \begin{cases} -\frac{1}{2}x^2, & x < 3 \\ -3x, & x \ge 3 \end{cases}$; $c = 3$.

Sketch the graph of the derivative of the function with the given graph.

37.

38.

39.

40.

41.

42.
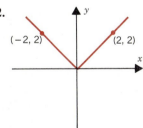

43. Show that

$$f(x) = \begin{cases} x^2, & x \le 1 \\ 2x, & x > 1 \end{cases}$$

is not differentiable at $x = 1$.   HINT: Consider Theorem 3.1.4.

44. Find A and B given that the function

$$f(x) = \begin{cases} x^3, & x \le 1 \\ Ax + B, & x > 1 \end{cases}$$

is differentiable at $x = 1$.

45. Let

$$f(x) = \begin{bmatrix} (x+1)^2, & x \le 0 \\ (x-1)^2, & x > 0 \end{bmatrix}.$$

(a) Compute $f'(x)$ for $x \ne 0$.   (b) Show that f is not differentiable at $x = 0$.

46. Given that f is differentiable at x_0, show that the function

$$g(x) = \begin{Bmatrix} f(x), & x \le x_0 \\ f'(x_0)(x - x_0) + f(x_0), & x > x_0 \end{Bmatrix}$$

is differentiable at x_0.   What is $g'(x_0)$?

Give an example of a function f that is defined for all real numbers and satisfies the given conditions.

47. $f'(x) = 0$ for all real x.   **48.** $f'(x) = 0$ for all $x \ne 0$; $f'(0)$ does not exist.

49. $f'(x)$ exists for all $x \ne -1$; $f'(-1)$ does not exist.

50. $f'(x)$ exists for all $x \ne \pm 1$;   neither $f'(1)$ nor $f'(-1)$ exists.

51. $f'(1) = 2$ and $f(1) = 7$.   **52.** $f'(2) = 5$ and $f(2) = 1$.

53. f is nowhere differentiable.   **54.** $f'(x) = 1$ for $x < 0$ and $f'(x) = -1$ for $x > 0$.

Program for computing difference quotients   (BASIC)

This program is meant to help you understand the way in which the difference quotient for a function converges to the derivative of the function at a given point.   The program prompts you to enter the function of your choice, calling if FNf.   [The specific example given in the program is

$$FNf = 3*x^2 \qquad (3x^2)$$

but you are free to make another choice, and this is part of the point: the function should be selected according to your interest.]   The program then asks for a point x at which the calculations are to be done, an initial increment h to be used in forming the difference quotient, and also a "number of times" n.   The computer first calculates the quotient

$$(FNf(x + h) - FNf(x))/(h).$$

Then it performs the same calculation with the increment h replaced by h/2, and iterates this procedure n times.

```
10 REM Compute value of difference quotient to approximate derivative
20 REM copyright © Colin C. Graham 1988-1989

30 REM change the line "def FNf = ..."
40 REM to fit your function

50 DEF FNf(x) = 3*x^2

100 INPUT "Enter x:"; x
110 INPUT "Enter h:"; h
130 INPUT "How many times?"; n

200 PRINT "h", "Difference Quotient"

300 FOR j = 1 TO n
310     PRINT h, (FNf(x + h) − FNf(x))/(h)
320     h = h/2
330 NEXT j

500 INPUT "Do another? (Y/N):"; a$
510 IF a$ = "Y" OR a$ = "y" THEN 100
520 END
```

3.2   SOME DIFFERENTIATION FORMULAS

Calculating the derivative of

$$f(x) = (x^3 - 2)(4x + 1) \qquad \text{or} \qquad f(x) = \frac{6x^2 - 1}{x^4 + 5x + 1}$$

by forming the proper difference quotient

$$\frac{f(x + h) - f(x)}{h}$$

and then taking the limit as h tends to 0 is somewhat laborious. Here we derive some general formulas that will enable us to calculate such derivatives quite quickly and easily.

We begin by pointing out that constant functions have derivative identically 0:

(3.2.1)
$$\text{if } \ f(x) = \alpha, \qquad \text{then } \ f'(x) = 0 \qquad \text{for all real } x$$

and the identity function has constant derivative 1:

(3.2.2)
$$\text{if } \ f(x) = x, \qquad \text{then } \ f'(x) = 1 \qquad \text{for all real } x.$$

Proof. For $f(x) = \alpha$,

$$f'(x) = \lim_{h \to 0} \frac{f(x + h) - f(x)}{h} = \lim_{h \to 0} \frac{\alpha - \alpha}{h} = \lim_{h \to 0} 0 = 0.$$

For $f(x) = x$,

$$f'(x) = \lim_{h \to 0} \frac{f(x + h) - f(x)}{h} = \lim_{h \to 0} \frac{(x + h) - x}{h} = \lim_{h \to 0} \frac{h}{h} = \lim_{h \to 0} 1 = 1. \quad \square$$

DEFINITION 3.2.3   DERIVATIVES OF SUMS AND SCALAR MULTIPLES

Let α be a real number. If f and g are differentiable at x, then $f + g$ and αf are differentiable at x. Moreover,

$$(f + g)'(x) = f'(x) + g'(x) \quad \text{and} \quad (\alpha f)'(x) = \alpha f'(x).$$

Proof. To verify the first formula note that

$$\frac{(f + g)(x + h) - (f + g)(x)}{h} = \frac{[f(x + h) + g(x + h)] - [f(x) + g(x)]}{h}$$

$$= \frac{f(x + h) - f(x)}{h} + \frac{g(x + h) - g(x)}{h}.$$

By definition

$$\lim_{h \to 0} \frac{f(x + h) - f(x)}{h} = f'(x), \qquad \lim_{h \to 0} \frac{g(x + h) - g(x)}{h} = g'(x).$$

Thus

$$\lim_{h\to 0} \frac{(f+g)(x+h) - (f+g)(x)}{h} = f'(x) + g'(x),$$

which means that

$$(f+g)'(x) = f'(x) + g'(x).$$

To verify the second formula we must show that

$$\lim_{h\to 0} \frac{(\alpha f)(x+h) - (\alpha f)(x)}{h} = \alpha f'(x).$$

This follows directly from the fact that

$$\frac{(\alpha f)(x+h) - (\alpha f)(x)}{h} = \alpha \left[\frac{f(x+h) - f(x)}{h}\right]. \quad \square$$

THEOREM 3.2.4   THE PRODUCT RULE

If f and g are differentiable at x, then so is their product, and

$$(fg)'(x) = f(x)g'(x) + g(x)f'(x).$$

Proof.   We form the difference quotient

$$\frac{(fg)(x+h) - (fg)(x)}{h} = \frac{f(x+h)\,g(x+h) - f(x)g(x)}{h}$$

and rewrite it as

$$f(x+h)\left[\frac{g(x+h) - g(x)}{h}\right] + g(x)\left[\frac{f(x+h) - f(x)}{h}\right].$$

(Here we have added and subtracted $f(x+h)g(x)$ in the numerator and then regrouped the terms so as to display difference quotients for f and g.)   Since f is differentiable at x, we know that f is continuous at x (Theorem 3.1.4) and thus

$$\lim_{h\to 0} f(x+h) = f(x).$$

Since

$$\lim_{h\to 0} \frac{g(x+h) - g(x)}{h} = g'(x) \quad \text{and} \quad \lim_{h\to 0} \frac{f(x+h) - f(x)}{h} = f'(x),$$

we obtain

$$\lim_{h\to 0} \frac{(fg)(x+h) - (fg)(x)}{h} = f(x)g'(x) + g(x)f'(x). \quad \square$$

Using the product rule, it is not hard to prove that

(3.2.5)

> for each positive integer n
>
> $p(x) = x^n$ has derivative $p'(x) = nx^{n-1}$.

In particular

$$p(x) = x \quad \text{has derivative} \quad p'(x) = 1 \cdot x^0 = 1,$$
$$p(x) = x^2 \quad \text{has derivative} \quad p'(x) = 2x,$$
$$p(x) = x^3 \quad \text{has derivative} \quad p'(x) = 3x^2,$$
$$p(x) = x^4 \quad \text{has derivative} \quad p'(x) = 4x^3,$$

and so on.

Proof of (3.2.5). We proceed by induction on n. If $n = 1$, then we have the identity function

$$p(x) = x,$$

which we know satisfies

$$p'(x) = 1 = 1 \cdot x^0.$$

This means that the formula holds for $n = 1$.

We suppose now that the result holds for $n = k$ and go on to show that it holds for $n = k + 1$. We let

$$p(x) = x^{k+1}$$

and note that

$$p(x) = x \cdot x^k.$$

Applying the product rule (3.2.4) and our inductive hypothesis, we obtain

$$p'(x) = x \cdot kx^{k-1} + 1 \cdot x^k = (k + 1)x^k.$$

This shows that the formula holds for $n = k + 1$.

By the axiom of induction the formula holds for all positive integers n. □

The formula for differentiating polynomials follows from (3.2.3) and (3.2.5):

(3.2.6)

> if $P(x) = a_n x^n + \cdots + a_2 x^2 + a_1 x + a_0,$
>
> then $P'(x) = na_n x^{n-1} + \cdots + 2a_2 x + a_1.$

For example,

$$P(x) = 12x^3 - 6x - 2 \quad \text{has derivative} \quad P'(x) = 36x^2 - 6$$

and

$$P(x) = \tfrac{1}{4}x^4 - 2x^2 + x + 5 \quad \text{has derivative} \quad P'(x) = x^3 - 4x + 1.$$

Problem 1. Differentiate

$$F(x) = (x^3 - 2)(4x + 1)$$

and find $F'(-1)$.

Solution. We have a product $F(x) = f(x)g(x)$ with

$$f(x) = x^3 - 2 \quad \text{and} \quad g(x) = 4x + 1.$$

The product rule gives

$$\begin{aligned}
F'(x) &= f(x)g'(x) + g(x)f'(x) \\
&= (x^3 - 2) \cdot 4 + (4x + 1)(3x^2) \\
&= 4x^3 - 8 + 12x^3 + 3x^2 \\
&= 16x^3 + 3x^2 - 8.
\end{aligned}$$

Setting $x = -1$, we have

$$F'(-1) = 16(-1)^3 + 3(-1)^2 - 8 = -16 + 3 - 8 = -21. \quad \square$$

Problem 2. Differentiate

$$F(x) = (ax + b)(cx + d).$$

Solution. We have a product $F(x) = f(x)g(x)$ with

$$f(x) = ax + b \quad \text{and} \quad g(x) = cx + d.$$

Again we use the product rule

$$F'(x) = f(x)g'(x) + g(x)f'(x).$$

In this case

$$F'(x) = (ax + b)c + (cx + d)a = 2acx + bc + ad.$$

We can also do the problem without using the product rule by first carrying out the multiplication:

$$F(x) = acx^2 + bcx + adx + bd$$

and then differentiating:

$$F'(x) = 2acx + bc + ad.$$

The result is the same. $\quad \square$

We come now to reciprocals.

> ### THEOREM 3.2.7 THE RECIPROCAL RULE
>
> If g is differentiable at x and $g(x) \neq 0$, then $1/g$ is differentiable at x and
>
> $$\left(\frac{1}{g}\right)'(x) = -\frac{g'(x)}{[g(x)]^2}.$$

Proof. Since g is differentiable at x, g is continuous at x (Theorem 3.1.4). Since $g(x) \neq 0$, we know that $1/g$ is continuous at x, and thus that

$$\lim_{h \to 0} \frac{1}{g(x+h)} = \frac{1}{g(x)}.$$

For h different from zero and sufficiently small, $g(x+h) \neq 0$ and

$$\frac{1}{h}\left[\frac{1}{g(x+h)} - \frac{1}{g(x)}\right] = -\left[\frac{g(x+h) - g(x)}{h}\right]\frac{1}{g(x+h)}\frac{1}{g(x)}.$$

As h tends to zero, the right-hand side (and thus the left) tends to

$$-\frac{g'(x)}{[g(x)]^2}. \quad \square$$

From this last result it is easy to see that the formula for the derivative of a power, x^n, also applies to negative powers; namely,

(3.2.8)
> for each negative integer n
>
> $p(x) = x^n$ has derivative $p'(x) = nx^{n-1}$.

This formula holds except, of course, at $x = 0$ where no negative power is even defined. In particular

$$p(x) = x^{-1} \quad \text{has derivative} \quad p'(x) = (-1)x^{-2} = -x^{-2},$$
$$p(x) = x^{-2} \quad \text{has derivative} \quad p'(x) = -2x^{-3},$$
$$p(x) = x^{-3} \quad \text{has derivative} \quad p'(x) = -3x^{-4},$$

and so on.

Proof of (3.2.8). Note that

$$p(x) = \frac{1}{g(x)} \quad \text{where} \quad g(x) = x^{-n} \text{ and } -n \text{ is a positive integer.}$$

The rule for reciprocals gives

$$p'(x) = -\frac{g'(x)}{[g(x)]^2} = -\frac{(-nx^{-n-1})}{x^{-2n}} = (nx^{-n-1})x^{2n} = nx^{n-1}. \quad \square$$

Problem 3. Differentiate

$$f(x) = \frac{5}{x^2} - \frac{6}{x}$$

and find $f'(\tfrac{1}{2})$.

Solution. To apply (3.2.8) we write

$$f(x) = 5x^{-2} - 6x^{-1}.$$

Differentiation gives

$$f'(x) = -10x^{-3} + 6x^{-2}.$$

Back in fractional notation

$$f'(x) = -\frac{10}{x^3} + \frac{6}{x^2}.$$

Setting $x = \tfrac{1}{2}$, we have

$$f'(\tfrac{1}{2}) = -\frac{10}{(\tfrac{1}{2})^3} + \frac{6}{(\tfrac{1}{2})^2} = -80 + 24 = -56. \quad \square$$

Problem 4. Differentiate

$$f(x) = \frac{1}{ax^2 + bx + c}.$$

Solution. Here we have a reciprocal $f(x) = 1/g(x)$ with

$$g(x) = ax^2 + bx + c.$$

The reciprocal rule (3.2.7) gives

$$f'(x) = -\frac{g'(x)}{[g(x)]^2} = -\frac{2ax + b}{[ax^2 + bx + c]^2}. \quad \square$$

Finally we come to quotients in general.

THEOREM 3.2.9 THE QUOTIENT RULE

If f and g are differentiable at x and $g(x) \neq 0$, then the quotient f/g is differentiable at x and

$$\left(\frac{f}{g}\right)'(x) = \frac{g(x)f'(x) - f(x)g'(x)}{[g(x)]^2}.$$

The proof is left to you.   HINT: $f/g = f \cdot 1/g$.

From the quotient rule you can see that all rational functi
are differentiable wherever they are defined.

Problem 5. Differentiate

$$F(x) = \frac{ax + b}{cx + d}.$$

Solution. We are dealing with a quotient $F(x) = f(x)/g(x)$. The quotient rule,

$$F'(x) = \frac{g(x)f'(x) - f(x)g'(x)}{[g(x)]^2},$$

gives

$$F'(x) = \frac{(cx + d) \cdot a - (ax + b) \cdot c}{(cx + d)^2} = \frac{ad - bc}{(cx + d)^2}. \quad \square$$

Problem 6. Differentiate

$$F(x) = \frac{6x^2 - 1}{x^4 + 5x + 1}.$$

Solution. Again we are dealing with a quotient $F(x) = f(x)/g(x)$. The quotient rule gives

$$F'(x) = \frac{(x^4 + 5x + 1)(12x) - (6x^2 - 1)(4x^3 + 5)}{(x^4 + 5x + 1)^2}. \quad \square$$

Problem 7. Find $f'(0)$, $f'(1)$, and $f'(2)$ for the function

$$f(x) = \frac{5x}{1 + x}.$$

Solution. First we find a general expression for $f'(x)$, and then we evaluate it at 0, 1, and 2. Using the quotient rule, we get

$$f'(x) = \frac{(1 + x)5 - 5x(1)}{(1 + x)^2} = \frac{5}{(1 + x)^2}.$$

This gives

$$f'(0) = \frac{5}{(1 + 0)^2} = 5, \quad f'(1) = \frac{5}{(1 + 1)^2} = \frac{5}{4}, \quad f'(2) = \frac{5}{(1 + 2)^2} = \frac{5}{9}. \quad \square$$

Problem 8. Find $f'(-1)$ for

$$f(x) = \frac{x^2}{ax^2 + b}.$$

Solution. We first find $f'(x)$ in general by the quotient rule:

$$f'(x) = \frac{(ax^2 + b)2x - x^2(2ax)}{(ax^2 + b)^2} = \frac{2bx}{(ax^2 + b)^2}.$$

Now we evaluate f' at -1:

$$f'(-1) = -\frac{2b}{(a + b)^2}. \quad \square$$

Remark. Some expressions are easier to differentiate if we rewrite them in more convenient form. For example, we can differentiate

$$f(x) = \frac{x^5 - 2x}{x^2} = \frac{x^4 - 2}{x}$$

by the quotient rule, or we can write

$$f(x) = (x^4 - 2)x^{-1}$$

and use the product rule; even better, we can notice that

$$f(x) = x^3 - 2x^{-1}$$

and proceed from there:

$$f'(x) = 3x^2 + 2x^{-2}. \quad \square$$

EXERCISES 3.2

Differentiate the following functions.

1. $F(x) = 1 - x$.

2. $F(x) = 2(1 + x)$.

3. $F(x) = 11x^5 - 6x^3 + 8$.

4. $F(x) = \dfrac{3}{x^2}$.

5. $F(x) = ax^2 + bx + c$.

6. $F(x) = \dfrac{x^4}{4} - \dfrac{x^3}{3} + \dfrac{x^2}{2} - \dfrac{x}{1}$.

7. $F(x) = -\dfrac{1}{x^2}$.

8. $F(x) = \dfrac{(x^2 + 2)}{x^3}$.

9. $G(x) = (x^2 - 1)(x - 3)$.

10. $F(x) = x - \dfrac{1}{x}$.

11. $G(x) = \dfrac{x^3}{1 - x}$.

12. $F(x) = \dfrac{ax - b}{cx - d}$.

13. $G(x) = \dfrac{x^2 - 1}{2x + 3}$.

14. $G(x) = \dfrac{7x^4 + 11}{x + 1}$.

15. $G(x) = (x - 1)(x - 2)$.

16. $G(x) = \dfrac{2x^2 + 1}{x + 2}$.

17. $G(x) = \dfrac{6 - 1/x}{x - 2}$.

18. $G(x) = \dfrac{1 + x^4}{x^2}$.

19. $G(x) = (9x^8 - 8x^9)\left(x + \dfrac{1}{x}\right)$.

20. $G(x) = \left(1 + \dfrac{1}{x}\right)\left(1 + \dfrac{1}{x^2}\right)$.

Find $f'(0)$ and $f'(1)$.

21. $f(x) = \dfrac{1}{x-2}.$

22. $f(x) = x^2(x+1).$

23. $f(x) = \dfrac{1-x^2}{1+x^2}.$

24. $f(x) = \dfrac{2x^2 + x + 1}{x^2 + 2x + 1}.$

25. $f(x) = \dfrac{ax+b}{cx+d}.$

26. $f(x) = \dfrac{ax^2 + bx + c}{cx^2 + bx + a}.$

Given that $h(0) = 3$ and $h'(0) = 2$, find $f'(0)$.

27. $f(x) = xh(x).$

28. $f(x) = 3x^2h(x) - 5x.$

29. $f(x) = h(x) - \dfrac{1}{h(x)}.$

30. $f(x) = h(x) + \dfrac{x}{h(x)}.$

Find an equation for the tangent to the graph of f at the point $(a, f(a))$.

31. $f(x) = \dfrac{x}{x+2};\quad a = -4.$

32. $f(x) = (x^3 - 2x + 1)(4x - 5);\quad a = 2.$

33. $f(x) = (x^2 - 3)(5x - x^3);\quad a = 1.$

34. $f(x) = x^2 - \dfrac{10}{x};\quad a = -2.$

35. $f(x) = 6/x^2;\quad a = 3.$

36. $f(x) = \dfrac{3x+5}{7x-3};\quad a = 1.$

Find the points where the tangent to the curve is horizontal.

37. $f(x) = (x-2)(x^2 - x - 11).$

38. $f(x) = x^2 - \dfrac{16}{x}.$

39. $f(x) = \dfrac{5x}{x^2 + 1}.$

40. $f(x) = (x+2)(x^2 - 2x - 8).$

41. $f(x) = x + \dfrac{4}{x^2}.$

42. $f(x) = \dfrac{x^2 - 2x + 4}{x^2 + 4}.$

Find the points where the tangent to the curve

43. $f(x) = -x^2 - 6$ is parallel to the line $y = 4x - 1$.

44. $f(x) = x^3 - 3x$ is perpendicular to the line $5y - 3x - 8 = 0$.

45. $f(x) = x^3 - x^2$ is perpendicular to the line $5y + x + 2 = 0$.

46. $f(x) = 4x - x^2$ is parallel to the line $2y = 3x - 5$.

47. Find the area of the triangle formed by the x-axis and the lines tangent and normal to the curve $f(x) = 6x - x^2$ at the point $(5, 5)$.

48. Find the area of the triangle formed by the x-axis and the lines tangent and normal to the curve $f(x) = 9 - x^2$ at the point $(2, 5)$.

49. Determine the coefficients A, B, C so that the curve $f(x) = Ax^2 + Bx + C$ will pass through the point $(1, 3)$ and be tangent to the line $4x + y = 8$ at the point $(2, 0)$.

50. Determine A, B, C, D so that the curve $f(x) = Ax^3 + Bx^2 + Cx + D$ will be tangent to the line $y = 3x - 3$ at the point $(1, 0)$ and tangent to the line $y = 18x - 27$ at the point $(2, 9)$.

51. Prove the validity of the quotient rule.

52. Verify that, if f, g, h are differentiable, then

$$(fgh)'(x) = f'(x)g(x)h(x) + f(x)g'(x)h(x) + f(x)g(x)h'(x).$$

HINT: Apply the product rule to $[f(x)g(x)]h(x)$.

53. Use the product rule (3.2.4) to show that, if f is differentiable, then

$$g(x) = [f(x)]^2 \quad \text{has derivative} \quad g'(x) = 2f(x)f'(x).$$

54. Show that, if f is differentiable, then

$$g(x) = [f(x)]^n \quad \text{has derivative} \quad g'(x) = n[f(x)]^{n-1}f'(x)$$

for each nonzero integer n.   HINT: Mimic the inductive proof of (3.2.5).

55. Find A and B given that the derivative of

$$f(x) = \begin{cases} Ax^3 + Bx + 2, & x \le 2 \\ Bx^2 - A, & x > 2 \end{cases}$$

is continuous for all real x.   HINT: f itself must be continuous.

56. Find A and B given that the derivative of

$$f(x) = \begin{cases} Ax^2 + B, & x < -1 \\ Bx^5 + Ax + 4, & x \ge -1 \end{cases}$$

is continuous for all real x.

3.3   THE d/dx NOTATION; DERIVATIVES OF HIGHER ORDER

The d/dx Notation

So far we have indicated the derivative by a prime, but there are other notations that are widely used, particularly in science and engineering.   The most popular of these is the double-d notation of Leibniz.   In the Leibniz notation the derivative of a function y is indicated by writing

$$\frac{dy}{dx}, \quad \frac{dy}{dt}, \quad \text{or} \quad \frac{dy}{dz}, \quad \text{etc.,}$$

depending upon whether the letter x, t, or z, etc., is being used for the elements of the domain of y.   For instance, if y is initially defined by

$$y(x) = x^3,$$

then the Leibniz notation gives

$$\frac{dy(x)}{dx} = 3x^2.$$

Usually writers drop the (x) and simply write

$$y = x^3 \quad \text{and} \quad \frac{dy}{dx} = 3x^2.$$

The symbols

$$\frac{d}{dx}, \quad \frac{d}{dt}, \quad \frac{d}{dz}, \quad \text{etc.,}$$

are also used as prefixes before expressions to be differentiated.   For example,

$$\frac{d}{dx}(x^3 - 4x) = 3x^2 - 4, \qquad \frac{d}{dt}(t^2 + 3t + 1) = 2t + 3, \qquad \frac{d}{dz}(z^5 - 1) = 5z^4.$$

In the Leibniz notation the differentiation formulas look like this:

$$\frac{d}{dx}[f(x) + g(x)] = \frac{d}{dx}[f(x)] + \frac{d}{dx}[g(x)], \qquad \frac{d}{dx}[\alpha f(x)] = \alpha \frac{d}{dx}[f(x)],$$

$$\frac{d}{dx}[f(x)g(x)] = f(x)\frac{d}{dx}[g(x)] + g(x)\frac{d}{dx}[f(x)],$$

$$\frac{d}{dx}\left[\frac{1}{g(x)}\right] = -\frac{1}{[g(x)]^2}\frac{d}{dx}[g(x)], \qquad \frac{d}{dx}\left[\frac{f(x)}{g(x)}\right] = \frac{g(x)\frac{d}{dx}[f(x)] - f(x)\frac{d}{dx}[g(x)]}{[g(x)]^2}.$$

Often functions f and g are replaced by u and v and the x is left out altogether. Then the formulas look like this:

$$\frac{d}{dx}(u+v) = \frac{du}{dx} + \frac{dv}{dx}, \qquad \frac{d}{dx}(\alpha u) = \alpha\frac{du}{dx},$$

$$\frac{d}{dx}(uv) = u\frac{dv}{dx} + v\frac{du}{dx},$$

$$\frac{d}{dx}\left(\frac{1}{v}\right) = -\frac{1}{v^2}\frac{dv}{dx}, \qquad \frac{d}{dx}\left(\frac{u}{v}\right) = \frac{v\frac{du}{dx} - u\frac{dv}{dx}}{v^2}.$$

The only way to develop a feeling for this notation is to use it. Below we work out some practice problems.

Problem 1. Find

$$\frac{dy}{dx} \quad \text{for} \quad y = \frac{3x-1}{5x+2}.$$

Solution. Here we can use the quotient rule:

$$\frac{dy}{dx} = \frac{(5x+2)\frac{d}{dx}(3x-1) - (3x-1)\frac{d}{dx}(5x+2)}{(5x+2)^2}$$

$$= \frac{(5x+2)(3) - (3x-1)(5)}{(5x+2)^2} = \frac{11}{(5x+2)^2}. \quad \square$$

Problem 2. Find

$$\frac{dy}{dx} \quad \text{for} \quad y = (x^3+1)(3x^5+2x-1).$$

Solution. Here we can use the product rule:

$$\frac{dy}{dx} = (x^3+1)\frac{d}{dx}(3x^5+2x-1) + (3x^5+2x-1)\frac{d}{dx}(x^3+1)$$

$$= (x^3+1)(15x^4+2) + (3x^5+2x-1)(3x^2)$$

$$= (15x^7+15x^4+2x^3+2) + (9x^7+6x^3-3x^2)$$

$$= 24x^7+15x^4+8x^3-3x^2+2. \quad \square$$

Problem 3. Find

$$\frac{d}{dt}\left(t^3 - \frac{t}{t^2 - 1}\right).$$

Solution

$$\frac{d}{dt}\left(t^3 - \frac{t}{t^2 - 1}\right) = \frac{d}{dt}(t^3) - \frac{d}{dt}\left(\frac{t}{t^2 - 1}\right)$$

$$= 3t^2 - \left[\frac{(t^2 - 1)(1) - t(2t)}{(t^2 - 1)^2}\right] = 3t^2 + \frac{t^2 + 1}{(t^2 - 1)^2}. \quad \square$$

Problem 4. Find

$$\frac{du}{dx} \quad \text{for} \quad u = x(x + 1)(x + 2).$$

Solution.   You can think of u as

$$[x(x + 1)](x + 2) \quad \text{or as} \quad x[(x + 1)(x + 2)].$$

From the first point of view,

$$\frac{du}{dx} = [x(x + 1)](1) + (x + 2)\frac{d}{dx}[x(x + 1)]$$

$$= x(x + 1) + (x + 2)[x(1) + (x + 1)(1)]$$

(∗) $$= x(x + 1) + (x + 2)(2x + 1).$$

From the second point of view,

$$\frac{du}{dx} = x\frac{d}{dx}[(x + 1)(x + 2)] + (x + 1)(x + 2)(1)$$

$$= x[(x + 1)(1) + (x + 2)(1)] + (x + 1)(x + 2)$$

(∗∗) $$= x(2x + 3) + (x + 1)(x + 2).$$

Both (∗) and (∗∗) can be multiplied out to give

$$\frac{du}{dx} = 3x^2 + 6x + 2.$$

This same result can be obtained by first carrying out the entire multiplication and then differentiating:

$$u = x(x + 1)(x + 2) = x(x^2 + 3x + 2) = x^3 + 3x^2 + 2x$$

so that

$$\frac{du}{dx} = 3x^2 + 6x + 2. \quad \square$$

Problem 5. Evaluate dy/dx at $x = 0$ and $x = 1$ given that

$$y = \frac{x^2}{x^2 - 4}.$$

Solution

$$\frac{dy}{dx} = \frac{(x^2 - 4)2x - x^2(2x)}{(x^2 - 4)^2} = -\frac{8x}{(x^2 - 4)^2}.$$

At $x = 0$, $dy/dx = 0$; at $x = 1$, $dy/dx = -\frac{8}{9}$. ☐

Derivatives of Higher Order

If a function f is differentiable, then we can form a new function f'. If f' is itself differentiable, then we can form its derivative, called the *second derivative of f*, and denoted by f''. So long as we have differentiability, we can continue in this manner, forming f''', etc. The prime notation is not used beyond the third order. For the fourth derivative we write $f^{(4)}$ and more generally, for the nth derivative, $f^{(n)}$.
   If $f(x) = x^5$, then

$$f'(x) = 5x^4, \quad f''(x) = 20x^3, \quad f'''(x) = 60x^2, \quad f^{(4)}(x) = 120x, \quad f^{(5)}(x) = 120.$$

All higher derivatives are identically zero. As a variant of this notation you can write $y = x^5$ and then

$$y' = 5x^4, \quad y'' = 20x^3, \quad y''' = 60x^2, \quad \text{etc.}$$

   Since each polynomial P has a derivative P' that is in turn a polynomial and each rational function Q has derivative Q' that is in turn a rational function, polynomials and rational functions have derivatives of all orders. In the case of a polynomial of degree n, derivatives of order greater than n are all identically zero. (Explain.)
   In the Leibniz notation the derivatives of higher order are written

$$\frac{d^2y}{dx^2} = \frac{d}{dx}\left(\frac{dy}{dx}\right), \quad \frac{d^3y}{dx^3} = \frac{d}{dx}\left(\frac{d^2y}{dx^2}\right), \quad \ldots, \quad \frac{d^ny}{dx^n} = \frac{d}{dx}\left(\frac{d^{n-1}y}{dx^{n-1}}\right), \quad \ldots$$

or

$$\frac{d^2}{dx^2}[f(x)] = \frac{d}{dx}\left[\frac{d}{dx}[f(x)]\right], \quad \frac{d^3}{dx^3}[f(x)] = \frac{d}{dx}\left[\frac{d^2}{dx^2}[f(x)]\right], \quad \ldots,$$

$$\frac{d^n}{dx^n}[f(x)] = \frac{d}{dx}\left[\frac{d^{n-1}}{dx^{n-1}}[f(x)]\right], \quad \ldots$$

Below we work out some examples.

Example 6. If

$$f(x) = x^4 - 3x^{-1} + 5,$$

then

$$f'(x) = 4x^3 + 3x^{-2} \quad \text{and} \quad f''(x) = 12x^2 - 6x^{-3}. \ \square$$

Example 7

$$\frac{d}{dx}(x^5 - 4x^3 + 7x) = 5x^4 - 12x^2 + 7$$

so that

$$\frac{d^2}{dx^2}(x^5 - 4x^3 + 7x) = \frac{d}{dx}(5x^4 - 12x^2 + 7) = 20x^3 - 24x$$

and

$$\frac{d^3}{dx^3}(x^5 - 4x^3 + 7x) = \frac{d}{dx}(20x^3 - 24x) = 60x^2 - 24. \quad \square$$

Example 8. Finally we consider $y = x^{-1}$. In the Leibniz notation

$$\frac{dy}{dx} = -x^{-2}, \quad \frac{d^2y}{dx^2} = 2x^{-3}, \quad \frac{d^3y}{dx^3} = -6x^{-4}, \quad \frac{d^4y}{dx^4} = 24x^{-5}$$

and, more generally,

$$\frac{d^n y}{dx^n} = (-1)^n n! \, x^{-n-1}.$$

In the prime notation

$$y' = -x^{-2}, \quad y'' = 2x^{-3}, \quad y''' = -6x^{-4}, \quad y^{(4)} = 24x^{-5}$$

and

$$y^{(n)} = (-1)^n n! \, x^{-n-1}. \quad \square$$

EXERCISES 3.3

Find dy/dx.

1. $y = 3x^4 - x^2 + 1$.   **2.** $y = x^2 + 2x^{-4}$.   **3.** $y = x - \dfrac{1}{x}$.

4. $y = \dfrac{2x}{1-x}$.   **5.** $y = \dfrac{x}{1+x^2}$.   **6.** $y = x(x-2)(x+1)$.

7. $y = \dfrac{x^2}{1-x}$.   **8.** $y = \left(\dfrac{x}{1+x}\right)\left(\dfrac{2-x}{3}\right)$.   **9.** $y = \dfrac{x^3+1}{x^3-1}$.   **10.** $y = \dfrac{x^2}{(1+x)}$.

Find the indicated derivative.

11. $\dfrac{d}{dx}(2x-5)$.   **12.** $\dfrac{d}{dx}(5x+2)$.

13. $\dfrac{d}{dx}[(3x^2 - x^{-1})(2x+5)]$.   **14.** $\dfrac{d}{dx}[(2x^2 + 3x^{-1})(2x - 3x^{-2})]$.

15. $\dfrac{d}{dt}\left(\dfrac{t^2+1}{t^2-1}\right)$.   **16.** $\dfrac{d}{dt}\left(\dfrac{2t^3+1}{t^4}\right)$.   **17.** $\dfrac{d}{dt}\left(\dfrac{t^4}{2t^3-1}\right)$.

18. $\dfrac{d}{dt}\left[\dfrac{t}{(1+t)^2}\right]$.   **19.** $\dfrac{d}{du}\left(\dfrac{2u}{1-2u}\right)$.   **20.** $\dfrac{d}{du}\left(\dfrac{u^2}{u^3+1}\right)$.

21. $\dfrac{d}{du}\left(\dfrac{u}{u-1} - \dfrac{u}{u+1}\right).$ **22.** $\dfrac{d}{du}[u^2(1-u^2)(1-u^3)].$ **23.** $\dfrac{d}{dx}\left(\dfrac{x^2}{1-x^2} - \dfrac{1-x^2}{x^2}\right).$

24. $\dfrac{d}{dx}\left(\dfrac{3x^4+2x+1}{x^4+x-1}\right).$ **25.** $\dfrac{d}{dx}\left(\dfrac{x^3+x^2+x+1}{x^3-x^2+x-1}\right).$ **26.** $\dfrac{d}{dx}\left(\dfrac{x^3+x^2+x-1}{x^3-x^2+x+1}\right).$

Evaluate dy/dx at $x = 2$.

27. $y = (x+1)(x+2)(x+3).$ **28.** $y = (x+1)(x^2+2)(x^3+3).$

29. $y = \dfrac{(x-1)(x-2)}{(x+2)}.$ **30.** $y = \dfrac{(x^2+1)(x^2-2)}{x^2+2}.$

Find the second derivative.

31. $f(x) = 7x^3 - 6x^5.$ **32.** $f(x) = 2x^5 - 6x^4 + 2x - 1.$

33. $f(x) = \dfrac{x^2-3}{x}.$ **34.** $f(x) = x^2 - \dfrac{1}{x^2}.$

35. $f(x) = (x^2-2)(x^{-2}+2).$ **36.** $f(x) = (2x-3)\left(\dfrac{2x+3}{x}\right).$

Find d^3y/dx^3.

37. $y = \frac{1}{3}x^3 + \frac{1}{2}x^2 + x + 1.$ **38.** $y = (1+5x)^2.$ **39.** $y = (2x-5)^2.$

40. $y = \frac{1}{6}x^3 - \frac{1}{4}x^2 + x - 3.$ **41.** $y = x^3 - \dfrac{1}{x^3}.$ **42.** $y = \dfrac{x^4+2}{x}.$

Find the indicated derivatives.

43. $\dfrac{d}{dx}\left[x\dfrac{d}{dx}(x-x^2)\right].$ **44.** $\dfrac{d^2}{dx^2}\left[(x^2-3x)\dfrac{d}{dx}(x+x^{-1})\right].$

45. $\dfrac{d^4}{dx^4}[3x-x^4].$ **46.** $\dfrac{d^5}{dx^5}[ax^4+bx^3+cx^2+dx+e].$

47. $\dfrac{d^2}{dx^2}\left[(1+2x)\dfrac{d^2}{dx^2}(5-x^3)\right].$ **48.** $\dfrac{d^3}{dx^3}\left[\dfrac{1}{x}\dfrac{d^2}{dx^2}(x^4-5x^2)\right].$

49. Show that in general

$$(fg)''(x) \neq f(x)g''(x) + f''(x)g(x).$$

50. Verify the identity

$$f(x)g''(x) - f''(x)g(x) = \dfrac{d}{dx}[f(x)g'(x) - f'(x)g(x)].$$

Determine the values of x for which (a) $f''(x) = 0$, (b) $f''(x) > 0$, (c) $f''(x) < 0$.

51. $f(x) = x^3.$ **52.** $f(x) = x^4.$

53. $f(x) = x^4 + 2x^3 - 12x^2 + 1.$ **54.** $f(x) = x^4 + 3x^3 - 6x^2 - x.$

55. Prove by induction that

$$\text{if}\quad y = x^{-1}, \qquad \text{then}\quad \dfrac{d^ny}{dx^n} = (-1)^n n! x^{-n-1}.$$

56. Let u, v, w be differentiable functions of x. Express the derivative of the product uvw in terms of the functions u, v, w and their derivatives.

3.4 THE DERIVATIVE AS A RATE OF CHANGE

In the case of a linear function $y = mx + b$ the graph is a straight line and the slope m measures the steepness of the line by giving the rate of climb of the line, *the rate of change of y with respect to x.*

As x changes from x_0 to x_1, y changes m times as much:

$$y_1 - y_0 = m(x_1 - x_0).$$ (Figure 3.4.1)

Thus the slope m gives the change in y per unit change in x.

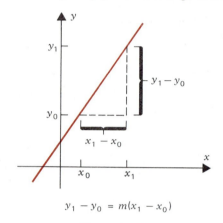

$$y_1 - y_0 = m(x_1 - x_0)$$

Figure 3.4.1

In the more general case of a differentiable function

$$y = f(x)$$

the graph is a curve.   The slope

$$\frac{dy}{dx} = f'(x)$$

still gives *the rate of change of y with respect to x,* but this rate can vary from point to point.   At $x = x_1$ (see Figure 3.4.2) the rate of change of y with respect to x is $f'(x_1)$; the steepness of the graph is that of a line of slope $f'(x_1)$.   At $x = x_2$ the rate of change of y with respect to x is $f'(x_2)$; the steepness of the graph is that of a line of slope $f'(x_2)$.   At $x = x_3$ the rate of change of y with respect to x is $f'(x_3)$; the steepness of the graph is that of a line of slope $f'(x_3)$.

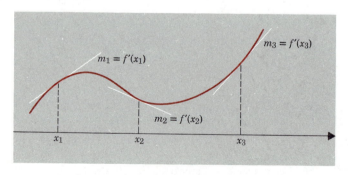

Figure 3.4.2

From your understanding of slope it should be apparent to you when examining Figure 3.4.2 that $f'(x_1) > 0$, $f'(x_2) < 0$, and $f'(x_3) > 0$, and that, in general, a function increases on any interval where the derivative remains positive and decreases on any interval where the derivative remains negative. We will take up this matter carefully in Chapter 4. Right now we look at rates of change in the context of some simple geometric figures.

Example 1. A square of side s has area $A = s^2$. Now suppose that s increases. As s increases from 1 to 2, A increases by 3 square units:

$$2^2 - 1^2 = 4 - 1 = 3. \qquad \text{(See Figure 3.4.3.)}$$

As s increases from 2 to 3, A increases by 5 square units:

$$3^2 - 2^2 = 9 - 4 = 5.$$

As s increases from 3 to 4, A increases by 7 square units:

$$4^2 - 3^2 = 16 - 9 = 7,$$

and so on. Continued unit changes in s cause ever-increasing changes in A.

In Figure 3.4.4 we have plotted A against s. The rate of change of A with respect to s is the derivative

$$\frac{dA}{ds} = 2s.$$

This appears in Figure 3.4.4 as the slope of the tangent line. At $s = 1$, $dA/ds = 2$; thus, at $s = 1$, A is increasing at the rate of a linear function of slope 2. At $s = 2$, $dA/ds = 4$; thus, at $s = 2$, A is increasing at the rate of a linear function of slope 4, and so on. At $s = k$, $dA/ds = 2k$ and A is increasing at the rate of a linear function of slope $2k$. $\square$

Figure 3.4.3

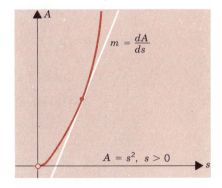

Figure 3.4.4

Example 2. Suppose that we have a circular cylinder of changing dimensions. When the base radius is r and the height is h, the cylinder has volume

$$V = \pi r^2 h.$$

If r remains constant while h changes, then V can be viewed as a function of h. The rate of change of V with respect to h is the derivative

$$\frac{dV}{dh} = \pi r^2.$$

(If h increases at the rate of α centimeters per minute, then V increases at the rate of $\pi r^2 \alpha$ cubic centimeters per minute.)

If h remains constant while r changes, then V can be viewed as a function of r. The rate of change of V with respect to r is the derivative

$$\frac{dV}{dr} = 2\pi rh.$$

(If r increases at the rate of α centimeters per minute, then V increases at the rate of $2\pi rh\alpha$ cubic centimeters per minute.)

Suppose now that r is changing but V is being kept constant. How is h changing with respect to r? To answer this we express h in terms of r and V:

$$h = \frac{V}{\pi r^2} = \frac{V}{\pi} r^{-2}.$$

Since V is being held constant, h is now a function of r. The rate of change of h with respect to r is the derivative

$$\frac{dh}{dr} = -\frac{2V}{\pi} r^{-3} = -\frac{2(\pi r^2 h)}{\pi} r^{-3} = -\frac{2h}{r}.$$

(If r increases at the rate of α centimeters per minute, then h *decreases* at the rate of $2h\alpha/r$ cubic centimeters per minute.) ▢

The derivative as a rate of change is one of the fundamental ideas of calculus. We will come back to it again and again. This brief section is only introductory.

EXERCISES 3.4   (Elementary area and volume formulas are listed in Section 1.2.)

1. Find the rate of change of the area of a circle with respect to the radius r. What is the rate when $r = 2$?

2. Find the rate of change of the volume of a cube with respect to the length s of a side. What is the rate when $s = 4$?

3. Find the rate of change of the area of a square with respect to the length z of a diagonal. What is the rate when $z = 4$?

4. Find the rate of change of $y = 1/x$ with respect to x at $x = -1$.

5. Find the rate of change of $y = [x(x + 1)]^{-1}$ with respect to x at $x = 2$.

6. Find the values of x for which the rate of change of $y = x^3 - 12x^2 + 45x - 1$ with respect to x is zero.

7. Find the rate of change of the volume of a ball with respect to the radius r.

8. Find the rate of change of the surface area of a ball with respect to the radius r. What is this rate of change when $r = r_0$? How must r_0 be chosen so that the rate of change is 1?

9. Find x_0 given that the rate of change of $y = 2x^2 + x - 1$ with respect to x at $x = x_0$ is 4.

10. Find the rate of change of the area A of a circle with respect to
    (a) the diameter d.   (b) the circumference C.

11. Find the rate of change of the volume V of a cube with respect to
    (a) the length w of a diagonal on one of the faces.
    (b) the length z of one of the diagonals of the cube.

12. The dimensions of a rectangle are changing in such a way that the area of the rectangle remains constant. Find the rate of change of the height h with respect to the base b.

13. The area of a sector in a circle is given by the formula $A = \frac{1}{2}r^2\theta$ where r is the radius and θ is the central angle measured in radians.
    (a) Find the rate of change of A with respect to θ if r remains constant.
    (b) Find the rate of change of A with respect to r if θ remains constant.
    (c) Find the rate of change of θ with respect to r if A remains constant.

14. The total surface area of a right circular cylinder is given by the formula $A = 2\pi r(r + h)$ where r is the radius and h is the height.
    (a) Find the rate of change of A with respect to h if r remains constant.
    (b) Find the rate of change of A with respect to r if h remains constant.
    (c) Find the rate of change of h with respect to r if A remains constant.

15. The edge of a cube is decreasing at the rate of 3 centimeters per second. How is the volume of the cube changing when the edge is 5 centimeters long?

16. The diameter of a sphere is increasing at the rate of 4 centimeters per minute. How is the volume of the sphere changing when the diameter is 10 centimeters long?

17. For what value of x is the rate of change of

$$y = ax^2 + bx + c \quad \text{with respect to } x$$

the same as the rate of change of

$$z = bx^2 + ax + c \quad \text{with respect to } x?$$

Assume $a \neq b$.

18. Find the rate of change of the product $f(x)g(x)h(x)$ with respect to x at $x = 1$ given that

$$f(1) = 0, \quad g(1) = 2, \quad h(1) = -2, \quad f'(1) = 1, \quad g'(1) = -1, \quad h'(1) = 0.$$

3.5   VELOCITY AND ACCELERATION; FREE FALL

Suppose that an object is moving along a straight line and that, for each time t during a certain time interval, the object has position (coordinate) $x(t)$.   If $x'(t)$ exists, then $x'(t)$ gives the *rate of change of position with respect to time*.   This rate of change of position is called the *velocity* of the object; in symbols,

(3.5.1)
$$v(t) = x'(t).$$

If the velocity function is itself differentiable, then its rate of change with respect to time is called the *acceleration*; in symbols,

(3.5.2)
$$a(t) = v'(t) = x''(t).$$

In the Leibniz notation,

(3.5.3)
$$v = \frac{dx}{dt} \quad \text{and} \quad a = \frac{dv}{dt} = \frac{d^2x}{dt^2}.$$

The *speed* v is by definition the absolute value of the velocity:

(3.5.4)
$$\text{speed at time } t = v(t) = |v(t)|.$$

1. Positive velocity indicates motion in the positive direction (x is increasing).   Negative velocity indicates motion in the negative direction (x is decreasing).

2. Positive acceleration indicates increasing velocity (increasing speed in the positive direction or decreasing speed in the negative direction).   Negative acceleration indicates decreasing velocity (decreasing speed in the positive direction or increasing speed in the negative direction).

3. It follows from (2) that, if the velocity and acceleration have the same sign, the object is speeding up, but if the velocity and acceleration have opposite signs, the object is slowing down.

Example 1.   An object moves along the x-axis, its position at each time t given by the function

$$x(t) = t^3 - 12t^2 + 36t - 27.$$

Let's study the motion from time $t = 0$ to time $t = 9$.
    The object starts out 27 units to the left of the origin:

$$x(0) = 0^3 - 12(0)^2 + 36(0) - 27 = -27$$

and ends up 54 units to the right of the origin:

$$x(9) = 9^3 - 12(9)^2 + 36(9) - 27 = 54.$$

We can find the velocity function by differentiating the position function:

$$v(t) = x'(t) = 3t^2 - 24t + 36 = 3(t-2)(t-6).$$

It is clear that

$$v(t) \text{ is } \begin{cases} \text{positive,} & \text{for } 0 \le t < 2 \\ 0, & \text{at } t = 2 \\ \text{negative,} & \text{for } 2 < t < 6 \\ 0, & \text{at } t = 6 \\ \text{positive,} & \text{for } 6 < t \le 9 \end{cases}.$$

We can interpret all this as follows: The object begins by moving to the right [$v(t)$ is positive for $0 \le t < 2$]; it comes to a stop at time $t = 2$ [$v(2) = 0$]; it then moves left [$v(t)$ is negative for $2 < t < 6$]; it stops at time $t = 6$ [$v(6) = 0$]; it then moves right and keeps going right [$v(t) > 0$ for $6 < t \le 9$].

We can find the acceleration by differentiating the velocity:

$$a(t) = v'(t) = 6t - 24 = 6(t-4).$$

Obviously

$$a(t) \text{ is } \begin{cases} \text{negative,} & \text{for } 0 \le t < 4 \\ 0, & \text{at } t = 4 \\ \text{positive,} & \text{for } 4 < t \le 9 \end{cases}.$$

At the beginning the velocity decreases, reaching a minimum at time $t = 4$. Then the velocity starts to increase and continues to increase.

Figure 3.5.1 shows a diagram for the sign of the velocity and a comparable diagram for the sign of the acceleration. Combining the two diagrams we have a brief description of the motion in convenient form.

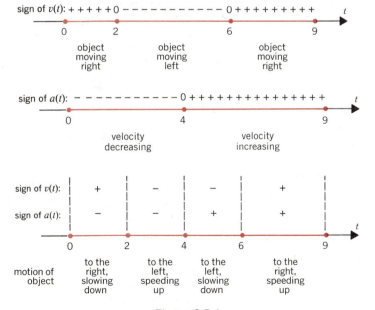

Figure 3.5.1

The direction of the motion at each time $t \in [1, 9]$ is represented schematically in Figure 3.5.2.

Figure 3.5.2

A better way to represent the motion is to graph x as a function of t, as we did in Figure 3.5.3. The velocity $v(t) = x'(t)$ then appears as the slope of the tangent to the curve. Note, for example, that at $t = 2$ and $t = 6$, the tangent is horizontal and the slope is 0. This reflects the fact that $v(2) = 0$ and $v(6) = 0$. $\square$

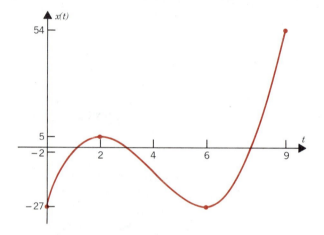

Figure 3.5.3

Before going on, a few words about units. The units of velocity and acceleration depend upon the units used to measure distance and the units used to measure time. The units of velocity are units of distance per unit time:

feet per second, meters per second, miles per hour, etc.

The units of acceleration are units of distance per unit time per unit time:

feet per second per second, meters per second per second,

miles per hour per hour, etc.

Free Fall   (Near the Surface of the Earth)

Imagine an object (for example, a rock or an apple) falling to the ground. We will assume that the object is in *free fall*: namely, that the gravitational pull on the object is constant throughout the fall and that there is no air resistance.†

---

† In practice, neither of these conditions is ever fully met. Gravitational attraction near the surface of the earth does vary somewhat with altitude, and there is always some air resistance. Nevertheless, in the setting within which we will be working the results that we obtain are correct within close approximation.

Galileo's formula for free fall gives the height of the object at each time t of the fall:

(3.5.5)
$$y(t) = -\tfrac{1}{2}gt^2 + v_0 t + y_0.†$$

Let's examine this formula. Since $y(0) = y_0$, the constant y_0 represents the height of the object at time $t = 0$. This is called *the initial position*. Differentiation gives

$$y'(t) = -gt + v_0.$$

Since $y'(0) = v_0$, the constant v_0 gives the velocity of the object at time $t = 0$. This is called *the initial velocity*. A second differentiation gives

$$y''(t) = -g.$$

This indicates that the object falls with constant negative acceleration $-g$. (Why negative?)

The constant g is a *gravitational constant*. If time is measured in seconds and distance in feet, then g is approximately 32 feet per second per second.†† In making numerical calculations we will take g as 32 feet per second per second. Equation 3.5.5 then reads

$$y(t) = -16t^2 + v_0 t + y_0.$$

Problem 2. A stone is dropped from a height of 1600 feet. In how many seconds does it hit the ground? What is the speed at impact?

Solution. Here $y_0 = 1600$ and $v_0 = 0$. Consequently, by (3.5.5),

$$y(t) = -16t^2 + 1600.$$

To find t at the moment of impact we set $y(t) = 0$. This gives

$$-16t^2 + 1600 = 0, \quad t^2 = 100, \quad t = \pm 10.$$

We can disregard the negative answer because the stone was not dropped before time $t = 0$. We conclude that it takes 10 seconds for the stone to hit the ground.

The velocity at impact is the velocity at time $t = 10$. Since

$$v(t) = y'(t) = -32t,$$

we have

$$v(10) = -320.$$

The speed at impact is 320 feet per second. □

---

† Galileo Galilei (1564–1642), a great Italian astronomer and mathematician, is popularly known today for his early experiments with falling objects. His astronomical observations led him to support the Copernican view of the solar system. For this he was brought before the Inquisition.

†† The value of this constant varies with latitude and elevation. It is approximately 32 feet per second per second at the equator at elevation zero. In Greenland it is about 32.23.

Problem 3.   An explosion causes debris to rise vertically with an initial velocity of 72 feet per second.

(a) In how many seconds does it attain maximum height?
(b) What is this maximum height?
(c) What is the speed of the debris as it reaches a height of 32 feet?   (i) going up? (ii) coming back down?

Solution.   The basic equation is again

$$y(t) = -16t^2 + v_0 t + y_0.$$

Here $y_0 = 0$ (it starts at ground level) and $v_0 = 72$ (the initial velocity is 72 feet per second).   The equation of motion is therefore

$$y(t) = -16t^2 + 72t.$$

Differentiation gives

$$v(t) = y'(t) = -32t + 72.$$

The maximum height is attained when the velocity is 0.   This occurs at time $t = \frac{72}{32} = \frac{9}{4}$.   Since $y(\frac{9}{4}) = 81$, the maximum height attained is 81 feet.

To answer part (c) we must first find those times t for which $y(t) = 32$.   Since

$$y(t) = -16t^2 + 72t,$$

the condition $y(t) = 32$ gives

$$16t^2 - 72t + 32 = 0.$$

This quadratic has two solutions, $t = \frac{1}{2}$ and $t = 4$.   Since $v(\frac{1}{2}) = 56$ and $v(4) = -56$, the velocity going up is 56 feet per second and the velocity coming down is -56 feet per second.   In each case the speed is 56 feet per second.   □

The Energy of a Falling Body   (Near the Surface of the Earth)

If we lift an object, we counteract the force of gravity.   In so doing we increase what the physicists call the *gravitational potential energy* of the object.   This can be defined as follows:

$$\text{gravitational potential energy} = \text{weight} \times \text{height}.$$

Since the weight of an object of mass m is mg (we take this from physics), we can write

$$\text{GPE} = mgy.$$

If we lift an object and release it, the object starts to drop.   As it drops, it loses height and therefore loses gravitational potential energy, but it picks up speed v.   The speed v with which the object falls gives the object another form of energy, called *kinetic energy,* the energy of motion.   This is defined as follows:

$$\text{KE} = \tfrac{1}{2}mv^2. \qquad \text{(general definition)}$$

This definition of kinetic energy applies to any kind of motion. For straight-line motion (which is the motion we are considering here), $v^2 = v^2$ and we can write

$$\text{KE} = \tfrac{1}{2}mv^2. \qquad \text{(for straight-line motion)}$$

Now let's go back to our falling object. As it drops, it loses potential energy GPE but gains kinetic energy KE. How much kinetic energy does it gain? Just enough to balance out what it loses in potential energy; namely, throughout the motion GPE + KE remains constant:

(3.5.6)
$$mgy + \tfrac{1}{2}mv^2 = C.$$

Proof. (You have seen that constant functions have derivative identically zero. The argument we give here is incomplete because it depends on the fact, so far unproven, that *only* constant functions have derivative identically zero. A proof of this will be given in Chapter 4.)

We will differentiate $mgy + \tfrac{1}{2}mv^2$ with respect to time t and show that the derivative is identically zero. It will follow then that the quantity remains constant. Observe first that

$$\frac{d}{dt}(v^2) = \frac{d}{dt}(vv) = v\frac{dv}{dt} + v\frac{dv}{dt} = 2v\frac{dv}{dt}.$$

Therefore

$$\frac{d}{dt}[mgy + \tfrac{1}{2}mv^2] = mg\frac{dy}{dt} + \tfrac{1}{2}m\frac{d}{dt}(v^2)$$

$$= mgv + \tfrac{1}{2}m\left[2v\frac{dv}{dt}\right]$$

$$= mgv + mv\frac{dv}{dt}$$

$$dv/dt = a = -g \longrightarrow$$

$$= mgv + mv(-g) = mgv - mgv = 0. \quad \square$$

Problem 4. An object at rest falls freely from height y_0. What is the speed of the object at height y?

Solution. We start with the energy equation

$$mgy + \tfrac{1}{2}mv^2 = C.$$

Since $v = 0$ at height $y = y_0$, it is clear that $C = mgy_0$. The energy equation can thus be written

$$mgy + \tfrac{1}{2}mv^2 = mgy_0.$$

Dividing by m we have

$$gy + \tfrac{1}{2}v^2 = gy_0$$
$$\tfrac{1}{2}v^2 = g(y_0 - y)$$
$$v^2 = 2g(y_0 - y).$$

It follows that $v = |v| = \sqrt{2g(y_0 - y)}$. $\quad \square$

EXERCISES 3.5

An object moves along a coordinate line, its position at each time $t \geq 0$ given by $x(t)$. Find the position, velocity, acceleration, and speed at time t_0.

1. $x(t) = 4 + 3t - t^2$; $t_0 = 5$.

2. $x(t) = 5t - t^3$; $t_0 = 3$.

3. $x(t) = t^3 - 6t$; $t_0 = 2$.

4. $x(t) = 4t^2 - 3t + 6$; $t_0 = 1$.

5. $x(t) = \dfrac{18}{t + 2}$; $t_0 = 1$.

6. $x(t) = \dfrac{2t}{t + 3}$; $t_0 = 3$.

7. $x(t) = (t^2 + 5t)(t^2 + t - 2)$; $t_0 = 1$.

8. $x(t) = (t^2 - 3t)(t^2 + 3t)$; $t_0 = 2$.

An object moves along a coordinate line, its position at each time $t \geq 0$ given by $x(t)$. Determine when, if ever, the object changes direction.

9. $x(t) = t^3 - 3t^2 + 3t$.

10. $x(t) = t + \dfrac{3}{t + 1}$.

11. $x(t) = t + \dfrac{5}{t + 2}$.

12. $x(t) = t^4 - 4t^3 + 6t^2 - 6t$.

13. $x(t) = \dfrac{t}{t^2 + 8}$.

14. $x(t) = 3t^5 + 10t^3 + 15t$.

Objects A, B, C move along the x-axis. Their positions from time $t = 0$ to time $t = t_3$ have been graphed in Figure 3.5.4.

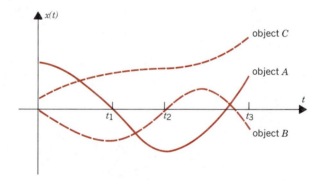

Figure 3.5.4

15. Which object begins furthest to the right?

16. Which object finishes furthest to the right?

17. Which object has the greatest speed at time t_1?

18. Which object maintains the same direction during the time interval $[t_1, t_3]$?

19. Which object begins by moving left?

20. Which object finishes moving left?

21. Which object changes direction at time t_2?

22. Which object speeds up during the time interval $[0, t_1]$?

23. Which object slows down during the time interval $[t_1, t_2]$?

24. Which object changes direction during the time interval $[t_2, t_3]$?

An object moves along the x-axis, its position at each time $t \geq 0$ given by $x(t)$. Determine the time interval(s), if any, during which the object satisfies the given condition.

25. $x(t) = t^4 - 12t^3 + 28t^2$; moving right.

26. $x(t) = t^3 - 12t^2 + 21t$; moving left.

27. $x(t) = 5t^4 - t^5$; speeding up.

28. $x(t) = 6t^2 - t^4$; slowing down.

29. $x(t) = t^3 - 6t^2 - 15t$; moving left and slowing down.

30. $x(t) = t^3 - 6t^2 - 15t$; moving right and slowing down.

31. $x(t) = t^4 - 8t^3 + 16t^2$; moving right and speeding up.

32. $x(t) = t^4 - 8t^3 + 16t^2$; moving left and speeding up.

(For the remaining exercises, neglect air resistance. For the numerical calculations, take g as 32 feet per second per second.)

33. An object is dropped and hits the ground 6 seconds later. From what height was it dropped?

34. Supplies are dropped from a helicopter and seconds later hit the ground at 384 feet per second. How high was the helicopter?

35. An object is projected vertically upward from ground level with velocity v_0. What is the height attained by the object?

36. An object projected vertically upward from ground level returns to earth in 8 seconds. What was the initial velocity?

37. An object projected vertically upward passes every height less than the maximum twice, once on the way up, once on the way down. Show that the speed is the same in each direction.

38. An object is projected vertically upward from the ground. Show that it takes the object the same amount of time to reach its maximum height as it takes for it to drop from that height back to the ground.

39. A rubber ball is thrown straight down from a height of 224 feet at a speed of 80 feet per second. If the ball always rebounds with one-fourth of its impact speed, what will be the speed of the ball the third time it hits the ground?

40. A ball is thrown straight up from ground level. How high will the ball go if it reaches a height of 64 feet in 2 seconds?

41. A stone is thrown upward from ground level. The initial speed is 32 feet per second. (a) In how many seconds will the stone hit the ground? (b) How high will it go? (c) With what minimum speed should the stone be thrown so as to reach a height of 36 feet?

42. To estimate the height of a bridge a man drops a stone into the water below. How high is the bridge (a) if the stone hits the water 3 seconds later? (b) if the man hears the splash 3 seconds later? (Use 1080 feet per second as the speed of sound.)

43. A falling stone is at a certain instant 100 feet above the ground. Two seconds later it is only 16 feet above the ground. (a) From what height was it dropped? (b) If it was thrown down with an initial speed of 5 feet per second, from what height was it thrown? (c) If it was thrown upward with an initial speed of 10 feet per second, from what height was it thrown?

44. A rubber ball is thrown straight down from a height of 4 feet. If the ball rebounds with one-half of its impact speed and returns exactly to its original height before falling again, how fast was it thrown originally?

3.6 THE CHAIN RULE

Here we take up the differentiation of composite functions. Until we get to Theorem 3.6.7, our approach is completely intuitive—no real definitions, no proofs, just informal discussion. Our purpose is to give you some experience with the standard computational procedures and some insight into why these procedures work. Theorem 3.6.7 puts it all on a sound footing.

Suppose that y is a differentiable function of u:

$$y = f(u)$$

and u in turn is a differentiable function of x:

$$u = g(x).$$

Then y is a composite function of x:

$$y = f(u) = f(g(x)).$$

Does y have a derivative with respect to x? Yes it does, and it's given by a formula that is easy to remember:

(3.6.1)

$$\frac{dy}{dx} = \frac{dy}{du}\frac{du}{dx}.$$

This formula, known as the *chain rule*, says that "the rate of change of y with respect to x is the rate of change of y with respect to u times the rate of change of u with respect to x." (Plausible as all this sounds, remember that we have proved nothing. All we have done is asserted that the composition of differentiable functions is differentiable and given you a formula, a formula that needs justification and is justified at the end of this section.)

Before using the chain rule in elaborate computations, let's confirm its validity in some simple instances.

If $y = 2u$ and $u = 3x$, then $y = 6x$. Clearly

$$\frac{dy}{dx} = 6, \qquad \frac{dy}{du} = 2, \qquad \frac{du}{dx} = 3$$

and so, in this case certainly, the chain rule is confirmed:

$$\frac{dy}{dx} = \frac{dy}{du}\frac{du}{dx}.$$

If $y = u^3$ and $u = x^2$, then $y = (x^2)^3 = x^6$. This time

$$\frac{dy}{dx} = 6x^5, \qquad \frac{dy}{du} = 3u^2 = 3x^4, \qquad \frac{du}{dx} = 2x$$

and once again

$$\frac{dy}{dx} = \frac{dy}{du}\frac{du}{dx}.$$

Problem 1. Find dy/dx by the chain rule given that

$$y = \frac{u-1}{u+1} \qquad \text{and} \qquad u = x^2.$$

Solution

$$\frac{dy}{du} = \frac{(u+1)1 - (u-1)1}{(u+1)^2} = \frac{2}{(u+1)^2} \qquad \text{and} \qquad \frac{du}{dx} = 2x$$

so that

$$\frac{dy}{dx} = \frac{dy}{du}\frac{du}{dx} = \left[\frac{2}{(u+1)^2}\right]2x = \frac{4x}{(x^2+1)^2}. \quad \square$$

Remark. We could have obtained the same result without the chain rule by first writing y as a function of x and then differentiating:

$$\text{with} \quad y = \frac{u-1}{u+1} \quad \text{and} \quad u = x^2 \quad \text{we have} \quad y = \frac{x^2-1}{x^2+1}$$

and therefore

$$\frac{dy}{dx} = \frac{(x^2+1)2x - (x^2-1)2x}{(x^2+1)^2} = \frac{4x}{(x^2+1)^2}. \quad \square$$

If f is a differentiable function of u and u is a differentiable function of x, then, according to (3.6.1),

(3.6.2)
$$\frac{d}{dx}[f(u)] = \frac{d}{du}[f(u)]\frac{du}{dx}.$$

[All we have done here is written y as $f(u)$.] This is the formulation of the chain rule that we will use most frequently in later work.

Suppose now that we were asked to calculate

$$\frac{d}{dx}[(x^2-1)^{100}].$$

We could expand $(x^2-1)^{100}$ into polynomial form by using the binomial theorem or, if we were masochistic, by repeated multiplication, but we would have a terrible mess on our hands and we would probably get a wrong answer. With (3.6.2) we can derive a formula that will render such calculations almost trivial.

Assuming (3.6.2), we can prove that, if u is a differentiable function of x and n is a positive or negative integer, then

(3.6.3)
$$\frac{d}{dx}(u^n) = nu^{n-1}\frac{du}{dx}.$$

Proof. Set $f(u) = u^n$. Then

$$\frac{d}{dx}(u^n) = \underset{\underset{\text{(3.6.2)}}{\uparrow}}{\frac{d}{du}}(u^n)\frac{du}{dx} = nu^{n-1}\frac{du}{dx}. \quad \square$$

To calculate

$$\frac{d}{dx}[(x^2-1)^{100}]$$

we set $u = x^2 - 1$. Then by our formula

$$\frac{d}{dx}[(x^2-1)^{100}] = 100(x^2-1)^{99}\frac{d}{dx}(x^2-1) = 100(x^2-1)^{99}2x = 200x(x^2-1)^{99}.$$

Here are more examples of a similar sort.

Example 2

$$\frac{d}{dx}\left[\left(x+\frac{1}{x}\right)^{-3}\right] = -3\left(x+\frac{1}{x}\right)^{-4}\frac{d}{dx}\left(x+\frac{1}{x}\right) = -3\left(x+\frac{1}{x}\right)^{-4}\left(1-\frac{1}{x^2}\right).$$ ☐

Example 3

$$\frac{d}{dx}[1-(2+3x)^2]^3 = 3[1-(2+3x)^2]^2\frac{d}{dx}[1-(2+3x)^2].$$

Since

$$\frac{d}{dx}[1-(2+3x)^2] = -2(2+3x)\frac{d}{dx}(3x) = -6(2+3x),$$

we have

$$\frac{d}{dx}[1-(2+3x)^2]^3 = -18(2+3x)[1-(2+3x)^2]^2.$$ ☐

Example 4

$$\frac{d}{dx}[4x(x^2+3)^2] = 4x\frac{d}{dx}(x^2+3)^2 + (x^2+3)^2\frac{d}{dx}(4x)$$

$$= 4x[2(x^2+3)2x] + 4(x^2+3)^2$$

$$= 16x^2(x^2+3) + 4(x^2+3)^2 = 4(x^2+3)(5x^2+3).$$ ☐

The formula

$$\frac{dy}{dx} = \frac{dy}{du}\frac{du}{dx}$$

can easily be extended to more variables. For example, if x itself depends on s, then we have

(3.6.4)
$$\frac{dy}{ds} = \frac{dy}{du}\frac{du}{dx}\frac{dx}{ds}.$$

If, in addition, s depends on t, then

(3.6.5)
$$\frac{dy}{dt} = \frac{dy}{du}\frac{du}{dx}\frac{dx}{ds}\frac{ds}{dt}$$

and so on. Each new dependence adds a new link to the chain.

Problem 5. Find dy/ds given that

$$y = 3u + 1, \qquad u = x^{-2}, \qquad x = 1 - s.$$

Solution

$$\frac{dy}{du} = 3, \qquad \frac{du}{dx} = -2x^{-3}, \qquad \frac{dx}{ds} = -1.$$

Therefore

$$\frac{dy}{ds} = \frac{dy}{du}\frac{du}{dx}\frac{dx}{ds} = (3)(-2x^{-3})(-1) = 6x^{-3} = 6(1-s)^{-3}. \quad \square$$

Problem 6. Find dy/dt at $t = 9$ given that

$$y = \frac{u+2}{u-1}, \qquad u = (3s-7)^2, \qquad s = \sqrt{t}.$$

Solution. As you can check,

$$\frac{dy}{du} = -\frac{3}{(u-1)^2}, \qquad \frac{du}{ds} = 6(3s-7), \qquad \frac{ds}{dt} = \frac{1}{2\sqrt{t}}.\dagger$$

At $t = 9$, we have $s = 3$ and $u = 4$ so that

$$\frac{dy}{du} = -\frac{3}{(4-1)^2} = -\frac{1}{3}, \qquad \frac{du}{ds} = 6(9-7) = 12, \qquad \frac{ds}{dt} = \frac{1}{2\sqrt{9}} = \frac{1}{6}.$$

Thus, at $t = 9$,

$$\frac{dy}{dt} = \frac{dy}{du}\frac{du}{ds}\frac{ds}{dt} = \left(-\frac{1}{3}\right)(12)\left(\frac{1}{6}\right) = -\frac{2}{3}. \quad \square$$

So far we have worked entirely in Leibniz's notation. What does the chain rule look like in prime notation?

Let's go back to the beginning. Once again let y be a differentiable function of u:

$$y = f(u),$$

and let u be a differentiable function of x:

$$u = g(x).$$

---

† It was shown in Section 3.1 that

$$\frac{d}{dx}(\sqrt{x}) = \frac{1}{2\sqrt{x}}.$$

Then

$$y = f(u) = f(g(x)) = (f \circ g)(x)$$

and, according to the chain rule (as yet unproved),

$$\frac{dy}{dx} = \frac{dy}{du}\frac{du}{dx}.$$

Since

$$\frac{dy}{dx} = \frac{d}{dx}[(f \circ g)(x)] = (f \circ g)'(x), \qquad \frac{dy}{du} = f'(u) = f'(g(x)), \qquad \frac{du}{dx} = g'(x),$$

the chain rule can be written

(3.6.6)
$$(f \circ g)'(x) = f'(g(x))\, g'(x).$$

The chain rule in prime notation says that "the derivative of a composition $f \circ g$ at x is the derivative of f at $g(x)$ times the derivative of g at x."

In Leibniz's notation the chain rule *appears* seductively simple, to some even obvious.   "After all, to prove it, all you have to do is cancel the du's":

$$\frac{dy}{dx} = \frac{dy}{du}\frac{du}{dx}.$$

Of course, this is just nonsense.   What would one cancel from

$$(f \circ g)'(x) = f'(g(x))\, g'(x)?$$

Although Leibniz's notation is useful for routine calculations, mathematicians generally turn to the prime notation when precision is required.

It is time for us to be precise.   How do we know that the composition of differentiable functions is differentiable?   What assumptions do we need?   Precisely under what circumstances is it true that

$$(f \circ g)'(x) = f'(g(x))\, g'(x)?$$

The following theorem provides the definitive answer.

THEOREM 3.6.7   THE CHAIN-RULE THEOREM

If g is differentiable at x and f is differentiable at $g(x)$, then the composition $f \circ g$ is differentiable at x and

$$(f \circ g)'(x) = f'(g(x))\, g'(x).$$

A proof of this theorem appears in the supplement to this section. The argument is not as easy as "canceling" the du's.

EXERCISES 3.6

Differentiate: (a) by expanding before differentiation, (b) by using the chain rule.   Then reconcile the results.

1. $f(x) = (x^2 + 1)^2$.

2. $f(x) = (x^3 - 1)^2$.

3. $f(x) = (2x + 1)^3$.

4. $f(x) = (x^2 + 1)^3$.

5. $f(x) = (x + x^{-1})^2$.

6. $f(x) = (3x^2 - 2x)^2$.

Differentiate.

7. $f(x) = (1 - 2x)^{-1}$.

8. $f(x) = (1 + 2x)^5$.

9. $f(x) = (x^5 - x^{10})^{20}$.

10. $f(x) = \left(x^2 + \dfrac{1}{x^2} \right)^3$.

11. $f(x) = \left(x - \dfrac{1}{x} \right)^4$.

12. $f(x) = \left(x + \dfrac{1}{x} \right)^3$.

13. $f(x) = (x - x^3 - x^5)^4$.

14. $f(t) = \left(\dfrac{1}{1 + t} \right)^4$.

15. $f(t) = (t^2 - 1)^{100}$.

16. $f(t) = (t - t^2)^3$.

17. $f(t) = (t^{-1} + t^{-2})^4$.

18. $f(x) = \left(\dfrac{4x + 3}{5x - 2} \right)^3$.

19. $f(x) = \left(\dfrac{3x}{x^2 + 1} \right)^4$.

20. $f(x) = [(2x + 1)^2 + (x + 1)^2]^3$.

21. $f(x) = (x^4 + x^2 + x)^2$.

22. $f(x) = (x^2 + 2x + 1)^3$.

23. $f(x) = \left(\dfrac{x^3}{3} + \dfrac{x^2}{2} + \dfrac{x}{1} \right)^{-1}$.

24. $f(x) = \left(\dfrac{x^2 + 2}{x^2 + 1} \right)^5$.

25. $f(x) = \left(\dfrac{1}{x + 2} - \dfrac{1}{x - 2} \right)^3$.

26. $f(x) = [(6x + x^5)^{-1} + x]^2$.

Find dy/dx at $x = 0$.

27. $y = \dfrac{1}{1 + u^2}$, $u = 2x + 1$.

28. $y = u + \dfrac{1}{u}$, $u = (3x + 1)^4$.

29. $y = \dfrac{2u}{1 - 4u}$, $u = (5x^2 + 1)^4$.

30. $y = u^3 - u + 1$, $u = \dfrac{1 - x}{1 + x}$.

Find dy/dt.

31. $y = \dfrac{1 - 7u}{1 + u^2}$, $u = 1 + x^2$, $x = 2t - 5$.

32. $y = 1 + u^2$, $u = \dfrac{1 - 7x}{1 + x^2}$, $x = 5t + 2$.

Find dy/dx at $x = 2$.

33. $y = (s + 3)^2$, $s = \sqrt{t - 3}$, $t = x^2$.

34. $y = \dfrac{1 + s}{1 - s}$, $s = t - \dfrac{1}{t}$, $t = \sqrt{x}$.

Given that

$$f(0) = 1, \quad f'(0) = 2, \quad f(1) = 0, \quad f'(1) = 1, \quad f(2) = 1, \quad f'(2) = 1,$$
$$g(0) = 2, \quad g'(0) = 1, \quad g(1) = 1, \quad g'(1) = 0, \quad g(2) = 2, \quad g'(2) = 1,$$
$$h(0) = 1, \quad h'(0) = 2, \quad h(1) = 2, \quad h'(1) = 1, \quad h(2) = 0, \quad h'(2) = 2,$$

evaluate the following.

35. $(f \circ g)'(0)$.

36. $(f \circ g)'(1)$.

37. $(f \circ g)'(2)$.

38. $(g \circ f)'(0)$.

39. $(g \circ f)'(1)$.

40. $(g \circ f)'(2)$.

41. $(f \circ h)'(0)$.

42. $(h \circ f)'(1)$.

43. $(h \circ f)'(0)$.

44. $(f \circ h \circ g)'(1)$.

45. $(g \circ f \circ h)'(2)$.

46. $(g \circ h \circ f)'(0)$.

Find the following derivatives.

47. $\dfrac{d}{dx}[f(x^2 + 1)].$ **48.** $\dfrac{d}{dx}\left[f\left(\dfrac{x-1}{x+1}\right)\right].$ **49.** $\dfrac{d}{dx}[[f(x)]^2 + 1].$ **50.** $\dfrac{d}{dx}\left[\dfrac{f(x)-1}{f(x)+1}\right].$

Determine the values of x for which

    (a) $f'(x) = 0.$ (b) $f'(x) > 0.$ (c) $f'(x) < 0.$

51. $f(x) = (1 + x^2)^{-2}.$ **52.** $f(x) = (1 - x^2)^2.$

53. $f(x) = x(1 + x^2)^{-1}.$ **54.** $f(x) = x(1 - x^2)^3.$

An object moves along a coordinate line, its position at each time $t \geq 0$ given by $x(t)$. Determine when the object changes direction.

55. $x(t) = (t + 1)^2(t - 9)^3.$ **56.** $x(t) = t(t - 8)^3.$

57. $x(t) = (t^3 - 12t)^4.$ **58.** $x(t) = (t^2 - 8t + 15)^3.$

59. Show that if $(x - a)^2$ is a factor of the polynomial $p(x)$, then $x - a$ is a factor of $p'(x)$.

60. Let f be a differentiable function. Show that
    (a) if f is even, then f' is odd.   (b) if f is odd, then f' is even.

Differentiate.

61. $f(x) = [(7x - x^{-1})^{-2} + 3x^2]^{-1}.$ **62.** $f(x) = [(x^2 - x^{-2})^3 - x]^5.$

63. $f(x) = [(x + x^{-1})^2 - (x^2 + x^{-2})^{-1}]^3.$ **64.** $f(x) = [(x^{-1} + 2x^{-2})^3 + 3x^{-3}]^4.$

65. (Calculator) Let $f(x) = (x^2 - 2)^{10}$. Determine

$$f'(1) = \lim_{h \to 0} \frac{f(1 + h) - f(1)}{1}$$

numerically by evaluating the quotient at $h = \pm 0.01, \pm 0.001, \pm 0.0001$. Compare your answer to what you obtain by using the chain rule.

66. (Calculator) Let $f(x) = \sqrt{x^5 - 16}$. Determine

$$f'(2) = \lim_{h \to 0} \frac{f(2 + h) - f(2)}{2}$$

numerically. Compare your answer to what you obtain by using the chain rule. [Recall that $d/dx\,(\sqrt{x}) = 1/2\sqrt{x}.$]

*SUPPLEMENT TO SECTION 3.6

To prove Theorem 3.6.7 it is convenient to use a slightly different formulation of derivative.

THEOREM 3.6.8

The function f is differentiable at x iff

$$\lim_{t \to x} \frac{f(t) - f(x)}{t - x} \quad \text{exists.}$$

If this limit exists, it is $f'(x)$.

Proof. For each t in the domain of f, $t \neq x$, define

$$G(t) = \frac{f(t) - f(x)}{t - x}.$$

The theorem follows from observing that f is differentiable at x iff

$$\lim_{h \to 0} G(x + h) \quad \text{exists},$$

and noting that

$$\lim_{h \to 0} G(x + h) = l \quad \text{iff} \quad \lim_{t \to x} G(t) = l. \quad \square$$

Proof of Theorem 3.6.7. By Theorem 3.6.8 it is enough to show that

$$\lim_{t \to x} \frac{f(g(t)) - f(g(x))}{t - x} = f'(g(x)) \, g'(x).$$

We begin by defining an auxiliary function F on the domain of f by setting

$$F(y) = \begin{cases} \dfrac{f(y) - f(g(x))}{y - g(x)}, & y \neq g(x) \\ f'(g(x)), & y = g(x) \end{cases}.$$

F is continuous at $g(x)$ since

$$\lim_{y \to g(x)} F(y) = \lim_{y \to g(x)} \frac{f(y) - f(g(x))}{y - g(x)},$$

and the right-hand side is (by Theorem 3.6.8) $f'(g(x))$, which is the value of F at $g(x)$. For $t \neq x$,

(1) $$\frac{f(g(t)) - f(g(x))}{t - x} = F(g(t)) \left[\frac{g(t) - g(x)}{t - x} \right].$$

To see this we note that, if $g(t) = g(x)$, then both sides are 0. If $g(t) \neq g(x)$, then

$$F(g(t)) = \frac{f(g(t)) - f(g(x))}{g(t) - g(x)},$$

so that again we have equality.

Since g, being differentiable at x, is continuous at x and since F is continuous at $g(x)$, we know that the composition $F \circ g$ is continuous at x. Thus

$$\lim_{t \to x} F(g(t)) = F(g(x)) = f'(g(x)).$$

$\underline{\uparrow}$ by our definition of F

This, together with Equation (1), gives

$$\lim_{t \to x} \frac{f(g(t)) - f(g(x))}{t - x} = f'(g(x)) \, g'(x). \quad \square$$

3.7 DIFFERENTIATING THE TRIGONOMETRIC FUNCTIONS

The calculus of the trigonometric functions is simplified by the use of radian measure. Throughout our work we will use radian measure and refer to degrees only in passing. An outline review of trigonometry—identities and graphs—appears in Section 1.2.

The derivative of the sine function is the cosine function:

(3.7.1)
$$\frac{d}{dx} (\sin x) = \cos x.$$

Proof. For $h \neq 0$,

$$\frac{\sin (x + h) - \sin x}{h} = \frac{[\sin x \cos h + \cos x \sin h] - [\sin x]}{h}$$

$$= \sin x \frac{\cos h - 1}{h} + \cos x \frac{\sin h}{h}.$$

As shown earlier

$$\lim_{h \to 0} \frac{\cos h - 1}{h} = 0 \quad \text{and} \quad \lim_{h \to 0} \frac{\sin h}{h} = 1. \tag{2.5.5}$$

Since $\sin x$ and $\cos x$ remain constant as h approaches zero,

$$\lim_{h \to 0} \frac{\sin (x + h) - \sin x}{h} = \sin x \left(\lim_{h \to 0} \frac{\cos h - 1}{h} \right) + \cos x \left(\lim_{h \to 0} \frac{\sin h}{h} \right)$$

$$= (\sin x)(0) + (\cos x)(1) = \cos x. \quad \blacksquare$$

The derivative of the cosine function is the negative of the sine function:

(3.7.2)

$$\frac{d}{dx} (\cos x) = - \sin x.$$

Proof

$$\cos (x + h) = \cos x \cos h - \sin x \sin h.$$

Therefore

$$\lim_{h \to 0} \frac{\cos (x + h) - \cos x}{h} = \lim_{h \to 0} \frac{[\cos x \cos h - \sin x \sin h] - [\cos x]}{h}$$

$$= \cos x \left(\lim_{h \to 0} \frac{\cos h - 1}{h} \right) - \sin x \left(\lim_{h \to 0} \frac{\sin h}{h} \right) = - \sin x. \quad \blacksquare$$

Example 1. To differentiate

$$f(x) = \cos x \sin x$$

we use the product rule:

$$f'(x) = \cos x \frac{d}{dx} (\sin x) + \sin x \frac{d}{dx} (\cos x)$$

$$= \cos x (\cos x) + \sin x (- \sin x) = \cos^2 x - \sin^2 x. \quad \blacksquare$$

We come now to the tangent function. Since $\tan x = \sin x / \cos x$, we have

$$\frac{d}{dx} (\tan x) = \frac{\cos x \dfrac{d}{dx} (\sin x) - \sin x \dfrac{d}{dx} (\cos x)}{\cos^2 x} = \frac{\cos^2 x + \sin^2 x}{\cos^2 x} = \frac{1}{\cos^2 x} = \sec^2 x.$$

The derivative of the tangent function is the secant squared:

(3.7.3)
$$\frac{d}{dx}(\tan x) = \sec^2 x.$$

The derivatives of the other trigonometric functions are as follows:

(3.7.4)
$$\frac{d}{dx}(\cot x) = -\csc^2 x,$$
$$\frac{d}{dx}(\sec x) = \sec x \tan x,$$
$$\frac{d}{dx}(\csc x) = -\csc x \cot x.$$

The verification of these formulas is left as an exercise.

It is time for some sample problems.

Problem 2. Find $f'(\pi/4)$ for

$$f(x) = x \cot x.$$

Solution. We first find $f'(x)$ by the product rule:

$$f'(x) = x\frac{d}{dx}(\cot x) + \cot x \frac{d}{dx}(x) = -x\csc^2 x + \cot x.$$

Now we evaluate f' at $\pi/4$:

$$f'(\pi/4) = -\frac{\pi}{4}(\sqrt{2})^2 + 1 = 1 - \frac{\pi}{2}. \quad \blacksquare$$

Problem 3. Find

$$\frac{d}{dx}\left[\frac{\sec x}{\tan x}\right].$$

Solution. By the quotient rule,

$$\frac{d}{dx}\left[\frac{\sec x}{\tan x}\right] = \frac{\tan x \dfrac{d}{dx}(\sec x) - \sec x \dfrac{d}{dx}(\tan x)}{\tan^2 x}$$

$$= \frac{\tan x(\sec x \tan x) - \sec x(\sec^2 x)}{\tan^2 x}$$

$$= \frac{\sec x[\tan^2 x - \sec^2 x]}{\tan^2 x} = -\frac{\sec x}{\tan^2 x}. \quad \blacksquare$$

Problem 4.   Find an equation for the tangent to the curve

$$y = \cos x$$

at the point where $x = \pi/3$.

Solution.   Since $\cos \pi/3 = \frac{1}{2}$, the point of tangency is $(\pi/3, \frac{1}{2})$.   To find the slope of the tangent, we evaluate the derivative

$$\frac{dy}{dx} = -\sin x$$

at $x = \pi/3$.   This gives $m = -\sqrt{3}/2$.   The equation for the tangent can be written

$$y - \frac{1}{2} = -\frac{\sqrt{3}}{2}\left(x - \frac{\pi}{3}\right). \quad \square$$

Problem 5.   An object moves along the x-axis, its position at each time t given by the function

$$x(t) = t + 2 \sin t.$$

Determine those times t from 0 to 2π when the object is moving to the left.

Solution.   The object is moving to the left only when $v(t) < 0$.   Since

$$v(t) = x'(t) = 1 + 2 \cos t,$$

the object is moving to the left only when $\cos t < -\frac{1}{2}$.   As you can check, the only t in $[0, 2\pi]$ for which $\cos t < -\frac{1}{2}$ are those t that lie between $2\pi/3$ and $4\pi/3$.   Thus from time $t = 0$ to $t = 2\pi$, the object is moving left only during the time interval $(2\pi/3, 4\pi/3)$. $\square$

The Chain Rule and the Trigonometric Functions

If f is a differentiable function of u and u is a differentiable function of x, then, as you saw in Section 3.6,

$$\frac{d}{dx}[f(u)] = \frac{d}{du}[f(u)]\frac{du}{dx}.$$

Written in this form, the derivatives of the six trigonometric functions appear as follows:

$$(3.7.5) \quad \begin{array}{ll} \dfrac{d}{dx}(\sin u) = \cos u\,\dfrac{du}{dx}. & \dfrac{d}{dx}(\cos u) = -\sin u\,\dfrac{du}{dx}. \\[2ex] \dfrac{d}{dx}(\tan u) = \sec^2 u\,\dfrac{du}{dx}. & \dfrac{d}{dx}(\cot u) = -\csc^2 u\,\dfrac{du}{dx}. \\[2ex] \dfrac{d}{dx}(\sec u) = \sec u \tan u\,\dfrac{du}{dx}. & \dfrac{d}{dx}(\csc u) = -\csc u \cot u\,\dfrac{du}{dx}. \end{array}$$

Example 6

$$\frac{d}{dx}(\cos 2x) = -\sin 2x\,\frac{d}{dx}(2x) = -2\sin 2x. \quad \blacksquare$$

Example 7

$$\frac{d}{dx}[\sec(x^2 + 1)] = \sec(x^2 + 1)\tan(x^2 + 1)\,\frac{d}{dx}(x^2 + 1)$$
$$= 2x\sec(x^2 + 1)\tan(x^2 + 1). \quad \blacksquare$$

Example 8

$$\frac{d}{dx}(\sin^3 \pi x) = 3\sin^2 \pi x\,\frac{d}{dx}(\sin \pi x)$$
$$= 3\sin^2 \pi x \cos \pi x\,\frac{d}{dx}(\pi x) = 3\pi \sin^2 \pi x \cos \pi x. \quad \blacksquare$$

Our treatment of the trigonometric functions has been based entirely on radian measure. When degrees are used, the derivatives of the trigonometric functions contain the extra factor $\frac{1}{180}\pi \cong 0.0175$.

Problem 9. Find

$$\frac{d}{dx}(\sin x°).$$

Solution. Since $x° = \frac{1}{180}\pi x$ radians,

$$\frac{d}{dx}(\sin x°) = \frac{d}{dx}\left(\sin \tfrac{1}{180}\pi x\right) = \tfrac{1}{180}\pi \cos \tfrac{1}{180}\pi x = \tfrac{1}{180}\pi \cos x°. \quad \blacksquare$$

The extra factor $\frac{1}{180}\pi$ is a disadvantage, particularly in problems where it occurs repeatedly. This tends to discourage the use of degree measure in theoretical work.

EXERCISES 3.7

Differentiate.

1. $y = 3 \cos x - 4 \sec x$.   **2.** $y = x^2 \sec x$.   **3.** $y = x^3 \csc x$.
4. $y = \sin^2 x$.   **5.** $y = \cos^2 t$.   **6.** $y = 3t^2 \tan t$.
7. $y = \sin^4 \sqrt{u}$.   **8.** $y = u \csc u^2$.   **9.** $y = \tan x^2$.
10. $y = \cos \sqrt{x}$.   **11.** $y = [x + \cot \pi x]^4$.   **12.** $y = [x^2 - \sec 2x]^3$.

Find the second derivative.

13. $y = \sin x$.   **14.** $y = \cos x$.   **15.** $y = \dfrac{\cos x}{1 + \sin x}$.

16. $y = \tan^3 2\pi x$.   **17.** $y = \cos^3 2u$.   **18.** $y = \sin^5 3t$.

19. $y = \tan 2t$.   **20.** $y = \cot 4u$.   **21.** $y = x^2 \sin 3x$.

22. $y = \dfrac{\sin x}{1 - \cos x}$.   **23.** $y = \sin^2 x + \cos^2 x$.   **24.** $y = \sec^2 x - \tan^2 x$.

Find the indicated derivative.

25. $\dfrac{d^4}{dx^4} (\sin x)$.   **26.** $\dfrac{d^4}{dx^4} (\cos x)$.   **27.** $\dfrac{d}{dt}\left[t^2 \dfrac{d^2}{dt^2} (t \cos 3t) \right]$.

28. $\dfrac{d}{dt}\left[t \dfrac{d}{dt} (\cos t^2) \right]$.   **29.** $\dfrac{d}{dx} [f(\sin 3x)]$.   **30.** $\dfrac{d}{dx} [\sin (f(3x))]$.

Find an equation for the tangent to the curve at $x = a$.

31. $y = \sin x$; $a = 0$.   **32.** $y = \tan x$; $a = \pi/6$.   **33.** $y = \cot x$; $a = \pi/6$.
34. $y = \cos x$; $a = 0$.   **35.** $y = \sec x$; $a = \pi/4$.   **36.** $y = \csc x$; $a = \pi/3$.

Determine the numbers x between 0 and 2π where the tangent to the curve is horizontal.

37. $y = \cos x$.   **38.** $y = \sin x$.   **39.** $y = \sin x + \sqrt{3} \cos x$.
40. $y = \cos x - \sqrt{3} \sin x$.   **41.** $y = \sin^2 x$.   **42.** $y = \cos^2 x$.
43. $y = \tan x - 2x$.   **44.** $y = 3 \cot x + 4x$.   **45.** $y = 2 \sec x + \tan x$.

46. $y = \cot x - 2 \csc x$.   **47.** $y = \dfrac{\csc x}{1 + \cot x}$.   **48.** $y = \dfrac{\sec x}{\tan x - 1}$.

An object moves along the x-axis, its position at each time t given by $x(t)$. Determine those times from $t = 0$ to $t = 2\pi$ when the object is moving to the right with increasing speed.

49. $x(t) = \sin 3t$.   **50.** $x(t) = \cos 2t$.   **51.** $x(t) = \sin t - \cos t$.
52. $x(t) = \sin t + \cos t$.   **53.** $x(t) = t + 2 \cos t$.   **54.** $x(t) = t - \sqrt{2} \sin t$.

Find dy/dt (a) by using (3.6.4) and (b) by writing y as a function of t and then differentiating.

55. $y = u^2 - 1$, $u = \sec x$, $x = \pi t$.   **56.** $y = [\frac{1}{2}(1 + u)]^3$, $u = \cos x$, $x = 2t$.
57. $y = [\frac{1}{2}(1 - u)]^4$, $u = \cos x$, $x = 2t$.   **58.** $y = 1 - u^2$, $u = \csc x$, $x = 3t$.

59. It can be shown by induction that the nth derivative of the sine function is given by the formula

$$\frac{d^n}{dx^n} (\sin x) = \left[\begin{cases} (-1)^{(n-1)/2} \cos x, & n \text{ odd} \\ (-1)^{n/2} \sin x, & n \text{ even} \end{cases} \right].$$

Persuade yourself that this formula is correct and obtain a similar formula for the nth derivative of the cosine function.

60. Verify the following differentiation formulas.

(a) $\dfrac{d}{dx}(\cot x) = -\csc^2 x.$ (b) $\dfrac{d}{dx}(\sec x) = \sec x \tan x.$ (c) $\dfrac{d}{dx}(\csc x) = -\csc x \cot x.$

61. (Calculator) Let $f(x) = \cos x^2$. Determine

$$f'(0) = \lim_{h \to 0} \frac{f(h) - f(0)}{h}$$

numerically. Compare your answer to what you obtain by the chain rule.

62. (Calculator) Let $f(x) = \cos^2 x$. Determine

$$f'(\tfrac{1}{4}\pi) = \lim_{h \to 0} \frac{f(\tfrac{1}{4}\pi + h) - f(\tfrac{1}{4}\pi)}{h}$$

numerically. Compare your answer to what you obtain by the chain rule.

3.8   DIFFERENTIATING INVERSES; RATIONAL POWERS

Differentiating Inverses

If f is a one-to-one function, then f has an inverse f^{-1}. This you already know. Suppose now that f is differentiable. Is f^{-1} necessarily differentiable? Yes, if f' does not take on the value 0. You will find a proof of this in Appendix B.3. Right now we assume this result and go on to describe how to calculate the derivative of f^{-1}.

To simplify notation we set $f^{-1} = g$. Then

$$f(g(x)) = x \quad \text{for all } x \text{ in the range of } f.$$

Differentiation gives

$$\frac{d}{dx}[f(g(x))] = 1,$$

$$f'(g(x))\, g'(x) = 1, \qquad \text{(we are assuming that } f \text{ and } g \text{ are differentiable)}$$

$$g'(x) = \frac{1}{f'(g(x))}. \qquad \text{(we are assuming that } f' \text{ is never 0)}$$

Substituting f^{-1} back in for g, we have

(3.8.1)
$$(f^{-1})'(x) = \frac{1}{f'(f^{-1}(x))}.$$

This is the standard formula for the derivative of an inverse.

In Leibniz's notation Formula 3.8.1 reads simply

(3.8.2)

$$\frac{dx}{dy} = \frac{1}{dy/dx}.$$

The rate of change of x with respect to y is the reciprocal of the rate of change of y with respect to x.

For some geometric reassurance of the validity of Formula 3.8.1 we refer you to Figure 3.8.1. The graphs of f and f^{-1} are reflections of one another in the line $y = x$. The tangent lines l_1 and l_2 are also reflections of one another. From the figure

$$(f^{-1})'(x) = \text{slope of } l_1 = \frac{f^{-1}(x) - b}{x - b}, \qquad f'(f^{-1}(x)) = \text{slope of } l_2 = \frac{x - b}{f^{-1}(x) - b};$$

namely, $(f^{-1})'(x)$ and $f'(f^{-1}(x))$ are indeed reciprocals.

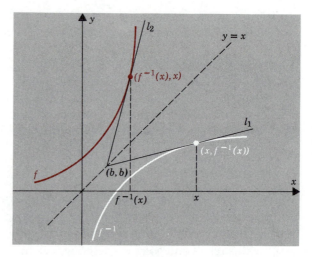

Figure 3.8.1

nth Roots

For $x > 0$, x^n is one-to-one and its derivative, nx^{n-1}, does not take on the value 0. The inverse $x^{1/n}$ is therefore differentiable and we can use the method just described to find its derivative.

$$(x^{1/n})^n = x,$$

$$\frac{d}{dx}[(x^{1/n})^n] = 1,$$

$$n(x^{1/n})^{n-1}\frac{d}{dx}(x^{1/n}) = 1,$$

$$\frac{d}{dx}(x^{1/n}) = \frac{1}{n}(x^{1/n})^{1-n}.$$

Since $(x^{1/n})^{1-n} = x^{(1/n)-1}$, we have the differentiation formula

(3.8.3)
$$\frac{d}{dx}(x^{1/n}) = \frac{1}{n}x^{(1/n)-1}.$$

We derived this formula for $x > 0$. If n is odd, then x^n is one-to-one everywhere and the formula also holds for $x < 0$. Same argument.

For suitable values of x,

$$\frac{d}{dx}(x^{1/2}) = \frac{1}{2}x^{-1/2}, \qquad \frac{d}{dx}(x^{1/3}) = \frac{1}{3}x^{-2/3}, \qquad \frac{d}{dx}(x^{1/4}) = \frac{1}{4}x^{-3/4}, \qquad \text{etc.}$$

Rational Powers

We come now to rational exponents p/q. We take q as positive and absorb the sign in p. You can see then that

$$\frac{d}{dx}(x^{p/q}) = \frac{d}{dx}[(x^{1/q})^p] = p(x^{1/q})^{p-1}\frac{d}{dx}(x^{1/q}) = p(x^{1/q})^{p-1}\frac{1}{q}x^{(1/q)-1} = \frac{p}{q}x^{(p/q)-1}.$$

power formula (3.6.3) ——   ——— root formula (3.8.3)

In short

(3.8.4)
$$\frac{d}{dx}(x^{p/q}) = \frac{p}{q}x^{(p/q)-1}.$$

Here are some simple examples:

$$\frac{d}{dx}(x^{2/3}) = \frac{2}{3}x^{-1/3}, \qquad \frac{d}{dx}(x^{5/2}) = \frac{5}{2}x^{3/2}, \qquad \frac{d}{dx}(x^{-7/9}) = -\frac{7}{9}x^{-16/9}. \quad \square$$

If u is a differentiable function of x, then, by the chain rule,

(3.8.5)
$$\frac{d}{dx}(u^{p/q}) = \frac{p}{q}u^{(p/q)-1}\frac{du}{dx}.$$

The verification of this is left to you. The result holds on every open x-interval where $u^{(p/q)-1}$ is defined.

Example 1

(a) $\dfrac{d}{dx}[(1 + x^2)^{1/5}] = \frac{1}{5}(1 + x^2)^{-4/5}(2x) = \frac{2}{5}x(1 + x^2)^{-4/5}$.

(b) $\dfrac{d}{dx}[(1 - x^2)^{1/5}] = \frac{1}{5}(1 - x^2)^{-4/5}(-2x) = -\frac{2}{5}x(1 - x^2)^{-4/5}$.

(c) $\dfrac{d}{dx}[(1 - x^2)^{1/4}] = \frac{1}{4}(1 - x^2)^{-3/4}(-2x) = -\frac{1}{2}x(1 - x^2)^{-3/4}$.

The first statement holds for all real x, the second for all $x \neq \pm 1$, the third only for $x \in (-1, 1)$. $\quad \square$

Example 2

$$\frac{d}{dx}\left[\left(\frac{x}{1+x^2}\right)^{1/2}\right] = \frac{1}{2}\left(\frac{x}{1+x^2}\right)^{-1/2}\frac{d}{dx}\left(\frac{x}{1+x^2}\right)$$

$$= \frac{1}{2}\left(\frac{x}{1+x^2}\right)^{-1/2}\frac{(1+x^2)(1)-x(2x)}{(1+x^2)^2}$$

$$= \frac{1}{2}\left(\frac{1+x^2}{x}\right)^{1/2}\frac{1-x^2}{(1+x^2)^2}$$

$$= \frac{1-x^2}{2x^{1/2}(1+x^2)^{3/2}}.$$

The result holds for all $x > 0$.   ☐

EXERCISES 3.8

Find dy/dx.

1. $y = (x^3 + 1)^{1/2}$.   **2.** $y = (x+1)^{1/3}$.   **3.** $y = x\sqrt{x^2+1}$.

4. $y = x^2\sqrt{x^2+1}$.   **5.** $y = \sqrt[4]{2x^2+1}$.   **6.** $y = (x+1)^{1/3}(x+2)^{2/3}$.

7. $y = \sqrt{2-x^2}\sqrt{3-x^2}$.   **8.** $y = \sqrt{(x^4-x+1)^3}$.

Compute.

9. $\dfrac{d}{dx}\left(\sqrt{x}+\dfrac{1}{\sqrt{x}}\right)$.   **10.** $\dfrac{d}{dx}\left(\sqrt{\dfrac{3x+1}{2x+5}}\right)$.   **11.** $\dfrac{d}{dx}\left(\dfrac{x}{\sqrt{x^2+1}}\right)$.

12. $\dfrac{d}{dx}\left(\dfrac{\sqrt{x^2+1}}{x}\right)$.   **13.** $\dfrac{d}{dx}\left(\sqrt[3]{x}+\dfrac{1}{\sqrt[3]{x}}\right)$.   **14.** $\dfrac{d}{dx}\left(\sqrt{\dfrac{ax+b}{cx+d}}\right)$.

15. (*Important*) Show the general form of the graph.
- (a) $f(x) = x^{1/n}$, n a positive even integer.
- (b) $f(x) = x^{1/n}$, n a positive odd integer.
- (c) $f(x) = x^{2/n}$, n an odd integer greater than 1.

Find the second derivative.

16. $y = \sqrt{a^2+x^2}$.   **17.** $y = \sqrt[3]{a+bx}$.   **18.** $y = x\sqrt{a^2-x^2}$.

19. $y = \sqrt{a^2-x^2}$.   **20.** $y = \sqrt{x}\tan\sqrt{x}$.   **21.** $y = \sqrt{x}\sin\sqrt{x}$.

22. Our geometric argument for the validity of Formula 3.8.1 collapses if the tangent marked l_1 in Figure 3.8.1 does not intersect the line $y = x$. Fill in this loophole.

Compute.

23. $\dfrac{d}{dx}[f(\sqrt{x}+1)]$.   **24.** $\dfrac{d}{dx}[\sqrt{[f(x)]^2+1}]$.

25. $\dfrac{d}{dx}[\sqrt{f(x^2+1)}]$.   **26.** $\dfrac{d}{dx}[f(\sqrt{x^2+1})]$.

Find a formula for $(f^{-1})'(x)$ given that f is one-to-one and its derivative satisfies the indicated equation.

27. $f'(x) = f(x)$.   **28.** $f'(x) = 1 + [f(x)]^2$.   **29.** $f'(x) = \sqrt{1-[f(x)]^2}$.

In economics the *elasticity* of demand is given by the formula

$$\epsilon = \frac{P}{Q}\left|\frac{dQ}{dP}\right|$$

where P is price and Q quantity.   The demand is said to be

$$\left\{\begin{array}{ll} \text{inelastic} & \text{where } \epsilon < 1 \\ \text{unitary} & \text{where } \epsilon = 1 \\ \text{elastic} & \text{where } \epsilon > 1 \end{array}\right\}.$$

Describe the elasticity of each of the following demand curves.   Keep in mind that Q and P must remain positive.

30. $Q = \dfrac{1}{\sqrt{P}}$.   **31.** $Q = \sqrt{300 - P}$.   **32.** $Q = (900 - P)^{4/5}$.

Given that f is one-to-one, $g = f \circ f$, and

$$f(1) = 2, \quad f'(1) = 6, \quad f(2) = 4, \quad f'(2) = 7,$$
$$f(3) = 5, \quad f'(3) = 8, \quad f(4) = 12, \quad f'(4) = 9,$$

deduce, if possible, the following values.

33. $(f^{-1})'(4)$.   **34.** $(f^{-1})'(2)$.   **35.** $(f^{-1})'(5)$.   **36.** $(f^{-1})'(6)$.

37. $g'(2)$.   **38.** $g'(1)$.   **39.** $(g^{-1})'(12)$.   **40.** $(g^{-1})'(4)$.

41. (Calculator) Let $f(x) = x^{3/4}$.   Determine

$$f'(16) = \lim_{h \to 0} \frac{f(16 + h) - f(16)}{h}$$

numerically.   Compare your answer to what you obtain by the methods of this section.

42. (Calculator) Determine

$$\lim_{h \to 0} \frac{(1 + h)^{98.6} - 1}{h}$$

numerically.   Confirm your answer by other means.

43. (Calculator) Determine

$$\lim_{x \to 32} \frac{x^{2/5} - 4}{x^{4/5} - 16}$$

numerically.   Confirm your answer by other means.

3.9 ADDITIONAL PRACTICE IN DIFFERENTIATION

Differentiate.

1. $y = x^{2/3} - a^{2/3}$.   **2.** $y = 2x^{3/4} + 4x^{-1/4}$.   **3.** $y = \dfrac{a + bx + cx^2}{x}$.

4. $y = \dfrac{\sqrt{x}}{2} - \dfrac{2}{\sqrt{x}}$.   **5.** $y = \sqrt{ax} + \dfrac{a}{\sqrt{ax}}$.   **6.** $s = \dfrac{a + bt + ct^2}{\sqrt{t}}$.

7. $r = \sqrt{1 - 2\theta}$.   **8.** $f(t) = (2 - 3t^2)^3$.   **9.** $f(x) = \dfrac{1}{\sqrt{a^2 - x^2}}$.

10. $y = \left(a - \dfrac{b}{x}\right)^2$.

11. $y = \left(a + \dfrac{b}{x^2}\right)^3$.

12. $y = x\sqrt{a + bx}$.

13. $s = t\sqrt{a^2 + t^2}$.

14. $y = \dfrac{a - x}{a + x}$.

15. $y = \dfrac{a^2 + x^2}{a^2 - x^2}$.

16. $y = \dfrac{\sqrt{a^2 + x^2}}{x}$.

17. $y = \dfrac{2 - x}{1 + 2x^2}$.

18. $y = \dfrac{x}{\sqrt{a^2 - x^2}}$.

19. $y = \dfrac{x}{\sqrt{a - bx}}$.

20. $r = \theta^2\sqrt{3 - 4\theta}$.

21. $y = \sqrt{\dfrac{1 - cx}{1 + cx}}$.

22. $f(x) = x\sqrt[3]{2 + 3x}$.

23. $y = \sqrt{\dfrac{a^2 + x^2}{a^2 - x^2}}$.

24. $s = \sqrt[3]{\dfrac{2 + 3t}{2 - 3t}}$.

25. $r = \dfrac{\sqrt[3]{a + b\theta}}{\theta}$.

26. $s = \sqrt{2t - \dfrac{1}{t^2}}$.

27. $y = \dfrac{b}{a}\sqrt{a^2 - x^2}$.

28. $y = (a^{3/5} - x^{3/5})^{5/3}$.

29. $y = (a^{2/3} - x^{2/3})^{3/2}$.

30. $y = (x + 2)^2\sqrt{x^2 + 2}$.

31. $y = \sec x^2$.

32. $y = \tan \sqrt{x}$.

33. $y = \cot^3 2x$.

34. $y = \csc^4 (3 - 2x)$.

35. $y = \sin (\cos x)$.

36. $y = \cos (\sin x)$.

Evaluate dy/dx at the given value of x.

37. $y = (x^2 - x)^3$, $\quad x = 3$.

38. $y = (4 - x^2)^3$, $\quad x = 3$.

39. $y = \sqrt[3]{x} + \sqrt{x}$, $\quad x = 64$.

40. $y = x\sqrt{3 + 2x}$, $\quad x = 3$.

41. $y = (2x)^{1/3} + (2x)^{2/3}$, $\quad x = 4$.

42. $y = \sqrt{9 + 4x^2}$, $\quad x = 2$.

43. $y = \dfrac{1}{\sqrt{25 - x^2}}$, $\quad x = 3$.

44. $y = \dfrac{x^2 + 2}{2 - x}$, $\quad x = 2$.

45. $y = \dfrac{\sqrt{16 + 3x}}{x}$, $\quad x = 3$.

46. $y = \dfrac{\sqrt{5 - 2x}}{2x + 1}$, $\quad x = \frac{1}{2}$.

47. $y = x\sqrt{8 - x^2}$, $\quad x = 2$.

48. $y = \sqrt{\dfrac{4x + 1}{5x - 1}}$, $\quad x = 2$.

49. $y = x^2\sqrt{1 + x^3}$, $\quad x = 2$.

50. $y = \sqrt{\dfrac{x^2 - 5}{10 - x^2}}$, $\quad x = 3$.

51. $y = \tan 2x$; $\quad x = \pi/6$.

52. $y = x^2 \sin^3 \pi x$; $\quad x = \frac{1}{6}$.

53. $y = \cos^3 4x$; $\quad x = \pi/12$.

54. $y = \cot 3x$; $\quad x = \pi/9$.

55. $y = x \csc \pi x$; $\quad x = \frac{1}{4}$.

56. $y = \sec^4 (x/4)$; $\quad x = \pi$.

3.10   IMPLICIT DIFFERENTIATION

Suppose you know that y is a function of x that satisfies the equation

$$3x^3y - 4y - 2x + 1 = 0.$$

One way to find dy/dx is to solve first for y:

$$(3x^3 - 4)y - 2x + 1 = 0,$$
$$(3x^3 - 4)y = 2x - 1,$$
$$y = \frac{2x - 1}{3x^3 - 4},$$

and then differentiate:

$$\frac{dy}{dx} = \frac{(3x^3 - 4)2 - (2x - 1)(9x^2)}{(3x^3 - 4)^2} = \frac{6x^3 - 8 - 18x^3 + 9x^2}{(3x^3 - 4)^2}$$

(1)
$$= -\frac{12x^3 - 9x^2 + 8}{(3x^3 - 4)^2}.$$

It is also possible to find dy/dx without first solving the equation for y. The technique is called *implicit differentiation*.

Let's return to the equation

$$3x^3y - 4y - 2x + 1 = 0.$$

Differentiating both sides of this equation with respect to x (remembering that y is a function of x), we have

$$\frac{d}{dx}(3x^3y - 4y - 2x + 1) = 0,$$

$$\left(\underbrace{3x^3\frac{dy}{dx} + 9x^2y}\right) - 4\frac{dy}{dx} - 2 = 0,$$

by the product rule

$$(3x^3 - 4)\frac{dy}{dx} = 2 - 9x^2y,$$

(2)
$$\frac{dy}{dx} = \frac{2 - 9x^2y}{3x^3 - 4}.$$

The answer looks different from what we obtained before because this time y appears on the right-hand side. This is generally no disadvantage. To satisfy yourself that the two answers are really the same all you have to do is substitute

$$y = \frac{2x - 1}{3x^3 - 4}$$

in (2). This gives

$$\frac{dy}{dx} = \frac{2 - 9x^2\left(\dfrac{2x - 1}{3x^3 - 4}\right)}{3x^3 - 4} = \frac{6x^3 - 8 - 18x^3 + 9x^2}{(3x^3 - 4)^2} = -\frac{12x^3 - 9x^2 + 8}{(3x^3 - 4)^2}$$

which is the answer (1) we obtained before.

Implicit differentiation is particularly useful where it is inconvenient (or impossible) to first solve the given equation for y.

Problem 1. Find the slope of the curve

$$x^3 - 3xy^2 + y^3 = 1 \qquad \text{at } (2, -1).$$

Solution. Differentiation gives

$$3x^2 - 3\left[x\left(2y\frac{dy}{dx}\right) + y^2\right] + 3y^2\frac{dy}{dx} = 0$$

(by the chain rule)

$$3x^2 - 6xy\frac{dy}{dx} - 3y^2 + 3y^2\frac{dy}{dx} = 0.$$

At $x = 2$ and $y = -1$, the equation becomes

$$12 + 12\frac{dy}{dx} - 3 + 3\frac{dy}{dx} = 0$$

$$15\frac{dy}{dx} = -9$$

$$\frac{dy}{dx} = -\frac{3}{5}.$$

The slope is $-\frac{3}{5}$. ■

We can also find higher derivatives by implicit differentiation.

Problem 2. Find d^2y/dx^2 given that

$$y^3 - x^2 = 4.$$

Solution. Differentiation with respect to x gives

$$(1) \qquad 3y^2\frac{dy}{dx} - 2x = 0.$$

Differentiating again we have

$$(2) \qquad 3y^2\left(\frac{d^2y}{dx^2}\right) + 6y\left(\frac{dy}{dx}\right)^2 - 2 = 0.$$

From (1) we know that

$$\frac{dy}{dx} = \frac{2x}{3y^2}.$$

Substituting this in (2) we have

$$3y^2\left(\frac{d^2y}{dx^2}\right) + 6y\left(\frac{2x}{3y^2}\right)^2 - 2 = 0$$

which, as you can check, gives

$$\frac{d^2y}{dx^2} = \frac{6y^3 - 8x^2}{9y^5}. \quad ■$$

A Theoretical Aside. If we differentiate $x^2 + y^2 = -1$ implicitly, we find that

$$2x + 2y \frac{dy}{dx} = 0 \quad \text{and therefore} \quad \frac{dy}{dx} = -\frac{x}{y}.$$

However, the result is meaningless. It is meaningless because there is no real-valued function y of x that satisfies the equation $x^2 + y^2 = -1$. Implicit differentiation can be applied meaningfully to an equation in x and y only if there is a function y of x that satisfies the equation. You can assume that such is the case for all the equations that appear in the exercises.

EXERCISES 3.10

Use implicit differentiation to obtain dy/dx in terms of x and y.

1. $x^2 + y^2 = 4.$
2. $x^3 + y^3 - 3xy = 0.$
3. $4x^2 + 9y^2 = 36.$
4. $\sqrt{x} + \sqrt{y} = 4.$
5. $x^4 + 4x^3y + y^4 = 1.$
6. $x^2 - x^2y + xy^2 + y^2 = 1.$
7. $(x - y)^2 - y = 0.$
8. $(y + 3x)^2 - 4x = 0.$
9. $\sin (x + y) = xy.$
10. $\tan xy = xy.$

Express d^2y/dx^2 in terms of x and y.

11. $y^2 + 2xy = 16.$
12. $x^2 - 2xy + 4y^2 = 3.$
13. $y^2 + xy - x^2 = 9.$
14. $x^2 - 3xy = 18.$

Find dy/dx and d^2y/dx^2 at the given point.

15. $x^2 - 4y^2 = 9;$ $(5, 2).$
16. $x^2 + 4xy + y^3 + 5 = 0;$ $(2, -1).$
17. $\cos (x + 2y) = 0;$ $(\pi/6, \pi/6).$
18. $x = \sin^2 y;$ $(\frac{1}{2}, \pi/4).$

Find equations for the tangent and normal at the indicated point.

19. $2x + 3y = 5;$ $(-2, 3).$
20. $9x^2 + 4y^2 = 72;$ $(2, 3).$
21. $x^2 + xy + 2y^2 = 28;$ $(-2, -3).$
22. $x^3 - axy + 3ay^2 = 3a^3;$ $(a, a).$

23. Show that all normals to the circle $x^2 + y^2 = r^2$ pass through the center of the circle.
24. Determine the x-intercept of the tangent to the parabola $y^2 = x$ at the point where $x = a$.

The angle between two curves is the angle between their tangents at the point of intersection. If the slopes are m_1 and m_2, then the angle of intersection α can be obtained from the formula

$$\tan \alpha = \left| \frac{m_2 - m_1}{1 + m_1 m_2} \right|. \tag{1.3.1}$$

25. At what angles do the parabolas $y^2 = 2px + p^2$ and $y^2 = p^2 - 2px$ intersect?
26. At what angle does the line $y = 2x$ intersect the curve $x^2 - xy + 2y^2 = 28$?
27. Show that the hyperbola $x^2 - y^2 = 5$ and the ellipse $4x^2 + 9y^2 = 72$ intersect at right angles.
28. Find the angles at which the circles $(x - 1)^2 + y^2 = 10$ and $x^2 + (y - 2)^2 = 5$ intersect.
29. Find equations for the tangents to the ellipse $4x^2 + y^2 = 72$ that are perpendicular to the line $2y + x + 3 = 0.$
30. Find equations for the normals to the hyperbola $4x^2 - y^2 = 36$ that are parallel to the line $2x + 5y - 4 = 0.$

3.11   RATES OF CHANGE PER UNIT TIME

Earlier you saw that velocity is the rate of change of position with respect to time t, and acceleration the rate of change of velocity with respect to time t.   In this section we work with other quantities that vary with time.   The fundamental point is this: *If Q is a quantity that varies with time, then the derivative dQ/dt gives the rate of change of that quantity per unit time.*

Problem 1.   A spherical balloon is expanding.   If the radius is increasing at the rate of 2 inches per minute, at what rate is the volume increasing when the radius is 5 inches?

Solution

Find $\dfrac{dV}{dt}$ when $r = 5$ inches

given that $\dfrac{dr}{dt} = 2$ in./min.

$V = \frac{4}{3}\pi r^3$ (volume of a sphere of radius r)

Differentiating $V = \frac{4}{3}\pi r^3$ with respect to t, we have

$$\frac{dV}{dt} = 4\pi r^2 \frac{dr}{dt}.$$

Substituting $r = 5$ and $dr/dt = 2$, we find that

$$\frac{dV}{dt} = 4\pi(5)^2(2) = 200\pi.$$

When the radius is 5 inches, the volume is increasing at the rate of 200π cubic inches per minute.   ◻

Problem 2.   A particle moves in circular orbit $x^2 + y^2 = 1$.   As it passes through the point $(\frac{1}{2}, \frac{1}{2}\sqrt{3})$, its y-coordinate decreases at the rate of 3 units per second.   At what rate is the x-coordinate changing?

Solution

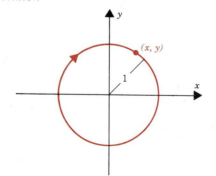

Find $\dfrac{dx}{dt}$ when $x = \frac{1}{2}$ and $y = \frac{1}{2}\sqrt{3}$

given that $\dfrac{dy}{dt} = -3$ units/sec.

$x^2 + y^2 = 1$ (equation of circle)

Differentiating $x^2 + y^2 = 1$ with respect to t, we have

$$2x\frac{dx}{dt} + 2y\frac{dy}{dt} = 0 \quad \text{and thus} \quad x\frac{dx}{dt} + y\frac{dy}{dt} = 0.$$

Substitution of $x = \frac{1}{2}$, $y = \frac{1}{2}\sqrt{3}$, and $dy/dt = -3$ gives

$$\frac{1}{2}\frac{dx}{dt} + \frac{1}{2}\sqrt{3}(-3) = 0 \quad \text{so that} \quad \frac{dx}{dt} = 3\sqrt{3}.$$

As the particle passes through the point $(\frac{1}{2}, \frac{1}{2}\sqrt{3})$, the x-coordinate is increasing at the rate of $3\sqrt{3}$ units per second. ☐

As you must have noticed, the first two sample problems were solved by the same general method, a method that we recommend to you for solving problems of this type:

Step 1. Draw a diagram, where relevant, and indicate the quantities that vary.

Step 2. Specify in mathematical form the rate of change you are looking for, and record all given information.

Step 3. Find an equation involving the variable whose rate of change is to be found.

Step 4. Differentiate with respect to time t the equation found in step 3.

Step 5. State the final answer in coherent form, specifying the units that you are using.

Problem 3. A water trough with vertical cross section in the shape of an equilateral triangle is being filled at a rate of 4 cubic feet per minute. Given that the trough is 12 feet long, how fast is the level of the water rising at the instant that the water reaches a depth of $1\frac{1}{2}$ feet?

Solution. Let x be the depth of the water in feet and V the volume of water in cubic feet.

cross section of trough

Find $\dfrac{dx}{dt}$ when $x = \frac{3}{2}$

given that $\dfrac{dV}{dt} = 4$ ft^3/min.

area of cross section $= \dfrac{\sqrt{3}}{3}x^2$

volume of water $= 12\left(\dfrac{\sqrt{3}}{3}x^2\right) = 4\sqrt{3}x^2$

Differentiation of $V = 4\sqrt{3}x^2$ with respect to t gives

$$\frac{dV}{dt} = 8\sqrt{3}x\,\frac{dx}{dt}.$$

Substituting $x = \frac{3}{2}$ and $dV/dt = 4$, we have

$$4 = 8\sqrt{3}\left(\frac{3}{2}\right)\frac{dx}{dt} \quad \text{and thus} \quad \frac{dx}{dt} = \frac{1}{9}\sqrt{3}.$$

At the instant that the water reaches a depth of $1\frac{1}{2}$ feet, the water level is rising at the rate of $\frac{1}{9}\sqrt{3}$ feet per second (about 0.19 feet per second).   ◻

Problem 4. Two ships, one heading west and the other east, approach each other on parallel courses 8 miles apart. Given that each ship is cruising at 20 miles per hour, at what rate is the distance between the ships diminishing when they are 10 miles apart?

Solution. Let y be the distance between the ships measured in miles.

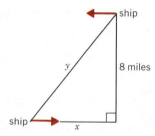

Find $\dfrac{dy}{dt}$ when $y = 10$

given that $\dfrac{dx}{dt} = -40$ mph.

$x^2 + 8^2 = y^2$ (Pythagorean theorem)

(Note that dx/dt is taken as negative since x gets smaller.)   Differentiating $x^2 + 8^2 = y^2$ with respect to t, we find that

$$2x\,\frac{dx}{dt} + 0 = 2y\,\frac{dy}{dt} \quad \text{and consequently} \quad x\,\frac{dx}{dt} = y\,\frac{dy}{dt}.$$

When $y = 10$, $x = 6$. (Why?) Substituting $x = 6$, $y = 10$, and $dx/dt = -40$, we have

$$6(-40) = 10\,\frac{dy}{dt} \quad \text{so that} \quad \frac{dy}{dt} = -24.$$

(Note that dy/dt is negative since y is decreasing.) When the two ships are 10 miles apart, the distance between them is diminishing at the rate of 24 miles per hour.   ◻

Problem 5. A conical paper cup 8 inches across the top and 6 inches deep is full of water. The cup springs a leak at the bottom and loses water at the rate of 2 cubic inches per minute. How fast is the water level dropping at the instant when the water is exactly 3 inches deep?

Solution. We begin with a diagram that represents the situation after the cup has been leaking for a while (Figure 3.11.1). We label the radius and height of the remaining "cone of water" *r* and *h*. We can relate *r* and *h* by similar triangles (Figure 3.11.2).

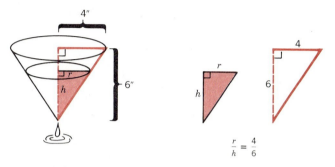

Figure 3.11.1   **Figure 3.11.2**

Let *h* be the depth of the water measured in inches.

$$\text{Find } \frac{dh}{dt} \text{ when } h = 3 \qquad \text{given that } \frac{dV}{dt} = -2 \text{ in.}^3/\text{min.}$$

$$V = \frac{1}{3}\pi r^2 h \quad \text{(volume of a cone)} \qquad \text{and} \qquad \frac{r}{h} = \frac{4}{6} \quad \text{(similar triangles)}$$

Using the second equation to eliminate *r* from the first equation, we have

$$V = \frac{1}{3}\pi \left(\frac{2h}{3}\right)^2 h = \frac{4}{27}\pi h^3.$$

Differentiation with respect to *t* gives

$$\frac{dV}{dt} = \frac{4}{9}\pi h^2 \frac{dh}{dt}.$$

Substituting *h* = 3 and *dV/dt* = −2, we have

$$-2 = \frac{4}{9}\pi(3)^2 \frac{dh}{dt} \quad \text{and thus} \quad \frac{dh}{dt} = -\frac{1}{2\pi}.$$

At the instant when the water is exactly 3 inches deep, the water level is dropping at the rate of $1/2\pi$ inches per minute (about 0.16 inches per minute). ▢

Problem 6. A balloon leaves the ground 500 feet away from an observer and rises vertically at the rate of 140 feet per minute. At what rate is the angle of inclination of the observer's line of sight increasing at the instant when the balloon is exactly 500 feet above the ground?

Solution. Let *x* be the altitude of the balloon and θ the angle of inclination of the observer's line of sight.

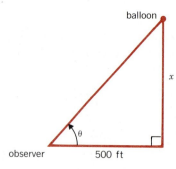

Find $\dfrac{d\theta}{dt}$ when $x = 500$ feet,

given that $\dfrac{dx}{dt} = 140$ ft/min.

$\tan \theta = \dfrac{x}{500}$

Differentiation with respect to t gives

$$\sec^2 \theta \, \frac{d\theta}{dt} = \frac{1}{500} \frac{dx}{dt}.$$

When $x = 500$, the triangle is isosceles, and therefore $\sec \theta = \sqrt{2}$. Substituting $\sec \theta = \sqrt{2}$ and $dx/dt = 140$, we have

$$(\sqrt{2})^2 \frac{d\theta}{dt} = \frac{1}{500}(140) \quad \text{and therefore} \quad \frac{d\theta}{dt} = 0.14.$$

At the instant when the balloon is exactly 500 feet above the ground, the inclination of the observer's line of sight is increasing at the rate of 0.14 radians per minute (about 8 degrees per minute). ▨

Problem 7. A ladder 13 feet long is leaning against the side of a building. If the foot of the ladder is pulled away from the building at the rate of 0.1 feet per second, how fast is the angle formed by the ladder and the ground changing at the instant when the top of the ladder is 12 feet above the ground?

Solution

Find $\dfrac{d\theta}{dt}$ when $y = 12$ feet

given that $\dfrac{dx}{dt} = 0.1$ ft/sec.

$\cos \theta = \dfrac{x}{13}$

Differentiation with respect to t gives

$$-\sin \theta \, \frac{d\theta}{dt} = \frac{1}{13} \frac{dx}{dt}.$$

When $y = 12$, $\sin \theta = \frac{12}{13}$. Substituting $\sin \theta = \frac{12}{13}$ and $dx/dt = 0.1$, we have

$$-\left(\frac{12}{13}\right)\frac{d\theta}{dt} = \frac{1}{13}(0.1) \quad \text{and thus} \quad \frac{d\theta}{dt} = -\frac{1}{120}.$$

At the instant that the top of the ladder is 12 feet above the ground, the angle formed by the ladder and the ground is decreasing at the rate of $\frac{1}{120}$ radians per second (about half a degree per second).   ▢

EXERCISES 3.11

1. A point moves along the straight line $x + 2y = 2$. Find: (a) the rate of change of the y-coordinate, given that the x-coordinate is increasing at a rate of 4 units per second; (b) the rate of change of the x-coordinate, given that the y-coordinate is decreasing at a rate of 2 units per second.

2. A particle is moving in circular orbit $x^2 + y^2 = 25$. As it passes through the point $(3, 4)$, its y-coordinate is decreasing at the rate of 2 units per second. How is the x-coordinate changing?

3. A heap of rubbish in the shape of a cube is being compacted. Given that the volume decreases at a rate of 2 cubic meters per minute, find the rate of change of an edge of the cube when the volume is exactly 27 cubic meters. What is the rate of change at that instant of the surface area of the cube?

4. At a certain instant the dimensions of a rectangle are a and b. These dimensions are changing at the rates m, n, respectively. Find the rate at which the area is changing.

5. The volume of a spherical balloon is increasing at a constant rate of 8 cubic feet per minute. How fast is the radius increasing when the radius is exactly 10 feet? How fast is the surface area increasing at that instant?

6. At a certain instant the side of an equilateral triangle is α centimeters long and increasing at the rate of k centimeters per minute. How fast is the area increasing?

7. The perimeter of a rectangle is fixed at 24 centimeters. If the length l of the rectangle is increasing at the rate of 1 centimeter per second, for what value of l does the area of the rectangle start to decrease?

8. The dimensions of a rectangle are changing in such a way that the perimeter is always 24 inches. Show that, at the instant when the area is 32 square inches, the area is either increasing or decreasing 4 times as fast as the length is increasing.

9. A rectangle is inscribed in a circle of radius 5 inches. If the length of the rectangle is decreasing at the rate of 2 inches per second, how fast is the area changing at the instant when the length is 6 inches?   (HINT: A diagonal of the rectangle is a diameter of the circle.)

10. A boat is held by a bow line that is wound about a windlass 6 feet higher than the bow of the boat. If the boat is drifting away at the rate of 8 feet per minute, how fast is the line unwinding when the bow is exactly 30 feet from the windlass?

11. Two boats are racing with constant speed toward a finish marker, boat A sailing from the south at 13 mph and boat B approaching from the east. When equidistant from the marker the boats are 16 miles apart and the distance between them is decreasing at the rate of 17 mph. Which boat will win the race?

12. A ladder 13 feet long is leaning against a wall. If the foot of the ladder is pulled away from the wall at the rate of 0.5 feet per second, how fast will the top of the ladder be dropping at the instant when the base is 5 feet from the wall?

13. A tank contains 1000 cubic feet of natural gas at a pressure of 5 pounds per square inch. Find the rate of change of the volume if the pressure decreases at a rate of 0.05 pounds per square inch per hour. (Assume Boyle's law: *pressure* × *volume* = *constant*.)

14. The adiabatic law for the expansion of air is $PV^{1.4} = C$. At a given instant the volume is 10 cubic feet and the pressure 50 pounds per square inch. At what rate is the pressure changing if the volume is decreasing at a rate of 1 cubic foot per second?

(*Important*) Angular Velocity; Uniform Circular Motion

As a particle moves along a circle of radius r, it effects a change in the central angle marked θ in Figure 3.11.3. The angle θ is measured in radians. The time rate of change of θ, $\omega = d\theta/dt$, is called the *angular velocity* of the particle. Circular motion with constant positive angular velocity is called *uniform circular motion.*

15. A particle in uniform circular motion traces out a circular arc. The time rate of change of the length of that arc is called the *speed* of the particle. What is the speed of a particle that moves along a circle of radius r with constant positive angular velocity ω?

16. What is the kinetic energy of the particle in Exercise 15 if the mass of the particle is m?

17. A point P moves uniformly along the circle $x^2 + y^2 = r^2$ with angular velocity ω. Find the xy-coordinates of P at time t given that the motion starts at time $t = 0$ with $\theta = \theta_0$.

18. With P as in Exercise 17 find the velocity and acceleration of the projection† of P onto (a) the x-axis. (b) the y-axis.

19. Figure 3.11.4 shows a sector in a circle of radius r. The sector is the union of triangle T and segment S. Suppose that the radius vector rotates counterclockwise with a constant angular velocity of ω radians per second. Show that the area of the sector changes at a constant rate, but that the area of T and the area of S do not change at a constant rate.

20. Take S and T as in Exercise 19. In general the area of S and the area of T change at different rates. There is, however, one value of θ between 0 and π at which both areas have the same instantaneous rate of change. Find this value of θ.

Figure 3.11.3

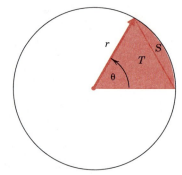

Figure 3.11.4

21. A man standing 3 feet from the base of a lamp post casts a shadow 4 feet long. If the man is 6 feet tall and walks away from the lamp post at a speed of 400 feet per minute, at what rate will his shadow lengthen?

_____

† The *projection* of $P(x, y)$ onto the x-axis is the point $Q(x, 0)$; the projection of $P(x, y)$ onto the y-axis is the point $R(0, y)$.

22. In the special theory of relativity the mass of a particle moving at velocity v is

$$\frac{m}{\sqrt{1 - v^2/c^2}}$$

    where m is the mass at rest and c is the speed of light. At what rate is the mass changing when the particle's velocity is $\frac{1}{2}c$ and the rate of change of the velocity is $0.01c$ per second?

23. Water is dripping through the bottom of a conical cup 4 inches across and 6 inches deep. Given that the cup loses half a cubic inch of water per minute, how fast is the water level dropping when the water is 3 inches deep?

24. Water flows from a faucet into a hemispherical basin 14 inches in diameter at a rate of 2 cubic inches per second. How fast does the water rise (a) when the water is exactly halfway to the top? (b) just as it runs over? (The volume of a spherical segment is given by $\pi r h^2 - \frac{1}{3}\pi h^3$ where r is the radius of the sphere and h is the depth of the segment.)

25. The base of an isosceles triangle is 6 feet. If the altitude is 4 feet and increasing at the rate of 2 inches per minute, at what rate is the vertex angle changing?

26. As a boy winds up the cord, his kite is moving horizontally at a height of 60 feet with a speed of 10 feet per minute. How fast is the inclination of the cord changing when its length is 100 feet?

27. A revolving searchlight $\frac{1}{2}$ mile from shore makes 1 revolution per minute. How fast is the light traveling along the straight beach at the instant it passes over a shorepoint 1 mile away from the shorepoint nearest to the searchlight?

28. Two cars, car A traveling east at 30 mph and car B traveling north at 22.5 mph, are heading toward an intersection I. At what rate is angle IAB changing at the instant when cars A and B are 300 feet and 400 feet, respectively, from the intersection?

29. A rope 32 feet long is attached to a weight and passed over a pulley 16 feet above the ground. The other end of the rope is pulled away along the ground at the rate of 3 feet per second. At what rate is the angle between the rope and the ground changing at the instant when the weight is exactly 4 feet off the ground?

30. A slingshot is made by fastening the two ends of a 10-inch rubber strip 6 inches apart. If the midpoint of the strip is drawn back at the rate of 1 inch per second, at what rate is the angle between the segments of the strip changing 8 seconds later?

31. A balloon is released 500 feet away from an observer. If the balloon rises vertically at the rate of 100 feet per minute and at the same time the wind is carrying it horizontally away from the observer at the rate of 75 feet per minute, at what rate is the angle of inclination of the observer's line of sight changing 6 minutes after the balloon has been released?

32. A searchlight is trained on a plane that flies directly above the light at an altitude of 2 miles at a speed of 400 miles per hour. How fast must the light be turning 2 seconds after the plane passes directly overhead?

3.12 DIFFERENTIALS; NEWTON-RAPHSON APPROXIMATIONS

Differentials

In Figure 3.12.1 you can see the graph of a function f and below it the graph of the tangent line at the point $(x, f(x))$. As the figure suggests, for small h, $f(x + h) - f(x)$,

Figure 3.12.1

the change in f from x to $x + h$, can be approximated by the product $f'(x)h$:

(3.12.1)

$$f(x + h) - f(x) \cong f'(x)h.$$

How good is this approximation? It is good in the sense that, for small h, the difference between the two quantities,

$$[f(x + h) - f(x)] - f'(x)h,$$

is small compared to h. How small? Small enough that the ratio

$$\frac{[f(x + h) - f(x)] - f'(x)h}{h}$$

tends to 0 as h tends to 0:

$$\lim_{h \to 0} \frac{f(x + h) - f(x) - f'(x)h}{h} = \lim_{h \to 0} \frac{f(x + h) - f(x)}{h} - \lim_{h \to 0} \frac{f'(x)h}{h} = f'(x) - f'(x) = 0.$$

The quantities $f(x + h) - f(x)$ and $f'(x)h$ have names:

DEFINITION 3.12.2

The difference $f(x + h) - f(x)$ is called the *increment* (of f from x to $x + h$) and is denoted by Δf:

$$\Delta f = f(x + h) - f(x).\dagger$$

The product $f'(x)h$ is called the *differential* (at x with increment h) and is denoted by df:

$$df = f'(x)h.$$

Display (3.12.1) says that, for small h, Δf and df are approximately equal:

---

† The symbol Δ is a capital δ; Δf is read "delta f."

$$\boxed{\Delta f \cong df.}$$

How approximately equal are they? Enough so that the ratio

$$\frac{\Delta f - df}{h}$$

tends to 0 as h tends to 0.

Let's see what all this amounts to in a very simple case. The area of a square of side x is given by the function

$$f(x) = x^2, \qquad x > 0.$$

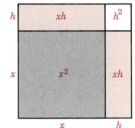

If the length of each side is increased from x to $x + h$, the area increases from $f(x)$ to $f(x + h)$. The change in area is the increment Δf:

$$\begin{aligned}
\Delta f &= f(x + h) - f(x) \\
&= (x + h)^2 - x^2 \\
&= (x^2 + 2xh + h^2) - x^2 \\
&= 2xh + h^2.
\end{aligned}$$

Figure 3.12.2

As an estimate for this change we can use the differential

$$df = f'(x)h = 2xh. \qquad \text{(Figure 3.12.2)}$$

The error of this estimate, the difference between the actual change Δf and the estimated change df, is the difference

$$\Delta f - df = h^2.$$

As promised, the error is small compared to h in the sense that

$$\frac{\Delta f - df}{h} = \frac{h^2}{h} = h$$

tends to 0 as h tends to 0.

Problem 1. Use a differential to estimate the change in $f(x) = x^{2/5}$,

(a) if x is increased from 32 to 34.   (b) if x is decreased from 1 to $\frac{9}{10}$.

Solution. Since $f'(x) = \frac{2}{5}(1/x)^{3/5}$, we have

$$df = f'(x)h = \frac{2}{5}(1/x)^{3/5}h.$$

For part (a) we set $x = 32$ and $h = 2$. The differential then becomes

$$\frac{2}{5}(\tfrac{1}{32})^{3/5}(2) = \frac{1}{10} = 0.10.$$

A change in x from 32 to 34 increases the value of f by approximately 0.10. (From our calculator the increment $\cong 0.098$.)

For part (b) we set $x = 1$ and $h = -\frac{1}{10}$. The differential then becomes

$$\tfrac{2}{5}(\tfrac{1}{1})^{3/5}(-\tfrac{1}{10}) = -\tfrac{2}{50} = -\tfrac{1}{25} = -0.04.$$

A change in x from 1 to $\frac{9}{10}$ decreases the value of f by approximately 0.04. (From our calculator the increment $\cong -0.041$.) ◻

Problem 2. Use a differential to estimate $\sqrt{104}$.

Solution. We know $\sqrt{100}$. What we need is an estimate for the increase of

$$f(x) = \sqrt{x}$$

from 100 to 104. Here

$$f'(x) = \frac{1}{2\sqrt{x}} \quad \text{and thus} \quad df = f'(x)h = \frac{h}{2\sqrt{x}}.$$

With $x = 100$ and $h = 4$, df becomes

$$\frac{4}{2\sqrt{100}} = \frac{1}{5} = 0.2.$$

A change in x from 100 to 104 increases the value of the square root by approximately 0.2. It follows that

$$\sqrt{104} \cong \sqrt{100} + 0.2 = 10 + 0.2 = 10.2.$$

As you can check, $(10.2)^2 = 104.04$, so that we are not far off. ◻

Newton-Raphson Method

Figure 3.12.3 shows the graph of a function f. We will assume that f is twice differentiable. Since the graph of f crosses the x-axis at $x = c$, the number c is a solution of the equation $f(x) = 0$. In the setup of Figure 3.12.3 we can approximate c as follows: Start at x_1 (see the figure). The tangent line at $(x_1, f(x_1))$ intersects the x-axis at some point x_2 which is closer to c than x_1. The tangent line at $(x_2, f(x_2))$ intersects the x-axis at some point x_3 which is closer to c than x_2. Continuing in this manner we obtain better and better approximations to c.

Figure 3.12.3

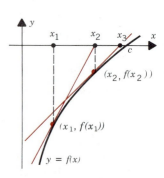

Figure 3.12.4

This method for locating a root of an equation $f(x) = 0$ is called the *Newton-Raphson method*. The method does not work in all cases, but it does work if x_1 can be joined to c by an interval I on the interior of which $f(x) f''(x) > 0$. For then, if f is positive on the interior of I, the graph of f is concave up on I and we have the situation pictured in Figure 3.12.3. On the other hand, if f is negative on the interior of I, then the graph is concave down on I and we have the situation pictured in Figure 3.12.4.

There is an algebraic connection between x_n and x_{n+1} that we now develop. The tangent line at $(x_n, f(x_n))$ has equation

$$y - f(x_n) = f'(x_n)(x - x_n).$$

The x-intercept of this line, x_{n+1}, can be found by setting $y = 0$:

$$0 - f(x_n) = f'(x_n)(x_{n+1} - x_n).$$

Solving this equation for x_{n+1}, we have

(3.12.3)

$$x_{n+1} = x_n - \frac{f(x_n)}{f'(x_n)}.$$

Example 3. The number $\sqrt{3}$ is a root of the equation $x^2 - 3 = 0$. We will estimate $\sqrt{3}$ by applying the Newton-Raphson method to the function $f(x) = x^2 - 3$ starting at $x_1 = 2$. (As you can check, $f(x) f''(x) > 0$ on $(\sqrt{3}, 2)$ and therefore we can be sure that the method applies.) Since $f'(x) = 2x$, the Newton-Raphson formula gives

$$x_{n+1} = x_n - \left(\frac{x_n^2 - 3}{2x_n} \right) = \frac{2x_n^2 + 3}{x_{n+1}}.$$

Successive calculations with this formula (here we are using a calculator) give

$$x_1 = 2, \qquad x_2 = 1.7500000, \qquad x_3 = 1.7321428, \qquad x_4 = 1.7320508.$$

Since $(1.7320508)^2 \cong 2.9999999$, the method has generated a very accurate estimate for $\sqrt{3}$ in only three steps. ■

EXERCISES 3.12

1. Use a differential to estimate the change in the volume of a cube caused by an increase h in the length of each side. Interpret geometrically the error of your estimate $\Delta V - dV$.
2. Use a differential to estimate the area of a ring of inner radius r and width h. What is the exact area?

Estimate the following by differentials.

3. $\sqrt[3]{1010}$.   4. $\sqrt{125}$.   5. $\sqrt[4]{15}$.   6. $1/\sqrt{24}$.

7. $\sqrt[5]{30}$.   8. $(26)^{2/3}$.   9. $(33)^{3/5}$.   10. $(33)^{-1/5}$.

Remembering that the domain of a trigonometric function is measured in radians rather than degrees, use differentials to estimate the following.

11. $\sin 46°$.   12. $\cos 62°$.   13. $\tan 28°$.   14. $\sin 43°$.

15. Estimate $f(2.8)$ given that $f(3) = 2$ and $f'(x) = (x^3 + 5)^{1/5}$.

16. Estimate $f(5.4)$ given that $f(5) = 1$ and $f'(x) = \sqrt[3]{x^2 + 2}$.

17. Find the approximate volume of a thin cylindrical sheet with open ends given that the inner radius is r, the height is h, and the thickness is t.

18. The diameter of a steel ball is measured to be 16 centimeters, with a maximum error of 0.3 centimeters. Estimate by differentials the maximum error in (a) the surface area when calculated by the formula $S = 4\pi r^2$; (b) the volume when calculated by the formula $V = \frac{4}{3}\pi r^3$.

19. A box is to be constructed in the form of a cube to hold 1000 cubic feet. Use a differential to estimate how accurately the inner edge must be made so that the volume will be correct to within 3 cubic feet.

20. Use differentials to estimate the values of x for which

  (a) $\sqrt{x+1} - \sqrt{x} < 0.01$. (b) $\sqrt[4]{x+1} - \sqrt[4]{x} < 0.002$.

21. The duration of one vibration of a pendulum of length l is given by the formula $t = \pi\sqrt{l/g}$. Here t is measured in seconds, and l is measured in feet. Take $\pi \cong 3.14$ and $g \cong 32$. Given that a pendulum of length 3.26 feet vibrates once a second, find the approximate change in t if the pendulum is lengthened by 0.01 feet.

22. As a metal cube is heated, the length of each edge increases $\frac{1}{10}\%$ per degree increase in temperature. Show by differentials that the surface area increases about $\frac{2}{10}\%$ per degree and the volume increases about $\frac{3}{10}\%$ per degree.

23. We are trying to determine the area of a circle by measuring the diameter. How accurately must we measure the diameter if our estimate is to be correct within 1%?

24. Estimate by differentials how precisely x must be determined if (a) x^n is to be accurate within 1%; (b) $x^{1/n}$ is to be accurate within 1%.

(Calculator) Use the Newton-Raphson method to estimate a root of the equation starting at the indicated value of x: (a) Express x_{n+1} in terms of x_n. (b) Give x_4 rounded off to five decimal places and evaluate f at that approximation.

25. $f(x) = x^2 - 24$; $x_1 = 5$.   **26.** $f(x) = x^2 - 17$; $x_1 = 5$.

27. $f(x) = x^3 - 25$; $x_1 = 3$.   **28.** $f(x) = x^5 - 30$; $x_1 = 2$.

29. $f(x) = \cos x - x$; $x_1 = 1$.   **30.** $f(x) = \sin x - x^2$; $x_1 = 1$.

31. $f(x) = 2 \sin x - x$; $x_1 = 2$.   **32.** $f(x) = x^3 - 4x + 1$; $x_1 = 2$.

Little-$o(h)$; The Tangent Line as Best Approximation

Let g be a function defined at least on an open interval containing the number 0. We say that $g(h)$ is *little-o(h)* and write $g(h) = o(h)$ iff, for small h, $g(h)$ is small enough compared to h that

$$\lim_{h \to 0} \frac{g(h)}{h} = 0, \quad \text{or equivalently,} \quad \lim_{h \to 0} \frac{g(h)}{|h|} = 0.$$

33. Which of the following statements is true?

  (a) $h^3 = o(h)$.   (b) $\dfrac{h^2}{h-1} = o(h)$.   (c) $h^{1/3} = o(h)$.

34. Show that, if $g(h) = o(h)$, then $\lim_{h \to 0} g(h) = 0$.

35. Show that, if $g_1(h) = o(h)$ and $g_2(h) = o(h)$, then

$$g_1(h) + g_2(h) = o(h) \quad \text{and} \quad g_1(h)g_2(h) = o(h).$$

36. Figure 3.12.5 shows the graph of a differentiable function f and a line with slope m that passes through the point $(x, f(x))$. The vertical separation at $x + h$ between the line with slope m and the graph of f has been labeled $g(h)$.

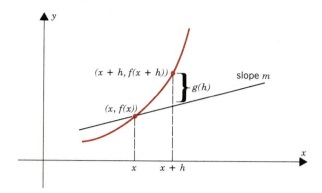

Figure 3.12.5

(a) Calculate $g(h)$.

(b) Show that, of all lines that pass through $(x, f(x))$, the tangent line is the line that best approximates the graph of f near the point $(x, f(x))$ by showing that

$$g(h) = o(h) \quad \text{iff} \quad m = f'(x).$$

Program for the Newton-Raphson method   (BASIC)

This program takes a lot of the work out of the process of applying the Newton-Raphson method for finding a root of an equation of the type FNf(x) = 0, where FNf is the computer's name for a function. As in previous programs, you are meant to replace the sample function FNf(x) by the function of your choice. You will also have to specify the correct derivative of your function, calling it FNfprime(x), and specify the value x that you consider to be a good starting point for finding a root of the equation FNf(x) = 0. After that the computer will take over.

```
10 REM Newton-Raphson method for finding roots
20 REM copyright © Colin C. Graham 1988-1989

30 REM change the lines "def FNf = ..." and "def FNfprime = ..."
40 REM    to fit your function

50 DEF FNf(x) = LOG(x) − 1
60 DEF FNfprime(x) = 1/x

70 REM warning: program does not avoid division by 0.

100 INPUT "Enter the number of iterations:"; ntimes
110 REM    13 gives a nice display on many computers

120 INPUT "Enter starting value:"; x

200    PRINT "x", "FNf(x)"
210    PRINT x, FNf(x)

300 FOR j = 0 TO ntimes
310       x = x − FNf(x)/FNfprime(x)
320       PRINT x, FNf(x)
330 NEXT j

500 INPUT "Do it again? (Y/N)"; a$
510 IF a$ = "Y" OR a$ = "y" THEN 100
520 END
```

3.13   ADDITIONAL EXERCISES

1. The rate of change of f at $x = x_0$ is twice as great as it is at $x = 1$. Find x_0 if
   (a) $f(x) = x^2$.   (b) $f(x) = 2x^3$.   (c) $f(x) = \sqrt{x}$.

2. Given that $y = 2x^3 - x^2$, determine the values of x for which $y = dy/dx$.

3. Find the points on the curve $y = \frac{2}{3}x^{3/2}$ where the inclination of the tangent line is
   (a) $45°$. (b) $60°$. (c) $30°$.

4. Find an equation for the tangent to the curve $y = x^3$ that passes through the point $(0, 2)$.

5. Find an equation for the line tangent to the graph of f^{-1} at the point (b, a) given that the slope of the graph of f at (a, b) is m.

6. Find an equation for the normal to the parabola $y = 5x + x^2$ that makes an angle of $45°$ with the x-axis.

Find a formula for the nth derivative.

7. $y = (a + bx)^n$.

8. $y = \dfrac{a}{bx + c}$.

9. If $y = 4x - x^3$ and x is increasing steadily at the rate of $\frac{1}{3}$ unit per second, how fast is the slope of the graph changing at the instant when $x = 2$?

10. A ball thrown straight up from the ground reaches a height of 24 feet in 1 second. How high will the ball go?

(*Important*) We have defined the derivative of f at c by

$$f'(c) = \lim_{h \to 0} \frac{f(c + h) - f(c)}{h}.$$

Setting $c + h = x$, we can write

(3.13.1)
$$f'(c) = \lim_{x \to c} \frac{f(x) - f(c)}{x - c}.$$

This alternative definition of derivative has advantages in certain situations. Convince yourself of the equivalence of both definitions by calculating $f'(c)$ by both methods.

11. $f(x) = x^2$; $c = 2$.

12. $f(x) = x^2 - 3x$; $c = 1$.

(*Important*) Find the following limits. [They are all derivatives set in the manner of (3.13.1).]

13. $\lim\limits_{x \to 1} \dfrac{x^{1/5} - 1}{x - 1}$.   [HINT: This is the derivative of $f(x) = x^{1/5}$ at $x = 1$.]

14. $\lim\limits_{x \to 2} \dfrac{x^5 - 32}{x - 2}$.

15. $\lim\limits_{x \to \pi} \dfrac{\sin x}{x - \pi}$.

16. $\lim\limits_{x \to 0} \dfrac{\cos x - 1}{x}$.

17. Find equations for all the tangents to the curve $y = x^3 - x$ that pass through the point $(-2, 2)$.

18. A particle moves along the parabola $y^2 = 12x$. At what point of the parabola do the x- and y-coordinates of the particle increase at the same rate?

19. The radius of a cone is increasing at the rate of 0.3 inches per minute but the volume remains constant. At what rate is the height of the cone changing at the instant when the radius is 4 inches and the height is 15 inches?

20. The area of a square is increasing at the rate of 6 square inches per minute when the side of the square measures 14 inches. How fast is the perimeter of the square changing at that instant?

21. The dimensions of a rectangle are changing in such a way that the perimeter remains 20 centimeters. How fast is the area of the rectangle changing at the instant when one side is 7 centimeters long and is decreasing at the rate of 0.4 centimeters per minute?

22. The dimensions of a rectangle are changing in such a way that the area remains 24 square inches. At what rate is the perimeter changing at the instant when one side is 6 inches long and is increasing at the rate of 0.3 inches per minute?

23. At what rate is the volume of a sphere changing at the instant when the surface area is increasing at the rate of 4 square inches per minute and the radius is increasing at the rate of 0.1 inches per minute?

24. At what rate is the surface area of a cube changing at the instant when the volume is decreasing at the rate of 9 cubic centimeters per minute and the length of a side is decreasing at the rate of 4 centimeters per minute?

25. An object moves along a coordinate line, its position at time t given by the function $x(t) = t + 2 \cos t$. Find those times t from 0 to 2π when the object is slowing down.

26. An object moves along a coordinate line, its position at time t given by the function $x(t) = \sqrt{t + 1}$. (a) Show that the acceleration is negative and proportional to the cube of the velocity. (b) Use differentials to find numerical estimates for the position, velocity, and acceleration of the object at time $t = 17$. Base your estimate on $t = 15$.

27. As a metal sphere is heated, its radius increases $\frac{1}{10}$% per degree increase in temperature. Show by differentials that the surface area increases about $\frac{2}{10}$% per degree and the volume increases about $\frac{3}{10}$% per degree.

28. An object moves along a coordinate line, its position at time t given by the function $x(t) = -t\sqrt{t + 1}$, $t \geq 0$. Find (a) the velocity at time $t = 1$. (b) the acceleration at time $t = 1$. (c) the speed at time $t = 1$. (d) the rate of change of the speed at time $t = 1$.

29. A horizontal trough 12 feet long has a vertical cross section in the shape of a trapezoid. The bottom is 3 feet wide, and the sides are inclined to the vertical at an angle whose sine is $\frac{4}{5}$. Given that water is poured into the trough at the rate of 10 cubic feet per minute, how fast is the water level rising when the water is exactly 2 feet deep?

30. In Exercise 29, at what rate is the water being drawn from the trough if the level is falling 0.1 feet per minute when the water is 3 feet deep?

31. A point P moves along the parabola $y = x^2$ so that its x-coordinate increases at the constant rate of k units per second. The projection of P on the x-axis is M. At what rate is the area of triangle OMP changing when P is at the point where $x = a$?

32. The diameter and height of a right circular cylinder are found at a certain instant to be 10 and 20 inches, respectively. If the diameter is increasing at the rate of 1 inch per minute, what change in the height will keep the volume constant?

33. The period (P seconds) of a complete oscillation of a pendulum of length l inches is given by the formula $P = 0.32 \sqrt{l}$. (a) Find the rate of change of the period with respect to the length when $l = 9$ inches. (b) Use a differential to estimate the change in P caused by an increase in l from 9 to 9.2 inches.

34. Show that an object projected vertically from the ground lands with the speed with which it was projected. Neglect air resistance.

35. Ballast dropped from a balloon that was rising at the rate of 5 feet per second reached the ground in 8 seconds. How high was the balloon when the ballast was dropped?

36. Had the balloon of Exercise 35 been falling at the rate of 5 feet per second, how long would it have taken for the ballast to reach the ground?

37. Isocost lines and indifference curves are studied in economics. For what value of C is the isocost line

$$Ax + By = 1 \qquad\qquad (A > 0, B > 0)$$

tangent to the indifference curve

$$y = \frac{1}{x} + C, \qquad x > 0?$$

Find the point of tangency (called the *equilibrium point*).

38. Let

$$P = f(Q) \qquad\qquad (P = \text{price}, Q = \text{output})$$

be the supply function of Figure 3.13.1. Compare angles θ and ϕ at points where the elasticity

$$\epsilon = \frac{f(Q)}{Q|f'(Q)|}$$

(a) is 1.   (b) is less than 1.   (c) is greater than 1.

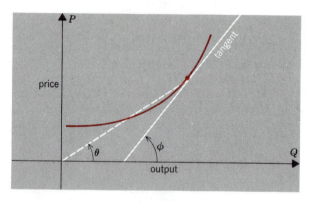

Figure 3.13.1

39. Determine the coefficients A, B, C so that the curve $y = Ax^2 + Bx + C$ will pass through the point $(1, 3)$ and be tangent to the line $x - y + 1 = 0$ at the point $(2, 3)$.

40. Find A, B, C, D given that the curve $y = Ax^3 + Bx^2 + Cx + D$ is tangent to the line $y = 5x - 4$ at the point $(1, 1)$ and tangent to the line $y = 9x$ at the point $(-1, -9)$.

41. An object moves along a coordinate line, its position at time t given by the function $x(t) = \cos 2t - 2 \cos t$. At what times t from 0 to 2π does the object change direction?

42. A boy walks on a straight horizontal path away from a light that hangs 12 feet above the path. How fast does his shadow lengthen if he is 5 feet tall and walks at a rate of 168 feet per minute?

43. A railroad track crosses a highway at an angle of $60°$. A locomotive is 500 feet from the intersection and moving away from it at the rate of 60 miles per hour. An automobile is 500 feet from the intersection and moving toward it at the rate of 30 miles per hour. What is the rate of change of the distance between them?   HINT: Use the law of cosines.

44. One ship was sailing south at 6 miles per hour; another east at 8 miles per hour. At 4 P.M. the second crossed the track of the first, where the first had been 2 hours before. (a) How was the distance between the ships changing at 3 P.M.? (b) At 5 P.M.?

45. A point P is moving uniformly along the circle $x^2 + y^2 = r^2$. (a) At what points of the circle is the projection of P onto the x-axis moving with maximum speed? (b) At what points is the projection of P onto the y-axis moving with maximum speed?   HINT: See Exercise 17, Section 3.11.

CHAPTER HIGHLIGHTS

3.1 The Derivative

If f is differentiable at x, then f is continuous at x. (p. 112) The converse is false.

3.2 Some Differentiation Formulas

$$(f + g)'(x) = f'(x) + g'(x)   (\alpha f)'(x) = \alpha f'(x) \quad \text{for each real } \alpha$$

$$(fg)'(x) = f(x)g'(x) + g(x)f'(x) \qquad \left(\frac{f}{g}\right)'(x) = \frac{g(x)f'(x) - f(x)g'(x)}{[g(x)]^2} \quad (g(x) \neq 0)$$

For each nonzero integer n, $p(x) = x^n$ has derivative $p'(x) = nx^{n-1}$.

3.3 The *d/dx* Notation; Derivatives of Higher Order

Another notation for the derivative is the *double-d* notation of Leibniz: if $y = f(x)$, then

$$\frac{dy}{dx} = f'(x).$$

Higher order derivatives are calculated by repeated differentiation. For example, the second derivative of $y = f(x)$ is the derivative of the derivative:

$$f''(x) = \frac{d^2y}{dx^2} = \frac{d}{dx}\left(\frac{dy}{dx}\right).$$

3.4 The Derivative as a Rate of Change

dy/dx gives the rate of change of y with respect to x.

3.5 Velocity and Acceleration; Free Fall

If the velocity and acceleration have the same sign, the object is *speeding up*. If the velocity and acceleration have opposite signs, the object is *slowing down*.

3.6 The Chain Rule

Various forms of the chain rule:

$$\frac{dy}{dx} = \frac{dy}{du}\frac{du}{dx}, \qquad \frac{d}{dx}[f(u)] = \frac{d}{du}[f(u)]\frac{du}{dx}, \qquad (f \circ g)'(x) = f'(g(x))\,g'(x).$$

If u is a differentiable function of x and n is a positive or negative integer, then

$$\frac{d}{dx}(u^n) = nu^{n-1}\frac{du}{dx}.$$

3.7 Differentiating the Trigonometric Functions

$$\frac{d}{dx}(\sin u) = \cos u \frac{du}{dx}, \qquad \frac{d}{dx}(\cos u) = -\sin u \frac{du}{dx},$$

$$\frac{d}{dx}(\tan u) = \sec^2 u \frac{du}{dx}, \qquad \frac{d}{dx}(\cot u) = -\csc^2 u \frac{du}{dx},$$

$$\frac{d}{dx}(\sec u) = \sec u \tan u \frac{du}{dx}, \qquad \frac{d}{dx}(\csc u) = -\csc u \cot u \frac{du}{dx}.$$

3.8 Differentiating Inverses; Rational Powers

derivative of an inverse: $(f^{-1})'(x) = \dfrac{1}{f'(f^{-1}(x))}, \qquad \dfrac{dx}{dy} = \dfrac{1}{dy/dx}$

derivative of a rational power: $\dfrac{d}{dx}(u^{p/q}) = \dfrac{p}{q} u^{(p/q)-1} \dfrac{du}{dx}$

3.9 Additional Practice in Differentiation

3.10 Implicit Differentiation

finding dy/dx from an equation in x and y without first solving the equation for y

3.11 Rates of Change per Unit Time

If a quantity Q varies with time, then the derivative dQ/dt gives the rate of change of that quantity per unit time.

A 5-step procedure for solving *related rates problems* is outlined on p. 165.

3.12 Differentials; Newton-Raphson Approximations

increment: $\Delta f = f(x + h) - f(x)$ differential: $df = f'(x)h$

$\Delta f \cong df$ in the sense that $\dfrac{\Delta f - df}{h}$ tends to 0 as $h \to 0$

Newton-Raphson method (p. 174)

3.13 Additional Exercises

alternative definition of derivative: $f'(c) = \lim\limits_{x \to c} \dfrac{f(x) - f(c)}{x - c}$

CHAPTER 4

THE MEAN-VALUE THEOREM AND APPLICATIONS

4.1 THE MEAN-VALUE THEOREM

In this section we prove a result known as *the mean-value theorem*. First stated by the French mathematician Joseph Louis Lagrange† (1736–1813), this theorem has come to permeate the theoretical structure of all calculus.

THEOREM 4.1.1 THE MEAN-VALUE THEOREM

If f is differentiable on (a, b) and continuous on $[a, b]$, then there is at least one number c in (a, b) for which

$$f'(c) = \frac{f(b) - f(a)}{b - a}.$$

The number

$$\frac{f(b) - f(a)}{b - a}$$

---

† Lagrange, whom you encountered earlier in connection with the prime notation for differentiation, was born in Turin, Italy. He spent twenty years as mathematician in residence at the court of Frederick the Great. Later he taught at the renowned École Polytechnique.

is the slope of the line l that passes through the points $(a, f(a))$ and $(b, f(b))$. To say that there is at least one number c for which

$$f'(c) = \frac{f(b) - f(a)}{b - a}$$

is to say that the graph of f has at least one point $(c, f(c))$ at which the tangent is parallel to l. See Figure 4.1.1.

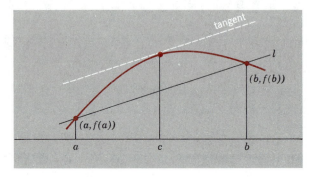

Figure 4.1.1

We will prove the mean-value theorem in steps. First we will show that, if a function f has a nonzero derivative at some point x_0, then, close to x_0, $f(x)$ is greater than $f(x_0)$ on one side of x_0 and less than $f(x_0)$ on the other side of x_0.

THEOREM 4.1.2

Let f be differentiable at x_0. If $f'(x_0) > 0$, then

$$f(x_0 - h) < f(x_0) < f(x_0 + h)$$

for all positive h sufficiently small. If $f'(x_0) < 0$, then

$$f(x_0 + h) < f(x_0) < f(x_0 - h)$$

for all positive h sufficiently small.

Proof. We take the case $f'(x_0) > 0$ and leave the other case to you. By the definition of derivative,

$$\lim_{k \to 0} \frac{f(x_0 + k) - f(x_0)}{k} = f'(x_0).$$

With $f'(x_0) > 0$ we can use $f'(x_0)$ itself as ϵ and conclude that there exists $\delta > 0$ such that

$$\text{if} \quad 0 < |k| < \delta, \quad \text{then} \quad \left| \frac{f(x_0 + k) - f(x_0)}{k} - f'(x_0) \right| < f'(x_0).$$

For such k we have

$$\frac{f(x_0 + k) - f(x_0)}{k} > 0. \qquad \text{(why?)}$$

If now $0 < h < \delta$, then

$$\frac{f(x_0 + h) - f(x_0)}{h} > 0 \quad \text{and} \quad \frac{f(x_0 - h) - f(x_0)}{-h} > 0,$$

and consequently

$$f(x_0) < f(x_0 + h) \quad \text{and} \quad f(x_0 - h) < f(x_0). \quad \square$$

Next we prove a special case of the mean-value theorem, known as Rolle's theorem (after the French mathematician Michel Rolle who first announced the result in 1691). In Rolle's theorem we make the additional assumption that $f(a)$ and $f(b)$ are both 0. (See Figure 4.1.2.) In this case the line joining $(a, f(a))$ to $(b, f(b))$ is horizontal. (It is the x-axis.) The conclusion is that there is a point $(c, f(c))$ at which the tangent is horizontal.

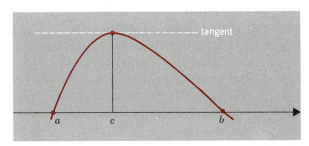

Figure 4.1.2

THEOREM 4.1.3 ROLLE'S THEOREM

Let f be differentiable on (a, b) and continuous on $[a, b]$. If $f(a)$ and $f(b)$ are both 0, then there is at least one number c in (a, b) for which

$$f'(c) = 0.$$

Proof. If f is constantly 0 on $[a, b]$, the result is obvious. If f is not constantly 0 on $[a, b]$, then either f takes on some positive values or some negative values. We assume the former and leave the other case to you.

Since f is continuous on $[a, b]$, at some point c of $[a, b]$, f must take on a maximum value. (Theorem 2.6.2.) This maximum value, $f(c)$, must be positive. Since $f(a)$ and $f(b)$ are both 0, c cannot be a and it cannot be b. This means that c must lie in the open interval (a, b) and therefore $f'(c)$ exists. Now $f'(c)$ cannot be greater than 0 and it cannot be less than 0 because each of these conditions would imply that f takes on values greater than $f(c)$. (This follows from Theorem 4.1.2.) We conclude therefore that $f'(c) = 0$. $\square$

We are now ready to give a proof of the mean-value theorem.

Proof of the Mean-Value Theorem.   We will create a function g that satisfies the conditions of Rolle's theorem and is so related to f that the conclusion $g'(c) = 0$ leads to the conclusion

$$f'(c) = \frac{f(b) - f(a)}{b - a}.$$

It is not hard to see that

$$g(x) = f(x) - \left[\frac{f(b) - f(a)}{b - a} (x - a) + f(a) \right]$$

is exactly such a function.   Geometrically $g(x)$ is represented in Figure 4.1.3.   The line that passes through $(a, f(a))$ and $(b, f(b))$ has the equation

$$y = \frac{f(b) - f(a)}{b - a} (x - a) + f(a).$$

[This is not hard to verify.   The slope is right, and, when $x = a$, $y = f(a)$.]   The difference

$$g(x) = f(x) - \left[\frac{f(b) - f(a)}{b - a} (x - a) + f(a) \right]$$

is simply the vertical separation between the graph of f and the line in question.

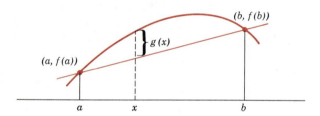

Figure 4.1.3

If f is differentiable on (a, b) and continuous on $[a, b]$, then so is g.   As you can check, $g(a)$ and $g(b)$ are both 0.   Therefore by Rolle's theorem there is at least one point c in (a, b) for which $g'(c) = 0$.   Since in general

$$g'(x) = f'(x) - \frac{f(b) - f(a)}{b - a},$$

in particular

$$g'(c) = f'(c) - \frac{f(b) - f(a)}{b - a}.$$

With $g'(c) = 0$, we have

$$f'(c) = \frac{f(b) - f(a)}{b - a}. \quad \square$$

[*Note*: The conclusion $f'(c) = \dfrac{f(b) - f(a)}{b - a}$ is often written $f(b) - f(a) = f'(c)(b - a)$.]

EXERCISES 4.1

After verifying that the function satisfies the conditions of the mean-value theorem on the indicated interval, find the admissible values of c.

1. $f(x) = x^2$; [1, 2].   **2.** $f(x) = 3\sqrt{x} - 4x$; [1, 4].   **3.** $f(x) = x^3$; [1, 3].

4. $f(x) = x^{2/3}$; [1, 8].   **5.** $f(x) = \sqrt{1 - x^2}$; [0, 1].   **6.** $f(x) = x^3 - 3x$; [−1, 1].

7. Determine whether the function $f(x) = \sqrt{1 - x^2}/(3 + x^2)$ satisfies the conditions of Rolle's theorem on the interval [−1, 1]. If so, find the admissible values of c.

8. Sketch the graph of

$$f(x) = \begin{cases} 2 + x^3, & x \le 1 \\ 3x, & x > 1 \end{cases}$$

and compute the derivative. Determine whether f satisfies the conditions of the mean-value theorem on the interval [−1, 2] and, if so, find the admissible values of c.

9. Sketch the graph of

$$f(x) = \begin{cases} 2x + 2, & x \le -1 \\ x^3 - x, & x > -1 \end{cases}$$

and compute the derivative. Determine whether f satisfies the conditions of the mean-value theorem on the interval [−3, 2] and, if so, find the admissible values of c.

10. Set $f(x) = x^{-1}$, $a = -1$, $b = 1$. Verify that there is no number c for which

$$f'(c) = \frac{f(b) - f(a)}{b - a}.$$

Explain how this does not violate the mean-value theorem.

11. Exercise 10 with $f(x) = |x|$.

12. Graph the function $f(x) = |2x - 1| - 3$ and compute the derivative. Verify that $f(-1) = 0 = f(2)$ and yet $f'(x)$ is never 0. Explain how this does not violate Rolle's theorem.

13. Show that the equation $6x^4 - 7x + 1 = 0$ does not have more than two distinct real roots. (Use Rolle's theorem.)

14. Show that the equation $6x^5 + 13x + 1 = 0$ has exactly one real root. (Use Rolle's theorem and the intermediate-value theorem.)

15. Show that the equation $x^3 + 9x^2 + 33x - 8 = 0$ has exactly one real root.

16. Let f be twice differentiable. Show that, if the equation $f(x) = 0$ has n distinct real roots, then the equation $f'(x) = 0$ has at least $n - 1$ distinct real roots and the equation $f''(x) = 0$ has at least $n - 2$ distinct real roots.

17. Let P be a nonconstant polynomial $P(x) = a_n x^n + \cdots + a_1 x + a_0$. Show that between any two consecutive roots of the equation $P'(x) = 0$ there is at most one root of the equation $P(x) = 0$.

18. Show that the equation $x^3 + ax + b = 0$ has exactly one real root if $a \ge 0$ and at most one real root between $-\frac{1}{3}\sqrt{3}|a|$ and $\frac{1}{3}\sqrt{3}|a|$ if $a < 0$.

19. Given that $|f'(x)| \le 1$ for all real numbers x, show that $|f(x_1) - f(x_2)| \le |x_1 - x_2|$ for all real numbers x_1 and x_2.

20. Let f be differentiable on an open interval I. Prove that, if $f'(x) = 0$ for all x in I, then f is constant on I.

21. Let f be differentiable on (a, b) with $f(a) = f(b) = 0$ and $f'(c) = 0$ for some c in (a, b). Show by example that f need not be continuous on $[a, b]$.

22. Prove that for all real x and y

   (a) $|\cos x - \cos y| \le |x - y|$.   (b) $|\sin x - \sin y| \le |x - y|$.

23. An object moves along a coordinate line, its position at time t given by $f(t)$ where f is a differentiable function. (a) What does Rolle's theorem say about the motion? (b) What does the mean-value theorem say about the motion?

24. (*Important*) Use the mean-value theorem to show that, if f is continuous at x and $x + h$ and differentiable in between, then

$$f(x + h) - f(x) = f'(x + \theta h)h$$

for some number θ between 0 and 1. (In some texts this is how the mean-value theorem is stated.)

25. Let $h > 0$. Suppose that f is continuous on $[x_0 - h, x_0 + h]$ and differentiable on $(x_0 - h, x_0) \cup (x_0, x_0 + h)$. Show that if

$$\lim_{x \to x_0^-} f'(x) = \lim_{x \to x_0^+} f'(x) = L,$$

then f is differentiable at x_0 and $f'(x_0) = L$.   HINT: Use Exercise 24.

26. (**Calculator**) Let $f(x) = x^4 + 7x^2 + 2$, $x \in [1, 3]$. According to the mean-value theorem there exists a number c in (1, 3) for which

$$f'(c) = \frac{f(3) - f(1)}{3 - 1}.$$

Give a numerical estimate for c correct to two decimal places.

27. (**Calculator**) Let $f(x) = x^2 + x^{-1}$, $x \in [2, 4]$. According to the mean-value theorem there exists a number c in (2, 4) for which

$$f'(c) = \frac{f(4) - f(2)}{4 - 2}.$$

Give a numerical estimate for c correct to two decimal places.

4.2   INCREASING AND DECREASING FUNCTIONS

To place our discussion on a solid footing we begin with a definition.

DEFINITION 4.2.1

A function f is said to

    (i) *increase* on the interval I iff for every two numbers x_1, x_2 in I

$$x_1 < x_2 \quad \text{implies} \quad f(x_1) < f(x_2).$$

    (ii) *decrease* on the interval I iff for every two numbers x_1, x_2 in I

$$x_1 < x_2 \quad \text{implies} \quad f(x_1) > f(x_2).$$

Preliminary Examples

 (a) The squaring function

$$f(x) = x^2 \qquad\qquad \text{(Figure 4.2.1)}$$

decreases on $(-\infty, 0]$ and increases on $[0, \infty)$.

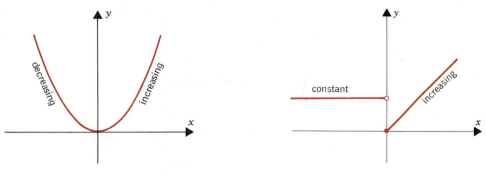

Figure 4.2.1　　　　　　　　　　**Figure 4.2.2**

(b) The function

$$f(x) = \begin{bmatrix} 1, & x < 0 \\ x, & x \geq 0 \end{bmatrix}$$ (Figure 4.2.2)

is constant on $(-\infty, 0)$ and increases on $[0, \infty)$.

(c) The cubing function

$$f(x) = x^3$$ (Figure 4.2.3)

is everywhere increasing.

(d) In the case of the Dirichlet function

$$g(x) = \begin{bmatrix} 1, & x \text{ rational} \\ 0, & x \text{ irrational} \end{bmatrix},$$ (Figure 4.2.4)

there is no interval on which the function increases and no interval on which the function decreases. On every interval the function jumps back and forth between 0 and 1 an infinite number of times. ☐

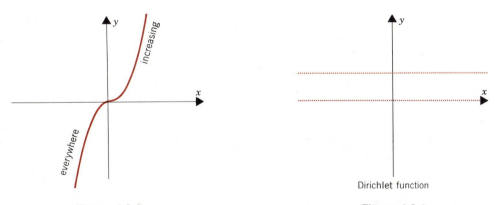

Figure 4.2.3　　　　　　　　　　**Figure 4.2.4**

If f is a differentiable function, then we can determine the intervals on which f increases and the intervals on which f decreases by examining the sign of the first derivative.

THEOREM 4.2.2

Let f be differentiable on an open interval I.

   (i) If $f'(x) > 0$ for all x in I, then f increases on I.

  (ii) If $f'(x) < 0$ for all x in I, then f decreases on I.

 (iii) If $f'(x) = 0$ for all x in I, then f is constant on I.

Proof.   We take x_1 and x_2 in I with $x_1 < x_2$. Since f is differentiable on (x_1, x_2) and continuous on $[x_1, x_2]$, we see from the mean-value theorem that there is a number c in (x_1, x_2) for which

$$f'(c) = \frac{f(x_2) - f(x_1)}{x_2 - x_1}.$$

In (i), we have $\dfrac{f(x_2) - f(x_1)}{x_2 - x_1} > 0$ and thus $f(x_1) < f(x_2)$.

In (ii), we have $\dfrac{f(x_2) - f(x_1)}{x_2 - x_1} < 0$ and thus $f(x_1) > f(x_2)$.

In (iii), we have $\dfrac{f(x_2) - f(x_1)}{x_2 - x_1} = 0$ and thus $f(x_1) = f(x_2)$.   ☐

    The theorem we just proved is useful, but it doesn't tell the complete story. Look, for example, at the squaring function $f(x) = x^2$. The derivative $f'(x) = 2x$ is negative for x in $(-\infty, 0)$, zero at $x = 0$, and positive for x in $(0, \infty)$. Theorem 4.2.2 assures us that

        f decreases on $(-\infty, 0)$ and increases on $(0, \infty)$,

but actually

        f decreases on $(-\infty, 0]$ and increases on $[0, \infty)$.

To get these stronger results we need a theorem that works for closed intervals too.

    To extend Theorem 4.2.2 so that it works for an arbitrary interval I, the only additional condition we need is continuity at the endpoint(s).

THEOREM 4.2.3

Let f be continuous on an arbitrary interval I and differentiable on the interior of I.

   (i) If $f'(x) > 0$ for all x in the interior of I, then f increases on all of I.

  (ii) If $f'(x) < 0$ for all x in the interior of I, then f decreases on all of I.

 (iii) If $f'(x) = 0$ for all x in the interior of I, then f is constant on all of I.

A proof of this result is not hard to construct. As you can verify, a word-for-word repetition of our proof of Theorem 4.2.2 also works here.

It is time for some examples.

Example 1. The function

$$f(x) = \sqrt{1 - x^2}$$

has derivative

$$f'(x) = -\frac{x}{\sqrt{1 - x^2}}.$$

Since $f'(x) > 0$ for all x in $(-1, 0)$ and f is continuous on $[-1, 0]$, f increases on $[-1, 0]$. Since $f'(x) < 0$ for all x in $(0, 1)$ and f is continuous on $[0, 1]$, f decreases on $[0, 1]$. The graph of f is a semicircle. (Figure 4.2.5) ☐

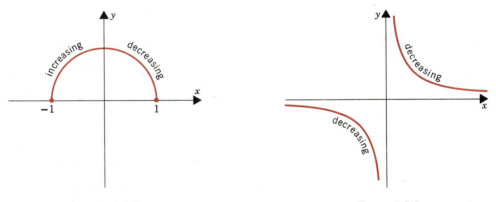

Figure 4.2.5   **Figure 4.2.6**

Example 2. The function

$$f(x) = \frac{1}{x}$$

is defined for all $x \neq 0$. The derivative

$$f'(x) = -\frac{1}{x^2}$$

is negative for all $x \neq 0$. Thus the function f decreases both on $(-\infty, 0)$ and on $(0, \infty)$. (Figure 4.2.6.) ☐

Example 3. The function

$$g(x) = 4x^5 - 15x^4 - 20x^3 + 110x^2 - 120x + 40$$

is a polynomial. It is therefore everywhere continuous and everywhere differentiable. Differentiation gives

$$g'(x) = 20x^4 - 60x^3 - 60x^2 + 220x - 120$$
$$= 20(x^4 - 3x^3 - 3x^2 + 11x - 6)$$
$$= 20(x + 2)(x - 1)^2(x - 3).$$

The derivative g' takes on the value 0 at -2, at 1 and at 3. These numbers determine four intervals on which g' keeps a constant sign:

$$(-\infty, -2), \qquad (-2, 1), \qquad (1, 3), \qquad (3, \infty).$$

The sign of g' on these intervals and the consequences for g are as follows:

Since g is everywhere continuous, g increases on $(-\infty, -2]$, decreases on $[-2, 3]$ and increases on $[3, \infty)$. See Figure 4.2.7. ☐

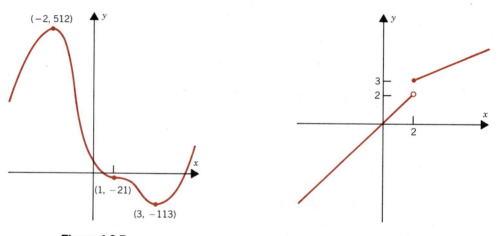

Figure 4.2.7

Figure 4.2.8

Theorems 4.2.2 and 4.2.3 have wide but limited applicability. If, for example, f is discontinuous at some point of its domain, then Theorems 4.2.2 and 4.2.3 do not tell the whole story.

Example 4. The function

$$f(x) = \begin{cases} x, & x < 2 \\ \frac{1}{2}x + 2, & x \ge 2 \end{cases}$$

is graphed in Figure 4.2.8. Obviously there is a discontinuity at $x = 2$. Differentiation gives

$$f'(x) = \begin{cases} 1, & x < 2 \\ \text{does not exist}, & x = 2 \\ \frac{1}{2}, & x > 2 \end{cases}.$$

Since $f'(x) > 0$ on $(-\infty, 2)$, we know from Theorem 4.2.2 that f increases on $(-\infty, 2)$. Since $f'(x) > 0$ on $(2, \infty)$ and f is continuous on $[2, \infty)$, we know from Theorem 4.2.3 that f increases on $[2, \infty)$. The obvious fact that f increases on all of $(-\infty, \infty)$ is not derivable from the theorems.† □

Equality of Derivatives

If two differentiable functions differ by a constant,

$$f(x) = g(x) + C,$$

then their derivatives are equal:

$$f'(x) = g'(x).$$

The converse is also true. In fact, we have the following theorem.

THEOREM 4.2.4

I. Let I be an open interval. If $f'(x) = g'(x)$ for all x in I, then f and g differ by a constant on I.

II. Let I be an arbitrary interval. If $f'(x) = g'(x)$ for all x in the interior of I and f and g are continuous on I, then f and g differ by a constant on I.

Proof. Set $h = f - g$. For the first assertion apply (iii) of Theorem 4.2.2 to h. For the second assertion apply (iii) of Theorem 4.2.3 to h. We leave the details as an exercise. □

We illustrate the theorem in Figure 4.2.9. At points with the same x-coordinate the slopes are equal, and thus the curves have the same steepness. The separation between the curves remains constant.

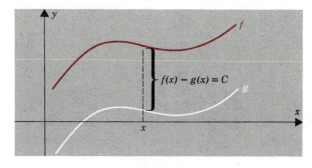

Figure 4.2.9

---

† The function

$$f(x) = \begin{cases} \tfrac{1}{2}x + 2, & x < 2 \\ x, & x \ge 2 \end{cases}$$

increases on $(-\infty, 2)$ and on $[2, \infty)$, but it does not increase on $(-\infty, \infty)$. Draw the graph.

Problem 5. Find f given that

$$f'(x) = 6x^2 - 7x - 5 \quad \text{for all real } x \quad \text{and} \quad f(2) = 1.$$

Solution. It is not hard to find a function with the right derivative:

$$\frac{d}{dx}(2x^3 - \tfrac{7}{2}x^2 - 5x) = 6x^2 - 7x - 5.$$

By Theorem 4.2.4 we know that $f(x)$ differs from $2x^3 - \tfrac{7}{2}x^2 - 5x$ only by some constant C. Thus we can write

$$f(x) = 2x^3 - \tfrac{7}{2}x^2 - 5x + C.$$

To evaluate C we use the fact that $f(2) = 1$. Since $f(2) = 1$ and also

$$f(2) = 2(2)^3 - \tfrac{7}{2}(2)^2 - 5(2) + C = 16 - 14 - 10 + C = -8 + C,$$

we have $-8 + C = 1$ and therefore $C = 9$. Thus

$$f(x) = 2x^3 - \tfrac{7}{2}x^2 - 5x + 9. \quad \square$$

EXERCISES 4.2

Find the intervals on which f increases and the intervals on which f decreases.

1. $f(x) = x^3 - 3x + 2.$ **2.** $f(x) = x^3 - 3x^2 + 6.$ **3.** $f(x) = x + \dfrac{1}{x}.$

4. $f(x) = (x - 3)^3.$ **5.** $f(x) = x^3(1 + x).$ **6.** $f(x) = x(x + 1)(x + 2).$

7. $f(x) = (x + 1)^4.$ **8.** $f(x) = 2x - \dfrac{1}{x^2}.$ **9.** $f(x) = \dfrac{1}{|x - 2|}.$

10. $f(x) = \dfrac{x}{1 + x^2}.$ **11.** $f(x) = \dfrac{x^2 + 1}{x^2 - 1}.$ **12.** $f(x) = \dfrac{x^2}{x^2 + 1}.$

13. $f(x) = |x^2 - 5|.$ **14.** $f(x) = x^2(1 + x)^2.$ **15.** $f(x) = \dfrac{x - 1}{x + 1}.$

16. $f(x) = x^2 + \dfrac{16}{x^2}.$ **17.** $f(x) = \left(\dfrac{1 - \sqrt{x}}{1 + \sqrt{x}}\right)^7.$ **18.** $f(x) = \sqrt{\dfrac{2 + x}{1 + x}}.$

19. $f(x) = \sqrt{\dfrac{1 + x^2}{2 + x^2}}.$ **20.** $f(x) = |x + 1||x - 2|.$ **21.** $f(x) = \sqrt{\dfrac{3 - x}{x}}.$

22. $f(x) = x + \sin x, \quad 0 \le x \le 2\pi.$ **23.** $f(x) = x - \cos x, \quad 0 \le x \le 2\pi.$
24. $f(x) = \cos^2 x, \quad 0 \le x \le \pi.$ **25.** $f(x) = \cos 2x + 2 \cos x, \quad 0 \le x \le \pi.$
26. $f(x) = \sin^2 x - \sqrt{3} \sin x, \quad 0 \le x \le \pi.$ **27.** $f(x) = \sqrt{3}\, x - \cos 2x, \quad 0 \le x \le \pi.$

Find f given the following information.

28. $f'(x) = x^2 - 1$ for all real x, $f(0) = 1.$ **29.** $f'(x) = x^2 - 1$ for all real x, $f(1) = 2.$
30. $f'(x) = 2x - 5$ for all real x, $f(2) = 4.$
31. $f'(x) = 5x^4 + 4x^3 + 3x^2 + 2x + 1$ for all real x, $f(0) = 5.$
32. $f'(x) = 4x^{-3}$ for $x > 0$, $f(1) = 0.$ **33.** $f'(x) = x^{1/3} - x^{1/2}$ for $x > 0$, $f(0) = 1.$
34. $f'(x) = x^{-5} - 5x^{-1/5}$ for $x > 0$, $f(1) = 0.$ **35.** $f'(x) = 2 + \sin x$ for all real x, $f(0) = 3.$

36. $f'(x) = 4x + \cos x$ for all real x, $f(0) = 1$.

Find the intervals on which f increases and the intervals on which f decreases.

37. $f(x) = \begin{cases} x + 7, & x < -3 \\ |x + 1|, & -3 \leq x < 1 \\ 5 - 2x, & 1 \leq x \end{cases}$.

38. $f(x) = \begin{cases} (x - 1)^2, & x < 1 \\ 5 - x, & 1 \leq x < 3 \\ 7 - 2x, & 3 \leq x \end{cases}$.

39. $f(x) = \begin{cases} 4 - x^2, & x < 1 \\ 7 - 2x, & 1 \leq x < 3 \\ 3x - 10, & 3 \leq x \end{cases}$.

40. $f(x) = \begin{cases} x + 2, & x < 0 \\ (x - 1)^2, & 0 < x < 3 \\ 8 - x, & 3 < x < 7 \\ 2x - 9, & 7 < x \\ 6, & x = 0, 3, 7 \end{cases}$.

41. Suppose that the function f increases on (a, b) and on (b, c). Suppose further that $\lim_{x \to b^-} f(x) = M$, $\lim_{x \to b^+} f(x) = N$ and $f(b) = L$. For which values of L, if any, can you conclude that f increases on (a, c) if (a) $M < N$? (b) $M > N$? (c) $M = N$?

42. Suppose that the function f increases on (a, b) and decreases on (b, c). Under what conditions can you conclude that f increases on $(a, b]$ and decreases on $[b, c)$?

43. Give an example of a function f that satisfies all of the following conditions: (i) f is defined for all real x, (ii) f' is positive wherever it exists, (iii) there is no interval on which f increases.

44. Prove Theorem 4.2.4.

45. Given that $f'(x) > g'(x)$ for all real x and $f(0) = g(0)$, compare $f(x)$ with $g(x)$ on $(-\infty, 0)$ and on $(0, \infty)$. Justify your answers.

46. Show, without making any assumptions of differentiability, that if f increases (decreases) on (a, b) and is continuous on $[a, b]$, then f increases (decreases) on $[a, b]$.

4.3   LOCAL EXTREME VALUES

In many problems of economics, engineering, and physics it is important to determine how large or how small a certain quantity can be. If the problem admits a mathematical formulation, it is often reducible to the problem of finding the maximum or minimum value of some function.

In this section we consider maximum and minimum values for functions defined on *an open interval* or on *the union of open intervals*. We begin with a definition.

DEFINITION 4.3.1   LOCAL EXTREME VALUES

A function f is said to have a *local maximum at c* iff

$$f(c) \geq f(x) \qquad \text{for all } x \text{ sufficiently close to } c.†$$

It is said to have a *local minimum at c* iff

$$f(c) \leq f(x) \qquad \text{for all } x \text{ sufficiently close to } c.$$

---

† What do we mean by saying that "such and such is true *for all x sufficiently close to c*"? We mean that it is true for all x in some open interval $(c - \delta, c + \delta)$ centered at c.

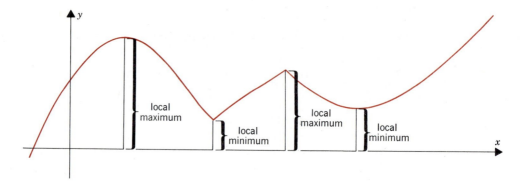

Figure 4.3.1

We illustrate these notions in Figure 4.3.1. A careful look at the figure suggests that local maxima and minima occur only at points where the tangent is horizontal [$f'(c) = 0$] or where there is no tangent line [$f'(c)$ does not exist]. This is indeed the case.

THEOREM 4.3.2

If f has a local maximum or minimum at c, then either

$$f'(c) = 0 \quad \text{or} \quad f'(c) \text{ does not exist.}$$

Proof. If $f'(c) > 0$ or $f'(c) < 0$, then, by Theorem 4.1.2, there must be numbers x_1 and x_2 that are arbitrarily close to c and yet satisfy

$$f(x_1) < f(c) < f(x_2).$$

This makes it impossible for a local maximum or minimum to occur at c. □

In view of this result, when searching for the local maxima and minima of a function f, the only points we need to test are *those points c in the domain of f for which $f'(c) = 0$ or $f'(c)$ does not exist.* Such points are called *critical points*.

We illustrate the technique for finding local maxima and minima by some examples. In each example the first step will be to find the critical points. We begin with very simple cases.

Example 1. For

$$f(x) = 3 - x^2 \qquad \text{(Figure 4.3.2)}$$

the derivative

$$f'(x) = -2x$$

exists everywhere. Since $f'(x) = 0$ only at $x = 0$, 0 is the only critical point. The number $f(0) = 3$ is obviously a local maximum. ☐

Figure 4.3.2   **Figure 4.3.3**

Example 2. In the case of

$$f(x) = |x + 1| + 2 \qquad \text{(Figure 4.3.3)}$$

differentiation gives

$$f'(x) = \begin{cases} -1, & x < -1 \\ 1, & x > -1 \end{cases}.$$

This derivative is never 0. It fails to exist only at -1. The number -1 is the only critical point. The value $f(-1) = 2$ is a local minimum. □

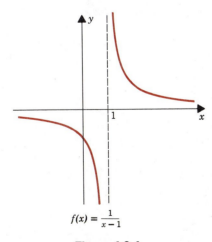

Figure 4.3.4

Example 3. Figure 4.3.4 shows the graph of the function

$$f(x) = \frac{1}{x - 1}.$$

The domain is $(-\infty, 1) \cup (1, \infty)$. The derivative

$$f'(x) = -\frac{1}{(x - 1)^2}$$

exists throughout the domain of f and is never 0. Thus there are no critical points and therefore no local extreme values. □

Caution.   The fact that c is a critical point of f does not guarantee that $f(c)$ is a local extreme value.   This is illustrated in the next two examples.   ☐

Example 4. In the case of the cubing function

$$f(x) = x^3 \qquad \text{(Figure 4.3.5)}$$

the derivative $f'(x) = 3x^2$ is 0 at 0, but $f(0) = 0$ is not a local extreme value.   The cubing function is everywhere increasing.   ☐

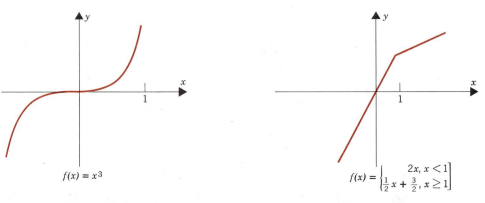

Figure 4.3.5	**Figure 4.3.6**

Example 5. The function

$$f(x) = \begin{Bmatrix} 2x, & x < 1 \\ \tfrac{1}{2}x + \tfrac{3}{2}, & x \geq 1 \end{Bmatrix}, \qquad \text{(Figure 4.3.6)}$$

is everywhere increasing.   Although 1 is a critical point [$f'(1)$ does not exist], $f(1) = 2$ is not a local extreme value.   ☐

There are two widely used tests for determining the behavior of a function at a critical point.   The first test (given in Theorem 4.3.3) requires that we examine the sign of the first derivative on both sides of the critical point.   The second test (given in Theorem 4.3.4) requires that we examine the sign of the second derivative at the critical point itself.

THEOREM 4.3.3   THE FIRST-DERIVATIVE TEST

Suppose that c is a critical point for f and f is continuous at c.   Now let δ be a positive number.

(i) If $f'(x) > 0$ for all x in $(c - \delta, c)$ and $f'(x) < 0$ for all x in $(c, c + \delta)$, then $f(c)$ is a local maximum.   (Figures 4.3.7 and 4.3.8.)

(ii) If $f'(x) < 0$ for all x in $(c - \delta, c)$ and $f'(x) > 0$ for all x in $(c, c + \delta)$, then $f(c)$ is a local minimum.   (Figures 4.3.9 and 4.3.10.)

(iii) If $f'(x)$ keeps a constant sign on $(c - \delta, c) \cup (c, c + \delta)$, then $f(c)$ is not a local extreme value.   (Figures 4.3.5 and 4.3.6.)

Proof. The result is a direct consequence of Theorem 4.2.3. The details of the proof are left to you as Exercise 31. ☐

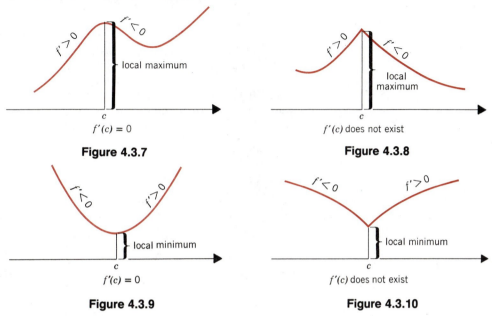

Figure 4.3.7

Figure 4.3.8

Figure 4.3.9

Figure 4.3.10

Example 6. The function

$$f(x) = x^4 - 2x^3, \qquad \text{all real } x$$

has derivative

$$f'(x) = 4x^3 - 6x^2 = 2x^2(2x - 3), \qquad \text{all real } x.$$

The only critical points are 0 and $\frac{3}{2}$. The sign of f' is recorded below.

Figure 4.3.11

Since f' keeps the same sign to both sides of 0, $f(0) = 0$ is not a local extreme value. However, $f(\frac{3}{2}) = -\frac{27}{16}$ is a local minimum. The graph of f appears in Figure 4.3.12. ☐

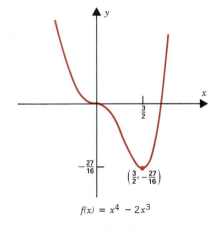

$$f(x) = x^4 - 2x^3$$

Figure 4.3.12

Example 7. The function

$$f(x) = 2x^{5/3} + 5x^{2/3}, \qquad \text{all real } x$$

has derivative

$$f'(x) = \frac{10}{3} x^{2/3} + \frac{10}{3} x^{-1/3} = \frac{10}{3} x^{-1/3}(x + 1), \qquad x \neq 0.$$

Since $f'(-1) = 0$ and $f'(0)$ does not exist, the critical points are -1 and 0. The sign of f' is recorded below. (To save space in the diagram we write "dne" for "does not exist.")

sign of f': $+ + + + + + + + + + + + 0 - - - - - - - - - - - - - - -$ dne $+ + + + + + + + +$ x

domain of f:

-1 0

Figure 4.3.13

In this case $f(-1) = 3$ is a local maximum and $f(0) = 0$ is a local minimum. The graph appears in Figure 4.3.14. ☐

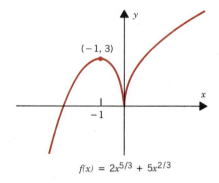

$$f(x) = 2x^{5/3} + 5x^{2/3}$$

Figure 4.3.14

In certain cases it may be difficult to determine the sign of the first derivative both to the left of c and to the right of c. If f is twice differentiable at c, the following test may be easier to apply.

THEOREM 4.3.4   THE SECOND-DERIVATIVE TEST

Suppose that $f'(c) = 0$.

If $f''(c) > 0$, then $f(c)$ is a local minimum value.

If $f''(c) < 0$, then $f(c)$ is a local maximum value.

Proof. We handle the case $f''(c) > 0$. The other is left as an exercise. Since f'' is the derivative of f', we see from Theorem 4.1.2 that there exists $\delta > 0$ such that, if

$$c - \delta < x_1 < c < x_2 < c + \delta,$$

then

$$f'(x_1) < f'(c) < f'(x_2).$$

Since $f'(c) = 0$, we have

$$f'(x) < 0 \quad \text{for } x \text{ in } (c - \delta, c) \qquad \text{and} \qquad f'(x) > 0 \quad \text{for } x \text{ in } (c, c + \delta).$$

By the first-derivative test $f(c)$ is a local minimum. ☐

Example 8. For

$$f(x) = x^3 - x$$

we have

$$f'(x) = 3x^2 - 1 \quad \text{and} \quad f''(x) = 6x.$$

The critical points are $-\frac{1}{3}\sqrt{3}$ and $\frac{1}{3}\sqrt{3}$. At each of these points the derivative is 0. Since $f''(-\frac{1}{3}\sqrt{3}) < 0$ and $f''(\frac{1}{3}\sqrt{3}) > 0$, we can conclude from the second-derivative test that $f(-\frac{1}{3}\sqrt{3}) = \frac{2}{9}\sqrt{3}$ is a local maximum and $f(\frac{1}{3}\sqrt{3}) = -\frac{2}{9}\sqrt{3}$ is a local minimum. ☐

We can apply the first-derivative test even at points where the function is not differentiable, provided that it is continuous there. On the other hand, the second-derivative test can only be applied at points where the function is twice differentiable and then only if the second derivative is different from zero. This is a drawback. Take, for example, the functions

$$f(x) = x^{4/3} \quad \text{and} \quad g(x) = x^4.$$

In the first case we have $f'(x) = \frac{4}{3}x^{1/3}$ so that

$$f'(0) = 0, \qquad f'(x) < 0 \quad \text{for } x < 0, \qquad f'(x) > 0 \quad \text{for } x > 0.$$

By the first-derivative test, $f(0) = 0$ is a local minimum. We cannot get this information from the second-derivative test because $f''(x) = \frac{4}{9}x^{-2/3}$ does not exist at $x = 0$. In the case of $g(x) = x^4$, we have $g'(x) = 4x^3$ so that

$$g'(0) = 0, \qquad g'(x) < 0 \quad \text{for } x < 0, \qquad g'(x) > 0 \quad \text{for } x > 0.$$

By the first-derivative test, $g(0) = 0$ is a local minimum. But here again the second-derivative test is of no avail: $g''(x) = 12x^2$ is 0 at $x = 0$.

EXERCISES 4.3

Find the critical points of f and the local extreme values.

1. $f(x) = x^3 + 3x - 2.$

2. $f(x) = 2x^4 - 4x^2 + 6.$

3. $f(x) = x + \dfrac{1}{x}.$

4. $f(x) = x^2 - \dfrac{3}{x^2}.$

5. $f(x) = x^2(1 - x).$

6. $f(x) = (1 - x)^2(1 + x).$

7. $f(x) = \dfrac{1 + x}{1 - x}.$

8. $f(x) = \dfrac{2 - 3x}{2 + x}.$

9. $f(x) = \dfrac{2}{x(x + 1)}.$

10. $f(x) = |x^2 - 16|.$

11. $f(x) = x^3(1 - x)^2.$

12. $f(x) = \left(\dfrac{x - 2}{x + 2}\right)^3.$

13. $f(x) = (1 - 2x)(x - 1)^3.$

14. $f(x) = (1 - x)(1 + x)^3.$

15. $f(x) = \dfrac{x^2}{1 + x}.$

16. $f(x) = \dfrac{|x|}{1 + |x|}.$

17. $f(x) = |x - 1||x + 2|.$

18. $f(x) = x\sqrt[3]{1 - x}.$

19. $f(x) = x^2\sqrt[3]{2 + x}.$

20. $f(x) = \dfrac{1}{x + 1} - \dfrac{1}{x - 2}.$

21. $f(x) = |x - 3| + |2x + 1|.$

22. $f(x) = x^{7/3} - 7x^{1/3}$.   **23.** $f(x) = x^{2/3} + 2x^{-1/3}$.   **24.** $f(x) = \dfrac{x^3}{x + 1}$

25. $f(x) = \sin x + \cos x$, $0 < x < 2\pi$.   **26.** $f(x) = x + \cos 2x$, $0 < x < \pi$.

27. $f(x) = \sin^2 x - \sqrt{3} \sin x$, $0 < x < \pi$.   **28.** $f(x) = \sin^2 x$, $0 < x < 2\pi$.

29. $f(x) = \sin x \cos x - 3 \sin x + 2x$, $0 < x < 2\pi$.

30. $f(x) = 2 \sin^3 x - 3 \sin x$, $0 < x < \pi$.

31. Prove Theorem 4.3.3 by applying Theorem 4.2.3.

32. Prove the validity of the second-derivative test in the case that $f''(c) < 0$.

33. Find the critical points and the local extreme values of the polynomial

$$P(x) = x^4 - 8x^3 + 22x^2 - 24x + 4.$$

Then show that the equation $P(x) = 0$ has exactly two real roots, both positive.

4.4   ENDPOINT AND ABSOLUTE EXTREME VALUES

Endpoint Maxima and Minima

For functions defined on an open interval or on the union of open intervals the critical points are those where the derivative is 0 or the derivative does not exist. For functions defined on a closed or half-closed interval

$$[a, b], \quad [a, b), \quad (a, b], \quad [a, \infty), \quad \text{or} \quad (-\infty, b]$$

the *endpoints* of the domain (a and b in the case of $[a, b]$, a in the case of $[a, b)$ or $[a, \infty)$, and b in the case of $(a, b]$ or $(-\infty, b]$) are also called *critical points*.

Endpoints can give rise to what are called *endpoint maxima* and *endpoint minima*. See, for example, Figures 4.4.1, 4.4.2, 4.4.3, and 4.4.4.

Figure 4.4.1

Figure 4.4.2

Figure 4.4.3

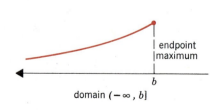

Figure 4.4.4

DEFINITION 4.4.1 ENDPOINT EXTREME VALUES

If c is an endpoint of the domain of f, then f is said to have an *endpoint maximum* at c iff

$$f(c) \geq f(x) \qquad \text{for all } x \text{ in the domain of } f \text{ sufficiently close to } c.$$

It is said to have an *endpoint minimum* at c iff

$$f(c) \leq f(x) \qquad \text{for all } x \text{ in the domain of } f \text{ sufficiently close to } c.$$

Endpoints are usually tested by examining the sign of the derivative at nearby points. Suppose, for example, that a is the left endpoint and f is continuous from the right at a. If $f'(x) < 0$ for x close to a, then f decreases on an interval of the form $[a, a + \delta]$ and therefore $f(a)$ must be an endpoint maximum. (Figure 4.4.1.) On the other hand, if $f'(x) > 0$ for x close to a, then f increases on an interval of the form $[a, a + \delta]$ and therefore $f(a)$ must be an endpoint minimum. (Figure 4.4.2.) Similar reasoning can be applied to right endpoints.

Absolute Maxima and Minima

Whether or not a function f has a local or endpoint extreme value at some point depends entirely on the behavior of f for x close to that point. Absolute extreme values, which we define below, depend on the behavior of the function on its entire domain.

We begin with a number d in the domain of f. Here d can be an interior point or an endpoint.

DEFINITION 4.4.2   ABSOLUTE EXTREME VALUES

$f(d)$ is called *the absolute maximum* of f iff

$$f(d) \geq f(x) \qquad \text{for all } x \text{ in the domain of } f;$$

$f(d)$ is called *the absolute minimum* of f iff

$$f(d) \leq f(x) \qquad \text{for all } x \text{ in the domain of } f.$$

A function can be continuous on an interval (or even differentiable there) without taking on an absolute maximum or an absolute minimum. (The functions depicted in Figures 4.4.2 and 4.4.3 have no absolute maximum. The function depicted in Figure 4.4.4 has no absolute minimum.) All we can say in general is that, if f takes on an absolute extreme value, then it does so at a critical point. If, however, the domain is a bounded closed interval $[a, b]$, then the continuity of the function guarantees the existence of both an absolute maximum and an absolute minimum (Theorem 2.6.2). In this case the most practical way to determine the absolute extremes is to gather together the local extremes and the endpoint extremes. The largest of these numbers is obviously the absolute maximum, and the smallest is the absolute minimum.

As $x \to \pm\infty$, $f(x) \to \pm\infty$

We now write down four definitions. Once you grasp the first one, the others become transparent.

To say that

$$\text{as } x \to \infty, \quad f(x) \to \infty$$

is to say that, *as x increases without bound, f(x) becomes arbitrarily large*: given any positive number M, there exists a positive number K such that,

$$\text{if } x \geq K, \quad \text{then } f(x) \geq M.$$

Thus, for example, as $x \to \infty$,

$$x^2 \to \infty, \quad \sqrt{1 + x^2} \to \infty, \quad \tan\left(\frac{\pi}{2} - \frac{1}{x^2}\right) \to \infty. \quad \square$$

To say that

$$\text{as } x \to \infty, \quad f(x) \to -\infty$$

is to say that, *as x increases without bound, f(x) becomes arbitrarily large negative*: given any negative number M, there exists a positive number K such that,

$$\text{if } x \geq K, \quad \text{then } f(x) \leq M.$$

Thus, for example, as $x \to \infty$,

$$-x^4 \to -\infty, \quad 1 - \sqrt{x} \to -\infty, \quad \tan\left(\frac{1}{x^2} - \frac{\pi}{2}\right) \to -\infty. \quad \square$$

To say that

$$\text{as } x \to -\infty, \quad f(x) \to \infty$$

is to say that, *as x decreases without bound, f(x) becomes arbitrarily large*: given any positive number M, there exists a negative number K such that,

$$\text{if } x \leq K, \quad \text{then } f(x) \geq M.$$

Thus, for example, as $x \to -\infty$,

$$x^2 \to \infty, \quad \sqrt{1 - x} \to \infty, \quad \tan\left(\frac{\pi}{2} - \frac{1}{x^2}\right) \to \infty. \quad \square$$

Finally, to say that

$$\text{as } x \to -\infty, \quad f(x) \to -\infty$$

is to say that, *as x decreases without bound, f(x) becomes arbitrarily large negative*: given any negative number M, there exists a negative number K such that,

$$\text{if } x \leq K, \quad \text{then } f(x) \leq M.$$

Thus, for example, as $x \to -\infty$,

$$x^3 \to -\infty, \qquad -\sqrt{1-x} \to -\infty, \qquad \tan\left(\frac{1}{x^2} - \frac{\pi}{2}\right) \to -\infty. \quad \square$$

Remark. As you can readily see, $f(x) \to -\infty$ iff $-f(x) \to \infty$. $\quad \square$

Suppose now that P is a nonconstant polynomial:

$$P(x) = a_0 x^n + a_1 x^{n-1} + \cdots + a_{n-1}x + a_n. \qquad (n \geq 1)$$

For values of x far from the origin, the leading term $a_0 x^n$ dominates. Thus what happens to $P(x)$ as $x \to \pm\infty$ depends entirely on what happens to $a_0 x^n$. (For confirmation, do Exercise 35.)

Example 1

(a) As $x \to \infty$, $3x^4 \to \infty$ and therefore $3x^4 - 100x^3 + 2x - 5 \to \infty$.

(b) As $x \to -\infty$, $5x^3 \to -\infty$ and therefore $5x^3 + 12x^2 + 80 \to -\infty$. $\quad \square$

Finally, we point out that, if $f(x) \to \infty$, then f cannot have an absolute maximum value and, if $f(x) \to -\infty$, then f cannot have an absolute minimum value.

A Summary on Finding All the Extreme Values (Local, Endpoint, and Absolute) of a Continuous Function f

1. Find the critical points. These are the endpoints of the domain and the interior points c at which $f'(c) = 0$ or $f'(c)$ does not exist.

2. Test each endpoint of the domain by examining the sign of the first derivative nearby.

3. Test each interior critical point c by examining the sign of the first derivative to both sides of c (first-derivative test) or by checking the sign of the second derivative at c itself (second-derivative test).

4. Determine whether any of the endpoint extremes and local extremes are absolute extremes.

Problem 2. Find the critical points and classify all the extreme values of

$$f(x) = \tfrac{1}{4}(x^3 - \tfrac{3}{2}x^2 - 6x + 2), \qquad x \in [-2, \infty).$$

Solution. The left endpoint -2 is a critical point. To find the interior critical points, we differentiate:

$$f'(x) = \tfrac{1}{4}(3x^2 - 3x - 6) = \tfrac{3}{4}(x+1)(x-2).$$

Since $f'(x) = 0$ at $x = -1$ and $x = 2$, the numbers -1 and 2 are interior critical points. The sign of f' is recorded in Figure 4.4.5:

Figure 4.4.5

We can see from the sign of f' that

$$f(-2) = \tfrac{1}{4}(-8 - 6 + 12 + 2) = 0 \quad \text{is an endpoint minimum;}$$
$$f(-1) = \tfrac{1}{4}(-1 - \tfrac{3}{2} + 6 + 2) = \tfrac{11}{8} \quad \text{is a local maximum;}$$
$$f(2) = \tfrac{1}{4}(8 - 6 - 12 + 2) = -2 \quad \text{is a local and absolute minimum.}$$

The function takes on no absolute maximum value since, as $x \to \infty$, $f(x) \to \infty$.   ☐

Problem 3.   Find the critical points and classify all the extreme values of

$$f(x) = x^2 - 2|x| + 2, \qquad x \in [-\tfrac{1}{2}, \tfrac{3}{2}].$$

Solution.   The endpoints $-\tfrac{1}{2}$ and $\tfrac{3}{2}$ are critical points.   On the open interval $(-\tfrac{1}{2}, \tfrac{3}{2})$ the function is differentiable except at 0:

$$f'(x) = \begin{cases} 2x + 2, & -\tfrac{1}{2} < x < 0 \\ 2x - 2, & 0 < x < \tfrac{3}{2} \end{cases}. \qquad \text{(check this out)}$$

This makes 0 a critical point.   Since $f'(x) = 0$ at $x = 1$, 1 is a critical point.   The endpoints $-\tfrac{1}{2}$ and $\tfrac{3}{2}$ are also critical points.
   The sign of f' is recorded in Figure 4.4.6:

Figure 4.4.6

Therefore

$$f(-\tfrac{1}{2}) = \tfrac{1}{4} - 1 + 2 = \tfrac{5}{4} \quad \text{is an endpoint minimum;}$$
$$f(0) = 2 \quad \text{is a local maximum;}$$
$$f(1) = 1 - 2 + 2 = 1 \quad \text{is a local minimum;}$$
$$f(\tfrac{3}{2}) = \tfrac{9}{4} - 3 + 2 = \tfrac{5}{4} \quad \text{is an endpoint maximum.}$$

Also $f(1)$ is the absolute minimum and $f(0) = 2$ is the absolute maximum.   The graph of the function is shown in Figure 4.4.7.   ☐

Figure 4.4.7

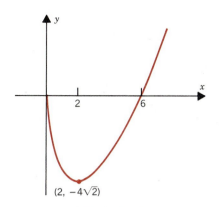

Figure 4.4.8

Problem 4. Find the critical points and classify all the extreme values of

$$f(x) = x\sqrt{x} - 6\sqrt{x}.$$

Solution. The domain is $[0, \infty)$. Therefore the endpoint 0 is a critical point. On $(0, \infty)$

$$f'(x) = \frac{3}{2}x^{1/2} - 3x^{-1/2} = \frac{3(x - 2)}{2\sqrt{x}}. \qquad \text{(check this out)}$$

Since $f'(x) = 0$ at $x = 2$, we see that 2 is an interior critical point.
The sign of f' is recorded below.

Figure 4.4.9

Therefore

$$f(0) = 0 \quad \text{is an endpoint maximum;}$$
$$f(2) = 2\sqrt{2} - 6\sqrt{2} = -4\sqrt{2} \quad \text{is a local and absolute minimum.}$$

Since $f(x) = \sqrt{x}(x - 6)$, it is easy to see that, as $x \to \infty$, $f(x) \to \infty$. Thus the function has no absolute maximum value. The graph of the function appears in Figure 4.4.8. ☐

EXERCISES 4.4

Find the critical points and classify the extreme values.

1. $f(x) = \sqrt{x + 2}$.

2. $f(x) = (x - 1)(x - 2)$.

3. $f(x) = x^2 - 4x + 1, \quad x \in [0, 3]$.

4. $f(x) = 2x^2 + 5x - 1, \quad x \in [-2, 0]$.

5. $f(x) = x^2 + \dfrac{1}{x}$.

6. $f(x) = x + \dfrac{1}{x^2}$.

7. $f(x) = x^2 + \dfrac{1}{x}, \quad x \in [\frac{1}{10}, 2]$.

8. $f(x) = x + \dfrac{1}{x^2}, \quad x \in [1, \sqrt{2}]$

9. $f(x) = (x - 1)(x - 2), \quad x \in [0, 2]$.

10. $f(x) = (x - 1)^2(x - 2)^2, \quad x \in [0, 4]$.

11. $f(x) = \dfrac{x}{4 + x^2}, \quad x \in [-3, 1]$.

12. $f(x) = \dfrac{x^2}{1 + x^2}, \quad x \in [-1, 2]$.

13. $f(x) = (x - \sqrt{x})^2$.

14. $f(x) = x\sqrt{4 - x^2}$.

15. $f(x) = x\sqrt{3 - x}$.

16. $f(x) = \sqrt{x} - \dfrac{1}{\sqrt{x}}$.

17. $f(x) = 1 - \sqrt[3]{x - 1}$.

18. $f(x) = (4x - 1)^{1/3}(2x - 1)^{2/3}$.

19. $f(x) = \sin^2 x - \sqrt{3} \cos x, \quad 0 \le x \le \pi$.

20. $f(x) = \cot x + x, \quad 0 < x \le \dfrac{2\pi}{3}$.

21. $f(x) = 2 \cos^3 x + 3 \cos x, \quad 0 \le x \le \pi$.

22. $f(x) = \sin 2x - x, \quad 0 \le x \le \pi$.

23. $f(x) = \tan x - x, \quad -\dfrac{\pi}{3} \le x < \dfrac{\pi}{2}$.

24. $f(x) = \sin^4 x - \sin^2 x, \quad 0 \le x \le \dfrac{2\pi}{3}$.

25. $f(x) = \begin{cases} -2x, & 0 \le x < 1 \\ x - 3, & 1 \le x \le 4 \\ 5 - x, & 4 < x \le 7 \end{cases}$.

26. $f(x) = \begin{cases} x + 9, & -8 \le x < -3 \\ x^2 + x, & -3 \le x \le 2 \\ 5x - 4, & 2 < x < 5 \end{cases}$.

27. $f(x) = \begin{cases} x^2 + 1, & -2 \le x < -1 \\ 5 + 2x - x^2, & -1 \le x \le 3 \\ x - 1, & 3 < x < 6 \end{cases}$.

28. $f(x) = \begin{cases} 2 - 2x - x^2, & -2 \le x \le 0 \\ |x - 2|, & 0 < x < 3 \\ \frac{1}{3}(x - 2)^3, & 3 \le x \le 4 \end{cases}$.

29. $f(x) = \begin{cases} |x + 1|, & -3 \le x < 0 \\ x^2 - 4x + 2, & 0 \le x < 3 \\ 2x - 7, & 3 \le x < 4 \end{cases}$.

30. $f(x) = \begin{cases} -x^2, & 0 \le x < 1 \\ -2x, & 1 < x < 2 \\ -\frac{1}{2}x^2, & 2 \le x \le 3 \end{cases}$.

31. Suppose that c is a critical point for f and $f'(x) > 0$ for $x \ne c$. Prove that, if $f(c)$ is a local maximum, then f is not continuous at c.

32. What can be said about the function f continuous on $[a, b]$, if for some c in (a, b), $f(c)$ is both a local maximum and a local minimum?

33. Suppose that f is continuous on its domain $[a, b]$ and $f(a) = f(b)$. Prove that f has at least one critical point in (a, b).

34. Suppose that $c_1 < c_2$ and that both $f(c_1)$ and $f(c_2)$ are local maximums. Prove that, if f is continuous on $[c_1, c_2]$, then, for some c in (c_1, c_2), $f(c)$ is a local minimum.

35. Let P be a nonconstant polynomial with positive leading coefficient:

$$P(x) = a_0 x^n + a_1 x^{n-1} + \cdots + a_{n-1} x + a_n. \qquad (n \ge 1, \ a_0 > 0)$$

Clearly, as $x \to \infty$, $a_0 x^n \to \infty$. Show that, as $x \to \infty$, $P(x) \to \infty$ by showing that, given any positive number M, there exists a positive number K such that, if $x \ge K$, then $f(x) \ge M$.

4.5 SOME MAX-MIN PROBLEMS

The techniques of the last two sections can be brought to bear on a wide variety of max-min problems. The *key step* in solving such problems is to express the quantity to be maximized or minimized as a function of one variable. If the function is differentiable, we can then differentiate and analyze the results.

Figure 4.5.1

Problem 1. A mirror in the shape of a rectangle capped by a semicircle is to have perimeter p. Choose the radius of the semicircular part so that the mirror has maximum area.

Solution. As in Figure 4.5.1 we let x be the radius of the semicircular part and y the height of the rectangular part. We want to express the area

$$A = \tfrac{1}{2}\pi x^2 + 2xy$$

as a function of x alone. To do this, we must express y in terms of x.

Since the perimeter is p, we have

$$p = 2x + 2y + \pi x$$

and thus

$$y = \tfrac{1}{2}[p - (2 + \pi)x].$$

Since y must be positive, x must remain between 0 and $p/(2 + \pi)$.

The area can now be expressed in terms of x alone:

$$\begin{aligned}
A(x) &= \tfrac{1}{2}\pi x^2 + 2xy \\
&= \tfrac{1}{2}\pi x^2 + 2x\{\tfrac{1}{2}[p - (2 + \pi)x]\} \\
&= \tfrac{1}{2}\pi x^2 + px - (2 + \pi)x^2 = px - (2 + \tfrac{1}{2}\pi)x^2.
\end{aligned}$$

We want to maximize the function

$$A(x) = px - (2 + \tfrac{1}{2}\pi)x^2, \qquad x \in \left(0, \frac{p}{2 + \pi}\right).$$

(*Key step completed.*)

The derivative

$$A'(x) = p - (4 + \pi)x$$

is 0 only at $x = p/(4 + \pi)$. Since $A'(x) > 0$ for $0 < x < p/(4 + \pi)$ and $A'(x) < 0$ for $p/(4 + \pi) < x < p/(2 + \pi)$, the function A is maximized by setting $x = p/(4 + \pi)$. Thus the mirror has maximum area when the radius of the semicircular part is $p/(4 + \pi)$. □

Problem 2. The sum of two nonnegative numbers is 3. Find the two numbers if the square of twice the first, diminished by twice the square of the second, is a minimum.

Solution. Denote the first number by x; the second number becomes $3 - x$. Since both x and $3 - x$ are to be nonnegative, we must have $0 \leq x \leq 3$. We want the minimum value of

$$f(x) = (2x)^2 - 2(3 - x)^2, \qquad 0 \leq x \leq 3.$$

(*Key step completed.*)
    Since

$$f'(x) = 8x + 4(3 - x) = 4x + 12 = 4(x + 3),$$

we have $f'(x) > 0$ on $(0, 3)$. Since f is continuous on $[0, 3]$, f increases on $[0, 3]$. The minimum value of f occurs at $x = 0$. The first number must be 0; the second must be 3. □

Problem 3. Determine the dimensions of the rectangle of greatest area that can be inscribed in a 3-4-5 right triangle with one side of the rectangle resting on the hypotenuse of the triangle.

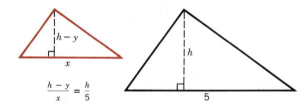

Figure 4.5.2   **Figure 4.5.3**

Solution. We label the dimensions as in Figure 4.5.2. To express the area $A = xy$ as a function of x alone, we need y in terms of x. We can get y in terms of x by similar triangles. The triangles of interest have been redrawn separately in Figure 4.5.3. From these triangles we see that

$$\frac{h - y}{x} = \frac{h}{5}$$

and therefore

(1) $y = h(1 - \tfrac{1}{5}x).$

We can determine h by calculating the area of the 3-4-5 triangle in two different ways:

$$\tfrac{1}{2}(5)(h) = \tfrac{5}{2}h \quad \text{and} \quad \tfrac{1}{2}(3)(4) = 6.$$

Therefore $\tfrac{5}{2}h = 6$ and thus $h = \tfrac{12}{5}$.

Substituting $h = \frac{12}{5}$ in equation (1) we have

(2) $$y = \tfrac{12}{5}(1 - \tfrac{1}{5}x) = \tfrac{12}{25}(5 - x).$$

Since both x and y must be positive, x must remain between 0 and 5. Since the area of the rectangle is the product

$$xy = x\tfrac{12}{25}(5 - x) = \tfrac{12}{5}x - \tfrac{12}{25}x^2,$$

the function we want to maximize is the function

$$A(x) = \tfrac{12}{5}x - \tfrac{12}{25}x^2, \qquad 0 < x < 5.$$

(*Key step completed.*)

As you can check by differentiation, this function takes on its absolute maximum value at $x = \frac{5}{2}$. Using (2) you can see that when $x = \frac{5}{2}$, $y = \frac{6}{5}$. The rectangle of greatest area that satisfies the conditions of this problem is a rectangle of length $\frac{5}{2}$ and width $\frac{6}{5}$. ☐

Problem 4. A lighthouse lies 3 miles offshore directly across from a point A of a straight coastline. Five miles down the coast from A there is a general store. The lighthouse keeper can row his boat at 4 miles per hour and he can walk at 6 miles per hour. To what point of the shore should he row so as to reach the store as quickly as possible?

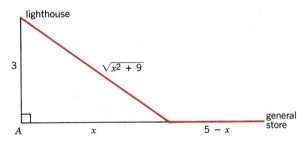

Figure 4.5.4

Solution. Figure 4.5.4 shows the geometry of the problem. If the lighthouse keeper lands x miles from A, he must row for $\sqrt{x^2 + 9}$ miles and walk for $5 - x$ miles. The total time required on that route is

$$\frac{\text{distance rowed}}{\text{rowing rate}} + \frac{\text{distance walked}}{\text{walking rate}} = \frac{\sqrt{x^2 + 9}}{4} + \frac{5 - x}{6}.$$

We want to find the value of x that minimizes the function

$$T(x) = \frac{\sqrt{x^2 + 9}}{4} + \frac{5 - x}{6}, \qquad 0 \le x \le 5.$$

(*Key step completed.*)

Differentiation gives

$$T'(x) = \frac{x}{4\sqrt{x^2 + 9}} - \frac{1}{6}, \qquad 0 < x < 5.$$

Setting $T'(x) = 0$, we find that

$$6x = 4\sqrt{x^2 + 9}$$
$$36x^2 = 16(x^2 + 9)$$
$$20x^2 = 144$$
$$x = \pm\tfrac{6}{5}\sqrt{5}.$$

The solution $x = -\tfrac{6}{5}\sqrt{5}$ is an extraneous root we introduced when we did the squaring. The value we want is $x = \tfrac{6}{5}\sqrt{5}$. As you can check by examining the sign of the derivative, T decreases for $0 < x < \tfrac{6}{5}\sqrt{5}$ and increases for $\tfrac{6}{5}\sqrt{5} < x < 5$. It follows that $x = \tfrac{6}{5}\sqrt{5}$ gives the absolute minimum value for T.

The lighthouse keeper can reach the general store most quickly by rowing to the point that lies $\tfrac{6}{5}\sqrt{5}$ miles (about 2.683 miles) down shore from A. ☐

EXERCISES 4.5

1. Find the greatest possible value for xy given that x and y are both positive and $x + y = 40$.
2. Find the dimensions of the rectangle of perimeter 24 that has the largest area.
3. A rectangular garden 200 square feet in area is to be fenced off against rabbits. Find the dimensions that will require the least amount of fencing if one side of the garden is already protected by a barn.
4. Find the largest possible area for a rectangle with base on the x-axis and upper vertices on the curve $y = 4 - x^2$.
5. Find the largest possible area for a rectangle inscribed in a circle of radius 4.
6. The cross section of a beam is in the form of a rectangle of length l and width w. (Figure 4.5.5.) Assuming that the strength of the beam varies directly with w^2l, what are the dimensions of the strongest beam that can be sawed from a round log of diameter 3 feet?

Figure 4.5.5

7. A rectangular playground is to be fenced off and divided in two by another fence parallel to one side of the playground. Six hundred feet of fencing is used. Find the dimensions of the playground that will enclose the greatest total area.
8. A Norman window is a window in the shape of a rectangle surmounted by a semicircle. Find the dimensions of the Norman window that admits the most light if the perimeter of the window is 30 feet.

9. Exercise 8, this time assuming that the semicircular portion of the window admits only one-third as much light per square foot as does the rectangular portion of the window.

10. One side of a rectangular field is bounded by a straight river. The other three sides are bounded by straight fences. The total length of the fence is 800 feet. Determine the dimensions of the field given that its area is a maximum.

11. Find the coordinates of P that maximize the area of the rectangle shown in Figure 4.5.6.

12. The base of a triangle is on the x-axis, one side lies along the line $y = 3x$, and the third side passes through the point $(1, 1)$. What is the slope of the third side if the area of the triangle is to be a minimum?

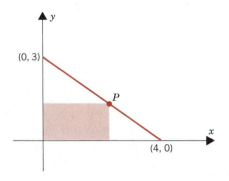

Figure 4.5.6

Figure 4.5.7

13. A triangle is formed by the coordinate axes and a line through the point $(2, 5)$ as in Figure 4.5.7. Determine the slope of this line if the area of the triangle is to be a minimum.

14. In the setting of Exercise 13 determine the slope of the line if the area is to be a maximum.

15. What are the dimensions of the base of the rectangular box of greatest volume that can be constructed from 100 square inches of cardboard if the base is to be twice as long as it is wide? Assume that the box has a top.

16. Exercise 15 under the assumption that the box has no top.

17. Find the dimensions of the isosceles triangle of largest area with perimeter 12.

18. Find the point(s) on the parabola $y = \frac{1}{8}x^2$ closest to the point $(0, 6)$.

19. Find the point(s) on the parabola $x = y^2$ closest to the point $(0, 3)$.

20. Find A and B given that the function $y = Ax^{-1/2} + Bx^{1/2}$ has a minimum value of 6 at $x = 9$.

21. A pentagon with a perimeter of 30 inches is to be constructed by adjoining an equilateral triangle to a rectangle. Find the dimensions of the rectangle and triangle that will maximize the area of the pentagon.

22. A 10-foot section of gutter is made from a 12-inch-wide strip of sheet metal by folding up 4-inch strips on each side so that they make the same angle with the bottom of the gutter. Determine the depth of the gutter that has the greatest carrying capacity. *Caution*: there are two ways to sketch the trapezoidal cross section, as shown in Figure 4.5.8.

Figure 4.5.8

23. From a rectangular piece of cardboard of dimensions 8×15 four congruent squares are to be cut out, one at each corner. See Figure 4.5.9. The remaining crosslike piece is then to be folded into an open box. What size squares should be cut out if the volume of the resulting box is to be a maximum?

15

Figure 4.5.9

24. A page is to contain 81 square centimeters of print. The margins at the top and bottom are to be 3 centimeters each and, at the sides, 2 centimeters each. Find the most economical dimensions given that the cost of a page varies directly with the perimeter of the page.

25. Let ABC be a triangle with vertices $A = (-3, 0)$, $B = (0, 6)$, $C = (3, 0)$. Let P be a point on the line segment that joins B to the origin. Find the position of P that minimizes the sum of the distances between P and the vertices.

26. Solve Exercise 25 with $A = (-6, 0)$, $B = (0, 3)$, $C = (6, 0)$.

27. An 8-foot-high fence is located 1 foot from a building. Determine the length of the shortest ladder that can be leaned against the building and touch the top of the fence.

28. Two hallways, one 8 feet wide and the other 6 feet wide, meet at right angles. Determine the length of the longest ladder that can be carried horizontally from one hallway into the other.

29. A rectangular banner has a red border and a white center. The width of the border at top and bottom is 8 inches and along the sides it is 6 inches. The total area is 27 square feet. What should be the dimensions of the banner if the area of the white center is to be a maximum?

30. Find the absolute maximum value of $y = x(r^2 + x^2)^{-3/2}$.

31. A string 28 inches long is to be cut into two pieces, one piece to form a square and the other to form a circle. How should the string be cut so as to (a) maximize the sum of the two areas? (b) minimize the sum of the two areas?

32. What is the maximum volume for a rectangular box (square base, no top) made from 12 square feet of cardboard?

33. Figure 4.5.10 shows a cylinder inscribed in a right circular cone of height 8 and base radius 5. Find the dimensions of the cylinder if its volume is to be a maximum.

34. A variant of Exercise 33. This time find the dimensions of the cylinder if the area of its curved surface is to be a maximum.

Figure 4.5.10

Figure 4.5.11

35. A rectangular box with square base and top is to be made to contain 1250 cubic feet. The material for the base costs 35 cents per square foot, for the top 15 cents per square foot, and for the sides 20 cents per square foot. Find the dimensions that will minimize the cost of the box.

36. What is the largest possible area for a parallelogram inscribed in a triangle ABC in the manner of Figure 4.5.11?

37. Find the dimensions of the isosceles triangle of least area that circumscribes a circle of radius r.

38. What is the maximal possible area for a triangle inscribed in a circle of radius r?

39. Figure 4.5.12 shows a cylinder inscribed in a sphere of radius R. Find the dimensions of the cylinder if its volume is to be a maximum.

40. A variant of Exercise 39. This time find the dimensions of the cylinder if the area of its curved surface is be a maximum.

Figure 4.5.12   **Figure 4.5.13**

41. A right circular cone is inscribed in a sphere of radius R as in Figure 4.5.13. Find the dimensions of the cone if its volume is to be a maximum.

42. What is the largest possible volume for a right circular cone of slant height a?

43. A power line is needed to connect a power station on the shore of a river to an island 4 kilometers downstream and 1 kilometer offshore. Find the minimum cost for such a line given that it costs $50,000 per kilometer to lay wire under water and $30,000 per kilometer to lay wire under ground.

44. A tapestry 7 feet high hangs on a wall. The lower edge is 9 feet above an observer's eye. How far from the wall should the observer stand to obtain the most favorable view? Namely, what distance from the wall maximizes the visual angle of the observer? HINT: Use the formula for $\tan (A - B)$.

45. A body of weight W is dragged along a horizontal plane by means of a force P whose line of action makes an angle θ with the plane. The magnitude of the force is given by the equation

$$P = \frac{mW}{m \sin \theta + \cos \theta}$$

where m denotes the coefficient of friction. For what value of θ is the pull a minimum?

46. If a projectile is fired from O so as to strike an inclined plane that makes a constant angle α with the horizontal, its range is given by the formula

$$R = \frac{2v^2 \cos \theta \sin (\theta - \alpha)}{g \cos^2 \alpha}$$

where v and g are constants and θ is the angle of elevation. Calculate θ for maximum range.

47. The lower edge of a movie theater screen 30 feet high is 6 feet above an observer's eye. How far from the screen should the observer sit to obtain the most favorable view? Namely, what distance from the screen maximizes the visual angle of the observer?

4.6 CONCAVITY AND POINTS OF INFLECTION

We begin with a picture, Figure 4.6.1. To the left of c_1 and between c_2 and c_3, the graph "curves up" (we call it *concave up*); between c_1 and c_2 and to the right of c_3, the graph "curves down" (we call it *concave down*).

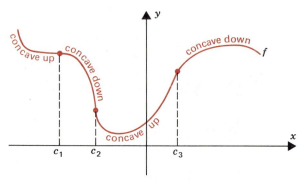

Figure 4.6.1

These terms deserve a precise definition.

DEFINITION 4.6.1 CONCAVITY

Let f be a function differentiable on an open interval I. The graph of f is said to be *concave up* on I iff f' increases on I; it is said to be *concave down* on I iff f' decreases on I.

In other words, the graph is concave up on an open interval where the slope increases and concave down on an open interval where the slope decreases.

Points that join arcs of opposite concavity are called *points of inflection*. The graph in Figure 4.6.1 has three of them: $(c_1, f(c_1))$, $(c_2, f(c_2))$, $(c_3, f(c_3))$.

Here is a formal definition.

DEFINITION 4.6.2 POINT OF INFLECTION

Let f be a function continuous at c. The point $(c, f(c))$ is called *a point of inflection* iff there exists $\delta > 0$ such that the graph of f is concave in one sense on $(c - \delta, c)$ and concave in the opposite sense on $(c, c + \delta)$.

Example 1. The graph of the squaring function $f(x) = x^2$ is concave up everywhere since the derivative $f'(x) = 2x$ is everywhere increasing. (See Figure 4.6.2.) ☐

Example 2. The graph of the cubing function $f(x) = x^3$ is concave down on $(-\infty, 0)$ and concave up on $(0, \infty)$:

$$f'(x) = 3x^2 \quad \text{decreases on } (-\infty, 0) \text{ and increases on } (0, \infty).$$

The origin is a point of inflection.   (See Figure 4.6.3.)   ☐

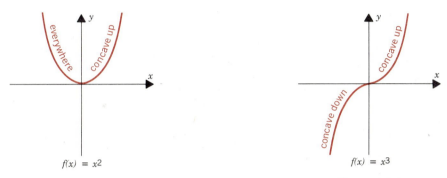

Figure 4.6.2   **Figure 4.6.3**

If f is twice differentiable, then, remembering that $f'' = (f')'$, we can determine the concavity of the graph by looking at the sign of the second derivative. If f'' is positive on I, then f' increases on I and the graph is concave up; if f'' is negative on I, then f' decreases on I and the graph is concave down.

The following result gives us a way of identifying possible points of inflection.

(4.6.3)

> If $(c, f(c))$ is a point of inflection, then either
> $$f''(c) = 0 \quad \text{or} \quad f''(c) \text{ does not exist.}$$

Proof. Suppose that $(c, f(c))$ is a point of inflection. Let's assume that the graph is concave up to the left of c and concave down to the right of c. The other case can be handled in a similar manner.

In this situation f' increases on an interval $(c - \delta, c)$ and decreases on an interval $(c, c + \delta)$.

Suppose now that $f''(c)$ exists. Then f' is continuous at c. It follows that f' increases on the half-open interval $(c - \delta, c]$ and decreases on the half-open interval $[c, c + \delta)$.† This says that f' has a local maximum at c. Since, by assumption, $f''(c)$ exists, $f''(c) = 0$. (Theorem 4.3.2 applied to f'.)

We have shown that, if $f''(c)$ exists, then $f''(c) = 0$. The other possibility, of course, is that $f''(c)$ does not exist.   ☐

---

† See Exercise 46, Section 4.2.

Example 3. For the function

$$f(x) = x^3 + \tfrac{1}{2}x^2 - 2x + 1 \qquad \text{(Figure 4.6.4)}$$

we have

$$f'(x) = 3x^2 + x - 2 \quad \text{and} \quad f''(x) = 6x + 1.$$

Since

$$f''(x) \text{ is } \begin{cases} \text{negative,} & \text{for} \quad x < -\tfrac{1}{6} \\ 0, & \text{at} \quad x = -\tfrac{1}{6} \\ \text{positive,} & \text{for} \quad x > -\tfrac{1}{6} \end{cases},$$

the graph of f is concave down on $(-\infty, -\tfrac{1}{6})$ and concave up on $(-\tfrac{1}{6}, \infty)$. The point $(-\tfrac{1}{6}, f(-\tfrac{1}{6}))$ is a point of inflection. ▢

Figure 4.6.4

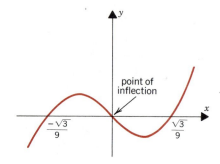

Figure 4.6.5

Example 4. For

$$f(x) = 3x^{5/3} - x \qquad \text{(Figure 4.6.5)}$$

we have

$$f'(x) = 5x^{2/3} - 1 \quad \text{and} \quad f''(x) = \tfrac{10}{3}x^{-1/3}.$$

The second derivative does not exist at $x = 0$. Since

$$f''(x) \text{ is } \begin{cases} \text{negative,} & \text{for} \quad x < 0 \\ \text{positive,} & \text{for} \quad x > 0 \end{cases},$$

the graph is concave down on $(-\infty, 0)$ and concave up on $(0, \infty)$. Since f is continuous at 0, the point $(0, f(0)) = (0, 0)$ is a point of inflection. ▢

Caution. The fact that

$$f''(c) = 0 \quad \text{or} \quad f''(c) \text{ does not exist}$$

does not guarantee that $(c, f(c))$ is a point of inflection. As you can verify, the function $f(x) = x^4$ satisfies $f''(0) = 0$, but the graph is always concave up and there are no points of inflection. If f is discontinuous at c, then $f''(c)$ does not exist but $(c, f(c))$ cannot be a point of inflection. A point of inflection occurs at c iff f is continuous at c and the point $(c, f(c))$ joins arcs of opposite concavity. ▢

EXERCISES 4.6

Describe the concavity of the graph of f and find the points of inflection (if any).

1. $f(x) = \dfrac{1}{x}$.

2. $f(x) = x + \dfrac{1}{x}$.

3. $f(x) = x^3 - 3x + 2$.

4. $f(x) = 2x^2 - 5x + 2$.

5. $f(x) = \frac{1}{4}x^4 - \frac{1}{2}x^2$.

6. $f(x) = x^3(1 - x)$.

7. $f(x) = \dfrac{x}{x^2 - 1}$.

8. $f(x) = \dfrac{x + 2}{x - 2}$.

9. $f(x) = (1 - x)^2(1 + x)^2$.

10. $f(x) = \dfrac{6x}{x^2 + 1}$.

11. $f(x) = \dfrac{1 - \sqrt{x}}{1 + \sqrt{x}}$.

12. $f(x) = (x - 3)^{1/5}$.

13. $f(x) = (x + 2)^{5/3}$.

14. $f(x) = x\sqrt{4 - x^2}$.

15. $f(x) = \sin^2 x, \quad x \in [0, \pi]$.

16. $f(x) = 2\cos^2 x - x^2, \quad x \in [0, \pi]$.

17. $f(x) = x^2 + \sin 2x, \quad x \in [0, \pi]$.

18. $f(x) = \sin^4 x, \quad x \in [0, \pi]$.

19. Find d given that $(d, f(d))$ is a point of inflection of the graph of

$$f(x) = (x - a)(x - b)(x - c).$$

20. Find c given that the graph of $f(x) = cx^2 + x^{-2}$ has a point of inflection at $(1, f(1))$.

21. Find a and b given that the graph of $f(x) = ax^3 + bx^2$ passes through $(-1, 1)$ and has a point of inflection at $x = \frac{1}{3}$.

22. Determine A and B so that the curve

$$y = Ax^{1/2} + Bx^{-1/2}$$

will have a point of inflection at $(1, 4)$.

23. Determine A and B so that the curve

$$y = A\cos 2x + B\sin 3x$$

will have a point of inflection at $(\pi/6, 5)$.

24. Find necessary and sufficient conditions on A and B for $f(x) = Ax^2 + Bx + C$
   (a) to decrease between A and B with graph concave up.
   (b) to increase between A and B with graph concave down.

4.7 SOME CURVE SKETCHING

During the course of the last few sections you have learned how to find the extreme values of a function, the intervals where it increases, the intervals where it decreases. You have also seen how to determine the concavity of the graph and how to find the points of inflection. This knowledge makes it possible to sketch the graph of a somewhat complicated function without having to plot point, after point, after point.

Before attempting to sketch a graph, we try to gather together the necessary information and record it in an organized form. Here is an outline of the procedure that we follow.

1. We look for special features (such as endpoints of the domain, clear behavior as $x \to \pm\infty$, and symmetry).

2. We calculate f' and f''.

3. We examine the sign of f' to determine extreme values and intervals of increase or decrease.

4. We examine the sign of f'' to determine points of inflection and intervals of different concavity.

5. We plot the points of interest in a preliminary sketch.

6. Finally we sketch the graph by connecting the points of our preliminary sketch, keeping an eye on all the information gathered earlier to make sure that the sketch "rises," "falls," and "bends" in the proper way.

Figure 4.7.1 shows some of the elements that we might include in a preliminary sketch.

local maximum: (2,4)   point of inflection: (3,2)   endpoint minimum: (4,1)

Figure 4.7.1

Example 1. We begin with the function

$$f(x) = \tfrac{1}{3}x^3 + \tfrac{1}{2}x^2 - 2x - 1, \qquad \text{all real } x.$$

This is a polynomial with leading term $\tfrac{1}{3}x^3$. Therefore, as $x \to -\infty$, $f(x) \to -\infty$ and, as $x \to \infty$, $f(x) \to \infty$. Since $f(0) = -1$, the graph intersects the y-axis at the point $(0, -1)$.

For all real x

$$f'(x) = x^2 + x - 2 = (x + 2)(x - 1)$$
$$f''(x) = 2x + 1 = 2(x + \tfrac{1}{2}).$$

The signs of f' and f'' are recorded in Figure 4.7.2:

Figure 4.7.2

Points of interest to be plotted. (See Figure 4.7.3 for preliminary sketch.)

$(-2, 2\frac{1}{3})$: $f(-2) = 2\frac{1}{3}$ is a local maximum.

$(-\frac{1}{2}, \frac{1}{12})$: point of inflection; slope $-\frac{9}{4}$.

$(0, -1)$: y-intercept -1.

$(1, -2\frac{1}{6})$: $f(1) = -2\frac{1}{6}$ is a local minimum.

The final graph is given in Figure 4.7.4. ◻

Figure 4.7.3

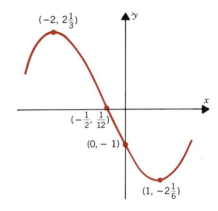

Figure 4.7.4

Example 2. Now we consider the function

$$f(x) = x^4 - 4x^3 + 1, \qquad x \in [-1, 5).$$

We note that $x = 5$ is a "missing endpoint." For $x \in (-1, 5)$

$$f'(x) = 4x^3 - 12x^2 = 4x^2(x - 3)$$
$$f''(x) = 12x^2 - 24x = 12x(x - 2).$$

The signs of f' and f'' are recorded in Figure 4.7.5:

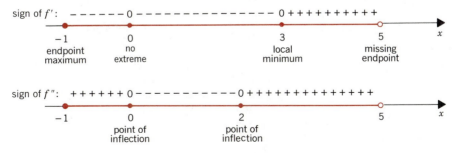

Figure 4.7.5

Points of interest to be plotted.   (See Figure 4.7.6 for preliminary sketch.)

$(-1, 6)$: $f(-1) = 6$ is an endpoint maximum.

$(0, 1)$:   no extreme here; point of inflection with horizontal tangent.

$(2, -15)$:   point of inflection; slope -16.

$(3, -26)$: $f(3) = -26$ is a local minimum.

As x approaches the missing endpoint 5 from the left, $f(x)$ increases toward a value of 126.

Since the range of f makes a scale drawing impractical, we must be content with a rough sketch as in Figure 4.7.7. In cases like this, it is particularly important to give the coordinates of the points of interest. ☐

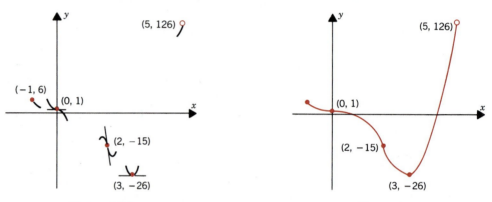

Figure 4.7.6   Figure 4.7.7

Example 3. The function

$$f(x) = 5x - 3x^{5/3}, \qquad \text{all real } x$$

is an odd function: $f(-x) = -f(x)$. This tells us that the graph of f is symmetric about the origin. Since we can write $f(x) = x(5 - 3x^{2/3})$, it is clear that, as $x \to \infty$, $f(x) \to -\infty$. By symmetry, as $x \to -\infty$, $f(x) \to \infty$.

Differentiation gives

$$f'(x) = 5 - 5x^{2/3} = 5(1 - x^{2/3}) = 5(1 + x^{1/3})(1 - x^{1/3}), \qquad \text{all real } x$$

$$f''(x) = -\tfrac{10}{3}x^{-1/3}, \qquad x \neq 0.$$

The signs of f' and f'' are recorded in Figure 4.7.8:

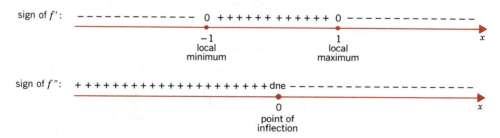

Figure 4.7.8

Points to be plotted.   (See Figure 4.7.9 for preliminary sketch.)

$$(-1, -2): \quad f(-1) = -2 \text{ is a local minimum.}$$
$$(0, 0): \quad \text{point of inflection; slope 5.}$$
$$(1, 2): \quad f(1) = 2 \text{ is a local maximum.}$$

The final graph appears in Figure 4.7.10. ☐

Figure 4.7.9

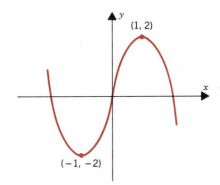

Figure 4.7.10

Remark. The most effective way to produce an accurate graph of a function is to take advantage of the graphic capabilities of some of today's computers. By the method outlined in this chapter we can produce a rough sketch that gives the general shape of the graph and shows its salient features. For our purposes this is sufficient. ☐

EXERCISES 4.7

Sketch the graph.

1. $f(x) = (x - 2)^2.$

2. $f(x) = 1 - (x - 2)^2.$

3. $f(x) = x^3 - 2x^2 + x + 1.$

4. $f(x) = x^3 - 9x^2 + 24x - 7.$

5. $f(x) = x^3 + 6x^2, \quad x \in [-4, 4].$

6. $f(x) = x^4 - 8x^2, \quad x \in [0, \infty).$

7. $f(x) = \frac{2}{3}x^3 - \frac{1}{2}x^2 - 10x - 1.$

8. $f(x) = 2 + (x - 4)^{1/3}, \quad x \in [4, \infty).$

9. $f(x) = 3x^4 - 4x^3 + 1.$

10. $f(x) = x(x^2 + 4)^2.$

11. $f(x) = 2\sqrt{x} - x, \quad x \in [0, 4].$

12. $f(x) = \frac{1}{4}x - \sqrt{x}, \quad x \in [0, 9].$

13. $f(x) = 2 + (x + 1)^{6/5}.$

14. $f(x) = 2 + (x + 1)^{7/5}.$

15. $f(x) = 3x^5 + 5x^3.$

16. $f(x) = 3x^4 + 4x^3.$

17. $f(x) = 1 + (x - 2)^{5/3}.$

18. $f(x) = 1 + (x - 2)^{4/3}.$

19. $f(x) = x^2(1 + x)^3.$

20. $f(x) = x^2(1 + x)^2.$

21. $f(x) = x\sqrt{1 - x}.$

22. $f(x) = \frac{1}{3}x^3 + x^2 - 8x + 4.$

23. $f(x) = x + \sin 2x, \quad x \in [0, \pi].$

24. $f(x) = \cos^3 x + 6 \cos x, \quad x \in [0, \pi].$

25. $f(x) = \cos^4 x, \quad x \in [0, \pi].$

26. $f(x) = \sqrt{3}\, x - \cos 2x, \quad x \in [0, \pi].$

27. $f(x) = 2 \sin^3 x + 3 \sin x, \quad x \in [0, \pi].$

28. $f(x) = \sin^4 x, \quad x \in [0, \pi].$

4.8 VERTICAL AND HORIZONTAL ASYMPTOTES; VERTICAL TANGENTS AND CUSPS

Vertical and Horizontal Asymptotes

In Figure 4.8.1 you can see the graph of the function

$$f(x) = \frac{1}{|x - c|} \qquad \text{for } x \text{ close to } c.$$

As $x \to c$, $f(x) \to \infty$: given any positive number M there exists a positive number δ such that

$$\text{if} \quad 0 < |x - c| < \delta, \qquad \text{then} \quad f(x) \geq M.$$

The line $x = c$ is called a *vertical asymptote*.

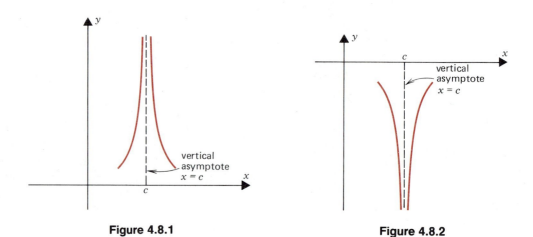

| Figure 4.8.1 | Figure 4.8.2 |

Figure 4.8.2 shows the graph of

$$g(x) = -\frac{1}{|x - c|} \qquad \text{for } x \text{ close to } c.$$

With obvious meaning, as $x \to c$, $g(x) \to -\infty$. Again the line $x = c$ is called a *vertical asymptote*.

Vertical asymptotes can also arise from one-sided behavior. With f and g as in Figure 4.8.3, we write

$$\text{as} \quad x \to c^-, \quad f(x) \to \infty \text{ and } g(x) \to -\infty.$$

With f and g as in Figure 4.8.4, we write

$$\text{as} \quad x \to c^+, \quad f(x) \to \infty \text{ and } g(x) \to -\infty.$$

In each case the vertical line $x = c$ is a vertical asymptote for both functions.

Figure 4.8.3

Figure 4.8.4

Figure 4.8.5

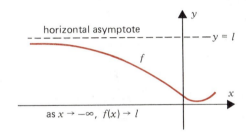

Figure 4.8.6

It is also possible for a function to have a *horizontal asymptote*. See Figures 4.8.5 and 4.8.6. The ϵ, K definition for these limits is left to you in the exercises. (See Exercise 35.)

Example 1. Figure 4.8.7 shows the graph of the function

$$f(x) = \frac{x}{x-2}.$$

As $x \to 2^-$, $f(x) \to -\infty$; as $x \to 2^+$, $f(x) \to \infty$. The line $x = 2$ is a vertical asymptote.

As $x \to \pm\infty$,

$$f(x) = \frac{x}{x-2} = \frac{1}{1-(2/x)} \to 1.$$

The line $y = 1$ is a horizontal asymptote. ☐

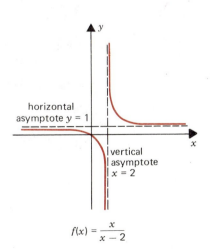

Figure 4.8.7

Example 2.   The function

$$g(x) = \frac{1}{1 + x^2}$$

is everywhere continuous, and thus there are no vertical asymptotes. As $x \to \pm\infty$, $g(x) \to 0$. The line $y = 0$ (the x-axis) is a horizontal asymptote. The graph of g is sketched in Figure 4.8.8.   ☐

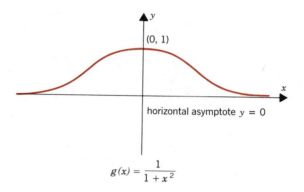

Figure 4.8.8

Example 3.   The function

$$f(x) = \frac{5 - 3x^2}{1 - x^2}$$

is continuous everywhere except at $x = \pm 1$. The asymptotic properties of f are made more apparent by writing

$$f(x) = \frac{3 - 3x^2 + 2}{1 - x^2} = 3 + \frac{2}{1 - x^2}.$$

The line $x = 1$ is a vertical asymptote:

as $x \to 1^-$, $1 - x^2$ decreases toward 0
and $f(x) \to \infty$;

as $x \to 1^+$, $1 - x^2$ increases toward 0
and $f(x) \to -\infty$.

The line $x = -1$ is also a vertical asymptote:

as $x \to -1^-$, $1 - x^2$ increases toward 0
and $f(x) \to -\infty$;

as $x \to -1^+$, $1 - x^2$ decreases toward 0
and $f(x) \to \infty$.

The line $y = 3$ is a horizontal asymptote:

as $x \to \pm\infty$, $\dfrac{2}{1 - x^2} \to 0$ and $f(x) \to 3$.

The graph appears in Figure 4.8.9.   ☐

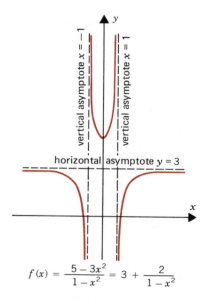

$$f(x) = \frac{5 - 3x^2}{1 - x^2} = 3 + \frac{2}{1 - x^2}$$

Figure 4.8.9

Vertical Tangents; Vertical Cusps

(For the remainder of this section assume that f is continuous at $x = c$ and differentiable for $x \neq c$.)

We say that the graph of f has a *vertical tangent* at the point $(c, f(c))$ iff

$$\text{as } x \to c, \quad f'(x) \to \infty \text{ or } f'(x) \to -\infty.$$

Figure 4.8.10 gives two examples.

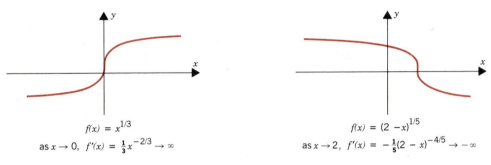

$$f(x) = x^{1/3}$$
$$\text{as } x \to 0, \quad f'(x) = \tfrac{1}{3}x^{-2/3} \to \infty$$

$$f(x) = (2 - x)^{1/5}$$
$$\text{as } x \to 2, \quad f'(x) = -\tfrac{1}{5}(2 - x)^{-4/5} \to -\infty$$

Figure 4.8.10

On occasion you will see a graph become almost vertical and then virtually double back on itself. Such a pattern signals the presence of what is known as a "vertical cusp." We say that the graph of f has a *vertical cusp* at $(c, f(c))$ iff

$$\text{as } x \to c^{-}, \quad f'(x) \to -\infty \qquad \text{and} \qquad \text{as } x \to c^{+}, \quad f'(x) \to \infty,$$

or

$$\text{as } x \to c^{-}, \quad f'(x) \to \infty \qquad \text{and} \qquad \text{as } x \to c^{+}, \quad f'(x) \to -\infty.$$

You can see some examples in Figure 4.8.11.

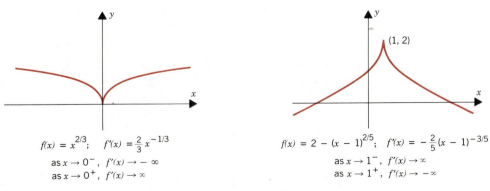

$$f(x) = x^{2/3}; \quad f'(x) = \tfrac{2}{3}x^{-1/3}$$
$$\text{as } x \to 0^{-}, \quad f'(x) \to -\infty$$
$$\text{as } x \to 0^{+}, \quad f'(x) \to \infty$$

$(1, 2)$

$$f(x) = 2 - (x - 1)^{2/5}; \quad f'(x) = -\tfrac{2}{5}(x - 1)^{-3/5}$$
$$\text{as } x \to 1^{-}, \quad f'(x) \to \infty$$
$$\text{as } x \to 1^{+}, \quad f'(x) \to -\infty$$

Figure 4.8.11

The fact that $f'(c)$ does not exist does *not* mean that the graph of f has either a vertical tangent or a vertical cusp at $(c, f(c))$. Unless the conditions spelled out above are met, the graph of f can simply be making a "corner" at $(c, f(c))$. For example, the function

$$f(x) = |x^3 - 1|$$

has derivative

$$f'(x) = \begin{bmatrix} -3x^2, & x < 1 \\ 3x^2, & x > 1 \end{bmatrix}.$$

At $x = 1$, $f'(x)$ does not exist. As $x \to 1^-$, $f'(x) \to -3$, and as $x \to 1^+$, $f'(x) \to 3$. There is no vertical tangent here and no vertical cusp. For the graph, see Figure 4.8.12.

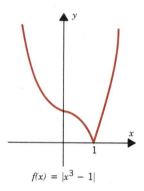

$f(x) = |x^3 - 1|$

Figure 4.8.12

EXERCISES 4.8

Find the vertical and horizontal asymptotes.

1. $f(x) = \dfrac{x}{3x - 1}.$

2. $f(x) = \dfrac{x^3}{x + 2}.$

3. $f(x) = \dfrac{x^2}{x - 2}.$

4. $f(x) = \dfrac{4x}{x^2 + 1}.$

5. $f(x) = \dfrac{2x}{x^2 - 9}.$

6. $f(x) = \dfrac{\sqrt{x}}{4\sqrt{x} - x}.$

7. $f(x) = \left(\dfrac{2x - 1}{4 + 3x}\right)^2.$

8. $f(x) = \dfrac{4x^2}{(3x - 1)^2}.$

9. $f(x) = \dfrac{3x}{(2x - 5)^2}.$

10. $f(x) = \left(\dfrac{x}{1 - 2x}\right)^3.$

11. $f(x) = \dfrac{3x}{\sqrt{4x^2 + 1}}.$

12. $f(x) = \dfrac{x^{1/3}}{x^{2/3} - 4}.$

13. $f(x) = \dfrac{\sqrt{x}}{2\sqrt{x} - x - 1}.$

14. $f(x) = \dfrac{2x}{\sqrt{x^2 - 1}}.$

15. $f(x) = \sqrt{x + 4} - \sqrt{x}.$

16. $f(x) = \sqrt{x} - \sqrt{x - 2}.$

17. $f(x) = \dfrac{\sin x}{\sin x - 1}.$

18. $f(x) = \dfrac{1}{\sec x - 1}.$

Determine whether or not the graph of f has a vertical tangent or a vertical cusp at c.

19. $f(x) = (x + 3)^{4/3}$; $c = -3$.

20. $f(x) = 3 + x^{2/5}$; $c = 0$.

21. $f(x) = (2 - x)^{4/5}$; $c = 2$.

22. $f(x) = (x + 1)^{-1/3}$; $c = -1$.

23. $f(x) = 2x^{3/5} - x^{6/5}$; $c = 0$.

24. $f(x) = (x - 5)^{7/5}$; $c = 5$.

25. $f(x) = (x + 2)^{-2/3}$; $c = -2$.

26. $f(x) = 4 - (2 - x)^{3/7}$; $c = 2$.

27. $f(x) = \sqrt{|x - 1|}$; $c = 1$.

28. $f(x) = x(x - 1)^{1/3}$; $c = 1$.

29. $f(x) = |(x + 8)^{1/3}|$; $c = -8$.

30. $f(x) = \sqrt{4 - x^2}$; $c = 2$.

31. $f(x) = \dfrac{x^{2/3} - 1}{|x^{1/3} - 1|}$; $c = 0$.

32. $f(x) = |(4 + x)^{3/5}|$; $c = -4$.

33. $f(x) = \begin{bmatrix} x^{1/3} + 2, & x \le 0 \\ 1 - x^{1/5}, & x > 0 \end{bmatrix}$; $c = 0$.

34. $f(x) = \begin{bmatrix} 1 + \sqrt{-x}, & x \le 0 \\ (4x - x^2)^{1/3}, & x > 0 \end{bmatrix}$; $c = 0$.

35. To say that, as $x \to \infty$, $f(x) \to l$ is to say that given a positive number ϵ, there exists a positive number K such that, if $x \geq K$, then $|f(x) - l| < \epsilon$. Construct a similar ϵ, K definition for

$$\text{as } x \to -\infty, \quad f(x) \to l.$$

4.9 ADDITIONAL EXERCISES ON GRAPHING

Sketch the graph

1. $f(x) = (x + 1)^3 - 3(x + 1)^2 + 3(x + 1)$.   **2.** $f(x) = (x - 1)^4 - 2(x - 1)^2$.

3. $f(x) = x^2(5 - x)^3$.   **4.** $f(x) = x^3(x + 5)^2$.   **5.** $f(x) = x^2 + \dfrac{2}{x}$.

6. $f(x) = x - \dfrac{1}{x}$.   **7.** $f(x) = \dfrac{x - 4}{x^2}$.   **8.** $f(x) = \dfrac{x + 2}{x^3}$.

9. $f(x) = 3 - |x^2 - 1|$.   **10.** $f(x) = 4 - |2x - x^2|$.   **11.** $f(x) = x^2 - 6x^{1/3}$.

12. $f(x) = x - x^{1/3}$.   **13.** $f(x) = x(x - 1)^{1/5}$.   **14.** $f(x) = x^2(x - 7)^{1/3}$.

15. $f(x) = \dfrac{2x}{4x - 3}$.   **16.** $f(x) = \dfrac{2x^2}{x + 1}$.   **17.** $f(x) = \dfrac{x}{(x + 3)^2}$.

18. $f(x) = \dfrac{x}{x^2 + 1}$.   **19.** $f(x) = \dfrac{x^2}{x^2 - 4}$.   **20.** $f(x) = \dfrac{2x}{x - 4}$.

21. $f(x) = \sqrt{\dfrac{x}{x - 2}}$.   **22.** $f(x) = \dfrac{2x}{\sqrt{x^2 + 1}}$.   **23.** $f(x) = \dfrac{x}{\sqrt{4x^2 + 1}}$.

24. $f(x) = \dfrac{x}{\sqrt{x^2 - 2}}$.   **25.** $f(x) = \dfrac{x^2}{\sqrt{x^2 - 2}}$.   **26.** $f(x) = \sqrt{\dfrac{x}{x + 4}}$.

27. $f(x) = 3 \sin 2x$, $\quad x \in [0, \pi]$.   **28.** $f(x) = 4 \cos 3x$, $\quad x \in [0, \pi]$.

29. $f(x) = 2 \sin 3x$, $\quad x \in [0, \pi]$.   **30.** $f(x) = 3 \cos 4x$, $\quad x \in [0, \pi]$.

31. $f(x) = (\sin x - \cos x)^2$, $\quad x \in [0, \pi]$.   **32.** $f(x) = 3 + 2 \cot x + \csc^2 x$, $\quad x \in (0, \tfrac{1}{2}\pi)$.

33. $f(x) = 2 \tan x - \sec^2 x$, $\quad x \in (0, \tfrac{1}{2}\pi)$.

34. Set

$$F(x) = \begin{cases} \sin (1/x), & x \neq 0 \\ 0, & x = 0 \end{cases}, \qquad G(x) = \begin{cases} x \sin (1/x), & x \neq 0 \\ 0, & x = 0 \end{cases},$$

$$H(x) = \begin{cases} x^2 \sin (1/x), & x \neq 0 \\ 0, & x = 0 \end{cases}.$$

(a) Sketch a figure displaying the general nature of the graph of F.
(b) Sketch a figure displaying the general nature of the graph of G.
(c) Sketch a figure displaying the general nature of the graph of H.
(d) Which of these functions is continuous at 0?
(e) Which of these functions is differentiable at 0?

4.10   ADDITIONAL MAX-MIN PROBLEMS

Problem 1. (*The angle of incidence equals the angle of reflection.*) Figure 4.10.1 depicts light from a point A reflected to a point B by a mirror. Two angles have been marked: the *angle of incidence*, θ_i, and the *angle of reflection*, θ_r. Experiment shows that $\theta_i = \theta_r$. Derive this result by postulating that the light that travels from A to the mirror and then to B follows the shortest possible path.†

Figure 4.10.1

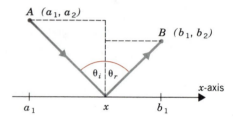

Figure 4.10.2

Solution. We can write the length of the path as a function of x: in the setup of Figure 4.10.2,

$$l(x) = \sqrt{(x - a_1)^2 + a_2^2} + \sqrt{(x - b_1)^2 + b_2^2}, \qquad x \in [a_1, b_1].$$

Differentiation gives

$$l'(x) = \frac{x - a_1}{\sqrt{(x - a_1)^2 + a_2^2}} + \frac{x - b_1}{\sqrt{(x - b_1)^2 + b_2^2}}.$$

We can therefore see that

$$l'(x) = 0 \quad \text{iff} \quad \frac{x - a_1}{\sqrt{(x - a_1)^2 + a_2^2}} = \frac{b_1 - x}{\sqrt{(x - b_1)^2 + b_2^2}}$$

$$\text{iff} \quad \sin \theta_i = \sin \theta_r \qquad\qquad \text{(see the figure)}$$

$$\text{iff} \quad \theta_i = \theta_r.$$

That $l(x)$ is minimal when $\theta_i = \theta_r$ can be seen from noting that $l''(x)$ is always positive:

$$l''(x) = \frac{a_2^2}{[(x - a_1)^2 + a_2^2]^{3/2}} + \frac{b_2^2}{[(x - b_1)^2 + b_2^2]^{3/2}} > 0. \quad \square$$

We must admit that there is a simpler way to do Problem 1 that requires no calculus. Can you find it?

Now we will work out a problem in which the function to be maximized is defined, not on an interval or union of intervals, but on a discrete set of points, in this case a finite collection of integers.

---

† This is a special case of Fermat's *Principle of Least Time*, which says that, of all (neighboring) paths, light chooses the one that demands the least time. If light passes from one medium to another, the geometrically shortest path is not necessarily the path of least time. (See Figure 7.11.1.)

Problem 2. A manufacturing plant has a capacity of 25 articles per week. Experience has shown that n articles per week can be sold at a price of p dollars each where $p = 110 - 2n$ and the cost of producing n articles is $600 + 10n + n^2$ dollars. How many articles should be made each week to give the largest profit?

Solution. The profit (P dollars) on the sale of n articles is

$$P = np - (600 + 10n + n^2).$$

With $p = 110 - 2n$, this simplifies to

$$P = 100n - 600 - 3n^2.$$

In this problem n must be an integer, and thus it makes no sense to differentiate P with respect to n. The formula shows that P is negative if n is less than 8 or greater than 25. By direct calculation we construct Table 4.10.1. The table shows that the largest profit comes from setting production at 17 articles per week.

TABLE 4.10.1

n	P	n	P	n	P
8	8	14	212	20	200
9	57	15	225	21	117
10	100	16	232	22	148
11	137	17	233	23	113
12	168	18	228	24	72
13	193	19	217	25	25

We can avoid such massive computation by considering the function

$$f(x) = 100x - 600 - 3x^2, \qquad 8 \le x \le 25. \qquad \text{(key step)}$$

This function is differentiable with respect to x, and for integral values of x it agrees with P. Differentiation of f gives

$$f'(x) = 100 - 6x.$$

Obviously, $f'(x) = 0$ at $x = \frac{100}{6} = 16\frac{2}{3}$. Since $f'(x) > 0$ on $(8, 16\frac{2}{3})$, f increases on $[8, 16\frac{2}{3}]$. Since $f'(x) < 0$ on $(16\frac{2}{3}, 25)$, f decreases on $[16\frac{2}{3}, 25]$. The largest value of f corresponding to an integral value of x will therefore occur at $x = 16$ or $x = 17$. Direct calculation of $f(16)$ and $f(17)$ shows that the choice of $x = 17$ is correct. ☐

EXERCISES 4.10

1. Davis Rent-A-TV derives an average net profit of $15 per customer if it services 1000 customers or less. If it services over 1000 customers, then the average profit decreases 1¢ for each customer above that number. How many customers give the maximum net profit?

2. A truck is to be driven 300 miles on a freeway at a constant speed of x miles per hour. Speed laws require that $35 \le x \le 55$. Assume that fuel costs $1.35 per gallon and is consumed at the rate of $2 + \frac{1}{600}x^2$ gallons per hour. Given that the driver's wages are $13 per hour, at what speed should the truck be driven to minimize the truck owner's expenses?

3. A triangle is formed by the coordinate axes and a line through the point (a, b) as in Figure 4.10.3. Determine the slope of this line if the area of the triangle is to be a minimum.

Figure 4.10.3

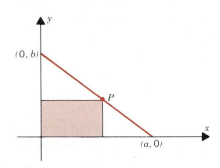

Figure 4.10.4

4. Find the coordinates of P that maximize the area of the rectangle shown in Figure 4.10.4.

5. Let x and y be positive numbers such that $x + y = 100$; let n be a positive integer. Find the values of x and y that minimize $x^n + y^n$.

6. In Exercise 5 find the values of x and y that maximize $x^n y^n$.

7. Find the dimensions of the cylinder of greatest volume that can be inscribed in a right circular cone of height H and base radius R.

8. Let Q denote output, R revenue, C cost, and P profit. Given that $P = R - C$, what is the relation between marginal revenue dR/dQ and marginal cost dC/dQ when profit is at a maximum?

9. A manufacturer finds that the total cost of producing Q tons is $aQ^2 + bQ + c$ dollars and the price at which each ton can be sold is $\beta - \alpha Q$. Assuming that a, b, c, α, β are all positive, what is the output for maximum profit?

10. Let $P = f(Q)$ be a demand curve. (P = price, Q = output) Show that at an output that maximizes total revenue (price × output) the elasticity

$$\epsilon = \frac{f(Q)}{Q|f'(Q)|} = 1.$$

11. A manufacturer of customized lighting fixtures finds that it can sell x standing lamps per week at p dollars each where $5x = 375 - 3p$. The cost of production is $500 + 15x + \frac{1}{5}x^2$ dollars. Show that maximum profit is obtained by setting production at about 30 lamps per week.

12. Suppose that the relation between x and p in Exercise 11 is given by the equation $x = 100 - 4\sqrt{5p}$. Show that for maximum profit the manufacturer should produce only about 25 lamps per week.

13. Suppose that the relation between x and p in Exercise 11 is $x^2 = 2500 - 20p$. What production produces maximum profit in this instance?

14. The total cost of producing Q units per week is

$$C(Q) = \tfrac{1}{3}Q^3 - 20Q^2 + 600Q + 1000 \quad \text{dollars.}$$

Given that the total revenues are

$$R(Q) = 420Q - 2Q^2 \quad \text{dollars,}$$

find the output that maximizes profit.

15. A steel plant is capable of producing Q_1 tons per day of low-grade steel and Q_2 tons per day of high-grade steel, where

$$Q_2 = \frac{40 - 5Q_1}{10 - Q_1}.$$

If the market price of low-grade steel is half that of high-grade steel, show that about $5\frac{1}{2}$ tons of low-grade steel should be produced per day for maximum receipts.

16. The total cost of producing Q articles per week is $aQ^2 + bQ + c$ dollars, and the price at which each can be sold is $p = \beta - \alpha Q^2$. Show that the output for maximum profit is

$$Q = \frac{\sqrt{a^2 + 3\alpha(\beta - b)} - a}{3\alpha}. \qquad \text{(take } a, b, c, \alpha, \beta \text{ as positive)}$$

17. In a certain industry an output of Q units can be obtained at a total cost of

$$C(Q) = aQ^2 + bQ + c \quad \text{dollars,}$$

bringing in total revenues of

$$R(Q) = \beta Q - \alpha Q^2 \quad \text{dollars.}$$

The government decides to impose an excise tax on the product. What tax rate (dollars per unit) will maximize the government's revenues from this tax? (Take a, b, c, α, β as positive.)

18. Find the dimensions of the rectangle of perimeter p that has the largest area.

19. Find the largest possible area for a rectangle inscribed in a circle of radius r.

20. Find the dimensions of the isosceles triangle of largest area with perimeter p.

21. A local bus company offers charter trips to Blue Mountain Museum at a fare of $37 per person if 16 to 35 passengers sign up for the trip. The company does not charter trips for fewer than 16 passengers. The bus has 48 seats. If more than 35 passengers sign up, then the fare for every passenger is reduced by 50 cents for each passenger in excess of 35 that signs up. Determine the number of passengers that generates the greatest revenue for the bus company.

22. The Hotwheels Rent-A-Car Co. derives an average net profit of $12 per customer if it services 50 customers or less. If it services over 50 customers, then the average net profit is decreased by 6 cents for each customer over 50. What number of customers produces the greatest total net profit for the company?

23. One side of a rectangular field is bounded by a straight river. The other three sides are bounded by straight fences. The total length of the fence is 800 feet. Each side of the field is at least 220 feet. Determine the dimensions of the field given that its area is a maximum.

24. An oil drum is to be made in the form of a right circular cylinder to contain 16π cubic feet. The upright drum is to be taller than it is wide but not more than 6 feet tall. Determine the dimensions of the drum with minimal surface area.

25. A closed rectangular box with a square base is to be built subject to the following specifications: the volume is to be 27 cubic feet, the area of the base may not exceed 18 square feet, the height of the box may not exceed 2 feet. Determine the dimensions for (a) minimal surface area; (b) maximal surface area.

26. Conical paper cups are usually made so that the depth is $\sqrt{2}$ times the radius of the rim. Show that this design requires the least amount of paper per unit volume.

27. An open box is made from a rectangular sheet of material measuring 8×15 inches by removing equal-sized squares from each corner of the sheet and folding up the sides. Each dimension of the resulting box is at least 2 inches. Determine the dimensions of the box of greatest volume.

28. A 28 inch piece of wire is to be cut into two pieces, one piece to form a circle and the other to form a square. Each piece is at least 9 inches long. How should the wire be cut so as to (a) maximize the sum of the two areas? (b) minimize the sum of the two areas?

29. Find the dimensions of the cylinder of maximum volume inscribable in a sphere of radius 6 inches. (Use the central angle θ subtended by the radius of the base.)

30. Solve Exercise 29 if the curved surface of the cylinder is to be a maximum.

31. The equation of the path of a ball is $y = mx - \frac{1}{800}(m^2 + 1)x^2$, where the origin is taken as the point from which the ball is thrown and m is the slope of the curve at the origin. For what value of m will the ball strike at the greatest distance along the same horizontal level?

32. In the setting of Exercise 31, for what value of m will the ball strike at the greatest height on a vertical wall 300 feet away?

33. The cost of erecting an office building is $1,000,000 for the first story, $1,100,000 for the second, $1,200,000 for the third, and so on. Other expenses (lot, basement, etc.) are $5,000,000. Assume that the net annual income is $200,000 per story. How many stories will provide the greatest rate of return on investment?

34. The sum of the surface areas of a sphere and a cube being given, show that the sum of the volumes will be least when the diameter of the sphere is equal to the edge of the cube. When will the sum of the volumes be greatest?

35. The distance from a point to a line is the distance from that point to the closest point of the line. (a) What point of the line $y = x$ is closest to the point (x_1, y_1)? (b) What is the distance from (x_1, y_1) to the line $y = x$?

36. Given that $\overline{PQ}$ is the longest or shortest line segment that can be drawn from $P(a, b)$ to the differentiable curve $y = f(x)$, show that $\overline{PQ}$ is perpendicular to the curve at Q.

37. What point on the curve $y = x^{3/2}$ is closest to $P(\frac{1}{2}, 0)$?

38. Let $P(x_0, y_0)$ be a point in the first quadrant. Draw a line through P that cuts the positive x-axis at $A(a, 0)$ and the positive y-axis at $B(0, b)$. Choose a and b so that (a) the area of $\triangle OAB$ is a minimum. (b) the length of $\overline{AB}$ is a minimum. (c) $a + b$ is a minimum. (d) the perpendicular distance from O to $\overline{AB}$ is a maximum.

39. Two sources of heat are placed s meters apart: a source of intensity a at A and a source of intensity b at B. The intensity of heat at a point P between A and B is then given by the formula

$$I = \frac{a}{x^2} + \frac{b}{(s - x)^2}$$

where x is the distance between P and A measured in meters. At what point P between A and B will the temperature be lowest?

40. Draw the parabola $y = x^2$. On the parabola mark a point $P \neq 0$. Through P draw the normal line. This normal line intersects the parabola at another point Q. Show that the distance between P and Q is minimized by setting $P = (\pm\frac{1}{2}\sqrt{2}, \frac{1}{2})$.

41. *The problem of Viviani.*† Two parallel lines are cut by a given line AB. See Figure 4.10.5. From the point C we draw a line to the point marked Q. How should Q be chosen so that the sum of the areas of the two triangles is a minimum?

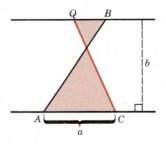

Figure 4.10.5

---

† Vicenzo Viviani (1622–1703), a pupil of Galileo.

Figure 4.10.6

Figure 4.10.7

42. An isosceles trapezoid is inscribed in a circle of radius r. (Figure 4.10.6) Given that one base has length $2r$, find the length of the other base if the area is a maximum.

43. Find the dimensions of the rectangular solid of maximum volume that can be cut from a solid sphere of radius r. (Figure 4.10.7) HINT: First do Exercise 19.

44. Find the base and altitude of the isosceles triangle of minimum area that circumscribes the ellipse $b^2x^2 + a^2y^2 = a^2b^2$ if the base is parallel to the x-axis.

45. A triangle is formed in the first quadrant by the coordinate axes and a tangent to the ellipse $b^2x^2 + a^2y^2 = a^2b^2$. Find the point of tangency if the triangle is to have minimum area.

CHAPTER HIGHLIGHTS

4.1 The Mean-Value Theorem

mean-value theorem (p. 183)   Rolle's theorem (p. 185)

4.2 Increasing and Decreasing Functions

f increases on an interval I (p. 188) f decreases on an interval I (p. 188)

One can find the intervals on which a differentiable function increases or decreases or is constant by examining the sign of the derivative.

If two functions have the same derivative on an interval, then they differ by a constant on that interval. (p. 193)

4.3 Local Extreme Values

local extreme: local maximum, local minimum (p. 195)
critical point (p. 196)   first-derivative test (p. 198)
second-derivative test (p. 200)

If f has a local extreme at c, then either $f'(c) = 0$ or $f'(c)$ does not exist; the converse is false.

The first-derivative test requires that we examine the sign of the first derivative on both sides of the critical point; the test is conclusive provided that the function is continuous at the critical point. The second-derivative test requires that we examine the sign of the second derivative at the critical point itself; the test is inconclusive if the second derivative is 0 at the critical point.

4.4 Endpoint and Absolute Extreme Values

endpoint maximum and endpoint minimum (p. 203)
absolute maximum and absolute minimum (p. 203)
as $x \to \pm\infty$, $f(x) \to \pm\infty$ (p. 204)

4.5 Some Max-Min Problems

The *key step* is to express the quantity to be maximized or minimized as a function of one variable.

4.6 Concavity and Points of Inflection

> concave up and concave down (p. 216)
> point of inflection (p. 216)

If $(c, f(c))$ is a point of inflection, then either $f''(c) = 0$ or $f''(c)$ does not exist; the converse is false.

4.7 Some Curve Sketching

A systematic procedure for sketching a graph is outlined on p. 220.

4.8 Vertical and Horizontal Asymptotes; Vertical Tangents and Cusps

> as $x \to c$, $f(x) \to \pm\infty$ (p. 224)   as $x \to \pm\infty$, $f(x) \to l$ (p. 229)
> vertical asymptote (p. 224)   horizontal asymptote (p. 225)
> vertical tangent (p. 227)   vertical cusp (p. 227)

4.9 Additional Exercises on Graphing

4.10 Additional Max-Min Problems

INTEGRATION

5.1 AN AREA PROBLEM; A SPEED-DISTANCE PROBLEM

An Area Problem

In Figure 5.1.1 you can see a region Ω bounded above by the graph of a continuous nonnegative function f, bounded below by the x-axis, bounded to the left by $x = a$, and bounded to the right by $x = b$. The question before us is this: what number, if any, should be called the area of Ω?

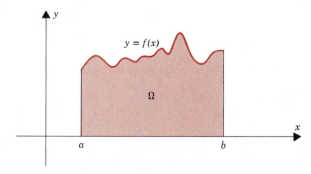

Figure 5.1.1

To begin to answer this question, we split up the interval $[a, b]$ into a finite number

of nonoverlapping intervals

$$[x_0, x_1], [x_1, x_2], \ldots, [x_{n-1}, x_n] \quad \text{with} \quad a = x_0 < x_1 < \cdots < x_n = b.$$

This breaks up the region Ω into n subregions:

$$\Omega_1, \Omega_2, \ldots, \Omega_n. \qquad \text{(Figure 5.1.2)}$$

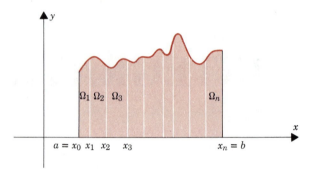

Figure 5.1.2

We can estimate the total area of Ω by estimating the area of each subregion Ω_i and adding up the results. Let's denote by M_i the maximum value of f on $[x_{i-1}, x_i]$ and by m_i the minimum value. Consider now the rectangles r_i and R_i of Figure 5.1.3.

Figure 5.1.3

Since

$$r_i \subseteq \Omega_i \subseteq R_i,$$

we must have

$$\text{area of } r_i \leq \text{area of } \Omega_i \leq \text{area of } R_i.$$

Since the area of a rectangle is the length times the width,

$$m_i(x_i - x_{i-1}) \leq \text{area of } \Omega_i \leq M_i(x_i - x_{i-1}).$$

Setting $x_i - x_{i-1} = \Delta x_i$ we have

$$m_i \, \Delta x_i \leq \text{area of } \Omega_i \leq M_i \, \Delta x_i.$$

This inequality holds for $i = 1, i = 2, \ldots, i = n$. Adding up these inequalities we get on the one hand

(5.1.1) $m_1 \Delta x_1 + m_2 \Delta x_2 + \cdots + m_n \Delta x_n \leq$ area of Ω

and on the other hand

(5.1.2)   area of $\Omega \leq M_1 \Delta x_1 + M_2 \Delta x_2 + \cdots + M_n \Delta x_n.$

A sum of the form

$$m_1 \Delta x_1 + m_2 \Delta x_2 + \cdots + m_n \Delta x_n \qquad \text{(Figure 5.1.4)}$$

is called a *lower sum for f.* A sum of the form

$$M_1 \Delta x_1 + M_2 \Delta x_2 + \cdots + M_n \Delta x_n \qquad \text{(Figure 5.1.5)}$$

is called an *upper sum for f.*

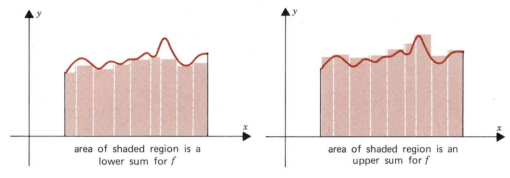

area of shaded region is a
lower sum for f

Figure 5.1.4

area of shaded region is an
upper sum for f

Figure 5.1.5

Inequalities (5.1.1) and (5.1.2) together tell us that for a number to be a candidate for the title "area of Ω" it must be greater than or equal to every lower sum for f, and it must be less than or equal to every upper sum. By an argument that we omit here, it can be proved that with f continuous on $[a, b]$ there is one and only one such number. This number we call *the area of* Ω.

Later we will return to the subject of area. At this point we turn to a speed-distance problem. As you will see, this new problem can be solved by the same technique that we just applied to the area problem.

A Speed-Distance Problem

If an object moves at a constant speed for a given period of time, then the total distance traveled is given by the familiar formula

$$\text{distance} = \text{speed} \times \text{time.}$$

Suppose now that during the course of the motion the speed does not remain constant but instead varies continuously. How can the total distance traveled be computed then?

To answer this question, we suppose that the motion begins at time a, ends at time b, and that during the time interval $[a, b]$ the speed varies continuously.

As in the case of the area problem we begin by breaking up the interval $[a, b]$ into a finite number of nonoverlapping intervals

$$[t_0, t_1], [t_1, t_2], \ldots, [t_{n-1}, t_n] \quad \text{with} \quad a = t_0 < t_1 < \cdots < t_n = b.$$

On each subinterval $[t_{i-1}, t_i]$ the object attains a certain maximum speed M_i and a certain minimum speed m_i.   (How do we know this?)   If throughout the time interval $[t_{i-1}, t_i]$ the object were to move constantly at its minimum speed, m_i, then it would cover a distance of $m_i \Delta t_i$ units.   If instead it were to move constantly at its maximum speed, M_i, then it would cover a distance of $M_i \Delta t_i$ units.   As it is, the actual distance traveled, call it s_i, must lie somewhere in between; namely, we must have

$$m_i \Delta t_i \leq s_i \leq M_i \Delta t_i.$$

The total distance traveled during the time interval $[a, b]$, call it s, must be the sum of the distances traveled during the subintervals $[t_{i-1}, t_i]$.   In other words we must have

$$s = s_1 + s_2 + \cdots + s_n.$$

Since

$$m_1 \Delta t_1 \leq s_1 \leq M_1 \Delta t_1$$
$$m_2 \Delta t_2 \leq s_2 \leq M_2 \Delta t_2$$
$$\vdots$$
$$m_n \Delta t_n \leq s_n \leq M_n \Delta t_n,$$

it follows by the addition of these inequalities that

$$m_1 \Delta t_1 + m_2 \Delta t_2 + \cdots + m_n \Delta t_n \leq s \leq M_1 \Delta t_1 + M_2 \Delta t_2 + \cdots + M_n \Delta t_n.$$

A sum of the form

$$m_1 \Delta t_1 + m_2 \Delta t_2 + \cdots + m_n \Delta t_n$$

is called a *lower sum* for the speed function.   A sum of the form

$$M_1 \Delta t_1 + M_2 \Delta t_2 + \cdots + M_n \Delta t_n$$

is called an *upper sum* for the speed function.   The inequality we just obtained for s tells us that s must be greater than or equal to every lower sum for the speed function, and it must be less than or equal to every upper sum.   As in the case of the area problem, it turns out that there is one and only one such number, and this is the total distance traveled.

5.2   THE DEFINITE INTEGRAL OF A CONTINUOUS FUNCTION

The process we used to solve the two problems of Section 5.1 is called *integration*, and the end results of this process are called *definite integrals*.   Our purpose here is to establish these notions more precisely.

(5.2.1) By a *partition* of the closed interval $[a, b]$ we mean a finite subset of $[a, b]$ which contains the points a and b.

We index the elements of a partition according to their natural order. Thus, if we write

$$P = \{x_0, x_1, \ldots, x_n\} \text{ is a partition of } [a, b],$$

you can conclude that

$$a = x_0 < x_1 < \cdots < x_n = b.$$

Example 1. The sets

$$\{0, 1\}, \quad \{0, \tfrac{1}{2}, 1\}, \quad \{0, \tfrac{1}{4}, \tfrac{1}{2}, 1\}, \quad \{0, \tfrac{1}{4}, \tfrac{1}{3}, \tfrac{1}{2}, \tfrac{5}{8}, 1\}$$

are all partitions of the interval $[0, 1]$. □

If $P = \{x_0, x_1, \ldots, x_n\}$ is a partition of $[a, b]$, then P breaks up $[a, b]$ into a finite number of nonoverlapping intervals

$$[x_0, x_1], [x_1, x_2], \ldots, [x_{n-1}, x_n]$$

of lengths $\Delta x_1, \Delta x_2, \ldots, \Delta x_n$ respectively.

Suppose now that f is continuous on $[a, b]$. Then on each interval $[x_{i-1}, x_i]$ the function f takes on a maximum value, M_i, and a minimum value, m_i.

(5.2.2) The number

$$U_f(P) = M_1 \Delta x_1 + M_2 \Delta x_2 + \cdots + M_n \Delta x_n$$

is called *the P upper sum for f*, and the number

$$L_f(P) = m_1 \Delta x_1 + m_2 \Delta x_2 + \cdots + m_n \Delta x_n$$

is called *the P lower sum for f*.

Example 2. The squaring function

$$f(x) = x^2$$

is continuous on $[0, 1]$. The partition $P = \{0, \tfrac{1}{4}, \tfrac{1}{2}, 1\}$ breaks up $[0, 1]$ into three subintervals:

$$[x_0, x_1] = [0, \tfrac{1}{4}], \qquad [x_1, x_2] = [\tfrac{1}{4}, \tfrac{1}{2}], \qquad [x_2, x_3] = [\tfrac{1}{2}, 1]$$

of lengths

$$\Delta x_1 = \tfrac{1}{4} - 0 = \tfrac{1}{4}, \qquad \Delta x_2 = \tfrac{1}{2} - \tfrac{1}{4} = \tfrac{1}{4}, \qquad \Delta x_3 = 1 - \tfrac{1}{2} = \tfrac{1}{2}$$

respectively. The maximum values taken on by f on these intervals are

$$M_1 = f(\tfrac{1}{4}) = \tfrac{1}{16}, \qquad M_2 = f(\tfrac{1}{2}) = \tfrac{1}{4}, \qquad M_3 = f(1) = 1.$$

The minimum values are

$$m_1 = f(0) = 0, \qquad m_2 = f(\tfrac{1}{4}) = \tfrac{1}{16}, \qquad m_3 = f(\tfrac{1}{2}) = \tfrac{1}{4}.$$

Thus

$$U_f(P) = M_1\Delta x_1 + M_2\Delta x_2 + M_3\Delta x_3 = \tfrac{1}{16}(\tfrac{1}{4}) + \tfrac{1}{4}(\tfrac{1}{4}) + 1(\tfrac{1}{2}) = \tfrac{37}{64}$$

and

$$L_f(P) = m_1\Delta x_1 + m_2\Delta x_2 + m_3\Delta x_3 = 0(\tfrac{1}{4}) + \tfrac{1}{16}(\tfrac{1}{4}) + \tfrac{1}{4}(\tfrac{1}{2}) = \tfrac{9}{64}. \quad \square$$

Example 3.   This time we take the function

$$f(x) = -x - 1$$

on the closed interval $[-1, 0]$.   The partition $P = \{-1, -\tfrac{1}{4}, 0\}$ breaks up $[-1, 0]$ into two intervals: $[-1, -\tfrac{1}{4}]$ and $[-\tfrac{1}{4}, 0]$.   As you can check,

$$U_f(P) = (0)(\tfrac{3}{4}) + (-\tfrac{3}{4})(\tfrac{1}{4}) = -\tfrac{3}{16} \quad \text{and} \quad L_f(P) = (-\tfrac{3}{4})(\tfrac{3}{4}) + (-1)(\tfrac{1}{4}) = -\tfrac{13}{16}. \quad \square$$

By an argument that we omit here (it appears in Appendix B.4) it can be proved that, with f continuous on $[a, b]$, there is one and only one number I that satisfies the inequality

$$L_f(P) \leq I \leq U_f(P) \qquad \text{for } all \text{ partitions } P \text{ of } [a, b].$$

This is the number we want.

DEFINITION 5.2.3   THE DEFINITE INTEGRAL

The unique number I that satisfies the inequality

$$L_f(P) \leq I \leq U_f(P) \qquad \text{for all partitions } P \text{ of } [a, b]$$

is called *the definite integral* (or more simply *the integral*) *of f from a to b* and is denoted by

$$\int_a^b f(x)\, dx.\dagger$$

The symbol $\int$ dates back to Leibniz and is called an *integral sign*.   It is really an elongated *S*—as in *Sum*.   The numbers a and b are called *the limits of integration*, and we'll speak of *integrating* a function from a to b.$\dagger\dagger$ The function being integrated is called *the integrand*.

$\dagger$ This is not the only notation.   Some mathematicians omit the dx and write simply

$$\int_a^b f.$$

We will keep the dx.   As we go on, you will see that it does serve a useful purpose.
$\dagger\dagger$ The word "limit" here has no connection with the limits discussed in Chapter 2.

In the expression

$$\int_a^b f(x)\, dx$$

the letter x is a "dummy variable"; in other words, it may be replaced by any other letter not already engaged. Thus, for example, there is no difference between

$$\int_a^b f(x)\, dx, \qquad \int_a^b f(t)\, dt, \qquad \text{and} \qquad \int_a^b f(z)\, dz.$$

All of these denote the definite integral of f from a to b.

Section 5.1 gives two immediate applications of the definite integral:

I. If f is nonnegative on $[a, b]$, then

$$A = \int_a^b f(x)\, dx$$

gives the area below the graph of f. See Figure 5.1.1.

II. If $|v(t)|$ is the speed of an object at time t, then

$$s = \int_a^b |v(t)|\, dt$$

gives the distance traveled from time a to time b.

We will come back to these ideas later. Right now, some simple computations.

Example 4. If $f(x) = \alpha$ for all x in $[a, b]$, then

$$\int_a^b f(x)\, dx = \alpha(b - a).$$

To see this, take $P = \{x_0, x_1, \ldots, x_n\}$ as an arbitrary partition of $[a, b]$. Since f is constantly α on $[a, b]$, it is constantly α on each subinterval $[x_{i-1}, x_i]$. Thus, M_i and m_i are both α. It follows that

$$U_f(P) = \alpha \Delta x_1 + \alpha \Delta x_2 + \cdots + \alpha \Delta x_n$$
$$= \alpha(\Delta x_1 + \Delta x_2 + \cdots + \Delta x_n) = \alpha(b - a)$$

and

$$L_f(P) = \alpha \Delta x_1 + \alpha \Delta x_2 + \cdots + \alpha \Delta x_n$$
$$= \alpha(\Delta x_1 + \Delta x_2 + \cdots + \Delta x_n) = \alpha(b - a).$$

Obviously then

$$L_f(P) \le \alpha(b - a) \le U_f(P).$$

Since this inequality holds for all partitions P of $[a, b]$, we can conclude that

$$\int_a^b f(x)\, dx = \alpha(b - a). \quad \square$$

This last result can be written

(5.2.4)

$$\int_a^b \alpha \, dx = \alpha(b - a).$$

Thus for example

$$\int_{-1}^1 4 \, dx = 4(2) = 8 \quad \text{and} \quad \int_4^{10} -2 \, dx = -2(6) = -12.$$

If $\alpha > 0$, the region below the graph is a rectangle of height α erected on the interval $[a, b]$. (Figure 5.2.1) The integral gives the area of this rectangle.

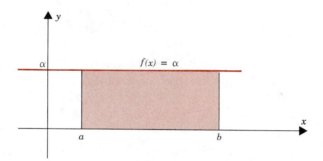

Figure 5.2.1

Example 5

(5.2.5)

$$\int_a^b x \, dx = \tfrac{1}{2}(b^2 - a^2).$$

To see this, take $P = \{x_0, x_1, \ldots, x_n\}$ as an arbitrary partition of $[a, b]$. On each subinterval $[x_{i-1}, x_i]$ the function $f(x) = x$ has a maximum $M_i = x_i$ and a minimum $m_i = x_{i-1}$. It follows that

$$U_f(P) = x_1 \Delta x_1 + x_2 \Delta x_2 + \cdots + x_n \Delta x_n$$

and

$$L_f(P) = x_0 \Delta x_1 + x_1 \Delta x_2 + \cdots + x_{n-1} \Delta x_n.$$

For each index i

$$x_{i-1} \le \tfrac{1}{2}(x_i + x_{i-1}) \le x_i. \qquad \text{(explain)}$$

Multiplication by $\Delta x_i = x_i - x_{i-1}$ gives

$$x_{i-1}\Delta x_i \le \tfrac{1}{2}(x_i^2 - x_{i-1}^2) \le x_i \Delta x_i.$$

Summing from $i = 1$ to $i = n$, we find that

$$L_f(P) \le \tfrac{1}{2}(x_1^2 - x_0^2) + \tfrac{1}{2}(x_2^2 - x_1^2) + \cdots + \tfrac{1}{2}(x_n^2 - x_{n-1}^2) \le U_f(P).$$

The middle collapses to

$$\tfrac{1}{2}(x_n^2 - x_0^2) = \tfrac{1}{2}(b^2 - a^2).$$

Consequently we have

$$L_f(P) \le \tfrac{1}{2}(b^2 - a^2) \le U_f(P).$$

Since P was chosen arbitrarily, we can conclude that this inequality holds for all partitions P of $[a, b]$. It follows then that

$$\int_a^b x \, dx = \tfrac{1}{2}(b^2 - a^2). \quad \square$$

If the interval $[a, b]$ lies to the right of the origin, then the region below the graph of

$$f(x) = x, \qquad x \in [a, b]$$

is the trapezoid of Figure 5.2.2. The integral

$$\int_a^b x \, dx$$

gives the area of this trapezoid: $A = (b - a)[\tfrac{1}{2}(a + b)] = \tfrac{1}{2}(b^2 - a^2). \quad \square$

Figure 5.2.2

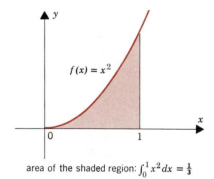

area of the shaded region: $\int_0^1 x^2 \, dx = \tfrac{1}{3}$

Figure 5.2.3

Example 6

$$\int_0^1 x^2 \, dx = \tfrac{1}{3}. \qquad\qquad \text{(Figure 5.2.3)}$$

This time take $P = \{x_0, x_1, \ldots, x_n\}$ as a partition of $[0, 1]$. On each subinterval $[x_{i-1}, x_i]$ the function $f(x) = x^2$ has a maximum $M_i = x_i^2$ and a minimum $m_i = x_{i-1}^2$. It follows that

$$U_f(P) = x_1^2 \Delta x_1 + \cdots + x_n^2 \Delta x_n$$

and

$$L_f(P) = x_0^2 \Delta x_1 + \cdots + x_{n-1}^2 \Delta x_n.$$

For each index i,

$$x_{i-1}^2 \le \tfrac{1}{3}(x_{i-1}^2 + x_{i-1}x_i + x_i^2) \le x_i^2. \qquad \text{(check this out)}$$

We now multiply this inequality by $\Delta x_i = x_i - x_{i-1}$. The middle term reduces to

$$\tfrac{1}{3}(x_i^3 - x_{i-1}^3), \qquad \text{(check this out)}$$

showing us that

$$x_{i-1}^2 \Delta x_i \le \tfrac{1}{3}(x_i^3 - x_{i-1}^3) \le x_i^2 \Delta x_i.$$

The sum of the terms on the left is $L_f(P)$. The sum of all the middle terms collapses to $\tfrac{1}{3}$:

$$\tfrac{1}{3}(x_1^3 - x_0^3 + x_2^3 - x_1^3 + \cdots + x_n^3 - x_{n-1}^3) = \tfrac{1}{3}(x_n^3 - x_0^3) = \tfrac{1}{3}(1^3 - 0^3) = \tfrac{1}{3}.$$

The sum of the terms on the right is $U_f(P)$. Clearly then

$$L_f(P) \le \tfrac{1}{3} \le U_f(P).$$

Since P was chosen arbitrarily, we can conclude that this inequality holds for all partitions P of $[0, 1]$. It follows therefore that

$$\int_0^1 x^2 \, dx = \tfrac{1}{3}. \quad \square$$

EXERCISES 5.2

Find $L_f(P)$ and $U_f(P)$.

1. $f(x) = 2x$, $\quad x \in [0, 1]$; $\quad P = \{0, \tfrac{1}{4}, \tfrac{1}{2}, 1\}$.
2. $f(x) = 1 - x$, $\quad x \in [0, 2]$; $\quad P = \{0, \tfrac{1}{3}, \tfrac{3}{4}, 1, 2\}$.
3. $f(x) = x^2$, $\quad x \in [-1, 0]$; $\quad P = \{-1, -\tfrac{1}{2}, -\tfrac{1}{4}, 0\}$.
4. $f(x) = 1 - x^2$, $\quad x \in [0, 1]$; $\quad P = \{0, \tfrac{1}{4}, \tfrac{1}{2}, 1\}$.
5. $f(x) = 1 + x^3$, $\quad x \in [0, 1]$; $\quad P = \{0, \tfrac{1}{2}, 1\}$.
6. $f(x) = \sqrt{x}$, $\quad x \in [0, 1]$; $\quad P = \{0, \tfrac{1}{25}, \tfrac{4}{25}, \tfrac{9}{25}, \tfrac{16}{25}, 1\}$.
7. $f(x) = |x|$, $\quad x \in [-1, 1]$; $\quad P = \{-1, -\tfrac{1}{2}, 0, \tfrac{1}{4}, 1\}$.
8. $f(x) = |x|$, $\quad x \in [-1, 1]$; $\quad P = \{-1, -\tfrac{1}{2}, -\tfrac{1}{4}, \tfrac{1}{2}, 1\}$.
9. $f(x) = x^2$, $\quad x \in [-1, 1]$; $\quad P = \{-1, -\tfrac{1}{4}, \tfrac{1}{4}, \tfrac{1}{2}, 1\}$.
10. $f(x) = x^2$, $\quad x \in [-1, 1]$; $\quad P = \{-1, -\tfrac{3}{4}, -\tfrac{1}{4}, \tfrac{1}{4}, \tfrac{1}{2}, 1\}$.
11. $f(x) = \sin x$, $\quad x \in [0, \pi]$; $\quad P = \{0, \tfrac{1}{6}\pi, \tfrac{1}{2}\pi, \pi\}$.
12. $f(x) = \cos x$, $\quad x \in [0, \pi]$; $\quad P = \{0, \tfrac{1}{3}\pi, \tfrac{1}{2}\pi, \pi\}$.
13. Explain why each of the following three statements is false. Take P as a partition of $[-1, 1]$.
    (a) $L_f(P) = 3$ and $U_f(P) = 2$.

    (b) $L_f(P) = 3$, $U_f(P) = 6$, and $\displaystyle\int_{-1}^1 f(x) \, dx = 2$.

    (c) $L_f(P) = 3$, $U_f(P) = 6$, and $\displaystyle\int_{-1}^1 f(x) \, dx = 10$.

14. (a) Given that $P = \{x_0, x_1, \ldots, x_n\}$ is an arbitrary partition of $[a, b]$, find $L_f(P)$ and $U_f(P)$ for $f(x) = x + 3$.
    (b) Use your answers to part (a) to evaluate

$$\int_a^b (x + 3)\, dx.$$

15. Exercise 14 with $f(x) = -3x$.

16. Exercise 14 with $f(x) = 1 + 2x$.

17. Evaluate

$$\int_0^1 x^3\, dx$$

by the methods of this section.   HINT: $b^4 - a^4 = (b^3 + b^2a + ba^2 + a^3)(b - a)$.

18. Consider the Dirichlet function f on $[0, 1]$:

$$f(x) = \begin{cases} 1, & x \text{ rational} \\ 0, & x \text{ irrational} \end{cases}.$$

    (a) Show that $L_f(P) \leq \tfrac{1}{2} \leq U_f(P)$ for all partitions P of $[0, 1]$.
    (b) Explain why we can *not* conclude that

$$\int_0^1 f(x)\, dx = \tfrac{1}{2}.$$

19. The definition of integral that we have given (Definition 5.2.3) can be applied to functions with a finite number of discontinuities.
    Suppose that f is continuous on $[a, b]$. If g is defined on $[a, b]$ and differs from f only at a finite number of points, then g can be integrated on $[a, b]$ and

$$\int_a^b g(x)\, dx = \int_a^b f(x)\, dx.$$

For example, the function

$$g(x) = \begin{cases} 2, & x \in [0, 3) \cup (3, 4) \\ 7, & x = 3 \end{cases}$$

differs from the constant function

$$f(x) = 2, \qquad x \in [0, 4]$$

only at $x = 3$. Clearly

$$\int_0^4 f(x)\, dx = 8.$$

Show that

$$\int_0^4 g(x)\, dx = 8$$

by showing that 8 is the unique number I that satisfies the inequality

$$L_g(P) \leq I \leq U_g(P) \qquad \text{for all partitions } P \text{ of } [0, 4].$$

5.3   THE FUNCTION $F(x) = \int_a^x f(t)\, dt$

The evaluation of a definite integral

$$\int_a^b f(x)\, dx$$

directly from its definition as the unique number I satisfying the inequality $L_f(P) \le I \le U_f(P)$ for all partitions P of $[a, b]$ is usually a laborious and difficult process.   Try, for example, to evaluate

$$\int_2^5 \left(x^3 + x^{5/2} - \frac{2x}{1 - x^2} \right) dx \quad \text{or} \quad \int_{-1/2}^{1/4} \frac{x}{1 - x^2}\, dx$$

by this method.   Theorem 5.4.2, called *the fundamental theorem of integral calculus,* gives us another way to evaluate such integrals.   This other way depends on a connection between integration and differentiation described in Theorem 5.3.5. The main purpose of this section is to prove Theorem 5.3.5.   Along the way we will pick up some information that is of interest in itself.

THEOREM 5.3.1

Let P and Q be partitions of the interval $[a, b]$.   If $P \subseteq Q$, then

$$L_f(P) \le L_f(Q) \quad \text{and} \quad U_f(Q) \le U_f(P).$$

This result is easy to see.   By adding points to a partition we tend to make the subintervals $[x_{i-1}, x_i]$ smaller.   This tends to make the minima, m_i, larger and the maxima, M_i, smaller.   Thus the lower sums are made bigger, and the upper sums are made smaller.   The idea is illustrated (for a positive function) in Figures 5.3.1 and 5.3.2.

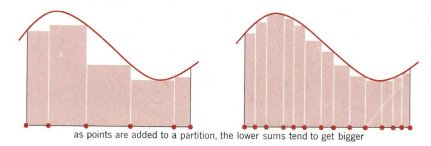

as points are added to a partition, the lower sums tend to get bigger

Figure 5.3.1

as points are added to a partition, the upper sums tend to get smaller

Figure 5.3.2

The next theorem says that the integral is *additive* on intervals.

THEOREM 5.3.2

If f is continuous on $[a, b]$ and $a < c < b$, then

$$\int_a^c f(t)\, dt + \int_c^b f(t)\, dt = \int_a^b f(t)\, dt.$$

For nonnegative functions f this theorem is easily understood in terms of area. The area of part I in Figure 5.3.3 is given by

$$\int_a^c f(t)\, dt;$$

the area of part II by

$$\int_c^b f(t)\, dt;$$

the area of the entire region by

$$\int_a^b f(t)\, dt.$$

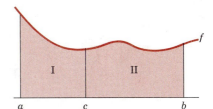

Figure 5.3.3

The theorem says that

---

the area of part I + the area of part II = the area of the entire region.

Theorem 5.3.2 can also be viewed in terms of speed and distance. With $f(t) \geq 0$, we can interpret $f(t)$ as the speed of an object at time t. With this interpretation,

$$\int_a^c f(t)\, dt = \text{distance traveled from time } a \text{ to time } c,$$

$$\int_c^b f(t)\, dt = \text{distance traveled from time } c \text{ to time } b,$$

$$\int_a^b f(t)\, dt = \text{distance traveled from time } a \text{ to time } b.$$

It is not surprising that the sum of the first two distances should equal the third.

The fact that the additivity theorem is so easy to understand does not relieve us of the necessity to prove it. Here is a proof.

Proof of Theorem 5.3.2.   To prove the theorem we need only show that for each partition P of $[a, b]$

$$L_f(P) \le \int_a^c f(t)\, dt + \int_c^b f(t)\, dt \le U_f(P). \qquad \text{(why?)}$$

We begin with an arbitrary partition of $[a, b]$:

$$P = \{x_0, x_1, \ldots, x_n\}.$$

Since the partition $Q = P \cup \{c\}$ contains P, we know from Theorem 5.3.1 that

(1) $$L_f(P) \le L_f(Q) \qquad \text{and} \qquad U_f(Q) \le U_f(P).$$

The sets

$$Q_1 = Q \cap [a, c] \qquad \text{and} \qquad Q_2 = Q \cap [c, b]$$

are partitions of $[a, c]$ and $[c, b]$, respectively.   Moreover

$$L_f(Q_1) + L_f(Q_2) = L_f(Q) \qquad \text{and} \qquad U_f(Q_1) + U_f(Q_2) = U_f(Q).$$

Since

$$L_f(Q_1) \le \int_a^c f(t)\, dt \le U_f(Q_1) \qquad \text{and} \qquad L_f(Q_2) \le \int_c^b f(t)\, dt \le U_f(Q_2),$$

we have

$$L_f(Q_1) + L_f(Q_2) \le \int_a^c f(t)\, dt + \int_c^b f(t)\, dt \le U_f(Q_1) + U_f(Q_2).$$

It follows that

$$L_f(Q) \le \int_a^c f(t)\, dt + \int_c^b f(t)\, dt \le U_f(Q)$$

and thus by (1) that

$$L_f(P) \le \int_a^c f(t)\, dt + \int_c^b f(t)\, dt \le U_f(P). \quad \square$$

Until now we have integrated only from left to right: from a number a to a number b greater than a.   We integrate in the other direction by defining

(5.3.3)
$$\int_b^a f(t)\, dt = -\int_a^b f(t)\, dt.$$

The integral from any number to itself is defined to be zero:

(5.3.4)
$$\int_c^c f(t)\, dt = 0$$

for all real c. With these additional conventions, the additivity condition

$$\int_a^c f(t)\, dt + \int_c^b f(t)\, dt = \int_a^b f(t)\, dt$$

holds for all choices of a, b, c from an interval on which f is continuous, no matter what the order of a, b, c happens to be. We have left the proof of this to you as an exercise.

We are now ready to state the all-important connection that exists between integration and differentiation. Our first step is to point out that, if f is continuous on $[a, b]$, then for each x in $[a, b]$, the integral

$$\int_a^x f(t)\, dt$$

is a number, and consequently we can define a function F on $[a, b]$ by setting

$$F(x) = \int_a^x f(t)\, dt.$$

THEOREM 5.3.5

If f is continuous on $[a, b]$, the function F defined on $[a, b]$ by setting

$$F(x) = \int_a^x f(t)\, dt$$

is continuous on $[a, b]$, differentiable on (a, b), and has derivative

$$F'(x) = f(x) \qquad \text{for all } x \text{ in } (a, b).$$

Proof. We begin with x in the half-open interval $[a, b)$ and show that

$$\lim_{h \to 0+} \frac{F(x + h) - F(x)}{h} = f(x).$$

(For a pictorial outline of the proof for f nonnegative see Figure 5.3.4.)

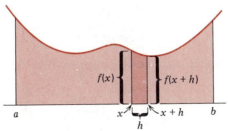

$F(x + h) = $ area from a to $x + h$ and $F(x) = $ area from a to x.

Therefore

$$F(x + h) - F(x) = \text{area from } x \text{ to } x + h,$$

$$\frac{F(x + h) - F(x)}{h} = \frac{\text{area from } x \text{ to } x + h}{h} \cong f(x) \text{ if } h \text{ is small}$$

Figure 5.3.4

If $x < x + h \le b$, then

$$F(x + h) - F(x) = \int_a^{x+h} f(t)\, dt - \int_a^x f(t)\, dt.$$

It follows that

$$F(x + h) - F(x) = \int_x^{x+h} f(t)\, dt. \qquad \text{(justify this step)}$$

Now set

$$M_h = \text{maximum value of } f \text{ on } [x, x + h]$$

and

$$m_h = \text{minimum value of } f \text{ on } [x, x + h].$$

Since

$$M_h[(x + h) - x] = M_h \cdot h$$

is an upper sum for f on $[x, x + h]$ and

$$m_h[(x + h) - x] = m_h \cdot h$$

is a lower sum for f on $[x, x + h]$,

$$m_h \cdot h \le \int_x^{x+h} f(t)\, dt \le M_h \cdot h$$

and, since $h > 0$,

$$m_h \le \frac{F(x + h) - F(x)}{h} \le M_h.$$

Since f is continuous on $[x, x + h]$,

(1)
$$\lim_{h \to 0^+} m_h = f(x) = \lim_{h \to 0^+} M_h$$

and thus

(2)
$$\lim_{h \to 0^+} \frac{F(x + h) - F(x)}{h} = f(x).$$

In a similar manner we can prove that, for x in the half-open interval $(a, b]$,

(3)
$$\lim_{h \to 0^-} \frac{F(x + h) - F(x)}{h} = f(x).$$

For x in the open interval (a, b), both (2) and (3) hold, and we have

$$F'(x) = \lim_{h \to 0} \frac{F(x + h) - F(x)}{h} = f(x).$$

This proves that F is differentiable on (a, b) and has derivative $F'(x) = f(x)$.

All that remains to be shown is that F is continuous from the right at a and continuous from the left at b.   Relation (2) at $x = a$ gives

$$\lim_{h \to 0^+} \frac{F(a + h) - F(a)}{h} = f(a).$$

It follows that

$$\lim_{h \to 0^+} [F(a + h) - F(a)] = 0 \qquad \text{(explain)}$$

and

$$\lim_{h \to 0^+} F(a + h) = F(a).$$

This shows that F is continuous from the right at a.   That F is continuous from the left at b can be shown by applying relation (3) at $x = b$.   ☐

EXERCISES 5.3

1. Given that

$$\int_0^1 f(x)\, dx = 6, \qquad \int_0^2 f(x)\, dx = 4, \qquad \int_2^5 f(x)\, dx = 1,$$

find each of the following:

(a) $\displaystyle\int_0^5 f(x)\, dx.$ (b) $\displaystyle\int_1^2 f(x)\, dx.$ (c) $\displaystyle\int_1^5 f(x)\, dx.$

(d) $\displaystyle\int_0^0 f(x)\, dx.$ (e) $\displaystyle\int_2^0 f(x)\, dx.$ (f) $\displaystyle\int_5^1 f(x)\, dx.$

2. Given that

$$\int_1^4 f(x)\, dx = 5, \qquad \int_3^4 f(x)\, dx = 7, \qquad \int_1^8 f(x)\, dx = 11,$$

find each of the following:

(a) $\displaystyle\int_4^8 f(x)\, dx.$ (b) $\displaystyle\int_4^3 f(x)\, dx.$ (c) $\displaystyle\int_1^3 f(x)\, dx.$

(d) $\displaystyle\int_3^8 f(x)\, dx.$ (e) $\displaystyle\int_8^4 f(x)\, dx.$ (f) $\displaystyle\int_4^4 f(x)\, dx.$

3. Use upper and lower sums to show that

$$0.5 < \int_1^2 \frac{dx}{x} < 1.$$

4. Use upper and lower sums to show that

$$0.6 < \int_0^1 \frac{dx}{1 + x^2} < 1.$$

5. For $x > -1$ set $F(x) = \displaystyle\int_0^x t \sqrt{t + 1}\, dt.$

(a) Find $F(0)$.   (b) Find $F'(x)$.   (c) Find $F'(2)$.
(d) Express $F(2)$ as an integral of $t \sqrt{t + 1}$.   (e) Express $-F(x)$ as an integral of $t \sqrt{t + 1}$.

6. For $x > 0$ set $F(x) = \int_1^x \dfrac{dt}{t}$.

   (a) Find $F(1)$.   (b) Find $F'(x)$.   (c) Find $F'(1)$.
   (d) Express $F(2)$ as an integral of $1/t$.   (e) Express $-F(x)$ as an integral of $1/t$.

For the given function F compute the following:

   (a) $F'(-1)$.   (b) $F'(0)$.   (c) $F'(\tfrac{1}{2})$.   (d) $F''(x)$.

7. $F(x) = \int_0^x \dfrac{dt}{t^2 + 9}$.   **8.** $F(x) = \int_x^0 \sqrt{t^2 + 1}\ dt$.   **9.** $F(x) = \int_x^1 t\sqrt{t^2 + 1}\ dt$.

10. $F(x) = \int_1^x \sin \pi t\ dt$.   **11.** $F(x) = \int_1^x \cos \pi t\ dt$.   **12.** $F(x) = \int_2^x (t + 1)^3\ dt$.

13. Explain why each of the following statements must be false.
   (a) $U_f(P_1) = 4$ for the partition $P_1 = \{0, 1, \tfrac{3}{2}, 2\}$ and
       $U_f(P_2) = 5$ for the partition $P_2 = \{0, \tfrac{1}{4}, 1, \tfrac{3}{2}, 2\}$.
   (b) $L_f(P_1) = 5$ for the partition $P_1 = \{0, 1, \tfrac{3}{2}, 2\}$ and
       $L_f(P_2) = 4$ for the partition $P_2 = \{0, \tfrac{1}{4}, 1, \tfrac{3}{2}, 2\}$.

14. Show that, if f is continuous on an interval I, then

$$\int_a^c f(t)\ dt + \int_c^b f(t)\ dt = \int_a^b f(t)\ dt$$

for *every* choice of a, b, c from I.   HINT: Assume $a < b$ and consider the four cases: $c = a, c = b, c < a, b < c$.   Then consider what happens if $a > b$ or $a = b$.

15. (*The First Mean-Value Theorem for Integrals.*)   Show that, if f is continuous on $[a, b]$, then there is at least one number c in (a, b) for which

$$\int_a^b f(x)\ dx = f(c)(b - a).$$

HINT: Apply the mean-value theorem to the function

$$F(x) = \int_a^x f(t)\ dt \quad \text{on } [a, b].$$

16. Show the validity of equation (1) in the proof of Theorem 5.3.5.

17. Extend Theorem 5.3.5 by showing that, if f is continuous on $[a, b]$ and c is *any* number in $[a, b]$, then the function

$$F(x) = \int_c^x f(t)\ dt$$

is continuous on $[a, b]$, differentiable on (a, b), and satisfies

$$F'(x) = f(x) \qquad \text{for all } x \text{ in } (a, b).$$

HINT: $\int_c^x f(t)\ dt = \int_c^a f(t)\ dt + \int_a^x f(t)\ dt$.

18. Complete the proof of Theorem 5.3.5 by showing that

$$\lim_{h \to 0^-} \frac{F(x + h) - F(x)}{h} = f(x) \qquad \text{for all } x \text{ in } (a, b].$$

5.4 THE FUNDAMENTAL THEOREM OF INTEGRAL CALCULUS

DEFINITION 5.4.1 ANTIDERIVATIVE

A function G is called an *antiderivative* for f on $[a, b]$ iff

G is continuous on $[a, b]$ and $G'(x) = f(x)$ for all $x \in (a, b)$.

Theorem 5.3.5 says that, if f is continuous on $[a, b]$, then

$$F(x) = \int_a^x f(t)\, dt$$

is an antiderivative for f on $[a, b]$. This gives us a prescription for constructing antiderivatives. It tells us that we can construct an antiderivative for f by integrating f.

The so-called "fundamental theorem" goes the other way. It gives us a prescription, not for finding antiderivatives, but for evaluating integrals. It tells us that we can evaluate

$$\int_a^b f(t)\, dt$$

by finding an antiderivative for f.

THEOREM 5.4.2 THE FUNDAMENTAL THEOREM OF INTEGRAL CALCULUS

Let f be continuous on $[a, b]$. If G is an antiderivative for f on $[a, b]$, then

$$\int_a^b f(t)\, dt = G(b) - G(a).$$

Proof. From Theorem 5.3.5 we know that the function

$$F(x) = \int_a^x f(t)\, dt$$

is an antiderivative for f on $[a, b]$. If G is also an antiderivative for f on $[a, b]$, then both F and G are continuous on $[a, b]$ and satisfy $F'(x) = G'(x)$ for all x in (a, b). From Theorem 4.2.4 we know that there exists a constant C such that

$$F(x) = G(x) + C \qquad \text{for all } x \text{ in } [a, b].$$

Since $F(a) = 0$,

$$G(a) + C = 0 \quad \text{and thus} \quad C = -G(a).$$

It follows that

$$F(x) = G(x) - G(a) \qquad \text{for all } x \text{ in } [a, b].$$

In particular,

$$F(b) = G(b) - G(a). \quad \square$$

We now evaluate some integrals by applying the fundamental theorem. In each case we use the simplest antiderivative we can think of.

Problem 1. Evaluate

$$\int_1^4 x \, dx.$$

Solution. As an antiderivative for $f(x) = x$ we can use the function

$$G(x) = \tfrac{1}{2}x^2. \qquad \text{(check this out)}$$

By the fundamental theorem

$$\int_1^4 x \, dx = G(4) - G(1) = \tfrac{1}{2}(4)^2 - \tfrac{1}{2}(1)^2 = 7\tfrac{1}{2}. \quad \square$$

Problem 2. Evaluate

$$\int_0^1 8x \, dx.$$

Solution. Here we can use the antiderivative $G(x) = 4x^2$:

$$\int_0^1 8x \, dx = G(1) - G(0) = 4(1)^2 - 4(0)^2 = 4. \quad \square$$

Expressions of the form $G(b) - G(a)$ are conveniently written

$$\left[G(x) \right]_a^b.$$

In this notation

$$\int_1^4 x \, dx = \left[\tfrac{1}{2}x^2 \right]_1^4 = \tfrac{1}{2}(4)^2 - \tfrac{1}{2}(1)^2 = 7\tfrac{1}{2},$$

and

$$\int_0^1 8x \, dx = \left[4x^2 \right]_0^1 = 4(1)^2 - 4(0)^2 = 4. \quad \square$$

From your study of differentiation you know that for all positive integers n

$$\frac{d}{dx}\left(\frac{1}{n+1}x^{n+1}\right) = x^n.$$

It follows that

(5.4.3)
$$\int_a^b x^n\, dx = \left[\frac{1}{n+1}x^{n+1}\right]_a^b = \frac{1}{n+1}(b^{n+1} - a^{n+1}).$$

Thus

$$\int_a^b x^2\, dx = \left[\tfrac{1}{3}x^3\right]_a^b = \tfrac{1}{3}(b^3 - a^3), \qquad \int_a^b x^3\, dx = \left[\tfrac{1}{4}x^4\right]_a^b = \tfrac{1}{4}(b^4 - a^4), \text{ etc. } \quad \square$$

Problem 3. Evaluate

$$\int_0^1 (2x - 6x^4 + 5)\, dx.$$

Solution. As an antiderivative we use $G(x) = x^2 - \tfrac{6}{5}x^5 + 5x$:

$$\int_0^1 (2x - 6x^4 + 5)\, dx = \left[x^2 - \tfrac{6}{5}x^5 + 5x\right]_0^1 = 1 - \tfrac{6}{5} + 5 = 4\tfrac{4}{5}. \quad \square$$

Problem 4. Evaluate

$$\int_{-1}^1 (x - 1)(x + 2)\, dx.$$

Solution. First we carry out the indicated multiplication:

$$(x - 1)(x + 2) = x^2 + x - 2.$$

As an antiderivative we use $G(x) = \tfrac{1}{3}x^3 + \tfrac{1}{2}x^2 - 2x$:

$$\int_{-1}^1 (x - 1)(x + 2)\, dx = \left[\tfrac{1}{3}x^3 + \tfrac{1}{2}x^2 - 2x\right]_{-1}^1 = -\tfrac{10}{3}. \quad \square$$

Formula 5.4.3 holds not only for positive integer exponents but for all rational exponents other than -1 provided that the *integrand*, the function being integrated, is defined over the interval of integration. Thus

$$\int_1^2 \frac{dx}{x^2} = \int_1^2 x^{-2}\, dx = \left[-x^{-1}\right]_1^2 = \left[-\frac{1}{x}\right]_1^2 = (-\tfrac{1}{2}) - (-1) = \tfrac{1}{2},$$

$$\int_4^9 \frac{dt}{\sqrt{t}} = \int_4^9 t^{-1/2}\, dt = \left[2t^{1/2}\right]_4^9 = 2(3) - 2(2) = 2,$$

$$\int_0^1 t^{5/3}\, dt = \left[\tfrac{3}{8}t^{8/3}\right]_0^1 = \tfrac{3}{8}(1)^{8/3} - \tfrac{3}{8}(0)^{8/3} = \tfrac{3}{8}. \quad \square$$

Now some slightly more complicated examples.   The essential step in each case is the determination of an antiderivative.   Check each computation in detail.

$$\int_1^2 \frac{x^4 + 1}{x^2}\, dx = \int_1^2 (x^2 + x^{-2})\, dx = \left[\tfrac{1}{3}x^3 - x^{-1} \right]_1^2 = \tfrac{17}{6}.$$

$$\int_1^5 \sqrt{x - 1}\, dx = \int_1^5 (x - 1)^{1/2}\, dx = \left[\tfrac{2}{3}(x - 1)^{3/2} \right]_1^5 = \tfrac{16}{3}.$$

$$\int_0^1 (4 - \sqrt{x})^2\, dx = \int_0^1 (16 - 8\sqrt{x} + x)\, dx = \left[16x - \tfrac{16}{3}x^{3/2} + \tfrac{1}{2}x^2 \right]_0^1 = \tfrac{67}{6}.$$

$$\int_1^2 -\frac{dt}{(t + 2)^2} = \int_1^2 -(t + 2)^{-2}\, dt = \left[(t + 2)^{-1} \right]_1^2 = -\tfrac{1}{12}. \quad \square$$

The Linearity of the Integral

Here are some simple properties of the integral that are often used in computation. Throughout take f and g as continuous functions.

I. Constants may be factored through the integral sign:

(5.4.4)
$$\int_a^b \alpha f(x)\, dx = \alpha \int_a^b f(x)\, dx.$$

For example,

$$\int_0^{10} \tfrac{3}{7}(x - 5)\, dx = \tfrac{3}{7} \int_0^{10} (x - 5)\, dx = \tfrac{3}{7} \left[\tfrac{1}{2}(x - 5)^2 \right]_0^{10} = 0,$$

$$\int_0^1 \frac{8}{(x + 1)^3}\, dx = 8 \int_0^1 \frac{dx}{(x + 1)^3} = 8 \left[-\tfrac{1}{2}(x + 1)^{-2} \right]_0^1 = 8(\tfrac{3}{8}) = 3. \quad \square$$

II. The integral of a sum is the sum of the integrals:

(5.4.5)
$$\int_a^b [f(x) + g(x)]\, dx = \int_a^b f(x)\, dx + \int_a^b g(x)\, dx.$$

For example,

$$\int_1^2 \left[(x - 1)^2 + \frac{1}{(x + 2)^2} \right] dx = \int_1^2 (x - 1)^2\, dx + \int_1^2 \frac{dx}{(x + 2)^2}$$

$$= \left[\tfrac{1}{3}(x - 1)^3 \right]_1^2 + \left[-(x + 2)^{-1} \right]_1^2$$

$$= \tfrac{1}{3} - \tfrac{1}{4} + \tfrac{1}{3} = \tfrac{5}{12}. \quad \square$$

III. The integral of a linear combination is the linear combination of the integrals:

(5.4.6)
$$\int_a^b [\alpha f(x) + \beta g(x)]\, dx = \alpha \int_a^b f(x)\, dx + \beta \int_a^b g(x)\, dx.$$

For example,

$$\int_2^{5/2} \left[\frac{3}{(x-1)^2} - 4x \right] dx = 3 \int_2^{5/2} \frac{dx}{(x-1)^2} - 2 \int_2^{5/2} 2x \, dx$$

$$= 3 \left[-(x-1)^{-1} \right]_2^{5/2} - 2 \left[x^2 \right]_2^{5/2}$$

$$= 3(-\tfrac{2}{3} + 1) - 2(\tfrac{25}{4} - 4) = -\tfrac{7}{2}. \quad \square$$

I and II are particular instances of III. To prove III, take F as antiderivative for f, G as antiderivative for g, and observe that $\alpha F + \beta G$ is an antiderivative for $\alpha f + \beta g$. The details are left to you.

EXERCISES 5.4

Evaluate the following integrals.

1. $\int_0^1 (2x - 3) \, dx$.

2. $\int_0^1 (3x + 2) \, dx$.

3. $\int_{-1}^0 5x^4 \, dx$.

4. $\int_1^2 (2x + x^2) \, dx$.

5. $\int_1^4 \sqrt{x} \, dx$.

6. $\int_0^4 \sqrt{x} \, dx$.

7. $\int_1^5 2\sqrt{x-1} \, dx$.

8. $\int_1^2 \left(\frac{3}{x^3} + 5x \right) dx$.

9. $\int_{-2}^0 (x+1)(x-2) \, dx$.

10. $\int_2^0 \frac{dx}{(x+1)^2}$.

11. $\int_3^3 \sqrt{x} \, dx$.

12. $\int_{-1}^0 (t-2)(t+1) \, dt$.

13. $\int_0^1 \frac{dt}{(t+2)^3}$.

14. $\int_1^0 (t^3 + t^2) \, dt$.

15. $\int_1^2 \left(3t + \frac{4}{t^2} \right) dt$.

16. $\int_{-1}^{-1} 7x^6 \, dx$.

17. $\int_0^1 (x^{3/2} - x^{1/2}) \, dx$.

18. $\int_0^1 (x^{3/4} - 2x^{1/2}) \, dx$.

19. $\int_0^1 (x+1)^{17} \, dx$.

20. $\int_0^a (a^2 x - x^3) \, dx$.

21. $\int_0^a (\sqrt{a} - \sqrt{x})^2 \, dx$.

22. $\int_{-1}^1 (x-2)^2 \, dx$.

23. $\int_1^2 \frac{6-t}{t^3} \, dt$.

24. $\int_1^2 \frac{2-t}{t^3} \, dt$.

25. $\int_0^1 x^2(x-1) \, dx$.

26. $\int_1^3 \left(x^2 - \frac{1}{x^2} \right) dx$.

27. $\int_1^2 2x(x^2 + 1) \, dx$.

28. $\int_0^1 3x^2(x^3 + 1) \, dx$.

29. $\int_0^{\pi/2} \cos x \, dx$.

30. $\int_0^\pi \sin x \, dx$.

31. $\int_0^{\pi/4} \sec^2 x \, dx$.

32. $\int_{\pi/6}^{\pi/3} \sec x \tan x \, dx$.

33. $\int_{\pi/6}^{\pi/4} \csc x \cot x \, dx$.

34. $\int_{\pi/4}^{\pi/3} \csc^2 x \, dx$.

35. $\int_0^{2\pi} \sin x \, dx$.

36. $\int_0^\pi \cos x \, dx$.

37. Define a function F on $[1, 8]$ such that $F'(x) = 1/x$ and (a) $F(2) = 0$; (b) $F(2) = -3$.

38. Define a function F on $[0, 4]$ such that $F'(x) = \sqrt{1 + x^2}$ and (a) $F(3) = 0$; (b) $F(3) = 1$.

39. Compare $\dfrac{d}{dx} \left[\displaystyle\int_a^x f(t) \, dt \right]$ to $\displaystyle\int_a^x \dfrac{d}{dt} [f(t)] \, dt$.

5.5   SOME AREA PROBLEMS

In Section 5.2 we noted that, if f is nonnegative and continuous on $[a, b]$, then the integral of f from a to b gives the area of the region below the graph of f. With Ω as in Figure 5.5.1

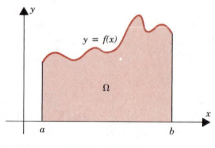

(5.5.1) $\quad$ area of $\Omega = \displaystyle\int_a^b f(x)\, dx.$

Figure 5.5.1

Problem 1. Find the area below the graph of the square-root function from $x = 0$ to $x = 1$.

Solution. The graph is pictured in Figure 5.5.2. The area below the graph is $\frac{2}{3}$:

$$\int_0^1 \sqrt{x}\, dx = \int_0^1 x^{1/2}\, dx = \left[\tfrac{2}{3}x^{3/2}\right]_0^1 = \tfrac{2}{3}. \quad \square$$

Figure 5.5.2

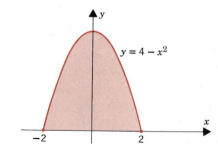

Figure 5.5.3

Problem 2. Find the area of the region bounded above by the curve $y = 4 - x^2$ and below by the x-axis.

Solution. The curve intersects the x-axis at $x = -2$ and $x = 2$. See Figure 5.5.3. The area of the region is $\frac{32}{3}$:

$$\int_{-2}^2 (4 - x^2)\, dx = \left[4x - \tfrac{1}{3}x^3\right]_{-2}^2 = \tfrac{32}{3}. \quad \square$$

Now we will calculate the areas of somewhat more complicated regions such as region Ω shown in Figure 5.5.4. To avoid excessive repetition, let's agree at the outset that throughout this section the symbols f, g, h represent continuous functions.

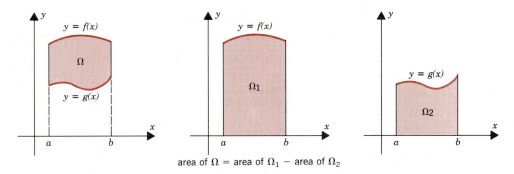

area of Ω = area of Ω_1 − area of Ω_2

Figure 5.5.4

Look at the region Ω shown in Figure 5.5.4. The upper boundary of Ω is the graph of a nonnegative function f and the lower boundary is the graph of a nonnegative function g. We can obtain the area of Ω by calculating the area of Ω_1 and subtracting off the area of Ω_2. Since

$$\text{area of } \Omega_1 = \int_a^b f(x)\,dx \quad \text{and} \quad \text{area of } \Omega_2 = \int_a^b g(x)\,dx,$$

we have

$$\text{area of } \Omega = \int_a^b f(x)\,dx - \int_a^b g(x)\,dx.$$

We can combine the two integrals and write

(5.5.2)
$$\text{area of } \Omega = \int_a^b [f(x) - g(x)]\,dx.$$

Problem 3. Find the area of the region bounded above by $y = x + 2$ and below by $y = x^2$.

Solution. The two curves intersect at $x = -1$ and $x = 2$:

$$x + 2 = x^2 \quad \text{iff} \quad x^2 - x - 2 = 0$$
$$\text{iff} \quad (x + 1)(x - 2) = 0.$$

The region, sketched in Figure 5.5.5, has area $\frac{9}{2}$:

$$\int_{-1}^{2} [(x + 2) - x^2]\,dx = \left[\tfrac{1}{2}x^2 + 2x - \tfrac{1}{3}x^3 \right]_{-1}^{2}$$
$$= (2 + 4 - \tfrac{8}{3}) - (\tfrac{1}{2} - 2 + \tfrac{1}{3})$$
$$= \tfrac{9}{2}. \quad \square$$

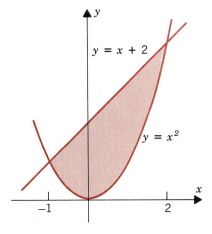

Figure 5.5.5

We derived Formula 5.5.2 under the assumption that f and g were both nonnegative, but that assumption is unnecessary. The formula holds for any region Ω that has

an upper boundary of the form $y = f(x), \quad x \in [a, b]$

and

a lower boundary of the form $y = g(x), \quad x \in [a, b]$.

Figure 5.5.6

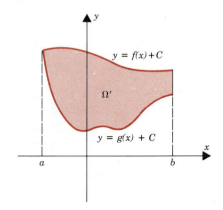

Figure 5.5.7

To see this, take Ω as in Figure 5.5.6. Obviously, Ω is congruent to the region marked Ω' in Figure 5.5.7. Ω' is Ω raised C units. Since Ω' lies entirely above the x-axis, the area of Ω' is given by the integral

$$\int_a^b \{[f(x) + C] - [g(x) + C]\}\, dx = \int_a^b [f(x) - g(x)]\, dx.$$

Since area of Ω = area of Ω',

$$\text{area of } \Omega = \int_a^b [f(x) - g(x)]\, dx$$

as asserted.

Problem 4. Find the area of the region shown in Figure 5.5.8.

Solution

$$\begin{aligned}
\text{area of } \Omega &= \int_{\pi/4}^{5\pi/4} [\sin x - \cos x]\, dx \\
&= \Big[-\cos x - \sin x\Big]_{\pi/4}^{5\pi/4} \\
&= 2\sqrt{2}. \quad \square
\end{aligned}$$

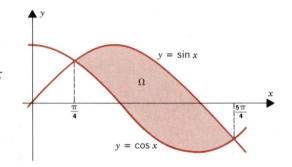

Figure 5.5.8

Problem 5.   Find the area between the curves

$$y = 4x \quad \text{and} \quad y = x^3$$

from $x = -2$ to $x = 2$.   See Figure 5.5.9.

Solution.   Notice that $y = x^3$ is the upper boundary from $x = -2$ to $x = 0$, but it is the lower boundary from $x = 0$ to $x = 2$.   Thus

$$\text{area} = \int_{-2}^{0} [x^3 - 4x]\, dx + \int_{0}^{2} [4x - x^3]\, dx$$

$$= \left[\tfrac{1}{4}x^4 - 2x^2 \right]_{-2}^{0} + \left[2x^2 - \tfrac{1}{4}x^4 \right]_{0}^{2}$$

$$= 8. \quad \square$$

Figure 5.5.9

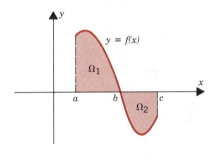

Figure 5.5.10

Problem 6.   Use integrals to represent the area of the region $\Omega = \Omega_1 \cup \Omega_2$ shaded in Figure 5.5.10.

Solution.   From $x = a$ to $x = b$, the curve $y = f(x)$ is above the x-axis.   Therefore

$$\text{area of } \Omega_1 = \int_{a}^{b} f(x)\, dx.$$

From $x = b$ to $x = c$, the curve $y = f(x)$ is below the x-axis.   The upper boundary for Ω_1 is the curve $y = 0$ (the x-axis) and the lower boundary is the curve $y = f(x)$.   Thus

$$\text{area of } \Omega_2 = \int_{b}^{c} [0 - f(x)]\, dx = - \int_{b}^{c} f(x)\, dx.$$

The area of Ω is the sum of these two areas:

$$\text{area of } \Omega = \int_{a}^{b} f(x)\, dx - \int_{b}^{c} f(x)\, dx. \quad \blacksquare$$

Problem 7. Use integrals to represent the area of the region shaded in Figure 5.5.11.

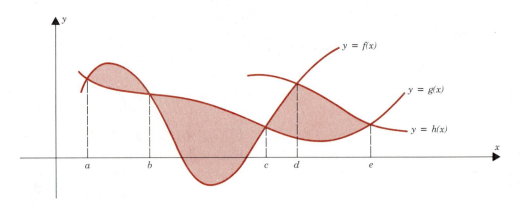

Figure 5.5.11

Solution

$$\text{Area} = \int_{a}^{b} [f(x) - g(x)]\, dx + \int_{b}^{c} [g(x) - f(x)]\, dx$$
$$+ \int_{c}^{d} [f(x) - g(x)]\, dx + \int_{d}^{e} [h(x) - g(x)]\, dx. \quad \square$$

EXERCISES 5.5

Find the area between the graph of f and the x-axis.

1. $f(x) = 2 + x^3, \quad x \in [0, 1].$

2. $f(x) = (x + 2)^{-2}, \quad x \in [0, 2].$

3. $f(x) = \sqrt{x + 1}, \quad x \in [3, 8].$

4. $f(x) = x^2(3 + x), \quad x \in [0, 8].$

5. $f(x) = (2x^2 + 1)^2, \quad x \in [0, 1].$

6. $f(x) = \frac{1}{2}(x + 1)^{-1/2}, \quad x \in [0, 8].$

7. $f(x) = x^2 - 4, \quad x \in [1, 2].$

8. $f(x) = \cos x, \quad x \in [\frac{1}{6}\pi, \frac{1}{3}\pi].$

9. $f(x) = \sin x, \quad x \in [\frac{1}{3}\pi, \frac{1}{2}\pi].$

10. $f(x) = x^3 + 1, \quad x \in [-2, -1].$

Sketch the region bounded by the curves and find its area.

11. $y = \sqrt{x}, \quad y = x^2.$

12. $y = 6x - x^2, \quad y = 2x.$

13. $y = 5 - x^2, \quad y = 3 - x.$

14. $y = 8, \quad y = x^2 + 2x.$

15. $y = 8 - x^2, \quad y = x^2.$

16. $y = \sqrt{x}, \quad y = \frac{1}{4}x.$

17. $x^3 - 10y^2 = 0, \quad x - y = 0.$

18. $y^2 - 27x = 0, \quad x + y = 0.$

19. $x - y^2 + 3 = 0, \quad x - 2y = 0.$

20. $y^2 = 2x, \quad x - y = 4.$

21. $y = x, \quad y = 2x, \quad y = 4.$

22. $y = x^2, \quad y = -\sqrt{x}, \quad x = 4.$

23. $y = \cos x, \quad y = 4x^2 - \pi^2.$

24. $y = \sin x, \quad y = \pi x - x^2.$

25. $y = x, \quad y = \sin x, \quad x = \pi/2.$

26. $y = x + 1, \quad y = \cos x, \quad x = \pi.$

5.6   INDEFINITE INTEGRALS

We begin with a continuous function f. If F is an antiderivative for f on $[a, b]$, then

(1)
$$\int_a^b f(x)\, dx = \left[F(x) \right]_a^b.$$

If C is a constant, then

$$\left[F(x) + C \right]_a^b = [F(b) + C] - [F(a) + C] = F(b) - F(a) = \left[F(x) \right]_a^b.$$

Thus we can replace (1) by writing

$$\int_a^b f(x)\, dx = \left[F(x) + C \right]_a^b.$$

If we have no particular interest in the interval $[a, b]$ but wish instead to emphasize that F is an antiderivative for f on *some* interval, then we can omit the a and the b and simply write

$$\int f(x)\, dx = F(x) + C.$$

Antiderivatives expressed in this manner are called *indefinite integrals*. The constant C is called the *constant of integration*.

Thus, for example,

$$\int x^2\, dx = \tfrac{1}{3}x^3 + C \qquad \text{and} \qquad \int \sqrt{s}\, ds = \tfrac{2}{3}s^{3/2} + C.$$

The following problems will serve to illustrate the notation further.

Problem 1. Find f given that

$$f'(x) = x^3 + 2 \quad \text{and} \quad f(0) = 1.$$

Solution. Since f' is the derivative of f, f is an antiderivative for f'. Thus

$$f(x) = \int (x^3 + 2)\, dx = \tfrac{1}{4}x^4 + 2x + C.$$

To evaluate C we use the fact that $f(0) = 1$. Since

$$f(0) = 1 \qquad \text{and} \qquad f(0) = \tfrac{1}{4}(0)^4 + 2(0) + C = C,$$

we see that $C = 1$. Therefore

$$f(x) = \tfrac{1}{4}x^4 + 2x + 1. \quad \square$$

Problem 2. Find f given that

$$f''(x) = 6x - 2, \quad f'(1) = -5, \quad \text{and} \quad f(1) = 3.$$

Solution. First we get f' by integrating f'':

$$f'(x) = \int (6x - 2) \, dx = 3x^2 - 2x + C.$$

Since

$$f'(1) = -5 \quad \text{and} \quad f'(1) = 3(1)^2 - 2(1) + C = 1 + C,$$

we have

$$-5 = 1 + C \quad \text{and thus} \quad C = -6.$$

Therefore

$$f'(x) = 3x^2 - 2x - 6.$$

Now we get to f by integrating f':

$$f(x) = \int (3x^2 - 2x - 6) \, dx = x^3 - x^2 - 6x + K.$$

(We are writing the constant of integration as K because we used C before and it would be confusing to have C with two different values in the same problem.) Since

$$f(1) = 3 \quad \text{and} \quad f(1) = (1)^3 - (1)^2 - 6(1) + K = -6 + K,$$

we have

$$3 = -6 + K \quad \text{and thus} \quad K = 9.$$

Therefore

$$f(x) = x^3 - x^2 - 6x + 9. \quad \square$$

EXERCISES 5.6

Calculate the following.

1. $\displaystyle\int \frac{dx}{x^4}.$

2. $\displaystyle\int (x - 1)^2 \, dx.$

3. $\displaystyle\int (ax + b) \, dx.$

4. $\displaystyle\int (ax^2 + b) \, dx.$

5. $\displaystyle\int \frac{dx}{\sqrt{1 + x}}.$

6. $\displaystyle\int \left(\frac{x^3 + 1}{x^5}\right) dx.$

7. $\displaystyle\int \left(\frac{x^3 - 1}{x^2}\right) dx.$

8. $\displaystyle\int \left(\sqrt{x} - \frac{1}{\sqrt{x}}\right) dx.$

9. $\displaystyle\int (t - a)(t - b) \, dt.$

10. $\displaystyle\int (t^2 - a)(t^2 - b) \, dt.$

11. $\displaystyle\int \frac{(t^2 - a)(t^2 - b)}{\sqrt{t}} \, dt.$

12. $\displaystyle\int (2 - \sqrt{x})(2 + \sqrt{x}) \, dx.$

13. $\displaystyle\int g(x)g'(x) \, dx.$

14. $\displaystyle\int \sin x \cos x \, dx.$

15. $\displaystyle\int \tan x \sec^2 x \, dx.$

16. $\displaystyle\int \frac{g'(x)}{[g(x)]^2} \, dx.$

17. $\displaystyle\int \frac{4}{(4x + 1)^2} \, dx.$

18. $\displaystyle\int \frac{3x^2}{(x^3 + 1)^2} \, dx.$

Find f from the information given.

19. $f'(x) = 2x - 1$, $f(3) = 4$.

20. $f'(x) = 3 - 4x$, $f(1) = 6$.

21. $f'(x) = ax + b$, $f(2) = 0$.

22. $f'(x) = ax^2 + bx + c$, $f(0) = 0$.

23. $f'(x) = \sin x$, $f(0) = 2$.

24. $f'(x) = \cos x$, $f(\pi) = 3$.

25. $f''(x) = 6x - 2$, $f'(0) = 1$, $f(0) = 2$.

26. $f''(x) = -12x^2$, $f'(0) = 1$, $f(0) = 2$.

27. $f''(x) = x^2 - x$, $f'(1) = 0$, $f(1) = 2$.

28. $f''(x) = 1 - x$, $f'(2) = 1$, $f(2) = 0$.

29. $f''(x) = \cos x$, $f'(0) = 1$, $f(0) = 2$.

30. $f''(x) = \sin x$, $f'(0) = -2$, $f(0) = 1$.

31. $f''(x) = 2x - 3$, $f(2) = -1$, $f(0) = 3$.

32. $f''(x) = 5 - 4x$, $f(1) = 1$, $f(0) = -2$.

33. Compare $\dfrac{d}{dx}\left[\displaystyle\int f(x)\,dx\right]$ to $\displaystyle\int \dfrac{d}{dx}[f(x)]\,dx$.

34. Calculate

$$\int [f(x)g''(x) - g(x)f''(x)]\,dx.$$

5.7   SOME MOTION PROBLEMS

Problem 1.   An object moves along a coordinate line with velocity

$$v(t) = 2 - 3t + t^2 \quad \text{units per second.}$$

Its initial position (position at time $t = 0$) is 2 units to the right of the origin. Find the position of the object 4 seconds later.

Solution.   Let $x(t)$ be the position (coordinate) of the object at time t. We are given that $x(0) = 2$. Since $x'(t) = v(t)$,

$$x(t) = \int v(t)\,dt = \int (2 - 3t + t^2)\,dt = 2t - \tfrac{3}{2}t^2 + \tfrac{1}{3}t^3 + C.$$

Since $x(0) = 2$ and $x(0) = 2(0) - \tfrac{3}{2}(0)^2 + \tfrac{1}{3}(0)^3 + C = C$, we have $C = 2$ and

$$x(t) = 2t - \tfrac{3}{2}t^2 + \tfrac{1}{3}t^3 + 2.$$

The position of the object at time $t = 4$ is the value of this function at $t = 4$:

$$x(4) = 2(4) - \tfrac{3}{2}(4)^2 + \tfrac{1}{3}(4)^3 + 2 = 7\tfrac{1}{3}.$$

At the end of 4 seconds the object is $7\tfrac{1}{3}$ units to the right of the origin. $\square$

Recall that speed is the absolute value of velocity:

$$\text{speed at time } t = |v(t)|$$

and the integral of the speed function gives the distance traveled:

$$\int_a^b |v(t)|\,dt = \text{distance traveled from time } t = a \text{ to time } t = b.$$

Problem 2. An object moves along the x-axis with acceleration $a(t) = 2t - 2$ units per second per second. Its initial position (position at time $t = 0$) is 5 units right of the origin. One second later the object is moving left at the rate of 4 units per second.

    (a) Find the position of the object at time $t = 4$ seconds.

    (b) How far does the object travel during these 4 seconds?

Solution. (a) Let $x(t)$ and $v(t)$ denote the position and velocity of the object at time t. We are given that $x(0) = 5$ and $v(1) = -4$. Since $v'(t) = a(t)$,

$$v(t) = \int a(t)\, dt = \int (2t - 2)\, dt = t^2 - 2t + C.$$

Since

$$v(1) = -4 \qquad \text{and} \qquad v(1) = (1)^2 - 2(1) + C = -1 + C,$$

we have $C = -3$ and

$$v(t) = t^2 - 2t - 3.$$

Since $x'(t) = v(t)$,

$$x(t) = \int v(t)\, dt = \int (t^2 - 2t - 3)\, dt = \tfrac{1}{3}t^3 - t^2 - 3t + K.$$

Since

$$x(0) = 5 \qquad \text{and} \qquad x(0) = \tfrac{1}{3}(0)^3 - (0)^2 - 3(0) + K = K,$$

we have $K = 5$. Therefore

$$x(t) = \tfrac{1}{3}t^3 - t^2 - 3t + 5.$$

As you can check, $x(4) = -\tfrac{5}{3}$. At time $t = 4$ the object is $\tfrac{5}{3}$ units to the left of the origin.

(b) The distance traveled from time $t = 0$ to $t = 4$ is given by the integral

$$s = \int_0^4 |v(t)|\, dt = \int_0^4 |t^2 - 2t - 3|\, dt.$$

To evaluate this integral we first remove the absolute-value sign. As you can verify,

$$|t^2 - 2t - 3| = \begin{cases} -(t^2 - 2t - 3), & 0 \le t \le 3 \\ t^2 - 2t - 3, & 3 \le t \le 4 \end{cases}.$$

Thus

$$s = \int_0^3 (3 + 2t - t^2)\, dt + \int_3^4 (t^2 - 2t - 3)\, dt$$

$$= \left[3t + t^2 - \tfrac{1}{3}t^3 \right]_0^3 + \left[\tfrac{1}{3}t^3 - t^2 - 3t \right]_3^4 = \tfrac{34}{3}.$$

During the 4 seconds the object travels a distance of $\tfrac{34}{3}$ units. $\square$

Question: The object in Problem 2 leaves $x = -5$ at time $t = 0$ and arrives at $x = -\frac{5}{3}$ at time $t = 4$. The separation between $x = -5$ and $x = -\frac{5}{3}$ is only $|-5 - (-\frac{5}{3})| = \frac{10}{3}$. How is it possible that the object travels a distance of $\frac{34}{3}$ units?

Answer: The object does not maintain a fixed direction. It changes direction at time $t = 3$. You can see this by noting that the velocity function

$$v(t) = t^2 - 2t - 3 = (t - 3)(t + 1)$$

changes signs at $t = 3$.

Problem 3. Find the equation of motion for an object that moves along a straight line with constant acceleration a from an initial position x_0 with initial velocity v_0.

Solution. Call the line of the motion the x-axis. Here $a(t) = a$ at all times t. To find the velocity we integrate the acceleration:

$$v(t) = \int a \, dt = at + C.$$

The constant C is the initial velocity v_0:

$$v_0 = v(0) = a \cdot 0 + C = C.$$

We see therefore that

$$v(t) = at + v_0.$$

To find the position function we integrate the velocity:

$$x(t) = \int v(t) \, dt = \int (at + v_0) \, dt = \tfrac{1}{2}at^2 + v_0 t + K.$$

The constant K is the initial position x_0:

$$x_0 = x(0) = \tfrac{1}{2}a \cdot 0^2 + v_0 \cdot 0 + K = K.$$

The equation of motion is

(5.7.1) $\qquad\qquad \boxed{x(t) = \tfrac{1}{2}at^2 + v_0 t + x_0.}$ † ☐

EXERCISES 5.7

1. An object moves along a coordinate line with velocity $v(t) = 6t^2 - 6$ units per second. Its initial position (position at time $t = 0$) is 2 units to the left of the origin. (a) Find the position of the object 3 seconds later. (b) Find the total distance traveled by the object during those 3 seconds.

---

† In the case of a free-falling body, $a = -g$ and we have Galileo's equation for free fall. (3.5.5)

2. An object moves along a coordinate line with acceleration $a(t) = (t + 2)^3$ units per second per second. (a) Find the velocity function given that the initial velocity is 3 units per second. (b) Find the position function given that the initial velocity is 3 units per second and the initial position is the origin.

3. An object moves along a coordinate line with acceleration $a(t) = (t + 1)^{-1/2}$ units per second per second. (a) Find the velocity function given that the initial velocity is 1 unit per second. (b) Find the position function given that the initial velocity is 1 unit per second and the initial position is the origin.

4. An object moves along a coordinate line with velocity $v(t) = t(1 - t)$ units per second. Its initial position is 2 units to the left of the origin. (a) Find the position of the object 10 seconds later. (b) Find the total distance traveled by the object during those 10 seconds.

5. A car traveling at 60 mph decelerates at 20 feet per second per second. (a) How long does it take for the car to come to a complete stop? (b) What distance is required to bring the car to a complete stop?

6. An object moves along the x-axis with constant acceleration. Express the position $x(t)$ in terms of the initial position x_0, the initial velocity v_0, the velocity $v(t)$, and the elapsed time t.

7. An object moves along the x-axis with constant acceleration a. Verify that

$$[v(t)]^2 = v_0^2 + 2a[x(t) - x_0].$$

8. A bobsled moving at 60 mph decelerates at a constant rate to 40 mph over a distance of 264 feet and continues to decelerate at that same rate until it comes to a full stop. (a) What is the acceleration of the sled in feet per second per second? (b) How long does it take to reduce the speed to 40 mph? (c) How long does it take to bring the sled to a complete stop from 60 mph? (d) Over what distance does the sled come to a complete stop from 60 mph?

9. In the AB-run, minicars start from a standstill at point A, race along a straight track, and come to a full stop at point B one-half mile away. Given that the cars can accelerate uniformly to a maximum speed of 60 mph in 20 seconds and can brake at a maximum rate of 22 feet per second per second, what is the best possible time for the completion of the AB-run?

Find the general law of motion of an object that moves in a straight line with acceleration $a(t)$. Write x_0 for initial position and v_0 for initial velocity.

10. $a(t) = \sin t$.   11. $a(t) = 2A + 6Bt$.   12. $a(t) = \cos t$.

13. As a particle moves about the plane, its x-coordinate changes at the rate of $t^2 - 5$ units per second and its y-coordinate changes at the rate of $3t$ units per second. If the particle is at the point $(4, 2)$ when $t = 2$ seconds, where is the particle 4 seconds later?

14. As a particle moves about the plane, its x-coordinate changes at the rate of $t - 2$ units per second and its y-coordinate changes at the rate of $\sqrt{t}$ units per second. If the particle is at the point $(3, 1)$ when $t = 4$ seconds, where is the particle 5 seconds later?

15. A particle moves along the x-axis with velocity $v(t) = At + B$. Determine A and B given that the initial velocity of the particle is 2 units per second and the position of the particle after 2 seconds of motion is 1 unit to the left of the initial position.

16. A particle moves along the x-axis with velocity $v(t) = At^2 + 1$. Determine A given that $x(1) = x(0)$. Compute the total distance traveled by the particle during the first second.

17. An object moves along a coordinate line with velocity $v(t) = \sin t$ units per second. The object passes through the origin at time $t = \pi/6$ seconds. When is the next time (a) that the object passes through the origin? (b) that the object passes through the origin moving from left to right?

18. Exercise 17 with $v(t) = \cos t$.

19. An automobile with varying velocity $v(t)$ moves in a fixed direction for 5 minutes and covers a distance of 4 miles. What theorem would you invoke to argue that for at least one instant the speedometer must have read 48 miles per hour?

20. A speeding motorcyclist sees his way blocked by a haywagon some distance s ahead and slams on his brakes. Given that the brakes impart to the motorcycle a constant negative acceleration a and that the haywagon is moving with speed v_1 in the same direction as the motorcycle, show that the motorcyclist can avoid collision only if he is traveling at a speed less than $v_1 + \sqrt{2|a|s}$.

21. Find the velocity $v(t)$ given that $a(t) = 2[v(t)]^2$ and $v_0 \neq 0$.

5.8 THE u-SUBSTITUTION; CHANGE OF VARIABLES

When we differentiate a composite function, we do so by the chain rule. In trying to calculate an indefinite integral, we are often called upon to apply the chain rule in reverse. This process is usually facilitated by making what we call a "u-substitution."

An integral of the form

$$\int f(g(x)) \, g'(x) \, dx$$

can be written

$$\int f(u) \, du$$

by setting

$$u = g(x), \qquad du = g'(x) \, dx.\dagger$$

If F is an antiderivative for f, then

$$\int f(g(x)) \, g'(x) \, dx = \int F'(g(x)) \, g'(x) \, dx = F(g(x)) + C.$$

by the chain rule ⬈

We can obtain the same result by calculating

$$\int f(u) \, du$$

and then substituting $g(x)$ back in for u:

$$\int f(u) \, du = F(u) + C = F(g(x)) + C.$$

---

† Think of $du = g'(x) \, dx$ as a "formal differential," writing dx for h.

Problem 1. Calculate

$$\int (x^2 - 1)^4 \, 2x \, dx$$

and then check the result by differentiation.

Solution. Set

$$u = x^2 - 1 \qquad \text{so that} \qquad du = 2x \, dx.$$

Then

$$\int (x^2 - 1)^4 \, 2x \, dx = \int u^4 \, du = \tfrac{1}{5} u^5 + C = \tfrac{1}{5}(x^2 - 1)^5 + C.$$

Checking:

$$\frac{d}{dx} [\tfrac{1}{5}(x^2 - 1)^5 + C] = \tfrac{5}{5}(x^2 - 1)^4 \frac{d}{dx}(x^2 - 1) \overset{\checkmark}{=} (x^2 - 1)^4 \, 2x. \quad \square$$

Problem 2. Calculate

$$\int \frac{dx}{(3 + 5x)^2}$$

and then check the result by differentiation.

Solution. Set

$$u = 3 + 5x \qquad \text{so that} \qquad du = 5 \, dx.$$

Then

$$\frac{dx}{(3 + 5x)^2} = \frac{\tfrac{1}{5} du}{u^2} = \frac{1}{5} \frac{du}{u^2}$$

and

$$\int \frac{dx}{(3 + 5x)^2} = \frac{1}{5} \int \frac{du}{u^2} = -\frac{1}{5u} + C\dagger = -\frac{1}{5(3 + 5x)} + C.$$

Checking:

$$\frac{d}{dx} \left[-\frac{1}{5(3 + 5x)} + C \right] = \frac{d}{dx} [-\tfrac{1}{5}(3 + 5x)^{-1}]$$

$$= (-\tfrac{1}{5})(-1)(3 + 5x)^{-2}(5) \overset{\checkmark}{=} \frac{1}{(3 + 5x)^2}. \quad \square$$

---

† One could write

$$\frac{1}{5} \int \frac{du}{u^2} = -\frac{1}{5} \left[\frac{1}{u} + C \right] = -\frac{1}{5u} + \frac{C}{5},$$

but, since C is arbitrary, $C/5$ is arbitrary, and we can therefore write C instead.

In the remaining problems we will leave the checking to you.

Problem 3. Calculate

$$\int x(2x^2 - 1)^3 \, dx.$$

Solution. Set

$$u = 2x^2 - 1, \qquad du = 4x \, dx.$$

Then

$$x(2x^2 - 1)^3 \, dx = \underbrace{(2x^2 - 1)^3}_{u^3} \underbrace{x \, dx}_{\frac{1}{4} \, du} = \tfrac{1}{4} u^3 \, du$$

and

$$\int x(2x^2 - 1)^3 \, dx = \tfrac{1}{4} \int u^3 \, du = \tfrac{1}{16} u^4 + C = \tfrac{1}{16} (2x^2 - 1)^4 + C. \quad \square$$

The key step in making a *u*-substitution is to find a substitution $u = g(x)$ such that the expression $du = g'(x) \, dx$ appears in the original integral (at least up to a constant factor) and the new integral

$$\int f(u) \, du$$

is easier to calculate than the original integral. In most cases the form of the original integrand will suggest a good choice for *u*.

Problem 4. Calculate

$$\int x^2 \sqrt{4 + x^3} \, dx.$$

Solution. Set

$$u = 4 + x^3, \qquad du = 3x^2 \, dx.$$

Then

$$x^2 \sqrt{4 + x^3} \, dx = \underbrace{\sqrt{4 + x^3}}_{\sqrt{u}} \underbrace{x^2 \, dx}_{\frac{1}{3} \, du} = \tfrac{1}{3} \sqrt{u} \, du$$

and

$$\int x^2 \sqrt{4 + x^3} \, dx = \tfrac{1}{3} \int \sqrt{u} \, du$$

$$= \tfrac{2}{9} u^{3/2} + C = \tfrac{2}{9} (4 + x^3)^{3/2} + C. \quad \square$$

Problem 5. Calculate

$$\int x(x-3)^5 \, dx.$$

Solution. Set

$$u = x - 3, \qquad du = dx.$$

Then

$$x(x-3)^5 \, dx = (u+3)u^5 \, du = (u^6 + 3u^5) \, du$$

and

$$\int x(x-3)^5 \, dx = \int (u^6 + 3u^5) \, du$$

$$= \tfrac{1}{7}u^7 + \tfrac{1}{2}u^6 + C = \tfrac{1}{7}(x-3)^7 + \tfrac{1}{2}(x-3)^6 + C. \quad \square$$

For definite integrals we have the following *change of variables formula*:

(5.8.1)
$$\int_a^b f(g(x)) \, g'(x) \, dx = \int_{g(a)}^{g(b)} f(u) \, du.$$

The formula holds provided that f and g' are both continuous. More precisely, g' must be continuous on an interval that joins a and b, and f must be continuous on the set of values taken on by g.

Proof. Let F be an antiderivative for f. Then $F' = f$ and

$$\int_a^b f(g(x)) \, g'(x) \, dx = \int_a^b F'(g(x)) \, g'(x) \, dx$$

$$= \left[F(g(x)) \right]_a^b = F(g(b)) - F(g(a)) = \int_{g(a)}^{g(b)} f(u) \, du. \quad \square$$

Problem 6. Evaluate

$$\int_1^2 \frac{10x^2}{(x^3+1)^2} \, dx.$$

Solution. Set

$$u = x^3 + 1, \qquad du = 3x^2 \, dx.$$

Then

$$\frac{10x^2}{(x^3+1)^2} \, dx = 10 \underbrace{\frac{\overbrace{x^2 \, dx}^{\frac{1}{3} du}}{(x^3+1)^2}}_{u^2} = \frac{10}{3} \frac{du}{u^2}.$$

At $x = 1$, $u = 2$. At $x = 2$, $u = 9$. Therefore

$$\int_1^2 \frac{10x^2}{(x^3 + 1)^2}\, dx = \frac{10}{3} \int_2^9 \frac{du}{u^2} = \frac{10}{3} \left[-\frac{1}{u} \right]_2^9 = \frac{35}{27}. \quad \square$$

Problem 7. Evaluate

$$\int_0^2 x\sqrt{4x^2 + 9}\, dx.$$

Solution. Set

$$u = 4x^2 + 9, \qquad du = 8x\, dx.$$

Then

$$x\sqrt{4x^2 + 9}\, dx = \underbrace{\sqrt{4x^2 + 9}}_{\sqrt{u}}\, \underbrace{x\, dx}_{\frac{1}{8}\, du} = \tfrac{1}{8}\, \sqrt{u}\, du.$$

At $x = 0$, $u = 9$. At $x = 2$, $u = 25$. Thus

$$\int_0^2 x\sqrt{4x^2 + 9}\, dx = \tfrac{1}{8} \int_9^{25} \sqrt{u}\, du = \tfrac{1}{12} \left[u^{3/2} \right]_9^{25} = \tfrac{49}{6}. \quad \square$$

Problem 8. Evaluate

$$\int_0^{\sqrt{3}} x^5 \sqrt{x^2 + 1}\, dx.$$

Solution. Set

$$u = x^2 + 1, \qquad du = 2x\, dx.$$

Then

$$x^5 \sqrt{x^2 + 1}\, dx = \underbrace{x^4}_{(u-1)^2}\, \underbrace{\sqrt{x^2 + 1}}_{\sqrt{u}}\, \underbrace{x\, dx}_{\frac{1}{2}\, du} = \tfrac{1}{2}(u - 1)^2 \sqrt{u}\, du.$$

At $x = 0$, $u = 1$. At $x = \sqrt{3}$, $u = 4$. Thus

$$\int_0^{\sqrt{3}} x^5 \sqrt{x^2 + 1}\, dx = \tfrac{1}{2} \int_1^4 (u - 1)^2 \sqrt{u}\, du$$

$$= \tfrac{1}{2} \int_1^4 (u^{5/2} - 2u^{3/2} + u^{1/2})\, du$$

$$= \tfrac{1}{2} \left[\tfrac{2}{7}u^{7/2} - \tfrac{4}{5}u^{5/2} + \tfrac{2}{3}u^{3/2} \right]_1^4 = \left[u^{3/2}(\tfrac{1}{7}u^2 - \tfrac{2}{5}u + \tfrac{1}{3}) \right]_1^4 = \tfrac{848}{105}. \quad \square$$

Now a few words about the role of symmetry in integration. Suppose that f is a continuous function defined on some interval of the form $[-a, a]$, some closed interval that is symmetric about the origin.

(5.8.2)

(a) If f is odd on $[-a, a]$ (i.e., if $f(-x) = -f(x)$ for all $x \in [-a, a]$), then

$$\int_{-a}^{a} f(x)\, dx = 0.$$

(b) If f is even on $[-a, a]$ (i.e., if $f(-x) = f(x)$ for all $x \in [-a, a]$), then

$$\int_{-a}^{a} f(x)\, dx = 2\int_{0}^{a} f(x)\, dx.$$

These assertions can be verified by a simple change of variables. (Exercise 34) Here we look at these assertions from the standpoint of area, referring to Figures 5.8.1 and 5.8.2.

odd f

Figure 5.8.1

even f

Figure 5.8.2

For the odd function,

$$\int_{-a}^{a} f(x)\, dx = \int_{-a}^{0} f(x)\, dx + \int_{0}^{a} f(x)\, dx = \text{area of } \Omega_1 - \text{area of } \Omega_2 = 0.$$

For the even function,

$$\int_{-a}^{a} f(x)\, dx = \text{area of } \Omega_1 + \text{area of } \Omega_2 = 2(\text{area of } \Omega_2) = 2\int_{0}^{a} f(x)\, dx.$$

Suppose we were asked to evaluate

$$\int_{-\pi}^{\pi} (\sin x - x \cos x)^3\, dx.$$

A laborious calculation would show that this integral is zero. We don't have to carry out that calculation. The integrand is an odd function, and thus we can tell immediately that the integral is zero.

EXERCISES 5.8

Work out the following integrals by a *u*-substitution.

1. $\displaystyle\int \frac{dx}{(2 - 3x)^2}.$ **2.** $\displaystyle\int \frac{dx}{\sqrt{2x + 1}}.$ **3.** $\displaystyle\int \sqrt{2x + 1}\, dx.$

4. $\displaystyle\int \sqrt{ax + b}\, dx.$ **5.** $\displaystyle\int (ax + b)^{3/4}\, dx.$ **6.** $\displaystyle\int 2ax(ax^2 + b)^4\, dx.$

7. $\displaystyle\int \frac{t}{(4t^2 + 9)^2}\, dt.$ **8.** $\displaystyle\int \frac{3t}{(t^2 + 1)^2}\, dt.$ **9.** $\displaystyle\int t^2(5t^3 + 9)^4\, dt.$

10. $\displaystyle\int t(1 + t^2)^3\, dt.$ **11.** $\displaystyle\int x^2(1 + x^3)^{1/4}\, dx.$ **12.** $\displaystyle\int x^{n-1}\sqrt{a + bx^n}\, dx.$

13. $\displaystyle\int \frac{s}{(1 + s^2)^3}\, ds.$ **14.** $\displaystyle\int \frac{2s}{\sqrt[3]{6 - 5s^2}}\, ds.$ **15.** $\displaystyle\int \frac{x}{\sqrt{x^2 + 1}}\, dx.$

16. $\displaystyle\int \frac{3ax^2 - 2bx}{\sqrt{ax^3 - bx^2}}\, dx.$ **17.** $\displaystyle\int x^2(1 - x^3)^{2/3}\, dx.$ **18.** $\displaystyle\int \frac{x^2}{(1 - x^3)^{2/3}}\, dx.$

19. $\displaystyle\int 5x(x^2 + 1)^{-3}\, dx.$ **20.** $\displaystyle\int 2x^3(1 - x^4)^{-1/4}\, dx.$ **21.** $\displaystyle\int x^{-3/4}(x^{1/4} + 1)^{-2}\, dx.$

22. $\displaystyle\int \frac{4x + 6}{\sqrt{x^2 + 3x + 1}}\, dx.$ **23.** $\displaystyle\int \frac{b^3x^3}{\sqrt{1 - a^4x^4}}\, dx.$ **24.** $\displaystyle\int \frac{x^{n-1}}{\sqrt{a + bx^n}}\, dx.$

Evaluate the following integrals by a *u*-substitution.

25. $\displaystyle\int_0^1 x(x^2 + 1)^3\, dx.$ **26.** $\displaystyle\int_{-1}^0 3x^2(4 + 2x^3)^2\, dx.$ **27.** $\displaystyle\int_0^1 5x(1 + x^2)^4\, dx.$

28. $\displaystyle\int_1^2 (6 - x)^{-3}\, dx.$ **29.** $\displaystyle\int_{-1}^1 \frac{r}{(1 + r^2)^4}\, dr.$ **30.** $\displaystyle\int_0^3 \frac{r}{\sqrt{r^2 + 16}}\, dr.$

31. $\displaystyle\int_0^a y\sqrt{a^2 - y^2}\, dy.$ **32.** $\displaystyle\int_0^a y\sqrt{a^2 + y^2}\, dy.$ **33.** $\displaystyle\int_{-a}^0 y^2\left(1 - \frac{y^3}{a^3}\right)^{-2} dy.$

34. Let *f* be a continuous function. Show by a change of variables that

    (a) if *f* is odd, then $\displaystyle\int_{-a}^a f(x)\, dx = 0;$ (b) if *f* is even, then $\displaystyle\int_{-a}^a f(x)\, dx = 2\int_0^a f(x)\, dx.$

Work out the following integrals by a *u*-substitution.

35. $\displaystyle\int x\sqrt{x + 1}\, dx.$ [set $u = x + 1$]   **36.** $\displaystyle\int 2x\sqrt{x - 1}\, dx.$

37. $\displaystyle\int x\sqrt{2x - 1}\, dx.$ **38.** $\displaystyle\int x^2\sqrt{x + 1}\, dx.$ **39.** $\displaystyle\int y(y + 1)^{12}\, dy.$

40. $\displaystyle\int y(y - 1)^{-7}\, dy.$ **41.** $\displaystyle\int t^2(t - 2)^5\, dt.$ **42.** $\displaystyle\int t(2t + 3)^8\, dt.$

43. $\displaystyle\int t^2(t - 2)^{-5}\, dt.$ **44.** $\displaystyle\int t(2t + 3)^{-8}\, dt.$ **45.** $\displaystyle\int x(x + 1)^{-1/5}\, dx.$

46. $\displaystyle\int x^2(2x-1)^{-2/3}\,dx.$ **47.** $\displaystyle\int_0^1 \frac{x+3}{\sqrt{x+1}}\,dx.$ **48.** $\displaystyle\int_0^1 \frac{x^2}{\sqrt{x+1}}\,dx.$

49. $\displaystyle\int_{-1}^0 x^3(x^2+1)^6\,dx.$ [set $u = x^2 + 1$]   **50.** $\displaystyle\int_1^{\sqrt{2}} x^3(x^2-1)^7\,dx.$

5.9   INTEGRATING THE TRIGONOMETRIC FUNCTIONS

Here and there you have been integrating some trigonometric functions, but we have not really focused on them. Now we do that.

Applying the Fundamental Theorem of Integral Calculus to the basic differentiation formulas we have

(5.9.1)
$$\int \cos x\,dx = \sin x + C, \qquad \int \sin x\,dx = -\cos x + C,$$
$$\int \sec^2 x\,dx = \tan x + C, \qquad \int \csc^2 x\,dx = -\cot x + C,$$
$$\int \sec x \tan x\,dx = \sec x + C, \qquad \int \csc x \cot x\,dx = -\csc x + C.$$

Problem 1. Find
$$\int \sin x \cos x\,dx.$$

Solution. Set
$$u = \sin x, \qquad du = \cos x\,dx.$$

Then
$$\int \sin x \cos x\,dx = \int u\,du = \tfrac{1}{2}u^2 + C = \tfrac{1}{2}\sin^2 x + C. \quad \square$$

Problem 2. Find
$$\int \sec^3 x \tan x\,dx.$$

Solution. Set
$$u = \sec x, \qquad du = \sec x \tan x\,dx.$$

Then

$$\sec^3 x \tan x \, dx = \underbrace{\sec^2 x}_{u^2} \underbrace{(\sec x \tan x) \, dx}_{du} = u^2 \, du$$

and

$$\int \sec^3 x \tan x \, dx = \int u^2 \, du = \tfrac{1}{3}u^3 + C = \tfrac{1}{3} \sec^3 x + C. \quad \square$$

Where an antiderivative is obvious, there is no need to make a substitution:

(1) $$\int \cos 3x \, dx = \tfrac{1}{3} \sin 3x + C$$

(2) $$\int \sec^2 \frac{\pi}{2} x \, dx = \frac{2}{\pi} \tan \frac{\pi}{2} x + C$$

(3) $$\int \sec (\pi - t) \tan (\pi - t) \, dt = - \sec (\pi - t) + C.$$

We can, of course, derive these results by substitution.

For (1) set

$$u = 3x, \qquad du = 3 \, dx.$$

Then

$$\int \cos 3x \, dx = \tfrac{1}{3} \int \cos u \, du = \tfrac{1}{3} \sin u + C = \tfrac{1}{3} \sin 3x + C.$$

For (2) set

$$u = \frac{\pi}{2} x, \qquad du = \frac{\pi}{2} \, dx.$$

Then

$$\int \sec^2 \frac{\pi}{2} x \, dx = \frac{2}{\pi} \int \sec^2 u \, du = \frac{2}{\pi} \tan u + C = \frac{2}{\pi} \tan \frac{\pi}{2} x + C.$$

For (3) set

$$u = \pi - t, \qquad du = - dt.$$

Then

$$\int \sec (\pi - t) \tan (\pi - t) \, dt = - \int \sec u \tan u \, du$$

$$= - \sec u + C = - \sec (\pi - t) + C. \quad \square$$

Problem 3. Find

$$\int x \cos \pi x^2 \, dx.$$

Solution. Set

$$u = \pi x^2, \qquad du = 2\pi x \, dx.$$

Then

$$x \cos \pi x^2 \, dx = \underbrace{\cos \pi x^2}_{\cos u} \underbrace{x \, dx}_{\frac{1}{2\pi} du} = \frac{1}{2\pi} \cos u \, du$$

and

$$\int x \cos \pi x^2 \, dx = \frac{1}{2\pi} \int \cos u \, du = \frac{1}{2\pi} \sin u + C = \frac{1}{2\pi} \sin \pi x^2 + C. \quad \square$$

Problem 4. Find

$$\int x^2 \csc^2 x^3 \cot^4 x^3 \, dx.$$

Solution. We can effect some simplification by setting

$$u = x^3, \qquad du = 3x^2 \, dx.$$

Then

$$x^2 \csc^2 x^3 \cot^4 x^3 \, dx = \underbrace{\cot^4 x^3}_{\cot^4 u} \underbrace{\csc^2 x^3}_{\csc^2 u} \underbrace{x^2 \, dx}_{\frac{1}{3} du} = \tfrac{1}{3} \cot^4 u \csc^2 u \, du$$

and

$$\int x^2 \csc^2 x^3 \cot^4 x^3 \, dx = \tfrac{1}{3} \int \cot^4 u \csc^2 u \, du.$$

We can calculate the integral on the right by setting

$$t = \cot u, \qquad dt = -\csc^2 u \, du.$$

Then

$$\tfrac{1}{3} \int \cot^4 u \csc^2 u \, du = -\tfrac{1}{3} \int t^4 \, dt = -\tfrac{1}{15} t^5 + C = -\tfrac{1}{15} \cot^5 u + C,$$

and thus

$$\int x^2 \csc^2 x^3 \cot^4 x^3 \, dx = -\tfrac{1}{15} \cot^5 u + C = -\tfrac{1}{15} \cot^5 x^3 + C.$$

We arrived at this by making two consecutive substitutions. First we set $u = x^3$

and then we set $t = \cot u$. We could have saved ourselves some work by setting $u = \cot x^3$ at the very beginning. With

$$u = \cot x^3, \qquad du = -\csc^2 x^3 \cdot 3x^2 \, dx$$

we have

$$x^2 \csc^2 x^3 \cot^4 x^3 \, dx = \underbrace{\cot^4 x^3}_{u^4} \underbrace{\csc^2 x^3 \, x^2 \, dx}_{-\frac{1}{3} \, du} = -\tfrac{1}{3} u^4 \, du$$

and

$$\int x^2 \csc^2 x^3 \cot^4 x^3 \, dx = -\tfrac{1}{3} \int u^4 \, du = -\tfrac{1}{15} u^5 + C = -\tfrac{1}{15} \cot^5 x^3 + C. \quad \square$$

Remark. All the integrals that we calculated in this section by substitution can be calculated without substitution. All that is required is a good sense of the chain rule.

1. $\int \sin x \cos x \, dx$. The cosine is the derivative of the sine. Thus

$$\int \sin x \cos x \, dx = \int \sin x \frac{d}{dx} (\sin x) \, dx = \tfrac{1}{2} \sin^2 x + C.$$

2. $\int \sec^3 x \tan x \, dx$. Write the integrand as

$$\sec^2 x (\sec x \tan x) = \sec^2 x \frac{d}{dx} (\sec x).$$

Then

$$\int \sec^3 x \tan x \, dx = \int \sec^2 x \frac{d}{dx} (\sec x) \, dx = \tfrac{1}{3} \sec^3 x + C.$$

3. $\int x \cos \pi x^2 \, dx$. The cosine is the derivative of the sine. Therefore

$$\frac{d}{dx} (\sin \pi x^2) = \cos \pi x^2 \cdot 2\pi x \qquad \text{and} \qquad x \cos \pi x^2 = \frac{1}{2\pi} \frac{d}{dx} (\sin \pi x^2).$$

Thus

$$\int x \cos \pi x^2 \, dx = \frac{1}{2\pi} \int \frac{d}{dx} (\sin \pi x^2) \, dx = \frac{1}{2\pi} \sin \pi x^2 + C.$$

4. $\int x^2 \csc^2 x^3 \cot^4 x^3 \, dx$. This integral may look complicated, but looked at in the right way, it becomes very simple. You know that the derivative of the cotangent is the negative of the cosecant squared. Therefore, by the chain rule,

$$\frac{d}{dx} (\cot x^3) = -\csc^2 x^3 \cdot 3x^2$$

and

$$x^2 \csc^2 x^3 = -\frac{1}{3} \frac{d}{dx} (\cot x^3).$$

Thus

$$\int x^2 \csc^2 x^3 \cot^4 x^3 \, dx = -\frac{1}{3} \int \cot^4 x^3 \cdot \frac{d}{dx} (\cot x^3) \, dx = -\frac{1}{15} \cot^5 x^3 + C.$$

There is nothing wrong with calculating integrals by substitution. All we are saying is that, with some experience, you will be able to calculate many integrals without it. ◻

We close this section by giving two important formulas:

(5.9.2) $$\int \cos^2 x \, dx = \tfrac{1}{2}x + \tfrac{1}{4} \sin 2x + C, \qquad \int \sin^2 x \, dx = \tfrac{1}{2}x - \tfrac{1}{4} \sin 2x + C.$$

You can derive these formulas by recalling that

$$\cos^2 x = \tfrac{1}{2}(1 + \cos 2x) \qquad \text{and} \qquad \sin^2 x = \tfrac{1}{2}(1 - \cos 2x).$$

We leave the details to you.

EXERCISES 5.9

Calculate these indefinite integrals.

1. $\displaystyle\int \cos (3x - 1) \, dx.$ 2. $\displaystyle\int \sin 2\pi x \, dx.$ 3. $\displaystyle\int \csc^2 \pi x \, dx.$

4. $\displaystyle\int \sec 2x \tan 2x \, dx.$ 5. $\displaystyle\int \sin (3 - 2x) \, dx.$ 6. $\displaystyle\int \sin^2 x \cos x \, dx.$

7. $\displaystyle\int \cos^4 x \sin x \, dx.$ 8. $\displaystyle\int x \sec^2 x^2 \, dx.$ 9. $\displaystyle\int x^{-1/2} \sin x^{1/2} \, dx.$

10. $\displaystyle\int \csc (1 - 2x) \cot (1 - 2x) \, dx.$ 11. $\displaystyle\int \sqrt{1 + \sin x} \cos x \, dx.$

12. $\displaystyle\int \frac{\sin x}{\sqrt{1 + \cos x}} \, dx.$ 13. $\displaystyle\int \frac{1}{\cos^2 x} \, dx.$

14. $\displaystyle\int (1 + \tan^2 x) \sec^2 x \, dx.$ 15. $\displaystyle\int x \sin^3 x^2 \cos x^2 \, dx.$

16. $\displaystyle\int \sqrt{1 + \cot x} \csc^2 x \, dx.$ 17. $\displaystyle\int (1 + \cot^2 x) \csc^2 x \, dx.$

18. $\displaystyle\int \cos^2 5x \, dx.$ [use (5.9.2)]   19. $\displaystyle\int \sin^2 3x \, dx.$ [use (5.9.2)]

20. $\displaystyle\int \frac{1}{\sin^2 x} \, dx.$ 21. $\displaystyle\int \frac{\sec^2 x}{\sqrt{1 + \tan x}} \, dx.$ 22. $\displaystyle\int x^2 \sin (4x^3 - 7) \, dx.$

Evaluate these integrals.

23. $\int_{-\pi}^{\pi} \sin x \, dx.$ **24.** $\int_{-\pi/3}^{\pi/3} \sec x \tan x \, dx.$ **25.** $\int_{1/4}^{1/3} \sec^2 \pi x \, dx.$

26. $\int_{0}^{1} \cos^2 \frac{\pi}{2} x \sin \frac{\pi}{2} x \, dx.$ **27.** $\int_{0}^{\pi/2} \sin^3 x \cos x \, dx.$ **28.** $\int_{0}^{\pi} \cos x \, dx.$

29. $\int_{\pi/6}^{\pi/4} \csc x \cot x \, dx.$ **30.** $\int_{0}^{2\pi} \sin^2 x \, dx.$ **31.** $\int_{0}^{2\pi} \cos^2 x \, dx.$

Find the area bounded by the following curves.

32. $y = \cos \pi x, \quad y = \sin \pi x, \quad x = 0, \quad x = \frac{1}{4}.$
33. $y = \cos^2 \pi x, \quad y = \sin^2 \pi x, \quad x = 0, \quad x = \frac{1}{4}.$
34. $y = \cos^2 \pi x, \quad y = -\sin^2 \pi x, \quad x = 0, \quad x = \frac{1}{4}.$
35. $y = \csc^2 \pi x, \quad y = \sec^2 \pi x, \quad x = \frac{1}{6}, \quad x = \frac{1}{4}.$

36. In the text we found that

$$\int \sin x \cos x \, dx = \tfrac{1}{2} \sin^2 x + C$$

by setting $u = \sin x$. Calculate the integral by setting $u = \cos x$ and then reconcile the two answers.

37. Calculate

$$\int \sec^2 x \tan x \, dx$$

(a) setting $u = \sec x$; (b) setting $u = \tan x$. (c) Reconcile your answers to (a) and (b).

38. (*The area of a circular region*) The circle $x^2 + y^2 = r^2$ encloses a circular disc of radius r. Justify the familiar formula $A = \pi r^2$ by integration.   HINT: The quarter disc in the first quadrant is the region below the curve $y = \sqrt{r^2 - x^2}, \quad x \in [0, r]$. Therefore

$$A = 4 \int_{0}^{r} \sqrt{r^2 - x^2} \, dx.$$

Evaluate the integral by setting $x = r \sin u, \quad dx = r \cos u \, du.$

39. Find the area enclosed by the ellipse $b^2 x^2 + a^2 y^2 = a^2 b^2$.

5.10   SOME FURTHER PROPERTIES OF THE DEFINITE INTEGRAL

In this section we feature some important general properties of the integral. The proofs are left mostly to you. You can assume throughout that the functions involved are continuous and that $a < b$.

I. The integral of a nonnegative function is nonnegative:

(5.10.1)   if $f(x) \geq 0$ for all $x \in [a, b]$,   then $\int_{a}^{b} f(x) \, dx \geq 0.$

The integral of a positive function is positive:

(5.10.2)

$$\text{if} \quad f(x) > 0 \text{ for all } x \in [a, b], \qquad \text{then} \qquad \int_a^b f(x)\, dx > 0.$$

The next property is an immediate consequence of Property I and linearity (5.4.6).

II. The integral is order-preserving:

(5.10.3)

$$\text{if} \quad f(x) \le g(x) \text{ for all } x \in [a, b], \qquad \text{then} \qquad \int_a^b f(x)\, dx \le \int_a^b g(x)\, dx$$

and

(5.10.4)

$$\text{if} \quad f(x) < g(x) \text{ for all } x \in [a, b], \qquad \text{then} \qquad \int_a^b f(x)\, dx < \int_a^b g(x)\, dx.$$

III. Just as the absolute value of a sum is less than or equal to the sum of the absolute values,

$$|x_1 + x_2 + \cdots + x_n| \le |x_1| + |x_2| + \cdots + |x_n|,$$

the absolute value of an integral is less than or equal to the integral of the absolute value:

(5.10.5)

$$\left| \int_a^b f(x)\, dx \right| \le \int_a^b |f(x)|\, dx.$$

Hint for Proof of (5.10.5). Show that

$$\int_a^b f(x)\, dx \qquad \text{and} \qquad -\int_a^b f(x)\, dx$$

are both less than or equal to

$$\int_a^b |f(x)|\, dx. \quad \square$$

The next property is one with which you are already familiar.

IV. If m is the minimum value of f on $[a, b]$ and M is the maximum value, then

(5.10.6)

$$m(b - a) \le \int_a^b f(x)\, dx \le M(b - a).$$

We come now to a generalization of Theorem 5.3.5. Its importance will become apparent in Chapter 7.

V. If u is a differentiable function of x and f is continuous, then

(5.10.7)
$$\frac{d}{dx}\left(\int_a^u f(t)\,dt\right) = f(u)\frac{du}{dx}.$$

Proof. The function

$$F(u) = \int_a^u f(t)\,dt$$

is differentiable with respect to u and

$$\frac{d}{du}[F(u)] = f(u). \qquad \text{(Theorem 5.3.5)}$$

Thus

$$\frac{d}{dx}\left(\int_a^u f(t)\,dt\right) = \frac{d}{dx}[F(u)] = \frac{d}{du}[F(u)]\frac{du}{dx} = f(u)\frac{du}{dx}. \quad \square$$

chain rule

Problem 1. Find

$$\frac{d}{dx}\left(\int_0^{x^3}\frac{dt}{1+t}\right).$$

Solution. At this stage you would probably be hard put to carry out the integration; it requires the logarithm function, not introduced in this text until Chapter 7. But, for our purposes, that doesn't matter. By (5.10.7) you know that

$$\frac{d}{dx}\left(\int_0^{x^3}\frac{dt}{1+t}\right) = \frac{1}{1+x^3}\,3x^2 = \frac{3x^2}{1+x^3}$$

without carrying out the integration. $\square$

Problem 2. Find

$$\frac{d}{dx}\left(\int_x^{2x}\frac{dt}{1+t^2}\right).$$

Solution. The idea is to express the integral in terms of integrals that have constant lower limits. Once we have done that, we can differentiate as before. By the additivity of the integral

$$\int_0^x\frac{dt}{1+t^2} + \int_x^{2x}\frac{dt}{1+t^2} = \int_0^{2x}\frac{dt}{1+t^2}.$$

Thus

$$\int_x^{2x}\frac{dt}{1+t^2} = \int_0^{2x}\frac{dt}{1+t^2} - \int_0^x\frac{dt}{1+t^2}.$$

Differentiation gives

$$\frac{d}{dx}\left(\int_x^{2x} \frac{dt}{1+t^2}\right) = \frac{d}{dx}\left(\int_0^{2x} \frac{dt}{1+t^2}\right) - \frac{d}{dx}\left(\int_0^x \frac{dt}{1+t^2}\right)$$

$$= \frac{1}{1+(2x)^2}(2) - \frac{1}{1+x^2}(1) = \frac{2}{1+4x^2} - \frac{1}{1+x^2}. \quad \square$$

$(5.10.7)$ ⟶

EXERCISES 5.10

Assume: f and g continuous, $a < b$, and

$$\int_a^b f(x)\, dx > \int_a^b g(x)\, dx.$$

Answer questions 1–6 giving supporting reasons.

1. Does it necessarily follow that $\int_a^b [f(x) - g(x)]\, dx > 0$?

2. Does it necessarily follow that $f(x) > g(x)$ for all $x \in [a, b]$?
3. Does it necessarily follow that $f(x) > g(x)$ for at least some $x \in [a, b]$?

4. Does it necessarily follow that $\left|\int_a^b f(x)\, dx\right| > \left|\int_a^b g(x)\, dx\right|$?

5. Does it necessarily follow that $\int_a^b |f(x)|\, dx > \int_a^b |g(x)|\, dx$?

6. Does it necessarily follow that $\int_a^b |f(x)|\, dx > \int_a^b g(x)\, dx$?

Assume: f continuous, $a < b$, and

$$\int_a^b f(x)\, dx = 0.$$

Answer questions 7–15 giving supporting reasons.

7. Does it necessarily follow that $f(x) = 0$ for all $x \in [a, b]$?
8. Does it necessarily follow that $f(x) = 0$ for at least some $x \in [a, b]$?

9. Does it necessarily follow that $\int_a^b |f(x)|\, dx = 0$?

10. Does it necessarily follow that $\left|\int_a^b f(x)\, dx\right| = 0$?

11. Must all upper sums $U_f(P)$ be nonnegative?
12. Must all upper sums $U_f(P)$ be positive?
13. Can a lower sum, $L_f(P)$, be positive?

14. Does it necessarily follow that $\int_a^b [f(x)]^2\, dx = 0$?

15. Does it necessarily follow that $\int_a^b [f(x) + 1]\, dx = b - a$?

Calculate the following derivatives.

16. $\dfrac{d}{dx}\left(\displaystyle\int_{1}^{x^2} \dfrac{dt}{t} \right).$

17. $\dfrac{d}{dx}\left(\displaystyle\int_{0}^{1+x^2} \dfrac{dt}{\sqrt{2t+5}} \right).$

18. $\dfrac{d}{dx}\left(\displaystyle\int_{0}^{x^3} \dfrac{dt}{\sqrt{1+t^2}} \right).$

19. $\dfrac{d}{dx}\left(\displaystyle\int_{x}^{a} f(t)\,dt \right).$

20. Show that, if u and v are differentiable functions of x and f is continuous, then

$$\frac{d}{dx}\left(\int_{u}^{v} f(t)\,dt \right) = f(v)\frac{dv}{dx} - f(u)\frac{du}{dx}.$$

HINT: Take a number a from the domain of f. Express the integral as the difference of two integrals, each with lower limit a.

Calculate the following derivatives using Exercise 20.

21. $\dfrac{d}{dx}\left(\displaystyle\int_{x}^{x^2} \dfrac{dt}{t} \right).$

22. $\dfrac{d}{dx}\left(\displaystyle\int_{1-x}^{1+x} \dfrac{t-1}{t}\,dt \right).$

23. $\dfrac{d}{dx}\left(\displaystyle\int_{x^{1/3}}^{2+3x} \dfrac{dt}{1+t^{3/2}} \right).$

24. $\dfrac{d}{dx}\left(\displaystyle\int_{\sqrt{x}}^{x^2+x} \dfrac{dt}{2+\sqrt{t}} \right).$

25. Verify Property I: (a) by considering lower sums $L_f(P)$; (b) by using an antiderivative.

26. Verify Property II.   **27.** Verify Property III.

28. (*Important*) Prove that, if f is continuous on $[a, b]$ and

$$\int_{a}^{b} |f(x)|\,dx = 0,$$

then $f(x) = 0$ for all x in $[a, b]$.   HINT: Exercise 48, Section 2.4.

29. Find $H'(2)$ given that

$$H(x) = \int_{2x}^{x^3-4} \frac{x}{1+\sqrt{t}}\,dt.$$

30. Find $H'(3)$ given that

$$H(x) = \frac{1}{x}\int_{3}^{x} [2t - 3H'(t)]\,dt.$$

5.11   MEAN-VALUE THEOREMS FOR INTEGRALS; AVERAGE VALUES

We begin with a result that we asked you to prove in Exercise 15, Section 5.3.

THEOREM 5.11.1   THE FIRST MEAN-VALUE THEOREM FOR INTEGRALS

If f is continuous on $[a, b]$, then there is a number c in (a, b) for which

$$\int_{a}^{b} f(x)\,dx = f(c)(b - a).$$

This number $f(c)$ is called *the average value of f on $[a, b]$.*

We then have the following identity:

(5.11.2)
$$\int_a^b f(x)\, dx = (\text{the average value of } f \text{ on } [a, b]) \cdot (b - a).$$

This identity provides a powerful and intuitive way of viewing the definite integral. Think for a moment about area. If f is constant and positive on $[a, b]$, then Ω, the region below the graph, is a rectangle. Its area is given by the formula

$$\text{area of } \Omega = (\text{the constant value of } f \text{ on } [a, b]) \cdot (b - a). \qquad \text{(Figure 5.11.1)}$$

If f is now allowed to vary continuously on $[a, b]$, then we have

$$\text{area of } \Omega = \int_a^b f(x)\, dx,$$

and the area formula reads

$$\text{area of } \Omega = (\text{the average value of } f \text{ on } [a, b]) \cdot (b - a). \qquad \text{(Figure 5.11.2)}$$

area = (the constant value of f) $\cdot$ $(b - a)$

Figure 5.11.1

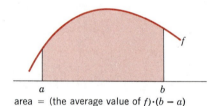

area = (the average value of f)$\cdot(b - a)$

Figure 5.11.2

Think now about motion. If an object moves along a line with constant speed $|v|$ during the time interval $[a, b]$, then

$$\text{the distance traveled} = (\text{the constant value of } |v| \text{ on } [a, b]) \cdot (b - a).$$

If the speed $|v|$ varies, then we have

$$\text{distance traveled} = \int_a^b |v(t)|\, dt$$

and the formula reads

$$\text{distance traveled} = (\text{the average value of } |v| \text{ on } [a, b]) \cdot (b - a).$$

We will take an interval $[a, b]$ and average out on that interval the simplest possible functions. The average of a constant k is of course k:

$$\int_a^b k\, dx = k(b - a).$$

The average of x is $\frac{1}{2}(b + a)$, the center of the interval:

$$\int_a^b x \, dx = \left[\frac{1}{2}x^2\right]_a^b = \frac{1}{2}(b^2 - a^2) = [\frac{1}{2}(b + a)](b - a).$$

What is the average of x^2?

$$\int_a^b x^2 \, dx = \left[\frac{1}{3}x^3\right]_a^b = \frac{1}{3}(b^3 - a^3) = [\frac{1}{3}(b^2 + ab + a^2)](b - a).$$

The average of x^2 on $[a, b]$ is $\frac{1}{3}(b^2 + ab + a^2)$.

There is an extension of Theorem 5.11.1 that, as you will see, is useful in applications.

THEOREM 5.11.3   THE SECOND MEAN-VALUE THEOREM FOR INTEGRALS

If f and g are continuous on $[a, b]$ and g is nonnegative, then there is a number c in (a, b) for which

$$\int_a^b f(x)g(x) \, dx = f(c) \int_a^b g(x) \, dx.$$

This number $f(c)$ is called *the g-weighted average of f on* $[a, b]$.

We will prove this theorem (and thereby have a proof of Theorem 5.11.1) at the end of the section.   First a brief excursion into physics.

The Mass of a Rod.   Imagine a thin rod (a straight material wire of negligible thickness) lying on the x-axis from $x = a$ to $x = b$.   If the *mass density* of the rod (the mass per unit length) is constant, then the total mass M of the rod is simply the density λ times the length of the rod: $M = \lambda(b - a)$.†   If the density λ varies continuously from point to point, say $\lambda = \lambda(x)$, then the mass of the rod is the average density of the rod times the length of the rod:

$$M = (\text{average density}) \times (\text{length}).$$

This is an integral:

(5.11.4)
$$M = \int_a^b \lambda(x) \, dx.$$

The Center of Mass of a Rod.   Continue with that same rod.   If the rod is homogeneous (constant density), then the center of mass x_M of the rod is simply the midpoint of the rod:

$$x_M = \frac{1}{2}(a + b). \qquad \text{(the average of x from a to b)}$$

---

† The symbol λ is the Greek letter "lambda."

If the rod is not homogeneous, the center of mass is still an average, but now a weighted average, *the density-weighted average of x from a to b*; namely, x_M is the point for which

$$x_M \int_a^b \lambda(x)\, dx = \int_a^b x\, \lambda(x)\, dx.$$

Since the integral on the left is simply M, we have

(5.11.5)
$$x_M M = \int_a^b x\, \lambda(x)\, dx.$$

Problem 1. A rod of length L is placed on the x-axis from $x = 0$ to $x = L$. Find the mass of the rod and the center of mass if the density of the rod varies directly as the distance from the $x = 0$ endpoint of the rod.

Solution. Here $\lambda(x) = kx$ where k is some positive constant. Therefore

$$M = \int_0^L kx\, dx = \left[\tfrac{1}{2}kx^2\right]_0^L = \tfrac{1}{2}kL^2$$

and

$$x_M M = \int_0^L x(kx)\, dx = \int_0^L kx^2\, dx = \left[\tfrac{1}{3}kx^3\right]_0^L = \tfrac{1}{3}kL^3.$$

Division by M gives $x_M = \tfrac{2}{3}L$.

In this instance the center of mass is to the right of the midpoint. This makes sense. After all, the density increases from left to right. Thus there is more mass to the right of the midpoint than there is to the left of it. ☐

We go back now to Theorem 5.11.3 and prove it. [There is no reason to worry about Theorem 5.11.1. That is just Theorem 5.11.3 with $g(x)$ identically 1.]

Proof of Theorem 5.11.3. Since f is continuous on $[a, b]$, f takes on a minimum value m on $[a, b]$ and a maximum value M. Since g is nonnegative on $[a, b]$,

$$mg(x) \le f(x)g(x) \le Mg(x) \qquad \text{for all } x \text{ in } [a, b].$$

Therefore

$$\int_a^b mg(x)\, dx \le \int_a^b f(x)g(x)\, dx \le \int_a^b Mg(x)\, dx$$

and

$$m \int_a^b g(x)\, dx \le \int_a^b f(x)g(x)\, dx \le M \int_a^b g(x)\, dx.$$

We know that $\int_a^b g(x)\, dx \geq 0$. If $\int_a^b g(x)\, dx = 0$, then $\int_a^b f(x)g(x)\, dx = 0$ and the theorem holds for all choices of c in (a, b). If $\int_a^b g(x)\, dx > 0$, then

$$m \leq \frac{\int_a^b f(x)g(x)\, dx}{\int_a^b g(x)\, dx} \leq M$$

and by the intermediate-value theorem (Theorem 2.6.1) there exists c in (a, b) for which

$$f(c) = \frac{\int_a^b f(x)g(x)\, dx}{\int_a^b g(x)\, dx}.$$

Obviously then

$$f(c) \int_a^b g(x)\, dx = \int_a^b f(x)g(x)\, dx. \quad \square$$

EXERCISES 5.11

Determine the average value on the indicated interval and find a point in this interval at which the function takes on this average value.

1. $f(x) = mx + b, \quad x \in [0, c]$.
2. $f(x) = x^2, \quad x \in [-1, 1]$.
3. $f(x) = x^3, \quad x \in [-1, 1]$.
4. $f(x) = x^{-2}, \quad x \in [1, 4]$.
5. $f(x) = |x|, \quad x \in [-2, 2]$.
6. $f(x) = x^{1/3}, \quad x \in [-8, 8]$.
7. $f(x) = 2x - x^2, \quad x \in [0, 2]$.
8. $f(x) = 3 - 2x, \quad x \in [0, 3]$.
9. $f(x) = \sqrt{x}, \quad x \in [0, 9]$.
10. $f(x) = 4 - x^2, \quad x \in [-2, 2]$.
11. $f(x) = \sin x, \quad x \in [0, 2\pi]$.
12. $f(x) = \cos x, \quad x \in [0, \pi]$.

13. Solve the following equation for A:

$$\int_a^b [f(x) - A]\, dx = 0.$$

14. Given that f is continuous on $[a, b]$, compare

$$f(b)(b - a) \quad \text{and} \quad \int_a^b f(x)\, dx$$

if f is (a) constant on $[a, b]$; (b) increasing on $[a, b]$; (c) decreasing on $[a, b]$.

15. In Chapter 4, we viewed $[f(b) - f(a)]/(b - a)$ as the average rate of change of f on $[a, b]$ and $f'(t)$ as the instantaneous rate of change at time t. If our new sense of average is to be consistent with the old one, we must have

$$\frac{f(b) - f(a)}{b - a} = \text{average of } f' \text{ on } [a, b].$$

Prove that this is the case.

16. Determine whether the assertion is true or false.

(a) $\left(\begin{array}{c} \text{the average of } f + g \\ \text{on } [a, b] \end{array}\right) = \left(\begin{array}{c} \text{the average of } f \\ \text{on } [a, b] \end{array}\right) + \left(\begin{array}{c} \text{the average of } g \\ \text{on } [a, b] \end{array}\right)$.

(b) $\left(\begin{array}{c} \text{the average of } \alpha f \\ \text{on } [a, b] \end{array}\right) = \alpha \left(\begin{array}{c} \text{the average of } f \\ \text{on } [a, b] \end{array}\right)$.

(c) $\left(\begin{array}{c} \text{the average of } fg \\ \text{on } [a, b] \end{array}\right) = \left(\begin{array}{c} \text{the average of } f \\ \text{on } [a, b] \end{array}\right)\left(\begin{array}{c} \text{the average of } g \\ \text{on } [a, b] \end{array}\right)$.

17. Find the average distance of the parabolic arc

$$y = x^2, \qquad x \in [0, \sqrt{3}]$$

from (a) the x-axis; (b) the y-axis; (c) the origin.

18. Find the average distance of the line segment

$$y = mx, \qquad x \in [0, 1]$$

from (a) the x-axis; (b) the y-axis; (c) the origin.

19. A stone falls from rest in a vacuum for x seconds. (a) Compare its terminal velocity to its average velocity; (b) compare its average velocity during the first $\frac{1}{2}x$ seconds to its average velocity during the next $\frac{1}{2}x$ seconds.

20. Let f be continuous. Show that, if f is an odd function, then its average value on every interval of the form $[-a, a]$ is zero.

21. A rod of length L is placed on the x-axis from $x = 0$ to $x = L$. Find the mass of the rod and the center of mass if the mass density of the rod varies directly (a) as the square root of the distance from $x = 0$. (b) as the square of the distance from $x = L$.

22. A rod of varying density, mass M, and center of mass x_M extends from $x = a$ to $x = b$. A partition $P = \{x_0, x_1, \ldots, x_n\}$ of $[a, b]$ decomposes the rod into n pieces in the obvious way. Show that, if the n pieces have masses $M_1, M_2, \ldots, M_n$ and centers of mass x_{M_1}, $x_{M_2}, \ldots, x_{M_n}$, then

$$x_M M = x_{M_1} M_1 + x_{M_2} M_2 + \cdots + x_{M_n} M_n.$$

23. A rod that has mass M and extends from $x = 0$ to $x = L$ consists of two pieces with masses M_1, M_2. Given that the center of mass of the entire rod is at $x = \frac{1}{4}L$ and the center of mass of the first piece is at $x = \frac{1}{8}L$, determine the center of mass of the second piece.

24. A rod that has mass M and extends from $x = 0$ to $x = L$ consists of two pieces. Find the mass of each piece given that the center of mass of the entire rod is at $x = \frac{2}{3}L$, the center of mass of the first piece is at $x = \frac{1}{4}L$, and the center of mass of the second piece is at $x = \frac{7}{8}L$.

25. Prove Theorem 5.11.1 by applying the mean-value theorem of differential calculus to the function

$$F(x) = \int_a^x f(t) \, dt, \qquad x \in [a, b].$$

(This exercise was given before.)

26. Let f be continuous on $[a, b]$. Let $a < c < b$. Prove that

$$f(c) = \lim_{h \to 0^+} (\text{average value of } f \text{ on } [c - h, c + h]).$$

27. Prove that two distinct continuous functions cannot have the same average on every interval.

5.12 ADDITIONAL EXERCISES

Work out the following integrals.

1. $\displaystyle\int (\sqrt{x - a} - \sqrt{x - b}) \, dx.$ **2.** $\displaystyle\int ax\sqrt{1 + bx^2} \, dx.$ **3.** $\displaystyle\int t^{-1/3}(t^{2/3} - 1)^2 \, dt.$

4. $\displaystyle\int t^2(1 + t^3)^{10} \, dt.$ **5.** $\displaystyle\int (1 + 2\sqrt{x})^2 \, dx.$ **6.** $\displaystyle\int \frac{1}{\sqrt{x}} (1 + 2\sqrt{x})^5 \, dx.$

7. $\displaystyle\int (x^{1/5} - x^{-1/5})^2 \, dx.$

8. $\displaystyle\int x\sqrt{2 - x^2} \, dx.$

9. $\displaystyle\int x\sqrt{2 - x} \, dx.$

10. $\displaystyle\int \frac{(1 + \sqrt{x})^5}{\sqrt{x}} \, dx.$

11. $\displaystyle\int \frac{(a + b\sqrt{y + 1})^2}{\sqrt{y + 1}} \, dy.$

12. $\displaystyle\int y\sqrt{y}(1 + y^2\sqrt{y})^2 \, dy.$

13. $\displaystyle\int \frac{g'(x)}{[g(x)]^3} \, dx.$

14. $\displaystyle\int [f'(x) + \cdots + f^{(n)}(x)] \, dx.$

15. $\displaystyle\int \frac{g(x)g'(x)}{\sqrt{1 + [g(x)]^2}} \, dx.$

16. $\displaystyle\int x \sin^3 x^2 \cos x^2 \, dx.$

17. $\displaystyle\int (\sec \theta - \tan \theta)^2 \, d\theta.$

18. $\displaystyle\int (\tan 3\theta + \cot 3\theta)^2 \, d\theta.$

19. $\displaystyle\int \frac{dx}{1 + \cos 2x}.$ HINT: Multiply numerator and denominator by $1 - \cos 2x$.

20. $\displaystyle\int \frac{dx}{1 + \sin 2x}.$

21. $\displaystyle\int \sec^4 \pi x \tan \pi x \, dx.$

22. $\displaystyle\int ax\sqrt{1 + bx} \, dx.$

23. $\displaystyle\int (1 + \sin^2 \pi x)^{-3} \sin 2\pi x \, dx.$

24. $\displaystyle\int ax^2\sqrt{1 + bx} \, dx.$

Find the area below the graph.

25. $y = x\sqrt{2x^2 + 1}, \quad x \in [0, 2].$

26. $y = \dfrac{x}{(2x^2 + 1)^2}, \quad x \in [0, 2].$

27. $y = x^{-3}(1 + x^{-2})^{-3}, \quad x \in [1, 2].$

28. $y = \dfrac{2x + 5}{(x + 2)^2(x + 3)^2}, \quad x \in [0, 1].$

29. At every point of the curve the slope is given by the equation

$$\frac{dy}{dx} = x\sqrt{x^2 + 1}.$$

Find an equation for the curve given that it passes through the point $(0, 1)$.

30. Find the area of the portion of the first quadrant that is bounded

above by $y = 2x$ and   below by $y = x\sqrt{3x^2 + 1}.$

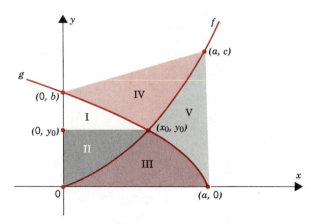

Figure 5.12.1

31. Find a formula for the area of each region drawn in Figure 5.12.1:
   (a) I.   (b) II.   (c) III.   (d) IV.   (e) V.

Sketch the region bounded by the curves and find its area.

32. $y^2 = 2x$, $x^2 = 3y$.

33. $y^2 = 6x$, $x^2 + 4y = 0$.

34. $y = 4 - x^2$, $y = x + 2$.

35. $y = 4 - x^2$, $x + y + 2 = 0$.

36. $y = \sqrt{x}$, the x-axis, $y = 6 - x$.

37. $y = x^3$, the x-axis, $x + y = 2$.

38. $4y = x^2 - x^4$, $x + y + 1 = 0$.

39. $x^2y = a^2$, the x-axis, $x = a$, $x = 2a$.

40. Show that, if f', g', and h' are continuous and $h(a) = h(b)$, then

$$\int_a^b f'(g(h(x)))g'(h(x))h'(x)\, dx = 0.$$

Assume that f and g are continuous, that $a < b$, and that

$$\int_a^b f(x)\, dx > \int_a^b g(x)\, dx.$$

Which of the following statements necessarily hold for all partitions P of $[a, b]$?   Support your answers.

41. $L_g(P) < U_f(P)$.

42. $L_g(P) < L_f(P)$.

43. $L_g(P) < \int_a^b f(x)\, dx$.

44. $U_g(P) < U_f(P)$.

45. $U_f(P) > \int_a^b g(x)\, dx$.

46. $U_g(P) < \int_a^b f(x)\, dx$.

In Exercises 47 and 48 find the general law of motion of an object that moves along the x-axis with the given acceleration $a(t)$.   Write x_0 for initial position and v_0 for initial velocity.   Here ω, A, and ϕ_0 are constants.

47. $a(t) = -\omega^2 A \sin(\omega t + \phi_0)$.

48. $a(t) = -\omega^2 A \cos(\omega t + \phi_0)$.

49. Find an equation $y = f(x)$ for the curve that passes through the point $(4, \tfrac{1}{3})$ and has varying slope

$$\frac{dy}{dx} = -\frac{1}{2\sqrt{x}(1 + \sqrt{x})^2}.$$

50. Sketch the curve of Exercise 49.

51. (Calculator)   Estimate the area under the curve of Exercise 49 from $x = 0$ to $x = 1$ by means of the partition

$$P = \{0, \tfrac{1}{100}, \tfrac{4}{100}, \tfrac{9}{100}, \tfrac{16}{100}, \tfrac{25}{100}, \tfrac{36}{100}, \tfrac{49}{100}, \tfrac{64}{100}, \tfrac{81}{100}, 1\}.$$

Base your estimate on $\tfrac{1}{2}[L_f(P) + U_f(P)]$.

Carry out the differentiation.

52. $\dfrac{d}{dx}\left(\displaystyle\int_0^x \dfrac{dt}{1 + t^2}\right).$

53. $\dfrac{d}{dx}\left(\displaystyle\int_0^{x^2} \dfrac{dt}{1 + t^2}\right).$

54. $\dfrac{d}{dx}\left(\displaystyle\int_x^{x^2} \dfrac{dt}{1 + t^2}\right).$

55. $\dfrac{d}{dx}\left(\displaystyle\int_0^{\sin x} \dfrac{dt}{1 - t^2}\right).$

56. $\dfrac{d}{dx}\left(\displaystyle\int_0^{\cos x} \dfrac{dt}{1 - t^2}\right).$

57. $\dfrac{d}{dx}\left(\displaystyle\int_0^{\tan x} \dfrac{dt}{1 + t^2}\right).$

58. Define a function L by setting

$$L(x) = \int_1^x \frac{dt}{t} \qquad \text{for all } x > 0.$$

(a) Show that L is an increasing function.   (b) Where does L take on the value 0?
(c) Show that $L(xy) = L(x) + L(y)$ for all positive x and y.

59. A rod extends from $x = 0$ to $x = a$. Find the center of mass if the density of the rod varies directly as the distance from $x = 2a$.

60. A rod of mass M and length L is to be cut out from a long piece that extends to the right from $x = 0$. Where should the cuts be made if the density of the long piece varies directly as the distance from $x = 0$? (Assume that $M \geq \frac{1}{2}kL^2$ where k is the constant of proportionality in the density function.)

61. Where does the function

$$f(x) = \int_0^x \frac{t-1}{1+t^2}\, dt$$

take on its minimum value?

62. Show that

(a)
$$\int_{a+c}^{b+c} f(x-c)\, dx = \int_a^b f(x)\, dx,$$

and, if $c \neq 0$,

(b)
$$\frac{1}{c} \int_{ac}^{bc} f\left(\frac{x}{c}\right) dx = \int_a^b f(x)\, dx.$$

Assume that f is everywhere continuous.

5.13   THE INTEGRAL AS THE LIMIT OF RIEMANN SUMS

Let f be some function continuous on a closed interval $[a, b]$. With our approach to integration (at this point we ask you to review Section 5.2) the definite integral

$$\int_a^b f(x)\, dx$$

is the unique number that satisfies the inequality

$$L_f(P) \leq \int_a^b f(x)\, dx \leq U_f(P)$$

for all partitions P of $[a, b]$. This method of obtaining the definite integral by *squeezing* toward it with upper and lower sums is called the *Darboux method.*†

There is another way to obtain the integral that is frequently used. Take a partition $P = \{x_0, x_1, \ldots, x_n\}$ of $[a, b]$. P breaks up $[a, b]$ into n subintervals

$$[x_0, x_1], [x_1, x_2], \ldots, [x_{n-1}, x_n]$$

of lengths

$$\Delta x_1, \Delta x_2, \ldots, \Delta x_n.$$

Now pick a point x_1^* from $[x_0, x_1]$ and form the product $f(x_1^*)\, \Delta x_1$; pick a point x_2^* from $[x_1, x_2]$ and form the product $f(x_2^*)\, \Delta x_2$; go on in this manner until you have

---

† After the French mathematician J. G. Darboux (1842–1917).

formed the products

$$f(x_1^*)\,\Delta x_1, f(x_2^*)\,\Delta x_2, \ \ldots \ , f(x_n^*)\,\Delta x_n.$$

The sum of these products

$$S^*(P) = f(x_1^*)\,\Delta x_1 + f(x_2^*)\,\Delta x_2 + \cdots + f(x_n^*)\,\Delta x_n$$

is called a *Riemann sum.*†

The definite integral can be viewed as the *limit* of such Riemann sums in the following sense: define $\|P\|$, the *norm* of P, by setting

$$\|P\| = \max \Delta x_i \qquad i = 1, 2, \ \ldots \ , n;$$

given any $\epsilon > 0$, there exists a $\delta > 0$ such that

$$\text{if} \quad \|P\| < \delta \quad \text{then} \quad \left| S^*(P) - \int_a^b f(x)\,dx \right| < \epsilon,$$

no matter how the x_i^* are chosen within $[x_{i-1}, x_i]$.

In symbols we write

(5.13.1) $$\int_a^b f(x)\,dx = \lim_{\|P\|\to 0} [f(x_1^*)\,\Delta x_1 + f(x_2^*)\,\Delta x_2 + \cdots + f(x_n^*)\,\Delta x_n].$$

A proof of this assertion appears in Appendix B.5. Figure 5.13.1 illustrates the idea. Here the base interval is broken up into 8 subintervals. The point x_1^* is chosen from $[x_0, x_1]$, x_2^* from $[x_1, x_2]$, and so on. While the integral represents the area under the curve, the Riemann sum represents the sum of the areas of the shaded rectangles. The difference between the two can be made as small as we wish (less than ϵ) simply by making the maximum length of the base subintervals sufficiently small — that is, by making $\|P\|$ sufficiently small.

Figure 5.13.1

---

† After the German mathematician G. F. B. Riemann (1826–1866).

EXERCISES 5.13

1. Let Ω be the region below the graph of $f(x) = x^2$, $x \in [0,1]$. Draw a figure showing the Riemann sum $S^*(P)$ as an estimate for this area. Take

$$P = \{0, \tfrac{1}{4}, \tfrac{1}{2}, \tfrac{3}{4}, 1\} \quad \text{and} \quad x_1^* = \tfrac{1}{8}, \quad x_2^* = \tfrac{3}{8}, \quad x_3^* = \tfrac{5}{8}, \quad x_4^* = \tfrac{7}{8}.$$

2. Let Ω be the region below the graph of $f(x) = \tfrac{3}{2}x + 1$, $x \in [0, 2]$. Draw a figure showing the Riemann sum $S^*(P)$ as an estimate for this area. Take

$$P = \{0, \tfrac{1}{4}, \tfrac{3}{4}, 1, \tfrac{3}{2}, 2\} \quad \text{and let the } x_i^* \text{ be the midpoints of the subintervals.}$$

3. Set $f(x) = 2x$, $x \in [0, 1]$. Take $P = \{0, \tfrac{1}{8}, \tfrac{1}{4}, \tfrac{1}{2}, \tfrac{3}{4}, 1\}$ and set

$$x_1^* = \tfrac{1}{16}, \quad x_2^* = \tfrac{3}{16}, \quad x_3^* = \tfrac{3}{8}, \quad x_4^* = \tfrac{5}{8}, \quad x_5^* = \tfrac{3}{4}.$$

Calculate the following:

(a) $\Delta x_1, \Delta x_2, \Delta x_3, \Delta x_4, \Delta x_5$.   (b) $\|P\|$.   (c) m_1, m_2, m_3, m_4, m_5.

(d) $f(x_1^*), f(x_2^*), f(x_3^*), f(x_4^*), f(x_5^*)$.   (e) M_1, M_2, M_3, M_4, M_5.

(f) $L_f(P)$.   (g) $S^*(P)$.   (h) $U_f(P)$.   (i) $\displaystyle\int_a^b f(x)\, dx$.

4. Let f be continuous on $[a, b]$, let $P = \{x_0, x_1, \ldots, x_n\}$ be a partition of $[a, b]$, and let $S^*(P)$ be any Riemann sum generated by P. Show that

$$L_f(P) \le S^*(P) \le U_f(P).$$

In numerical computations the base interval $[a, b]$ is usually broken up into n subintervals each of length $(b - a)/n$. The Riemann sums then take the form

$$S_n^* = \frac{b - a}{n}[f(x_1^*) + f(x_2^*) + \cdots + f(x_n^*)].$$

Choose $\epsilon > 0$. Break up the interval $[0, 1]$ into n subintervals each of length $1/n$. Take $x_1^* = 1/n, x_2^* = 2/n, \ldots, x_n^* = n/n$.

5. (a) Determine the Riemann sum S_n^* for

$$\int_0^1 x\, dx.$$

(b) Show that

$$\left| S_n^* - \int_0^1 x\, dx \right| < \epsilon \quad \text{if } n > 1/\epsilon.$$

HINT: $1 + 2 + \cdots + n = \tfrac{1}{2}n(n + 1)$.

6. (a) Determine the Riemann sum S_n^* for

$$\int_0^1 x^2\, dx.$$

(b) Show that

$$\left| S_n^* - \int_0^1 x^2\, dx \right| < \epsilon \quad \text{if } n > 1/\epsilon.$$

HINT: $1^2 + 2^2 + \cdots + n^2 = \tfrac{1}{6}n(n + 1)(2n + 1)$.

7. (a) Determine the Riemann sum S_n^* for

$$\int_0^1 x^3\, dx.$$

(b) Show that

$$\left| S_n^* - \int_0^1 x^3 \, dx \right| < \epsilon \qquad \text{if } n > 1/\epsilon.$$

HINT: $1^3 + 2^3 + \cdots + n^3 = (1 + 2 + \cdots + n)^2$.

8. Let f be a function continuous on $[a, b]$. Show that, if P is a partition of $[a, b]$, then $L_f(P)$, $U_f(P)$, and $\frac{1}{2}[L_f(P) + U_f(P)]$ are all Riemann sums.

9. (Calculator) Set $f(x) = x \cos x^2$, $x \in [0, 2]$. Take $P = \{0, \frac{1}{3}, \frac{2}{3}, 1, \frac{4}{3}, \frac{5}{3}, 2\}$ and set

$$x_1^* = \tfrac{1}{6}, \quad x_2^* = \tfrac{3}{6}, \quad x_3^* = \tfrac{5}{6}, \quad x_4^* = \tfrac{7}{6}, \quad x_5^* = \tfrac{9}{6}, \quad x_6^* = \tfrac{11}{6}.$$

Calculate $S^*(P)$ and compare this to the value of the integral

$$\int_0^2 x \cos x^2 \, dx.$$

10. (Calculator) Set $f(x) = \sec^2 x$, $x \in [0, 1]$. Take $P = \{0, \frac{2}{10}, \frac{4}{10}, \frac{6}{10}, \frac{8}{10}, 1\}$ and set

$$x_1^* = \tfrac{1}{10}, \quad x_2^* = \tfrac{3}{10}, \quad x_3^* = \tfrac{5}{10}, \quad x_4^* = \tfrac{7}{10}, \quad x_5^* = \tfrac{9}{10}.$$

Calculate $S^*(P)$ and compare this to the value of the integral

$$\int_0^1 \sec^2 x \, dx.$$

 Program for numerical integration via Riemann sums   (BASIC)

This program emphasizes the important fact that one need not always try to compute the value of a definite integral by finding an explicit antiderivative for the integrand and evaluating it between the limits of integration. Instead one can try to estimate the value directly by using Riemann sums. The computer makes this a practical and potentially very accurate method for finding answers. The program is straightforward and as usual works with a sample function that is to be replaced by the function of your choice. The interval in question and the size of the subintervals to be used are also at your disposal.

```
10 REM Numerical integration via Riemann sums
20 REM copyright © Colin C. Graham 1988-1989

30 REM change the line "def FNf(x) = ..."
40 REM to fit your function

50 DEF FNf(x) = 4/(1 + x*x)

100 INPUT "Enter left endpoint"; a
110 INPUT "Enter right endpoint"; b
120 INPUT "Enter number of divisions"; n

200 delta = (b − a)/n
210 integral = 0

300 REM this is left endpt estimate
310 REM for right endpt estimate use j = 1 to n
320 FOR j = 0 TO n − 1
330     integral = integral + FNf(a + j*delta)*delta
340 NEXT j

400 PRINT integral

500 INPUT "Do again? (Y/N)"; a$
510 IF a$ = "Y" OR a$ = "y" THEN 100
520 END
```

CHAPTER HIGHLIGHTS

5.1 An Area Problem; A Speed-Distance Problem

5.2 The Definite Integral of a Continuous Function

partition (p. 241)   upper sum; lower sum (p. 241)
definite integral (p. 242)   limits of integration; integrand (p. 242)

5.3 The Function $F(x) = \displaystyle\int_a^x f(t)\, dt$

additivity of the integral (p. 249)

$$\frac{d}{dx}\left[\int_a^x f(t)\, dt\right] = f(x) \quad \text{provided } f \text{ is continuous (p. 251)}$$

5.4 The Fundamental Theorem of Integral Calculus

antiderivative (p. 255)   fundamental theorem (p. 255)

linearity of the integral (p. 258)

5.5 Some Area Problems

If f and g are continuous and $f(x) \geq g(x)$ for all x in $[a, b]$, then

$$\int_a^b [f(x) - g(x)]\, dx$$

gives the area between the graph of f and the graph of g over $[a, b]$.

5.6 Indefinite Integrals

indefinite integral; constant of integration (p. 265)

5.7 Some Motion Problems

For an object that moves along a coordinate line with velocity $v(t)$

$$\int_a^b |v(t)|\, dt = \text{distance traveled from time } t = a \text{ to time } t = b$$

whereas

$$\int_a^b v(t)\, dt = \text{net displacement from time } t = a \text{ to time } t = b.$$

equation for linear motion with constant acceleration (p. 269)

5.8 The u-Substitution; Change of Variables

An integral of the form $\displaystyle\int f(g(x))\, g'(x)\, dx$ can be written $\displaystyle\int f(u)\, du$ by setting

$$u = g(x), \qquad du = g'(x)\, dx.$$

For definite integrals

$$\int_a^b f(g(x))\, g'(x)\, dx = \int_{g(a)}^{g(b)} f(u)\, du.$$

5.9 Integrating the Trigonometric Functions

$$\int \cos x \, dx = \sin x + C,$$

$$\int \sin x \, dx = -\cos x + C,$$

$$\int \sec^2 x \, dx = \tan x + C,$$

$$\int \csc^2 x \, dx = -\cot x + C,$$

$$\int \sec x \tan x \, dx = \sec x + C,$$

$$\int \csc x \cot x \, dx = -\csc x + C,$$

$$\int \cos^2 x \, dx = \tfrac{1}{2}x + \tfrac{1}{4}(\sin 2x) + C,$$

$$\int \sin^2 x \, dx = \tfrac{1}{2}x - \tfrac{1}{4}(\sin 2x) + C.$$

5.10 Some Further Properties of the Definite Integral

The integral of a nonnegative function is nonnegative; the integral of a positive function is positive; the integral is order-preserving. (pp. 283–284)

$$\left| \int_a^b f(x) \, dx \right| \le \int_a^b |f(x)| \, dx, \qquad \frac{d}{dx}\left[\int_a^u f(t) \, dt \right] = f(u) \frac{du}{dx}$$

5.11 Mean-Value Theorems for Integrals; Average Values

first mean-value theorem for integrals (p. 287)
second mean-value theorem for integrals (p. 289)
mass of a rod (p. 289)

average value (p. 287)
weighted average (p. 289)
center of mass of a rod (p. 290)

5.12 Additional Exercises

5.13 The Integral as the Limit of Riemann Sums

Darboux method (p. 295)   Riemann sum (p. 296)

$$\int_a^b f(x) \, dx = \lim_{\|P\| \to 0} [f(x_1^*) \, \Delta x_1 + f(x_2^*) \, \Delta x_2 + \cdots + f(x_n^*) \, \Delta x_n]$$

SOME APPLICATIONS OF THE INTEGRAL

6.1 MORE ON AREA

Representative Rectangles

You have seen that the definite integral can be viewed as the limit of Riemann sums:

$$(1) \qquad \int_a^b f(x)\, dx = \lim_{\|P\| \to 0} [f(x_1^*)\Delta x_1 + f(x_2^*)\Delta x_2 + \cdots + f(x_n^*)\Delta x_n].$$

With x_i^* chosen arbitrarily from $[x_{i-1}, x_i]$, you can think of $f(x_i^*)$ as a *representative* value of f for that interval. If f is positive, then the product

$$f(x_i^*)\Delta x_i$$

gives the area of the *representative rectangle* shown in Figure 6.1.1. Formula (1) tells us that we can approximate the area under the curve as closely as we wish by adding up the areas of representative rectangles. (Figure 6.1.2)

Figure 6.1.1

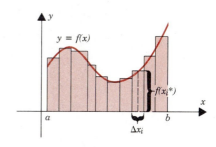

Figure 6.1.2

Figure 6.1.3 shows a region Ω bounded above by the graph of a function f and bounded below by the graph of a function g. As you have seen before, we can calculate the area of Ω by integrating with respect to x the *vertical separation*

$$f(x) - g(x)$$

from $x = a$ to $x = b$:

$$\text{area }(\Omega) = \int_a^b [f(x) - g(x)]\, dx.$$

In this case the approximating Riemann sums are of the form

$$[f(x_1^*) - g(x_1^*)]\, \Delta x_1 + [f(x_2^*) - g(x_2^*)]\, \Delta x_2 + \cdots + [f(x_n^*) - g(x_n^*)]\, \Delta x_n.$$

The dimensions of a representative rectangle are now

$$f(x_i^*) - g(x_i^*) \qquad \text{and} \qquad \Delta x_i.$$

 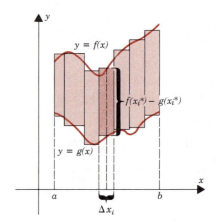

Figure 6.1.3

Areas by Integration with Respect to *y*

In Figure 6.1.4 you can see a region the boundaries of which are not functions of x but functions of y instead. In such a case we set the representative rectangles horizontally and we calculate the area of the region as the limit of Riemann sums of the form

$$[F(y_1^*) - G(y_1^*)]\, \Delta y_1 + [F(y_2^*) - G(y_2^*)]\, \Delta y_2 + \cdots + [F(y_n^*) - G(y_n^*)]\, \Delta y_n.$$

The area of the region is thus given by the integral

$$\int_c^d [F(y) - G(y)]\, dy.$$

Here we are integrating the *horizontal separation*

$$F(y) - G(y)$$

with respect to y.

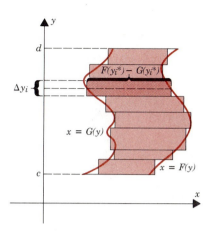

Figure 6.1.4

Problem 1. Find the area of the region bounded on the left by $x = y^2$ and on the right by $x = 3 - 2y^2$.

Solution. The region is sketched in Figure 6.1.5. The easiest way to calculate the area is to set our representative rectangles horizontally and integrate with respect to y.

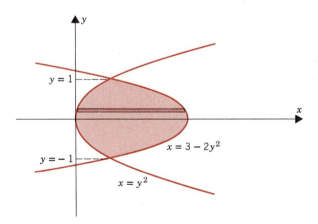

Figure 6.1.5

The two curves intersect at $y = -1$ and $y = 1$. We can find the area of the region by integrating the horizontal separation

$$(3 - 2y^2) - y^2 = 3 - 3y^2$$

from $y = -1$ to $y = 1$:

$$A = \int_{-1}^{1} (3 - 3y^2) \, dy = 4. \quad \square$$

Problem 2. Calculate the area of the region bounded by the curves $x = y^2$ and $x - y = 2$ by integrating (a) with respect to x, (b) with respect to y.

Solution. (a) To integrate with respect to x we set the representative rectangles vertically. See Figure 6.1.6. The upper boundary of the region is the curve $y = \sqrt{x}$. However, the lower boundary is described by two different equations: $y = -\sqrt{x}$ from $x = 0$ to $x = 1$ and $y = x - 2$ from $x = 1$ to $x = 4$. Thus, we need two integrals:

$$A = \int_0^1 [\sqrt{x} - (-\sqrt{x})]\, dx + \int_1^4 [\sqrt{x} - (x - 2)]\, dx$$

$$= 2\int_0^1 \sqrt{x}\, dx + \int_1^4 (\sqrt{x} - x + 2)\, dx = \left[\tfrac{4}{3}x^{3/2}\right]_0^1 + \left[\tfrac{2}{3}x^{3/2} - \tfrac{1}{2}x^2 + 2x\right]_1^4 = \tfrac{9}{2}.$$

(b) See Figure 6.1.7. To integrate with respect to y we set the representative rectangles horizontally. Now the right boundary is the line $x = y + 2$ and the left boundary is the curve $x = y^2$. Since y ranges from -1 to 2,

$$A = \int_{-1}^2 [(y + 2) - y^2]\, dy = \left[\tfrac{1}{2}y^2 + 2y - \tfrac{1}{3}y^3\right]_{-1}^2 = \tfrac{9}{2}. \quad \square$$

Figure 6.1.6

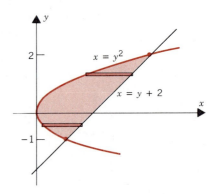

Figure 6.1.7

EXERCISES 6.1

Sketch the region bounded by the curves. Represent the area of the region by one or more integrals (a) in terms of x; (b) in terms of y.

1. $y = x^2,\quad y = x + 2.$

2. $y = x^2,\quad y = -4x.$

3. $y = x^3,\quad y = 2x^2.$

4. $y = \sqrt{x},\quad y = x^3.$

5. $y = -\sqrt{x},\quad y = x - 6,\quad y = 0.$

6. $x = y^3,\quad x = 3y + 2.$

7. $y = |x|,\quad 3y - x = 8.$

8. $y = x,\quad y = 2x,\quad y = 3.$

9. $x + 4 = y^2,\quad x = 5.$

10. $x = |y|,\quad x = 2.$

11. $y = 2x,\quad x + y = 9,\quad y = x - 1.$

12. $y = x^3,\quad y = x^2 + x - 1.$

13. $y = x^{1/3},\quad y = x^2 + x - 1.$

14. $y = x + 1,\quad y + 3x = 13,\quad 3y + x + 1 = 0.$

Sketch the region bounded by the curves and find its area.

15. $4x = 4y - y^2$, $4x - y = 0$.

16. $x + y^2 - 4 = 0$, $x + y = 2$.

17. $x = y^2$, $x = 3 - 2y^2$.

18. $x + y = 2y^2$, $y = x^3$.

19. $x + y - y^3 = 0$, $x - y + y^2 = 0$.

20. $8x = y^3$, $8x = 2y^3 + y^2 - 2y$.

6.2 VOLUME BY PARALLEL CROSS SECTIONS; DISCS AND WASHERS

One way to calculate the volume of a solid is to introduce a coordinate axis and then examine the cross sections of the solid that are perpendicular to that axis. In Figure 6.2.1 we picture a solid and a coordinate axis that we label the x-axis. As in the figure, we suppose that the solid lies entirely between $x = a$ and $x = b$. The figure shows an arbitrary cross section perpendicular to the x-axis. By $A(x)$ we mean the area of the cross section that has coordinate x.

Figure 6.2.1

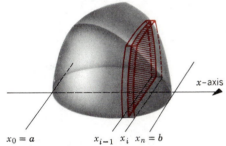

Figure 6.2.2

If the cross-sectional area, $A(x)$, varies continuously with x, then we can find the volume of the solid by integrating $A(x)$ from a to b:

(6.2.1)
$$V = \int_a^b A(x)\, dx.$$

To see this, let $P = \{x_0, x_1, \ldots, x_n\}$ be a partition of $[a, b]$. The volume of the solid from x_{i-1} to x_i can be approximated by a product of the form

$$A(x_i^*)\Delta x_i \qquad \text{(Figure 6.2.2)}$$

with x_i^* chosen arbitrarily from $[x_{i-1}, x_i]$. The volume of the entire solid can therefore be approximated by sums of the form

$$A(x_1^*)\Delta x_1 + A(x_2^*)\Delta x_2 + \cdots + A(x_n^*)\Delta x_n.$$

These are Riemann sums which, as $\|P\| \to 0$, converge to

$$\int_a^b A(x)\, dx. \quad \square$$

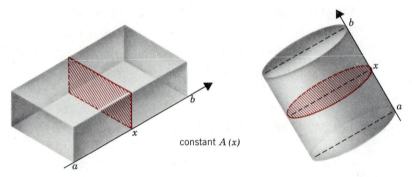

Figure 6.2.3

For solids of constant cross-sectional area (such as those of Figure 6.2.3), the volume is simply the cross-sectional area times the thickness:

$$V = \text{(the constant cross-sectional area)} \cdot (b - a).$$

For solids with varying cross-sectional area

$$V = \int_a^b A(x)\, dx,$$

which is to say

(6.2.2) $V = \text{(the average cross-sectional area)} \cdot (b - a).$ (5.11.2)

The calculation of volumes of solids of arbitrary shape is left for Chapter 17. Here we restrict our attention to solids with very simple cross sections.

Problem 1.   Find the volume of a triangular pyramid with base area B and height h.

Solution.   In Figure 6.2.4 you can see a coordinate line that passes through the apex and is perpendicular to the base. To the apex we assign coordinate 0, and to the base point coordinate h. The cross section at a distance of x units from the apex ($0 \le x \le h$) is a triangle similar to the base triangle. The corresponding sides of these triangles are proportional, the factor of proportionality being x/h. This means that the areas are also proportional, but now the factor of proportionality is $(x/h)^2$.†
Thus we have

$$A(x) = B\,\frac{x^2}{h^2} \quad \text{and} \quad V = \int_0^h A(x)\, dx = \frac{B}{h^2} \int_0^h x^2\, dx = \frac{B}{h^2}\left(\tfrac{1}{3}h^3\right) = \tfrac{1}{3}Bh.$$

This formula was discovered in ancient times. It was known to Eudoxus of Cnidos (circa 400 B.C.).   ☐

---

† If the base triangle has dimensions a, b, c, then its area is $ab \sin \theta$ where θ is the angle opposite the side of length c. For the x-level triangle the dimensions are $a(x/h)$, $b(x/h)$, $c(x/h)$, and the area is $a(x/h)b(x/h) \sin \theta = ab \sin \theta\,(x/h)^2$.

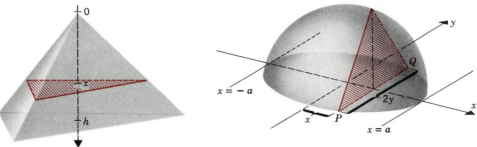

Figure 6.2.4

Figure 6.2.5

Problem 2. The base of a solid is the region bounded by the ellipse

$$\frac{x^2}{a^2} + \frac{y^2}{b^2} = 1.$$

Find the volume of the solid given that the cross sections perpendicular to the x-axis are equilateral triangles.

Solution. Take x as in Figure 6.2.5. The cross section with coordinate x is an equilateral triangle with side $\overline{PQ}$. The equation of the ellipse can be written

$$y^2 = \frac{b^2}{a^2}(a^2 - x^2).$$

Since

$$\text{length of } \overline{PQ} = 2y = \frac{2b}{a}\sqrt{a^2 - x^2},$$

the equilateral triangle has area

$$A(x) = \frac{\sqrt{3}b^2}{a^2}(a^2 - x^2).\dagger$$

We can find the volume of the solid by integrating $A(x)$ from $x = -a$ to $x = a$:

$$V = \underbrace{\int_{-a}^{a} A(x)\, dx}_{\text{by symmetry}} = 2\int_{0}^{a} A(x)\, dx$$

$$= \frac{2\sqrt{3}b^2}{a^2}\int_{0}^{a}(a^2 - x^2)\, dx$$

$$= \frac{2\sqrt{3}b^2}{a^2}\left[a^2 x - \frac{x^3}{3}\right]_{0}^{a} = \tfrac{4}{3}\sqrt{3}\, ab^2. \quad \square$$

$\dagger$ In general, the area of an equilateral triangle is $\tfrac{1}{4}\sqrt{3}s^2$, where s is the length of a side.

Problem 3. The base of a solid is the region between the parabolas

$$x = y^2 \qquad \text{and} \qquad x = 3 - 2y^2.$$

Find the volume of the solid given that the cross sections perpendicular to the x-axis are squares.

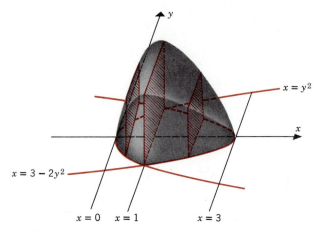

Figure 6.2.6

Solution. The two parabolas intersect at $x = 1$. (Figure 6.2.6) From $x = 0$ to $x = 1$, the cross section with coordinate x has area

$$A(x) = (2y)^2 = 4y^2 = 4x.$$

(Here we are measuring the span across the first parabola $x = y^2$.) The volume of the solid from $x = 0$ to $x = 1$ is

$$V_1 = \int_0^1 4x \, dx = 4 \left[\tfrac{1}{2}x^2 \right]_0^1 = 2.$$

From $x = 1$ to $x = 3$, the cross section with coordinate x has area

$$A(x) = (2y)^2 = 4y^2 = 2(3 - x).$$

(Here we are measuring the span across the second parabola $x = 3 - 2y^2$.) The volume of the solid from $x = 1$ to $x = 3$ is

$$V_2 = \int_1^3 2(3 - x) \, dx = 2 \left[-\tfrac{1}{2}(3 - x)^2 \right]_1^3 = 4.$$

The total volume is

$$V_1 + V_2 = 6. \quad \square$$

Solids of Revolution: Disc Method

Suppose that f is nonnegative and continuous on $[a, b]$. (See Figure 6.2.7.) If we revolve the region below the graph of f about the x-axis, we obtain a solid. The volume

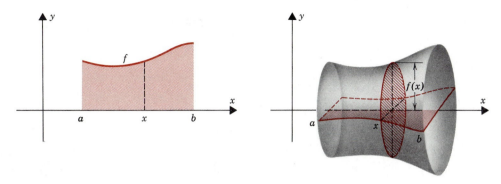

Figure 6.2.7

of this solid is given by the formula

(6.2.3)
$$V = \int_a^b \pi[f(x)]^2 \, dx.$$

Proof. The cross section with coordinate x is a circular *disc* of radius $f(x)$. The cross-sectional area is thus $\pi[f(x)]^2$. ☐

Among the simplest solids of revolution are the cone and the sphere.

Example 4. We can generate a cone of base radius r and height h by revolving about the x-axis the region below the graph of

$$f(x) = \frac{r}{h}x, \qquad 0 \le x \le h. \qquad \text{(draw a figure)}$$

By Formula 6.2.3

$$\text{volume of cone} = \int_0^h \pi \frac{r^2}{h^2} x^2 \, dx = \tfrac{1}{3}\pi r^2 h. \quad ☐$$

Example 5. A sphere of radius r can be obtained by revolving about the x-axis the region below the graph of

$$f(x) = \sqrt{r^2 - x^2}, \qquad -r \le x \le r. \qquad \text{(draw a figure)}$$

Therefore

$$\text{volume of sphere} = \int_{-r}^r \pi(r^2 - x^2) \, dx = \pi \left[r^2 x - \tfrac{1}{3}x^3 \right]_{-r}^r = \tfrac{4}{3}\pi r^3.$$

This result was obtained by Archimedes in the third century B.C. ☐

Solids of Revolution: Washer Method

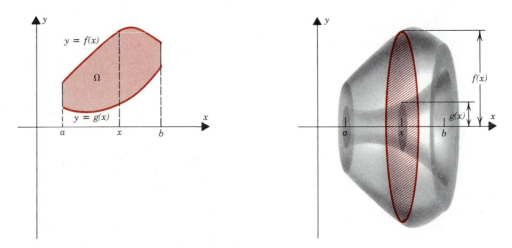

Figure 6.2.8

The washer method is a slight generalization of the disc method. Suppose that f and g are nonnegative continuous functions with $g(x) \leq f(x)$ for all x in $[a, b]$. (See Figure 6.2.8.) If we revolve the region Ω about the x-axis, we obtain a solid. The volume of this solid is given by the formula

(6.2.4)
$$V = \int_a^b \pi([f(x)]^2 - [g(x)]^2)\, dx.$$
washer method
about x-axis

Proof. The cross section with coordinate x is a *washer* of outer radius $f(x)$, inner radius $g(x)$, and area

$$A(x) = \pi[f(x)]^2 - \pi[g(x)]^2 = \pi([f(x)]^2 - [g(x)]^2).$$

We can get the volume of the solid by integrating this function from a to b.   ☐

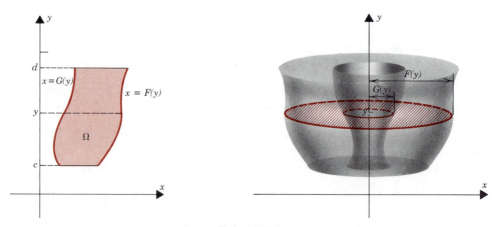

Figure 6.2.9

Suppose now that the boundaries are functions of y rather than x. (See Figure 6.2.9.) By revolving Ω *about the y-axis,* we obtain a solid. It is clear from (6.2.4) that in this case

(6.2.5)

$$V = \int_c^d \pi([F(y)]^2 - [G(y)]^2)\, dy.$$

washer method
about y-axis

Problem 6. Find the volume of the solid generated by revolving the region between $y = x^2$ and $y = 2x$ about the x-axis.

Solution. We refer to Figure 6.2.10. For each x from 0 to 2, the x cross section is a washer of outer radius $2x$ and inner radius x^2. By (6.2.4)

$$V = \int_0^2 \pi[(2x)^2 - (x^2)^2]\, dx = \pi \int_0^2 (4x^2 - x^4)\, dx = \pi\left[\tfrac{4}{3}x^3 - \tfrac{1}{5}x^5\right]_0^2 = \tfrac{64}{15}\pi. \quad \square$$

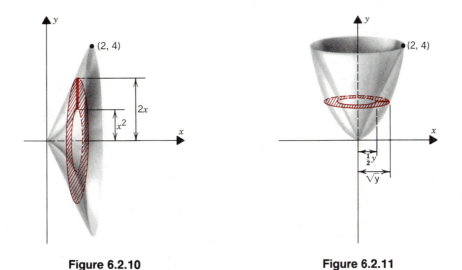

Figure 6.2.10　　　　　　　　**Figure 6.2.11**

Problem 7. Find the volume of the solid generated by revolving the region between $y = x^2$ and $y = 2x$ about the y-axis.

Solution. The solid is depicted in Figure 6.2.11. For each y from 0 to 4, the y cross section is a washer of outer radius $\sqrt{y}$ and inner radius $\tfrac{1}{2}y$. By (6.2.5)

$$V = \int_0^4 \pi[(\sqrt{y})^2 - (\tfrac{1}{2}y)^2]\, dy = \pi \int_0^4 (y - \tfrac{1}{4}y^2)\, dy = \pi\left[\tfrac{1}{2}y^2 - \tfrac{1}{12}y^3\right]_0^4 = \tfrac{8}{3}\pi. \quad \square$$

Remark.   These last two problems concerned solids generated by revolving the *same* region about *different* axes.   Notice that the solids are different and have different volumes.   ☐

EXERCISES 6.2

Sketch the region Ω bounded by the curves and find the volume of the solid generated by revolving this region about the x-axis.

1. $y = x$, $y = 0$, $x = 1$.

2. $x + y = 3$, $y = 0$, $x = 0$.

3. $y = x^2$, $y = 9$.

4. $y = x^3$, $y = 8$, $x = 0$.

5. $y = \sqrt{x}$, $y = x^3$.

6. $y = x^2$, $y = x^{1/3}$.

7. $y = x^3$, $x + y = 10$, $y = 1$.

8. $y = \sqrt{x}$, $x + y = 6$, $y = 1$.

9. $y = x^2$, $y = x + 2$.

10. $y = x^2$, $y = 2 - x$.

11. $y = \sqrt{4 - x^2}$, $y = 0$.

12. $y = 1 - |x|$, $y = 0$.

13. $y = \sec x$, $x = 0$, $x = \frac{1}{4}\pi$, $y = 0$.

14. $y = \csc x$, $x = \frac{1}{4}\pi$, $x = \frac{3}{4}\pi$, $y = 0$.

15. $y = \cos x$, $y = x + 1$, $x = \frac{1}{2}\pi$.

16. $y = \sin x$, $x = \frac{1}{4}\pi$, $x = \frac{1}{2}\pi$, $y = 0$.

Sketch the region Ω bounded by the curves and find the volume of the solid generated by revolving this region about the y-axis.

17. $y = 2x$, $y = 4$, $x = 0$.

18. $x + 3y = 6$, $x = 0$, $y = 0$.

19. $x = y^3$, $x = 8$, $y = 0$.

20. $x = y^2$, $x = 4$.

21. $y = \sqrt{x}$, $y = x^3$.

22. $y = x^2$, $y = x^{1/3}$.

23. $y = x$, $y = 2x$, $x = 4$.

24. $x + y = 3$, $2x + y = 6$, $x = 0$.

25. $x = y^2$, $x = 2 - y^2$.

26. $x = \sqrt{9 - y^2}$, $x = 0$.

27. The base of a solid is the circle $x^2 + y^2 = r^2$.   Find the volume of the solid given that the cross sections perpendicular to the x-axis are (a) squares, (b) equilateral triangles.

28. The base of a solid is the region bounded by the ellipse $b^2x^2 + a^2y^2 = a^2b^2$.   Find the volume of the solid given that the cross sections perpendicular to the x-axis are (a) isosceles right triangles each with hypotenuse on the xy-plane, (b) squares, (c) isosceles triangles of height 2.

29. The base of a solid is the region bounded by $y = x^2$ and $y = 4$.   Find the volume of the solid given that the cross sections perpendicular to the x-axis are (a) squares, (b) semicircles, (c) equilateral triangles.

30. The base of a solid is the region between the parabolas $x = y^2$ and $x = 3 - 2y^2$.   Find the volume of the solid given that the cross sections perpendicular to the x-axis are (a) rectangles of height h, (b) equilateral triangles, (c) isosceles right triangles each with hypotenuse on the xy-plane.

31. Exercise 29 with the cross sections perpendicular to the y-axis.

32. Exercise 30 with the cross sections perpendicular to the y-axis.

33. Find the volume generated by revolving the ellipse $b^2x^2 + a^2y^2 = a^2b^2$ about the x-axis.

34. Exercise 33 with the ellipse revolved about the y-axis.

35. Derive a formula for the volume of the frustum of a cone in terms of the height h, the lower base radius R, and the upper base radius r.   (See Figure 6.2.12.)

36. Find the volume enclosed by the surface that is generated by revolving the equilateral triangle with vertices $(0, 0)$, $(a, 0)$, $(\frac{1}{2}a, \frac{1}{2}\sqrt{3}a)$ about the x-axis.

37. A hemispheric basin of radius r feet is being used to store water.   To what percent of capacity is it filled when the water is (a) $\frac{1}{2}r$ feet deep?   (b) $\frac{1}{3}r$ feet deep?

Figure 6.2.12

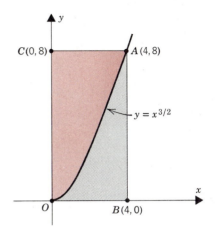

Figure 6.2.13

38. A sphere of radius r is cut by two parallel planes: one, a units above the equator; the other, b units above the equator. Find the volume of the portion of the sphere that lies between the two planes. Assume that $a < b$.

39. Find the volume when the region OAB of Figure 6.2.13 is revolved about
(a) the x-axis.   (b) AB.   (c) CA.   (d) the y-axis.

40. Find the volume when the region OAC of Figure 6.2.13 is revolved about
(a) the y-axis.   (b) CA.   (c) AB.   (d) the x-axis.

6.3   VOLUME BY THE SHELL METHOD

Figure 6.3.1

To describe the shell method of computing volumes, we begin with a solid cylinder of radius R and height h, and from it we cut out a cylindrical core of radius r.   (Figure 6.3.1)

Since the original cylinder had volume $\pi R^2 h$ and the piece removed had volume $\pi r^2 h$, the solid cylindrical shell that remains has volume

(6.3.1) $$\pi R^2 h - \pi r^2 h = \pi h(R + r)(R - r).$$

We will use this shortly.

Consider now a function f that is nonnegative and continuous on an interval $[a, b]$ that does not contain the origin in its interior.   For convenience we assume that $a \geq 0$.   If the region below the graph is revolved about the y-axis, then a solid T is

Figure 6.3.2

generated (Figure 6.3.2). The volume of this solid is given by the *shell method formula*:

(6.3.2)
$$V = \int_a^b 2\pi x\, f(x)\, dx.$$

To see this, we take a partition $P = \{x_0, x_1, \ldots, x_n\}$ of $[a, b]$ and concentrate on what's happening on the ith subinterval $[x_{i-1}, x_i]$. Recall that, when we form a Riemann sum, we are free to choose x_i^* as any point from $[x_{i-1}, x_i]$. For convenience we take x_i^* as the midpoint $\frac{1}{2}(x_{i-1} + x_i)$. The representative rectangle of height $f(x_i^*)$ and base Δx_i (see Figure 6.3.2) generates a cylindrical shell of height $h = f(x_i^*)$, inner radius $r = x_{i-1}$, and outer radius $R = x_i$. We can calculate the volume of this shell by (6.3.1). Since

$$h = f(x_i^*) \quad \text{and} \quad R + r = x_i + x_{i-1} = 2x_i^* \quad \text{and} \quad R - r = \Delta x_i,$$

the volume of this shell is

$$\pi h(R + r)(R - r) = 2\pi x_i^*\, f(x_i^*)\Delta x_i.$$

The volume of the entire solid can be approximated by adding up the volumes of these shells:

$$V \cong 2\pi x_1^* f(x_1^*)\Delta x_1 + 2\pi x_2^* f(x_2^*)\Delta x_2 + \cdots + 2\pi x_n^* f(x_n^*)\Delta x_n.$$

The sums on the right are Riemann sums which, as $\|P\| \to 0$, converge to

$$\int_a^b 2\pi x\, f(x)\, dx. \quad \blacksquare$$

For a simple interpretation of Formula 6.3.2 we refer to Figure 6.3.3.

As the region below the graph of f is revolved about the y-axis, the line segment x units from the y-axis generates a cylinder of radius x, height $f(x)$, and lateral area $2\pi x\, f(x)$. The shell method formula expresses the volume of a solid of revolution as the average lateral area of these cylinders times the length of the base interval:

(6.3.3)
$$V = (\text{average lateral area of the cylinders}) \cdot (b - a).$$

Figure 6.3.3

The shell method formula can be generalized. With Ω as in Figure 6.3.4, the volume generated by revolving Ω about the y-axis is given by the formula

(6.3.4)
$$V = \int_a^b 2\pi x [f(x) - g(x)] \, dx.$$

shell method
about y-axis

The integrand $2\pi x [f(x) - g(x)]$ is the lateral area of the cylinder in Figure 6.3.4.

Figure 6.3.4   **Figure 6.3.5**

We can also apply the shell method to solids generated by revolving a region about the x-axis. See Figure 6.3.5. In this instance the curved boundaries are functions of y rather than x and we have

(6.3.5)
$$V = \int_c^d 2\pi y [F(y) - G(y)] \, dy.$$

shell method
about x-axis

The integrand $2\pi y [F(y) - G(y)]$ is the lateral area of the cylinder in Figure 6.3.5.

Problem 1. Find the volume of the solid generated by revolving the region between

$$y = x^2 \quad \text{and} \quad y = 2x$$

about the y-axis.

Solution. We refer to Figure 6.3.6. For each x from 0 to 2 the line segment x units from the y-axis generates a cylinder of radius x, height $(2x - x^2)$, and lateral area $2\pi x(2x - x^2)$. By (6.3.4)

$$V = \int_0^2 2\pi x(2x - x^2)\, dx = 2\pi \int_0^2 (2x^2 - x^3)\, dx = 2\pi \left[\tfrac{2}{3}x^3 - \tfrac{1}{4}x^4 \right]_0^2 = \tfrac{8}{3}\pi. \quad \square$$

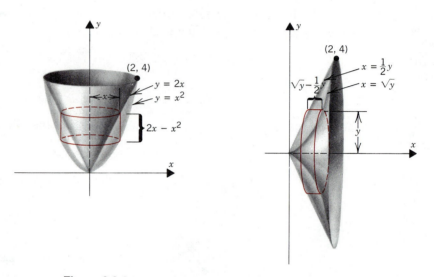

Figure 6.3.6   **Figure 6.3.7**

Problem 2. Find the volume of the solid generated by revolving the region between

$$y = x^2 \quad \text{and} \quad y = 2x$$

about the x-axis.

Solution. We begin by expressing these boundaries as functions of y. We write $x = \sqrt{y}$ for the right boundary and $x = \tfrac{1}{2}y$ for the left boundary. (See Figure 6.3.7.) The shell of radius y has height $(\sqrt{y} - \tfrac{1}{2}y)$. Thus, by (6.3.5)

$$V = \int_0^4 2\pi y(\sqrt{y} - \tfrac{1}{2}y)\, dy = \pi \int_0^4 (2y^{3/2} - y^2)\, dy = \pi \left[\tfrac{4}{5}y^{5/2} - \tfrac{1}{3}y^3 \right]_0^4 = \tfrac{64}{15}\pi. \quad \square$$

Remark. In Section 6.2 we calculated the volumes of the same solids (and got the same answers) by the washer method. ◻

EXERCISES 6.3

Sketch the region Ω bounded by the curves and use the shell method to find the volume of the solid generated by revolving Ω about the y-axis.

1. $y = x$, $y = 0$, $x = 1$.
2. $x + y = 3$, $y = 0$, $x = 0$.
3. $y = \sqrt{x}$, $x = 4$, $y = 0$.
4. $y = x^3$, $x = 2$, $y = 0$.
5. $y = \sqrt{x}$, $y = x^3$.
6. $y = x^2$, $y = x^{1/3}$.
7. $y = x$, $y = 2x$, $y = 4$.
8. $y = x$, $y = 1$, $x + y = 6$.
9. $x = y^2$, $x = y + 2$.
10. $x = y^2$, $x = 2 - y$.
11. $x = \sqrt{9 - y^2}$, $x = 0$.
12. $x = |y|$, $x = 2 - y^2$.

Sketch the region Ω bounded by the curves and use the shell method to find the volume of the solid generated by revolving Ω about the x-axis.

13. $x + 3y = 6$, $y = 0$, $x = 0$.
14. $y = x$, $y = 5$, $x = 0$.
15. $y = x^2$, $y = 9$.
16. $y = x^3$, $y = 8$, $x = 0$.
17. $y = \sqrt{x}$, $y = x^3$.
18. $y = x^2$, $y = x^{1/3}$.
19. $y = x^2$, $y = x + 2$.
20. $y = x^2$, $y = 2 - x$.
21. $y = x$, $y = 2x$, $x = 4$.
22. $y = x$, $x + y = 8$, $x = 1$.
23. $y = \sqrt{1 - x^2}$, $x + y = 1$.
24. $y = x^2$, $y = 2 - |x|$.

25. Use the shell method to find the volume enclosed by the surface obtained by revolving the ellipse $b^2x^2 + a^2y^2 = a^2b^2$ about the y-axis.

26. Exercise 25 with the ellipse revolved about the x-axis.

27. Find the volume enclosed by the surface generated by revolving the equilateral triangle with vertices $(0, 0)$, $(a, 0)$, $(\frac{1}{2}a, \frac{1}{2}\sqrt{3}a)$ about the y-axis.

28. A ball of radius r is cut into two pieces by a horizontal plane a units above the center of the ball. Determine the volume of the upper piece by using the shell method.

29. Carry out Exercise 39 of Section 6.2, this time using the shell method.

30. Carry out Exercise 40 of Section 6.2, this time using the shell method.

6.4   THE CENTROID OF A REGION; PAPPUS'S THEOREM ON VOLUMES

The Centroid of a Region

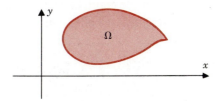

Figure 6.4.1

In Section 5.11 you saw how to calculate the center of mass of a thin rod. Suppose now that we have a thin distribution of matter, a *plate,* laid out in the xy-plane in the shape of some region Ω. (Figure 6.4.1) If the mass density of the plate varies from point to point, then the determination of the center of mass of the plate requires double integration. (Chapter 17) If, however, the mass density of the plate is constant throughout Ω, then the center of mass of the plate depends only on the shape of Ω and falls on a point that we call the *centroid* of Ω. Unless Ω has a very complicated shape, we can calculate the centroid of Ω by ordinary one-variable integration.

We will use two guiding principles to find the centroid of a region Ω. The first is obvious. The second we take from physics.†

Principle 1: Symmetry. If the region has an axis of symmetry, then the centroid $(\bar{x}, \bar{y})$ lies somewhere along that axis. (It follows from Principle 1 that, if the region has a center, then that center is the centroid.)

Principle 2: Additivity. If the region, having area A, consists of a finite number of pieces with areas $A_1, \ldots, A_n$ and centroids $(\bar{x}_1, \bar{y}_1), \ldots, (\bar{x}_n, \bar{y}_n)$, then

(6.4.1)
$$\bar{x}A = \bar{x}_1 A_1 + \cdots + \bar{x}_n A_n \quad \text{and} \quad \bar{y}A = \bar{y}_1 A_1 + \cdots + \bar{y}_n A_n.$$

Figure 6.4.2

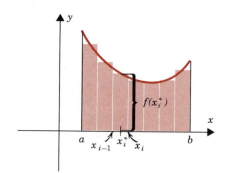

Figure 6.4.3

We are now ready to bring the techniques of calculus into play. Figure 6.4.2 shows the region Ω under the graph of a continuous function f. Let's denote the area of Ω by A. The centroid $(\bar{x}, \bar{y})$ of Ω can be obtained from the following formulas:

(6.4.2)
$$\bar{x}A = \int_a^b x f(x)\, dx, \qquad \bar{y}A = \int_a^b \tfrac{1}{2}[f(x)]^2\, dx.$$

To derive these formulas we choose a partition $P = \{x_0, x_1, \ldots, x_n\}$ of $[a, b]$. This breaks up $[a, b]$ into n subintervals $[x_{i-1}, x_i]$. Choosing x_i^* as the midpoint of $[x_{i-1}, x_i]$, we form the midpoint rectangles R_i shown in Figure 6.4.3. The area of R_i is $f(x_i^*)\Delta x_i$ and the centroid of R_i is its center $(x_i^*, \tfrac{1}{2}f(x_i^*))$. By (6.4.1), the centroid

---

† For now on faith. It is easily justified by double integration. (Chapter 17)

$(\bar{x}_P, \bar{y}_P)$ of the union of all these rectangles satisfies the following equations:

$$\bar{x}_P A_P = x_1^* f(x_1^*)\, \Delta x_1 + \cdots + x_n^* f(x_n^*)\, \Delta x_n,$$

$$\bar{y}_P A_P = \tfrac{1}{2}[f(x_1^*)]^2\, \Delta x_1 + \cdots + \tfrac{1}{2}[f(x_n^*)]^2\, \Delta x_n.$$

(Here A_P represents the area of the union of the n rectangles.)   As $\|P\| \to 0$, the union of rectangles tends to the shape of Ω and the equations we just derived tend to the formulas given in (6.4.2).   ☐

Problem 1. Find the centroid of the quarter-disc shown in Figure 6.4.4.

Solution. The quarter-disc is symmetric about the line $x = y$. We know therefore that $\bar{x} = \bar{y}$. Here

$$\bar{y}A = \int_0^r \tfrac{1}{2}[f(x)]^2\, dx = \int_0^r \tfrac{1}{2}(r^2 - x^2)\, dx = \tfrac{1}{3}r^3.$$
$$f(x) = \sqrt{r^2 - x^2} \longrightarrow$$

Since $A = \tfrac{1}{4}\pi r^2$,

$$\bar{y} = \frac{\tfrac{1}{3}r^3}{\tfrac{1}{4}\pi r^2} = \frac{4r}{3\pi}.$$

The centroid of the quarter-disc is the point

$$\left(\frac{4r}{3\pi}, \frac{4r}{3\pi}\right). \quad ☐$$

Figure 6.4.4

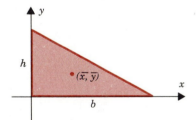

Figure 6.4.5

Problem 2. Find the centroid of the right triangle shown in Figure 6.4.5.

Solution. There is no symmetry that we can use here. The hypotenuse lies on the line

$$y = -\frac{h}{b}x + h.$$

Hence

$$\bar{x}A = \int_0^b x f(x)\, dx = \int_0^b \left(-\frac{h}{b}x^2 + hx\right) dx = \tfrac{1}{6}b^2 h$$

and

$$\bar{y}A = \int_0^b \tfrac{1}{2}[f(x)]^2 \, dx = \tfrac{1}{2}\int_0^b \left(\frac{h^2}{b^2}x^2 - \frac{2h^2}{b}x + h^2\right) dx = \tfrac{1}{6}bh^2.$$

Since $A = \tfrac{1}{2}bh$, we have

$$\bar{x} = \frac{\tfrac{1}{6}b^2 h}{\tfrac{1}{2}bh} = \tfrac{1}{3}b \quad \text{and} \quad \bar{y} = \frac{\tfrac{1}{6}bh^2}{\tfrac{1}{2}bh} = \tfrac{1}{3}h.$$

The centroid is the point $(\tfrac{1}{3}b, \tfrac{1}{3}h)$. ☐

Figure 6.4.6 shows the region Ω between the graphs of two continuous functions f and g. In this case, if Ω has area A and centroid $(\bar{x}, \bar{y})$, then

(6.4.3) $\qquad \bar{x}A = \int_a^b x[f(x) - g(x)] \, dx, \qquad \bar{y}A = \int_a^b \tfrac{1}{2}([f(x)]^2 - [g(x)]^2) \, dx.$

Proof. Let A_f be the area below the graph of f and let A_g be the area below the graph of g. Then in obvious notation

$$\bar{x}A + \bar{x}_g A_g = \bar{x}_f A_f \quad \text{and} \quad \bar{y}A + \bar{y}_g A_g = \bar{y}_f A_f. \tag{6.4.1}$$

Therefore

$$\bar{x}A = \bar{x}_f A_f - \bar{x}_g A_g = \int_a^b x f(x) \, dx - \int_a^b x g(x) \, dx = \int_a^b x[f(x) - g(x)] \, dx$$

and

$$\bar{y}A = \bar{y}_f A_f - \bar{y}_g A_g = \int_a^b \tfrac{1}{2}[f(x)]^2 \, dx - \int_a^b \tfrac{1}{2}[g(x)]^2 \, dx = \int_a^b \tfrac{1}{2}([f(x)]^2 - [g(x)]^2) \, dx. \; ☐$$

Figure 6.4.6

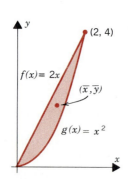

Figure 6.4.7

Problem 3. Find the centroid of the region shown in Figure 6.4.7.

Solution. Here there is no symmetry we can appeal to. We must carry out the calculations.

$$A = \int_0^2 [f(x) - g(x)] \, dx = \int_0^2 (2x - x^2) \, dx = \tfrac{4}{3},$$

$$\bar{x}A = \int_0^2 x[f(x) - g(x)] \, dx = \int_0^2 (2x^2 - x^3) \, dx = \tfrac{4}{3},$$

$$\bar{y}A = \int_0^2 \tfrac{1}{2}([f(x)]^2 - [g(x)]^2) \, dx = \tfrac{1}{2} \int_0^2 (4x^2 - x^4) \, dx = \tfrac{32}{15}.$$

Therefore $\bar{x} = \tfrac{4}{3}/\tfrac{4}{3} = 1$ and $\bar{y} = \tfrac{32}{15}/\tfrac{4}{3} = \tfrac{8}{5}$. ☐

Pappus's Theorem on Volumes

All the formulas that we have derived for volumes of solids of revolution are simple corollaries to an observation made by a brilliant, ancient Greek, Pappus of Alexandria (circa 300 A.D.).

THEOREM 6.4.4 PAPPUS'S THEOREM ON VOLUMES†

A plane region is revolved about an axis that lies in its plane. If the region does not cross the axis, then the volume of the resulting solid of revolution is the area of the region multiplied by the circumference of the circle described by the centroid of the region:

$$V = 2\pi \bar{R} A$$

where A is the area of the region and $\bar{R}$ is the distance from the axis to the centroid of the region.

Basically we have derived only two formulas for the volumes of solids of revolution:

1. *The Washer Method Formula.* If the region Ω of Figure 6.4.6 is revolved about the x-axis, the resulting solid has volume

$$V_x = \int_a^b \pi([f(x)]^2 - [g(x)]^2) \, dx.$$

2. *The Shell Method Formula.* If the region Ω of Figure 6.4.6 is revolved about the y-axis, the resulting solid has volume

$$V_y = \int_a^b 2\pi x[f(x) - g(x)] \, dx.$$

Note that

$$V_x = \int_a^b \pi([f(x)]^2 - [g(x)]^2) \, dx = 2\pi \int_a^b \tfrac{1}{2}([f(x)]^2 - [g(x)]^2) \, dx = 2\pi \bar{y} A = 2\pi \bar{R} A$$

---

† This theorem is found in Book VII of Pappus's "Mathematical Collection," largely a survey of ancient geometry to which Pappus made many original contributions (among them this theorem). Much of what we know today of Greek geometry we owe to Pappus.

and

$$V_y = \int_a^b 2\pi x [f(x) - g(x)] \, dx = 2\pi \bar{x} A = 2\pi \bar{R} A,$$

exactly as predicted by Pappus.

Remark. In stating Pappus's theorem we assumed a complete revolution. If Ω is only partially revolved about the given axis, then the volume of the resulting solid is simply the area of Ω multiplied by the length of the circular arc described by the centroid of Ω. ◻

Applications of Pappus's Theorem

Problem 4. Earlier we saw that the region shown in Figure 6.4.7 has area $\frac{4}{3}$ and centroid $(1, \frac{8}{5})$. Find the volume of the solid generated by revolving this region about the line $y = 5$.

Solution. Here $\bar{R} = 5 - \frac{8}{5} = \frac{17}{5}$ and $A = \frac{4}{3}$. Hence

$$V = 2\pi(\tfrac{17}{5})(\tfrac{4}{3}) = \tfrac{136}{15}\pi. \quad ◻$$

Problem 5. Find the volume of the doughnut (called *torus* in mathematics) generated by revolving the circular disc

$$(x - a)^2 + (y - b)^2 \leq c^2, \qquad a > c, \ b > c$$

about (a) the *x*-axis, (b) the *y*-axis.

Solution. The centroid of the disc is the center (a, b). This lies b units from the *x*-axis and a units from the *y*-axis. The area of the disc is πc^2. Therefore

$$\text{(a)} \quad V_x = 2\pi(b)(\pi c^2) = 2\pi^2 bc^2, \qquad \text{(b)} \quad V_y = 2\pi(a)(\pi c^2) = 2\pi^2 ac^2. \quad ◻$$

Problem 6. Find the centroid of the half-disc

$$x^2 + y^2 \leq r^2, \qquad y \geq 0$$

by appealing to Pappus's theorem.

Solution. Since the half-disc is symmetric about the *y*-axis, we know that $\bar{x} = 0$. All we need to find is $\bar{y}$.

If we revolve the half-disc about the *x*-axis, we obtain a solid ball of volume $\frac{4}{3}\pi r^3$. The area of the half-disc is $\frac{1}{2}\pi r^2$. By Pappus's theorem

$$\tfrac{4}{3}\pi r^3 = 2\pi \bar{y}(\tfrac{1}{2}\pi r^2).$$

Simple division gives $\bar{y} = 4r/3\pi$. ◻

Remark. Centroids of solids of revolution are discussed in the exercises. ◻

EXERCISES 6.4

Sketch the region bounded by the curves. Determine the centroid of the region and the volume generated by revolving the region about each of the coordinate axes.

1. $y = \sqrt{x}$, $y = 0$, $x = 4$.
2. $y = x^3$, $y = 0$, $x = 2$.
3. $y = x^2$, $y = x^{1/3}$.
4. $y = x^3$, $y = \sqrt{x}$.
5. $y = 2x$, $y = 2$, $x = 3$.
6. $y = 3x$, $y = 6$, $x = 1$.
7. $y = x^2 + 2$, $y = 6$, $x = 0$.
8. $y = x^2 + 1$, $y = 1$, $x = 3$.
9. $\sqrt{x} + \sqrt{y} = 1$, $x + y = 1$.
10. $y = \sqrt{1 - x^2}$, $x + y = 1$.
11. $y = x^2$, $y = 0$, $x = 1$, $x = 2$.
12. $y = x^{1/3}$, $y = 1$, $x = 8$.
13. $y = x$, $x + y = 6$, $y = 1$.
14. $y = x$, $y = 2x$, $x = 3$.

Find the centroid of the bounded region determined by the following curves.

15. $y = 6x - x^2$, $y = x$.
16. $y = 4x - x^2$, $y = 2x - 3$.
17. $x^2 = 4y$, $x - 2y + 4 = 0$.
18. $y = x^2$, $2x - y + 3 = 0$.
19. $y^3 = x^2$, $2y = x$.
20. $y^2 = 2x$, $y = x - x^2$.
21. $y = x^2 - 2x$, $y = 6x - x^2$.
22. $y = 6x - x^2$, $x + y = 6$.
23. $x + 1 = 0$, $x + y^2 = 0$.
24. $\sqrt{x} + \sqrt{y} = \sqrt{a}$, $x = 0$, $y = 0$.

25. Let Ω be the annular region (ring) formed by the circles

$$x^2 + y^2 = \tfrac{1}{4} \quad \text{and} \quad x^2 + y^2 = 4.$$

(a) Locate the centroid of Ω. (b) Locate the centroid of the first-quadrant part of Ω. (c) Locate the centroid of the upper half of Ω.

26. The ellipse $b^2x^2 + a^2y^2 = a^2b^2$ encloses a region of area πab. Locate the centroid of the upper half of the region.

27. The rectangle of Figure 6.4.8 is revolved about the line marked l. Find the volume of the resulting solid.

Figure 6.4.8   **Figure 6.4.9**

28. In Problem 2 of this section you saw that the centroid of the triangle in Figure 6.4.5 is the point $(\tfrac{1}{3}b, \tfrac{1}{3}h)$.
    (a) Verify that the line segments that join the centroid to the vertices divide the triangle into three triangles of equal area.
    (b) Find the distance d from the centroid of the triangle to the hypotenuse.
    (c) Find the volume generated by revolving the triangle about the hypotenuse.

29. The triangular region of Figure 6.4.9 is the union of two right triangles Ω_1, Ω_2. Find the centroid (a) of Ω_1. (b) of Ω_2. (c) of the entire region.

30. Find the volume of the solid generated by revolving the entire triangular region of Figure 6.4.9 about (a) the x-axis. (b) the y-axis.

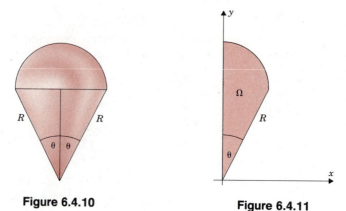

| Figure 6.4.10 | Figure 6.4.11 |

31. (a) Find the volume of the ice-cream cone shown in Figure 6.4.10.   (a right circular cone topped by a solid hemisphere)

(b) Find $\bar{x}$ for the region Ω of Figure 6.4.11.

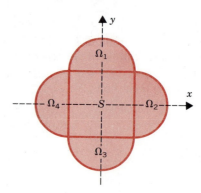

Figure 6.4.12

32. The region Ω of Figure 6.4.12 consists of a square S of side $2r$ and four semidiscs of radius r. Find the centroid of each of the following.

(a) Ω.   (b) Ω_1.   (c) $S \cup \Omega_1$.   (d) $S \cup \Omega_3$.   (e) $S \cup \Omega_1 \cup \Omega_3$.   (f) $S \cup \Omega_1 \cup \Omega_2$.
(g) $S \cup \Omega_1 \cup \Omega_2 \cup \Omega_3$.

33. (a) In Section 5.11 we stated that the mass of a rod that extends from $x = a$ to $x = b$ is given by the formula

$$M = \int_a^b \lambda(x)\, dx$$

where λ is the mass density function.   Derive this formula by taking an arbitrary partition $P = \{x_0, x_1, \ldots, x_n\}$ of $[a, b]$, estimating the mass contributed by $[x_{i-1}, x_i]$, adding up these contributions, and then taking the limit as $\|P\| \to 0$.

(b) In Section 5.11 we also gave the center of mass formula

$$x_M M = \int_a^b x\, \lambda(x)\, dx.$$

This formula can be derived from the physical premise that, if the rod is broken up into a finite number of pieces of masses $M_1, \ldots, M_n$ and centers of mass $x_{M_1}, \ldots, x_{M_n}$, then

$$x_M M = x_{M_1} M_1 + \cdots + x_{M_n} M_n.$$

Carry out the derivation.

34. (*The Centroid of a Solid of Revolution.*) If a solid is *homogeneous* (constant mass density), then the center of mass of the solid depends only on the shape of the solid and is called the *centroid*. The determination of the centroid of a solid of arbitrary shape requires triple integration. (Chapter 17) However, if the solid is a solid of revolution, then the centroid can be found by ordinary one-variable integration.

(a) Let Ω be the region of Figure 6.4.2 and let T be the solid generated by revolving Ω about the x-axis. By symmetry the centroid of T lies on the x-axis. Thus the centroid of T is completely determined by its x-coordinate $\bar{x}$. Show that

(6.4.5)
$$\bar{x} V_x = \int_a^b \pi x [f(x)]^2 \, dx$$

basing your argument on the following additivity principle: If a solid of volume V consists of a finite number of pieces with volumes $V_1, \ldots, V_n$ and the centroids of the pieces have x-coordinates $\bar{x}_1, \ldots, \bar{x}_n$, then

$$\bar{x} V = \bar{x}_1 V_1 + \cdots + \bar{x}_n V_n.$$

(b) Let Ω be the region of Figure 6.4.2 and let T be the solid generated by revolving Ω about the y-axis. By symmetry the centroid of T lies on the y-axis. Thus the centroid of T is completely determined by its y-coordinate $\bar{y}$. Show that

(6.4.6)
$$\bar{y} V_y = \int_a^b \pi x [f(x)]^2 \, dx$$

basing your argument on the following additivity principle: If a solid of volume V consists of a finite number of pieces with volumes $V_1, \ldots, V_n$ and the centroids of the pieces have y-coordinates $\bar{y}_1, \ldots, \bar{y}_n$, then

$$\bar{y} V = \bar{y}_1 V_1 + \cdots + \bar{y}_n V_n.$$

Locate the centroid of each of the following configurations.

35. A solid cone of base radius r and height h.

36. A solid hemisphere of radius r.

37. The solid generated by revolving the region below the graph of $f(x) = \sqrt{x}, \quad x \in [0, 1]$
(a) about the x-axis. (b) about the y-axis.

38. The solid generated by revolving the region below the graph of $f(x) = 4 - x^2, \quad x \in [0, 2]$
(a) about the x-axis. (b) about the y-axis.

39. The solid generated by revolving about the x-axis the first-quadrant part of the region enclosed by the ellipse $b^2 x^2 + a^2 y^2 = a^2 b^2$.

40. A segment of height h cut from a ball of radius r. Take $h < r$.

6.5   THE NOTION OF WORK

We begin with a constant force F that acts along some line that we call the x-axis. By convention F is positive if it acts in the direction of increasing x and negative if it acts in the direction of decreasing x.   (Figure 6.5.1)

Figure 6.5.1

Suppose now that an object moves along the x-axis from $x = a$ to $x = b$ subject to this constant force F.   The *work* done by F during the displacement is by definition the *force times the displacement*:

(6.5.1)
$$W = F \times (b - a).$$

It is not hard to see that, if F acts in the direction of the motion, then $W > 0$, but, if F acts against the motion, then $W < 0$.   Thus, for example, if an object slides off a table and falls to the floor, then the work done by gravity is positive (after all, Earth's gravity points down).   But if an object is lifted from the floor and raised to tabletop level, then the work done by gravity is by definition negative.†

Figure 6.5.2

To repeat, if an object moves from $x = a$ to $x = b$ subject to a constant force F, then the work done by F is the constant value of F times $b - a$.   What is the work done by F if F does not remain constant but instead varies continuously as a function of x? As you would expect, we then define the work done by F as the *average value* of F times $b - a$:

(6.5.2)
$$W = \int_a^b F(x) \, dx.$$
(Figure 6.5.2)

---

† However, the work done by the hand that lifts the object is positive.

Hooke's Law

You can sense a variable force in the action of a steel spring. Stretch a spring within its elastic limit and you feel a pull in the opposite direction. The greater the stretching, the harder the pull of the spring. Compress a spring within its elastic limit and you feel a push against you. The greater the compression, the harder the push. According to Hooke's law (Robert Hooke, 1635–1703) the force exerted by the spring can be written

$$F(x) = -kx$$

where k is a positive number, called *the spring constant,* and x is the displacement from the equilibrium configuration. The minus sign indicates that the spring force always acts in the direction opposite to the direction in which the spring has been deformed (the force always acts so as to restore the spring to its equilibrium state).

Remark. Hooke's law is only an approximation, but it is a good approximation for small displacements. In the problems that follow we assume that the restoring force of the spring is given by Hooke's law. □

Problem 1. A spring of natural length L compressed to length $\frac{7}{8} L$ exerts a force F_0. Find the work done by the spring in restoring itself to natural length.

Solution. Place the spring on the x-axis so that the equilibrium point falls at the origin. View compression as a move to the left.

Compressed $\frac{1}{8} L$ units to the left, the spring exerts a force F_0. Thus by Hooke's law

$$F_0 = F(-\tfrac{1}{8} L) = -k(-\tfrac{1}{8} L) = \tfrac{1}{8} kL.$$

This tells us that $k = 8F_0/L$. Therefore the force law for this spring reads

$$F(x) = -\left(\frac{8F_0}{L}\right) x.$$

To find the work done by this spring in restoring itself to equilibrium, we integrate $F(x)$ from $x = -\frac{1}{8} L$ to $x = 0$:

$$W = \int_{-L/8}^{0} F(x)\, dx = \int_{-L/8}^{0} -\left(\frac{8F_0}{L}\right) x\, dx = -\frac{8F_0}{L}\left[\frac{x^2}{2}\right]_{-L/8}^{0} = \frac{LF_0}{16}. \quad \square$$

Problem 2. For the spring in Problem 1, what work must we do to stretch the spring to length $\frac{11}{10} L$?

Solution. To stretch the spring we must counteract the force of the spring. The force exerted by the spring when stretched x units is

$$F(x) = -\left(\frac{8F_0}{L}\right) x. \qquad \text{(Problem 1)}$$

To counter this force we must apply the opposite force

$$-F(x) = \left(\frac{8F_0}{L}\right) x.$$

The work we must do to stretch the spring to length $\frac{11}{10} L$ can be found by integrating $-F(x)$ from $x = 0$ to $x = \frac{1}{10} L$:

$$W = \int_0^{L/10} -F(x) \, dx = \int_0^{L/10} \left(\frac{8F_0}{L}\right) x \, dx = \frac{8F_0}{L}\left[\frac{x^2}{2}\right]_0^{L/10} = \frac{LF_0}{25}. \quad \square$$

If force is measured in pounds and distance in feet, the units of work are foot-pounds; thus if a force of 500 pounds pushes a car for 60 feet, the work done is 30,000 foot-pounds.†

Problem 3. Stretched $\frac{1}{3}$ foot beyond its natural length, a certain spring exerts a restoring force of 10 pounds. What work must we do to stretch the spring another $\frac{1}{3}$ foot?

Solution. Place the spring on the x-axis so that the equilibrium point falls at the origin. View stretching as a move to the right. As usual, assume Hooke's law: $F(x) = -kx$.

When the spring is stretched $\frac{1}{3}$ foot, it exerts a force of -10 pounds (10 pounds to the left). Therefore $-10 = -k(\frac{1}{3})$ and $k = 30$ (i.e., 30 pounds per foot).

To find the work we must do to stretch the spring another $\frac{1}{3}$ foot, we integrate the opposite force $-F(x) = 30x$ from $x = \frac{1}{3}$ to $x = \frac{2}{3}$:

$$W = \int_{1/3}^{2/3} 30x \, dx = 30\left[\frac{1}{2}x^2\right]_{1/3}^{2/3} = 5 \text{ foot-pounds.} \quad \square$$

Pumping Out a Tank

To lift an object we must counteract the force of gravity. Consequently, the work done in lifting an object is given by the formula

work = (weight of the object) × (distance lifted).

If we lift a leaking sand bag or pump out a water tank from above, the calculation of work is more complicated. In the first instance the weight varies during the motion. (There is less sand in the bag as we keep lifting.) In the second instance the distance varies. (Water at the top of the tank does not have to be pumped as far as water at the bottom.) The sand bag problem is left to the exercises. Here we take up the second problem.

---

† There are other units of work. If force is given in Newtons and distance measured in meters, then work is given in *Newton-meters,* also called *Joules.* If force is given in dynes and distance in centimeters, then work is given in *dyne-centimeters,* called *ergs.* For our purposes foot-pounds will suffice.

Figure 6.5.3 depicts a storage tank filled to within a feet of the top with some liquid. Assume that the liquid is homogeneous and weighs σ pounds per cubic foot. Suppose now that this storage tank is pumped out from above until the level of the liquid drops to b feet below the top of the tank. How much work has been done?

Figure 6.5.3

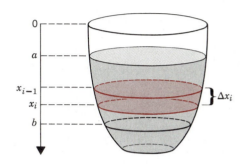

Figure 6.5.4

We can answer this question by the methods of integral calculus. For each $x \in [a, b]$, we let

$A(x) = $ cross-sectional area x feet below the top of the tank,

$s(x) = $ distance that the x-level must be lifted.

We let $P = \{x_0, x_1, \ldots, x_n\}$ be an arbitrary partition of $[a, b]$ and focus our attention on the ith subinterval $[x_{i-1}, x_i]$. (Figure 6.5.4.) Taking x_i^* as an arbitrary point in the ith subinterval, we have

$A(x_i^*)\Delta x_i = $ approximate volume of the ith layer of liquid,

$\sigma A(x_i^*)\Delta x_i = $ approximate weight of this volume,

$s(x_i^*) = $ approximate distance this weight is to be moved,

and therefore

$\sigma s(x_i^*)A(x_i^*)\Delta x_i = $ approximate work (weight × distance) required to pump this layer of liquid to the top of the tank.

The work required to pump out all the liquid can be approximated by adding up all these last terms:

$$W \cong \sigma s(x_1^*)A(x_1^*)\Delta x_1 + \sigma s(x_2^*)A(x_2^*)\Delta x_2 + \cdots + \sigma s(x_n^*)A(x_n^*)\Delta x_n.$$

The sums on the right are Riemann sums which, as $\|P\| \to 0$, converge to give

(6.5.3)
$$W = \int_a^b \sigma s(x)A(x)\, dx.$$
□

We use this formula in the next problem.

Problem 4. A hemispherical water tank of radius 10 feet is being pumped out. Find the work done in lowering the water level from 2 feet below the top of the tank to 4 feet below the top of the tank given that the pump is placed (a) at the top of the tank, (b) 3 feet above the top of the tank.

Solution. As the weight of water take 62.5 pounds per cubic foot. It is not hard to see that the cross section x feet from the top of the tank is a disc of radius $\sqrt{100 - x^2}$. Its area is therefore

$$A(x) = \pi(100 - x^2).$$

For part (a) we have $s(x) = x$, so that

$$W = \int_2^4 62.5\pi x(100 - x^2)\, dx = 33{,}750\pi \text{ foot-pounds.}$$

For part (b) we have $s(x) = x + 3$, so that

$$W = \int_2^4 62.5\pi(x + 3)(100 - x^2)\, dx = 67{,}750\pi \text{ foot-pounds.} \quad \square$$

EXERCISES 6.5

1. Stretched 4 feet beyond its natural length, a certain spring exerts a restoring force of 200 pounds. What work must we do to stretch the spring (a) 1 foot beyond its natural length? (b) $1\frac{1}{2}$ feet beyond its natural length?

2. A certain spring has natural length L. Given that W is the work done in stretching the spring from L feet to $L + a$ feet, find the work done in stretching the spring (a) from L feet to $L + 2a$ feet, (b) from L feet to $L + na$ feet, (c) from $L + a$ feet to $L + 2a$ feet, (d) from $L + a$ feet to $L + na$ feet.

3. Find the natural length of a heavy metal spring, given that the work done in stretching it from a length of 2 feet to a length of 2.1 feet is one-half of the work done in stretching it from a length of 2.1 feet to a length of 2.2 feet.

4. A vertical cylindrical tank of radius 2 feet and height 6 feet is full of water. Find the work done in pumping out the water (a) to an outlet at the top of the tank, (b) to a level 5 feet above the top of the tank. (Assume that the water weighs 62.5 pounds per cubic foot.)

5. A horizontal cylindrical tank of radius 3 feet and length 8 feet is half full of oil weighing 60 pounds per cubic foot. What is the work done in pumping out the oil (a) to the top of the tank, (b) to a level 4 feet above the top of the tank?

6. What is the work done by gravity if the tank of Exercise 5 is completely drained through an opening at the bottom?

7. A conical container (vertex down) of radius r feet and height h feet is full of a liquid weighing σ pounds per cubic foot. Find the work done in pumping out the top $\frac{1}{2}h$ feet of liquid (a) to the top of the tank, (b) to a level k feet above the top of the tank.

8. What is the work done by gravity if the tank of Exercise 7 is completely drained through an opening at the bottom?

9. The force of gravity exerted by the earth on a mass m at a distance r from the center of the earth is given by Newton's formula

$$F = -G\frac{mM}{r^2}$$

where M is the mass of the earth and G is a gravitational constant. Find the work done by gravity in pulling a mass m from $r = r_1$ to $r = r_2$.

10. A box that weighs w pounds is dropped to the floor from a height of d feet. (a) What is the work done by gravity? (b) Show that the work is the same if the box slides to the floor along a smooth inclined plane.

11. A 100-pound bag of sand is lifted for 2 seconds at the rate of 4 feet per second. Find the work done in lifting the bag if the sand leaks out at the rate of $1\frac{1}{2}$ pounds per second.

12. A water container initially weighing w pounds is hoisted by a crane at the rate of n feet per second. What is the work done if the tank is raised m feet and the water leaks out constantly at the rate of p gallons per second? (Assume that the water weighs 8.3 pounds per gallon.)

13. A rope of length l feet that weighs σ pounds per foot is lying on the ground. What is the work done in lifting the rope so that it hangs from a beam (a) l feet high, (b) $2l$ feet high?

14. A load of weight w is lifted from the bottom of a shaft h feet deep. Find the work done given that the rope used to hoist the load weighs σ pounds per foot.

15. An 800-pound steel beam hangs from a 50-foot cable which weighs 6 pounds per foot. Find the work done in winding 20 feet of the cable about a steel drum.

*6.6 THE FORCE AGAINST A DAM

When an object is submerged in a liquid, it experiences a force from the liquid around it. This force is perpendicular to the surface of the solid at each of its points.

A surface patch of area A at depth h experiences a force

(6.6.1)
$$F = \sigma h A$$

where σ is the *weight density* of the liquid (the weight per unit volume). This is determined experimentally. In this discussion we will measure area in square feet, depth in feet, and σ in pounds per cubic foot. The force against a surface is then measured in pounds.

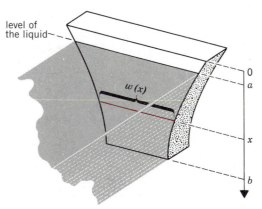

Figure 6.6.1

In Figure 6.6.1 we have depicted a vertical wall standing against a body of liquid. (Think of it as a dam or part of a container.) We want to calculate the force exerted by the liquid on this wall.

As in the figure we assume that the liquid sits from depth a to depth b and we let $w(x)$ denote the width of the wall at depth x. A partition $P = \{x_0, x_1, \ldots, x_n\}$ of

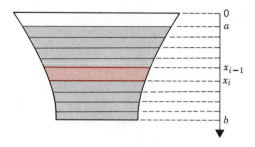

Figure 6.6.2

$[a, b]$ of small norm subdivides the wall into n narrow horizontal strips.   (Figure 6.6.2.)

We can estimate the force on the ith strip by taking x_i^* as the midpoint of $[x_{i-1}, x_i]$.   Then

$$w(x_i^*) = \text{the approximate width of the } i\text{th strip}$$

and

$$w(x_i^*) \, \Delta x_i = \text{the approximate area of the } i\text{th strip}.$$

Since this strip is narrow, all the points of the strip are approximately at depth x_i^*. Thus, using (6.6.1), we can estimate the force on the ith strip by the product

$$\sigma x_i^* \, w(x_i^*) \, \Delta x_i.$$

Adding up all these estimates we have an estimate for the force on the entire wall:

$$F \cong \sigma x_1^* \, w(x_1^*) \, \Delta x_1 + \sigma x_2^* \, w(x_2^*) \, \Delta x_2 + \cdots + \sigma x_n^* \, w(x_n^*) \, \Delta x_n.$$

The sum on the right is a Riemann sum for the integral

$$\int_a^b \sigma x \, w(x) \, dx$$

and as such converges to that integral as $\|P\| \to 0$.   Thus we have

(6.6.2)

$$\text{Force against wall} = \int_a^b \sigma x \, w(x) \, dx.$$

Figure 6.6.3

Problem 1. A horizontal circular water main (Figure 6.6.3) 6 feet in diameter is capped when half full of water. What is the force of the water against the cap?

Solution. Here

$$w(x) = 2\sqrt{9 - x^2} \quad \text{and} \quad \sigma = 62.5 \text{ pounds per cubic foot.}$$

The force exerted against the cap is

$$F = \int_0^3 (62.5)x(2\sqrt{9 - x^2})\,dx = 62.5 \int_0^3 2x\sqrt{9 - x^2}\,dx = 1125 \text{ pounds.} \quad \square$$

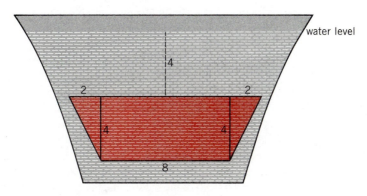

Figure 6.6.4

Problem 2. A metal plate in the form of a trapezoid is affixed to a vertical dam as in Figure 6.6.4. The dimensions shown are given in feet. Find the hydrostatic force on the plate.

Solution. First we find the width of the plate x feet below the water level. By similar triangles (see Figure 6.6.5)

$$t = \tfrac{1}{2}(8 - x) \quad \text{so that} \quad w(x) = 8 + 2t = 16 - x.$$

The force against the plate is

$$\int_4^8 62.5x(16 - x)\,dx = 14{,}666\tfrac{2}{3} \text{ pounds.} \quad \square$$

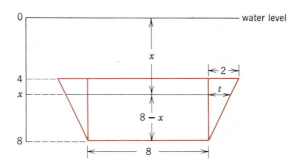

Figure 6.6.5

EXERCISES *6.6

1. Each end of a horizontal oil tank is an ellipse with horizontal axis 12 feet long and vertical axis 6 feet long. Calculate the force on one end when the tank is half full of oil weighing 60 pounds per cubic foot.

2. Each vertical end of a vat is a segment of a parabola (vertex down) 8 feet across the top and 16 feet deep. Calculate the force on an end when the vat is full of liquid weighing 70 pounds per cubic foot.

3. The vertical ends of a water trough are isosceles right triangles with the 90° angle at the bottom. Calculate the force on each triangle when the trough is full of water given that the legs of the triangle are 8 feet long.

4. The vertical ends of a water trough are isosceles triangles 5 feet across the top and 5 feet deep. Calculate the force on an end when the trough is full of water.

5. A horizontal cylindrical tank of diameter 8 feet is half full of oil weighing 60 pounds per cubic foot. Calculate the force on one end.

6. Calculate the force on one end if the tank of Exercise 5 is full.

7. A rectangular metal plate 10 feet by 6 feet is affixed to a vertical dam with the center of the plate 11 feet below water level. Find the hydrostatic force on the plate (a) if the 10-foot sides are horizontal. (b) if the 6-foot sides are horizontal.

8. A vertical cylindrical tank of diameter 30 feet and height 50 feet is full of oil weighing 60 pounds per cubic foot. Find the force on the curved surface.

9. Relate the force on a dam to the centroid of the submerged surface of the dam.

10. Two identical metal plates are affixed to a vertical dam. The centroid of the first plate is at depth h_1, the centroid of the second plate is at depth h_2. Compare the forces on the two plates.   HINT: Exercise 9.

*6.7   NEWTON'S LAW OF GRAVITATIONAL ATTRACTION

Newton's Law of Gravitational Attraction states that a point mass M attracts a point mass m with a force that varies in magnitude directly as the product of the two masses and inversely as the square of the distance between them:

(6.7.1)
$$F = G\,\frac{mM}{r^2}.$$

Here G is a constant that depends on the units employed.

To calculate the gravitational effect of an extended body (an object that is not a particle) we have to use integration.

Problem 1. A point mass m is placed on the perpendicular bisector of a thin uniform rod at a distance h from the rod. The rod exerts gravitational attraction on m. Find the magnitude of the attraction given that the rod has mass M and length L.

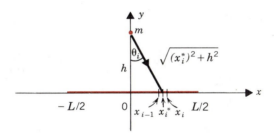

Figure 6.7.1

Solution. Since the rod is uniform, it has constant mass density $\lambda = M/L$. Think of the rod as lying on the x-axis and the point mass m on the y-axis. (Figure 6.7.1)

It is clear that any pull to the right exerted by the right half of the rod is counterbalanced by an equal pull to the left by the left half of the rod. Therefore the net force on m is down, toward $x = 0$. By symmetry the force on m toward $x = 0$ exerted by the entire rod is exactly twice the force exerted by the right half of the rod. It will be enough therefore to work with the right half of the rod.

We begin by partitioning the interval $[0, L/2]$ and focus our attention on the ith interval $[x_{i-1}, x_i]$. The mass of this little piece of rod is $\lambda \, \Delta x_i$. Its force on m is less than it would be if its entire mass were concentrated at the nearest point x_{i-1} but greater than it would be if its entire mass were concentrated at the furthest point x_i. (See the figure.) We can therefore view all the mass of the ith piece as concentrated at some point x_i^* within the interval $[x_{i-1}, x_i]$. Following this line of reasoning we can think of the ith piece as attracting m toward x_i^* with a force of magnitude

$$G \frac{m\lambda \Delta x_i}{(x_i^*)^2 + h^2}.$$

But part of the force is to the right. The downward part, the part toward $x = 0$, is only

$$\left(G \frac{m\lambda \Delta x_i}{(x_i^*)^2 + h^2}\right) \cos \theta_i = \left(G \frac{m\lambda \Delta x_i}{(x_i^*)^2 + h^2}\right)\left(\frac{h}{\sqrt{(x_i^*)^2 + h^2}}\right) = G \frac{m\lambda h \Delta x_i}{[(x_i^*)^2 + h^2]^{3/2}}.$$

see the figure

By adding up all these terms we get an estimate for the downward force exerted on m by the right half of the rod. Doubling this we get an estimate for the downward force exerted on m by the entire rod:

$$F \cong 2Gm\lambda h \left[\frac{\Delta x_1}{[(x_1^*)^2 + h^2]^{3/2}} + \frac{\Delta x_2}{[(x_2^*)^2 + h^2]^{3/2}} + \cdots + \frac{\Delta x_n}{[(x_n^*)^2 + h^2]^{3/2}}\right].$$

The bracketed term is a Riemann sum, which, as the norm of the partition tends to zero, tends to

$$\int_0^{L/2} \frac{dx}{(x^2 + h^2)^{3/2}} = \left[\frac{x}{h^2 \sqrt{x^2 + h^2}}\right]_0^{L/2} = \frac{L/2}{h^2 \sqrt{(L/2)^2 + h^2}}.$$

check by differentiation

Remembering the factor $2Gm\lambda h$, we have

$$F = (2Gm\lambda h)\left(\frac{L/2}{h^2\sqrt{(L/2)^2 + h^2}}\right) = \frac{Gm\lambda L}{h\sqrt{(L/2)^2 + h^2}} = \frac{GmM}{h\sqrt{(L/2)^2 + h^2}}.\dagger \quad \square$$

$$\lambda L = M$$

Problem 2. A point mass m is placed halfway between the endpoints of a uniform semicircular wire of radius R and mass M. The wire exerts a gravitational force on m that (by symmetry) is directed to the midpoint of the wire. Find the magnitude of this force.

Figure 6.7.2

Solution. The angle marked θ in Figure 6.7.2 ranges from 0 to π. A partition

$$P = \{0 = \theta_0, \theta_1, \ldots, \theta_{i-1}, \theta_i, \ldots, \theta_n = \pi\}$$

of $[0, \pi]$ induces a decomposition of the wire into n little arcs of length $R\,\Delta\theta_i$ and mass $\lambda R\,\Delta\theta_i$ where $\lambda = M/\pi R$. (λ is the mass density of the wire, the mass per unit length of wire.) This ith little piece attracts m with a force of magnitude

$$G\frac{m\lambda R\,\Delta\theta_i}{R^2}.$$

But part of this force is to the side. The upward part (see the figure) is approximately

$$\left(G\frac{m\lambda R\,\Delta\theta_i}{R^2}\right)\sin\theta_i^* = \left(\frac{Gm\lambda}{R}\right)\sin\theta_i^*\,\Delta\theta_i$$

where θ_i^* is some angle between θ_{i-1} and θ_i. Adding up these terms we get

$$\frac{Gm\lambda}{R}[\sin\theta_1^*\,\Delta\theta_1 + \sin\theta_2^*\,\Delta\theta_2 + \cdots + \sin\theta_n^*\,\Delta\theta_n].$$

---

† We can write this force as

$$G\frac{mM}{r^2}$$

where r is the geometric mean of the two distances: the distance from m to the center of the rod and the distance from m to the endpoints of the rod. (The geometric mean of two positive numbers a and b is by definition $\sqrt{ab}$.)

The bracketed term is a Riemann sum, which, as the norm of the partition tends to 0, tends to

$$\int_0^\pi \sin \theta \, d\theta = \left[-\cos \theta \right]_0^\pi = 2.$$

The semicircular wire attracts m to its midpoint with a force of magnitude

$$\frac{2Gm\lambda}{R} \underset{\lambda = M/\pi R}{=} \frac{2GmM}{\pi R^2}. \qquad \square$$

EXERCISES *6.7

1. A uniform rod of length L and mass M exerts an attractive gravitational force an a point mass m placed on the line of the rod at a distance h from an end of the rod. Find the magnitude of the attractive force.

2. Four uniform rods each of mass M and length L are laid out in the form of a square in the xy-plane. A point mass m is then placed on the z-axis at a distance h above the center of the square. The material square exerts a gravitational force on m that (by symmetry) is directed toward the center of the square. Find the magnitude of this force.

3. A uniform circular wire of radius R and total mass M exerts a gravitational force on the point mass m shown in Figure 6.7.3. By symmetry this force is directed toward the center of the circle. Find the magnitude of the force.

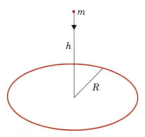

Figure 6.7.3

4. Problem 1 of this section given that the density of the rod varies directly as the distance from the center of the rod.

5. Suppose an object of mass m is in free fall near the surface of the earth. What is the magnitude of the gravitational force on the object at height h? It can be shown from Newton's Law of Gravitation that the magnitude of the force on m is approximately

$$G \frac{mM}{(R + h)^2}$$

where M is the mass of the earth and R is the radius of the earth. According to Galilean ideas, the force on m has magnitude mg where g is a constant. Reconcile these two ideas.

6. A Tinkertoy-like object consists of two identical bobs, each of mass m, held together by a stick of length L and negligible mass. The object is suspended, stick vertical. Show that the *difference* in the pull of the earth's gravity on the two ends of the object is given within good approximation by the expression

$$F = \left(\frac{2GmM}{R^3} \right) L$$

with M and R as in Exercise 5. (This is sometimes referred to as a "tidal force," because an analogous force, caused by a *difference* in the gravitational pull of the moon on those parts of the ocean that are closest to the moon and those that are farthest away, is largely responsible for the earth's tides.)

CHAPTER HIGHLIGHTS

6.1 More on Area

representative rectangle (p. 301)
area by integration with respect to y (p. 302)

6.2 Volume by Parallel Cross Sections; Discs and Washers

volume by parallel cross sections (p. 305)
disc method (p. 309)
washer method (x-axis) (p. 310)
washer method (y-axis) (p. 311)

6.3 Volume by the Shell Method

shell method (y-axis) (p. 315)   shell method (x-axis) (p. 315)

6.4 The Centroid of a Region; Pappus's Theorem on Volumes

principles for finding the centroid of a region (p. 318)

$$\bar{x}A = \int_a^b x[f(x) - g(x)]\, dx, \qquad \bar{y}A = \int_a^b \tfrac{1}{2}([f(x)]^2 - [g(x)]^2)\, dx$$

Pappus's theorem on volumes (p. 321)

centroid of a solid of revolution: $\bar{x}\, V_x = \int_a^b \pi x[f(x)]^2\, dx, \quad \bar{y}\, V_y = \int_a^b \pi x[f(x)]^2\, dx$

6.5 The Notion of Work

Hooke's law (p. 327)   pumping out a tank (p. 328)

*6.6 The Force Against a Dam

Force against a vertical wall $= \int_a^b \sigma x\, w(x)\, dx$

*6.7 Newton's Law of Gravitational Attraction

$$F = G\, \frac{mM}{r^2}$$

CHAPTER

7

THE TRANSCENDENTAL FUNCTIONS

Some real numbers satisfy polynomial equations with integer coefficients:

$$\tfrac{3}{5} \text{ satisfies the equation } 5x - 3 = 0,$$
$$\sqrt{2} \text{ satisfies the equation } x^2 - 2 = 0.$$

Such numbers are called *algebraic.* There are, however, numbers that are not algebraic, among them π. Such numbers are called *transcendental.*

Some functions f satisfy polynomial equations with polynomial coefficients:

$$f(x) = \frac{x}{\pi x + \sqrt{2}} \text{ satisfies the equation } (\pi x + \sqrt{2})f(x) - x = 0,$$

$$f(x) = 2\sqrt{x} - 3x^2 \text{ satisfies the equation } [f(x)]^2 + 6x^2 f(x) + (9x^4 - 4x) = 0.$$

Such functions are called *algebraic.* There are, however, functions that are not algebraic; these are called *transcendental.* The functions that we study in this chapter (the logarithm, the exponential, and the trigonometric functions and their inverses) are all transcendental.

7.1 IN SEARCH OF A NOTION OF LOGARITHM

If B is a positive number different from 1, the logarithm to the base B is defined in elementary mathematics by setting

$$C = \log_B A \quad \text{iff} \quad B^C = A.$$

Generally the base 10 is chosen and the defining relation becomes

$$C = \log_{10} A \quad \text{iff} \quad 10^C = A.$$

The basic properties of $\log_{10}$ can then be summarized as follows: with $X, Y > 0$,

$$\log_{10}(XY) = \log_{10} X + \log_{10} Y, \qquad \log_{10} 1 = 0,$$
$$\log_{10}(1/Y) = -\log_{10} Y, \qquad \log_{10}(X/Y) = \log_{10} X - \log_{10} Y,$$
$$\log_{10} X^Y = Y \log_{10} X, \qquad \log_{10} 10 = 1.$$

This elementary notion of logarithm is inadequate for calculus. It is unclear: what is meant by 10^C if C is irrational? It does not lend itself well to the methods of calculus: how would you differentiate $Y = \log_{10} X$ knowing only that $10^Y = X$?

Here we take an entirely different approach to logarithms. Instead of trying to tamper with the elementary definition, we discard it altogether. From our point of view the fundamental property of logarithms is that they transform multiplication into addition:

the log of a product = the sum of the logs.

Taking this as the central idea we are led to a general notion of logarithm that encompasses the elementary notion, lends itself well to the methods of calculus, and leads us naturally to a choice of base that simplifies many calculations.

DEFINITION 7.1.1

A *logarithm* function is a nonconstant differentiable function f defined on the set of positive numbers such that for all $x > 0$ and $y > 0$

$$f(xy) = f(x) + f(y).$$

Let's assume for the time being that such logarithm functions exist, and let's see what we can find out about them. In the first place, if f is such a function, then

$$f(1) = f(1 \cdot 1) = f(1) + f(1) = 2f(1) \quad \text{and so} \quad f(1) = 0.$$

Taking $y > 0$, we have

$$0 = f(1) = f(y \cdot 1/y) = f(y) + f(1/y)$$

and therefore

$$f(1/y) = -f(y).$$

Taking $x > 0$ and $y > 0$, we have

$$f(x/y) = f(x \cdot 1/y) = f(x) + f(1/y),$$

which, in view of the previous result, means that

$$f(x/y) = f(x) - f(y).$$

We are now ready to look for the derivative. (Remember, we are *assuming* that f is differentiable.) We begin by forming the difference quotient

$$\frac{f(x+h) - f(x)}{h}.$$

From what we have discovered about f,

$$f(x + h) - f(x) = f\left(\frac{x + h}{x}\right) = f(1 + h/x),$$

and therefore

$$\frac{f(x + h) - f(x)}{h} = \frac{f(1 + h/x)}{h}.$$

Remembering that $f(1) = 0$ and multiplying the denominator by x/x, we have

$$\frac{f(x + h) - f(x)}{h} = \frac{1}{x}\left[\frac{f(1 + h/x) - f(1)}{h/x}\right].$$

As h tends to 0, the left side tends to $f'(x)$, and, while $1/x$ remains fixed, h/x tends to 0 and thus the bracketed expression tends to $f'(1)$. In short,

(7.1.2)
$$\boxed{f'(x) = \frac{1}{x}f'(1).}$$

We have now established that, if f is a logarithm, then

$$f(1) = 0 \quad \text{and} \quad f'(x) = \frac{1}{x}f'(1).$$

We can't have $f'(1) = 0$ for that would make f constant. (Explain.) The most natural alternative is to set $f'(1) = 1$.† The derivative is then $1/x$.

This function, which takes on the value 0 at 1 and has derivative $1/x$ for $x > 0$, must, by the fundamental theorem of integral calculus, take the form

$$\int_1^x \frac{dt}{t}. \qquad \text{(check this out)}$$

7.2 THE LOGARITHM FUNCTION, PART I

DEFINITION 7.2.1

The function

$$L(x) = \int_1^x \frac{dt}{t}, \qquad x > 0$$

is called the (*natural*) *logarithm function*.

---

† This, as you will see later, is tantamount to a choice of base.

Here are some obvious properties of L:

(1) L is defined on $(0, \infty)$ with derivative

$$L'(x) = \frac{1}{x} \qquad \text{for all } x > 0.$$

L' is positive on $(0, \infty)$, and therefore L is an increasing function.

(2) L is continuous on $(0, \infty)$ since it is there differentiable.

(3) For $x > 1$, $L(x)$ gives the area of the region shaded in Figure 7.2.1.

(4) $L(x)$ is negative if $0 < x < 1$, zero at $x = 1$, positive if $x > 1$.

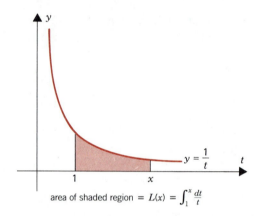

area of shaded region $= L(x) = \int_1^x \frac{dt}{t}$

Figure 7.2.1

The following result is fundamental.

THEOREM 7.2.2

If x and y are positive, then

$$L(xy) = L(x) + L(y).$$

Proof. Since

$$\frac{d}{dx}[L(x)] = \frac{1}{x} \qquad \text{and} \qquad \frac{d}{dx}[L(xy)] = \frac{y}{xy} = \frac{1}{x},$$

$$\text{chain rule} \longrightarrow$$

we can be sure that

$$L(xy) = L(x) + C.$$

We can evaluate the constant by taking $x = 1$:

$$L(y) = L(1 \cdot y) = L(1) + C = C.$$

$$L(1) = 0 \longrightarrow$$

Therefore $L(xy) = L(x) + L(y)$ as asserted. ▫

We come now to another important result.

THEOREM 7.2.3

If x is positive and p/q is rational, then

$$L(x^{p/q}) = \frac{p}{q} L(x).$$

Proof. You have seen that $d/dx[L(x)] = 1/x$. By the chain rule

$$\frac{d}{dx}[L(x^{p/q})] = \frac{1}{x^{p/q}} \frac{d}{dx}(x^{p/q}) = \frac{1}{x^{p/q}}\left(\frac{p}{q}\right)x^{(p/q)-1} = \frac{p}{q}\left(\frac{1}{x}\right) = \frac{d}{dx}\left[\frac{p}{q}L(x)\right].$$

$$\underset{(3.8.4)}{\uparrow}$$

Since both functions have the same derivative, they differ by a constant:

$$L(x^{p/q}) = \frac{p}{q} L(x) + C.$$

Since both functions are zero at $x = 1$, $C = 0$ and the theorem is proved. ☐

The domain of L is $(0, \infty)$. What is the range of L?

THEOREM 7.2.4

The range of L is $(-\infty, \infty)$.

Proof. Since L is continuous on $(0, \infty)$, we know from the intermediate-value theorem that it "skips" no values. Thus, its range is an interval. To show that the interval is $(-\infty, \infty)$, we need only show that it is unbounded above and unbounded below. We can do this by taking M as an arbitrary positive number and showing that L takes on values greater than M and values less than $-M$.

Since

$$L(2) = \int_1^2 \frac{dt}{t}$$

is positive (explain), we know that some positive multiple of $L(2)$ must be greater than M; namely, we know that there exists a positive integer n such that

$$nL(2) > M.$$

Multiplying this equation by -1 we have

$$-nL(2) < -M.$$

Since

$$nL(2) = L(2^n) \quad \text{and} \quad -nL(2) = L(2^{-n}),$$

we have

$$L(2^n) > M \quad \text{and} \quad L(2^{-n}) < -M.$$

This proves the unboundedness. ☐

The Number *e*

Since the range of L is $(-\infty, \infty)$ and L is an increasing function, we know that L takes on every value and does so only once. In particular, there is one and only one number at which the function L takes on the value 1. *This unique number is denoted by the letter e.*†

Since

$$(7.2.5) \qquad L(e) = \int_1^e \frac{dt}{t} = 1,$$

it follows from Theorem 7.2.3 that

$$(7.2.6) \qquad L(e^{p/q}) = \frac{p}{q} \qquad \text{for all rational numbers } \frac{p}{q}.$$

Because of this relation, we call L *the logarithm to the base e* and sometimes write

$$L(x) = \log_e x.$$

The number e arises naturally in many settings. Accordingly, we call $L(x)$ the natural logarithm and write

$$(7.2.7) \qquad L(x) = \ln x. \quad ††$$

Here are the basic properties we have established for $\ln x$:

$$(7.2.8)$$

$$\ln 1 = 0, \qquad \ln e = 1.$$
$$\ln xy = \ln x + \ln y. \qquad\qquad (x > 0, \ y > 0)$$
$$\ln 1/x = -\ln x. \qquad\qquad\qquad (x > 0)$$
$$\ln x/y = \ln x - \ln y. \qquad\qquad (x > 0, \ y > 0)$$
$$\ln x^r = r \ln x. \qquad\qquad (x > 0, \ r \text{ rational})$$

Notice how closely these rules parallel the familiar rules for common logarithms (base 10). Later we will show that the last of these rules also holds for irrational exponents.

---

† After the Swiss mathematician Leonhard Euler (1707–1783), considered by many the greatest mathematician of the eighteenth century.

†† Logarithms to bases other than e will be taken up later [they arise by other choices of $f'(1)$], but by far the most important logarithm in calculus is the logarithm to the base e. So much so, that when we speak of the logarithm of a number x and don't specify the base, you can be sure that we are talking about the *natural logarithm* $\ln x$.

The Graph of the Logarithm Function

You know that the logarithm function

$$\ln x = \int_1^x \frac{dt}{t}$$

has domain $(0, \infty)$, range $(-\infty, \infty)$, and derivative

$$\frac{d}{dx}(\ln x) = \frac{1}{x} > 0.$$

For small x the derivative is large (near 0 the curve is steep); for large x the derivative is small (far out the curve flattens out). At $x = 1$ the logarithm is 0 and its derivative $1/x$ is 1. (The graph crosses the x-axis at the point (1, 0), and the tangent line at that point is parallel to the line $y = x$.) The second derivative

$$\frac{d^2}{dx^2}(\ln x) = -\frac{1}{x^2}$$

is negative on $(0, \infty)$. (The graph is concave down throughout.) We have sketched the graph in Figure 7.2.2. The y-axis is a vertical asymptote:

$$\text{as } x \to 0^+, \quad \ln x \to -\infty.$$

Figure 7.2.2

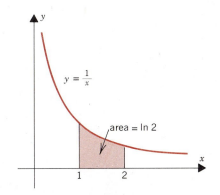

Figure 7.2.3

Problem 1. Estimate

$$\ln 2 = \int_1^2 \frac{dt}{t} \qquad \text{(Figure 7.2.3)}$$

from the partition

$$P = \{1 = \tfrac{10}{10}, \tfrac{11}{10}, \tfrac{12}{10}, \tfrac{13}{10}, \tfrac{14}{10}, \tfrac{15}{10}, \tfrac{16}{10}, \tfrac{17}{10}, \tfrac{18}{10}, \tfrac{19}{10}, \tfrac{20}{10} = 2\}.$$

Solution. Using a calculator we find that

$$L_f(P) = \tfrac{1}{10}\left(\tfrac{10}{11} + \tfrac{10}{12} + \tfrac{10}{13} + \tfrac{10}{14} + \tfrac{10}{15} + \tfrac{10}{16} + \tfrac{10}{17} + \tfrac{10}{18} + \tfrac{10}{19} + \tfrac{10}{20}\right)$$

$$= \tfrac{1}{11} + \tfrac{1}{12} + \tfrac{1}{13} + \tfrac{1}{14} + \tfrac{1}{15} + \tfrac{1}{16} + \tfrac{1}{17} + \tfrac{1}{18} + \tfrac{1}{19} + \tfrac{1}{20} > 0.668$$

and

$$U_f(P) = \tfrac{1}{10}(\tfrac{10}{10} + \tfrac{10}{11} + \tfrac{10}{12} + \tfrac{10}{13} + \tfrac{10}{14} + \tfrac{10}{15} + \tfrac{10}{16} + \tfrac{10}{17} + \tfrac{10}{18} + \tfrac{10}{19})$$
$$= \tfrac{1}{10} + \tfrac{1}{11} + \tfrac{1}{12} + \tfrac{1}{13} + \tfrac{1}{14} + \tfrac{1}{15} + \tfrac{1}{16} + \tfrac{1}{17} + \tfrac{1}{18} + \tfrac{1}{19} < 0.719.$$

Thus we have

$$0.668 < L_f(P) < \ln 2 < U_f(P) < 0.719.$$

The average of these two estimates is

$$\tfrac{1}{2}(0.668 + 0.719) = 0.6935.$$

We are not far off. Three-place tables carry the estimate 0.693. ☐

Table 7.2.1 gives the natural logarithms of the integers 2 through 10 rounded off to the nearest hundredth. A more extended table appears in Appendix C.

TABLE 7.2.1

n	$\ln n$	n	$\ln n$
1	0.00	6	1.79
2	0.69	7	1.95
3	1.10	8	2.08
4	1.39	9	2.20
5	1.61	10	2.30

Problem 2. Estimate the following logarithms on the basis of Table 7.2.1.

(a) $\ln 0.2$.   (b) $\ln 0.25$.   (c) $\ln 2.4$.   (d) $\ln 90$.

Solution

(a) $\ln 0.2 = \ln \tfrac{1}{5} = -\ln 5 \cong -1.61$.   (b) $\ln 0.25 = \ln \tfrac{1}{4} = -\ln 4 \cong -1.39$.

(c) $\ln 2.4 = \ln \dfrac{12}{5} = \ln \dfrac{(3)(4)}{5} = \ln 3 + \ln 4 - \ln 5 \cong 0.88$.

(d) $\ln 90 = \ln [(9)(10)] = \ln 9 + \ln 10 \cong 4.50$. ☐

Problem 3. Estimate e on the basis of Table 7.2.1.

Solution. So far all we know is that $\ln e = 1$. From the table you can see that

$$3 \ln 3 - \ln 10 \cong 1.$$

The expression on the left can be written

$$\ln 3^3 - \ln 10 = \ln 27 - \ln 10 = \ln \tfrac{27}{10} = \ln 2.7.$$

Thus $\ln 2.7 \cong 1$ and $e \cong 2.7$. (To eight decimal places $e \cong 2.71828182$. For most purposes we take $e \cong 2.72$.) ☐

EXERCISES 7.2

Estimate the following natural logarithms on the basis of Table 7.2.1.

1. ln 20.

2. ln 16.

3. ln 1.6.

4. ln 3^4.

5. ln 0.1.

6. ln 2.5.

7. ln 7.2.

8. ln $\sqrt{630}$.

9. ln $\sqrt{2}$.

10. ln 0.4.

11. ln 2^5.

12. ln 1.8.

13. Interpret the equation ln n = ln mn − ln m in terms of area under the curve $y = 1/x$. Draw a figure.

14. Given that $0 < x < 1$, express as a logarithm the area under the curve $y = 1/t$ from $t = x$ to $t = 1$.

15. Estimate

$$\ln 1.5 = \int_1^{1.5} \frac{dt}{t}$$

using the approximation $\frac{1}{2}[L_f(P) + U_f(P)]$ with $P = \{1 = \frac{8}{8}, \frac{9}{8}, \frac{10}{8}, \frac{11}{8}, \frac{12}{8} = 1.5\}$.

16. Estimate

$$\ln 2.5 = \int_1^{2.5} \frac{dt}{t}$$

using the approximation $\frac{1}{2}[L_f(P) + U_f(P)]$ with $P = \{1 = \frac{4}{4}, \frac{5}{4}, \frac{6}{4}, \frac{7}{4}, \frac{8}{4}, \frac{9}{4}, \frac{10}{4} = \frac{5}{2}\}$.

17. Taking ln 5 ≅ 1.61, use differentials to estimate (a) ln 5.2.   (b) ln 4.8.   (c) ln 5.5.

18. Taking ln 10 ≅ 2.30, use differentials to estimate (a) ln 10.3.   (b) ln 9.6.   (c) ln 11.

Solve the following equations for x.

19. ln $x = 2$.

20. ln $x = -1$.

21. $(2 - \ln x) \ln x = 0$.

22. $\frac{1}{2}$ ln $x = $ ln $(2x - 1)$.

23. ln $[(2x + 1)(x + 2)] = 2$ ln $(x + 2)$.

24. 2 ln $(x + 2) - \frac{1}{2}$ ln $x^4 = 1$.

(Calculator)   Evaluate numerically.

25. $\lim\limits_{x \to 1} \dfrac{\ln x}{x - 1}$.

26. $\lim\limits_{x \to 0^+} x \ln x$.

27. $\lim\limits_{x \to 0^+} \sqrt{x} \ln x$.

7.3   THE LOGARITHM FUNCTION, PART II

Differentiation and Graphing

We know that for $x > 0$

$$\frac{d}{dx} (\ln x) = \frac{1}{x}.$$

If u is a positive differentiable function of x, then by the chain rule

$$\frac{d}{dx} (\ln u) = \frac{d}{du} (\ln u) \frac{du}{dx} = \frac{1}{u} \frac{du}{dx}.$$

Thus, for example,

$$\frac{d}{dx} [\ln (1 + x^2)] = \frac{1}{1 + x^2} \cdot 2x = \frac{2x}{1 + x^2} \qquad \text{for all real } x$$

and

$$\frac{d}{dx}[\ln(1+3x)] = \frac{1}{1+3x} \cdot 3 = \frac{3}{1+3x} \qquad \text{for all } x > -\tfrac{1}{3}.$$

Problem 1. Find the domain of f and find $f'(x)$ if

$$f(x) = \ln(x\sqrt{1+3x}).$$

Solution. For x to be in the domain of f, we must have $x\sqrt{1+3x} > 0$, and consequently, $x > 0$. The domain is the set of positive numbers.

Before differentiating f, we make use of the special properties of the logarithm:

$$f(x) = \ln(x\sqrt{1+3x}) = \ln x + \ln[(1+3x)^{1/2}] = \ln x + \tfrac{1}{2}\ln(1+3x).$$

From this we have

$$f'(x) = \frac{1}{x} + \frac{1}{2}\left(\frac{3}{1+3x}\right). \quad \square$$

Problem 2. Sketch the graph of

$$f(x) = \ln|x|.$$

Solution. This is an even function, $f(-x) = f(x)$, defined for all numbers $x \neq 0$. The graph has two branches:

$$y = \ln(-x), \quad x < 0 \qquad \text{and} \qquad y = \ln x, \quad x > 0.$$

Each branch is the mirror image of the other. (Figure 7.3.1) $\square$

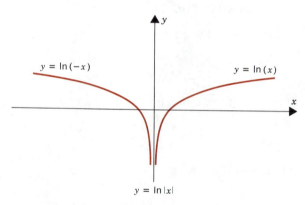

Figure 7.3.1

Problem 3. Show that

(7.3.1)

$$\frac{d}{dx}(\ln|x|) = \frac{1}{x} \qquad \text{for all } x \neq 0.$$

Solution. For $x > 0$,

$$\frac{d}{dx}(\ln|x|) = \frac{d}{dx}(\ln x) = \frac{1}{x}.$$

For $x < 0$, we have $|x| = -x > 0$, so that

$$\frac{d}{dx}(\ln|x|) = \frac{d}{dx}[\ln(-x)] = \frac{1}{-x}\frac{d}{dx}(-x) = \left(\frac{1}{-x}\right)(-1) = \frac{1}{x}. \quad \square$$

Problem 4. Find the domain of f and find $f'(x)$ given that

$$f(x) = (\ln x^2)^3.$$

Solution. Since the logarithm function is defined only for positive numbers, we must have $x^2 > 0$. The domain of f consists of all $x \neq 0$. Before differentiating note that

$$f(x) = (\ln x^2)^3 = (2\ln|x|)^3 = 8(\ln|x|)^3.$$

It follows that

$$f'(x) = 24(\ln|x|)^2\frac{d}{dx}(\ln|x|) = 24(\ln|x|)^2\frac{1}{x} = \frac{24}{x}(\ln|x|)^2. \quad \square$$

Problem 5. Show that, if u is a differentiable function of x, then for $u \neq 0$

(7.3.2)

$$\frac{d}{dx}(\ln|u|) = \frac{1}{u}\frac{du}{dx}.$$

Solution

$$\frac{d}{dx}(\ln|u|) = \frac{d}{du}(\ln|u|)\frac{du}{dx} = \frac{1}{u}\frac{du}{dx}. \quad \square$$

Here are two examples:

$$\frac{d}{dx}(\ln|1 - x^3|) = \frac{1}{1 - x^3}\frac{d}{dx}(1 - x^3) = \frac{-3x^2}{1 - x^3} = \frac{3x^2}{x^3 - 1}.$$

$$\frac{d}{dx}\left(\ln\left|\frac{x - 1}{x - 2}\right|\right) = \frac{d}{dx}(\ln|x - 1|) - \frac{d}{dx}(\ln|x - 2|) = \frac{1}{x - 1} - \frac{1}{x - 2}. \quad \square$$

Problem 6. Let

$$f(x) = \ln\left(\frac{x^4}{x-1}\right).$$

(a) Specify the domain of f. (b) On what intervals does f increase? Decrease? (c) Find the extreme values of f. (d) Determine the concavity of the graph and find the points of inflection. (e) Sketch the graph specifying the asymptotes if any.

Solution. Since the logarithm function is defined only for positive numbers, the domain of f is the open interval $(1, \infty)$.

Making use of the special properties of the logarithm we write

$$f(x) = \ln x^4 - \ln(x-1) = 4\ln x - \ln(x-1).$$

Differentiation gives

$$f'(x) = \frac{4}{x} - \frac{1}{x-1} = \frac{3x-4}{x(x-1)}$$

and

$$f''(x) = -\frac{4}{x^2} + \frac{1}{(x-1)^2} = -\frac{(x-2)(3x-2)}{x^2(x-1)^2}.$$

Since the domain of f is $(1, \infty)$, we consider only the values of x greater than 1.
  It is not hard to see that

$$f'(x) \text{ is } \begin{cases} \text{negative,} & \text{for} \quad 1 < x < \frac{4}{3} \\ \qquad 0, & \text{at} \qquad x = \frac{4}{3} \\ \text{positive,} & \text{for} \qquad x > \frac{4}{3} \end{cases}.$$

Thus f decreases on $(1, \frac{4}{3}]$ and increases on $[\frac{4}{3}, \infty)$. By the first derivative test

$$f(\tfrac{4}{3}) = 4\ln 4 - 3\ln 3 \cong 2.25$$

is a local and absolute minimum. There are no other extreme values.
  Since

$$f''(x) \text{ is } \begin{cases} \text{positive,} & \text{for} \quad 1 < x < 2 \\ \qquad 0, & \text{at} \qquad x = 2 \\ \text{negative,} & \text{for} \qquad x > 2 \end{cases},$$

the graph is concave up on $(1, 2)$ and concave down on $(2, \infty)$. The point

$$(2, f(2)) = (2, 4\ln 2) \cong (2, 2.77)$$

is the only point of inflection.
  Before sketching the graph, note that the derivative

$$f'(x) = \frac{4}{x} - \frac{1}{x-1}$$

is very large negative for x close to 1 and very close to 0 for x large. This means that the graph is very steep for x close to 1 and very flat for x large.   See Figure 7.3.2.   The line $x = 1$ is a vertical asymptote: as $x \to 1^+$, $f(x) \to \infty$.   ☐

Figure 7.3.2

Integration

The integral counterpart of (7.3.1) takes the form

(7.3.3)
$$\int \frac{dx}{x} = \ln |x| + C.$$

   In practice

$$\int \frac{g'(x)}{g(x)} \, dx \quad \text{is reduced to} \quad \int \frac{du}{u}$$

by setting

$$u = g(x), \qquad du = g'(x) \, dx.$$

Problem 7. Calculate

$$\int \frac{8x}{x^2 - 1} \, dx.$$

Solution. Set

$$u = x^2 - 1, \qquad du = 2x \, dx.$$

$$\int \frac{8x}{x^2 - 1} \, dx = 4 \int \frac{du}{u} = 4 \ln |u| + C = 4 \ln |x^2 - 1| + C. \quad ☐$$

Problem 8. Calculate

$$\int \frac{x^2}{1 - 4x^3} \, dx.$$

Solution. Set

$$u = 1 - 4x^3, \qquad du = -12x^2\, dx.$$

$$\int \frac{x^2}{1 - 4x^3}\, dx = -\frac{1}{12} \int \frac{du}{u} = -\frac{1}{12} \ln |u| + C = -\frac{1}{12} \ln |1 - 4x^3| + C. \quad \square$$

Problem 9. Calculate

$$\int \frac{\ln x}{x}\, dx.$$

Solution. Set

$$u = \ln x, \qquad du = \frac{1}{x}\, dx.$$

$$\int \frac{\ln x}{x}\, dx = \int u\, du = \tfrac{1}{2}u^2 + C = \tfrac{1}{2}(\ln x)^2 + C. \quad \square$$

Problem 10. Evaluate

$$\int_1^2 \frac{6x^2 - 2}{x^3 - x + 1}\, dx.$$

Solution. Set

$$u = x^3 - x + 1, \qquad du = (3x^2 - 1)\, dx.$$

At $x = 1$, $u = 1$; at $x = 2$, $u = 7$.

$$\int_1^2 \frac{6x^2 - 2}{x^3 - x + 1}\, dx = 2 \int_1^7 \frac{du}{u} = 2 \left[\ln |u| \right]_1^7 = 2(\ln 7 - \ln 1) = 2 \ln 7. \quad \square$$

Logarithmic Differentiation

We can differentiate a lengthy product

$$g(x) = g_1(x)g_2(x) \cdots g_n(x)$$

by first writing

$$\ln |g(x)| = \ln \left(|g_1(x)| |g_2(x)| \cdots |g_n(x)| \right)$$
$$= \ln |g_1(x)| + \ln |g_2(x)| + \cdots + \ln |g_n(x)|$$

and then differentiating:

$$\frac{g'(x)}{g(x)} = \frac{g_1'(x)}{g_1(x)} + \frac{g_2'(x)}{g_2(x)} + \cdots + \frac{g_n'(x)}{g_n(x)}.$$

Multiplication by $g(x)$ gives

(7.3.4)
$$g'(x) = g(x) \left(\frac{g_1'(x)}{g_1(x)} + \frac{g_2'(x)}{g_2(x)} + \cdots + \frac{g_n'(x)}{g_n(x)} \right).$$

The process by which $g'(x)$ was obtained is called *logarithmic differentiation*. Logarithmic differentiation is valid at all points x where $g(x) \neq 0$. At points x where $g(x) = 0$, none of it makes sense.

A product of n factors

$$g(x) = g_1(x)g_2(x) \cdots g_n(x)$$

can, of course, also be differentiated by repeated applications of the product rule (3.2.4). The great advantage of logarithmic differentiation is that it gives us an explicit formula for the derivative, a formula that's easy to remember and easy to work with.

Problem 11. Given that

$$g(x) = x(x - 1)(x - 2)(x - 3)$$

find $g'(x)$ for $x \neq 0, 1, 2, 3$.

Solution. We can write down $g'(x)$ directly from Formula 7.3.4.

$$g'(x) = x(x - 1)(x - 2)(x - 3) \left(\frac{1}{x} + \frac{1}{x - 1} + \frac{1}{x - 2} + \frac{1}{x - 3} \right).$$

Or, we can go through the process by which we derived Formula 7.3.4.

$$\ln |g(x)| = \ln |x| + \ln |x - 1| + \ln |x - 2| + \ln |x - 3|,$$

$$\frac{g'(x)}{g(x)} = \frac{1}{x} + \frac{1}{x - 1} + \frac{1}{x - 2} + \frac{1}{x - 3},$$

$$g'(x) = x(x - 1)(x - 2)(x - 3) \left(\frac{1}{x} + \frac{1}{x - 1} + \frac{1}{x - 2} + \frac{1}{x - 3} \right). \quad \square$$

Logarithmic differentiation can also be used with quotients.

Problem 12. Given that

$$g(x) = \frac{(x^2 + 1)^3(2x - 5)^2}{(x^2 + 5)^2},$$

find $g'(x)$ for $x \neq \frac{5}{2}$.

Solution. Our first step is to write

$$g(x) = (x^2 + 1)^3(2x - 5)^2(x^2 + 5)^{-2}.$$

Then according to Formula 7.3.4

$$g'(x) = \frac{(x^2 + 1)^3(2x - 5)^2}{(x^2 + 5)^2} \left[\frac{3(x^2 + 1)^2(2x)}{(x^2 + 1)^3} + \frac{2(2x - 5)(2)}{(2x - 5)^2} + \frac{(-2)(x^2 + 5)^{-3}(2x)}{(x^2 + 5)^{-2}} \right]$$

$$= \frac{(x^2 + 1)^3(2x - 5)^2}{(x^2 + 5)^2} \left(\frac{6x}{x^2 + 1} + \frac{4}{2x - 5} - \frac{4x}{x^2 + 5} \right). \quad \square$$

EXERCISES 7.3

Determine the domain and find the derivative.

1. $f(x) = \ln 4x$.

2. $f(x) = \ln (2x + 1)$.

3. $f(x) = \ln (x^3 + 1)$.

4. $f(x) = \ln [(x + 1)^3]$.

5. $f(x) = \ln \sqrt{1 + x^2}$.

6. $f(x) = (\ln x)^3$.

7. $f(x) = \ln |x^4 - 1|$.

8. $f(x) = \ln (\ln x)$.

9. $f(x) = x \ln x$.

10. $f(x) = \ln \left| \dfrac{x + 2}{x^3 - 1} \right|$.

11. $f(x) = \dfrac{1}{\ln x}$.

12. $f(x) = \ln \sqrt[4]{x^2 + 1}$.

13. $f(x) = \dfrac{\ln (x + 1)}{x + 1}$.

14. $f(x) = \ln \sqrt{\dfrac{1 - x}{2 - x}}$.

Calculate the following integrals.

15. $\displaystyle\int \dfrac{dx}{x + 1}$.

16. $\displaystyle\int \dfrac{dx}{3 - x}$.

17. $\displaystyle\int \dfrac{x}{3 - x^2} \, dx$.

18. $\displaystyle\int \dfrac{x + 1}{x^2} \, dx$.

19. $\displaystyle\int \dfrac{x}{(3 - x^2)^2} \, dx$.

20. $\displaystyle\int \dfrac{\ln (x + a)}{x + a} \, dx$.

21. $\displaystyle\int \left(\dfrac{1}{x + 2} - \dfrac{1}{x - 2} \right) dx$.

22. $\displaystyle\int \dfrac{x^2}{2x^3 - 1} \, dx$.

23. $\displaystyle\int \dfrac{dx}{x(\ln x)^2}$.

24. $\displaystyle\int \left(\dfrac{1}{x - a} - \dfrac{1}{x - b} \right) dx$.

25. $\displaystyle\int \dfrac{\sqrt{x}}{1 + x\sqrt{x}} \, dx$.

26. $\displaystyle\int x \left(\dfrac{1}{x^2 - a^2} - \dfrac{1}{x^2 - b^2} \right) dx$.

Evaluate.

27. $\displaystyle\int_1^e \dfrac{dx}{x}$.

28. $\displaystyle\int_1^{e^2} \dfrac{dx}{x}$.

29. $\displaystyle\int_e^{e^2} \dfrac{dx}{x}$.

30. $\displaystyle\int_3^4 \dfrac{dx}{1 - x}$.

31. $\displaystyle\int_4^5 \dfrac{x}{x^2 - 1} \, dx$.

32. $\displaystyle\int_1^e \dfrac{\ln x}{x} \, dx$.

33. $\displaystyle\int_0^1 \dfrac{\ln (x + 1)}{x + 1} \, dx$.

34. $\displaystyle\int_0^1 \left(\dfrac{1}{x + 1} - \dfrac{1}{x + 2} \right) dx$.

Calculate the derivative by logarithmic differentiation.

35. $g(x) = (x^2 + 1)^2 (x - 1)^5 x^3$.

36. $g(x) = x(x + a)(x + b)(x + c)$.

37. $g(x) = \dfrac{x^4(x - 1)}{(x + 2)(x^2 + 1)}$.

38. $g(x) = \dfrac{(1 + x)(2 + x)x}{(4 + x)(2 - x)}$.

39. $g(x) = \sqrt{\dfrac{(x - 1)(x - 2)}{(x - 3)(x - 4)}}$.

40. $g(x) = \dfrac{x^2(x^2 + 1)(x^2 + 2)}{(x^2 - 1)(x^2 - 5)}$.

Find the area of the part of the first quadrant that lies between

41. $x + 4y - 5 = 0$ and $xy = 1$.

42. $x + y - 3 = 0$ and $xy = 2$.

43. A particle moves along a line with acceleration $a(t) = -(t + 1)^{-2}$ feet per second per second. Find the distance traveled by the particle during the time interval $[0, 4]$, given that the initial velocity $v(0)$ is 1 foot per second.

44. Exercise 43 with $v(0)$ as 2 feet per second.

Find a formula for the nth derivative.

45. $\dfrac{d^n}{dx^n} (\ln x)$.

46. $\dfrac{d^n}{dx^n} [\ln (1 - x)]$.

47. $\dfrac{d^n}{dx^n} (\ln 2x)$.

48. $\dfrac{d^n}{dx^n} \left(\ln \dfrac{1}{x} \right)$.

49. (a) Verify that

$$\ln x = \int_1^x \frac{dt}{t} < \int_1^x \frac{dt}{\sqrt{t}} \qquad \text{for } x > 1.$$

(b) Use (a) to show that

$$0 < \ln x < 2(\sqrt{x} - 1) \qquad \text{for } x > 1.$$

(c) Use (b) to show that

$$2x\left(1 - \frac{1}{\sqrt{x}}\right) < x \ln x < 0 \qquad \text{for } 0 < x < 1.$$

(d) Use (c) to show that

$$\lim_{x \to 0^+} (x \ln x) = 0.$$

50. (a) Show that for $n = 2$ Formula 7.3.4 reduces to the product rule (3.2.4).

(b) Show that Formula 7.3.4 applied to

$$g(x) = \frac{g_1(x)}{g_2(x)}$$

reduces to the quotient rule (3.2.9).

For each of the functions below, (i) find the domain, (ii) find the intervals where the function increases and the intervals where it decreases, (iii) find the extreme values, (iv) determine the concavity of the graph and find the points of inflection, and finally, (v) sketch the graph, indicating asymptotes.

51. $f(x) = \ln 2x.$ **52.** $f(x) = x - \ln x.$ **53.** $f(x) = \ln (4 - x).$

54. $f(x) = \ln (4 - x^2).$ **55.** $f(x) = x \ln x.$ **56.** $f(x) = \ln (8x - x^2).$

57. $f(x) = \ln \left(\frac{x}{1 + x^2}\right).$ **58.** $f(x) = \ln \left(\frac{x^3}{x - 1}\right).$

7.4   THE EXPONENTIAL FUNCTION

Rational powers of e already have an established meaning: by $e^{p/q}$ we mean the qth root of e raised to the pth power. But what is meant by $e^{\sqrt{2}}$ or e^{π}?

Earlier we proved that each rational power $e^{p/q}$ had logarithm p/q:

(7.4.1) $$\ln e^{p/q} = \frac{p}{q}.$$

The definition of e^z for z irrational is patterned after this relation.

> **DEFINITION 7.4.2**
>
> If z is irrational, then by e^z we mean the unique number which has logarithm z:
>
> $$\ln e^z = z.$$

What is $e^{\sqrt{2}}$? It is the unique number that has logarithm $\sqrt{2}$. What is e^{π}? It is the unique number that has logarithm π. Note that e^x now has meaning for every real value of x: it is the unique number with logarithm x.

DEFINITION 7.4.3

The function

$$E(x) = e^x \qquad \text{for all real } x$$

is called the *exponential function*.

Some properties of the exponential function follow.

(1) In the first place

(7.4.4) $\qquad\boxed{\ln e^x = x \qquad \text{for all real } x.}$

Writing $L(x) = \ln x$ and $E(x) = e^x$, we have

$$L(E(x)) = x \qquad \text{for all real } x.$$

This says that *the exponential function is the inverse of the logarithm function.*

(2) The graph of the exponential function appears in Figure 7.4.1. It can be obtained from the graph of the logarithm by reflection in the line $y = x$.

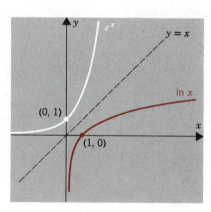

Figure 7.4.1

(3) Since the graph of the logarithm remains to the right of the y-axis, the graph of the exponential function remains above the x-axis:

(7.4.5) $\qquad\boxed{e^x > 0 \qquad \text{for all real } x.}$

(4) Since the graph of the logarithm crosses the x-axis at $(1, 0)$, the graph of the exponential function crosses the y-axis at $(0, 1)$:

$$\ln 1 = 0 \quad \text{gives} \quad e^0 = 1.$$

(5) Since the y-axis is a vertical asymptote for the graph of the logarithm function, the x-axis is a horizontal asymptote for the graph of the exponential function:

$$\text{as } x \to -\infty, \quad e^x \to 0.$$

(6) Since the exponential function is the inverse of the logarithm function, the logarithm function is the inverse of the exponential function; namely, we must have

(7.4.6)
$$e^{\ln x} = x \qquad \text{for all } x > 0.$$

You can verify this equation directly by observing that both sides have the same logarithm:

$$\ln (e^{\ln x}) = \ln x$$

since, for all real t, $\ln e^t = t$. □

You know that for rational exponents

$$e^{(p/q + r/s)} = e^{p/q} \cdot e^{r/s}.$$

This property holds for all exponents, including irrational exponents.

THEOREM 7.4.7

$$e^{x+y} = e^x \cdot e^y \qquad \text{for all real } x \text{ and } y.$$

Proof

$$\ln (e^x \cdot e^y) = \ln e^x + \ln e^y = x + y = \ln e^{x+y}.$$

Since the logarithm is one-to-one, we must have

$$e^{x+y} = e^x \cdot e^y. \quad □$$

We leave it to you to show that

(7.4.8)
$$e^{-y} = \frac{1}{e^y} \qquad \text{and} \qquad e^{x-y} = \frac{e^x}{e^y}.$$

Table 7.4.1 gives some values of the exponential function rounded off to the nearest hundredth. More extended tables appear in Appendix C.

TABLE 7.4.1

x	e^x	x	e^x
0.1	1.11	1.1	3.00
0.2	1.22	1.2	3.32
0.3	1.35	1.3	3.67
0.4	1.49	1.4	4.06
0.5	1.65	1.5	4.48
0.6	1.82	1.6	4.95
0.7	2.01	1.7	5.47
0.8	2.23	1.8	6.05
0.9	2.46	1.9	6.68
1.0	2.72	2.0	7.39

Problem 1.   Estimate the following powers of e on the basis of Table 7.4.1.

(a) $e^{-0.2}$.   (b) $e^{2.4}$.   (c) $e^{3.1}$.

Solution.   The idea, of course, is to use the laws of exponents.

(a) $e^{-0.2} = 1/e^{0.2} \cong 1/1.22 \cong 0.82$.

(b) $e^{2.4} = e^{2+0.4} = (e^2)(e^{0.4}) \cong (7.39)(1.49) \cong 11.01$.

(c) $e^{3.1} = e^{1.7+1.4} = (e^{1.7})(e^{1.4}) \cong (5.47)(4.06) \cong 22.21$.   □

We come now to one of the most important results in calculus.   It is marvelously simple.

THEOREM 7.4.9

The exponential function is its own derivative: for all real x

$$\frac{d}{dx}(e^x) = e^x.$$

Proof.   The logarithm function is differentiable, and its derivative is never 0.   It follows from our discussion in Section 3.8 that its inverse, the exponential function, is also differentiable.   Knowing this, we can show that

$$\frac{d}{dx}(e^x) = e^x$$

by differentiating the identity

$$\ln e^x = x.$$

On the left-hand side, the chain rule gives

$$\frac{d}{dx}(\ln e^x) = \frac{1}{e^x}\frac{d}{dx}(e^x).$$

On the right-hand side, the derivative is 1:

$$\frac{d}{dx}(x) = 1.$$

Equating these derivatives, we have

$$\frac{1}{e^x}\frac{d}{dx}(e^x) = 1 \quad \text{and thus} \quad \frac{d}{dx}(e^x) = e^x.$$ □

We frequently run across expressions of the form e^u where u is a function of x.   If u is differentiable, then the chain rule gives

(7.4.10)

$$\frac{d}{dx}(e^u) = e^u\frac{du}{dx}.$$

Proof

$$\frac{d}{dx}(e^u) = \frac{d}{du}(e^u)\frac{du}{dx} = e^u \frac{du}{dx}. \quad \square$$

Example 2

(a) $\dfrac{d}{dx}(e^{kx}) = e^{kx} \cdot k = ke^{kx}.$

(b) $\dfrac{d}{dx}(e^{\sqrt{x}}) = e^{\sqrt{x}}\dfrac{d}{dx}(\sqrt{x}) = e^{\sqrt{x}}\left(\dfrac{1}{2\sqrt{x}}\right) = \dfrac{1}{2\sqrt{x}}e^{\sqrt{x}}.$

(c) $\dfrac{d}{dx}(e^{-x^2}) = e^{-x^2}\dfrac{d}{dx}(-x^2) = e^{-x^2}(-2x) = -2xe^{-x^2}. \quad \square$

The relation

$$\frac{d}{dx}(e^x) = e^x \quad \text{and its corollary} \quad \frac{d}{dx}(e^{kx}) = ke^{kx}$$

have important applications to engineering, physics, chemistry, biology, and economics. We will discuss some of these applications in Section 7.6.

Problem 3. Let

$$f(x) = xe^{-x} \qquad \text{for all real } x.$$

(a) On what intervals does f increase? Decrease? (b) Find the extreme values of f. (c) Determine the concavity of the graph and find the points of inflection. (d) Sketch the graph indicating the asymptotes if any.

Solution. We have

$$f(x) = xe^{-x},$$
$$f'(x) = xe^{-x}(-1) + e^{-x} = (1 - x)e^{-x},$$
$$f''(x) = (1 - x)e^{-x}(-1) - e^{-x} = (x - 2)e^{-x}.$$

Since $e^x > 0$ for all real x,

$$f'(x) \text{ is } \begin{cases} \text{positive,} & \text{for} \quad x < 1 \\ 0, & \text{at} \quad x = 1 \\ \text{negative,} & \text{for} \quad x > 1 \end{cases}.$$

The function f increases on $(-\infty, 1]$ and decreases on $[1, \infty)$. The number

$$f(1) = \frac{1}{e} \cong \frac{1}{2.72} \cong 0.37$$

is a local and absolute maximum. There are no other extreme values. Since

$$f''(x) \text{ is } \begin{cases} \text{negative,} & \text{for} \quad x < 2 \\ 0, & \text{at} \quad x = 2 \\ \text{positive,} & \text{for} \quad x > 2 \end{cases},$$

the graph is concave down on $(-\infty, 2)$ and concave up on $(2, \infty)$. The point

$$(2, f(2)) = (2, 2e^{-2}) \cong \left(2, \frac{2}{(2.72)^2}\right) \cong (2, 0.27)$$

is the only point of inflection. The graph is given in Figure 7.4.2. The x-axis is a horizontal asymptote: as $x \to \infty$, $f(x) \to 0$. $\square$

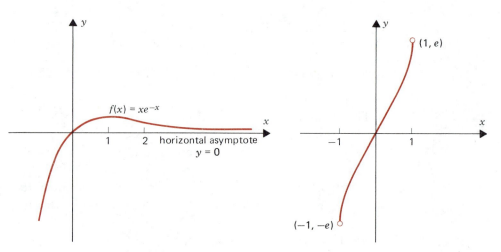

$f(x) = xe^{-x}$

horizontal asymptote
$y = 0$

$(1, e)$

$(-1, -e)$

Figure 7.4.2   **Figure 7.4.3**

Problem 4.   Let

$$f(x) = xe^{x^2}   \qquad \text{for all } x \in (-1, 1).$$

(a) On what intervals is f increasing? Decreasing? (b) Find the extreme values of f.
(c) Determine the concavity of the graph and find the points of inflection. (d) Sketch the graph.

Solution.   Here

$$f(x) = xe^{x^2},$$
$$f'(x) = xe^{x^2}(2x) + e^{x^2} = (1 + 2x^2)e^{x^2},$$
$$f''(x) = (1 + 2x^2)e^{x^2}(2x) + 4xe^{x^2} = 2x(2x^2 + 3)e^{x^2}.$$

Since $f'(x) > 0$ for all $x \in (-1, 1)$, f increases on $(-1, 1)$, and there are no extreme values. Since

$$f''(x) \text{ is } \begin{cases} \text{negative,} & \text{for} \quad x \in (-1, 0) \\ \quad\quad 0, & \text{at} \quad\quad x = 0 \\ \text{positive,} & \text{for} \quad x \in (0, 1) \end{cases},$$

the graph is concave down on $(-1, 0)$ and concave up on $(0, 1)$. The point $(0, f(0)) = (0, 0)$ is the only point of inflection. Since $f(-x) = -f(x)$, f is an odd function, and thus its graph (Figure 7.4.3) is symmetric with respect to the origin. $\square$

The integral counterpart of (7.4.9) takes the form

(7.4.11)
$$\int e^x \, dx = e^x + C.$$

In practice

$$\int e^{g(x)} \, g'(x) \, dx \quad \text{is reduced to} \quad \int e^u \, du$$

by setting

$$u = g(x), \quad du = g'(x) \, dx.$$

Problem 5. Find

$$\int 9e^{3x} \, dx.$$

Solution. Set

$$u = 3x, \quad du = 3 \, dx.$$

$$\int 9e^{3x} \, dx = 3 \int e^u \, du = 3e^u + C = 3e^{3x} + C.$$

If you recognize at the very beginning that

$$3e^{3x} = \frac{d}{dx} (e^{3x}),$$

then you can dispense with the u-substitution and simply write

$$\int 9e^{3x} \, dx = 3 \int 3e^{3x} \, dx = 3e^{3x} + C. \quad \blacksquare$$

Problem 6. Find

$$\int \frac{e^{\sqrt{x}}}{\sqrt{x}} \, dx.$$

Solution. Set

$$u = \sqrt{x}, \quad du = \frac{1}{2\sqrt{x}} \, dx.$$

$$\int \frac{e^{\sqrt{x}}}{\sqrt{x}} \, dx = 2 \int e^u \, du = 2e^u + C = 2e^{\sqrt{x}} + C.$$

If you recognize immediately that

$$\frac{1}{2} \left(\frac{e^{\sqrt{x}}}{\sqrt{x}} \right) = \frac{d}{dx} (e^{\sqrt{x}}),$$

then you can dispense with the u-substitution and integrate directly:

$$\int \frac{e^{\sqrt{x}}}{\sqrt{x}}\,dx = 2 \int \frac{1}{2}\!\left(\frac{e^{\sqrt{x}}}{\sqrt{x}}\right) dx = 2e^{\sqrt{x}} + C. \quad \square$$

Problem 7. Find

$$\int \frac{e^{3x}}{e^{3x} + 1}\,dx.$$

Solution. We can put this integral in the form

$$\int \frac{du}{u}$$

by setting

$$u = e^{3x} + 1, \qquad du = 3e^{3x}\,dx.$$

Then

$$\int \frac{e^{3x}}{e^{3x} + 1}\,dx = \frac{1}{3} \int \frac{du}{u} = \frac{1}{3}\ln|u| + C = \frac{1}{3}\ln\,(e^{3x} + 1) + C. \quad \square$$

Problem 8. Evaluate

$$\int_0^{\ln 2} e^x\,dx.$$

Solution

$$\int_0^{\ln 2} e^x\,dx = \left[\,e^x\,\right]_0^{\ln 2} = e^{\ln 2} - e^0 = 2 - 1 = 1. \quad \square$$

Problem 9. Evaluate

$$\int_0^1 e^x(e^x + 1)^{1/5}\,dx.$$

Solution. Set

$$u = e^x + 1, \qquad du = e^x\,dx.$$

At $x = 0$, $u = 2$; at $x = 1$, $u = e + 1$. Thus

$$\int_0^1 e^x(e^x + 1)^{1/5}\,dx = \int_2^{e+1} u^{1/5}\,du = \left[\tfrac{5}{6}u^{6/5}\right]_2^{e+1} = \tfrac{5}{6}[(e + 1)^{6/5} - 2^{6/5}]. \quad \square$$

Remark. The u-substitution simplifies many calculations, but you will find with experience that in many cases you can carry out the integration more quickly without it. ▪

EXERCISES 7.4

Differentiate the following functions.

1. $y = e^{-2x}$.

2. $y = 3e^{2x+1}$.

3. $y = e^{x^2-1}$.

4. $y = 2e^{-4x}$.

5. $y = e^x \ln x$.

6. $y = x^2 e^x$.

7. $y = x^{-1} e^{-x}$.

8. $y = e^{\sqrt{x}+1}$.

9. $y = \frac{1}{2}(e^x + e^{-x})$.

10. $y = \frac{1}{2}(e^x - e^{-x})$.

11. $y = e^{\sqrt{x}} \ln \sqrt{x}$.

12. $y = (1 - e^{4x})^2$.

13. $y = (e^x + e^{-x})^2$.

14. $y = (3 - 2e^{-x})^3$.

15. $y = (e^{x^2} + 1)^2$.

16. $y = (e^{2x} - e^{-2x})^2$.

17. $y = (x^2 - 2x + 2)e^x$.

18. $y = x^2 e^x - xe^{x^2}$.

19. $y = \dfrac{e^x - 1}{e^x + 1}$.

20. $y = \dfrac{e^{2x} - 1}{e^{2x} + 1}$.

21. $y = \dfrac{e^{ax} - e^{bx}}{e^{ax} + e^{bx}}$.

22. $y = \ln e^{3x}$.

23. $y = e^{4 \ln x}$.

24. $y = e^{\sqrt{1-x^2}}$.

Calculate the following indefinite integrals.

25. $\displaystyle\int e^{2x}\, dx$.

26. $\displaystyle\int e^{-2x}\, dx$.

27. $\displaystyle\int e^{kx}\, dx$.

28. $\displaystyle\int e^{ax+b}\, dx$.

29. $\displaystyle\int xe^{x^2}\, dx$.

30. $\displaystyle\int xe^{-x^2}\, dx$.

31. $\displaystyle\int \dfrac{e^{1/x}}{x^2}\, dx$.

32. $\displaystyle\int \dfrac{e^{2\sqrt{x}}}{\sqrt{x}}\, dx$.

33. $\displaystyle\int (e^x + e^{-x})^2\, dx$.

34. $\displaystyle\int e^{\ln x}\, dx$.

35. $\displaystyle\int \ln e^x\, dx$.

36. $\displaystyle\int (e^{-x} - 1)^2\, dx$.

37. $\displaystyle\int \dfrac{4}{\sqrt{e^x}}\, dx$.

38. $\displaystyle\int \dfrac{e^x}{e^x + 1}\, dx$.

39. $\displaystyle\int \dfrac{e^x}{\sqrt{e^x + 1}}\, dx$.

40. $\displaystyle\int \dfrac{2e^x}{\sqrt[3]{e^x + 1}}\, dx$.

41. $\displaystyle\int \dfrac{e^{2x}}{2e^{2x} + 3}\, dx$.

42. $\displaystyle\int \dfrac{xe^{ax^2}}{e^{ax^2} + 1}\, dx$.

Evaluate the following definite integrals.

43. $\displaystyle\int_0^1 e^x\, dx$.

44. $\displaystyle\int_0^1 e^{-kx}\, dx$.

45. $\displaystyle\int_0^{\ln \pi} e^{-6x}\, dx$.

46. $\displaystyle\int_0^1 xe^{-x^2}\, dx$.

47. $\displaystyle\int_0^1 \dfrac{e^x + 1}{e^x}\, dx$.

48. $\displaystyle\int_0^1 \dfrac{4 - e^x}{e^x}\, dx$.

49. $\displaystyle\int_0^{\ln 2} \dfrac{e^x}{e^x + 1}\, dx$.

50. $\displaystyle\int_0^1 \dfrac{e^x + e^{-x}}{2}\, dx$.

51. $\displaystyle\int_{\ln 2}^{\ln 3} (e^x - e^{-x})^2\, dx$.

52. $\displaystyle\int_0^1 \dfrac{e^x}{4 - e^x}\, dx$.

53. $\displaystyle\int_0^1 x(e^{x^2} + 2)\, dx$.

54. $\displaystyle\int_1^2 (2 - e^{-x})^2\, dx$.

Estimate on the basis of Table 7.4.1.

55. $e^{-0.4}$.

56. $e^{2.6}$.

57. $e^{2.8}$.

58. $e^{-2.1}$.

Estimate using differentials.

59. $e^{2.03}$. $(e^2 \cong 7.39)$

60. $e^{-0.15}$. $(e^0 = 1)$

61. $e^{2.85}$. $(e^3 \cong 20.09)$

62. e^{2a+h}. $(e^a = A)$

63. A particle moves on a coordinate line with its position at time t given by the function $x(t) = Ae^{ct} + Be^{-ct}$. Show that the acceleration of the particle is proportional to its position.

Sketch the region bounded by the curves and find its area.

64. $x = e^{2y}$, $\quad x = e^{-y}$, $\quad x = 4$. **65.** $y = e^x$, $\quad y = e^{2x}$, $\quad y = e^4$.

66. $y = e^x$, $\quad y = e$, $\quad y = x$, $\quad x = 0$. **67.** $x = e^y$, $\quad y = 1$, $\quad y = 2$, $\quad x = 2$.

68. The function $f(x) = e^{-x^2}$ is very important in statistics. (a) What is the symmetry of the graph? (b) On what intervals does the function increase? decrease? (c) The function has only one extreme value. Where is that value taken on and what is it? (d) Determine the concavity of the graph and find the points of inflection. (e) The graph has a horizontal asymptote. What is it? (f) Sketch the graph.

For each of the functions below, (i) find the domain, (ii) find the intervals where the function increases and the intervals where it decreases, (iii) find the extreme values, (iv) determine the concavity and find the points of inflection, and finally, (v) sketch the graph, indicating all asymptotes.

69. $f(x) = \frac{1}{2}(e^x + e^{-x})$. **70.** $f(x) = \frac{1}{2}(e^x - e^{-x})$. **71.** $f(x) = xe^x$.

72. $f(x) = (1 - x)e^x$. **73.** $f(x) = e^{(1/x)^2}$. **74.** $f(x) = x^2 e^{-x}$.

75. Take $a > 0$ and refer to Figure 7.4.4. (a) Find the points of tangency, marked A and B. (b) Find the area of region I. (c) Find the area of region II.

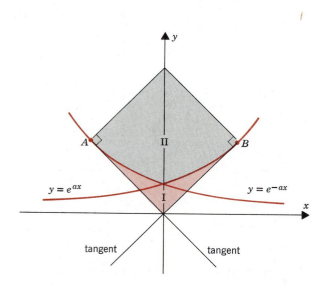

Figure 7.4.4

76. Prove that for all $x > 0$ and all positive integers n

$$e^x > 1 + x + \frac{x^2}{2!} + \frac{x^3}{3!} + \cdots + \frac{x^n}{n!},$$

where $n!$, read "n factorial," is shorthand for

$$n(n - 1)(n - 2) \cdots 2 \cdot 1.$$

HINT: $\quad e^x = 1 + \int_0^x e^t\, dt > 1 + \int_0^x dt = 1 + x,$

$$e^x = 1 + \int_0^x e^t\, dt > 1 + \int_0^x (1 + t)\, dt = 1 + x + \frac{x^2}{2}, \quad \text{etc.}$$

77. Prove that, if n is a positive integer, then

$$e^x > x^n \qquad \text{for all } x \text{ sufficiently large.}$$

HINT: Use Exercise 76.

(Calculator) Evaluate the following limits numerically and justify your answers by other means.

78. $\lim\limits_{x \to 0} \dfrac{e^{10x} - 1}{x}$.

79. $\lim\limits_{x \to 1} \dfrac{e^{x^3} - e}{x - 1}$.

80. $\lim\limits_{x \to 1} \dfrac{e^x - e}{\ln x}$.

HINT: Remember the alternative definition of derivative

$$f'(c) = \lim_{x \to c} \frac{f(x) - f(c)}{x - c}. \tag{3.13.1}$$

7.5 ARBITRARY POWERS; OTHER BASES; ESTIMATING e

Arbitrary Powers: the Function $f(x) = x^r$

The elementary notion of exponent applies only to rational numbers. Expressions such as

$$10^5, \quad 2^{1/3}, \quad 7^{-4/5}, \quad \pi^{-1/2}$$

make sense, but so far we have attached no meaning to expressions such as

$$10^{\sqrt{2}}, \quad 2^\pi, \quad 7^{-\sqrt{3}}, \quad \pi^e.$$

The extension of our sense of exponent to allow for irrational exponents is conveniently done by making use of the logarithm function and the exponential function. The heart of the matter is to observe that for $x > 0$ and p/q rational

$$x^{p/q} = e^{(p/q)\ln x}.$$

(To verify this, take the log of both sides.) We *define* x^z for irrational z by setting

$$x^z = e^{z \ln x}.$$

We then have the following result:

(7.5.1)
> if $x > 0$, then
> $$x^r = e^{r \ln x} \quad \text{for all real numbers } r.$$

In particular

$$10^{\sqrt{2}} = e^{\sqrt{2}\ln 10}, \quad 2^\pi = e^{\pi \ln 2}, \quad 7^{-\sqrt{3}} = e^{-\sqrt{3}\ln 7}, \quad \pi^e = e^{e \ln \pi}. \quad \square$$

With this extended sense of exponent the usual laws of exponents

(7.5.2)
> $$x^{r+s} = x^r x^s, \qquad x^{r-s} = \frac{x^r}{x^s}, \qquad (x^r)^s = x^{rs}$$

still hold:

$$x^{r+s} = e^{(r+s)\ln x} = e^{r\ln x} \cdot e^{s\ln x} = x^r x^s,$$

$$x^{r-s} = e^{(r-s)\ln x} = e^{r\ln x} \cdot e^{-s\ln x} = \frac{e^{r\ln x}}{e^{s\ln x}} = \frac{x^r}{x^s},$$

$$(x^r)^s = e^{s\ln x^r} = e^{rs\ln x} = x^{rs}. \quad \square$$

The differentiation of arbitrary powers has a familiar look:

(7.5.3)

$$\frac{d}{dx}(x^r) = rx^{r-1} \qquad \text{for all } x > 0.$$

Proof

$$\frac{d}{dx}(x^r) = \frac{d}{dx}(e^{r\ln x}) = e^{r\ln x}\frac{d}{dx}(r\ln x) = x^r\frac{r}{x} = rx^{r-1}.$$

You can also write $f(x) = x^r$ and use logarithmic differentiation:

$$\ln f(x) = r\ln x$$

$$\frac{f'(x)}{f(x)} = \frac{r}{x}$$

$$f'(x) = \frac{rf(x)}{x} = \frac{rx^r}{x} = rx^{r-1}. \quad \square$$

Thus

$$\frac{d}{dx}(x^{\sqrt{2}}) = \sqrt{2}\,x^{\sqrt{2}-1}, \qquad \frac{d}{dx}(x^\pi) = \pi x^{\pi-1}.$$

If u is a positive differentiable function of x, then, by the chain rule,

(7.5.4)

$$\frac{d}{dx}(u^r) = ru^{r-1}\frac{du}{dx}.$$

Proof

$$\frac{d}{dx}(u^r) = \frac{d}{du}(u^r)\frac{du}{dx} = ru^{r-1}\frac{du}{dx}. \quad \square$$

For example,

$$\frac{d}{dx}[(x^2+5)^{\sqrt{3}}] = \sqrt{3}(x^2+5)^{\sqrt{3}-1}(2x) = 2\sqrt{3}\,x(x^2+5)^{\sqrt{3}-1}.$$

Problem 1. Find

$$\int \frac{x^3}{(2x^4+1)^\pi}\,dx.$$

Solution. Set

$$u = 2x^4 + 1, \qquad du = 8x^3 \, dx.$$

$$\int \frac{x^3}{(2x^4 + 1)^\pi} \, dx = \frac{1}{8} \int \frac{du}{u^\pi} = \frac{1}{8} \left(\frac{u^{1-\pi}}{1 - \pi} \right) + C = \frac{(2x^4 + 1)^{1-\pi}}{8(1 - \pi)} + C. \quad \square$$

Problem 2. Find

$$\frac{d}{dx} (x^x).$$

Solution. One way to find this derivative is to observe that $x^x = e^{x \ln x}$ and then differentiate:

$$\frac{d}{dx} (x^x) = \frac{d}{dx} (e^{x \ln x}) = e^{x \ln x} \left(x \cdot \frac{1}{x} + \ln x \right) = x^x (1 + \ln x).$$

Another way to find the derivative of x^x is to set $f(x) = x^x$ and use logarithmic differentiation:

$$\ln f(x) = x \ln x$$

$$\frac{f'(x)}{f(x)} = x \cdot \frac{1}{x} + \ln x = 1 + \ln x$$

$$f'(x) = f(x)(1 + \ln x) = x^x (1 + \ln x). \quad \square$$

Base p: the Function $f(x) = p^x$

To form the function $f(x) = x^r$ we take a positive variable x and raise it to a constant power r. To form the function $f(x) = p^x$ we take a positive constant p and raise it to a variable power x. Since $1^x = 1$ for all x, the function is of interest only if $p \neq 1$.

The high status enjoyed by Euler's number e comes from the fact that

$$\frac{d}{dx} (e^x) = e^x.$$

For other bases the derivative has an extra factor:

(7.5.5)
$$\frac{d}{dx} (p^x) = p^x \ln p.$$

Proof

$$\frac{d}{dx} (p^x) = \frac{d}{dx} (e^{x \ln p}) = e^{x \ln p} \ln p = p^x \ln p. \quad \square$$

Thus, for example,

$$\frac{d}{dx} (2^x) = 2^x \ln 2 \qquad \text{and} \qquad \frac{d}{dx} (10^x) = 10^x \ln 10.$$

If u is a differentiable function of x, then, by the chain rule,

(7.5.6)
$$\frac{d}{dx}(p^u) = p^u \ln p \, \frac{du}{dx}.$$

Proof

$$\frac{d}{dx}(p^u) = \frac{d}{du}(p^u)\frac{du}{dx} = p^u \ln p \, \frac{du}{dx}. \quad \square$$

For example,

$$\frac{d}{dx}(2^{3x^2}) = 2^{3x^2}(\ln 2)(6x) = 6x \, 2^{3x^2} \ln 2.$$

Problem 3. Find

$$\int 2^x \, dx.$$

Solution. Set

$$u = 2^x, \qquad du = 2^x \ln 2 \, dx.$$

$$\int 2^x \, dx = \frac{1}{\ln 2}\int du = \frac{1}{\ln 2}u + C = \frac{1}{\ln 2}2^x + C. \quad \square$$

Problem 4. Evaluate

$$\int_1^2 3^{-x} \, dx.$$

Solution. Set

$$u = 3^{-x}, \qquad du = -3^{-x}\ln 3 \, dx.$$

At $x = 1$, $u = \frac{1}{3}$; at $x = 2$, $u = \frac{1}{9}$. Thus

$$\int_1^2 3^{-x} \, dx = -\frac{1}{\ln 3}\int_{1/3}^{1/9} du = -\frac{1}{\ln 3}\left(\frac{1}{9} - \frac{1}{3}\right) = \frac{2}{9 \ln 3}. \quad \square$$

Base p: the Function $f(x) = \log_p x$

If $p > 0$, then

$$\ln p^t = t \ln p \qquad \text{for all } t.$$

If p is also different from 1, then $\ln p \neq 0$, and we have

$$\frac{\ln p^t}{\ln p} = t.$$

This indicates that the function

$$f(x) = \frac{\ln x}{\ln p}$$

satisfies the relation

$$f(p^t) = t \qquad \text{for all real } t.$$

In view of this we call

$$\frac{\ln x}{\ln p}$$

the logarithm of x to the base p and write

(7.5.7)
$$\log_p x = \frac{\ln x}{\ln p}.$$

For example,

$$\log_2 32 = \frac{\ln 32}{\ln 2} = \frac{\ln 2^5}{\ln 2} = \frac{5 \ln 2}{\ln 2} = 5$$

and

$$\log_{100} \left(\tfrac{1}{10}\right) = \frac{\ln \left(\tfrac{1}{10}\right)}{\ln 100} = \frac{\ln 10^{-1}}{\ln 10^2} = \frac{-\ln 10}{2 \ln 10} = -\frac{1}{2}. \quad \square$$

We can obtain these same results more directly from the relation

(7.5.8)
$$\log_p p^t = t.$$

Accordingly

$$\log_2 32 = \log_2 2^5 = 5 \qquad \text{and} \qquad \log_{100} \left(\tfrac{1}{10}\right) = \log_{100} \left(100^{-1/2}\right) = -\tfrac{1}{2}.$$

Differentiating (7.5.7) we have

$$\frac{d}{dx} (\log_p x) = \frac{1}{x \ln p}.\dagger$$

---

† The function $f(x) = \log_p x$ satisfies

$$f'(x) = \frac{1}{x \ln p}, \qquad f'(1) = \frac{1}{\ln p}.$$

This means that in general

$$f'(x) = \frac{1}{x} f'(1).$$

We predicted this from general considerations in Section 7.1. (See Formula 7.1.2.)

When p is e, the factor $\ln p$ is 1 and we have

$$\frac{d}{dx}(\log_e x) = \frac{1}{x}.$$

The logarithm to the base e, $\ln = \log_e$, is called the "*natural logarithm*" because it is the logarithm with the simplest derivative.

Example 5

(a) $\dfrac{d}{dx}(\log_5 |x|) = \dfrac{d}{dx}\left(\dfrac{\ln |x|}{\ln 5}\right) = \dfrac{1}{x \ln 5}.$

(b) $\dfrac{d}{dx}[\log_2 (3x^2 + 1)] = \dfrac{d}{dx}\left[\dfrac{\ln (3x^2 + 1)}{\ln 2}\right] = \dfrac{6x}{(3x^2 + 1) \ln 2}.$

(c) $\displaystyle\int \dfrac{dx}{x \ln 2} = \dfrac{\ln |x|}{\ln 2} + C = \log_2 |x| + C.$ $\square$

Estimating the Number *e*

We defined the logarithm function by setting

$$\ln x = \int_1^x \frac{dt}{t}, \qquad x > 0.$$

We can derive a numerical estimate for e from this integral representation of the logarithm.

Since e is irrational (Exercise 36, Section 12.6), we cannot hope to express e as a terminating decimal. (Nor as a repeating decimal.) What we can do is describe a simple technique by which one can compute the value of e to any desired degree of accuracy.

THEOREM 7.5.9

For each positive integer n,

$$\left(1 + \frac{1}{n}\right)^n \le e \le \left(1 + \frac{1}{n}\right)^{n+1}.$$

Proof. Refer throughout to Figure 7.5.1.

$$\ln\left(1 + \frac{1}{n}\right) = \int_1^{1+1/n} \frac{dt}{t} \le \int_1^{1+1/n} 1\, dt = \frac{1}{n}.$$

— since $\dfrac{1}{t} \le 1$ throughout the interval of integration

$$\ln\left(1+\frac{1}{n}\right) = \int_{1}^{1+1/n} \frac{dt}{t} \geq \int_{1}^{1+1/n} \frac{dt}{1+1/n} = \frac{1}{1+1/n} \cdot \frac{1}{n} = \frac{1}{n+1}.$$

since $\frac{1}{t} \geq \frac{1}{1+1/n}$ throughout the interval of integration

We have now shown that

$$\frac{1}{n+1} \leq \ln\left(1+\frac{1}{n}\right) \leq \frac{1}{n}.$$

From the inequality on the right

$$1 + \frac{1}{n} \leq e^{1/n} \qquad \text{and thus} \qquad \left(1+\frac{1}{n}\right)^{n} \leq e.$$

From the inequality on the left

$$e^{1/(n+1)} \leq 1 + \frac{1}{n} \qquad \text{and thus} \qquad e \leq \left(1+\frac{1}{n}\right)^{n+1}. \quad \square$$

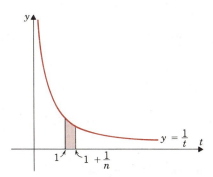

Figure 7.5.1

The inequality

$$\left(1+\frac{1}{n}\right)^{n} \leq e \leq \left(1+\frac{1}{n}\right)^{n+1}$$

is an elegant characterization of e but not a very efficient tool for calculating e. Reading from our calculator we find that

$$(1+\tfrac{1}{100})^{100} \cong 2.7048138 \qquad \text{and} \qquad (1+\tfrac{1}{100})^{101} \cong 2.7318619.$$

This gives e rounded off to one decimal place: $e \cong 2.7$. Going on to $n = 1000$, we find that

$$(1+\tfrac{1}{1000})^{1000} \cong 2.7169239 \qquad \text{and} \qquad (1+\tfrac{1}{1000})^{1001} \cong 2.7196409.$$

This gives a two-place estimate for e: $e \cong 2.72$.   To get a five-place estimate for e, we have to go to about $n = 1,000,000$:

$$(1 + \tfrac{1}{1,000,000})^{1,000,000} \cong 2.7182805 \qquad \text{and} \qquad (1 + \tfrac{1}{1,000,000})^{1,000,001} \cong 2.7182832$$

and therefore, to five decimal places, $e \cong 2.71828$.

Remark.   A more efficient way of estimating e is provided by Taylor series.   (Chapter 12) ◻

EXERCISES 7.5

Find.

1. $\log_2 64$.

2. $\log_2 \frac{1}{64}$.

3. $\log_{64} \frac{1}{2}$.

4. $\log_{10} 0.01$.

5. $\log_5 1$.

6. $\log_5 0.2$.

7. $\log_5 125$.

8. $\log_2 4^3$.

9. $\log_{32} 8$.

10. $\log_{100} 10^{-4/5}$.

11. $\log_{10} 100^{-4/5}$.

12. $\log_9 \sqrt{3}$.

Show that

13. $\log_p xy = \log_p x + \log_p y$.

14. $\log_p \dfrac{1}{x} = -\log_p x$.

15. $\log_p x^y = y \log_p x$.

16. $\log_p \dfrac{x}{y} = \log_p x - \log_p y$.

Find those numbers x, if any, for which

17. $10^x = e^x$.

18. $\log_5 x = 0.04$.

19. $\log_x 10 = \log_4 100$.

20. $\log_x 2 = \log_3 x$.

21. $\log_2 x = \displaystyle\int_2^x \dfrac{dt}{t}$.

22. $\log_x 10 = \log_2 (\tfrac{1}{10})$.

23. Estimate $\ln a$ given that $e^{t_1} < a < e^{t_2}$.   **24.** Estimate e^b given that $\ln x_1 < b < \ln x_2$.

Calculate the following integrals.

25. $\displaystyle\int 3^x \, dx$.

26. $\displaystyle\int 2^{-x} \, dx$.

27. $\displaystyle\int (x^3 + 3^{-x}) \, dx$.

28. $\displaystyle\int x10^{-x^2} \, dx$.

29. $\displaystyle\int \dfrac{dx}{x \ln 5}$.

30. $\displaystyle\int \dfrac{\log_5 x}{x} \, dx$.

31. $\displaystyle\int \dfrac{\log_2 x^3}{x} \, dx$.

32. $\displaystyle\int \dfrac{\log_3 \sqrt{x} - 1}{x} \, dx$.

33. $\displaystyle\int x10^{x^2} \, dx$.

34. Show that, if a, b, c are positive, then

$$\log_a c = \log_a b \, \log_b c$$

provided that a and b are both different from 1.

Find $f'(e)$.

35. $f(x) = \log_3 x$.

36. $f(x) = x \log_3 x$.

37. $f(x) = \ln (\ln x)$.

38. $f(x) = \log_3 (\log_2 x)$.

Derive the formula for $f'(x)$ by logarithmic differentiation.

39. $f(x) = p^x$.

40. $f(x) = p^{g(x)}$.

Find by logarithmic differentiation.

41. $\dfrac{d}{dx}[(x+1)^x]$.

42. $\dfrac{d}{dx}[(\ln x)^x]$.

43. $\dfrac{d}{dx}[(\ln x)^{\ln x}]$.

44. $\dfrac{d}{dx}\left[\left(\dfrac{1}{x}\right)^x\right]$.

45. $\dfrac{d}{dx}[(x^2+2)^{\ln x}]$.

46. $\dfrac{d}{dx}[(\ln x)^{x^2+2}]$.

Sketch figures in which you compare the following pairs of graphs.

47. $f(x)=e^x$ and $g(x)=2^x$.

48. $f(x)=e^x$ and $g(x)=3^x$.

49. $f(x)=e^x$ and $g(x)=e^{-x}$.

50. $f(x)=2^x$ and $g(x)=2^{-x}$.

51. $f(x)=\ln x$ and $g(x)=\log_3 x$.

52. $f(x)=\ln x$ and $g(x)=\log_2 x$.

53. $f(x)=2^x$ and $g(x)=\log_2 x$.

54. $f(x)=10^x$ and $g(x)=\log_{10} x$.

For each of the functions below (i) specify the domain, (ii) find the intervals where the function increases and those where it decreases, (iii) find the extreme values.

55. $f(x)=10^{1-x^2}$.

56. $f(x)=10^{1/(1-x^2)}$.

57. $f(x)=10^{\sqrt{1-x^2}}$.

58. $f(x)=\log_{10}\sqrt{1-x^2}$.

Evaluate.

59. $\displaystyle\int_1^2 2^{-x}\,dx$.

60. $\displaystyle\int_0^1 4^x\,dx$.

61. $\displaystyle\int_1^4 \dfrac{dx}{x\ln 2}$.

62. $\displaystyle\int_0^2 p^{x/2}\,dx$.

63. $\displaystyle\int_0^1 x10^{1+x^2}\,dx$.

64. $\displaystyle\int_1^3 \dfrac{\log_3 x}{x}\,dx$.

65. $\displaystyle\int_{10}^{100} \dfrac{dx}{x\log_{10} x}$.

66. $\displaystyle\int_0^1 \dfrac{5p^{\sqrt{x+1}}}{\sqrt{x+1}}\,dx$.

67. $\displaystyle\int_0^1 (2^x+x^2)\,dx$.

Estimate the following logarithms on the basis of Table 7.2.1.

68. $\log_{10} 4$.

69. $\log_{10} 7$.

70. $\log_{10} 12$.

71. $\log_{10} 45$.

72. Prove that

$$\lim_{h\to 0}(1+h)^{1/h}=e.$$

HINT: At $x=1$ the logarithm function has derivative 1:

$$\lim_{h\to 0}\dfrac{\ln(1+h)-\ln 1}{h}=1.$$

(Calculator)   Evaluate and then explain the result.

73. $5^{(\ln 17)/(\ln 5)}$.

74. $7^{1/\ln 7}$.

75. $16^{1/\ln 2}$.

Program for computing e as the limit of a product   (BASIC)

This program computes the value of the product

$$\left(1+\dfrac{1}{n}\right)^n,$$

which approximates the value of e. It is instructive to have the computer make the calculation for several fairly large values of n to see the speed with which the correct value of e is approached.

10 **REM** Computation of e as the limit of a product
20 **REM** copyright © Colin C. Graham 1988-1989

```
100 INPUT "Enter n:"; n
200 term = 1 + (1/n)
210 e = 1

300 FOR j = 1 TO n
310    e = e*term
330    PRINT e
340 NEXT j

500 INPUT "Do it again? (Y/N)"; a$
510 IF a$ = "Y" OR a$ = "y" THEN 100
520 END
```

7.6   EXPONENTIAL GROWTH AND DECLINE; COMPOUND INTEREST

We begin by comparing exponential change to linear change. Let y be a quantity that changes with time: $y = y(t)$.

If y is a linear function,

$$y(t) = kt + C,$$

then y changes by the *same amount during all periods of the same duration*:

$$y(t + \Delta t) = k[t + \Delta t] + C = (kt + C) + k\,\Delta t = y(t) + k\,\Delta t.$$

During every period of length Δt, y changes by the amount $k\,\Delta t$.

If, on the other hand, y is an exponential function,

$$y(t) = Ce^{kt},$$

then y changes by *the same factor during all periods of the same duration*:

$$y(t + \Delta t) = Ce^{k[t + \Delta t]} = Ce^{kt + k\Delta t} = Ce^{kt} \cdot e^{k\Delta t} = e^{k\Delta t}y(t).$$

During every period of length Δt, y changes by a factor of $e^{k\Delta t}$.

An exponential

$$f(t) = Ce^{kt}$$

has the property that its derivative $f'(t)$ is proportional to $f(t)$:

$$f'(t) = Cke^{kt} = kCe^{kt} = k\,f(t).$$

Moreover it is the only such function:

THEOREM 7.6.1

If

$$f'(t) = k\,f(t) \qquad \text{for all } t \text{ in some interval } I,$$

then f is of the form

$$f(t) = Ce^{kt} \qquad \text{for all } t \text{ in } I.$$

An equation that involves derivatives is called a *differential equation,* and a function that satisfies the equation is called a *solution of the equation.* According to Theorem 7.6.1 every solution of the differential equation $f'(t) = kf(t)$ can be written $f(t) = Ce^{kt}$.

$$f'(t) = k\,f(t)$$
$$f'(t) - k\,f(t) = 0$$
$$e^{-kt}f'(t) - ke^{-kt}f(t) = 0 \qquad \text{(we mutiplied by } e^{-kt})$$
$$\frac{d}{dt}[e^{-kt}f(t)] = 0$$
$$e^{-kt}f(t) = C$$
$$f(t) = Ce^{kt}. \quad \square$$

The constant C is the value of f at 0: $f(0) = Ce^0 = C$. This is usually called *the initial value of f.*

Problem 1. Find $f(t)$ given that $f'(t) = 2f(t)$ for all t and $f(0) = \sqrt{2}$.

Solution. The fact that $f'(t) = 2f(t)$ tells us that $f(t) = Ce^{2t}$ where C is some constant. The fact that $f(0) = \sqrt{2}$ tells us that $C = \sqrt{2}$. Thus $f(t) = \sqrt{2}\,e^{2t}$. $\square$

Remark. For the numerical calculations in this section, use a calculator or refer to the tables in Appendix C. $\square$

Problem 2. Population tends to grow with time at a rate roughly proportional to the population present. According to the Bureau of the Census, the population of the United States in 1970 was approximately 203 million and in 1980, 227 million. Use this information to estimate the year in which the population of the United States may be expected to match the current world population of approximately 5 billion.

Solution. Let $P(t)$ be the population (in millions) of the United States t years after 1970. The basic equation $P'(t) = kP(t)$ gives

$$P(t) = Ce^{kt}.$$

Since $P(0) = 203$, we have

(1) $$P(t) = 203e^{kt}.$$

We can use the fact that $P(10) = 227$ to eliminate k from (1):

(∗) $$227 = 203e^{10k}, \qquad e^{10k} = \frac{227}{203}, \qquad e^k = \left(\frac{227}{203}\right)^{1/10},$$

and, thus, Equation (1) can be written

$$P(t) = 203\left(\frac{227}{203}\right)^{t/10}.$$

We want to find the value of t for which $P(t) = 5000$. Thus we set

$$203 \left(\frac{227}{203}\right)^{t/10} = 5000$$

$$\ln 203 + \frac{t}{10} \ln \left(\frac{227}{203}\right) = \ln 5000$$

$$t = \frac{10 \ln (5000/203)}{\ln (227/203)} \cong 287. \qquad \text{\textemdash\, calculator}$$

According to our calculations, the population of the United States may be expected to match the current world population in the year $1970 + 287 = 2257$. $\quad\square$

Problem 3.   The rate of decay of a radioactive substance is proportional to the amount of such substance present.

(a) Today we have A_0 grams of a radioactive substance. Given that one-third of the substance decays every 5 years, how much will be left t years from today?

(b) What is the *half-life* of the substance?   (That is, how long does it take for half of the substance to decay?)

Solution.   (a) Let $A(t)$ be the amount of radioactive substance present t years from today. Since the rate of decay is proportional to the amount present, we know that

$$A'(t) = kA(t) \qquad \text{and thus} \qquad A(t) = Ce^{kt}.$$

The constant C represents the amount at time $t = 0$. In this case $C = A_0$ and we have

$$(1) \hspace{6cm} A(t) = A_0 e^{kt}.$$

At the end of 5 years, one-third of A_0 will have decayed and therefore two-thirds of A_0 will have remained:

$$A(5) = \tfrac{2}{3}A_0.$$

We can use this relation to eliminate k from (1):

$$A(5) = A_0 e^{5k} = \tfrac{2}{3}A_0, \qquad e^{5k} = \tfrac{2}{3}, \qquad e^k = (\tfrac{2}{3})^{1/5},$$

and consequently (1) can be written

$$A(t) = A_0(\tfrac{2}{3})^{t/5}.$$

(b) To find the half-life of the substance we must find the value of t for which

$$A_0(\tfrac{2}{3})^{t/5} = \tfrac{1}{2}A_0.$$

Dividing by A_0, we have

$$(\tfrac{2}{3})^{t/5} = \tfrac{1}{2}$$

so that

$$(\tfrac{2}{3})^t = (\tfrac{1}{2})^5, \qquad t \ln \tfrac{2}{3} = 5 \ln \tfrac{1}{2}, \qquad t(\ln 2 - \ln 3) = -5 \ln 2,$$

and

calculator

$$t = \frac{5 \ln 2}{\ln 3 - \ln 2} = \frac{5 \ln 2}{\ln 1.5} \cong 8.55.$$

The half-life of the substance is about $8\frac{1}{2}$ years. □

Compound Interest

Consider money invested at interest rate r. If the accumulated interest is credited once a year, then the interest is said to be compounded annually; if twice a year, then semiannually; if four times a year, then quarterly. The idea can be pursued further. Interest can be credited every day, every hour, every second, every half-second, and so on. In the limiting case, interest is credited instantaneously. Economists call this *continuous compounding.*

The economists' formula for continuous compounding is a simple exponential:

(7.6.2)
$$A(t) = A_0 e^{rt}.$$

Here t is measured in years,

$A(t)$ = the principal in dollars at time t,

$A_0 = A(0)$ = the initial investment,

r = the annual interest rate.

The rate r is called the *nominal* interest rate. Compounding makes the *effective* rate higher.

A Derivation of the Compound Interest Formula. We take h as a small time increment and observe that

$$A(t + h) - A(t) = \text{interest earned from time } t \text{ to time } t + h.$$

Had the principal remained $A(t)$ from time t to time $t + h$, the interest earned during this time period would have been

$$rhA(t).$$

Had the principal been $A(t + h)$ throughout the time interval, the interest earned would have been

$$rhA(t + h).$$

The actual interest earned must be somewhere in between:

$$rhA(t) \leq A(t + h) - A(t) \leq rhA(t + h).$$

Dividing by h, we get

$$rA(t) \leq \frac{A(t+h) - A(t)}{h} \leq rA(t+h).$$

If A varies continuously, then, as h tends to zero, $rA(t+h)$ tends to $rA(t)$ and (by the pinching theorem) the difference quotient in the middle must also tend to $rA(t)$:

$$\lim_{h \to 0} \frac{A(t+h) - A(t)}{h} = rA(t).$$

This says that

$$A'(t) = rA(t),$$

from which it follows that

$$A(t) = Ce^{rt}.$$

With A_0 as the initial investment, we have $C = A_0$ and

$$A(t) = A_0 e^{rt}. \quad \square$$

Problem 4. Find the amount of interest earned by \$100 compounded continuously at 6% for 5 years.

Solution

$$A(5) = 100e^{(0.06)5} = 100e^{0.30} \overset{\text{calculator}}{\cong} 134.99.$$

The interest earned is approximately \$35. $\quad \square$

Problem 5. A sum of money is earning interest at the rate of 10% compounded continuously. What is the effective interest rate?

Solution. At the end of one year each dollar grows to

$$e^{0.10} \overset{\text{calculator}}{\cong} 1.105.$$

The interest earned is $10\frac{1}{2}$ cents. The effective interest rate is $10\frac{1}{2}\%$. $\quad \square$

Problem 6. How long does it take a sum of money to double at 5% compounded continuously?

Solution. In general

$$A(t) = A_0 e^{0.05t}.$$

We set

$$2A_0 = A_0 e^{0.05t}$$

and solve for t:

calculator

$$2 = e^{0.05t}, \qquad \ln 2 = 0.05t = \tfrac{1}{20}t, \qquad t = 20 \ln 2 \cong 13.86.$$

It takes about 13 years, 10 months, and 14 days. □

A dollar today is better than a dollar tomorrow because during that time we could be earning interest.

Problem 7. What is the present value of $1000 forty months from now? Assume continuous compounding at 6%.

Solution. We begin with

$$A(t) = A_0 e^{0.06t}$$

and solve for A_0:

$$A_0 = A(t)e^{-0.06t}.$$

Since forty months is $\tfrac{10}{3}$ years, $t = \tfrac{10}{3}$ and

calculator

$$A_0 = A(\tfrac{10}{3})e^{(-0.06)(10/3)} = 1000e^{-0.20} \cong 818.73.$$

The present value of $1000 forty months from now is about $818.73. □

Problem 8. Mr. A earns $28,000 a year. How much will he have to be earning 10 years from today to retain his present purchasing power? Assume a rate of inflation of 7%.

Solution. The purchasing power of a dollar t years from now will have been eroded by a factor of $e^{-0.07t}$. Earnings of S dollars a year 10 years from now will have the purchasing power of $Se^{-0.07(10)} = Se^{-0.7}$ dollars today. Setting

$$28,000 = Se^{-0.7}$$

we have

calculator

$$S = (28,000)e^{0.7} \cong 56,385.$$

Mr. A will have to be earning about $56,385 a year to maintain his present purchasing power. □

Problem 9. Producers of brandy find that the value of their inventory (the price at which they can sell it) increases with time according to the formula

$$V(t) = V_0 e^{\sqrt{t}/2}. \qquad\qquad (t \text{ measured in years})$$

How long should they plan to keep the brandy in storage? Neglect storage costs and assume an 8% annual rate of inflation.

Solution. Future dollars should be discounted by a factor of $e^{-0.08t}$. What we want to maximize is not $V(t)$ but the present value of the brandy

$$f(t) = V(t)e^{-0.08t} = V_0 e^{(\sqrt{t}/2)-0.08t}.$$

Differentiation gives

$$f'(t) = V_0 e^{(\sqrt{t}/2)-0.08t}\left(\frac{1}{4\sqrt{t}} - 0.08\right).$$

Setting $f'(t) = 0$, we find that

$$\frac{1}{4\sqrt{t}} = 0.08, \qquad \sqrt{t} = \frac{1}{0.32}, \qquad t = \left(\frac{1}{0.32}\right)^2 \overset{\text{calculator}}{\cong} 9.77.$$

They should plan to store it for about 9 years and 9 months. ☐

A Mixing Problem

In our approach to mixing problems of the sort presented below we will assume that uniform concentration is maintained by some kind of "stirring."

Problem 10. Water from a polluted reservoir (pollution level p_0 grams per liter) is being drawn off at the rate of n liters per hour and being replaced by less polluted water (pollution level p_1 grams per liter). Given that the reservoir contains M liters, how long will it take to reduce the pollution to p grams per liter?

Solution. Let $A(t)$ be the total number of grams of pollution in the reservoir at time t. We want to find the time t at which

$$\frac{A(t)}{M} = p.$$

At each time t, pollutants leave the reservoir at the rate of

$$n\frac{A(t)}{M} \quad \text{grams per hour}$$

and enter the reservoir at the rate of

$$np_1 \quad \text{grams per hour.}$$

It follows that

$$A'(t) = \text{rate in} - \text{rate out} = np_1 - n\frac{A(t)}{M}$$

and we have

$$A'(t) = -\frac{n}{M}[A(t) - Mp_1].$$

Since constants have derivative 0, we can write

$$\frac{d}{dt}[A(t) - Mp_1] = -\frac{n}{M}[A(t) - Mp_1]$$

and apply Theorem 7.6.1. By the theorem

$$A(t) - Mp_1 = Ce^{-nt/M}$$

so that

$$\frac{A(t)}{M} = p_1 + \frac{C}{M}e^{-nt/M}.$$

The constant C is determined by the initial pollution level p_0:

$$p_0 = \frac{A(0)}{M} = p_1 + \frac{C}{M} \qquad \text{so that} \qquad C = M(p_0 - p_1).$$

Using this value for C, we have

$$\frac{A(t)}{M} = p_1 + (p_0 - p_1)e^{-nt/M}.$$

This equation gives the pollution level at each time t. To find the time at which the pollution level has been reduced to p, we set

$$p_1 + (p_0 - p_1)e^{-nt/M} = p$$

and solve for t:

$$e^{-nt/M} = \frac{p - p_1}{p_0 - p_1}, \qquad -\frac{n}{M}t = \ln\frac{p - p_1}{p_0 - p_1}, \qquad t = \frac{M}{n}\ln\frac{p_0 - p_1}{p - p_1}. \quad \square$$

Problem 11. In the problem just considered, how long would it take to reduce the pollution by 50% if the reservoir contained 50,000,000 liters, the water were replaced at the rate of 25,000 liters per hour, and the replacement water were completely pollution free?

Solution. In general

$$t = \frac{M}{n}\ln\frac{p_0 - p_1}{p - p_1}.$$

Here

$$M = 50{,}000{,}000, \quad n = 25{,}000, \quad p_1 = 0, \quad p = \tfrac{1}{2}p_0,$$

so that

$$t = 2000\ln 2 \cong 1386.$$

It would take about 1386 hours. $\quad \square$

EXERCISES 7.6

1. **(Calculator)** Find the amount of interest earned by $500 compounded continuously for 10 years (a) at 6%. (b) at 8%. (c) at 10%.

2. **(Calculator)** How long does it take for a sum of money to double when compounded continuously (a) at 6%? (b) at 8%? (c) at 10%?

3. **(Calculator)** What is the present value of a $1000 bond that matures in 5 years? Assume continuous compounding (a) at 6%. (b) at 8%. (c) at 10%.

4. At what rate r of continuous compounding does a sum of money increase by a factor of e (a) in one year? (b) in n years?

5. **(Calculator)** At what rate r of continuous compounding does a sum of money triple in 20 years?

6. **(Calculator)** At what rate r of continuous compounding does a sum of money double in 10 years?

7. At a certain moment a 100-gallon mixing tank is full of brine containing 0.25 pounds of salt per gallon. Find the amount of salt present t minutes later if the brine is being continuously drawn off at the rate of 3 gallons per minute and replaced by brine containing 0.2 pounds of salt per gallon.

8. **(Calculator)** Water is pumped into a tank to dilute a saline solution. The volume of the solution, call it V, is kept constant by continuous outflow. The amount of salt in the tank, call it s, depends on the amount of water that has been pumped in, call this x. Given that

$$\frac{ds}{dx} = -\frac{s}{V},$$

find the amount of water that must be pumped into the tank to eliminate 50% of the salt. Take V as 10,000 gallons.

9. A 200-liter tank initially full of water develops a leak at the bottom. Given that 20% of the water leaks out in the first 5 minutes, find the amount of water left in the tank t minutes after the leak develops if the water drains off at a rate that is proportional to the product of the time elapsed and the amount of water present.

10. In Exercise 9 assume that the water drains off at a rate that is proportional to the amount of water present.

In Exercises 11–16 remember that the rate of decay of a radioactive substance is proportional to the amount of substance present.

11. **(Calculator)** What is the half-life of a radioactive substance if it takes 4 years for a quarter of the substance to decay?

12. **(Calculator)** A year ago there were 4 grams of a radioactive substance. Now there are 3 grams. How much was there 10 years ago?

13. **(Calculator)** Two years ago there were 5 grams of a radioactive substance. Now there are 4 grams. How much will remain 3 years from now?

14. **(Calculator)** What is the half-life of a radioactive substance if it takes 5 years for a third of the substance to decay?

15. Suppose the half-life of a radioactive substance is n years. What percentage of the substance present at the start of a year will decay during the ensuing year?

16. A radioactive substance weighed n grams at time $t = 0$. Today, 5 years later, the substance weighs m grams. How much will it weigh 5 years from now?

17. (*The power of exponential growth*) Imagine two racers competing on the x-axis (which has been calibrated in meters), a linear racer LIN [position function of the form $x_1(t) = kt + C$] and an exponential racer EXP [position function of the form $x_2(t) = e^{kt} + C$]. Suppose that both racers start out simultaneously from the origin, LIN at one million meters per second, EXP at only one meter per second. In the early stages of the race, fast-starting

LIN will move far ahead of EXP, but in time EXP will catch up to LIN, pass her, and leave her hopelessly behind. Show that this is true as follows:

(a) Express the position of each racer as a function of time, measuring t in seconds.

(b) Show that LIN's lead over EXP starts to decline about 13.8 seconds into the race.

(c) Show that LIN is still ahead of EXP some 15 seconds into the race but far behind 3 seconds later. (Use $e^3 \cong 20$.)

(d) Show that, once EXP passes LIN, LIN can never catch up.

18. (*The weakness of logarithmic growth*) Having been soundly beaten in the race in Exercise 17, LIN finds an opponent she can beat, LOG, the logarithmic racer [position function $x_3(t) = k \ln (t + 1) + C$]. Once again the racetrack is the x-axis calibrated in meters. Both racers start out at the origin, LOG at one million meters per second, LIN at only one meter per second. (LIN is tired from the previous race.) In this race LOG will shoot ahead and remain ahead for a long time, but eventually LIN will catch up to LOG, pass her, and leave her permanently behind. Show that this is true as follows:

(a) Express the position of each racer as a function of time t, measuring t in seconds.

(b) Show that LOG's lead over LIN starts to decline $10^6 - 1$ seconds into the race.

(c) Show that LOG is still ahead of LIN $10^7 - 1$ seconds into the race but behind LIN $10^8 - 1$ seconds into the race.

(d) Show that, once LIN passes LOG, LOG can never catch up.

19. (**Calculator**) Atmospheric pressure p varies with altitude h according to the equation

$$\frac{dp}{dh} = kp, \qquad \text{where } k \text{ is a constant.}$$

Given that p is 15 pounds per square inch at sea level and 10 pounds per square inch at 10,000 feet, find p at (a) 5000 feet. (b) 15,000 feet.

20. (**Calculator**) You are 45 years old and are looking toward an annual pension of $30,000 per year at age 65. What will be the present-day purchasing power of your pension if inflation over this period runs at (a) 6%? (b) 9%? (c) 10%?

21. (**Calculator**) Tuition at XYZ College is currently $6000. What would you expect to pay as tuition 3 years from now if inflation runs at (a) 5%? (b) 8%? (c) 12%?

22. A boat moving in still water is subject to a retardation proportional to its velocity. Show that the velocity t seconds after the power is shut off is given by the formula $v = ce^{-kt}$ where c is the velocity at the instant the power is shut off.

23. (**Calculator**) A boat is drifting in still water at 4 miles per hour; 1 minute later, at 2 miles per hour. How far has the boat drifted in that 1 minute? (See Exercise 22.)

24. (**Calculator**) Population tends to grow with time at a rate roughly proportional to the population present. According to the Bureau of the Census, the population of the United States in 1970 was approximately 203 million and in 1980, 227 million. Use this information to estimate the population of 1960. (The actual figure was about 179 million.)

25. (**Calculator**) Use the data of Exercise 24 to predict the population for the year 2000.

26. (**Calculator**) Use the data of Exercise 24 to estimate how long it takes for the population to double.

27. (**Calculator**) A lumber company finds from experience that the value of its standing timber increases with time according to the formula

$$V(t) = V_0(\tfrac{3}{2})^{\sqrt{t}}.$$

Here t is measured in years and V_0 is the value at planting time. How long should the company wait before cutting the timber? Neglect costs and assume continuous compounding at 5%.

28. Determine the time period during which $y = Ce^{kt}$ changes by a factor of q.

29. During the process of inversion, the amount A of raw sugar present decreases at a rate proportional to A. During the first 10 hours, 1000 pounds of raw sugar have been reduced to 800 pounds. How many pounds will remain after 10 more hours of inversion?

30. The number of bacteria present in a given culture increases at a rate proportional to the number present. When first observed, the culture contained n_0 bacteria, an hour later, n_1.
(a) Find the number present t hours after the observations began.
(b) How long did it take for the number of bacteria to double?

31. (*Important*) Let k_1, k_2 be constants, $k_1 \neq 0$. Show that

$$\text{if} \quad f'(t) = k_1 f(t) + k_2 \quad \text{then} \quad f(t) = Ce^{k_1 t} - k_2/k_1$$

where C is an arbitrary constant.

HINT: By adding a constant to a function we don't change its derivative. Therefore we can write

$$f'(t) = k_1 f(t) + k_2 \quad \text{as} \quad d/dt\,[f(t) + k_2/k_1] = k_1[f(t) + k_2/k_1].$$

Proceed from there.

The remaining exercises are based on Exercise 31.

32. Find $f(t)$ given that $f'(t) = k[2 - f(t)]$ and $f(0) = 0$.

33. An object falling in air is subject not only to gravitational force but also to air resistance. Find $v(t)$, the velocity of the object at time t, given that

$$v'(t) + Kv(t) = 32 \quad \text{and} \quad v(0) = 0. \qquad \text{(take } K > 0)$$

Show that $v(t)$ cannot exceed $32/K$. ($32/K$ is called the *terminal velocity*.)

34. *Newton's Law of Cooling* states that the rate of change of the temperature T of an object is proportional to the difference between T and the temperature τ of the surrounding medium:

$$\frac{dT}{dt} = -k(T - \tau) \qquad \text{where } k \text{ is a positive constant.}$$

(a) Solve this equation for T given that $T(0) = T_0$.
(b) Take $T_0 > \tau$. What is the limiting temperature to which the object cools as t increases? What happens if $T_0 < \tau$?

35. **(Calculator)** A cup of coffee is served to you at $185°$ F in a room where the temperature is $65°$ F. Two minutes later the temperature of the coffee has dropped to $155°$ F. How many more minutes would you expect to wait for the coffee to cool to $105°$F? HINT: See Exercise 34.

36. The current i in an electric circuit varies with time t according to the formula

$$L\frac{di}{dt} + Ri = E$$

where E (the voltage), L (the inductance), and R (the resistance) are positive constants. Measure time in seconds, current in amperes, and suppose that the initial current is 0.
(a) Find a formula for the current at each subsequent time t.
(b) What upper limit does the current approach as t increases?
(c) In how many seconds will the current reach 90% of its upper limit?

7.7 INTEGRATION BY PARTS

We begin with the differentiation formula

$$f(x)g'(x) + f'(x)g(x) = (f \cdot g)'(x).$$

Integrating both sides, we get

$$\int f(x)g'(x)\,dx + \int f'(x)g(x)\,dx = \int (f \cdot g)'(x)\,dx.$$

Since

$$\int (f \cdot g)'(x)\,dx = f(x)g(x) + C,$$

we have

$$\int f(x)g'(x)\,dx + \int f'(x)g(x)\,dx = f(x)g(x) + C$$

and therefore

$$\int f(x)g'(x)\,dx = f(x)g(x) - \int f'(x)g(x)\,dx + C.$$

Since the computation of

$$\int f'(x)g(x)\,dx$$

will yield its own arbitrary constant, there is no reason to keep the constant C. We therefore drop it and write

(7.7.1)
$$\int f(x)g'(x)\,dx = f(x)g(x) - \int f'(x)g(x)\,dx.$$

This formula, called the formula for *integration by parts,* enables us to find

$$\int f(x)g'(x)\,dx$$

by computing

$$\int f'(x)g(x)\,dx$$

instead. It is, of course, of practical use only if the second integral is easier to compute than the first.

In practice we usually set

$$u = f(x), \qquad dv = g'(x)\,dx$$

and find

$$du = f'(x)\,dx, \qquad v = g(x).$$

The formula for integration by parts can then be written

(7.7.2)
$$\int u\,dv = uv - \int v\,du.$$

Success with this formula depends on choosing u and dv so that

$$\int v\,du \quad \text{is easier to calculate than} \quad \int u\,dv.$$

Problem 1. Calculate

$$\int xe^x\,dx.$$

Solution. We want to separate x from e^x. Setting

$$u = x, \qquad dv = e^x\,dx,$$

we have

$$du = dx, \qquad v = e^x.$$

Accordingly,

$$\int xe^x\,dx = \int u\,dv = uv - \int v\,du = xe^x - \int e^x\,dx = xe^x - e^x + C.$$

Our choice of u and dv worked out very well. Had we set

$$u = e^x, \qquad dv = x\,dx,$$

we would have had

$$du = e^x, \qquad v = \tfrac{1}{2}x^2.$$

Integration by parts would then have given

$$\int xe^x\,dx = \int u\,dv = uv - \int v\,du = \tfrac{1}{2}x^2 e^x - \tfrac{1}{2}\int x^2 e^x\,dx,$$

an integral more complicated than the one we started with. A good choice of u and dv is crucial. ☐

As the next problem shows, effective choices of u and dv are not unique.

Problem 2. Calculate

$$\int x \ln x\,dx.$$

Solution. Setting

$$u = \ln x, \qquad dv = x\,dx,$$

we have

$$du = \frac{1}{x}\,dx, \qquad v = \frac{x^2}{2}.$$

Then

$$\int x \ln x \, dx = \int u \, dv = uv - \int v \, du$$

$$= \frac{x^2}{2} \ln x - \int \frac{1}{x} \frac{x^2}{2} \, dx = \frac{1}{2} x^2 \ln x - \frac{1}{4} x^2 + C.$$

Alternative Solution.   This time we set

$$u = x \ln x, \qquad\qquad dv = dx,$$

so that

$$du = (1 + \ln x) \, dx, \qquad v = x.$$

Substituting these selections in

$$\int u \, dv = uv - \int v \, du,$$

we find that

(1) $$\int x \ln x \, dx = x^2 \ln x - \int x(1 + \ln x) \, dx.$$

It may seem that the integral on the right is more complicated than the one we started with.   However, we can rewrite equation (1) as

$$\int x \ln x \, dx = x^2 \ln x - \int x \, dx - \int x \ln x \, dx.$$

It follows that

$$2 \int x \ln x \, dx = x^2 \ln x - \int x \, dx$$

and thus

$$\int x \ln x \, dx = \frac{1}{2} x^2 \ln x - \frac{1}{4} x^2 + C$$

as before.   ☐

Problem 3.   Calculate

$$\int x \sin x \, dx.$$

Solution.   Setting

$$u = x, \qquad dv = \sin x \, dx,$$

we have

$$du = dx, \qquad v = -\cos x.$$

Therefore

$$\int x \sin x \, dx = -x \cos x - \int -\cos x \, dx = -x \cos x + \sin x + C. \quad \Box$$

Integration by parts is often used to calculate integrals where the integrand is a mixture of function types; for example, polynomials and exponentials, or polynomials and trigonometric functions, etc. Some integrands, however, are better left as mixtures; for example, it is easy to see that

$$\int 2xe^{x^2} \, dx = e^{x^2} + C \qquad \text{and} \qquad \int 3x^2 \cos x^3 \, dx = \sin x^3 + C.$$

Any attempt to separate these integrands by integration by parts is counterproductive. The mixtures in these integrands arise from the chain rule and we need these mixtures to calculate the integrals.

Problem 4. Calculate

$$\int x^5 e^{x^3} \, dx.$$

Solution. To integrate e^{x^3} we need an x^2 factor. So we will keep x^2 together with e^{x^3}. Setting

$$u = x^3, \qquad dv = x^2 e^{x^3} \, dx,$$

we have

$$du = 3x^2 \, dx, \qquad v = \tfrac{1}{3} e^{x^3}.$$

Thus

$$\int x^5 e^{x^3} \, dx = \int u \, dv = uv - \int v \, du$$

$$= \tfrac{1}{3} x^3 e^{x^3} - \int x^2 e^{x^3} \, dx$$

$$= \tfrac{1}{3} x^3 e^{x^3} - \tfrac{1}{3} e^{x^3} + C = \tfrac{1}{3} (x^3 - 1) e^{x^3} + C. \quad \Box$$

To calculate some integrals you may have to integrate by parts more than once.

Problem 5. Evaluate

$$\int_0^1 x^2 e^x \, dx.$$

Solution. First we calculate the indefinite integral

$$\int x^2 e^x \, dx$$

by separating x^2 from e^x. Setting

$$u = x^2, \qquad dv = e^x \, dx,$$

we have

$$du = 2x \, dx, \qquad v = e^x,$$

and thus

$$\int x^2 e^x \, dx = \int u \, dv = uv - \int v \, du = x^2 e^x - \int 2x e^x \, dx.$$

We now compute the integral on the right again by parts. This time we set

$$u = 2x, \qquad dv = e^x \, dx.$$

This gives

$$du = 2 \, dx, \qquad v = e^x$$

and thus

$$\int 2x e^x \, dx = \int u \, dv = uv - \int v \, du = 2x e^x - \int 2e^x \, dx = 2x e^x - 2e^x + C.$$

This together with our earlier calculations gives

$$\int x^2 e^x \, dx = x^2 e^x - 2x e^x + 2e^x + C.$$

It follows now that

$$\int_0^1 x^2 e^x \, dx = \left[x^2 e^x - 2x e^x + 2e^x \right]_0^1 = (e - 2e + 2e) - 2 = e - 2. \quad \square$$

Problem 6. Find

$$\int e^x \cos x \, dx.$$

Solution. Here we integrate by parts twice. First we write

$$u = e^x, \qquad dv = \cos x \, dx,$$
$$du = e^x \, dx, \qquad v = \sin x.$$

This gives

$$(1) \qquad \int e^x \cos x \, dx = \int u \, dv = uv - \int v \, du = e^x \sin x - \int e^x \sin x \, dx.$$

Now we work with the integral on the right. Setting

$$u = e^x, \qquad dv = \sin x \, dx,$$
$$du = e^x \, dx, \qquad v = -\cos x,$$

we have

$$(2) \quad \int e^x \sin x \, dx = \int u \, dv = uv - \int v \, du = -e^x \cos x + \int e^x \cos x \, dx.$$

Substituting (2) in (1), we get

$$\int e^x \cos x \, dx = e^x \sin x + e^x \cos x - \int e^x \cos x \, dx,$$

$$2 \int e^x \cos x \, dx = e^x(\sin x + \cos x),$$

$$\int e^x \cos x \, dx = \tfrac{1}{2} e^x(\sin x + \cos x).$$

Since this integral is an indefinite integral, we add an arbitrary constant C:

$$\int e^x \cos x \, dx = \tfrac{1}{2} e^x(\sin x + \cos x) + C. \quad \square$$

Finally, the techniques of integration by parts enable us to integrate the logarithm function:

(7.7.3)
$$\int \ln x \, dx = x \ln x - x + C.$$

The derivation of this important formula is left to you as an exercise.

EXERCISES 7.7

Work out the following integrals and check your results by differentiation.

1. $\displaystyle\int xe^{-x} \, dx.$

2. $\displaystyle\int \ln(-x) \, dx.$

3. $\displaystyle\int x^2 \ln x \, dx.$

4. $\displaystyle\int x 2^x \, dx.$

5. $\displaystyle\int x^2 e^{-x^3} \, dx.$

6. $\displaystyle\int x \ln x^2 \, dx.$

7. $\displaystyle\int x^2 e^{-x} \, dx.$

8. $\displaystyle\int x^3 e^{-x^2} \, dx.$

9. $\displaystyle\int \frac{x^2}{\sqrt{1-x}} \, dx.$

10. $\displaystyle\int \frac{dx}{x(\ln x)^3}.$

11. $\displaystyle\int x \ln \sqrt{x} \, dx.$

12. $\displaystyle\int x\sqrt{x+1} \, dx.$

13. $\displaystyle\int \frac{\ln(x+1)}{\sqrt{x+1}} \, dx.$

14. $\displaystyle\int x^2(e^x - 1) \, dx.$

15. $\displaystyle\int (\ln x)^2 \, dx.$

16. $\displaystyle\int x(x+5)^{-14} \, dx.$

17. $\displaystyle\int x^3 3^x \, dx.$

18. $\displaystyle\int \sqrt{x} \ln x \, dx.$

19. $\displaystyle\int x(x+5)^{14} \, dx.$

20. $\displaystyle\int (2^x + x^2)^2 \, dx.$

21. $\displaystyle\int x \cos x \, dx.$

22. $\int x^2 \sin x \, dx.$ **23.** $\int x^2(x+1)^9 \, dx.$ **24.** $\int x^2(2x-1)^{-7} \, dx.$

25. $\int e^x \sin x \, dx.$ **26.** $\int (e^x + 2x)^2 \, dx.$ **27.** $\int \ln(1+x^2) \, dx.$

28. $\int x \ln(x+1) \, dx.$ **29.** $\int x^n \ln x \, dx. \ (n \neq -1)$ **30.** $\int e^{3x} \cos 2x \, dx.$

31. $\int x^3 \sin x^2 \, dx.$ **32.** $\int x^3 \sin x \, dx.$ **33.** $\int x^4 e^x \, dx.$

34. Derive (7.7.3) by integrating by parts.

Find the centroid of the region under the graph.

35. $f(x) = e^x, \quad x \in [0, 1].$ **36.** $f(x) = e^{-x}, \quad x \in [0, 1].$
37. $f(x) = \sin x, \quad x \in [0, \pi].$ **38.** $f(x) = \cos x, \quad x \in [0, \tfrac{1}{2}\pi].$

39. The mass density of a rod that extends from $x = 0$ to $x = 1$ is given by the function $\lambda(x) = e^{kx}$ where k is a constant. (a) Calculate the mass of the rod. (b) Find the center of mass of the rod.

40. The mass density of a rod that extends from $x = 2$ to $x = 3$ is given by the logarithm function $f(x) = \ln x$. (a) Calculate the mass of the rod. (b) Find the center of mass of the rod.

Find the volume generated by revolving the region under the graph about the *y*-axis.

41. $f(x) = e^{\alpha x}, \quad x \in [0, 1].$ **42.** $f(x) = \sin \pi x, \quad x \in [0, 1].$
43. $f(x) = \cos \tfrac{1}{2} \pi x, \quad x \in [0, 1].$ **44.** $f(x) = x \sin x, \quad x \in [0, \pi].$
45. $f(x) = xe^x, \quad x \in [0, 1].$ **46.** $f(x) = x \cos x, \quad x \in [0, \tfrac{1}{2}\pi].$

47. Let Ω be the region under the curve $y = e^x$, $x \in [0, 1]$. Find the centroid of the solid generated by revolving Ω about the *x*-axis. (6.4.5)

48. Let Ω be the region under the curve $y = \sin x$, $x \in [0, \tfrac{1}{2}\pi]$. Find the centroid of the solid generated by revolving Ω about the *x*-axis. (6.4.5)

7.8 SIMPLE HARMONIC MOTION

An object moves along a straight line. Instead of continuing in one direction, it moves back and forth, oscillating about a central point. Now call the central point $x = 0$ and denote by $x(t)$ the displacement at time t. If the acceleration is a constant negative multiple of the displacement,

$$a(t) = -kx(t) \qquad \text{with } k > 0,$$

then the object is said to be in *simple harmonic motion*. The purpose of this section is to analyze such motion.

Since, by definition,

$$a(t) = x''(t),$$

in simple harmonic motion

$$x''(t) = -kx(t).$$

This gives

$$x''(t) + kx(t) = 0.$$

To emphasize that k is positive, we set $k = \omega^2$. The equation of motion then takes the form

(7.8.1) $\boxed{x''(t) + \omega^2 x(t) = 0.}$

It is easy to show that any function of the form

$$x(t) = C_1 \sin(\omega t + C_2)$$

satisfies this equation:

$$x'(t) = C_1 \omega \cos(\omega t + C_2)$$
$$x''(t) = -C_1 \omega^2 \sin(\omega t + C_2) = -\omega^2 x(t)$$

and thus

$$x''(t) + \omega^2 x(t) = 0.$$

What is not easy to see, but equally true, is that every solution of Equation 7.8.1 can be written in the form

(7.8.2) $\boxed{x(t) = C_1 \sin(\omega t + C_2) \quad \text{with} \quad C_1 \geq 0 \quad \text{and} \quad C_2 \in [0, 2\pi).}$

Proof. Suppose that f is some function that satisfies Equation 7.8.1 on some open interval I. Now let t_0 be any number in I. The function has a value at t_0, $f(t_0)$, and a derivative at t_0, $f'(t_0)$. Now write

$$g(t) = C_1 \sin(\omega t + C_2).$$

An elementary calculation that we leave to you shows that we can adjust the constants C_1, C_2 taking $C_1 \geq 0$ and $C_2 \in [0, 2\pi)$ so that $g(t_0) = f(t_0)$ and $g'(t_0) = f'(t_0)$. Let's suppose that this has been done. Since

$$f''(t) + \omega^2 f(t) = 0 \quad \text{and} \quad g''(t) + \omega^2 g(t) = 0,$$

we have

$$f''(t) - g''(t) + \omega^2[f(t) - g(t)] = 0.$$

Multiplication by $f'(t) - g'(t)$ gives

$$[f'(t) - g'(t)][f''(t) - g''(t)] + \omega^2[f(t) - g(t)][f'(t) - g'(t)] = 0.$$

Integrating both sides of the equation and clearing fractions, we have

$$[f'(t) - g'(t)]^2 + \omega^2[f(t) - g(t)]^2 = C.$$

Since $g(t_0) = f(t_0)$ and $g'(t_0) = f'(t_0)$, the constant C must be zero. Thus for all t in I

$$[f'(t) - g'(t)]^2 + \omega^2[f(t) - g(t)]^2 = 0.$$

This, of course, implies that $f(t) = g(t)$ for all t in I and proves the assertion.† ☐

---

† We are indebted to Professor Edwin Hewitt of the University of Washington for suggesting this little argument.

In practice, the first constant $C_1 \geq 0$ is written A and the second constant $C_2 \in [0, 2\pi)$ is written ϕ_0. The solutions of (7.8.1) then read

(7.8.3)
$$x(t) = A \sin(\omega t + \phi_0).$$

Now let's analyze the motion measuring t in seconds. By adding $2\pi/\omega$ to t we increase $\omega t + \phi_0$ by 2π:

$$\omega\left(t + \frac{2\pi}{\omega}\right) + \phi_0 = \omega t + \phi_0 + 2\pi.$$

This means that the motion is *periodic* with *period* $2\pi/\omega$:

$$T = \frac{2\pi}{\omega}.$$

A complete oscillation takes $2\pi/\omega$ seconds. The reciprocal of the period gives the number of complete oscillations per second. This is called the *frequency*:

$$f = \frac{\omega}{2\pi}.$$

The number ω is called the *angular frequency*. Since $\sin(\omega t + \phi_0)$ oscillates between -1 and 1,

$$x(t) = A \sin(\omega t + \phi_0)$$

oscillates between $-A$ and A. The number A is called the *amplitude* of the motion.

In Figure 7.8.1 we have plotted x against t. The oscillations along the x-axis are now waves in the xt-plane. The period of the motion, $2\pi/\omega$, is the t distance (the time separation) between consecutive wave crests. The amplitude of the motion, A, is the height of the waves measured in x units from $x = 0$. The number ϕ_0 is known as the *phase constant*. The phase constant determines the initial displacement (in the xt-plane the height of the wave at time $t = 0$). If $\phi_0 = 0$, the object starts at the center of the interval of motion (the wave starts at the origin of the xt-plane).

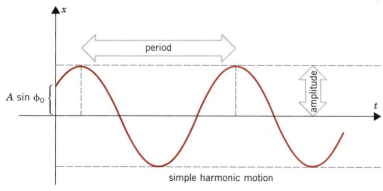

Figure 7.8.1

Problem 1. Find an equation for the oscillatory motion given that the period is $2\pi/3$ and, at time $t = 0$, $x = 1$ and $v = 3$.

Solution. We begin by setting

$$x(t) = A \sin (\omega t + \phi_0).$$

In general the period is $2\pi/\omega$, so that here

$$\frac{2\pi}{\omega} = \frac{2\pi}{3} \quad \text{and thus} \quad \omega = 3.$$

The equation of motion takes the form

$$x(t) = A \sin (3t + \phi_0).$$

By differentiation

$$v(t) = 3A \cos (3t + \phi_0).$$

The conditions at $t = 0$ give

$$1 = x(0) = A \sin \phi_0, \qquad 3 = v(0) = 3A \cos \phi_0$$

and therefore

$$1 = A \sin \phi_0, \qquad 1 = A \cos \phi_0.$$

Adding the squares, we have

$$2 = A^2 \sin^2 \phi_0 + A^2 \cos^2 \phi_0 = A^2.$$

Since $A > 0$, $A = \sqrt{2}$.

To find ϕ_2 we note that

$$1 = \sqrt{2} \sin \phi_0, \qquad 1 = \sqrt{2} \cos \phi_0.$$

These equations are satisfied by setting $\phi_0 = \frac{1}{4}\pi$. The equation of motion can be written

$$x(t) = \sqrt{2} \sin (3t + \tfrac{1}{4}\pi). \quad \square$$

Problem 2. An object in simple harmonic motion passes through the central point $x = 0$ at time 0 and every second thereafter. Find an equation for the motion if $v(0) = -4$.

Solution

$$x(t) = A \sin (\omega t + \phi_0).$$

On each complete cycle the object must pass through the central point twice — once going one way, once going the other way. The period is thus 2 seconds:

$$\frac{2\pi}{\omega} = 2 \quad \text{and therefore} \quad \omega = \pi.$$

We now know that

$$x(t) = A \sin (\pi t + \phi_0)$$

and consequently

$$v(t) = \pi A \cos (\pi t + \phi_0).$$

The initial conditions can be written

$$0 = x(0) = A \sin \phi_0, \qquad -4 = v(0) = \pi A \cos \phi_0$$

so that

$$0 = A \sin \phi_0, \qquad -\frac{4}{\pi} = A \cos \phi_0.$$

These equations give

$$A = \frac{4}{\pi}, \qquad \phi_0 = \pi. \qquad \text{(check this out)}$$

The equation of motion can be written

$$x(t) = \frac{4}{\pi} \sin (\pi t + \pi). \quad \square$$

Problem 3. A coil spring hangs naturally to a length l_0. When a bob of mass m is attached to it, the spring stretches l_1 inches. The bob is later pulled down an additional x_0 inches and then released.   What is the resulting motion?

Solution.   Throughout we refer to Figure 7.8.2, taking the downward direction as positive.

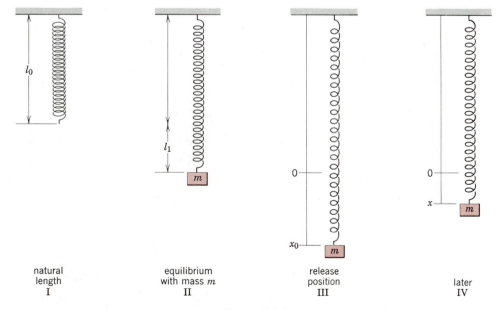

natural
length
I

equilibrium
with mass m
II

release
position
III

later
IV

Figure 7.8.2

We begin by analyzing the forces acting on the bob at general position x (stage IV). First there is the weight of the bob:

$$F_1 = mg.$$

This is a downward force, and by our choice of coordinate system, positive. Then there is the restoring force of the spring. This force, by Hooke's law, is proportional to the total displacement $l_1 + x$ and acts in the opposite direction:

$$F_2 = -k(l_1 + x) \qquad \text{with } k > 0.$$

If we neglect resistance, then these are the only forces acting on the bob. Under these conditions the total force is

$$F = F_1 + F_2 = mg - k(l_1 + x),$$

which we rewrite as

(1) $$F = (mg - kl_1) - kx.$$

At stage II (Figure 7.8.2) there was equilibrium. The force of gravity, mg, plus the force of the spring, $-kl_1$, must have been 0:

$$mg - kl_1 = 0.$$

Equation (1) can therefore be simplified to

$$F = -kx.$$

Using Newton's

$$F = ma \qquad\qquad \text{(force = mass × acceleration)}$$

we have

$$ma = -kx \quad \text{and thus} \quad a = -\frac{k}{m}x.$$

At any time t,

$$x''(t) = -\frac{k}{m}x(t).$$

Since $k/m > 0$, we can set $\omega = \sqrt{k/m}$ and write

$$x''(t) = -\omega^2 x(t).$$

The motion of the bob is simple harmonic motion with period $T = 2\pi/\omega$. ☐

There is something remarkable about harmonic motion that we have not yet specifically pointed out; namely, that the frequency $f = \omega/2\pi$ is completely independent of the amplitude of the motion. The oscillations of the bob in Problem 3 occur with frequency

$$f = \frac{\sqrt{k/m}}{2\pi}. \qquad\qquad \text{(There } \omega = \sqrt{k/m}.)$$

By adjusting the spring constant k and the mass of the bob m, we can calibrate the spring–bob system so that the oscillations take place exactly once a second (at least almost exactly). We then have a primitive timepiece (a first cousin of the windup clock).

With the passing of time, friction and air resistance reduce the amplitude of the oscillations but not their frequency. By giving the bob a little push or pull once in a while (by rewinding our clock), we can restore the amplitude of the oscillations and thus maintain the steady "ticking."

In this section we examined the simplest form of oscillation, and everywhere there were sines and cosines. Sines and cosines play a central role in the study of all oscillations. Wherever there are waves to analyze (sound waves, radio waves, light waves, etc.) sines and cosines hold the center stage.

EXERCISES 7.8

1. An object is in simple harmonic motion. Find an equation for the motion given that the period is $\frac{1}{4}\pi$ and, at time $t = 0$, $x = 1$ and $v = 0$. What is the amplitude? What is the frequency?

2. An object is in simple harmonic motion. Find an equation for the motion given that the frequency is $1/\pi$ and, at time $t = 0$, $x = 0$ and $v = -2$. What is the amplitude? What is the period?

3. An object is in simple harmonic motion with period T and amplitude A. What is the velocity at the central point $x = 0$?

4. An object is in simple harmonic motion with period T. Find the amplitude given that $v = \pm v_0$ at $x = x_0$.

5. An object in simple harmonic motion passes through the central point $x = 0$ at time $t = 0$ and every 3 seconds thereafter. Find the equation of motion given that $v(0) = 5$.

6. Show that simple harmonic motion $x(t) = A \sin(\omega t + \phi_0)$ can just as well be written (a) $x(t) = A \cos(\omega t + \phi_1)$. (b) $x(t) = B \sin \omega t + C \cos \omega t$.

7. What is $x(t)$ for the bob of mass m in Problem 3?

8. Find the positions of the bob in Problem 3 where the bob attains (a) maximum speed. (b) zero speed. (c) maximum acceleration. (d) zero acceleration.

9. Where does the bob in Problem 3 take on half of its maximum speed?

10. Find the maximal kinetic energy obtained by the bob in Problem 3. (Remember: KE $= \frac{1}{2}mv^2$ where m is the mass of the object and v is the speed.)

11. Find the time average of the kinetic energy of the bob in Problem 3 during one period T.

12. Express the velocity of the bob in Problem 3 in terms of k, m, x_0, and $x(t)$.

13. Given that $x''(t) = 8 - 4x(t)$ with $x(0) = 0$ and $x'(0) = 0$, show that the motion is simple harmonic motion centered at $x = 2$. Find the amplitude and the period.

14. Figure 7.8.3 shows a pendulum of mass m swinging on an arm of length L. The angle θ is measured counterclockwise. Neglecting friction and the weight of the arm, we can describe the motion by the equation

$$mL\theta''(t) = -mg \sin \theta(t)$$

which reduces to

$$\theta''(t) = -\frac{g}{L} \sin \theta(t).$$

(a) For small angles we replace $\sin \theta$ by θ and write

$$\theta''(t) \cong -\frac{g}{L} \theta(t).$$

Justify this step.

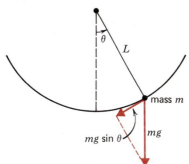

Figure 7.8.3

(b) Solve the approximate equation in (a) given that the pendulum
   (i) is held at an angle $\theta_0 > 0$ and released at time $t = 0$.
   (ii) passes through the vertical position at time $t = 0$ and $\theta'(0) = -\sqrt{g/L}\,\theta_0$.
(c) Find L given that the motion repeats itself every 2 seconds.

Figure 7.8.4

15. A cylindrical buoy of mass m and radius r centimeters floats with its axis vertical in a liquid of density ρ kilograms per cubic centimeter. (Figure 7.8.4) Suppose that the buoy is pushed x_0 centimeters down into the liquid.
   (a) Neglecting friction and given that the buoyancy force is equal to the weight of the liquid displaced, show that the buoy bobs up and down in simple harmonic motion by finding the equation of motion.
   (b) Solve the equation obtained in (a). Specify the amplitude and the period.
16. Explain in detail the connection between uniform circular motion and simple harmonic motion.

7.9   MORE ON THE INTEGRATION OF THE TRIGONOMETRIC FUNCTIONS

In Section 5.9 you saw that

(7.9.1)

$$\int \cos x \, dx = \sin x + C, \qquad \int \sin x \, dx = -\cos x + C,$$

$$\int \sec^2 x \, dx = \tan x + C, \qquad \int \csc^2 x \, dx = -\cot x + C,$$

$$\int \sec x \tan x \, dx = \sec x + C, \qquad \int \csc x \cot x \, dx = -\csc x + C,$$

$$\int \cos^2 x \, dx = \tfrac{1}{2}x + \tfrac{1}{4}\sin 2x + C, \qquad \int \sin^2 x \, dx = \tfrac{1}{2}x - \tfrac{1}{4}\sin 2x + C.$$

Now that you are familiar with the logarithm function, we can add four more basic formulas to the list.

$$\text{(i)} \quad \int \tan x \, dx = \ln |\sec x| + C.$$

$$\text{(ii)} \quad \int \cot x \, dx = \ln |\sin x| + C.$$

(7.9.2)

$$\text{(iii)} \quad \int \sec x \, dx = \ln |\sec x + \tan x| + C.$$

$$\text{(iv)} \quad \int \csc x \, dx = \ln |\csc x - \cot x| + C.$$

Each of these formulas can be derived as follows:

$$\int \frac{g'(x)}{g(x)} \, dx = \int \frac{du}{u} = \ln |u| + C = \ln |g(x)| + C.$$

$$\text{set } u = g(x), \quad du = g'(x) \, dx$$

Derivation of Formulas (i)–(iii)

$$\text{(i)} \quad \int \tan x \, dx = \int \frac{\sin x}{\cos x} \, dx \qquad\qquad [\text{set } u = \cos x, \quad du = -\sin x \, dx]$$

$$= -\int \frac{du}{u} = -\ln |u| + C$$

$$= -\ln |\cos x| + C = \ln \left| \frac{1}{\cos x} \right| + C = \ln |\sec x| + C. \quad \square$$

$$\text{(ii)} \quad \int \cot x \, dx = \int \frac{\cos x}{\sin x} \, dx \qquad\qquad [\text{set } u = \sin x, \quad du = \cos x \, dx]$$

$$= \int \frac{du}{u} = \ln |u| + C = \ln |\sin x| + C. \quad \square$$

$$\text{(iii)} \quad \int \sec x \, dx \overset{\dagger}{=} \int \sec x \frac{\sec x + \tan x}{\sec x + \tan x} \, dx$$

$$= \int \frac{\sec x \tan x + \sec^2 x}{\sec x + \tan x} \, dx$$

$$[\text{set } u = \sec x + \tan x, \quad du = (\sec x \tan x + \sec^2 x) \, dx]$$

$$= \int \frac{du}{u} = \ln |u| + C = \ln |\sec x + \tan x| + C. \quad \square$$

The derivation of formula (iv) is left to you.

---

† Only experience prompts us to multiply numerator and denominator by $\sec x + \tan x$.

Problem 1. Calculate

$$\int \cot \pi x \, dx.$$

Solution. Set

$$u = \pi x, \qquad du = \pi \, dx.$$

$$\int \cot \pi x \, dx = \frac{1}{\pi} \int \cot u \, du = \frac{1}{\pi} \ln |\sin u| + C = \frac{1}{\pi} \ln |\sin \pi x| + C. \quad \square$$

Remark. The u-substitution simplifies many calculations, but you will find with experience that you can carry out many integrations without it. $\square$

Problem 2. Evaluate

$$\int_0^{\pi/8} \sec 2x \, dx.$$

Solution

$$\int_0^{\pi/8} \sec 2x \, dx = \tfrac{1}{2} \left[\ln |\sec 2x + \tan 2x| \right]_0^{\pi/8}$$

$$= \tfrac{1}{2} [\ln (\sqrt{2} + 1) - \ln 1] = \tfrac{1}{2} \ln (\sqrt{2} + 1) \cong 0.44. \quad \square$$

Whenever the integrand is a quotient, a useful tactic is to see if the integral can be written in the form $\int \frac{du}{u}$.

Problem 3. Calculate

$$\int \frac{\cos 3x}{2 + \sin 3x} \, dx.$$

Solution. Set

$$u = 2 + \sin 3x, \qquad du = 3 \cos 3x \, dx.$$

$$\int \frac{\cos 3x}{2 + \sin 3x} \, dx = \frac{1}{3} \int \frac{du}{u} = \frac{1}{3} \ln |u| + C = \frac{1}{3} \ln |2 + \sin 3x| + C. \quad \square$$

Problem 4. Evaluate

$$\int_{\pi/4}^{\pi/3} \frac{\sec^2 x}{1 + \tan x} \, dx.$$

Solution

$$\int_{\pi/4}^{\pi/3} \frac{\sec^2 x}{1 + \tan x} \, dx = \left[\ln (1 + \tan x) \right]_{\pi/4}^{\pi/3}$$

$$= \ln (1 + \sqrt{3}) - \ln (1 + 1) = \ln \left(\frac{1 + \sqrt{3}}{2} \right) \cong 0.31. \quad \square$$

EXERCISES 7.9

Calculate these indefinite integrals.

1. $\int \tan 3x \, dx.$

2. $\int \sec \tfrac{1}{2}\pi x \, dx.$

3. $\int \csc \pi x \, dx.$

4. $\int \cot (\pi - x) \, dx.$

5. $\int e^x \cot e^x \, dx.$

6. $\int \dfrac{\csc^2 x}{2 + \cot x} \, dx.$

7. $\int \dfrac{\sin 2x}{3 - 2 \cos 2x} \, dx.$

8. $\int e^{\csc x} \csc x \cot x \, dx.$

9. $\int e^{\tan 3x} \sec^2 3x \, dx.$

10. $\int e^x \cos e^x \, dx.$

11. $\int x \sec x^2 \, dx.$

12. $\int \dfrac{\sec e^{-2x}}{e^{2x}} \, dx.$

13. $\int \cot x \ln (\sin x) \, dx.$

14. $\int \dfrac{\tan (\ln x)}{x} \, dx.$

15. $\int (1 + \sec x)^2 \, dx.$

16. $\int \tan x \ln (\sec x) \, dx.$

17. $\int \left(\dfrac{\csc x}{1 + \cot x}\right)^2 dx.$

18. $\int (3 - \csc x)^2 \, dx.$

Evaluate these definite integrals.

19. $\int_{\pi/6}^{\pi/2} \dfrac{\cos x}{1 + \sin x} \, dx.$

20. $\int_{\pi/4}^{\pi/2} (1 + \csc x)^2 \, dx.$

21. $\int_{\pi/4}^{\pi/2} \cot x \, dx.$

22. $\int_{1/4}^{1/3} \tan \pi x \, dx.$

23. $\int_{0}^{\ln \pi/4} e^x \sec e^x \, dx.$

24. $\int_{\pi/4}^{\pi/2} \dfrac{\csc^2 x}{3 + \cot x} \, dx.$

Sketch the region bounded by the curves and find its area.

25. $y = \sec x, \quad y = 2, \quad x = 0, \quad x = \pi/6.$
27. $y = \tan x, \quad y = 1, \quad x = 0.$

26. $y = \csc \tfrac{1}{2}\pi x, \quad y = x, \quad x = \tfrac{1}{2}.$
28. $y = \sec x, \quad y = \cos x, \quad x = \pi/4.$

7.10   THE INVERSE TRIGONOMETRIC FUNCTIONS

Since none of the trigonometric functions are one-to-one, none of them have inverses. What then are the inverse trigonometric functions?

The Inverse Sine

The range of the sine function is the closed interval $[-1, 1]$. Although not one-to-one on its full domain, the sine function is one-to-one on the closed interval $[-\tfrac{1}{2}\pi, \tfrac{1}{2}\pi]$, and on that interval it takes on as a value every number in $[-1, 1]$. Thus, if $x \in [-1, 1]$, there is one and only one number in the interval $[-\tfrac{1}{2}\pi, \tfrac{1}{2}\pi]$ at which the sine function takes on the value x. This number is called the *inverse sine of x* and is written $\sin^{-1} x$.†

The *inverse sine function*

$$y = \sin^{-1} x, \qquad x \in [-1, 1]$$

___

† *Caution*: The -1 is not an exponent. Do not confuse $\sin^{-1} x$ with the reciprocal $1/(\sin x)$.

is the inverse of the function

$$y = \sin x, \quad x \in [-\tfrac{1}{2}\pi, \tfrac{1}{2}\pi].$$

The graphs of these functions are pictured in Figure 7.10.1. Each curve is the reflection of the other in the line $y = x$.

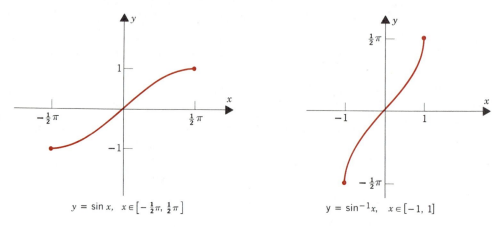

$$y = \sin x, \quad x \in \left[-\tfrac{1}{2}\pi, \tfrac{1}{2}\pi\right] \qquad\qquad y = \sin^{-1}x, \quad x \in [-1, 1]$$

Figure 7.10.1

Because these functions are inverses,

(7.10.1)   for all $x \in [-1, 1]$, $\sin(\sin^{-1} x) = x$

and

(7.10.2)   for all $x \in [-\tfrac{1}{2}\pi, \tfrac{1}{2}\pi]$, $\sin^{-1}(\sin x) = x$.

Table 7.10.1 gives some representative values of the sine function from $x = -\tfrac{1}{2}\pi$ to $x = \tfrac{1}{2}\pi$. Reversing the order of the columns we have a table for the inverse sine (Table 7.10.2).

TABLE 7.10.1			TABLE 7.10.2	
x	$\sin x$		x	$\sin^{-1} x$
$-\tfrac{1}{2}\pi$	-1		-1	$-\tfrac{1}{2}\pi$
$-\tfrac{1}{3}\pi$	$-\tfrac{1}{2}\sqrt{3}$		$-\tfrac{1}{2}\sqrt{3}$	$-\tfrac{1}{3}\pi$
$-\tfrac{1}{4}\pi$	$-\tfrac{1}{2}\sqrt{2}$		$-\tfrac{1}{2}\sqrt{2}$	$-\tfrac{1}{4}\pi$
$-\tfrac{1}{6}\pi$	$-\tfrac{1}{2}$		$-\tfrac{1}{2}$	$-\tfrac{1}{6}\pi$
0	0		0	0
$\tfrac{1}{6}\pi$	$\tfrac{1}{2}$		$\tfrac{1}{2}$	$\tfrac{1}{6}\pi$
$\tfrac{1}{4}\pi$	$\tfrac{1}{2}\sqrt{2}$		$\tfrac{1}{2}\sqrt{2}$	$\tfrac{1}{4}\pi$
$\tfrac{1}{3}\pi$	$\tfrac{1}{2}\sqrt{3}$		$\tfrac{1}{2}\sqrt{3}$	$\tfrac{1}{3}\pi$
$\tfrac{1}{2}\pi$	1		1	$\tfrac{1}{2}\pi$

On the basis of Table 7.10.2 one could guess that for all $x \in [-1, 1]$

$$\sin^{-1}(-x) = -\sin^{-1} x.$$

This is indeed the case. Being the inverse of an odd function (sin $(-x) = -\sin x$ for all $x \in [-\frac{1}{2}\pi, \frac{1}{2}\pi]$), the inverse sine is itself an odd function. (Check that out.)

Table 4 at the back of the book gives the values of the principal trigonometric functions from 0 radians to 1.58 radians (approximately $\frac{1}{2}\pi$ radians). According to that table, for example,

$$\sin 0.32 \cong 0.315 \quad \text{and} \quad \sin 0.81 \cong 0.724.$$

Reading backward from that table and using the fact that the inverse sine is an odd function, we have

$$\sin^{-1} 0.315 \cong 0.32 \quad \text{and} \quad \sin^{-1}(-0.724) = -\sin^{-1} 0.724 \cong -0.81.$$

Problem 1. Calculate if defined:

      (a) $\sin^{-1}(\sin \frac{1}{16}\pi)$.   (b) $\sin^{-1}(\sin \frac{5}{2}\pi)$.   (c) $\sin(\sin^{-1} \frac{1}{3})$.

      (d) $\sin^{-1}(\sin \frac{9}{5}\pi)$.   (e) $\sin(\sin^{-1} 2)$.

Solution. (a) Since $\frac{1}{16}\pi$ is within the interval $[-\frac{1}{2}\pi, \frac{1}{2}\pi]$, we know by (7.10.2) that

$$\sin^{-1}(\sin \tfrac{1}{16}\pi) = \tfrac{1}{16}\pi.$$

(b) Since $\frac{5}{2}\pi$ is not within the interval $[-\frac{1}{2}\pi, \frac{1}{2}\pi]$, we cannot apply (7.10.2) directly. However, $\frac{5}{2}\pi = \frac{1}{2}\pi + 2\pi$. Thus

$$\sin^{-1}(\sin \tfrac{5}{2}\pi) = \sin^{-1}(\sin (\tfrac{1}{2}\pi + 2\pi)) = \sin^{-1}(\sin \tfrac{1}{2}\pi) = \tfrac{1}{2}\pi.$$

by (7.10.2) ⬏

(c) By (7.10.1)

$$\sin(\sin^{-1} \tfrac{1}{3}) = \tfrac{1}{3}.$$

(d) Since $\frac{9}{5}\pi$ is not within the interval $[-\frac{1}{2}\pi, \frac{1}{2}\pi]$, we cannot apply (7.10.2) directly. However, $\frac{9}{5}\pi = 2\pi - \frac{1}{5}\pi$. Thus

$$\sin^{-1}(\sin \tfrac{9}{5}\pi) = \sin^{-1}(\sin (2\pi - \tfrac{1}{5}\pi)) = \sin^{-1}(\sin (-\tfrac{1}{5}\pi)) = -\tfrac{1}{5}\pi.$$

by (7.10.2) ⬏

(e) The expression $\sin(\sin^{-1} 2)$ makes no sense since 2 is not within the domain of the inverse sine. The inverse sine is defined only on $[-1, 1]$. ◻

If $0 < x < 1$, then $\sin^{-1} x$ is the radian measure of the acute angle that has sine x. We can construct an angle of radian measure $\sin^{-1} x$ by drawing a right triangle with a leg of length x and a hypotenuse of length 1. (Figure 7.10.2)

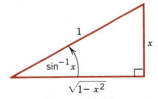

Figure 7.10.2

Reading from the figure we have

$$\sin(\sin^{-1} x) = x, \qquad\qquad \cos(\sin^{-1} x) = \sqrt{1 - x^2}$$

$$\tan(\sin^{-1} x) = \frac{x}{\sqrt{1 - x^2}}, \qquad \cot(\sin^{-1} x) = \frac{\sqrt{1 - x^2}}{x}$$

$$\sec(\sin^{-1} x) = \frac{1}{\sqrt{1 - x^2}}, \qquad \csc(\sin^{-1} x) = \frac{1}{x}.$$

Since the derivative of the sine function,

$$\frac{d}{dx}(\sin x) = \cos x,$$

does not take on the value 0 on the *open* interval $(-\tfrac{1}{2}\pi, \tfrac{1}{2}\pi)$, the inverse sine function is differentiable on the *open* interval $(-1, 1)$.† We can find the derivative as follows:

$$y = \sin^{-1} x,$$

$$\sin y = x,$$

$$\cos y \frac{dy}{dx} = 1,$$

$$\frac{dy}{dx} = \frac{1}{\cos y} = \frac{1}{\sqrt{1 - x^2}}. \quad \square$$

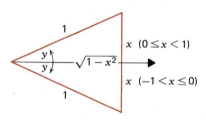

In short

(7.10.3)

$$\frac{d}{dx}(\sin^{-1} x) = \frac{1}{\sqrt{1 - x^2}}.$$

Problem 2. Find

$$\frac{d}{dx}[\sin^{-1}(3x^2)].$$

Solution. In general, by the chain rule,

$$\frac{d}{dx}[\sin^{-1} u] = \frac{d}{du}[\sin^{-1} u]\frac{du}{dx} = \frac{1}{\sqrt{1 - u^2}}\frac{du}{dx}.$$

Thus

$$\frac{d}{dx}[\sin^{-1}(3x^2)] = \frac{1}{\sqrt{1 - (3x^2)^2}}\frac{d}{dx}(3x^2) = \frac{6x}{\sqrt{1 - 9x^4}}. \quad \blacksquare$$

---

† See the first paragraph of Section 3.8.

Problem 3.   Show that for $a > 0$

(7.10.4)

$$\int \frac{dx}{\sqrt{a^2 - x^2}} = \sin^{-1}\left(\frac{x}{a}\right) + C.$$

Solution.   We change variables so that the a^2 becomes 1 and we can use (7.10.3).   We set

$$au = x, \qquad a\,du = dx.$$

Then

$$\int \frac{dx}{\sqrt{a^2 - x^2}} = \int \frac{a\,du}{\sqrt{a^2 - a^2u^2}} = \int \frac{du}{\sqrt{1 - u^2}} = \sin^{-1} u + C = \sin^{-1}\left(\frac{x}{a}\right) + C. \quad \square$$

since $a > 0$

Problem 4.   Evaluate

$$\int_0^{\sqrt{3}} \frac{dx}{\sqrt{4 - x^2}}.$$

Solution.   By (7.10.4)

$$\int \frac{dx}{\sqrt{4 - x^2}} = \sin^{-1}\left(\frac{x}{2}\right) + C.$$

It follows that

$$\int_0^{\sqrt{3}} \frac{dx}{\sqrt{4 - x^2}} = \left[\sin^{-1}\left(\frac{x}{2}\right)\right]_0^{\sqrt{3}} = \sin^{-1}\frac{\sqrt{3}}{2} - \sin^{-1} 0 = \frac{\pi}{3} - 0 = \frac{\pi}{3}. \quad \square$$

We have not talked about integrating the inverse sine.   An integration by parts that we leave to you shows that

(7.10.5)

$$\int \sin^{-1} x\,dx = x \sin^{-1} x + \sqrt{1 - x^2} + C.$$

The Inverse Tangent

Although not one-to-one on its full domain, the tangent function is one-to-one on the open interval $(-\frac{1}{2}\pi, \frac{1}{2}\pi)$ and on that interval it takes on as a value every real number. Thus, if x is real, there is one and only one number in the open interval $(-\frac{1}{2}\pi, \frac{1}{2}\pi)$ at which the tangent function takes on the value x.   This number is called the *inverse tangent of x* and is written $\tan^{-1} x$.

The *inverse tangent function*

$$y = \tan^{-1} x, \qquad x \text{ real}$$

is the inverse of the function

$$y = \tan x, \qquad x \in (-\tfrac{1}{2}\pi, \tfrac{1}{2}\pi).$$

The graphs of these two functions are given in Figure 7.10.3. Each curve is a reflection of the other in the line $y = x$. While the tangent has vertical asymptotes, the inverse tangent has horizontal asymptotes. Both functions are odd functions.

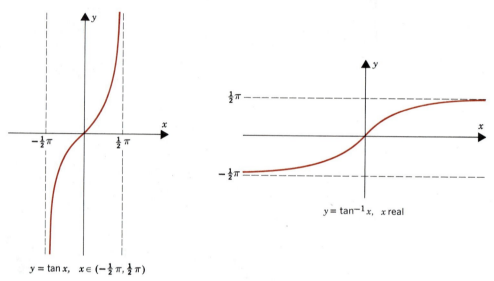

$y = \tan x, \quad x \in (-\tfrac{1}{2}\pi, \tfrac{1}{2}\pi)$

$y = \tan^{-1} x, \quad x \text{ real}$

Figure 7.10.3

Because these functions are inverses,

(7.10.6)   for all real numbers x, $\tan (\tan^{-1} x) = x$

and

(7.10.7)   for all $x \in (-\tfrac{1}{2}\pi, \tfrac{1}{2}\pi)$, $\tan^{-1} (\tan x) = x.$

It is hard to make a mistake with the first relation since it applies for all real numbers. The second relation requires the usual care:

$$\tan^{-1} (\tan \tfrac{1}{4}\pi) = \tfrac{1}{4}\pi \qquad \text{but} \qquad \tan^{-1} (\tan \tfrac{7}{5}\pi) \neq \tfrac{7}{5}\pi.$$

We can calculate $\tan^{-1} (\tan \tfrac{7}{5}\pi)$ as follows:

$$\tan^{-1} (\tan \tfrac{7}{5}\pi) = \tan^{-1} (\tan (\tfrac{2}{5}\pi + \pi)) = \tan^{-1} (\tan \tfrac{2}{5}\pi) = \tfrac{2}{5}\pi.$$

The relation $\tan^{-1} (\tan \tfrac{2}{5}\pi) = \tfrac{2}{5}\pi$ is valid because $\tfrac{2}{5}\pi$ is within the interval $(-\tfrac{1}{2}\pi, \tfrac{1}{2}\pi)$.

If $x > 0$, then $\tan^{-1} x$ is the radian measure of the acute angle that has tangent x. We can construct an angle of radian measure $\tan^{-1} x$ by drawing a right triangle with legs of length x and 1 (Figure 7.10.4). The values of

$$\tan (\tan^{-1} x), \qquad \cot (\tan^{-1} x),$$

$$\sin (\tan^{-1} x), \qquad \cos (\tan^{-1} x),$$

$$\sec (\tan^{-1} x), \qquad \csc (\tan^{-1} x)$$

can all be read from this triangle.

Figure 7.10.4

Since the derivative of the tangent function

$$\frac{d}{dx} (\tan x) = \sec^2 x = \frac{1}{\cos^2 x},$$

is never 0, the inverse tangent function is everywhere differentiable (Section 3.8). We can find the derivative as we did for the inverse sine:

$$y = \tan^{-1} x,$$

$$\tan y = x,$$

$$\sec^2 y \frac{dy}{dx} = 1,$$

$$\frac{dy}{dx} = \frac{1}{\sec^2 y} = \cos^2 y = \frac{1}{1 + x^2}. \quad \blacksquare$$

We have found that

(7.10.8)
$$\frac{d}{dx} (\tan^{-1} x) = \frac{1}{1 + x^2}.$$

Problem 5. Calculate

$$\frac{d}{dx} [\tan^{-1} (ax^2 + bx + c)].$$

Solution. In general, by the chain rule,

$$\frac{d}{dx} [\tan^{-1} u] = \frac{d}{du} [\tan^{-1} u] \frac{du}{dx} = \frac{1}{1 + u^2} \frac{du}{dx}.$$

Thus

$$\frac{d}{dx} [\tan^{-1} (ax^2 + bx + c)] = \frac{1}{1 + (ax^2 + bx + c)^2} \frac{d}{dx} (ax^2 + bx + c)$$

$$= \frac{2ax + b}{1 + (ax^2 + bx + c)^2}. \quad \blacksquare$$

Problem 6. Show that

(7.10.9)
$$\int \frac{dx}{a^2 + x^2} = \frac{1}{a} \tan^{-1}\left(\frac{x}{a}\right) + C.$$

Solution. We change variables so that a^2 is replaced by 1 and we can use (7.10.8). We set

$$au = x, \qquad a\, du = dx.$$

Then

$$\int \frac{dx}{a^2 + x^2} = \int \frac{a\, du}{a^2 + a^2 u^2} = \frac{1}{a} \int \frac{du}{1 + u^2}$$

$$= \frac{1}{a} \tan^{-1} u + C = \frac{1}{a} \tan^{-1}\left(\frac{x}{a}\right) + C. \quad \square$$

$$(7.10.8) \longrightarrow$$

Problem 7. Evaluate

$$\int_0^2 \frac{dx}{4 + x^2}.$$

Solution. By (7.10.9)

$$\int \frac{dx}{4 + x^2} = \int \frac{dx}{2^2 + x^2} = \frac{1}{2} \tan^{-1}\left(\frac{x}{2}\right) + C$$

so that

$$\int_0^2 \frac{dx}{4 + x^2} = \left[\frac{1}{2} \tan^{-1}\left(\frac{x}{2}\right)\right]_0^2 = \frac{1}{2} \tan^{-1} 1 = \frac{\pi}{8}. \quad \square$$

Finally, an integration by parts that we leave to you shows that

(7.10.10)
$$\int \tan^{-1} x \, dx = x \tan^{-1} x - \tfrac{1}{2} \ln (1 + x^2) + C.$$

The Other Trigonometric Inverses

There are four other trigonometric inverses:

the inverse cosine, $y = \cos^{-1} x$, is the inverse of $y = \cos x, \ x \in [0, \pi]$;

the inverse cotangent, $y = \cot^{-1} x$, is the inverse of $y = \cot x, \ x \in (0, \pi)$;

the inverse secant, $y = \sec^{-1} x$, is the inverse of $y = \sec x$, $x \in [0, \frac{1}{2}\pi) \cup (\frac{1}{2}\pi, \pi]$;

the inverse cosecant, $y = \csc^{-1} x$, is the inverse of $y = \csc x$, $x \in [-\frac{1}{2}\pi, 0) \cup (0, \frac{1}{2}\pi]$.

Of these four functions only the first two are much used. Figure 7.10.5 shows them all between 0 and $\frac{1}{2}\pi$ in terms of right triangles.

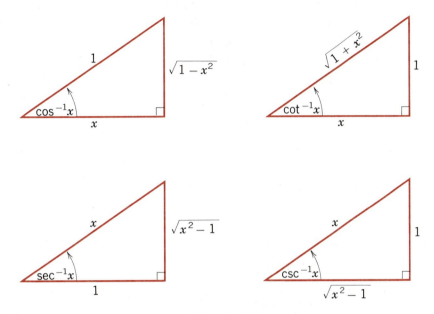

Figure 7.10.5

A Word on Notation

The $(\ \ )^{-1}$ notation that we have been using for the trigonometric inverses is not the only notation in general use: the inverses

$$\sin^{-1} t, \quad \tan^{-1} t, \quad \cos^{-1} t, \quad \text{and so on}$$

are often written

$$\text{arc } \sin t, \quad \text{arc } \tan t, \quad \text{arc } \cos t, \quad \text{etc.}$$

To familiarize you further with the notation, we use it in some of the exercises.

EXERCISES 7.10

Determine the exact value:

1. $\tan^{-1} 0$.

2. $\cos^{-1} 0$.

3. $\cos^{-1} \frac{1}{2}$.

4. $\tan^{-1} \sqrt{3}$.

5. $\tan^{-1} (-\sqrt{3})$.

6. $\cos^{-1} (-\frac{1}{2})$.

7. $\text{arc } \cos 1$.

8. $\text{arc } \tan (-1)$.

9. $\text{arc } \sin (-\frac{1}{2})$.

10. $\text{arc } \cos (-\frac{1}{2}\sqrt{3})$.

11. $\cos (\cos^{-1} \frac{1}{2})$.

12. $\sin^{-1} (\sin \frac{7}{4}\pi)$.

13. $\tan^{-1} (\tan \frac{7}{4}\pi)$.

14. $\text{arc } \tan (\tan \frac{5}{6}\pi)$.

15. $\cos (\text{arc } \sin \frac{1}{2})$.

16. $\sin (\text{arc } \cos \frac{1}{2})$.

17. $\tan^{-1} (\cos 0)$.

18. $\sin^{-1} (\sin (-\frac{7}{4}\pi))$.

19. $\sin (\cos^{-1} (-\frac{1}{2}))$.

20. $\text{arc } \tan (\cos \frac{3}{2}\pi)$.

Find the approximate value by using Table 4 at the back of the book.

21. $\sin^{-1} 0.918$. **22.** arc $\sin(-0.795)$. **23.** $\tan^{-1}(-0.493)$.

24. arc tan 3.111. **25.** arc cos (0.960). **26.** $\cos^{-1}(-0.142)$.

Taking $x > 0$, calculate the following from Figure 7.10.4.

27. $\cos(\tan^{-1} x)$. **28.** $\sin(\tan^{-1} x)$. **29.** $\tan(\tan^{-1} x)$.

30. $\cot(\tan^{-1} x)$. **31.** $\sec(\tan^{-1} x)$. **32.** $\csc(\tan^{-1} x)$.

Taking $0 < x < 1$, calculate the following from an appropriate right triangle.

33. $\sin(\cos^{-1} x)$. **34.** $\cos(\cot^{-1} x)$. **35.** $\sec(\cot^{-1} x)$.

36. $\tan(\cos^{-1} x)$. **37.** $\cot(\cos^{-1} x)$. **38.** $\csc(\cot^{-1} x)$.

Differentiate.

39. $y = \tan^{-1}(x + 1)$. **40.** $y = \tan^{-1} \sqrt{x}$. **41.** $y = \sin^{-1} x^2$.

42. $f(x) = e^x \sin^{-1} x$. **43.** $f(x) = x \sin^{-1} 2x$. **44.** $f(x) = e^{\tan^{-1} x}$.

45. $u = (\sin^{-1} x)^2$. **46.** $v = \tan^{-1}(e^x)$. **47.** $y = \dfrac{\tan^{-1} x}{x}$.

48. $y = \tan^{-1}\left(\dfrac{2}{x}\right)$. **49.** $f(x) = \sqrt{\tan^{-1} 2x}$. **50.** $f(x) = \ln(\tan^{-1} x)$.

51. $y = \tan^{-1}(\ln x)$. **52.** $y = \tan^{-1}(\sin x)$. **53.** $\theta = \sin^{-1}(\sqrt{1 - r^2})$.

54. $\theta = \sin^{-1}\left(\dfrac{r}{r + 1}\right)$. **55.** $\theta = \tan^{-1}\left(\dfrac{c + r}{1 - cr}\right)$. **56.** $\theta = \tan^{-1}\left(\dfrac{1}{1 + r^2}\right)$.

57. $f(x) = \sqrt{c^2 - x^2} + c \sin^{-1}\left(\dfrac{x}{c}\right)$.† **58.** $f(x) = \frac{1}{3} \sin^{-1}(3x - 4x^2)$.

59. $y = \dfrac{x}{\sqrt{c^2 - x^2}} - \sin^{-1}\left(\dfrac{x}{c}\right)$.† **60.** $y = x\sqrt{c^2 - x^2} + c^2 \sin^{-1}\left(\dfrac{x}{c}\right)$.†

61. Show that for $a > 0$

(7.10.11)
$$\int \frac{dx}{\sqrt{a^2 - (x + b)^2}} = \sin^{-1}\left(\frac{x + b}{a}\right) + C.$$

62. Show that for $a \neq 0$

(7.10.12)
$$\int \frac{dx}{a^2 + (x + b)^2} = \frac{1}{a} \tan^{-1}\left(\frac{x + b}{a}\right) + C.$$

Evaluate.

63. $\displaystyle\int_0^1 \frac{dx}{1 + x^2}$. **64.** $\displaystyle\int_{-1}^1 \frac{dx}{1 + x^2}$. **65.** $\displaystyle\int_0^{1/\sqrt{2}} \frac{dx}{\sqrt{1 - x^2}}$.

66. $\displaystyle\int_0^1 \frac{dx}{\sqrt{4 - x^2}}$. **67.** $\displaystyle\int_0^5 \frac{dx}{25 + x^2}$. **68.** $\displaystyle\int_{-4}^4 \frac{dx}{16 + x^2}$.

69. $\displaystyle\int_0^{3/2} \frac{dx}{9 + 4x^2}$. **70.** $\displaystyle\int_2^5 \frac{dx}{9 + (x - 2)^2}$. **71.** $\displaystyle\int_{-3}^{-2} \frac{dx}{\sqrt{4 - (x + 3)^2}}$.

---

† Take $c > 0$.

72. $\displaystyle\int_{\ln 2}^{\ln 3} \frac{e^{-x}}{\sqrt{1 - e^{-2x}}}\, dx.$ **73.** $\displaystyle\int_{0}^{\ln 2} \frac{e^{x}}{1 + e^{2x}}\, dx.$ **74.** $\displaystyle\int_{0}^{1/2} \frac{1}{\sqrt{3 - 4x^2}}\, dx.$

75. Derive Formula 7.10.5.

76. Derive Formula 7.10.10.

Calculate these indefinite integrals.

77. $\displaystyle\int \frac{x}{\sqrt{1 - x^4}}\, dx.$ **78.** $\displaystyle\int \frac{\sec^2 x}{\sqrt{9 - \tan^2 x}}\, dx.$ **79.** $\displaystyle\int \frac{x}{1 + x^4}\, dx.$

80. $\displaystyle\int x \tan^{-1} x\, dx.$ **81.** $\displaystyle\int \frac{\sec^2 x}{9 + \tan^2 x}\, dx.$ **82.** $\displaystyle\int x \sin^{-1} 2x^2\, dx.$

83. Show that $\sin^{-1} x + \cos^{-1} x = \frac{1}{2}\pi$ for all $x \in [-1, 1]$.

84. Show that $\tan^{-1} x + \cot^{-1} x = \frac{1}{2}\pi$ for all real x.

Determine the following derivatives.

85. $\dfrac{d}{dx}(\cos^{-1} x)$ for $-1 < x < 1$.   **86.** $\dfrac{d}{dx}(\cot^{-1} x)$ for all real x.

87. Show that for $x < -1$ and $x > 1$

(7.10.13)
$$\frac{d}{dx}(\sec^{-1} x) = \frac{1}{|x|\sqrt{x^2 - 1}}.$$

88. (Calculator) Evaluate

$$\lim_{x \to 0} \frac{\sin^{-1} x}{x}$$

numerically. Justify your answer by other means.

89. (Calculator) Estimate the integral

$$\int_{0}^{0.5} \frac{1}{\sqrt{1 - x^2}}\, dx$$

by using the partition $\{0, 0.1, 0.2, 0.3, 0.4, 0.5\}$ and the intermediate points

$$x_1^* = 0.05, \quad x_2^* = 0.15, \quad x_3^* = 0.25, \quad x_4^* = 0.35, \quad x_5^* = 0.45.$$

Note that the sine of your estimate is close to 0.5. Explain the reason for this.

7.11 THE HYPERBOLIC SINE AND COSINE

The *hyperbolic sine* and *hyperbolic cosine* are the functions

(7.11.1)
$$\sinh x = \tfrac{1}{2}(e^x - e^{-x}), \qquad \cosh x = \tfrac{1}{2}(e^x + e^{-x}).$$

The reasons for these names will become apparent as we go on.
Since

$$\frac{d}{dx}(\sinh x) = \frac{d}{dx}[\tfrac{1}{2}(e^x - e^{-x})] = \tfrac{1}{2}(e^x + e^{-x})$$

and

$$\frac{d}{dx}(\cosh x) = \frac{d}{dx}[\tfrac{1}{2}(e^x + e^{-x})] = \tfrac{1}{2}(e^x - e^{-x}),$$

we have

(7.11.2)
$$\frac{d}{dx}(\sinh x) = \cosh x, \qquad \frac{d}{dx}(\cosh x) = \sinh x.$$

In short, each of these functions is the derivative of the other.

The Graphs

We begin with the hyperbolic sine. Since

$$\sinh(-x) = \tfrac{1}{2}(e^{-x} - e^x) = -\tfrac{1}{2}(e^x - e^{-x}) = -\sinh x,$$

the hyperbolic sine is an odd function. The graph is therefore symmetric about the origin. Since

$$\frac{d}{dx}(\sinh x) = \cosh x = \tfrac{1}{2}(e^x + e^{-x}) > 0 \qquad \text{for all real } x,$$

the hyperbolic sine increases everywhere. Since

$$\frac{d^2}{dx^2}(\sinh x) = \frac{d}{dx}(\cosh x) = \sinh x = \tfrac{1}{2}(e^x - e^{-x}),$$

you can see that

$$\frac{d^2}{dx^2}(\sinh x) \quad \text{is} \quad \begin{cases} \text{negative,} & \text{for} \quad x < 0 \\ 0, & \text{at} \quad x = 0 \\ \text{positive,} & \text{for} \quad x > 0 \end{cases}.$$

The graph is therefore concave down on $(-\infty, 0)$ and concave up on $(0, \infty)$. The point $(0, \sinh 0) = (0, 0)$ is the only point of inflection. The slope at the origin is $\cosh 0 = 1$. A sketch of the graph appears in Figure 7.11.1. ◻

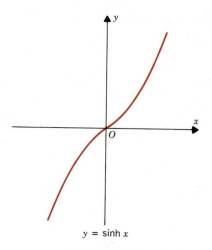

$y = \sinh x$

Figure 7.11.1

We turn now to the hyperbolic cosine. Since

$$\cosh(-x) = \tfrac{1}{2}(e^{-x} + e^{x}) = \tfrac{1}{2}(e^{x} + e^{-x}) = \cosh x,$$

the hyperbolic cosine is an even function. The graph is therefore symmetric about the y-axis. Since

$$\frac{d}{dx}(\cosh x) = \sinh x,$$

you can see that

$$\frac{d}{dx}(\cosh x) \quad \text{is} \quad \left\{ \begin{array}{rll} \text{negative,} & \text{for} & x < 0 \\ 0, & \text{at} & x = 0 \\ \text{positive,} & \text{for} & x > 0 \end{array} \right\}.$$

The function therefore decreases on $(-\infty, 0]$ and increases on $[0, \infty)$. The number

$$\cosh 0 = \tfrac{1}{2}(e^0 + e^{-0}) = \tfrac{1}{2}(1 + 1) = 1$$

is a local and absolute minimum. There are no other extreme values. Since

$$\frac{d^2}{dx^2}(\cosh x) = \frac{d}{dx}(\sinh x) = \cosh x > 0 \qquad \text{for all real } x,$$

the graph is everywhere concave up.   (Figure 7.11.2)   ☐

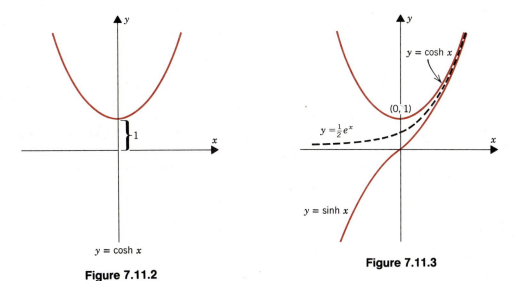

$y = \cosh x$

Figure 7.11.2

Figure 7.11.3

Figure 7.11.3 shows the graphs of three functions

$$y = \sinh x = \tfrac{1}{2}(e^x - e^{-x}), \qquad y = \tfrac{1}{2}e^x, \qquad y = \cosh x = \tfrac{1}{2}(e^x + e^{-x}).$$

For all real x

$$\sinh x < \tfrac{1}{2}e^x < \cosh x. \qquad\qquad (e^{-x} > 0)$$

Although markedly different for negative x, these functions are almost indistinguishable for large positive x. Reason: as $x \to \infty$, $e^{-x} \to 0$.

Identities

The hyperbolic sine and cosine satisfy identities similar to those satisfied by the "circular" sine and cosine.

(7.11.3)

$$\cosh^2 t - \sinh^2 t = 1,$$
$$\sinh(t + s) = \sinh t \cosh s + \cosh t \sinh s,$$
$$\cosh(t + s) = \cosh t \cosh s + \sinh t \sinh s,$$
$$\sinh 2t = 2 \sinh t \cosh t,$$
$$\cosh 2t = \cosh^2 t + \sinh^2 t.$$

The verification of these identities is left to you as an exercise.

Relation to the Hyperbola $x^2 - y^2 = 1$

The hyperbolic sine and cosine are related to the hyperbola $x^2 - y^2 = 1$ much as the "circular" sine and cosine are related to the circle $x^2 + y^2 = 1$:

1. For each real t

$$\cos^2 t + \sin^2 t = 1,$$

and thus the point $(\cos t, \sin t)$ lies on the circle $x^2 + y^2 = 1$. For each real t

$$\cosh^2 t - \sinh^2 t = 1,$$

and thus the point $(\cosh t, \sinh t)$ lies on the hyperbola $x^2 - y^2 = 1$.

2. For each t in $[0, 2\pi]$ (see Figure 7.11.4), the number $\frac{1}{2}t$ gives the area of the circular sector generated by the circular arc that begins at $(1, 0)$ and ends at $(\cos t, \sin t)$. As we prove below, for each $t > 0$ (see Figure 7.11.5), the number $\frac{1}{2}t$ gives the area of the hyperbolic sector generated by the hyperbolic arc that begins at $(1, 0)$ and ends at $(\cosh t, \sinh t)$.

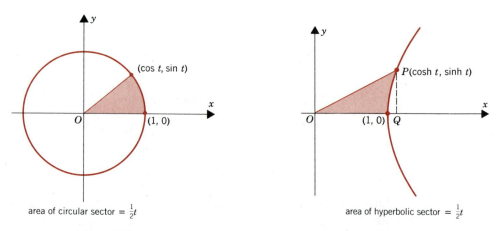

area of circular sector $= \frac{1}{2}t$

Figure 7.11.4

area of hyperbolic sector $= \frac{1}{2}t$

Figure 7.11.5

Proof. Let's call the area of the hyperbolic sector $A(t)$. It is not hard to see that

$$A(t) = \tfrac{1}{2} \cosh t \sinh t - \int_1^{\cosh t} \sqrt{x^2 - 1} \, dx.$$

The first term, $\tfrac{1}{2} \cosh t \sinh t$, gives the area of the triangle OPQ, and the integral

$$\int_1^{\cosh t} \sqrt{x^2 - 1} \, dx$$

gives the area of the unshaded portion of the triangle. We wish to show that

$$A(t) = \tfrac{1}{2}t \qquad \text{for all } t \geq 0.$$

We will do so by showing that

$$A'(t) = \tfrac{1}{2} \quad \text{for all } t > 0 \qquad \text{and} \qquad A(0) = 0.$$

Differentiating $A(t)$, we have

$$A'(t) = \tfrac{1}{2} \left[\cosh t \frac{d}{dt} (\sinh t) + \sinh t \frac{d}{dt} (\cosh t) \right] - \frac{d}{dt} \left(\int_1^{\cosh t} \sqrt{x^2 - 1} \, dx \right)$$

and therefore

(1) $$A'(t) = \tfrac{1}{2}(\cosh^2 t + \sinh^2 t) - \frac{d}{dt} \left(\int_1^{\cosh t} \sqrt{x^2 - 1} \, dx \right).$$

Now we differentiate the integral:

$$\frac{d}{dt} \left(\int_1^{\cosh t} \sqrt{x^2 - 1} \, dx \right) = \sqrt{\cosh^2 t - 1} \, \frac{d}{dt} (\cosh t) = \sinh t \cdot \sinh t = \sinh^2 t.$$

$$\underset{\displaystyle (5.10.7)}{\big\uparrow}$$

Substituting this last expression into equation (1), we have

$$A'(t) = \tfrac{1}{2}(\cosh^2 t + \sinh^2 t) - \sinh^2 t = \tfrac{1}{2}(\cosh^2 t - \sinh^2 t) = \tfrac{1}{2}.$$

It is not hard to see that $A(0) = 0$:

$$A(0) = \tfrac{1}{2} \cosh 0 \sinh 0 - \int_1^{\cosh 0} \sqrt{x^2 - 1} \, dx = 0. \quad \square$$

EXERCISES 7.11

Differentiate.

1. $y = \sinh x^2$.

2. $y = \cosh (x + a)$.

3. $y = \sqrt{\cosh ax}$.

4. $y = (\sinh ax)(\cosh ax)$.

5. $y = \dfrac{\sinh x}{\cosh x - 1}$.

6. $y = \dfrac{\sinh x}{x}$.

7. $y = a \sinh bx - b \cosh ax$.

8. $y = e^x (\cosh x + \sinh x)$.

9. $y = \ln |\sinh ax|$.

10. $y = \ln |1 - \cosh ax|$.

Verify the following identities.

11. $\cosh^2 t - \sinh^2 t = 1$. **12.** $\sinh(t+s) = \sinh t \cosh s + \cosh t \sinh s$.

13. $\cosh(t+s) = \cosh t \cosh s + \sinh t \sinh s$.

14. $\sinh 2t = 2 \sinh t \cosh t$. **15.** $\cosh 2t = \cosh^2 t + \sinh^2 t$.

Find the absolute extreme values.

16. $y = 5 \cosh x + 4 \sinh x$. **17.** $y = -5 \cosh x + 4 \sinh x$.

18. $y = 4 \cosh x + 5 \sinh x$.

19. Show that for each positive integer n

$$(\cosh x + \sinh x)^n = \cosh nx + \sinh nx.$$

20. Verify that $y = A \cosh cx + B \sinh cx$ satisfies the equation $y'' - c^2 y = 0$.

21. Determine A, B, and c so that $y = A \cosh cx + B \sinh cx$ satisfies the conditions $y'' - 9y = 0$, $y(0) = 2$, $y'(0) = 1$. Take $c > 0$.

22. Determine A, B, and c so that $y = A \cosh cx + B \sinh cx$ satisfies the conditions $4y'' - y = 0$, $y(0) = 1$, $y'(0) = 2$. Take $c > 0$.

Calculate the following indefinite integrals.

23. $\displaystyle\int \cosh ax \, dx$. **24.** $\displaystyle\int \sinh ax \, dx$. **25.** $\displaystyle\int \sinh^2 ax \cosh ax \, dx$.

26. $\displaystyle\int \sinh ax \cosh^2 ax \, dx$. **27.** $\displaystyle\int \frac{\sinh ax}{\cosh ax} \, dx$. **28.** $\displaystyle\int \frac{\cosh ax}{\sinh ax} \, dx$.

29. $\displaystyle\int \frac{\sinh ax}{\cosh^2 ax} \, dx$. **30.** $\displaystyle\int x \cosh x \, dx$. **31.** $\displaystyle\int x \sinh x \, dx$.

32. $\displaystyle\int \sinh^2 x \, dx$. **33.** $\displaystyle\int \cosh^2 x \, dx$. **34.** $\displaystyle\int x^2 \sinh x \, dx$.

Find the area of the region under the curve and determine the centroid.

35. $y = \cosh x$, $x \in [0, 1]$. **36.** $y = \sinh x$, $x \in [0, 1]$.

37. Find the volume generated by revolving the region between

$$y = \cosh x, \quad x \in [0, 1] \quad \text{and} \quad y = \sinh x, \quad x \in [0, 1]$$

(a) about the x-axis. (b) about the y-axis.

38. Let Ω be the region under the curve $y = \cosh x$, $x \in [0, 1]$. Find the centroid of the solid generated by revolving Ω (a) about the x-axis. (b) about the y-axis.

*7.12 THE OTHER HYPERBOLIC FUNCTIONS

The hyperbolic tangent is defined by setting

$$\tanh x = \frac{\sinh x}{\cosh x} = \frac{e^x - e^{-x}}{e^x + e^{-x}}.$$

There is also a *hyperbolic cotangent*, a *hyperbolic secant*, and a *hyperbolic cosecant*:

$$\coth x = \frac{\cosh x}{\sinh x}, \qquad \text{sech } x = \frac{1}{\cosh x}, \qquad \text{csch } x = \frac{1}{\sinh x}.$$

The derivatives are as follows:

(7.12.1)

$$\frac{d}{dx}(\tanh x) = \text{sech}^2 x, \qquad \frac{d}{dx}(\coth x) = -\text{csch}^2 x,$$

$$\frac{d}{dx}(\text{sech } x) = -\text{sech } x \tanh x, \qquad \frac{d}{dx}(\text{csch } x) = -\text{csch } x \coth x.$$

These formulas are easy to verify. For instance,

$$\frac{d}{dx}(\tanh x) = \frac{d}{dx}\left(\frac{\sinh x}{\cosh x}\right) = \frac{\cosh x \frac{d}{dx}(\sinh x) - \sinh x \frac{d}{dx}(\cosh x)}{\cosh^2 x}$$

$$= \frac{\cosh^2 x - \sinh^2 x}{\cosh^2 x} = \frac{1}{\cosh^2 x} = \text{sech}^2 x.$$

We leave it to you to verify the other formulas. ☐

Let's examine the hyperbolic tangent a little further. Since

$$\tanh(-x) = \frac{\sinh(-x)}{\cosh(-x)} = \frac{-\sinh x}{\cosh x} = -\tanh x,$$

the hyperbolic tangent is an odd function and thus the graph is symmetric about the origin. Since

$$\frac{d}{dx}(\tanh x) = \text{sech}^2 x > 0 \qquad \text{for all real } x,$$

the function is everywhere increasing. From the relation

$$\tanh x = \frac{e^x - e^{-x}}{e^x + e^{-x}} = 1 - \frac{2}{e^{2x} + 1}$$

you can see that $\tanh x$ always remains between -1 and 1. Moreover,

$$\text{as } x \to \infty, \quad \tanh x \to 1 \qquad \text{and} \qquad \text{as } x \to -\infty, \quad \tanh x \to -1.$$

The lines $y = 1$ and $y = -1$ are horizontal asymptotes. To check on the concavity of the graph, we take the second derivative:

$$\frac{d^2}{dx^2}(\tanh x) = \frac{d}{dx}(\text{sech}^2 x) = 2 \text{ sech } x \frac{d}{dx}(\text{sech } x)$$

$$= 2 \text{ sech } x (-\text{sech } x \tanh x)$$

$$= -2 \text{ sech}^2 x \tanh x.$$

Since

$$\tanh x = \frac{e^x - e^{-x}}{e^x + e^{-x}} \quad \text{is} \quad \begin{cases} \text{negative,} & \text{for} \quad x < 0 \\ 0, & \text{at} \quad x = 0 \\ \text{positive,} & \text{for} \quad x > 0 \end{cases},$$

you can see that

$$\frac{d^2}{dx^2}(\tanh x) \quad \text{is} \quad \begin{cases} \text{positive,} & \text{for} \quad x < 0 \\ 0, & \text{at} \quad x = 0 \\ \text{negative,} & \text{for} \quad x > 0 \end{cases}.$$

The graph is therefore concave up on $(-\infty, 0)$ and concave down on $(0, \infty)$. The point $(0, \tanh 0) = (0, 0)$ is a point of inflection. At the origin the slope is

$$\text{sech}^2\, 0 = \frac{1}{\cosh^2 0} = 1.$$

For a picture of the graph see Figure 7.12.1.

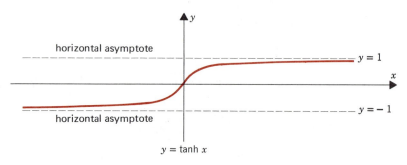

Figure 7.12.1

The Hyperbolic Inverses

The hyperbolic inverses that are important to us are the *inverse hyperbolic sine*, the *inverse hyperbolic cosine*, and the *inverse hyperbolic tangent*. These functions

$$y = \sinh^{-1} x, \qquad y = \cosh^{-1} x, \qquad y = \tanh^{-1} x$$

are the inverses of

$$y = \sinh x, \qquad y = \cosh x \quad (x \geq 0), \qquad y = \tanh x,$$

respectively.

THEOREM 7.12.2

(i) $\sinh^{-1} x = \ln (x + \sqrt{x^2 + 1})$, $\qquad\qquad$ x real

(ii) $\cosh^{-1} x = \ln (x + \sqrt{x^2 - 1})$, $\qquad\qquad$ $x \geq 1$

(iii) $\tanh^{-1} x = \frac{1}{2} \ln \left(\frac{1 + x}{1 - x} \right)$, $\qquad\qquad$ $-1 < x < 1$.

Proof. To prove (i), we set $y = \sinh^{-1} x$ and note that

$$\sinh y = x.$$

This gives in sequence

$$\tfrac{1}{2}(e^y - e^{-y}) = x, \qquad e^y - e^{-y} = 2x, \qquad e^{2y} - 2xe^y - 1 = 0.$$

This last equation is a quadratic equation in e^y. From the general quadratic formula we find that

$$e^y = \tfrac{1}{2}(2x \pm \sqrt{4x^2 + 4}) = x \pm \sqrt{x^2 + 1}.$$

Since $e^y > 0$, the minus sign on the right is impossible. Consequently, we have

$$e^y = x + \sqrt{x^2 + 1}$$

and, taking logs,

$$y = \ln (x + \sqrt{x^2 + 1}).$$

To prove (ii), we set

$$y = \cosh^{-1} x, \qquad x \geq 1$$

and note that

$$\cosh y = x \quad \text{and} \quad y \geq 0.$$

This gives in sequence

$$\tfrac{1}{2}(e^y + e^{-y}) = x, \qquad e^y + e^{-y} = 2x, \qquad e^{2y} - 2xe^y + 1 = 0.$$

Again we have a quadratic in e^y. Here the general quadratic formula gives

$$e^y = \tfrac{1}{2}(2x \pm \sqrt{4x^2 - 4}) = x \pm \sqrt{x^2 - 1}.$$

Since y is nonnegative,

$$e^y = x \pm \sqrt{x^2 - 1}$$

cannot be less than 1. This renders the negative sign impossible (check this out) and leaves

$$e^y = x + \sqrt{x^2 - 1}$$

as the only possibility. Taking logs, we get

$$y = \ln (x + \sqrt{x^2 - 1}).$$

The proof of (iii) is left as an exercise. ∎

EXERCISES *7.12

Differentiate.

1. $y = \tanh^2 x$.
2. $y = \tanh^2 3x$.
3. $y = \ln (\tanh x)$.
4. $y = \tanh (\ln x)$.
5. $y = \sinh (\tan^{-1} e^{2x})$.
6. $y = \operatorname{sech} (3x^2 + 1)$.
7. $y = \coth (\sqrt{x^2 + 1})$.
8. $y = \ln (\operatorname{sech} x)$.
9. $y = \dfrac{\operatorname{sech} x}{1 + \cosh x}$.
10. $y = \dfrac{\cosh x}{1 + \operatorname{sech} x}$.

Verify the following differentiation formulas.

11. $\dfrac{d}{dx}(\coth x) = -\operatorname{csch}^2 x.$

12. $\dfrac{d}{dx}(\operatorname{sech} x) = -\operatorname{sech} x \tanh x.$

13. $\dfrac{d}{dx}(\operatorname{csch} x) = -\operatorname{csch} x \coth x.$

14. Show that

$$\tanh (t + s) = \frac{\tanh t + \tanh s}{1 + \tanh t \tanh s}.$$

15. Given that $\tanh x_0 = \frac{4}{5}$, find (a) $\operatorname{sech} x_0$.   HINT: $1 - \tanh^2 x = \operatorname{sech}^2 x$.   Then find (b) $\cosh x_0$.   (c) $\sinh x_0$.   (d) $\coth x_0$.   (e) $\operatorname{csch} x_0$.

16. Given that $\tanh t_0 = -\frac{5}{12}$, evaluate the remaining hyperbolic functions at t_0.

17. Show that, if $x^2 \geq 1$, then $x - \sqrt{x^2 - 1} \leq 1$.

18. Show that

$$\tanh^{-1} x = \frac{1}{2} \ln \left(\frac{1 + x}{1 - x}\right), \qquad -1 < x < 1.$$

19. Show that

(7.12.3)
$$\frac{d}{dx}(\sinh^{-1} x) = \frac{1}{\sqrt{x^2 + 1}}, \qquad x \text{ real.}$$

20. Show that

(7.12.4)
$$\frac{d}{dx}(\cosh^{-1} x) = \frac{1}{\sqrt{x^2 - 1}}, \qquad x > 1.$$

21. Show that

(7.12.5)
$$\frac{d}{dx}(\tanh^{-1} x) = \frac{1}{1 - x^2}, \qquad -1 < x < 1.$$

22. Sketch the graphs of (a) $y = \sinh^{-1} x$.   (b) $y = \cosh^{-1} x$.   (c) $y = \tanh^{-1} x$.

23. Given that $\tan \phi = \sinh x$, show that

(a) $\dfrac{d\phi}{dx} = \operatorname{sech} x.$ (b) $x = \ln (\sec \phi + \tan \phi).$ (c) $\dfrac{dx}{d\phi} = \sec \phi.$

7.13   ADDITIONAL EXERCISES

1. Find the minimum value of $y = ae^{kx} + be^{-kx}$, $a > 0$, $b > 0$.

2. Show that, if $y = \frac{1}{2}a(e^{x/a} + e^{-x/a})$, then $y'' = y/a^2$.

3. Find the points of inflection of the graph of $y = e^{-x^2}$.

4. A rectangle has one side on the x-axis and the two upper vertices on the graph of $y = e^{-x^2}$. Where should the upper vertices be placed so as to maximize the area of the rectangle?

5. A telegraph cable consists of a core of copper wires with a covering made of nonconducting material. If x is the ratio of the radius of the core to the thickness of the covering, the speed of signaling is proportional to the product $x^2 \ln (1/x)$. For what value of x is the speed a maximum?

6. A rectangle has two sides along the coordinate axes and one vertex at a point P that moves along the curve $y = e^x$ in such a way that y increases at the rate of $\frac{1}{2}$ unit per minute. How fast is the area of the rectangle increasing when $y = 3$?

7. Find

$$\lim_{h \to 0} \frac{1}{h} (e^h - 1).$$

8. Find the extreme values and points of inflection of $y = x/\ln x$.

Find the area below the graph.

9. $y = \dfrac{x}{x^2 + 1}, \quad x \in [0, 1].$

10. $y = \dfrac{1}{x^2 + 1}, \quad x \in [0, 1].$

11. $y = \dfrac{1}{\sqrt{1 - x^2}}, \quad x \in [0, \frac{1}{2}].$

12. $y = \dfrac{x}{\sqrt{1 - x^2}}, \quad x \in [0, \frac{1}{2}].$

A particle moves along a coordinate line so that at time t it has position $x(t)$, velocity $v(t)$, and acceleration $a(t)$.

13. Find $x(1)$ given that $x(0) = 0$ and $v(t) = 1/(t^2 + 1)$.

14. Find $a(1)$ given that $v(t) = 1/(t^2 + 1)$.

15. Find $a(0)$ given that $x(t) = \tan^{-1} (1 + t)$.

16. Find $v(0)$ given that $v(1) = 1$ and $a(t) = (\tan^{-1} t)/(t^2 + 1)$.

17. Show that the average slope of the logarithm curve from $x = a$ to $x = b$ is

$$\frac{1}{b - a} \ln \frac{b}{a}.$$

18. Find the average value of $\sin^2 x$ from $x = 0$ to $x = \pi$. (This average is used in the theory of alternating currents.)

19. Find the area of the region bounded by the curve $xy = a^2$, the x-axis, and the vertical lines $x = a, x = 2a$.

20. What are the averages of the sine and cosine on an interval of length 2π?

21. A ball is thrown from ground level. Take the origin as the point of release, take the x-axis as ground level, measure distances in feet, and take the path of the ball as given by the equation $y = x - \frac{1}{100}x^2$. (a) At what angle is the ball released? (b) At what angle will it strike a wall 75 feet away? (c) At what angle will it fall on a horizontal roof 16 feet high?

22. An object moves with acceleration $a(t) = g - kv(t)$, initial velocity $v = 0$, and initial position $x = 0$. Show that the following relations hold.

(a) $v(t) = \dfrac{g}{k} (1 - e^{-kt}).$ (b) $x(t) = \dfrac{g}{k^2} (kt + e^{-kt} - 1).$

(c) $kx(t) + v(t) + \dfrac{g}{k} \ln \left(1 - \dfrac{kv(t)}{g} \right) = 0.$

23. Let Ω be the region below the graph of $f(x) = (1 + x^2)^{-1/2}, \quad x \in [0, \sqrt{3}]$. Find the volume generated by revolving Ω about (a) the x-axis. (b) the y-axis.

24. Let Ω be the region below the graph of $f(x) = (1 - x^2)^{-1/4}$, $x \in [0, \frac{1}{2}]$.   Find the volume generated by revolving Ω about (a) the x-axis.   (b) the y-axis.

25. Find the centroid of the region under the curve $y = \sec \frac{1}{2} \pi x$, $x \in [-\frac{1}{2}, \frac{1}{2}]$.

26. Given that $|a| < 1$, find the value of b for which

$$\int_0^1 \frac{b}{\sqrt{1 - b^2 x^2}} \, dx = \int_0^a \frac{dx}{\sqrt{1 - x^2}}.$$

27. Show that for all real numbers a,

$$\int_0^1 \frac{a}{1 + a^2 x^2} \, dx = \int_0^a \frac{dx}{1 + x^2}.$$

28. Find the centroid of the region under the curve $y = (1 - x^2)^{-1/2}$, $x \in [0, \frac{1}{2}]$.

29. Show that for all positive integers $m < n$

$$\ln \frac{n+1}{m} < \frac{1}{m} + \frac{1}{m+1} + \cdots + \frac{1}{n} < \ln \frac{n}{m-1}.$$

HINT:

$$\frac{1}{k+1} < \int_k^{k+1} \frac{dx}{x} < \frac{1}{k}.$$

30. Sketch a figure displaying the three graphs
(a) $y = e^{x/4}$; $y = e^{-x/4}$; $y = e^{-x/4} \sin \frac{1}{2} \pi x$, $x \geq 0$.
(b) $y = e^{x/4}$; $y = e^{-x/4}$; $y = e^{-x/4} \cos \frac{1}{2} \pi x$, $x \geq 0$.

31. Let $p(x)$ be a polynomial of degree n and set $P(x) = p(x) + p'(x) + \cdots + p^{(n)}(x)$.   Show that

$$\frac{d}{dx} [e^{-x} P(x)] = - e^{-x} p(x).$$

32. Verify the following formulas:

(a) $\quad \int x \sin^{-1} x \, dx = \frac{1}{4}[(2x^2 - 1) \sin^{-1} x + x\sqrt{1 - x^2}] + C.$

(b) $\quad \int (\sin^{-1} x)^2 \, dx = x(\sin^{-1} x)^2 + 2(\sin^{-1} x)\sqrt{1 - x^2} - 2x + C.$

33. Find the centroid of the region under the curve $y = \sin^{-1} x$, $x \in [0, 1]$.

34. Let Ω be the region under the curve $y = e^{-x}$, $x \in [0, 1]$.   Find the centroid of the solid generated by revolving Ω (a) about the x-axis.   (b) about the y-axis.

35. Let Ω be the region under the graph of the logarithm function from $x = 1$ to $x = e$.   (a) Find the area of Ω.   (b) Find the centroid of Ω.   (c) Find the volume generated by revolving Ω about each of the coordinate axes.   (d) Find the distance of the centroid of Ω from the line $y = x$.   (e) Find the volume generated by revolving Ω about the line $y = x$.

36. Let Ω be the region in Exercise 35.   Find the centroid of the solid generated by revolving Ω (a) about the x-axis.   (b) about the y-axis.

Revenue Streams

In Section 7.6 we talked about continuous compounding and you saw that the present value of A dollars t years from now is given by the formula

$$P.V. = Ae^{-rt}$$

where r is the annual rate of continuous compounding.

Suppose now that revenue flows continually at a constant rate of R dollars per year for n years. What is the present value of such a revenue stream? It's not hard to show that the present value of such a revenue stream is given by the formula

(7.13.1)
$$P.V. = \int_0^n Re^{-rt}\,dt.$$

Let S be a constant revenue stream of 1000 dollars per year.

37. (Calculator) What is the present value of the first 4 years of revenue? Assume continuous compounding (a) at 4%. (b) at 8%.

38. (Calculator) What is the present value of the fifth year of revenue with continuous compounding (a) at 4%? (b) at 8%?

Revenues usually do not flow at a constant rate R but rather at a time-dependent rate $R(t)$. In general, $R(t)$ tends to increase when business is good and tends to decrease when business is poor. "Growth companies" are so dubbed because of their continually increasing $R(t)$. The "cyclical companies" owe their appellation to a fluctuating $R(t)$; $R(t)$ increases for awhile, then decreases for awhile, and then the cycle is supposedly repeated.

If we postulate a varying $R(t)$, then the present value of an n-year stream is given by the formula

(7.13.2)
$$P.V. = \int_0^n R(t)e^{-rt}\,dt.$$

Here, as before, r is the rate of continuous compounding used to discount future dollars.

Let S be a revenue stream with $R(t) = 1000 + 60t$ dollars per year.

39. (Calculator) What is the present value of the first 2 years of revenue? Assume continuous compounding (a) at 5%. (b) at 10%.

40. (Calculator) What is the present value of the third year of revenue with continuous compounding (a) at 5%? (b) at 10%?

Let S be a revenue stream with $R(t) = 1000 + 80t$ dollars per year.

41. (Calculator) What is the present value of the fourth year of revenue with continuous compounding (a) at 6%? (b) at 8%?

42. (Calculator) What is the present value of the first 4 years of revenue? Assume continuous compounding (a) at 6%. (b) at 8%.

43. Let Ω be the region under the curve $y = \sinh x$, $x \in [0, 1]$. Find the volume of the solid generated by revolving Ω (a) about the x-axis. (b) about the y-axis.

44. Locate the centroid of the solids in Exercise 43.

Refraction

(Dip a straight stick in a pool of water and it appears to bend.)

Only in a vacuum does light travel at speed c.   (the famous c of $E = mc^2$)   Light does not travel as fast through a material medium.   The *index of refraction n* of a medium relates the speed of light in that medium to c:

(7.13.3)

$$n = \frac{c}{\text{speed of light in medium}}.$$

When light travels from one medium to another, it changes direction.   We say that the light is *refracted.*   Experiment shows that *the angle of refraction θ_r* is related to *the angle of incidence θ_i* by Snell's law:

(7.13.4) $$n_i \sin \theta_i = n_r \sin \theta_r. \quad †$$ (Figure 7.13.1)

Figure 7.13.1

45. A light beam passes from a medium with index of refraction n_1 through a plane sheet of material whose top and bottom faces are parallel and then out into some other medium with index of refraction n_2.   (Figure 7.13.2)

Figure 7.13.2

---

† Like the Law of Reflection, Snell's Law of Refraction can be derived from Fermat's Principle of Least Time.   (Exercise 26, Section 16.8)

(a) Show that Snell's law implies that $n_1 \sin \theta_1 = n_2 \sin \theta_2$ regardless of the thickness of the sheet or its index of refraction.

(b) (*A star is not where it appears to be.*) Consider a beam of light traveling through an atmosphere whose index of refraction varies with height, $n = n(y)$. The light follows some curved path $y = y(x)$. Think of the atmosphere as a succession of parallel slabs. When a light ray strikes a slab at height y, it is traveling at some angle θ to the vertical; when it emerges, at height $y = \Delta y$, it is traveling at a slightly different angle, $\theta + \Delta\theta$. Using the result in part (a), show that

$$\frac{1}{n}\frac{dn}{dy} = -\cot\theta\,\frac{d\theta}{dy} = \frac{d^2y/dx^2}{1 + (dy/dx)^2}.$$

[Note that θ is the angle that the tangent to the curve makes with the *vertical*. Therefore $\cot\theta = dy/dx$.]

(c) Verify that the slope of the light path must vary in such a way that

$$1 + (dy/dx)^2 = (\text{constant})[n(y)]^2.$$

(d) How must n vary with height y for the light to travel along a circular arc?

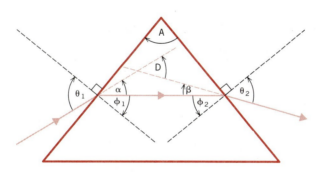

Figure 7.13.3

46. A beam of light passes through a glass prism with index of refraction n. (See Figure 7.13.3.) Setting the index of refraction of air equal to 1 (it is very close to 1), Snell's law gives

$$\sin \theta_1 = n \sin \phi_1 \quad \text{and} \quad \sin \theta_2 = n \sin \phi_2.$$

(a) The difference in directions of the incoming and outgoing rays, marked D in the figure, is called the *angle of deviation*. Show from the geometry of the figure that

$$D = (\theta_1 - \phi_1) + (\theta_2 - \phi_2) \quad \text{and} \quad \phi_1 + \phi_2 = A.$$

Then use these relations together with Snell's law to derive the following result:

$$D = \sin^{-1}[n \sin \phi_1] + \sin^{-1}[n \sin(A - \phi_1)] - A.$$

(b) Show that minimum deviation occurs when θ_1 is such that

$$\sin \theta_1 = n \sin \frac{A}{2} = \sin\left(\frac{D + A}{2}\right).$$

(Since angles A and D can be measured to very high precision, this result provides a simple and accurate way to measure the index of refraction of many materials.)

CHAPTER HIGHLIGHTS

7.1 In Search of a Notion of Logarithm

definition of a logarithm function (p. 340)

7.2 The Logarithm Function, Part I

natural logarithm: $\ln x = \displaystyle\int_1^x \frac{dt}{t}$, $x > 0$;   domain $(0, \infty)$, range $(-\infty, \infty)$

$\ln 1 = 0$, $\ln e = 1$

for $x, y > 0$ and r rational

$$\ln xy = \ln x + \ln y, \quad \ln \frac{1}{x} = -\ln x, \quad \ln \frac{x}{y} = \ln x - \ln y, \quad \ln x^r = r \ln x.$$

graph of $y = \ln x$ (p. 345)

7.3 The Logarithm Function, Part II

$$\frac{d}{dx}(\ln |u|) = \frac{1}{u}\frac{du}{dx} \qquad\qquad \int \frac{g'(x)}{g(x)}\,dx = \ln |g(x)| + C$$

logarithmic differentiation (p. 352)

7.4 The Exponential Function

The exponential function $y = e^x$ is the inverse of the logarithm function $y = \ln x$.

graph of $y = e^x$ (p. 356);   domain $(-\infty, \infty)$, range $(0, \infty)$

$$\ln (e^x) = x, \qquad e^{\ln x} = x, \qquad e^0 = 1, \qquad e^{-x} = \frac{1}{e^x}, \qquad e^{x+y} = e^x e^y, \qquad e^{x-y} = \frac{e^x}{e^y}$$

$$\frac{d}{dx}(e^u) = e^u \frac{du}{dx} \qquad\qquad \int e^{g(x)}g'(x)\,dx = e^{g(x)} + C$$

7.5 Arbitrary Powers; Other Bases; Estimating e

$x^r = e^{r \ln x}$ for all $x > 0$, all real r

$$\frac{d}{dx}(p^u) = p^u \ln p \frac{du}{dx} \quad (p \text{ a positive constant}) \qquad\qquad \log_p x = \frac{\ln x}{\ln p}$$

$$\frac{d}{dx}(\log_p u) = \frac{1}{u \ln p}\frac{du}{dx}$$

$$\left(1 + \frac{1}{n}\right)^n \le e \le \left(1 + \frac{1}{n}\right)^{n+1} \qquad e \cong 2.71828$$

7.6 Exponential Growth and Decline; Compound Interest

Linear functions of time change by the same *amount* during all periods of the same duration; exponential functions of time change by the same *factor* during all periods of the same duration.

All the functions that satisfy the equation $f'(t) = k f(t)$ are of the form $f(t) = Ce^{kt}$.

population growth (p. 375)
radioactive decay, half-life (p. 376)
compound interest, continuous compounding (p. 377)
power of exponential growth (p. 382)
weakness of logarithmic growth (p. 383)

7.7 Integration by Parts

$$\int u\, dv = uv - \int v\, du \qquad \int \ln x\, dx = x \ln x - x + C$$

Success with the technique depends on choosing u and dv so that $\int v\, du$ is easier to integrate than $\int u\, dv$.

Integration by parts is often used to calculate integrals when the integrand is a mixture of function types; for example, polynomials and exponentials, or polynomials and trigonometric functions.

To calculate some integrals you may have to integrate by parts more than once.

7.8 Simple Harmonic Motion

Every solution of the equation $x''(t) + \omega^2 x(t) = 0$ can be written

$$x(t) = A \sin (\omega t + \phi_0) \qquad \text{with } A \geq 0, \ \phi_0 \in [0, 2\pi).$$

simple harmonic motion (p. 391)
period, frequency, angular frequency (p. 393)
amplitude (p. 393)
phase constant (p. 393)

Other ways to write simple harmonic motion:

$$x(t) = A \cos (\omega t + \phi_1), \qquad x(t) = A \sin \omega t + B \cos \omega t.$$

7.9 More on the Integration of the Trigonometric Functions

$$\int \tan x\, dx = \ln |\sec x| + C, \qquad \int \sec x\, dx = \ln |\sec x + \tan x| + C,$$

$$\int \cot x\, dx = \ln |\sin x| + C, \qquad \int \csc x\, dx = \ln |\csc x - \cot x| + C.$$

7.10 The Inverse Trigonometric Functions

the inverse sine, $y = \sin^{-1} x$, is the inverse of $y = \sin x$, $x \in [-\frac{1}{2}\pi, \frac{1}{2}\pi]$
the inverse tangent, $y = \tan^{-1} x$, is the inverse of $y = \tan x$, $x \in (-\frac{1}{2}\pi, \frac{1}{2}\pi)$

graph of $y = \sin^{-1} x$ (p. 402)   graph of $y = \tan^{-1} x$ (p. 406)

$$\frac{d}{dx}(\sin^{-1} x) = \frac{1}{\sqrt{1 - x^2}} \qquad \int \frac{dx}{\sqrt{a^2 - x^2}} = \sin^{-1}\left(\frac{x}{a}\right) + C \quad (a > 0)$$

$$\frac{d}{dx}(\tan^{-1} x) = \frac{1}{1 + x^2} \qquad \int \frac{dx}{a^2 + x^2} = \frac{1}{a}\tan^{-1}\left(\frac{x}{a}\right) + C \quad (a \neq 0)$$

$\int \sin^{-1} x \, dx$ and $\int \tan^{-1} x \, dx$ can be calculated by integration by parts.

definition of the remaining inverse trigonometric functions (p. 408)

7.11 The Hyperbolic Sine and Cosine

$$\sinh x = \tfrac{1}{2}(e^x - e^{-x}), \qquad \cosh x = \tfrac{1}{2}(e^x + e^{-x}),$$

$$\frac{d}{dx}(\sinh x) = \cosh x, \qquad \frac{d}{dx}(\cosh x) = \sinh x.$$

graphs (pp. 412–413)   basic identities (p. 414)

*7.12 The Other Hyperbolic Functions

$$\tanh x = \frac{\sinh x}{\cosh x}, \qquad \coth x = \frac{\cosh x}{\sinh x},$$

$$\operatorname{sech} x = \frac{1}{\cosh x}, \qquad \operatorname{csch} x = \frac{1}{\sinh x}.$$

derivatives (p. 417)   hyperbolic inverses (p. 418)
derivatives of hyperbolic inverses (p. 420)

7.13 Additional Exercises

revenue streams (p. 423)

CHAPTER

8

TECHNIQUES OF INTEGRATION

8.1   REVIEW

We begin by listing the more important integrals with which you are already familiar. A more extended table of integrals appears on the inside covers of the book.

1. $\displaystyle\int \alpha\, dx = \alpha x + C.$

2. $\displaystyle\int x^n\, dx = \frac{1}{n+1}x^{n+1} + C, \qquad n \neq -1.$

3. $\displaystyle\int \frac{dx}{x} = \ln|x| + C.$

4. $\displaystyle\int e^x\, dx = e^x + C.$

5. $\displaystyle\int p^x\, dx = \frac{p^x}{\ln p} + C.$

6. $\displaystyle\int \ln x\, dx = x \ln x - x + C.$

7. $\displaystyle\int xe^x\, dx = xe^x - e^x + C.$

8. $\displaystyle\int x \ln x\, dx = \tfrac{1}{2}x^2 \ln x - \tfrac{1}{4}x^2 + C.$

9. $\displaystyle\int \sin x\, dx = -\cos x + C.$

10. $\displaystyle\int \cos x\, dx = \sin x + C.$

11. $\displaystyle\int x \sin x\, dx = -x \cos x + \sin x + C.$

12. $\displaystyle\int x \cos x\, dx = x \sin x + \cos x + C.$

13. $\displaystyle\int \sin^2 x\, dx = \tfrac{1}{2}x - \tfrac{1}{4}\sin 2x + C.$

14. $\displaystyle\int \cos^2 x\, dx = \tfrac{1}{2}x + \tfrac{1}{4}\sin 2x + C.$

15. $\displaystyle\int \tan x\, dx = \ln |\sec x| + C.$

16. $\displaystyle\int \cot x\, dx = \ln |\sin x| + C.$

17. $\displaystyle\int \sec x\, dx = \ln |\sec x + \tan x| + C.$

18. $\displaystyle\int \csc x\, dx = \ln |\csc x - \cot x| + C.$

19. $\displaystyle\int \sec x \tan x\, dx = \sec x + C.$

20. $\displaystyle\int \csc x \cot x\, dx = -\csc x + C.$

21. $\displaystyle\int \sec^2 x\, dx = \tan x + C.$

22. $\displaystyle\int \csc^2 x\, dx = -\cot x + C.$

23. $\displaystyle\int \frac{dx}{\sqrt{1 - x^2}} = \sin^{-1} x + C.$

24. $\displaystyle\int \frac{dx}{1 + x^2} = \tan^{-1} x + C.$

25. $\displaystyle\int \sin^{-1} x\, dx = x \sin^{-1} x + \sqrt{1 - x^2} + C.$

26. $\displaystyle\int \tan^{-1} x\, dx = x \tan^{-1} x - \tfrac{1}{2} \ln (1 + x^2)$

27. $\displaystyle\int \frac{dx}{|x|\sqrt{x^2 - 1}} = \sec^{-1} x + C.$

28. $\displaystyle\int \sinh x\, dx = \cosh x + C.$

29. $\displaystyle\int \cosh x\, dx = \sinh x + C.$

30. $\displaystyle\int \sinh^2 x\, dx = \tfrac{1}{2}[\sinh x \cosh x - x] + C.$

31. $\displaystyle\int \cosh^2 x\, dx = \tfrac{1}{2}[\sinh x \cosh x + x] + C.$

32. $\displaystyle\int x \sinh x\, dx = x \cosh x - \sinh x + C.$

33. $\displaystyle\int x \cosh x\, dx = x \sinh x - \cosh x + C.$

In the spirit of review we work out a few integrals.

Problem 1. Find

$$\int x \tan x^2\, dx.$$

Solution. Set

$$u = x^2, \qquad du = 2x\, dx.$$

Then

$$\int x \tan x^2\, dx = \tfrac{1}{2}\int \tan u\, du = \tfrac{1}{2} \ln |\sec u| + C = \tfrac{1}{2} \ln |\sec x^2| + C. \quad \square$$

Problem 2. Compute

$$\int_0^1 \frac{e^x}{e^x + 2}\, dx.$$

Solution. Set

$$u = e^x + 2, \qquad du = e^x\, dx.$$

At $x = 0$, $u = 3$; at $x = 1$, $u = e + 2$. Thus

$$\int_0^1 \frac{e^x}{e^x + 2}\, dx = \int_3^{e+2} \frac{du}{u} = \left[\ln |u|\right]_3^{e+2}$$
$$= \ln (e + 2) - \ln 3 = \ln \left[\tfrac{1}{3}(e + 2)\right] \cong 0.45. \quad \square$$

Problem 3. Find

$$\int x \sec^2 x\, dx.$$

Solution. We integrate by parts. Setting

$$u = x, \qquad dv = \sec^2 x$$

we have

$$du = dx, \qquad v = \tan x.$$

Therefore

$$\int x \sec^2 x\, dx = x \tan x - \int \tan x\, dx = x \tan x - \ln |\sec x| + C. \quad \square$$

The final problem involves a little algebra.

Problem 4. Find

$$\int \frac{dx}{x^2 + 2x + 5}.$$

Solution. First we complete the square in the denominator:

$$\int \frac{dx}{x^2 + 2x + 5} = \int \frac{dx}{(x^2 + 2x + 1) + 4} = \int \frac{dx}{(x + 1)^2 + 2^2} = (*).$$

We know that

$$\int \frac{du}{u^2 + 1} = \tan^{-1} u + C.$$

Setting

$$2u = x + 1, \qquad 2du = dx,$$

we have

$$(*) = \int \frac{2du}{4u^2 + 4} = \tfrac{1}{2} \int \frac{du}{u^2 + 1} = \tfrac{1}{2} \tan^{-1} u + C = \tfrac{1}{2} \tan^{-1} \left[\tfrac{1}{2}(x + 1)\right] + C. \quad \square$$

EXERCISES 8.1

Work out the following integrals.

1. $\int e^{2-x}\, dx.$

2. $\int \cos \tfrac{2}{3}x\, dx.$

3. $\int_0^1 \sin \pi x\, dx.$

4. $\int_0^t \sec \pi x \tan \pi x\, dx.$

5. $\int \sec^2 (1 - x)\, dx.$

6. $\int \dfrac{dx}{5^x}.$

7. $\int_{\pi/6}^{\pi/3} \cot x\, dx.$

8. $\int_0^1 \dfrac{x^3}{1 + x^4}\, dx.$

9. $\int \dfrac{x}{\sqrt{1 - x^2}}\, dx.$

10. $\int_{-\pi/4}^{\pi/4} \dfrac{dx}{\cos^2 x}.$

11. $\int_{-\pi/4}^{\pi/4} \dfrac{\sin x}{\cos^2 x}\, dx.$

12. $\int \dfrac{e^{\sqrt{x}}}{\sqrt{x}}\, dx.$

13. $\int_1^2 \dfrac{e^{1/x}}{x^2}\, dx.$

14. $\int_{\pi/6}^{\pi/3} \csc x\, dx.$

15. $\int \dfrac{x}{x^2 + 1}\, dx.$

16. $\int \dfrac{x^3}{\sqrt{1 - x^4}}\, dx.$

17. $\int_0^c \dfrac{dx}{x^2 + c^2}.$

18. $\int a^x e^x\, dx.$

19. $\int \dfrac{\sec^2 \theta}{\sqrt{3} \tan \theta + 1}\, d\theta.$

20. $\int \dfrac{\sin \phi}{3 - 2 \cos \phi}\, d\phi.$

21. $\int \dfrac{e^x}{ae^x - b}\, dx.$

22. $\int_0^{\pi/4} \dfrac{\sec^2 x \tan x}{\sqrt{2 + \sec^2 x}}\, dx.$

23. $\int \dfrac{1 + \cos 2x}{\sin^2 2x}\, dx.$

24. $\int \dfrac{dx}{x^2 - 4x + 13}.$

25. $\int \dfrac{x}{(x + 1)^2 + 4}\, dx.$

26. $\int \dfrac{\ln x}{x}\, dx.$

27. $\int \dfrac{x}{\sqrt{1 - x^4}}\, dx.$

28. $\int \dfrac{e^x}{1 + e^{2x}}\, dx.$

29. $\int \dfrac{dx}{x^2 + 6x + 10}.$

30. $\int e^x \tan e^x\, dx.$

31. $\int x \sin x^2\, dx.$

32. $\int \dfrac{x + 1}{\sqrt{1 - x^2}}\, dx.$

33. $\int \tan^2 x\, dx.$

34. $\int x \sin 2x\, dx.$

35. $\int x^3 e^{-x^2}\, dx.$

36. $\int e^{2x} \sin 3x\, dx.$

8.2   PARTIAL FRACTIONS

A rational function is by definition the quotient of two polynomials. Thus, for example,

$$\frac{1}{x^2 - 4}, \quad \frac{2x^2 + 3}{x(x - 1)^2}, \quad \frac{-2x}{(x + 1)(x^2 + 1)}, \quad \frac{1}{x(x^2 + x + 1)},$$

$$\frac{3x^4 + x^3 + 20x^2 + 3x + 31}{(x + 1)(x^2 + 4)^2}, \quad \frac{x^5}{x^2 - 1}$$

are all rational functions, but

$$\frac{1}{\sqrt{x}}, \quad \ln x, \quad \frac{|x - 2|}{x^2}$$

are not rational functions.

To integrate a rational function it is usually necessary to first rewrite it as a polynomial (which may be identically 0) plus fractions of the form

(8.2.1)
$$\frac{A}{(x - \alpha)^k} \quad \text{and} \quad \frac{Bx + C}{(x^2 + \beta x + \gamma)^k}$$

with the quadratic $x^2 + \beta x + \gamma$ irreducible (i.e., not factorable into linear terms with real coefficients).† Such fractions are called *partial fractions*.

It is shown in algebra that every rational function can be written in such a way. Here are some examples.

Example 1. (*The denominator splits into distinct linear factors.*) For

$$\frac{1}{x^2 - 4} = \frac{1}{(x - 2)(x + 2)},$$

we write

$$\frac{1}{x^2 - 4} = \frac{A}{x - 2} + \frac{B}{x + 2}.$$

Clearing fractions, we have

(1)
$$1 = A(x + 2) + B(x - 2).$$

To find A and B we substitute numbers for x:

setting $x = 2$, we get $1 = 4A$, which gives $A = \frac{1}{4}$;

setting $x = -2$, we get $1 = -4B$, which gives $B = -\frac{1}{4}$.††

The desired decomposition reads

$$\frac{1}{x^2 - 4} = \frac{1}{4(x - 2)} - \frac{1}{4(x + 2)}.$$

You can check this by carrying out the subtraction on the right. □

> In general, each distinct linear factor $x - c$ in the denominator gives rise to a term of the form
> $$\frac{A}{x - c}.$$

---

† $\beta^2 - 4\gamma < 0$.

†† Another way of finding A and B consists of rewriting (1) as
$$1 = (A + B)x + 2A - 2B$$
and then equating coefficients to produce the system of equations
$$A + B = 0$$
$$2A - 2B = 1.$$

We can then get A and B by solving these equations simultaneously. Usually this approach involves more algebra. We will use the method of substituting *well-chosen* values of x.

Example 2. (*The denominator has a repeated linear factor.*) For

$$\frac{2x^2 + 3}{x(x - 1)^2},$$

we write

$$\frac{2x^2 + 3}{x(x - 1)^2} = \frac{A}{x} + \frac{B}{x - 1} + \frac{C}{(x - 1)^2}.$$

This leads to

$$2x^2 + 3 = A(x - 1)^2 + Bx(x - 1) + Cx.$$

To determine the three coefficients A, B, C we need to substitute three values for x. We select 0 and 1 because for those values of x several terms on the right side will drop out. As a third value of x, any other number will do; we select 2 just to keep the arithmetic simple.

Setting $x = 0$, we get $3 = A$.

Setting $x = 1$, we get $5 = C$.

Setting $x = 2$, we get $11 = A + 2B + 2C$,

which, with $A = 3$ and $C = 5$, gives $B = -1$.

The decomposition is therefore

$$\frac{2x^2 + 3}{x(x - 1)^2} = \frac{3}{x} - \frac{1}{x - 1} + \frac{5}{(x - 1)^2}. \quad \square$$

In general, each factor of the form $(x - c)^k$ in the denominator gives rise to an expression of the form

$$\frac{A_1}{x - c} + \frac{A_2}{(x - c)^2} + \cdots + \frac{A_k}{(x - c)^k}.$$

Example 3. (*The denominator has an irreducible quadratic factor.*) For

$$\frac{-2x}{(x + 1)(x^2 + 1)},$$

we write

$$\frac{-2x}{(x + 1)(x^2 + 1)} = \frac{A}{x + 1} + \frac{Bx + C}{x^2 + 1}$$

and obtain

$$-2x = A(x^2 + 1) + (Bx + C)(x + 1).$$

This time we use -1, 0, and 1.

Setting $x = -1$, we get $2 = 2A$, which gives $A = 1$.

Setting $x = 0$, we get $0 = A + C$, which gives $C = -1$.

Setting $x = 1$, we get $-2 = 2A + 2B + 2C$,

which, with $A = 1$ and $C = -1$, gives $B = -1$.

The decomposition reads

$$\frac{-2x}{(x+1)(x^2+1)} = \frac{1}{x+1} - \frac{x+1}{x^2+1}. \quad \square$$

Example 4. (*The denominator has an irreducible quadratic factor.*) For

$$\frac{1}{x(x^2+x+1)}$$

we write

$$\frac{1}{x(x^2+x+1)} = \frac{A}{x} + \frac{Bx+C}{x^2+x+1}$$

and obtain

$$1 = A(x^2+x+1) + (Bx+C)x.$$

Again we select values of x that produce zeros or simple arithmetic on the right side.

$$
\begin{array}{ll}
1 = A & (x = 0), \\
1 = 3A + B + C & (x = 1), \\
1 = A + B - C & (x = -1).
\end{array}
$$

From this we find that

$$A = 1, \quad B = -1, \quad C = -1,$$

and therefore

$$\frac{1}{x(x^2+x+1)} = \frac{1}{x} - \frac{x+1}{x^2+x+1}. \quad \blacksquare$$

In general, each irreducible quadratic factor $x^2 + \beta x + \gamma$ in the denominator gives rise to a term of the form

$$\frac{Ax+B}{x^2+\beta x+\gamma}.$$

Example 5. (*The denominator has a repeated irreducible quadratic factor.*) For

$$\frac{3x^4 + x^3 + 20x^2 + 3x + 31}{(x + 1)(x^2 + 4)^2}$$

we write

$$\frac{3x^4 + x^3 + 20x^2 + 3x + 31}{(x + 1)(x^2 + 4)^2} = \frac{A}{x + 1} + \frac{Bx + C}{x^2 + 4} + \frac{Dx + E}{(x^2 + 4)^2}.$$

This gives

$$3x^4 + x^3 + 20x^2 + 3x + 31$$
$$= A(x^2 + 4)^2 + (Bx + C)(x + 1)(x^2 + 4) + (Dx + E)(x + 1).$$

This time we use $-1, 0, 1, 2,$ and -2.

$$
\begin{aligned}
50 &= 25A && (x = -1), \\
31 &= 16A \quad\quad\; + 4C \quad\quad + E && (x = 0), \\
58 &= 25A + 10B + 10C + 2D + 2E && (x = 1), \\
173 &= 64A + 48B + 24C + 6D + 3E && (x = 2), \\
145 &= 64A + 16B - \;\; 8C + 2D - \;\; E && (x = -2).
\end{aligned}
$$

With a little patience you can see that

$$A = 2, \quad B = 1, \quad C = 0, \quad D = 0, \quad \text{and} \quad E = -1.$$

This gives the decomposition

$$\frac{3x^4 + x^3 + 20x^2 + 3x + 31}{(x + 1)(x^2 + 4)^2} = \frac{2}{x + 1} + \frac{x}{x^2 + 4} - \frac{1}{(x^2 + 4)^2}. \quad \square$$

In general, each multiple irreducible quadratic factor $(x^2 + \beta x + \gamma)^k$ in the denominator gives rise to an expression of the form

$$\frac{A_1 x + B_1}{x^2 + \beta x + \gamma} + \frac{A_2 x + B_2}{(x^2 + \beta x + \gamma)^2} + \cdots + \frac{A_k x + B_k}{(x^2 + \beta x + \gamma)^k}.$$

A polynomial appears in the decomposition when the degree of the numerator is greater than or equal to that of the denominator.

Example 6. (*The decomposition contains a polynomial.*) For

$$\frac{x^5 + 2}{x^2 - 1},$$

first carry out the suggested division:

$$
x^2 - 1 \overline{)\begin{array}{c} x^3 + x \\ x^5 \qquad\qquad + 2 \\ \underline{x^5 - x^3} \\ x^3 \\ \underline{x^3 - x} \\ x + 2. \end{array}}
$$

This gives

$$
\frac{x^5 + 2}{x^2 - 1} = x^3 + x + \frac{x + 2}{x^2 - 1}.
$$

Since the denominator of the fraction splits into linear factors, we write

$$
\frac{x + 2}{x^2 - 1} = \frac{A}{x + 1} + \frac{B}{x - 1}.
$$

This gives

$$
x + 2 = A(x - 1) + B(x + 1).
$$

Substitution of $x = 1$ gives $B = \frac{3}{2}$; substitution of $x = -1$ gives $A = -\frac{1}{2}$. The decomposition takes the form

$$
\frac{x^5 + 2}{x^2 - 1} = x^3 + x - \frac{1}{2(x + 1)} + \frac{3}{2(x - 1)}. \quad \square
$$

We have been decomposing rational functions into partial fractions so as to be able to integrate them. Here we carry out the integrations leaving some of the details to you.

Example 1′

$$
\int \frac{dx}{x^2 - 4} = \frac{1}{4} \int \left(\frac{1}{x - 2} - \frac{1}{x + 2} \right) dx
$$

$$
= \frac{1}{4} \left(\ln |x - 2| - \ln |x + 2| \right) + C = \frac{1}{4} \ln \left| \frac{x - 2}{x + 2} \right| + C. \quad \square
$$

Example 2′

$$
\int \frac{2x^2 + 3}{x(x - 1)^2} \, dx = \int \left[\frac{3}{x} - \frac{1}{x - 1} + \frac{5}{(x - 1)^2} \right] dx
$$

$$
= 3 \ln |x| - \ln |x - 1| - \frac{5}{x - 1} + C. \quad \square
$$

Example 3′

$$
\int \frac{-2x}{(x + 1)(x^2 + 1)} \, dx = \int \left(\frac{1}{x + 1} - \frac{x + 1}{x^2 + 1} \right) dx.
$$

Since

$$\int \frac{dx}{x+1} = \ln|x+1| + C_1$$

and

$$\int \frac{x+1}{x^2+1}\,dx = \frac{1}{2}\int \frac{2x}{x^2+1}\,dx + \int \frac{dx}{x^2+1} = \frac{1}{2}\ln(x^2+1) + \tan^{-1}x + C_2,$$

we have

$$\int \frac{-2x}{(x+1)(x^2+1)}\,dx = \ln|x+1| - \frac{1}{2}\ln(x^2+1) - \tan^{-1}x + C. \quad \square$$

Example 4′

$$\int \frac{dx}{x(x^2+x+1)} = \int \left(\frac{1}{x} - \frac{x+1}{x^2+x+1}\right) dx = \ln|x| - \int \frac{x+1}{x^2+x+1}\,dx = (*).$$

To compute the remaining integral, note that

$$\int \frac{x+1}{x^2+x+1}\,dx = \frac{1}{2}\int \frac{2x+1}{x^2+x+1}\,dx + \frac{1}{2}\int \frac{dx}{x^2+x+1}.$$

The first integral is a logarithm:

$$\frac{1}{2}\int \frac{2x+1}{x^2+x+1}\,dx = \frac{1}{2}\ln(x^2+x+1) + C_1;$$

the second integral is an inverse tangent:

$$\frac{1}{2}\int \frac{dx}{x^2+x+1} = \frac{1}{2}\int \frac{dx}{(x+\frac{1}{2})^2 + (\sqrt{3}/2)^2} = \frac{1}{\sqrt{3}}\tan^{-1}\left[\frac{2}{\sqrt{3}}\left(x+\frac{1}{2}\right)\right] + C_2.$$

Combining results, we have

$$(*) = \ln|x| - \frac{1}{2}\ln(x^2+x+1) - \frac{1}{\sqrt{3}}\tan^{-1}\left[\frac{2}{\sqrt{3}}\left(x+\frac{1}{2}\right)\right] + C. \quad \square$$

Example 5′

$$\int \frac{3x^4 + x^3 + 20x^2 + 3x + 31}{(x+1)(x^2+4)^2}\,dx = \int \left[\frac{2}{x+1} + \frac{x}{x^2+4} - \frac{1}{(x^2+4)^2}\right] dx.$$

The first two fractions are easy to integrate:

$$\int \frac{2}{x+1}\,dx = 2\ln|x+1| + C_1,$$

$$\int \frac{x}{x^2+4}\,dx = \frac{1}{2}\int \frac{2x}{x^2+4}\,dx = \frac{1}{2}\ln(x^2+4) + C_2.$$

The integral of the last fraction is of the form

$$\int \frac{dx}{(x^2 + c^2)^n}.$$

All such integrals can be calculated by setting $c \tan u = x$:

(8.2.2)
$$\int \frac{dx}{(x^2 + c^2)^n} = \frac{1}{c^{2n-1}} \int \cos^{2(n-1)} u \; du. \qquad †$$
$$c \tan u = x$$

The integration of such trigonometric powers is discussed in detail in the next section. Here we have

$$\int \frac{dx}{(x^2 + 4)^2} = \frac{1}{8} \int \cos^2 u \; du$$
$$2 \tan u = x$$

$$= \frac{1}{16} \int (1 + \cos 2u) \; du$$

half-angle formula
$$= \frac{1}{16} u + \frac{1}{32} \sin 2u + C_3$$

$$= \frac{1}{16} u + \frac{1}{16} \sin u \cos u + C_3$$

$\sin 2u = 2 \sin u \cos u$
$$= \frac{1}{16} \tan^{-1} \frac{x}{2} + \frac{1}{16} \left(\frac{x}{\sqrt{x^2 + 4}} \right) \left(\frac{2}{\sqrt{x^2 + 4}} \right) + C_3$$

$$= \frac{1}{16} \tan^{-1} \frac{x}{2} + \frac{1}{8} \left(\frac{x}{x^2 + 4} \right) + C_3.$$

The integral we want is therefore

$$2 \ln |x + 1| + \frac{1}{2} \ln (x^2 + 4) - \frac{1}{8} \left(\frac{x}{x^2 + 4} \right) - \frac{1}{16} \tan^{-1} \frac{x}{2} + C. \quad \square$$

Example 6′

$$\int \frac{x^5 + 2}{x^2 - 1} dx = \int \left[x^3 + x - \frac{1}{2(x + 1)} + \frac{3}{2(x - 1)} \right] dx$$

$$= \tfrac{1}{4} x^4 + \tfrac{1}{2} x^2 - \tfrac{1}{2} \ln |x + 1| + \tfrac{3}{2} \ln |x - 1| + C. \quad \square$$

---

† The proof of this reduction formula is left to you in the exercises.

Partial fractions come up in a variety of physical problems. Here is an example.

Example 7. (*Velocity-dependent forces*) When an object of mass m is moving through air or any viscous medium, it is acted upon by a frictional force opposite to its direction of motion. This frictional force depends on the velocity of the object and (within close approximation) is given by a formula of the form

$$F(v) = -\alpha v - \beta v^2.$$

Here α and β are positive constants that depend on properties of the object (its size and shape) and on properties of the medium (its density and viscosity). For very low speeds, the term βv^2 is small compared to the first and can be ignored. For ordinary speeds, βv^2 is usually the dominant term and then the linear term is dropped. We will keep both terms, assume that the frictional force is the only force acting, and derive a formula that shows how the velocity v varies with time. We will take v as positive.

From Newton's $F = ma$ we get

$$m \frac{dv}{dt} = -\alpha v - \beta v^2, \quad \text{which gives} \quad 1 = -\frac{m \, dv/dt}{\alpha v + \beta v^2}.$$

We now integrate with respect to t:

$$t = -m \int \frac{dv/dt}{v(\alpha + \beta v)} \, dt = -\frac{m}{\alpha} \int \left(\frac{dv/dt}{v} - \frac{\beta \, dv/dt}{\alpha + \beta v} \right) dt$$

$$= -\frac{m}{\alpha} [\ln v - \ln (\alpha + \beta v)] + C.$$

At $t = 0$ the object has some initial velocity $v(0) = v_0$. We evaluate the constant C from that and find that

$$C = \frac{m}{\alpha} [\ln v_0 - \ln (\alpha + \beta v_0)].$$

Thus

$$t = -\frac{m}{\alpha} [\ln v - \ln (\alpha + \beta v)] + \frac{m}{\alpha} [\ln v_0 - \ln (\alpha + \beta v_0)]$$

$$= \frac{m}{\alpha} \ln \left[\frac{v_0(\alpha + \beta v)}{v(\alpha + \beta v_0)} \right].$$

Multiplying through by α/m and then exponentiating both sides of the equation, we find that

$$e^{\alpha t/m} = \frac{v_0(\alpha + \beta v)}{v(\alpha + \beta v_0)}.$$

A bit of algebra that we leave to you shows that

$$v = \frac{\alpha v_0 e^{-\alpha t/m}}{\alpha + \beta v_0(1 - e^{-\alpha t/m})}. \quad \square$$

EXERCISES 8.2

Work out the following integrals.

1. $\int \dfrac{7}{(x-2)(x+5)}\,dx.$ **2.** $\int \dfrac{x}{(x+1)(x+2)(x+3)}\,dx.$ **3.** $\int \dfrac{2x^2+3}{x^2(x-1)}\,dx.$

4. $\int \dfrac{x^2+1}{x(x^2-1)}\,dx.$ **5.** $\int \dfrac{x^5}{(x-2)^2}\,dx.$ **6.** $\int \dfrac{x^5}{x-2}\,dx.$

7. $\int \dfrac{x+3}{x^2-3x+2}\,dx.$ **8.** $\int \dfrac{x^2+3}{x^2-3x+2}\,dx.$ **9.** $\int \dfrac{dx}{(x-1)^3}.$

10. $\int \dfrac{dx}{x^2+2x+2}.$ **11.** $\int \dfrac{x^2}{(x-1)^2(x+1)}\,dx.$ **12.** $\int \dfrac{2x-1}{(x+1)^2(x-2)^2}\,dx.$

13. $\int \dfrac{dx}{x^4-16}.$ **14.** $\int \dfrac{x}{x^3-1}\,dx.$ **15.** $\int \dfrac{x^3+4x^2-4x-1}{(x^2+1)^2}\,dx.$

16. $\int \dfrac{dx}{(x^2+16)^2}.$ **17.** $\int \dfrac{dx}{x^4+4}.^\dagger$ **18.** $\int \dfrac{dx}{x^4+16}.^\dagger$

19. Show that

$$\text{if} \quad y=\frac{1}{x^2-1} \quad \text{then} \quad \frac{d^n y}{dx^n}=\frac{1}{2}(-1)^n\, n!\left[\frac{1}{(x-1)^{n+1}}-\frac{1}{(x+1)^{n+1}}\right].$$

20. Verify reduction formula 8.2.2.

21. Find the centroid of the region under the curve $y=(x^2+1)^{-1},\ x\in[0,1].$

22. Find the centroid of the solid generated by revolving the region of Exercise 21 (a) about the x-axis. (b) about the y-axis.

23. It is known that m parts of chemical A combine with n parts of chemical B to produce a compound C. Suppose that the rate at which C is produced varies directly with the product of the amounts of A and B present at that instant. Find the amount of C produced in t minutes from an initial mixing of A_0 pounds of A with B_0 pounds of B, given that
(a) $n=m,\ A_0=B_0,$ and A_0 pounds of C are produced in the first minute.
(b) $n=m,\ A_0=\frac{1}{2}B_0,$ and A_0 pounds of C are produced in the first minute.
(c) $n\neq m,\ A_0=B_0,$ and A_0 pounds of C are produced in the first minute.
HINT: Denote by $A(t)$, $B(t)$, and $C(t)$ the amounts of A, B, and C present at time t. Observe that $C'(t)=kA(t)B(t)$. Then note that

$$A_0-A(t)=\frac{m}{m+n}\,C(t) \quad \text{and} \quad B_0-B(t)=\frac{n}{m+n}\,C(t)$$

and thus

$$C'(t)=k\left[A_0-\frac{m}{m+n}\,C(t)\right]\left[B_0-\frac{n}{m+n}\,C(t)\right].$$

24. A descending parachutist is acted on by two forces: a constant downward force mg and the upward force of air resistance, which (within close approximation) is of the form $-\beta v^2$ where β is a positive constant. (In this problem we are taking the downward direction as positive.)
(a) Express t in terms of the velocity v, the initial velocity v_0, and the constant $v_c=\sqrt{mg/\beta}.$

$\dagger$ HINT: With $a>0,\ x^4+a^2=(x^2+\sqrt{2a}\,x+a)(x^2-\sqrt{2a}\,x+a).$

(b) Express v as a function of t.

(c) Express the acceleration a as a function of t. Verify that the acceleration never changes sign and in time tends to zero.

(d) Show that in time v tends to v_c. (This number v_c is called the terminal velocity.)

Exercises *25 and *26 presuppose some familiarity with Section *6.7.

*25. A rod of length L and mass M exerts a gravitational force on a point mass m placed on the line of the rod at a distance h from the rod. Find the magnitude of the attractive force given that the density of the rod varies directly as the distance from the point of the rod farthest from m.

*26. Exercise 25 given that the density varies directly as the distance from the point of the rod closest to m.

8.3   POWERS AND PRODUCTS OF SINES AND COSINES

I. We begin by explaining how to calculate integrals of the form

$$\int \sin^m x \cos^n x \, dx \qquad \text{with } m \text{ or } n \text{ odd.}$$

Suppose that n is odd. If $n = 1$, the integral is of the form

(1)
$$\int \sin^m x \cos x \, dx = \frac{1}{m+1} \sin^{m+1} x + C.$$

If $n > 1$, write

$$\cos^n x = \cos^{n-1} x \cos x.$$

Since $n - 1$ is even, $\cos^{n-1} x$ can be expressed in powers of $\sin^2 x$ by noting that $\cos^2 x = 1 - \sin^2 x$. The integral then takes the form

$$\int (\text{sum of powers of } \sin x) \cdot \cos x \, dx,$$

which can be broken up into integrals of form (1).

Similarly if m is odd, write

$$\sin^m x = \sin^{m-1} x \sin x$$

and use the substitution $\sin^2 x = 1 - \cos^2 x$.

Example 1

$$\int \sin^2 x \cos^5 x \, dx = \int \sin^2 x \cos^4 x \cos x \, dx$$

$$= \int \sin^2 x \, (1 - \sin^2 x)^2 \cos x \, dx$$

$$= \int (\sin^2 x - 2\sin^4 x + \sin^6 x)\cos x \, dx$$

$$= \int \sin^2 x \cos x \, dx - 2\int \sin^4 x \cos x \, dx + \int \sin^6 x \cos x \, dx$$

$$= \tfrac{1}{3}\sin^3 x - \tfrac{2}{5}\sin^5 x + \tfrac{1}{7}\sin^7 x + C. \quad \square$$

Example 2

$$\int \sin^5 x \, dx = \int \sin^4 x \sin x \, dx$$

$$= \int (1 - \cos^2 x)^2 \sin x \, dx$$

$$= \int (1 - 2\cos^2 x + \cos^4 x) \sin x \, dx$$

$$= \int \sin x \, dx - 2\int \cos^2 x \sin x \, dx + \int \cos^4 x \sin x \, dx$$

$$= -\cos x + \tfrac{2}{3}\cos^3 x - \tfrac{1}{5}\cos^5 x + C. \quad \square$$

II. To calculate integrals of the form

$$\int \sin^m x \cos^n x \, dx \qquad \text{with } m \text{ and } n \text{ both even}$$

use the following identities:

$$\sin x \cos x = \tfrac{1}{2}\sin 2x, \qquad \sin^2 x = \tfrac{1}{2} - \tfrac{1}{2}\cos 2x, \qquad \cos^2 x = \tfrac{1}{2} + \tfrac{1}{2}\cos 2x.$$

Example 3

$$\int \cos^2 x \, dx = \int (\tfrac{1}{2} + \tfrac{1}{2}\cos 2x) \, dx$$

$$= \tfrac{1}{2}\int dx + \tfrac{1}{2}\int \cos 2x \, dx = \tfrac{1}{2}x + \tfrac{1}{4}\sin 2x + C. \quad \square$$

Example 4

$$\int \sin^2 x \cos^2 x \, dx = \tfrac{1}{4}\int \sin^2 2x \, dx$$

$$= \tfrac{1}{4}\int (\tfrac{1}{2} - \tfrac{1}{2}\cos 4x) \, dx$$

$$= \tfrac{1}{8}\int dx - \tfrac{1}{8}\int \cos 4x \, dx = \tfrac{1}{8}x - \tfrac{1}{32}\sin 4x + C. \quad \square$$

Example 5

$$\int \sin^4 x \cos^2 x \, dx = \int (\sin x \cos x)^2 \sin^2 x \, dx$$

$$= \int \tfrac{1}{4} \sin^2 2x \, (\tfrac{1}{2} - \tfrac{1}{2} \cos 2x) \, dx$$

$$= \tfrac{1}{8} \int \sin^2 2x \, dx - \tfrac{1}{8} \int \sin^2 2x \cos 2x \, dx$$

$$= \tfrac{1}{8} \int (\tfrac{1}{2} - \tfrac{1}{2} \cos 4x) \, dx - \tfrac{1}{8} \int \sin^2 2x \cos 2x \, dx$$

$$= \tfrac{1}{16} x - \tfrac{1}{64} \sin 4x - \tfrac{1}{48} \sin^3 2x + C. \quad \square$$

III. Finally we come to integrals of the form

$$\int \sin mx \cos nx \, dx, \qquad \int \sin mx \sin nx \, dx, \qquad \int \cos mx \cos nx \, dx.$$

If $m = n$, there is no difficulty. For $m \neq n$ use the identities

$$\sin A \cos B = \tfrac{1}{2}[\sin (A - B) + \sin (A + B)],$$
$$\sin A \sin B = \tfrac{1}{2}[\cos (A - B) - \cos (A + B)],$$
$$\cos A \cos B = \tfrac{1}{2}[\cos (A - B) + \cos (A + B)].$$

These identities follow readily from the familiar addition formulas:

$$\sin (A + B) = \sin A \cos B + \cos A \sin B,$$
$$\sin (A - B) = \sin A \cos B - \cos A \sin B,$$
$$\cos (A + B) = \cos A \cos B - \sin A \sin B,$$
$$\cos (A - B) = \cos A \cos B + \sin A \sin B.$$

Example 6

$$\int \sin 5x \sin 3x \, dx = \int \tfrac{1}{2} (\cos 2x - \cos 8x) \, dx = \tfrac{1}{4} \sin 2x - \tfrac{1}{16} \sin 8x + C. \quad \square$$

EXERCISES 8.3

Work out the following integrals.

1. $\int \sin^3 x \, dx.$ **2.** $\int \cos^2 4x \, dx.$ **3.** $\int \sin^2 3x \, dx.$

4. $\int \cos^3 x \, dx.$ **5.** $\int \cos^4 x \sin^3 x \, dx.$ **6.** $\int \sin^3 x \cos^2 x \, dx.$

7. $\int \sin^3 x \cos^3 x \, dx.$ **8.** $\int \sin^2 x \cos^4 x \, dx.$ **9.** $\int \sin^2 x \cos^3 x \, dx.$

10. $\int \sin^4 x \cos^3 x \, dx.$ **11.** $\int \sin^4 x \, dx.$ **12.** $\int \cos^3 x \cos 2x \, dx.$

13. $\int \sin 2x \cos 3x \, dx.$ **14.** $\int \cos 2x \sin 3x \, dx.$ **15.** $\int \sin^2 x \sin 2x \, dx.$

16. $\int \cos^4 x \, dx.$ **17.** $\int \sin^4 x \cos^4 x \, dx.$ **18.** $\int \sin^7 x \, dx.$

19. $\int \sin^6 x \, dx.$ **20.** $\int \cos^5 x \sin^5 x \, dx.$ **21.** $\int \cos^7 x \, dx.$

22. $\int \cos^6 x \, dx.$ **23.** $\int \cos 3x \cos 2x \, dx.$ **24.** $\int \sin 3x \sin 2x \, dx.$

Calculate the following integrals by a trigonometric substitution.

25. $\int \dfrac{dx}{(x^2 + 1)^3}.$ [set $x = \tan u$]   **26.** $\int \dfrac{dx}{(x^2 + 4)^3}.$

27. $\int \dfrac{dx}{[(x + 1)^2 + 1]^2}.$ **28.** $\int \dfrac{dx}{[(2x + 1)^2 + 9]^2}.$

Evaluate the following integrals. They are important in applied mathematics.

29. $\dfrac{1}{\pi} \displaystyle\int_0^{2\pi} \cos^2 nx \, dx.$ **30.** $\dfrac{1}{\pi} \displaystyle\int_0^{2\pi} \sin^2 nx \, dx.$

31. $\displaystyle\int_0^{2\pi} \sin mx \cos nx \, dx, \quad m \neq n.$ **32.** $\displaystyle\int_0^{2\pi} \sin mx \sin nx \, dx, \quad m \neq n.$

33. $\displaystyle\int_0^{2\pi} \cos mx \cos nx \, dx, \quad m \neq n.$

34. (a) Use integration by parts to derive the reduction formula:

$$\int \sin^{n+2} x \, dx = -\frac{\sin^{n+1} x \cos x}{n + 2} + \frac{n + 1}{n + 2} \int \sin^n x \, dx.$$

(b) Using (a), verify that

$$\int_0^{\pi/2} \sin^{n+2} x \, dx = \frac{n + 1}{n + 2} \int_0^{\pi/2} \sin^n x \, dx.$$

(c) Show that

$$\int_0^{\pi/2} \sin^m x \, dx = \begin{cases} \left(\dfrac{(m - 1) \cdots 5 \cdot 3 \cdot 1}{m \cdots 6 \cdot 4 \cdot 2}\right) \dfrac{\pi}{2}, & m \text{ even}, \, m \geq 2 \\ \dfrac{(m - 1) \cdots 4 \cdot 2}{m \cdots 5 \cdot 3}, & m \text{ odd}, \, m \geq 3 \end{cases}.$$

(d) Show that

$$\int_0^{\pi/2} \cos^m x \, dx = \int_0^{\pi/2} \sin^m x \, dx.$$

8.4 OTHER TRIGONOMETRIC POWERS

I. First we take up integrals of the form

$$\int \tan^n x \, dx, \qquad \int \cot^n x \, dx.$$

To integrate $\tan^n x$, set

$$\tan^n x = \tan^{n-2} x \tan^2 x = (\tan^{n-2} x)(\sec^2 x - 1) = \tan^{n-2} x \sec^2 x - \tan^{n-2} x.$$

To integrate $\cot^n x$, set

$$\cot^n x = \cot^{n-2} x \cot^2 x = (\cot^{n-2} x)(\csc^2 x - 1) = \cot^{n-2} x \csc^2 x - \cot^{n-2} x.$$

Example 1

$$\int \tan^6 x \, dx = \int (\tan^4 x \sec^2 x - \tan^4 x) \, dx$$

$$= \int (\tan^4 x \sec^2 x - \tan^2 x \sec^2 x + \tan^2 x) \, dx$$

$$= \int (\tan^4 x \sec^2 x - \tan^2 x \sec^2 x + \sec^2 x - 1) \, dx$$

$$= \int \tan^4 x \sec^2 x \, dx - \int \tan^2 x \sec^2 x \, dx + \int \sec^2 x \, dx - \int dx$$

$$= \tfrac{1}{5} \tan^5 x - \tfrac{1}{3} \tan^3 x + \tan x - x + C. \quad \square$$

II. Now consider the integrals

$$\int \sec^n x \, dx, \qquad \int \csc^n x \, dx.$$

For even powers write

$$\sec^n x = \sec^{n-2} x \sec^2 x = (\tan^2 x + 1)^{(n-2)/2} \sec^2 x$$

and

$$\csc^n x = \csc^{n-2} x \csc^2 x = (\cot^2 x + 1)^{(n-2)/2} \csc^2 x.$$

Example 2

$$\int \sec^4 x \, dx = \int \sec^2 x \sec^2 x \, dx$$

$$= \int (\tan^2 x + 1) \sec^2 x \, dx$$

$$= \int \tan^2 x \sec^2 x \, dx + \int \sec^2 x \, dx = \tfrac{1}{3} \tan^3 x + \tan x + C. \quad \square$$

The odd powers you can integrate by parts. For $\sec^n x$ set

$$u = \sec^{n-2} x, \quad dv = \sec^2 x \, dx$$

and, when $\tan^2 x$ appears, use the identity $\tan^2 x = \sec^2 x - 1$. You can handle $\csc^n x$ in a similar manner.

Example 3. For

$$\int \sec^3 x \, dx$$

write

$$u = \sec x, \qquad\qquad dv = \sec^2 x \, dx,$$
$$du = \sec x \tan x \, dx, \qquad v = \tan x.$$

$$\int \sec^3 x \, dx = \int \sec x \sec^2 x \, dx$$

$$= \int u \, dv = uv - \int v \, du$$

$$= \sec x \tan x - \int \tan^2 x \sec x \, dx$$

$$= \sec x \tan x - \int (\sec^2 x - 1) \sec x \, dx$$

$$= \sec x \tan x - \int \sec^3 x \, dx + \int \sec x \, dx$$

$$= \sec x \tan x - \int \sec^3 x \, dx + \ln |\sec x + \tan x|$$

$$2 \int \sec^3 x \, dx = \sec x \tan x + \ln |\sec x + \tan x|$$

$$\int \sec^3 x \, dx = \tfrac{1}{2} \sec x \tan x + \tfrac{1}{2} \ln |\sec x + \tan x|.$$

Now add the arbitrary constant:

$$\int \sec^3 x \, dx = \tfrac{1}{2} \sec x \tan x + \tfrac{1}{2} \ln |\sec x + \tan x| + C.$$

This integral occurs so frequently in applications that you will find it listed on the inside covers of this text. □

III. Finally we come to integrals of the form

$$\int \tan^m x \sec^n x \, dx, \qquad \int \cot^m x \csc^n x \, dx.$$

When n is even, write

$$\tan^m x \sec^n x = \tan^m x \sec^{n-2} x \sec^2 x$$

and express $\sec^{n-2} x$ entirely in terms of $\tan^2 x$ using $\sec^2 x = \tan^2 x + 1$.

Example 4

$$\int \tan^5 x \sec^4 x \, dx = \int \tan^5 x \sec^2 x \sec^2 x \, dx$$

$$= \int \tan^5 x (\tan^2 x + 1) \sec^2 x \, dx$$

$$= \int \tan^7 x \sec^2 x \, dx + \int \tan^5 x \sec^2 x \, dx$$

$$= \tfrac{1}{8} \tan^8 x + \tfrac{1}{6} \tan^6 x + C. \quad \square$$

When n and m are both odd, write

$$\tan^m x \sec^n x = \tan^{m-1} x \sec^{n-1} x \sec x \tan x$$

and express $\tan^{m-1} x$ entirely in terms of $\sec^2 x$ using $\tan^2 x = \sec^2 x - 1$.

Example 5

$$\int \tan^5 x \sec^3 x \, dx = \int \tan^4 x \sec^2 x \sec x \tan x \, dx$$

$$= \int (\sec^2 x - 1)^2 \sec^2 x \sec x \tan x \, dx$$

$$= \int (\sec^6 x - 2 \sec^4 x + \sec^2 x) \sec x \tan x \, dx$$

$$= \tfrac{1}{7} \sec^7 x - \tfrac{2}{5} \sec^5 x + \tfrac{1}{3} \sec^3 x + C. \quad \square$$

Finally, if n is odd and m is even, write the product entirely in terms of $\sec x$ and integrate by parts.

Example 6

$$\int \tan^2 x \sec x \, dx = \int (\sec^2 x - 1) \sec x \, dx$$

$$= \int (\sec^3 x - \sec x) \, dx = \int \sec^3 x \, dx - \int \sec x \, dx.$$

We have calculated each of these integrals before:

$$\int \sec^3 x \, dx = \tfrac{1}{2} \sec x \tan x + \tfrac{1}{2} \ln |\sec x + \tan x| + C$$

and

$$\int \sec x \, dx = \ln|\sec x + \tan x| + C.$$

It follows that

$$\int \tan^2 x \sec x \, dx = \tfrac{1}{2} \sec x \tan x - \tfrac{1}{2} \ln|\sec x + \tan x| + C. \quad \square$$

You can handle integrals of $\cot^m x \csc^n x$ in a similar manner.

EXERCISES 8.4

Calculate.

1. $\displaystyle\int \tan^2 3x \, dx.$ 2. $\displaystyle\int \cot^2 5x \, dx.$ 3. $\displaystyle\int \sec^2 \pi x \, dx.$

4. $\displaystyle\int \csc^2 2x \, dx.$ 5. $\displaystyle\int \tan^3 x \, dx.$ 6. $\displaystyle\int \cot^3 x \, dx.$

7. $\displaystyle\int \tan^2 x \sec^2 x \, dx.$ 8. $\displaystyle\int \cot^2 x \csc^2 x \, dx.$ 9. $\displaystyle\int \csc^3 x \, dx.$

10. $\displaystyle\int \sec^3 \pi x \, dx.$ 11. $\displaystyle\int \cot^4 x \, dx.$ 12. $\displaystyle\int \tan^4 x \, dx.$

13. $\displaystyle\int \cot^3 x \csc^3 x \, dx.$ 14. $\displaystyle\int \tan^3 x \sec^3 x \, dx.$ 15. $\displaystyle\int \csc^4 2x \, dx.$

16. $\displaystyle\int \sec^4 3x \, dx.$ 17. $\displaystyle\int \cot^2 x \csc x \, dx.$ 18. $\displaystyle\int \csc^3 \tfrac{1}{2}x \, dx.$

19. $\displaystyle\int \tan^5 3x \, dx.$ 20. $\displaystyle\int \cot^5 2x \, dx.$ 21. $\displaystyle\int \sec^5 x \, dx.$

22. $\displaystyle\int \csc^5 x \, dx.$ 23. $\displaystyle\int \tan^4 x \sec^4 x \, dx.$ 24. $\displaystyle\int \cot^4 x \csc^4 x \, dx.$

8.5   INTEGRALS INVOLVING $\sqrt{a^2 \pm x^2}$ AND $\sqrt{x^2 \pm a^2}$

Such integrals can usually be calculated by a trigonometric substitution:

$$\text{for} \quad \sqrt{a^2 - x^2} \quad \text{set} \quad a \sin u = x,$$
$$\text{for} \quad \sqrt{a^2 + x^2} \quad \text{set} \quad a \tan u = x,$$
$$\text{for} \quad \sqrt{x^2 - a^2} \quad \text{set} \quad a \sec u = x.$$

In each case take $a > 0$.

Problem 1. Find

$$\int \frac{dx}{(a^2 - x^2)^{3/2}}.$$

Solution. Set

$$a \sin u = x, \qquad a \cos u \, du = dx.$$

$$\int \frac{dx}{(a^2 - x^2)^{3/2}} = \int \frac{a \cos u}{(a^2 - a^2 \sin^2 u)^{3/2}} \, du$$

$$= \frac{1}{a^2} \int \frac{\cos u}{\cos^3 u} \, du$$

$$= \frac{1}{a^2} \int \sec^2 u \, du$$

$$= \frac{1}{a^2} \tan u + C = \frac{x}{a^2 \sqrt{a^2 - x^2}} + C. \quad \square$$

Problem 2. Find

$$\int \sqrt{a^2 + x^2} \, dx.$$

Solution. Set

$$a \tan u = x, \qquad a \sec^2 u \, du = dx.$$

$$\int \sqrt{a^2 + x^2} \, dx = \int \sqrt{a^2 + a^2 \tan^2 u} \, a \sec^2 u \, du$$

$$= a^2 \int \sqrt{1 + \tan^2 u} \, \sec^2 u \, du$$

$$= a^2 \int \sec u \cdot \sec^2 u \, du$$

$$= a^2 \int \sec^3 u \, du$$

by Section 8.4

$$= \frac{a^2}{2} (\sec u \tan u + \ln |\sec u + \tan u|) + C$$

$$= \frac{a^2}{2} \left[\frac{\sqrt{a^2 + x^2}}{a} \left(\frac{x}{a} \right) + \ln \left| \frac{\sqrt{a^2 + x^2}}{a} + \frac{x}{a} \right| \right] + C$$

$$= \tfrac{1}{2} x \sqrt{a^2 + x^2} + \tfrac{1}{2} a^2 \ln (x + \sqrt{a^2 + x^2}) - \tfrac{1}{2} a^2 \ln a + C.$$

We can absorb the constant $-\tfrac{1}{2} a^2 \ln a$ in C and write

(8.5.1) $$\int \sqrt{a^2 + x^2} \, dx = \tfrac{1}{2} x \sqrt{a^2 + x^2} + \tfrac{1}{2} a^2 \ln (x + \sqrt{a^2 + x^2}) + C.$$

This is one of the standard formulas. $\square$

Now a slight variation.

Problem 3. Find

$$\int \frac{dx}{x\sqrt{4x^2 + 9}}.$$

Solution. Set

$$3 \tan u = 2x, \qquad 3 \sec^2 u \, du = 2 \, dx.$$

$$\int \frac{dx}{x\sqrt{4x^2 + 9}} = \int \frac{\frac{3}{2} \sec^2 u}{\frac{3}{2} \tan u \cdot 3 \sec u} \, du$$

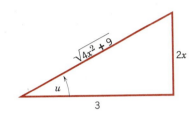

$$= \frac{1}{3} \int \frac{\sec u}{\tan u} \, du$$

$$= \frac{1}{3} \int \csc u \, du$$

$$= \frac{1}{3} \ln |\csc u - \cot u| + C$$

$$= \frac{1}{3} \ln \left| \frac{\sqrt{4x^2 + 9} - 3}{2x} \right| + C. \quad \square$$

The next problem requires that we first complete the square under the radical.

Problem 4. Find

$$\int \frac{x}{\sqrt{x^2 + 2x - 3}} \, dx.$$

Solution. First note that

$$\int \frac{x}{\sqrt{x^2 + 2x - 3}} \, dx = \int \frac{x}{\sqrt{(x + 1)^2 - 4}} \, dx.$$

Now set

$$2 \sec u = x + 1, \qquad 2 \sec u \tan u = dx.$$

Then

$$\int \frac{x}{\sqrt{(x + 1)^2 - 4}} \, dx = \int \frac{(2 \sec u - 1)2 \sec u \tan u}{2 \tan u} \, du$$

$$= \int (2 \sec^2 u - \sec u) \, du$$

$$= 2 \tan u - \ln |\sec u + \tan u| + C$$

$$= \sqrt{x^2 + 2x - 3} - \ln \left| \frac{x + 1 + \sqrt{x^2 + 2x - 3}}{2} \right| + C. \quad \square$$

Finally, we note that a trigonometric substitution may also be effective in cases where the quadratic in the integrand is not under a radical. For example, see reduction formula (8.2.2).

EXERCISES 8.5

Work out the following integrals.

1. $\displaystyle\int \frac{dx}{\sqrt{a^2 - x^2}}.$

2. $\displaystyle\int \frac{dx}{(x^2 + 2)^{3/2}}.$

3. $\displaystyle\int \frac{dx}{(5 - x^2)^{3/2}}.$

4. $\displaystyle\int \frac{x}{\sqrt{x^2 - 4}}\, dx.$

5. $\displaystyle\int \sqrt{x^2 - 1}\, dx.$

6. $\displaystyle\int \frac{x}{\sqrt{4 - x^2}}\, dx.$

7. $\displaystyle\int \frac{x^2}{\sqrt{4 - x^2}}\, dx.$

8. $\displaystyle\int \frac{x^2}{\sqrt{x^2 - 4}}\, dx.$

9. $\displaystyle\int \frac{x}{(1 - x^2)^{3/2}}\, dx.$

10. $\displaystyle\int \frac{x^2}{\sqrt{4 + x^2}}\, dx.$

11. $\displaystyle\int \frac{x^2}{(1 - x^2)^{3/2}}\, dx.$

12. $\displaystyle\int \frac{x}{a^2 + x^2}\, dx.$

13. $\displaystyle\int x \sqrt{4 - x^2}\, dx.$

14. $\displaystyle\int \frac{e^x}{\sqrt{9 - e^{2x}}}\, dx.$

15. $\displaystyle\int \frac{x^2}{(x^2 + 8)^{3/2}}\, dx.$

16. $\displaystyle\int \frac{\sqrt{1 - x^2}}{x^4}\, dx.$

17. $\displaystyle\int \frac{dx}{x\sqrt{a^2 - x^2}}.$

18. $\displaystyle\int \frac{dx}{\sqrt{x^2 + a^2}}.$

19. $\displaystyle\int \frac{dx}{\sqrt{x^2 - a^2}}.$

20. $\displaystyle\int \sqrt{a^2 - x^2}\, dx.$

21. $\displaystyle\int e^x \sqrt{e^{2x} - 1}\, dx.$

22. $\displaystyle\int \frac{\sqrt{x^2 - 1}}{x}\, dx.$

23. $\displaystyle\int \frac{dx}{x^2\sqrt{a^2 + x^2}}.$

24. $\displaystyle\int \frac{dx}{x^2\sqrt{a^2 - x^2}}.$

25. $\displaystyle\int \frac{dx}{x^2\sqrt{x^2 - a^2}}.$

26. $\displaystyle\int \frac{dx}{e^x\sqrt{4 + e^{2x}}}.$

27. $\displaystyle\int \frac{dx}{e^x\sqrt{e^{2x} - 9}}.$

28. $\displaystyle\int \frac{dx}{\sqrt{x^2 - 2x - 3}}.$

29. $\displaystyle\int \frac{dx}{(x^2 - 4x + 4)^{3/2}}.$

30. $\displaystyle\int \frac{x}{\sqrt{6x - x^2}}\, dx.$

31. $\displaystyle\int x\sqrt{6x - x^2 - 8}\, dx.$

32. $\displaystyle\int \frac{x + 2}{\sqrt{x^2 + 4x + 13}}\, dx.$

33. $\displaystyle\int \frac{x}{(x^2 + 2x + 5)^2}\, dx.$

34. $\displaystyle\int \frac{x}{\sqrt{x^2 - 2x - 3}}\, dx.$

35. $\displaystyle\int \frac{x + 3}{\sqrt{x^2 + 4x + 13}}\, dx.$

36. $\displaystyle\int x\sqrt{x^2 + 6x}\, dx.$

37. $\displaystyle\int \sqrt{6x - x^2 - 8}\, dx.$

38. $\displaystyle\int \frac{dx}{(4 - x^2)^2}.$

39. $\displaystyle\int x(8 - 2x - x^2)^{3/2}\, dx.$

40. Calculate

$$\int x \sin^{-1} x\, dx.$$

41. Calculate the mass and the center of mass of a rod that extends from $x = 0$ to $x = a > 0$ and has mass density $\lambda(x) = (x^2 + a^2)^{-1/2}$.

42. Calculate the mass and the center of mass of the rod in Exercise 41 if the mass density is given by $\lambda(x) = (x^2 + a^2)^{-3/2}$.

For Exercises 43–45 let Ω be the region under the curve $y = \sqrt{x^2 - a^2}$, $x \in [a, \sqrt{2}\, a]$.

43. Sketch Ω, find its area, and locate the centroid.

44. Find the volume of the solid generated by revolving Ω about the x-axis and determine the centroid of that solid.

45. Find the volume of the solid generated by revolving Ω about the y-axis and determine the centroid of that solid.

8.6   SOME RATIONALIZING SUBSTITUTIONS

Problem 1. Find

$$\int \frac{dx}{1 + \sqrt{x}}.$$

Solution. To rationalize the integrand we set

$$u^2 = x, \qquad 2u \, du = dx.$$

With $u = \sqrt{x}$,

$$\int \frac{dx}{1 + \sqrt{x}} = \int \frac{2u}{1 + u} \, du = \int \left(2 - \frac{2}{1 + u}\right) du$$
$$= 2u - 2 \ln|1 + u| + C$$
$$= 2\sqrt{x} - 2 \ln|1 + \sqrt{x}| + C. \quad \square$$

Problem 2. Find

$$\int \frac{x^{1/2}}{4(1 + x^{3/4})} \, dx.$$

Solution. Here we set

$$u^4 = x, \qquad 4u^3 \, du = dx.$$

With $u = x^{1/4}$,

$$\int \frac{x^{1/2}}{4(1 + x^{3/4})} \, dx = \int \frac{(u^2)(4u^3)}{4(1 + u^3)} \, du$$
$$= \int \frac{u^5}{1 + u^3} \, du = \int \left(u^2 - \frac{u^2}{1 + u^3}\right) du$$
$$= \tfrac{1}{3}u^3 - \tfrac{1}{3} \ln|1 + u^3| + C$$
$$= \tfrac{1}{3}x^{3/4} - \tfrac{1}{3} \ln|1 + x^{3/4}| + C. \quad \square$$

Problem 3. Find

$$\int \sqrt{1 - e^x} \, dx.$$

Solution. To rationalize the integrand we set

$$u^2 = 1 - e^x.$$

To find dx in terms of u and du we solve the equation for x:

$$1 - u^2 = e^x, \qquad \ln(1 - u^2) = x, \qquad -\frac{2u}{1 - u^2} \, du = dx.$$

The rest is straightforward:

$$\int \sqrt{1 - e^x}\, dx = \int u\left(-\frac{2u}{1 - u^2}\right) du$$

$$= \int \frac{2u^2}{u^2 - 1}\, du = \int \left(2 + \frac{1}{u - 1} - \frac{1}{u + 1}\right) du$$

$$= 2u + \ln|u - 1| - \ln|u + 1| + C$$

$$= 2u + \ln\left|\frac{u - 1}{u + 1}\right| + C$$

$$= 2\sqrt{1 - e^x} + \ln\left|\frac{\sqrt{1 - e^x} - 1}{\sqrt{1 - e^x} + 1}\right| + C. \quad \square$$

EXERCISES 8.6

Work out the following integrals.

1. $\displaystyle\int \frac{dx}{1 - \sqrt{x}}.$ **2.** $\displaystyle\int \frac{\sqrt{x}}{1 + x}\, dx.$ **3.** $\displaystyle\int \sqrt{1 + e^x}\, dx.$

4. $\displaystyle\int \frac{dx}{x(x^{1/3} - 1)}.$ **5.** $\displaystyle\int x\sqrt{1 + x}\, dx.$ [(a) set $u^2 = 1 + x$; (b) set $u = 1 + x$]

6. $\displaystyle\int x^2\sqrt{1 + x}\, dx.$ [(a) set $u^2 = 1 + x$; (b) set $u = 1 + x$]   **7.** $\displaystyle\int (x + 2)\sqrt{x - 1}\, dx.$

8. $\displaystyle\int (x - 1)\sqrt{x + 2}\, dx.$ **9.** $\displaystyle\int \frac{x^3}{(1 + x^2)^3}\, dx.$ **10.** $\displaystyle\int x(1 + x)^{1/3}\, dx.$

11. $\displaystyle\int \frac{\sqrt{x}}{\sqrt{x} - 1}\, dx.$ **12.** $\displaystyle\int \frac{x}{\sqrt{x} + 1}\, dx.$ **13.** $\displaystyle\int \frac{\sqrt{x - 1} + 1}{\sqrt{x - 1} - 1}\, dx.$

14. $\displaystyle\int \frac{1 - e^x}{1 + e^x}\, dx.$ **15.** $\displaystyle\int \frac{dx}{\sqrt{1 + e^x}}.$ **16.** $\displaystyle\int \frac{dx}{1 + e^{-x}}.$

17. $\displaystyle\int \frac{x}{\sqrt{x} + 4}\, dx.$ **18.** $\displaystyle\int \frac{x + 1}{x\sqrt{x - 2}}\, dx.$ **19.** $\displaystyle\int 2x^2(4x + 1)^{-5/2}\, dx.$

20. $\displaystyle\int x^2\sqrt{x - 1}\, dx.$ **21.** $\displaystyle\int \frac{x}{(ax + b)^{3/2}}\, dx.$ **22.** $\displaystyle\int \frac{x}{\sqrt{ax + b}}\, dx.$

8.7   NUMERICAL INTEGRATION

To evaluate a definite integral by the formula

$$\int_a^b f(x)\, dx = F(b) - F(a)$$

we must be able to find an antiderivative F and we must be able to evaluate this antiderivative both at a and at b. When this is not possible, the method fails.

The method fails even for such simple-looking integrals as

$$\int_0^1 \sqrt{x}\, \sin x \, dx \qquad \text{and} \qquad \int_0^1 e^{-x^2} \, dx.$$

There are no *elementary functions* with derivatives $\sqrt{x} \sin x$ and e^{-x^2}.

Here we take up some simple numerical methods for estimating definite integrals — methods that you can use whether or not you can find an antiderivative. All the methods we describe involve only simple arithmetic and are ideally suited to the computer.

We focus now on

$$\int_a^b f(x) \, dx.$$

As usual, we suppose that f is continuous on $[a, b]$ and, for pictorial convenience, assume that f is positive. We begin by subdividing $[a, b]$ into n subintervals each of length $(b - a)/n$:

$$[a, b] = [x_0, x_1] \cup \cdots \cup [x_{i-1}, x_i] \cup \cdots \cup [x_{n-1}, x_n],$$

with

$$\Delta x_i = \frac{b - a}{n}.$$

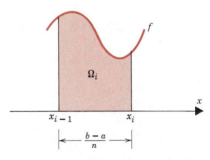

Figure 8.7.1

The region Ω_i pictured in Figure 8.7.1 can be approximated in many ways.

(1) By the left-endpoint rectangle (Figure 8.7.2):

$$\text{area} = f(x_{i-1}) \, \Delta x_i$$
$$= f(x_{i-1}) \left(\frac{b - a}{n} \right).$$

Figure 8.7.2

(2) By the right-endpoint rectangle (Figure 8.7.3):

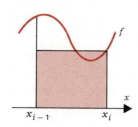

$$\text{area} = f(x_i)\, \Delta x_i$$
$$= f(x_i) \left(\frac{b-a}{n} \right).$$

Figure 8.7.3

(3) By the midpoint rectangle (Figure 8.7.4):

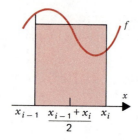

$$\text{area} = f\left(\frac{x_{i-1} + x_i}{2} \right) \Delta x_i$$
$$= f\left(\frac{x_{i-1} + x_i}{2} \right) \left(\frac{b-a}{n} \right).$$

Figure 8.7.4

(4) By a trapezoid (Figure 8.7.5):

$$\text{area} = \frac{1}{2}\left[f(x_{i-1}) + f(x_i) \right] \Delta x_i$$
$$= \frac{1}{2}\left[f(x_{i-1}) + f(x_i) \right] \left(\frac{b-a}{n} \right).$$

Figure 8.7.5

(5) By a parabolic region (Figure 8.7.6): take the parabola $y = Ax^2 + Bx + C$ that passes through the three points indicated.

$$\text{area} = \frac{1}{6}\left[f(x_{i-1}) + 4f\left(\frac{x_{i-1} + x_i}{2} \right) + f(x_i) \right] \Delta x_i$$
$$= \left[f(x_{i-1}) + 4f\left(\frac{x_{i-1} + x_i}{2} \right) + f(x_i) \right] \left(\frac{b-a}{6n} \right).$$

Figure 8.7.6

You can verify this formula by doing Exercises 7 and 8.   (If the three points are collinear, the parabola degenerates to a straight line and the parabolic region becomes a trapezoid.   The formula then gives the area of the trapezoid.)

The approximations to Ω_i just considered yield the following estimates for

$$\int_a^b f(x)\, dx.$$

(1) The left-endpoint estimate:

$$L_n = \frac{b-a}{n}[f(x_0) + f(x_1) + \cdots + f(x_{n-1})].$$

(2) The right-endpoint estimate:

$$R_n = \frac{b-a}{n}[f(x_1) + f(x_2) + \cdots + f(x_n)].$$

(3) The midpoint estimate:

$$M_n = \frac{b-a}{n}\left[f\left(\frac{x_0 + x_1}{2}\right) + \cdots + f\left(\frac{x_{n-1} + x_n}{2}\right)\right].$$

(4) The trapezoidal estimate (*trapezoidal rule*):

$$T_n = \frac{b-a}{n}\left[\frac{f(x_0) + f(x_1)}{2} + \frac{f(x_1) + f(x_2)}{2} + \cdots + \frac{f(x_{n-1}) + f(x_n)}{2}\right]$$

$$= \frac{b-a}{2n}[f(x_0) + 2f(x_1) + \cdots + 2f(x_{n-1}) + f(x_n)].$$

(5) The parabolic estimate (*Simpson's rule*):

$$S_n = \frac{b-a}{6n}\left\{ f(x_0) + f(x_n) + 2[f(x_1) + \cdots + f(x_{n-1})] \right.$$

$$\left. + 4\left[f\left(\frac{x_0 + x_1}{2}\right) + \cdots + f\left(\frac{x_{n-1} + x_n}{2}\right)\right]\right\}.$$

The first three estimates, L_n, R_n, M_n, are Riemann sums (Section 5.13). T_n and S_n, although not explicitly defined as Riemann sums, can be written as Riemann sums. (See Exercise 18.)   It follows from (5.13.1) that any one of these estimates can be used to approximate the integral as closely as we may wish.   All we have to do is take n sufficiently large.

As an example, we will find the approximate value of

$$\ln 2 = \int_1^2 \frac{dx}{x}$$

by applying each of the five estimates.   Here

$$f(x) = \frac{1}{x}, \qquad [a, b] = [1, 2].$$

Taking $n = 5$, we have

$$\frac{b-a}{n} = \frac{2-1}{5} = \frac{1}{5}.$$

The partition points are

$$x_0 = \tfrac{5}{5}, \quad x_1 = \tfrac{6}{5}, \quad x_2 = \tfrac{7}{5}, \quad x_3 = \tfrac{8}{5}, \quad x_4 = \tfrac{9}{5}, \quad x_5 = \tfrac{10}{5}.$$

The five estimates then run as follows:

$$L_5 = \tfrac{1}{5}(\tfrac{5}{5} + \tfrac{5}{6} + \tfrac{5}{7} + \tfrac{5}{8} + \tfrac{5}{9}) = (\tfrac{1}{5} + \tfrac{1}{6} + \tfrac{1}{7} + \tfrac{1}{8} + \tfrac{1}{9}) \cong 0.75,$$

$$R_5 = \tfrac{1}{5}(\tfrac{5}{6} + \tfrac{5}{7} + \tfrac{5}{8} + \tfrac{5}{9} + \tfrac{5}{10}) = (\tfrac{1}{6} + \tfrac{1}{7} + \tfrac{1}{8} + \tfrac{1}{9} + \tfrac{1}{10}) \cong 0.65,$$

$$M_5 = \tfrac{1}{5}(\tfrac{10}{11} + \tfrac{10}{13} + \tfrac{10}{15} + \tfrac{10}{17} + \tfrac{10}{19}) = 2(\tfrac{1}{11} + \tfrac{1}{13} + \tfrac{1}{15} + \tfrac{1}{17} + \tfrac{1}{19}) \cong 0.69,$$

$$T_5 = \tfrac{1}{10}(\tfrac{5}{5} + \tfrac{10}{6} + \tfrac{10}{7} + \tfrac{10}{8} + \tfrac{10}{9} + \tfrac{5}{10}) = (\tfrac{1}{10} + \tfrac{1}{6} + \tfrac{1}{7} + \tfrac{1}{8} + \tfrac{1}{9} + \tfrac{1}{20}) \cong 0.70,$$

$$S_5 = \tfrac{1}{30}[\tfrac{5}{5} + \tfrac{5}{10} + 2(\tfrac{5}{6} + \tfrac{5}{7} + \tfrac{5}{8} + \tfrac{5}{9}) + 4(\tfrac{10}{11} + \tfrac{10}{13} + \tfrac{10}{15} + \tfrac{10}{17} + \tfrac{10}{19})] \cong 0.69.$$

Since the integrand $1/x$ decreases throughout the interval $[1, 2]$, you can expect the left-endpoint estimate, 0.75, to be too large and you can expect the right-endpoint estimate, 0.65, to be too small.   The other estimates should be better.

Table 1 at the end of the book gives $\ln 2 \cong 0.693$.   The estimate 0.69 is correct to the nearest hundredth.

Problem 1.   Find the approximate value of

$$\int_0^2 \sqrt{4 + x^3}\, dx$$

by the trapezoidal rule.   Take $n = 4$.

Solution.   Each subinterval has length

$$\frac{b-a}{n} = \frac{2-0}{4} = \frac{1}{2}.$$

The partition points are

$$x_0 = 0, \quad x_1 = \tfrac{1}{2}, \quad x_2 = 1, \quad x_3 = \tfrac{3}{2}, \quad x_4 = 2.$$

Consequently

$$T_4 = \tfrac{1}{4}[f(0) + 2f(\tfrac{1}{2}) + 2f(1) + 2f(\tfrac{3}{2}) + f(2)].$$

Using a calculator and rounding off to three decimal places, we have

$$f(0) = 2.000, \quad f(\tfrac{1}{2}) \cong 2.031, \quad f(1) \cong 2.236, \quad f(\tfrac{3}{2}) \cong 2.716, \quad f(2) \cong 3.464.$$

Thus

$$T_4 \cong \tfrac{1}{4}(2.000 + 4.062 + 4.472 + 5.432 + 3.464) \cong 4.858. \quad \square$$

Problem 2. Find the approximate value of

$$\int_0^2 \sqrt{4 + x^3}\, dx$$

by Simpson's rule. Take $n = 2$.

Solution. There are two intervals each of length

$$\frac{b - a}{n} = \frac{2 - 0}{2} = 1.$$

Here

$$x_0 = 0, \quad x_1 = 1, \quad x_2 = 2 \quad \text{and} \quad \frac{x_0 + x_1}{2} = \frac{1}{2}, \quad \frac{x_1 + x_2}{2} = \frac{3}{2}.$$

Simpson's rule yields

$$S_2 = \tfrac{1}{6}[f(0) + f(2) + 2f(1) + 4f(\tfrac{1}{2}) + 4f(\tfrac{3}{2})],$$

which, in light of the square-root estimates given in the last problem, gives

$$S_2 \cong \tfrac{1}{6}(2.000 + 3.464 + 4.472 + 8.124 + 10.864) \cong 4.821. \quad \square$$

Error Estimates

A numerical estimate is useful only to the extent that we can gauge its accuracy. When we use any kind of approximation method, we face two forms of error: the error inherent in the method we use (we call this the *theoretical error*) and the error that accumulates from rounding off the decimals that arise during the course of computation (we call this the *round-off error*). The nature of round-off error is obvious. We will speak first about theoretical error.

We begin with a function f continuous and increasing on $[a, b]$. We subdivide $[a, b]$ into n nonoverlapping intervals, each of length $(b - a)/n$. We want to estimate

$$\int_a^b f(x)\, dx$$

by the left-endpoint method. What is the theoretical error? It should be clear from Figure 8.7.7 that the theoretical error does not exceed

(1) $$[f(b) - f(a)]\left(\frac{b - a}{n}\right).$$

The error is represented by the sum of the areas of the shaded regions. These regions, when shifted to the right, all fit together within a rectangle of height $f(b) - f(a)$ and base $(b - a)/n$.

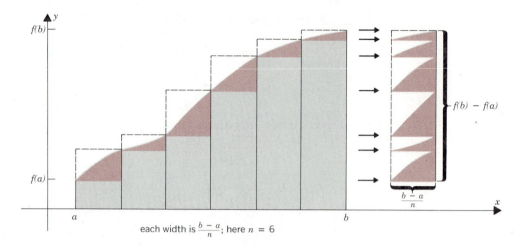

Figure 8.7.7

Similar reasoning shows that, under the same circumstances, the theoretical error associated with the trapezoidal method does not exceed

(2) $$\frac{1}{2}[f(b) - f(a)]\left(\frac{b - a}{n}\right).$$

In this setting at least, the trapezoidal estimate does a better job.

We can improve on (2) considerably, even dropping the requirement that f be increasing. It is shown in texts on numerical analysis that, if f is twice differentiable, then the theoretical error of the trapezoidal rule can be written

(8.7.1) $$\frac{(b - a)^3}{12n^2} f''(c)$$

where c is some number between a and b. Usually we cannot pinpoint c any further. Nevertheless, as we show below, (8.7.1) is a powerful tool.

Recall the trapezoidal-rule estimate

$$\int_1^2 \frac{dx}{x} \cong 0.70.$$

We arrived at this taking $n = 5$. We will apply (8.7.1) to find the theoretical error. Here

$$f(x) = \frac{1}{x}, \qquad f'(x) = -\frac{1}{x^2}, \qquad f''(x) = \frac{2}{x^3}.$$

With $a = 1$ and $b = 2$, we have c between 1 and 2. Since $n = 5$,

$$\left|\frac{(b - a)^3}{12n^2} f''(c)\right| = \frac{1}{300} \frac{2}{c^3} < \frac{1}{150} < 0.007.$$

$$\underset{c > 1}{}$$

The estimate 0.70 is in theoretical error by less than 0.007.

To obtain an estimate for

$$\int_1^2 \frac{dx}{x}$$

that is accurate to four decimal places, we need

(3) $$\left| \frac{(b-a)^3}{12n^2} f''(c) \right| < 0.00005.$$

Since

$$\left| \frac{(b-a)^3}{12n^2} f''(c) \right| \le \frac{1}{12n^2} \frac{2}{c^3} < \frac{1}{6n^2},$$

we can guarantee that (3) holds by having

$$\frac{1}{6n^2} < 0.00005.$$

As you can check, this requires that n be greater than 57.

Even more effective than the trapezoidal rule is Simpson's rule. The theoretical error for Simpson's rule can be written

(8.7.2)
$$\frac{(b-a)^5}{180n^4} f^{(4)}(c)$$

where, as before, c is some number between a and b. Whereas (8.7.1) varies as $1/n^2$, this quantity varies as $1/n^4$. Thus, for comparable n, we can expect greater accuracy from Simpson's rule.

Finally, a word about round-off error. Any numerical procedure requires careful consideration of round-off error. To illustrate this point, we rework our trapezoidal estimate for

$$\int_1^2 \frac{dx}{x},$$

again taking $n = 5$, but this time assuming that our computer or calculator can store only two significant digits. As before,

(3) $$\int_1^2 \frac{dx}{x} \cong \tfrac{1}{10}[\tfrac{1}{1} + 2(\tfrac{5}{6}) + 2(\tfrac{5}{7}) + 2(\tfrac{5}{8}) + 2(\tfrac{5}{9}) + (\tfrac{1}{2})].$$

Now our limited round-off machine goes to work:

$$\int_1^2 \frac{dx}{x} \cong (0.10)[(1.0) + 2(0.83) + 2(0.71) + 2(0.62) + 2(0.44) + (0.50)]$$

$$\text{``=''} (0.10)[(1.0) + (1.7) + (1.4) + (1.2) + (0.88) + (0.50)]$$

$$= (0.10)[6.7] = 0.67.$$

Earlier we used (8.7.1) to show that estimate (3) is in error by no more than 0.007 and found 0.70 as the approximation. Now with our limited round-off machine, we simplified (3) in a different way and obtained 0.67 as the approximation. The apparent error due to crude round off, $0.70 - 0.67 = 0.03$, exceeds the error of the approximation method itself. The lesson should be clear: round-off error is important.

EXERCISES 8.7

1. **(Calculator)** Estimate

$$\int_0^{12} x^2 \, dx$$

using (a) the left-endpoint estimate, $n = 12$. (b) the right-endpoint estimate, $n = 12$. (c) the midpoint estimate, $n = 6$. (d) the trapezoidal rule, $n = 12$. (e) Simpson's rule, $n = 6$. Check your results by performing the integration.

2. **(Calculator)** Estimate

$$\int_0^1 \sin^2 \pi x \, dx$$

using (a) the midpoint estimate, $n = 3$. (b) the trapezoidal rule, $n = 6$. (c) Simpson's rule, $n = 3$. Check your results by performing the integration.

3. **(Calculator)** Estimate

$$\int_0^3 \frac{dx}{1 + x^3}$$

using (a) the left-endpoint estimate, $n = 6$. (b) the right-endpoint estimate, $n = 6$. (c) the midpoint estimate, $n = 3$. (d) the trapezoidal rule, $n = 6$. (e) Simpson's rule, $n = 3$.

4. **(Calculator)** Estimate

$$\int_0^\pi \frac{\sin x}{\pi + x} \, dx$$

using (a) the trapezoidal rule, $n = 6$. (b) Simpson's rule, $n = 3$.

5. **(Calculator)** Find the approximate value of π by estimating the integral

$$\frac{\pi}{4} = \tan^{-1} 1 = \int_0^1 \frac{dx}{1 + x^2}$$

using (a) the trapezoidal rule, $n = 4$. (b) Simpson's rule, $n = 4$.

6. **(Calculator)** Estimate

$$\int_0^2 \frac{dx}{\sqrt{4 + x^3}}$$

using (a) the trapezoidal rule, $n = 4$. (b) Simpson's rule, $n = 2$.

7. Show that there is one and only one curve of the form $y = Ax^2 + Bx + C$ through three distinct points with different x-coordinates.

8. Show that the function $g(x) = Ax^2 + Bx + C$ satisfies the condition

$$\int_a^b g(x) \, dx = \frac{b - a}{6} \left[g(a) + 4g\left(\frac{a + b}{2}\right) + g(b) \right]$$

for every interval $[a, b]$.

(Calculator)   Determine the values of n for which a theoretical error less than ϵ can be guaranteed if the integral is estimated using (a) the trapezoidal rule.   (b) Simpson's rule.

9. $\displaystyle\int_1^4 \sqrt{x}\ dx;\quad \epsilon = 0.01.$

10. $\displaystyle\int_1^3 x^5\ dx;\quad \epsilon = 0.01.$

11. $\displaystyle\int_1^4 \sqrt{x}\ dx;\quad \epsilon = 0.00001.$

12. $\displaystyle\int_1^3 x^5\ dx;\quad \epsilon = 0.00001.$

13. $\displaystyle\int_0^\pi \sin x\ dx;\quad \epsilon = 0.001$

14. $\displaystyle\int_0^\pi \cos x\ dx;\quad \epsilon = 0.001.$

15. $\displaystyle\int_1^3 e^x\ dx;\quad \epsilon = 0.01.$

16. $\displaystyle\int_1^e \ln x\ dx;\quad \epsilon = 0.01.$

17. Show that Simpson's rule is exact (theoretical error zero) for every polynomial of degree 3 or less.

18. Show that, if f is continuous, then T_n and S_n can both be written as Riemann sums. HINT: Both

$$\frac{1}{2}[f(x_{i-1}) + f(x_i)] \quad\text{and}\quad \frac{1}{6}\left[f(x_{i-1}) + 4f\left(\frac{x_{i-1} + x_i}{2}\right) + f(x_i)\right]$$

lie between m_i and M_i, the minimum and maximum values of f on $[x_{i-1}, x_i]$.

Numerical integration via the trapezoidal rule and Simpson's rule   (BASIC)

Like the earlier program on numerical integration via Riemann sums, these programs offer a practical alternative to the method of computing definite integrals by explicitly finding antiderivatives. You may be interested to try the two methods on an integral whose value you know to see for yourself which gives the better results with the same size subinterval.

```
10 REM Numerical integration via the trapezoidal rule
20 REM copyright © Colin C. Graham 1988-1989
30 REM change the line "def FNf(x) = ..."
40 REM to fit your function
50 DEF FNf(x) = 4/(1 + x*x)
100 INPUT "Enter left endpoint"; a
120 INPUT "Enter right endpoint"; b
130 INPUT "Enter number of divisions"; n
200 delta = (b − a)/n
210 integral = 0
300 FOR j = 1 TO n − 1
310     integral = integral + FNf(a + j*delta)*delta
320 NEXT j
330 integral = integral + delta*(FNf(a) + FNf(b))/2
400 PRINT integral
510 INPUT "Do again? (Y/N)"; a$
520 IF a$ = "Y" OR a$ = "y" THEN 100
530 END
```

```
10 REM Numerical integration via Simpson's rule
20 REM copyright © Colin C. Graham 1988-1989

30 REM change the line "def FNf(x) = ..."
40 REM to fit your function

50 DEF FNf(x) = 4/(1 + x*x)

100 INPUT "Enter left endpoint"; a
110 INPUT "Enter right endpoint"; b
120 INPUT "Enter one-half number of divisions"; n

200 n = 2*n
210 delta = (b − a)/n
220 evenones = 0
230 oddones = 0

300 FOR j = 1 TO n − 1 STEP 2
310     oddones = oddones + FNf(a + j*delta)*delta
320 NEXT j
330 FOR j = 2 TO n − 2 STEP 2
340     evenones = evenones + FNf(a + j*delta)*delta
350 NEXT j
360 integral = 4*oddones + 2*evenones
370 integral = (integral + delta*(FNf(a) + FNf(b)))/3

400 PRINT integral

500 INPUT "Do again? (Y/N)"; a$
510 IF a$ = "Y" OR a$ = "y" THEN 100
520 END
```

8.8   ADDITIONAL EXERCISES

Work out the following integrals and check your answers by differentiation.

1. $\int 10^{nx}\,dx.$

2. $\int \tan\left(\dfrac{\pi}{n}x\right)dx.$

3. $\int \sqrt{2x+1}\,dx.$

4. $\int x\sqrt{2x+1}\,dx.$

5. $\int e^x \tan e^x\,dx.$

6. $\int \dfrac{\sin\sqrt{x}}{\sqrt{x}}\,dx.$

7. $\int \dfrac{dx}{a^2x^2+b^2}.$

8. $\int \sin 2x \cos x\,dx.$

9. $\int \ln(x\sqrt{x})\,dx.$

10. $\int \sec 2x\,dx.$

11. $\int \dfrac{x^3}{\sqrt{1+x^2}}\,dx.$

12. $\int x2^x\,dx.$

13. $\int \sin^2\left(\dfrac{\pi}{n}x\right)dx.$

14. $\int \dfrac{dx}{x^3-1}.$

15. $\int \dfrac{dx}{a\sqrt{x}+b}.$

16. $\int (1-\sec x)^2\,dx.$

17. $\int \dfrac{\sin 3x}{2+\cos 3x}\,dx.$

18. $\int \dfrac{\sqrt{a-x}}{\sqrt{a+x}}\,dx.\quad (a>0)$

19. $\int \dfrac{\sin x}{\cos^3 x}\, dx.$ **20.** $\int \dfrac{\cos x}{\sin^3 x}\, dx.$ **21.** $\int \dfrac{dx}{2x^2 - 2x + 1}.$

22. $\int \dfrac{dx}{\sin x \cos x}.$ **23.** $\int \dfrac{\sqrt{a + x}}{\sqrt{a - x}}\, dx.$ **24.** $\int a^{2x}\, dx.$

25. $\int \dfrac{\sqrt{a^2 - x^2}}{x^2}\, dx.$ **26.** $\int \ln \sqrt{x + 1}\, dx.$ **27.** $\int \dfrac{x}{(x + 1)^2}\, dx.$

28. $\int \dfrac{\sin^5 x}{\cos^7 x}\, dx.$ **29.** $\int \dfrac{1 - \sin 2x}{1 + \sin 2x}\, dx.$ **30.** $\int \ln (ax + b)\, dx.$

31. $\int x \ln (ax + b)\, dx.$ **32.** $\int (\tan x + \cot x)^2\, dx.$ **33.** $\int \dfrac{dx}{\sqrt{x + 1} - \sqrt{x}}.$

34. $\int \dfrac{e^{-\sqrt{x}}}{\sqrt{x}}\, dx.$ **35.** $\int \dfrac{x^2}{1 + x^2}\, dx.$ **36.** $\int \sqrt{\dfrac{x^2}{9} - 1}\, dx.$

37. $\int \dfrac{-x^2}{\sqrt{1 - x^2}}\, dx.$ **38.** $\int e^x \sin \pi x\, dx.$ **39.** $\int \sinh^2 x\, dx.$

40. $\int 2x \sinh x\, dx.$ **41.** $\int e^{-x} \cosh x\, dx.$ **42.** $\int x \ln \sqrt{x^2 + 1}\, dx.$

43. $\int x \tan^{-1} (x - 3)\, dx.$ **44.** $\int \dfrac{dx}{2 - \sqrt{x}}.$ **45.** $\int \dfrac{2}{x(1 + x^2)}\, dx.$

46. $\int \dfrac{dx}{\sqrt{2x - x^2}}.$ **47.** $\int \dfrac{\cos^4 x}{\sin^2 x}\, dx.$ **48.** $\int \dfrac{x - 3}{x^2(x + 1)}\, dx.$

49. $\int \dfrac{\sqrt{x^2 + 4}}{x}\, dx.$ **50.** $\int \dfrac{e^x}{\sqrt{e^x + 1}}\, dx.$ **51.** $\int \dfrac{dx}{x\sqrt{9 - x^2}}.$

52. $\int \dfrac{dx}{e^x - 2e^{-x}}.$ **53.** $\int \ln (1 - \sqrt{x})\, dx.$ **54.** $\int x \tan^2 \pi x\, dx.$

55. $\int \dfrac{dx}{\sqrt{1 - e^{2x}}}.$ **56.** $\int \dfrac{\cos 2x}{\cos x}\, dx.$ **57.** $\int \dfrac{\sec^3 x}{\tan x}\, dx.$

58. $\int \sin^3 x \sec x\, dx.$ **59.** $\int \dfrac{\sin 4x}{\sin x}\, dx.$ **60.** $\int \dfrac{x^2 + x}{\sqrt{1 - x^2}}\, dx.$

61. $\int \sin^5 \left(\dfrac{x}{2}\right) dx.$ **62.** $\int \csc x \tan x\, dx.$ **63.** $\int \cot^2 x \sec x\, dx.$

64. $\int \sec^3 x \sin x\, dx.$ **65.** $\int \dfrac{\sin x}{\sin 2x}\, dx.$ **66.** $\int \dfrac{x^2}{\sqrt{3 - 2x - x^2}}\, dx.$

67. $\int x^2 \sin^{-1} x\, dx.$ **68.** $\int \dfrac{x + 3}{\sqrt{x^2 + 2x - 8}}\, dx.$ **69.** $\int x\sqrt{x^2 + 2x - 8}\, dx.$

70. $\int x^2 \tan^{-1} x\, dx.$ **71.** $\int (\sin^2 x - \cos x)^2\, dx.$ **72.** $\int \sin 2x \cos 3x\, dx.$

73. $\int \dfrac{3}{\sqrt{2 - 3x - 4x^2}}\, dx.$ **74.** $\int \left(\dfrac{x}{\sqrt{a^2 - x^2}} - \dfrac{\sqrt{a^2 - x^2}}{x}\right) dx.$

CHAPTER HIGHLIGHTS

8.1 Review

A table of integrals appears on the inside covers.

8.2 Partial Fractions

We can integrate any rational function by writing it as a polynomial (that may be identically zero) plus fractions of the form

$$\frac{A}{(x - \alpha)^k} \quad \text{and} \quad \frac{Bx + C}{(x^2 + \beta x + \gamma)^k}.$$

velocity-dependent forces (p. 440)

8.3 Powers and Products of Sines and Cosines

Integrals of the form

$$\int \sin^m x \cos^n x \, dx$$

can be calculated by using the basic identity $\sin^2 x + \cos^2 x = 1$ and the double angle formulas

$$\sin x \cos x = \tfrac{1}{2} \sin 2x, \quad \sin^2 x = \tfrac{1}{2} - \tfrac{1}{2} \cos 2x, \quad \cos^2 x = \tfrac{1}{2} + \tfrac{1}{2} \cos 2x.$$

8.4 Other Trigonometric Powers

The main tools for calculating such integrals are the identities

$$1 + \tan^2 x = \sec^2 x, \quad 1 + \cot^2 x = \csc^2 x$$

and integration by parts.

8.5 Integrals Involving $\sqrt{a^2 \pm x^2}$ and $\sqrt{x^2 \pm a^2}$

Such integrals are usually calculated by a trigonometric substitution:

$$\text{for } \sqrt{a^2 - x^2} \quad \text{set} \quad a \sin u = x,$$
$$\text{for } \sqrt{a^2 + x^2} \quad \text{set} \quad a \tan u = x,$$
$$\text{for } \sqrt{x^2 - a^2} \quad \text{set} \quad a \sec u = x.$$

Sometimes it is necessary to first complete the square under the radical. (p. 451)

A trigonometric substitution may also be effective in cases where the quadratic in the integrand is not under a radical.

8.6 Some Rationalizing Substitutions

8.7 Numerical Integration

left-endpoint, right-endpoint, and midpoint estimates; trapezoidal rule; Simpson's rule (p. 457)

theoretical error in trapezoidal rule varies as $1/n^2$

theoretical error in Simpson's rule varies as $1/n^4$

round-off error (p. 459)

8.8 Additional Exercises

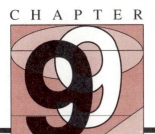

THE CONIC SECTIONS

9.1 INTRODUCTION

If a "double right circular cone" is cut by a plane, the resulting intersection is called a *conic section* or, more briefly, a *conic*. In Figure 9.1.1 we depict three important cases.

By choosing a plane perpendicular to the axis of the cone, we can obtain a circle. The other possibilities are a point, a line, or a pair of lines. Try to visualize this.

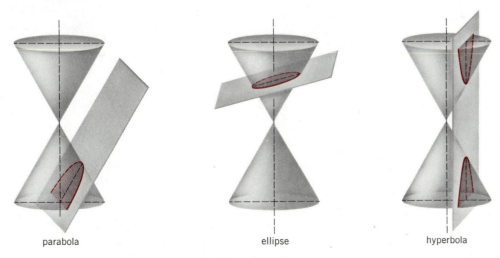

parabola   ellipse   hyperbola

Figure 9.1.1

This three-dimensional approach to the conic sections goes back to Apollonius of Perga, a Greek of the third century B.C.   He wrote eight books on the subject.

We will take a different approach.   We will define parabola, ellipse, and hyperbola entirely in terms of plane geometry.   But first some preliminary considerations.

9.2   TRANSLATIONS; THE DISTANCE BETWEEN A POINT AND A LINE

In Figure 9.2.1 we have drawn a rectangular coordinate system and marked a point $O'(x_0, y_0)$.   Think of the Oxy system as a rigid frame and in your mind slide it along the plane, without letting it turn, so that the origin O falls on the point O'.   (Figure 9.2.2)   Such a move, called a *translation*, produces a new coordinate system $O'XY$.

Figure 9.2.1

Figure 9.2.2

A point P now has two pairs of coordinates: a pair (x, y) with respect to the Oxy system and a pair (X, Y) with respect to the $O'XY$ system.   (Figure 9.2.3)   To see the relation between these coordinates, note that, starting at O, we can reach P by first going to O' and then going on to P; thus

$$x = x_0 + X, \qquad y = y_0 + Y.$$

Figure 9.2.3

Translations are often used to simplify geometric arguments.   If you carried out part (c) of Exercise 55 in Section 1.3, you found that the distance between a line l: $Ax + By + C = 0$ and the origin is given by the formula

(9.2.1)
$$d(O, l) = \frac{|C|}{\sqrt{A^2 + B^2}}.$$

We will derive the formula for the distance between a line and an arbitrary point by translating the coordinate system.

THEOREM 9.2.2

The distance between the line $l: Ax + By + C = 0$ and the point $P_1(x_1, y_1)$ is given by the formula

$$d(P_1, l) = \frac{|Ax_1 + By_1 + C|}{\sqrt{A^2 + B^2}}.$$

Proof. We translate the Oxy coordinate system to obtain a new system $O'XY$ with O' falling on the point P_1. The new coordinates are now related to the old coordinates as follows:

$$x = x_1 + X, \qquad y = y_1 + Y.$$

In the xy-system, l has equation

$$Ax + By + C = 0.$$

In the XY-system, l has equation

$$A(x_1 + X) + B(y_1 + Y) + C = 0.$$

We can write this last equation as

$$AX + BY + K = 0 \qquad \text{with} \quad K = Ax_1 + By_1 + C.$$

The distance we want is the distance between the line $AX + BY + K = 0$ and the new origin O'. By (9.2.1), this distance is

$$\frac{|K|}{\sqrt{A^2 + B^2}}.$$

Since $K = Ax_1 + By_1 + C$, we have

$$d(P_1, l) = \frac{|Ax_1 + By_1 + C|}{\sqrt{A^2 + B^2}}. \quad \square$$

Problem 1. Find the distance between the line

$$l: 3x + 4y - 5 = 0$$

and (a) the origin, (b) the point $P(-6, 2)$.

Solution

(a) $d(O, l) = \dfrac{|C|}{\sqrt{A^2 + B^2}} = \dfrac{|-5|}{\sqrt{3^2 + 4^2}} = \dfrac{5}{\sqrt{25}} = 1.$

(b) $d(P, l) = \dfrac{|Ax_1 + By_1 + C|}{\sqrt{A^2 + B^2}} = \dfrac{|3(-6) + 4(2) - 5|}{\sqrt{3^2 + 4^2}} = \dfrac{15}{\sqrt{25}} = 3. \quad \square$

If a curve has an equation in x and y, then that equation can be written in the form $f(x, y) = K$ with K a constant:

$$2y = 3x - 5 \quad \text{can be written} \quad 3x - 2y = 5,$$
$$y^2 = x^2 + x + 1 \quad \text{can be written} \quad x^2 - y^2 + x = -1,$$
$$y^2 = x^3 \quad \text{can be written} \quad x^3 - y^2 = 0, \text{ etc.}$$

Now let's start with a curve

$$C: f(x, y) = K.$$

The curve

$$C_1: f(x - 2, y - 3) = K$$

is a translation of C; set

$$x = 2 + X, \qquad y = 3 + Y$$

and the equation of C_1 takes the form

$$f(X, Y) = K.$$

The curve C_1 is the curve C displaced 2 units right and 3 units up. The curves

$$C_2: f(x + 2, y + 3) = K, \quad C_3: f(x - 2, y + 3) = K, \quad C_4: f(x + 2, y - 3) = K,$$

are also translations of C. C_2 is C displaced 2 units left and 3 units down; C_3 is C displaced 2 units right and 3 units down; C_4 is C displaced 2 units left, 3 units up.

EXERCISES 9.2

1. Find the distance between the line $5x + 12y + 2 = 0$ and
   (a) the origin.   (b) the point $P(1, -3)$.
2. Find the distance between the line $2x - 3y + 1 = 0$ and
   (a) the origin.   (b) the point $P(-2, 5)$.
3. Which of the points $(0, 1)$, $(1, 0)$, and $(-1, 1)$ is closest to $l: 8x + 7y - 6 = 0$?   Which is farthest from l?
4. Consider the triangle with vertices $A(2, 0)$, $B(4, 3)$, $C(5, -1)$. Which of these vertices is farthest from the opposite side?
5. Find the area of the triangle with vertices $(1, -2)$, $(-1, 3)$, $(2, 4)$.
6. Find the area of the triangle with vertices $(-1, 1)$, $(3, \sqrt{2})$, $(\sqrt{2}, -1)$.
7. Write the equation of a line $Ax + By + C = 0$ in *normal form*:

$$x \cos \alpha + y \sin \alpha = p \qquad \text{with } p \geq 0.$$

   What is the geometric significance of p? of α?
8. Find an expression for the distance between the parallel lines $Ax + By + C = 0$ and $Ax + By + C' = 0$.
9. Write an equation in x and y for the line $l: 4x + 5y + 3 = 0$ displaced
   (a) 1 unit left, 2 units up.   (b) 1 unit right, 2 units up.
   (c) 1 unit left, 2 units down.   (d) 1 unit right, 2 units down.
10. Exercise 9 with $l: 3x - 2y + 7 = 0$.

11. Write an equation in x and y for the curve C: $x^2 = y^3$ displaced
   (a) 3 units right, 4 units up.   (b) 3 units left, 4 units up.
   (c) 4 units right, 3 units down.   (d) 4 units left, 3 units down.

12. Exercise 11 with C: $x^2 + y^2 = 1$.

13. A ray l is rotating clockwise about the point $Q(-b^2, 0)$ at the rate of 1 revolution per minute. How fast is the distance between l and the origin changing at the moment that l has slope $\frac{3}{4}$?

14. A ray l is rotating counterclockwise about the point $Q(-b^2, 0)$ at the rate of $\frac{1}{2}$ revolution per minute. What is the slope of l at the moment that l is receding most quickly from the point $P(b^2, -1)$?

9.3 THE PARABOLA

Figure 9.3.1 shows a line l and a point F not on l.

(9.3.1)   The set of points P equidistant from F and l is called a *parabola*.

See Figure 9.3.2.

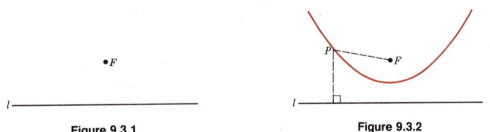

Figure 9.3.1   **Figure 9.3.2**

The line l is called the *directrix* of the parabola, and the point F is called the *focus*. (You will see why later on.) The line through F perpendicular to l is called the *axis* of the parabola. (It is the axis of symmetry.) The point at which the axis intersects the parabola is called the *vertex*. (See Figure 9.3.3.)

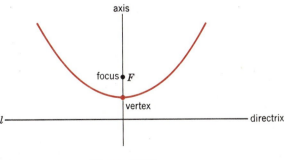

Figure 9.3.3

The equation of a parabola is particularly simple if we place the vertex at the origin and the focus on one of the coordinate axes. Suppose for the moment that the focus

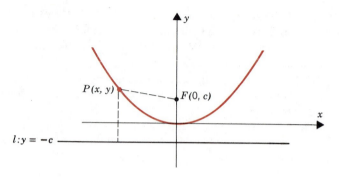

Figure 9.3.4

F is on the y-axis.   Then F has coordinates of the form $(0, c)$.   With the vertex at the origin, the directrix has equation $y = -c$.   (See Figure 9.3.4.)

Every point $P(x, y)$ that lies on this parabola has the property that

$$d(P, F) = d(P, l).$$

Since

$$d(P, F) = \sqrt{x^2 + (y - c)^2} \qquad \text{and} \qquad d(P, l) = |y + c|,$$

you can see that

$$\sqrt{x^2 + (y - c)^2} = |y + c|,$$
$$x^2 + (y - c)^2 = |y + c|^2 = (y + c)^2,$$
$$x^2 + y^2 - 2cy + c^2 = y^2 + 2cy + c^2,$$
$$x^2 = 4cy. \quad \square$$

You have just seen that the equation

(9.3.2) $\boxed{x^2 = 4cy}$ (Figure 9.3.5)

represents a parabola with vertex at the origin and focus at $(0, c)$.   By interchanging

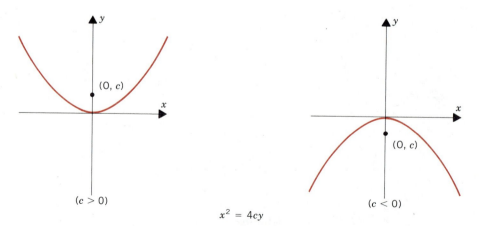

$$x^2 = 4cy$$

Figure 9.3.5

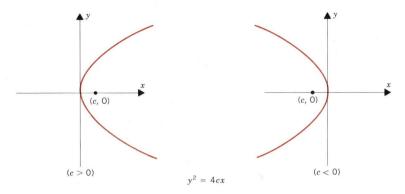

Figure 9.3.6

the roles of x and y, you can see that the equation

(9.3.3) $$y^2 = 4cx$$ (Figure 9.3.6)

represents a parabola with vertex at the origin and focus at $(c, 0)$.

Problem 1.   Sketch the parabola specifying the vertex, focus, directrix, and axis:

$$\text{(a) } x^2 = -4y. \qquad \text{(b) } y^2 = 3x.$$

Solution.   (a) The equation $x^2 = -4y$ has the form

$$x^2 = 4cy \qquad \text{with} \quad c = -1. \qquad \text{(Figure 9.3.7)}$$

The vertex is at the origin, and the focus is at $(0, -1)$; the directrix is the horizontal line $y = 1$; the axis of the parabola is the y-axis.
(b) The equation $y^2 = 3x$ has the form

$$y^2 = 4cx \qquad \text{with} \quad c = \tfrac{3}{4}. \qquad \text{(Figure 9.3.7)}$$

The vertex is at the origin, and the focus is at $(\tfrac{3}{4}, 0)$; the directrix is the vertical line $x = -\tfrac{3}{4}$; the axis of the parabola is the x-axis.   ☐

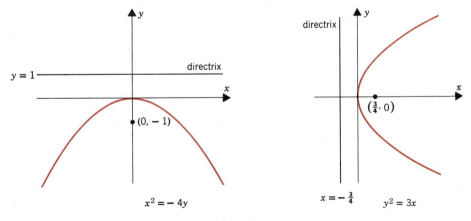

Figure 9.3.7

Every parabola with vertical axis is a translation of a parabola with equation of the form $x^2 = 4cy$, and every parabola with horizontal axis is a translation of a parabola with equation of the form $y^2 = 4cx$.

Problem 2. Identify the curve

$$(x - 4)^2 = 8(y + 3).$$

Solution. The curve is the parabola

$$x^2 = 8y \qquad\qquad \text{(here } c = 2\text{)}$$

displaced 4 units right and 3 units down. The parabola $x^2 = 8y$ has vertex at the origin; the focus is at $(0, 2)$, and the directrix is the line $y = -2$. Thus, the parabola $(x - 4)^2 = 8(y + 3)$ has vertex at $(4, -3)$; the focus is at $(4, -1)$, and the directrix is the line $y = -5$. ☐

Problem 3. Identify the curve

$$(y - 1)^2 = 8(x + 3).$$

Solution. This is the parabola

$$y^2 = 8x \qquad\qquad \text{(here } c = 2\text{)}$$

displaced 3 units left, 1 unit up. The parabola $y^2 = 8x$ has vertex at the origin; the focus is at $(2, 0)$, and the directrix is the vertical line $x = -2$. Thus, the parabola $(y - 1)^2 = 8(x + 3)$ has vertex at $(-3, 1)$; the focus is at $(-1, 1)$, and the directrix is the line $x = -5$. ☐

Problem 4. Identify the curve

$$y = x^2 + 2x - 2.$$

Solution. We first complete the square on the right by adding 3 to both sides of the equation:

$$y + 3 = x^2 + 2x + 1 = (x + 1)^2.$$

This gives

$$(x + 1)^2 = y + 3.$$

This is the parabola

$$x^2 = y \qquad\qquad \text{(here } c = \tfrac{1}{4}\text{)}$$

displaced 1 unit left, 3 units down. The parabola $x^2 = y$ has vertex at the origin; the focus is at the point $(0, \tfrac{1}{4})$ and the directrix is the line $y = -\tfrac{1}{4}$. Thus, the parabola $(x + 1)^2 = y + 3$ has vertex at $(-1, -3)$; the focus is at $(-1, -\tfrac{11}{4})$ and the directrix is the line $y = -\tfrac{13}{4}$. ☐

By the method of the last problem one can show that every quadratic

$$y = Ax^2 + Bx + C$$

represents a parabola with vertical axis. It looks like ∪ if A is positive and like ∩ if A is negative.

Parabolic Mirrors

You are already familiar with the geometric principle of reflected light: that the angle of reflection equals the angle of incidence. (Problem 1, Section 4.10)

Now take a parabola and revolve it about its axis. This gives you a parabolic surface. A curved mirror of that form is called a *parabolic mirror.* Such mirrors are used in searchlights and telescopes. Our purpose here is to explain why.

We begin with a parabola and choose the coordinate system so that the equation takes the form $x^2 = 4cy$ with $c > 0$. We can express y in terms of x by writing

$$y = \frac{x^2}{4c}.$$

Since

$$\frac{dy}{dx} = \frac{2x}{4c} = \frac{x}{2c},$$

the tangent line at the point $P(x_0, y_0)$ has slope $m = x_0/2c$ and equation

(1) $$(y - y_0) = \frac{x_0}{2c}(x - x_0).$$

For the rest we refer to Figure 9.3.8.

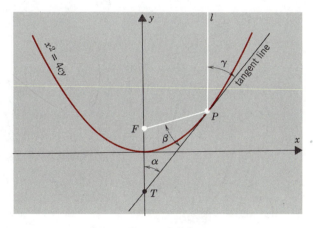

Figure 9.3.8

In the figure we have drawn a ray (a half-line) l parallel to the axis of the parabola, the y-axis. We want to show that the angles marked β and γ are equal.

Setting $x = 0$ in equation (1), we find that

$$y = y_0 - \frac{x_0^2}{2c}.$$

Since the point (x_0, y_0) lies on the parabola, we have $x_0^2 = 4cy_0$, and thus

$$y_0 - \frac{x_0^2}{2c} = y_0 - \frac{4cy_0}{2c} = -y_0.$$

The y-coordinate of the point marked T is $-y_0$. Since the focus F is at $(0, c)$,

$$d(F, T) = y_0 + c.$$

The distance between F and P is also $y_0 + c$:

$$d(F, P) = \sqrt{x_0^2 + (y_0 - c)^2} = \sqrt{4cy_0 + (y_0 - c)^2} = \sqrt{(y_0 + c)^2} = y_0 + c.$$
$$x_0^2 = 4cy_0 \overset{\textstyle\uparrow}{}$$

Since $d(F, T) = d(F, P)$, the triangle TFP is isosceles and the angles marked α and β are equal. Since l is parallel to the y-axis, $\alpha = \gamma$ and thus (and this is what we wanted to show)

$$\beta = \gamma.$$

The fact that $\beta = \gamma$ has important optical consequences. It means (Figure 9.3.9) that light emitted from a source at the focus of a parabolic mirror is reflected in a beam parallel to the axis of that mirror; this is the principle of the searchlight. It also means that light coming to a parabolic mirror in a beam parallel to the axis of the mirror is reflected entirely to the focus; this is the principle of the reflecting telescope.

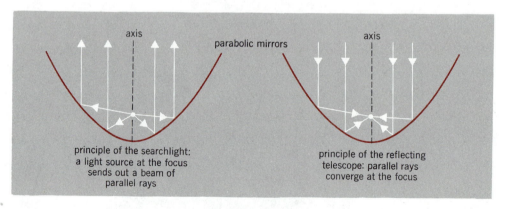

Figure 9.3.9

Parabolic Trajectories

In the early part of the seventeenth century Galileo Galilei observed the motion of stones projected from the tower of Pisa and observed that their trajectory was parabolic. By simple calculus, together with some simplifying physical assumptions, we obtain results that agree with Galileo's observations.

Figure 9.3.10

Figure 9.3.11

Consider a projectile fired at angle θ from a point (x_0, y_0) with initial velocity v_0 (Figure 9.3.10). The horizontal component of v_0 is $v_0 \cos \theta$, and the vertical component is $v_0 \sin \theta$ (Figure 9.3.11).

Let's neglect air resistance and the curvature of the earth. Under these circumstances there is no horizontal acceleration:

$$x''(t) = 0.$$

The only vertical acceleration is due to gravity:

$$y''(t) = -g.$$

From the first equation, we have

$$x'(t) = C$$

and, since $x'(0) = v_0 \cos \theta$,

$$x'(t) = v_0 \cos \theta.$$

Integrating again, we have

$$x(t) = (v_0 \cos \theta)t + C$$

and, since $x(0) = x_0$,

(1) $$x(t) = (v_0 \cos \theta)t + x_0.$$

The relation $y''(t) = -g$ gives

$$y'(t) = -gt + C$$

and, since $y'(0) = v_0 \sin \theta$,

$$y'(t) = -gt + v_0 \sin \theta.$$

Integrating again, we find that

$$y(t) = -\tfrac{1}{2}gt^2 + (v_0 \sin \theta)t + C.$$

Since $y(0) = y_0$, we have

(2) $$y(t) = -\tfrac{1}{2}gt^2 + (v_0 \sin \theta)t + y_0.$$

From (1)

$$t = \frac{1}{v_0 \cos \theta} [x(t) - x_0].$$

Substitute this expression for t in (2) and you will find that

$$y(t) = -\frac{g}{2v_0^2} \sec^2 \theta \, [x(t) - x_0]^2 + \tan \theta \, [x(t) - x_0] + y_0.$$

The trajectory (the path followed by the projectile) is the curve

(9.3.4) $$y = -\frac{g}{2v_0^2} \sec^2 \theta \, [x - x_0]^2 + \tan \theta \, [x - x_0] + y_0.$$

This is a quadratic in x and therefore a parabola. ☐

EXERCISES 9.3

Sketch the parabola and give an equation for it.

1. vertex $(0, 0)$, focus $(2, 0)$.
2. vertex $(0, 0)$, focus $(-2, 0)$.
3. vertex $(-1, 3)$, focus $(-1, 0)$.
4. vertex $(1, 2)$, focus $(1, 3)$.
5. focus $(1, 1)$, directrix $y = -1$.
6. focus $(2, -2)$, directrix $x = -5$.
7. focus $(1, 1)$, directrix $x = 2$.
8. focus $(2, 0)$, directrix $y = 3$.

Find the vertex, focus, axis, and directrix; then sketch the parabola.

9. $y^2 = 2x$.
10. $x^2 = -5y$.
11. $2y = 4x^2 - 1$.
12. $y^2 = 2(x - 1)$.
13. $(x + 2)^2 = 12 - 8y$.
14. $y - 3 = 2(x - 1)^2$.
15. $x = y^2 + y + 1$.
16. $y = x^2 + x + 1$.

Find an equation for the indicated parabola.

17. focus $(1, 2)$, directrix $x + y + 1 = 0$.
18. vertex $(2, 0)$, directrix $2x - y = 0$.
19. vertex $(2, 0)$, focus $(0, 2)$.
20. vertex $(3, 0)$, focus $(0, 1)$.

21. Show that every parabola has an equation of the form

$$(\alpha x + \beta y)^2 = \gamma x + \delta y + \epsilon \quad \text{with} \quad \alpha^2 + \beta^2 \neq 0.$$

   HINT: Take l: $Ax + By + C = 0$ as the directrix and $F(a, b)$ as the focus.

22. A line through the focus of a parabola intersects the parabola at two points P and Q. Show that the tangent line through P meets the tangent line through Q at right angles.

23. A parabola intersects a rectangle of area A at two opposite vertices. Show that, if one side of the rectangle falls on the axis of the parabola, then the parabola subdivides the rectangle into two pieces, one of area $\frac{1}{3}A$, the other of area $\frac{2}{3}A$.

24. (a) Show that every parabola with axis parallel to the y-axis has an equation of the form $y = Ax^2 + Bx + C$ with $A \neq 0$. (b) Find the vertex, the focus, and the directrix of the parabola $y = Ax^2 + Bx + C$.

25. Find equations for all the parabolas that pass through the point $(5, 6)$ and have directrix $y = 1$, axis $x = 2$.

26. Find an equation for the parabola that has horizontal axis, vertex $(-1, 1)$, and passes through the point $(-6, 13)$.

The line that passes through the focus of a parabola and is parallel to the directrix intersects the parabola at two points A and B. The line segment $\overline{AB}$ is called the *latus rectum* of the parabola. In Exercises 27 – 30 we work with the parabola $x^2 = 4cy$, $c > 0$. By Ω we mean the region bounded below by the parabola and above by the latus rectum.

27. Find the length of the latus rectum.

28. What is the slope of the parabola at the endpoints of the latus rectum?

29. Determine the area of Ω and locate the centroid.

30. Find the volume of the solid generated by revolving Ω about the y-axis and determine the centroid of the solid (6.4.5).

In Exercises 31 – 36 we neglect air resistance and the curvature of the earth. We measure distance in feet, time in seconds, and set $g = 32$ feet per second per second. We take O as the origin, the x-axis as ground level, and consider a projectile fired from O at an angle θ with initial velocity v_0.

31. Find an equation for the trajectory.

32. What is the maximum height attained by the projectile?

33. Find the range of the projectile.

34. How many seconds after firing does the impact take place?

35. How should θ be chosen so as to maximize the range?

36. How should θ be chosen so that the range becomes r?

37. Suppose that a flexible inelastic cable (Figure 9.3.12) fixed at the ends supports a horizontal load. (Imagine a suspension bridge and think of the load on the cable as the roadway.) Show that, if the load has constant weight per unit length, then the cable hangs in the form of a parabola.

HINT: The part of the cable that supports the load from 0 to x is subject to the following forces:

      (1) the weight of the load, which in this case is proportional to x

      (2) the horizontal pull at 0: $p(0)$

      (3) the tangential pull at x: $p(x)$.

Balancing the vertical forces we have

$$kx = p(x) \sin \theta. \qquad \text{[weight = vertical pull at } x\text{]}$$

Balancing the horizontal forces we have

$$p(0) = p(x) \cos \theta. \qquad \text{[pull at 0 = horizontal pull at } x\text{]}$$

Figure 9.3.12

38. The parabolic mirror of a telescope gathers parallel light rays from a distant star and directs them all to the focus. Show that all the light paths to the focus are of the same length.

39. All equilateral triangles are similar; they differ only in scale. Show that the same is true of all parabolas.

9.4 THE ELLIPSE

Start with two points F_1, F_2 and a number k greater than the distance between them.

(9.4.1)

> The set of all points P for which
>
> $$d(P, F_1) + d(P, F_2) = k$$
>
> is called an *ellipse*. F_1 and F_2 are called the *foci*.

The idea is illustrated in Figure 9.4.1. A string is looped over tacks placed at the foci. The pencil placed in the loop traces out an ellipse.

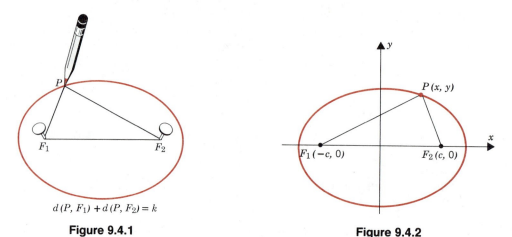

$$d(P, F_1) + d(P, F_2) = k$$

Figure 9.4.1   **Figure 9.4.2**

Figure 9.4.2 shows an ellipse in what is called *standard position*: foci along the x-axis at equal distances from the origin. We will derive an equation for this ellipse setting $k = 2a$.

A point $P(x, y)$ lies on the ellipse iff

$$d(P, F_1) + d(P, F_2) = 2a.$$

With F_1 at $(-c, 0)$ and F_2 at $(c, 0)$ we have

$$\sqrt{(x + c)^2 + y^2} + \sqrt{(x - c)^2 + y^2} = 2a.$$

Transferring the second term to the right-hand side and squaring both sides, we get

$$(x + c)^2 + y^2 = 4a^2 + (x - c)^2 + y^2 - 4a\sqrt{(x - c)^2 + y^2}.$$

This reduces to

$$4a\sqrt{(x - c)^2 + y^2} = 4(a^2 - cx).$$

Canceling the factor 4 and squaring again, we obtain

$$a^2(x^2 - 2cx + c^2 + y^2) = a^4 - 2a^2cx + c^2x^2.$$

This in turn reduces to

$$(a^2 - c^2)x^2 + a^2y^2 = a^2(a^2 - c^2),$$

which we write as

(9.4.2)
$$\frac{x^2}{a^2} + \frac{y^2}{a^2 - c^2} = 1. \qquad \square$$

Usually we set $b = \sqrt{a^2 - c^2}$. The equation for an ellipse in standard position then takes the form

(9.4.3)
$$\frac{x^2}{a^2} + \frac{y^2}{b^2} = 1 \qquad \text{with} \quad a > b.$$

The roles played by a, b, c can be read from Figure 9.4.3.

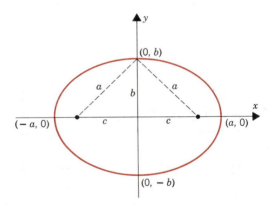

Figure 9.4.3

Every ellipse has four *vertices*. In Figure 9.4.3 these are marked $(a, 0)$, $(-a, 0)$, $(0, b)$, $(0, -b)$. The line segments that join opposite vertices are called the *axes* of the ellipse. The axis that contains the foci is called the *major axis*, the other the *minor axis*. In standard position the major axis is horizontal and has length $2a$; the minor axis is vertical and has length $2b$. The point at which the axes intersect is called the *center* of the ellipse. In standard position the center is at the origin.

Example 1. The equation $16x^2 + 25y^2 = 400$ can be written

$$\frac{x^2}{25} + \frac{y^2}{16} = 1. \qquad \text{(divide by 400)}$$

Here $a = 5$, $b = 4$, and $c = \sqrt{a^2 - b^2} = \sqrt{9} = 3$. The equation is in the form of (9.4.2). It is an ellipse in standard position with foci at $(-3, 0)$ and $(3, 0)$. The major axis has length $2a = 10$, and the minor axis has length $2b = 8$. The center is at the origin. The ellipse is sketched in Figure 9.4.4. □

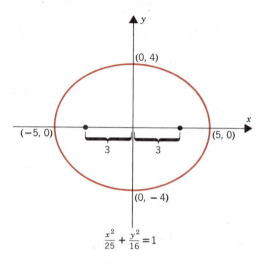

$$\frac{x^2}{25} + \frac{y^2}{16} = 1$$

Figure 9.4.4

Example 2. The equation

$$\frac{x^2}{16} + \frac{y^2}{25} = 1$$

does not represent an ellipse in standard position because $25 > 16$. This equation is the equation of Example 1 with x and y interchanged. It represents the ellipse of Example 1 reflected in the line $y = x$. (See Figure 9.4.5.) The foci are now on the y-axis, at $(0, -3)$ and $(0, 3)$. The major axis, now vertical, has length 10, and the minor axis has length 8. The center remains at the origin. □

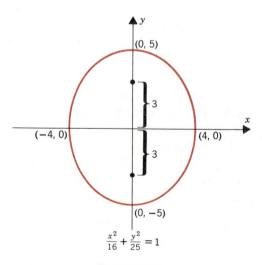

$$\frac{x^2}{16} + \frac{y^2}{25} = 1$$

Figure 9.4.5

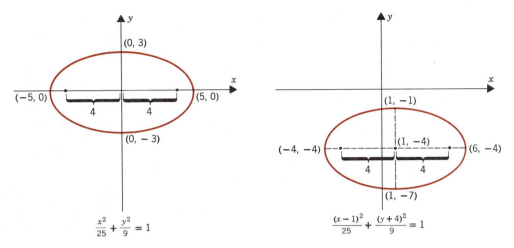

Figure 9.4.6

Example 3. Figure 9.4.6 shows two ellipses:

$$\frac{x^2}{25} + \frac{y^2}{9} = 1 \qquad \text{and} \qquad \frac{(x-1)^2}{25} + \frac{(y+4)^2}{9} = 1.$$

The first ellipse is in standard position. Here $a = 5$, $b = 3$, $c = \sqrt{a^2 - b^2} = 4$. The foci are at $(-4, 0)$ and $(4, 0)$. The major axis has length 10, and the minor axis has length 6.

The second ellipse is the first ellipse displaced 1 unit right and 4 units down. The center is now at the point $(1, -4)$. The foci are at $(-3, -4)$ and $(5, -4)$. ◻

Example 4. To identify the curve

$$4x^2 - 8x + y^2 + 4y - 8 = 0.$$

we write

$$4(x^2 - 2x + \quad) + (y^2 + 4y + \quad) = 8.$$

By completing the squares within the parentheses, we get

$$4(x^2 - 2x + 1) + (y^2 + 4y + 4) = 16,$$
$$4(x - 1)^2 + (y + 2)^2 = 16,$$

(∗) $$\frac{(x-1)^2}{4} + \frac{(y+2)^2}{16} = 1.$$

This is the ellipse

(∗∗) $$\frac{x^2}{16} + \frac{y^2}{4} = 1 \qquad (a = 4, b = 2, c = \sqrt{16 - 4} = 2\sqrt{3})$$

reflected in the line $y = x$ and then displaced 1 unit right and 2 units down. Since the foci of (∗∗) are at $(-2\sqrt{3}, 0)$ and $(2\sqrt{3}, 0)$, the foci of (∗) are at $(1, -2 - 2\sqrt{3})$ and $(1, -2 + 2\sqrt{3})$. The major axis, now vertical, has length 8; the minor axis has length 4. ◻

Elliptical Reflectors

Like the parabola, the ellipse has an interesting reflecting property. To derive it, we consider the ellipse

$$\frac{x^2}{a^2} + \frac{y^2}{b^2} = 1.$$

Differentiation with respect to x gives

$$\frac{2x}{a^2} + \frac{2y}{b^2}\frac{dy}{dx} = 0 \qquad \text{and thus} \qquad \frac{dy}{dx} = -\frac{b^2x}{a^2y}.$$

The slope at the point (x_0, y_0) is therefore

$$-\frac{b^2x_0}{a^2y_0}$$

and the tangent line has equation

$$y - y_0 = -\frac{b^2x_0}{a^2y_0}(x - x_0).$$

We can rewrite this last equation as

$$(b^2x_0)x + (a^2y_0)y - a^2b^2 = 0.$$

We can now show the following:

(9.4.4)   At each point P of the ellipse, the focal radii $\overline{F_1P}$ and $\overline{F_2P}$ make equal angles with the tangent.

Proof. If P lies on the x-axis, the focal radii are coincident and there is nothing to show. To visualize the argument for a point $P = P(x_0, y_0)$ not on the x-axis, see Figure 9.4.7.

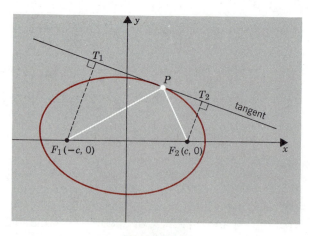

Figure 9.4.7

To show that $\overline{F_1P}$ and $\overline{F_2P}$ make equal angles with the tangent we need only show that the triangles PT_1F_1 and PT_2F_2 are similar. We can do this by showing that

$$\frac{d(T_1, F_1)}{d(F_1, P)} = \frac{d(T_2, F_2)}{d(F_2, P)}$$

or, equivalently, by showing that

$$\frac{|-b^2x_0c - a^2b^2|}{\sqrt{(x_0 + c)^2 + y_0^2}} = \frac{|b^2x_0c - a^2b^2|}{\sqrt{(x_0 - c)^2 + y_0^2}}.$$

The validity of this last equation can be seen by canceling the factor b^2 and then squaring. This gives

$$\frac{(x_0c + a^2)^2}{(x_0 + c)^2 + y_0^2} = \frac{(x_0c - a^2)^2}{(x_0 - c)^2 + y_0^2},$$

which can be simplified to

$$(a^2 - c^2)x_0^2 + a^2y_0^2 = a^2(a^2 - c^2) \quad \text{and thus to} \quad \frac{x_0^2}{a^2} + \frac{y_0^2}{b^2} = 1.$$

This last equation holds since the point $P(x_0, y_0)$ is on the ellipse. $\quad\square$

The result we just proved has the following physical consequence:

(9.4.5) | An elliptical reflector takes light or sound originating at one focus and converges it at the other focus.

In elliptical rooms called "whispering chambers," a whisper at one focus, inaudible nearby, is easily heard at the other focus. You will experience this phenomenon if you visit the Statuary Room in the Capitol.

EXERCISES 9.4

For each of the following ellipses find (a) the center, (b) the foci, (c) the length of the major axis, and (d) the length of the minor axis. Then sketch the figure.

1. $x^2/9 + y^2/4 = 1$.
2. $x^2/4 + y^2/9 = 1$.
3. $3x^2 + 2y^2 = 12$.
4. $3x^2 + 4y^2 - 12 = 0$.
5. $4x^2 + 9y^2 - 18y = 27$.
6. $4x^2 + y^2 - 6y + 5 = 0$.
7. $4(x - 1)^2 + y^2 = 64$.
8. $16(x - 2)^2 + 25(y - 3)^2 = 400$.

Find an equation for the ellipse that satisfies the given conditions.

9. Foci at $(-1, 0)$, $(1, 0)$; major axis 6.
10. Foci at $(0, -1)$, $(0, 1)$; major axis 6.
11. Foci at $(1, 3)$, $(1, 9)$; minor axis 8.
12. Foci at $(3, 1)$, $(9, 1)$; minor axis 10.
13. Focus at $(1, 1)$; center at $(1, 3)$; major axis 10.
14. Center at $(2, 1)$; vertices at $(2, 6)$, $(1, 1)$.
15. Major axis 10; vertices at $(3, 2)$, $(3, -4)$.
16. Focus at $(6, 2)$; vertices at $(1, 7)$, $(1, -3)$.

17. What is the length of the string in Figure 9.4.1?

18. Show that the set of all points ($a \cos t$, $b \sin t$) with t real lie on an ellipse.

19. Find the distance between the foci of an ellipse of area A if the length of the major axis is $2a$.

20. Show that in an ellipse the product of the distances between the foci and a tangent to the ellipse [$d(F_1, T_1)\, d(F_2, T_2)$ in Figure 9.4.7] is always the square of one-half the length of the minor axis.

21. Locate the foci of the ellipse given that the point (3, 4) lies on the ellipse and the ends of the major axis are at (0, 0) and (10, 0).

22. Find the centroid of the first-quadrant portion of the elliptical region $b^2x^2 + a^2y^2 \leq a^2b^2$.

Although all parabolas have exactly the same shape (Exercise 39, Section 9.3), ellipses come in different shapes. The shape of an ellipse depends on its *eccentricity e*. This is half the distance between the foci divided by half the length of the major axis:

(9.4.6)
$$e = c/a.$$

For every ellipse $0 < e < 1$.
Determine the eccentricity of the ellipse.

23. $x^2/25 + y^2/16 = 1$.　　　　　　　　**24.** $x^2/16 + y^2/25 = 1$.

25. $(x - 1)^2/25 + (y + 2)^2/9 = 1$.　　　**26.** $(x + 1)^2/169 + (y - 1)^2/144 = 1$.

27. Suppose that E_1 and E_2 are both ellipses with the same major axis. Compare the shape of E_1 to the shape of E_2 if $e_1 < e_2$.

28. What happens to an ellipse with major axis $2a$ if e tends to zero?

29. What happens to an ellipse with major axis $2a$ if e tends to 1?

Write an equation for the ellipse.

30. Major axis from $(-3, 0)$ to $(3, 0)$, eccentricity $\frac{1}{3}$.

31. Major axis from $(-3, 0)$ to $(3, 0)$, eccentricity $\frac{2}{3}\sqrt{2}$.

32. Let l be a line and let F be a point not on l. You have seen that the set of points P for which

$$d(F, P) = d(l, P)$$

is a parabola. Show that, if $0 < e < 1$, then the set of all points P for which

$$d(F, P) = e\, d(l, P)$$

is an ellipse of eccentricity e. HINT: Begin by choosing a coordinate system whereby F falls on the origin and l is a vertical line $x = d$.

9.5　THE HYPERBOLA

Start with two points F_1, F_2 and take a positive number k less than the distance between them.

(9.5.1)

The set of all points P for which

$$|d(P, F_1) - d(P, F_2)| = k$$

is called a *hyperbola*. F_1 and F_2 are called the *foci*.

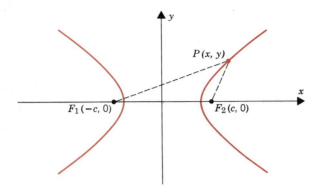

Figure 9.5.1

Figure 9.5.1 shows a hyperbola in what is called *standard position*: foci along the x-axis at equal distances from the origin. We will derive an equation for this hyperbola setting $k = 2a$.

A point $P(x, y)$ lies on the hyperbola iff

$$|d(P, F_1) - d(P, F_2)| = 2a.$$

With F_1 at $(-c, 0)$ and F_2 at $(c, 0)$ we have

$$\sqrt{(x + c)^2 + y^2} - \sqrt{(x - c)^2 + y^2} = \pm 2a. \qquad \text{(explain)}$$

Transferring the second term to the right and squaring both sides, we obtain

$$(x + c)^2 + y^2 = 4a^2 \pm 4a\sqrt{(x - c)^2 + y^2} + (x - c)^2 + y^2.$$

This equation reduces to

$$xc - a^2 = \pm a\sqrt{(x - c)^2 + y^2}.$$

Squaring once more, we find that

$$x^2c^2 - 2a^2xc + a^4 = a^2(x^2 - 2xc + c^2 + y^2),$$

which reduces to

$$(c^2 - a^2)x^2 - a^2y^2 = a^2(c^2 - a^2),$$

and thus to

(9.5.2) $$\frac{x^2}{a^2} - \frac{y^2}{c^2 - a^2} = 1. \qquad \square$$

Usually we set $b = \sqrt{c^2 - a^2}$. The equation for a hyperbola in standard position then takes the form

(9.5.3) $$\frac{x^2}{a^2} - \frac{y^2}{b^2} = 1.$$

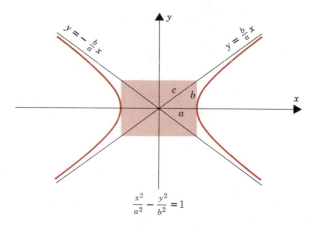

Figure 9.5.2

The roles played by a, b, c can be read from Figure 9.5.2.
As the figure suggests, the hyperbola remains between the lines

$$y = \frac{b}{a}x \quad \text{and} \quad y = -\frac{b}{a}x.$$

These lines are called the *asymptotes* of the hyperbola. They can be obtained from Equation 9.5.2 by replacing the 1 on the right-hand side by 0:

$$\frac{x^2}{a^2} - \frac{y^2}{b^2} = 0 \qquad \text{gives} \qquad y = \pm\frac{b}{a}x.$$

As $x \to \pm\infty$, the vertical separation between the hyperbola and the asymptotes tends to zero. To see this, solve the equation

$$\frac{x^2}{a^2} - \frac{y^2}{b^2} = 1$$

for y. This gives

$$y = \pm\sqrt{\frac{b^2}{a^2}x^2 - b^2}.$$

In all four quadrants the vertical separation between the hyperbola and the asymptotes can be written

$$\left| \frac{b}{a}|x| - \sqrt{\frac{b^2}{a^2}x^2 - b^2} \right|. \qquad \text{(check this)}$$

As $x \to \pm\infty$,

$$\left| \frac{b}{a}|x| - \sqrt{\frac{b^2}{a^2}x^2 - b^2} \right| = \left| \frac{b}{a}|x| - \sqrt{\frac{b^2}{a^2}x^2 - b^2} \right| \frac{\left| \frac{b}{a}|x| + \sqrt{\frac{b^2}{a^2}x^2 - b^2} \right|}{\left| \frac{b}{a}|x| + \sqrt{\frac{b^2}{a^2}x^2 - b^2} \right|}$$

$$= \frac{b^2}{\left| \frac{b}{a}|x| + \sqrt{\frac{b^2}{a^2}x^2 - b^2} \right|} \longrightarrow 0. \quad \square$$

The line determined by the foci of a hyperbola intersects the hyperbola at two points, called the *vertices*. The line segment that joins the vertices is called the *transverse axis*. The midpoint of the transverse axis is called the *center* of the hyperbola.

In standard position (Figure 9.5.2), the vertices are $(\pm a, 0)$, the transverse axis has length $2a$, and the center is at the origin.

Example 1. The equation

$$\frac{x^2}{1} - \frac{y^2}{3} = 1 \qquad \text{(Figure 9.5.3)}$$

represents a hyperbola in standard position; here $a = 1, b = \sqrt{3}, c = \sqrt{1 + 3} = 2$. The center is at the origin. The foci are at $(-2, 0)$ and $(2, 0)$. The vertices are at $(-1, 0)$ and $(1, 0)$. The transverse axis has length 2. We can obtain the asymptotes by setting

$$\frac{x^2}{1} - \frac{y^2}{3} = 0.$$

The asymptotes are the lines $y = \pm\sqrt{3}x$. □

Figure 9.5.3

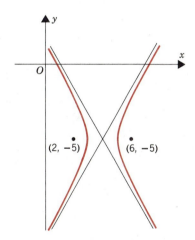

Figure 9.5.4

Example 2. The hyperbola

$$\frac{(x - 4)^2}{1} - \frac{(y + 5)^2}{3} = 1 \qquad \text{(Figure 9.5.4)}$$

is the hyperbola

$$\frac{x^2}{1} - \frac{y^2}{3} = 1$$

of Example 1 displaced 4 units right and 5 units down. The center of the hyperbola is now at the point $(4, -5)$. The foci are at $(2, -5)$ and $(6, -5)$. The vertices are at $(3, -5)$ and $(5, -5)$. The new asymptotes are the lines $y + 5 = \pm\sqrt{3}(x - 4)$. □

Example 3.　The hyperbola

$$\frac{y^2}{1} - \frac{x^2}{3} = 1 \qquad \text{(Figure 9.5.5)}$$

is the hyperbola

$$\frac{x^2}{1} - \frac{y^2}{3} = 1$$

of Example 1 reflected in the line $y = x$. The center is still at the origin. The foci are now at $(0, -2)$ and $(0, 2)$. The vertices are at $(0, -1)$ and $(0, 1)$. The asymptotes are the lines $x = \pm\sqrt{3}y$. ☐

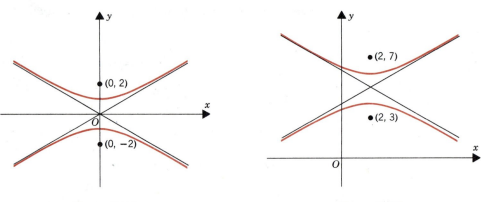

Figure 9.5.5　　　　　　　　**Figure 9.5.6**

Example 4.　The hyperbola

$$\frac{(y - 5)^2}{1} - \frac{(x - 2)^2}{3} = 1 \qquad \text{(Figure 9.5.6)}$$

is the hyperbola of Example 3 displaced 2 units right and 5 units up. ☐

Hyperbolic Reflectors

A straightforward calculation that you are asked to carry out in the exercises shows that

(9.5.4)　　*at each point P of a hyperbola, the tangent line bisects the angle between the focal radii $\overline{F_1P}$ and $\overline{F_2P}$.*

The optical consequences of this are illustrated in Figure 9.5.7. There you see the right branch of a hyperbola with foci F_1, F_2. Light or sound aimed at F_2 from any point to the left of the reflector is beamed to F_1.

Figure 9.5.7

An Application to Range Finding

If observers, located at two listening posts at a known distance apart, time the firing of a cannon, the time difference multiplied by the velocity of sound gives the value of $2a$ and hence determines a hyperbola on which the cannon must be located. A third listening post gives two more hyperbolas. The cannon is found where the hyperbolas intersect.

(Simple variants of this appear in the exercises.)

EXERCISES 9.5

Find an equation for the indicated hyperbola.

1. Foci at $(-5, 0)$, $(5, 0)$; transverse axis 6.
2. Foci at $(-13, 0)$, $(13, 0)$; transverse axis 10.
3. Foci at $(0, -13)$, $(0, 13)$; transverse axis 10.
4. Foci at $(0, -13)$, $(0, 13)$; transverse axis 24.
5. Foci at $(-5, 1)$, $(5, 1)$; transverse axis 6.
6. Foci at $(-3, 1)$, $(7, 1)$; transverse axis 6.
7. Foci at $(-1, -1)$, $(-1, 1)$; transverse axis $\frac{1}{2}$.
8. Foci at $(2, 1)$, $(2, 5)$; transverse axis 3.

For each of the following hyperbolas find the center, the vertices, the foci, the asymptotes, and the length of the transverse axis. Then sketch the figure.

9. $x^2 - y^2 = 1$.   10. $y^2 - x^2 = 1$.   11. $x^2/9 - y^2/16 = 1$.

12. $x^2/16 - y^2/9 = 1$.   13. $y^2/16 - x^2/9 = 1$.   14. $y^2/9 - x^2/16 = 1$.

15. $(x - 1)^2/9 - (y - 3)^2/16 = 1$.   16. $(x - 1)^2/16 - (y - 3)^2/9 = 1$.

17. $4x^2 - 8x - y^2 + 6y - 1 = 0$.   18. $-3x^2 + y^2 - 6x = 0$.

19. Find the center, the vertices, the foci, the asymptotes, and the length of the transverse axis of the hyperbola with equation $xy = 1$.   HINT: Define new XY-coordinates by setting $x = X + Y$ and $y = X - Y$.

For Exercises 20–22 we refer to the hyperbola in Figure 9.5.2.

20. Find functions $x = x(t)$, $y = y(t)$ such that, as t ranges over the set of real numbers, the points $(x(t), y(t))$ traverse (a) the right branch of the hyperbola. (b) the left branch of the hyperbola.

21. Find the area of the region between the right branch of the hyperbola and the vertical line $x = 2a$.

22. Show that at each point P of the hyperbola the tangent at P bisects the angle between the focal radii $\overline{F_1 P}$ and $\overline{F_2 P}$.

The shape of a hyperbola is determined by its *eccentricity e*. This is half the distance between the foci divided by half the length of the transverse axis:

(9.5.5)

$$e = c/a.$$

For all hyperbolas $e > 1$.

Determine the eccentricity of the hyperbola.

23. $x^2/9 - y^2/16 = 1$.   **24.** $x^2/16 - y^2/9 = 1$.

25. $x^2 - y^2 = 1$.   **26.** $x^2/25 - y^2/144 = 1$.

27. Suppose H_1 and H_2 are both hyperbolas with the same transverse axis. Compare the shape of H_1 to the shape of H_2 if $e_1 < e_2$.

28. What happens to a hyperbola if e tends to 1?

29. What happens to a hyperbola if e increases without bound?

30. (Compare to Exercise 32 of Section 9.4.) Let l be a line and let F be a point not on l. Show that, if $e > 1$, then the set of all points P for which

$$d(F, P) = e \, d(l, P)$$

is a hyperbola of eccentricity e. HINT: Begin by choosing a coordinate system whereby F falls on the origin and l is a vertical line $x = d$.

31. (**Calculator**) A meteor crashes somewhere in the hills that lie north of point A. The impact is heard at point A and four seconds later it is heard at point B. Two seconds still later it is heard at point C. Locate the point of impact given that A lies two miles due east of B and two miles due west of C. (Take 0.20 miles per second as the speed of sound.)

32. (**Calculator**) A radio signal is received at the points marked P_1, P_2, P_3, P_4 in Figure 9.5.8. Suppose that the signal arrives at P_1 six hundred microseconds after it arrives at P_2 and arrives at P_4 eight hundred microseconds after it arrives at P_3. Locate the source of the signal given that radio waves travel at the speed of light, 186,000 miles per second. (A microsecond is a millionth of a second.)

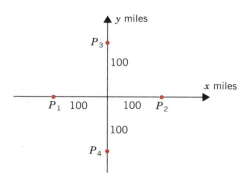

Figure 9.5.8

*9.6   ROTATIONS; ELIMINATING THE xy-TERM

We begin by referring to Figure 9.6.1. From the figure

$$\cos \theta = \frac{x}{r}, \qquad \sin \theta = \frac{y}{r}.$$

Thus

(9.6.1) $\boxed{x = r \cos \theta, \qquad y = r \sin \theta.}$

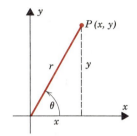

Figure 9.6.1

These equations come up repeatedly in calculus. We will consider them in more detail in Section 10.1 when we take up polar coordinates.

Figure 9.6.2

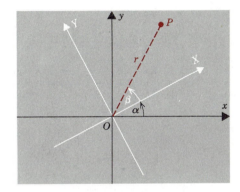

Figure 9.6.3

Consider now a rectangular coordinate system Oxy. If we rotate this system counterclockwise α radians about the origin, we obtain a new coordinate system OXY. See Figure 9.6.2.

A point P will now have two pairs of rectangular coordinates:

(x, y) in the Oxy system   and (X, Y) in the OXY system.

Here we investigate the relation between (x, y) and (X, Y). With P as in Figure 9.6.3,

$$x = r \cos (\alpha + \beta), \qquad y = r \sin (\alpha + \beta)$$

and

$$X = r \cos \beta, \qquad Y = r \sin \beta.$$

Since

$$\cos (\alpha + \beta) = \cos \alpha \cos \beta - \sin \alpha \sin \beta,$$
$$\sin (\alpha + \beta) = \sin \alpha \cos \beta + \cos \alpha \sin \beta,$$

we have

$$x = r \cos(\alpha + \beta) = (\cos \alpha)\, r \cos \beta - (\sin \alpha)\, r \sin \beta,$$
$$y = r \sin(\alpha + \beta) = (\sin \alpha)\, r \cos \beta + (\cos \alpha)\, r \sin \beta,$$

and therefore

(9.6.2) $\boxed{x = (\cos \alpha)X - (\sin \alpha)Y, \qquad y = (\sin \alpha)X + (\cos \alpha)Y.}$

These formulas give the algebraic consequences of a counterclockwise rotation of α radians.

Eliminating the *xy*-Term

Rotations of the coordinate system enable us to simplify equations of the second degree by eliminating the *xy*-term; that is, if in the *Oxy* system, *S* has an equation of the form

(1) $\qquad\qquad\qquad ax^2 + bxy + cy^2 + dx + ey + f = 0 \qquad$ with $\quad b \neq 0,$

then there exists a coordinate system *OXY*, differing from *Oxy* by a rotation, such that in the *OXY* system *S* has an equation of the form

$$AX^2 + CY^2 + DX + EY + F = 0.$$

To see this, substitute

$$x = (\cos \alpha)X - (\sin \alpha)Y, \qquad y = (\sin \alpha)X + (\cos \alpha)Y$$

in equation (1). This will give you a second-degree equation in *X* and *Y* in which the coefficient of *XY* is

$$-2a \cos \alpha \sin \alpha + b(\cos^2 \alpha - \sin^2 \alpha) + 2c \cos \alpha \sin \alpha.$$

This can be simplified to

$$(c - a) \sin 2\alpha + b \cos 2\alpha.$$

To eliminate the *XY* term we must have this coefficient equal to zero, that is, we must have

$$(a - c) \sin 2\alpha = b \cos 2\alpha.$$

There are two possibilities here: $a = c$, $a \neq c$.

I. If $a = c$, then

$$\cos 2\alpha = 0,$$
$$2\alpha = \tfrac{1}{2}\pi + n\pi, \qquad\qquad (\text{n an arbitrary integer})$$
$$\alpha = \tfrac{1}{4}\pi + \tfrac{1}{2}n\pi.$$

Thus, if $a = c$, we can eliminate the *XY* term by setting

(9.6.3) $\qquad\qquad\qquad\qquad \boxed{\alpha = \tfrac{1}{4}\pi.} \qquad\qquad\qquad (\text{choose } n = 0)$

II. If $a \neq c$, then

$$\tan 2\alpha = \frac{b}{a - c},$$

$$2\alpha = \tan^{-1}\left(\frac{b}{a - c}\right) + n\pi,$$

$$\alpha = \tfrac{1}{2}\tan^{-1}\left(\frac{b}{a - c}\right) + \tfrac{1}{2}n\pi.$$

Thus, if $a \neq c$, we can eliminate the XY term by setting

(9.6.4)
$$\alpha = \tfrac{1}{2}\tan^{-1}\left(\frac{b}{a - c}\right).$$
(choose $n = 0$)

Example 1. In the case of

$$xy - 2 = 0,$$

we have $a = c$ and thus can choose $\alpha = \tfrac{1}{4}\pi$. Setting

$$x = (\cos\tfrac{1}{4}\pi)X - (\sin\tfrac{1}{4}\pi)Y = \tfrac{1}{2}\sqrt{2}\,(X - Y),$$
$$y = (\sin\tfrac{1}{4}\pi)X + (\cos\tfrac{1}{4}\pi)Y = \tfrac{1}{2}\sqrt{2}\,(X + Y),$$

we find that $xy - 2 = 0$ becomes

$$\tfrac{1}{2}(X^2 - Y^2) - 2 = 0,$$

which can be written

$$\frac{X^2}{4} - \frac{Y^2}{4} = 1.$$

This is the equation of a hyperbola in standard position in the OXY system. The hyperbola is shown in Figure 9.6.4. □

Figure 9.6.4

Example 2. In the case of

$$11x^2 + 4\sqrt{3}xy + 7y^2 - 1 = 0,$$

we have $a = 11$, $b = 4\sqrt{3}$, and $c = 7$. Thus we can choose

$$\alpha = \tfrac{1}{2}\tan^{-1}\left(\frac{b}{a-c}\right) = \tfrac{1}{2}\tan^{-1}\sqrt{3} = \tfrac{1}{6}\pi.$$

Setting

$$x = (\cos\tfrac{1}{6}\pi)X - (\sin\tfrac{1}{6}\pi)Y = \tfrac{1}{2}(\sqrt{3}X - Y),$$
$$y = (\sin\tfrac{1}{6}\pi)X + (\cos\tfrac{1}{6}\pi)Y = \tfrac{1}{2}(X + \sqrt{3}Y),$$

we find that our initial equation simplifies to $13X^2 + 5Y^2 - 1 = 0$, which we can write as

$$\frac{X^2}{(1/\sqrt{13})^2} + \frac{Y^2}{(1/\sqrt{5})^2} = 1.$$

This is the equation of an ellipse. The ellipse is pictured in Figure 9.6.5 ☐

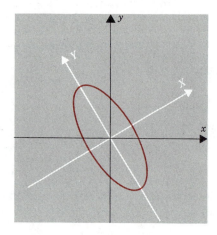

Figure 9.6.5

EXERCISES *9.6

For each of the equations below, (a) find a rotation $\alpha \in (-\tfrac{1}{4}\pi, \tfrac{1}{4}\pi]$ that eliminates the xy-term; (b) rewrite the equation in terms of the new coordinate system; (c) sketch the curve displaying both coordinate systems.

1. $xy = 1$.

2. $xy - y + x = 1$.

3. $11x^2 + 10\sqrt{3}xy + y^2 - 4 = 0$.

4. $52x^2 - 72xy + 73y^2 - 100 = 0$.

5. $x^2 - 2xy + y^2 + x + y = 0$.

6. $3x^2 + 2\sqrt{3}xy + y^2 - 2x + 2\sqrt{3}y = 0$.

7. $x^2 + 2\sqrt{3}xy + 3y^2 + 2\sqrt{3}x - 2y = 0$.

8. $2x^2 + 4\sqrt{3}xy + 6y^2 + (8 - \sqrt{3})x + (8\sqrt{3} + 1)y + 8 = 0$.

Find a rotation $\alpha \in (-\tfrac{1}{4}\pi, \tfrac{1}{4}\pi]$ that eliminates the xy-term. Then find $\cos\alpha$ and $\sin\alpha$.

9. $x^2 + xy + Kx + Ly + M = 0$.

10. $5x^2 + 24xy + 12y^2 + Kx + Ly + M = 0$.

The Second-Degree Equation

It is possible to draw general conclusions about the graph of a second-degree equation

$$ax^2 + bxy + cy^2 + dx + ey + f = 0, \qquad a, b, c \text{ not all } 0,$$

just from the *discriminant* $\Delta = b^2 - 4ac$. There are three cases:

Case 1 If $\Delta < 0$, the graph is an ellipse, a circle, a point, or empty.

Case 2 If $\Delta > 0$, the graph is a hyperbola or a pair of intersecting lines.

Case 3 If $\Delta = 0$, the graph is a parabola, a line, a pair of lines, or empty.

Below we outline how these assertions can be verified. A useful first step is to rotate the coordinate system so that the equation takes the form

$$(1) \qquad AX^2 + CY^2 + DX + EY + F = 0.$$

An elementary but time-consuming computation shows that the discriminant is unchanged by a rotation, so that in this instance we have

$$\Delta = b^2 - 4ac = -4AC.$$

Moreover, A and C cannot both be zero. If $\Delta < 0$, then $AC > 0$ and we can rewrite (1) as

$$\frac{X^2}{C} + \frac{D}{AC} X + \frac{Y^2}{A} + \frac{EY}{AC} + \frac{F}{AC} = 0.$$

By completing the squares, we obtain an equation of the form

$$\frac{(X - \alpha)^2}{(\sqrt{|C|})^2} + \frac{(Y - \beta)^2}{(\sqrt{|A|})^2} = K.$$

If $K > 0$, we have an ellipse or a circle. If $K = 0$, we have the point (α, β). If $K < 0$, the set is empty.

If $\Delta > 0$, then $AC < 0$. Proceeding as before, we obtain an equation of the form

$$\frac{(X - \alpha)^2}{(\sqrt{|C|})^2} - \frac{(Y - \beta)^2}{(\sqrt{|A|})^2} = K.$$

If $K \neq 0$, we have a hyperbola. If $K = 0$, the equation becomes

$$\left(\frac{X - \alpha}{\sqrt{|C|}} - \frac{Y - \beta}{\sqrt{|A|}} \right)\left(\frac{X - \alpha}{\sqrt{|C|}} + \frac{Y - \beta}{\sqrt{|A|}} \right) = 0,$$

so that we have a pair of lines intersecting at the point (α, β).

If $\Delta = 0$, then $AC = 0$, so that either $A = 0$ or $C = 0$. Since A and C are not both zero, there is no loss in generality in assuming that $A \neq 0$ and $C = 0$. In this case equation (1) reduces to

$$AX^2 + DX + EY + F = 0.$$

Dividing by A and completing the square we have an equation of the form

$$(X - \alpha)^2 = \beta Y + K.$$

If $\beta \neq 0$, we have a parabola. If $\beta = 0$ and $K = 0$, we have a line. If $\beta = 0$ and $K > 0$, we have a pair of parallel lines. If $\beta = 0$ and $K < 0$, the set is empty. ▢

9.7   ADDITIONAL EXERCISES

Describe the following curves in detail.

1. $x^2 - 4y - 4 = 0$.
2. $3x^2 + 2y^2 - 6 = 0$.
3. $x^2 - 4y^2 - 10x + 41 = 0$.
4. $9x^2 - 4y^2 - 18x - 8y - 31 = 0$.
5. $x^2 + 3y^2 + 6x + 8 = 0$.
6. $x^2 - 10x - 8y + 41 = 0$.
7. $y^2 + 4y + 2x + 1 = 0$.
8. $9x^2 + 4y^2 - 18x - 8y - 23 = 0$.
9. $9x^2 + 25y^2 + 100y + 99 = 0$.
10. $7x^2 - y^2 + 42x + 14y + 21 = 0$.
11. $7x^2 - 5y^2 + 14x - 40y = 118$.
12. $2x^2 - 3y^2 + 4\sqrt{3}x - 6\sqrt{3}y = 9$.
13. $(x^2 - 4y)(4x^2 + 9y^2 - 36) = 0$.
14. $(x^2 - 4y)(x^2 - 4y^2) = 0$.

CHAPTER HIGHLIGHTS

9.1 Introduction

conic section (p. 467)

9.2 Translations; The Distance Between a Point and a Line

translation (p. 468) $d(P_1, l) = \dfrac{|Ax_1 + By_1 + C|}{\sqrt{A^2 + B^2}}$

9.3 The Parabola

directrix, focus, axis, vertex (p. 471)   latus rectum (p. 479)
reflecting property (p. 475)   parabolic trajectories (p. 476)

A parabola is the set of points equidistant from a fixed line and a fixed point not on that line.

9.4 The Ellipse

foci, standard position (p. 480)   vertices, axes, center (p. 481)
elliptical reflectors (p. 484)   eccentricity (p. 486)

An ellipse is the set of points the sum of whose distances from two fixed points is constant.

9.5 The Hyperbola

foci, standard position (pp. 486–487)   asymptotes (p. 488)
vertices, transverse axis, center (p. 489)   hyperbolic reflectors (p. 490)
eccentricity (p. 492)

A hyperbola is the set of points the difference of whose distances from two fixed points is constant.

*9.6 Rotations; Eliminating the xy-Term

rotation formulas: $x = (\cos \alpha)X - (\sin \alpha)Y$, $y = (\sin \alpha)X + (\cos \alpha)Y$
discriminant of a second degree equation in x and y (p. 497)

9.7 Additional Exercises

C H A P T E R

10

POLAR COORDINATES; PARAMETRIC EQUATIONS

10.1 POLAR COORDINATES

The purpose of coordinates is to fix position with respect to a frame of reference. When we use rectangular coordinates, our frame of reference is a pair of lines that intersect at right angles. For a *polar coordinate system*, the frame of reference is a point O that we call the *pole* and a ray that emanates from it that we call the *polar axis*. (Figure 10.1.1)

In Figure 10.1.2 we have drawn two more rays from the pole. One lies at an angle of θ radians from the polar axis; we call it *ray* θ. The opposite ray lies at an angle of $\theta + \pi$ radians; we call it *ray* $\theta + \pi$.

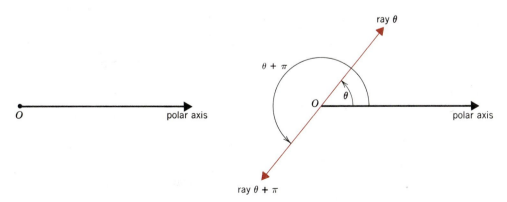

Figure 10.1.1   **Figure 10.1.2**

Figure 10.1.3 shows some points along these same rays, labeled with *polar coordinates*.

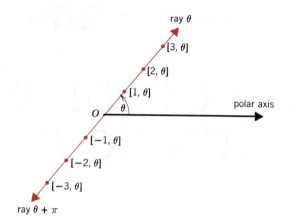

Figure 10.1.3

(10.1.1)	In general, a point is given *polar coordinates* $[r, \theta]$ iff it lies at a distance $\lvert r \rvert$ from the pole
	along the ray θ, if $r \geq 0$, and along the ray $\theta + \pi$, if $r < 0$.

Figure 10.1.4 shows the point $[2, \frac{2}{3}\pi]$ at a distance of 2 units from the pole along the ray $\frac{2}{3}\pi$. The point $[-2, \frac{2}{3}\pi]$ also lies 2 units from the pole, not along the ray $\frac{2}{3}\pi$, but along the opposite ray.

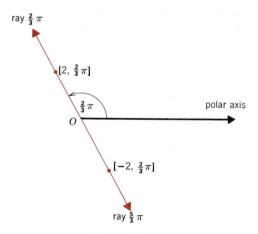

Figure 10.1.4

Polar coordinates are not unique. Many pairs $[r, \theta]$ can represent the same point.

1. If $r = 0$, it does not matter how we choose θ. The resulting point is still the pole:

(10.1.2) $O = [0, \theta]$ for all θ.

2. Geometrically there is no distinction between angles that differ by an integral multiple of 2π. Consequently, as suggested in Figure 10.1.5,

(10.1.3) $[r, \theta] = [r, \theta + 2n\pi]$ for all integers n.

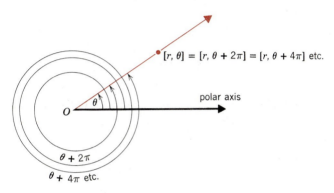

Figure 10.1.5

3. Adding π to the second coordinate is equivalent to changing the sign of the first coordinate:

(10.1.4) $[r, \theta + \pi] = [-r, \theta].$ (Figure 10.1.6)

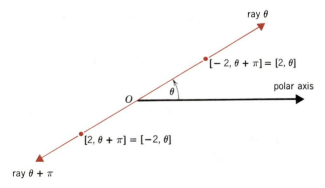

Figure 10.1.6

Relation to Rectangular Coordinates

In Figure 10.1.7 we have superimposed a polar coordinate system on a rectangular coordinate system. We have placed the pole at the origin and the polar axis along the positive x-axis.

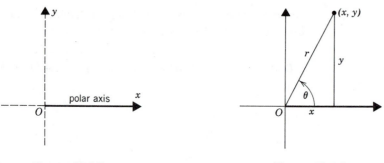

Figure 10.1.7   **Figure 10.1.8**

The relation between polar coordinates $[r, \theta]$ and rectangular coordinates (x, y) is given by the following equations:

(10.1.5) $$x = r \cos \theta, \qquad y = r \sin \theta.$$

Proof. If $r = 0$, the formulas hold, since the point $[r, \theta]$ is then the origin and both x and y are 0:

$$0 = 0 \cos \theta, \qquad 0 = 0 \sin \theta.$$

For $r > 0$, we refer to Figure 10.1.8.†   From the figure,

$$\cos \theta = \frac{x}{r}, \qquad \sin \theta = \frac{y}{r}$$

and therefore

$$x = r \cos \theta, \qquad y = r \sin \theta.$$

Suppose now that $r < 0$. Since $[r, \theta] = [-r, \theta + \pi]$ and $-r > 0$, we know from the previous case that

$$x = -r \cos (\theta + \pi), \qquad y = -r \sin (\theta + \pi).$$

Since

$$\cos (\theta + \pi) = -\cos \theta \qquad \text{and} \qquad \sin (\theta + \pi) = -\sin \theta,$$

once again we have

$$x = r \cos \theta, \qquad y = r \sin \theta. \quad \square$$

---

† For simplicity we have placed (x, y) in the first quadrant. A similar argument works in each of the other quadrants.

From the relations we just proved you can see that, unless $x = 0$,

(10.1.6)
$$\tan \theta = \frac{y}{x}$$

and, under all circumstances,

(10.1.7)
$$x^2 + y^2 = r^2.$$
(check this out)

Problem 1. Find the rectangular coordinates of the point P with polar coordinates $[-2, \frac{1}{3}\pi]$.

Solution. The relations

$$x = r \cos \theta, \qquad y = r \sin \theta$$

give

$$x = -2 \cos \tfrac{1}{3}\pi = -2(\tfrac{1}{2}) = -1, \qquad y = -2 \sin \tfrac{1}{3}\pi = -2(\tfrac{1}{2}\sqrt{3}) = -\sqrt{3}.$$

The point P has rectangular coordinates $(-1, -\sqrt{3})$. ▢

Problem 2. Find all possible polar coordinates for the point P that has rectangular coordinates $(-2, 2\sqrt{3})$.

Solution. We know that

$$r \cos \theta = -2, \qquad r \sin \theta = 2\sqrt{3}.$$

We can get the possible values of r by squaring these expressions and then adding them:

$$r^2 = r^2 \cos^2 \theta + r^2 \sin^2 \theta = (-2)^2 + (2\sqrt{3})^2 = 16,$$

so that $r = \pm 4$.

Taking $r = 4$, we have

$$4 \cos \theta = -2, \qquad 4 \sin \theta = 2\sqrt{3}$$
$$\cos \theta = -\tfrac{1}{2}, \qquad \sin \theta = \tfrac{1}{2}\sqrt{3}.$$

These equations are satisfied by setting $\theta = \frac{2}{3}\pi$, or more generally, by setting

$$\theta = \tfrac{2}{3}\pi + 2n\pi.$$

The polar coordinates of P with first coordinate $r = 4$ are all pairs of the form

$$[4, \tfrac{2}{3}\pi + 2n\pi]$$

where n ranges over the set of all integers.

We could go through the same process again, this time taking $r = -4$, but there is no need to do so. Since $[r, \theta] = [-r, \theta + \pi]$, we know that

$$[4, \tfrac{2}{3}\pi + 2n\pi] = [-4, (\tfrac{2}{3}\pi + \pi) + 2n\pi].$$

The polar coordinates of P with first coordinate $r = -4$ are thus all pairs of the form

$$[-4, \tfrac{5}{3}\pi + 2n\pi]$$

where n again ranges over the set of all integers. ▢

Let's specify some simple sets in polar coordinates.

1. In rectangular coordinates the circle of radius a centered at the origin has equation

$$x^2 + y^2 = a^2.$$

The equation for this circle in polar coordinates is simply

$$r = a.$$

The interior of the circle is given by $0 \leq r < a$ and the exterior by $r > a$.

2. The line through the origin with inclination α is given by the polar equation

$$\theta = \alpha.$$

3. The vertical line $x = a$ becomes

$$r \cos \theta = a$$

and the horizontal line $y = b$ becomes

$$r \sin \theta = b.$$

4. The line $Ax + By + C = 0$ can be written

$$r(A \cos \theta + B \sin \theta) + C = 0.$$

Problem 3. Find an equation in polar coordinates for the hyperbola $x^2 - y^2 = a^2$.

Solution. Setting $x = r \cos \theta$ and $y = r \sin \theta$, we have

$$r^2 \cos^2 \theta - r^2 \sin^2 \theta = a^2,$$
$$r^2(\cos^2 \theta - \sin^2 \theta) = a^2,$$
$$r^2 \cos 2\theta = a^2. \quad \square$$

Problem 4. Show that the equation $r = 2a \cos \theta$ represents a circle.

Solution. Multiplication by r gives

$$r^2 = 2ar \cos \theta,$$
$$x^2 + y^2 = 2ax,$$
$$x^2 - 2ax + y^2 = 0,$$
$$x^2 - 2ax + a^2 + y^2 = a^2,$$
$$(x - a)^2 + y^2 = a^2.$$

This is a circle of radius a centered at the point with rectangular coordinates $(a, 0)$. $\square$

Symmetry

Symmetry with respect to each of the coordinate axes and with respect to the origin is illustrated in Figure 10.1.9. The coordinates marked are, of course, not the only ones possible. (The difficulties that this can cause are explained in Section 10.4.)

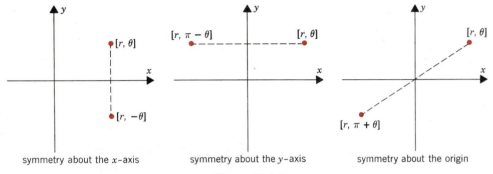

symmetry about the x-axis   symmetry about the y-axis   symmetry about the origin

Figure 10.1.9

Problem 5. Test the lemniscate $r^2 = \cos 2\theta$ for symmetry.

Solution. Since

$$\cos [2(-\theta)] = \cos (-2\theta) = \cos 2\theta,$$

you can see that, if $[r, \theta]$ is on the curve, then so is $[r, -\theta]$. This says that the curve is symmetric about the x-axis. Since

$$\cos [2(\pi - \theta)] = \cos (2\pi - 2\theta) = \cos (-2\theta) = \cos 2\theta,$$

you can see that, if $[r, \theta]$ is on the curve, then so is $[r, \pi - \theta]$. The curve is therefore symmetric about the y-axis.

Being symmetric about both axes, the curve must also be symmetric about the origin. You can verify this directly by noting that

$$\cos [2(\pi + \theta)] = \cos (2\pi + 2\theta) = \cos 2\theta,$$

so that, if $[r, \theta]$ lies on the curve, then so does $[r, \pi + \theta]$. A sketch of the lemniscate appears in Figure 10.1.10. ☐

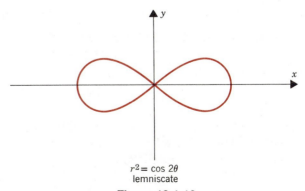

$r^2 = \cos 2\theta$
lemniscate

Figure 10.1.10

EXERCISES 10.1

Plot the following points:

1. $[1, \frac{1}{3}\pi]$.   **2.** $[1, \frac{1}{2}\pi]$.   **3.** $[-1, \frac{1}{3}\pi]$.   **4.** $[-1, -\frac{1}{3}\pi]$.
5. $[4, \frac{5}{4}\pi]$.   **6.** $[-2, 0]$.   **7.** $[-\frac{1}{2}, \pi]$.   **8.** $[\frac{1}{3}, \frac{2}{3}\pi]$.

Find the rectangular coordinates of each of the following points:

9. $[3, \frac{1}{2}\pi]$.   **10.** $[4, \frac{1}{6}\pi]$.   **11.** $[-1, -\pi]$.   **12.** $[-1, \frac{1}{4}\pi]$.
13. $[-3, -\frac{1}{3}\pi]$.   **14.** $[2, 0]$.   **15.** $[3, -\frac{1}{2}\pi]$.   **16.** $[2, 3\pi]$.

The following points are given in rectangular coordinates. Find all possible polar coordinates for each point.

17. $(0, 1)$.   **18.** $(1, 0)$.   **19.** $(-3, 0)$.   **20.** $(4, 4)$.
21. $(2, -2)$.   **22.** $(3, -3\sqrt{3})$.   **23.** $(4\sqrt{3}, 4)$.   **24.** $(\sqrt{3}, -1)$.

25. Find a formula for the distance between $[r_1, \theta_1]$ and $[r_2, \theta_2]$.

26. Show that for $r_1 > 0, r_2 > 0, |\theta_1 - \theta_2| < \pi$ the distance formula you found in Exercise 25 is just the law of cosines.

Find the point $[r, \theta]$ symmetric to the given point about (a) the x-axis, (b) the y-axis, (c) the origin. Express your answer with $r > 0$ and $\theta \in [0, 2\pi)$.

27. $[\frac{1}{2}, \frac{1}{6}\pi]$.   **28.** $[3, -\frac{5}{4}\pi]$.   **29.** $[-2, \frac{1}{3}\pi]$.   **30.** $[-3, -\frac{7}{4}\pi]$.

Test these curves for symmetry about the coordinate axes and the origin.

31. $r = 2 + \cos\theta$.   **32.** $r = \cos 2\theta$.   **33.** $r(\sin\theta + \cos\theta) = 1$.
34. $r\sin\theta = 1$.   **35.** $r^2 \sin 2\theta = 1$.   **36.** $r^2 \cos 2\theta = 1$.

Write the equation in polar coordinates.

37. $x = 2$.   **38.** $y = 3$.   **39.** $2xy = 1$.
40. $x^2 + y^2 = 9$.   **41.** $x^2 + (y - 2)^2 = 4$.   **42.** $(x - a)^2 + y^2 = a^2$.
43. $y = x$.   **44.** $x^2 - y^2 = 4$.   **45.** $x^2 + y^2 + x = \sqrt{x^2 + y^2}$.
46. $y = mx$.   **47.** $(x^2 + y^2)^2 = 2xy$.   **48.** $(x^2 + y^2)^2 = x^2 - y^2$.

Identify the curve and write the equation in rectangular coordinates.

49. $r\sin\theta = 4$.   **50.** $r\cos\theta = 4$.   **51.** $\theta = \frac{1}{3}\pi$.
52. $\theta^2 = \frac{1}{9}\pi^2$.   **53.** $r = 2(1 - \cos\theta)^{-1}$.   **54.** $r = 4\sin(\theta + \pi)$.
55. $r = 3\cos\theta$.   **56.** $\theta = -\frac{1}{2}\pi$.   **57.** $\tan\theta = 2$.
58. $r = 2\sin\theta$.   **59.** $\theta^2 = \frac{1}{4}\pi^2$.   **60.** $r = (2 - \cos\theta)^{-1}$.

10.2 GRAPHING IN POLAR COORDINATES

We begin with the equation

$$r = \theta, \qquad \theta \geq 0.$$

The graph is a nonending spiral, part of the famous *spiral of Archimedes*. The curve is shown in detail from $\theta = 0$ to $\theta = 2\pi$ in Figure 10.2.1. At $\theta = 0$, $r = 0$; at $\theta = \frac{1}{4}\pi$, $r = \frac{1}{4}\pi$; at $\theta = \frac{1}{2}\pi$, $r = \frac{1}{2}\pi$; etc. ☐

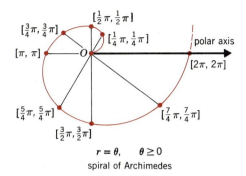

$$r = \theta, \qquad \theta \geq 0$$

spiral of Archimedes

Figure 10.2.1

The next examples involve trigonometric functions.

Example 1. To sketch the curve

$$r = 1 - 2 \cos \theta$$

we refer to the graph of the cosine function in rectangular coordinates. (Figure 10.2.2)

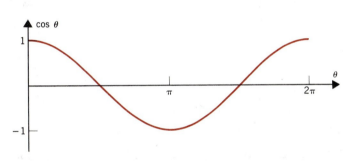

Figure 10.2.2

Since the cosine function is periodic, the curve $r = 1 - 2 \cos \theta$ is a closed curve. We will draw it from 0 to 2π. Outside that interval the curve repeats itself.

To begin the sketch of

$$r = 1 - 2 \cos \theta,$$

we first find the values of θ for which $r = 0$ or $|r|$ is a local maximum:

$r = 0$ at $\theta = \frac{1}{3}\pi, \frac{5}{3}\pi$ for then $\cos \theta = \frac{1}{2}$; $\qquad |r|$ is a local maximum at $\theta = 0, \pi, 2\pi$.

These five values of θ generate four intervals:

$$[0, \tfrac{1}{3}\pi], \quad [\tfrac{1}{3}\pi, \pi], \quad [\pi, \tfrac{5}{3}\pi], \quad [\tfrac{5}{3}\pi, 2\pi].$$

We sketch the curve in four stages. These stages are shown in Figure 10.2.3.

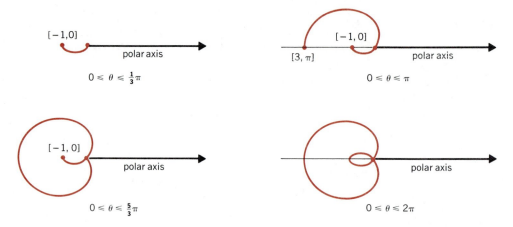

Figure 10.2.3

As θ increases from 0 to $\frac{1}{3}\pi$, cos θ decreases from 1 to $\frac{1}{2}$ and $r = 1 - 2\cos\theta$ increases from -1 to 0.

As θ increases from $\frac{1}{3}\pi$ to π, cos θ decreases from $\frac{1}{2}$ to -1 and r increases from 0 to 3.

As θ increases from π to $\frac{5}{3}\pi$, cos θ increases from -1 to $\frac{1}{2}$ and r decreases from 3 to 0.

Finally, as θ increases from $\frac{5}{3}\pi$ to 2π, cos θ increases from $\frac{1}{2}$ to 1 and r decreases from 0 to -1.

As we could have read from the equation, the curve is symmetric about the x-axis.   ◻

Example 2.   To sketch the curve

$$r = \cos 2\theta$$

we refer to the graph of cos 2θ in rectangular coordinates.   (Figure 10.2.4)

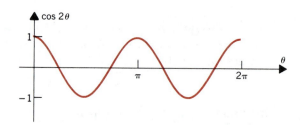

Figure 10.2.4

The values of θ for which r is zero or has an extreme value are as follows:

$$\theta = \tfrac{1}{4}n\pi, \qquad n = 0, 1, \ldots, 8.$$

The curve drawn in the corresponding eight stages is shown in Figure 10.2.5.   ◻

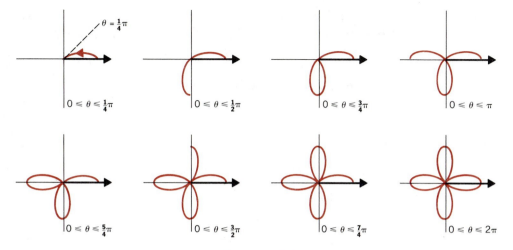

Figure 10.2.5

Example 3. Figure 10.2.6 shows four *cardioids,* heart-shaped curves.

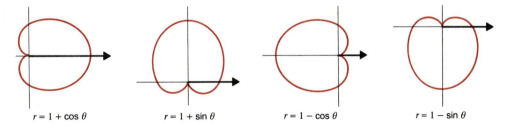

Figure 10.2.6

Rotation of $r = 1 + \cos\theta$ by $\frac{1}{2}\pi$ radians gives

$$r = 1 + \cos(\theta - \tfrac{1}{2}\pi) = 1 + \sin\theta.$$

Rotation by another $\frac{1}{2}\pi$ radians gives

$$r = 1 + \cos(\theta - \pi) = 1 - \cos\theta.$$

Rotation by yet another $\frac{1}{2}\pi$ radians gives

$$r = 1 + \cos(\theta - \tfrac{3}{2}\pi) = 1 - \sin\theta.$$

Notice how easy it is to rotate axes in polar coordinates: each change

$$\cos\theta \to \sin\theta \to -\cos\theta \to -\sin\theta$$

represents a rotation by $\frac{1}{2}\pi$ radians. ☐

At this point we will try to give you a brief survey of some of the basic polar curves, leaving the parabola, the ellipse, and the hyperbola to Section *10.3. (The numbers a and b that appear below are to be interpreted as nonzero constants.)

Lines: $\theta = a$, $r = a \sec \theta$, $r = a \csc \theta$.   (Figure 10.2.7)

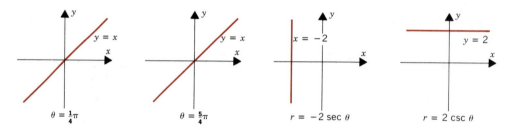

Figure 10.2.7

Circles: $r = a$, $r = a \sin \theta$, $r = a \cos \theta$.   (Figure 10.2.8)

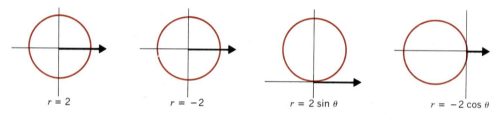

Figure 10.2.8

Limaçons:† $r = a + b \sin \theta$, $r = a + b \cos \theta$.   (Figure 10.2.9)

The general shape of the curve depends on the relative magnitude of $|a|$ and $|b|$.

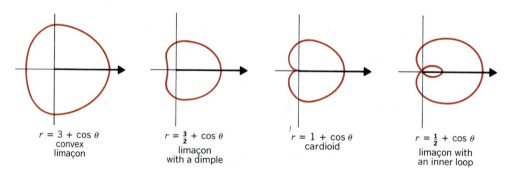

Figure 10.2.9

---

† From the French term for snail.   The word is pronounced with a soft c.

***Lemniscates:*† $r^2 = a \sin 2\theta$, $\qquad r^2 = a \cos 2\theta$. (Figure 10.2.10)

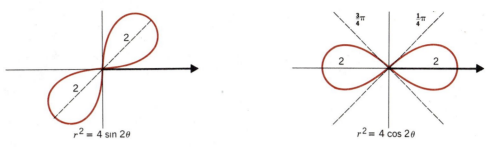

$r^2 = 4 \sin 2\theta$ $\qquad$ $r^2 = 4 \cos 2\theta$

Figure 10.2.10

Petal Curves: $r = a \sin n\theta$, $\qquad r = a \cos n\theta$ $\qquad$ integer n. (Figure 10.2.11)

If n is *odd*, there are n petals. If n is *even,* there are $2n$ petals.

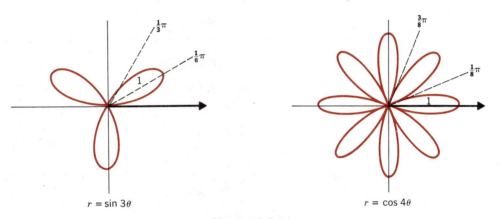

$r = \sin 3\theta$ $\qquad\qquad\qquad$ $r = \cos 4\theta$

Figure 10.2.11

Spirals: $r = a\theta$, spiral of Archimedes. (Figure 10.2.12)

$\qquad\qquad$ $r = e^{a\theta}$, $\;$ $\ln r = a\theta$ logarithmic spiral. (Figure 10.2.13)

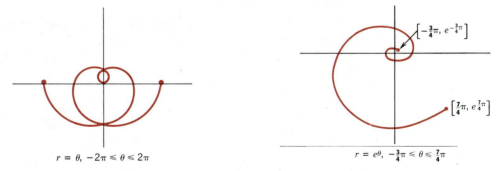

$r = \theta$, $-2\pi \leq \theta \leq 2\pi$ $\qquad\qquad$ $r = e^{\theta}$, $-\frac{3}{4}\pi \leq \theta \leq \frac{7}{4}\pi$

Figure 10.2.12 $\qquad\qquad\qquad\qquad$ **Figure 10.2.13**

---

† From the Latin *lemniscatus,* meaning "adorned with pendent ribbons."

EXERCISES 10.2

Sketch the following polar curves.

1. $\theta = -\frac{1}{4}\pi$.
2. $r = -3$.
3. $r = 4$.
4. $r = 3\cos\theta$.
5. $r = -2\sin\theta$.
6. $\theta = \frac{2}{3}\pi$.
7. $r\csc\theta = 3$.
8. $r = 1 - \cos\theta$.
9. $r = \theta$, $-\frac{1}{2}\pi \le \theta \le \pi$.
10. $r\sec\theta = -2$.
11. $r = \sin 3\theta$.
12. $r^2 = \cos 2\theta$.
13. $r^2 = \sin 2\theta$.
14. $r = \cos 2\theta$.
15. $r^2 = 4$, $0 \le \theta \le \frac{3}{4}\pi$.
16. $r = \sin\theta$.
17. $r^3 = 9r$.
18. $\theta = -\frac{1}{4}\pi$, $1 \le r < 2$.
19. $r = -1 + \sin\theta$.
20. $r^2 = 4r$.
21. $r = \sin 2\theta$.
22. $r = \cos 3\theta$, $0 \le \theta \le \frac{1}{2}\pi$.
23. $r = \cos 5\theta$, $0 \le \theta \le \frac{1}{2}\pi$.
24. $r = e^\theta$, $-\pi \le \theta \le \pi$.
25. $r = 2 + \sin\theta$.
26. $r = \cot\theta$.
27. $r = \tan\theta$.
28. $r = 2 - \cos\theta$.
29. $r = 2 + \sec\theta$.
30. $r = 3 - \csc\theta$.
31. $r = -1 + 2\cos\theta$.

*10.3   THE CONIC SECTIONS IN POLAR COORDINATES

(This material is used later in this text only in Section *14.6 where we examine planetary motion.)

We begin with a theorem. In the proof of the theorem we use both polar and rectangular coordinates. For convenience we refer to the pole as the origin.

THEOREM 10.3.1

Choose positive numbers e and d and write the polar equation

$$r = \frac{ed}{1 + e\cos\theta}.$$

There are three possibilities. (Figure 10.3.1)

I. If $0 < e < 1$, the equation represents an ellipse of eccentricity e with right focus at the origin, major axis horizontal.†

II. If $e = 1$, the equation represents a parabola with focus at the origin, directrix $x = d$.††

III. If $e > 1$, the equation represents a hyperbola of eccentricity e with left focus at the origin, transverse axis horizontal.†††

---

† The ellipse $\dfrac{(x + c)^2}{a^2} + \dfrac{y^2}{a^2 - c^2} = 1$ with $a = \dfrac{ed}{1 - e^2}$, $c = ea$.

†† The parabola $y^2 = -4\dfrac{d}{2}\left(x - \dfrac{d}{2}\right)$.

††† The hyperbola $\dfrac{(x - c)^2}{a^2} - \dfrac{y^2}{c^2 - a^2} = 1$ with $a = \dfrac{ed}{e^2 - 1}$, $c = ea$.

To verify this, we begin by clearing the denominator. Multiplication by $1 + e \cos \theta$ gives

$$r + er \cos \theta = ed.$$

Therefore

$$r = ed - er \cos \theta$$
$$r^2 = e^2 d^2 - 2e^2 dr \cos \theta + e^2 r^2 \cos^2 \theta$$

(*)
$$x^2 + y^2 = e^2 d^2 - 2e^2 dx + e^2 x^2.$$

If $0 < e < 1$, equation (*) can be written

$$(1 - e^2)x^2 + 2e^2 dx + y^2 = e^2 d^2.$$

By completing the square for x we can rearrange this equation to read

$$\left(x + \frac{e^2 d}{1 - e^2}\right)^2 + \frac{y^2}{1 - e^2} = \frac{e^2 d^2}{(1 - e^2)^2}. \qquad \text{(carry out the details)}$$

Setting $a = ed/(1 - e^2)$ and $c = ea$, we have

$$(x + c)^2 + \frac{y^2}{1 - e^2} = a^2$$

$$\frac{(x + c)^2}{a^2} + \frac{y^2}{(1 - e^2)a^2} = 1$$

$$\frac{(x + c)^2}{a^2} + \frac{y^2}{a^2 - c^2} = 1.$$

This is an equation of an ellipse of eccentricity e with right focus at the origin, major axis horizontal.

If $e = 1$, equation (*) reads $y^2 = d^2 - 2dx$. This can be written

$$y^2 = -4\frac{d}{2}\left(x - \frac{d}{2}\right).$$

As you can check, this is the equation of a parabola with focus at the origin and directrix $x = d$.

The case $e > 1$ is left to you. ◻

$$r = \frac{ed}{1 + e \cos \theta} \qquad (e, d > 0)$$

$0 < e < 1,$ an ellipse

$e = 1,$ a parabola

$e > 1,$ a hyperbola

Figure 10.3.1

Problem 1. The ellipse

$$r = \frac{8}{4 + 3 \cos \theta}$$

has right focus at the pole, major axis horizontal. Without resorting to *xy*-coordinates, (a) find the eccentricity of the ellipse, (b) locate the ends of the major axis, (c) locate the center of the ellipse, (d) locate the second focus, (e) determine the length of the minor axis, (f) determine the width of the ellipse at the foci, and finally (g) sketch the ellipse.

Solution. (a) Dividing numerator and denominator by 4, we have

$$r = \frac{2}{1 + \frac{3}{4} \cos \theta}.$$

The eccentricity of the ellipse is $\frac{3}{4}$.

(b) At the right end of the major axis, $\theta = 0$, $\cos \theta = 1$, and $r = \frac{8}{7}$. At the left end, $\theta = \pi$, $\cos \theta = -1$, and $r = 8$. One end of the major axis lies 8 units to the left of the pole, the other end lies $\frac{8}{7}$ units to the right of the pole.

(c) The center of the ellipse lies halfway between the endpoints of the major axis, in this case $\frac{24}{7}$ units to the left of the pole.

(d) In general the focal separation $2c$ divided by the length of the major axis $2a$ gives the eccentricity. Here $2c/\frac{64}{7} = \frac{3}{4}$ and therefore $2c = \frac{48}{7}$. The second focus lies $\frac{48}{7}$ units to the left of the pole.

(You can get the same result by symmetry: since the right focus lies $\frac{8}{7}$ units from the right end of the major axis, the other focus lies $\frac{8}{7}$ units from the left end of the major axis and thus $8 - \frac{8}{7} = \frac{48}{7}$ units to the left of the pole.)

(e) In general the length of the minor axis is $2b = 2\sqrt{a^2 - c^2}$. Here

$$2b = 2\sqrt{(\tfrac{32}{7})^2 - (\tfrac{24}{7})^2} = \tfrac{16}{7}\sqrt{7}.$$

The minor axis has length $\frac{16}{7}\sqrt{7} \cong 6.05$.

(f) The width of the ellipse at the foci is $2r$ where $\theta = \frac{1}{2}\pi$. The width at the foci is 4.

(g) A sketch of the ellipse appears in Figure 10.3.2. ☐

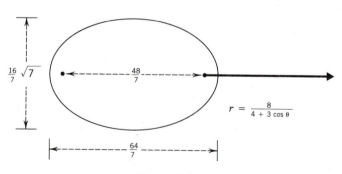

Figure 10.3.2

Problem 2. Sketch the ellipse

$$r = \frac{8}{4 + 3 \sin \theta}.$$

Solution. No need to do much here. This ellipse is the ellipse in Problem 1 rotated by $\frac{1}{2}\pi$ radians. See Figure 10.3.3. ☐

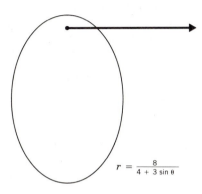

$r = \frac{8}{4 + 3 \sin \theta}$

Figure 10.3.3

EXERCISES *10.3

1. The parabola

$$r = \frac{1}{1 + \cos \theta}$$

has focus at the pole and directrix $x = 1$. Without resorting to xy-coordinates, (a) locate the vertex of the parabola, (b) find the length of the latus rectum (the width of the parabola at the focus), and (c) sketch the parabola.

Sketch the following parabolas.

2. $r = \dfrac{1}{1 + \sin \theta}.$ 3. $r = \dfrac{1}{1 - \cos \theta}.$ 4. $r = \dfrac{1}{1 - \sin \theta}.$

5. The ellipse

$$r = \frac{2}{3 + 2 \cos \theta}$$

has right focus at the pole, major axis horizontal. Without resorting to xy-coordinates, (a) find the eccentricity of the ellipse, (b) locate the ends of the major axis, (c) locate the center of the ellipse, (d) locate the second focus, (e) determine the length of the minor axis, (f) determine the width of the ellipse at the foci, and finally (g) sketch the ellipse.

Sketch the following ellipses.

6. $r = \dfrac{2}{3 - 2 \sin \theta}.$ 7. $r = \dfrac{2}{3 - 2 \cos \theta}.$ 8. $r = \dfrac{2}{3 + 2 \sin \theta}.$

9. The hyperbola

$$r = \frac{6}{1 + 2 \cos \theta}$$

has left focus at the pole, transverse axis horizontal. Without resorting to xy-coordinates, (a) find the eccentricity of the hyperbola, (b) locate the ends of the transverse axes (for one of these you need $r < 0$), (c) locate the center of the hyperbola, (d) locate the second focus, (e) determine the width of the hyperbola at the foci, and (f) sketch the hyperbola.

Sketch the following hyperbolas.

10. $r = \dfrac{6}{1 + 2 \sin \theta}.$ **11.** $r = \dfrac{6}{1 - 2 \sin \theta}.$ **12.** $r = \dfrac{6}{1 - 2 \cos \theta}.$

Identify the conic section and write an equation for it in rectangular coordinates.

13. $r = \dfrac{12}{2 + \sin \theta}.$ **14.** $r = \dfrac{4}{1 + 3 \cos \theta}.$ **15.** $r = \dfrac{9}{5 - 4 \sin \theta}.$

16. Let e and d be positive numbers and let l be the vertical line $x = d$. Show that the equation

$$r = \frac{ed}{1 + e \cos \theta}$$

gives the set of all points P for which $d(P, O) = e\, d(P, l)$.

10.4   THE INTERSECTION OF POLAR CURVES

The fact that a single point has many pairs of polar coordinates can cause complications. In particular, it means that a point $[r_1, \theta_1]$ can lie on a curve given by a polar equation although the coordinates r_1 and θ_1 do not satisfy the equation. For example, the coordinates of $[2, \pi]$ do not satisfy the equation $r^2 = 4 \cos \theta$:

$$r^2 = 2^2 = 4 \quad \text{but} \quad 4 \cos \theta = 4 \cos \pi = -4.$$

Nevertheless the point $[2, \pi]$ does lie on the curve $r^2 = 4 \cos \theta$. It lies on the curve because $[2, \pi] = [-2, 0]$ and the coordinates of $[-2, 0]$ satisfy the equation:

$$r^2 = (-2)^2 = 4, \qquad 4 \cos \theta = 4 \cos 0 = 4.$$

The difficulties are compounded when we deal with two or more curves. Here is an example.

Problem 1. Find the points where the cardioids

$$r = a(1 - \cos \theta) \quad \text{and} \quad r = a(1 + \cos \theta) \qquad (a > 0)$$

intersect.

Solution. We begin by solving the two equations simultaneously. Adding these equations, we get $2r = 2a$ and thus $r = a$. This tells us that $\cos \theta = 0$ and therefore $\theta = \frac{1}{2}\pi + n\pi$. The points $[a, \frac{1}{2}\pi + n\pi]$ all lie on both curves. Not all of these points are distinct:

for n even, $[a, \frac{1}{2}\pi + n\pi] = [a, \frac{1}{2}\pi]$;   for n odd, $[a, \frac{1}{2}\pi + n\pi] = [a, \frac{3}{2}\pi]$.

In short, by solving the two equations simultaneously we have arrived at two common points:

$$[a, \tfrac{1}{2}\pi] = (0, a) \quad \text{and} \quad [a, \tfrac{3}{2}\pi] = (0, -a).$$

There is, however, a third point at which the curves intersect, and that is the origin O. (Figure 10.4.1)

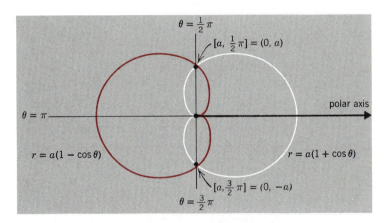

Figure 10.4.1

The origin clearly lies on both curves:

for $r = a(1 - \cos \theta)$ take $\theta = 0, 2\pi$, etc.

for $r = a(1 + \cos \theta)$ take $\theta = \pi, 3\pi$, etc.

The reason that the origin does not appear when we solve the two equations simultaneously is that the curves do not pass through the origin "simultaneously"; that is, they do not pass through the origin for the same values of θ. Think of each of the equations

$$r = a(1 - \cos \theta) \qquad \text{and} \qquad r = a(1 + \cos \theta)$$

as giving the position of an object at time θ. At the points we found by solving the two equations simultaneously, the objects collide. (They both arrive there at the same time.) At the origin the situation is different. Both objects pass through the origin, but no collision takes place because the objects pass through the origin at *different* times. $\square$

Remark. Problems of incidence (does such and such a point lie on the curve with the following polar equation?) and problems of intersection (where do such and such polar curves intersect?) are usually handled more easily by first changing to rectangular coordinates. So it is with symmetry. For example, the curve

$$C: \quad r^2 = \sin \theta$$

is symmetric about the *x*-axis:

(1) $$\text{if } [r, \theta] \in C, \quad \text{then } [r, -\theta] \in C.$$

But this is not easy to see from the polar equation because, in general, if the coordinates of the first point satisfy the equation, the coordinates of the second point do not. One way to see that (1) is valid is to note that

$$[r, -\theta] = [-r, \pi - \theta]$$

and then verify that, if the coordinates of $[r, \theta]$ satisfy the equation, then so do the coordinates of $[-r, \pi - \theta]$. But all this is very cumbersome. The easiest way to see that the curve $r^2 = \sin \theta$ is symmetric about the *x*-axis is to write it as

$$(x^2 + y^2)^3 = y^2.$$

The other symmetries are then also clear. ☐

EXERCISES 10.4

Determine whether the point lies on the curve.

1. $r^2 \cos \theta = 1$; $[1, \pi]$.
2. $r^2 = \cos 2\theta$; $[1, \frac{1}{4}\pi]$.
3. $r = \sin \frac{1}{3}\theta$; $[\frac{1}{2}, \frac{1}{2}\pi]$.
4. $r^2 = \sin 3\theta$; $[1, -\frac{5}{6}\pi]$.

5. Show that the point $[2, \pi]$ lies on both $r^2 = 4 \cos \theta$ and $r = 3 + \cos \theta$.
6. Show that the point $[2, \frac{1}{2}\pi]$ lies on both $r^2 \sin 2\theta = 4$ and $r = 2 \cos \theta$.

Find the points at which the curves intersect. Express your answers in rectangular coordinates.

7. $r = \sin \theta$, $r = -\cos \theta$.
8. $r^2 = \sin \theta$, $r = 2 - \sin \theta$.
9. $r = \cos^2 \theta$, $r = -1$.
10. $r = 2 \sin \theta$, $r = 2 \cos \theta$.
11. $r = 1 - \cos \theta$, $r = \cos \theta$.
12. $r = 1 - \cos \theta$, $r = \sin \theta$.

13. $r = \dfrac{1}{1 - \cos \theta}$, $r \sin \theta = 2$.
14. $r = \dfrac{1}{2 - \cos \theta}$, $r \cos \theta = 1$.

15. $r = \cos 3\theta$, $r = \cos \theta$.
16. $r = \sin 2\theta$, $r = \sin \theta$.

10.5 AREA IN POLAR COORDINATES

Here we develop a technique for calculating the area of a region the boundary of which is given in polar coordinates.

As a start, we suppose that α and β are two real numbers with $\alpha < \beta \leq \alpha + 2\pi$. We take ρ as a function that is continuous on $[\alpha, \beta]$ and keeps a constant sign on that interval. We want the area of the polar region Γ generated by the curve

$$r = \rho(\theta), \quad \alpha \leq \theta \leq \beta.$$

Such a region is portrayed in Figure 10.5.1.

Figure 10.5.1

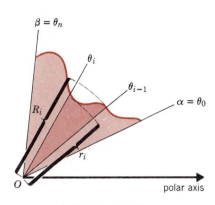

Figure 10.5.2

In the figure $\rho(\theta)$ remains nonnegative. If $\rho(\theta)$ were negative, the region Γ would appear on the opposite side of the pole. In either case, the area of Γ is given by the formula

(10.5.1)
$$A = \int_{\alpha}^{\beta} \tfrac{1}{2}[\rho(\theta)]^2 \, d\theta.$$

Proof. We consider the case where $\rho(\theta) \geq 0$. We take $P = \{\theta_0, \theta_1, \ldots, \theta_n\}$ as a partition of $[\alpha, \beta]$ and direct our attention to what happens between θ_{i-1} and θ_i. We set

$$r_i = \text{min value of } \rho \text{ on } [\theta_{i-1}, \theta_i] \quad \text{and} \quad R_i = \text{max value of } \rho \text{ on } [\theta_{i-1}, \theta_i].$$

The part of Γ that lies between θ_{i-1} and θ_i contains a circular sector of radius r_i and central angle $\Delta\theta_i = \theta_i - \theta_{i-1}$ and is contained in a circular sector of radius R_i and central angle $\Delta\theta_i = \theta_i - \theta_{i-1}$. (See Figure 10.5.2.) Its area A_i must therefore satisfy the inequality

$$\tfrac{1}{2}r_i^2 \, \Delta\theta_i \leq A_i \leq \tfrac{1}{2}R_i^2 \, \Delta\theta_i.\dagger$$

By summing these inequalities from $i = 1$ to $i = n$, you can see that the total area A must satisfy the inequality

(1)
$$L_f(P) \leq A \leq U_f(P)$$

where

$$f(\theta) = \tfrac{1}{2}[\rho(\theta)]^2.$$

Since f is continuous and (1) holds for every partition P of $[a, b]$, we must have

$$A = \int_{\alpha}^{\beta} f(\theta) \, d\theta = \int_{\alpha}^{\beta} \tfrac{1}{2}[\rho(\theta)]^2 \, d\theta. \quad \square$$

$\dagger$ The area of a circular sector of radius r and central angle α is $\tfrac{1}{2}r^2\alpha$.

Remark. If ρ is constant, Γ is a circular sector and

the area of Γ = (the constant value of $\frac{1}{2}\rho^2$ on $[\alpha, \beta]$) $\cdot$ $(\beta - \alpha)$.

If ρ varies continuously,

(10.5.2)   the area of Γ = (the average value of $\frac{1}{2}\rho^2$ on $[\alpha, \beta]$) $\cdot$ $(\beta - \alpha)$. $\square$

Problem 1. Calculate the area enclosed by the cardioid

$$r = 1 - \cos \theta. \qquad \text{(Figure 10.2.6)}$$

Solution

$$\text{area} = \int_0^{2\pi} \tfrac{1}{2}(1 - \cos \theta)^2 \, d\theta = \tfrac{1}{2} \int_0^{2\pi} (1 - 2 \cos \theta + \cos^2 \theta) \, d\theta$$

$$= \tfrac{1}{2} \int_0^{2\pi} (\tfrac{3}{2} - 2 \cos \theta + \cos 2\theta) \, d\theta.$$

$\underset{\text{half-angle formula: } \cos^2 \theta = \frac{1}{2} + \frac{1}{2} \cos 2\theta}{\underline{}}$

Since

$$\int_0^{2\pi} \cos \theta \, d\theta = 0 \qquad \text{and} \qquad \int_0^{2\pi} \cos 2\theta \, d\theta = 0,$$

we have

$$\text{area} = \tfrac{1}{2} \int_0^{2\pi} \tfrac{3}{2} \, d\theta = \tfrac{3}{4} \int_0^{2\pi} d\theta = \tfrac{3}{2}\pi. \quad \square$$

A slightly more complicated type of region is pictured in Figure 10.5.3. We can calculate the area of such a region Ω by taking the area out to $r = \rho_2(\theta)$ and subtracting from it the area out to $r = \rho_1(\theta)$. This gives the formula

(10.5.3) $$\text{area of } \Omega = \int_\alpha^\beta \tfrac{1}{2}([\rho_2(\theta)]^2 - [\rho_1(\theta)]^2) \, d\theta.$$

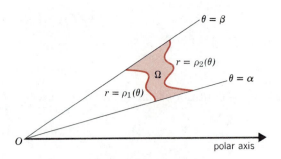

Figure 10.5.3

Problem 2. Find the area of the region that consists of all points that lie within the circle $r = 2 \cos \theta$ but outside the circle $r = 1$.

Solution. The region is shown in Figure 10.5.4. Our first step is to find values of θ for the two points where the circles intersect:

$$2 \cos \theta = 1, \qquad \cos \theta = \tfrac{1}{2}, \qquad \theta = \tfrac{1}{3}\pi, \tfrac{5}{3}\pi.$$

Since the region is symmetric about the polar axis, the area below the polar axis equals the area above the polar axis. Thus

$$A = 2 \int_0^{\pi/3} \tfrac{1}{2}([2 \cos \theta]^2 - [1]^2) \, d\theta.$$

Carry out the integration and you will see that $A = \tfrac{1}{3}\pi + \tfrac{1}{2}\sqrt{3} \cong 1.91$. ☐

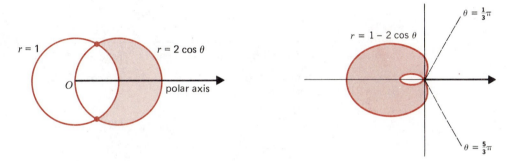

Figure 10.5.4   **Figure 10.5.5**

To find the area between two polar curves, we first determine the curves that serve as outer and inner boundaries of the region and the intervals of θ values over which these boundaries are traced out. Since the polar coordinates of a point are not unique, extra care is needed to determine these intervals of θ values.

Example 3. The area of the region between the inner and outer loops of the limaçon

$$r = 1 - 2 \cos \theta \qquad\qquad \text{(Figure 10.5.5)}$$

can be computed as the difference between

$$\text{area within outer loop} = \int_{\pi/3}^{5\pi/3} \tfrac{1}{2}[1 - 2 \cos \theta]^2 \, d\theta$$

and

$$\text{area within inner loop} = \int_0^{\pi/3} \tfrac{1}{2}[1 - 2 \cos \theta]^2 \, d\theta + \int_{5\pi/3}^{2\pi} \tfrac{1}{2}.1 - 2 \cos \theta]^2 \, d\theta \quad ☐$$

Example 4. The area of the region marked Ω in Figure 10.5.6 can be represented as follows:

$$\text{area of region } \Omega = \int_{\pi/3}^{\pi} \tfrac{1}{2}[2 - \cos\theta]^2 \, d\theta - \int_{\pi/3}^{\pi/2} \tfrac{1}{2}[3\cos\theta]^2 \, d\theta. \quad \square$$

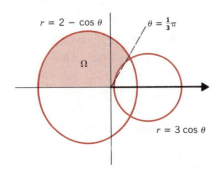

Figure 10.5.6

EXERCISES 10.5

Calculate the area enclosed by the following curves. Take $a > 0$.

1. $r = a\cos\theta$ from $\theta = -\tfrac{1}{2}\pi$ to $\theta = \tfrac{1}{2}\pi$.   **2.** $r = a\cos 3\theta$ from $\theta = -\tfrac{1}{6}\pi$ to $\theta = \tfrac{1}{6}\pi$.

3. $r = a\sqrt{\cos 2\theta}$ from $\theta = -\tfrac{1}{4}\pi$ to $\theta = \tfrac{1}{4}\pi$.

4. $r = a(1 + \cos 3\theta)$ from $\theta = -\tfrac{1}{3}\pi$ to $\theta = \tfrac{1}{3}\pi$.

5. $r^2 = a^2\sin^2\theta$.   **6.** $r^2 = a^2\sin^2 2\theta$.

Calculate the area of the region bounded by the following.

7. $r = \tan 2\theta$ and the rays $\theta = 0, \quad \theta = \tfrac{1}{8}\pi$.

8. $r = \cos\theta, \quad r = \sin\theta,$ and the rays $\theta = 0, \quad \theta = \tfrac{1}{4}\pi$.

9. $r = 2\cos\theta, \quad r = \cos\theta,$ and the rays $\theta = 0, \quad \theta = \tfrac{1}{4}\pi$.

10. $r = 1 + \cos\theta, \quad r = \cos\theta,$ and the rays $\theta = 0, \quad \theta = \tfrac{1}{2}\pi$.

11. $r = a(4\cos\theta - \sec\theta)$ and the rays $\theta = 0, \quad \theta = \tfrac{1}{4}\pi$.

12. $r = \tfrac{1}{2}\sec^2\tfrac{1}{2}\theta$ and the vertical line through the origin.

Find the area of the region bounded by the following.

13. $r = e^\theta, \quad 0 \le \theta \le \pi; \qquad r = \theta, \quad 0 \le \theta \le \pi; \qquad$ the rays $\theta = 0, \quad \theta = \pi.$

14. $r = e^\theta, \quad 2\pi \le \theta \le 3\pi; \qquad r = \theta, \quad 0 \le \theta \le \pi; \qquad$ the rays $\theta = 0, \quad \theta = \pi.$

15. $r = e^\theta, \quad 0 \le \theta \le \pi; \qquad r = e^{\theta/2}, \quad 0 \le \theta \le \pi; \qquad$ the rays $\theta = 2\pi, \quad \theta = 3\pi.$

16. $r = e^\theta, \quad 0 \le \theta \le \pi; \qquad r = e^\theta, \quad 2\pi \le \theta \le 3\pi; \qquad$ the rays $\theta = 0, \quad \theta = \pi.$

Represent the area by one or more integrals.

17. Outside $r = 2$, but inside $r = 4\sin\theta$.

18. Outside $r = 1 - \cos\theta$, but inside $r = 1 + \cos\theta$.

19. Inside $r = 4$, and to the right of $r = 2\sec\theta$.   **20.** Inside $r = 2$, but outside $r = 4\cos\theta$.

21. Inside $r = 4$, and between the lines $\theta = \tfrac{1}{2}\pi$ and $r = 2\sec\theta$.

22. Inside the inner loop of $r = 1 - 2\sin\theta$.   **23.** Inside one petal of $r = 2\sin 3\theta$.

24. Outside $r = 1 + \cos\theta$, but inside $r = 2 - \cos\theta$.

25. Interior to both $r = 1 - \sin\theta$ and $r = \sin\theta$.   **26.** Inside one petal of $r = 5\cos 6\theta$.

10.6 CURVES GIVEN PARAMETRICALLY

We begin with a pair of functions $x = x(t)$, $y = y(t)$ differentiable on the interior of an interval I. At the endpoints of I (if any) we require only continuity.

For each number t in I we can interpret $(x(t), y(t))$ as the point with x-coordinate $x(t)$ and y-coordinate $y(t)$. Then, as t ranges over I, the point $(x(t), y(t))$ traces out a path in the xy-plane. (Figure 10.6.1) We call such a path a *parametrized curve* and refer to t as the *parameter*.

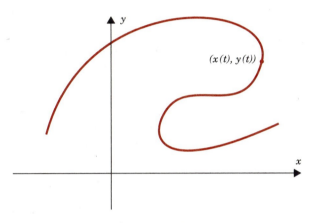

Figure 10.6.1

Problem 1. Identify the curve parametrized by the functions

$$x(t) = t + 1, \quad y(t) = 2t - 5 \qquad t \text{ real.}$$

Solution. We can express $y(t)$ in terms of $x(t)$:

$$y(t) = 2[x(t) - 1] - 5 = 2x(t) - 7.$$

The functions parametrize the line $y = 2x - 7$: as t ranges over the set of real numbers, the point $(x(t), y(t))$ traces out the line $y = 2x - 7$. ☐

Problem 2. Identify the curve parametrized by the functions

$$x(t) = 2t, \quad y(t) = t^2 \qquad t \text{ real.}$$

Solution. You can see that for all t

$$y(t) = \tfrac{1}{4}[x(t)]^2.$$

The functions parametrize the parabola $y = \tfrac{1}{4}x^2$: as t ranges over the set of real numbers, the point $(x(t), y(t))$ traces out the parabola $y = \tfrac{1}{4}x^2$. ☐

Problem 3. Identify the curve parametrized by the functions

$$x(t) = \sin^2 t, \quad y(t) = \cos t \qquad t \in [0, \pi].$$

Solution. Note first that

$$x(t) = \sin^2 t = 1 - \cos^2 t = 1 - [y(t)]^2.$$

The points $(x(t), y(t))$ all lie on the parabola

$$x = 1 - y^2. \qquad \qquad \text{(Figure 10.6.2)}$$

At $t = 0$, $x = 0$ and $y = 1$; at $t = \pi$, $x = 0$ and $y = -1$. As t ranges from 0 to π, the point $(x(t), y(t))$ traverses the parabolic arc

$$x = 1 - y^2, \qquad -1 \le y \le 1$$

from the point $(0, 1)$ to the point $(0, -1)$. ☐

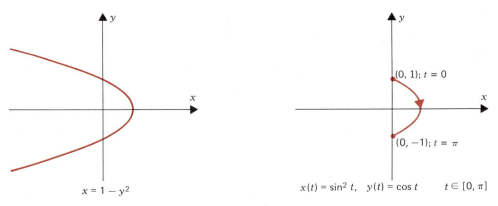

$$x = 1 - y^2 \qquad\qquad\qquad x(t) = \sin^2 t, \quad y(t) = \cos t \qquad t \in [0, \pi]$$

Figure 10.6.2

Remark. Changing the domain in the previous problem to all real t does *not* give us any more of the parabola. For any given t we still have

$$0 \le x(t) \le 1 \qquad \text{and} \qquad -1 \le y(t) \le 1.$$

As t ranges over the set of real numbers, the point $(x(t), y(t))$ traces out that same parabolic arc back and forth an infinite number of times. ☐

The functions

(10.6.1) $\boxed{x(t) = x_0 + t(x_1 - x_0), \quad y(t) = y_0 + t(y_1 - y_0) \qquad t \text{ real}}$

parametrize the line that passes through (x_0, y_0) and (x_1, y_1).

Proof. If $x_1 = x_0$, then we have

$$x(t) = x_0, \quad y(t) = y_0 + t(y_1 - y_0).$$

As t ranges over the set of real numbers, $x(t)$ remains constantly x_0 and $y(t)$ ranges over the set of real numbers. The functions parametrize the vertical line $x = x_0$. Since $x_1 = x_0$, both (x_0, y_0) and (x_1, y_1) lie on this vertical line.

If $x_1 \neq x_0$, then we can solve the first equation for t:

$$t = \frac{x(t) - x_0}{x_1 - x_0}.$$

Substituting this into the second equation we obtain the identity

$$y(t) - y_0 = \frac{y_1 - y_0}{x_1 - x_0} [x(t) - x_0].$$

The functions parametrize the line with equation

$$y - y_0 = \frac{y_1 - y_0}{x_1 - x_0} (x - x_0).$$

This is the line determined by (x_0, y_0) and (x_1, y_1). ☐

The functions $x(t) = a \cos t, \ y(t) = b \sin t$ satisfy the identity

$$\frac{[x(t)]^2}{a^2} + \frac{[y(t)]^2}{b^2} = 1.$$

As t ranges over any interval of length 2π, the point $(x(t), y(t))$ traces out the ellipse

$$\frac{x^2}{a^2} + \frac{y^2}{b^2} = 1.$$

Usually we let t range from 0 to 2π and parametrize the ellipse by setting

(10.6.2) $\boxed{x(t) = a \cos t, \quad y(t) = b \sin t \qquad t \in [0, 2\pi].}$

If $b = a$, we have a circle. We can parametrize the circle

$$x^2 + y^2 = a^2$$

by setting

(10.6.3) $\boxed{x(t) = a \cos t, \quad y(t) = a \sin t \qquad t \in [0, 2\pi].}$

Interpreting the parameter t as time measured in seconds, we can think of a pair of parametric equations, $x = x(t)$ and $y = y(t)$, as describing the motion of a particle in the xy-plane. Different parametrizations of the same curve represent different ways of traversing that curve.

Example 4. The line that passes through the points (1, 2) and (3, 6) has equation $y = 2x$. The line segment that joins these same points is given by

$$y = 2x, \quad 1 \leq x \leq 3.$$

We will parametrize this line segment in different ways and interpret each parametrization as the motion of a particle.

We begin by setting

$$x(t) = t, \quad y(t) = 2t \quad t \in [1, 3].$$

At time $t = 1$, the particle is at the point (1, 2). It traverses the line segment and arrives at the point (3, 6) at time $t = 3$.

Now we set

$$x(t) = t + 1, \quad y(t) = 2t + 2 \quad t \in [0, 2].$$

At time $t = 0$, the particle is at the point (1, 2). It traverses the line segment and arrives at the point (3, 6) at time $t = 2$.

The equations

$$x(t) = 3 - t, \quad y(t) = 6 - 2t \quad t \in [0, 2]$$

represent a traversal of that same line segment in the opposite direction. At time $t = 0$, the particle is at (3, 6). It arrives at (1, 2) at time $t = 2$.

Set

$$x(t) = 3 - 4t, \quad y(t) = 6 - 8t \quad t \in [0, \tfrac{1}{2}].$$

Now the particle traverses the same line segment in only half a second. At time $t = 0$, the particle is at (3, 6). It arrives at (1, 2) at time $t = \tfrac{1}{2}$.

Finally we set

$$x(t) = 2 - \cos t, \quad y(t) = 4 - 2 \cos t \quad t \in [0, 4\pi].$$

In this instance the particle begins and ends its motion at the point (1, 2), having traced and retraced the line segment twice during a span of 4π seconds. See Figure 10.6.3. ☐

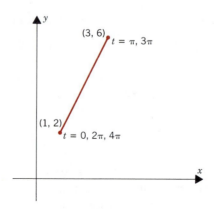

Figure 10.6.3

Remark. If the path of an object is given by a time parameter t and we eliminate the parameter to obtain an equation in x and y, it may be that we obtain a clearer view of the path, but we do so at considerable expense. The equation in x and y does not tell us where the particle is at any time t. The parametric equations do. ☐

Example 5. We return to the ellipse

$$\frac{x^2}{a^2} + \frac{y^2}{b^2} = 1.$$

A particle with position given by the equations

$$x(t) = a \cos t, \quad y(t) = b \sin t \qquad \text{with } t \in [0, 2\pi]$$

traverses the ellipse in a counterclockwise manner. It begins at the point $(a, 0)$ and makes a full circuit in 2π seconds. If the equations of motion are

$$x(t) = a \cos 2\pi t, \quad y(t) = -b \sin 2\pi t \qquad \text{with } t \in [0, 1],$$

the particle still travels the same ellipse, but in a different manner. Once again it starts at $(a, 0)$, but this time it moves clockwise and makes the full circuit in only one second. If the equations of motion are

$$x(t) = a \sin 4\pi t, \quad y(t) = b \cos 4\pi t \qquad \text{with } t \in [0, \infty),$$

the motion begins at $(0, b)$ and goes on in perpetuity. The motion is clockwise, a complete circuit taking place every half second. ☐

Intersections and Collisions

Problem 6. Two particles start at the same instant, the first along the linear path

$$x_1(t) = \tfrac{16}{3} - \tfrac{8}{3}t, \quad y_1(t) = 4t - 5 \qquad t \geq 0$$

and the second along the elliptical path

$$x_2(t) = 2 \sin \tfrac{1}{2}\pi t, \quad y_2(t) = -3 \cos \tfrac{1}{2}\pi t \qquad t \geq 0.$$

(a) At what points, if any, do the paths intersect?

(b) At what points, if any, do the particles collide?

Solution. To see where the paths intersect, we find equations for them in x and y. The linear path can be written

$$3x + 2y - 6 = 0, \qquad x \leq \tfrac{16}{3}$$

and the elliptical path

$$\frac{x^2}{4} + \frac{y^2}{9} = 1.$$

Solving the two equations simultaneously, we get

$$x = 2, \quad y = 0 \quad \text{and} \quad x = 0, \quad y = 3.$$

This means that the paths intersect at the points $(2, 0)$ and $(0, 3)$. This answers part (a).

Now for part (b). The first particle passes through $(2, 0)$ only when

$$x_1(t) = \tfrac{16}{3} - \tfrac{8}{3}t = 2 \quad \text{and} \quad y_1(t) = 4t - 5 = 0.$$

As you can check, this happens only when $t = \tfrac{5}{4}$. When $t = \tfrac{5}{4}$, the second particle is elsewhere. Hence no collision takes place at $(2, 0)$. There is however a collision at $(0, 3)$ because both particles get there at exactly the same time, $t = 2$:

$$x_1(2) = 0 = x_2(2), \qquad y_1(2) = 3 = y_2(2). \quad \square$$

EXERCISES 10.6

Express the curve by an equation in x and y.

1. $x(t) = t^2, \quad y(t) = 2t + 1.$ **2.** $x(t) = 3t - 1, \quad y(t) = 5 - 2t.$
3. $x(t) = t^2, \quad y(t) = 4t^4 + 1.$ **4.** $x(t) = 2t - 1, \quad y(t) = 8t^3 - 5.$
5. $x(t) = 2 \cos t, \quad y(t) = 3 \sin t.$ **6.** $x(t) = \sec^2 t, \quad y(t) = 2 + \tan t.$
7. $x(t) = \tan t, \quad y(t) = \sec t.$ **8.** $x(t) = 2 - \sin t, \quad y(t) = \cos t.$
9. $x(t) = \sin t, \quad y(t) = 1 + \cos^2 t.$ **10.** $x(t) = e^t, \quad y(t) = 4 - e^{2t}.$
11. $x(t) = 4 \sin t, \quad y(t) = 3 + 2 \sin t.$ **12.** $x(t) = \csc t, \quad y(t) = \cot t.$

Express the curve by an equation in x and y; then sketch the curve.

13. $x(t) = e^{2t}, \quad y(t) = e^{2t} - 1, \quad t \leq 0.$
14. $x(t) = 3 \cos t, \quad y(t) = 2 - \cos t, \quad 0 \leq t \leq \pi.$
15. $x(t) = \sin t, \quad y(t) = \csc t, \quad 0 < t \leq \tfrac{1}{4}\pi.$ **16.** $x(t) = 1/t, \quad y(t) = 1/t^2, \quad 0 < t < 3.$
17. $x(t) = 3 + 2t, \quad y(t) = 5 - 4t, \quad -1 \leq t \leq 2.$
18. $x(t) = \sec t, \quad y(t) = \tan t, \quad 0 \leq t \leq \tfrac{1}{4}\pi.$ **19.** $x(t) = \sin \pi t, \quad y(t) = 2t, \quad 0 \leq t \leq 4.$
20. $x(t) = 2 \sin t, \quad y(t) = \cos t, \quad 0 \leq t \leq \tfrac{1}{2}\pi.$
21. $x(t) = \cot t, \quad y(t) = \csc t, \quad \tfrac{1}{4}\pi \leq t < \tfrac{1}{2}\pi.$

22. (*Important*) Parametrize (a) the curve $y = f(x)$, $x \in [a, b]$. (b) the polar curve $r = f(\theta)$, $\theta \in [\alpha, \beta]$.

23. A particle with position given by the equations

$$x(t) = \sin 2\pi t, \quad y(t) = \cos 2\pi t \qquad t \in [0, 1]$$

starts at the point $(0, 1)$ and traverses the unit circle $x^2 + y^2 = 1$ once in a clockwise manner. Write equations of the form

$$x(t) = f(t), \quad y(t) = g(t) \qquad t \in [0, 1]$$

so that the particle
(a) begins at $(0, 1)$ and traverses the circle once in a counterclockwise manner;
(b) begins at $(0, 1)$ and traverses the circle twice in a clockwise manner;
(c) traverses the quarter circle from $(1, 0)$ to $(0, 1)$;
(d) traverses the three-quarter circle from $(1, 0)$ to $(0, 1)$.

24. A particle with position given by the equations

$$x(t) = 3 \cos 2\pi t, \quad y(t) = 4 \sin 2\pi t \qquad t \in [0, 1]$$

starts at the point $(3, 0)$ and traverses the ellipse $16x^2 + 9y^2 = 144$ once in a counterclockwise manner. Write equations of the form

$$x(t) = f(t), \quad y(t) = g(t) \qquad t \in [0, 1]$$

so that the particle
(a) begins at $(3, 0)$ and traverses the ellipse once in a clockwise manner;
(b) begins at $(0, 4)$ and traverses the ellipse once in a clockwise manner;
(c) begins at $(-3, 0)$ and traverses the ellipse twice in a counterclockwise manner;
(d) traverses the upper half of the ellipse from $(3, 0)$ to $(0, 3)$.

25. Find a parametrization

$$x = x(t), \quad y = y(t) \qquad t \in (-1, 1)$$

for the horizontal line $y = 2$.

26. Find a parametrization

$$x(t) = \sin f(t), \quad y(t) = \cos f(t) \qquad t \in (0, 1)$$

which traces out the unit circle infinitely often.

Find a parametrization

$$x = x(t), \quad y = y(t) \qquad t \in [0, 1]$$

for the given curve.

27. The line segment from $(3, 7)$ to $(8, 5)$.　　　**28.** The line segment from $(2, 6)$ to $(6, 3)$.

29. The parabolic arc $x = 1 - y^2$ from $(0, -1)$ to $(0, 1)$.

30. The parabolic arc $x = y^2$ from $(4, 2)$ to $(0, 0)$.

31. The curve $y^2 = x^3$ from $(4, 8)$ to $(1, 1)$.　　　**32.** The curve $y^3 = x^2$ from $(1, 1)$ to $(8, 4)$.

33. The curve $y = f(x), \quad x \in [a, b]$.

34. (*Important*) Suppose that the curve

$$C: \quad x = x(t), \quad y = y(t) \qquad t \in [c, d]$$

is the graph of a nonnegative function f over an interval $[a, b]$. Suppose that x' and y are continuous, $x(c) = a$, and $x(d) = b$.
(a) (*The area under a parametrized curve*) Show that

(10.6.4) 　　　the area below $C = \displaystyle\int_c^d y(t)x'(t)\, dt.$

HINT: Since C is the graph of f, we know that $y(t) = f(x(t))$.

(b) (*The centroid of a region under a parametrized curve*) Show that, if the region under C has area A and centroid $(\bar{x}, \bar{y})$, then

(10.6.5) 　　　$\bar{x}A = \displaystyle\int_c^d x(t)y(t)x'(t)\, dt, \qquad \bar{y}A = \int_c^d \tfrac{1}{2}[y(t)]^2 x'(t)\, dt.$

(c) (*The volume of the solid generated by revolving about a coordinate axis the region under a parametrized curve*) Show that

(10.6.6)

$$V_x = \int_c^d \pi [y(t)]^2 x'(t)\, dt, \qquad V_y = \int_c^d 2\pi x(t) y(t) x'(t)\, dt.$$

provided $x(c) \geq 0$

(d) (*The centroid of the solid generated by revolving about a coordinate axis the region under a parametrized curve*) Show that

(10.6.7)

$$\bar{x} V_x = \int_c^d \pi x(t) [y(t)]^2 x'(t)\, dt, \qquad \bar{y} V_y = \int_c^d \pi x(t) [y(t)]^2 x'(t)\, dt.$$

provided $x(c) \geq 0$

35. Sketch the curve

$$x(t) = rt, \quad y(t) = r(1 - \cos t) \qquad t \in [0, 2\pi].$$

and find the area below it.

36. Determine the centroid of the region under the curve in Exercise 35.

37. Find the volume generated by revolving the region in Exercise 36 (a) about the x-axis. (b) about the y-axis.

38. Find the centroid of the solid generated by revolving the region of Exercise 36 (a) about the x-axis. (b) about the y-axis.

39. Give a parametrization for the upper half of the ellipse $b^2x^2 + a^2y^2 = a^2b^2$ that satisfies the requirements of (10.6.4).

40. Use the parametrization you chose for Exercise 39 to find (a) the area of the region enclosed by the ellipse. (b) the centroid of the upper half of that region.

41. Two particles start at the same instant, the first along the ray

$$x(t) = 2t + 6, \quad y(t) = 5 - 4t \qquad t \geq 0$$

and the second along the circular path

$$x(t) = 3 - 5 \cos \pi t, \quad y(t) = 1 + 5 \sin \pi t \qquad t \geq 0.$$

(a) At what points, if any, do these paths intersect?
(b) At what points, if any, will the particles collide?

42. Two particles start at the same instant, the first along the elliptical path

$$x_1(t) = 2 - 3 \cos \pi t, \quad y_1(t) = 3 + 7 \sin \pi t \qquad t \geq 0$$

and the second along the parabolic path

$$x_2(t) = 3t + 2, \quad y_2(t) = -\tfrac{7}{15}(3t + 1)^2 + \tfrac{157}{15} \qquad t \geq 0.$$

(a) At what points, if any, do these paths intersect?
(b) At what points, if any, will the particles collide?

We can determine the points where a parametrized curve

$$C: \ x = x(t), \quad y = y(t) \qquad t \in I$$

intersects itself by finding the numbers r and s in I $(r \neq s)$ for which

$$x(r) = x(s) \quad \text{and} \quad y(r) = y(s).$$

Use this method to find the point(s) of self-intersection for each of the following curves.

43. $x(t) = t^2 - 2t$, $y(t) = t^3 - 3t^2 + 2t$,   real t.
44. $x(t) = \cos t\,(1 - 2\sin t)$, $y(t) = \sin t\,(1 - 2\sin t)$, $t \in [0, \pi]$.
45. $x(t) = \sin 2\pi t$, $y(t) = 2t - t^2$, $t \in [0, 4]$.
46. $x(t) = t^3 - 4t$, $y(t) = t^3 - 3t^2 + 2t$,   real t.

Graphing Programs (BASIC)

These programs make graphing curves simpler by computing the rectangular coordinates for a variety of points on the curve you have chosen. In each case, the curve is presumed to be given parametrically, with parameter t. In the program "Graphing polar coordinate curves," you are to enter the polar coordinates r and θ as functions of t. The result of computation, however, will be the *rectangular* coordinates of a sequence of points on the curve. In the program "Graph a parametric curve," everything is in rectangular coordinates to begin with: you enter x and y as functions of t, and the program returns the rectangular coordinates of a sequence of points on the curve.

```
10 REM Graphing polar coordinate curves
20 REM copyright © Colin C. Graham 1988-1989

30 REM change the two lines "def FNr(t)..." and
40 REM "def FNtheta(t)..." to fit your functions

50 DEF FNr(t) = 1 + COS(t)
60 DEF FNtheta(t) = t

100 INPUT "Enter starting value:"; a
110     REM (a will usually be 0), though exactly 0 will be bad for curves such as "rsin(theta) = 1"
120 INPUT "Enter ending value:"; b
130     REM (b will usually be 2Pi) - again 2Pi bad for curves such as "rsin(theta) = 1"
140 INPUT "Enter number of points:"; n

300 FOR j = 0 TO n
310     t = a + (b − a)*j/n
320     PRINT FNr(t)*COS(FNtheta(t)), FNr(t)*SIN(FNtheta(t))
330 NEXT j

500 INPUT "Do it again? (Y/N)"; a$
510 IF a$ = "Y" OR a$ = "y" THEN 100
520 END
```

```
10 REM Graph a parametric curve
20 REM copyright © Colin C. Graham 1988-1989

30 REM change the two lines "def FNx(t) =..." and
40 REM "def FNy(t)..." to fit your functions

50 DEF FNx(t) = COS(t)
60 DEF FNy(t) = SIN(t)

100 INPUT "Enter starting value:"; a
120 INPUT "Enter ending value:"; b
140 INPUT "Enter number of points:"; n

300 FOR j = 0 TO n
310     t = a + (b − a)*j/n
320     PRINT FNx(t), FNy(t)
330 NEXT j

500 INPUT "Do it again? (Y/N)"; a$
510 IF a$ = "Y" OR a$ = "y" THEN 100
520 END
```

10.7   TANGENTS TO CURVES GIVEN PARAMETRICALLY

A parametrized curve

$$C: \quad x = x(t), \quad y = y(t) \qquad \text{with } t \in I$$

can intersect itself.   Thus, at any given point, it can have

(i) one tangent,   (ii) two or more tangents,   or   (iii) no tangent at all.

We illustrate these possibilities in Figure 10.7.1.

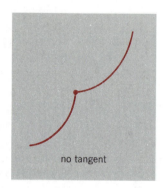

one tangent   two tangents   no tangent

Figure 10.7.1

As before, we are assuming that $x'(t)$ and $y'(t)$ exist, at least on the interior of I. To make sure that at least one tangent line exists at each point of c, we will make the additional *assumption* that

(10.7.1)
$$[x'(t)]^2 + [y'(t)]^2 \neq 0.$$

(Without this assumption most anything can happen.   See Exercises 31–35.)

Now choose a point (x_0, y_0) on the curve C and a time t_0 at which

$$x(t_0) = x_0 \quad \text{and} \quad y(t_0) = y_0.$$

We want the slope of the curve as it passes through the point (x_0, y_0) at time t_0.† To find this slope, we assume that $x'(t_0) \neq 0$.   With $x'(t_0) \neq 0$, we can be sure that, for h sufficiently small,

$$x(t_0 + h) - x(t_0) \neq 0. \qquad \text{(explain)}$$

For such h we can form the quotient

$$\frac{y(t_0 + h) - y(t_0)}{x(t_0 + h) - x(t_0)}.$$

---

† It could pass through the point (x_0, y_0) at other times also.

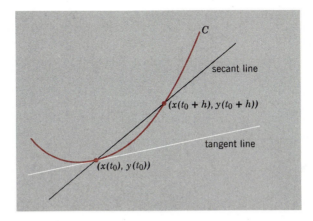

Figure 10.7.2

This quotient is the slope of the secant line pictured in Figure 10.7.2. The limit of this quotient as h tends to zero is the slope of the tangent line and thus the slope of the curve. Since

$$\frac{y(t_0 + h) - y(t_0)}{x(t_0 + h) - x(t_0)} = \frac{(1/h)[y(t_0 + h) - y(t_0)]}{(1/h)[x(t_0 + h) - x(t_0)]} \to \frac{y'(t_0)}{x'(t_0)} \qquad \text{as } h \to 0,$$

you can see that

(10.7.2)
$$m = \frac{y'(t_0)}{x'(t_0)}.$$

The equation for the tangent line can be written

$$y - y(t_0) = \frac{y'(t_0)}{x'(t_0)} [x - x(t_0)].$$

Multiplication by $x'(t_0)$ gives

$$y'(t_0)[x - x(t_0)] - x'(t_0)[y - y(t_0)] = 0,$$

and thus

(10.7.3)
$$y'(t_0)[x - x_0] - x'(t_0)[y - y_0] = 0.$$

We derived this equation under the assumption that $x'(t_0) \neq 0$. If $x'(t_0) = 0$, Equation 10.7.3 still makes sense. It is simply $y'(t_0)[x - x_0] = 0$, which, since $y'(t_0) \neq 0$,† can be simplified to read

(10.7.4)
$$x = x_0.$$

In this instance the line is vertical, and we say that the curve has a *vertical tangent*.

---

† We are assuming that $[x'(t)]^2 + [y'(t)]^2$ is never 0. Since $x'(t_0) = 0$, $y'(t_0) \neq 0$.

Problem 1. Find the tangent(s) to the curve

$$x(t) = t^3, \quad y(t) = 1 - t$$

at the point $(8, -1)$.

Solution. Since the curve passes through the point $(8, -1)$ only when $t = 2$, there can be only one tangent line at that point. With

$$x(t) = t^3, \quad y(t) = 1 - t$$

we have

$$x'(t) = 3t^2, \quad y'(t) = -1$$

and therefore

$$x'(2) = 12, \quad y'(2) = -1.$$

The tangent line has equation

$$(-1)[x - 8] - (12)[y - (-1)] = 0. \tag{10.7.3}$$

This reduces to

$$x + 12y + 4 = 0. \quad \square$$

Problem 2. Find the points of the curve

$$x(t) = 3 - 4 \sin t, \quad y(t) = 4 + 3 \cos t$$

at which there is (i) a horizontal tangent, (ii) a vertical tangent.

Solution. Observe first of all that the derivatives

$$x'(t) = -4 \cos t \quad \text{and} \quad y'(t) = -3 \sin t$$

are never 0 simultaneously.

To find the points at which there is a horizontal tangent, we set $y'(t) = 0$. This gives $t = n\pi$. Horizontal tangents occur at all points of the form $(x(n\pi), y(n\pi))$. Since

$$x(n\pi) = 3 - 4 \sin n\pi = 3 \quad \text{and} \quad y(n\pi) = 4 + 3 \cos n\pi = \begin{cases} 7, & n \text{ even} \\ 1, & n \text{ odd} \end{cases},$$

there are horizontal tangents at $(3, 7)$ and $(3, 1)$.

To find the vertical tangents, we set $x'(t) = 0$. This gives $t = \frac{1}{2}\pi + n\pi$. Vertical tangents occur at all points of the form $(x(\frac{1}{2}\pi + n\pi), y(\frac{1}{2}\pi + n\pi))$. Since

$$x(\tfrac{1}{2}\pi + n\pi) = 3 - 4 \sin (\tfrac{1}{2}\pi + n\pi) = \begin{cases} -1, & n \text{ even} \\ 7, & n \text{ odd} \end{cases}$$

and

$$y(\tfrac{1}{2}\pi + n\pi) = 4 + 3 \cos (\tfrac{1}{2}\pi + n\pi) = 4,$$

there are vertical tangents at $(-1, 4)$ and $(7, 4)$. $\square$

Problem 3. Find the tangent(s) to the curve

$$x(t) = t^2 - 2t + 1, \quad y(t) = t^4 - 4t^2 + 4$$

at the point $(1, 4)$.

Solution. The curve passes through the point $(1, 4)$ when $t = 0$ and when $t = 2$. (Check this out.) Differentiation gives

$$x'(t) = 2t - 2, \quad y'(t) = 4t^3 - 8t.$$

When $t = 0$,

$$x'(t) \neq 0 \quad \text{and} \quad y'(t) = 0;$$

the tangent line is horizontal and its equation can be written $y = 4$. When $t = 2$,

$$x'(t) = 2 \quad \text{and} \quad y'(t) = 16;$$

the equation for the tangent line can therefore be written

$$16(x - 1) - 2(y - 4) = 0, \qquad\qquad (10.7.3)$$

or, more simply,

$$y - 4 = 8(x - 1). \quad \square$$

We can apply these ideas to a curve given in polar coordinates by an equation of the form $r = f(\theta)$. The coordinate transformations

$$x = r \cos \theta, \quad y = r \sin \theta$$

enable us to parametrize such a curve by setting

$$x(\theta) = f(\theta) \cos \theta, \quad y(\theta) = f(\theta) \sin \theta.$$

Problem 4. Find the slope of the spiral

$$r = a\theta \qquad \text{at } \theta = \tfrac{1}{2}\pi.$$

Solution. We write

$$x(\theta) = r \cos \theta = a\theta \cos \theta, \quad y(\theta) = r \sin \theta = a\theta \sin \theta.$$

Now we differentiate:

$$x'(\theta) = -a\theta \sin \theta + a \cos \theta, \quad y'(\theta) = a\theta \cos \theta + a \sin \theta.$$

Since

$$x'(\tfrac{1}{2}\pi) = -\tfrac{1}{2}\pi a \quad \text{and} \quad y'(\tfrac{1}{2}\pi) = a,$$

the slope of the curve at $\theta = \tfrac{1}{2}\pi$ is

$$\frac{y'(\tfrac{1}{2}\pi)}{x'(\tfrac{1}{2}\pi)} = -\frac{2}{\pi}. \quad \square$$

Problem 5. Find the points of the cardioid

$$r = 1 - \cos \theta$$

at which the tangent line is vertical.

Solution. Since the cosine function has period 2π, we need only concern ourselves with θ in $[0, 2\pi)$. Parametrically we have

$$x(\theta) = (1 - \cos \theta) \cos \theta, \quad y(\theta) = (1 - \cos \theta) \sin \theta.$$

Differentiating and simplifying, we find that

$$x'(\theta) = (2 \cos \theta - 1) \sin \theta, \quad y'(\theta) = (1 - \cos \theta)(1 + 2 \cos \theta).$$

The only numbers in the interval $[0, 2\pi)$ at which x' is zero and y' is not zero are $\frac{1}{3}\pi$, π, and $\frac{5}{3}\pi$. The tangent line is vertical at

$$[\tfrac{1}{2}, \tfrac{1}{3}\pi], \quad [2, \pi], \quad [\tfrac{1}{2}, \tfrac{5}{3}\pi].$$

These points have rectangular coordinates

$$(\tfrac{1}{4}, \tfrac{1}{4}\sqrt{3}), \quad (-2, 0), \quad (\tfrac{1}{4}, -\tfrac{1}{4}\sqrt{3}). \quad \square$$

EXERCISES 10.7

Find an equation in x and y for the line tangent to the curve.

1. $x(t) = t, \quad y(t) = t^3 - 1; \quad t = 1.$
2. $x(t) = t^2, \quad y(t) = t + 5; \quad t = 2.$
3. $x(t) = 2t, \quad y(t) = \cos \pi t; \quad t = 0.$
4. $x(t) = 2t - 1, \quad y(t) = t^4; \quad t = 1.$
5. $x(t) = t^2, \quad y(t) = (2 - t)^2; \quad t = \frac{1}{2}.$
6. $x(t) = 1/t, \quad y(t) = t^2 + 1; \quad t = 1.$
7. $x(t) = \cos^3 t, \quad y(t) = \sin^3 t; \quad t = \frac{1}{4}\pi.$
8. $x(t) = e^t, \quad y(t) = 3e^{-t}; \quad t = 0.$

Find an equation in x and y for the line tangent to the polar curve.

9. $r = 4 - 2 \sin \theta, \quad \theta = 0.$
10. $r = 4 \cos 2\theta, \quad \theta = \frac{1}{2}\pi.$

11. $r = \dfrac{4}{5 - \cos \theta}, \quad \theta = \frac{1}{2}\pi.$
12. $r = \dfrac{5}{4 - \cos \theta}, \quad \theta = \frac{1}{6}\pi.$

13. $r = \dfrac{\sin \theta - \cos \theta}{\sin \theta + \cos \theta}, \quad \theta = 0.$
14. $r = \dfrac{\sin \theta + \cos \theta}{\sin \theta - \cos \theta}, \quad \theta = \frac{1}{2}\pi.$

Parametrize the curve by a pair of differentiable functions

$$x = x(t), \quad y = y(t) \qquad \text{with} \quad [x'(t)]^2 + [y'(t)]^2 \neq 0.$$

Sketch the curve and determine the tangent line at the origin by the method of this section.

15. $y = x^3.$ **16.** $x = y^3.$ **17.** $y^5 = x^3.$ **18.** $y^3 = x^5.$

Find the points (x, y) at which the curve has (a) a horizontal tangent, (b) a vertical tangent. Then sketch the curve.

19. $x(t) = 3t - t^3, \quad y(t) = t + 1.$
20. $x(t) = t^2 - 2t, \quad y(t) = t^3 - 12t.$
21. $x(t) = 3 - 4 \sin t, \quad y(t) = 4 + 3 \cos t.$
22. $x(t) = \sin 2t, \quad y(t) = \sin t.$
23. $x(t) = t^2 - 2t, \quad y(t) = t^3 - 3t^2 + 2t.$
24. $x(t) = 2 - 5 \cos t, \quad y(t) = 3 + \sin t.$
25. $x(t) = \cos t, \quad y(t) = \sin 2t.$
26. $x(t) = 3 + 2 \sin t, \quad y(t) = 2 + 5 \sin t.$

27. Find the tangent(s) to the curve

$$x(t) = -t + 2 \cos \tfrac{1}{4}\pi t, \quad y(t) = t^4 - 4t^2$$

at the point $(2, 0)$.

28. Find the tangent(s) to the curve.

$$x(t) = t^3 - t, \quad y(t) = t \sin \tfrac{1}{2}\pi t$$

at the point $(0, 1)$.

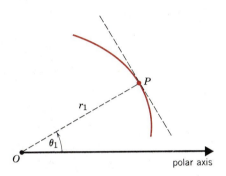

Figure 10.7.3

29. Let $P = [r_1, \theta_1]$ be a point on a polar curve $r = f(\theta)$ as in Figure 10.7.3. Show that, if $f'(\theta_1) = 0$ but $f(\theta_1) \neq 0$, then the tangent line at P is perpendicular to the line segment $\overline{OP}$.

30. If $0 < a < 1$, the polar curve $r = a - \cos \theta$ is a limaçon with an inner loop. Choose a so that the curve will intersect itself at the pole in a right angle.

Verify that $x'(0) = y'(0) = 0$ and that the given description holds at the point where $t = 0$. Sketch the graph.

31. $x(t) = t^3, \quad y(t) = t^2$; cusp.

32. $x(t) = t^3, \quad y(t) = t^5$; horizontal tangent.

33. $x(t) = t^5, \quad y(t) = t^3$; vertical tangent.

34. $x(t) = t^3 - 1, \quad y(t) = 2t^3$; tangent with slope 2.

35. $x(t) = t^2, \quad y(t) = t^2 + 1$; no tangent line.

36. Suppose that $x = x(t)$, $y = y(t)$ are twice differentiable functions that parametrize a curve. Take a point on the curve at which $x'(t) \neq 0$ and d^2y/dx^2 exists. Show that

(10.7.5)

$$\frac{d^2y}{dx^2} = \frac{x'(t)y''(t) - y'(t)x''(t)}{[x'(t)]^3}.$$

Calculate d^2y/dx^2 at the indicated point without eliminating the parameter.

37. $x(t) = \cos t, \quad y(t) = \sin t; \quad t = \tfrac{1}{6}\pi.$ **38.** $x(t) = t^3, \quad y(t) = t - 2; \quad t = 1.$

39. $x(t) = e^t, \quad y(t) = e^{-t}; \quad t = 0.$ **40.** $x(t) = \sin^2 t, \quad y(t) = \cos t; \quad t = \tfrac{1}{4}\pi.$

10.8 THE LEAST UPPER BOUND AXIOM

In Section 10.9 we will be talking about the *length* of a curve. To be able to do that with precision we first need to look a little deeper into the real number system.

We begin with a set S of real numbers. As indicated in Chapter 1, a number M is called an *upper bound* for S iff

$$x \leq M \qquad \text{for all } x \in S.$$

Of course, not all sets of real numbers have upper bounds. Those that do are said to be *bounded above*.

It is clear that every set that has a greatest element has an upper bound: if b is the greatest element of S, then

$$x \leq b \qquad \text{for all } x \in S;$$

this makes b an upper bound for S. The converse is false: the sets

$$(-\infty, 0) \quad \text{and} \quad \left\{ \frac{1}{2}, \frac{2}{3}, \frac{3}{4}, \cdots, \frac{n}{n+1}, \cdots \right\}$$

both have upper bounds (2 for instance), but neither has a greatest element.

Let's return to the first set, $(-\infty, 0)$. While $(-\infty, 0)$ does not have a greatest element, the set of its upper bounds, $[0, \infty)$, does have a least element, 0. We call 0 the *least upper bound of* $(-\infty, 0)$.

Now let's reexamine the second set. While the set of quotients

$$\frac{n}{n+1} = 1 - \frac{1}{n+1}$$

does not have a greatest element, the set of its upper bounds, $[1, \infty)$, does have a least element, 1. We call 1 the *least upper bound* of that set of quotients.

We are ready now to make explicit one of the key *assumptions* that we make about the real number system. It is called the *least upper bound axiom*.

AXIOM 10.8.1   THE LEAST UPPER BOUND AXIOM

Every nonempty set of real numbers that has an upper bound has a *least* upper bound.

To indicate the least upper bound of a set S, we will write lub S. Here are some examples:

1. lub $(-\infty, 0) = 0$,   lub $(-\infty, 0] = 0$.

2. lub $(-4, -1) = -1$,   lub $(-4, -1] = -1$.

3. lub $\left\{ \dfrac{1}{2}, \dfrac{2}{3}, \dfrac{3}{4}, \cdots, \dfrac{n}{n+1}, \cdots \right\} = 1$.

4. lub $\left\{ -\dfrac{1}{2}, -\dfrac{1}{8}, -\dfrac{1}{27}, \cdots, -\dfrac{1}{n^3}, \cdots \right\} = 0$.

5. lub $\{x : x^2 < 3\} = $ lub $\{x : -\sqrt{3} < x < \sqrt{3}\} = \sqrt{3}$.   □

The least upper bound of a set has a special property that deserves particular attention. The idea is this: the fact that M is the least upper bound of the set S does not tell us that M is in S (it need not be), but it does tell us that we can approximate M as closely as we may wish by elements of S.

THEOREM 10.8.2

If M is the least upper bound of the set S and ϵ is a positive number, then there is at least one number s in S such that

$$M - \epsilon < s \le M.$$

Proof. Let $\epsilon > 0$. Since M is an upper bound for S, the condition $s \le M$ is satisfied by all numbers s in S. All we have to show therefore is that there is some number s in S such that

$$M - \epsilon < s.$$

Suppose on the contrary that there is no such number in S. We then have

$$x \le M - \epsilon \qquad \text{for all } x \in S.$$

This makes $M - \epsilon$ an upper bound for S. But this cannot be, for then $M - \epsilon$ is an upper bound for S that is *less* than M, and by assumption, M is the *least* upper bound. ☐

The theorem we just proved is illustrated in Figure 10.8.1. Take S as the set of points marked in the figure. If $M = \mathrm{lub}\ S$, then S has at least one element in every half-open interval of the form $(M - \epsilon, M]$.

$M - \epsilon$ M

Figure 10.8.1

Example 1

(a) Let

$$S = \left\{ \frac{1}{2}, \frac{2}{3}, \frac{3}{4}, \cdots, \frac{n}{n+1}, \cdots \right\}$$

and take $\epsilon = 0.0001$. Since 1 is the least upper bound of S, there must be a number s in S such that

$$1 - 0.0001 < s \le 1.$$

There is: take, for example, $s = \frac{99999}{100000}$. ☐

(b) Let

$$S = \{1, 2, 3\}$$

and take $\epsilon = $ | |||||||| S, there must be a number s in S such that

$$3 - 0.000001 < s \leq 3.$$

There is: $s = 3$. □

We come now to lower bounds. In the first place, a number m is called a *lower bound* for S iff

$$m \leq x \qquad \text{for all } x \in S.$$

Sets that have lower bounds are said to be *bounded below*. Not all sets have lower bounds; those that do have *greatest lower bounds*. This need not be taken as an axiom. Using the least upper bound axiom, we can prove it as a theorem.

THEOREM 10.8.3

Every nonempty set of real numbers that has a lower bound has a *greatest lower bound*.

Proof. Suppose that S is nonempty and that it has a lower bound x. Then

$$x \leq s \qquad \text{for all } s \in S.$$

It follows that $-s \leq -x$ for all $s \in S$; that is,

$$\{-s : s \in S\} \quad \text{has an upper bound } -x.$$

From the least upper bound axiom we conclude that $\{-s : s \in S\}$ has a least upper bound; call it x_0. Since $-s \leq x_0$ for all $s \in S$, you can see that

$$-x_0 \leq s \qquad \text{for all } s \in S,$$

and thus $-x_0$ is a lower bound for S. We now assert that $-x_0$ is the greatest lower bound of the set S. To see this, note that, if there existed a number x_1 satisfying

$$-x_0 < x_1 \leq s \qquad \text{for all } s \in S,$$

then we would have

$$-s \leq -x_1 < x_0 \qquad \text{for all } s \in S,$$

and thus x_0 would not be the *least* upper bound of $\{-s : s \in S\}$.† □

---

† We proved Theorem 10.8.3 by assuming the least upper bound axiom. We could have proceeded the other way. We could have set Theorem 10.8.3 as an axiom, and then proved the least upper bound axiom as a theorem.

As in the case of the least upper bound, the greatest lower bound of a set need not be in the set but can be approximated as closely as we wish by members of the set. In short, we have the following theorem, the proof of which is left as an exercise.

THEOREM 10.8.4

If m is the greatest lower bound of the set S and ϵ is a positive number, then there is at least one number s in S such that

$$m \leq s < m + \epsilon.$$

The theorem is illustrated in Figure 10.8.2. If $m = \text{glb } S$ (that is, if m is the greatest lower bound of the set S), then S has at least one element in every half-open interval of the form $[m, m + \epsilon)$.

m $m + \epsilon$

Figure 10.8.2

Remark. Remember the intermediate-value theorem? It states that a continuous function skips no values. Remember the maximum-minimum theorem? It states that on a bounded closed interval a continuous function takes on both a maximum and a minimum value. We have been using these two results right along but we have not proved them. Now that you understand least upper bounds and greatest lower bounds, you are in a position to follow proofs of both of these theorems. (See Appendix B.) Better still, try to prove the theorems yourself. ☐

EXERCISES 10.8

Find the least upper bound (if it exists) and the greatest lower bound (if it exists) for each of the following sets.

1. $(0, 2)$.   **2.** $[0, 2]$.   **3.** $(0, \infty)$.   **4.** $(-\infty, 1)$.

5. $\{x : x^2 < 4\}$.   **6.** $\{x : |x - 1| < 2\}$.   **7.** $\{x : x^3 \geq 8\}$.   **8.** $\{x : x^4 \leq 16\}$.

9. $\{2\frac{1}{2}, 2\frac{1}{3}, 2\frac{1}{4}, \ldots\}$.   **10.** $\{-1, -\frac{1}{2}, -\frac{1}{3}, -\frac{1}{4}, \ldots\}$.

11. $\{0.9, 0.99, 0.999, \ldots\}$.   **12.** $\{-2, 2, -2.1, 2.1, -2.11, 2.11, \ldots\}$.

13. $\{x : \ln x < 1\}$.   **14.** $\{x : \ln x > 0\}$.   **15.** $\{x : x^2 + x - 1 < 0\}$.

16. $\{x : x^2 + x + 2 \geq 0\}$.   **17.** $\{x : x^2 > 4\}$.   **18.** $\{x : |x - 1| > 2\}$.

19. $\{x : \sin x \geq -1\}$.   **20.** $\{x : e^x < 1\}$.

Illustrate the validity of Theorem 10.8.4 taking S and ϵ as given below.

21. $S = \{\frac{1}{11}, (\frac{1}{11})^2, (\frac{1}{11})^3, \ldots, (\frac{1}{11})^n, \ldots\}$, $\epsilon = 0.001$.

22. $S = \{1, 2, 3, 4\}$, $\epsilon = 0.0001$.

23. $S = \{\frac{1}{10}, \frac{1}{1000}, \frac{1}{100000}, \ldots, (\frac{1}{10})^{2n-1}, \ldots\}$, $\epsilon = (\frac{1}{10})^k$ $(k \geq 1)$.

24. $S = \{\frac{1}{2}, \frac{1}{4}, \frac{1}{8}, \ldots, (\frac{1}{2})^n, \ldots\}$, $\epsilon = (\frac{1}{4})^k$ $(k \geq 1)$.

25. Prove Theorem 10.8.4 by imitating the proof of Theorem 10.8.2.

10.9   ARC LENGTH AND SPEED

Definition of Arc Length

We come now to the notion of arc length.   In Figure 10.9.1 we have sketched a curve C, which we assume is parametrized by a pair of *continuously differentiable* functions†

$$x = x(t), \quad y = y(t) \qquad t \in [a, b].$$

What we want to do is assign a length to this curve C.

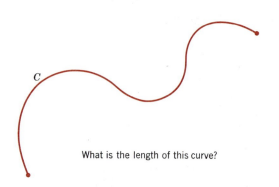

What is the length of this curve?

Figure 10.9.1

a polygonal path inscribed in the curve C

Figure 10.9.2

Here our experience in Chapter 5 can be used as a model.   To decide what should be meant by the area of a region Ω, we approximated Ω by the union of a finite number of rectangles.   To decide what should be meant by the length of C, we will approximate C by the union of a finite number of line segments.

Each point t in $[a, b]$ gives rise to a point $P = P(x(t), y(t))$ that lies on C.   By choosing a finite number of points in $[a, b]$,

$$a = t_0 < t_1 < \ \cdots \ < t_{i-1} < t_i < \ \cdots \ < t_{n-1} < t_n = b,$$

we obtain a finite number of points on C,

$$P_0, P_1, \ldots, P_{i-1}, P_i, \ldots, P_{n-1}, P_n.$$

We join these points consecutively by line segments and call the resulting path,

$$\gamma = \overline{P_0 P_1} \cup \ \cdots \ \cup \overline{P_{i-1} P_i} \cup \ \cdots \ \cup \overline{P_{n-1} P_n},$$

a *polygonal path* inscribed in C.   (See Figure 10.9.2.)

The length of such a polygonal path is the sum of the distances between consecutive vertices:

$$\text{length of } \gamma = L(\gamma) = d(P_0, P_1) + \ \cdots \ + d(P_{i-1}, P_i) + \ \cdots \ + d(P_{n-1}, P_n).$$

---

† By this we mean functions that have continuous first derivatives.

The path γ serves as an approximation to the curve C, but obviously a better approximation can be obtained by adding more vertices to γ. At this point let's ask ourselves exactly what it is we require of the number that we shall call the length of C. Certainly we require that

$$L(\gamma) \leq \text{the length of } C \qquad \text{for each } \gamma \text{ inscribed in } C.$$

But that is not enough. There is another requirement that seems reasonable. If we can choose γ to approximate C as closely as we wish, then we should be able to choose γ so that $L(\gamma)$ approximates the length of C as closely as we wish; namely, for each positive number ϵ there should exist a polygonal path γ such that

$$(\text{length of } C) - \epsilon < L(\gamma) \leq \text{length of } C.$$

Theorem 10.8.2 tells us that we can achieve this result by defining the length of C as the least upper bound of all the $L(\gamma)$. This in fact is what we do.

DEFINITION 10.9.1

$$\text{The length of } C = \begin{cases} \text{the least upper bound of the set of all} \\ \text{lengths of polygonal paths inscribed in } C \end{cases}.$$

Arc-Length Formulas

Now that we have explained what we mean by the length of a parametrized curve, it is time to describe a practical way to compute it. The basic result is easy to state. If C is parametrized by a pair of continuously differentiable functions

$$x = x(t), \quad y = y(t) \qquad t \in [a, b],$$

then

(10.9.2)
$$\text{the length of } C = \int_a^b \sqrt{[x'(t)]^2 + [y'(t)]^2} \, dt.$$

Obviously, this is not something to be taken on faith. It has to be proved. We will do so, but not until Chapter 14. For now we assume the result and carry out some computations.

Example 1. The functions

$$x(t) = a \cos t, \quad y(t) = a \sin t \qquad t \in [0, 2\pi]$$

parametrize a circle C of radius a. Formula 10.9.2 gives

$$L(C) = \int_0^{2\pi} \sqrt{a^2 \sin^2 t + a^2 \cos^2 t} \, dt = \int_0^{2\pi} a \, dt = 2\pi a,$$

which is reassuring. ◻

Suppose now that C is the graph of a continuously differentiable function

$$y = f(x), \qquad x \in [a, b].$$

Then we can parametrize C by setting

$$x(t) = t, \quad y(t) = f(t) \qquad t \in [a, b].$$

Since

$$x'(t) = 1 \qquad \text{and} \qquad y'(t) = f'(t),$$

Formula 10.9.2 gives

$$L(C) = \int_a^b \sqrt{1 + [f'(t)]^2} \, dt.$$

Replacing t by x we can write

(10.9.3) $\qquad$ the length of the graph of $f = \displaystyle\int_a^b \sqrt{1 + [f'(x)]^2} \, dx.$

A direct derivation of this formula is outlined in Exercise 38.

Example 2. If

$$f(x) = \tfrac{1}{6}x^3 + \tfrac{1}{2}x^{-1},$$

then

$$f'(x) = \tfrac{3}{6}x^2 - \tfrac{1}{2}x^{-2} = \tfrac{1}{2}x^2 - \tfrac{1}{2}x^{-2}.$$

Therefore

$$1 + [f'(x)]^2 = 1 + (\tfrac{1}{4}x^4 - \tfrac{1}{2} + \tfrac{1}{4}x^{-4}) = \tfrac{1}{4}x^4 + \tfrac{1}{2} + \tfrac{1}{4}x^{-4} = (\tfrac{1}{2}x^2 + \tfrac{1}{2}x^{-2})^2.$$

The length of the graph from $x = 1$ to $x = 3$ is

$$\int_1^3 \sqrt{1 + [f'(x)]^2} \, dx = \int_1^3 (\tfrac{1}{2}x^2 + \tfrac{1}{2}x^{-2}) \, dx = \left[\tfrac{1}{6}x^3 - \tfrac{1}{2}x^{-1} \right]_1^3 = \tfrac{14}{3}. \quad \square$$

Example 3. The graph of the function

$$f(x) = x^2, \qquad x \in [0, 1]$$

is a parabolic arc. The length of this arc is given by

$$\int_0^1 \sqrt{1 + [f'(x)]^2} \, dx = \int_0^1 \sqrt{1 + 4x^2} \, dx = 2 \int_0^1 \sqrt{(\tfrac{1}{2})^2 + x^2} \, dx$$

$$= \left[x\sqrt{(\tfrac{1}{2})^2 + x^2} + (\tfrac{1}{2})^2 \ln \left(x + \sqrt{(\tfrac{1}{2})^2 + x^2} \right) \right]_0^1$$

by 8.5.1 $\longrightarrow$

$$= \tfrac{1}{2}\sqrt{5} + \tfrac{1}{4} \ln (2 + \sqrt{5}) \approx 1.48. \quad \square$$

Suppose now that C is the graph of a polar function

$$r = \rho(\theta), \qquad \alpha \leq \theta \leq \beta.$$

We can parametrize C by setting

$$x(\theta) = \rho(\theta) \cos \theta, \quad y(\theta) = \rho(\theta) \sin \theta \qquad \theta \in [\alpha, \beta].$$

A straightforward calculation that we leave to you shows that

$$[x'(\theta)]^2 + [y'(\theta)]^2 = [\rho(\theta)]^2 + [\rho'(\theta)]^2.$$

The arc-length formula now reads

(10.9.4)
$$L(C) = \int_\alpha^\beta \sqrt{[\rho(\theta)]^2 + [\rho'(\theta)]^2} \, d\theta.$$

Example 4. For fixed $a > 0$ the equation $r = a$ represents a circle of radius a. Here

$$\rho(\theta) = a \quad \text{and} \quad \rho'(\theta) = 0.$$

The circumference of this circle is

$$\int_0^{2\pi} \sqrt{[\rho(\theta)]^2 + [\rho'(\theta)]^2} \, d\theta = \int_0^{2\pi} \sqrt{a^2 + 0^2} \, d\theta = \int_0^{2\pi} a \, d\theta = 2\pi a. \quad \square$$

Example 5. In the case of the cardioid $r = a(1 - \cos \theta)$, we have

$$\rho(\theta) = a(1 - \cos \theta), \qquad \rho'(\theta) = a \sin \theta.$$

Here

$$[\rho(\theta)]^2 + [\rho'(\theta)]^2 = a^2(1 - 2 \cos \theta + \cos^2 \theta) + a^2 \sin^2 \theta = 2a^2(1 - \cos \theta).$$

The identity

$$\tfrac{1}{2}(1 - \cos \theta) = \sin^2 \tfrac{1}{2}\theta$$

gives

$$[\rho(\theta)]^2 + [\rho'(\theta)]^2 = 4a^2 \sin^2 \tfrac{1}{2}\theta.$$

The length of the cardioid is $8a$:

$$\int_0^{2\pi} \sqrt{[\rho(\theta)]^2 + [\rho'(\theta)]^2} \, d\theta = \int_0^{2\pi} 2a \sin \tfrac{1}{2}\theta \, d\theta = 4a \left[-\cos \tfrac{1}{2}\theta \right]_0^{2\pi} = 8a. \quad \square$$

The Geometric Significance of *dx/ds* and *dy/ds*.

Figure 10.9.3 shows the graph of a function $y = f(x)$ which we assume to be continuously differentiable. At the point (x, y) the tangent line has an inclination marked α.

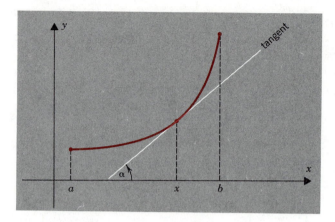

Figure 10.9.3

The length of the graph from a to x can be written

$$s(x) = \int_a^x \sqrt{1 + [f'(t)]^2}\, dt.$$

Differentiation with respect to x gives $s'(x) = \sqrt{1 + [f'(x)]^2}$. Using the Leibniz notation we have

$$\frac{ds}{dx} = \sqrt{1 + \left(\frac{dy}{dx}\right)^2} = \sqrt{1 + \tan^2 \alpha} = \sec \alpha.$$

$$\underset{\sec \alpha > 0 \quad \text{for } \alpha \in (-\tfrac{1}{2}\pi, \tfrac{1}{2}\pi)}{\uparrow}$$

By (3.8.2)

$$\frac{dx}{ds} = \frac{1}{\sec \alpha} = \cos \alpha.$$

To find dy/ds we note that

$$\tan \alpha = \frac{dy}{dx} = \frac{dy}{ds}\frac{ds}{dx} = \frac{dy}{ds}\sec \alpha.$$

$$\underset{\text{chain rule}}{\uparrow}$$

Multiplication by $\sec \alpha$ gives

$$\frac{dy}{ds} = \sin \alpha.$$

For the record

(10.9.5) $\dfrac{dx}{ds} = \cos \alpha$ and $\dfrac{dy}{ds} = \sin \alpha$ where α is the inclination of the tangent line.

(We will be using this in a moment.)

Speed Along a Plane Curve

So far we have talked about speed only in connection with straight-line motion. How can we calculate the speed of an object that moves along a curve? Imagine an object moving along some curved path. Suppose that $(x(t), y(t))$ gives the position of the object at time t. The distance traveled by the object from time zero to any later time t is simply the length of the path up to time t:

$$s(t) = \int_0^t \sqrt{[x'(u)]^2 + [y'(u)]^2} \, du.$$

The time rate of change of this distance is what we call the *speed* of the object. Denoting the speed of the object at time t by $v(t)$ we have

(10.9.6) $$v(t) = s'(t) = \sqrt{[x'(t)]^2 + [y'(t)]^2}.$$

Problem 6. The path of the projectile in Exercises 31 – 36 in Section 9.3 is given in terms of the time parameter t by the following equations:

$$x(t) = (v_0 \cos \theta)t, \qquad y(t) = -16t^2 + (v_0 \sin \theta)t.$$

From those exercises we know that the projectile impacts at time $t = \frac{1}{16} v_0 \sin \theta$. Calculate the speed at impact.

Solution. Since

$$x'(t) = v_0 \cos \theta \quad \text{and} \quad y'(t) = -32t + v_0 \sin \theta,$$

we have

$$v(t) = \sqrt{v_0^2 \cos^2 \theta + (-32t + v_0 \sin \theta)^2}.$$

The speed at impact is therefore

$$\sqrt{v_0^2 \cos^2 \theta + (-2v_0 \sin \theta + v_0 \sin \theta)^2} = \sqrt{v_0^2 \cos^2 \theta + v_0^2 \sin^2 \theta} = |v_0|,$$

which is exactly the speed with which the projectile was fired.† □

In the Leibniz notation the equation for speed reads

(10.9.7) $$v = \frac{ds}{dt} = \sqrt{\left(\frac{dx}{dt}\right)^2 + \left(\frac{dy}{dt}\right)^2}.$$

If we know the speed of an object and we know its mass, then we can calculate its kinetic energy.

---

† This can be obtained from energy considerations. See Exercise 39.

Problem 7.   A particle of mass m slides down a frictionless curve (see Figure 10.9.4) from a point (x_0, y_0) to a point (x_1, y_1) under the force of gravity.   As discussed in Section 3.5, the particle has two forms of energy during the motion: gravitational potential energy mgy and kinetic energy $\frac{1}{2}mv^2$.   Show that the sum of these two quantities remains constant:

$$\overset{\text{GPE}}{mgy} + \overset{\text{KE}}{\tfrac{1}{2}mv^2} = C.$$

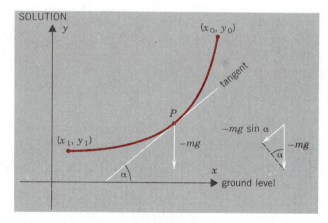

Figure 10.9.4

The particle is subjected to a vertical force $-mg$ (a downward force of magnitude mg).   Since the particle is constrained to remain on the curve, the effective force on the particle is tangential.   The tangential component of the vertical force is $-mg\sin\alpha$.   (See Figure 10.9.4.)   The speed of the particle is ds/dt and the tangential acceleration is d^2s/dt^2.   (It is as if the particle were moving along the tangent line.)   Therefore, by Newton's $F = ma$, we have

$$m\frac{d^2s}{dt^2} = \underset{\text{by (10.9.5)}}{-mg\sin\alpha} = -mg\frac{dy}{ds}.$$

We can therefore write

$$mg\frac{dy}{ds} + m\frac{d^2s}{dt^2} = 0$$

$$mg\frac{dy}{ds}\frac{ds}{dt} + m\frac{ds}{dt}\frac{d^2s}{dt^2} = 0 \qquad \left(\text{multiplied by } \tfrac{ds}{dt}\right)$$

$$mg\frac{dy}{dt} + mv\frac{dv}{dt} = 0. \qquad \text{(chain rule)}$$

Integrating with respect to t we have

$$mgy + \tfrac{1}{2}mv^2 = C$$

as asserted.   □

EXERCISES 10.9

Find the length of the graph and compare it to the straight-line distance between the endpoints of the graph.

1. $f(x) = 2x + 3$, $x \in [0, 1]$.
2. $f(x) = 3x + 2$, $x \in [0, 1]$.
3. $f(x) = (x - \frac{4}{9})^{3/2}$, $x \in [1, 4]$.
4. $f(x) = x^{3/2}$, $x \in [0, 44]$.
5. $f(x) = \frac{1}{3}\sqrt{x}(x - 3)$, $x \in [0, 3]$.
6. $f(x) = \frac{2}{3}(x - 1)^{3/2}$, $x \in [1, 2]$.
7. $f(x) = \frac{1}{3}(x^2 + 2)^{3/2}$, $x \in [0, 1]$.
8. $f(x) = \frac{1}{3}(x^2 - 2)^{3/2}$, $x \in [2, 4]$.
9. $f(x) = \frac{1}{4}x^2 - \frac{1}{2}\ln x$, $x \in [1, 5]$.
10. $f(x) = \frac{1}{8}x^2 - \ln x$, $x \in [1, 4]$.
11. $f(x) = \frac{3}{8}x^{4/3} - \frac{3}{4}x^{2/3}$, $x \in [1, 8]$.
12. $f(x) = \frac{1}{10}x^5 + \frac{1}{6}x^{-3}$, $x \in [1, 2]$.
13. $f(x) = \ln (\sec x)$, $x \in [0, \frac{1}{4}\pi]$.
14. $f(x) = \frac{1}{2}x^2$, $x \in [0, 1]$.

15. $f(x) = \frac{1}{2}x\sqrt{x^2 - 1} - \frac{1}{2}\ln (x + \sqrt{x^2 - 1})$, $x \in [1, 2]$.

16. $f(x) = \cosh x$, $x \in [0, \ln 2]$.

17. $f(x) = \frac{1}{2}x\sqrt{3 - x^2} + \frac{3}{2}\sin^{-1}(\frac{1}{3}\sqrt{3}x)$, $x \in [0, 1]$.

18. $f(x) = \ln (\sin x)$, $x \in [\frac{1}{6}\pi, \frac{1}{2}\pi]$.

The equations below give the position of a particle at each time t during the time interval specified. Find the initial speed of the particle, the terminal speed, and the distance traveled.

19. $x(t) = t^2$, $y(t) = 2t$ from $t = 0$ to $t = \sqrt{3}$.

20. $x(t) = t - 1$, $y(t) = \frac{1}{2}t^2$ from $t = 0$ to $t = 1$.

21. $x(t) = t^2$, $y(t) = t^3$ from $t = 0$ to $t = 1$.

22. $x(t) = a \cos^3 t$, $y(t) = a \sin^3 t$ from $t = 0$ to $t = \frac{1}{2}\pi$.

23. $x(t) = e^t \sin t$, $y(t) = e^t \cos t$ from $t = 0$ to $t = \pi$.

24. $x(t) = \cos t + t \sin t$, $y(t) = \sin t - t \cos t$ from $t = 0$ to $t = \pi$.

Find the length of the polar curve.

25. $r = 1$ from $\theta = 0$ to $\theta = 2\pi$.
26. $r = 3$ from $\theta = 0$ to $\theta = \pi$.

27. $r = e^\theta$ from $\theta = 0$ to $\theta = 4\pi$.   (logarithmic spiral)

28. $r = ae^\theta$ from $\theta = -2\pi$ to $\theta = 2\pi$.
29. $r = e^{2\theta}$ from $\theta = 0$ to $\theta = 2\pi$.

30. $r = 1 + \cos \theta$ from $\theta = 0$ to $\theta = 2\pi$.
31. $r = 1 - \cos \theta$ from $\theta = 0$ to $\theta = \frac{1}{2}\pi$.

32. $r = 2a \sec \theta$ from $\theta = 0$ to $\theta = \frac{1}{4}\pi$.

33. At time t a particle has position

$$x(t) = 1 + \tan^{-1} t, \quad y(t) = 1 - \ln \sqrt{1 + t^2}.$$

Find the total distance traveled from time $t = 0$ to time $t = 1$. Give the initial speed and the terminal speed.

34. At time t a particle has position

$$x(t) = 1 - \cos t, \quad y(t) = t - \sin t.$$

Find the total distance traveled from time $t = 0$ to time $t = 2\pi$. Give the initial speed and the terminal speed.

35. Find c given that the length of the curve $y = \ln x$ from $x = 1$ to $x = e$ equals the length of the curve $y = e^x$ from $x = 0$ to $x = c$.

36. Find the length of the curve $y = x^{2/3}$, $x \in [1, 8]$. HINT: Work with the mirror image $y = x^{3/2}$, $x \in [1, 4]$.

37. Show that the function $f(x) = \cosh x$ has the following property: for every interval $[a, b]$ the length of the graph equals the area under the graph.

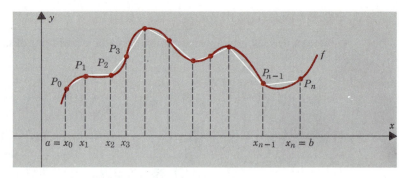

Figure 10.9.5

38. Figure 10.9.5 shows the graph of a continuously differentiable function f from $x = a$ to $x = b$ together with a polygonal approximation. Show that the length of this polygonal approximation can be written as the following Riemann sum:

$$\sqrt{1 + [f'(x_1^*)]^2}\,\Delta x_1 + \sqrt{1 + [f'(x_2^*)]^2}\,\Delta x_2 + \cdots + \sqrt{1 + [f'(x_n^*)]^2}\,\Delta x_n.$$

As $\|P\| = \max \Delta x_i$ tends to 0, such Riemann sums tend to

$$\int_a^b \sqrt{1 + [f'(x)]^2}\,dx.$$

39. Rework Problem 6, this time from energy considerations. (a) Verify that GPE + KE remains constant. (b) Then derive the speed at impact from an energy equation.

40. The path of the projectile discussed under the heading "Parabolic Trajectories" in Section 9.3 is given in terms of the time parameter t by the following equations:

$$x(t) = (v_0 \cos \theta)t + x_0, \qquad y(t) = -\tfrac{1}{2}gt^2 + (v_0 \sin \theta)t + y_0.$$

The projectile impacts at ground level, $y = 0$. Find the speed at impact.

41. Suppose that f is continuously differentiable from $x = a$ to $x = b$. Show that the

(10.9.8)   length of the graph of $f = \displaystyle\int_a^b |\sec [\alpha(x)]|\,dx$ where $\alpha(x)$ is the inclination of the tangent line at $(x, f(x))$.

42. Show that a homogeneous, flexible, inelastic rope hanging from two fixed points assumes the shape of a *catenary*:

$$f(x) = a \cosh\left(\frac{x}{a}\right) = \frac{a}{2}\,(e^{x/a} + e^{-x/a}).$$

HINT: Refer to Figure 10.9.6. The part of the rope that corresponds to the interval $[0, x]$ is subject to the following forces:

(1) its weight, which is proportional to its length;
(2) a horizontal pull at 0, $p(0)$;
(3) a tangential pull at x, $p(x)$.

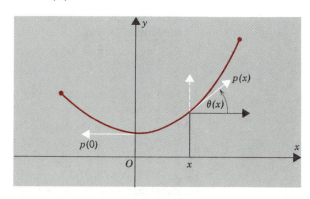

Figure 10.9.6

10.10 THE AREA OF A SURFACE OF REVOLUTION; THE CENTROID OF A CURVE; PAPPUS'S THEOREM ON SURFACE AREA

The Area of a Surface of Revolution

In Figure 10.10.1 you can see the frustum of a cone; one radius is marked r, the other R, and the slant height is marked s.

Figure 10.10.1

An interesting elementary calculation that we leave to you shows that the area of this slanted surface is given by the formula

(10.10.1) $$A = \pi(r + R)s.$$ (Exercise 19)

This little formula forms the basis for all that follows.

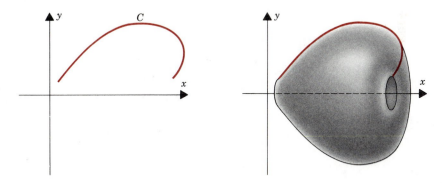

Figure 10.10.2

Let C be a curve in the upper half-plane. (Figure 10.10.2) The curve can meet the x-axis but only at a finite number of points. We will assume that C is parametrized by a pair of continuously differentiable functions

$$x = x(t), \quad y = y(t) \qquad t \in [c, d].$$

Furthermore, we will assume that C is *simple*: no two values of t between c and d give rise to the same point of C.

If we revolve C about the x-axis, we obtain a surface of revolution. The area of that surface is given by the formula

(10.10.2)

$$A = \int_c^d 2\pi y(t) \sqrt{[x'(t)]^2 + [y'(t)]^2}\ dt.$$

We will try to outline how this formula comes about.

Each partition $P = \{c = x_0 < x_1 < \cdots < x_n = d\}$ of $[c, d]$ generates a polygonal approximation to C. (Figure 10.10.3) Call this polygonal approximation C_p. By revolving C_p about the x-axis we get a surface made up of n conical frustums.

Figure 10.10.3

Figure 10.10.4

The ith frustum (Figure 10.10.4) has slant height

$$s_i = \sqrt{[x(t_i) - x(t_{i-1})]^2 + [y(t_i) - y(t_{i-1})]^2}$$

$$= \sqrt{\left[\frac{x(t_i) - x(t_{i-1})}{t_i - t_{i-1}}\right]^2 + \left[\frac{y(t_i) - y(t_{i-1})}{t_i - t_{i-1}}\right]^2}\ (t_i - t_{i-1})$$

and the lateral area $\pi[y(t_{i-1}) + y(t_i)]s_i$ (see Formula 10.10.1) can be written

$$\pi[y(t_{i-1}) + y(t_i)] \sqrt{\left[\frac{x(t_i) - x(t_{i-1})}{t_i - t_{i-1}}\right]^2 + \left[\frac{y(t_i) - y(t_{i-1})}{t_i - t_{i-1}}\right]^2}\ (t_i - t_{i-1}).$$

There exist points t_i^*, t_i^{**}, t_i^{***} all in $[t_{i-1}, t_i]$ such that

$$y(t_i) + y(t_{i-1}) \underset{\uparrow}{=} 2y(t_i^*), \qquad \frac{x(t_i) - x(t_{i-1})}{t_i - t_{i-1}} \underset{\uparrow}{=} x'(t_i^{**}), \qquad \frac{y(t_i) - y(t_{i-1})}{t_i - t_{i-1}} = y'(t_i^{***}).$$

intermediate-value theorem   mean-value theorem

Since $t_i - t_{i-1} = \Delta t_i$, we can write the lateral area of the ith frustum as

$$2\pi y(t_i^*) \sqrt{[x'(t_i^{**})]^2 + [y'(t_i^{***})]^2}\ \Delta t_i.$$

The area generated by revolving all of C_p is the sum of these terms:

$$2\pi y(t_1^*)\sqrt{[x'(t_1^{**})]^2 + [y'(t_1^{***})]^2}\,\Delta t_1 + \cdots + 2\pi y(t_n^*)\sqrt{[x'(t_n^{**})]^2 + [y'(t_n^{***})]^2}\,\Delta t_n.$$

This is not a Riemann sum: we don't know that $t_i^* = t_i^{**} = t_i^{***}$. But it is "close" to a Riemann sum. Close enough that, as $\|P\| \to 0$, this "almost" Riemann sum tends to the integral

$$\int_c^d 2\pi y(t)\sqrt{[x'(t)]^2 + [y'(t)]^2}\,dt.$$

That this is so follows from a theorem of advanced calculus known as Duhamel's principle. We will not attempt to fill in the details. ☐

Problem 1. Derive a formula for the surface area of a sphere from (10.10.2).

Solution. We can generate a sphere of radius r by revolving the arc

$$x(t) = r\cos t, \quad y(t) = r\sin t \qquad t \in [0, \pi]$$

about the x-axis. Differentiation gives

$$x'(t) = -r\sin t, \quad y'(t) = r\cos t.$$

By Formula 10.10.2

$$A = 2\pi \int_0^\pi r\sin t\,\sqrt{r^2(\sin^2 t + \cos^2 t)}\,dt$$

$$= 2\pi r^2 \int_0^\pi \sin t\,dt = 2\pi r^2 \left[-\cos t\right]_0^\pi = 4\pi r^2. \quad ☐$$

Problem 2. Find the area of the surface generated by revolving about the x-axis the curve

$$y^2 - 2\ln y = 4x \qquad \text{from } y = 1 \text{ to } y = 2.$$

Solution. We can represent the curve parametrically by setting

$$x(t) = \tfrac{1}{4}(t^2 - 2\ln t), \quad y(t) = t \qquad t \in [1, 2].$$

Here

$$x'(t) = \tfrac{1}{2}(t - t^{-1}), \quad y'(t) = 1$$

and

$$[x'(t)]^2 + [y'(t)]^2 = [\tfrac{1}{2}(t + t^{-1})]^2. \qquad \text{(check this)}$$

It follows that

$$A = \int_1^2 2\pi t[\tfrac{1}{2}(t + t^{-1})]\,dt = \int_1^2 \pi(t^2 + 1)\,dt = \pi\left[\tfrac{1}{3}t^3 + t\right]_1^2 = \tfrac{10}{3}\pi. \quad ☐$$

Suppose now that f is some nonnegative function defined from $x = a$ to $x = b$. If f' is continuous, then the graph of f is a continuously differentiable curve in the upper half-plane. The area of the surface generated by revolving this graph about the x-axis is given by the formula

(10.10.3)
$$A = \int_a^b 2\pi f(x)\sqrt{1 + [f'(x)]^2}\ dx.$$

This follows readily from (10.10.2). Set

$$x = t, \quad y(t) = f(t) \qquad t \in [a, b].$$

Apply (10.10.2) and then replace the dummy variable t by x. ☐

Problem 3. Find the area of the surface generated by revolving about the x-axis the graph of the sine function from $x = 0$ to $x = \frac{1}{2}\pi$.

Solution. Setting $f(x) = \sin x$, we have $f'(x) = \cos x$ and therefore

$$A = \int_0^{\pi/2} 2\pi \sin x \sqrt{1 + \cos^2 x}\ dx.$$

To calculate this integral, we set

$$u = \cos x, \qquad du = -\sin x\ dx.$$

At $x = 0$, $u = 1$; at $x = \frac{1}{2}\pi$, $u = 0$. Therefore

$$A = -2\pi \int_1^0 \sqrt{1 + u^2}\ du = 2\pi \int_0^1 \sqrt{1 + u^2}\ du$$

$$= 2\pi \left[\tfrac{1}{2} u \sqrt{1 + u^2} + \tfrac{1}{2} \ln\left(u + \sqrt{1 + u^2}\right) \right]_0^1$$

by (8.5.1) ——

$$= \pi[\sqrt{2} + \ln(1 + \sqrt{2})] \cong 2.3\pi. \quad ☐$$

The Centroid of a Curve

The centroid of a plane region Ω is the center of mass of a homogeneous plate in the shape of Ω. Likewise, the centroid of a solid of revolution T is the center of mass of a homogeneous solid in the shape of T. All this you know from Section 6.4.

What do we mean by the centroid of a plane curve C? Exactly what you would expect. By the *centroid* of a plane curve C, we mean the center of mass of a homogeneous wire in the shape of C. We can calculate the centroid of a curve from the following principles, which we take from physics.

Principle 1: Symmetry If a curve has an axis of symmetry, then the centroid $(\bar{x}, \bar{y})$ lies somewhere along that axis.

Principle 2: Additivity If a curve with length L is broken up into a finite number of pieces with arc lengths $\Delta s_1, \ldots, \Delta s_n$ and centroids $(\bar{x}_1, \bar{y}_1), \ldots, (\bar{x}_n, \bar{y}_n)$, then

$$\bar{x}L = \bar{x}_1 \Delta s_1 + \cdots + \bar{x}_n \Delta s_n \quad \text{and} \quad \bar{y}L = \bar{y}_1 \Delta s_1 + \cdots + \bar{y}_n \Delta s_n.$$

Figure 10.10.5 shows a curve C that begins at A and ends at B. Let's suppose that the curve is continuously differentiable and that the length of the curve is L. We want a formula for the centroid $(\bar{x}, \bar{y})$.

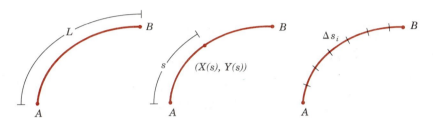

Figure 10.10.5

Let $(X(s), Y(s))$ be the point on C that is at an arc distance s from the initial point A. (What we are doing here is called *parametrizing C by arc length*.) A partition $P = \{0 = s_0 < s_1 < \cdots < s_n = L\}$ of $[0, L]$ breaks up C into n little pieces of lengths $\Delta s_1, \ldots, \Delta s_n$ and centroids $(\bar{x}_1, \bar{y}_1), \ldots, (\bar{x}_n, \bar{y}_n)$. From Principle 2 we know that

$$\bar{x}L = \bar{x}_1 \Delta s_1 + \cdots + \bar{x}_n \Delta s_n \quad \text{and} \quad \bar{y}L = \bar{y}_1 \Delta s_1 + \cdots + \bar{y}_n \Delta s_n.$$

Since $(\bar{x}_i, \bar{y}_i)$ lies on the ith piece, there exists s_i^* in $[s_{i-1}, s_i]$ for which $\bar{x}_i = X(s_i^*)$ and there exists s_i^{**} in $[s_{i-1}, s_i]$ for which $\bar{y}_i = Y(s_i^{**})$. We can therefore write

$$\bar{x}L = X(s_1^*) \Delta s_1 + \cdots + X(s_n^*) \Delta s_n, \qquad \bar{y}L = Y(s_1^{**}) \Delta s_1 + \cdots + Y(s_n^{**}) \Delta s_n.$$

The sums on the right are Riemann sums tending to easily recognizable limits: letting $\|P\| \to 0$ we have

(10.10.4)
$$\bar{x}L = \int_0^L X(s)\, ds \quad \text{and} \quad \bar{y}L = \int_0^L Y(s)\, ds.$$

These formulas give the centroid of a curve in terms of the parameter arc length. It is but a short step from here to formulas of more ready applicability.

Suppose that the curve C is given parametrically by the functions

$$x = x(t), \quad y = y(t) \qquad t \in [c, d]$$

where t is now an arbitrary parameter. Then

$$s(t) = \int_c^t \sqrt{[x'(u)]^2 + [y'(u)]^2}\, du, \qquad ds = s'(t)\, dt = \sqrt{[x'(t)]^2 + [y'(t)]^2}.$$

At $s = 0$, $t = c$; at $s = L$, $t = d$. Changing variables in (10.10.4) from s to t, we have

$$\bar{x}L = \int_c^d X(s(t)) s'(t)\, dt = \int_c^d X(s(t)) \sqrt{[x'(t)]^2 + [y'(t)]^2}\, dt$$

and

$$\bar{y}L = \int_c^d Y(s(t))s'(t)\,dt = \int_c^d Y(s(t))\sqrt{[x'(t)]^2 + [y'(t)]^2}\,dt.$$

A moment's reflection shows that

$$X(s(t)) = x(t) \quad \text{and} \quad Y(s(t)) = y(t).$$

We can then write

(10.10.5)

$$\bar{x}L = \int_c^d x(t)\sqrt{[x'(t)]^2 + [y'(t)]^2}\,dt,$$

$$\bar{y}L = \int_c^d y(t)\sqrt{[x'(t)]^2 + [y'(t)]^2}\,dt.$$

These are the centroid formulas in their most useful form.

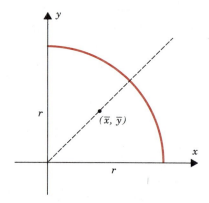

Figure 10.10.6

Problem 4. Find the centroid of the quarter-circle shown in Figure 10.10.6.

Solution. We can parametrize that quarter-circle by setting

$$x(t) = r\cos t, \quad y(t) = r\sin t \qquad t \in [0, \pi/2].$$

Since the curve is symmetric about the line $x = y$, we know that $\bar{x} = \bar{y}$. Here $x'(t) = -r\sin t$ and $y'(t) = r\cos t$. Therefore

$$\sqrt{[x'(t)]^2 + [y'(t)]^2} = \sqrt{r^2\sin^2 t + r^2\cos^2 t} = r.$$

By (10.10.5)

$$\bar{y}L = \int_0^{\pi/2} (r\sin t)r\,dt = r^2\int_0^{\pi/2}\sin t\,dt = r^2\left[-\cos t\right]_0^{\pi/2} = r^2.$$

Note that $L = \pi r/2$. Therefore $\bar{y} = r^2/L = 2r/\pi$. The centroid of the quarter-circle is the point $(2r/\pi, 2r/\pi)$. [Note that this point is closer to the curve than the centroid of the quarter-disc. (Problem 1, Section 6.4)] ☐

Problem 5. Find the centroid of the cardioid $r = a(1 - \cos \theta)$.

Solution. The curve (see Figure 10.4.1) is symmetric about the *x*-axis. Thus $\bar{y} = 0$. To find $\bar{x}$ we parametrize the curve as follows:

$$x(\theta) = r \cos \theta = a(1 - \cos \theta) \cos \theta,$$
$$y(\theta) = r \sin \theta = a(1 - \cos \theta) \sin \theta \qquad \theta \in [0, 2\pi].$$

A straightforward calculation shows that

$$[x'(\theta)]^2 + [y'(\theta)]^2 = 4a^2 \sin^2 \tfrac{1}{2}\theta.$$

Applying (10.10.5) we have

$$\bar{x}L = \int_0^{2\pi} [a(1 - \cos \theta) \cos \theta][2a \sin \tfrac{1}{2}\theta]\, d\theta = -\tfrac{32}{5}a^2.$$

check this out ⟶

By Example 5 in Section 10.9, $L = 8a$. Thus $\bar{x} = (-\tfrac{32}{5}a^2)/8a = -\tfrac{4}{5}a$. The centroid of the curve is the point $(-\tfrac{4}{5}a, 0)$. ☐

If C is a curve of the form

$$y = f(x), \qquad x \in [a, b],$$

then the formulas in (10.10.5) give

(10.10.6)
$$\bar{x}L = \int_a^b x\sqrt{1 + [f'(x)]^2}\, dx, \qquad \bar{y}L = \int_a^b f(x)\sqrt{1 + [f'(x)]^2}\, dx.$$

The details are left to you. ☐

Pappus's Theorem on Surface Area

That same Pappus who gave us that wonderful theorem on volumes of solids of revolution (Theorem 6.4.4) gave us the following equally marvelous result on surface area:

THEOREM 10.10.7 PAPPUS'S THEOREM ON SURFACE AREA

A plane curve is revolved about an axis that lies in its plane. The curve may meet the axis but, if so, only at a finite number of points. If the curve does not cross the axis, then the area of the resulting surface of revolution is the length of the curve multiplied by the circumference of the circle described by the centroid of the curve:

$$A = 2\pi \bar{R}L$$

where L is the length of the curve and $\bar{R}$ is the distance from the axis to the centroid of the curve.

Pappus did not have calculus to help him when he made his inspired guesses: he did his work 13 centuries before Newton or Leibniz were born. With the formulas that we have developed through calculus (through Newton and Leibniz, that is) Pappus's theorem is easily verified. Call the plane of the curve the xy-plane and call the axis of rotation the x-axis. Then $\overline{R} = \overline{y}$ and

$$A = \int_c^d 2\pi y(t)\sqrt{[x'(t)]^2 + [y'(t)]^2}\, dt$$

$$= 2\pi \int_c^d y(t)\sqrt{[x'(t)]^2 + [y'(t)]^2}\, dt = 2\pi \overline{y} L = 2\pi \overline{R} L. \quad \square$$

EXERCISES 10.10

Find the length of the curve, locate the centroid, and determine the area of the surface generated by revolving the curve about the x-axis.

1. $f(x) = 4, \quad x \in [0, 1]$.

2. $f(x) = 2x, \quad x \in [0, 1]$.

3. $y = \frac{4}{3}x, \quad x \in [0, 3]$.

4. $-\frac{12}{5}x + 12, \quad x \in [0, 5]$.

5. $x(t) = 3t, \quad y(t) = 4t; \quad t \in [0, 2]$.

6. $r = 5, \quad \theta \in [0, \frac{1}{4}\pi]$.

7. $x(t) = 2\cos t, \quad y(t) = 2\sin t; \quad t \in [0, \frac{1}{6}\pi]$.

8. $x(t) = \cos^3 t, \quad y(t) = \sin^3 t; \quad t \in [0, \frac{1}{2}\pi]$.

9. $x^2 + y^2 = a^2, \quad x \in [-\frac{1}{2}a, \frac{1}{2}a]$.

10. $r = 1 + \cos\theta, \quad \theta \in [0, \pi]$.

Find the area of the surface generated by revolving the curve about the x-axis.

11. $f(x) = \frac{1}{3}x^3, \quad x \in [0, 2]$.

12. $f(x) = \sqrt{x}, \quad x \in [1, 2]$.

13. $4y = x^3, \quad x \in [0, 1]$.

14. $y^2 = 9x, \quad x \in [0, 4]$.

15. $y = \cos x, \quad x \in [0, \frac{1}{2}\pi]$.

16. $f(x) = 2\sqrt{1 - x}, \quad x \in [-1, 0]$.

17. $r = e^\theta, \quad \theta \in [0, \frac{1}{2}\pi]$.

18. $y = \cosh x, \quad x \in [0, \ln 2]$.

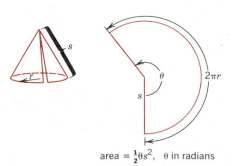

area $= \frac{1}{2}\theta s^2$, θ in radians

Figure 10.10.7

19. By cutting a cone of slant height s and base radius r along a lateral edge and laying the surface flat, we can form a sector of a circle of radius s. (See Figure 10.10.7.) Use this idea to verify Formula 10.10.1.

20. Figure 10.10.8 shows a ring formed by two quarter-circles. Call the corresponding quarter-discs Ω_a and Ω_n. By Section 6.4, Ω_a has centroid $(4a/3\pi, 4a/3\pi)$ and Ω_r has centroid $(4r/3\pi, 4r/3\pi)$.

(a) Without integration calculate the centroid of the ring.

(b) Find the centroid of the outer arc from your answer to (a) by letting a tend to r.

Figure 10.10.8

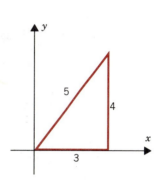

Figure 10.10.9

21. (a) Find the centroid of each side of the triangle in Figure 10.10.9.

(b) Use your answers to (a) to calculate the centroid of the triangle.

(c) What is the centroid of the triangular region?

(d) What is the centroid of the curve consisting of sides 4 and 5?

(e) Use Pappus's theorem to find the slanted surface area of a cone of base radius 4 and height 3.

(f) Use Pappus's theorem to find the total surface area of the cone in (e). (This time include the base.)

22. Find the area of the surface generated by revolving about the x-axis the curve

(a) $2x = y\sqrt{y^2 - 1} + \ln|y - \sqrt{y^2 - 1}|$, $y \in [2, 5]$.   (b) $6a^2xy = y^4 + 3a^4$, $y \in [a, 3a]$.

23. Use Pappus's theorem to find the surface area of the *torus* generated by revolving about the x-axis the circle $x^2 + (y - b)^2 = a^2$. $(0 < a \le b)$

24. (a) We calculated the total surface area of a sphere from (10.10.2) not (10.10.3). Could we just as well have used (10.10.3)? Explain.

(b) Verify that Formula 10.10.2 applied to

$$C: \quad x(t) = \cos t, \quad y(t) = r \quad \text{with } t \in [0, 2\pi]$$

gives $A = 8\pi r$. Note that the surface obtained by revolving C about the x-axis is a cylinder of base radius r and height h, and therefore A should be $4\pi r$. What's wrong?

25. (*An interesting property of the sphere*) Slice a sphere along two parallel planes that are a fixed distance apart. Show that the surface area of the band so obtained is independent of where the cuts are made.

26. Locate the centroid of a first-quadrant circular arc

$$C: \quad x(t) = r \cos t, \quad y(t) = r \sin t \quad t \in [\theta_1, \theta_2].$$

27. Find the surface area of the ellipsoid obtained by revolving the ellipse

$$\frac{x^2}{a^2} + \frac{y^2}{b^2} = 1 \qquad\qquad (0 < b < a)$$

(a) about its major axis.   (b) about its minor axis.

The Centroid of a Surface of Revolution   If a material surface of revolution is homogeneous (constant mass density), then the center of mass of that material surface is called the *centroid*. The determination of the centroid of a surface of arbitrary shape requires surface integration. (Chapter 18)   However, if the surface is a surface of revolution, then the centroid can be found by ordinary one-variable integration.

28. Let C be a simple curve in the upper half-plane parametrized by a pair of continuously differentiable functions

$$x = x(t), \quad y = y(t) \qquad t \in [c, d].$$

By revolving C about the x-axis we obtain a surface of revolution, the area of which we denote by A.   By symmetry the centroid of the surface lies on the x-axis.   Thus the centroid is completely determined by its x-coordinate $\bar{x}$.   Show that

(10.10.8)

$$\bar{x}A = \int_c^d 2\pi x(t) y(t) \sqrt{[x'(t)]^2 + [y'(t)]^2} \, dt$$

by assuming the following additivity principle: If the surface is broken up into n surfaces of revolution with areas $A_1, \ldots, A_n$ and the centroids of the surfaces have x-coordinates $\bar{x}_1, \ldots, \bar{x}_n$, then

$$\bar{x}A = \bar{x}_1 A_1 + \cdots + \bar{x}_n A_n.$$

29. Locate the centroid of a hemisphere of radius r.
30. Locate the centroid of a conical surface of base radius r and height h.
31. Where is the centroid of the lateral surface of the frustum of a cone of height h with base radii r and R?

---

*SUPPLEMENT TO SECTION 10.10   THE GRAVITATIONAL FORCE EXERTED BY A SPHERICAL SHELL

(This material presupposes some familiarity with Section *6.7.)

Problem 6.   A homogeneous spherical shell of radius R and total mass M attracts a particle that is outside the shell toward the center of the shell.   That is clear by symmetry.   Find the magnitude of the force if the particle has mass m and lies at a distance a from the center of the shell.

Solution

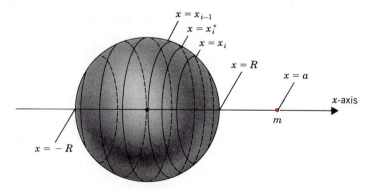

Figure 10.10.10

As in Figure 10.10.10 we place the center of the sphere at $x = 0$ and the particle at $x = a > 0$. We then partition the base interval $[-R, R]$ by equally spaced points $-R = x_0 < x_1 < \cdots < x_n = R$. These points decompose the sphere into n vertical bands of equal width. These bands have equal surface area (Exercise 25) and therefore equal mass $M_i = M/n$. Since $n \, \Delta x_n = 2R$, we have

$$M_i = (M/2R) \, \Delta x_i.$$

The gravitational effect of this ith band is, within close approximation, the same as that of a circular wire of the same mass M_i centered at $x = x_i^*$ where x_i^* is the midpoint of $[x_{i-1}, x_i]$. By Exercise 3 in Section *6.7 this mass configuration attracts m along the x-axis with a force of magnitude

$$F_i = \frac{GmM_i h_i}{(h_i^2 + R_i^2)^{3/2}} = \frac{GmM}{2R} \left[\frac{h_i}{(h_i^2 + R_i^2)^{3/2}} \right] \Delta x_i$$

where h_i is the distance from the center of the wire to the mass m and R_i is the radius of the wire. Since $h_i = a - x_i^*$ and $R_i^2 = R^2 - (x_i^*)^2$, we have

$$h_i^2 + R_i^2 = (a - x_i^*)^2 + R^2 - (x_i^*)^2 = a^2 + R^2 - 2ax_i^*.$$

It follows that

$$F_i = \frac{GmM}{2R} \left[\frac{a - x_i^*}{(a^2 + R^2 - 2ax_i^*)^{3/2}} \right] \Delta x_i.$$

Adding up these terms we have

$$F = \frac{GmM}{2R} \left[\frac{a - x_1^*}{(a^2 + R^2 - 2ax_1^*)^{3/2}} \Delta x_1 + \cdots + \frac{a - x_n^*}{(a^2 + R^2 - 2ax_n^*)^{3/2}} \Delta x_n \right].$$

As max $\Delta x_i \to 0$, the sum converges to an integral and we have

(*) $$F = \frac{GmM}{2R} \int_{-R}^{R} \frac{a - x}{(a^2 + R^2 - 2ax)^{3/2}} \, dx.$$

You can calculate this integral using integration by parts. Set

$$u = a - x, \qquad dv = \frac{dx}{(a^2 + R^2 - 2ax)^{3/2}}$$

and (with a little perseverance) you will see that

$$F = \frac{GmM}{a^2}.$$

The spherical shell attracts the particle as if all of its mass were concentrated at its center.† ☐

The particle in Problem 6 was outside the shell. Suppose the particle is inside the shell. What is the force on the particle then? Answer: The force is zero. To see this, go back to equation (*) and calculate the integral as before using the fact that now $a < R$.

---

† The same holds true for a homogeneous solid ball. (Think of the ball as being made up of a large number of thin concentric shells.)

*10.11   THE CYCLOID

(In this section the exercises are intertwined with the text.)

Take a wheel (a roll of tape will do) and fix your eyes on some point of the rim.   Call that point
P.   Now roll the wheel slowly, keeping your eyes on P.   The jumping-kangaroo path described
by P is called a *cycloid*.

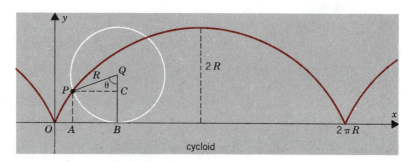

Figure 10.11.1

To obtain a mathematical characterization of the cycloid, call the radius of the wheel R and
set the wheel on the x-axis so that the point P starts out at the origin.   Figure 10.11.1 shows P
after a turn of θ radians.

1. Show that the cycloid can be parametrized by the functions

(10.11.1) $$x(\theta) = R(\theta - \sin \theta), \quad y(\theta) = R(1 - \cos \theta).$$

HINT: Length of $\overline{OB}$ = length of $\overparen{PB}$ = $R\theta$.

2. At the end of each arch the cycloid comes to a cusp.   Verify that x' and y' are both 0 at the
end of each arch.

3. Show that the line tangent to the cycloid at P (a) passes through the top of the circle and
(b) intersects the x-axis at an angle of $\frac{1}{2}(\pi - \theta)$ radians.

4. Express the cycloid from $\theta = 0$ to $\theta = \pi$ by an equation in x and y.   (The equation is quite
complicated and not easy to work with.   The properties of the cycloid are much easier to fathom from
the parametric equations.)

5. Find the length of an arch of the cycloid.   (10.9.2)

6. Show that the area under an arch of the cycloid is three times the area of the rolling circle.
(10.6.4)

7. Locate the centroid of the region under the first arch of the cycloid.   (10.6.5)

8. Find the volume of the solid generated by revolving about the x-axis the region under an
arch of the cycloid.   (10.6.6)

9. Find the volume of the solid generated by revolving about the y-axis the region under the
first arch of the cycloid.   (10.6.6)

10. Determine the centroid of the first arch of the cycloid.   (10.10.5)

11. Find the area of the surface generated by revolving about the x-axis an arch of the cycloid.
(10.10.2)

The Inverted Cycloid

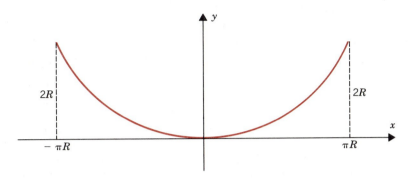

Figure 10.11.2

12. Show that the arch of the inverted cycloid shown in Figure 10.11.2 can be parametrized by setting

$$x = R(\phi + \sin \phi), \quad y = R(1 - \cos \phi) \qquad \phi \in [-\pi, \pi].$$

Proceed as follows:

(a) Reflect the arch of Figure 10.11.1 in the x-axis and find the θ parametrization for that.
(b) Raise the arch obtained in part (a) so that the low point rests on the x-axis and find the θ parametrization for that.
(c) Obtain the arch of Figure 10.11.2 by translating πR units to the left the arch obtained in part (b). Write down the θ parametrization for this arch and then set $\phi = \theta - \pi$.

13. Find the inclination α of the line tangent to the inverted arch at the point $(x(\phi), y(\phi))$.

14. Let s be the arc distance from the low point of the inverted arch to the point $(x(\phi), y(\phi))$ of that same arch. Show that $s = 4R \sin \frac{1}{2}\phi = 4r \sin \alpha$ where α is the inclination of the tangent line at $(x(\phi), y(\phi))$.

The Tautochrone

Visualize two particles (beads if you prefer) sliding without friction down an arch of an inverted cycloid. (Figure 10.11.3) If the two particles are released at the same time from different positions, which will reach the bottom first? Neither: They will both get there at exactly the same time.† Being the only curve that produces this effect, the inverted arch of a cycloid is known as *the tautochrone*, the *same-time* curve.

Figure 10.11.3

---

† This was understood in 1673 by Christian Huygens, the inventor of the pendulum clock. By tracking the bob of the pendulum along the arch of an inverted cycloid, Huygens was able to stabilize the frequency of the oscillations of the pendulum and thereby improve the accuracy of his clock.

15. Verify that the inverted arch of a cycloid has the tautochrone property by taking the following steps:

   (a) Show that the effective gravitational force on a particle of mass m is $-mg \sin \alpha$ where α is the inclination of the tangent line at the position of the particle. Conclude then that

   (∗)
   $$\frac{d^2s}{dt^2} = -g \sin \alpha.$$

   (b) Combine (∗) with Exercise 14 to show that the particle is in simple harmonic motion with period

   $$T = 4\pi \sqrt{R/g}.$$

   (Thus, while the amplitude of the motion depends on the point of release, the frequency does not. Two particles released simultaneously from different points of the curve will reach the low point of the curve in exactly the same amount of time: $T/4 = \pi \sqrt{R/g}$.)

The Brachystochrone

A particle is to descend without friction along a curve from some point A to a point B not directly below it. (See Figure 10.11.4.) What should be the shape of the curve so that the particle descends from A to B in the least possible time?

Figure 10.11.4

This question was first formulated by Johann Bernoulli and posed by him as a challenge to the scientific community in 1696. The challenge was readily accepted and within months the answer was found—by Johann Bernoulli himself, by his brother Jacob, by Newton, by Leibniz, and by L'Hospital. The answer? Part of an inverted cycloid. Because of this, the inverted cycloid is heralded as the *brachystochrone,* the *least-time* curve.

A proof that the inverted cycloid is the least time curve, the curve of quickest descent, is beyond our reach. The argument requires a sophisticated variant of calculus known as *the calculus of variations.* We can, however, compare the time of descent along a cycloid to the time of descent along a straight-line path.

16. You have seen that a particle descends along the inverted arch of a cycloid from $(\pi R, 2R)$ to $(0, 0)$ in time $t = T/4 = \pi \sqrt{R/g}$. What is the time of descent along a straight-line path?

10.12 ADDITIONAL EXERCISES

1. Sketch the curve $r^2 = 4 \sin 2\theta$ and find the area it encloses.

2. Sketch the curve $r = 2 + \sin \theta$ and find the area it encloses.

3. Sketch the curve $r = 2 + \sin 3\theta$ and find the area it encloses.

4. Sketch the region bounded by the curve $r = \sec \theta + \tan \theta$, the polar axis, and the ray $\theta = \frac{1}{4}\pi$. Then find the area of the region.

Find the length of the curve.

5. $y^2 = x^3$ from $x = 0$ to $x = \frac{5}{9}$. (upper branch)

6. $9y^2 = 4(1 + x^2)^3$ from $x = 0$ to $x = 3$. (upper branch)

7. $ay^2 = x^3$ from $x = 0$ to $x = 5a$. (upper branch)

8. $y = \ln(1 - x^2)$ from $x = 0$ to $x = \frac{1}{2}$.

9. $r = 2(1 + \cos \theta)^{-1}$ from $\theta = 0$ to $\theta = \frac{1}{2}\pi$.

10. $x(t) = \cos t,\quad y(t) = \sin^2 t$ from $t = 0$ to $t = \pi$.

11. $r = a \sin^3 \frac{1}{3}\theta$ from $\theta = 0$ to $\theta = 3\pi$.

12. An object moves in a plane so that dx/dt and d^2y/dt^2 are constant but not zero. Identify the path of the object.

Find the area of the surface generated by revolving the curve about the x-axis.

13. $y^2 = 2px$ from $x = 0$ to $x = 4p$.

14. $y^2 = 4x$ from $x = 0$ to $x = 24$.

15. $6a^2xy = x^4 + 3a^4$ from $x = a$ to $x = 2a$.

16. $3x^2 + 4y^2 = 3a^2$.

17. $x(t) = \frac{2}{3}t^{3/2},\quad y(t) = t\quad\quad t \in [3, 8]$.

18. Show that, if f is continuously differentiable on $[a, b]$ and f' is never zero, then

$$\text{the length of the graph of } f^{-1} = \text{the length of the graph of } f.$$

19. Locate the centroid of the catenary $y = a \cosh(x/a),\ x \in [-a, a]$.

Exercises 20–23 concern the *astroid*: the curve

$$x^{2/3} + y^{2/3} = r^{2/3}. \qquad\qquad (r > 0)$$

20. Sketch the astroid and show that the curve can be parametrized by setting

$$x(\theta) = r \cos^3 \theta,\quad y(\theta) = r \sin^3 \theta \qquad \theta \in [0, 2\pi].$$

21. Find the arc length of the astroid.

22. Locate the centroid of the first-quadrant part of the astroid.

23. Find the area enclosed by the astroid.   HINT: Parametrize the upper half so that you can apply (10.6.4).

24. A particle starting at time $t = 0$ at the point $(4, 2)$ moves until time $t = 1$ with $x'(t) = x(t)$ and $y'(t) = 2y(t)$. Find the initial speed v_0, the terminal speed v_1, and the distance traveled s.

25. Locate the centroid of the cardioid $r = a(1 - \cos \theta),\ a > 0$.

26. A particle moves from time $t = 0$ to time $t = 1$ so that

$$x(t) = 4t - \sin \pi t,\quad y(t) = 4t + \cos \pi t.$$

(a) When does the particle have minimum speed? When does it have maximum speed?

(b) What is the area under the curved path?

(c) What is the slope of the tangent line at the point $(1 - \frac{1}{2}\sqrt{2},\ 1 + \frac{1}{2}\sqrt{2})$?

(The remaining exercises presuppose some familiarity with Section *10.11.)

27. (*The trochoid*) We return to the wheel of radius R shown in Figure 10.11.1. Choose a point S between P and Q (a point along the spoke $\overline{QP}$). As the wheel rolls along the axis, the point S describes a curve. Such a curve is called a *trochoid*. Figure 10.12.1 shows the position of S after a turn of θ radians.

(a) Parametrize the trochoid in terms of θ, taking b as the length of $\overline{QS}$.
(b) Draw a figure showing the path traced by a point T that is outside the wheel along an extension of the spoke $\overline{QP}$.

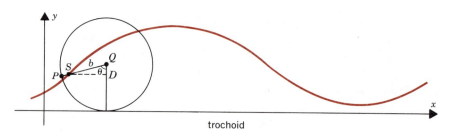

trochoid

Figure 10.12.1

28. (*The epicycloid and the hypocycloid*) Figure 10.12.2 shows the curve traced out by a point P of a circle of radius a as this circle rolls around a circle of radius R. Such a curve is called an *epicycloid*. Figure 10.12.3 shows the curve traced out by P if the rolling circle lies inside the other circle. Such a curve is called a *hypocycloid*.

(a) Parametrize the epicycloid in Figure 10.12.2 in terms of the angle θ formed by the positive x-axis and the line through the center of the circles.
(b) Sketch the epicycloid for the case $a = R$ and identify this familiar curve.
(c) Parametrize the hypocycloid in Figure 10.12.3 in terms of the angle θ formed by the positive x-axis and the line through the center of the circles.

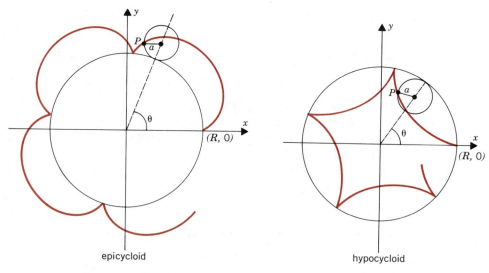

epicycloid

Figure 10.12.2

hypocycloid

Figure 10.12.3

CHAPTER HIGHLIGHTS

10.1 Polar Coordinates

relation between rectangular coordinates (x, y) and polar coordinates $[r, \theta]$:

$$x = r \cos \theta, \quad y = r \sin \theta \qquad \tan \theta = \frac{y}{x}, \qquad x^2 + y^2 = r^2.$$

10.2 Graphing in Polar Coordinates

cardioids (p. 509)   lines (p. 510)   circles (p. 510)
limaçons (p. 510)   spirals (p. 511)   lemniscates (p. 511)
petal curves (p. 511)

*10.3 The Conic Sections in Polar Coordinates

$$r = \frac{ed}{1 + e \cos \theta}$$

ellipse $0 < e < 1$
parabola $e = 1$
hyperbola $e > 1$

10.4 The Intersection of Polar Curves

If we think of each of two polar equations as giving the position of an object at time θ, then simultaneous solution of the two equations gives us the points where the objects collide. To identify the points where two polar curves intersect, but do not collide, we usually need to convert the equations to rectangular coordinates.

10.5 Area in Polar Coordinates

Let ρ_1 and ρ_2 be positive continuous functions defined on a closed interval $[\alpha, \beta]$ of length at most 2π. If $\rho_1(\theta) \leq \rho_2(\theta)$ for all θ in $[\alpha, \beta]$, then the area of the region between the polar curves $r = \rho_1(\theta)$ and $r = \rho_2(\theta)$ is given by the formula

$$A = \int_{\alpha}^{\beta} \tfrac{1}{2}([\rho_2(\theta)]^2 - [\rho_1(\theta)]^2) \, d\theta.$$

10.6 Curves Given Parametrically

Let $x = x(t)$, $y = y(t)$ be a pair of functions differentiable on the interior of some interval I. At the endpoints of I (if any) we require only continuity. For each t in I we can interpret $(x(t), y(t))$ as the point with x-coordinate $x(t)$ and y-coordinate $y(t)$. Then as t ranges over I, the point $(x(t), y(t))$ traces a path in the xy-plane. We call such a path a *parametrized curve* and refer to t as the *parameter*.

line: $x(t) = x_0 + t(x_1 - x_0), \quad y(t) = y_0 + t(y_1 - y_0)$ t real
circle: $x(t) = a \cos t, \quad y(t) = a \sin t$ $t \in [0, 2\pi]$
ellipse: $x(t) = a \cos t, \quad y(t) = b \sin t$ $t \in [0, 2\pi]$

10.7 Tangents to Curves Given Parametrically

tangent line at (x_0, y_0): $y'(t_0)[x - x_0] - x'(t_0)[y - y_0] = 0$
provided that $[x'(t_0)]^2 + [y'(t_0)]^2 \neq 0$

10.8 The Least Upper Bound Axiom

10.9 Arc Length and Speed

length of a parametrized curve: $L = \int_a^b \sqrt{[x'(t)]^2 + [y'(t)]^2}\, dt$

length of a graph: $L = \int_a^b \sqrt{1 + [f'(x)]^2}\, dx$

speed along a curve parametrized by time t: $v(t) = \sqrt{[x'(t)]^2 + [y'(t)]^2}$

10.10 The Area of a Surface of Revolution; The Centroid of a Curve; Pappus's Theorem on Surface Area

revolution of curve about x-axis: $A = \int_c^d 2\pi y(t)\sqrt{[x'(t)]^2 + [y'(t)]^2}\, dt$

revolution of graph about x-axis: $A = \int_a^b 2\pi f(x)\sqrt{1 + [f'(x)]^2}\, dx$

parametrizing a curve by arc length (p. 555)
principles for finding the centroid of a plane curve (pp. 554–555)

centroid of a plane curve: $\bar{x}L = \int_c^d x(t)\sqrt{[x'(t)]^2 + [y'(t)]^2}\, dt$,

$\bar{y}L = \int_c^d y(t)\sqrt{[x'(t)]^2 + [y'(t)]^2}\, dt$

Pappus's theorem on surface area: $A = 2\pi \bar{R}L$
*gravitational force exerted by a spherical shell (p. 560)

*10.11 The Cycloid

10.12 Additional Exercises

11

SEQUENCES; INDETERMINATE FORMS; IMPROPER INTEGRALS

11.1 SEQUENCES OF REAL NUMBERS

So far our attention has been fixed on functions defined on an interval or on the union of intervals. Here we study functions defined on the set of positive integers.

DEFINITION 11.1.1

A function that takes on real values and is defined on the set of positive integers is called a *sequence of real numbers*.

The functions defined on the set of positive integers by setting

$$a(n) = n^2, \qquad b(n) = \frac{n}{n+1}, \qquad c(n) = \sqrt{\ln n}, \qquad d(n) = \frac{e^n}{n}$$

are all sequences of real numbers.

The notions developed for functions in general carry over to sequences. For example, a sequence is said to be bounded (bounded above, bounded below) iff its range is bounded (bounded above, bounded below). If a and b are sequences, then their linear combination $\alpha a + \beta b$ and their product ab are also sequences:

$$(\alpha a + \beta b)(n) = \alpha a(n) + \beta b(n) \qquad \text{and} \qquad (ab)(n) = a(n) \cdot b(n).$$

If the sequence b does not take on the value 0, then the reciprocal $1/b$ is a sequence and so is the quotient a/b:

$$\frac{1}{b}(n) = \frac{1}{b(n)} \qquad \text{and} \qquad \frac{a}{b}(n) = \frac{a(n)}{b(n)}.$$

The number $a(n)$, called *the nth term of the sequence a,* is usually written a_n and the sequence itself is written

$$\{a_1, a_2, a_3, \ldots\}$$

or, even more simply,

$$\{a_n\}.$$

For example, the sequence of reciprocals defined by setting

$$a_n = 1/n \qquad \text{for all } n$$

can be written

$$\{1, \tfrac{1}{2}, \tfrac{1}{3}, \ldots\} \quad \text{or} \quad \{1/n\}.$$

The sequence defined by setting

$$a_n = 10^{1/n} \qquad \text{for all } n$$

can be written

$$\{10, 10^{1/2}, 10^{1/3}, \ldots\} \quad \text{or} \quad \{10^{1/n}\}.$$

In this notation, we can write

$$\alpha\{a_n\} + \beta\{b_n\} = \{\alpha a_n + \beta b_n\}, \qquad \{a_n\}\{b_n\} = \{a_n b_n\}$$

and, provided that none of the b_n are zero,

$$\frac{1}{\{b_n\}} = \{1/b_n\} \qquad \text{and} \qquad \frac{\{a_n\}}{\{b_n\}} = \{a_n/b_n\}.$$

The following assertions will serve to illustrate the notation further.

(1) The sequence $\{1/2^n\}$ multiplied by 5 is the sequence $\{5/2^n\}$:

$$5\{1/2^n\} = \{5/2^n\}.$$

We can also write

$$5\{\tfrac{1}{2}, \tfrac{1}{4}, \tfrac{1}{8}, \tfrac{1}{16}, \ldots\} = \{\tfrac{5}{2}, \tfrac{5}{4}, \tfrac{5}{8}, \tfrac{5}{16}, \ldots\}.$$

(2) The sequence $\{n\}$ plus the sequence $\{1/n\}$ is the sequence $\{n + 1/n\}$:

$$\{n\} + \{1/n\} = \{n + 1/n\}.$$

In expanded form

$$\{1, 2, 3, \ldots\} + \{1, \tfrac{1}{2}, \tfrac{1}{3}, \ldots\} = \{2, 2\tfrac{1}{2}, 3\tfrac{1}{3}, \ldots\}.$$

(3) The sequence $\{n\}$ times the sequence $\{\sqrt{n}\}$ is the sequence $\{n\sqrt{n}\}$:

$$\{n\}\{\sqrt{n}\} = \{n\sqrt{n}\};$$

the sequence $\{2^n\}$ divided by the sequence $\{n\}$ is the sequence $\{2^n/n\}$:

$$\frac{\{2^n\}}{\{n\}} = \{2^n/n\}.$$

(4) The sequence $\{1/n\}$ is bounded below by 0 and bounded above by 1:

$$0 \le 1/n \le 1 \qquad \text{for all } n.$$

(5) The sequence $\{2^n\}$ is bounded below by 2:

$$2 \le 2^n \qquad \text{for all } n.$$

It is not bounded above: there is no fixed number M that satisfies

$$2^n \le M \qquad \text{for all } n.$$

(6) The sequence $a_n = (-1)^n 2^n$ can be written

$$\{-2, 4, -8, 16, -32, 64, \ldots\}.$$

It is unbounded below and unbounded above. $\square$

DEFINITION 11.1.2

A sequence $\{a_n\}$ is said to be

(i) *increasing*   iff $a_n < a_{n+1}$ for each positive integer n,

(ii) *nondecreasing*   iff $a_n \le a_{n+1}$ for each positive integer n,

(iii) *decreasing*   iff $a_n > a_{n+1}$ for each positive integer n,

(iv) *nonincreasing*   iff $a_n \ge a_{n+1}$ for each positive integer n.

If any of these four properties holds, the sequence is said to be *monotonic*.

The sequences

$$\{1, \tfrac{1}{2}, \tfrac{1}{3}, \ldots\}, \qquad \{2, 4, 8, 16, \ldots\}, \qquad \{2, 2, 4, 4, 8, 8, 16, 16, \ldots\}$$

are all monotonic, but the sequence

$$\{1, \tfrac{1}{2}, 1, \tfrac{1}{3}, 1, \tfrac{1}{4}, \ldots\}$$

is not monotonic.

The following examples are less trivial.

Example 1. The sequence

$$a_n = \frac{n}{n+1}$$

is increasing. It is bounded below by $\frac{1}{2}$ and above by 1.

Proof. Since

$$\frac{a_{n+1}}{a_n} = \frac{n+1}{n+2} \cdot \frac{n+1}{n} = \frac{n^2 + 2n + 1}{n^2 + 2n} > 1,$$

we have $a_n < a_{n+1}$. This confirms that the sequence is increasing. Since the sequence can be written

$$\{\tfrac{1}{2}, \tfrac{2}{3}, \tfrac{3}{4}, \tfrac{4}{5}, \tfrac{5}{6}, \ldots\},$$

the bounds are obvious. $\square$

Example 2. The sequence

$$a_n = \frac{2^n}{n!}$$

is nonincreasing. It decreases for $n \geq 2$.

Proof. The first two terms are equal:

$$a_1 = \frac{2^1}{1!} = 2 = \frac{2^2}{2!} = a_2.$$

For $n \geq 2$ the sequence decreases:

$$\frac{a_{n+1}}{a_n} = \frac{2^{n+1}}{(n+1)!} \cdot \frac{n!}{2^n} = \frac{2}{n+1} < 1 \qquad \text{if } n \geq 2. \quad \square$$

Example 3. If $|c| > 1$, the sequence

$$a_n = |c|^n$$

increases without bound.

Proof. Assume that $|c| > 1$. Then

$$\frac{a_{n+1}}{a_n} = \frac{|c|^{n+1}}{|c|^n} = |c| > 1.$$

This shows that the sequence increases. To show the unboundedness, we take an arbitrary positive number M and show that there exists a positive integer k for which

$$|c|^k \geq M.$$

A suitable k is one such that

$$k \geq \frac{\ln M}{\ln |c|},$$

for then

$$k \ln |c| \geq \ln M, \qquad \ln |c|^k \geq \ln M, \qquad \text{and thus} \qquad |c|^k \geq M. \quad \square$$

Since sequences are defined on the set of positive integers and not on an interval, they are not directly susceptible to the methods of calculus. Fortunately, we can often circumvent this difficulty by dealing initially, not with the sequence itself, but with a function of a real variable x that agrees with the given sequence for all positive integers n.

Example 4. The sequence

$$a_n = \frac{n}{e^n}$$

decreases. It is bounded above by $1/e$ and below by 0.

Proof. We will work with the function

$$f(x) = \frac{x}{e^x}.$$

Differentiation gives

$$f'(x) = \frac{e^x - xe^x}{e^{2x}} = \frac{1 - x}{e^x}.$$

Since $f'(x)$ is 0 at $x = 1$ and negative for $x > 1$, f decreases on $[1, \infty)$. Obviously, then, the sequence $a_n = n/e^n$ decreases.

Since the sequence decreases, its first term $a_1 = 1/e$ is the greatest term. Thus $1/e$ is certainly an upper bound. Since all the terms of the sequence are positive, the number 0 is a lower bound. $\quad \square$

Example 5. The sequence

$$a_n = n^{1/n}$$

decreases for $n \geq 3$.

Proof. We could compare a_n with a_{n+1} directly, but it is easier to consider the function

$$f(x) = x^{1/x}$$

instead. Since

$$f(x) = e^{(1/x)\ln x},$$

we have

$$f'(x) = e^{(1/x)\ln x} \frac{d}{dx}\left(\frac{1}{x}\ln x\right) = x^{1/x}\left(\frac{1 - \ln x}{x^2}\right).$$

For $x > e$, $f'(x) < 0$. This shows that f decreases on $[e, \infty)$. Since $3 > e$, f decreases on $[3, \infty)$. It follows that $\{a_n\}$ decreases for $n \geq 3$. $\square$

Remark. We must be careful when we examine a function of a real variable x in order to analyze the behavior of a sequence. The function $y = f(x)$ and the sequence $y_n = f(n)$ may behave differently. For example, the sequence

$$f(n) = \frac{1}{n - 11.5}$$

is bounded (by 2 and -2) even though the function

$$f(x) = \frac{1}{x - 11.5}$$

is not bounded. Also, the sequence $f(n) = \sin n\pi$ is identically zero and thus monotonic even though the function $f(x) = \sin \pi x$ is not monotonic. $\square$

EXERCISES 11.1

Determine the boundedness and monotonicity of the indicated sequence.

1. $\dfrac{2}{n}$. **2.** $\dfrac{(-1)^n}{n}$. **3.** $\sqrt{n}$. **4.** $(1.001)^n$.

5. $\dfrac{n + (-1)^n}{n}$. **6.** $\dfrac{n - 1}{n}$. **7.** $(0.9)^n$. **8.** $\sqrt{n^2 + 1}$.

9. $\dfrac{n^2}{n + 1}$. **10.** $\dfrac{2^n}{4^n + 1}$. **11.** $\dfrac{4n}{\sqrt{4n^2 + 1}}$. **12.** $\dfrac{n + 1}{n^2}$.

13. $\dfrac{4^n}{2^n + 100}$. **14.** $\dfrac{n^2}{\sqrt{n^3 + 1}}$. **15.** $\dfrac{10^{10}\sqrt{n}}{n + 1}$. **16.** $\dfrac{n^2 + 1}{3n + 2}$.

17. $\ln\left(\dfrac{2n}{n + 1}\right)$. **18.** $\dfrac{n + 2}{3^{10}\sqrt{n}}$. **19.** $\dfrac{(n + 1)^2}{n^2}$. **20.** $(-1)^n\sqrt{n}$.

21. $\sqrt{4 - \dfrac{1}{n}}$. **22.** $\ln\left(\dfrac{n + 1}{n}\right)$. **23.** $(-1)^{2n+1}\sqrt{n}$. **24.** $\dfrac{\sqrt{n + 1}}{\sqrt{n}}$.

25. $\dfrac{2^n - 1}{2^n}$. **26.** $\dfrac{1}{2n} - \dfrac{1}{2n + 3}$. **27.** $\sin\dfrac{\pi}{n + 1}$. **28.** $\left(-\tfrac{1}{2}\right)^n$.

29. $(1.2)^{-n}$. **30.** $\dfrac{n + 3}{\ln(n + 3)}$. **31.** $\dfrac{1}{n} - \dfrac{1}{n + 1}$. **32.** $\cos n\pi$.

33. $\dfrac{\ln(n + 2)}{n + 2}$. **34.** $\dfrac{(-2)^n}{n^{10}}$. **35.** $\dfrac{3^n}{(n + 1)^2}$. **36.** $\dfrac{1 - \left(\tfrac{1}{2}\right)^n}{\left(\tfrac{1}{2}\right)^n}$.

37. Show that the sequence $\{5^n/n!\}$ decreases for $n \geq 5$. Is the sequence nonincreasing?

38. Let M be a positive integer. Show that $\{M^n/n!\}$ decreases for $n \geq M$.

39. Show that, if $0 < c < d$, then the sequence

$$a_n = (c^n + d^n)^{1/n}$$

is bounded and monotonic.

40. Show that linear combinations and products of bounded sequences are bounded.

A sequence $\{a_n\}$ is said to be defined *recursively* if, for some $k \geq 1$, the terms $a_1, a_2, \ldots, a_k$ are given and a_n is specified in terms of $a_1, a_2, \ldots, a_{n-1}$ for each $n \geq k$. The formula specifying a_n for $n \geq k$ in terms of some (or all) of its predecessors is called a *recurrence relation*. Write down the first six terms of the sequence and then give the general term.

41. $a_1 = 1$; $a_{n+1} = \dfrac{1}{n+1} a_n$.

42. $a_1 = 1$; $a_{n+1} = a_n + 3n(n+1) + 1$.

43. $a_1 = 1$; $a_{n+1} = \frac{1}{2}(a_n + 1)$.

44. $a_1 = 1$; $a_{n+1} = \frac{1}{2}a_n + 1$.

45. $a_1 = 1$; $a_{n+1} = a_n + 2$.

46. $a_1 = 1$; $a_{n+1} = \dfrac{n}{n+1} a_n$.

47. $a_1 = 1$; $a_{n+1} = 3a_n + 1$.

48. $a_1 = 1$; $a_{n+1} = 4a_n + 3$.

49. $a_1 = 1$; $a_{n+1} = a_n + 2n + 1$.

50. $a_1 = 1$; $a_{n+1} = 2a_n + 1$.

51. $a_1 = 1, a_2 = 3$; $a_{n+1} = a_n + \cdots + a_1$.

52. $a_1 = 3$; $a_{n+1} = 4 - a_n$.

53. $a_1 = 2, a_2 = 1, a_3 = 2$; $a_{n+1} = 6 - (a_n + a_{n-1} + a_{n-2})$.

54. $a_1 = 1, a_2 = 2$; $a_{n+1} = 2a_n - a_{n-1}$.

55. $a_1 = 1, a_2 = 3$; $a_{n+1} = 2a_n - a_{n-1}$.

56. $a_1 = 1, a_2 = 3$; $a_{n+1} = 3a_n - 2n - 1$.

Use mathematical induction to prove the following assertions for all $n \geq 1$.

57. If $a_1 = 1$ and $a_{n+1} = 2a_n + 1$, then $a_n = 2^n - 1$.

58. If $a_1 = 3$ and $a_{n+1} = a_n + 5$, then $a_n = 5n - 2$.

59. If $a_1 = 1$ and $a_{n+1} = \dfrac{n+1}{2n} a_n$, then $a_n = \dfrac{n}{2^{n-1}}$.

60. If $a_1 = 1$ and $a_{n+1} = a_n - \dfrac{1}{n(n+1)}$, then $a_n = \dfrac{1}{n}$.

11.2   THE LIMIT OF A SEQUENCE

The gist of

$$\lim_{x \to c} f(x) = l$$

is that we can make $f(x)$ as close as we wish to the number l simply by requiring that x be sufficiently close to c. The gist of

$$\lim_{n \to \infty} a_n = l$$

(read "the limit of a_n as n tends to infinity is l") is that we can make a_n as close as we wish to the number l simply by requiring that n be sufficiently large.

DEFINITION 11.2.1   LIMIT

$$\lim_{n \to \infty} a_n = l \quad \text{iff} \quad \begin{cases} \text{for each } \epsilon > 0 \text{ there exists an integer } k > 0 \text{ such that} \\ \text{if } \ n \geq k, \quad \text{then} \quad |a_n - l| < \epsilon. \end{cases}$$

Example 1

$$\lim_{n \to \infty} \frac{1}{n} = 0.$$

Proof. Take $\epsilon > 0$ and choose an integer $k > 1/\epsilon$. Then $1/k < \epsilon$, and, if $n \geq k$,

$$0 < \frac{1}{n} \leq \frac{1}{k} < \epsilon \quad \text{and thus} \quad \left| \frac{1}{n} - 0 \right| < \epsilon. \quad \square$$

Example 2

$$\lim_{n \to \infty} \frac{2n - 1}{n} = 2.$$

Proof. Let $\epsilon > 0$. We must show that there exists an integer k such that

$$\left| \frac{2n - 1}{n} - 2 \right| < \epsilon \quad \text{for all } n \geq k.$$

Since

$$\left| \frac{2n - 1}{n} - 2 \right| = \left| \frac{2n - 1 - 2n}{n} \right| = \left| -\frac{1}{n} \right| = \frac{1}{n},$$

again we need only choose $k > 1/\epsilon$. $\square$

The next example justifies the familiar statement

$$\tfrac{1}{3} = 0.333 \ \cdots .$$

Example 3.   The decimal fractions

$$a_n = 0.\overset{n}{\overbrace{33 \ \cdots \ 3}}$$

tend to $\tfrac{1}{3}$ as a limit:

$$\lim_{n \to \infty} a_n = \tfrac{1}{3}.$$

Proof. Let $\epsilon > 0$. In the first place

$$(1) \qquad \left| a_n - \frac{1}{3} \right| = \left| 0.\overset{n}{\overbrace{33 \cdots 3}} - \frac{1}{3} \right| = \left| \frac{0.\overset{n}{\overbrace{99 \cdots 9}} - 1}{3} \right| = \frac{1}{3} \cdot \frac{1}{10^n} < \frac{1}{10^n}.$$

Now choose k so that $1/10^k < \epsilon$. If $n \geq k$, then by (1)

$$\left| a_n - \frac{1}{3} \right| < \frac{1}{10^n} \leq \frac{1}{10^k} < \epsilon. \quad \square$$

Limit Theorems

The limit process for sequences is so similar to the limit process you have already studied that you may find you can prove many of the limit theorems yourself. In any case, try to come up with your own proofs and refer to these only if necessary.

THEOREM 11.2.2 UNIQUENESS OF LIMIT

If $\lim_{n \to \infty} a_n = l$ and $\lim_{n \to \infty} a_n = m$, then $l = m$.

A proof, similar to the proof of Theorem 2.3.1, is given in the supplement at the end of this section.

DEFINITION 11.2.3

A sequence that has a limit is said to be *convergent*. A sequence that has no limit is said to be *divergent*.

Instead of writing

$$\lim_{n \to \infty} a_n = l,$$

we will often write

$$a_n \to l \qquad \text{(read ``}a_n \text{ converges to } l\text{'')}$$

or more fully

$$a_n \to l \quad \text{as} \quad n \to \infty.$$

THEOREM 11.2.4

Every convergent sequence is bounded.

Proof. Assume that $a_n \to l$ and choose any positive number: 1, for instance. Using 1 as ϵ, you can see that there must exist k such that

$$|a_n - l| < 1 \qquad \text{for all } n \geq k.$$

This means that

$$|a_n| < 1 + |l| \qquad \text{for all } n \geq k$$

and, consequently,

$$|a_n| \leq \max \{|a_1|, |a_2|, \ldots, |a_{k-1}|, 1 + |l|\} \qquad \text{for all } n.$$

This proves that $\{a_n\}$ is bounded. ◻

Since every convergent sequence is bounded, a sequence that is not bounded cannot be convergent; namely,

(11.2.5)
> every unbounded sequence is divergent.

The sequences

$$a_n = \tfrac{1}{2}n, \qquad b_n = \frac{n^2}{n+1}, \qquad c_n = n \ln n$$

are all unbounded. Each of these sequences is therefore divergent.

Boundedness does not imply convergence. As a counterexample, consider the oscillating sequence

$$\{1, 0, 1, 0, \ldots\}.$$

This sequence is certainly bounded (above by 1 and below by 0), but obviously it does not converge: the limit would have to be arbitrarily close to both 0 and 1 simultaneously.

Boundedness together with monotonicity does imply convergence.

> **THEOREM 11.2.6**
>
> A bounded nondecreasing sequence converges to the least upper bound of its range; a bounded nonincreasing sequence converges to the greatest lower bound of its range.

Proof. Suppose that $\{a_n\}$ is bounded and nondecreasing. If l is the least upper bound of the range of this sequence, then

$$a_n \leq l \qquad \text{for all } n.$$

Now let ϵ be an arbitrary positive number. By Theorem 10.8.2 there exists a_k such that

$$l - \epsilon < a_k.$$

Since the sequence is nondecreasing,

$$a_k \leq a_n \qquad \text{for all } n \geq k.$$

It follows that

$$l - \epsilon < a_n \leq l \qquad \text{for all } n \geq k.$$

This shows that

$$|a_n - l| < \epsilon \qquad \text{for all } n \geq k$$

and proves that

$$a_n \to l.$$

The nonincreasing case can be handled in a similar manner. ☐

Example 4. Take the sequence

$$\{(3^n + 4^n)^{1/n}\}.$$

Since

$$3 = (3^n)^{1/n} < (3^n + 4^n)^{1/n} < (2 \cdot 4^n)^{1/n} = 2^{1/n} \cdot 4 \leq 8,$$

the sequence is bounded. Note that

$$\begin{aligned}
(3^n + 4^n)^{(n+1)/n} &= (3^n + 4^n)^{1/n}(3^n + 4^n) \\
&= (3^n + 4^n)^{1/n}3^n + (3^n + 4^n)^{1/n}4^n \\
&> (3^n)^{1/n}3^n + (4^n)^{1/n}4^n = 3 \cdot 3^n + 4 \cdot 4^n = 3^{n+1} + 4^{n+1}.
\end{aligned}$$

Taking the $(n + 1)$st root of both extremes, we find that

$$(3^n + 4^n)^{1/n} > (3^{n+1} + 4^{n+1})^{1/(n+1)}.$$

The sequence is decreasing. Being bounded, it must be convergent. (Later you will be asked to show that the limit is 4.) ☐

THEOREM 11.2.7

Let α be a real number. If $a_n \to l$ and $b_n \to m$, then

    (i) $a_n + b_n \to l + m$,   (ii) $\alpha a_n \to \alpha l$,   (iii) $a_n b_n \to lm$.

If, in addition, $m \neq 0$ and b_n is never 0, then

        (iv) $\dfrac{1}{b_n} \to \dfrac{1}{m}$ and   (v) $\dfrac{a_n}{b_n} \to \dfrac{l}{m}$.

Proofs of parts (i) and (ii) are left as exercises. For proofs of parts (iii)–(v), see the supplement at the end of this section.

We are now in a position to handle any rational sequence

$$a_n = \frac{\alpha_k n^k + \alpha_{k-1} n^{k-1} + \cdots + \alpha_0}{\beta_j n^j + \beta_{j-1} n^{j-1} + \cdots + \beta_0}.$$

To determine the behavior of such a sequence we need only divide both numerator and denominator by the highest power of n that occurs.

Example 5

$$\frac{3n^4 - 2n^2 + 1}{n^5 - 3n^3} = \frac{3/n - 2/n^3 + 1/n^5}{1 - 3/n^2} \rightarrow \frac{0}{1} = 0. \quad \square$$

Example 6

$$\frac{1 - 4n^7}{n^7 + 12n} = \frac{1/n^7 - 4}{1 + 12/n^6} \rightarrow \frac{-4}{1} = -4. \quad \square$$

Example 7

$$\frac{n^4 - 3n^2 + n + 2}{n^3 + 7n} = \frac{1 - 3/n^2 + 1/n^3 + 2/n^4}{1/n + 7/n^3}.$$

Since the numerator tends to 1 and the denominator tends to 0, the sequence is unbounded. Therefore it cannot converge. $\square$

THEOREM 11.2.8

$$a_n \rightarrow l \quad \text{iff} \quad a_n - l \rightarrow 0 \quad \text{iff} \quad |a_n - l| \rightarrow 0.$$

We leave the proof to you.

THEOREM 11.2.9   THE PINCHING THEOREM FOR SEQUENCES

Suppose that for all n sufficiently large

$$a_n \le b_n \le c_n.$$

If $a_n \rightarrow l$ and $c_n \rightarrow l$, then $b_n \rightarrow l$.

Once again the proof is left to you.

As an immediate and obvious consequence of the pinching theorem we have the following corollary.

(11.2.10)

Suppose that for all n sufficiently large

$$|b_n| \leq |c_n|.$$

If $|c_n| \to 0$, then $|b_n| \to 0$.

Example 8

$$\frac{\cos n}{n} \to 0 \quad \text{since} \quad \left| \frac{\cos n}{n} \right| \leq \frac{1}{n} \quad \text{and} \quad \frac{1}{n} \to 0. \quad \square$$

Example 9

$$\sqrt{4 + \left(\frac{1}{n} \right)^2} \to 2$$

since

$$2 \leq \sqrt{4 + \left(\frac{1}{n} \right)^2} \leq \sqrt{4 + 4 \left(\frac{1}{n} \right) + \left(\frac{1}{n} \right)^2} = 2 + \frac{1}{n} \quad \text{and} \quad 2 + \frac{1}{n} \to 2. \quad \square$$

Example 10

(11.2.11)

$$\lim_{n \to \infty} \left(1 + \frac{1}{n} \right)^n = e.$$

Proof. You have already seen that, for all positive integers n,

$$\left(1 + \frac{1}{n} \right)^n \leq e \leq \left(1 + \frac{1}{n} \right)^{n+1}. \qquad \text{(Theorem 7.5.9)}$$

Dividing the right-hand inequality by $1 + 1/n$, we have

$$\frac{e}{1 + 1/n} \leq \left(1 + \frac{1}{n} \right)^n.$$

Combining this with the left-hand inequality, we can write

$$\frac{e}{1 + 1/n} \leq \left(1 + \frac{1}{n} \right)^n \leq e.$$

Since

$$\frac{e}{1 + 1/n} \to \frac{e}{1} = e,$$

we can conclude from the pinching theorem that

$$\left(1 + \frac{1}{n} \right)^n \to e. \quad \blacksquare$$

The sequences

$$\left\{ \cos \frac{\pi}{n} \right\}, \quad \left\{ \ln \left(\frac{n}{n+1} \right) \right\}, \quad \{e^{1/n}\}, \quad \left\{ \tan \left(\sqrt{\frac{\pi^2 n^2 - 8}{16 n^2}} \right) \right\}$$

are all of the form $\{f(c_n)\}$ with f a continuous function. Such sequences are frequently easy to deal with. The basic idea is this: when a continuous function is applied to a convergent sequence, the result is itself a convergent sequence. More precisely, we have the following theorem.

THEOREM 11.2.12

Suppose that

$$c_n \to c$$

and that, for each n, c_n is in the domain of f. If f is continuous at c, then

$$f(c_n) \to f(c).$$

Proof. We assume that f is continuous at c and take $\epsilon > 0$. From the continuity of f at c we know that there exists $\delta > 0$ such that

$$\text{if} \quad |x - c| < \delta, \quad \text{then} \quad |f(x) - f(c)| < \epsilon.$$

Since $c_n \to c$, we know that there exists a positive integer k such that

$$\text{if} \quad n \geq k, \quad \text{then} \quad |c_n - c| < \delta.$$

It follows therefore that

$$\text{if} \quad n \geq k, \quad \text{then} \quad |f(c_n) - f(c)| < \epsilon. \quad \square$$

Example 11. Since $\pi/n \to 0$ and the cosine function is continuous at 0,

$$\cos(\pi/n) \to \cos 0 = 1. \quad \square$$

Example 12. Since

$$\frac{n}{n+1} = \frac{1}{1 + 1/n} \to 1$$

and the logarithm function is continuous at 1,

$$\ln \left(\frac{n}{n+1} \right) \to \ln 1 = 0. \quad \square$$

Example 13. Since $1/n \to 0$ and the exponential function is continuous at 0,

$$e^{1/n} \to e^0 = 1. \quad \square$$

Example 14. Since

$$\frac{\pi^2 n^2 - 8}{16n^2} = \frac{\pi^2 - 8/n^2}{16} \to \frac{\pi^2}{16}$$

and the function $f(x) = \tan \sqrt{x}$ is continuous at $\pi^2/16$,

$$\tan\left(\sqrt{\frac{\pi^2 n^2 - 8}{16n^2}}\right) \to \tan\left(\sqrt{\frac{\pi^2}{16}}\right) = \tan\frac{\pi}{4} = 1. \quad \square$$

Example 15. Since

$$\frac{2n+1}{n} + \left(5 - \frac{1}{n^2}\right) \to 7$$

and the square-root function is continuous at 7,

$$\sqrt{\frac{2n+1}{n} + \left(5 - \frac{1}{n^2}\right)} \to \sqrt{7}. \quad \square$$

Example 16. Since the absolute-value function is everywhere continuous,

$$a_n \to l \quad \text{implies} \quad |a_n| \to |l|. \quad \square$$

Remark. For some time now we have asked you to take on faith two fundamentals of integration: that continuous functions do have definite integrals and that these integrals can be expressed as limits of Riemann sums. We could not give you proofs of these assertions because we did not have the necessary tools. Now we do. Proofs are given in Appendix B. □

EXERCISES 11.2

State whether or not the sequence converges and, if it does, find the limit.

1. 2^n. **2.** $\dfrac{2}{n}$. **3.** $\dfrac{(-1)^n}{n}$. **4.** $\sqrt{n}$.

5. $\dfrac{n-1}{n}$. **6.** $\dfrac{n+(-1)^n}{n}$. **7.** $\dfrac{n+1}{n^2}$. **8.** $\sin\dfrac{\pi}{2n}$.

9. $\dfrac{2^n}{4^n+1}$. **10.** $\dfrac{n^2}{n+1}$. **11.** $(-1)^n\sqrt{n}$. **12.** $\dfrac{4n}{\sqrt{n^2+1}}$.

13. $(-\tfrac{1}{2})^n$. **14.** $\dfrac{4^n}{2^n+10^6}$. **15.** $\tan\dfrac{n\pi}{4n+1}$. **16.** $\dfrac{10^{10}\sqrt{n}}{n+1}$.

17. $\dfrac{(2n+1)^2}{(3n-1)^2}$. **18.** $\ln\left(\dfrac{2n}{n+1}\right)$. **19.** $\dfrac{n^2}{\sqrt{2n^4+1}}$. **20.** $\dfrac{n^4-1}{n^4+n-6}$.

21. $\cos n\pi$. **22.** $\dfrac{n^5}{17n^4+12}$. **23.** $e^{1/\sqrt{n}}$. **24.** $\sqrt{4-\dfrac{1}{n}}$.

25. $(0.9)^{-n}$.

26. $\dfrac{2^n - 1}{2^n}$.

27. $\ln n - \ln (n+1)$.

28. $\dfrac{1}{n} - \dfrac{1}{n+1}$.

29. $\dfrac{\sqrt{n+1}}{2\sqrt{n}}$.

30. $(0.9)^n$.

31. $\left(1 + \dfrac{1}{n}\right)^{2n}$.

32. $\left(1 + \dfrac{1}{n}\right)^{n/2}$.

33. $\dfrac{2^n}{n^2}$.

34. $\dfrac{(n+1)\cos\sqrt{n}}{n(1+\sqrt{n})}$.

35. $\dfrac{\sqrt{n}\sin(e^n\pi)}{n+1}$.

36. $2\ln 3n - \ln(n^2 + 1)$.

37. Prove that, if $a_n \to l$ and $b_n \to m$, then $a_n + b_n \to l + m$.

38. Let α be a real number. Prove that, if $a_n \to l$, then $\alpha a_n \to \alpha l$.

39. Prove that

$$\left(1 + \frac{1}{n}\right)^{n+1} \to e \quad \text{given that} \quad \left(1 + \frac{1}{n}\right)^n \to e.$$

40. Determine the convergence or divergence of a rational sequence

$$a_n = \frac{\alpha_k n^k + \alpha_{k-1} n^{k-1} + \cdots + \alpha_0}{\beta_j n^j + \beta_{j-1} n^{j-1} + \cdots + \beta_0} \quad \text{with } \alpha_k \neq 0, \beta_j \neq 0,$$

given that (a) $k = j$. (b) $k < j$. (c) $k > j$.

41. Prove that a bounded nonincreasing sequence converges to the greatest lower bound of its range.

42. Let $\{a_n\}$ be a sequence of real numbers. Let $\{e_n\}$ be the sequence of even terms:

$$e_n = a_{2n}$$

and let $\{o_n\}$ be the sequence of odd terms:

$$o_n = a_{2n-1}.$$

Show that

$$a_n \to l \quad \text{iff} \quad e_n \to l \quad \text{and} \quad o_n \to l.$$

43. Prove the pinching theorem for sequences.

44. Show that

$$\frac{2^n}{n!} \to 0.$$

HINT: First show that

$$\frac{2^n}{n!} = \frac{2}{1} \cdot \frac{2}{2} \cdot \frac{2}{3} \cdot \; \cdots \; \cdot \frac{2}{n} \le \frac{4}{n}.$$

45. Prove that $(1/n)^{1/p} \to 0$ for all odd positive integers p.

46. Prove Theorem 11.2.8.

The following sequences are defined recursively.† Determine in each case whether the sequence converges and, if so, find the limit. Start each sequence with $a_1 = 1$.

47. $a_{n+1} = \dfrac{1}{e} a_n$.

48. $a_{n+1} = 2^{n+1} a_n$.

49. $a_{n+1} = \dfrac{1}{n+1} a_n$.

50. $a_{n+1} = \dfrac{n}{n+1} a_n$.

51. $a_{n+1} = 1 - a_n$.

52. $a_{n+1} = -a_n$.

53. $a_{n+1} = \tfrac{1}{2} a_n + 1$.

54. $a_{n+1} = \tfrac{1}{3} a_n + 1$.

---

† The notion was introduced in the exercises at the end of Section 11.1.

(Calculator) Evaluate numerically the limit of each sequence as $n \to \infty$. Some of these sequences converge more rapidly than others. Determine for each sequence the least value of n for which the nth term differs from the limit by less than 0.001.

55. $\dfrac{1}{n^2}$. **56.** $\dfrac{1}{\sqrt{n}}$. **57.** $\dfrac{n}{10^n}$. **58.** $\dfrac{n^{10}}{10^n}$.

59. $\dfrac{1}{n!}$. **60.** $\dfrac{2^n}{n!}$. **61.** $\dfrac{\ln n}{n^2}$. **62.** $\dfrac{\ln n}{n}$.

*SUPPLEMENT TO SECTION 11.2

Proof of Theorem 11.2.2. If $l \neq m$, then

$$\tfrac{1}{2}|l - m| > 0.$$

The assumption that $\lim\limits_{n \to \infty} a_n = l$ and $\lim\limits_{n \to \infty} a_n = m$ gives the existence of k_1 such that

$$\text{if} \quad n \geq k_1, \quad \text{then} \quad |a_n - l| < \tfrac{1}{2}|l - m|$$

and the existence of k_2 such that

$$\text{if} \quad n \geq k_2, \quad \text{then} \quad |a_n - m| < \tfrac{1}{2}|l - m|.\dagger$$

For $n \geq \max\{k_1, k_2\}$ we have

$$|a_n - l| + |a_n - m| < |l - m|.$$

By the triangle inequality we have

$$|l - m| = |(l - a_n) + (a_n - m)| \leq |l - a_n| + |a_n - m| = |a_n - l| + |a_n - m|.$$

Combining the last two statements, we have

$$|l - m| < |l - m|.$$

The hypothesis $l \neq m$ has led to an absurdity. We conclude that $l = m$. □

Proof of Theorem 11.2.7(iii)–(v). To prove (iii), we set $\epsilon > 0$. For each n,

$$|a_n b_n - lm| = |(a_n b_n - a_n m) + (a_n m - lm)|$$
$$\leq |a_n||b_n - m| + |m||a_n - l|.$$

Since $\{a_n\}$ is convergent, $\{a_n\}$ is bounded; that is, there exists $M > 0$ such that

$$|a_n| \leq M \quad \text{for all } n.$$

Since $|m| < |m| + 1$, we have

(1) $$|a_n b_n - lm| \leq M|b_n - m| + (|m| + 1)|a_n - l|.\dagger\dagger$$

Since $b_n \to m$, we know that there exists k_1 such that

$$\text{if} \quad n \geq k_1, \quad \text{then} \quad |b_n - m| < \frac{\epsilon}{2M}.$$

$\dagger$ We can reach these conclusions from Definition 11.2.1 by taking $\tfrac{1}{2}|l - m|$ as ϵ.

$\dagger\dagger$ Soon we will want to divide by the coefficient of $|a_n - l|$. We have replaced $|m|$ by $|m| + 1$ because $|m|$ can be zero.

Since $a_n \to l$, we know that there exists k_2 such that

$$\text{if} \quad n \geq k_2, \quad \text{then} \quad |a_n - l| < \frac{\epsilon}{2(|m| + 1)}.$$

For $n \geq \max\{k_1, k_2\}$ both conditions hold, and consequently

$$M|b_n - m| + (|m| + 1)|a_n - l| < \frac{\epsilon}{2} + \frac{\epsilon}{2} = \epsilon.$$

In view of (1), we can conclude that

$$\text{if} \quad n \geq \max\{k_1, k_2\}, \quad \text{then} \quad |a_n b_n - lm| < \epsilon.$$

This proves that

$$a_n b_n \to lm. \quad \square$$

To prove (iv), once again we set $\epsilon > 0$. In the first place

$$\left| \frac{1}{b_n} - \frac{1}{m} \right| = \left| \frac{m - b_n}{b_n m} \right| = \frac{|b_n - m|}{|b_n||m|}.$$

Since $b_n \to m$ and $|m|/2 > 0$, there exists k_1 such that

$$\text{if} \quad n \geq k_1, \quad \text{then} \quad |b_n - m| < \frac{|m|}{2}.$$

This tells us that for $n \geq k_1$ we have

$$|b_n| > \frac{|m|}{2} \quad \text{and thus} \quad \frac{1}{|b_n|} < \frac{2}{|m|}.$$

Thus for $n \geq k_1$ we have

$$(2) \qquad\qquad \left| \frac{1}{b_n} - \frac{1}{m} \right| \leq \frac{2}{|m|^2} |b_n - m|.$$

Since $b_n \to m$, there exists k_2 such that

$$\text{if} \quad n \geq k_2, \quad \text{then} \quad |b_n - m| < \frac{\epsilon|m|^2}{2}.$$

Thus for $n \geq k_2$ we have

$$\frac{2}{|m|^2} |b_n - m| < \epsilon.$$

In view of (2), we can be sure that

$$\text{if} \quad n \geq \max\{k_1, k_2\}, \quad \text{then} \quad \left| \frac{1}{b_n} - \frac{1}{m} \right| < \epsilon.$$

This proves that

$$\frac{1}{b_n} \to \frac{1}{m}. \quad \square$$

The proof of (v) is now easy:

$$\frac{a_n}{b_n} = a_n \cdot \frac{1}{b_n} \to l \cdot \frac{1}{m} = \frac{l}{m}. \quad \square$$

11.3   SOME IMPORTANT LIMITS

Our purpose here is to familiarize you with some limits that are particularly important in calculus and to give you more experience with limit arguments.

(11.3.1)

> If $x > 0$, then
>
> $x^{1/n} \to 1$ as $n \to \infty$.

Proof.   Note that

$$\ln (x^{1/n}) = \frac{1}{n} \ln x \to 0.$$

Since the exponential function is continuous at 0, it follows from Theorem 11.2.12 that

$$x^{1/n} = e^{(1/n)\ln x} \to e^0 = 1. \quad \square$$

(11.3.2)

> If $|x| < 1$, then
>
> $x^n \to 0$ as $n \to \infty$.

Proof.   Take $|x| < 1$ and observe that $\{|x|^n\}$ is a decreasing sequence:

$$|x|^{n+1} = |x|\,|x|^n < |x|^n.$$

Now let $\epsilon > 0$.   By (11.3.1)

$$\epsilon^{1/n} \to 1 \quad \text{as} \quad n \to \infty.$$

Thus there exists an integer $k > 0$ such that

$$|x| < \epsilon^{1/k}. \hspace{3cm} \text{(explain)}$$

Obviously, then, $|x|^k < \epsilon$.   Since $\{|x|^n\}$ is a decreasing sequence,

$$|x^n| = |x|^n < \epsilon \quad \text{for all } n \geq k. \quad \square$$

(11.3.3)

> For each real x
>
> $\dfrac{x^n}{n!} \to 0$ as $n \to \infty$.

Proof.   Choose an integer k such that $k > |x|$.   For $n > k$,

$$\frac{k^n}{n!} = \left(\frac{k^k}{k!}\right)\left[\frac{k}{k+1}\,\frac{k}{k+2} \cdots \frac{k}{n-1}\right]\left(\frac{k}{n}\right) < \left(\frac{k^{k+1}}{k!}\right)\left(\frac{1}{n}\right).$$

the middle term is less than 1

Since $k > |x|$, we have

$$0 < \frac{|x|^n}{n!} < \frac{k^n}{n!} < \left(\frac{k^{k+1}}{k!}\right)\left(\frac{1}{n}\right).$$

Since k is fixed and $1/n \to 0$, it follows from the pinching theorem that

$$\frac{|x|^n}{n!} \to 0 \quad \text{and thus} \quad \frac{x^n}{n!} \to 0. \quad \square$$

(11.3.4)

> For each $\alpha > 0$
>
> $$\frac{1}{n^\alpha} \to 0 \quad \text{as} \quad n \to \infty.$$

Proof. Since $\alpha > 0$, there exists an odd positive integer p such that $1/p < \alpha$. Then

$$0 < \frac{1}{n^\alpha} = \left(\frac{1}{n}\right)^\alpha \le \left(\frac{1}{n}\right)^{1/p}.$$

Since $1/n \to 0$ and $f(x) = x^{1/p}$ is continuous at 0, we have

$$\left(\frac{1}{n}\right)^{1/p} \to 0 \quad \text{and thus by the pinching theorem} \quad \frac{1}{n^\alpha} \to 0. \quad \square$$

(11.3.5)

> $$\frac{\ln n}{n} \to 0 \quad \text{as} \quad n \to \infty.$$

Proof. A routine proof can be based on L'Hospital's rule (11.5.1), but that is not available to us yet. We will appeal to the pinching theorem and base our argument on the integral representation of the logarithm:

$$0 \le \frac{\ln n}{n} = \frac{1}{n} \int_1^n \frac{dt}{t} \le \frac{1}{n} \int_1^n \frac{dt}{\sqrt{t}} = \frac{2}{n}(\sqrt{n} - 1) = 2\left(\frac{1}{\sqrt{n}} - \frac{1}{n}\right) \to 0. \quad \square$$

(11.3.6)

> $$n^{1/n} \to 1 \quad \text{as} \quad n \to \infty.$$

Proof. We know that

$$n^{1/n} = e^{(1/n)\ln n}.$$

Since

$$(1/n) \ln n \to 0 \tag{11.3.5}$$

and the exponential function is continuous at 0, it follows from (11.2.12) that

$$n^{1/n} \to e^0 = 1. \quad \square$$

(11.3.7)

For each real x

$$\left(1 + \frac{x}{n}\right)^n \to e^x \quad \text{as} \quad n \to \infty.$$

Proof. For $x = 0$, the result is obvious. For $x \neq 0$,

$$\ln\left(1 + \frac{x}{n}\right)^n = n \ln\left(1 + \frac{x}{n}\right) = x\left[\frac{\ln(1 + x/n) - \ln 1}{x/n}\right].$$

The crux here is to recognize that the bracketed expression is a difference quotient for the logarithm function. Once we see this, we can write

$$\lim_{n \to \infty}\left[\frac{\ln(1 + x/n) - \ln 1}{x/n}\right] = \lim_{h \to 0}\left[\frac{\ln(1 + h) - \ln 1}{h}\right] = 1.\dagger$$

It follows that

$$\ln\left(1 + \frac{x}{n}\right)^n \to x \quad \text{and therefore} \quad \left(1 + \frac{x}{n}\right)^n = e^{\ln(1 + x/n)^n} \to e^x. \quad \square$$

EXERCISES 11.3

State whether or not the sequence converges as $n \to \infty$; if it does, find the limit.

1. $2^{2/n}$.

2. $e^{-\alpha/n}$.

3. $\left(\dfrac{2}{n}\right)^n$.

4. $\dfrac{\log_{10} n}{n}$.

5. $\dfrac{\ln(n + 1)}{n}$.

6. $\dfrac{3^n}{4^n}$.

7. $\dfrac{x^{100n}}{n!}$.

8. $n^{1/(n+2)}$.

9. $n^{\alpha/n}, \quad \alpha > 0$.

10. $\ln\left(\dfrac{n + 1}{n}\right)$.

11. $\dfrac{3^{n+1}}{4^{n-1}}$.

12. $\displaystyle\int_{-n}^{0} e^{2x}\, dx$.

13. $(n + 2)^{1/n}$.

14. $\left(1 - \dfrac{1}{n}\right)^n$.

15. $\displaystyle\int_{0}^{n} e^{-x}\, dx$.

16. $\dfrac{2^{3n-1}}{7^{n+2}}$.

17. $\displaystyle\int_{-n}^{n} \dfrac{dx}{1 + x^2}$.

18. $\displaystyle\int_{0}^{n} e^{-nx}\, dx$.

19. $(n + 2)^{1/(n+2)}$.

20. $n^2 \sin n\pi$.

21. $\dfrac{\ln n^2}{n}$.

22. $\displaystyle\int_{-1+1/n}^{1-1/n} \dfrac{dx}{\sqrt{1 - x^2}}$.

23. $n^2 \sin \dfrac{\pi}{n}$.

24. $\dfrac{n!}{2^n}$.

---

† For each $t > 0$

$$\lim_{h \to 0} \frac{\ln(t + h) - \ln t}{h} = \frac{d}{dt}(\ln t) = \frac{1}{t}.$$

25. $\dfrac{5^{n+1}}{4^{2n-1}}.$

26. $\left(1 + \dfrac{x}{n}\right)^{3n}.$

27. $\left(\dfrac{n+1}{n+2}\right)^n.$

28. $\displaystyle\int_{1/n}^{1} \dfrac{dx}{\sqrt{x}}.$

29. $\displaystyle\int_{n}^{n+1} e^{-x^2}\, dx.$

30. $\left(1 + \dfrac{1}{n^2}\right)^n.$

31. $\dfrac{n^n}{2^{n^2}}.$

32. $\displaystyle\int_{0}^{1/n} \cos e^x\, dx.$

33. $\left(1 + \dfrac{x}{2n}\right)^{2n}.$

34. $\left(1 + \dfrac{1}{n}\right)^{n^2}.$

35. $\displaystyle\int_{-1/n}^{1/n} \sin x^2\, dx.$

36. $\left(t + \dfrac{x}{n}\right)^n, \quad t > 0, \quad x > 0.$

37. Show that $\lim\limits_{n\to\infty} (\sqrt{n+1} - \sqrt{n}) = 0.$
38. Show that $\lim\limits_{n\to\infty} (\sqrt{n^2+n} - n) = \frac{1}{2}.$

39. Find $\lim\limits_{n\to\infty} [2n \sin (\pi/n)].$ What is the geometric significance of this limit?

HINT: Think of a regular polygon of n sides inscribed in the unit circle.

40. Show that

$$\text{if} \quad 0 < c < d, \quad \text{then} \quad (c^n + d^n)^{1/n} \to d.$$

Find the following limits.

41. $\lim\limits_{n\to\infty} \dfrac{1 + 2 + \cdots + n}{n^2}.$ HINT: $1 + 2 + \cdots + n = \dfrac{n(n+1)}{2}.$

42. $\lim\limits_{n\to\infty} \dfrac{1^2 + 2^2 + \cdots + n^2}{(1+n)(2+n)}.$ HINT: $1^2 + 2^2 + \cdots + n^2 = \dfrac{n(n+1)(2n+1)}{6}.$

43. $\lim\limits_{n\to\infty} \dfrac{1^3 + 2^3 + \cdots + n^3}{2n^4 + n - 1}.$ HINT: $1^3 + 2^3 + \cdots + n^3 = \dfrac{n^2(n+1)^2}{4}.$

44. A sequence $\{a_n\}$ is said to be a *Cauchy sequence*† iff

(11.3.8)
> for each $\epsilon > 0$ there exists a positive integer k such that
> $$|a_n - a_m| < \epsilon \qquad \text{for all } m, n \geq k.$$

Show that

(11.3.9)
> every convergent sequence is a Cauchy sequence.

It is also true that every Cauchy sequence is convergent, but this is more difficult to prove.

45. (*Arithmetic means*) Let

$$m_n = \frac{1}{n}(a_1 + a_2 + \cdots + a_n).$$

(a) Prove that if $\{a_n\}$ is increasing, then $\{m_n\}$ is increasing.

(b) Prove that if $a_n \to 0$, then $m_n \to 0$.

HINT: Choose an integer $j > 0$ such that, if $n \geq j$, then $a_n < \epsilon/2$. Then for $n \geq j$,

$$|m_n| < \frac{|a_1 + a_2 + \cdots + a_j|}{n} + \frac{\epsilon}{2}\left(\frac{n-j}{n}\right).$$

---

† After the French baron Augustin Louis Cauchy (1789–1857), one of the most prolific mathematicians of all time.

46. (Calculator) You have seen that for all real x

$$\lim_{n \to \infty} \left(1 + \frac{x}{n}\right)^n = e^x.$$

However, the rate of convergence is different for different x. Verify that at $n = 100$, $(1 + 1/n)^n$ is within 1% of its limit, while $(1 + 5/n)^n$ is still about 12% from its limit. Give comparable accuracy estimates for these two sequences at $n = 1000$.

47. (Calculator) Evaluate

$$\lim_{n \to \infty} \left(\sin \frac{1}{n}\right)^{1/n}$$

numerically and justify your answer by other means.

48. (Calculator) We have stated that

$$\lim_{n \to \infty} (\sqrt{n^2 + n} - n) = \tfrac{1}{2}. \qquad \text{(Exercise 38)}$$

Evaluate numerically

$$\lim_{n \to \infty} [(n^3 + n^2)^{1/3} - n].$$

Formulate a conjecture about

$$\lim_{n \to \infty} [(n^k + n^{k-1})^{1/k} - n], \qquad k = 1, 2, 3, \ldots$$

and prove that your conjecture is valid.

11.4 THE INDETERMINATE FORM (0/0)

Here we are concerned with limits of quotients $f(x)/g(x)$ where the numerator and denominator both tend to 0 and elementary methods fail or are difficult to apply.

THEOREM 11.4.1 L'HOSPITAL'S RULE (0/0)†

Suppose that

$$f(x) \to 0 \quad \text{and} \quad g(x) \to 0$$

as $x \to c^+, x \to c^-, x \to c, x \to \infty,$ or $x \to -\infty.$

If $\dfrac{f'(x)}{g'(x)} \to l,$ then $\dfrac{f(x)}{g(x)} \to l.$

We will prove the validity of L'Hospital's rule later in the section. First we demonstrate its usefulness.

---

† Named after a Frenchman G. F. A. L'Hospital (1661–1704). The result was actually discovered by his teacher, Jakob Bernoulli (1654–1705).

Problem 1. Find

$$\lim_{x \to \pi/2} \frac{\cos x}{\pi - 2x}.$$

Solution.   As $x \to \pi/2$, both the numerator $f(x) = \cos x$ and the denominator $g(x) = \pi - 2x$ tend to zero, but it is not at all obvious what happens to the quotient

$$\frac{f(x)}{g(x)} = \frac{\cos x}{\pi - 2x}.$$

Therefore we test the quotient of derivatives:

$$\frac{f'(x)}{g'(x)} = \frac{-\sin x}{-2} = \frac{\sin x}{2} \to \frac{1}{2} \quad \text{as} \quad x \to \pi/2.$$

It follows from L'Hospital's rule that

$$\frac{\cos x}{\pi - 2x} \to \frac{1}{2} \quad \text{as} \quad x \to \pi/2.$$

We can express all this on just one line using * to indicate the differentiation of numerator and denominator:

$$\lim_{x \to \pi/2} \frac{\cos x}{\pi - 2x} \overset{*}{=} \lim_{x \to \pi/2} \frac{-\sin x}{-2} = \lim_{x \to \pi/2} \frac{\sin x}{2} = \frac{1}{2}. \quad \square$$

Problem 2. Find

$$\lim_{x \to 0^+} \frac{x}{\sin \sqrt{x}}.$$

Solution.   As $x \to 0^+$, both numerator and denominator tend to 0. Since

$$\frac{f'(x)}{g'(x)} = \frac{1}{(\cos \sqrt{x})(1/2\sqrt{x})} = \frac{2\sqrt{x}}{\cos \sqrt{x}} \to 0 \quad \text{as} \quad x \to 0^+,$$

it follows from L'Hospital's rule that

$$\frac{x}{\sin \sqrt{x}} \to 0 \quad \text{as} \quad x \to 0^+.$$

For short we can write

$$\lim_{x \to 0^+} \frac{x}{\sin \sqrt{x}} \overset{*}{=} \lim_{x \to 0^+} \frac{2\sqrt{x}}{\cos \sqrt{x}} = 0. \quad \square$$

Sometimes it is necessary to differentiate numerator and denominator more than once. The next problem gives such an instance.

Problem 3. Find

$$\lim_{x \to 0} \frac{e^x - x - 1}{x^2}.$$

Solution. As $x \to 0$, both numerator and denominator tend to 0. Here

$$\frac{f'(x)}{g'(x)} = \frac{e^x - 1}{2x}.$$

Since both numerator and denominator still tend to 0, we differentiate again:

$$\frac{f''(x)}{g''(x)} = \frac{e^x}{2}.$$

Since this last quotient tends to $\frac{1}{2}$, we can conclude that

$$\frac{e^x - 1}{2x} \to \frac{1}{2} \quad \text{and therefore} \quad \frac{e^x - x - 1}{x^2} \to \frac{1}{2}.$$

For short we can write

$$\lim_{x \to 0} \frac{e^x - x - 1}{x^2} \overset{*}{=} \lim_{x \to 0} \frac{e^x - 1}{2x} \overset{*}{=} \lim_{x \to 0} \frac{e^x}{2} = \frac{1}{2}. \quad \square$$

Problem 4. Find

$$\lim_{x \to 0^+} (1 + x)^{1/x}.$$

Solution. Here we are dealing with an indeterminate of the form 1^∞: as $x \to 0^+$, $1 + x \to 0$ and $1/x$ increases without bound.

   As the expression stands, it is not amenable to L'Hospital's rule. To make it so, we take its logarithm. This gives

$$\lim_{x \to 0^+} \ln (1 + x)^{1/x} = \lim_{x \to 0^+} \frac{\ln (1 + x)}{x} \overset{*}{=} \lim_{x \to 0^+} \frac{1}{1 + x} = 1.$$

Since $\ln (1 + x)^{1/x}$ tends to 1, $(1 + x)^{1/x}$ itself tends to e. $\square$†

---

† Setting $x = 1/n$, we have the familiar result

$$\left(1 + \frac{1}{n}\right)^n \to e.$$

Problem 5. Find

$$\lim_{n \to \infty} (\cos \pi/n)^n.$$

Solution. This again is an indeterminate of the form 1^∞. To apply the methods of this section we replace the integer variable n by the real variable x and examine the behavior of

$$(\cos \pi/x)^x \quad \text{as} \quad x \to \infty.$$

Taking the logarithm of this expression, we have

$$\ln (\cos \pi/x)^x = x \ln (\cos \pi/x).$$

The right side is an indeterminate of the form $\infty \cdot 0$. To apply L'Hospital's rule, we express it as an indeterminate of the form 0/0:

$$x \ln (\cos \pi/x) = \frac{\ln (\cos \pi/x)}{1/x}.$$

We then have

$$\lim_{x \to \infty} \ln (\cos \pi/x)^x = \lim_{x \to \infty} \frac{\ln (\cos \pi/x)}{1/x}$$

$$\overset{*}{=} \lim_{x \to \infty} \frac{(-\sin \pi/x)(-\pi/x^2)}{(\cos \pi/x)(-1/x^2)} = \lim_{x \to \infty} (-\pi \tan \pi/x) = 0.$$

This shows that $(\cos \pi/x)^x$ tends to 1 and thus that $(\cos \pi/n)^n$ tends to 1. $\quad\square$

To derive L'Hospital's rule, we need a generalization of the mean-value theorem.

THEOREM 11.4.2 THE CAUCHY MEAN-VALUE THEOREM†

Suppose that f and g are differentiable on (a, b) and continuous on $[a, b]$. If g' is never 0 in (a, b), then there is a number c in (a, b) for which

$$\frac{f'(c)}{g'(c)} = \frac{f(b) - f(a)}{g(b) - g(a)}.$$

Proof. We can prove this by applying the mean-value theorem (4.1.1) to the function

$$G(x) = [g(b) - g(a)][f(x) - f(a)] - [g(x) - g(a)][f(b) - f(a)].$$

---

† Another contribution of A. L. Cauchy, after whom Cauchy sequences were named.

Since

$$G(a) = 0 \quad \text{and} \quad G(b) = 0,$$

there exists (by the mean-value theorem) a number c in (a, b) for which

$$G'(c) = 0.$$

Since, in general,

$$G'(x) = [g(b) - g(a)]f'(x) - g'(x)[f(b) - f(a)],$$

we must have

$$[g(b) - g(a)]f'(c) - g'(c)[f(b) - f(a)] = 0,$$

and thus

$$[g(b) - g(a)]f'(c) = g'(c)[f(b) - f(a)].$$

Since g' is never 0 in (a, b),

$$g'(c) \neq 0 \quad \text{and} \quad g(b) - g(a) \neq 0.$$

$\uparrow$ —— explain

We can therefore divide by these numbers and obtain

$$\frac{f'(c)}{g'(c)} = \frac{f(b) - f(a)}{g(b) - g(a)}. \quad \square$$

Now we prove L'Hospital's rule for the case $x \to c^+$. We assume that, as $x \to c^+$,

$$f(x) \to 0, \quad g(x) \to 0, \quad \text{and} \quad \frac{f'(x)}{g'(x)} \to l$$

and show that

$$\frac{f(x)}{g(x)} \to l.$$

Proof. The fact that

$$\frac{f'(x)}{g'(x)} \to l \quad \text{as} \quad x \to c^+$$

assures us that both f' and g' exist on a set of the form $(c, c + h]$ and that g' is not zero there. By setting $f(c) = 0$ and $g(c) = 0$, we ensure that f and g are both continuous on $[c, c + h]$. We can now apply the Cauchy mean-value theorem and conclude that there exists a number c_h between c and $c + h$ such that

$$\frac{f'(c_h)}{g'(c_h)} = \frac{f(c + h) - f(c)}{g(c + h) - g(c)} = \frac{f(c + h)}{g(c + h)}.$$

The result is now obtained by letting $h \to 0^+$. Since the left side tends to l, the right side tends to l. $\square$

Here is a proof of L'Hospital's rule for the case $x \to \infty$.

Proof. The key here is to set $x = 1/t$:

$$\lim_{x \to \infty} \frac{f'(x)}{g'(x)} = \lim_{t \to 0^+} \frac{f'(1/t)}{g'(1/t)} = \lim_{t \to 0^+} \frac{-t^{-2}f'(1/t)}{-t^{-2}g'(1/t)} = \lim_{t \to 0^+} \frac{f(1/t)}{g(1/t)} = \lim_{x \to \infty} \frac{f(x)}{g(x)}. \quad \square$$

by L'Hospital's rule for the case $t \to 0^+$

Caution. L'Hospital's rule does not apply when the numerator or the denominator has a finite nonzero limit. For example,

$$\lim_{x \to 0} \frac{x}{x + \cos x} = \frac{0}{1} = 0.$$

A blind application of L'Hospital's rule would lead to

$$\lim_{x \to 0} \frac{x}{x + \cos x} \overset{*}{=} \lim_{x \to 0} \frac{1}{1 - \sin x} = 1.$$

This is wrong. $\quad \square$

EXERCISES 11.4

Find the indicated limit.

1. $\lim\limits_{x \to 0^+} \dfrac{\sin x}{\sqrt{x}}$.

2. $\lim\limits_{x \to 1} \dfrac{\ln x}{1 - x}$.

3. $\lim\limits_{x \to 0} \dfrac{e^x - 1}{\ln(1 + x)}$.

4. $\lim\limits_{x \to 4} \dfrac{\sqrt{x} - 2}{x - 4}$.

5. $\lim\limits_{x \to \pi/2} \dfrac{\cos x}{\sin 2x}$.

6. $\lim\limits_{x \to a} \dfrac{x - a}{x^n - a^n}$.

7. $\lim\limits_{x \to 0} \dfrac{2^x - 1}{x}$.

8. $\lim\limits_{x \to 0} \dfrac{\tan^{-1} x}{x}$.

9. $\lim\limits_{x \to 1} \dfrac{x^{1/2} - x^{1/4}}{x - 1}$.

10. $\lim\limits_{x \to 0} \dfrac{e^x - 1}{x(1 + x)}$.

11. $\lim\limits_{x \to 0} \dfrac{e^x - e^{-x}}{\sin x}$.

12. $\lim\limits_{x \to 0} \dfrac{1 - \cos x}{3x}$.

13. $\lim\limits_{x \to 0} \dfrac{x + \sin \pi x}{x - \sin \pi x}$.

14. $\lim\limits_{x \to 0} \dfrac{a^x - (a + 1)^x}{x}$.

15. $\lim\limits_{x \to 0} (e^x + x)^{1/x}$.

16. $\lim\limits_{x \to \infty} \left(1 + \dfrac{1}{x}\right)^x$.

17. $\lim\limits_{x \to 0} \dfrac{\tan \pi x}{e^x - 1}$.

18. $\lim\limits_{x \to 0} (e^x + 3x)^{1/x}$.

19. $\lim\limits_{x \to 0} \dfrac{1 + x - e^x}{x(e^x - 1)}$.

20. $\lim\limits_{x \to 0} \dfrac{\ln(\sec x)}{x^2}$.

21. $\lim\limits_{x \to 0} \dfrac{x - \tan x}{x - \sin x}$.

22. $\lim\limits_{x \to 0} \dfrac{xe^{nx} - x}{1 - \cos nx}$.

23. $\lim\limits_{x \to 1^-} \dfrac{\sqrt{1 - x^2}}{\sqrt{1 - x^3}}$.

24. $\lim\limits_{x \to 0} \left(\dfrac{1}{\sin x} - \dfrac{1}{x}\right)$.

25. $\lim\limits_{x \to \pi/2} \dfrac{\ln(\sin x)}{(\pi - 2x)^2}$.

26. $\lim\limits_{x \to 0^+} \dfrac{\sqrt{x}}{\sqrt{x} + \sin \sqrt{x}}$.

27. $\lim\limits_{x \to 1} \left(\dfrac{1}{\ln x} - \dfrac{x}{x - 1}\right)$.

28. $\lim\limits_{x \to 0} \dfrac{\sqrt{a + x} - \sqrt{a - x}}{x}$.

29. $\lim\limits_{x \to \pi/4} \dfrac{\sec^2 x - 2\tan x}{1 + \cos 4x}$.

30. $\lim\limits_{x \to 0} \dfrac{x - \sin^{-1} x}{\sin^3 x}$.

Find the limit of the sequence.

31. $\lim\limits_{n\to\infty} n(\pi/2 - \tan^{-1} n).$

32. $\lim\limits_{n\to\infty} [\ln (1 - 1/n) \csc 1/n].$

33. $\lim\limits_{n\to\infty} \dfrac{1}{n[\ln (n + 1) - \ln n]}.$

34. $\lim\limits_{n\to\infty} \dfrac{\sinh \pi/n - \sin \pi/n}{\sin^3 \pi/n}.$

35. Find the fallacy:

$$\lim_{x\to 0} \frac{2 + x + \sin x}{x^3 + x - \cos x} \overset{*}{=} \lim_{x\to 0} \frac{1 + \cos x}{3x^2 + 1 + \sin x} \overset{*}{=} \lim_{x\to 0} \frac{-\sin x}{6x + \cos x} = \frac{0}{1} = 0.$$

36. Show that, if $a > 0$, then

$$\lim_{n\to\infty} n(a^{1/n} - 1) = \ln a.$$

37. Given that f is continuous, use L'Hospital's rule to determine

$$\lim_{x\to 0} \left(\frac{1}{x} \int_0^x f(t)\, dt \right).$$

(Calculator) Evaluate the limit numerically and prove that your answer is correct.

38. $\lim\limits_{x\to\infty} \left[x^2 \left(\cos \dfrac{1}{x} - 1 \right) \right].$

39. $\lim\limits_{x\to\infty} \left[x^3 \left(\sin \dfrac{1}{x} - \dfrac{1}{x} \right) \right].$

11.5 THE INDETERMINATE FORM (∞/∞)

We come now to limits of quotients $f(x)/g(x)$ where numerator and denominator both tend to ∞. We take l as a fixed real number.

THEOREM 11.5.1 L'HOSPITAL'S RULE (∞/∞)

Suppose that

$$f(x) \to \pm\infty \quad \text{and} \quad g(x) \to \pm\infty$$

as $x \to c^-, x \to c^+, x \to c, x \to \infty$, or $x \to -\infty$.

If $\dfrac{f'(x)}{g'(x)} \to l,$ then $\dfrac{f(x)}{g(x)} \to l.$

While the proof of L'Hospital's rule in this setting is a little more complicated than it was in the (0/0) case,† the application of the rule is much the same.

---

† We omit the proof.

Problem 1. Show that

(11.5.2)

$$\lim_{x \to \infty} \frac{\ln x}{x} = 0.$$

Solution. Both numerator and denominator tend to ∞ with x. L'Hospital's rule gives

$$\lim_{x \to \infty} \frac{\ln x}{x} \overset{*}{=} \lim_{x \to \infty} \frac{1}{x} = 0. \quad \Box$$

Problem 2. Show that

(11.5.3)

$$\lim_{x \to \infty} \frac{x^k}{e^x} = 0.$$

Solution. Here we differentiate numerator and denominator k times:

$$\lim_{x \to \infty} \frac{x^k}{e^x} \overset{*}{=} \lim_{x \to \infty} \frac{kx^{k-1}}{e^x} \overset{*}{=} \lim_{x \to \infty} \frac{k(k-1)x^{k-2}}{e^x} \overset{*}{=} \cdots \overset{*}{=} \lim_{x \to \infty} \frac{k!}{e^x} = 0. \quad \Box$$

Problem 3. Show that

(11.5.4)

$$\lim_{x \to 0^+} x^x = 1.$$

Solution. Here we are dealing with an indeterminate of the form 0^0. Our first step is to take the logarithm of x^x. Then we apply L'Hospital's rule:

$$\lim_{x \to 0^+} \ln x^x = \lim_{x \to 0^+} (x \ln x) = \lim_{x \to 0^+} \frac{\ln x}{1/x} \overset{*}{=} \lim_{x \to 0^+} \frac{1/x}{-1/x^2} = \lim_{x \to 0^+} (-x) = 0.$$

Since $\ln x^x$ tends to 0, x^x must tend to 1. $\quad \Box$

Problem 4. Show that

$$\lim_{x \to \infty} (x^2 + 1)^{1/\ln x} = e^2.$$

Solution. Here we have an indeterminate of the form ∞^0. Taking the logarithm and then applying L'Hospital's rule, we find that

$$\lim_{x \to \infty} \ln (x^2 + 1)^{1/\ln x} = \lim_{x \to \infty} \frac{\ln (x^2 + 1)}{\ln x} \overset{*}{=} \lim_{x \to \infty} \frac{2x/(x^2 + 1)}{1/x} = \lim_{x \to \infty} \frac{2x^2}{x^2 + 1} = 2.$$

It follows that

$$\lim_{x \to \infty} (x^2 + 1)^{1/\ln x} = e^2. \quad \Box$$

Problem 5. Find the limit as $n \to \infty$ of the sequence

$$a_n = \frac{\ln n}{\sqrt{n}}.$$

Solution. To bring into play the methods of calculus we investigate

$$\lim_{x \to \infty} \frac{\ln x}{\sqrt{x}}$$

instead. Since both numerator and denominator tend to ∞ with x, we try L'Hospital's rule:

$$\lim_{x \to \infty} \frac{\ln x}{\sqrt{x}} \overset{*}{=} \lim_{x \to \infty} \frac{1/x}{1/2\sqrt{x}} = \lim_{x \to \infty} \frac{2}{\sqrt{x}} = 0.$$

The limit of the sequence must also be 0. ☐

Caution. L'Hospital's rule does not apply when the numerator or denominator has a finite nonzero limit. While

$$\lim_{x \to 0^+} \frac{1 + x}{\sin x} = \infty,$$

a misapplication of L'Hospital's rule can lead to the *incorrect conclusion* that

$$\lim_{x \to 0^+} \frac{1 + x}{\sin x} \overset{*}{=} \lim_{x \to 0^+} \frac{1}{-\cos x} = -1. \quad ☐$$

EXERCISES 11.5

Find the indicated limit.

1. $\displaystyle \lim_{x \to -\infty} \frac{x^2 + 1}{1 - x}$.

2. $\displaystyle \lim_{x \to \infty} \frac{20x}{x^2 + 1}$.

3. $\displaystyle \lim_{x \to \infty} \frac{x^3}{1 - x^3}$.

4. $\displaystyle \lim_{x \to \infty} \frac{x^3 - 1}{2 - x}$.

5. $\displaystyle \lim_{x \to \infty} \left(x^2 \sin \frac{1}{x} \right)$.

6. $\displaystyle \lim_{x \to \infty} \frac{\ln x^k}{x}$.

7. $\displaystyle \lim_{x \to \pi/2^-} \frac{\tan 5x}{\tan x}$.

8. $\displaystyle \lim_{x \to 0} (x \ln |\sin x|)$.

9. $\displaystyle \lim_{x \to 0^+} x^{2x}$.

10. $\displaystyle \lim_{x \to \infty} \left(x \sin \frac{\pi}{x} \right)$.

11. $\displaystyle \lim_{x \to 0} [x (\ln |x|)^2]$.

11. $\displaystyle \lim_{x \to 0^+} \frac{\ln x}{\cot x}$.

13. $\displaystyle \lim_{x \to \infty} \left(\frac{1}{x} \int_0^x e^{t^2} \, dt \right)$.

14. $\displaystyle \lim_{x \to \infty} \frac{\sqrt{1 + x^2}}{x}$.

15. $\displaystyle \lim_{x \to 0} \left[\frac{1}{\sin^2 x} - \frac{1}{x^2} \right]$.

16. $\displaystyle \lim_{x \to 0} |\sin x|^x$.

17. $\displaystyle \lim_{x \to 1} x^{1/(x-1)}$.

18. $\displaystyle \lim_{x \to 0^+} x^{\sin x}$.

19. $\displaystyle \lim_{x \to \infty} \left(\cos \frac{1}{x} \right)^x$.

20. $\displaystyle \lim_{x \to \pi/2} |\sec x|^{\cos x}$.

21. $\displaystyle \lim_{x \to 0} \left[\frac{1}{\ln (1 + x)} - \frac{1}{x} \right]$.

22. $\lim\limits_{x\to\infty} (x^2 + a^2)^{(1/x)^2}.$

23. $\lim\limits_{x\to 1} \left[\dfrac{x}{x-1} - \dfrac{1}{\ln x}\right].$

24. $\lim\limits_{x\to\infty} \ln \left(\dfrac{x^2-1}{x^2+1}\right)^3.$

25. $\lim\limits_{x\to\infty} (\sqrt{x^2 + 2x} - x).$

26. $\lim\limits_{x\to\infty} \dfrac{1}{x} \displaystyle\int_0^x \sin\left(\dfrac{1}{t+1}\right) dt.$

27. $\lim\limits_{x\to\infty} (x^3 + 1)^{1/\ln x}.$

28. $\lim\limits_{x\to\infty} (e^x + 1)^{1/x}.$

29. $\lim\limits_{x\to\infty} (\cosh x)^{1/x}.$

30. $\lim\limits_{x\to\infty} (x^4 + 1)^{1/\ln x}.$

Find the limit of the sequence.

31. $\lim\limits_{n\to\infty} \left(\dfrac{1}{n} \ln \dfrac{1}{n}\right).$

32. $\lim\limits_{n\to\infty} \dfrac{n^k}{2^n}.$

33. $\lim\limits_{n\to\infty} (\ln n)^{1/n}.$

34. $\lim\limits_{n\to\infty} \dfrac{\ln n}{n^p}, \quad (p > 0).$

35. $\lim\limits_{n\to\infty} (n^2 + n)^{1/n}.$

36. $\lim\limits_{n\to\infty} n^{\sin(\pi/n)}.$

37. $\lim\limits_{n\to\infty} \dfrac{n^2 \ln n}{e^n}.$

38. $\lim\limits_{n\to\infty} (\sqrt{n} - 1)^{1/\sqrt{n}}.$

Sketch the following curves specifying all vertical and horizontal asymptotes.

39. $y = x^2 - \dfrac{1}{x^3}.$

40. $y = \sqrt{\dfrac{x}{x-1}}.$

41. $y = xe^x.$

42. $y = xe^{-x}.$

43. $y = x^2 e^{-x}.$

44. $y = \dfrac{\ln x}{x}.$

The graphs of two functions $y = f(x)$ and $y = g(x)$ are said to be *asymptotic as* $x \to \infty$ iff

$$\lim_{x\to\infty} [f(x) - g(x)] = 0;$$

they are said to be *asymptotic as* $x \to -\infty$ iff

$$\lim_{x\to-\infty} [f(x) - g(x)] = 0.$$

45. Show that the hyperbolic arc $y = (b/a)\sqrt{x^2 - a^2}$ is asymptotic to the line $y = (b/a)x$ as $x \to \infty$.

46. Show that the graphs of $y = \cosh x$ and $y = \sinh x$ are asymptotic.

47. Give an example of a function the graph of which is asymptotic to the cubing function $y = x^3$ as $x \to \infty$.

48. Give an example of a function the graph of which is asymptotic to the parabola $y = x^2$ as $x \to \infty$ and crosses the graph of the parabola twice.

49. Give an example of a function the graph of which is asymptotic to the line $y = x$ as $x \to \infty$ and crosses the graph of the line infinitely often.

50. Let P be a polynomial of degree n: $P(x) = a_n x^n + a_{n-1}x^{n-1} + \cdots + a_0$. Let Q be a polynomial of degree $m < n$: $Q(x) = b_m x^m + b_{m-1}x^{m-1} + \cdots + b_0$. Find

$$\lim_{x\to\infty} \frac{P(x)}{Q(x)}$$

given that a_n and b_m have (a) the same sign, (b) opposite signs.

51. Find the fallacy:

$$\lim_{x\to 0^+} \frac{x^2}{\sin x} \overset{*}{=} \lim_{x\to 0^+} \frac{2x}{\cos x} \overset{*}{=} \lim_{x\to 0^+} \frac{2}{-\sin x} = -\infty.$$

52. Show by induction that, for each positive integer k,

$$\lim_{n\to\infty} \frac{(\ln n)^k}{n} = 0.$$

11.6   IMPROPER INTEGRALS

We begin with a function f continuous on an unbounded interval $[a, \infty)$. For each number $b > a$ we can form the integral

$$\int_a^b f(x)\, dx.$$

If, as b tends to ∞, this integral tends to a finite limit l,

$$\lim_{b \to \infty} \int_a^b f(x)\, dx = l,$$

then we write

$$\int_a^\infty f(x)\, dx = l$$

and say that

the improper integral $\int_a^\infty f(x)\, dx$ *converges to l.*

Otherwise, we say that

the improper integral $\int_a^\infty f(x)\, dx$ *diverges.*

In a similar manner,

improper integrals $\int_{-\infty}^b f(x)\, dx$ arise as limits of the form $\lim_{a \to -\infty} \int_a^b f(x)\, dx$.

Example 1

(a) $\int_1^\infty e^{-x}\, dx = \dfrac{1}{e}.$ (b) $\int_1^\infty \dfrac{dx}{x}$ diverges.

(c) $\int_1^\infty \dfrac{dx}{x^2} = 1.$ (d) $\int_{-\infty}^1 \cos \pi x\, dx$ diverges.

Verification

(a) $\displaystyle \int_1^\infty e^{-x}\, dx = \lim_{b \to \infty} \int_1^b e^{-x}\, dx = \lim_{b \to \infty} \left[-e^{-x} \right]_1^b = \lim_{b \to \infty} \left(\frac{1}{e} - \frac{1}{e^b} \right) = \frac{1}{e}.$

(b) $\displaystyle \int_1^\infty \frac{dx}{x} = \lim_{b \to \infty} \int_1^b \frac{dx}{x} = \lim_{b \to \infty} \ln b = \infty.$

(c) $\displaystyle \int_1^\infty \frac{dx}{x^2} = \lim_{b \to \infty} \int_1^b \frac{dx}{x^2} = \lim_{b \to \infty} \left[-\frac{1}{x} \right]_1^b = \lim_{b \to \infty} \left(1 - \frac{1}{b} \right) = 1.$

(d)   Note first that

$$\int_a^1 \cos \pi x \, dx = \left[\frac{1}{\pi} \sin \pi x \right]_a^1 = -\frac{1}{\pi} \sin \pi a.$$

As a tends to $-\infty$, $\sin \pi a$ oscillates between -1 and 1. Therefore the integral oscillates between $1/\pi$ and $-1/\pi$ and does not converge. ☐

The usual formulas for area and volume are extended to the unbounded case by means of improper integrals.

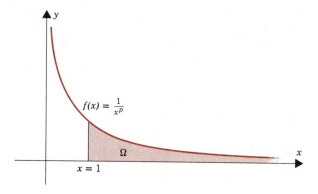

Figure 11.6.1

Example 2. If Ω is the region below the graph of

$$f(x) = \frac{1}{x^p}, \qquad x \geq 1 \qquad \text{(Figure 11.6.1)}$$

then

$$\text{the area of } \Omega = \left\{ \begin{array}{ll} \dfrac{1}{p-1}, & \text{if } p > 1 \\ \infty, & \text{if } p \leq 1 \end{array} \right].$$

This comes about from setting

$$\text{area of } \Omega = \lim_{b \to \infty} \int_1^b \frac{dx}{x^p} = \int_1^\infty \frac{dx}{x^p}.$$

For $p \neq 1$,

$$\int_1^\infty \frac{dx}{x^p} = \lim_{b \to \infty} \int_1^b \frac{dx}{x^p} = \lim_{b \to \infty} \frac{1}{1-p}(b^{1-p} - 1) = \left\{ \begin{array}{ll} \dfrac{1}{p-1}, & \text{if } p > 1 \\ \infty, & \text{if } p < 1 \end{array} \right].$$

For $p = 1$,

$$\int_1^\infty \frac{dx}{x^p} = \int_1^\infty \frac{dx}{x} = \infty,$$

as you have seen already. ☐

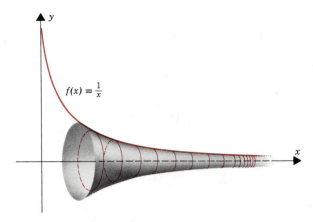

Figure 11.6.2

Example 3. From the last example you know that the region below the graph of

$$f(x) = \frac{1}{x}, \qquad x \geq 1$$

has infinite area. Suppose that this region with infinite area is revolved about the x-axis (see Figure 11.6.2). What is the volume of the resulting solid? It may surprise you somewhat, but the volume is not infinite. In fact, it is π. We can see this by the disc method:

$$V_x = \pi \int_1^\infty \frac{dx}{x^2} = \pi \cdot 1 = \pi. \quad \square$$

For future reference we record the following:

(11.6.1) $\int_1^\infty \frac{dx}{x^p}$ converges for $p > 1$ and diverges for $p \leq 1$.

It is often difficult to determine the convergence or divergence of a given integral by direct methods, but we can usually gain some information by comparison with integrals of known behavior.

(11.6.2)

(*A comparison test*) Suppose that f and g are continuous and

$$0 \leq f(x) \leq g(x) \qquad \text{for all } x \in [a, \infty).$$

(i) If $\int_a^\infty g(x)\, dx$ converges, then $\int_a^\infty f(x)\, dx$ converges.

(ii) If $\int_a^\infty f(x)\, dx$ diverges, then $\int_a^\infty g(x)\, dx$ diverges.

A similar test holds for integrals from $-\infty$ to b. The proof of (11.6.2) is left as an exercise.

Example 4. The improper integral

$$\int_1^\infty \frac{dx}{\sqrt{1+x^3}}$$

converges since

$$\frac{1}{\sqrt{1+x^3}} < \frac{1}{x^{3/2}} \quad \text{for } x \in [1, \infty) \quad \text{and} \quad \int_1^\infty \frac{dx}{x^{3/2}} \text{ converges.} \quad \square$$

Example 5. The improper integral

$$\int_1^\infty \frac{dx}{\sqrt{1+x^2}}$$

diverges since

$$\frac{1}{1+x} \le \frac{1}{\sqrt{1+x^2}} \quad \text{for } x \in [1, \infty) \quad \text{and} \quad \int_1^\infty \frac{dx}{1+x} \text{ diverges.} \quad \square$$

Suppose now that f is continuous on $(-\infty, \infty)$. The *improper integral*

$$\int_{-\infty}^\infty f(x)\, dx$$

is said to *converge* iff

$$\int_{-\infty}^0 f(x)\, dx \quad \text{and} \quad \int_0^\infty f(x)\, dx$$

both converge. We then set

$$\int_{-\infty}^\infty f(x)\, dx = l + m$$

where

$$\int_{-\infty}^0 f(x)\, dx = l \quad \text{and} \quad \int_0^\infty f(x)\, dx = m.$$

Improper integrals can also arise on bounded intervals. Suppose that f is continuous on the half-open interval $[a, b)$ but unbounded there. See Figure 11.6.3. If

$$\lim_{c \to b^-} \int_a^c f(x)\, dx = l \qquad (l \text{ finite}),$$

then we say that the *improper integral*

$$\int_a^b f(x)\, dx$$

converges to l. Otherwise, we say that the *improper integral diverges.* Similarly, functions continuous but unbounded on intervals of the form $(a, b]$ lead to limits of the form

$$\lim_{c \to a^+} \int_c^b f(x)\, dx$$

and thus to improper integrals

$$\int_a^b f(x)\, dx.$$

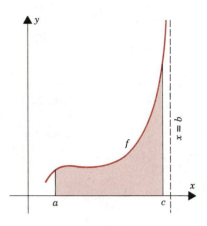

Figure 11.6.3

Example 6

(a) $\displaystyle\int_0^1 (1 - x)^{-2/3}\, dx = 3.$

(b) $\displaystyle\int_0^1 \frac{dx}{x}$ diverges.

Verification

(a) $\displaystyle\int_0^1 (1 - x)^{-2/3}\, dx = \lim_{c \to 1^-} \int_0^c (1 - x)^{-2/3}\, dx$

$$= \lim_{c \to 1^-} \left[-3(1 - x)^{1/3} \right]_0^c = \lim_{c \to 1^-} [-3(1 - c)^{1/3} + 3] = 3.$$

(b) $\displaystyle\int_0^1 \frac{dx}{x} = \lim_{c \to 0^+} \int_c^1 \frac{dx}{x} = \lim_{c \to 0^+} (-\ln c) = \infty.$ ▢

Now suppose that f is continuous on an interval $[a, b]$ except at some point c in (a, b) and that $f(x) \to \pm\infty$ as $x \to c^-$ or as $x \to c^+$. We say that the improper integral

$$\int_a^b f(x)\, dx$$

converges iff *both* of the improper integrals

$$\int_a^c f(x)\, dx \quad \text{and} \quad \int_c^b f(x)\, dx \quad \text{converge.}$$

Example 7. To evaluate

(*)
$$\int_1^4 \frac{dx}{(x-2)^2}$$

we need to calculate

$$\lim_{c \to 2^-} \int_1^c \frac{dx}{(x-2)^2} \quad \text{and} \quad \lim_{c \to 2^+} \int_c^4 \frac{dx}{(x-2)^2}.$$

As you can verify, neither of these limits exists and thus improper integral (*) diverges.

Notice that, if we ignore the fact that integral (*) is improper, then we are led to the *incorrect conclusion* that

$$\int_1^4 \frac{dx}{(x-2)^2} = \left[\frac{-1}{x-2}\right]_1^4 = -\frac{3}{2}. \quad \square$$

EXERCISES 11.6

Evaluate the improper integrals that converge.

1. $\int_1^\infty \frac{dx}{x^2}.$

2. $\int_0^\infty \frac{dx}{1+x^2}.$

3. $\int_0^\infty \frac{dx}{4+x^2}.$

4. $\int_0^\infty e^{-px}\,dx, \quad p > 0.$

5. $\int_0^\infty e^{px}\,dx, \quad p > 0.$

6. $\int_0^1 \frac{dx}{\sqrt{x}}.$

7. $\int_0^8 \frac{dx}{x^{2/3}}.$

8. $\int_0^1 \frac{dx}{x^2}.$

9. $\int_0^1 \frac{dx}{\sqrt{1-x^2}}.$

10. $\int_0^1 \frac{dx}{\sqrt{1-x}}.$

11. $\int_0^2 \frac{x}{\sqrt{4-x^2}}\,dx.$

12. $\int_0^a \frac{dx}{\sqrt{a^2-x^2}}.$

13. $\int_e^\infty \frac{\ln x}{x}\,dx.$

14. $\int_e^\infty \frac{dx}{x\ln x}.$

15. $\int_0^1 x\ln x\,dx.$

16. $\int_e^\infty \frac{dx}{x(\ln x)^2}.$

17. $\int_{-\infty}^\infty \frac{dx}{1+x^2}.$

18. $\int_2^\infty \frac{dx}{x^2-1}.$

19. $\int_{-\infty}^\infty \frac{dx}{x^2}.$

20. $\int_{1/3}^3 \frac{dx}{\sqrt[3]{3x-1}}.$

21. $\int_1^\infty \frac{dx}{x(x+1)}.$

22. $\int_{-\infty}^0 xe^x\,dx.$

23. $\int_3^5 \frac{x}{\sqrt{x^2-9}}\,dx.$

24. $\int_1^4 \frac{dx}{x^2-4}.$

25. $\int_{-3}^3 \frac{dx}{x(x+1)}.$

26. $\int_1^\infty \frac{x}{(1+x^2)^2}\,dx.$

27. $\int_{-3}^1 \frac{dx}{x^2-4}.$

28. $\int_0^\infty \sinh x\,dx.$

29. $\int_0^\infty \cosh x\,dx.$

30. $\int_1^4 \frac{dx}{x^2-5x+6}.$

31. Let Ω be the region under the curve $y = e^{-x}, x \geq 0$. (a) Sketch Ω. (b) Find the area of Ω. (c) Find the volume of the solid obtained by revolving Ω about the x-axis. (d) Find the volume obtained by revolving Ω about the y-axis. (e) Find the lateral surface area of the solid in part (c).

32. What point would you call the centroid of the region in Exercise 31? Does Pappus's theorem work in this instance?

33. Let Ω be the region under the curve $y = e^{-x^2}$, $x \geq 0$. (a) Show that Ω has finite area. (The area is actually $\frac{1}{2}\sqrt{\pi}$, as you will see in Chapter 17.) (b) Calculate the volume generated by revolving Ω about the y-axis.

34. Let Ω be the region bounded below by $y(x^2 + 1) = x$, above by $xy = 1$, and to the left by $x = 1$. (a) Find the area of Ω. (b) Show that the solid generated by revolving Ω about the x-axis has finite volume. (c) Calculate the volume generated by revolving Ω about the y-axis.

35. Let Ω be the region below the curve $y = x^{-1/4}, 0 < x \leq 1$. (a) Sketch Ω. (b) Find the area of Ω. (c) Find the volume of the solid obtained by revolving Ω about the x-axis. (d) Find the volume of the solid obtained by revolving Ω about the y-axis.

36. Prove the validity of the comparison test (11.6.2).

Use the comparison test (11.6.2) to determine which of the following integrals converge.

37. $\displaystyle\int_1^\infty \frac{x}{\sqrt{1 + x^5}}\, dx.$

38. $\displaystyle\int_1^\infty 2^{-x^2}\, dx.$

39. $\displaystyle\int_0^\infty (1 + x^5)^{-1/6}\, dx.$

40. $\displaystyle\int_\pi^\infty \frac{\sin^2 2x}{x^2}\, dx.$

41. $\displaystyle\int_1^\infty \frac{\ln x}{x^2}\, dx.$

42. $\displaystyle\int_e^\infty \frac{dx}{\sqrt{x + 1}\, \ln x}.$

43. Calculate the arc distance from the origin to the point $(x(\theta_1), y(\theta_1))$ along the exponential spiral $r = ae^{c\theta}$. (Take $a > 0, c > 0$.)

44. The function

$$f(x) = \frac{1}{\sqrt{2\pi}} \int_{-\infty}^{x} e^{-t^2/2}\, dt$$

is important in statistics. Prove that the integral on the right converges for all real x.

11.7 ADDITIONAL EXERCISES

State whether the indicated sequence converges, and if it does, find the limit.

1. $3^{1/n}.$

2. $n2^{1/n}.$

3. $\cos n\pi \sin n\pi.$

4. $\dfrac{(n + 1)(n + 2)}{(n + 3)(n + 4)}.$

5. $\left(\dfrac{n}{1 + n}\right)^{1/n}.$

6. $\dfrac{n^2 + 5n + 1}{n^3 + 1}.$

7. $\cos \dfrac{\pi}{n} \sin \dfrac{\pi}{n}.$

8. $\dfrac{n^\pi}{1 - n^\pi}.$

9. $\left(2 + \dfrac{1}{n}\right)^n.$

10. $\dfrac{\pi}{n} \cos \dfrac{\pi}{n}.$

11. $\dfrac{\ln [n(n + 1)]}{n}.$

12. $\left[\ln \left(1 + \dfrac{1}{n}\right)\right]^n.$

13. $\dfrac{\pi}{n} \ln \dfrac{n}{\pi}.$

14. $\dfrac{\pi}{n} e^{\pi/n}.$

15. $\dfrac{n}{\pi} \sin n\pi.$

16. $3 \ln 2n - \ln (n^3 + 1).$

17. $\displaystyle\int_n^{n+1} e^{-x}\, dx.$

18. $\displaystyle\int_1^n \frac{dx}{\sqrt{x}}.$

19. Show that, if $a_n \to l$, then $b_n = a_{n+1} \to l$.

20. If f is a function continuous for all real x, then, for each real number c, we can form the sequence

$$f(c), \quad f^2(c) = f(f(c)), \quad \ldots, \quad f^n(f^{n-1}(c)), \quad \ldots.$$

Show that, if $f^n(c) \to l$, then $f(l) = l$. [Thus l is a solution of the equation $f(x) = x$.]

21. (Calculator) Choose any real number a and form the sequence

$$\cos a, \quad \cos(\cos a), \quad \cos(\cos(\cos a)), \quad \ldots \ .$$

Convince yourself numerically that this sequence converges to some number l. Determine l and verify that $\cos l = l$. (This is an effective numerical method for solving the equation $\cos x = x$.)

22. (Calculator) Find a numerical solution to the equation $\sin(\cos x) = x$.   HINT: Use the method of Exercise 21.

23. (Calculator) Find a numerical solution to the equation $\cos(\sin x) = x$.

24. Evaluate

$$\int_0^a \ln \frac{1}{x}\,dx \qquad \text{for } a > 0.$$

25. (*Useful later*) Let f be a function, continuous, decreasing, and positive on $[1, \infty)$. Show that

$$\int_1^\infty f(x)\,dx \quad \text{converges} \qquad \text{iff} \qquad \text{the sequence } \left\{ \int_1^n f(x)\,dx \right\} \quad \text{converges.}$$

26. Find the length of the curve $y = (a^{2/3} - x^{2/3})^{3/2}$ from $x = 0$ to $x = a > 0$.

27. Let f be a function continuous on $(-\infty, \infty)$ and L a real number.
(a) Show that

$$\text{if} \quad \int_{-\infty}^\infty f(x)\,dx = L \qquad \text{then} \qquad \lim_{c \to \infty} \int_{-c}^c f(x)\,dx = L.$$

(b) Show that the converse of (a) is false by showing that

$$\lim_{c \to \infty} \int_{-c}^c x\,dx = 0 \qquad \text{but} \qquad \int_{-\infty}^\infty x\,dx \quad \text{diverges.}$$

28. Show that

$$\int_{-\infty}^\infty f(x)\,dx = L \qquad \text{iff} \qquad \lim_{c \to \infty} \int_{-c}^c f(x)\,dx = L$$

in the event that f is (a) nonnegative or (b) even.

The present value of a perpetual stream of revenues that flows continually at the rate of $R(t)$ dollars per year is given by the formula

$$P.V. = \int_0^\infty R(t)\,e^{-rt}\,dt,$$

where r is the rate of continuous compounding.

29. What is the present value of such a stream if $R(t)$ is constantly \$100 per year and r is 5%?

30. What is the present value if $R(t) = 100 + 60t$ dollars per year and r is 10%?

(Calculator) Evaluate the limit numerically and then prove that your answer is correct.

31. $\displaystyle \lim_{x \to \infty} \left\{ x^2 \left[\ln\left(1 + \frac{5}{x}\right) - \frac{5}{x} \right] \right\}.$ **32.** $\displaystyle \lim_{x \to \infty} \left[x^3 \left(e^{1/x} - 1 - \frac{1}{x} - \frac{1}{2x^2} \right) \right].$

33. A straight wire carrying a current I extends along the full length of the x-axis. P is an arbitrary point at perpendicular distance a from the wire. The contribution ΔB_i to the magnetic field B at P arising from a very short wire segment of length Δx_i is given approximately by

$$k_M I \frac{\sin \theta_i^*}{(r_i^*)^2}$$

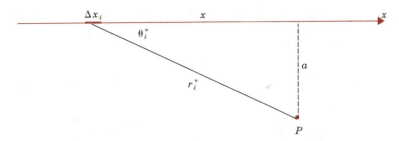

Figure 11.7.1

where θ_i^* and r_i^* are as in Figure 11.7.1. Calculate B at P.   HINT: Place the origin of the x-axis opposite to P. Express as an integral the contribution to B of an interval $[-c, c]$, and then take the limit as $c \to \infty$.

34. In any sample of gas molecules, some move faster than average, others slower than average. The number $N(v_1, v_2)$ of molecules that have their speeds within the interval $v_1 \le v \le v_2$ can be calculated from the Maxwell-Boltzmann formula:

$$N(v_1, v_2) = C \int_{v_1}^{v_2} v^2 e^{-\alpha v^2} \, dv.\dagger$$

Here C and α are positive constants. (The constant α can be expressed in terms of the mass of a molecule and the temperature of the sample by the equation $\alpha = m/2kT$; k is Boltzmann's constant.)

(a) $N = N(0, \infty)$ is the total number of molecules in the sample. Express C in terms of N and α.   HINT: Use the fact, proven in Section 17.5, that

$$\int_0^\infty e^{-x^2} \, dx = \tfrac{1}{2}\sqrt{\pi}.$$

(b) The *average speed* of a molecule, $\bar{v}$, can be found by forming a weighted average of all the possible speeds v using the "Maxwell-Boltzmann factor" $v^2 e^{-\alpha v^2}$ as a weighting factor:

$$N\bar{v} = C \int_0^\infty v \cdot v^2 e^{-\alpha v^2} \, dv.$$

Show that this implies that $\bar{v} = \sqrt{4kT/\pi m}$.

(c) The *average kinetic energy* of a molecule, $\overline{\tfrac{1}{2}mv^2}$, is given by a similar weighted average:

$$\overline{\tfrac{1}{2}mv^2} = \frac{C}{N} \int_0^\infty \tfrac{1}{2}mv^2 \cdot v^2 e^{-\alpha v^2} \, dv.$$

Show that $\overline{\tfrac{1}{2}mv^2} = \tfrac{3}{2}kT$. (Note that $\overline{\tfrac{1}{2}mv^2} = \tfrac{1}{2}m\overline{v^2}$ but this is not the same as the quantity $\tfrac{1}{2}m(\bar{v})^2$; they differ by a factor of $3\pi/8$.)

CHAPTER HIGHLIGHTS

11.1 Sequences of Real Numbers

---

† $N(v_1, v_2)$ must of course be an integer, but the integral on the right is seldom an integer. This formula (an important one) cannot be interpreted as a strict equality; it is only an approximation that holds extremely well for the large number of molecules in a typical gas sample.

It is often possible to obtain useful information about a sequence $y_n = f(n)$ by applying the techniques of calculus to the function $y = f(x)$.

11.2 The Limit of a Sequence

limit of a sequence (p. 576)   convergent, divergent (p. 577)

pinching theorem (p. 580) $\displaystyle\lim_{n\to\infty}\left(1+\frac{1}{n}\right)^n = e$

Every convergent sequence is bounded (p. 577); thus, every unbounded sequence is divergent.

A bounded, monotonic sequence converges. (p. 578)

Suppose that $c_n \to c$ and all the c_n are in the domain of f. If f is continuous at c, then $f(c_n) \to f(c)$. (p. 582)

11.3 Some Important Limits

for $x > 0$, $\displaystyle\lim_{n\to\infty} x^{1/n} = 1$ for $|x| < 1$, $\displaystyle\lim_{n\to\infty} x^n = 0$

for each real x, $\displaystyle\lim_{n\to\infty}\frac{x^n}{n!} = 0$ for $\alpha > 0$, $\displaystyle\lim_{n\to\infty}\frac{1}{n^\alpha} = 0$

$\displaystyle\lim_{n\to\infty}\frac{\ln n}{n} = 0$ $\displaystyle\lim_{n\to\infty} n^{1/n} = 1$

for each real x, $\displaystyle\lim_{n\to\infty}\left(1+\frac{x}{n}\right)^n = e^x$ Cauchy sequence (p. 590)

11.4 The Indeterminate Form (0/0)

L'Hospital's rule (0/0) (p. 591)   Cauchy mean-value theorem (p. 594)

If a product $f(x)g(x)$ gives rise to the indeterminate form $0 \cdot \infty$, then that same product written as $f(x)/(1/g(x))$ gives rise to the indeterminate form 0/0.

11.5 The Indeterminate Form (∞/∞)

L'Hospital's rule (∞/∞) (p. 597)   asymptotic graphs (p. 600)

$\displaystyle\lim_{x\to\infty}\frac{\ln x}{x} = 0$ $\displaystyle\lim_{x\to\infty}\frac{x^k}{e^x} = 0$ $\displaystyle\lim_{x\to 0^+} x^x = 1$

11.6 Improper Integrals

improper integrals (p. 601)   a comparison test (p. 603)

$\displaystyle\int_1^\infty \frac{dx}{x^p}$ converges for $p > 1$ and diverges for $p \le 1$

11.7 Additional Exercises

CHAPTER

12

INFINITE SERIES

12.1 SIGMA NOTATION

To indicate the sequence

$$\{1, \tfrac{1}{2}, \tfrac{1}{4}, \ldots\}$$

we can set $a_n = (\tfrac{1}{2})^{n-1}$ and write

$$\{a_1, a_2, a_3, \ldots\}.$$

We can also set $b_n = (\tfrac{1}{2})^n$ and write

$$\{b_0, b_1, b_2, \ldots\},$$

thereby beginning with the index 0. More generally, we can set $c_n = (\tfrac{1}{2})^{n-p}$ and write

$$\{c_p, c_{p+1}, c_{p+2}, \ldots\},$$

thereby beginning with the index p. In this chapter we often begin with an index other than 1.

The symbol Σ is the capital Greek letter "sigma." We write

(1)
$$\sum_{k=0}^{n} a_k$$

(read "the sum of the a sub k from k equals 0 to k equals n") to indicate the sum

$$a_0 + a_1 + \cdots + a_n.$$

More generally, if $n \geq m$, we write

(2)
$$\sum_{k=m}^{n} a_k$$

to indicate the sum

$$a_m + a_{m+1} + \cdots + a_n.$$

In (1) and (2) the letter "k" is being used as a "dummy" variable. That is, it can be replaced by any letter not already engaged. For instance,

$$\sum_{i=3}^{7} a_i, \qquad \sum_{j=3}^{7} a_j, \qquad \sum_{k=3}^{7} a_k$$

can all be used to indicate the sum

$$a_3 + a_4 + a_5 + a_6 + a_7.$$

Translating

$$(a_0 + \cdots + a_n) + (b_0 + \cdots + b_n) = (a_0 + b_0) + \cdots + (a_n + b_n),$$
$$\alpha(a_0 + \cdots + a_n) = \alpha a_0 + \cdots + \alpha a_n,$$
$$(a_0 + \cdots + a_m) + (a_{m+1} + \cdots + a_n) = a_0 + \cdots + a_n$$

into the Σ-notation, we have

$$\sum_{k=0}^{n} a_k + \sum_{k=0}^{n} b_k = \sum_{k=0}^{n} (a_k + b_k), \qquad \alpha \sum_{k=0}^{n} a_k = \sum_{k=0}^{n} \alpha a_k,$$
$$\sum_{k=0}^{m} a_k + \sum_{k=m+1}^{n} a_k = \sum_{k=0}^{n} a_k.$$

At times it is convenient to change indices. In this connection note that

$$\sum_{k=j}^{n} a_k = \sum_{i=0}^{n-j} a_{i+j}. \qquad \text{(set } i = k - j\text{)}$$

Both expressions are abbreviations for $a_j + a_{j+1} + \cdots + a_n$.

You can familiarize yourself further with this notation by doing the exercises below, but first one more remark. If all the a_k are equal to some fixed number x, then

$$\sum_{k=0}^{n} a_k \quad \text{can be written} \quad \sum_{k=0}^{n} x.$$

Obviously then

$$\sum_{k=0}^{n} x = \overbrace{x + x + \cdots + x}^{n+1} = (n+1)x.$$

In particular

$$\sum_{k=0}^{n} 1 = n + 1.$$

EXERCISES 12.1

Evaluate the following expressions.

1. $\sum_{k=0}^{2} (3k + 1).$ **2.** $\sum_{k=1}^{4} (3k - 1).$ **3.** $\sum_{k=0}^{3} 2^k.$ **4.** $\sum_{k=0}^{3} (-1)^k 2^k.$

5. $\sum_{k=0}^{3} (-1)^k 2^{k+1}.$ **6.** $\sum_{k=2}^{5} (-1)^{k+1} 2^{k-1}.$ **7.** $\sum_{k=1}^{4} \frac{1}{2^k}.$ **8.** $\sum_{k=2}^{5} \frac{1}{k!}.$

9. $\sum_{k=3}^{5} \frac{(-1)^k}{k!}.$ **10.** $\sum_{k=2}^{4} \frac{1}{3^{k-1}}.$ **11.** $\sum_{k=0}^{3} (\tfrac{1}{2})^{2k}.$ **12.** $\sum_{k=1}^{3} (-1)^{k+1}(\tfrac{1}{2})^{2k-1}.$

Express in sigma notation.

13. the lower sum $m_1 \Delta x_1 + m_2 \Delta x_2 + \cdots + m_n \Delta x_n.$
14. the upper sum $M_1 \Delta x_1 + M_2 \Delta x_2 + \cdots + M_n \Delta x_n.$
15. the Riemann sum $f(x_1^*) \Delta x_1 + f(x_2^*) \Delta x_2 + \cdots + f(x_n^*) \Delta x_n.$

16. $a^5 + a^4 b + a^3 b^2 + a^2 b^3 + ab^4 + b^5.$ **17.** $a^5 - a^4 b + a^3 b^2 - a^2 b^3 + ab^4 - b^5.$
18. $a^n + a^{n-1} b + \cdots + ab^{n-1} + b^n.$ **19.** $a_0 x^4 + a_1 x^3 + a_2 x^2 + a_3 x + a_4.$
20. $a_0 x^n + a_1 x^{n-1} + \cdots + a_{n-1} x + a_n.$ **21.** $1 - 2x + 3x^2 - 4x^3 + 5x^4.$
22. $3x - 4x^2 + 5x^3 - 6x^4.$

Write the following sums as $\sum_{k=3}^{10} a_k$ and $\sum_{i=0}^{7} a_{i+3}$:

23. $\dfrac{1}{2^3} + \dfrac{1}{2^4} + \cdots + \dfrac{1}{2^{10}}.$ **24.** $\dfrac{3^3}{3!} + \dfrac{4^4}{4!} + \cdots + \dfrac{10^{10}}{10!}.$

25. $\dfrac{3}{4} - \dfrac{4}{5} + \cdots - \dfrac{10}{11}.$ **26.** $\dfrac{1}{3} + \dfrac{1}{5} + \dfrac{1}{7} + \cdots + \dfrac{1}{17}.$

27. (a) (*Important*) Show that for $x \neq 1$

$$\sum_{k=0}^{n} x^k = \frac{1 - x^{n+1}}{1 - x}.$$

(b) Determine whether the sequence $a_n = \sum_{k=0}^{n} \dfrac{1}{3^k}$ converges and, if it does, find the limit.

28. Express $\sum_{k=1}^{n} \dfrac{a_k}{10^k}$ as a decimal fraction, given that each a_k is an integer from 0 to 9.

29. Let p be a positive integer. Show that, as $n \to \infty,$

$$a_n \to l \quad \text{iff} \quad a_{n-p} \to l.$$

30. Show that

$$\sum_{k=1}^{n} \frac{1}{\sqrt{k}} \geq \sqrt{n}.$$

Verify by induction.

31. $\sum_{k=1}^{n} k = \tfrac{1}{2}(n)(n + 1).$ **32.** $\sum_{k=1}^{n} (2k - 1) = n^2.$

33. $\sum_{k=1}^{n} k^2 = \tfrac{1}{6}(n)(n + 1)(2n + 1).$ **34.** $\sum_{k=1}^{n} k^3 = \left(\sum_{k=1}^{n} k \right)^2.$

12.2 INFINITE SERIES

Introduction; Definitions

While it is possible to add two numbers, three numbers, a hundred numbers, or even a million numbers, it is impossible to add an infinite number of numbers. The theory of *infinite series* arose from attempts to circumvent this impossibility.

To form an infinite series we begin with an infinite sequence of real numbers: a_0, $a_1, a_2, \ldots$. We can't form the sum of all the a_k (there is an infinite number of them), but we can form the *partial sums*

$$s_0 = a_0 = \sum_{k=0}^{0} a_k,$$

$$s_1 = a_0 + a_1 = \sum_{k=0}^{1} a_k,$$

$$s_2 = a_0 + a_1 + a_2 = \sum_{k=0}^{2} a_k,$$

$$s_3 = a_0 + a_1 + a_2 + a_3 = \sum_{k=0}^{3} a_k,$$

$$\vdots$$

$$s_n = a_0 + a_1 + a_2 + a_3 + \cdots + a_n = \sum_{k=0}^{n} a_k$$

$$\vdots$$

If the sequence $\{s_n\}$ of partial sums converges to a finite limit l, we write

$$\sum_{k=0}^{\infty} a_k = l$$

and say that

$$\text{the } \textit{series} \quad \sum_{k=0}^{\infty} a_k \quad \textit{converges to } l.$$

We call l the *sum* of the series. If the sequence of partial sums diverges, we say that

$$\text{the } \textit{series} \quad \sum_{k=0}^{\infty} a_k \quad \textit{diverges}.$$

Remark. It is important to note that the sum of a series is not a sum in the ordinary sense. It is a limit. ☐

Here are some examples.

Example 1. We begin with the series

$$\sum_{k=0}^{\infty} \frac{1}{(k+1)(k+2)}.$$

To determine whether or not this series converges we must examine the partial sums. Since

$$\frac{1}{(k+1)(k+2)} = \frac{1}{k+1} - \frac{1}{k+2},$$

you can see that

$$s_n = \frac{1}{1 \cdot 2} + \frac{1}{2 \cdot 3} + \cdots + \frac{1}{n(n+1)} + \frac{1}{(n+1)(n+2)}$$

$$= \left(\frac{1}{1} - \frac{1}{2}\right) + \left(\frac{1}{2} - \frac{1}{3}\right) + \cdots + \left(\frac{1}{n} - \frac{1}{n+1}\right) + \left(\frac{1}{n+1} - \frac{1}{n+2}\right)$$

$$= 1 - \frac{1}{2} + \frac{1}{2} - \frac{1}{3} + \cdots + \frac{1}{n} - \frac{1}{n+1} + \frac{1}{n+1} - \frac{1}{n+2}.$$

Since all but the first and last terms occur in pairs with opposite signs, the sum collapses to give

$$s_n = 1 - \frac{1}{n+2}.$$

Obviously, as $n \to \infty$, $s_n \to 1$. This means that the series converges to 1:

$$\sum_{k=0}^{\infty} \frac{1}{(k+1)(k+2)} = 1. \quad \square$$

Example 2. Here we examine two divergent series

$$\sum_{k=0}^{\infty} 2^k \quad \text{and} \quad \sum_{k=1}^{\infty} (-1)^k.$$

The partial sums of the first series take the form

$$s_n = \sum_{k=0}^{n} 2^k = 1 + 2 + \cdots + 2^n.$$

The sequence $\{s_n\}$ is unbounded and therefore divergent (11.2.5). This means that the series diverges.

For the second series we have

$$s_n = -1 \quad \text{if } n \text{ is odd} \quad \text{and} \quad s_n = 0 \quad \text{if } n \text{ is even}.$$

The sequence of partial sums looks like this:

$$-1, \quad 0, \quad -1, \quad 0, \quad -1, \quad 0, \quad \cdots.$$

The series diverges since the sequence of partial sums diverges. ▪

The Geometric Series

In school you studied geometric progression: $1,\ x,\ x^2,\ x^3,\ \cdots$. The sums

$$1,\quad 1 + x,\quad 1 + x + x^2,\quad 1 + x + x^2 + x^3,\ \cdots$$

generated by numbers in geometric progression are the partial sums of what is known as the *geometric series*:

$$\sum_{k=0}^{\infty} x^k.$$

The geometric series arises in so many contexts that it merits special attention.

The following result is fundamental:

(12.2.1)

> (i) if $|x| < 1$,   then $\displaystyle\sum_{k=0}^{\infty} x^k = \frac{1}{1-x}$;
>
> (ii) if $|x| \geq 1$,   then $\displaystyle\sum_{k=0}^{\infty} x^k$ diverges.

Proof.   The nth partial sum of the geometric series

$$\sum_{k=0}^{\infty} x^k$$

takes the form

$$(1) \qquad\qquad s_n = 1 + x + \cdots + x^n.$$

Multiplication by x gives

$$x s_n = x + x^2 + \cdots + x^{n+1}.$$

Subtracting the second equation from the first, we find that

$$(1 - x)s_n = 1 - x^{n+1}.$$

For $x \neq 1$, this gives

$$(2) \qquad\qquad s_n = \frac{1 - x^{n+1}}{1 - x}.$$

If $|x| < 1$, then $x^{n+1} \to 0$ and thus, by equation (2),

$$s_n \to \frac{1}{1 - x}.$$

This proves (i).

Now let's prove (ii).   For $x = 1$, we use equation (1) and deduce that $s_n = n + 1$. Obviously, $\{s_n\}$ diverges.   For $x \neq 1$ with $|x| \geq 1$, we use equation (2).   Since in this instance $\{x^{n+1}\}$ diverges, $\{s_n\}$ diverges. $\blacksquare$

You may have seen (12.2.1) before written as

$$a + ar + ar^2 + \cdots + ar^n + \cdots = \left\{ \begin{array}{ll} \dfrac{a}{1-r}, & |r| < 1 \\[2mm] \text{diverges}, & |r| \geq 1 \end{array} \right].$$

Taking $a = 1$ and $r = \frac{1}{2}$, we have

$$\sum_{k=0}^{\infty} \frac{1}{2^k} = \frac{1}{1 - \frac{1}{2}} = 2.$$

Begin the summation at $k = 1$ instead of at $k = 0$, and you see that

(12.2.2)
$$\sum_{k=1}^{\infty} \frac{1}{2^k} = 1.$$

The partial sums of this series

$$s_1 = \tfrac{1}{2},$$
$$s_2 = \tfrac{1}{2} + \tfrac{1}{4} = \tfrac{3}{4},$$
$$s_3 = \tfrac{1}{2} + \tfrac{1}{4} + \tfrac{1}{8} = \tfrac{7}{8},$$
$$s_4 = \tfrac{1}{2} + \tfrac{1}{4} + \tfrac{1}{8} + \tfrac{1}{16} = \tfrac{15}{16},$$
$$s_5 = \tfrac{1}{2} + \tfrac{1}{4} + \tfrac{1}{8} + \tfrac{1}{16} + \tfrac{1}{32} = \tfrac{31}{32},$$
etc.

are illustrated in Figure 12.2.1. Each new partial sum lies halfway between the previous partial sum and the number 1.

Figure 12.2.1

The convergence of the geometric series at $x = \frac{1}{10}$ enables us to assign a precise meaning to infinite decimals. Begin with the fact that

$$\sum_{k=0}^{\infty} \left(\frac{1}{10} \right)^k = \sum_{k=0}^{\infty} \frac{1}{10^k} = \frac{1}{1 - \frac{1}{10}} = \frac{10}{9}.$$

This gives

$$\sum_{k=1}^{\infty} \frac{1}{10^k} = \frac{1}{9}$$

and shows that the partial sums

$$s_n = \frac{1}{10} + \frac{1}{10^2} + \cdots + \frac{1}{10^n}$$

are all less than $\frac{1}{9}$. Now take a series of the form

$$\sum_{k=1}^{\infty} \frac{a_k}{10^k} \qquad \text{with} \quad a_k = 0, 1, \ldots, \text{or } 9.$$

Its partial sums

$$t_n = \frac{a_1}{10} + \frac{a_2}{10^2} + \cdots + \frac{a_n}{10^n}$$

are all less than 1:

$$t_n = \frac{a_1}{10} + \frac{a_2}{10^2} + \cdots + \frac{a_n}{10^n} \leq 9 \left(\frac{1}{10} + \frac{1}{10^2} + \cdots + \frac{1}{10^n} \right) = 9s_n < 9 \left(\frac{1}{9} \right) = 1.$$

Since $\{t_n\}$ is nondecreasing, as well as bounded above, $\{t_n\}$ is convergent; this means that the series

$$\sum_{k=1}^{\infty} \frac{a_k}{10^k}$$

is convergent.   The sum of this series is what we mean by the infinite decimal

$$0.a_1 a_2 a_3 \cdots a_n \cdots .$$

Below are two simple problems that lead naturally to the geometric series.   You will find more in the exercises.

Problem 3.   An electric fan is turned off, and the blades begin to lose speed.   Given that the blades turn N times during the first second of no power and lose at least $\sigma\%$ of their speed with the passing of each ensuing second, show that the blades cannot turn more than $100N\sigma^{-1}$ times after power shutdown.

Solution.   The number of turns during the first second of no power is

$$N.$$

The number of turns during the first 2 seconds is at most

$$N + \left(1 - \frac{1}{100} \sigma \right) N;$$

during the first 3 seconds, at most

$$N + \left(1 - \frac{1}{100} \sigma \right) N + \left(1 - \frac{1}{100} \sigma \right)^2 N;$$

and, during the first $n + 1$ seconds, at most

$$N \sum_{k=0}^{n} \left(1 - \frac{1}{100} \sigma \right)^k.$$

The total number of turns after power shutdown cannot exceed the limiting value

$$N \sum_{k=0}^{\infty} \left(1 - \frac{1}{100} \sigma \right)^k = N \left[\frac{1}{1 - (1 - \frac{1}{100}\sigma)} \right] = \frac{100N}{\sigma} = 100N\sigma^{-1}. \quad \square$$

Figure 12.2.2

Problem 4. According to Figure 12.2.2, it is 2 o'clock. At what time between 2 and 3 o'clock will the two hands coincide?

Solution. We will solve the problem by setting up a geometric series. We will then confirm our answer by approaching the problem from a different perspective.

The hour hand travels one-twelfth as fast as the minute hand. At 2 o'clock the minute hand points to 12 and the hour hand points to 2. By the time the minute hand reaches 2, the hour hand points to $2 + \frac{1}{6}$. By the time the minute hand reaches $2 + \frac{1}{6}$, the hour hand points to

$$2 + \frac{1}{6} + \frac{1}{6 \cdot 12}.$$

By the time the minute hand reaches

$$2 + \frac{1}{6} + \frac{1}{6 \cdot 12},$$

the hour hand points to

$$2 + \frac{1}{6} + \frac{1}{6 \cdot 12} + \frac{1}{6 \cdot 12^2}$$

and so on. In general, by the time the minute hand reaches

$$2 + \frac{1}{6} + \frac{1}{6 \cdot 12} + \cdots + \frac{1}{6 \cdot 12^{n-1}} = 2 + \frac{1}{6} \sum_{k=0}^{n-1} \frac{1}{12^k},$$

the hour hand points to

$$2 + \frac{1}{6} + \frac{1}{6 \cdot 12} + \cdots + \frac{1}{6 \cdot 12^{n-1}} + \frac{1}{6 \cdot 12^n} = 2 + \frac{1}{6} \sum_{k=0}^{n} \frac{1}{12^k}.$$

The two hands coincide when they both point to the limiting value

$$2 + \frac{1}{6} \sum_{k=0}^{\infty} \frac{1}{12^k} = 2 + \frac{1}{6} \left(\frac{1}{1 - \frac{1}{12}} \right) = 2 + \frac{2}{11}.$$

This happens at $2 + \frac{2}{11}$ o'clock, approximately 10 minutes and 55 seconds after 2.

We can confirm this as follows. Suppose that the two hands meet at hour $2 + x$. The hour hand will have moved a distance x and the minute hand a distance $2 + x$. Since the minute hand moves 12 times as fast as the hour hand, we have

$$12x = 2 + x, \qquad 11x = 2, \qquad \text{and thus} \qquad x = \frac{2}{11}. \quad \square$$

We will return to the geometric series later. Right now we turn our attention to series in general.

Some Basic Results

THEOREM 12.2.3

1. If $\displaystyle\sum_{k=0}^{\infty} a_k$ converges and $\displaystyle\sum_{k=0}^{\infty} b_k$ converges, then $\displaystyle\sum_{k=0}^{\infty} (a_k + b_k)$ converges.

    Moreover, if $\displaystyle\sum_{k=0}^{\infty} a_k = l$ and $\displaystyle\sum_{k=0}^{\infty} b_k = m$, then $\displaystyle\sum_{k=0}^{\infty} (a_k + b_k) = l + m$.

2. If $\displaystyle\sum_{k=0}^{\infty} a_k$ converges, then $\displaystyle\sum_{k=0}^{\infty} \alpha a_k$ converges for each real number α. More-

    over, if $\displaystyle\sum_{k=0}^{\infty} a_k = l$, then $\displaystyle\sum_{k=0}^{\infty} \alpha a_k = \alpha l$.

Proof. Let

$$s_n = \sum_{k=0}^{n} a_k, \qquad t_n = \sum_{k=0}^{n} b_k, \qquad u_n = \sum_{k=0}^{n} (a_k + b_k), \qquad v_n = \sum_{k=0}^{n} \alpha a_k.$$

Note that

$$u_n = s_n + t_n \qquad \text{and} \qquad v_n = \alpha s_n.$$

If $s_n \to l$ and $t_n \to m$, then

$$u_n \to l + m \qquad \text{and} \qquad v_n \to \alpha l. \quad \square$$

THEOREM 12.2.4

The *kth term* of a convergent series tends to 0; namely,

$$\text{if } \sum_{k=0}^{\infty} a_k \text{ converges}, \quad \text{then} \quad a_k \to 0.$$

Proof. To say that the series converges is to say that the sequence of partial sums converges to some number l:

$$s_n = \sum_{k=0}^{n} a_k \to l.$$

Obviously, then, $s_{n-1} \to l$. Since $a_n = s_n - s_{n-1}$, we have $a_n \to l - l = 0$. A change in notation gives $a_k \to 0$. $\quad \square$

The next result is an obvious, but important, consequence of Theorem 12.2.4.

THEOREM 12.2.5 A DIVERGENCE TEST

If $a_k \not\to 0$, then $\displaystyle\sum_{k=0}^{\infty} a_k$ diverges.

Example 5

(a) Since $\dfrac{k}{k+1} \not\to 0$, the series

$$\sum_{k=0}^{\infty} \frac{k}{k+1} = 0 + \frac{1}{2} + \frac{2}{3} + \frac{3}{4} + \frac{4}{5} + \cdots \quad \text{diverges.}$$

(b) Since $\sin k \not\to 0$, the series

$$\sum_{k=0}^{\infty} \sin k = \sin 0 + \sin 1 + \sin 2 + \sin 3 + \cdots \quad \text{diverges.} \quad \square$$

Caution. Theorem 12.2.4 does *not* say that, if $a_k \to 0$, then $\displaystyle\sum_{k=0}^{\infty} a_k$ converges. There are divergent series for which $a_k \to 0$.

Example 6. In the case of

$$\sum_{k=1}^{\infty} \frac{1}{\sqrt{k}} = \frac{1}{\sqrt{1}} + \frac{1}{\sqrt{2}} + \frac{1}{\sqrt{3}} + \frac{1}{\sqrt{4}} + \cdots$$

we have

$$a_k = \frac{1}{\sqrt{k}} \to 0,$$

but, since

$$s_n = \frac{1}{\sqrt{1}} + \frac{1}{\sqrt{2}} + \cdots + \frac{1}{\sqrt{n}} > \frac{n}{\sqrt{n}} = \sqrt{n},$$

the sequence of partial sums is unbounded, and therefore the series diverges. $\square$

EXERCISES 12.2

Find the sum of the series.

1. $\displaystyle\sum_{k=3}^{\infty} \frac{1}{(k+1)(k+2)}.$

2. $\displaystyle\sum_{k=0}^{\infty} \frac{1}{(k+3)(k+4)}.$

3. $\displaystyle\sum_{k=1}^{\infty} \frac{1}{2k(k+1)}.$

4. $\displaystyle\sum_{k=3}^{\infty} \frac{1}{k^2 - k}.$

5. $\displaystyle\sum_{k=1}^{\infty} \frac{1}{k(k+3)}.$

6. $\displaystyle\sum_{k=0}^{\infty} \frac{1}{(k+1)(k+3)}.$

7. $\displaystyle\sum_{k=0}^{\infty} \frac{3}{10^k}.$

8. $\displaystyle\sum_{k=0}^{\infty} \frac{12}{100^k}.$

9. $\displaystyle\sum_{k=0}^{\infty} \frac{67}{1000^k}.$

10. $\displaystyle\sum_{k=0}^{\infty} \frac{(-1)^k}{5^k}.$

11. $\displaystyle\sum_{k=0}^{\infty} \left(\frac{3}{4}\right)^k.$

12. $\displaystyle\sum_{k=0}^{\infty} \frac{3^k + 4^k}{5^k}.$

13. $\displaystyle\sum_{k=0}^{\infty} \frac{1 - 2^k}{3^k}.$

14. $\displaystyle\sum_{k=0}^{\infty} \left(\frac{25}{10^k} - \frac{6}{100^k}\right).$

15. $\displaystyle\sum_{k=3}^{\infty} \frac{1}{2^{k-1}}.$

16. $\displaystyle\sum_{k=0}^{\infty} \frac{1}{2^{k+3}}.$

17. $\displaystyle\sum_{k=0}^{\infty} \frac{2^{k+3}}{3^k}.$

18. $\displaystyle\sum_{k=2}^{\infty} \frac{3^{k-1}}{4^{3k+1}}.$

Write the decimal fraction as an infinite series and express the sum as the quotient of two integers.

19. $0.777\cdots.$

20. $0.999\cdots.$

21. $0.\overline{24}\overline{24}\cdots.$

22. $0.\overline{89}\overline{89}\cdots.$

23. $0.\overline{112}\overline{112}\overline{112}\cdots.$

24. $0.\overline{315}\overline{315}\overline{315}\cdots.$

25. $0.62\overline{45}\overline{45}\cdots.$

26. $0.112\overline{019}\overline{019}\cdots.$

27. Using series, show that every repeating decimal represents a rational number (the quotient of two integers).

28. Show that
$$\sum_{k=0}^{\infty} a_k = l \quad \text{iff} \quad \sum_{k=j+1}^{\infty} a_k = l - (a_0 + \cdots + a_j).$$

Derive these results from the geometric series.

29. $\displaystyle\sum_{k=0}^{\infty} (-1)^k x^k = \frac{1}{1+x}, \quad |x| < 1.$

30. $\displaystyle\sum_{k=0}^{\infty} (-1)^k x^{2k} = \frac{1}{1+x^2}, \quad |x| < 1.$

Find a series expansion for the given expression.

31. $\dfrac{x}{1-x}$ for $|x| < 1.$

32. $\dfrac{x}{1+x}$ for $|x| < 1.$

33. $\dfrac{x}{1+x^2}$ for $|x| < 1.$

34. $\dfrac{1}{4-x^2}$ for $|x| < 2.$

35. $\dfrac{1}{1+4x^2}$ for $|x| < \dfrac{1}{2}.$

36. $\dfrac{x^2}{1-x}$ for $|x| < 1.$

37. At some time between 4 and 5 o'clock the minute hand is directly above the hour hand. Express this time as a geometric series. What is the sum of this series?

38. Given that a ball dropped to the floor rebounds to a height proportional to the height from which it is dropped, find the total distance traveled by a ball dropped from a height of 6 feet if it rebounds initially to a height of 3 feet.

39. Exercise 38 under the supposition that the ball rebounds initially to a height of 2 feet.

40. In the setting of Exercise 38, to what height does the ball rebound initially if the total distance traveled by the ball is 21 feet?

41. How much money must you deposit at $r\%$ interest compounded annually to enable your descendants to withdraw n_1 dollars at the end of the first year, n_2 dollars at the end of the second year, n_3 dollars at the end of the third year, and so on in perpetuity? Assume that the sequence $\{n_k\}$ is bounded, $n_k \leq N$ for all k, and express your answer as an infinite series.

42. Sum the series you obtained in Exercise 41 setting
(a) $r = 5, \ n_k = 5000(\tfrac{1}{2})^{k-1}.$ (b) $r = 6, \ n_k = 1000(0.8)^{k-1}.$ (c) $r = 5, \ n_k = N.$

43. Show that
$$\sum_{k=1}^{\infty} \ln\left(\frac{k+1}{k}\right) \quad \text{diverges} \quad \text{even though} \quad \lim_{k\to\infty} \ln\left(\frac{k+1}{k}\right) = 0.$$

44. Show that

$$\sum_{k=1}^{\infty} \left(\frac{k+1}{k}\right)^k \quad \text{diverges.}$$

45. Let $\{d_k\}$ be a sequence of real numbers that converges to 0.
   (a) Show that

$$\sum_{k=1}^{\infty} (d_k - d_{k+1}) = d_1.$$

   (b) Sum the following series:

   (i) $\displaystyle\sum_{k=1}^{\infty} \frac{\sqrt{k+1} - \sqrt{k}}{\sqrt{k(k+1)}}.$ (ii) $\displaystyle\sum_{k=1}^{\infty} \frac{2k+1}{2k^2(k+1)^2}.$

46. Show that

$$\sum_{k=1}^{\infty} kx^{k-1} = \frac{1}{(1-x)^2} \quad \text{for } |x| < 1.$$

   HINT: Verify that s_n, the nth partial sum, satisfies the identity

$$(1-x)^2 s_n = 1 - (n+1)x^n + nx^{n+1}.$$

Program for summing a series   (BASIC)

This program can be used to sum a series in which there is a simple relation between successive terms, (as often happens in practice). In this instance, each term is half of the preceding one. By making a simple modification in the program you can sum other series. Try, for example, to generate a program for each of the following series:

$$\sum_{k=0}^{\infty} \left(\frac{3}{4}\right)^k, \quad \sum_{k=0}^{\infty} \frac{1}{k!}, \quad \sum_{k=1}^{\infty} \frac{1}{k}.$$

```
10 REM Sum a series
20 REM copyright © Colin C. Graham 1988-1989

30 REM change the three lines "firstterm = ...", "nextterm = ..."
40 REM    and "thisterm = ..." to fit your particular series

50 firstterm = 1

100 INPUT "Enter number of terms:"; n

200 sum = firstterm
210 thisterm = firstterm

300 FOR j = 1 TO n
310    PRINT  sum: REM omit this line if you don't want to see intermediate result
320    nextterm = thisterm/2
330    sum = sum + nextterm
340    thisterm = nextterm
350 NEXT j

400 PRINT sum

500 INPUT "Do it again? (Y/N)"; a$
510 IF a$ = "Y" OR a$ = "y" THEN 100
520 END
```

12.3   THE INTEGRAL TEST; COMPARISON THEOREMS

Here and in the next section we direct our attention to *series with nonnegative terms*: $a_k \geq 0$ for all k. For such series the following theorem is fundamental.

THEOREM 12.3.1

A series with nonnegative terms converges iff the sequence of partial sums is bounded.

Proof.   Assume that the series converges. Then the sequence of partial sums is convergent and therefore bounded (Theorem 11.2.4).

Suppose now that the sequence of partial sums is bounded. Since the terms are nonnegative, the sequence is nondecreasing. By being bounded and nondecreasing, the sequence of partial sums converges (Theorem 11.2.6). This means that the series converges. ☐

The convergence or divergence of a series can sometimes be deduced from the convergence or divergence of a closely related improper integral.

THEOREM 12.3.2   THE INTEGRAL TEST

If f is continuous, decreasing, and positive on $[1, \infty)$, then

$$\sum_{k=1}^{\infty} f(k) \quad \text{converges} \quad \text{iff} \quad \int_{1}^{\infty} f(x)\, dx \quad \text{converges.}$$

Proof.   In Exercise 25, Section 11.7, you were asked to show that with f continuous, decreasing, and positive on $[1, \infty)$

$$\int_{1}^{\infty} f(x)\, dx \quad \text{converges} \quad \text{iff} \quad \text{the sequence} \quad \left\{ \int_{1}^{n} f(x)\, dx \right\} \quad \text{converges.}$$

We assume this result and base our proof on the behavior of the sequence of integrals. To visualize our argument see Figure 12.3.1.

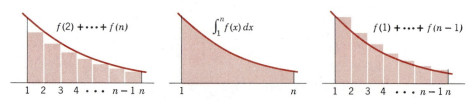

Figure 12.3.1

Since f decreases on the interval $[1, n]$,

$$f(2) + \cdots + f(n) \quad \text{is a lower sum for } f \text{ on } [1, n]$$

and

$$f(1) + \cdots + f(n - 1) \quad \text{is an upper sum for } f \text{ on } [1, n].$$

Consequently

$$f(2) + \cdots + f(n) \leq \int_1^n f(x)\, dx \quad \text{and} \quad \int_1^n f(x)\, dx \leq f(1) + \cdots + f(n - 1).$$

If the sequence of integrals converges, it is bounded. By the first inequality the sequence of partial sums is bounded and the series is therefore convergent.

Suppose now that the sequence of integrals diverges. Since f is positive, the sequence of integrals increases:

$$\int_1^n f(x)\, dx < \int_1^{n+1} f(x)\, dx.$$

Since this sequence diverges, it must be unbounded. By the second inequality, the sequence of partial sums must be unbounded and the series divergent. ◻

Applying the Integral Test

Example 1. (*The Harmonic Series*)

(12.3.3)

$$\sum_{k=1}^{\infty} \frac{1}{k} = 1 + \frac{1}{2} + \frac{1}{3} + \frac{1}{4} + \cdots \quad \text{diverges.}$$

Proof. The function $f(x) = 1/x$ is continuous, decreasing, and positive on $[1, \infty)$. We know that

$$\int_1^{\infty} \frac{dx}{x} \quad \text{diverges.} \tag{11.6.1}$$

By the integral test

$$\sum_{k=1}^{\infty} \frac{1}{k} \quad \text{diverges.} \quad ◻$$

The next example gives a more general result.

Example 2. (*The p-series*)

(12.3.4)

$$\sum_{k=1}^{\infty} \frac{1}{k^p} = 1 + \frac{1}{2^p} + \frac{1}{3^p} + \frac{1}{4^p} + \cdots \quad \text{converges} \quad \text{iff} \quad p > 1.$$

Proof. If $p \leq 0$, then each term of the series is greater than or equal to 1 and, by the divergence test (12.2.5), the series cannot converge. We assume therefore that $p > 0$. The function $f(x) = 1/x^p$ is then continuous, decreasing, and positive on $[1, \infty)$. Thus by the integral test

$$\sum_{k=1}^{\infty} \frac{1}{k^p} \quad \text{converges} \quad \text{iff} \quad \int_{1}^{\infty} \frac{dx}{x^p} \quad \text{converges}.$$

Earlier you saw that

$$\int_{1}^{\infty} \frac{dx}{x^p} \quad \text{converges} \quad \text{iff} \quad p > 1. \qquad (11.6.1)$$

It follows that

$$\sum_{k=1}^{\infty} \frac{1}{k^p} \quad \text{converges} \quad \text{iff} \quad p > 1. \quad \square$$

Example 3.   Here we show that the series

$$\sum_{k=1}^{\infty} \frac{1}{k \ln (k + 1)} = \frac{1}{\ln 2} + \frac{1}{2 \ln 3} + \frac{1}{3 \ln 4} + \cdots$$

diverges. We begin by setting

$$f(x) = \frac{1}{x \ln (x + 1)}.$$

Since f is continuous, decreasing, and positive on $[1, \infty)$, we can use the integral test. Note that

$$\int_{1}^{b} \frac{dx}{x \ln (x + 1)} > \int_{1}^{b} \frac{dx}{(x + 1) \ln (x + 1)}$$

$$= \left[\ln (\ln (x + 1)) \right]_{1}^{b} = \ln (\ln (b + 1)) - \ln (\ln 2).$$

As $b \to \infty$, $\ln (\ln (b + 1)) \to \infty$. This shows that

$$\int_{1}^{\infty} \frac{dx}{x \ln (x + 1)} \quad \text{diverges}.$$

It follows that the series diverges. $\square$

Remark on Notation.   You have seen that for each $j \geq 0$

$$\sum_{k=0}^{\infty} a_k \quad \text{converges} \quad \text{iff} \quad \sum_{k=j+1}^{\infty} a_k \quad \text{converges}$$

(Exercise 28, Section 12.2). This tells you that, in determining whether or not a series converges, it does not matter where we begin the summation.† Where detailed indexing would contribute nothing, we will omit it and write $\Sigma\, a_k$ without specifying where the summation begins. For instance, it makes sense to say that

$$\Sigma\, \frac{1}{k^2} \quad \text{converges} \quad \text{and} \quad \Sigma\, \frac{1}{k} \quad \text{diverges}$$

without specifying where we begin the summation. ☐

The convergence or divergence of a series with nonnegative terms is usually deduced by comparison with a series of known behavior.

THEOREM 12.3.5 THE BASIC COMPARISON THEOREM

Let $\Sigma\, a_k$ be a series with nonnegative terms.

  (i) $\Sigma\, a_k$ converges if there exists a convergent series $\Sigma\, c_k$ with nonnegative terms such that $a_k \leq c_k$ for all k sufficiently large;

 (ii) $\Sigma\, a_k$ diverges if there exists a divergent series $\Sigma\, d_k$ with nonnegative terms such that $d_k \leq a_k$ for all k sufficiently large.

Proof. The proof is just a matter of noting that, in the first instance, the partial sums of $\Sigma\, a_k$ are bounded and, in the second instance, unbounded. The details are left to you. ☐

Example 4

(a) $\Sigma\, \dfrac{1}{2k^3 + 1}$ converges by comparison with $\Sigma\, \dfrac{1}{k^3}$:

$$\frac{1}{2k^3 + 1} < \frac{1}{k^3} \quad \text{and} \quad \Sigma\, \frac{1}{k^3} \quad \text{converges.} \quad ☐$$

(b) $\Sigma\, \dfrac{1}{3k + 1}$ diverges by comparison with $\Sigma\, \dfrac{1}{3(k + 1)}$:

$$\frac{1}{3(k + 1)} < \frac{1}{3k + 1} \quad \text{and} \quad \Sigma\, \frac{1}{3(k + 1)} = \frac{1}{3} \Sigma\, \frac{1}{k + 1} \quad \text{diverges.} \quad ☐$$

(c) $\Sigma\, \dfrac{k^3}{k^5 + 5k^4 + 7}$ converges by comparison with $\Sigma\, \dfrac{1}{k^2}$:

$$\frac{k^3}{k^5 + 5k^4 + 7} < \frac{k^3}{k^5} = \frac{1}{k^2} \quad \text{and} \quad \Sigma\, \frac{1}{k^2} \quad \text{converges.} \quad ☐$$

---

† In the convergent case it does, however, affect the sum. Thus, for example,

$$\sum_{k=0}^{\infty} (\tfrac{1}{2})^k = 2, \quad \sum_{k=1}^{\infty} (\tfrac{1}{2})^k = 1, \quad \sum_{k=2}^{\infty} (\tfrac{1}{2})^k = \tfrac{1}{2}, \quad \text{and so forth.}$$

Problem 5. Show that

$$\Sigma \frac{1}{\ln{(k + 6)}} \quad \text{diverges.}$$

Solution. We know that as $k \to \infty$

$$\frac{\ln k}{k} \to 0. \qquad\qquad (11.3.5)$$

It follows that

$$\frac{\ln{(k + 6)}}{k + 6} \to 0$$

and thus that

$$\frac{\ln{(k + 6)}}{k} = \frac{\ln{(k + 6)}}{k + 6}\left(\frac{k + 6}{k}\right) \to 0.$$

Thus, for k sufficiently large,

$$\ln{(k + 6)} < k \quad\text{and}\quad \frac{1}{k} < \frac{1}{\ln{(k + 6)}}.$$

Since

$$\Sigma \frac{1}{k} \quad \text{diverges,}$$

we can conclude that

$$\Sigma \frac{1}{\ln{(k + 6)}} \quad \text{diverges.} \quad \square$$

Remark. Another way to show that $\ln{(k + 6)} < k$ for large k is to examine the function $f(x) = x - \ln{(x + 6)}$. At $x = 3$ the function is positive:

$$f(3) = 3 - \ln 9 \cong 3 - 2.197 > 0.$$

Since

$$f'(x) = 1 - \frac{1}{x + 6} > 0 \qquad \text{for all } x > 0,$$

$f(x) > 0$ for all $x \geq 3$. It follows that

$$\ln{(x + 6)} < x \qquad \text{for all } x \geq 3. \quad \square$$

We come now to a somewhat more sophisticated comparison theorem. Our proof relies on the basic comparison theorem.

THEOREM 12.3.6   THE LIMIT COMPARISON THEOREM

Let $\Sigma\, a_k$ and $\Sigma\, b_k$ be series with positive terms. If $a_k/b_k \to L$, where L is some positive number, then

$$\Sigma\, a_k \text{ converges} \qquad \text{iff} \qquad \Sigma\, b_k \text{ converges.}$$

Proof.   Choose ϵ between 0 and L. Since $a_k/b_k \to L$, we know that for all k sufficiently large (for all k greater than some k_0)

$$\left| \frac{a_k}{b_k} - L \right| < \epsilon.$$

For such k we have

$$L - \epsilon < \frac{a_k}{b_k} < L + \epsilon$$

and thus

$$(L - \epsilon)b_k < a_k < (L + \epsilon)b_k.$$

This last inequality is what we needed:

if $\Sigma\, a_k$ converges, then $\Sigma\, (L - \epsilon)b_k$ converges, and thus $\Sigma\, b_k$ converges;

if $\Sigma\, b_k$ converges, then $\Sigma\, (L + \epsilon)b_k$ converges, and thus $\Sigma\, a_k$ converges. ☐

To apply the limit comparison theorem to a series $\Sigma\, a_k$, we must first find a series $\Sigma\, b_k$ of known behavior for which a_k/b_k converges to a positive number.

Problem 6.   Determine whether the series

$$\Sigma\, \frac{3k^2 + 2k + 1}{k^3 + 1}$$

converges or diverges.

Solution.   For large values of k, the terms with the highest powers of k dominate. Here $3k^2$ dominates the numerator and k^3 dominates the denominator. Thus

$$\frac{3k^2 + 2k + 1}{k^3 + 1} \quad \text{differs little from} \quad \frac{3k^2}{k^3} = \frac{3}{k}.$$

Since

$$\frac{3k^2 + 2k + 1}{k^3 + 1} \div \frac{3}{k} = \frac{3k^3 + 2k^2 + k}{3k^3 + 3} = \frac{1 + 2/3k + 1/3k^2}{1 + 1/k^3} \to 1$$

and

$$\Sigma\, \frac{3}{k} = 3\Sigma\, \frac{1}{k} \quad \text{diverges,}$$

we know that the series diverges. ☐

Problem 7. Determine whether the series

$$\Sigma \; \frac{5\sqrt{k} + 100}{2k^2\sqrt{k} + 9\sqrt{k}}$$

converges or diverges.

Solution. For large values of k, $5\sqrt{k}$ dominates the numerator and $2k^2\sqrt{k}$ dominates the denominator. Thus, for such k,

$$\frac{5\sqrt{k} + 100}{2k^2\sqrt{k} + 9\sqrt{k}} \quad \text{differs little from} \quad \frac{5\sqrt{k}}{2k^2\sqrt{k}} = \frac{5}{2k^2}.$$

Since

$$\frac{5\sqrt{k} + 100}{2k^2\sqrt{k} + 9\sqrt{k}} \div \frac{5}{2k^2} = \frac{10k^2\sqrt{k} + 200k^2}{10k^2\sqrt{k} + 45\sqrt{k}} = \frac{1 + 20/\sqrt{k}}{1 + 9/2k^2} \to 1$$

and

$$\Sigma \; \frac{5}{2k^2} = \frac{5}{2} \, \Sigma \; \frac{1}{k^2} \quad \text{converges,}$$

the series converges.   ☐

Problem 8. Determine whether the series

$$\Sigma \; \sin \frac{\pi}{k}$$

converges or diverges.

Solution. Recall that

$$\text{as } x \to 0, \quad \frac{\sin x}{x} \to 1. \tag{2.5.5}$$

As $k \to \infty$, $\pi/k \to 0$ and thus

$$\frac{\sin \pi/k}{\pi/k} \to 1.$$

Since $\Sigma \, \pi/k$ diverges, $\Sigma \, \sin \pi/k$ diverges.   ☐

Remark. The question of what we can and cannot conclude by limit comparison if $a_k/b_k \to 0$ is taken up in Exercise 31.   ☐

EXERCISES 12.3

Determine whether the series converges or diverges.

1. $\Sigma \dfrac{k}{k^3 + 1}$.

2. $\Sigma \dfrac{1}{3k + 2}$.

3. $\Sigma \dfrac{1}{(2k + 1)^2}$.

4. $\Sigma \dfrac{\ln k}{k}$.

5. $\Sigma \dfrac{1}{\sqrt{k + 1}}$.

6. $\Sigma \dfrac{1}{k^2 + 1}$.

7. $\Sigma \dfrac{1}{\sqrt{2k^2 - k}}$.

8. $\Sigma \left(\dfrac{5}{2}\right)^{-k}$.

9. $\Sigma \dfrac{\tan^{-1} k}{1 + k^2}$.

10. $\Sigma \dfrac{\ln k}{k^3}$.

11. $\Sigma \dfrac{1}{k^{2/3}}$.

12. $\Sigma \dfrac{1}{(k + 1)(k + 2)(k + 3)}$.

13. $\Sigma \left(\dfrac{3}{4}\right)^{-k}$.

14. $\Sigma \dfrac{1}{1 + 2 \ln k}$.

15. $\Sigma \dfrac{\ln \sqrt{k}}{k}$.

16. $\Sigma \dfrac{2}{k \, (\ln k)^2}$.

17. $\Sigma \dfrac{1}{2 + 3^{-k}}$.

18. $\Sigma \dfrac{7k + 2}{2k^5 + 7}$.

19. $\Sigma \dfrac{2k + 5}{5k^3 + 3k^2}$.

20. $\Sigma \dfrac{k^4 - 1}{3k^2 + 5}$.

21. $\Sigma \dfrac{1}{k \ln k}$.

22. $\Sigma \dfrac{1}{2^{k+1} - 1}$.

23. $\Sigma \dfrac{k^2}{k^4 - k^3 + 1}$.

24. $\Sigma \dfrac{k^{3/2}}{k^{5/2} + 2k - 1}$.

25. $\Sigma \dfrac{2k + 1}{\sqrt{k^4 + 1}}$.

26. $\Sigma \dfrac{2k + 1}{\sqrt{k^3 + 1}}$.

27. $\Sigma \dfrac{2k + 1}{\sqrt{k^5 + 1}}$.

28. $\Sigma \dfrac{1}{\sqrt{2k(k + 1)}}$.

29. $\Sigma \, ke^{-k^2}$.

30. $\Sigma \, k^2 2^{-k^3}$.

31. Let $\Sigma \, a_k$ and $\Sigma \, b_k$ be series with positive terms and suppose that $a_k/b_k \to 0$.
   (a) Show that, if $\Sigma \, b_k$ converges, then $\Sigma \, a_k$ converges.
   (b) Show that, if $\Sigma \, a_k$ diverges, then $\Sigma \, b_k$ diverges.
   (c) Show by example that, if $\Sigma \, a_k$ converges, then $\Sigma \, b_k$ may converge or diverge.
   (d) Show by example that, if $\Sigma \, b_k$ diverges, then $\Sigma \, a_k$ may converge or diverge.
   [Parts (c) and (d) explain why we stipulated $L > 0$ in Theorem 12.3.6.]

32. All the results of this section were stated for series with nonnegative terms. Corresponding results hold for *series with nonpositive terms*: $a_k \le 0$ for all k.
   (a) State a comparison theorem analogous to Theorem 12.3.5, this time for series with nonpositive terms.
   (b) As stated, the integral test (Theorem 12.3.2) applies only to series with positive terms. State the equivalent result for series with negative terms.

33. This exercise demonstrates that we cannot always use the same testing series for both the basic comparison test and the limit comparison test.
   (a) Show that

$$\Sigma \, \frac{\ln n}{n \sqrt{n}} \quad \text{converges by comparison with} \quad \Sigma \, \frac{1}{n^{5/4}}.$$

   (b) Show that the limit comparison test does not apply.

12.4   THE ROOT TEST; THE RATIO TEST

Comparison with geometric series

$$\Sigma \, x^k$$

and with the *p*-series

$$\Sigma \, \frac{1}{k^p}$$

leads to two important tests for convergence: the root test and the ratio test.

THEOREM 12.4.1   THE ROOT TEST

Let $\Sigma \, a_k$ be a series with nonnegative terms and suppose that

$$(a_k)^{1/k} \to \rho.$$

If $\rho < 1$, $\Sigma \, a_k$ converges.   If $\rho > 1$, $\Sigma \, a_k$ diverges.   If $\rho = 1$, the test is inconclusive.

Proof. We suppose first that $\rho < 1$ and choose μ so that

$$\rho < \mu < 1.$$

Since $(a_k)^{1/k} \to \rho$, we have

$$(a_k)^{1/k} < \mu \qquad \text{for all } k \text{ sufficiently large.} \qquad \text{(explain)}$$

Thus

$$a_k < \mu^k \qquad \text{for all } k \text{ sufficiently large.}$$

Since $\Sigma \, \mu^k$ converges (a geometric series with $0 < \mu < 1$), we know by the basic comparison theorem that $\Sigma \, a_k$ converges.

We suppose now that $\rho > 1$ and choose μ so that

$$\rho > \mu > 1.$$

Since $(a_k)^{1/k} \to \rho$, we have

$$(a_k)^{1/k} > \mu \qquad \text{for all } k \text{ sufficiently large.} \qquad \text{(explain)}$$

Thus

$$a_k > \mu^k \qquad \text{for all } k \text{ sufficiently large.}$$

Since $\Sigma \, \mu^k$ diverges (a geometric series with $\mu > 1$), the basic comparison theorem tells us that $\Sigma \, a_k$ diverges.

To see the inconclusiveness of the root test when $\rho = 1$, note that $(a_k)^{1/k} \to 1$ for both $\Sigma\, 1/k^2$ and $\Sigma\, 1/k$.† The first series converges, but the second diverges.   ☐

Applying the Root Test

Example 1. For the series

$$\Sigma \frac{1}{(\ln k)^k}$$

we have

$$(a_k)^{1/k} = \frac{1}{\ln k} \to 0.$$

The series converges.   ☐

Example 2. For the series

$$\Sigma \frac{2^k}{k^3}$$

we have

$$(a_k)^{1/k} = 2 \left(\frac{1}{k}\right)^{3/k} = 2 \left[\left(\frac{1}{k}\right)^{1/k}\right]^3 \to 2 \cdot 1^3 = 2.$$

The series diverges.   ☐

Example 3. In the case of

$$\Sigma \left(1 - \frac{1}{k}\right)^k,$$

we have

$$(a_k)^{1/k} = 1 - \frac{1}{k} \to 1.$$

Here the root test is inconclusive. It is also unnecessary: since $a_k = (1 - 1/k)^k$ converges to $1/e$ and not to 0 (11.3.7), the series diverges (12.2.5).   ☐

----

† In the first instance

$$(a_k)^{1/k} = \left(\frac{1}{k^2}\right)^{1/k} = \left(\frac{1}{k^{1/k}}\right)^2 \to 1^2 = 1;$$

in the second instance

$$(a_k)^{1/k} = \left(\frac{1}{k}\right)^{1/k} = \frac{1}{k^{1/k}} \to 1.$$

THEOREM 12.4.2   THE RATIO TEST

Let $\Sigma \, a_k$ be a series with positive terms and suppose that

$$\frac{a_{k+1}}{a_k} \to \lambda.$$

If $\lambda < 1$, $\Sigma \, a_k$ converges.   If $\lambda > 1$, $\Sigma \, a_k$ diverges.   If $\lambda = 1$, the test is inconclusive.

Proof. We suppose first that $\lambda < 1$ and choose μ so that $\lambda < \mu < 1$.   Since

$$\frac{a_{k+1}}{a_k} \to \lambda,$$

we know that there exists $k_0 > 0$ such that

$$\text{if } k \geq k_0, \quad \text{then} \quad \frac{a_{k+1}}{a_k} < \mu. \qquad \text{(explain)}$$

This gives

$$a_{k_0+1} < \mu a_{k_0}, \qquad a_{k_0+2} < \mu a_{k_0+1} < \mu^2 a_{k_0},$$

and more generally,

$$a_{k_0+j} < \mu^j a_{k_0}.$$

For $k > k_0$ we have

(1)
$$a_k < \mu^{k-k_0} a_{k_0} = \frac{a_{k_0}}{\mu^{k_0}} \mu^k.$$

$$\underline{\qquad} \text{ set } j = k - k_0$$

Since $\mu < 1$,

$$\Sigma \frac{a_{k_0}}{\mu^{k_0}} \mu^k = \frac{a_{k_0}}{\mu^{k_0}} \Sigma \mu^k \quad \text{converges.}$$

Recalling (1), you can see by the basic comparison theorem that $\Sigma \, a_k$ converges.   The proof of the rest of the theorem is left to the exercises.   ☐

Remark. Contrary to some people's intuition the root and ratio tests are *not* equivalent.   See Exercise 36.   ☐

Applying the Ratio Test

Example 4.   The ratio test shows that the series

$$\Sigma \frac{1}{k!}$$

converges:

$$\frac{a_{k+1}}{a_k} = \frac{1}{(k+1)!} \cdot \frac{k!}{1} = \frac{1}{k+1} \to 0. \quad \square$$

Example 5. For the series

$$\Sigma \frac{k}{10^k}$$

we have

$$\frac{a_{k+1}}{a_k} = \frac{k+1}{10^{k+1}} \cdot \frac{10^k}{k} = \frac{1}{10} \frac{k+1}{k} \to \frac{1}{10}.$$

The series converges.† □

Example 6. For the series

$$\Sigma \frac{k^k}{k!}$$

we have

$$\frac{a_{k+1}}{a_k} = \frac{(k+1)^{k+1}}{(k+1)!} \cdot \frac{k!}{k^k} = \left(\frac{k+1}{k}\right)^k = \left(1 + \frac{1}{k}\right)^k \to e.$$

Since $e > 1$, the series diverges. □

Example 7. For the series

$$\Sigma \frac{1}{2k+1}$$

the ratio test is inconclusive:

$$\frac{a_{k+1}}{a_k} = \frac{1}{2(k+1)+1} \cdot \frac{2k+1}{1} = \frac{2k+1}{2k+3} = \frac{2+1/k}{2+3/k} \to 1.$$

Therefore, we have to look further. Comparison with the harmonic series shows that the series diverges:

$$\frac{1}{2k+1} > \frac{1}{2(k+1)} \quad \text{and} \quad \Sigma \frac{1}{2(k+1)} \quad \text{diverges.} \quad \square$$

---

† This series can be summed explicitly. See Exercise 33.

Summary on Convergence Tests

In general, the root test is used only if powers are involved. The ratio test is particularly effective with factorials and with combinations of powers and factorials. If the terms are rational functions of k, the ratio test is inconclusive and the root test is difficult to apply. Rational terms are most easily handled by comparison or limit comparison with a p-series, $\Sigma\, 1/k^p$. If the terms have the configuration of a derivative, you may be able to apply the integral test. Finally, keep in mind that, if $a_k \not\to 0$, then there is no reason to apply any special convergence test. The series diverges. (12.2.5)

EXERCISES 12.4

Determine whether the series converges or diverges.

1. $\Sigma\, \dfrac{10^k}{k!}$.

2. $\Sigma\, \dfrac{1}{k2^k}$.

3. $\Sigma\, \dfrac{1}{k^k}$.

4. $\Sigma\, \left(\dfrac{k}{2k+1}\right)^k$.

5. $\Sigma\, \dfrac{k!}{100^k}$.

6. $\Sigma\, \dfrac{(\ln k)^2}{k}$.

7. $\Sigma\, \dfrac{k^2+2}{k^3+6k}$.

8. $\Sigma\, \dfrac{1}{(\ln k)^k}$.

9. $\Sigma\, k\left(\dfrac{2}{3}\right)^k$.

10. $\Sigma\, \dfrac{1}{(\ln k)^{10}}$.

11. $\Sigma\, \dfrac{1}{1+\sqrt{k}}$.

12. $\Sigma\, \dfrac{2k+\sqrt{k}}{k^3+\sqrt{k}}$.

13. $\Sigma\, \dfrac{k!}{10^{4k}}$.

14. $\Sigma\, \dfrac{k^2}{e^k}$.

15. $\Sigma\, \dfrac{\sqrt{k}}{k^2+1}$.

16. $\Sigma\, \dfrac{2^k k!}{k^k}$.

17. $\Sigma\, \dfrac{k!}{(k+2)!}$.

18. $\Sigma\, \dfrac{1}{k}\left(\dfrac{1}{\ln k}\right)^{3/2}$.

19. $\Sigma\, \dfrac{1}{k}\left(\dfrac{1}{\ln k}\right)^{1/2}$.

20. $\Sigma\, \dfrac{1}{\sqrt{k^3-1}}$.

21. $\Sigma\, \left(\dfrac{k}{k+100}\right)^k$.

22. $\Sigma\, \dfrac{(k!)^2}{(2k)!}$.

23. $\Sigma\, k^{-(1+1/k)}$.

24. $\Sigma\, \dfrac{11}{1+100^{-k}}$.

25. $\Sigma\, \dfrac{\ln k}{e^k}$.

26. $\Sigma\, \dfrac{k!}{k^k}$.

27. $\Sigma\, \dfrac{\ln k}{k^2}$.

28. $\Sigma\, \dfrac{k!}{1\cdot 3\cdot\,\cdots\,\cdot(2k-1)}$.

29. $\Sigma\, \dfrac{2\cdot 4\cdot\,\cdots\,\cdot 2k}{(2k)!}$.

30. $\Sigma\, \dfrac{(2k+1)^{2k}}{(5k^2+1)^k}$.

31. $\Sigma\, \dfrac{k!(2k)!}{(3k)!}$.

32. $\Sigma\, \dfrac{\ln k}{k^{5/4}}$.

33. Find the sum of the series $\frac{1}{10} + \frac{2}{100} + \frac{3}{1000} + \frac{4}{10000} + \cdots$. HINT: Exercise 46 of Section 12.2.

34. Complete the proof of the ratio test.
    (a) Prove that, if $\lambda > 1$, then $\Sigma\, a_k$ diverges.
    (b) Prove that, if $\lambda = 1$, the ratio test is inconclusive.
        HINT: Consider $\Sigma\, 1/k$ and $\Sigma\, 1/k^2$.

35. Let $\{a_k\}$ be a sequence of positive numbers, and take $r > 0$. Use the root test to show that, if $(a_k)^{1/k} \to \rho$ and $\rho < 1/r$, then $\Sigma\, a_k r^k$ converges.

36. Consider the series $\frac{1}{2} + 1 + \frac{1}{8} + \frac{1}{4} + \frac{1}{32} + \frac{1}{16} + \cdots$ formed by rearranging a convergent geometric series. (a) Use the root test to show that the series converges. (b) Show that the ratio test does not apply.

12.5   ABSOLUTE AND CONDITIONAL CONVERGENCE; ALTERNATING SERIES

In this section we consider series that have both positive and negative terms.

Absolute and Conditional Convergence

Let $\Sigma\, a_k$ be a series with both positive and negative terms. One way to show that $\Sigma\, a_k$ converges is to show that $\Sigma\, |a_k|$ converges.

THEOREM 12.5.1

$$\text{If } \Sigma\, |a_k| \text{ converges, } \quad \text{then } \Sigma\, a_k \text{ converges.}$$

Proof. For each k,

$$-|a_k| \le a_k \le |a_k| \quad \text{and therefore} \quad 0 \le a_k + |a_k| \le 2|a_k|.$$

If $\Sigma\, |a_k|$ converges, then $\Sigma\, 2|a_k| = 2\,\Sigma\, |a_k|$ converges, and therefore, by the basic comparison theorem, $\Sigma\, (a_k + |a_k|)$ converges. Since

$$a_k = (a_k + |a_k|) - |a_k|,$$

we can conclude that $\Sigma\, a_k$ converges. $\square$

Series $\Sigma\, a_k$ for which $\Sigma\, |a_k|$ converge are called *absolutely convergent*. The theorem we just proved says that

(12.5.2)   absolutely convergent series are convergent.

As we will show presently, the converse is false. There are convergent series that are not absolutely convergent. Such series are called *conditionally convergent*.

Example 1

$$1 - \frac{1}{2^2} + \frac{1}{3^2} - \frac{1}{4^2} + \frac{1}{5^2} - \frac{1}{6^2} + \cdots .$$

If we replace each term by its absolute value, we obtain the series

$$1 + \frac{1}{2^2} + \frac{1}{3^2} + \frac{1}{4^2} + \frac{1}{5^2} + \frac{1}{6^2} + \cdots .$$

This is a p-series with $p = 2$. It is therefore convergent. This means that the initial series is absolutely convergent. $\square$

Example 2

$$1 - \frac{1}{2} - \frac{1}{2^2} + \frac{1}{2^3} - \frac{1}{2^4} - \frac{1}{2^5} + \frac{1}{2^6} - \frac{1}{2^7} - \frac{1}{2^8} + \cdots .$$

If we replace each term by its absolute value, we obtain the series

$$1 + \frac{1}{2} + \frac{1}{2^2} + \frac{1}{2^3} + \frac{1}{2^4} + \frac{1}{2^5} + \frac{1}{2^6} + \frac{1}{2^7} + \frac{1}{2^8} + \cdots .$$

This is a convergent geometric series.   The initial series is therefore absolutely convergent.   ☐

Example 3

$$1 - \frac{1}{2} + \frac{1}{3} - \frac{1}{4} + \frac{1}{5} - \frac{1}{6} + \cdots .$$

This series is only conditionally convergent.   It is convergent (see the next theorem), but it is not absolutely convergent: if we replace each term by its absolute value, we obtain the divergent harmonic series

$$1 + \frac{1}{2} + \frac{1}{3} + \frac{1}{4} + \frac{1}{5} + \frac{1}{6} + \cdots .\dagger \quad ☐$$

Example 4

$$\frac{1}{2} - 1 + \frac{1}{3} - 1 + \frac{1}{4} - 1 + \frac{1}{5} - 1 + \cdots .$$

Here the terms do not tend to zero.   The series is divergent.   ☐

Alternating Series

Series in which the consecutive terms have opposite signs are called *alternating series*.   Here are two examples:

$$1 - \frac{1}{2} + \frac{1}{3} - \frac{1}{4} + \frac{1}{5} - \frac{1}{6} + \cdots , \qquad 1 - \frac{1}{\sqrt{2}} + \frac{1}{\sqrt{3}} - \frac{1}{\sqrt{4}} + \frac{1}{\sqrt{5}} - \frac{1}{\sqrt{6}} + \cdots .$$

―――――

† In Section 12.6 we will show that the original series,

$$1 - \tfrac{1}{2} + \tfrac{1}{3} - \tfrac{1}{4} + \tfrac{1}{5} - \tfrac{1}{6} + \cdots ,$$

converges to ln 2.

The series

$$1 - \frac{1}{2} - \frac{1}{3} + \frac{1}{4} - \frac{1}{5} - \frac{1}{6} + \cdots$$

is not an alternating series because there are consecutive terms with the same sign.

THEOREM 12.5.3 ALTERNATING SERIES†

Let $\{a_k\}$ be a decreasing sequence of positive numbers.

$$\text{If} \quad a_k \to 0, \quad \text{then} \quad \sum_{k=0}^{\infty} (-1)^k a_k \quad \text{converges.}$$

Proof. First we look at the even partial sums, s_{2m}. Since

$$s_{2m} = (a_0 - a_1) + (a_2 - a_3) + \cdots + (a_{2m-2} - a_{2m-1}) + a_{2m}$$

is the sum of positive numbers, the even partial sums are all positive. Since

$$s_{2m+2} = s_{2m} - (a_{2m+1} - a_{2m+2}) \quad \text{and} \quad a_{2m+1} - a_{2m+2} > 0,$$

we have

$$s_{2m+2} < s_{2m}.$$

This means that the sequence of even partial sums is decreasing. Being bounded below by 0, it is convergent; say,

$$s_{2m} \to l.$$

Now

$$s_{2m+1} = s_{2m} - a_{2m+1}.$$

Since $a_{2m+1} \to 0$, we also have

$$s_{2m+1} \to l.$$

Since both the even and the odd partial sums tend to l, the sequence of all partial sums tends to l (Exercise 42, Section 11.2). ☐

From this theorem you can see that the following series all converge:

$$1 - \frac{1}{2} + \frac{1}{3} - \frac{1}{4} + \frac{1}{5} - \frac{1}{6} + \cdots, \qquad 1 - \frac{1}{\sqrt{2}} + \frac{1}{\sqrt{3}} - \frac{1}{\sqrt{4}} + \frac{1}{\sqrt{5}} - \frac{1}{\sqrt{6}} + \cdots,$$

$$1 - \frac{1}{2!} + \frac{1}{3!} - \frac{1}{4!} + \frac{1}{5!} - \frac{1}{6!} + \cdots.$$

The first two series converge only conditionally; the third converges absolutely.

---

† This theorem dates back to Leibniz. He observed the result in 1705.

An Estimate for Alternating Series You have seen that if $\{a_k\}$ is a decreasing sequence of positive numbers that tends to 0, then

$$\sum_{k=0}^{\infty} (-1)^k a_k \quad \text{converges to a sum } l.$$

(12.5.4)
> This sum l of an alternating series lies between consecutive partial sums s_n, s_{n+1}, and thus s_n approximates l to within a_{n+1}:
>
> $$|s_n - l| < a_{n+1}.$$

Proof. For all n

$$a_{n+1} > a_{n+2}.$$

If n is odd,

$$s_{n+2} = s_n + a_{n+1} - a_{n+2} > s_n;$$

if n is even,

$$s_{n+2} = s_n - a_{n+1} + a_{n+2} < s_n.$$

The odd partial sums increase toward l; the even partial sums decrease toward l.
For odd n

$$s_n < l < s_{n+1} = s_n + a_{n+1},$$

and for even n

$$s_n - a_{n+1} = s_{n+1} < l < s_n.$$

Thus, for all n, l lies between s_n and s_{n+1}, and s_n approximates l to within a_{n+1}. ☐

Example 5. Both

$$1 - \frac{1}{2} + \frac{1}{3} - \frac{1}{4} + \frac{1}{5} - \frac{1}{6} + \cdots \quad \text{and} \quad 1 - \frac{1}{2^2} + \frac{1}{3^2} - \frac{1}{4^2} + \frac{1}{5^2} - \frac{1}{6^2} + \cdots$$

are convergent alternating series. The nth partial sum of the first series approximates the sum of that series within $1/(n + 1)$; the nth partial sum of the second series approximates the sum of the second series within $1/(n + 1)^2$. The second series converges more rapidly than the first series. ☐

Rearrangements

A *rearrangement* of a series Σa_k is a series that has exactly the same terms but in a different order. Thus, for example,

$$1 + \frac{1}{3^3} - \frac{1}{2^2} + \frac{1}{5^5} - \frac{1}{4^4} + \frac{1}{7^7} - \frac{1}{6^6} + \cdots$$

and

$$1 + \frac{1}{3^3} + \frac{1}{5^5} - \frac{1}{2^2} - \frac{1}{4^4} + \frac{1}{7^7} + \frac{1}{9^9} - \cdots$$

are both rearrangements of

$$1 - \frac{1}{2^2} + \frac{1}{3^3} - \frac{1}{4^4} + \frac{1}{5^5} - \frac{1}{6^6} + \frac{1}{7^7} - \cdots .$$

In 1867 Riemann published a theorem on rearrangements of series that underscores the importance of distinguishing between absolute convergence and conditional convergence. According to this theorem all rearrangements of an absolutely convergent series converge absolutely to the same sum. In sharp contrast, a series that is only conditionally convergent can be rearranged to converge to any number we please. It can also be arranged to diverge to $+\infty$, or to diverge to $-\infty$, or even to oscillate between any two bounds we choose.†

EXERCISES 12.5

Test these series for (a) absolute convergence, (b) conditional convergence.

1. $1 + (-1) + 1 + \cdots + (-1)^k + \cdots .$

2. $\dfrac{1}{4} - \dfrac{1}{6} + \dfrac{1}{8} - \dfrac{1}{10} + \cdots + \dfrac{(-1)^k}{2k} + \cdots .$

3. $\dfrac{1}{2} - \dfrac{2}{3} + \dfrac{3}{4} - \dfrac{4}{5} + \cdots + (-1)^k \dfrac{k}{k+1} + \cdots .$

4. $\dfrac{1}{2\ln 2} - \dfrac{1}{3\ln 3} + \dfrac{1}{4\ln 4} - \dfrac{1}{5\ln 5} + \cdots + (-1)^k \dfrac{1}{k\ln k} + \cdots .$

5. $\Sigma (-1)^k \dfrac{\ln k}{k}.$ **6.** $\Sigma (-1)^k \dfrac{k}{\ln k}.$ **7.** $\Sigma \left(\dfrac{1}{k} - \dfrac{1}{k!} \right).$

8. $\Sigma \dfrac{k^3}{2^k}.$ **9.** $\Sigma (-1)^k \dfrac{1}{2k+1}.$ **10.** $\Sigma (-1)^k \dfrac{(k!)^2}{(2k)!}.$

11. $\Sigma \dfrac{k!}{(-2)^k}.$ **12.** $\Sigma \sin\left(\dfrac{k\pi}{4} \right).$ **13.** $\Sigma (-1)^k (\sqrt{k+1} - \sqrt{k}).$

14. $\Sigma (-1)^k \dfrac{k}{k^2+1}.$ **15.** $\Sigma \sin\left(\dfrac{\pi}{4k^2} \right).$ **16.** $\Sigma \dfrac{(-1)^k}{\sqrt{k(k+1)}}.$

17. $\Sigma (-1)^k \dfrac{k}{2^k}.$ **18.** $\Sigma \left(\dfrac{1}{\sqrt{k}} - \dfrac{1}{\sqrt{k+1}} \right).$ **19.** $\Sigma \dfrac{(-1)^k}{k - 2\sqrt{k}}.$

20. $\dfrac{1}{2} - \dfrac{1}{3} - \dfrac{1}{4} + \dfrac{1}{5} - \dfrac{1}{6} - \dfrac{1}{7} + \cdots + \dfrac{1}{3k+2} - \dfrac{1}{3k+3} - \dfrac{1}{3k+4} + \cdots .$

21. $\dfrac{2 \cdot 3}{4 \cdot 5} - \dfrac{5 \cdot 6}{7 \cdot 8} + \cdots + (-1)^k \dfrac{(3k+2)(3k+3)}{(3k+4)(3k+5)} + \cdots .$

---

† For a complete proof see pp. 138–139, 318–320 in Konrad Knopp's *Theory and Applications of Infinite Series* (Second English Edition), Blackie & Son Limited, London, 1951.

22. Let s_n be the nth partial sum of the series

$$\sum_{k=0}^{\infty} (-1)^k \frac{1}{10^k}.$$

Find the least value of n for which s_n approximates the sum of the series within
(a) 0.001.   (b) 0.0001.

23. Find the sum of the series in Exercise 22.

24. Verify that the series

$$1 - \frac{1}{2} + \frac{1}{2} - \frac{1}{3} + \frac{1}{2} - \frac{1}{3} + \frac{1}{3} - \frac{1}{4} + \frac{1}{3} - \frac{1}{4} + \frac{1}{3} - \frac{1}{4} + \cdots$$

diverges and explain how this does not violate the theorem on alternating series.

25. Let l be the sum of the series

$$\sum_{k=0}^{\infty} (-1)^k \frac{1}{k!}$$

and let s_n be the nth partial sum. Find the least value of n for which s_n approximates l within (a) 0.01.   (b) 0.001.

26. Let $\{a_k\}$ be a nonincreasing sequence of positive numbers that converges to 0. Does the alternating series $\Sigma\,(-1)^k a_k$ necessarily converge?

27. Can the hypothesis of Theorem 12.5.3 be relaxed to require only that $\{a_{2k}\}$ and $\{a_{2k+1}\}$ be decreasing sequences of positive numbers with limit zero?

28. Indicate how a conditionally convergent series can be rearranged (a) to converge to an arbitrary real number l; (b) to diverge to $+\infty$; (c) to diverge to $-\infty$. HINT: Collect the positive terms $p_1, p_2, p_3, \ldots$ and also the negative terms $n_1, n_2, n_3, \ldots$ in the order in which they appear in the original series.

29. In Section 12.8 we prove that, if $\Sigma\,a_k x_1^k$ converges, then $\Sigma\,a_k x^k$ converges absolutely for $|x| < |x_1|$. Try to prove this now.

12.6   TAYLOR POLYNOMIALS IN x; TAYLOR SERIES IN x

Taylor Polynomials in x

We begin with a function f continuous at 0 and set $P_0(x) = f(0)$. If f is differentiable at 0, the linear function that best approximates f at points close to 0 is the linear function

$$P_1(x) = f(0) + f'(0)x;$$

P_1 has the same value as f at 0 and also the same first derivative (the same rate of change):

$$P_1(0) = f(0), \quad P_1'(0) = f'(0).$$

If f has two derivatives at 0, then we can get a better approximation to f by using the quadratic polynomial

$$P_2(x) = f(0) + f'(0)x + \frac{f''(0)}{2!}\, x^2;$$

P_2 has the same value as f at 0 and the same first two derivatives:

$$P_2(0) = f(0), \quad P_2'(0) = f'(0), \quad P_2''(0) = f''(0).$$

If f has three derivatives at 0, we can form the cubic polynomial

$$P_3(x) = f(0) + f'(0)x + \frac{f''(0)}{2!} x^2 + \frac{f'''(0)}{3!} x^3;$$

P_3 has the same value as f at 0 and the same first three derivatives:

$$P_3(0) = f(0), \quad P_3'(0) = f'(0), \quad P_3''(0) = f''(0), \quad P_3'''(0) = f'''(0).$$

More generally, if f has n derivatives at 0, we can form the polynomial

$$P_n(x) = f(0) + f'(0)x + \frac{f''(0)}{2!} x^2 + \cdots + \frac{f^{(n)}(0)}{n!} x^n;$$

P_n is the polynomial of degree n that has the same value as f at 0 and the same first n derivatives:

$$P_n(0) = f(0), \quad P_n'(0) = f'(0), \quad P_n''(0) = f''(0), \quad \ldots, \quad P_n^{(n)}(0) = f^{(n)}(0).$$

These approximating polynomials $P_0(x), P_1(x), P_2(x), \ldots, P_n(x)$ are called *Taylor polynomials* after the English mathematician Brook Taylor (1685–1731). Taylor introduced these polynomials in the year 1712.

Example 1. The exponential function

$$f(x) = e^x$$

has derivatives

$$f'(x) = e^x, \quad f''(x) = e^x, \quad f'''(x) = e^x, \quad \text{etc.}$$

Thus

$$f(0) = 1, \quad f'(0) = 1, \quad f''(0) = 1, \quad f'''(0) = 1, \quad \ldots, \quad f^{(n)}(0) = 1.$$

The nth Taylor polynomial takes the form

$$P_n(x) = 1 + x + \frac{x^2}{2!} + \frac{x^3}{3!} + \cdots + \frac{x^n}{n!}. \quad \square$$

Example 2. To find the Taylor polynomials that approximate the sine function we write

$$f(x) = \sin x, \quad f'(x) = \cos x, \quad f''(x) = -\sin x, \quad f'''(x) = -\cos x.$$

The pattern now repeats itself:

$$f^{(4)}(x) = \sin x, \quad f^{(5)}(x) = \cos x, \quad f^{(6)}(x) = -\sin x, \quad f^{(7)}(x) = -\cos x.$$

At 0, the sine function and all its even derivatives are 0. The odd derivatives are alternately 1 and −1:

$$f'(0) = 1, \quad f'''(0) = -1, \quad f^{(5)}(0) = 1, \quad f^{(7)}(0) = -1, \quad \text{etc.}$$

The Taylor polynomials are therefore as follows:

$$P_0(x) = 0$$

$$P_1(x) = P_2(x) = x$$

$$P_3(x) = P_4(x) = x - \frac{x^3}{3!}$$

$$P_5(x) = P_6(x) = x - \frac{x^3}{3!} + \frac{x^5}{5!}$$

$$P_7(x) = P_8(x) = x - \frac{x^3}{3!} + \frac{x^5}{5!} - \frac{x^7}{7!}, \quad \text{etc.}$$

Only odd powers appear. □

It is not enough to say that the Taylor polynomials

$$P_n(x) = f(0) + f'(0)x + \frac{f''(0)}{2!}x^2 + \cdots + \frac{f^{(n)}(0)}{n!}x^n$$

approximate $f(x)$. We must describe the accuracy of the approximation. Our first step is to prove a result known as Taylor's theorem.

THEOREM 12.6.1   TAYLOR'S THEOREM

If f has $n + 1$ continuous derivatives on the interval I that joins 0 to x, then

$$f(x) = f(0) + f'(0)x + \frac{f''(0)}{2!}x^2 + \cdots + \frac{f^{(n)}(0)}{n!}x^n + R_{n+1}(x)$$

where the *remainder* $R_{n+1}(x)$ is given by the formula

$$R_{n+1}(x) = \frac{1}{n!}\int_0^x f^{(n+1)}(t)(x - t)^n \, dt.$$

Proof.   Integration by parts (see Exercise 33) gives

$$f'(0)x = \int_0^x f'(t)\, dt - \int_0^x f''(t)(x - t)\, dt,$$

$$\frac{f''(0)}{2!}x^2 = \int_0^x f''(t)(x - t)\, dt - \frac{1}{2!}\int_0^x f'''(t)(x - t)^2 \, dt,$$

$$\frac{f'''(0)}{3!}x^3 = \frac{1}{2!}\int_0^x f'''(t)(x - t)^2 \, dt - \frac{1}{3!}\int_0^x f^{(4)}(t)(x - t)^3 \, dt,$$

$$\vdots$$

$$\frac{f^{(n)}(0)}{n!}x^n = \frac{1}{(n - 1)!}\int_0^x f^{(n)}(t)(x - t)^{n-1} \, dt - \frac{1}{n!}\int_0^x f^{(n+1)}(t)(x - t)^n \, dt.$$

We now add these equations. The sum on the left is simply

$$f'(0)x + \frac{f''(0)}{2!} x^2 + \frac{f'''(0)}{3!} x^3 + \cdots + \frac{f^{(n)}(0)}{n!} x^n = P_n(x) - f(0).$$

The sum on the right telescopes to give

$$\int_0^x f'(t)\, dt - \frac{1}{n!} \int_0^x f^{(n+1)}(t)(x-t)^n\, dt = f(x) - f(0) - \frac{1}{n!} \int_0^x f^{(n+1)}(t)(x-t)^n\, dt.$$

It follows that

$$f(x) = P_n(x) + \frac{1}{n!} \int_0^x f^{(n+1)}(t)(x-t)^n\, dt,$$

and thus

$$R_{n+1}(x) = \frac{1}{n!} \int_0^x f^{(n+1)}(t)(x-t)^n\, dt. \quad \square$$

We can now describe how closely

$$P_n(x) = f(0) + f'(0)x + \frac{f''(0)}{2!} x^2 + \cdots + \frac{f^{(n)}(0)}{n!} x^n$$

approximates $f(x)$ by giving an estimate for the remainder

$$R_{n+1}(x) = \frac{1}{n!} \int_0^x f^{(n+1)}(t)(x-t)^n\, dt.$$

Working with the integral on the right you can see that

(12.6.2)
$$|R_{n+1}(x)| \leq \left(\max_{t \in I} |f^{(n+1)}(t)| \right) \frac{|x|^{n+1}}{(n+1)!}.$$

The derivation of this *remainder estimate* is left to you as an exercise.

Example 3. The Taylor polynomials of the exponential function

$$f(x) = e^x$$

take the form

$$P_n(x) = 1 + x + \frac{x^2}{2!} + \cdots + \frac{x^n}{n!}. \qquad \text{(Example 1)}$$

We will show with our remainder estimate that for all real x

$$R_{n+1}(x) \to 0,$$

and therefore we can approximate e^x as closely as we wish by Taylor polynomials.

We begin by fixing x and letting M be the maximum value of the exponential function on the interval I that joins 0 to x. (If $x > 0$, then $M = e^x$ but, if $x < 0$, $M = e^0 = 1$.) Since

$$f^{(n+1)}(t) = e^t \qquad \text{for all } n,$$

we have

$$\max_{t \in I} |f^{(n+1)}(t)| = M \qquad \text{for all } n.$$

Thus by (12.6.2)

$$|R_{n+1}(x)| \leq M \frac{|x|^{n+1}}{(n+1)!}.$$

By (11.3.3) we know that

$$\frac{|x|^{n+1}}{(n+1)!} \to 0.$$

It follows then that $R_{n+1}(x) \to 0$ as asserted. ☐

Example 4.   We return to the sine function

$$f(x) = \sin x$$

and its Taylor polynomials

$$P_1(x) = P_2(x) = x$$

$$P_3(x) = P_4(x) = x - \frac{x^3}{3!}$$

$$P_5(x) = P_6(x) = x - \frac{x^3}{3!} + \frac{x^5}{5!}, \quad \text{etc.}$$

The pattern of derivatives was established in Example 2; namely, for all k,

$$f^{(4k)}(x) = \sin x, \quad f^{(4k+1)}(x) = \cos x, \quad f^{(4k+2)}(x) = -\sin x, \quad f^{(4k+3)}(x) = -\cos x.$$

Thus, for all n and all real t,

$$|f^{(n+1)}(t)| \leq 1.$$

It follows from our remainder estimate (12.6.2) that

$$|R_{n+1}(x)| \leq \frac{|x|^{n+1}}{(n+1)!}.$$

Since

$$\frac{|x|^{n+1}}{(n+1)!} \to 0 \qquad \text{for all real } x,$$

we see that $R_{n+1}(x) \to 0$ for all real x. Thus the sequence of Taylor polynomials converges to the sine function and therefore can be used to approximate $\sin x$ as closely as we may wish. ☐

Taylor Series in *x*

By definition $0! = 1$. By adopting the convention that $f^{(0)} = f$, we can write Taylor polynomials

$$P_n(x) = f(0) + f'(0)x + \frac{f''(0)}{2!} x^2 + \cdots + \frac{f^{(n)}(0)}{n!} x^n$$

in Σ notation:

$$P_n(x) = \sum_{k=0}^{n} \frac{f^{(k)}(0)}{k!} x^k.$$

If f is infinitely differentiable at $x = 0$, then we have

$$f(x) = \sum_{k=0}^{n} \frac{f^{(k)}(0)}{k!} x^k + R_{n+1}(x)$$

for all positive integers n. If, as in the case of the exponential function and the sine function, $R_{n+1}(x) \to 0$, then

$$\sum_{k=0}^{n} \frac{f^{(k)}(0)}{k!} x^k \to f(x).$$

In this case, we say that $f(x)$ can be expanded as a *Taylor series in x* and write

(12.6.3)
$$f(x) = \sum_{k=0}^{\infty} \frac{f^{(k)}(0)}{k!} x^k.$$

[Taylor series in *x* are sometimes called Maclaurin series after Colin Maclaurin, a Scottish mathematician (1698–1746). In some circles the name Maclaurin remains attached to these series, although Taylor considered them some twenty years before Maclaurin.]

From Example 3 it is clear that

(12.6.4)
$$e^x = \sum_{k=0}^{\infty} \frac{x^k}{k!} = 1 + x + \frac{x^2}{2!} + \frac{x^3}{3!} + \cdots \quad \text{for all real } x.$$

From Example 4 we have

(12.6.5)
$$\sin x = \sum_{k=0}^{\infty} \frac{(-1)^k}{(2k+1)!} x^{2k+1} = x - \frac{x^3}{3!} + \frac{x^5}{5!} - \frac{x^7}{7!} + \cdots \quad \text{for all real } x.$$

We leave it to you as an exercise to show that

(12.6.6)
$$\cos x = \sum_{k=0}^{\infty} \frac{(-1)^k}{(2k)!} x^{2k} = 1 - \frac{x^2}{2!} + \frac{x^4}{4!} - \frac{x^6}{6!} + \cdots \quad \text{for all real } x.$$

We come now to the logarithm function. Since $\ln x$ is not defined at $x = 0$, we cannot expand $\ln x$ in powers of x. We will work instead with $\ln (1 + x)$.

(12.6.7)
$$\ln (1 + x) = \sum_{k=1}^{\infty} \frac{(-1)^{k+1}}{k} x^k = x - \frac{x^2}{2} + \frac{x^3}{3} - \cdots \quad \text{for } -1 < x \le 1.$$

Proof.† The function
$$f(x) = \ln (1 + x)$$

has derivatives

$$f'(x) = \frac{1}{1 + x}, \quad f''(x) = -\frac{1}{(1 + x)^2}, \quad f'''(x) = \frac{2}{(1 + x)^3},$$

$$f^{(4)}(x) = -\frac{3!}{(1 + x)^4}, \quad f^{(5)}(x) = \frac{4!}{(1 + x)^5}, \quad \text{and so on.}$$

For $k \ge 1$

$$f^{(k)}(x) = (-1)^{k+1} \frac{(k - 1)!}{(1 + x)^k}, \quad f^{(k)}(0) = (-1)^{k+1}(k - 1)!, \quad \frac{f^{(k)}(0)}{k!} = \frac{(-1)^{k+1}}{k}.$$

Since $f(0) = 0$, the nth Taylor polynomial takes the form

$$P_n(x) = \sum_{k=1}^{n} (-1)^{k+1} \frac{x^k}{k} = x - \frac{x^2}{2} + \cdots + (-1)^{n+1} \frac{x^n}{n}.$$

All we have to show therefore is that

$$R_{n+1}(x) \to 0 \qquad \text{for } -1 < x \le 1.$$

Instead of trying to apply our usual remainder estimate [in this case, that estimate is not delicate enough to show that $R_{n+1}(x) \to 0$ for $-1 < x < -\frac{1}{2}$], we write the remainder in its integral form. From Taylor's theorem

$$R_{n+1}(x) = \frac{1}{n!} \int_0^x f^{(n+1)}(t)(x - t)^n \, dt,$$

so that in this case

$$R_{n+1}(x) = \frac{1}{n!} \int_0^x (-1)^{n+2} \frac{n!}{(1 + t)^{n+1}} (x - t)^n \, dt = (-1)^n \int_0^x \frac{(x - t)^n}{(1 + t)^{n+1}} \, dt.$$

For $0 \le x \le 1$ we have

$$|R_{n+1}(x)| = \int_0^x \frac{(x - t)^n}{(1 + t)^{n+1}} \, dt \le \int_0^x (x - t)^n \, dt = \frac{x^{n+1}}{n + 1} \to 0.$$

$$\underset{\text{explain}}{\uparrow}$$

---

† The proof we give here illustrates the methods of this section. Another, much simpler way of obtaining this series expansion is given in Section 12.9.

For $-1 < x < 0$ we have

$$|R_{n+1}(x)| = \left| \int_0^x \frac{(x-t)^n}{(1+t)^{n+1}} \, dt \right| = \int_x^0 \left(\frac{t-x}{1+t} \right)^n \frac{1}{1+t} \, dt.$$

By the First Mean-Value Theorem for Integrals (5.11.1) there exists a number x_n between x and 0 such that

$$\int_x^0 \left(\frac{t-x}{1+t} \right)^n \frac{1}{1+t} \, dt = \left(\frac{x_n - x}{1+x_n} \right)^n \left(\frac{1}{1+x_n} \right)(-x).$$

Since $-x = |x|$ and $0 < 1 + x < 1 + x_n$, we can conclude that

$$|R_{n+1}(x)| < \left(\frac{x_n + |x|}{1+x_n} \right)^n \left(\frac{|x|}{1+x} \right).$$

Since $|x| < 1$ and $x_n < 0$, we have

$$x_n < |x|x_n, \qquad x_n + |x| < |x|x_n + |x| = |x|(1 + x_n)$$

and thus

$$\frac{x_n + |x|}{1+x_n} < |x|.$$

It now follows that

$$|R_{n+1}(x)| < |x|^n \left(\frac{|x|}{1+x} \right)$$

and, since $|x| < 1$, that $R_{n+1}(x) \to 0$. $\square$

Remark. The series expansion for $\ln(1+x)$ that we just verified for $-1 < x \le 1$ cannot be extended to other values of x. For $x \le -1$ neither side makes sense: $\ln(1+x)$ is not defined, and the series on the right diverges. For $x \ge 1$, $\ln(1+x)$ is defined, but the series on the right diverges and hence does not represent the function. At $x = 1$ the series gives an intriguing result:

$$\ln 2 = 1 - \tfrac{1}{2} + \tfrac{1}{3} - \tfrac{1}{4} + \cdots. \quad \square$$

We want to emphasize again the role played by the remainder term $R_{n+1}(x)$. We can form a Taylor series

$$\sum_{k=0}^{\infty} \frac{f^{(k)}(0)}{k!} x^k$$

for any function f with derivatives of all orders at $x = 0$, but such a series need not converge at any number $x \neq 0$. Even if it does converge, the sum need not be $f(x)$. (See Exercise 35.) The Taylor series converges to $f(x)$ if and only if the remainder term $R_{n+1}(x)$ tends to 0.

Some Numerical Calculations

If the Taylor series converges to $f(x)$, we can use the partial sums (the Taylor polynomials) to calculate $f(x)$ as accurately as we wish. In what follows we show some sample calculations. For ready reference we list some values of $k!$ and $1/k!$

TABLE 12.6.1
$k!$

| $2! = 2$ |
| $3! = 6$ |
| $4! = 24$ |
| $5! = 120$ |
| $6! = 720$ |
| $7! = 5,040$ |
| $8! = 40,320$ |

TABLE 12.6.2

$1/k!$

$$0.16666 < \frac{1}{3!} < 0.16667 \qquad 0.00138 < \frac{1}{6!} < 0.00139$$

$$0.04166 < \frac{1}{4!} < 0.04167 \qquad 0.00019 < \frac{1}{7!} < 0.00020$$

$$0.00833 < \frac{1}{5!} < 0.00834 \qquad 0.00002 < \frac{1}{8!} < 0.00003$$

Problem 5. Estimate e within 0.001.

Solution. For all x

$$e^x = 1 + x + \frac{x^2}{2!} + \cdots + \frac{x^n}{n!} + \cdots .$$

Taking $x = 1$ we have

$$e = 1 + 1 + \frac{1}{2!} + \cdots + \frac{1}{n!} + \cdots .$$

From Example 3 we know that the nth partial sum of this series, the Taylor polynomial

$$P_n(1) = 1 + 1 + \frac{1}{2!} + \cdots + \frac{1}{n!},$$

approximates e within

$$|R_{n+1}(1)| \le e \frac{|1|^{n+1}}{(n+1)!} < \frac{3}{(n+1)!}.$$

here $M = e^1 = e$ ⎯⎯⎯⎯ $e < 3$

Since

$$\frac{3}{7!} = \frac{3}{5040} = \frac{1}{1680} < 0.001,$$

we can take $n = 6$ and be sure that

$$P_6(1) = 1 + 1 + \frac{1}{2!} + \frac{1}{3!} + \frac{1}{4!} + \frac{1}{5!} + \frac{1}{6!} = \frac{1957}{720}$$

differs from e by less than 0.001.

Our calculator gives

$$\frac{1957}{720} \cong 2.7180556 \quad \text{and} \quad e \cong 2.7182818. \quad \square$$

Problem 6. Estimate $e^{0.2}$ within 0.001.

Solution. The exponential series at $x = 0.2$ gives

$$e^{0.2} = 1 + 0.2 + \frac{(0.2)^2}{2!} + \cdots + \frac{(0.2)^n}{n!} + \cdots .$$

From Example 3 we know that the nth partial sum of this series, the Taylor polynomial

$$P_n(0.2) = 1 + 0.2 + \frac{(0.2)^2}{2!} + \cdots + \frac{(0.2)^n}{n!},$$

approximates $e^{0.2}$ within

$$|R_{n+1}(0.2)| \le e^{0.2} \frac{|0.2|^{n+1}}{(n+1)!} < 3 \frac{(0.2)^{n+1}}{(n+1)!}.$$
$$\uparrow\_\_\_\_ \text{ here } M = e^{0.2}$$

Since

$$3 \frac{(0.2)^4}{4!} = \frac{(3)(16)}{240,000} < 0.001,$$

we can take $n = 3$ and be sure that

$$P_3(0.2) = 1 + 0.2 + \frac{(0.2)^2}{2!} + \frac{(0.2)^3}{3!} = \frac{7326}{6000} = 1.221$$

differs from $e^{0.2}$ by less than 0.001.

Our calculator gives

$$e^{0.2} \cong 1.2214028. \quad \square$$

Problem 7. Estimate $\sin 0.5$ within 0.001.

Solution. At $x = 0.5$ the sine series gives

$$\sin 0.5 = 0.5 - \frac{(0.5)^3}{3!} + \frac{(0.5)^5}{5!} - \frac{(0.5)^7}{7!} + \cdots .$$

From Example 4 we know that the nth partial sum, the nth Taylor polynomial

$P_n(0.5)$, approximates sin 0.5 within

$$|R_{n+1}(0.5)| \le \frac{(0.5)^{n+1}}{(n+1)!}.$$

Since

$$\frac{(0.5)^5}{5!} = \frac{1}{(2^5)(5!)} = \frac{1}{(32)(120)} = \frac{1}{3840} < 0.001,$$

we can be sure that

$$P_4(0.5) = P_3(0.5) = 0.5 - \frac{(0.5)^3}{3!} = \frac{23}{48}$$

the coefficient of x^4 is 0

approximates sin 0.5 within 0.001.

Our calculator gives

$$\frac{23}{48} \cong 0.4791666 \quad \text{and} \quad \sin 0.5 \cong 0.4794255. \quad \square$$

Remark. We could have solved the last problem without reference to the remainder estimate derived in Example 4. The series for sin 0.5 is a convergent alternating series with decreasing terms. By (12.5.4) we can conclude immediately that sin 0.5 lies between every two consecutive partial sums. In particular

$$0.5 - \frac{(0.5)^3}{3!} < \sin 0.5 < 0.5 - \frac{(0.5)^3}{3!} + \frac{(0.5)^5}{5!}. \quad \square$$

Problem 8. Estimate ln 1.4 within 0.01.

Solution. By (12.6.7)

$$\ln 1.4 = \ln(1 + 0.4) = 0.4 - \tfrac{1}{2}(0.4)^2 + \tfrac{1}{3}(0.4)^3 - \tfrac{1}{4}(0.4)^4 + \cdots .$$

This is a convergent alternating series with decreasing terms. Therefore ln 1.4 lies between every two consecutive partial sums.

The first term less than 0.01 is

$$\tfrac{1}{4}(0.4)^4 = \tfrac{1}{4}(0.0256) = 0.0064.$$

The relation

$$0.4 - \tfrac{1}{2}(0.4)^2 + \tfrac{1}{3}(0.4)^3 - \tfrac{1}{4}(0.4)^4 < \ln 1.4 < 0.4 - \tfrac{1}{2}(0.4)^2 + \tfrac{1}{3}(0.4)^3$$

gives

$$0.335 < \ln 1.4 < 0.341.$$

Within the prescribed limits of accuracy we can take ln 1.4 $\cong$ 0.34. $\square$†

———

† A much more effective tool for computing logarithms is given in the exercises.

EXERCISES 12.6

Find the Taylor polynomial $P_4(x)$ for each of the following functions.

1. $x - \cos x$.   **2.** $\sqrt{1 + x}$.   **3.** $\ln \cos x$.   **4.** $\sec x$.

Find the Taylor polynomial $P_5(x)$ for each of the following functions.

5. $(1 + x)^{-1}$.   **6.** $e^x \sin x$.   **7.** $\tan x$.   **8.** $x \cos x^2$.

9. Determine $P_0(x)$, $P_1(x)$, $P_2(x)$, $P_3(x)$ for $1 - x + 3x^2 + 5x^3$.

10. Determine $P_0(x)$, $P_1(x)$, $P_2(x)$, $P_3(x)$ for $(x + 1)^3$.

Determine the nth Taylor polynomial $P_n(x)$.

11. e^{-x}.   **12.** $\sinh x$.   **13.** $\cosh x$.   **14.** $\ln (1 - x)$.

Use Taylor polynomials to estimate the following within 0.01.

15. $\sqrt{e}$.   **16.** $\sin 0.3$.   **17.** $\sin 1$.   **18.** $\ln 1.2$.   **19.** $\cos 1$.   **20.** $e^{0.8}$.

21. Let $P_n(x)$ be the nth Taylor polynomial of

$$f(x) = \ln (1 + x).$$

Find the least integer n for which (a) $P_n(0.5)$ approximates $\ln 1.5$ within 0.01.
(b) $P_n(0.3)$ approximates $\ln 1.3$ within 0.01.   (c) $P_n(1)$ approximates $\ln 2$ within 0.001.

22. Let $P_n(x)$ be the nth Taylor polynomial of

$$f(x) = \sin x.$$

Find the least integer n for which (a) $P_n(1)$ approximates $\sin 1$ within 0.001.
(b) $P_n(2)$ approximates $\sin 2$ within 0.001.   (c) $P_n(3)$ approximates $\sin 3$ within 0.001.

23. Show that a polynomial $P(x) = a_0 + a_1 x + \cdots + a_n x^n$ is its own Taylor series.

24. Show that

$$\cos x = \sum_{k=0}^{\infty} \frac{(-1)^k}{(2k)!} x^{2k} \qquad \text{for all real } x.$$

Derive a series expansion in x for each of the following and specify the numbers x for which the expansion is valid. Take $a > 0$.

25. e^{ax}.   HINT: Set $t = ax$ and expand e^t in powers of t.

26. $\sin ax$.   **27.** $\cos ax$.   **28.** $\ln (1 - ax)$.

29. $\ln (a + x)$.   HINT: $\ln (a + x) = \ln [a(1 + x/a)]$.

30. The series we derived for $\ln (1 + x)$ converges too slowly to be of much practical use. The following logarithm series converges much more quickly:

(12.6.8) $$\ln \left(\frac{1 + x}{1 - x} \right) = 2 \left(x + \frac{x^3}{3} + \frac{x^5}{5} + \cdots \right) \qquad \text{for } -1 < x < 1.$$

Derive this series expansion.

31. Set $x = \frac{1}{3}$ and use the first three nonzero terms of (12.6.8) to estimate $\ln 2$.

32. Use the first two nonzero terms of (12.6.8) to estimate $\ln 1.4$.

33. Verify the identity

$$\frac{f^{(k)}(0)}{k!} x^k = \frac{1}{(k-1)!} \int_0^x f^{(k)}(t)(x-t)^{k-1}\, dt - \frac{1}{k!} \int_0^x f^{(k+1)}(t)(x-t)^k\, dt$$

by computing the second integral by parts.

34. Derive the remainder estimate (12.6.2).

35. Show that for the function

$$f(x) = \begin{cases} e^{-1/x^2}, & x \neq 0 \\ 0, & x = 0 \end{cases},$$

$f^{(k)}(0) = 0$, $k = 0, 1, 2$. Similar arguments show that $f^{(k)}(0) = 0$ for all k. The Taylor series in x is therefore identically 0 and does not represent the function except at $x = 0$.

36. (*Important*) Show that e is irrational by following these steps.
    (1) Take the expansion

$$e = \sum_{k=0}^{\infty} \frac{1}{k!}$$

and show that the qth partial sum

$$s_q = \sum_{k=0}^{q} \frac{1}{k!}$$

satisfies the inequality

$$0 < q!\,(e - s_q) < \frac{1}{q}.$$

(2) Show that $q!s_q$ is an integer and argue that, if e were of the form p/q, then $q!\,(e - s_q)$ would be a positive integer less than 1.

Program for computing e via sums (BASIC)

This program makes use of the formula

$$e = \sum_{j=1}^{\infty} \frac{1}{j!}$$

for computing the value of e. You need only set the number of terms to be used by selecting a value for n.

```
10 REM Computation of e via sums
20 REM copyright © Colin C. Graham 1988-1989

100 INPUT "Enter n:"; n

200 term = 1
210 e = 1

300 FOR j = 1 TO n
310    term = term/j
320    e = e + term
330    PRINT e
340 NEXT j

500 INPUT "Do it again? (Y/N)"; a$
510 IF a$ = "Y" OR a$ = "y" THEN 100
520 END
```

12.7   TAYLOR POLYNOMIALS IN $x - a$; TAYLOR SERIES IN $x - a$

So far we have considered series expansions only in powers of x. Here we generalize to expansions in powers of $x - a$ where a is an arbitrary real number. We begin with a more general version of Taylor's theorem.

THEOREM 12.7.1   TAYLOR'S THEOREM

If g has $n + 1$ continuous derivatives on the interval I that joins a to x, then

$$g(x) = g(a) + g'(a)(x - a) + \frac{g''(a)}{2!}(x - a)^2 + \cdots + \frac{g^{(n)}(a)}{n!}(x - a)^n + R_{n+1}(x)$$

where

$$R_{n+1}(x) = \frac{1}{n!} \int_a^x g^{(n+1)}(s)(x - s)^n \, ds.$$

In this more general setting the remainder estimate can be written

(12.7.2)
$$|R_{n+1}(x)| \leq \left(\max_{s \in I} |g^{(n+1)}(s)| \right) \frac{|x - a|^{n+1}}{(n + 1)!}.$$

If $R_{n+1}(x) \to 0$, then we have

$$g(x) = g(a) + g'(a)(x - a) + \frac{g''(a)}{2!}(x - a)^2 + \cdots + \frac{g^{(n)}(a)}{n!}(x - a)^n + \cdots.$$

In sigma notation we have

(12.7.3)
$$g(x) = \sum_{k=0}^{\infty} \frac{g^{(k)}(a)}{k!}(x - a)^k.$$

This is known as the Taylor expansion of $g(x)$ in powers of $x - a$. The series on the right is called a *Taylor series in* $x - a$.

All this differs from what you saw before only by a translation. Define

$$f(t) = g(t + a).$$

Then obviously

$$f^{(k)}(t) = g^{(k)}(t + a) \qquad \text{and} \qquad f^{(k)}(0) = g^{(k)}(a).$$

The results of this section as stated for g can be derived by applying the results of Section 12.6 to the function f.

Problem 1. Expand $g(x) = 4x^3 - 3x^2 + 5x - 1$ in powers of $x - 2$.

Solution. We want to evaluate g and its derivatives at $x = 2$.

$$g(x) = 4x^3 - 3x^2 + 5x - 1$$
$$g'(x) = 12x^2 - 6x + 5$$
$$g''(x) = 24x - 6$$
$$g'''(x) = 24.$$

All higher derivatives are identically 0.

Substitution gives $g(2) = 29$, $g'(2) = 41$, $g''(2) = 42$, $g'''(2) = 24$ and $g^{(k)}(2) = 0$ for all $k \geq 4$. Thus from (12.7.3)

$$g(x) = 29 + 41(x - 2) + \frac{42}{2!}(x - 2)^2 + \frac{24}{3!}(x - 2)^3$$
$$= 29 + 41(x - 2) + 21(x - 2)^2 + 4(x - 2)^3. \quad \square$$

Problem 2. Expand $g(x) = x^2 \ln x$ in powers of $x - 1$.

Solution. We want to evaluate g and its derivatives at $x = 1$.

$$g(x) = x^2 \ln x$$
$$g'(x) = x + 2x \ln x$$
$$g''(x) = 3 + 2 \ln x$$
$$g'''(x) = 2x^{-1}$$
$$g^{(4)}(x) = -2x^{-2}$$
$$g^{(5)}(x) = (2)(2)x^{-3}$$
$$g^{(6)}(x) = -(2)(2)(3)x^{-4} = -2(3!)x^{-4}$$
$$g^{(7)}(x) = (2)(2)(3)(4)x^{-5} = (2)(4!)x^{-5}, \quad \text{etc.}$$

The pattern is now clear: for $k \geq 3$

$$g^{(k)}(x) = (-1)^{k+1}2(k - 3)!\, x^{-k+2}.$$

Evaluation at $x = 1$ gives $g(1) = 0$, $g'(1) = 1$, $g''(1) = 3$ and, for $k \geq 3$,

$$g^{(k)}(1) = (-1)^{k+1}2(k - 3)!.$$

The expansion in powers of $x - 1$ can be written

$$g(x) = (x - 1) + \frac{3}{2!}(x - 1)^2 + \sum_{k=3}^{\infty} \frac{(-1)^{k+1}2(k - 3)!}{k!}(x - 1)^k$$
$$= (x - 1) + \frac{3}{2}(x - 1)^2 + \sum_{k=3}^{\infty} \frac{(-1)^{k+1}2}{k(k - 1)(k - 2)}(x - 1)^k. \quad \square$$

Another way to expand $g(x)$ in powers of $x - a$ is to expand $g(t + a)$ in powers of t and then set $t = x - a$. This is the approach we take when the expansion in t is known to us, or, if not known to us, easily available.

Example 3. We can expand $g(x) = e^{x/2}$ in powers of $x - 3$ by expanding

$$g(t + 3) = e^{(t+3)/2} \qquad \text{in powers of } t$$

and then setting $t = x - 3$.

Note that

$$g(t + 3) = e^{3/2}e^{t/2} = e^{3/2} \sum_{k=0}^{\infty} \frac{(t/2)^k}{k!} = e^{3/2} \sum_{k=0}^{\infty} \frac{1}{2^k k!} t^k.$$

exponential series ⎯⎯⎯⎯⎯↗

Setting $t = x - 3$, we have

$$g(x) = e^{3/2} \sum_{k=0}^{\infty} \frac{1}{2^k k!} (x - 3)^k.$$

Since the expansion of $g(t + 3)$ is valid for all real t, the expansion of $g(x)$ is valid for all real x. ☐

Taking this same approach, we can easily prove that

(12.7.4)

> for $0 < x \leq 2a$
>
> $$\ln x = \ln a + \frac{1}{a}(x - a) - \frac{1}{2a^2}(x - a)^2 + \frac{1}{3a^3}(x - a)^3 - \cdots.$$

Proof. We will expand $\ln(a + t)$ in powers of t and then set $t = x - a$.

In the first place

$$\ln(a + t) = \ln\left[a\left(1 + \frac{t}{a}\right)\right] = \ln a + \ln\left(1 + \frac{t}{a}\right).$$

From (12.6.7) it is clear that

$$\ln\left(1 + \frac{t}{a}\right) = \frac{t}{a} - \frac{1}{2}\left(\frac{t}{a}\right)^2 + \frac{1}{3}\left(\frac{t}{a}\right)^3 - \cdots \qquad \text{for } -a < t \leq a.$$

Adding $\ln a$ to both sides, we have

$$\ln(a + t) = \ln a + \frac{1}{a}t - \frac{1}{2a^2}t^2 + \frac{1}{3a^3}t^3 - \cdots \qquad \text{for } -a < t \leq a.$$

Setting $t = x - a$, we find that

$$\ln x = \ln a + \frac{1}{a}(x - a) - \frac{1}{2a^2}(x - a)^2 + \frac{1}{3a^3}(x - a)^3 - \cdots$$

for all x such that $-a < x - a \leq a$; that is, for all x such that $0 < x \leq 2a$. ☐

EXERCISES 12.7

Expand $g(x)$ as indicated and specify the values of x for which the expansion is valid.

1. $g(x) = 3x^3 - 2x^2 + 4x + 1$ in powers of $x - 1$.
2. $g(x) = x^4 - x^3 + x^2 - x + 1$ in powers of $x - 2$.
3. $g(x) = 2x^5 + x^2 - 3x - 5$ in powers of $x + 1$.
4. $g(x) = x^{-1}$ in powers of $x - 1$.   **5.** $g(x) = (1 + x)^{-1}$ in powers of $x - 1$.
6. $g(x) = (b + x)^{-1}$ in powers of $x - a$, $a \neq -b$.
7. $g(x) = (1 - 2x)^{-1}$ in powers of $x + 2$.   **8.** $g(x) = e^{-4x}$ in powers of $x + 1$.
9. $g(x) = \sin x$ in powers of $x - \pi$.   **10.** $g(x) = \sin x$ in powers of $x - \frac{1}{2}\pi$.
11. $g(x) = \cos x$ in powers of $x - \pi$.   **12.** $g(x) = \cos x$ in powers of $x - \frac{1}{2}\pi$.
13. $g(x) = \sin \frac{1}{2}\pi x$ in powers of $x - 1$.   **14.** $g(x) = \sin \pi x$ in powers of $x - 1$.
15. $g(x) = \ln(1 + 2x)$ in powers of $x - 1$.   **16.** $g(x) = \ln(2 + 3x)$ in powers of $x - 4$.

Expand $g(x)$ as indicated.

17. $g(x) = x \ln x$ in powers of $x - 2$.   **18.** $g(x) = x^2 + e^{3x}$ in powers of $x - 2$.
19. $g(x) = x \sin x$ in powers of x.   **20.** $g(x) = \ln(x^2)$ in powers of $x - 1$.
21. $g(x) = (1 - 2x)^{-3}$ in powers of $x + 2$.   **22.** $g(x) = \sin^2 x$ in powers of $x - \frac{1}{2}\pi$.
23. $g(x) = \cos^2 x$ in powers of $x - \pi$.   **24.** $g(x) = (1 + 2x)^{-4}$ in powers of $x - 2$.
25. $g(x) = x^n$ in powers of $x - 1$.   **26.** $g(x) = (x - 1)^n$ in powers of x.

27. (a) Expand e^x in powers of $x - a$.
   (b) Use the expansion to show that $e^{x_1 + x_2} = e^{x_1} e^{x_2}$.
   (c) Expand e^{-x} in powers of $x - a$.
28. (a) Expand $\sin x$ and $\cos x$ in powers of $x - a$.
   (b) Show that both series are absolutely convergent for all real x.
   (c) As noted earlier (Section 12.5), Riemann proved that the order of the terms of an absolutely convergent series may be changed without altering the sum of the series. Use Riemann's discovery and the Taylor expansions of part (a) to derive the addition formulas

$$\sin(x_1 + x_2) = \sin x_1 \cos x_2 + \cos x_1 \sin x_2,$$
$$\cos(x_1 + x_2) = \cos x_1 \cos x_2 - \sin x_1 \sin x_2.$$

12.8   POWER SERIES

You have become familiar with Taylor series

$$\sum_{k=0}^{\infty} \frac{f^{(k)}(a)}{k!} (x - a)^k \quad \text{and} \quad \sum_{k=0}^{\infty} \frac{f^{(k)}(0)}{k!} x^k.$$

Here we study series of the form

$$\sum_{k=0}^{\infty} a_k (x - a)^k \quad \text{and} \quad \sum_{k=0}^{\infty} a_k x^k$$

without regard to how the coefficients have been generated.   Such series are called *power series*.

Math 342: Analysis - Writing Assignment - 2000

General Information on Content and Procedure:

This semester you will write one paper. The paper will require both expository and technical writing.

1. Part of your grade is determined by the depth of coverage of the material. A paper should show outside research (for example, into the history of the problem). It is equally important to show your understanding of the material, evidenced by original examples, substantive transitions, interpretation of definitions and theorems, and overall cohesiveness of the ideas and concepts presented. In fact, one cannot get an A without these aspects.

2. Make a SENTENCE outline and submit this with the first version of the paper.

3. The paper is to be typed using a word processor.

4. Include a cover page and a bibliography. **Number the pages.**

5. The audience is a classmate.

6. The final form of the paper is to be your own work, not lifted from notes, someone else's paper, or another text.

7. Your paper should look like a textbook in terms of the formatting. See your pamphlet on the Advanced W in the Mathematics Department.

8. Your paper will be graded according to the Guidelines for Writing papers in Mathematics found in the pamphlet on the Advanced W in the Mathematics Department.

9. The exercises are stated in the form of "show that" or "prove that." In the paper, each exercise will be a theorem, an example, or a proposition. Statements of theorems and propositions do NOT include the words "show that" or "prove that."

10. A GOOD draft of the paper is due by February 23. The instructor will critique the paper. Following the instructor's critique, you will revise the paper. The revised paper and the original are to be turned in together by March 9.

Comment: The instructor's comments will be most helpful if you have given your best effort to _____ something clearly in need of major work.

Since a simple translation converts

$$\sum_{k=0}^{\infty} a_k(x-a)^k \quad \text{into} \quad \sum_{k=0}^{\infty} a_k x^k,$$

we can focus our attention on power series of the form

$$\sum_{k=0}^{\infty} a_k x^k.$$

When detailed indexing is unnecessary, we will omit it and write

$$\Sigma \, a_k x^k.$$

We begin the discussion with a definition.

DEFINITION 12.8.1

A power series $\Sigma \, a_k x^k$ is said to converge

   (i) at x_1 iff $\Sigma \, a_k x_1^k$ converges.

   (ii) on the set S iff $\Sigma \, a_k x^k$ converges for each x in S.

The following result is fundamental.

THEOREM 12.8.2

If $\Sigma \, a_k x^k$ converges at $x_1 \neq 0$, then it converges absolutely for $|x| < |x_1|$.

If $\Sigma \, a_k x^k$ diverges at x_1, then it diverges for $|x| > |x_1|$.

Proof.   If $\Sigma \, a_k x_1^k$ converges, then $a_k x_1^k \to 0$.   In particular, for k sufficiently large,

$$|a_k x_1^k| \le 1$$

and thus

$$|a_k x^k| = |a_k x_1^k| \left| \frac{x}{x_1} \right|^k \le \left| \frac{x}{x_1} \right|^k.$$

For $|x| < |x_1|$, we have

$$\left| \frac{x}{x_1} \right| < 1.$$

The convergence of $\Sigma \, |a_k x^k|$ follows by comparison with the geometric series. This proves the first statement.

   Suppose now that $\Sigma \, a_k x_1^k$ diverges. By the previous argument, there cannot exist x with $|x| > |x_1|$ such that $\Sigma \, a_k x^k$ converges. The existence of such an x would imply the absolute convergence of $\Sigma \, a_k x_1^k$. This proves the second statement. $\square$

From the theorem we just proved you can see that there are exactly three possibilities for a power series:

Case I.   *The series converges only at 0.*   This is what happens with

$$\Sigma\ k^k x^k.$$

As you can tell from the root test, the kth term, $k^k x^k$, tends to 0 only if $x = 0$.

Case II.   *The series converges everywhere absolutely.*   This is what happens with the exponential series

$$\Sigma\ \frac{x^k}{k!}.$$

Case III.   *There exists a positive number r such that the series converges absolutely for $|x| < r$ and diverges for $|x| > r$.*   This is what happens with the geometric series

$$\Sigma\ x^k.$$

In this instance, there is absolute convergence for $|x| < 1$, divergence for $|x| > 1$.

Associated with each case is a *radius of convergence*:

> In Case I, we say that the radius of convergence is 0.
> In Case II, we say that the radius of convergence is ∞.
> In Case III, we say that the radius of convergence is r.

The three cases are pictured in Figure 12.8.1.

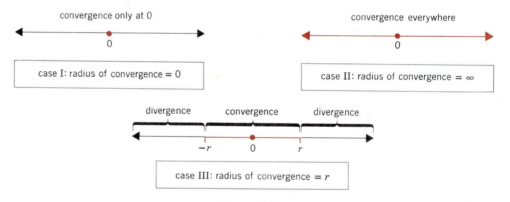

Figure 12.8.1

In general, the behavior of a power series at $-r$ and at r is not predictable. The series

$$\Sigma\ x^k, \qquad \Sigma\ \frac{1}{k}\ x^k, \qquad \Sigma\ \frac{(-1)^k}{k}\ x^k, \qquad \Sigma\ \frac{1}{k^2}\ x^k,$$

all have radius of convergence 1, but while the first series converges only on $(-1, 1)$, the second series converges on $[-1, 1)$, the third on $(-1, 1]$, and the fourth on $[-1, 1]$.

The maximal interval on which a power series converges is called the *interval of convergence*. For a series with infinite radius of convergence, the interval of convergence is $(-\infty, \infty)$. For a series with radius of convergence r, the interval of convergence can be $[-r, r]$, $(-r, r]$, $[-r, r)$, or $(-r, r)$. For a series with radius of convergence 0, the interval of convergence reduces to a point, $\{0\}$.

Problem 1. Verify that the series

(1)
$$\Sigma \frac{(-1)^k}{k} x^k$$

has interval of convergence $(-1, 1]$.

Solution. First we show that the radius of convergence is 1 (that the series converges absolutely for $|x| < 1$ and diverges for $|x| > 1$). We do this by forming the series

(2)
$$\Sigma \left| \frac{(-1)^k}{k} x^k \right| = \Sigma \frac{1}{k} |x|^k$$

and applying the ratio test.
  We set

$$b_k = \frac{1}{k} |x|^k$$

and note that

$$\frac{b_{k+1}}{b_k} = \frac{k}{k+1} \frac{|x|^{k+1}}{|x|^k} = \frac{k}{k+1} |x| \to |x|.$$

By the ratio test, series (2) converges for $|x| < 1$ and diverges for $|x| > 1$.† It follows that series (1) converges absolutely for $|x| < 1$ and diverges for $|x| > 1$. The radius of convergence is therefore 1.
  Now we test the endpoints $x = -1$ and $x = 1$. At $x = -1$

$$\Sigma \frac{(-1)^k}{k} x^k \quad \text{becomes} \quad \Sigma \frac{(-1)^k}{k} (-1)^k = \Sigma \frac{1}{k}.$$

This is the harmonic series which, as you know, diverges. At $x = 1$

$$\Sigma \frac{(-1)^k}{k} x^k \quad \text{becomes} \quad \Sigma \frac{(-1)^k}{k}.$$

This is a convergent alternating series.
  We have shown that series (1) converges absolutely for $|x| < 1$, diverges at -1, and converges at 1. The interval of convergence is $(-1, 1]$. ☐

---

† We could also have used the root test:

$$(b_k)^{1/k} = \left| \frac{1}{k} \right|^{1/k} |x| = \frac{1}{k^{1/k}} |x| \to |x|.$$

Problem 2. Verify that the series

(1)
$$\Sigma \frac{1}{k^2} x^k$$

has interval of convergence $[-1, 1]$.

Solution. First we examine the series

(2)
$$\Sigma \left| \frac{1}{k^2} x^k \right| = \Sigma \frac{1}{k^2} |x|^k.$$

Here again we use the ratio test. We set

$$b_k = \frac{1}{k^2} |x|^k$$

and note that

$$\frac{b_{k+1}}{b_k} = \frac{k^2}{(k+1)^2} \frac{|x|^{k+1}}{|x|^k} = \left(\frac{k}{k+1} \right)^2 |x| \rightarrow |x|.$$

By the ratio test, (2) converges for $|x| < 1$ and diverges for $|x| > 1$.† This shows that (1) converges absolutely for $|x| < 1$ and diverges for $|x| > 1$. The radius of convergence is therefore 1.

Now for the endpoints. At $x = -1$,

$$\Sigma \frac{1}{k^2} x^k \quad \text{takes the form} \quad \Sigma \frac{(-1)^k}{k^2} = -1 + \tfrac{1}{4} - \tfrac{1}{9} + \tfrac{1}{16} - \cdots.$$

This is a convergent alternating series. At $x = 1$,

$$\Sigma \frac{1}{k^2} x^k \quad \text{becomes} \quad \Sigma \frac{1}{k^2}.$$

This is a convergent *p*-series. The interval of convergence is therefore the entire closed interval $[-1, 1]$. ☐

Problem 3. Find the interval of convergence of

(1)
$$\Sigma \frac{k}{6^k} x^k.$$

Solution. We begin by examining the series

(2)
$$\Sigma \left| \frac{k}{6^k} x^k \right| = \Sigma \frac{k}{6^k} |x|^k.$$

---

† Once again we could have used the root test:

$$(b_k)^{1/k} = \frac{1}{k^{2/k}} |x| \rightarrow |x|.$$

We set

$$b_k = \frac{k}{6^k} |x|^k$$

and apply the root test.   (The ratio test will also work.)   Since

$$b^{1/k} = \tfrac{1}{6} k^{1/k} |x| \rightarrow \tfrac{1}{6} |x|,$$

you can see that (2) converges

$$\text{for} \quad \tfrac{1}{6} |x| < 1 \qquad\qquad \text{(for } |x| < 6)$$

and diverges

$$\text{for} \quad \tfrac{1}{6} |x| > 1. \qquad\qquad \text{(for } |x| > 6)$$

This shows that (1) converges absolutely for $|x| < 6$ and diverges for $|x| > 6$.   The radius of convergence is 6.

It is easy to see that (1) diverges both at -6 and at 6.   The interval of convergence is therefore $(-6, 6)$.   ☐

Problem 4.   Find the interval of convergence of

(1)
$$\Sigma \frac{(2k)!}{(3k)!} x^k.$$

Solution.   We begin by examining the series

(2)
$$\Sigma \left| \frac{(2k)!}{(3k)!} x^k \right| = \Sigma \frac{(2k)!}{(3k)!} |x|^k.$$

Set

$$b_k = \frac{(2k)!}{(3k)!} |x|^k.$$

Since factorials are involved, we will use the ratio test.   Note that

$$\frac{b_{k+1}}{b_k} = \frac{[2(k+1)]!}{[3(k+1)]!} \frac{(3k)!}{(2k)!} \frac{|x|^{k+1}}{|x|^k} = \frac{(2k+2)(2k+1)}{(3k+3)(3k+2)(3k+1)} |x|.$$

Since

$$\frac{(2k+2)(2k+1)}{(3k+3)(3k+2)(3k+1)} \rightarrow 0 \qquad \text{as } k \rightarrow \infty$$

(the numerator is a quadratic in k, the denominator a cubic), the ratio b_{k+1}/b_k tends to 0 no matter what x is.   By the ratio test, (2) converges for all x and therefore (1) converges absolutely for all x.   The radius of convergence is ∞ and the interval of convergence is $(-\infty, \infty)$.   ☐

Problem 5. Find the interval of convergence of

$$\Sigma \left(\tfrac{1}{2}k\right)^k x^k.$$

Solution. Since $\left(\tfrac{1}{2}k\right)^k x^k \to 0$ only if $x = 0$ (explain), there is no need to invoke the ratio test or the root test. By (12.2.5) the series can converge only at $x = 0$. There it converges trivially. □

EXERCISES 12.8

Find the interval of convergence.

1. $\Sigma\ kx^k.$ **2.** $\Sigma\ \dfrac{1}{k}\,x^k.$ **3.** $\Sigma\ \dfrac{1}{(2k)!}\,x^k.$ **4.** $\Sigma\ \dfrac{2^k}{k^2}\,x^k.$

5. $\Sigma\ (-k)^{2k}x^{2k}.$ **6.** $\Sigma\ \dfrac{(-1)^k}{\sqrt{k}}\,x^k.$ **7.** $\Sigma\ \dfrac{1}{k2^k}\,x^k.$ **8.** $\Sigma\ \dfrac{1}{k^2 2^k}\,x^k.$

9. $\Sigma\ \left(\dfrac{k}{100}\right)^k x^k.$ **10.** $\Sigma\ \dfrac{k^2}{1+k^2}\,x^k.$ **11.** $\Sigma\ \dfrac{2^k}{\sqrt{k}}\,x^k.$ **12.** $\Sigma\ \dfrac{1}{\ln k}\,x^k.$

13. $\Sigma\ \dfrac{k-1}{k}\,x^k.$ **14.** $\Sigma\ ka^k x^k.$ **15.** $\Sigma\ \dfrac{k}{10^k}\,x^k.$ **16.** $\Sigma\ \dfrac{3^{k^2}}{e^k}\,x^k.$

17. $\Sigma\ \dfrac{x^k}{k^k}.$ **18.** $\Sigma\ \dfrac{7^k}{k!}\,x^k.$ **19.** $\Sigma\ \dfrac{(-1)^k}{k^k}\,(x-2)^k.$

20. $\Sigma\ \dfrac{(-1)^k a^k}{k^2}\,(x-a)^k.$ **21.** $\Sigma\ \dfrac{\ln k}{2^k}\,(x-2)^k.$ **22.** $\Sigma\ \dfrac{1}{(\ln k)^k}\,(x-1)^k.$

23. $\Sigma\ (-1)^k (\tfrac{2}{3})^k\,(x+1)^k.$ **24.** $\Sigma\ \dfrac{2^{1/k}\pi^k}{k(k+1)(k+2)}\,(x-2)^k.$

25. Let $\Sigma\ a_k x^k$ be a power series, and let r be its radius of convergence.
(a) Given that $|a_k|^{1/k} \to \rho$, show that, if $\rho \neq 0$, then $r = 1/\rho$ and, if $\rho = 0$, then $r = \infty$.
(b) Given that $|a_{k+1}/a_k| \to \lambda$, show that, if $\lambda \neq 0$, then $r = 1/\lambda$ and, if $\lambda = 0$, then $r = \infty$.

26. Find the interval of convergence of the series $\Sigma\ s_k x^k$ where s_k is the kth partial sum of the series

$$\sum_{n=1}^{\infty} 1/n.$$

12.9   DIFFERENTIATION AND INTEGRATION OF POWER SERIES

We begin with a simple but important result.

THEOREM 12.9.1

If

$$\sum_{k=0}^{\infty} a_k x^k = a_0 + a_1 x + a_2 x^2 + \cdots + a_n x^n + \cdots$$

converges on $(-c, c)$, then

$$\sum_{k=0}^{\infty} \frac{d}{dx}(a_k x^k) = \sum_{k=1}^{\infty} k a_k x^{k-1} = a_1 + 2a_2 x + \cdots + na_n x^{n-1} + \cdots$$

also converges on $(-c, c)$.

Proof. Let's assume that

$$\sum_{k=0}^{\infty} a_k x^k \quad \text{converges on } (-c, c).$$

By Theorem 12.8.2 it converges there absolutely.

Now let x be some fixed number in $(-c, c)$ and choose $\epsilon > 0$ such that

$$|x| < |x| + \epsilon < |c|.$$

Since $|x| + \epsilon$ lies within the interval of convergence,

$$\sum_{k=0}^{\infty} |a_k(|x| + \epsilon)^k| \quad \text{converges.}$$

In Exercise 34 you are asked to show that, for all k sufficiently large,

$$|kx^{k-1}| \le (|x| + \epsilon)^k.$$

It follows that for all such k

$$|ka_k x^{k-1}| \le |a_k(|x| + \epsilon)^k|.$$

Since

$$\sum_{k=0}^{\infty} |a_k(|x| + \epsilon)^k| \quad \text{converges,}$$

we can conclude that

$$\sum_{k=0}^{\infty} \left| \frac{d}{dx}(a_k x^k) \right| = \sum_{k=1}^{\infty} |ka_k x^{k-1}| \quad \text{converges,}$$

and thus that

$$\sum_{k=0}^{\infty} \frac{d}{dx}(a_k x^k) = \sum_{k=1}^{\infty} k a_k x^{k-1} \quad \text{converges.} \quad \square$$

Repeated application of the theorem shows that

$$\sum_{k=0}^{\infty} \frac{d^2}{dx^2}(a_k x^k), \qquad \sum_{k=0}^{\infty} \frac{d^3}{dx^3}(a_k x^k), \qquad \sum_{k=0}^{\infty} \frac{d^4}{dx^4}(a_k x^k), \qquad \text{etc.}$$

all converge on $(-c, c)$.

Example 1. Since the geometric series

$$\sum_{k=0}^{\infty} x^k = 1 + x + x^2 + x^3 + x^4 + x^5 + x^6 + \cdots$$

converges on $(-1, 1)$, the series

$$\sum_{k=0}^{\infty} \frac{d}{dx}(x^k) = \sum_{k=1}^{\infty} kx^{k-1} = 1 + 2x + 3x^2 + 4x^3 + 5x^4 + 6x^5 + \cdots,$$

$$\sum_{k=0}^{\infty} \frac{d^2}{dx^2}(x^k) = \sum_{k=2}^{\infty} k(k-1)x^{k-2} = 2 + 6x + 12x^2 + 20x^3 + 30x^4 + \cdots,$$

$$\sum_{k=0}^{\infty} \frac{d^3}{dx^3}(x^k) = \sum_{k=3}^{\infty} k(k-1)(k-2)x^{k-3} = 6 + 24x + 60x^2 + 120x^3 + \cdots,$$

$$\vdots$$

all converge on $(-1, 1)$. □

Suppose now that

$$\sum_{k=0}^{\infty} a_k x^k \quad \text{converges on } (-c, c).$$

Then, as you just saw,

$$\sum_{k=0}^{\infty} \frac{d}{dx}(a_k x^k) \quad \text{also converges on } (-c, c).$$

Using the first series, we can define a function f on $(-c, c)$ by setting

$$f(x) = \sum_{k=0}^{\infty} a_k x^k.$$

Using the second series, we can define a function g on $(-c, c)$ by setting

$$g(x) = \sum_{k=0}^{\infty} \frac{d}{dx}(a_k x^k).$$

The crucial point is that

$$f'(x) = g(x).$$

THEOREM 12.9.2   THE DIFFERENTIABILITY THEOREM

If

$$f(x) = \sum_{k=0}^{\infty} a_k x^k \qquad \text{for all } x \text{ in } (-c, c),$$

then f is differentiable on $(-c, c)$ and

$$f'(x) = \sum_{k=0}^{\infty} \frac{d}{dx}(a_k x^k) \qquad \text{for all } x \text{ in } (-c, c).$$

By applying this theorem to f', you can see that f' is itself differentiable. This in turn implies that f'' is differentiable, and so on. In short, f has derivatives of all orders.

We can summarize as follows:

---

in the interior of its interval of convergence a power series defines an infinitely differentiable function, the derivatives of which can be obtained by differentiating term by term:

$$\frac{d^n}{dx^n}\left(\sum_{k=0}^{\infty} a_k x^k\right) = \sum_{k=0}^{\infty} \frac{d^n}{dx^n}(a_k x^k) \qquad \text{for all } n.$$

For a detailed proof of the differentiability theorem see the supplement at the end of this section. We go on to examples.

Example 2. You know that

$$\frac{d}{dx}(e^x) = e^x.$$

You can see this directly by differentiating the exponential series:

$$\frac{d}{dx}(e^x) = \frac{d}{dx}\left(\sum_{k=0}^{\infty} \frac{x^k}{k!}\right) = \sum_{k=0}^{\infty} \frac{d}{dx}\left(\frac{x^k}{k!}\right) = \sum_{k=1}^{\infty} \frac{x^{k-1}}{(k-1)!} = \sum_{n=0}^{\infty} \frac{x^n}{n!} = e^x. \quad \square$$

$$\text{set } n = k - 1 \longrightarrow$$

Example 3. You have seen that

$$\sin x = x - \frac{x^3}{3!} + \frac{x^5}{5!} - \frac{x^7}{7!} + \frac{x^9}{9!} - \cdots$$

and

$$\cos x = 1 - \frac{x^2}{2!} + \frac{x^4}{4!} - \frac{x^6}{6!} + \frac{x^8}{8!} - \cdots .$$

The relations

$$\frac{d}{dx}(\sin x) = \cos x, \qquad \frac{d}{dx}(\cos x) = -\sin x$$

can be confirmed by differentiating the series term by term:

$$\frac{d}{dx}(\sin x) = 1 - \frac{3x^2}{3!} + \frac{5x^4}{5!} - \frac{7x^6}{7!} + \frac{9x^8}{9!} - \cdots$$

$$= 1 - \frac{x^2}{2!} + \frac{x^4}{4!} - \frac{x^6}{6!} + \frac{x^8}{8!} - \cdots = \cos x,$$

$$\frac{d}{dx}(\cos x) = -\frac{2x}{2!} + \frac{4x^3}{4!} - \frac{6x^5}{6!} + \frac{8x^7}{8!} - \cdots$$

$$= -x + \frac{x^3}{3!} - \frac{x^5}{5!} + \frac{x^7}{7!} - \cdots$$

$$= -\left(x - \frac{x^3}{3!} + \frac{x^5}{5!} - \frac{x^7}{7!} + \cdots\right) = -\sin x. \quad \square$$

Example 4. We can sum the series

$$\sum_{k=1}^{\infty} \frac{x^k}{k} \qquad \text{for all } x \text{ in } (-1, 1)$$

by setting

$$g(x) = \sum_{k=1}^{\infty} \frac{x^k}{k} \qquad \text{for all } x \text{ in } (-1, 1)$$

and noting that

$$g'(x) = \sum_{k=1}^{\infty} \frac{kx^{k-1}}{k} = \sum_{k=1}^{\infty} x^{k-1} = \sum_{n=0}^{\infty} x^n = \frac{1}{1-x}.$$

the geometric series

With

$$g'(x) = \frac{1}{1-x} \qquad \text{and} \qquad g(0) = 0,$$

we can conclude that

$$g(x) = -\ln(1-x) = \ln\left(\frac{1}{1-x}\right).$$

It follows that

$$\sum_{k=1}^{\infty} \frac{x^k}{k} = \ln\left(\frac{1}{1-x}\right) \qquad \text{for all } x \text{ in } (-1, 1). \quad \square$$

Power series can also be integrated term by term.

THEOREM 12.9.3 TERM-BY-TERM INTEGRATION

If $f(x) = \sum\limits_{k=0}^{\infty} a_k x^k$ converges on $(-c, c)$, then

$$g(x) = \sum_{k=0}^{\infty} \frac{a_k}{k+1} x^{k+1} \quad \text{converges on } (-c, c) \text{ and } \int f(x)\, dx = g(x) + C.$$

Proof. If $\sum\limits_{k=0}^{\infty} a_k x^k$ converges on $(-c, c)$, then $\sum\limits_{k=0}^{\infty} |a_k x^k|$ converges on $(-c, c)$. Since

$$\left| \frac{a_k}{k+1} x^k \right| \le |a_k x^k| \qquad \text{for all } k,$$

we know by comparison that

$$\sum_{k=0}^{\infty} \left| \frac{a_k}{k+1} x^k \right| \quad \text{also converges on } (-c, c).$$

It follows that

$$x \sum_{k=0}^{\infty} \frac{a_k}{k+1} x^k = \sum_{k=0}^{\infty} \frac{a_k}{k+1} x^{k+1} \quad \text{converges on } (-c, c).$$

With

$$f(x) = \sum_{k=0}^{\infty} a_k x^k \qquad \text{and} \qquad g(x) = \sum_{k=0}^{\infty} \frac{a_k}{k+1} x^{k+1},$$

we know from the differentiability theorem that

$$g'(x) = f(x) \qquad \text{and therefore} \qquad \int f(x)\, dx = g(x) + C. \quad \square$$

Term-by-term integration can be expressed as follows:

(12.9.4)
$$\int \left(\sum_{k=0}^{\infty} a_k x^k \right) dx = \left(\sum_{k=0}^{\infty} \frac{a_k}{k+1} x^{k+1} \right) + C.$$

If a power series converges at c and converges at d, then it converges at all numbers in between and

(12.9.5)
$$\int_c^d \left(\sum_{k=0}^{\infty} a_k x^k \right) dx = \sum_{k=0}^{\infty} \left(\int_c^d a_k x^k\, dx \right) = \sum_{k=0}^{\infty} \frac{a_k}{k+1} (d^{k+1} - c^{k+1}).$$

Example 5.   You are familiar with the series expansion

$$\frac{1}{1 + x} = \frac{1}{1 - (-x)} = \sum_{k=0}^{\infty} (-1)^k x^k.$$

It is valid for all x in $(-1, 1)$ and for no other x.   Integrating term by term we have

$$\ln(1 + x) = \int \left(\sum_{k=0}^{\infty} (-1)^k x^k \right) dx = \left(\sum_{k=0}^{\infty} \frac{(-1)^k}{k + 1} x^{k+1} \right) + C$$

for all x in $(-1, 1)$.   At $x = 0$ both $\ln(1 + x)$ and the series on the right are 0.   It follows that $C = 0$ and thus

$$\ln(1 + x) = \sum_{k=0}^{\infty} \frac{(-1)^k}{k + 1} x^{k+1} = x - \frac{x^2}{2} + \frac{x^3}{3} - \frac{x^4}{4} + \cdots$$

for all x in $(-1, 1)$.   ☐

In Section 12.6 we were able to prove that this expansion for $\ln(1 + x)$ was valid on the half-closed interval $(-1, 1]$; this gave us an expansion for $\ln 2$.   Term-by-term integration gives us only the open interval $(-1, 1)$.   Well, you may say, it's easy to see that the logarithm series also converges at $x = 1$.†   True enough, but why to $\ln 2$? This takes us back to consideration of the remainder term, the method of Section 12.6.

There is, however, another way to proceed.   The great Norwegian mathematician Niels Henrik Abel (1802–1829) proved the following result: suppose that

$$\sum_{k=0}^{\infty} a_k x^k \quad \text{converges on } (-c, c) \text{ and there represents } f(x).$$

If f is continuous at one of the endpoints (c or $-c$) and the series converges there, then the series represents the function at that point.   Using Abel's theorem it is evident that the series for $\ln(1 + x)$ does represent the function at $x = 1$.

We come now to another important series expansion:

(12.9.6) $$\tan^{-1} x = x - \frac{x^3}{3} + \frac{x^5}{5} - \frac{x^7}{7} + \cdots \qquad \text{for } -1 \le x \le 1.$$

Proof.   For x in $(-1, 1)$

$$\frac{1}{1 + x^2} = \frac{1}{1 - (-x^2)} = \sum_{k=0}^{\infty} (-1)^k x^{2k}$$

so that, by integration,

---

† An alternating series with $a_k \to 0$.

$$\tan^{-1} x = \int \left(\sum_{k=0}^{\infty} (-1)^k x^{2k} \right) dx = \left(\sum_{k=0}^{\infty} \frac{(-1)^k}{2k+1} x^{2k+1} \right) + C.$$

The constant C is 0 because the series on the right and the inverse tangent are both 0 at $x = 0$. Thus, for all x in $(-1, 1)$, we have

$$\tan^{-1} x = \sum_{k=0}^{\infty} \frac{(-1)^k}{2k+1} x^{2k+1} = x - \frac{x^3}{3} + \frac{x^5}{5} - \frac{x^7}{7} + \cdots.$$

That the series also represents the function at $x = -1$ and $x = 1$ follows directly from Abel's theorem: at both these points $\tan^{-1} x$ is continuous, and at both of these points the series converges. ☐

Since $\tan^{-1} 1 = \frac{1}{4}\pi$, we have

$$\tfrac{1}{4}\pi = 1 - \tfrac{1}{3} + \tfrac{1}{5} - \tfrac{1}{7} + \tfrac{1}{9} - \cdots.$$

This series was known to the Scottish mathematician James Gregory in 1671. It is an elegant formula for π, but it converges too slowly for computational purposes. A much more effective way of computing π is outlined in the supplement at the end of this section.

Term-by-term integration provides a method of calculating some (otherwise rather intractable) definite integrals. Suppose that you are trying to evaluate

$$\int_a^b f(x) \, dx$$

but cannot find an antiderivative. If you can expand $f(x)$ in a convergent power series, then you can estimate the integral by forming the series and integrating term by term.

Example 6.   We will estimate

$$\int_0^1 e^{-x^2} \, dx$$

by expanding the integral in a power series and integrating term by term. Our starting point is the expansion

$$e^x = 1 + x + \frac{x^2}{2!} + \frac{x^3}{3!} + \frac{x^4}{4!} + \frac{x^5}{5!} + \frac{x^6}{6!} + \cdots.$$

From this we see that

$$e^{-x^2} = 1 - x^2 + \frac{x^4}{2!} - \frac{x^6}{3!} + \frac{x^8}{4!} - \frac{x^{10}}{5!} + \frac{x^{12}}{6!} - \cdots$$

and therefore

$$\int_0^1 e^{-x^2} \, dx = \left[x - \frac{x^3}{3} + \frac{x^5}{5(2!)} - \frac{x^7}{7(3!)} + \frac{x^9}{9(4!)} - \frac{x^{11}}{11(5!)} + \frac{x^{13}}{13(6!)} - \cdots \right]_0^1$$

$$= 1 - \frac{1}{3} + \frac{1}{5(2!)} - \frac{1}{7(3!)} + \frac{1}{9(4!)} - \frac{1}{11(5!)} + \frac{1}{13(6!)} - \cdots.$$

This is an alternating series with declining terms. Therefore we know that the integral lies between consecutive partial sums. In particular it lies between

$$1 - \frac{1}{3} + \frac{1}{5(2!)} - \frac{1}{7(3!)} + \frac{1}{9(4!)} - \frac{1}{11(5!)}$$

and

$$\left[1 - \frac{1}{3} + \frac{1}{5(2!)} - \frac{1}{7(3!)} + \frac{1}{9(4!)} - \frac{1}{11(5!)}\right] + \frac{1}{13(6!)}.$$

As you can check, the first sum is greater than 0.7458 and the second one is less than 0.7466. It follows that

$$0.7458 < \int_0^1 e^{-x^2}\, dx < 0.7466.$$

The estimate 0.746 approximates the integral within 0.001.   ☐

The integral of Example 6 was easy to estimate numerically because it could be expressed as an alternating series with declining terms. The next example requires more subtlety and illustrates a method more general than that used in Example 6.

Example 7. We want to estimate

$$\int_0^1 e^{x^2}\, dx.$$

If we proceed exactly as in Example 6, we find that

$$\int_0^1 e^{x^2}\, dx = 1 + \frac{1}{3} + \frac{1}{5(2!)} + \frac{1}{7(3!)} + \frac{1}{9(4!)} + \frac{1}{11(5!)} + \frac{1}{13(6!)} + \cdots .$$

We now have a series expansion for the integral, but that expansion does not guide us directly to a numerical estimate for the integral. We know that s_n, the nth partial sum of the series, approximates the integral, but we don't know the accuracy of the approximation. We have no handle on the remainder left by s_n.

We start again, this time keeping track of the remainder. For $x \in [0, 1]$

$$0 \le e^x - \left(1 + x + \frac{x^2}{2!} + \cdots + \frac{x^n}{n!}\right) = R_{n+1}(x) \le e\left[\frac{x^{n+1}}{(n+1)!}\right] \le \frac{3}{(n+1)!}.$$

$$\text{(12.6.2)}$$

If $x \in [0, 1]$, then $x^2 \in [0, 1]$ and therefore

$$0 \le e^{x^2} - \left(1 + x^2 + \frac{x^4}{2!} + \cdots + \frac{x^{2n}}{n!}\right) \le \frac{3}{(n+1)!}.$$

Integrating this inequality from $x = 0$ to $x = 1$, we have

$$0 \leq \int_0^1 \left[e^{x^2} - \left(1 + x^2 + \frac{x^4}{2!} + \cdots + \frac{x^{2n}}{n!} \right) \right] dx \leq \int_0^1 \frac{3}{(n+1)!} \, dx.$$

Carrying out the integration where possible, we see that

$$0 \leq \int_0^1 e^{x^2} \, dx - \left[1 + \frac{1}{3} + \frac{1}{5(2!)} + \cdots + \frac{1}{(2n+1)(n!)} \right] \leq \frac{3}{(n+1)!}.$$

We can use this inequality to estimate the integral as closely as we wish. Since

$$\frac{3}{7!} = \frac{1}{1680} < 0.0006,$$

we see that

$$\alpha = 1 + \frac{1}{3} + \frac{1}{5(2!)} + \frac{1}{7(3!)} + \frac{1}{9(4!)} + \frac{1}{11(5!)} + \frac{1}{13(6!)}$$

approximates the integral within 0.0006. Arithmetical computation shows that

$$1.4626 \leq \alpha \leq 1.4627.$$

It follows that

$$1.4626 \leq \int_0^1 e^{x^2} \, dx \leq 1.4627 + 0.0006 = 1.4633.$$

The estimate 1.463 approximates the integral within 0.0004. □

It is time to relate Taylor series

$$\sum_{k=0}^{\infty} \frac{f^{(k)}(0)}{k!} x^k$$

to power series in general. The relation is very simple:

> On its interval of convergence a power series is the Taylor series of its sum.

To see this, all you have to do is differentiate

$$f(x) = a_0 + a_1 x + a_2 x^2 + \cdots + a_k x^k + \cdots$$

term by term. Do this and you will find that $f^{(k)}(0) = k! \, a_k$ and therefore

$$a_k = \frac{f^{(k)}(0)}{k!}.$$

The a_k are the Taylor coefficients of f.

We end this section by carrying out a few simple expansions.

Problem 8. Expand cosh x and sinh x in powers of x.

Solution. There is no need to go through the labor of computing the Taylor coefficients

$$\frac{f^{(k)}(0)}{k!}$$

by differentiation. We know that

$$\cosh x = \tfrac{1}{2}(e^x + e^{-x}) \quad \text{and} \quad \sinh x = \tfrac{1}{2}(e^x - e^{-x}). \tag{7.11.1}$$

Since

$$e^x = 1 + x + \frac{x^2}{2!} + \frac{x^3}{3!} + \frac{x^4}{4!} + \frac{x^5}{5!} + \cdots ,$$

we have

$$e^{-x} = 1 - x + \frac{x^2}{2!} - \frac{x^3}{3!} + \frac{x^4}{4!} - \frac{x^5}{5!} + \cdots .$$

Thus

$$\cosh x = \frac{1}{2}\left(2 + 2\frac{x^2}{2!} + 2\frac{x^4}{4!} + \cdots\right) = 1 + \frac{x^2}{2!} + \frac{x^4}{4!} + \cdots = \sum_{k=0}^{\infty} \frac{x^{2k}}{(2k)!}$$

and

$$\sinh x = \frac{1}{2}\left(2x + 2\frac{x^3}{3!} + 2\frac{x^5}{5!} + \cdots\right) = x + \frac{x^3}{3!} + \frac{x^5}{5!} + \cdots = \sum_{k=0}^{\infty} \frac{x^{2k+1}}{(2k+1)!}.$$

Both expansions are valid for all real x, since the exponential expansions are valid for all real x. □

Problem 9. Expand $x^2 \cos x^3$ in powers of x.

Solution

$$\cos x = 1 - \frac{x^2}{2!} + \frac{x^4}{4!} - \frac{x^6}{6!} + \cdots .$$

Thus

$$\cos x^3 = 1 - \frac{(x^3)^2}{2!} + \frac{(x^3)^4}{4!} - \frac{(x^3)^6}{6!} + \cdots = 1 - \frac{x^6}{2!} + \frac{x^{12}}{4!} - \frac{x^{18}}{6!} + \cdots ,$$

and

$$x^2 \cos x^3 = x^2 - \frac{x^8}{2!} + \frac{x^{14}}{4!} - \frac{x^{20}}{6!} + \cdots .$$

This expansion is valid for all real x, since the expansion for $\cos x$ is valid for all real x. □

Alternative Solution. Since

$$x^2 \cos x^3 = \frac{d}{dx}\left(\frac{1}{3}\sin x^3\right),$$

we can derive the expansion for $x^2 \cos x^3$ by expanding $\frac{1}{3}\sin x^3$ and then differentiating term by term. $\square$

EXERCISES 12.9

Expand in powers of x, basing your calculations on the geometric series

$$\frac{1}{1-x} = 1 + x + x^2 + \cdots + x^n + \cdots.$$

1. $\dfrac{1}{(1-x)^2}$.

2. $\dfrac{1}{(1-x)^3}$.

3. $\dfrac{1}{(1-x)^k}$.

4. $\ln(1-x)$.

5. $\ln(1-x^2)$.

6. $\ln(2-3x)$.

Expand in powers of x, basing your calculations on the tangent series:

$$\tan x = x + \tfrac{1}{3}x^3 + \tfrac{2}{15}x^5 + \tfrac{17}{315}x^7 + \cdots.$$

7. $\sec^2 x$.

8. $\ln\cos x$.

Find $f^{(9)}(0)$.

9. $f(x) = x^2 \sin x$.

10. $f(x) = x\cos x^2$.

Expand in powers of x.

11. $\sin x^2$.

12. $x^2 \tan^{-1} x$.

13. e^{3x^3}.

14. $\dfrac{1-x}{1+x}$.

15. $\dfrac{2x}{1-x^2}$.

16. $x\sinh x^2$.

17. $\dfrac{1}{1-x} + e^x$.

18. $\cosh x \sinh x$.

19. $x\ln(1+x^3)$.

20. $(x^2 + x)\ln(1+x)$.

21. $x^3 e^{-x^3}$.

22. $x^5(\sin x + \cos 2x)$.

(Calculator) Estimate within 0.01.

23. $\displaystyle\int_0^1 e^{-x^3}\,dx$.

24. $\displaystyle\int_0^1 \sin x^2\,dx$.

25. $\displaystyle\int_0^1 \sin\sqrt{x}\,dx$.

26. $\displaystyle\int_0^1 x^4 e^{-x^2}\,dx$.

27. $\displaystyle\int_0^1 \tan^{-1} x^2\,dx$.

28. $\displaystyle\int_1^2 \frac{1-\cos x}{x}\,dx$.

Sum the following series.

29. $\displaystyle\sum_{k=0}^{\infty} \frac{1}{k!}x^{3k}$.

30. $\displaystyle\sum_{k=0}^{\infty} \frac{1}{k!}x^{3k+1}$.

31. $\displaystyle\sum_{k=1}^{\infty} \frac{3k}{k!}x^{3k-1}$.

32. Deduce the differentiation formulas

$$\frac{d}{dx}(\sinh x) = \cosh x, \qquad \frac{d}{dx}(\cosh x) = \sinh x$$

from the expansions of $\sinh x$ and $\cosh x$ in powers of x.

33. Show that, if $\Sigma\, a_k x^k$ and $\Sigma\, b_k x^k$ both converge to the same sum on some interval, then $a_k = b_k$ for each k.

34. Show that, if $\epsilon > 0$, then

$$|kx^{k-1}| < (|x| + \epsilon)^k \qquad \text{for all } k \text{ sufficiently large.}$$

HINT: Take the kth root of the left side and let $k \to \infty$.

(**Calculator**) Estimate within 0.001 by the method of this section and check your result by carrying out the integration directly.

35. $\displaystyle\int_0^{1/2} x \ln (1 + x)\, dx.$ 36. $\displaystyle\int_0^1 x \sin x\, dx.$ 37. $\displaystyle\int_0^1 xe^{-x}\, dx.$

38. Show that

$$0 \le \int_0^2 e^{x^2}\, dx - \left[2 + \frac{2^3}{3} + \frac{2^5}{5(2!)} + \cdots + \frac{2^{2n+1}}{(2n + 1)n!} \right] < \frac{e^4 2^{2n+3}}{(n + 1)!}.$$

Program for estimating the integral of $f(x) = e^{-x^2}$ (BASIC)

The integral

$$\int_0^b e^{-x^2}\, dx$$

occurs frequently in practice, because it is related to the normal distribution ("bell curve") in probability theory. The unfortunate thing about this integral is that there is no way to find an antiderivative *in closed form* for the function

$$f(x) = e^{-x^2}.$$

(That is, if you start from the elementary functions, like polynomials, trigonometric functions, logarithms, and exponentials, and perform any compositions you like and any arithmetic operations, you will never be able to write down an antiderivative for the function above. There *is* an antiderivative, of course—the fundamental theorem of calculus tells us that—it is just that the antiderivative can't be conveniently written down in the way in which we are accustomed to doing it.) In these circumstances, the value of the integral must be computed numerically. This integral uses the series for the function and integrates the series term by term.

```
10 REM Estimate Integral exp(-x^2)
20 REM Copyright © Colin C. Graham 1988-1989

100 INPUT "Enter number of iterations:"; n
110 INPUT "Enter right endpoint:"; b

200 Integral = b
210 oneovernfactorial = 1
220 numerator = -1
230 PRINT "GaussInt from 0 to"; b

300 FOR j = 1 TO n
310     PRINT Integral
320     oneovernfactorial = oneovernfactorial/j
330     Integral = Integral + (numerator*oneovernfactorial/(2*j + 1))
340     numerator = -b*b*numerator
350 NEXT j
```

400 **PRINT** Integral

500 **INPUT** "Do it again? (Y/N)"; a$
510 **IF** a$ = "Y" **OR** a$ = "y" **THEN** 100
520 **END**

*SUPPLEMENT TO SECTION 12.9

Proof of Theorem 12.9.2

Set

$$f(x) = \sum_{k=0}^{\infty} a_k x^k \quad \text{and} \quad g(x) = \sum_{k=0}^{\infty} \frac{d}{dx}(a_k x^k) = \sum_{k=1}^{\infty} ka_k x^{k-1}.$$

Select x from $(-c, c)$. We want to show that

$$\lim_{h \to 0} \frac{f(x+h) - f(x)}{h} = g(x).$$

For $x + h$ in $(-c, c)$, $h \neq 0$, we have

$$\left| g(x) - \frac{f(x+h) - f(x)}{h} \right| = \left| \sum_{k=1}^{\infty} ka_k x^{k-1} - \sum_{k=0}^{\infty} \frac{a_k(x+h)^k - a_k x^k}{h} \right|$$

$$= \left| \sum_{k=1}^{\infty} ka_k x^{k-1} - \sum_{k=1}^{\infty} a_k \left[\frac{(x+h)^k - x^k}{h} \right] \right|.$$

By the mean-value theorem

$$\frac{(x+h)^k - x^k}{h} = kt_k^{k-1}$$

for some number t_k between x and $x + h$. Thus we can write

$$\left| g(x) - \frac{f(x+h) - f(x)}{h} \right| = \left| \sum_{k=1}^{\infty} ka_k x^{k-1} - \sum_{k=1}^{\infty} ka_k t_k^{k-1} \right|$$

$$= \left| \sum_{k=1}^{\infty} ka_k(x^{k-1} - t_k^{k-1}) \right|$$

$$= \left| \sum_{k=2}^{\infty} ka_k(x^{k-1} - t_k^{k-1}) \right|.$$

By the mean-value theorem

$$\frac{x^{k-1} - t_k^{k-1}}{x - t_k} = (k-1)p_{k-1}^{k-2}$$

for some number p_{k-1} between x and t_k. Obviously, then,

$$|x^{k-1} - t_k^{k-1}| = |x - t_k||(k-1)p_{k-1}^{k-2}|.$$

Since $|x - t_k| < |h|$ and $|p_{k-1}| \leq |\alpha|$ where $|\alpha| = \max \{|x|, |x + h|\}$,

$$|x^{k-1} - t_k^{k-1}| \leq |h||(k-1)\alpha^{k-2}|.$$

Thus

$$\left| g(x) - \frac{f(x+h) - f(x)}{h} \right| \leq |h| \sum_{k=2}^{\infty} |k(k-1)a_k \alpha^{k-2}|.$$

Since the series converges,

$$\lim_{h \to 0} \left(|h| \sum_{k=2}^{\infty} |k(k-1)a_k \alpha^{k-2}| \right) = 0.$$

This gives

$$\lim_{h \to 0} \left| g(x) - \frac{f(x+h) - f(x)}{h} \right| = 0 \quad \text{and thus} \quad \lim_{h \to 0} \frac{f(x+h) - f(x)}{h} = g(x). \quad \square$$

Calculating π

We base our computation of π on the inverse tangent series

$$\tan^{-1} x = x - \frac{x^3}{3} + \frac{x^5}{5} - \frac{x^7}{7} + \cdots \qquad \text{for } -1 \le x \le 1$$

and the relation

(12.9.7)
$$\tfrac{1}{4}\pi = 4 \tan^{-1} \tfrac{1}{5} - \tan^{-1} \tfrac{1}{239}.\dagger$$

The inverse tangent series gives

$$\tan^{-1} \tfrac{1}{5} = \tfrac{1}{5} - \tfrac{1}{3}(\tfrac{1}{5})^3 + \tfrac{1}{5}(\tfrac{1}{5})^5 - \tfrac{1}{7}(\tfrac{1}{5})^7 + \cdots$$

and

$$\tan^{-1} \tfrac{1}{239} = \tfrac{1}{239} - \tfrac{1}{3}(\tfrac{1}{239})^3 + \tfrac{1}{5}(\tfrac{1}{239})^5 - \tfrac{1}{7}(\tfrac{1}{239})^7 + \cdots .$$

These are alternating series $\Sigma (-1)^k a_k$ with a_k decreasing toward 0. Thus we know that

$$\tfrac{1}{5} - \tfrac{1}{3}(\tfrac{1}{5})^3 \le \tan^{-1} \tfrac{1}{5} \le \tfrac{1}{5} - \tfrac{1}{3}(\tfrac{1}{5})^3 + \tfrac{1}{5}(\tfrac{1}{5})^5$$

and

$$\tfrac{1}{239} - \tfrac{1}{3}(\tfrac{1}{239})^3 \le \tan^{-1} \tfrac{1}{239} \le \tfrac{1}{239}.$$

With these inequalities together with relation (12.9.7), we can show that

$$3.14 < \pi < 3.147.$$

By using six terms of the series for $\tan^{-1} \tfrac{1}{5}$ and still only two of the series for $\tan^{-1} \tfrac{1}{239}$, we can show that

$$3.14159262 < \pi < 3.14159267.$$

Greater accuracy can be obtained by taking more terms into account. For instance, fifteen terms of the series for $\tan^{-1} \tfrac{1}{5}$ and just four terms of the series for $\tan^{-1} \tfrac{1}{239}$ determine π to twenty decimal places:

$$\pi \cong 3.14159\ 26535\ 89793\ 23846.$$

---

† This relation was discovered in 1706 by John Machin, a Scotsman. It can be verified by repeated applications of the addition formula

$$\tan(A + B) = \frac{\tan A + \tan B}{1 - \tan A \tan B}.$$

First calculate $\tan(2 \tan^{-1} \tfrac{1}{5})$, then $\tan(4 \tan^{-1} \tfrac{1}{5})$, and finally $\tan(4 \tan^{-1} \tfrac{1}{5} - \tan^{-1} \tfrac{1}{239})$.

12.10   THE BINOMIAL SERIES

Through a collection of problems we invite you to derive for yourself the basic properties of one of the most celebrated series of all—*the binomial series.*

Start with the binomial $1 + x$ (2 terms). Choose a real number $\alpha \neq 0$ and form the function

$$f(x) = (1 + x)^{\alpha}.$$

Problem 1. Show that

$$\frac{f^{(k)}(0)}{k!} = \frac{\alpha[\alpha - 1][\alpha - 2] \cdots [\alpha - (k - 1)]}{k!}.$$

The number you just obtained is the coefficient of x^k in the expansion of $(1 + x)^{\alpha}$. It is called *the kth binomial coefficient* and is usually denoted by $\binom{\alpha}{k}$:

$$\binom{\alpha}{k} = \frac{\alpha[\alpha - 1][\alpha - 2] \cdots [\alpha - (k - 1)]}{k!}.$$

Problem 2. Show that the binomial series

$$\Sigma \binom{\alpha}{k} x^k$$

has radius of convergence 1.   HINT: Use the ratio test.

From Problem 2 you know that the binomial series converges on the open interval $(-1, 1)$ and defines there an infinitely differentiable function. The next thing to show is that this function (the one defined by the series) is actually $(1 + x)^{\alpha}$. To do this, you first need some other results.

Problem 3. Verify the identity

$$(k + 1) \binom{\alpha}{k + 1} + k \binom{\alpha}{k} = \alpha \binom{\alpha}{k}.$$

Problem 4. Use the identity of Problem 3 to show that the sum of the binomial series

$$\phi(x) = \sum_{k=0}^{\infty} \binom{\alpha}{k} x^k$$

satisfies the differential equation

$$(1 + x)\phi'(x) = \alpha\phi(x) \qquad \text{for all } x \text{ in } (-1, 1)$$

together with the side condition $\phi(0) = 1$.

You are now in a position to prove the main result.

Problem 5. Show that

(12.10.1)

$$(1 + x)^\alpha = \sum_{k=0}^{\infty} \binom{\alpha}{k} x^k \qquad \text{for all } x \text{ in } (-1, 1).$$

You can probably get a better feeling for the series by writing out the first few terms:

(12.10.2)

$$(1 + x)^\alpha = 1 + \alpha x + \frac{\alpha(\alpha - 1)}{2!} x^2 + \frac{\alpha(\alpha - 1)(\alpha - 2)}{3!} x^3 + \cdots.$$

EXERCISES 12.10

Expand in powers of x up to x^4.

1. $\sqrt{1 + x}.$ **2.** $\sqrt{1 - x}.$ **3.** $\sqrt{1 + x^2}.$ **4.** $\sqrt{1 - x^2}.$

5. $\dfrac{1}{\sqrt{1 + x}}.$ **6.** $\dfrac{1}{\sqrt[3]{1 + x}}.$ **7.** $\sqrt[4]{1 - x}.$ **8.** $\dfrac{1}{\sqrt[4]{1 + x}}.$

(Calculator) Estimate by using the first three terms of a binomial expansion rounding off your answer to four decimal places.

9. $\sqrt{98}.$ HINT: $\sqrt{98} = (100 - 2)^{1/2} = 10(1 - \frac{1}{50})^{1/2}.$ **10.** $\sqrt[5]{36}.$

11. $\sqrt[3]{9}.$ **12.** $\sqrt[4]{620}.$ **13.** $17^{-1/4}.$ **14.** $9^{-1/3}.$

12.11 ADDITIONAL EXERCISES

Sum the following series.

1. $\displaystyle\sum_{k=0}^{\infty} \left(\frac{1}{4}\right)^k.$ **2.** $\displaystyle\sum_{k=0}^{\infty} \left(\frac{3}{4}\right)^{k+1}.$ **3.** $\displaystyle\sum_{k=0}^{\infty} (-1)^k \left(\frac{1}{2}\right)^k.$

4. $\displaystyle\sum_{k=0}^{\infty} \frac{(\ln 2)^k}{k!}.$ **5.** $\displaystyle\sum_{k=1}^{\infty} \left(\frac{1}{k} - \frac{1}{k+1}\right).$ **6.** $\displaystyle\sum_{k=2}^{\infty} \left(\frac{1}{k^2} - \frac{1}{(k+1)^2}\right).$

7. $\displaystyle\sum_{k=0}^{\infty} x^{5k+1}.$ **8.** $\displaystyle\sum_{k=0}^{\infty} 2x^{3k+2}.$ **9.** $\displaystyle\sum_{k=1}^{\infty} \tfrac{3}{2} x^{2k-1}.$

10. $\displaystyle\sum_{k=1}^{\infty} \frac{1}{(k-1)!} x^k.$ **11.** $\displaystyle\sum_{k=1}^{\infty} \frac{k^2}{k!}.$ **12.** $\displaystyle\sum_{k=1}^{\infty} \frac{1}{k(k+1)(k+2)}.$

Test for (a) absolute convergence, (b) conditional convergence.

13. $\displaystyle\sum_{k=0}^{\infty} \frac{1}{2k+1} = 1 + \frac{1}{3} + \frac{1}{5} + \cdots.$

14. $\displaystyle\sum_{k=0}^{\infty} \frac{1}{(2k+1)(2k+3)} = \frac{1}{1 \cdot 3} + \frac{1}{3 \cdot 5} + \frac{1}{5 \cdot 7} + \cdots.$

15. $\displaystyle\sum_{k=1}^{\infty} \frac{(-1)^{k+1}}{(k+1)(k+2)} = \frac{1}{2\cdot 3} - \frac{1}{3\cdot 4} + \frac{1}{4\cdot 5} - \cdots .$

16. $\displaystyle\sum_{k=2}^{\infty} \frac{1}{k\ln k} = \frac{1}{2\ln 2} + \frac{1}{3\ln 3} + \frac{1}{4\ln 4} + \cdots .$

17. $\displaystyle\sum_{k=0}^{\infty} \frac{(-1)^k}{2k+1} = 1 - \frac{1}{3} + \frac{1}{5} - \cdots .$

18. $\displaystyle\sum_{k=1}^{\infty} (-1)^{k+1} \frac{100^k}{k!} = 100 - \frac{100^2}{2!} + \frac{100^3}{3!} - \cdots .$

19. $\displaystyle\sum_{k=1}^{\infty} (-1)^{k-1} \frac{k}{3^{k-1}} = 1 - \frac{2}{3} + \frac{3}{3^2} - \cdots .$

20. $\displaystyle\sum_{k=1}^{\infty} k\left(\frac{3}{4}\right)^k = \frac{3}{4} + 2\left(\frac{3}{4}\right)^2 + 3\left(\frac{3}{4}\right)^3 + \cdots .$

21. $\displaystyle\sum_{k=1}^{\infty} \frac{(-1)^{k-1}}{\sqrt{(k+1)(k+2)}} = \frac{1}{\sqrt{2\cdot 3}} - \frac{1}{\sqrt{3\cdot 4}} + \frac{1}{\sqrt{4\cdot 5}} - \cdots .$

22. $\displaystyle\sum_{k=1}^{\infty} \frac{(-1)^{k-1}}{\sqrt[k]{5}} = \frac{1}{5} - \frac{1}{\sqrt{5}} + \frac{1}{\sqrt[3]{5}} - \cdots .$

Find the interval of convergence.

23. $\Sigma \dfrac{5^k}{k}(x-2)^k.$

24. $\Sigma \dfrac{(-1)^k}{3^k} x^{k+1}.$

25. $\Sigma (k+1)k(x-1)^{2k}.$

26. $\Sigma (-1)^k 4^k x^{2k}.$

27. $\Sigma \dfrac{k}{2k+1} x^{2k+1}.$

28. $\Sigma \dfrac{1}{2^{k!}}(x-2)^k.$

29. $\Sigma \dfrac{k!}{2}(x+1)^k.$

30. $\Sigma \dfrac{(-1)^k}{\sqrt{k}}(x+3)^k.$

31. $\Sigma \dfrac{(-1)^k k}{3^{2k}} x^k.$

32. $\Sigma \ln k\, (x-2)^k.$

33. $\Sigma \dfrac{(-1)^k}{5^{k+1}}(x-2)^k.$

34. $\Sigma \dfrac{2^k}{(2k)!}(x-1)^{2k}.$

Expand in powers of x.

35. $xe^{5x^2}.$

36. $\ln(1+x^2).$

37. $\sqrt{x}\tan^{-1}\sqrt{x}.$

38. $a^x.$

39. $(x+x^2)(\sin x^2).$

40. $x\ln\left(\dfrac{1+x^2}{1-x^2}\right).$

41. $e^{\sin x}$ up to x^4.

42. $e^{\sin x}\cos x$ up to x^3.

43. $(1-x^2)^{-1/2}$ up to x^4.

44. $\sin^{-1} x$ up to x^5.

Estimate within 0.01 from a series expansion.

45. $\displaystyle\int_0^{1/2} \frac{dx}{1+x^4}.$

46. $e^{2/3}.$

47. $\sqrt[3]{68}.$

48. $\displaystyle\int_0^1 x\sin x^4\, dx.$

49. Find the sum of the series

$$\sum_{k=1}^{\infty} a_k \quad \text{given that} \quad a_k = \int_k^{k+1} xe^{-x}\, dx.$$

50. Show that

$$\sum_{k=1}^{\infty} \frac{1}{k^2} = 1 + \sum_{k=1}^{\infty} \frac{1}{k^2(k+1)}.$$

51. Determine whether or not the series

$$\sum_{k=2}^{\infty} a_k$$

converges or diverges.   If it converges, find the sum.

(a) $a_k = \sum_{n=0}^{\infty} \left(\frac{1}{k}\right)^n.$ (b) $a_k = \sum_{n=1}^{\infty} \left(\frac{1}{k}\right)^n.$ (c) $a_k = \sum_{n=2}^{\infty} \left(\frac{1}{k}\right)^n.$

52. (*The Lagrange form of the remainder*) Show that the remainder in Taylor's theorem (Theorem 12.6.1) can be written

$$R_{n+1}(x) = f^{(n+1)}(c_{n+1}) \frac{x^{n+1}}{(n+1)!}$$

with c_{n+1} between 0 and x.   HINT: For $x > 0$

$$\frac{1}{n!} \int_0^x m_{n+1}(x-t)^n \, dt \le R_{n+1}(x) \le \frac{1}{n!} \int_0^x M_{n+1}(x-t)^n \, dt$$

where

$$m_{n+1} = \min_{t \in I} f^{(n+1)}(t) \quad \text{and} \quad M_{n+1} = \max_{t \in I} f^{(n+1)}(t).$$

53. Show that every sequence of real numbers can be covered by a sequence of open intervals of arbitrarily small total length; namely, show that if $\{x_1, x_2, x_3, \ldots\}$ is a sequence of real numbers and ϵ is positive, then there exists a sequence of open intervals (a_n, b_n) with $a_n < x_n < b_n$ such that

$$\sum_{n=1}^{\infty} (b_n - a_n) < \epsilon.$$

CHAPTER HIGHLIGHTS

12.1 Sigma Notation

12.2 Infinite Series

partial sums (p. 614)   convergence, divergence (p. 614)
sum of a series (p. 614)   a divergence test (p. 621)

geometric series: $\sum_{k=0}^{\infty} x^k = \left\{ \begin{array}{l} \dfrac{1}{1-x}, \quad |x| < 1 \\ \text{diverges}, \quad |x| \ge 1 \end{array} \right.$

If $\sum_{k=0}^{\infty} a_k$ converges, then $a_k \to 0$.   The converse is false.

12.3 The Integral Test; Comparison Theorems

integral test (p. 624)   basic comparison (p. 627)
limit comparison (p. 629)

harmonic series: $\sum_{k=1}^{\infty} \frac{1}{k}$ diverges p-series: $\sum_{k=1}^{\infty} \frac{1}{k^p}$ converges iff $p > 1$

12.4 The Root Test; The Ratio Test

root test (p. 632)   ratio test (p. 634)
summary on convergence tests (p. 636)

12.5 Absolute and Conditional Convergence; Alternating Series

absolutely convergent, conditionally convergent (p. 637)
convergence theorem for alternating series (p. 639)
an estimate for alternating series (p. 640)
rearrangements (p. 640)

12.6 Taylor Polynomials in x; Taylor Series in x

Taylor polynomials in x (p. 642)   remainder term $R_{n+1}(x)$ (p. 645)
remainder estimate (p. 645)

Taylor series in x (Maclaurin series): $\displaystyle\sum_{k=0}^{\infty} \frac{f^{(k)}(0)}{k!} x^k$

$$e^x = \sum_{k=0}^{\infty} \frac{x^k}{k!}, \quad \text{all real } x \qquad\qquad \ln(1+x) = \sum_{k=1}^{\infty} \frac{(-1)^{k+1}}{k} x^k, \quad -1 < x \le 1$$

$$\sin x = \sum_{k=0}^{\infty} \frac{(-1)^k}{(2k+1)!} x^{2k+1}, \quad \text{all real } x \qquad \cos x = \sum_{k=0}^{\infty} \frac{(-1)^k}{(2k)!} x^{2k}, \quad \text{all real } x$$

12.7 Taylor Polynomials in $x - a$; Taylor Series in $x - a$

Taylor series in $x - a$: $\displaystyle\sum_{k=0}^{\infty} \frac{g^{(k)}(a)}{k!} (x-a)^k$

12.8 Power Series

power series (p. 658)   radius of convergence (p. 660)
interval of convergence (p. 661)

If a power series converges at $x_1 \ne 0$, then it converges absolutely for $|x| < |x_1|$; if it diverges at x_1, then it diverges for $|x| > |x_1|$.

12.9 Differentiation and Integration of Power Series

$$\tan^{-1} x = \sum_{k=0}^{\infty} \frac{(-1)^k}{2k+1} x^{2k+1}, \quad -1 \le x \le 1$$

$$\cosh x = \sum_{k=0}^{\infty} \frac{x^{2k}}{(2k)!}, \quad \text{all real } x \qquad\qquad \sinh x = \sum_{k=0}^{\infty} \frac{x^{2k+1}}{(2k+1)!}, \quad \text{all real } x$$

On the interior of its interval of convergence, a power series can be differentiated and integrated term by term.

On its interval of convergence a power series is the Taylor series of its sum.

12.10 The Binomial Series

$$(1+x)^\alpha = \sum_{k=0}^{\infty} \binom{\alpha}{k} x^k = 1 + \alpha x + \frac{\alpha(\alpha-1)}{2!} x^2 + \cdots, \quad -1 < x < 1$$

12.11 Additional Exercises

13

VECTORS

13.1 CARTESIAN SPACE COORDINATES

To introduce a Cartesian coordinate system in three-dimensional space we begin
with a plane Cartesian coordinate system O-xy. Through the point O, which we
continue to call the origin, we pass a third line, perpendicular to the other two. This
third line we call the z-axis. We coordinatize it using the same scale as before,
assigning z-coordinate 0 to the origin O.

For later convenience we orient the z-axis so that O-xyz forms a "right-handed"
system. That is, if the index finger of the right hand points along the positive x-axis
and the middle finger along the positive y-axis, then the thumb will point along the
positive z-axis. (See Figure 13.1.1.)

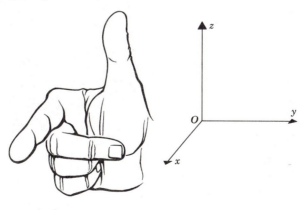

Figure 13.1.1

The point on the x-axis with x-coordinate x_0 has space coordinates $(x_0, 0, 0)$; the point on the y-axis with y-coordinate y_0 has space coordinates $(0, y_0, 0)$; the point on the z-axis with z-coordinate z_0 has space coordinates $(0, 0, z_0)$.

The point P with coordinates $x = x_0$, $y = y_0$, $z = z_0$ is denoted by $P(x_0, y_0, z_0)$. The plane through P perpendicular to the x-axis will intersect the x-axis at the point $(x_0, 0, 0)$; the plane through P perpendicular to the y-axis will intersect the y-axis at $(0, y_0, 0)$; the plane through P perpendicular to the z-axis will intersect the z-axis at $(0, 0, z_0)$. (See, for example, Figure 13.1.2.)

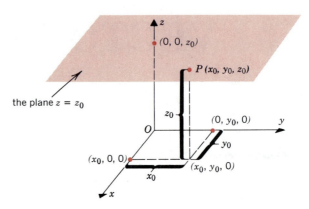

Figure 13.1.2

A point is in the xy-plane iff it is of the form $(x, y, 0)$. Thus, an equation for the xy-plane is $z = 0$. The equation

$$z = z_0$$

represents a plane parallel to the xy-plane, z_0 units above the xy-plane when z_0 is positive, $-z_0$ units below the xy-plane when z_0 is negative. Similarly, equations of the form $x = $ a constant represent planes parallel to the yz-plane, and equations of the form $y = $ a constant represent planes parallel to xz-plane. (See, for example, Figure 13.1.3. There we have drawn the planes $x = 1$ and $y = 3$.)

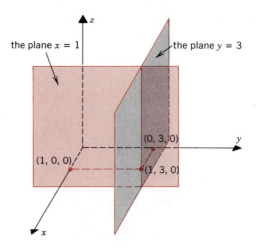

Figure 13.1.3

The Distance Formula

The distance $d(P_1, P_2)$ between two points $P_1(x_1, y_1, z_1)$ and $P_2(x_2, y_2, z_2)$ can be found by applying the Pythagorean theorem twice.

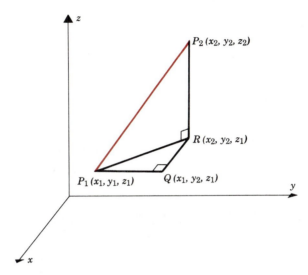

Figure 13.1.4

With Q and R as in Figure 13.1.4, P_1P_2R and P_1RQ are both right triangles. From the first triangle

$$[d(P_1, P_2)]^2 = [d(P_1, R)]^2 + [d(R, P_2)]^2,$$

and from the second triangle

$$[d(P_1, R)]^2 = [d(Q, R)]^2 + [d(P_1, Q)]^2.$$

Combining equations,

$$[d(P_1, P_2)]^2 = [d(Q, R)]^2 + [d(P_1, Q)]^2 + [d(R, P_2)]^2$$
$$= (x_2 - x_1)^2 + (y_2 - y_1)^2 + (z_2 - z_1)^2.$$

Taking square roots, we have the distance formula:

(13.1.1)
$$d(P_1, P_2) = \sqrt{(x_2 - x_1)^2 + (y_2 - y_1)^2 + (z_2 - z_1)^2}.$$

The *sphere* of radius r centered at $P_0(a, b, c)$ is the set of all points $P(x, y, z)$ with $d(P, P_0) = r$. We can obtain the equation of this sphere by using (13.1.1):

(13.1.2)
$$\text{Equation of Sphere}$$
$$(x - a)^2 + (y - b)^2 + (z - c)^2 = r^2.$$

Symmetry

You are already familiar with two kinds of symmetry: symmetry about a point and symmetry about a line. In space we can also speak of symmetry about a plane. The idea is obvious. (See Figure 13.1.5.)

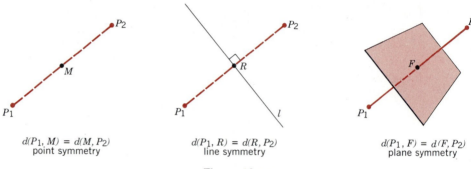

$$d(P_1, M) = d(M, P_2)$$
point symmetry

$$d(P_1, R) = d(R, P_2)$$
line symmetry

$$d(P_1, F) = d(F, P_2)$$
plane symmetry

Figure 13.1.5

In particular the endpoints $P_1(x_1, y_1, z_1)$ and $P_2(x_2, y_2, z_2)$ of line segment $\overline{P_1 P_2}$ are symmetric about the midpoint of the segment:

(13.1.3)

> Midpoint Formula
> $$\left(\frac{x_1 + x_2}{2}, \frac{y_1 + y_2}{2}, \frac{z_1 + z_2}{2} \right).$$

EXERCISES 13.1

Plot points A and B on a right-handed coordinate system. Then calculate the length of the line segment $\overline{AB}$ and find the midpoint.

1. $A(2, 0, 0)$, $B(0, 0, -4)$.   **2.** $A(0, -2, 0)$, $B(0, 0, 6)$.
3. $A(0, -2, 5)$, $B(4, 1, 0)$.   **4.** $A(4, 3, 0)$, $B(-2, 0, 6)$.

Find an equation for the plane through $(3, 1, -2)$ that satisfies the given condition.

5. Parallel to xy-plane.   **6.** Parallel to xz-plane.
7. Perpendicular to y-axis.   **8.** Perpendicular to z-axis.
9. Parallel to yz-plane.   **10.** Perpendicular to x-axis.

Find an equation for the sphere that satisfies the given conditions.

11. Centered at $(0, 2, -1)$ with radius 3.
12. Centered at $(1, 0, -2)$ with radius 4.
13. Centered at $(2, 4, -4)$ and passes through the origin.
14. Centered at the origin and passes through $(1, -2, 2)$.
15. The line segment joining $(0, 4, 2)$ and $(6, 0, 2)$ is a diameter.
16. Centered at $(2, 3, -4)$ and tangent to the xy-plane.

17. Centered at $(2, 3, -4)$ and tangent to the plane $x = 7$.

18. Centered at $(2, 3, -4)$ and tangent to the plane $y = 1$.

The points $P(a, b, c)$ and $Q(2, 3, 5)$ are symmetric in the sense given below. Find a, b, c.

19. About the xy-plane.　　　　**20.** About the xz-plane.

21. About the yz-plane.　　　　**22.** About the x-axis.

23. About the y-axis.　　　　**24.** About the z-axis.

25. About the origin.　　　　**26.** About the plane $x = 1$.

27. About the plane $y = -1$.　　**28.** About the plane $z = 4$.

29. About the point $(0, 2, 1)$.　　**30.** About the point $(4, 0, 1)$.

31. Find an equation for each sphere that passes through the point $(5, 1, 4)$ and is tangent to all three coordinate planes.

32. Find an equation for the largest sphere that is centered at $(2, 1, -2)$ and intersects the sphere $x^2 + y^2 + z^2 = 1$.

The point R lies on the line segment that joins $P(a_1, a_2, a_3)$ and $Q(b_1, b_2, b_3)$.

33. (a) Find the coordinates of R given that

$$d(P, R) = t \, d(P, Q) \qquad \text{where } 0 \le t \le 1.$$

(b) Determine the value of t for which R is the midpoint of $\overline{PQ}$.

34. (a) Find the coordinates of R given that

$$d(P, R) = r \, d(R, Q) \qquad \text{where } r > 0.$$

(b) Determine the value of r for which R is the midpoint of $\overline{PQ}$.

13.2 DISPLACEMENTS AND FORCES

Our purpose here is to motivate the notion of vector. A quick reading of this section will do.

A displacement along a coordinate line can be specified by a real number and depicted by an arrow. For a displacement of a_1 units we can use the number a_1 and an arrow that begins at any number x and ends at $x + a_1$. If $a_1 > 0$ (and we follow convention), the arrow will point to the right, and, if $a_1 < 0$, the arrow will point to the left. (Figure 13.2.1)

displacement a_1

Figure 13.2.1

Displacements in the xy-plane are more interesting. Instead of having two possible directions, there are now an infinite number of directions.   (Figure 13.2.2)

a displacement (a_1, a_2) in the xy-plane

Figure 13.2.2

The most general displacements take place in space. Here ordered triples of real numbers come into play. A displacement of a_1 units in x-coordinate, a_2 units in y-coordinate, and a_3 units in z-coordinate can be indicated by an arrow that begins at any point (x, y, z) and ends at $(x + a_1, y + a_2, z + a_3)$.   (Figure 13.2.3)

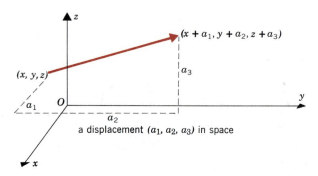

a displacement (a_1, a_2, a_3) in space

Figure 13.2.3

We can follow one displacement by another. A displacement (a_1, a_2, a_3) followed by (b_1, b_2, b_3) results in a total displacement $(a_1 + b_1, a_2 + b_2, a_3 + b_3)$. We can express this by writing

$$(a_1, a_2, a_3) + (b_1, b_2, b_3) = (a_1 + b_1, a_2 + b_2, a_3 + b_3).$$

We can picture the first displacement by an arrow from some point $P(x, y, z)$ to

$$Q(x + a_1, y + a_2, z + a_3),$$

the second displacement by an arrow from $Q(x + a_1, y + a_2, z + a_3)$ to

$$R(x + a_1 + b_1, y + a_2 + b_2, z + a_3 + b_3),$$

and then the resultant displacement by an arrow from $P(x, y, z)$ to

$$R(x + a_1 + b_1, y + a_2 + b_2, z + a_3 + b_3).$$

The three arrows then form a triangular pattern that is easy to remember (Figure 13.2.4).

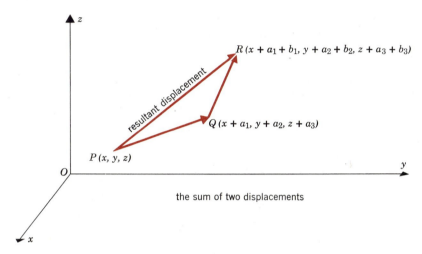

the sum of two displacements

Figure 13.2.4

From a displacement (a_1, a_2, a_3) and a real number α, we can form a new displacement $(\alpha a_1, \alpha a_2, \alpha a_3)$. We view this new displacement as α times the initial displacement and write

$$\alpha(a_1, a_2, a_3) = (\alpha a_1, \alpha a_2, \alpha a_3).$$

The effect of multiplying a displacement by a number α is to change the magnitude of the displacement by a factor of $|\alpha|$, keeping the same direction if $\alpha > 0$, but reversing the direction if $\alpha < 0$.

The displacement

$$2(a_1, a_2, a_3) = (2a_1, 2a_2, 2a_3)$$

is twice as long as (a_1, a_2, a_3) and has the same direction; the displacement

$$-(a_1, a_2, a_3) = (-a_1, -a_2, -a_3)$$

has the same length as (a_1, a_2, a_3) but the opposite direction; the displacement

$$-2(a_1, a_2, a_3) = (-2a_1, -2a_2, -2a_3)$$

is twice as long as (a_1, a_2, a_3) and has the opposite direction.   (Figure 13.2.5)

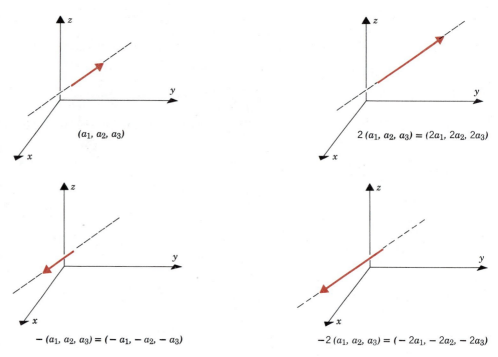

Figure 13.2.5

The algebraic pattern

$$(a_1, a_2, a_3) + (b_1, b_2, b_3) = (a_1 + b_1, a_2 + b_2, a_3 + b_3)$$
$$\alpha(a_1, a_2, a_3) = (\alpha a_1, \alpha a_2, \alpha a_3)$$

arises naturally in other settings; for example, in the analysis of forces. A force is completely determined by its components along the x, y, and z axes. If these components are a_1, a_2, a_3, respectively, then the force can be represented by the ordered triple (a_1, a_2, a_3). See Figure 13.2.6.

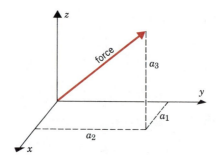

Figure 13.2.6

If two forces $\mathbf{F}_1 = (a_1, a_2, a_3)$ and $\mathbf{F}_2 = (b_1, b_2, b_3)$ are applied simultaneously at the same point, the effect is the same as that produced by the single force

$$\mathbf{F}_3 = (a_1 + b_1, a_2 + b_2, a_3 + b_3).$$

We call $\mathbf{F}_3$ the *resultant* or *total force* and write $\mathbf{F}_1 + \mathbf{F}_2 = \mathbf{F}_3$. For the ordered triples,

$$(a_1, a_2, a_3) + (b_1, b_2, b_3) = (a_1 + b_1, a_2 + b_2, a_3 + b_3).$$

Pictorially we have the usual force diagram, Figure 13.2.7. It is a parallelogram with the sides representing $\mathbf{F}_1 = (a_1, a_2, a_3)$ and $\mathbf{F}_2 = (b_1, b_2, b_3)$ and the diagonal representing

$$\mathbf{F}_3 = \mathbf{F}_1 + \mathbf{F}_2 = (a_1 + b_1, a_2 + b_2, a_3 + b_3).$$

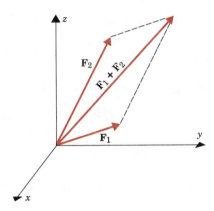

Figure 13.2.7

For any force $\mathbf{F} = (a_1, a_2, a_3)$ and any real number α, the force of $\alpha\mathbf{F}$ is defined by the equation

$$\alpha\mathbf{F} = (\alpha a_1, \alpha a_2, \alpha a_3).$$

Thus, once again we have

$$\alpha(a_1, a_2, a_3) = (\alpha a_1, \alpha a_2, \alpha a_3).$$

13.3   VECTORS

The algebra of number triples that we introduced in the last section to discuss displacements and forces is so prodigiously rich in applications that it has found a firm place in the world of science and engineering and has generated much mathematics. It is to this mathematics that we now turn.

DEFINITION 13.3.1

Ordered triples of real numbers subject to addition

$$(a_1, a_2, a_3) + (b_1, b_2, b_3) = (a_1 + b_1, a_2 + b_2, a_3 + b_3)$$

and multiplication by scalars (real numbers)

$$\alpha(a_1, a_2, a_3) = (\alpha a_1, \alpha a_2, \alpha a_3)$$

are called *vectors*.†

Two vectors are called *equal* iff they have exactly the same *components*:

$$(a_1, a_2, a_3) = (b_1, b_2, b_3) \quad \text{iff} \quad a_1 = b_1, a_2 = b_2, a_3 = b_3.$$

Geometrically, we can depict vectors by arrows. To depict the vector (a_1, a_2, a_3) we can choose any initial point $Q = Q(x, y, z)$ and use the arrow

$$\overrightarrow{QR} \quad \text{with} \quad R = R(x + a_1, y + a_2, z + a_3). \qquad \text{(Figure 13.3.1)}$$

Usually we choose the origin as the initial point and use the arrow

$$\overrightarrow{OP} \quad \text{with} \quad P = P(a_1, a_2, a_3).††$$

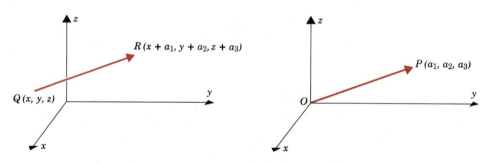

representations of (a_1, a_2, a_3)

Figure 13.3.1

We will use boldface letters to denote vectors. Thus for

$$\mathbf{a} = (a_1, a_2, a_3) \quad \text{and} \quad \mathbf{b} = (b_1, b_2, b_3)$$

we have

$$\mathbf{a} + \mathbf{b} = (a_1 + b_1, a_2 + b_2, a_3 + b_3),$$

---

† Strictly speaking, these are the vectors in three-dimensional Euclidean space. There are also four-dimensional vectors, five-dimensional vectors, etc. We will be working only in three dimensions.
†† The ordered triple (0, 0, 0) will not be represented by an arrow but simply by the origin.

and, if α is a *scalar* (a real number),

$$\alpha\mathbf{a} = (\alpha a_1, \alpha a_2, \alpha a_3).$$

For the *zero vector* $(0, 0, 0)$ we will use the symbol $\mathbf{0}$. Obviously

$$0\mathbf{a} = \mathbf{0} \qquad \text{for all vectors } \mathbf{a}.$$

By the vector $-\mathbf{b}$ we mean $(-1)\mathbf{b}$; that is,

$$-(b_1, b_2, b_3) = (-b_1, -b_2, -b_3).$$

By $\mathbf{a} - \mathbf{b}$ we mean $\mathbf{a} + (-\mathbf{b})$; that is,

$$(a_1, a_2, a_3) - (b_1, b_2, b_3) = (a_1 - b_1, a_2 - b_2, a_3 - b_3).$$

Problem 1. Given that

$$\mathbf{a} = (1, -1, 2), \qquad \mathbf{b} = (2, 3, -1), \qquad \mathbf{c} = (8, 7, 1),$$

find (a) $\mathbf{a} - \mathbf{b}$. (b) $2\mathbf{a} + \mathbf{b}$. (c) $3\mathbf{a} - 7\mathbf{b}$. (d) $2\mathbf{a} + 3\mathbf{b} - \mathbf{c}$.

Solution

(a) $\mathbf{a} - \mathbf{b} = (1, -1, 2) - (2, 3, -1) = (1 - 2, -1 - 3, 2 + 1) = (-1, -4, 3).$

(b) $2\mathbf{a} + \mathbf{b} = 2(1, -1, 2) + (2, 3, -1) = (2, -2, 4) + (2, 3, -1) = (4, 1, 3).$

(c) $3\mathbf{a} - 7\mathbf{b} = 3(1, -1, 2) - 7(2, 3, -1)$

$$= (3, -3, 6) - (14, 21, -7) = (-11, -24, 13).$$

(d) $2\mathbf{a} + 3\mathbf{b} - \mathbf{c} = 2(1, -1, 2) + 3(2, 3, -1) - (8, 7, 1)$

$$= (2, -2, 4) + (6, 9, -3) - (8, 7, 1) = (0, 0, 0) = \mathbf{0}. \quad \square$$

The addition of vectors can be visualized as the "addition" of arrows by the parallelogram law (as in the case of forces). If you picture $\mathbf{a}$, $\mathbf{b}$, and $\mathbf{a} + \mathbf{b}$ all as arrows emanating from the same point, say the origin, then $\mathbf{a} + \mathbf{b}$ acts as the diagonal of the parallelogram generated by $\mathbf{a}$ and $\mathbf{b}$. (See Figure 13.3.2.)

Figure 13.3.2

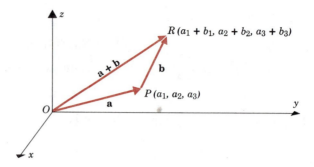

Figure 13.3.3

The addition of vectors can also be visualized as the "tail-to-head" addition of arrows (as in the case of displacements). If, instead of starting **b** at the origin, you start **b** at the tip of **a**, then **a** + **b** goes from the tail of **a** to the tip of **b**. (Figure 13.3.3) These two pictorial representations of vector addition lead to the same result.

DEFINITION 13.3.2

Two nonzero vectors **a** and **b** are said to be *parallel* provided that

$$\mathbf{a} = \alpha\mathbf{b} \qquad \text{for some real number } \alpha.$$

If $\alpha > 0$, **a** and **b** are said to have the *same direction*; if $\alpha < 0$, they are said to have *opposite directions*.

In the case of $\mathbf{a} = (2, -2, 6)$, $\mathbf{b} = (1, -1, 3)$, $\mathbf{c} = (-1, 1, -3)$ we have

$$\mathbf{a} = 2\mathbf{b} \quad \text{and} \quad \mathbf{a} = -2\mathbf{c}.$$

This tells us that **a** and **b** are parallel and have the same direction, whereas **a** and **c**, though parallel, have opposite directions. (See Figure 13.3.4.)

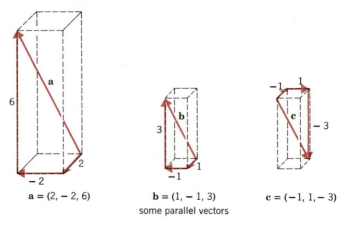

Figure 13.3.4

Definition 13.3.2 did not include the zero vector **0**. By special convention, **0** is said to be *parallel to every vector*.

(Since **0** is represented geometrically by a point, there is no geometric meaning to saying that **0** is parallel to another vector. However, it simplifies the statement of certain results to maintain that **0** is parallel to every vector. Algebraically this is warranted by the fact that **0** is a scalar multiple of every vector **b**: **0** = 0**b**.)

Problem 2. Show that, if **a** and **b** are parallel to **c**, then every linear combination $\alpha\mathbf{a} + \beta\mathbf{b}$ is also parallel to **c**.

Solution. Suppose that **a** and **b** are parallel to **c**. If **c** = **0**, then every vector is parallel to **c** and there is nothing to prove. If **c** ≠ **0**, then **a** and **b** are scalar multiples of **c**:

$$\mathbf{a} = \alpha_1\mathbf{c}, \qquad \mathbf{b} = \beta_1\mathbf{c}.$$

Then

$$\alpha\mathbf{a} + \beta\mathbf{b} = \alpha(\alpha_1\mathbf{c}) + \beta(\beta_1\mathbf{c}) = (\alpha\alpha_1 + \beta\beta_1)\mathbf{c}$$

is also parallel to **c**. □

DEFINITION 13.3.3

The *norm* of a vector $\mathbf{a} = (a_1, a_2, a_3)$ is the number

$$\|\mathbf{a}\| = \sqrt{a_1^2 + a_2^2 + a_3^2}.$$

The norm of **a** is also called the *length* or *magnitude* of **a**. The reason for this is simple: if we represent the vector **a** by an arrow, say

$$\overrightarrow{QR} \quad \text{with} \quad Q = Q(x, y, z) \quad \text{and} \quad R = R(x + a_1, y + a_2, z + a_3),$$

then $\|\mathbf{a}\|$ gives the length of $\overrightarrow{QR}$, or, what amounts to the same thing, the distance between Q and R.

The norm properties of vectors are very similar to the absolute-value properties of real numbers. In particular

(13.3.4)

$$(1)\ \|\mathbf{a}\| \geq 0 \quad \text{and} \quad \|\mathbf{a}\| = 0 \quad \text{iff} \quad \mathbf{a} = \mathbf{0}.$$
$$(2)\ \|\alpha\mathbf{a}\| = |\alpha|\,\|\mathbf{a}\|.$$
$$(3)\ \|\mathbf{a} + \mathbf{b}\| \leq \|\mathbf{a}\| + \|\mathbf{b}\|. \qquad \text{(the triangle inequality)}$$

Property (1) is obvious. Property (2) is easy to verify:

$$\|\alpha\mathbf{a}\| = \sqrt{(\alpha a_1)^2 + (\alpha a_2)^2 + (\alpha a_3)^2} = |\alpha|\sqrt{a_1^2 + a_2^2 + a_3^2} = |\alpha|\,\|\mathbf{a}\|.$$

We prove Property (3), the triangle inequality, in the next section, where we have "dot products" at our disposal. A proof at this time would be laborious.

Figure 13.3.5

You can interpret the triangle inequality as saying that the length of a side of a triangle cannot exceed the sum of the lengths of the other two sides. (See Figure 13.3.5.)

Problem 3. Given that $\mathbf{a} = (1, -2, 3)$ and $\mathbf{b} = (-4, 1, 0)$, calculate

(a) $\|\mathbf{a}\|$.   (b) $\|\mathbf{b}\|$.   (c) $\|\mathbf{a} + \mathbf{b}\|$.   (d) $\|\mathbf{a} - \mathbf{b}\|$.   (e) $\|-7\mathbf{a}\|$.   (f) $\|2\mathbf{a} - 3\mathbf{b}\|$.

Solution

(a) $\|\mathbf{a}\| = \sqrt{1^2 + (-2)^2 + (3)^2} = \sqrt{1 + 4 + 9} = \sqrt{14}$.
(b) $\|\mathbf{b}\| = \sqrt{(-4)^2 + 1^2 + 0^2} = \sqrt{16 + 1 + 0} = \sqrt{17}$.
(c) $\|\mathbf{a} + \mathbf{b}\| = \|(-3, -1, 3)\| = \sqrt{(-3)^2 + (-1)^2 + (3)^2} = \sqrt{9 + 1 + 9} = \sqrt{19}$.
(d) $\|\mathbf{a} - \mathbf{b}\| = \|(5, -3, 3)\| = \sqrt{(5)^2 + (-3)^2 + (3)^2} = \sqrt{25 + 9 + 9} = \sqrt{43}$.
(e) $\|-7\mathbf{a}\| = |-7|\|\mathbf{a}\| = 7\sqrt{14}$.
(f) $\|2\mathbf{a} - 3\mathbf{b}\| = \|2(1, -2, 3) - 3(-4, 1, 0)\|$
$\qquad = \|(14, -7, 6)\| = \sqrt{(14)^2 + (-7)^2 + (6)^2} = \sqrt{196 + 49 + 36} = \sqrt{281}$. ☐

To multiply a nonzero vector by a nonzero scalar α is to change its length by a factor of $|\alpha|$,

$$\|\alpha\mathbf{a}\| = |\alpha|\|\mathbf{a}\|,$$

keeping the same direction if $\alpha > 0$ and reversing the direction if $\alpha < 0$. The vector obtained from $\mathbf{a}$ simply by reversing its direction is the vector $(-1)\mathbf{a} = -\mathbf{a}$. (See Figure 13.3.6.)

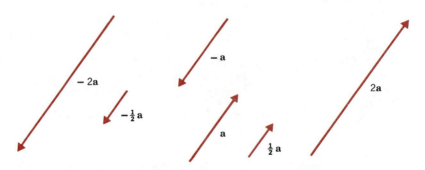

Figure 13.3.6

Since $\mathbf{a} - \mathbf{b} = \mathbf{a} + (-\mathbf{b})$, we can draw the vector $\mathbf{a} - \mathbf{b}$ by drawing $-\mathbf{b}$ and adding it to the vector $\mathbf{a}$. (Figure 13.3.7)

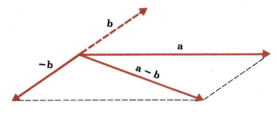

Figure 13.3.7

We can obtain the same result more easily by noting that $\mathbf{a} - \mathbf{b}$ is the vector that we must add to $\mathbf{b}$ to obtain $\mathbf{a}$. (Figure 13.3.8)

Figure 13.3.8

Vectors of norm 1 are called *unit vectors*. If $\mathbf{b}$ is a nonzero vector, then there is a unit vector $\mathbf{u_b}$ that has the direction of $\mathbf{b}$. To find $\mathbf{u_b}$, note that

$$\|\mathbf{u_b}\| = 1 \quad \text{and} \quad \mathbf{u_b} = \alpha\mathbf{b} \quad \text{for some } \alpha > 0.$$

It follows that

$$1 = \|\mathbf{u_b}\| = \|\alpha\mathbf{b}\| = |\alpha|\,\|\mathbf{b}\| = \alpha\|\mathbf{b}\|.$$

Thus

$$\alpha = \frac{1}{\|\mathbf{b}\|} \quad \text{and consequently} \quad \mathbf{u_b} = \frac{1}{\|\mathbf{b}\|}\mathbf{b} = \frac{\mathbf{b}}{\|\mathbf{b}\|}.$$

While

$$\mathbf{u_b} = \frac{\mathbf{b}}{\|\mathbf{b}\|}$$

is the unit vector in the direction of $\mathbf{b}$,

$$-\mathbf{u_b} = -\frac{\mathbf{b}}{\|\mathbf{b}\|}$$

is the unit vector in the opposite direction.

We single out for special attention the vectors

$$\mathbf{i} = (1, 0, 0), \qquad \mathbf{j} = (0, 1, 0), \qquad \mathbf{k} = (0, 0, 1).$$

These vectors all have norm 1 and, if pictured as emanating from the origin, lie along the positive coordinate axes. They are called *the unit coordinate vectors.* (Figure 13.3.9)

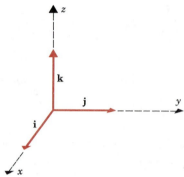

Figure 13.3.9

Every vector can be expressed as a linear combination of the unit coordinate vectors:

(13.3.5)   for $\mathbf{a} = (a_1, a_2, a_3)$ we have $\mathbf{a} = a_1\mathbf{i} + a_2\mathbf{j} + a_3\mathbf{k}.$

Proof

$$(a_1, a_2, a_3) = (a_1, 0, 0) + (0, a_2, 0) + (0, 0, a_3)$$
$$= a_1(1, 0, 0) + a_2(0, 1, 0) + a_3(0, 0, 1)$$
$$= a_1\mathbf{i} + a_2\mathbf{j} + a_3\mathbf{k}. \quad \square$$

The numbers a_1, a_2, a_3 are called the $\mathbf{i}$, $\mathbf{j}$, $\mathbf{k}$ components of the vector $\mathbf{a}$.

Problem 4. Given that

$$\mathbf{a} = 3\mathbf{i} - \mathbf{j} + \mathbf{k} \quad \text{and} \quad \mathbf{b} = 2\mathbf{i} + 3\mathbf{j} - \mathbf{k},$$

(1) express $2\mathbf{a} - \mathbf{b}$ as a linear combination of $\mathbf{i}$, $\mathbf{j}$, $\mathbf{k}$.
(2) calculate $\|2\mathbf{a} - \mathbf{b}\|$.
(3) find the unit vector $\mathbf{u}_c$ in the direction of $\mathbf{c} = 2\mathbf{a} - \mathbf{b}$.

Solution

(1) $2\mathbf{a} - \mathbf{b} = 2(3\mathbf{i} - \mathbf{j} + \mathbf{k}) - (2\mathbf{i} + 3\mathbf{j} - \mathbf{k})$
$$= 6\mathbf{i} - 2\mathbf{j} + 2\mathbf{k} - 2\mathbf{i} - 3\mathbf{j} + \mathbf{k} = 4\mathbf{i} - 5\mathbf{j} + 3\mathbf{k}.$$

(2) $\|2\mathbf{a} - \mathbf{b}\| = \|4\mathbf{i} - 5\mathbf{j} + 3\mathbf{k}\| = \sqrt{16 + 25 + 9} = \sqrt{50} = 5\sqrt{2}.$

(3) $\mathbf{u}_c = \dfrac{2\mathbf{a} - \mathbf{b}}{\|2\mathbf{a} - \mathbf{b}\|} = \dfrac{1}{5\sqrt{2}}(4\mathbf{i} - 5\mathbf{j} + 3\mathbf{k}). \quad \square$

EXERCISES 13.3

Simplify the following linear combinations.

1. $(2\mathbf{i} - \mathbf{j} + \mathbf{k}) + (\mathbf{i} - 3\mathbf{j} + 5\mathbf{k})$. **2.** $(6\mathbf{j} - \mathbf{k}) + (3\mathbf{i} - \mathbf{j} + 2\mathbf{k})$.

3. $2(\mathbf{j} + \mathbf{k}) - 3(\mathbf{i} + \mathbf{j} - 2\mathbf{k})$. **4.** $2(\mathbf{i} - \mathbf{j}) + 6(2\mathbf{i} + \mathbf{j} - 2\mathbf{k})$.

Calculate the norm of the following vectors.

5. $3\mathbf{i} + 4\mathbf{j}$. **6.** $\mathbf{i} - \mathbf{j}$. **7.** $2\mathbf{i} + \mathbf{j} - 2\mathbf{k}$.

8. $6\mathbf{i} + 2\mathbf{j} - \mathbf{k}$. **9.** $\frac{1}{2}(\mathbf{i} + 4\mathbf{j}) - (\frac{3}{2}\mathbf{i} + \mathbf{k})$. **10.** $(\mathbf{i} - \mathbf{j}) + 2(\mathbf{j} - \mathbf{i}) + (\mathbf{k} - \mathbf{j})$.

11. Let

$$\mathbf{a} = \mathbf{i} - \mathbf{j} + 2\mathbf{k}, \quad \mathbf{b} = 2\mathbf{i} - \mathbf{j} + 2\mathbf{k}, \quad \mathbf{c} = 3\mathbf{i} - 3\mathbf{j} + 6\mathbf{k}, \quad \mathbf{d} = -2\mathbf{i} + 2\mathbf{j} - 4\mathbf{k}.$$

(a) Which vectors are parallel? (b) Which vectors have the same direction?
(c) Which vectors have opposite directions?

12. (*Important*) Prove the following version of the triangle inequality:

$$\left| \|\mathbf{a}\| - \|\mathbf{b}\| \right| \le \|\mathbf{a} - \mathbf{b}\|.$$

HINT: Apply the previously stated triangle inequality to $\mathbf{a} = (\mathbf{a} - \mathbf{b}) + \mathbf{b}$.

13. Label the vectors.

(i)

(ii)

(iii)

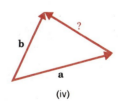

(iv)

14. Let $\mathbf{a} = (1, 1, 1)$, $\mathbf{b} = (-1, 3, 2)$, $\mathbf{c} = (-3, 0, 1)$, $\mathbf{d} = (4, -1, 1)$.
(a) Express $\mathbf{a} + 2\mathbf{b} + 3\mathbf{c} + 4\mathbf{d}$ as a linear combination of $\mathbf{i}, \mathbf{j}, \mathbf{k}$.
(b) Find scalars A, B, C such that $\mathbf{d} = A\mathbf{a} + B\mathbf{b} + C\mathbf{c}$.

15. Let $\mathbf{a} = (2, 0, -1)$, $\mathbf{b} = (1, 3, 5)$, $\mathbf{c} = (-1, 1, 1)$, $\mathbf{d} = (1, 1, 6)$.
(a) Express $\mathbf{a} - 3\mathbf{b} + 2\mathbf{c} + 4\mathbf{d}$ as a linear combination of $\mathbf{i}, \mathbf{j}, \mathbf{k}$.
(b) Find scalars A, B, C such that $\mathbf{d} = A\mathbf{a} + B\mathbf{b} + C\mathbf{c}$.

16. Find α given that $3\mathbf{i} + \mathbf{j} - \mathbf{k}$ and $\alpha\mathbf{i} - 4\mathbf{j} + 4\mathbf{k}$ are parallel.

17. Find α given that $3\mathbf{i} + \mathbf{j}$ and $\alpha\mathbf{j} - \mathbf{k}$ have the same length.

18. Find the unit vector in the direction of $\mathbf{i} - 2\mathbf{j} + 2\mathbf{k}$.

19. Find α given that $\|\alpha\mathbf{i} + (\alpha - 1)\mathbf{j} + (\alpha + 1)\mathbf{k}\| = 2$.

20. Find the vector of norm 2 in the direction of $\mathbf{i} + 2\mathbf{j} - \mathbf{k}$.

21. Find the vectors of norm 2 parallel to $3\mathbf{j} + 2\mathbf{k}$.

22. Express **c** in terms of **a** and **b**, given that the tip of **c** bisects the line segment.

(i)   (ii)

23. Let **a** and **b** be nonzero vectors such that $\|\mathbf{a} - \mathbf{b}\| = \|\mathbf{a} + \mathbf{b}\|$.
(a) What can you conclude about the parallelogram generated by **a** and **b**?
(b) Show that, if $\mathbf{a} = a_1\mathbf{i} + a_2\mathbf{j} + a_3\mathbf{k}$ and $\mathbf{b} = b_1\mathbf{i} + b_2\mathbf{j} + b_3\mathbf{k}$, then

$$a_1b_1 + a_2b_2 + a_3b_3 = 0.$$

24. (a) Show that, if **a** and **b** have the same direction, then $\|\mathbf{a} + \mathbf{b}\| = \|\mathbf{a}\| + \|\mathbf{b}\|$.
(b) Does this equation necessarily hold if **a** and **b** are only parallel?

A vector **r** emanating from the origin is called a *radius vector*. Each radius vector determines a unique point of space: the point at the tip of the vector. Conversely, each point of space determines a unique radius vector: the radius vector whose tip falls on that point. This one-to-one correspondence between the set of all radius vectors and the set of all points in three-dimensional space enables us to use radius vectors to specify sets in space. Thus, for example, the radius-vector equation $\|\mathbf{r}\| = 3$ can be used to represent the sphere of radius 3 centered at the origin: that sphere consists of the tips of all the radius vectors **r** that satisfy that equation.

25. Write a radius-vector equation or inequality for each of the following sets.
(a) The sphere of radius 3 centered at $P(a_1, a_2, a_3)$.
(b) The set of all points on or inside the sphere of radius 2 centered at the origin. (This set is called the *ball* of radius 2 about the origin.)
(c) The ball of radius 1 about the point $P(a_1, a_2, a_3)$.
(d) The set of all points equidistant from $P(a_1, a_2, a_3)$ and $Q(b_1, b_2, b_3)$. (This set forms a plane.)
(e) The set of all points the sum of whose distances from $P(a_1, a_2, a_3)$ and $Q(b_1, b_2, b_3)$ is a constant $k > d(P, Q)$. [Such a set is called an *ellipsoid*. The ellipsoid is a three-dimensional analogue of the ellipse. What happens if $k = d(P, Q)$? if $k < d(P, Q)$?]

13.4   THE DOT PRODUCT

(In this section we introduce the first of the two products we will define for vectors.)

Introduction

We begin with two nonzero vectors

$$\mathbf{a} = a_1\mathbf{i} + a_2\mathbf{j} + a_3\mathbf{k}, \qquad \mathbf{b} = b_1\mathbf{i} + b_2\mathbf{j} + b_3\mathbf{k}.$$

How can we tell from the components of these vectors whether these vectors meet at right angles? To explore this question we draw Figure 13.4.1.

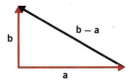

Figure 13.4.1

By the Pythagorean Theorem **a** and **b** meet at right angles iff

$$\|\mathbf{a}\|^2 + \|\mathbf{b}\|^2 = \|\mathbf{b} - \mathbf{a}\|^2.$$

In terms of components this equation reads

$$(a_1^2 + a_2^2 + a_3^2) + (b_1^2 + b_2^2 + b_3^2) = (b_1 - a_1)^2 + (b_2 - a_2)^2 + (b_3 - a_3)^2,$$

which, as you can readily check, simplifies to

$$a_1 b_1 + a_2 b_2 + a_3 b_3 = 0. \quad \square$$

The expression $a_1 b_1 + a_2 b_2 + a_3 b_3$ is widely used in geometry and in physics. It has a name, the *dot product* of **a** and **b**, and there is a special notation for it, **a** · **b**. The notion is so important that it deserves a formal definition.

Definition of the Dot Product

> **DEFINITION 13.4.1**
>
> For any two vectors
>
> $$\mathbf{a} = a_1\mathbf{i} + a_2\mathbf{j} + a_3\mathbf{k} \quad \text{and} \quad \mathbf{b} = b_1\mathbf{i} + b_2\mathbf{j} + b_3\mathbf{k}$$
>
> we define the *dot product* **a** · **b** by setting
>
> $$\mathbf{a} \cdot \mathbf{b} = a_1 b_1 + a_2 b_2 + a_3 b_3.$$

Example 1. For

$$\mathbf{a} = 2\mathbf{i} - \mathbf{j} + 3\mathbf{k}, \qquad \mathbf{b} = -3\mathbf{i} + \mathbf{j} + 4\mathbf{k}, \qquad \mathbf{c} = \mathbf{i} + 3\mathbf{j}$$

we have

$$\mathbf{a} \cdot \mathbf{b} = (2)(-3) + (-1)(1) + (3)(4) = -6 - 1 + 12 = 5,$$
$$\mathbf{a} \cdot \mathbf{c} = (2)(1) + (-1)(3) + (3)(0) = 2 - 3 = -1,$$
$$\mathbf{b} \cdot \mathbf{c} = (-3)(1) + (1)(3) + (4)(0) = -3 + 3 = 0.$$

The last equation tells us that **b** and **c** meet at right angles. (Verify this by drawing a figure.) $\square$

Because **a** · **b** is not a vector, but a scalar, it is sometimes called the *scalar product* of **a** and **b**. We will continue to call it the dot product and speak of dotting **a** with **b**.

Properties of the Dot Product

If we dot a vector with itself, we obtain the square of its norm:

(13.4.2)
$$\mathbf{a} \cdot \mathbf{a} = \|\mathbf{a}\|^2.$$

Proof. $\mathbf{a} \cdot \mathbf{a} = a_1 a_1 + a_2 a_2 + a_3 a_3 = a_1^2 + a_2^2 + a_3^2 = \|\mathbf{a}\|^2.$ □

The dot product of any vector with the zero vector is zero:

(13.4.3)
$$\mathbf{a} \cdot \mathbf{0} = 0, \qquad \mathbf{0} \cdot \mathbf{a} = 0.$$

Proof. $(a_1)(0) + (a_2)(0) + (a_3)(0) = 0, \qquad (0)(a_1) + (0)(a_2) + (0)(a_3) = 0.$ □

The dot product is commutative:

(13.4.4)
$$\mathbf{a} \cdot \mathbf{b} = \mathbf{b} \cdot \mathbf{a},$$

and scalars can be factored:

(13.4.5)
$$\alpha\mathbf{a} \cdot \beta\mathbf{b} = \alpha\beta(\mathbf{a} \cdot \mathbf{b}).$$

Proof
$$\mathbf{a} \cdot \mathbf{b} = a_1 b_1 + a_2 b_2 + a_3 b_3 = b_1 a_1 + b_2 a_2 + b_3 a_3 = \mathbf{b} \cdot \mathbf{a},$$

and
$$\alpha\mathbf{a} \cdot \beta\mathbf{b} = (\alpha a_1)(\beta b_1) + (\alpha a_2)(\beta b_2) + (\alpha a_3)(\beta b_3)$$
$$= \alpha\beta(a_1 b_1 + a_2 b_2 + a_3 b_3) = \alpha\beta(\mathbf{a} \cdot \mathbf{b}). \quad \square$$

The dot distributes over addition:

(13.4.6)
$$\mathbf{a} \cdot (\mathbf{b} + \mathbf{c}) = \mathbf{a} \cdot \mathbf{b} + \mathbf{a} \cdot \mathbf{c}, \qquad (\mathbf{a} + \mathbf{b}) \cdot \mathbf{c} = \mathbf{a} \cdot \mathbf{c} + \mathbf{b} \cdot \mathbf{c}.$$

Proof
$$\mathbf{a} \cdot (\mathbf{b} + \mathbf{c}) = a_1(b_1 + c_1) + a_2(b_2 + c_2) + a_3(b_3 + c_3)$$
$$= a_1 b_1 + a_1 c_1 + a_2 b_2 + a_2 c_2 + a_3 b_3 + a_3 c_3$$
$$= (a_1 b_1 + a_2 b_2 + a_3 b_3) + (a_1 c_1 + a_2 c_2 + a_3 c_3)$$
$$= \mathbf{a} \cdot \mathbf{b} + \mathbf{a} \cdot \mathbf{c}.$$

The second equation can be verified in a similar manner. □

Problem 2. Given that

$$\|\mathbf{a}\| = 1, \quad \|\mathbf{b}\| = 3, \quad \|\mathbf{c}\| = 4, \quad \mathbf{a} \cdot \mathbf{b} = 0, \quad \mathbf{a} \cdot \mathbf{c} = 1, \quad \mathbf{b} \cdot \mathbf{c} = -2,$$

find   (a) $3\mathbf{a} \cdot (\mathbf{b} + 4\mathbf{c})$.   (b) $(\mathbf{a} - \mathbf{b}) \cdot (2\mathbf{a} + \mathbf{b})$.   (c) $[(\mathbf{b} \cdot \mathbf{c})\mathbf{a} - (\mathbf{a} \cdot \mathbf{c})\mathbf{b}] \cdot \mathbf{c}$.

Solution

(a) $3\mathbf{a} \cdot (\mathbf{b} + 4\mathbf{c}) = (3\mathbf{a} \cdot \mathbf{b}) + (3\mathbf{a} \cdot 4\mathbf{c}) = 3(\mathbf{a} \cdot \mathbf{b}) + 12(\mathbf{a} \cdot \mathbf{c}) = 12.$

(b) $(\mathbf{a} - \mathbf{b}) \cdot (2\mathbf{a} + \mathbf{b}) = (\mathbf{a} \cdot 2\mathbf{a}) + (\mathbf{a} \cdot \mathbf{b}) + (-\mathbf{b} \cdot 2\mathbf{a}) + (-\mathbf{b} \cdot \mathbf{b})$

$$= 2(\mathbf{a} \cdot \mathbf{a}) + (\mathbf{a} \cdot \mathbf{b}) - 2(\mathbf{b} \cdot \mathbf{a}) - (\mathbf{b} \cdot \mathbf{b})$$

$$= 2\|\mathbf{a}\|^2 + (\mathbf{a} \cdot \mathbf{b}) - 2(\mathbf{a} \cdot \mathbf{b}) - \|\mathbf{b}\|^2$$

$$= 2 + 0 - 2(0) - 9 = -7.$$

(c) $[(\mathbf{b} \cdot \mathbf{c})\mathbf{a} - (\mathbf{a} \cdot \mathbf{c})\mathbf{b}] \cdot \mathbf{c} = [(\mathbf{b} \cdot \mathbf{c})\mathbf{a} \cdot \mathbf{c}] - [(\mathbf{a} \cdot \mathbf{c})\mathbf{b} \cdot \mathbf{c}]$

$$= (\mathbf{b} \cdot \mathbf{c})(\mathbf{a} \cdot \mathbf{c}) - (\mathbf{a} \cdot \mathbf{c})(\mathbf{b} \cdot \mathbf{c}) = 0. \quad \square$$

Geometric Interpretation of the Dot Product

We begin with a triangle with sides a, b, c.   (Figure 13.4.2)

Figure 13.4.2

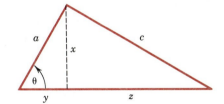

Figure 13.4.3

If θ were $\frac{1}{2}\pi$, the Pythagorean Theorem would tell us that $c^2 = a^2 + b^2$. The Law of Cosines,

$$c^2 = a^2 + b^2 - 2ab \cos \theta,$$

generalizes this rule to any angle θ.

To derive the Law of Cosines we drop a perpendicular to side b. (Figure 13.4.3) We then have

$$c^2 = z^2 + x^2 = (b - y)^2 + x^2 = b^2 - 2by + y^2 + x^2.$$

From the figure $y^2 + x^2 = a^2$ and $y = a \cos \theta$. Therefore

$$c^2 = a^2 + b^2 - 2ab \cos \theta,$$

as asserted. (What if the angle θ is obtuse? We leave that case to you.) $\square$

Now back to dot products. If neither **a** nor **b** is zero, we can interpret **a** · **b** from the triangle of Figure 13.4.4. The lengths of the sides are $\|\mathbf{a}\|$, $\|\mathbf{b}\|$, $\|\mathbf{a} - \mathbf{b}\|$. By the law of cosines

Figure 13.4.4

$$\|\mathbf{a} - \mathbf{b}\|^2 = \|\mathbf{a}\|^2 + \|\mathbf{b}\|^2 - 2\|\mathbf{a}\|\,\|\mathbf{b}\| \cos \theta.$$

This gives

$$2\|\mathbf{a}\|\,\|\mathbf{b}\| \cos \theta = \|\mathbf{a}\|^2 + \|\mathbf{b}\|^2 - \|\mathbf{a} - \mathbf{b}\|^2$$
$$= a_1^2 + a_2^2 + a_3^2 + b_1^2 + b_2^2 + b_3^2 - (a_1 - b_1)^2 - (a_2 - b_2)^2 - (a_3 - b_3)^2$$
$$= 2(a_1 b_1 + a_2 b_2 + a_3 b_3) = 2(\mathbf{a} \cdot \mathbf{b}),$$

and thus

(13.4.7) $\boxed{\mathbf{a} \cdot \mathbf{b} = \|\mathbf{a}\|\,\|\mathbf{b}\| \cos \theta.}$

By convention, θ, the angle between **a** and **b**, is measured in radians and taken from 0 to π. (no negative angles)

From (13.4.7) you can see that the dot product of two vectors depends on the norms of the vectors and on the angle between them. For vectors of a given norm, the dot product measures the extent to which the vectors agree in direction. As the difference in direction increases, the dot product decreases:

If **a** and **b** have the same direction, then $\theta = 0$ and

$$\mathbf{a} \cdot \mathbf{b} = \|\mathbf{a}\|\,\|\mathbf{b}\|; \qquad\qquad (\cos 0 = 1)$$

this is the largest possible value for **a** · **b**.
If **a** and **b** meet at right angles, then $\theta = \frac{1}{2}\pi$ and

$$\mathbf{a} \cdot \mathbf{b} = 0. \qquad\qquad (\cos \tfrac{1}{2}\pi = 0)$$

If **a** and **b** have opposite directions, then $\theta = \pi$ and

$$\mathbf{a} \cdot \mathbf{b} = -\|\mathbf{a}\|\,\|\mathbf{b}\|; \qquad\qquad (\cos \pi = -1)$$

this is the least possible value for **a** · **b**.

Two vectors are said to be *perpendicular* iff they lie at right angles or one of the vectors is the zero vector; in other words, two vectors are said to be perpendicular iff their dot product is zero.† In symbols

(13.4.8) $\boxed{\mathbf{a} \perp \mathbf{b} \quad \text{iff} \quad \mathbf{a} \cdot \mathbf{b} = 0.}$ (Figure 13.4.5)

The unit coordinate vectors are obviously mutually perpendicular:

$$\mathbf{i} \cdot \mathbf{j} = 0, \quad \mathbf{i} \cdot \mathbf{k} = 0, \quad \mathbf{j} \cdot \mathbf{k} = 0.$$

---

† This makes the zero vector both parallel and perpendicular to every vector. There is, however, no contradiction since we never apply the notion of angle to the zero vector.

$$\mathbf{a} \cdot \mathbf{b} = 0$$

Figure 13.4.5

Problem 3. Verify that the vectors

$$\mathbf{a} = 2\mathbf{i} + \mathbf{j} + \mathbf{k} \qquad \text{and} \qquad \mathbf{b} = \mathbf{i} + \mathbf{j} - 3\mathbf{k}$$

are perpendicular.

Solution

$$\mathbf{a} \cdot \mathbf{b} = (2)(1) + (1)(1) + (1)(-3) = 2 + 1 - 3 = 0. \quad \square$$

Problem 4. Find the value of α for which

$$3\mathbf{i} - \alpha\mathbf{j} + \mathbf{k} \perp \mathbf{i} + 2\mathbf{j}.$$

Solution. For the two vectors to be perpendicular their dot product must be zero. Since

$$(3\mathbf{i} - \alpha\mathbf{j} + \mathbf{k}) \cdot (\mathbf{i} + 2\mathbf{j}) = (3)(1) + (-\alpha)2 + (1)(0) = 3 - 2\alpha,$$

α must be $\frac{3}{2}$. $\quad \square$

With $\mathbf{a}$ and $\mathbf{b}$ both different from zero, we can divide the equation

$$\mathbf{a} \cdot \mathbf{b} = \|\mathbf{a}\| \, \|\mathbf{b}\| \cos \theta$$

by $\|\mathbf{a}\| \, \|\mathbf{b}\|$ to obtain

$$\cos \theta = \frac{\mathbf{a} \cdot \mathbf{b}}{\|\mathbf{a}\| \, \|\mathbf{b}\|} = \left(\frac{\mathbf{a}}{\|\mathbf{a}\|} \cdot \frac{\mathbf{b}}{\|\mathbf{b}\|} \right).$$

Writing

$$\mathbf{u_a} = \frac{\mathbf{a}}{\|\mathbf{a}\|} \qquad \text{and} \qquad \mathbf{u_b} = \frac{\mathbf{b}}{\|\mathbf{b}\|},$$

we have

(13.4.9)
$$\boxed{\cos \theta = (\mathbf{u_a} \cdot \mathbf{u_b}).}$$

The cosine of the angle between two vectors is the dot product of the corresponding unit vectors.

Problem 5. Calculate the angle between

$$\mathbf{a} = 2\mathbf{i} + 3\mathbf{j} + 2\mathbf{k} \qquad \text{and} \qquad \mathbf{b} = \mathbf{i} + 2\mathbf{j} - \mathbf{k}.$$

Solution

$$\mathbf{u_a} = \frac{2\mathbf{i} + 3\mathbf{j} + 2\mathbf{k}}{\|2\mathbf{i} + 3\mathbf{j} + 2\mathbf{k}\|} = \frac{1}{\sqrt{17}}(2\mathbf{i} + 3\mathbf{j} + 2\mathbf{k}), \qquad \mathbf{u_b} = \frac{\mathbf{i} + 2\mathbf{j} - \mathbf{k}}{\|\mathbf{i} + 2\mathbf{j} - \mathbf{k}\|} = \frac{1}{\sqrt{6}}(\mathbf{i} + 2\mathbf{j} - \mathbf{k}).$$

Therefore

$$\cos\theta = \mathbf{u_a} \cdot \mathbf{u_b} = \frac{1}{\sqrt{17}}\frac{1}{\sqrt{6}}[(2\mathbf{i} + 3\mathbf{j} + 2\mathbf{k}) \cdot (\mathbf{i} + 2\mathbf{j} - \mathbf{k})].$$

Since

$$(2\mathbf{i} + 3\mathbf{j} + 2\mathbf{k}) \cdot (\mathbf{i} + 2\mathbf{j} - \mathbf{k}) = (2)(1) + (3)(2) + (2)(-1) = 6,$$

we have

$$\cos\theta = \frac{6}{\sqrt{17}\sqrt{6}} = \frac{1}{17}\sqrt{102} \cong \frac{10.1}{17} \cong 0.594.$$

Table 4 (Appendix C) gives $\theta \cong 0.935$ radians, which is about 54 degrees. $\square$

Problem 6. Show that, if **a** and **b** are both perpendicular to **c**, then every linear combination $\alpha\mathbf{a} + \beta\mathbf{b}$ is also perpendicular to **c**.

Solution. Suppose that **a** and **b** are both perpendicular to **c**. Then

$$\mathbf{a} \cdot \mathbf{c} = 0 \qquad \text{and} \qquad \mathbf{b} \cdot \mathbf{c} = 0.$$

It follows that

$$(\alpha\mathbf{a} + \beta\mathbf{b}) \cdot \mathbf{c} = \alpha\underbrace{(\mathbf{a} \cdot \mathbf{c})}_{0} + \beta\underbrace{(\mathbf{b} \cdot \mathbf{c})}_{0} = 0$$

and therefore $\alpha\mathbf{a} + \beta\mathbf{b}$ is perpendicular to **c**. $\square$

Projections and Components

If $\mathbf{b} \neq \mathbf{0}$, then every vector **a** can be written in a unique manner as the sum of a vector $\mathbf{a}_\parallel$ parallel to **b** and a vector $\mathbf{a}_\perp$ perpendicular to **b**:

$$\mathbf{a} = \mathbf{a}_\parallel + \mathbf{a}_\perp. \qquad\qquad \text{(Exercise 23)}$$

The idea is illustrated in Figure 13.4.6.

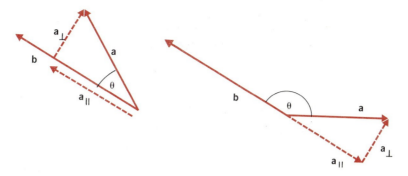

Figure 13.4.6

If **a** is parallel to **b**, then $\mathbf{a}_\parallel = \mathbf{a}$ and $\mathbf{a}_\perp = \mathbf{0}$. If **b** is perpendicular to **a**, then $\mathbf{a}_\parallel = \mathbf{0}$ and $\mathbf{a}_\perp = \mathbf{a}$.

The vector $\mathbf{a}_\parallel$ is called *the projection of* **a** *on* **b** and is denoted by $\mathbf{proj_b a}$. Since $\mathbf{proj_b a}$ is parallel to **b**, it is a scalar multiple of the unit vector in the direction of **b**:

$$\mathbf{proj_b a} = \lambda \mathbf{u_b}.$$

The scalar λ is called *the component of* **a** *in the direction of* **b** (or more briefly, the **b**-*component* of **a**) and is denoted by $\mathrm{comp_b a}$. In symbols,

(13.4.10) $\mathbf{proj_b a} = (\mathrm{comp_b a})\mathbf{u_b}.$

The component of **a** in the direction of **b** measures the "advance" of **a** in the direction of **b**. In Figure 13.4.6 we used θ to indicate the angle between **a** and **b**. If $0 \leq \theta < \frac{1}{2}\pi$, the projection and **b** have the same direction and the component is positive. If $\theta = \frac{1}{2}\pi$, the projection is **0** and the component is 0. If $\frac{1}{2}\pi < \theta \leq \pi$, the projection and **b** have opposite directions and, consequently, the component is negative.

Projections and components are closely related to dot products: if $\mathbf{b} \neq \mathbf{0}$, then

(13.4.11) $\mathbf{proj_b a} = (\mathbf{a \cdot u_b})\mathbf{u_b}$ and $\mathrm{comp_b a} = \mathbf{a \cdot u_b}.$

Proof. The second assertion follows trivially from the first. We will prove the first.
We begin with the identity

$$\mathbf{a} = (\mathbf{a \cdot u_b})\mathbf{u_b} + [\mathbf{a} - (\mathbf{a \cdot u_b})\mathbf{u_b}].$$

Since the first vector $(\mathbf{a \cdot u_b})\mathbf{u_b}$ is a scalar multiple of **b**, it is parallel to **b**. All we have to show now is that the second vector is perpendicular to **b**. We do this by showing that its dot product with $\mathbf{u_b}$ is zero:

$$[\mathbf{a} - (\mathbf{a \cdot u_b})\mathbf{u_b}] \cdot \mathbf{u_b} = (\mathbf{a \cdot u_b}) - (\mathbf{a \cdot u_b})(\mathbf{u_b \cdot u_b}) = 0. \quad \square$$

$$\underset{\mathbf{u_b \cdot u_b} = \|\mathbf{u_b}\|^2 = 1}{\uparrow}$$

Problem 7. Find comp$_b$**a** and **proj**$_b$**a** given that

$$\mathbf{a} = -2\mathbf{i} + \mathbf{j} + \mathbf{k} \qquad \text{and} \qquad \mathbf{b} = 4\mathbf{i} - 3\mathbf{j} + \mathbf{k}.$$

Solution. Since

$$\|\mathbf{b}\| = \sqrt{(4)^2 + (-3)^2 + (1)^2} = \sqrt{26},$$

we have

$$\mathbf{u_b} = \frac{\mathbf{b}}{\|\mathbf{b}\|} = \frac{1}{\sqrt{26}}(4\mathbf{i} - 3\mathbf{j} + \mathbf{k}).$$

Thus

$$\text{comp}_b\mathbf{a} = \mathbf{a} \cdot \mathbf{u_b} = (-2\mathbf{i} + \mathbf{j} + \mathbf{k}) \cdot \frac{1}{\sqrt{26}}(4\mathbf{i} - 3\mathbf{j} + \mathbf{k})$$

$$= \frac{1}{\sqrt{26}}[(-2)(4) + (1)(-3) + (1)(1)] = -\frac{10}{\sqrt{26}} = -\frac{5}{13}\sqrt{26}$$

and

$$\mathbf{proj}_b\mathbf{a} = (\text{comp}_b\mathbf{a})\,\mathbf{u_b} = -\frac{5}{13}(4\mathbf{i} - 3\mathbf{j} + \mathbf{k}). \quad \square$$

The following characterization of $\mathbf{a} \cdot \mathbf{b}$ is frequently used in physical applications:

(13.4.12)

$$\text{if } \mathbf{b} \neq 0, \quad \mathbf{a} \cdot \mathbf{b} = (\text{comp}_b\mathbf{a})\|\mathbf{b}\|.$$

Proof

$$\mathbf{a} \cdot \mathbf{b} = \left(\mathbf{a} \cdot \frac{\mathbf{b}}{\|\mathbf{b}\|}\right)\|\mathbf{b}\| = (\mathbf{a} \cdot \mathbf{u_b})\|\mathbf{b}\| = (\text{comp}_b\mathbf{a})\|\mathbf{b}\|. \quad \square$$

For an arbitrary vector $\mathbf{a} = a_1\mathbf{i} + a_2\mathbf{j} + a_3\mathbf{k}$ we have

$$\text{comp}_i\mathbf{a} = \mathbf{a} \cdot \mathbf{i} = a_1, \qquad \text{comp}_j\mathbf{a} = \mathbf{a} \cdot \mathbf{j} = a_2, \qquad \text{comp}_k\mathbf{a} = \mathbf{a} \cdot \mathbf{k} = a_3.$$

This agrees with our previous use of the term "component" (Section 13.3) and gives the identity

(13.4.13)

$$\mathbf{a} = (\mathbf{a} \cdot \mathbf{i})\mathbf{i} + (\mathbf{a} \cdot \mathbf{j})\mathbf{j} + (\mathbf{a} \cdot \mathbf{k})\mathbf{k}.$$

Direction Angles, Direction Cosines

In Figure 13.4.7 we have portrayed a nonzero vector **a**. The angles α, β, γ that the vector makes with the unit coordinate vectors are called the *direction angles* of **a**, and

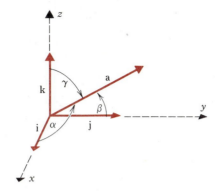

Figure 13.4.7

$\cos \alpha$, $\cos \beta$, $\cos \gamma$ are called the *direction cosines.* By (13.4.7)

$$\mathbf{a} \cdot \mathbf{i} = \|\mathbf{a}\| \cos \alpha, \qquad \mathbf{a} \cdot \mathbf{j} = \|\mathbf{a}\| \cos \beta, \qquad \mathbf{a} \cdot \mathbf{k} = \|\mathbf{a}\| \cos \gamma.$$

Thus by (13.4.13)

(13.4.14)
$$\mathbf{a} = \|\mathbf{a}\| (\cos \alpha \, \mathbf{i} + \cos \beta \, \mathbf{j} + \cos \gamma \, \mathbf{k}).$$

Taking the norm of both sides, we have

$$\|\mathbf{a}\| = \|\mathbf{a}\| \sqrt{\cos^2 \alpha + \cos^2 \beta + \cos^2 \gamma}$$

and therefore

(13.4.15)
$$\cos^2 \alpha + \cos^2 \beta + \cos^2 \gamma = 1.$$

The sum of the squares of the direction cosines is always 1.
For a unit vector $\mathbf{u}$, Equation 13.4.14 takes the form

(13.4.16)
$$\mathbf{u} = \cos \alpha \, \mathbf{i} + \cos \beta \, \mathbf{j} + \cos \gamma \, \mathbf{k}.$$

Thus, for a unit vector, the $\mathbf{i}, \mathbf{j}, \mathbf{k}$ components are simply the direction cosines.

Problem 8. Find the unit vector with direction angles

$$\alpha = \tfrac{1}{4}\pi, \qquad \beta = \tfrac{2}{3}\pi, \qquad \gamma = \tfrac{1}{3}\pi.$$

What is the vector of norm 4 with these same direction angles?

Solution. The unit vector with these direction angles is

$$\cos \tfrac{1}{4}\pi \, \mathbf{i} + \cos \tfrac{2}{3}\pi \, \mathbf{j} + \cos \tfrac{1}{3}\pi \, \mathbf{k} = \tfrac{1}{2}\sqrt{2} \, \mathbf{i} - \tfrac{1}{2}\mathbf{j} + \tfrac{1}{2}\mathbf{k}.$$

The vector of norm 4 with these direction angles is

$$4(\tfrac{1}{2}\sqrt{2} \, \mathbf{i} - \tfrac{1}{2}\mathbf{j} + \tfrac{1}{2}\mathbf{k}) = 2\sqrt{2} \, \mathbf{i} - 2\mathbf{j} + 2\mathbf{k}. \quad \square$$

Problem 9. Find the direction cosines of

$$\mathbf{a} = 2\mathbf{i} + 3\mathbf{j} - 6\mathbf{k}.$$

What are the direction angles?

Solution. Here

$$\|\mathbf{a}\| = \sqrt{2^2 + 3^2 + (-6)^2} = 7$$

so that

$$2 = 7 \cos \alpha, \qquad 3 = 7 \cos \beta, \qquad -6 = 7 \cos \gamma$$

and

$$\cos \alpha = \tfrac{2}{7}, \qquad \cos \beta = \tfrac{3}{7}, \qquad \cos \gamma = -\tfrac{6}{7}.$$

Since angles between vectors are measured in radians and taken from 0 to π,

$$\alpha = \cos^{-1} \tfrac{2}{7} \cong 1.28 \text{ radians}, \qquad \beta = \cos^{-1} \tfrac{3}{7} \cong 1.13 \text{ radians},$$
$$\gamma = \cos^{-1}(-\tfrac{6}{7}) \cong 2.60 \text{ radians}. \quad \square$$

Proving the Triangle Inequality

The relation

$$\mathbf{a} \cdot \mathbf{b} = \|\mathbf{a}\| \|\mathbf{b}\| \cos \theta$$

gives

(13.4.17) $\boxed{|\mathbf{a} \cdot \mathbf{b}| \le \|\mathbf{a}\| \|\mathbf{b}\|.}$ $(|\cos \theta| \le 1)$

This inequality, called *Schwarz's inequality*, enables us to give a simple proof of the triangle inequality

$$\|\mathbf{a} + \mathbf{b}\| \le \|\mathbf{a}\| + \|\mathbf{b}\|.$$

Proof

$$\|\mathbf{a} + \mathbf{b}\|^2 = (\mathbf{a} + \mathbf{b}) \cdot (\mathbf{a} + \mathbf{b})$$
$$= (\mathbf{a} \cdot \mathbf{a}) + (\mathbf{b} \cdot \mathbf{a}) + (\mathbf{a} \cdot \mathbf{b}) + (\mathbf{b} \cdot \mathbf{b})$$
$$= \|\mathbf{a}\|^2 + 2(\mathbf{a} \cdot \mathbf{b}) + \|\mathbf{b}\|^2$$
$$\le \|\mathbf{a}\|^2 + 2\|\mathbf{a}\| \|\mathbf{b}\| + \|\mathbf{b}\|^2 = (\|\mathbf{a}\| + \|\mathbf{b}\|)^2$$

by Schwarz's inequality

and therefore

$$\|\mathbf{a} + \mathbf{b}\| \le \|\mathbf{a}\| + \|\mathbf{b}\|. \quad \square$$

It is worth remarking that Schwarz's inequality, and hence the triangle inequality, can be proved purely algebraically. (See Exercise 34.)

EXERCISES 13.4

Simplify.

1. $(3\mathbf{a} \cdot \mathbf{b}) - (\mathbf{a} \cdot 2\mathbf{b})$.   **2.** $\mathbf{a} \cdot (\mathbf{a} - \mathbf{b}) + \mathbf{b} \cdot (\mathbf{b} + \mathbf{a})$.   **3.** $(\mathbf{a} - \mathbf{b}) \cdot \mathbf{c} + \mathbf{b} \cdot (\mathbf{c} + \mathbf{a})$.

4. $\mathbf{a} \cdot (\mathbf{a} + 2\mathbf{c}) + (2\mathbf{b} - \mathbf{a}) \cdot (\mathbf{a} + 2\mathbf{c}) - 2\mathbf{b} \cdot (\mathbf{a} + 2\mathbf{c})$.

5. Taking

$$\mathbf{a} = 2\mathbf{i} + \mathbf{j}, \qquad \mathbf{b} = 3\mathbf{i} - \mathbf{j} + 2\mathbf{k}, \qquad \mathbf{c} = 4\mathbf{i} + 3\mathbf{k}$$

calculate
(a) the three dot products, $\mathbf{a} \cdot \mathbf{b}, \mathbf{a} \cdot \mathbf{c}, \mathbf{b} \cdot \mathbf{c}$;
(b) the cosines of the angles between these vectors;
(c) the component of $\mathbf{a}$ (i) in the $\mathbf{b}$ direction, (ii) in the $\mathbf{c}$ direction;
(d) the projection of $\mathbf{a}$ (i) in the $\mathbf{b}$ direction, (ii) in the $\mathbf{c}$ direction.

6. Exercise 5 with $\mathbf{a} = \mathbf{j} + 3\mathbf{k}, \ \mathbf{b} = 2\mathbf{i} - \mathbf{j} + 2\mathbf{k}, \ \mathbf{c} = 3\mathbf{i} - \mathbf{k}$.

7. Find the unit vector with direction angles $\frac{1}{3}\pi, \frac{1}{4}\pi, \frac{2}{3}\pi$.

8. Find the vector of norm 2 with direction angles $\frac{1}{4}\pi, \frac{1}{4}\pi, \frac{1}{2}\pi$.

9. Find the angle between the vectors $3\mathbf{i} - \mathbf{j} - 2\mathbf{k}$ and $\mathbf{i} + 2\mathbf{j} - 3\mathbf{k}$.

10. Find the angle between the vectors $2\mathbf{i} - 3\mathbf{j} + \mathbf{k}$ and $-3\mathbf{i} + \mathbf{j} + 9\mathbf{k}$.

11. Find the direction angles of the vector $\mathbf{i} - \mathbf{j} + \sqrt{2}\,\mathbf{k}$.

12. Find the direction angles of the vector $\mathbf{i} - \sqrt{3}\,\mathbf{k}$.

13. Show that
(a) $\mathbf{proj_b}\alpha\mathbf{a} = \alpha\,\mathbf{proj_b a}$ for all real α, and (b) $\mathbf{proj_b}(\mathbf{a} + \mathbf{c}) = \mathbf{proj_b a} + \mathbf{proj_b c}$.

14. Show that

(a) $\mathbf{proj_{\beta b} a} = \mathbf{proj_b a}$ for all real $\beta \neq 0$, but (b) $\mathrm{comp}_{\beta\mathbf{b}}\mathbf{a} = \left\{ \begin{array}{l} \mathrm{comp_b a}, \quad \text{for } \beta > 0 \\ -\mathrm{comp_b a}, \quad \text{for } \beta < 0 \end{array} \right]$.

15. (a) (*Important*) Let $\mathbf{a} \neq \mathbf{0}$. Show that $\mathbf{a} \cdot \mathbf{b} = \mathbf{a} \cdot \mathbf{c}$ does not necessarily imply that $\mathbf{b} = \mathbf{c}$, but only that $\mathbf{b}$ and $\mathbf{c}$ have the same projection on $\mathbf{a}$. Draw a figure illustrating this for $\mathbf{b}$ and $\mathbf{c}$ different from $\mathbf{0}$.
(b) Show that if $\mathbf{u} \cdot \mathbf{b} = \mathbf{u} \cdot \mathbf{c}$ for all unit vectors $\mathbf{u}$, then $\mathbf{b} = \mathbf{c}$.   HINT: Consider the unit coordinate vectors.

16. What can you conclude about $\mathbf{a}$ and $\mathbf{b}$ given that
(a) $\|\mathbf{a}\|^2 + \|\mathbf{b}\|^2 = \|\mathbf{a} + \mathbf{b}\|^2$?   (b) $\|\mathbf{a}\|^2 + \|\mathbf{b}\|^2 = \|\mathbf{a} - \mathbf{b}\|^2$?
HINT: Draw figures.

17. (a) Show that

$$4(\mathbf{a} \cdot \mathbf{b}) = \|\mathbf{a} + \mathbf{b}\|^2 - \|\mathbf{a} - \mathbf{b}\|^2.$$

(b) Use (a) to verify that

$$\mathbf{a} \perp \mathbf{b} \qquad \text{iff} \qquad \|\mathbf{a} + \mathbf{b}\| = \|\mathbf{a} - \mathbf{b}\|.$$

(c) Show that, if $\mathbf{a}$ and $\mathbf{b}$ are nonzero vectors such that

$$\mathbf{a} + \mathbf{b} \perp \mathbf{a} - \mathbf{b} \qquad \text{and} \qquad \|\mathbf{a} + \mathbf{b}\| = \|\mathbf{a} - \mathbf{b}\|,$$

then the parallelogram generated by $\mathbf{a}$ and $\mathbf{b}$ is a square.

18. Under what conditions does $|\mathbf{a} \cdot \mathbf{b}| = \|\mathbf{a}\|\,\|\mathbf{b}\|$?

19. What are the direction angles of $-\mathbf{a}$ if the direction angles of $\mathbf{a}$ are α, β, γ?

20. Let θ be the angle between $\mathbf{a}$ and $\mathbf{b}$, and let β be a negative number. Use the dot product to compute the angle between $\mathbf{a}$ and $\beta\mathbf{b}$ in terms of θ. Draw a figure to verify your answer geometrically.

21. Show that, if $\mathbf{a} \perp \mathbf{b}$ and $\mathbf{a} \perp \mathbf{c}$, then $\mathbf{a} \perp (\alpha\mathbf{b} + \beta\mathbf{c})$ for all real α, β.

22. Show that, if $\mathbf{a} \parallel \mathbf{b}$ and $\mathbf{a} \parallel \mathbf{c}$, then $\mathbf{a} \parallel (\alpha\mathbf{b} + \beta\mathbf{c})$ for all real α, β.

23. (*Important*) Show that, if **b** is a nonzero vector, then every vector **a** can be written in a unique manner as the sum of a vector $\mathbf{a}_\parallel$ parallel to **b** and a vector $\mathbf{a}_\perp$ perpendicular to **b**:

$$\mathbf{a} = \mathbf{a}_\parallel + \mathbf{a}_\perp .$$

24. Find all the numbers x for which $x\mathbf{i} + 11\mathbf{j} - 3\mathbf{k} \perp 2x\mathbf{i} - x\mathbf{j} - 5\mathbf{k}$.
25. Find all the numbers x for which the angle between $\mathbf{c} = x\mathbf{i} + \mathbf{j} + \mathbf{k}$ and $\mathbf{d} = \mathbf{i} + x\mathbf{j} + \mathbf{k}$ is $\frac{1}{3}\pi$.
26. Set $\mathbf{a} = \mathbf{i} + x\mathbf{j} + \mathbf{k}$ and $\mathbf{b} = 2\mathbf{i} - \mathbf{j} + y\mathbf{k}$. Compute all values of x and y for which $\mathbf{a} \perp \mathbf{b}$ and $\|\mathbf{a}\| = \|\mathbf{b}\|$.
27. Find the angle between the diagonal of a cube and one of its edges.
28. Find the angle between the diagonal of a cube and the diagonal of one of its faces.
29. (a) Show that $\frac{1}{4}\pi, \frac{1}{6}\pi, \frac{2}{3}\pi$ cannot be the direction angles of a vector.
    (b) Show that, if $\mathbf{a} = a_1\mathbf{i} + a_2\mathbf{j} + a_3\mathbf{k}$ has direction angles $\alpha, \frac{1}{4}\pi, \frac{1}{4}\pi$, then $a_1 = 0$.
30. If a force **F** is applied throughout a displacement **r**, then the work done by **F** is given by the formula

$$W = (\text{comp}_\mathbf{r}\, \mathbf{F})\|\mathbf{r}\|.$$

    (a) Express the work W as a dot product.   (b) What is the work done if $\mathbf{F} \perp \mathbf{r}$?
    (c) Two forces of the same magnitude $\mathbf{F}_1$ and $\mathbf{F}_2$ are applied throughout a displacement **r** at angles θ_1 and θ_2, respectively. Compare the work done by $\mathbf{F}_1$ to that done by $\mathbf{F}_2$ if (i) $\theta_1 = -\theta_2$. (ii) $\theta_1 = \frac{1}{3}\pi$, $\theta_2 = \frac{1}{6}\pi$.
    (d) What is the work done by **F** if the object moves completely around a triangle? Justify your answer.
31. Find the unit vectors **u** that are perpendicular to both $\mathbf{i} + 2\mathbf{j} + \mathbf{k}$ and $3\mathbf{i} - 4\mathbf{j} + 2\mathbf{k}$.
32. Find two mutually perpendicular unit vectors that are perpendicular to $2\mathbf{i} + 3\mathbf{j}$.
33. Two points on a sphere are called *antipodal* iff they are opposite endpoints of a diameter. Show that, if P_1 and P_2 are antipodal points on a sphere and Q is any other point on the sphere, then $\overline{P_1 Q} \perp \overline{P_2 Q}$.
34. Give an algebraic proof of Schwarz's inequality $|\mathbf{a} \cdot \mathbf{b}| \le \|\mathbf{a}\|\|\mathbf{b}\|$.   HINT: If $\mathbf{b} = \mathbf{0}$, the inequality is trivial, so assume $\mathbf{b} \ne \mathbf{0}$. Note that for any number λ we have $\|\mathbf{a} - \lambda\mathbf{b}\|^2 \ge 0$. First expand this inequality using the fact that $\|\mathbf{a} - \lambda\mathbf{b}\|^2 = (\mathbf{a} - \lambda\mathbf{b}) \cdot (\mathbf{a} - \lambda\mathbf{b})$. After collecting terms, make the special choice $\lambda = (\mathbf{a} \cdot \mathbf{b})/\|\mathbf{b}\|^2$ and see what happens.

13.5   MATRICES AND DETERMINANTS

(In this section we introduce determinants. They will be useful to us in the next section where we define the cross product of two vectors.)

By a *matrix* we mean a rectangular arrangement of numbers enclosed in parentheses. For example,

$$\begin{pmatrix} 2 & 4 \\ 3 & 1 \end{pmatrix} \quad \begin{pmatrix} 1 & 6 & 3 \\ 5 & 2 & 2 \end{pmatrix} \quad \begin{pmatrix} 2 & 4 & 0 \\ 4 & 7 & 1 \\ 0 & 1 & 1 \end{pmatrix}$$

are all matrices. The numbers occurring in a matrix are called the *entries*.

Matrices are useful for representing certain kinds of data in a compact way. The first matrix above might represent the coefficients of x and y in a pair of simultaneous linear equations

$$2x + 4y = A$$
$$3x + y = B.$$

The second matrix might represent the results of three successive throws of two dice. The third matrix might represent the components of three forces acting at a certain point.

Matrices have become an indispensable scientific tool. They are the subject of a beautiful algebraic theory that has been elaborated in many directions to serve the purposes of various fields. The study of matrix theory is outside the scope of this book. We will confine ourselves to what we need to introduce determinants.

Each matrix has a certain number of rows and a certain number of columns. A matrix with m rows and n columns is called an $m \times n$ *matrix*. Thus the first matrix above is a 2×2 matrix, the second a 2×3 matrix, the third a 3×3 matrix. The first and third matrices are called *square*: they have the same number of rows as columns. Here we will be working with square matrices as these are the only ones that have determinants.

We could give a definition of determinant that is applicable to all square matrices, but the definition is complicated and would serve little purpose at this point. Our interest here is in the 2×2 case and the 3×3 case. We begin with the 2×2 case.

(13.5.1)

The *determinant* of the matrix

$$\begin{pmatrix} a_1 & a_2 \\ b_1 & b_2 \end{pmatrix}$$

is the number $a_1 b_2 - a_2 b_1$.

We have a special notation for the determinant. We change the parentheses of the matrix to vertical bars:

$$\text{determinant of } \begin{pmatrix} a_1 & a_2 \\ b_1 & b_2 \end{pmatrix} = \begin{vmatrix} a_1 & a_2 \\ b_1 & b_2 \end{vmatrix} = a_1 b_2 - a_2 b_1.$$

Thus, for example,

$$\begin{vmatrix} 5 & 8 \\ 4 & 2 \end{vmatrix} = (5 \cdot 2) - (8 \cdot 4) = 10 - 32 = -22 \quad \text{and} \quad \begin{vmatrix} 4 & 0 \\ 0 & \frac{1}{4} \end{vmatrix} = (4 \cdot \tfrac{1}{4}) - (0 \cdot 0) = 1.$$

We remark on three properties of 2×2 determinants:

(1) If the rows or columns of a 2×2 determinant are interchanged, the determinant changes sign:

$$\begin{vmatrix} b_1 & b_2 \\ a_1 & a_2 \end{vmatrix} = - \begin{vmatrix} a_1 & a_2 \\ b_1 & b_2 \end{vmatrix}, \qquad \begin{vmatrix} a_2 & a_1 \\ b_2 & b_1 \end{vmatrix} = - \begin{vmatrix} a_1 & a_2 \\ b_1 & b_2 \end{vmatrix}.$$

Proof. Just note that

$$b_1 a_2 - b_2 a_1 = -(a_1 b_2 - a_2 b_1) \quad \text{and} \quad a_2 b_1 - a_1 b_2 = -(a_1 b_2 - a_2 b_1). \quad \square$$

(2) A common factor can be removed from any row or column and placed as a multiplier in front of the determinant:

$$\begin{vmatrix} \lambda a_1 & \lambda a_2 \\ b_1 & b_2 \end{vmatrix} = \lambda \begin{vmatrix} a_1 & a_2 \\ b_1 & b_2 \end{vmatrix}, \qquad \begin{vmatrix} \lambda a_1 & a_2 \\ \lambda b_1 & b_2 \end{vmatrix} = \lambda \begin{vmatrix} a_1 & a_2 \\ b_1 & b_2 \end{vmatrix}.$$

Proof.   Just note that

$$(\lambda a_1)b_2 - (\lambda a_2)b_1 = \lambda(a_1 b_2 - a_2 b_1) \quad \text{and} \quad (\lambda a_1)b_2 - a_2(\lambda b_1) = \lambda(a_1 b_2 - a_2 b_1). \quad \square$$

(3) If the rows or columns of a 2 × 2 determinant are the same, the determinant is 0.

Proof

$$\begin{vmatrix} a_1 & a_2 \\ a_1 & a_2 \end{vmatrix} = a_1 a_2 - a_2 a_1 = 0, \qquad \begin{vmatrix} a_1 & a_1 \\ b_1 & b_1 \end{vmatrix} = a_1 b_1 - a_1 b_1 = 0. \quad \square$$

The determinant of a 3 × 3 matrix is harder to define.   One definition is this:

$$\begin{vmatrix} a_1 & a_2 & a_3 \\ b_1 & b_2 & b_3 \\ c_1 & c_2 & c_3 \end{vmatrix} = a_1 b_2 c_3 - a_1 b_3 c_2 + a_2 b_3 c_1 - a_2 b_1 c_3 + a_3 b_1 c_2 - a_3 b_2 c_1.$$

The problem with this definition is that it is hard to remember.   What saves us is that the expansion on the right can be conveniently written in terms of 2 × 2 determinants; namely, the expression on the right can be written

$$a_1(b_2 c_3 - b_3 c_2) - a_2(b_1 c_3 - b_3 c_1) + a_3(b_1 c_2 - b_2 c_1),$$

which turns into

$$a_1 \begin{vmatrix} b_2 & b_3 \\ c_2 & c_3 \end{vmatrix} - a_2 \begin{vmatrix} b_1 & b_3 \\ c_1 & c_3 \end{vmatrix} + a_3 \begin{vmatrix} b_1 & b_2 \\ c_1 & c_2 \end{vmatrix}.$$

We then have

(13.5.2)
$$\begin{vmatrix} a_1 & a_2 & a_3 \\ b_1 & b_2 & b_3 \\ c_1 & c_2 & c_3 \end{vmatrix} = a_1 \begin{vmatrix} b_2 & b_3 \\ c_2 & c_3 \end{vmatrix} - a_2 \begin{vmatrix} b_1 & b_3 \\ c_1 & c_3 \end{vmatrix} + a_3 \begin{vmatrix} b_1 & b_2 \\ c_1 & c_2 \end{vmatrix}.$$

We will take this as our definition.   It is called the *expansion of the determinant along the first row*.   Note that the coefficients are the entries a_1, a_2, a_3 of the first row, that they occur alternately with + and − signs, and that each is multiplied by a determinant.   You can remember which determinant goes with which entry a_i as follows: in the original matrix, mentally cross out the row and column in which the entry a_i is found, and take the determinant of the remaining 2 × 2 matrix.   For example, the determinant that goes with a_3 is

$$\begin{vmatrix} a_1 & a_2 & a_3 \\ b_1 & b_2 & b_3 \\ c_1 & c_2 & c_3 \end{vmatrix} = \begin{vmatrix} b_1 & b_2 \\ c_1 & c_2 \end{vmatrix}.$$

When first starting to work with specific 3×3 determinants, it is a good idea to set up the formula with blank 2×2 determinants:

$$\begin{vmatrix} a_1 & a_2 & a_3 \\ b_1 & b_2 & b_3 \\ c_1 & c_2 & c_3 \end{vmatrix} = a_1 \begin{vmatrix} & \\ & \end{vmatrix} - a_2 \begin{vmatrix} & \\ & \end{vmatrix} + a_3 \begin{vmatrix} & \\ & \end{vmatrix}$$

and then fill in the 2×2 determinants by the "crossing out" rule explained above.

Example 1

$$\begin{vmatrix} 1 & 2 & 1 \\ 0 & 3 & 4 \\ 6 & 2 & 5 \end{vmatrix} = 1 \begin{vmatrix} 3 & 4 \\ 2 & 5 \end{vmatrix} - 2 \begin{vmatrix} 0 & 4 \\ 6 & 5 \end{vmatrix} + 1 \begin{vmatrix} 0 & 3 \\ 6 & 2 \end{vmatrix}$$

$$= 1(15 - 8) - 2(0 - 24) + 1(0 - 18)$$

$$= 7 + 48 - 18 = 37. \quad \square$$

A straightforward (but somewhat laborious) calculation shows that 3×3 determinants have the same three properties we proved earlier for 2×2 determinants:

(1) If two rows or columns are interchanged, the determinant changes sign.

(2) A common factor can be removed from any row or column and placed as a multiplier in front of the determinant.

(3) If two rows or columns are the same, the determinant is 0.

EXERCISES 13.5

Evaluate the following determinants.

1. $\begin{vmatrix} 1 & 2 \\ 3 & 4 \end{vmatrix}.$ **2.** $\begin{vmatrix} 1 & -1 \\ -1 & 1 \end{vmatrix}.$ **3.** $\begin{vmatrix} 1 & 1 \\ a & a \end{vmatrix}.$ **4.** $\begin{vmatrix} a & b \\ b & d \end{vmatrix}.$

5. $\begin{vmatrix} 1 & 0 & 3 \\ 2 & 4 & 1 \\ 0 & 1 & 0 \end{vmatrix}.$ **6.** $\begin{vmatrix} 1 & 0 & 0 \\ 0 & 2 & 0 \\ 0 & 0 & 3 \end{vmatrix}.$ **7.** $\begin{vmatrix} 0 & 0 & 1 \\ 0 & 2 & 0 \\ 3 & 0 & 0 \end{vmatrix}.$ **8.** $\begin{vmatrix} a & 0 & 0 \\ b & c & 0 \\ d & e & f \end{vmatrix}.$

A matrix is *unimodular* iff it has determinant 1. Determine whether the matrix is unimodular.

9. $\begin{pmatrix} 1 & 0 \\ 0 & -1 \end{pmatrix}.$ **10.** $\begin{pmatrix} 1 & 1 \\ 1 & 0 \end{pmatrix}.$ **11.** $\begin{pmatrix} 1 & 0 \\ 1 & 1 \end{pmatrix}.$ **12.** $\begin{pmatrix} \cos\theta & \sin\theta \\ -\sin\theta & \cos\theta \end{pmatrix}.$

13. $\begin{pmatrix} 1 & 2 & 3 \\ 0 & \cosh x & \sinh x \\ 0 & \sinh x & \cosh x \end{pmatrix}.$ **14.** $\begin{pmatrix} 1 & 0 & 1 \\ 0 & 1 & 1 \\ 1 & 0 & 0 \end{pmatrix}.$

Find the values of x that make the matrix unimodular.

15. $\begin{pmatrix} a & b \\ x & c \end{pmatrix}.$ **16.** $\begin{pmatrix} x & x \\ x & 2 \end{pmatrix}.$ **17.** $\begin{pmatrix} 1 & 1 & 1 \\ 1 & x & 0 \\ 1 & 1 & x^2 \end{pmatrix}.$

18. If A is a matrix, its *transpose* A^T is obtained by interchanging the rows and columns. Thus

$$\begin{pmatrix} a_1 & a_2 \\ b_1 & b_2 \end{pmatrix}^T = \begin{pmatrix} a_1 & b_1 \\ a_2 & b_2 \end{pmatrix} \quad \text{and} \quad \begin{pmatrix} a_1 & a_2 & a_3 \\ b_1 & b_2 & b_3 \\ c_1 & c_2 & c_3 \end{pmatrix}^T = \begin{pmatrix} a_1 & b_1 & c_1 \\ a_2 & b_2 & c_2 \\ a_3 & b_3 & c_3 \end{pmatrix}.$$

Show that the determinant of a matrix equals the determinant of its transpose: (a) the 2×2 case. (b) the 3×3 case.

Verify the following assertions.

19. $\begin{vmatrix} 1 & 2 & 3 \\ 4 & 5 & 6 \\ 7 & 8 & 9 \end{vmatrix} + \begin{vmatrix} 4 & 5 & 6 \\ 1 & 2 & 3 \\ 7 & 8 & 9 \end{vmatrix} = 0.$

20. $\begin{vmatrix} 1 & 2 & 3 \\ 4 & 5 & 6 \\ 7 & 8 & 9 \end{vmatrix} = \begin{vmatrix} 4 & 5 & 6 \\ 7 & 8 & 9 \\ 1 & 2 & 3 \end{vmatrix}.$

21. $\begin{vmatrix} 1 & 2 & 3 \\ 4 & 5 & 6 \\ 7 & 8 & 9 \end{vmatrix} + \begin{vmatrix} 1 & 2 & 3 \\ 1 & 2 & 3 \\ 7 & 8 & 9 \end{vmatrix} = \begin{vmatrix} 1 & 2 & 3 \\ 4 & 5 & 6 \\ 7 & 8 & 9 \end{vmatrix}.$

22. $\frac{1}{2} \begin{vmatrix} 1 & 0 & 7 \\ 3 & 4 & 5 \\ 2 & 4 & 6 \end{vmatrix} = \begin{vmatrix} 1 & 0 & 7 \\ 3 & 4 & 5 \\ 1 & 2 & 3 \end{vmatrix}.$

23. $\begin{vmatrix} 1 & 2 & 3 \\ x & 2x & 3x \\ 4 & 5 & 6 \end{vmatrix} = 0.$

24. $\begin{vmatrix} 2 & 4 & 6 \\ 1 & 0 & 5 \\ 3 & 3 & 1 \end{vmatrix} = -2 \begin{vmatrix} 1 & 0 & 5 \\ 1 & 2 & 3 \\ 3 & 3 & 1 \end{vmatrix}.$

25. $\begin{vmatrix} \lambda^2 & \lambda^2 \\ a & b \end{vmatrix} - \begin{vmatrix} 2\lambda & 2\lambda \\ a & b \end{vmatrix} + \begin{vmatrix} 1 & 1 \\ a & b \end{vmatrix} = (\lambda - 1)^2 \begin{vmatrix} 1 & 1 \\ a & b \end{vmatrix}.$

26. (a) Verify that the equations

$$3x + 4y = 6$$
$$2x - 3y = 7$$

are solved by the prescription

$$x = \frac{\begin{vmatrix} 6 & 4 \\ 7 & -3 \end{vmatrix}}{\begin{vmatrix} 3 & 4 \\ 2 & -3 \end{vmatrix}}, \qquad y = \frac{\begin{vmatrix} 3 & 6 \\ 2 & 7 \end{vmatrix}}{\begin{vmatrix} 3 & 4 \\ 2 & -3 \end{vmatrix}}.$$

(b) More generally, verify that the equations

$$a_1 x + a_2 y = d$$
$$b_1 x + b_2 y = e$$

are solved by the prescription

$$x = \frac{\begin{vmatrix} d & a_2 \\ e & b_2 \end{vmatrix}}{\begin{vmatrix} a_1 & a_2 \\ b_1 & b_2 \end{vmatrix}}, \qquad y = \frac{\begin{vmatrix} a_1 & d \\ b_1 & e \end{vmatrix}}{\begin{vmatrix} a_1 & a_2 \\ b_1 & b_2 \end{vmatrix}}$$

provided that the determinant in the denominator is different from 0.

(c) Conjecture an analogous rule for solving three linear equations in three unknowns.

27. Show that a 3×3 determinant can be "expanded along the bottom row" as follows:

$$\begin{vmatrix} a_1 & a_2 & a_3 \\ b_1 & b_2 & b_3 \\ c_1 & c_2 & c_3 \end{vmatrix} = c_1 \begin{vmatrix} a_2 & a_3 \\ b_2 & b_3 \end{vmatrix} - c_2 \begin{vmatrix} a_1 & a_3 \\ b_1 & b_3 \end{vmatrix} + c_3 \begin{vmatrix} a_1 & a_2 \\ b_1 & b_2 \end{vmatrix}.$$

HINT: You can check this directly by writing out the values of the determinants on the right, or you can interchange rows twice to bring the bottom row to the top and then use expansion along the top row.

28. Verify the following additivity property for determinants differing only in the first row:

$$\begin{vmatrix} a_1 & a_2 & a_3 \\ b_1 & b_2 & b_3 \\ c_1 & c_2 & c_3 \end{vmatrix} + \begin{vmatrix} a_1' & a_2' & a_3' \\ b_1 & b_2 & b_3 \\ c_1 & c_2 & c_3 \end{vmatrix} = \begin{vmatrix} a_1 + a_1' & a_2 + a_2' & a_3 + a_3' \\ b_1 & b_2 & b_3 \\ c_1 & c_2 & c_3 \end{vmatrix}.$$

Similar rules hold for the other rows and all columns.

Verify the following assertions.

29. $\begin{vmatrix} a & a^2 \\ b & b^2 \end{vmatrix} = ab \begin{vmatrix} 1 & a \\ 1 & b \end{vmatrix}.$

30. $\begin{vmatrix} 1 & x & x^2 \\ x & x^2 & x^3 \\ x^2 & x^3 & x^4 \end{vmatrix} = 0.$

31. $\begin{vmatrix} a_1 + b_1 & a_2 + b_2 & a_3 + b_3 \\ b_1 & b_2 & b_3 \\ c_1 & c_2 & c_3 \end{vmatrix} = \begin{vmatrix} a_1 & a_2 & a_3 \\ b_1 & b_2 & b_3 \\ c_1 & c_2 & c_3 \end{vmatrix}.$

13.6   THE CROSS PRODUCT

Definition

While the dot product $\mathbf{a} \cdot \mathbf{b}$ is a scalar (and as such is sometimes called the scalar product of $\mathbf{a}$ and $\mathbf{b}$), the cross product $\mathbf{a} \times \mathbf{b}$ is a vector (sometimes called the *vector product* of $\mathbf{a}$ and $\mathbf{b}$). What is the vector $\mathbf{a} \times \mathbf{b}$? We could directly write down a formula that gives the components of $\mathbf{a} \times \mathbf{b}$ in terms of the components of $\mathbf{a}$ and $\mathbf{b}$, but at this stage that would reveal little. Instead we will begin geometrically. We will define $\mathbf{a} \times \mathbf{b}$ by giving its direction and its magnitude.

The definition of $\mathbf{a} \times \mathbf{b}$. If the vectors $\mathbf{a}$ and $\mathbf{b}$ are not parallel, they determine a plane. The vector $\mathbf{a} \times \mathbf{b}$ is perpendicular to this plane and directed in such a way that (like $\mathbf{i}, \mathbf{j}, \mathbf{k}$) the vectors $\mathbf{a}, \mathbf{b}, \mathbf{a} \times \mathbf{b}$ form a right-handed triple. (Figure 13.6.1) [If the index finger of the right hand points along $\mathbf{a}$ and the middle finger points along $\mathbf{b}$, then the thumb will point in the vicinity of $\mathbf{a} \times \mathbf{b}$.]

Figure 13.6.1

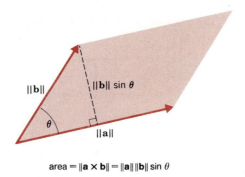

$$\text{area} = \|\mathbf{a} \times \mathbf{b}\| = \|\mathbf{a}\| \|\mathbf{b}\| \sin \theta$$

Figure 13.6.2

The magnitude of $\mathbf{a} \times \mathbf{b}$. If $\mathbf{a}$ and $\mathbf{b}$ are not parallel, they generate a parallelogram. (Figure 13.6.2) The magnitude of $\mathbf{a} \times \mathbf{b}$ is the area of this parallelogram: $\|\mathbf{a}\| \|\mathbf{b}\| \sin \theta$.

One more point. What if $\mathbf{a}$ and $\mathbf{b}$ are parallel? Then there is no parallelogram and we define $\mathbf{a} \times \mathbf{b} = \mathbf{0}$.

We summarize all this below.

DEFINITION 13.6.1

If $\mathbf{a}$ and $\mathbf{b}$ are not parallel, then $\mathbf{a} \times \mathbf{b}$ is the unique vector with the following properties:

1. $\mathbf{a} \times \mathbf{b}$ is perpendicular to the plane of $\mathbf{a}$ and $\mathbf{b}$.

2. $\mathbf{a}, \mathbf{b}, \mathbf{a} \times \mathbf{b}$ form a right-handed triple.

3. $\|\mathbf{a} \times \mathbf{b}\| = \|\mathbf{a}\| \|\mathbf{b}\| \sin \theta$ where θ is the angle between $\mathbf{a}$ and $\mathbf{b}$.

If $\mathbf{a}$ and $\mathbf{b}$ are parallel, then $\mathbf{a} \times \mathbf{b} = \mathbf{0}$.

Properties of Right-Handed Triples

I. Note first of all that if $(\mathbf{a}, \mathbf{b}, \mathbf{c})$ form a right-handed triple, then $(\mathbf{b}, \mathbf{c}, \mathbf{a})$ and $(\mathbf{c}, \mathbf{a}, \mathbf{b})$ also form right-handed triples. (Figure 13.6.3) To maintain right-handedness we don't have to keep the vectors in the same order but we do have to keep them in the *same cyclic order*:

$$\mathbf{a} \to \mathbf{b} \to \mathbf{c}$$

Alter the cyclic order and you reverse the orientation.

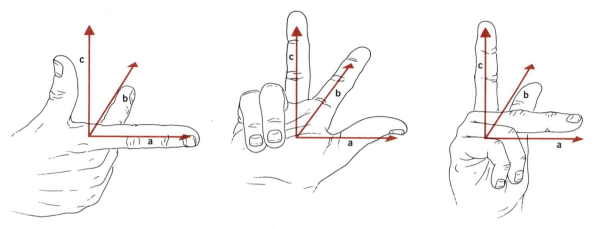

Figure 13.6.3

II. A triple $(\mathbf{a}, \mathbf{b}, \mathbf{c})$ is right-handed iff $\mathbf{c}$ and $\mathbf{a} \times \mathbf{b}$ lie on the same side of the plane determined by $\mathbf{a}$ and $\mathbf{b}$. (Figure 13.6.4) This means that $(\mathbf{a}, \mathbf{b}, \mathbf{c})$ is right-handed iff $(\mathbf{a} \times \mathbf{b}) \cdot \mathbf{c} > 0$. (Explain)

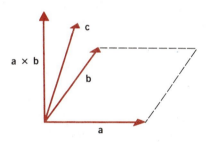

Figure 13.6.4

III. It follows from II that, if $(\mathbf{a}, \mathbf{b}, \mathbf{c})$ is right-handed, then $(\mathbf{a}, \mathbf{b}, -\mathbf{c})$ is not right-handed. Similarly $(-\mathbf{a}, \mathbf{b}, \mathbf{c})$ and $(\mathbf{a}, -\mathbf{b}, \mathbf{c})$ are also not right-handed. However, multiplication by positive scalars does maintain right-handedness: if $(\mathbf{a}, \mathbf{b}, \mathbf{c})$ is right-handed and α, β, γ are positive, then $(\alpha\mathbf{a}, \beta\mathbf{b}, \gamma\mathbf{c})$ is also right-handed.

Properties of the Cross Product

The cross product is *anticommutative*:

(13.6.2)
$$\mathbf{b} \times \mathbf{a} = -(\mathbf{a} \times \mathbf{b}).$$

To see this, note that both vectors are perpendicular to $\mathbf{a}$ and $\mathbf{b}$ and both have the same norm. Thus $\mathbf{b} \times \mathbf{a} = \pm(\mathbf{a} \times \mathbf{b})$. That the minus sign holds and not the plus sign follows from observing that $\mathbf{b}, \mathbf{a}, \mathbf{b} \times \mathbf{a}$ must be right-handed and that $\mathbf{b}, \mathbf{a}, \mathbf{a} \times \mathbf{b}$ is not right-handed since $\mathbf{a}, \mathbf{b}, \mathbf{a} \times \mathbf{b}$ is right-handed. ☐

Scalars can be factored:

(13.6.3)
$$\alpha \mathbf{a} \times \beta \mathbf{b} = \alpha\beta(\mathbf{a} \times \mathbf{b}).$$

If α or β is zero, the result is obvious. We will assume that α and β are both nonzero. In this case the two vectors are perpendicular to $\mathbf{a}$, perpendicular to $\mathbf{b}$, and have the same norm. Thus $\alpha \mathbf{a} \times \beta \mathbf{b} = \pm\alpha\beta(\mathbf{a} \times \mathbf{b})$. That the positive sign holds comes from noting that $\alpha \mathbf{a}$, $\beta \mathbf{b}$, $\alpha\beta(\mathbf{a} \times \mathbf{b})$ is a right-handed triple. This is obvious if α and β are both positive. If not, two of the three coefficients α, β, $\alpha\beta$ are negative and the other is positive. In this case the first minus sign reverses the orientation but the second one restores it. ☐

Finally, there are two distributive laws, the verification of which we postpone for a moment.

(13.6.4)
$$\mathbf{a} \times (\mathbf{b} + \mathbf{c}) = (\mathbf{a} \times \mathbf{b}) + (\mathbf{a} \times \mathbf{c}), \qquad (\mathbf{a} + \mathbf{b}) \times \mathbf{c} = (\mathbf{a} \times \mathbf{c}) + (\mathbf{b} \times \mathbf{c}).$$

The Scalar Triple Product

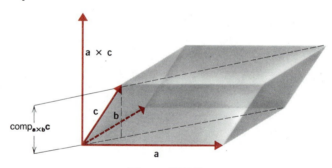

Figure 13.6.5

Earlier we saw that $\mathbf{a}$, $\mathbf{b}$, $\mathbf{c}$ is a right-handed triple iff $(\mathbf{a} \times \mathbf{b}) \cdot \mathbf{c} > 0$. The expression $(\mathbf{a} \times \mathbf{b}) \cdot \mathbf{c}$ is called a *scalar triple product*. The absolute value of this number (it is a number and not a vector) has geometric significance. To describe it we refer to Figure 13.6.5. There you see a parallelepiped with edges $\mathbf{a}$, $\mathbf{b}$, $\mathbf{c}$. The absolute value of the scalar triple product gives the volume of that parallelepiped:

(13.6.5)
$$V = |(\mathbf{a} \times \mathbf{b}) \cdot \mathbf{c}|.$$

Proof. The area of the base is $\|\mathbf{a} \times \mathbf{b}\|$. The height is $|\text{comp}_{\mathbf{a} \times \mathbf{b}}\,\mathbf{c}|$. Therefore

$$V = |\text{comp}_{\mathbf{a} \times \mathbf{b}}\,\mathbf{c}|\,\|\mathbf{a} \times \mathbf{b}\| = |(\mathbf{a} \times \mathbf{b}) \cdot \mathbf{c}|. \quad ☐$$

⌐——— (13.4.12)

Of course we could have chosen a different base and a different height. Therefore

$$|(\mathbf{a} \times \mathbf{b}) \cdot \mathbf{c}| = |(\mathbf{c} \times \mathbf{a}) \cdot \mathbf{b}| = |(\mathbf{b} \times \mathbf{c}) \cdot \mathbf{a}|.$$

Since the **a**, **b**, **c** appear above in the same cyclic order, the expressions inside the absolute value signs all have the same sign. (Property II of right-handed triples) Therefore

(13.6.6)
$$(\mathbf{a} \times \mathbf{b}) \cdot \mathbf{c} = (\mathbf{c} \times \mathbf{a}) \cdot \mathbf{b} = (\mathbf{b} \times \mathbf{c}) \cdot \mathbf{a}.$$

Verification of the Distributive Laws

We will verify the first distributive law,

$$\mathbf{a} \times (\mathbf{b} + \mathbf{c}) = (\mathbf{a} \times \mathbf{b}) + (\mathbf{a} \times \mathbf{c}).$$

The second follows readily from this one. The argument is left to you as an exercise.

Take an arbitrary vector **r** and form the dot product $[\mathbf{a} \times (\mathbf{b} + \mathbf{c})] \cdot \mathbf{r}$. We can then write

$$[\mathbf{a} \times (\mathbf{b} + \mathbf{c})] \cdot \mathbf{r} = (\mathbf{r} \times \mathbf{a}) \cdot (\mathbf{b} + \mathbf{c}) \qquad (13.6.6)$$

$$= [(\mathbf{r} \times \mathbf{a}) \cdot \mathbf{b}] + [(\mathbf{r} \times \mathbf{a}) \cdot \mathbf{c}] \qquad (13.4.6)$$

$$= [(\mathbf{a} \times \mathbf{b}) \cdot \mathbf{r}] + [(\mathbf{a} \times \mathbf{c}) \cdot \mathbf{r}] \qquad (13.6.6)$$

$$= [(\mathbf{a} \times \mathbf{b}) + (\mathbf{a} \times \mathbf{c})] \cdot \mathbf{r}. \qquad (13.4.6)$$

Since this holds true for all vectors **r**, it holds true for **i**, **j**, **k** and proves that

$$\mathbf{a} \times (\mathbf{b} + \mathbf{c}) = (\mathbf{a} \times \mathbf{b}) + (\mathbf{a} \times \mathbf{c}). \quad \square$$

The Components of $a \times b$

You have learned a lot about cross products, but you still have not seen $\mathbf{a} \times \mathbf{b}$ expressed in terms of the components of **a** and **b**. To derive the formula that does this, we need to observe one more fact, one which is obvious from the very definition of cross product:

(13.6.7)
$$\mathbf{i} \times \mathbf{j} = \mathbf{k}, \qquad \mathbf{j} \times \mathbf{k} = \mathbf{i}, \qquad \mathbf{k} \times \mathbf{i} = \mathbf{j}.$$
(Figure 13.6.6)

One way to remember these products is to arrange **i**, **j**, **k** in cyclic order, $\mathbf{i} \to \mathbf{j} \to \mathbf{k}$, and note that

[each coordinate unit vector] $\times$ [the next one] = [the third one].

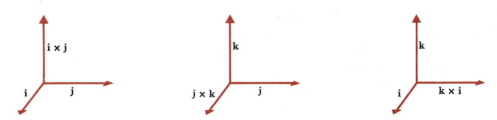

Figure 13.6.6

THEOREM 13.6.8

For vectors $\mathbf{a} = a_1\mathbf{i} + a_2\mathbf{j} + a_3\mathbf{k}$ and $\mathbf{b} = b_1\mathbf{i} + b_2\mathbf{j} + b_3\mathbf{k}$

$$\mathbf{a} \times \mathbf{b} = \begin{vmatrix} \mathbf{i} & \mathbf{j} & \mathbf{k} \\ a_1 & a_2 & a_3 \\ b_1 & b_2 & b_3 \end{vmatrix} = \begin{vmatrix} a_2 & a_3 \\ b_2 & b_3 \end{vmatrix}\mathbf{i} - \begin{vmatrix} a_1 & a_3 \\ b_1 & b_3 \end{vmatrix}\mathbf{j} + \begin{vmatrix} a_1 & a_2 \\ b_1 & b_2 \end{vmatrix}\mathbf{k}.$$

(The 3×3 determinant with $\mathbf{i}, \mathbf{j}, \mathbf{k}$ in the top row is there only as a mnemonic device.)

Proof. The hard work has all been done. With what you know about cross products now, the proof is just a matter of simple algebra:

$$\mathbf{a} \times \mathbf{b} = (a_1\mathbf{i} + a_2\mathbf{j} + a_3\mathbf{k}) \times (b_1\mathbf{i} + b_2\mathbf{j} + b_3\mathbf{k})$$

$$= a_1b_2(\mathbf{i} \times \mathbf{j}) + a_1b_3(\mathbf{i} \times \mathbf{k}) + a_2b_1(\mathbf{j} \times \mathbf{i}) + a_2b_3(\mathbf{j} \times \mathbf{k}) + a_3b_1(\mathbf{k} \times \mathbf{i}) + a_3b_2(\mathbf{k} \times \mathbf{j})$$

$$\uparrow\!\!\!\!\rule{1cm}{0.4pt}\; \mathbf{i} \times \mathbf{i} = \mathbf{j} \times \mathbf{j} = \mathbf{k} \times \mathbf{k} = \mathbf{0}$$

$$= a_1b_2\mathbf{k} - a_1b_3\mathbf{j} - a_2b_1\mathbf{k} + a_2b_3\mathbf{i} + a_3b_1\mathbf{j} - a_3b_2\mathbf{i}$$

$$= (a_2b_3 - a_3b_2)\mathbf{i} - (a_1b_3 - a_3b_1)\mathbf{j} + (a_1b_2 - a_2b_1)\mathbf{k}$$

$$= \begin{vmatrix} a_2 & a_3 \\ b_2 & b_3 \end{vmatrix}\mathbf{i} - \begin{vmatrix} a_1 & a_3 \\ b_1 & b_3 \end{vmatrix}\mathbf{j} + \begin{vmatrix} a_1 & a_2 \\ b_1 & b_2 \end{vmatrix}\mathbf{k}. \quad \square$$

Problems

Problem 1. Calculate $\mathbf{a} \times \mathbf{b}$ given that

$$\mathbf{a} = \mathbf{i} - 2\mathbf{j} + 3\mathbf{k} \quad \text{and} \quad \mathbf{b} = 2\mathbf{i} + \mathbf{j} - \mathbf{k}.$$

Solution

$$\mathbf{a} \times \mathbf{b} = \begin{vmatrix} \mathbf{i} & \mathbf{j} & \mathbf{k} \\ 1 & -2 & 3 \\ 2 & 1 & -1 \end{vmatrix} = \begin{vmatrix} -2 & 3 \\ 1 & -1 \end{vmatrix}\mathbf{i} - \begin{vmatrix} 1 & 3 \\ 2 & -1 \end{vmatrix}\mathbf{j} + \begin{vmatrix} 1 & -2 \\ 2 & 1 \end{vmatrix}\mathbf{k} = -\mathbf{i} + 7\mathbf{j} + 5\mathbf{k}. \quad \square$$

Problem 2. Calculate $\mathbf{a} \times \mathbf{b}$ given that

$$\mathbf{a} = \mathbf{i} - \mathbf{j} \quad \text{and} \quad \mathbf{b} = \mathbf{i} + \mathbf{k}.$$

Solution

$$\mathbf{a} \times \mathbf{b} = \begin{vmatrix} \mathbf{i} & \mathbf{j} & \mathbf{k} \\ 1 & -1 & 0 \\ 1 & 0 & 1 \end{vmatrix} = \begin{vmatrix} -1 & 0 \\ 0 & 1 \end{vmatrix}\mathbf{i} - \begin{vmatrix} 1 & 0 \\ 1 & 1 \end{vmatrix}\mathbf{j} + \begin{vmatrix} 1 & -1 \\ 1 & 0 \end{vmatrix}\mathbf{k} = -\mathbf{i} - \mathbf{j} + \mathbf{k}. \quad \square$$

In Problems 1 and 2 we calculated some cross products using determinants. Of course, we can obtain the same results just by applying the distributive laws. For example, for $\mathbf{a} = \mathbf{i} - 2\mathbf{j} + 3\mathbf{k}$ and $\mathbf{b} = 2\mathbf{i} + \mathbf{j} - \mathbf{k}$, we have

$$\mathbf{a} \times \mathbf{b} = (\mathbf{i} - 2\mathbf{j} + 3\mathbf{k}) \times (2\mathbf{i} + \mathbf{j} - \mathbf{k})$$
$$= (\mathbf{i} \times \mathbf{j}) - (\mathbf{i} \times \mathbf{k}) - 4(\mathbf{j} \times \mathbf{i}) + 2(\mathbf{j} \times \mathbf{k}) + 6(\mathbf{k} \times \mathbf{i}) + 3(\mathbf{k} \times \mathbf{j})$$
$$= \mathbf{k} + \mathbf{j} + 4\mathbf{k} + 2\mathbf{i} + 6\mathbf{j} - 3\mathbf{i} = -\mathbf{i} + 7\mathbf{j} + 5\mathbf{k}.$$

Problem 3. Show that the scalar triple product can be written as a determinant:

(13.6.9)
$$(\mathbf{a} \times \mathbf{b}) \cdot \mathbf{c} = \begin{vmatrix} a_1 & a_2 & a_3 \\ b_1 & b_2 & b_3 \\ c_1 & c_2 & c_3 \end{vmatrix}.$$

Solution

$$(\mathbf{a} \times \mathbf{b}) \cdot \mathbf{c} = \mathbf{c} \cdot (\mathbf{a} \times \mathbf{b})$$
$$= (c_1\mathbf{i} + c_2\mathbf{j} + c_3\mathbf{k}) \cdot \left(\begin{vmatrix} a_2 & a_3 \\ b_2 & b_3 \end{vmatrix} \mathbf{i} - \begin{vmatrix} a_1 & a_3 \\ b_1 & b_3 \end{vmatrix} \mathbf{j} + \begin{vmatrix} a_1 & a_2 \\ b_1 & b_2 \end{vmatrix} \mathbf{k} \right)$$
$$= c_1 \begin{vmatrix} a_2 & a_3 \\ b_2 & b_3 \end{vmatrix} - c_2 \begin{vmatrix} a_1 & a_3 \\ b_1 & b_3 \end{vmatrix} + c_3 \begin{vmatrix} a_1 & a_2 \\ b_1 & b_2 \end{vmatrix}.$$

This is the expansion of

$$\begin{vmatrix} a_1 & a_2 & a_3 \\ b_1 & b_2 & b_3 \\ c_1 & c_2 & c_3 \end{vmatrix}$$

by the bottom row.   (Exercise 27, Section 13.5) □

A SUGGESTION: Vectors were defined as ordered triples, and many of the early proofs were done by "breaking up" vectors into their components. This may give you the impression that the method of "breakup" and working with the components is the first thing to try when confronted with a problem that involves vectors. If it is a *computational* problem, this method may give good results. But if you have to *analyze* a situation involving vectors, particularly one in which geometry plays a role, then the "breakup" strategy is seldom the best. Think instead of using the *operations* we have defined on vectors: addition, subtraction, scalar multiplication, dot product, cross product. Being geometrically motivated, these operations are likely to provide greater understanding than breaking up everything in sight into components.

Problem 4. Let $\mathbf{a}, \mathbf{b}, \mathbf{c}$ be nonzero vectors that do not lie in the same plane. Find all the vectors $\mathbf{d}$ for which

$(*)$ $$\mathbf{d} \cdot \mathbf{a} = \mathbf{d} \cdot \mathbf{b} = \mathbf{d} \cdot \mathbf{c}.$$

Solution. We could begin by writing

$$\mathbf{d} = d_1\mathbf{i} + d_2\mathbf{j} + d_3\mathbf{k}, \quad \mathbf{a} = a_1\mathbf{i} + a_2\mathbf{j} + a_3\mathbf{k}, \quad \mathbf{b} = \text{etc.}$$

Equation (∗) would then take the form

$$d_1a_1 + d_2a_2 + d_3a_3 = d_1b_1 + d_2b_2 + d_3b_3 = d_1c_1 + d_2c_2 + d_3c_3.$$

We would then be faced with finding all d_1, d_2, d_3 that satisfied these equations. This is a messy task.

Here is a better approach. The vectors **d** that satisfy (∗) are the vectors **d** for which

$$\mathbf{d} \cdot (\mathbf{a} - \mathbf{b}) = 0 \qquad \text{and} \qquad \mathbf{d} \cdot (\mathbf{b} - \mathbf{c}) = 0.$$

These are the vectors **d** that are perpendicular to both $\mathbf{a} - \mathbf{b}$ and $\mathbf{b} - \mathbf{c}$. One such vector is $(\mathbf{a} - \mathbf{b}) \times (\mathbf{b} - \mathbf{c})$. The vectors **d** that satisfy (∗) are the scalar multiples of that cross product. ☐

Problem 5. Verify that

$$\|\mathbf{a} \times \mathbf{b}\|^2 + (\mathbf{a} \cdot \mathbf{b})^2 = \|\mathbf{a}\|^2 \|\mathbf{b}\|^2.$$

Solution. We could begin by writing

$$\|\mathbf{a} \times \mathbf{b}\| = (a_2b_3 - a_3b_2)^2 + (a_1b_3 - a_3b_1)^2 + (a_1b_2 - a_2b_1)^2$$
$$(\mathbf{a} \cdot \mathbf{b})^2 = (a_1b_1 + a_2b_2 + a_3b_3)^2$$
$$\|\mathbf{a}\|^2 \|\mathbf{b}\|^2 = \text{etc.,}$$

but this would take us into a morass of arithmetic. It is much more fruitful to proceed as follows:

$$\|\mathbf{a} \times \mathbf{b}\| = \|\mathbf{a}\| \|\mathbf{b}\| \sin \theta \qquad \text{and} \qquad \mathbf{a} \cdot \mathbf{b} = \|\mathbf{a}\| \|\mathbf{b}\| \cos \theta.$$

Therefore

$$\|\mathbf{a} \times \mathbf{b}\|^2 + (\mathbf{a} \cdot \mathbf{b})^2 = \|\mathbf{a}\|^2 \|\mathbf{b}\|^2 \sin^2 \theta + \|\mathbf{a}\|^2 \|\mathbf{b}\|^2 \cos^2 \theta$$
$$= \|\mathbf{a}\|^2 \|\mathbf{b}\|^2 (\sin^2 \theta + \cos^2 \theta) = \|\mathbf{a}\|^2 \|\mathbf{b}\|^2. \quad ☐$$

Three Important Identities

It may be tempting to think that $\mathbf{a} \times (\mathbf{b} \times \mathbf{c}) = (\mathbf{a} \times \mathbf{b}) \times \mathbf{c}$. This is in general false:

$$\mathbf{i} \times (\mathbf{i} \times \mathbf{j}) = \mathbf{i} \times \mathbf{k} = -\mathbf{j} \qquad \text{but} \qquad (\mathbf{i} \times \mathbf{i}) \times \mathbf{j} = \mathbf{0} \times \mathbf{j} = \mathbf{0}.$$

What is true instead is that

(13.6.10)

$$\mathbf{a} \times (\mathbf{b} \times \mathbf{c}) = (\mathbf{a} \cdot \mathbf{c})\mathbf{b} - (\mathbf{a} \cdot \mathbf{b})\mathbf{c},$$
$$(\mathbf{a} \times \mathbf{b}) \times \mathbf{c} = (\mathbf{c} \cdot \mathbf{a})\mathbf{b} - (\mathbf{c} \cdot \mathbf{b})\mathbf{a}.$$

As pointed out years ago by Professor Wilfred Kaplan of the University of Michigan, both formulas can be remembered as follows:

triple cross product = (outer · remote)adjacent − (outer · adjacent)remote.†

---

† See W. Kaplan, *Advanced Calculus,* Addison-Wesley, Reading, Massachusetts, 1952.

There is one more identity that we want to mention:

(13.6.11) $\qquad \boxed{(\mathbf{a} \times \mathbf{b}) \cdot (\mathbf{c} \times \mathbf{d}) = (\mathbf{a} \cdot \mathbf{c})(\mathbf{b} \cdot \mathbf{d}) - (\mathbf{a} \cdot \mathbf{d})(\mathbf{b} \cdot \mathbf{c}).}$

The proof of this, as well as the proof of (13.6.10), is left to you in the exercises.

Remark. Dot products and cross products are almost ubiquitous in physics and engineering. Work is a dot product. So is the power expended by a force. Every line integral is a dot product. Torque and angular momentum are cross products. Turn on a television set and watch the dots on the screen. How they move is determined by a cross product. Take a look at any text on electromagnetism. It is all based on Maxwell's four equations. Two of them are dot products, two of them are cross products. ☐

EXERCISES 13.6

Calculate.

1. $(\mathbf{i} + \mathbf{j}) \times (\mathbf{i} - \mathbf{j})$.   **2.** $(\mathbf{i} - \mathbf{j}) \times (\mathbf{j} - \mathbf{i})$.   **3.** $(\mathbf{i} - \mathbf{j}) \times (\mathbf{j} - \mathbf{k})$.

4. $\mathbf{j} \times (2\mathbf{i} - \mathbf{k})$.   **5.** $(2\mathbf{j} - \mathbf{k}) \times (\mathbf{i} - 3\mathbf{j})$.   **6.** $\mathbf{i} \cdot (\mathbf{j} \times \mathbf{k})$.

7. $\mathbf{j} \cdot (\mathbf{i} \times \mathbf{k})$.   **8.** $(\mathbf{j} \times \mathbf{i}) \cdot (\mathbf{i} \times \mathbf{k})$.   **9.** $(\mathbf{i} \times \mathbf{j}) \times \mathbf{k}$.

10. $\mathbf{k} \cdot (\mathbf{j} \times \mathbf{i})$.   **11.** $\mathbf{j} \cdot (\mathbf{k} \times \mathbf{i})$.   **12.** $\mathbf{j} \times (\mathbf{k} \times \mathbf{i})$.

Calculate.

13. $(\mathbf{i} + \mathbf{j} + \mathbf{k}) \times (2\mathbf{i} + \mathbf{k})$.   **14.** $(2\mathbf{i} - \mathbf{k}) \times (\mathbf{i} - 2\mathbf{j} + 2\mathbf{k})$.

15. $[2\mathbf{i} + \mathbf{j}] \cdot [(\mathbf{i} - 3\mathbf{j} + \mathbf{k}) \times (4\mathbf{i} + \mathbf{k})]$.   **16.** $[(-2\mathbf{i} + \mathbf{j} - 3\mathbf{k}) \times \mathbf{i}] \times [\mathbf{i} + \mathbf{j}]$.

17. $[(\mathbf{i} - \mathbf{j}) \times (\mathbf{j} - \mathbf{k})] \times [\mathbf{i} + 5\mathbf{k}]$.   **18.** $[\mathbf{i} - \mathbf{j}] \times [(\mathbf{j} - \mathbf{k}) \times (\mathbf{j} + 5\mathbf{k})]$.

Use a cross product to find the area of triangle *PQR*.

19. $P(0, 1, 0)$, $Q(-1, 1, 2)$, $R(2, 1, -1)$.

20. $P(1, 2, 3)$, $Q(-1, 3, 2)$, $R(3, -1, 2)$.

Find the volume of the parallelepiped with the given edges.

21. $\mathbf{i} + \mathbf{j}$, $2\mathbf{i} - \mathbf{k}$, $3\mathbf{j} + \mathbf{k}$.   **22.** $\mathbf{i} - 3\mathbf{j} + \mathbf{k}$, $2\mathbf{j} - \mathbf{k}$, $\mathbf{i} + \mathbf{j} - 2\mathbf{k}$.

23. Express $(\mathbf{a} + \mathbf{b}) \times (\mathbf{a} - \mathbf{b})$ as a scalar multiple of $\mathbf{a} \times \mathbf{b}$.

24. Earlier we verified that $\mathbf{a} \times (\mathbf{b} + \mathbf{c}) = (\mathbf{a} \times \mathbf{b}) + (\mathbf{a} \times \mathbf{c})$. Show now that

$$(\mathbf{a} + \mathbf{b}) \times \mathbf{c} = (\mathbf{a} \times \mathbf{c}) + (\mathbf{b} \times \mathbf{c}).$$

25. Suppose that $\mathbf{a} \times \mathbf{i} = \mathbf{0}$ and $\mathbf{a} \times \mathbf{j} = \mathbf{0}$. What can you conclude about $\mathbf{a}$?

26. Let $\mathbf{a} = a_1\mathbf{i} + a_2\mathbf{j}$ and $\mathbf{b} = b_1\mathbf{i} + b_2\mathbf{j}$ be nonzero vectors in the *xy*-plane. Show that $\mathbf{a} \times \mathbf{b}$ is parallel to $\mathbf{k}$.

27. Express $(\alpha\mathbf{a} + \beta\mathbf{b}) \times (\gamma\mathbf{a} + \delta\mathbf{b})$ as a scalar multiple of $\mathbf{a} \times \mathbf{b}$.

28. (a) Let $\mathbf{a}, \mathbf{b}, \mathbf{c}$ be distinct nonzero vectors. Show that

$$\mathbf{a} \times \mathbf{b} = \mathbf{a} \times \mathbf{c} \qquad \text{iff} \qquad \mathbf{a} \text{ and } \mathbf{b} - \mathbf{c} \text{ are parallel.}$$

(b) Sketch a figure depicting all those vectors $\mathbf{c}$ that satisfy $\mathbf{a} \times \mathbf{b} = \mathbf{a} \times \mathbf{c}$.

29. Which of the following dot products are equal?

$$\mathbf{a} \cdot (\mathbf{b} \times \mathbf{c}) \qquad \mathbf{a} \cdot (\mathbf{c} \times \mathbf{b}) \qquad (\mathbf{a} \times \mathbf{b}) \cdot \mathbf{c} \qquad (\mathbf{c} \times \mathbf{a}) \cdot \mathbf{b}$$

$$(\mathbf{b} \times \mathbf{c}) \cdot \mathbf{a} \qquad \mathbf{c} \cdot (\mathbf{b} \times \mathbf{a}) \qquad (-\mathbf{a} \times \mathbf{b}) \cdot \mathbf{c} \qquad (\mathbf{a} \times -\mathbf{c}) \cdot \mathbf{b}$$

30. (a) Show that

$$\mathbf{a} \times (\mathbf{b} \times \mathbf{c}) = (\mathbf{a} \cdot \mathbf{c})\mathbf{b} - (\mathbf{a} \cdot \mathbf{b})\mathbf{c}.$$

HINT: Verify that the **i** components of the two sides agree.   By analogy, it will be clear that the same holds true for the **j** and **k** components.

(b) Show that

$$(\mathbf{a} \times \mathbf{b}) \times \mathbf{c} = (\mathbf{c} \cdot \mathbf{a})\mathbf{b} - (\mathbf{c} \cdot \mathbf{b})\mathbf{a}.$$

HINT: $(\mathbf{a} \times \mathbf{b}) \times \mathbf{c} = -[\mathbf{c} \times (\mathbf{a} \times \mathbf{b})]$.

(c) Finally, show that

$$(\mathbf{a} \times \mathbf{b}) \cdot (\mathbf{c} \times \mathbf{d}) = (\mathbf{a} \cdot \mathbf{c})(\mathbf{b} \cdot \mathbf{d}) - (\mathbf{a} \cdot \mathbf{d})(\mathbf{b} \cdot \mathbf{c}).$$

HINT: Set $\mathbf{c} \times \mathbf{d} = \mathbf{r}$ and use (13.6.6).

31. Let **a**, **b** be nonzero vectors with $\mathbf{a} \perp \mathbf{b}$, and set $\mathbf{c} = \mathbf{a} \times \mathbf{b}$.   Express $\mathbf{c} \times \mathbf{a}$ in terms of **b**.

32. Given that **u** is a unit vector, show that each vector **a** can be decomposed as follows into a part parallel to **u** and a part perpendicular to **u**:

$$\mathbf{a} = \underbrace{(\mathbf{a} \cdot \mathbf{u})\mathbf{u}}_{\substack{\text{parallel} \\ \text{to } \mathbf{u}}} + \underbrace{(\mathbf{u} \times \mathbf{a}) \times \mathbf{u}}_{\substack{\text{perpendicular} \\ \text{to } \mathbf{u}}}.$$

HINT: Use Exercise 30(b).

13.7   LINES

Vectors used to specify position are called *position vectors*.   Position vectors that emanate from the origin are known as *radius vectors*.   In this section we use radius vectors to characterize lines.

Vector Parametrizations

We begin with the idea that two distinct points determine a line.   In Figure 13.7.1 we have marked two points and the line *l* that they determine.   To obtain a vector characterization of *l*, we choose the vectors $\mathbf{r}_0$ and **d** as in Figure 13.7.2.   Since we began with two distinct points, the vector **d** is nonzero.

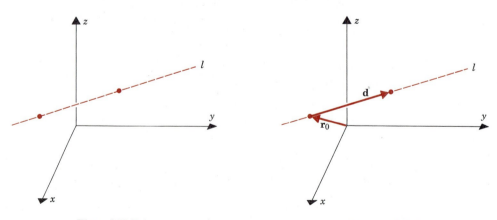

Figure 13.7.1   Figure 13.7.2

In Figure 13.7.3 we have drawn an additional vector $\mathbf{r}$. The vector that begins at the tip of $\mathbf{r}_0$ and ends at the tip of $\mathbf{r}$ is $\mathbf{r} - \mathbf{r}_0$. Therefore, the tip of $\mathbf{r}$ will fall on l iff

$$\mathbf{r} - \mathbf{r}_0 \quad \text{and} \quad \mathbf{d} \quad \text{are parallel;}$$

this in turn will happen iff

$$\mathbf{r} - \mathbf{r}_0 = t\mathbf{d} \qquad \text{for some real } t$$

or equivalently, iff

$$\mathbf{r} = \mathbf{r}_0 + t\mathbf{d} \qquad \text{for some real } t.$$

The vector equation

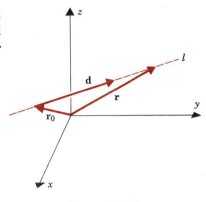

Figure 13.7.3

(13.7.1)

$$\boxed{\mathbf{r}(t) = \mathbf{r}_0 + t\mathbf{d}, \qquad t \text{ real}}$$

parametrizes the line l: by varying t, we vary the vector $\mathbf{r}(t)$, but its tip remains on l; as t ranges over the set of real numbers, the tip of $\mathbf{r}(t)$ traces out the line l.

Now set

$$\mathbf{r}_0 = x_0\mathbf{i} + y_0\mathbf{j} + z_0\mathbf{k}, \qquad \mathbf{d} = d_1\mathbf{i} + d_2\mathbf{j} + d_3\mathbf{k}$$

and denote the tip of $\mathbf{r}_0$ by $P(x_0, y_0, z_0)$. The line

$$\mathbf{r}(t) = \mathbf{r}_0 + t\mathbf{d}$$

passes through the point $P(x_0, y_0, z_0)$ and is parallel to $\mathbf{d}$. The vector $\mathbf{d}$, which by assumption is not zero, is called a *direction vector* for l, and its components d_1, d_2, d_3 are called *direction numbers*.

Problem 1. Find a vector equation that parametrizes the line that passes through the point $P(1, -1, 2)$ and is parallel to the vector $2\mathbf{i} - 3\mathbf{j} + \mathbf{k}$.

Solution. Here we can set

$$\mathbf{r}_0 = \mathbf{i} - \mathbf{j} + 2\mathbf{k} \quad \text{and} \quad \mathbf{d} = 2\mathbf{i} - 3\mathbf{j} + \mathbf{k}.$$

As a vector parametrization for the line we have

$$\mathbf{r}(t) = (\mathbf{i} - \mathbf{j} + 2\mathbf{k}) + t(2\mathbf{i} - 3\mathbf{j} + \mathbf{k}),$$

which we can rewrite as

$$\mathbf{r}(t) = (1 + 2t)\mathbf{i} - (1 + 3t)\mathbf{j} + (2 + t)\mathbf{k}. \quad \square$$

As a direction vector for a given line we can take any nonzero vector that is parallel to the line. Thus, if $\mathbf{d}$ is a direction vector for l, so is $\alpha\mathbf{d}$, provided that $\alpha \neq 0$. If d_1, d_2, d_3 are direction numbers for l, so are $\alpha d_1, \alpha d_2, \alpha d_3$, provided again that $\alpha \neq 0$.

The line that passes through the origin with direction vector **d** can be parametrized by setting

(13.7.2)
$$\boxed{\mathbf{r}(t) = t\mathbf{d}.}$$

There are, however, other ways to parametrize this line.

Problem 2. Find all vector parametrizations (13.7.1) for the line through the origin with direction vector **d**.

Solution. Since **d** is a direction vector, so is every vector $\alpha\mathbf{d}$ with $\alpha \neq 0$. We therefore write

(∗)
$$\mathbf{r}(t) = \mathbf{r}_0 + t\alpha\mathbf{d}.$$

Since the line passes through the origin, it may appear at first glance that $\mathbf{r}_0$ has to be **0**, but that is not true. From the fact that the line passes through the origin we can conclude only that $\mathbf{r}_0 + t\alpha\mathbf{d}$ is **0** for some value of t. Call this value t_0. Then

$$\mathbf{r}_0 + t_0\alpha\mathbf{d} = \mathbf{0} \quad \text{and thus} \quad \mathbf{r}_0 = -t_0\alpha\mathbf{d}.$$

Substitution in (∗) gives

$$\mathbf{r}(t) = -t_0\alpha\mathbf{d} + t\alpha\mathbf{d} = (t - t_0)\alpha\mathbf{d}.$$

All the desired parametrizations can be written

$$\mathbf{r}(t) = (t - t_0)\alpha\mathbf{d} \quad \text{with } \alpha \text{ and } t_0 \text{ real, } \alpha \neq 0,$$

and all equations of this form parametrize that same line. ☐

Scalar Parametric Equations

The line that passes through the point $P(x_0, y_0, z_0)$ with direction numbers d_1, d_2, d_3 can be parametrized by three scalar equations:

(13.7.3)
$$\boxed{x(t) = x_0 + d_1 t, \quad y(t) = y_0 + d_2 t, \quad z(t) = z_0 + d_3 t.}$$

These equations are the **i, j, k** components of the vector equation $\mathbf{r}(t) = \mathbf{r}_0 + t\mathbf{d}$.

Problem 3. Write scalar parametric equations for the line that passes through the point $P(-1, 4, 2)$ with direction numbers 1, 2, 3.

Solution. In this case the scalar equations

$$x(t) = x_0 + d_1 t, \quad y(t) = y_0 + d_2 t, \quad z(t) = z_0 + d_3 t$$

take the form

$$x(t) = -1 + t, \quad y(t) = 4 + 2t, \quad z(t) = 2 + 3t. \quad ☐$$

Problem 4. What direction numbers are displayed by the parametric equations

$$x(t) = 3 - t, \quad y(t) = 2 + 4t, \quad z(t) = 1 - 5t?$$

What other direction numbers could be used for the same line?

Solution. The direction numbers displayed are $-1, 4, -5$. Any triple of the form

$$-\alpha, \quad 4\alpha, \quad -5\alpha \qquad \text{with } \alpha \neq 0$$

could be used as a set of direction numbers for that same line. ☐

Symmetric Form

If the direction numbers are all different from zero, then all the scalar parametric equations can be solved for t:

$$t = \frac{x(t) - x_0}{d_1}, \quad t = \frac{y(t) - y_0}{d_2}, \quad t = \frac{z(t) - z_0}{d_3}.$$

Eliminating the parameter t, we obtain three equations:

$$\frac{x - x_0}{d_1} = \frac{y - y_0}{d_2}, \qquad \frac{y - y_0}{d_2} = \frac{z - z_0}{d_3}, \qquad \frac{x - x_0}{d_1} = \frac{z - z_0}{d_3}.$$

Obviously, any two of these equations suffice; the third is redundant and can be discarded. Rather than decide which equation to discard, we simply write

(13.7.4)
$$\frac{x - x_0}{d_1} = \frac{y - y_0}{d_2} = \frac{z - z_0}{d_3}.$$

These are the equations of a line written in *symmetric form*. They can be used only if d_1, d_2, d_3 are all different from zero.

Problem 5. Write equations in symmetric form for the line that passes through $P(x_0, y_0, z_0)$ and $Q(x_1, y_1, z_1)$. Under what conditions are the equations valid?

Solution. As direction numbers we can take the triple

$$x_1 - x_0, \quad y_1 - y_0, \quad z_1 - z_0.$$

We can base our calculations on $P(x_0, y_0, z_0)$ and write

$$\frac{x - x_0}{x_1 - x_0} = \frac{y - y_0}{y_1 - y_0} = \frac{z - z_0}{z_1 - z_0},$$

or we can base our calculations on $Q(x_1, y_1, z_1)$ and write

$$\frac{x - x_1}{x_1 - x_0} = \frac{y - y_1}{y_1 - y_0} = \frac{z - z_1}{z_1 - z_0}.$$

Both sets of equations are valid provided that $x_1 \neq x_0, y_1 \neq y_0, z_1 \neq z_0$. ☐

Equations 13.7.4 can be used only if the direction numbers are all different from zero. If one of the direction numbers is zero, then one of the coordinates is constant. As you will see, this simplifies the algebra. Geometrically, it means that the line lies on a plane that is parallel to one of the coordinate planes.

Suppose, for example, that $d_3 = 0$. Then the scalar parametric equations take the form

$$x(t) = x_0 + d_1 t, \quad y(t) = y_0 + d_2 t, \quad z(t) = z_0.$$

Eliminating t, we are left with two equations:

$$\frac{x - x_0}{d_1} = \frac{y - y_0}{d_2}, \qquad z = z_0.$$

The line lies on the horizontal plane $z = z_0$, and its projection onto the xy-plane (see Figure 13.7.4) is the line l' with equation

$$\frac{x - x_0}{d_1} = \frac{y - y_0}{d_2}.$$

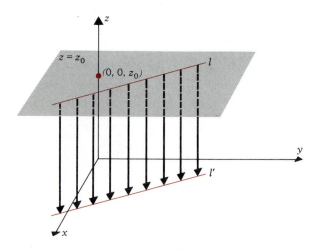

Figure 13.7.4

Intersecting Lines, Parallel Lines

Two distinct lines

$$l_1 : \mathbf{r}(t) = \mathbf{r}_0 + t\mathbf{d}, \qquad l_2 : \mathbf{R}(u) = \mathbf{R}_0 + u\mathbf{D}$$

intersect iff there are numbers t and u at which

$$\mathbf{r}(t) = \mathbf{R}(u).$$

Problem 6. Find the point at which the lines

$$l_1 : \mathbf{r}(t) = (\mathbf{i} - 6\mathbf{j} + 2\mathbf{k}) + t\,(\mathbf{i} + 2\mathbf{j} + \mathbf{k}), \qquad l_2 : \mathbf{R}(u) = (4\mathbf{j} + \mathbf{k}) + u(2\mathbf{i} + \mathbf{j} + 2\mathbf{k})$$

intersect.

Solution. We set

$$\mathbf{r}(t) = \mathbf{R}(u)$$

and solve for t and u:

$$(\mathbf{i} - 6\mathbf{j} + 2\mathbf{k}) + t(\mathbf{i} + 2\mathbf{j} + \mathbf{k}) = (4\mathbf{j} + \mathbf{k}) + u(2\mathbf{i} + \mathbf{j} + 2\mathbf{k}),$$
$$\mathbf{i} - 6\mathbf{j} + 2\mathbf{k} + t\mathbf{i} + 2t\mathbf{j} + t\mathbf{k} = 4\mathbf{j} + \mathbf{k} + 2u\mathbf{i} + u\mathbf{j} + 2u\mathbf{k},$$

and therefore

$$(1 + t - 2u)\mathbf{i} + (-10 + 2t - u)\mathbf{j} + (1 + t - 2u)\mathbf{k} = 0.$$

This tells us that

$$1 + t - 2u = 0,$$
$$-10 + 2t - u = 0,$$
$$1 + t - 2u = 0.$$

Note that the first and third equations are the same. Solving the first two equations simultaneously, we obtain $t = 7$, $u = 4$. As you can verify,

$$\mathbf{r}(7) = 8\mathbf{i} + 8\mathbf{j} + 9\mathbf{k} = \mathbf{R}(4).$$

The two lines intersect at the tip of this vector, which is the point $P(8, 8, 9)$.   ☐

In the setting of plane geometry we can think of two lines as parallel iff they do not intersect. This point of view is not satisfactory in three-dimensional space. (See Figure 13.7.5.)

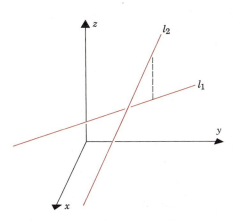

Figure 13.7.5

The lines l_1 and l_2 marked in Figure 13.7.5 do not intersect, and yet we would hesitate to call them parallel. We can avoid this difficulty by using direction vectors: we will call two lines *parallel* iff their direction vectors are parallel.†

---

† Nonparallel, nonintersecting lines are said to be *skew*.

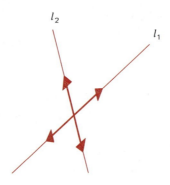

Figure 13.7.6

If two lines l_1, l_2 intersect, we can find the angle between them by finding the angle between their direction vectors, **d** and **D**. Depending on our choice of direction vectors, there are two such angles, each the supplement of the other (Figure 13.7.6). We choose the smaller of the two angles, the one with nonnegative cosine:

(13.7.5)
$$\cos \theta = |\mathbf{u_d} \cdot \mathbf{u_D}|.$$

Problem 7. Earlier we verified that the lines

$$l_1: \mathbf{r}(t) = (\mathbf{i} - 6\mathbf{j} + 2\mathbf{k}) + t(\mathbf{i} + 2\mathbf{j} + \mathbf{k}), \qquad l_2: \mathbf{R}(u) = (4\mathbf{j} + \mathbf{k}) + u(2\mathbf{i} + \mathbf{j} + 2\mathbf{k})$$

intersect at $P(8, 8, 9)$. What is the angle between these lines?

Solution. As direction vectors we can take

$$\mathbf{d} = \mathbf{i} + 2\mathbf{j} + \mathbf{k} \quad \text{and} \quad \mathbf{D} = 2\mathbf{i} + \mathbf{j} + 2\mathbf{k}.$$

Then, as you can check,

$$\mathbf{u_d} = \tfrac{1}{6}\sqrt{6}\,(\mathbf{i} + 2\mathbf{j} + \mathbf{k}) \quad \text{and} \quad \mathbf{u_D} = \tfrac{1}{3}(2\mathbf{i} + \mathbf{j} + 2\mathbf{k}).$$

It follows that

$$\cos \theta = |\mathbf{u_d} \cdot \mathbf{u_D}| = \tfrac{1}{3}\sqrt{6} \cong 0.816 \quad \text{and} \quad \theta \cong 0.62 \text{ radians.} \quad \square$$

Two intersecting lines are said to be *perpendicular* iff their direction vectors are perpendicular.

Problem 8. Let l_1 and l_2 be the lines of the last problem. These lines intersect at $P(8, 8, 9)$. Find a vector parametrization for the line l_3 that passes through $P(8, 8, 9)$ and is perpendicular to both l_1 and l_2.

Solution. We are given that l_3 passes through $P(8, 8, 9)$. All we need to parametrize this line is a direction vector $\mathbf{d_3}$. We require that $\mathbf{d_3}$ be perpendicular to the direction vectors of l_1 and l_2; namely, we require that

$$\mathbf{d}_3 \perp \mathbf{d} = \mathbf{i} + 2\mathbf{j} + \mathbf{k}, \qquad \mathbf{d}_3 \perp \mathbf{D} = 2\mathbf{i} + \mathbf{j} + 2\mathbf{k}.$$

Since $\mathbf{d} \times \mathbf{D}$ is perpendicular to both $\mathbf{d}$ and $\mathbf{D}$, we can set

$$\mathbf{d}_3 = \mathbf{d} \times \mathbf{D} = \begin{vmatrix} \mathbf{i} & \mathbf{j} & \mathbf{k} \\ 1 & 2 & 1 \\ 2 & 1 & 2 \end{vmatrix} = \begin{vmatrix} 2 & 1 \\ 1 & 2 \end{vmatrix} \mathbf{i} - \begin{vmatrix} 1 & 1 \\ 2 & 2 \end{vmatrix} \mathbf{j} + \begin{vmatrix} 1 & 2 \\ 2 & 1 \end{vmatrix} \mathbf{k} = 3\mathbf{i} - 3\mathbf{k}.$$

As a parametrization for l_3 we can write

$$\gamma_3(t) = (8\mathbf{i} + 8\mathbf{j} + 9\mathbf{k}) + t(3\mathbf{i} - 3\mathbf{k}). \quad \square$$

Problem 9. (a) Find a vector parametrization for the line

$$l: y = mx + b \qquad \text{in the } xy\text{-plane.}$$

(b) Show by vector methods that

$$l_1: y = m_1 x + b_1 \quad \perp \quad l_2: y = m_2 x + b_2 \qquad \text{iff} \qquad m_1 m_2 = -1.$$

Solution. (a) We seek a parametrization of the form

$$\mathbf{r}(t) = \mathbf{r}_0 + t\mathbf{d}.$$

Since $P(0, b)$ lies on l, we can set

$$\mathbf{r}_0 = 0\mathbf{i} + b\mathbf{j} = b\mathbf{j}.$$

To find a direction vector, we take $x_1 \neq 0$ and note that the point $Q(x_1, mx_1 + b)$ also lies on l. (See Figure 13.7.7.) As direction numbers we can take

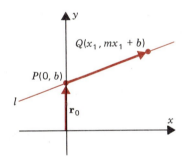

Figure 13.7.7

$$x_1 - 0 = x_1 \quad \text{and} \quad (mx_1 + b) - b = mx_1$$

or, more simply, 1 and m. This choice of direction numbers gives us the direction vector $\mathbf{d} = \mathbf{i} + m\mathbf{j}$. The vector equation

$$\mathbf{r}(t) = b\mathbf{j} + t(\mathbf{i} + m\mathbf{j})$$

parametrizes the line l.

(b) As direction vectors for l_1 and l_2 we have

$$\mathbf{d}_1 = \mathbf{i} + m_1\mathbf{j} \qquad \text{and} \qquad \mathbf{d}_2 = \mathbf{i} + m_2\mathbf{j}.$$

Since

$$\mathbf{d}_1 \cdot \mathbf{d}_2 = (\mathbf{i} + m_1\mathbf{j}) \cdot (\mathbf{i} + m_2\mathbf{j}) = 1 + m_1 m_2,$$

you can see that

$$\mathbf{d}_1 \cdot \mathbf{d}_2 = 0 \quad \text{iff} \quad m_1 m_2 = -1. \quad \square$$

Distance from a Point to a Line

In Figure 13.7.8 we have drawn a line l and a point P not on l. We are interested in finding the distance $d(P, l)$ between P and l.

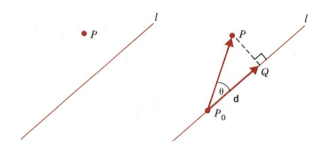

Figure 13.7.8

With P_0 and Q as in the figure we can take $\mathbf{d} = \overrightarrow{P_0 Q}$ as a direction vector for l. From the figure you can see that

$$d(P, l) = \|\overrightarrow{QP}\| = \|\overrightarrow{P_0 P}\| \sin \theta.$$

Since $\|\overrightarrow{P_0 P} \times \mathbf{d}\| = \|\overrightarrow{P_0 P}\| \|\mathbf{d}\| \sin \theta$, we have

(13.7.6)
$$d(P, l) = \frac{\|\overrightarrow{P_0 P} \times \mathbf{d}\|}{\|\mathbf{d}\|}.$$

This elegant little formula gives the distance from a point P to any line l in terms of any point P_0 on l and any direction vector $\mathbf{d}$ for l. ☐

Computations based on this formula are left to the exercises.

EXERCISES 13.7

1. Which of the points $P(1, 2, 0)$, $Q(-5, 1, 5)$, $R(-4, 2, 5)$ lie on the line

$$l: \mathbf{r}(t) = (\mathbf{i} + 2\mathbf{j}) + t(6\mathbf{i} + \mathbf{j} - 5\mathbf{k})?$$

2. Determine which of the lines are parallel:

$l_1: \mathbf{r}_1(t) = (\mathbf{i} + 2\mathbf{k}) + t(\mathbf{i} - 2\mathbf{j} + 3\mathbf{k})$, $l_2: \mathbf{r}_2(u) = (\mathbf{i} + 2\mathbf{k}) + u(\mathbf{i} + 2\mathbf{j} - 3\mathbf{k})$,

$l_3: \mathbf{r}_3(v) = (6\mathbf{i} - \mathbf{j}) - v(2\mathbf{i} - 4\mathbf{j} + 6\mathbf{k})$, $l_4: \mathbf{r}_4(w) = (\frac{1}{2} + \frac{1}{2}w)\mathbf{i} - w\mathbf{j} + (1 + \frac{3}{2}w)\mathbf{k}$.

Find a vector parametrization for the line that satisfies the given conditions.

3. Passes through $P(3, 1, 0)$ and is parallel to the line $\mathbf{r}(t) = (\mathbf{i} - \mathbf{j}) + t\mathbf{k}$.

4. Passes through $P(1, -1, 2)$ and is parallel to the line $\mathbf{r}(t) = t(3\mathbf{i} - \mathbf{j} + \mathbf{k})$.

5. Passes through the origin and $Q(x_1, y_1, z_1)$.

6. Passes through $P(x_0, y_0, z_0)$ and $Q(x_1, y_1, z_1)$.

Find a set of scalar parametric equations for the line that satisfies the given conditions.

7. Passes through $P(1, 0, 3)$ and $Q(2, -1, 4)$.

8. Passes through $P(x_0, y_0, z_0)$ and $Q(x_1, y_1, z_1)$.

9. Passes through $P(2, -2, 3)$ and is perpendicular to the xz-plane.

10. Passes through $(1, 4, -3)$ and is perpendicular to the yz-plane.

11. Give a vector parametrization for the line that passes through $P(-1, 2, -3)$ and is parallel to the line $2(x + 1) = 4(y - 3) = z$.

12. Write equations in symmetric form for the line that passes through the origin and the point $P(x_0, y_0, z_0)$.

Find the point where l_1 and l_2 intersect and give the angle of intersection.

13. $l_1: \mathbf{r}_1(t) = \mathbf{i} + t\mathbf{j}$, $l_2: \mathbf{r}_2(u) = \mathbf{j} + u(\mathbf{i} + \mathbf{j})$.

14. $l_1: \mathbf{r}_1(t) = (\mathbf{i} - 4\sqrt{3}\,\mathbf{j}) + t(\mathbf{i} + \sqrt{3}\,\mathbf{j})$, $l_2: \mathbf{r}_2(u) = (4\mathbf{i} + 3\sqrt{3}\,\mathbf{j}) + u(\mathbf{i} - \sqrt{3}\,\mathbf{j})$.

15. l_1: $x_1(t) = 3 + t,$ $y_1(t) = 1 - t,$ $z_1(t) = 5 + 2t.$
   l_2: $x_2(u) = 1,$ $y_2(u) = 4 + u,$ $z_2(u) = 2 + u.$

16. l_1: $x_1(t) = 1 + t,$ $y_1(t) = -1 - t,$ $z_1(t) = -4 + 2t.$
   l_2: $x_2(u) = 1 - u,$ $y_2(u) = 1 + 3u,$ $z_2(u) = 2u.$

17. Where does the line

$$\frac{x - x_0}{d_1} = \frac{y - y_0}{d_2} = \frac{z - z_0}{d_3}$$

intersect the xy-plane?

18. What can you conclude about the lines

$$\frac{x - x_0}{d_1} = \frac{y - y_0}{d_2} = \frac{z - z_0}{d_3}, \qquad \frac{x - x_0}{D_1} = \frac{y - y_0}{D_2} = \frac{z - z_0}{D_3}$$

given that $d_1 D_1 + d_2 D_2 + d_3 D_3 = 0$?

19. What can you conclude about the lines

$$\frac{x - x_0}{d_1} = \frac{y - y_0}{d_2} = \frac{z - z_0}{d_3}, \qquad \frac{x - x_1}{D_1} = \frac{y - y_1}{D_2} = \frac{z - z_1}{D_3}$$

given that $d_1/D_1 = d_2/D_2 = d_3/D_3$?

20. Let P_0, P_1 be two distinct points and let $\mathbf{r}_0, \mathbf{r}_1$ be the radius vectors that they determine:

$$\mathbf{r}_0 = \overrightarrow{OP_0}, \quad \mathbf{r}_1 = \overrightarrow{OP_1}.$$

As t ranges over the set of real numbers, $\mathbf{r}(t) = \mathbf{r}_0 + t(\mathbf{r}_1 - \mathbf{r}_0)$ traces out the line determined by P_0 and P_1. Restrict t so that $\mathbf{r}(t)$ traces out only the line segment $P_0 P_1$.

21. Find a vector parametrization for the line segment that begins at $(2, 7, -1)$ and ends at $(4, 2, 3)$.

22. Restrict t so that the equations

$$x(t) = 7 - 5t, \quad y(t) = -3 + 2t, \quad z(t) = 4 - t$$

parametrize the line segment that begins at $(12, -5, 5)$ and ends at $(-3, 1, 2)$.

23. Determine a unit vector $\mathbf{u}$ and the values of t for which the equation

$$\mathbf{r}(t) = (6\mathbf{i} - 5\mathbf{j} + \mathbf{k}) + t\mathbf{u}$$

parametrizes the line segment that begins at $P(0, -2, 7)$ and ends at $Q(-4, 0, 11)$.

24. Suppose that the lines

$$l_1: \mathbf{r}(t) = \mathbf{r}_0 + t\mathbf{d}, \qquad l_2: \mathbf{R}(u) = \mathbf{R}_0 + u\mathbf{D}$$

intersect at right angles. Show that the point of intersection is the origin iff $\mathbf{r}(t) \perp \mathbf{R}(u)$ for all real numbers t and u.

25. Find scalar parametric equations for all the lines that are perpendicular to the line

$$x(t) = 1 + 2t, \quad y(t) = 3 - 4t, \quad z(t) = 2 + 6t$$

and intersect that line at the point $P(3, -1, 8)$.

26. Suppose that $\mathbf{r}(t) = \mathbf{r}_0 + t\mathbf{d}$ and $\mathbf{R}(u) = \mathbf{R}_0 + u\mathbf{D}$ both parametrize the same line. (a) Show that $\mathbf{R}_0 = \mathbf{r}_0 + t_0\mathbf{d}$ for some real number t_0. (b) Then show that, for some real number α, $\mathbf{R}(u) = \mathbf{r}_0 + (t_0 + \alpha u)\mathbf{d}$ for all real u.

Find the distance from $P(1, 0, 2)$ to the indicated line.

27. The line through the origin parallel to $2\mathbf{i} - \mathbf{j} + 2\mathbf{k}$.
28. The line through $P_0(1, -1, 1)$ parallel to $\mathbf{i} - 2\mathbf{j} - 2\mathbf{k}$.

Find the distance from the point to the line.

29. $P(1, 2, 3), \quad l\colon \mathbf{r}(t) = \mathbf{i} + 2\mathbf{k} + t(\mathbf{i} - 2\mathbf{j} + 3\mathbf{k})$.
30. $P(0, 0, 0), \quad l\colon \mathbf{r}(t) = \mathbf{i} + t\mathbf{j}$.
31. $P(1, 0, 1), \quad l\colon \mathbf{r}(t) = 2\mathbf{i} - \mathbf{j} + t(\mathbf{i} + \mathbf{j})$.

32. Find the distance from the point $P(x_0, y_0, z_0)$ to the line $y = mx + b$ in the xy-plane.

33. What is the distance from the origin (a) to the line that joins $P(1, 1, 1)$ and $Q(2, 2, 1)$? (b) to the line segment that joins these same points? [For (b) find the point of the line segment $\overline{PQ}$ closest to the origin and calculate its distance from the origin.]

34. Let l be the line

$$\mathbf{r}(t) = \mathbf{r}_0 + t\mathbf{d}.$$

(a) Find the scalar t_0 for which $\mathbf{r}(t_0) \perp l$.
(b) Find the parametrizations $\mathbf{R}(t) = \mathbf{R}_0 + t\mathbf{D}$ for l in which $\mathbf{R}_0 \perp l$ and $\|\mathbf{D}\| = 1$. These are the *standard vector parametrizations* for l.

Find the standard vector parametrizations (Exercise 34) for the specified line.

35. The line through $P(0, 1, -2)$ parallel to $\mathbf{i} - \mathbf{j} + 3\mathbf{k}$.
36. The line through $P(\sqrt{3}, 0, 0)$ parallel to $\mathbf{i} + \mathbf{j} + \mathbf{k}$.

37. Let A, B, C be the vertices of a triangle in the xy-plane. Given that $0 < s < 1$, determine the values of t for which the tip of the radius vector

$$\overrightarrow{OA} + s\,\overrightarrow{AB} + t\,\overrightarrow{BC}$$

lies inside the triangle.   HINT: Draw a diagram.

13.8 PLANES

Ways of Specifying a Plane

How can we specify a plane? By giving three points on it, so long as they are not all on the same line. By giving two lines on it. By giving a line on it and a point on it, so long as the point does not lie on the line. There is still another way to specify a plane and that is to give a point from the plane and a nonzero vector perpendicular to the plane.

Scalar Equation of a Plane

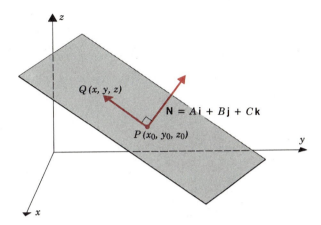

Figure 13.8.1

Figure 13.8.1 shows a plane. On it we have marked a point $P(x_0, y_0, z_0)$ and starting at that point a nonzero vector $\mathbf{N} = A\mathbf{i} + B\mathbf{j} + C\mathbf{k}$ perpendicular to the plane. We call $\mathbf{N}$ a *normal vector*. We can obtain an equation for the plane in terms of the coordinates of P and the components of $\mathbf{N}$.

To find such an equation we take an arbitrary point $Q(x, y, z)$ in space and form the vector

$$\overrightarrow{PQ} = (x - x_0)\mathbf{i} + (y - y_0)\mathbf{j} + (z - z_0)\mathbf{k}.$$

The point Q will lie on the given plane iff

$$\mathbf{N} \cdot \overrightarrow{PQ} = 0,$$

which is to say, iff

(13.8.1) $$\boxed{A(x - x_0) + B(y - y_0) + C(z - z_0) = 0.}$$

This is an equation in x, y, z for the plane that passes through $P(x_0, y_0, z_0)$ and has normal vector $\mathbf{N} = A\mathbf{i} + B\mathbf{j} + C\mathbf{k}$.

Remark. If $\mathbf{N}$ is normal to a given plane, then so is every nonzero scalar multiple of $\mathbf{N}$. Suppose we had chosen $-2\mathbf{N}$ as our normal. Then Equation 13.8.1 would have read

$$-2A(x - x_0) - 2B(y - y_0) - 2C(z - z_0) = 0.$$

Cancelling the -2, we would have the same equation we had before. It does not matter which normal we choose. They all give equivalent equations. □

We can write Equation 13.8.1 in the form

$$Ax + By + Cz + D = 0$$

simply by setting $D = -Ax_0 - By_0 - Cz_0$.

Problem 1.   Write an equation for the plane that passes through the point $P(1, 0, 2)$ and has normal vector $\mathbf{N} = 3\mathbf{i} - 2\mathbf{j} + \mathbf{k}$.

Solution.   The general equation

$$A(x - x_0) + B(y - y_0) + C(z - z_0) = 0$$

becomes

$$3(x - 1) + (-2)(y - 0) + (1)(z - 2) = 0,$$

which simplifies to

$$3x - 2y + z - 5 = 0. \quad \square$$

Problem 2.   Find an equation for the plane that passes through $P(1, 3, -2)$ and is perpendicular to the vector $\mathbf{i} - \mathbf{j}$.

Solution.   We can take $\mathbf{i} - \mathbf{j}$ as the normal:

$$\mathbf{N} = \mathbf{i} - \mathbf{j}.$$

An equation for the plane is

$$1(x - 1) + (-1)(y - 3) + 0(z + 2) = 0.$$

This simplifies to

$$x - y + 2 = 0.$$

The last equation looks very much like the equation of a line.   If our domain of discourse were the xy-plane, then the equation $x - y + 2 = 0$ would represent a line. In this case, however, our domain of discourse is three-dimensional space.   Here the equation $x - y + 2 = 0$ represents the set of all points $Q(x, y, z)$ where $x - y + 2 = 0$ and z is unrestricted.   This set forms a vertical plane that intersects the xy-plane in the line just mentioned. $\square$

Problem 3.   Show that every equation of the form

$$ax + by + cz + d = 0 \qquad \text{with} \quad \sqrt{a^2 + b^2 + c^2} \neq 0$$

represents a plane in space.

Solution.   Since $\sqrt{a^2 + b^2 + c^2} \neq 0$, the numbers a, b, c are not all zero, and therefore there exist numbers x_0, y_0, z_0 such that

$$ax_0 + by_0 + cz_0 + d = 0.\dagger$$

---

† Would such numbers necessarily exist if $\sqrt{a^2 + b^2 + c^2}$ were zero?

Equation

$$ax + by + cz + d = 0$$

can now be written

$$(ax + by + cz + d) - (ax_0 + by_0 + cz_0 + d) = 0,$$

and thus

$$a(x - x_0) + b(y - y_0) + c(z - z_0) = 0.$$

This equation (and hence the initial equation) represents the plane through the point $P(x_0, y_0, z_0)$ with normal $\mathbf{N} = a\mathbf{i} + b\mathbf{j} + c\mathbf{k}$. The initial assumption that $\sqrt{a^2 + b^2 + c^2} \neq 0$ guarantees that $\mathbf{N} \neq \mathbf{0}$. ☐

Vector Equation of a Plane

We can write the equation of a plane entirely in vector notation. With

$$\mathbf{N} = A\mathbf{i} + B\mathbf{j} + C\mathbf{k}$$

and

$$\mathbf{r}_0 = x_0\mathbf{i} + y_0\mathbf{j} + z_0\mathbf{k}, \qquad \mathbf{r} = x\mathbf{i} + y\mathbf{j} + z\mathbf{k},$$

Equation 13.8.1 reads

(13.8.2)
$$\boxed{\mathbf{N} \cdot (\mathbf{r} - \mathbf{r}_0) = 0.}$$

This vector equation represents the plane that passes through the tip of $\mathbf{r}_0$ and has normal $\mathbf{N}$. (See Figure 13.8.2.) If the plane passes through the origin, we can take $\mathbf{r}_0 = \mathbf{0}$. Equation 13.8.2 then takes the form

(13.8.3)
$$\boxed{\mathbf{N} \cdot \mathbf{r} = 0.}$$

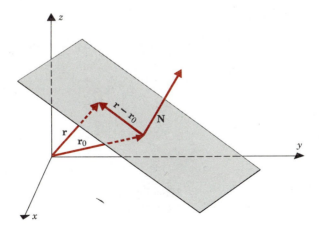

Figure 13.8.2

Collinear Vectors, Coplanar Vectors

Collinear points are points that lie on the same line; coplanar points are points that lie on the same plane. The terms "collinear" and "coplanar" are also applied to vectors: by definition, two vectors **a** and **b** are said to be *collinear* iff there exist scalars s and t not both 0 such that

$$s\mathbf{a} + t\mathbf{b} = \mathbf{0}.$$

If $s \neq 0$, then $\mathbf{a} = -(t/s)\mathbf{b}$; if $t \neq 0$, then $\mathbf{b} = -(s/t)\mathbf{a}$. Collinear vectors are thus parallel. If we set

$$\mathbf{a} = \overrightarrow{PA} \quad \text{and} \quad \mathbf{b} = \overrightarrow{PB},$$

then the points P, A, B all fall on the same line; hence the term "collinear vectors."

Three vectors **a**, **b**, **c** are said to be *coplanar* iff there exist scalars s, t, u not all zero such that

$$s\mathbf{a} + t\mathbf{b} + u\mathbf{c} = \mathbf{0}.$$

This term, too, is justified:

(13.8.4)

> $\mathbf{a} = \overrightarrow{PA}$, $\mathbf{b} = \overrightarrow{PB}$, $\mathbf{c} = \overrightarrow{PC}$ are coplanar vectors iff the points P, A, B, C all lie on the same plane.

Proof. Here we show that, if the three vectors are coplanar, then the four points all lie on the same plane. We leave the converse to you.

Suppose that the three vectors are coplanar. Then we can write

$$s\,\overrightarrow{PA} + t\,\overrightarrow{PB} + u\,\overrightarrow{PC} = \mathbf{0} \qquad \text{with } s, t, u \text{ not all zero.}$$

For definiteness, we assume that $s \neq 0$. Then

$$\overrightarrow{PA} = -\frac{t}{s}\,\overrightarrow{PB} - \frac{u}{s}\,\overrightarrow{PC}.$$

Since $\overrightarrow{PB} \times \overrightarrow{PC}$ is perpendicular to both $\overrightarrow{PB}$ and $\overrightarrow{PC}$, we see that

$$(\overrightarrow{PB} \times \overrightarrow{PC}) \cdot \overrightarrow{PA} = (\overrightarrow{PB} \times \overrightarrow{PC}) \cdot \left(-\frac{t}{s}\,\overrightarrow{PB} - \frac{u}{s}\,\overrightarrow{PC}\right)$$

$$= -\frac{t}{s}(\overrightarrow{PB} \times \overrightarrow{PC}) \cdot \overrightarrow{PB} - \frac{u}{s}(\overrightarrow{PB} \times \overrightarrow{PC}) \cdot \overrightarrow{PC}$$

$$= 0.$$

But $|(\overrightarrow{PB} \times \overrightarrow{PC}) \cdot \overrightarrow{PA}|$ gives the volume of the parallelepiped with edges $\overrightarrow{PA}$, $\overrightarrow{PB}$, $\overrightarrow{PC}$. This volume can be zero only if P, A, B, C all lie on the same plane. ∎

Unit Normals

If **N** is normal to a given plane, then all other normals to that plane are parallel to **N** and hence scalar multiples of **N**. In particular there are only two normals of length 1:

$$\mathbf{u_N} = \frac{\mathbf{N}}{\|\mathbf{N}\|} \quad \text{and} \quad -\mathbf{u_N} = -\frac{\mathbf{N}}{\|\mathbf{N}\|}.$$

These are called the *unit normals*.

Problem 4. Find the unit normals for the plane

$$3x - 4y + 12z + 8 = 0.$$

Solution. We can take

$$\mathbf{N} = 3\mathbf{i} - 4\mathbf{j} + 12\mathbf{k}.$$

Since

$$\|\mathbf{N}\| = \sqrt{(3)^2 + (-4)^2 + (12)^2} = \sqrt{169} = 13,$$

we have

$$\mathbf{u_N} = \tfrac{1}{13}(3\mathbf{i} - 4\mathbf{j} + 12\mathbf{k}) \quad \text{and} \quad -\mathbf{u_N} = -\tfrac{1}{13}(3\mathbf{i} - 4\mathbf{j} + 12\mathbf{k}). \quad \square$$

Parallel Planes, Intersecting Planes

Two planes are called *parallel* iff their normals are parallel. If two planes p_1, p_2 intersect, we can find the angle between them by finding the angle between their normals, $\mathbf{N_1}$, $\mathbf{N_2}$. (See Figure 13.8.3.) Depending on our choice of normals, there are two such angles, each the supplement of the other. We will choose the smaller angle, the one with the nonnegative cosine:

(13.8.5)
$$\cos \theta = |\mathbf{u_{N_1}} \cdot \mathbf{u_{N_2}}|.$$

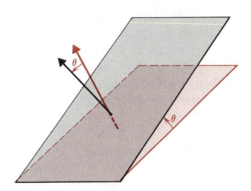

Figure 13.8.3

Problem 5.   Here are some planes:

$$p_1: 2(x-1) - 3y + 5(z-2) = 0, \qquad p_2: -4x + 6y + 10z + 24 = 0,$$
$$p_3: 4x - 6y - 10z + 1 = 0, \qquad p_4: 2x - 3y + 5z - 12 = 0.$$

(a) Indicate which planes are identical.
(b) Indicate which planes are distinct but parallel.
(c) Find the angle between p_1 and p_2.

Solution

(a) p_1 and p_4 are identical, as you can verify by simplifying the equation of p_1.
(b) p_2 and p_3 are distinct but parallel. The planes are distinct since $P(0, 0, \frac{1}{10})$ lies on p_3 but not on p_2; they are parallel since the normals

$$-4\mathbf{i} + 6\mathbf{j} + 10\mathbf{k} \qquad \text{and} \qquad 4\mathbf{i} - 6\mathbf{j} - 10\mathbf{k}$$

are parallel.
(c) Taking

$$\mathbf{N}_1 = 2\mathbf{i} - 3\mathbf{j} + 5\mathbf{k} \qquad \text{and} \qquad \mathbf{N}_2 = -4\mathbf{i} + 6\mathbf{j} + 10\mathbf{k},$$

we have

$$\mathbf{u}_{\mathbf{N}_1} = \frac{1}{\sqrt{38}}\,(2\mathbf{i} - 3\mathbf{j} + 5\mathbf{k}) \qquad \text{and} \qquad \mathbf{u}_{\mathbf{N}_2} = \frac{1}{\sqrt{152}}\,(-4\mathbf{i} + 6\mathbf{j} + 10\mathbf{k}).$$

As you can check,

$$\cos\theta = |\mathbf{u}_{\mathbf{N}_1} \cdot \mathbf{u}_{\mathbf{N}_2}| = \tfrac{6}{19} \cong 0.316 \quad \text{and thus} \quad \theta \cong 1.25 \text{ radians.} \quad \square$$

Problem 6.   The planes

$$p_1: A_1 x + B_1 y + C_1 z + D_1 = 0, \qquad p_2: A_2 x + B_2 y + C_2 z + D_2 = 0$$

intersect to form a line l. Find a vector parametrization for l.

Solution. We need to find a point P_0 on l and a direction vector for l. Finding P_0 is a matter of finding numbers x, y, z that simultaneously satisfy the equations given above for p_1 and p_2. In concrete cases this is not hard, and we will not try to give a general formula for such a P_0. We will just assume that P_0 has been found and focus on finding a direction vector for l.

Since p_1 and p_2 intersect in a line, the normals

$$\mathbf{N}_1 = A_1\mathbf{i} + B_1\mathbf{j} + C_1\mathbf{k}, \qquad \mathbf{N}_2 = A_2\mathbf{i} + B_2\mathbf{j} + C_2\mathbf{k}$$

are not parallel. This guarantees that $\mathbf{N}_1 \times \mathbf{N}_2$ is not $\mathbf{0}$. Since l lies on both p_1 and p_2, the line l, like the vector $\mathbf{N}_1 \times \mathbf{N}_2$, is perpendicular to both $\mathbf{N}_1$ and $\mathbf{N}_2$. This makes l parallel to $\mathbf{N}_1 \times \mathbf{N}_2$. (See Figure 13.8.4.) We can therefore take $\mathbf{N}_1 \times \mathbf{N}_2$ as a direction vector for l and write

$$l: \mathbf{r}(t) = \overrightarrow{OP_0} + t\,(\mathbf{N}_1 \times \mathbf{N}_2). \quad \square$$

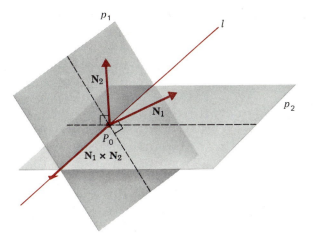

Figure 13.8.4

The Plane Determined by Three Noncollinear Points

Suppose now that we are given three noncollinear points P_1, P_2, P_3. These points determine a plane. How can we find an equation for that plane?

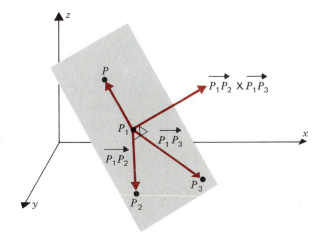

Figure 13.8.5

First we form the vectors $\overrightarrow{P_1P_2}$, $\overrightarrow{P_1P_3}$. Since P_1, P_2, P_3 are noncollinear, the vectors are not parallel. Therefore their cross product $\overrightarrow{P_1P_2} \times \overrightarrow{P_1P_3}$ can be used as a normal for the plane. We are back in a familiar situation. We have a point of the plane, say P_1, and we have a normal vector, $\overrightarrow{P_1P_2} \times \overrightarrow{P_1P_3}$. A point P will lie on the plane iff

(13.8.6)
$$\overrightarrow{P_1P} \cdot (\overrightarrow{P_1P_2} \times \overrightarrow{P_1P_3}) = 0.$$
(Figure 13.8.5)

Problem 7.   Find an equation in x, y, z for the plane that passes through the points $P_1(0, 1, 1)$, $P_2(1, 0, 1)$, $P_3(1, 1, 0)$.

Solution.   The point $P = P(x, y, z)$ will lie on this plane iff

$$\overrightarrow{P_1P} \cdot (\overrightarrow{P_1P_2} \times \overrightarrow{P_1P_3}) = 0.$$

Here

$$\overrightarrow{P_1P} = x\mathbf{i} + (y - 1)\mathbf{j} + (z - 1)\mathbf{k}, \qquad \overrightarrow{P_1P_2} = \mathbf{i} - \mathbf{j}, \qquad \overrightarrow{P_1P_3} = \mathbf{i} - \mathbf{k}.$$

As you can check,

$$\overrightarrow{P_1P_2} \times \overrightarrow{P_1P_3} = \mathbf{i} + \mathbf{j} + \mathbf{k}$$

and thus

$$\overrightarrow{P_1P} \cdot (\overrightarrow{P_1P_2} \times \overrightarrow{P_1P_3}) = [x\mathbf{i} + (y - 1)\mathbf{j} + (z - 1)\mathbf{k}] \cdot [\mathbf{i} + \mathbf{j} + \mathbf{k}]$$
$$= x + (y - 1) + (z - 1) = x + y + z - 2.$$

The equation for the plane reads

$$x + y + z - 2 = 0. \quad \square$$

The Distance from a Point to a Plane

In Figure 13.8.6 we have drawn a plane p: $Ax + By + Cz + D = 0$ and a point $P_1(x_1, y_1, z_1)$ not on p.

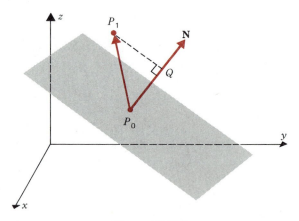

Figure 13.8.6

The distance between the point P_1 and the plane p is given by the formula

(13.8.7)
$$d(P_1, p) = \frac{|Ax_1 + By_1 + Cz_1 + D|}{\sqrt{A^2 + B^2 + C^2}}.$$

Proof. Pick any point $P_0(x_0, y_0, z_0)$ in the plane. As a normal to p we can take the vector

$$\mathbf{N} = A\mathbf{i} + B\mathbf{j} + C\mathbf{k}.$$

Then

$$\mathbf{u_N} = \frac{A\mathbf{i} + B\mathbf{j} + C\mathbf{k}}{\sqrt{A^2 + B^2 + C^2}}$$

is the corresponding unit normal. From the figure

$$d(P_1, p) = d(P_0, Q) = |\text{comp}_{\mathbf{N}} \overrightarrow{P_0 P_1}|$$
$$= |\overrightarrow{P_0 P_1} \cdot \mathbf{u_N}|$$
$$= \frac{|(x_1 - x_0)A + (y_1 - y_0)B + (z_1 - z_0)C|}{\sqrt{A^2 + B^2 + C^2}}$$
$$= \frac{|Ax_1 + By_1 + Cz_1 - (Ax_0 + By_0 + Cz_0)|}{\sqrt{A^2 + B^2 + C^2}}.$$

Since $P_0(x_0, y_0, z_0)$ lies on p,

$$Ax_0 + By_0 + Cz_0 = -D$$

and

$$d(P_1, p) = \frac{|Ax_1 + By_1 + Cz_1 + D|}{\sqrt{A^2 + B^2 + C^2}}. \quad \square$$

EXERCISES 13.8

1. Which of the points $P(3, 2, 1)$, $Q(2, 3, 1)$, $R(1, 4, 1)$ lie on the plane
$$3(x - 1) + 4y - 5(z + 2) = 0?$$

2. Which of the points $P(2, 1, -2)$, $Q(2, 0, 0)$, $R(4, 1, -1)$, $S(0, -1, -3)$ lie on the plane $\mathbf{N} \cdot (\mathbf{r} - \mathbf{r_0}) = 0$ if $\mathbf{N} = \mathbf{i} - 3\mathbf{j} + \mathbf{k}$ and $\mathbf{r_0} = 4\mathbf{i} + \mathbf{j} - \mathbf{k}$?

3. Write an equation for the plane that passes through the point $P(2, 3, 4)$ and is perpendicular to $\mathbf{i} - 4\mathbf{j} + 3\mathbf{k}$.

4. Write an equation for the plane that passes through the point $P(1, -2, 3)$ and is perpendicular to $\mathbf{j} + 2\mathbf{k}$.

5. Write an equation for the plane that passes through the point $P(2, 1, 1)$ and is parallel to the plane $3x - 2y + 5z - 2 = 0$.

6. Write an equation for the plane that passes through the point $P_0(x_0, y_0, z_0)$ and is perpendicular to $\overrightarrow{OP_0}$.

7. Find the unit normals for the plane $2x - 3y + 7z - 3 = 0$.

8. Show that the plane $x/a + y/b + z/c = 1$ intersects the coordinate axes at $x = a$, $y = b$, $z = c$. This is the equation of a plane in *intercept form*.

9. Write the equation of the plane $4x + 5y - 6z = 60$ in intercept form.

10. Where does the plane $3x - y + 4z + 2 = 0$ intersect the coordinate axes?

Find the angle between the planes:

11. $5(x - 1) - 3(y + 2) + 2z = 0,$ $x + 3(y - 1) + 2(z + 4) = 0.$
12. $2x - y + 3z = 5,$ $5x + 5y - z = 1.$
13. $x - y + z - 1 = 0,$ $2x + y + 3z + 5 = 0.$

Determine whether or not the vectors are coplanar.

14. $\mathbf{i},$ $\mathbf{i} - 2\mathbf{j},$ $3\mathbf{j} + \mathbf{k}.$ 　　　　　　　 **15.** $4\mathbf{j} - \mathbf{k},$ $3\mathbf{i} + \mathbf{j} + 2\mathbf{k},$ $\mathbf{0}.$
16. $\mathbf{j} - \mathbf{k},$ $3\mathbf{i} - \mathbf{j} + 2\mathbf{k},$ $3\mathbf{i} - 2\mathbf{j} + 3\mathbf{k}.$ 　　 **17.** $\mathbf{i} + \mathbf{j} + \mathbf{k},$ $2\mathbf{i} - \mathbf{j},$ $3\mathbf{i} - \mathbf{j} - \mathbf{k}.$

18. Find the distance between the parallel planes

$$Ax + By + Cz + D_1 = 0 \quad \text{and} \quad Ax + By + Cz + D_2 = 0.$$

19. Find the distance from $P(2, -1, 3)$ to the plane $2x + 4y - z + 1 = 0.$

20. Write equations in symmetric form for the line that passes through $P_0(x_0, y_0, z_0)$ and is perpendicular to the plane $Ax + By + Cz + D = 0.$

Find an equation in x, y, z for the plane that passes through the given points.

21. $P_1(1, 0, 1),$ $P_2(2, 1, 0),$ $P_3(1, 1, 1).$
22. $P_1(1, 1, 1),$ $P_2(2, -2, -1),$ $P_3(0, 2, 1).$

23. Show that the equations of a line in symmetric form

$$\frac{x - x_0}{d_1} = \frac{y - y_0}{d_2} = \frac{z - z_0}{d_3}$$

express the line as an intersection of two planes by finding equations for two such planes.

24. Suppose that we are given equations in x, y, z for two planes that intersect in a line. How can we determine a triple x_0, y_0, z_0 that places $P(x_0, y_0, z_0)$ on the line? One possible procedure is this: select a value for x, say x_0; substitute x_0 in the two equations given; we now have two linear equations in y and z that we can solve simultaneously. Is this procedure foolproof? If not, identify the circumstances under which the procedure will fail and indicate how the procedure can be made airtight.

Find a set of scalar parametric equations for the line formed by the two intersecting planes.   HINT: To find a point P_0 on the line, see Exercise 24.

25. $p_1: x + 2y + 3z = 0,$ $p_2: -3x + 4y + z = 0.$
26. $p_1: x + y + z + 1 = 0,$ $p_2: x - y + z + 2 = 0.$

Let l be the line determined by $P_1, P_2,$ and let p be the plane determined by $Q_1, Q_2, Q_3.$ Where, if anywhere, does l intersect p?

27. $P_1(1, -1, 2),$ $P_2(-2, 3, 1);$ $Q_1(2, 0, -4),$ $Q_2(1, 2, 3),$ $Q_3(-1, 2, 1).$
28. $P_1(4, -3, 1),$ $P_2(2, -2, 3);$ $Q_1(2, 0, -4),$ $Q_2(1, 2, 3),$ $Q_3(-1, 2, 1).$
29. Let l_1, l_2 be lines that pass through the origin and have direction vectors

$$\mathbf{d} = \mathbf{i} + 2\mathbf{j} + 4\mathbf{k}, \qquad \mathbf{D} = -\mathbf{i} - \mathbf{j} + 3\mathbf{k}.$$

Find an equation for the plane that contains l_1 and $l_2.$

30. Show that two nonparallel lines $\mathbf{r}(t) = \mathbf{r}_0 + t\mathbf{d}$ and $\mathbf{R}(t) = \mathbf{R}_0 + t\mathbf{D}$ intersect iff the vectors $\mathbf{r}_0 - \mathbf{R}_0, \mathbf{d},$ and $\mathbf{D}$ are coplanar.

31. Starting with the plane that contains the point P and has normal $\mathbf{N}$, describe the set of points Q on the plane for which $\mathbf{N} + \overrightarrow{PQ} \perp \mathbf{N} - \overrightarrow{PQ}.$ HINT: Draw a figure.

32. Let $\mathbf{a}, \mathbf{b}, \mathbf{c}$ be three nonzero vectors such that the angle between any pair of them is $\frac{1}{2}\pi.$ Can these vectors be coplanar?

33. Suppose that $\mathbf{N} = A\mathbf{i} + B\mathbf{j} + C\mathbf{k}$ is a nonzero vector, tail placed on the plane $Ax + By + Cz + D = 0$. Take $P_1(x_1, y_1, z_1)$ as a point of space and set $\alpha = Ax_1 + By_1 + Cz_1 + D$. If $\alpha = 0$, then P_1 lies on the plane. What can you conclude about P_1 if α is positive? if α is negative?

34. Suppose that $\mathbf{a}, \mathbf{b}, \mathbf{c}$ are radius vectors the tips of which are not collinear. Give a geometric interpretation of the equation

$$\begin{vmatrix} x - a_1 & y - a_2 & z - a_3 \\ x - b_1 & y - b_2 & z - b_3 \\ x - c_1 & y - c_2 & z - c_3 \end{vmatrix} = 0.$$

13.9   SOME GEOMETRY BY VECTOR METHODS

Try your hand at proving the following theorems by vector methods. Follow the hints if you like, but you may find it more interesting to disregard them and come up with proofs that are entirely your own.

Proposition 1. The diagonals of a parallelogram are perpendicular iff the parallelogram is a rhombus.

Hint for Proof. With $\mathbf{a}$ and $\mathbf{b}$ as in Figure 13.9.1 the diagonals are $\mathbf{a} + \mathbf{b}$ and $\mathbf{a} - \mathbf{b}$. Show that $(\mathbf{a} + \mathbf{b}) \cdot (\mathbf{a} - \mathbf{b}) = 0$ iff $\|\mathbf{a}\| = \|\mathbf{b}\|$. ☐

Proposition 2. Every angle inscribed in a semicircle is a right angle.

Hint for Proof. Take $\mathbf{c}$ and $\mathbf{d}$ as in Figure 13.9.2; express $\mathbf{c}$ and $\mathbf{d}$ in terms of $\mathbf{a}$ and $\mathbf{b}$; then show that $\mathbf{c} \cdot \mathbf{d} = 0$. ☐

Figure 13.9.1

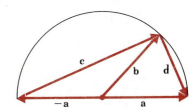

Figure 13.9.2

Proposition 3. In a parallelogram the sum of the squares of the lengths of the diagonals equals the sum of the squares of the lengths of the sides.

Hint for Proof. With $\mathbf{a}$ and $\mathbf{b}$ as in Figure 13.9.1 the diagonals are $\mathbf{a} + \mathbf{b}$ and $\mathbf{a} - \mathbf{b}$. Show that $\|\mathbf{a} + \mathbf{b}\|^2 + \|\mathbf{a} - \mathbf{b}\|^2 = 2\|\mathbf{a}\|^2 + 2\|\mathbf{b}\|^2$. ☐

Proposition 4. The three altitudes of a triangle meet at one point.

Hint for Proof. As in Figure 13.9.3 assume that the altitudes from P_1 and P_2 intersect at Q. Use the fact that $\overrightarrow{P_1Q} \perp \overrightarrow{OP_2}$ and $\overrightarrow{P_2Q} \perp \overrightarrow{OP_1}$ to show that $\overrightarrow{OQ} \perp \overrightarrow{P_1P_2}$. □

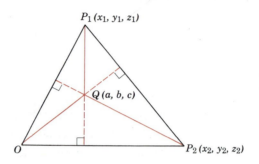

Figure 13.9.3

Proposition 5. The three medians of a triangle meet at one point.

Hint for Proof. With l_1, l_2, l_3 as in Figure 13.9.4,

$$l_1 : \mathbf{r}_1(t) = t(\mathbf{a} + \mathbf{b}), \qquad l_2 : \mathbf{r}_2(u) = \tfrac{1}{2}\mathbf{b} + u(\mathbf{a} - \tfrac{1}{2}\mathbf{b}), \qquad l_3 : \mathbf{r}_3(v) = \tfrac{1}{2}\mathbf{a} + v(\mathbf{b} - \tfrac{1}{2}\mathbf{a}).$$

Show that l_1 intersects both l_2 and l_3 at the same point. □

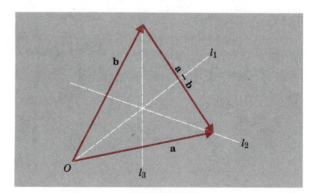

Figure 13.9.4

Proposition 6 (The law of sines). If a triangle has sides $\mathbf{a}$, $\mathbf{b}$, $\mathbf{c}$ and opposite angles A, B, C, then

$$\frac{\sin A}{\|\mathbf{a}\|} = \frac{\sin B}{\|\mathbf{b}\|} = \frac{\sin C}{\|\mathbf{c}\|}.$$

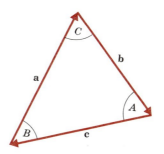

Figure 13.9.5

Hint for Proof. With **a**, **b**, **c** as in Figure 13.9.5, $\mathbf{a} + \mathbf{b} + \mathbf{c} = \mathbf{0}$. Observe then that $\mathbf{a} \times [\mathbf{a} + \mathbf{b} + \mathbf{c}] = \mathbf{0}$ and $\mathbf{b} \times [\mathbf{a} + \mathbf{b} + \mathbf{c}] = \mathbf{0}$. ☐

Proposition 7. If two planes have a point in common, then they have a line in common.

Hint for Proof. As equations for the two planes, take $\mathbf{n} \cdot (\mathbf{r} - \mathbf{r}_0) = 0$ and $\mathbf{N} \cdot (\mathbf{R} - \mathbf{R}_0) = 0$. If the point $P(a_1, a_2, a_3)$ lies on both planes, then the vector $\mathbf{a} = a_1\mathbf{i} + a_2\mathbf{j} + a_3\mathbf{k}$ satisfies both equations. In that case we have $\mathbf{n} \cdot (\mathbf{a} - \mathbf{r}_0) = 0$ and $\mathbf{N} \cdot (\mathbf{a} - \mathbf{R}_0) = 0$. Consider the line $\mathbf{r}(t) = \mathbf{a} + t(\mathbf{n} \times \mathbf{N})$. ☐

13.10 ADDITIONAL EXERCISES

1. Throughout this exercise let

$$P_1 = P_1(3, 2, -1), \qquad P_2 = P_2(7, -5, 4), \qquad P_3 = P_3(6, 6, -4).$$

(a) Find the length of the line segment that joins P_1 to P_2.

(b) What are the coordinates of the midpoint of this line segment?

(c) If one of the points in (a) is the endpoint of a line segment and the other the midpoint, what are the coordinates of the other endpoint?

(d) Find an equation in x, y, z for the set of all points equidistant from P_1 and P_2.

(e) Find scalar parametric equations for the line that passes through P_3 and is parallel to the line through P_1, P_2.

(f) Find scalar parametric equations for the line that passes through P_3 and is perpendicular to the line through P_1, P_2.

(g) The lines in (f) determine a plane. Find an equation for this plane.

2. Exercise 1 with $P_1 = P_1(4, 2, -3)$, $P_2 = P_2(-2, 1, 4)$, $P_3 = P_3(1, -1, -6)$.

3. Let l be the line $x + 1 = y + 2 = z + 1$ and let $P = P(3, 1, -2)$. Find the point Q on l such that $\overrightarrow{PQ} \perp l$.

4. (a) Are the points $P(3, 2, -1)$, $Q(7, -5, 4)$, $R(-2, 5, 0)$ collinear?

(b) Are the points $P(3, 2, -1)$, $Q(7, -5, 4)$, $R(-2, 5, 0)$, $S(2, 1, 6)$ coplanar?

5. Find the distance from $P(2, 1, -1)$ to the plane $x - 2y + 2z + 5 = 0$.

Write an equation for the plane that satisfies the given conditions.

6. Contains the points $P(1, -2, 1)$, $Q(2, 0, 3)$, $R(0, 1, -1)$.

7. Contains the point $P(2, 1, -3)$ and is perpendicular to the line

$$\frac{x+1}{2} = \frac{y-1}{3} = -\frac{z}{4}.$$

8. Contains the point $P(1, -2, -1)$ and is parallel to the plane $3x + 2y - z = -4$.

9. Contains the lines

$$x = -1 + 2t, \quad y = 2 + 3t, \quad z = 1 + t \quad \text{and} \quad x = -1 + u, \quad y = 2 - u, \quad z = 1 + 2u.$$

10. Contains the point $P(3, -1, 2)$ and the line $x = 2 + 2t$, $y = -1 + 3t$, $z = -2t$.

11. The planes $2x + y - 3z + 6 = 0$ and $x + 4y - 5z - 7 = 0$ intersect in a line. Find scalar parametric equations for this line.

12. Verify that the plane $5x - 3y - z - 6 = 0$ contains the line

$$x = 1 + 2t, \quad y = -1 + 3t, \quad z = 2 + t.$$

13. Where does the line

$$\frac{x+2}{3} = \frac{y-1}{2} = \frac{z+6}{1}$$

intersect the plane $2x + y - 3z + 6 = 0$?

14. Write an equation for the plane that contains the line with scalar parametric equations $x = -1 + 3t, y = 1 + 2t, z = 2 + 4t$ and is perpendicular to the plane $2x + y - 3z + 4 = 0$.

15. Find an equation for the plane that passes through the point $P(2, 1, -3)$ and contains the line determined by the planes $3x + y - z = 2$ and $2x + y + 4z = 1$.

16. Given that $\mathbf{a} = 2\mathbf{i} - \mathbf{j} + 3\mathbf{k}$ and $\mathbf{b} = 3\mathbf{i} + 2\mathbf{j} - \mathbf{k}$, find all vectors of length 4 that are perpendicular to both $\mathbf{a}$ and $\mathbf{b}$.

17. Find the distance from the point $P(4, 6, -4)$ to the line joining $Q(2, 2, 1)$ to $R(4, 3, -1)$.

18. The setting is the xy-plane.
    (a) Verify that the line l: $Ax + By + C = 0$ can be parametrized by setting

$$\mathbf{r}(t) = (-C/A)\mathbf{i} + (B\mathbf{i} - A\mathbf{j})t.$$

    (b) Show that $A\mathbf{i} + B\mathbf{j}$ is normal to l.
    (c) Show that

$$d(O, l) = \frac{|C|}{\sqrt{A^2 + B^2}}$$

    by vector methods.

19. Show that the midpoints of the sides of a quadrilateral are the four vertices of a parallelogram.

20. Use vector methods to show that the line segment that joins the midpoints of two sides of a triangle is parallel to the third side. Show that the length of this line segment is one half the length of the third side.

21. Determine α so that $(\mathbf{a} + \mathbf{b}) \times (\mathbf{a} - \mathbf{b}) = \alpha(\mathbf{a} \times \mathbf{b})$ for all $\mathbf{a}$ and $\mathbf{b}$.

22. Show that

$$(\mathbf{a} \times \mathbf{b}) \times (\mathbf{c} \times \mathbf{d}) = (\mathbf{a} \times \mathbf{b} \cdot \mathbf{d})\mathbf{c} - (\mathbf{a} \times \mathbf{b} \cdot \mathbf{c})\mathbf{d}.$$

23. Show that, for all vectors $\mathbf{a}$ and $\mathbf{b}$,

$$(\|\mathbf{b}\|\mathbf{a} - \|\mathbf{a}\|\mathbf{b}) \perp (\|\mathbf{b}\|\mathbf{a} + \|\mathbf{a}\|\mathbf{b}).$$

24. Show that, if two planes

$$p_1: A_1x + B_1y + C_1z + D_1 = 0 \qquad \text{and} \qquad p_2: A_2x + B_2y + C_2z + D_2 = 0$$

intersect in a line l, then every plane p that contains the line l has an equation of the form

$$\alpha(A_1x + B_1y + C_1z + D_1) + \beta(A_2x + B_2y + C_2z + D_2) = 0.$$

25. Find an equation for the plane that passes through the point $(1, 4, -3)$ and contains the line of intersection of the planes

$$3x - 2y + 4z - 7 = 0 \qquad \text{and} \qquad x + 5y - 2z + 9 = 0.$$

CHAPTER HIGHLIGHTS

13.1 Cartesian Space Coordinates

distance formula (p. 687)   equation of sphere (p. 687)
midpoint formula (p. 688)

13.2 Displacements and Forces

13.3 Vectors

addition, multiplication by scalars (p. 694)   component (p. 694)
parallel vectors (p. 696)   norm (p. 697)   unit vector (p. 699)

The zero vector **0** is parallel to every vector.

13.4 The Dot Product

$\mathbf{a} \cdot \mathbf{b} = a_1b_1 + a_2b_2 + a_3b_3 = \|\mathbf{a}\|\,\|\mathbf{b}\| \cos\theta$ $\mathbf{a} \perp \mathbf{b}$ iff $\mathbf{a} \cdot \mathbf{b} = 0$ $\mathbf{a} = \mathbf{a}_\| + \mathbf{a}_\perp$
projections and components (p. 709)   direction angles, direction cosines
Schwarz's inequality (p. 712)   (pp. 710–711)

13.5 Matrices and Determinants

matrix (p. 714)   determinant (pp. 715, 716)

13.6 The Cross Product

definition of $\mathbf{a} \times \mathbf{b}$ (p. 720)   properties of right-handed triples (p. 720)
properties of the cross product (p. 721)   distributive laws (p. 722)
scalar triple product (p. 722)   components of $\mathbf{a} \times \mathbf{b}$ (p. 724)
identities (pp. 726, 727)

13.7 Lines

position vector, radius vector (p. 728)
direction vector, direction numbers (p. 729)
vector parametrization: $\mathbf{r}(t) = \mathbf{r}_0 + t\mathbf{d}$
scalar parametric equations: $x(t) = x_0 + d_1t, \; y(t) = y_0 + d_2t, \; z(t) = z_0 + d_3t$
symmetric form (p. 731)
distance from a point to a line (p. 736)

Two lines are parallel iff their direction vectors are parallel; two intersecting lines are perpendicular iff their direction vectors are perpendicular.

13.8 Planes

13.9 Some Geometry by Vector Methods

13.10 Additional Exercises

CHAPTER 14

VECTOR CALCULUS

14.1 VECTOR FUNCTIONS

Introduction

If f_1, f_2, f_3 are real-valued functions defined on some interval I, then for each $t \in I$ we can form the vector

$$\mathbf{f}(t) = f_1(t)\mathbf{i} + f_2(t)\mathbf{j} + f_3(t)\mathbf{k}$$

and thereby create a *vector-valued function* $\mathbf{f}$. For short we will call such a function a *vector function*.

Example 1. Taking

$$f_1(t) = x_0 + d_1 t, \quad f_2(t) = y_0 + d_2 t, \quad f_3(t) = z_0 + d_3 t$$

we can form the vector function

$$\mathbf{f}(t) = (x_0 + d_1 t)\mathbf{i} + (y_0 + d_2 t)\mathbf{j} + (z_0 + d_3 t)\mathbf{k}.$$

If d_1, d_2, d_3 are not all 0, then the radius vector $\mathbf{f}(t)$ traces out the line that passes through the point $P(x_0, y_0, z_0)$ and has direction numbers d_1, d_2, d_3. If d_1, d_2, d_3 are all 0, then we have the constant function

$$\mathbf{f}(t) = x_0\mathbf{i} + y_0\mathbf{j} + z_0\mathbf{k}. \quad \square$$

Example 2. From the functions

$$f_1(t) = \cos t, \quad f_2(t) = \sin t, \quad f_3(t) = 0$$

we can form the function

$$\mathbf{f}(t) = \cos t\, \mathbf{i} + \sin t\, \mathbf{j}.$$

For each t

$$\|\mathbf{f}(t)\| = \sqrt{\cos^2 t + \sin^2 t} = 1.$$

Since the third component is zero, the radius vector $\mathbf{f}(t)$ lies in the xy-plane. As t increases, the tip of $\mathbf{f}(t)$ traces out the unit circle in a counterclockwise manner, effecting a complete revolution as t increases by 2π.   (Figure 14.1.1)   ☐

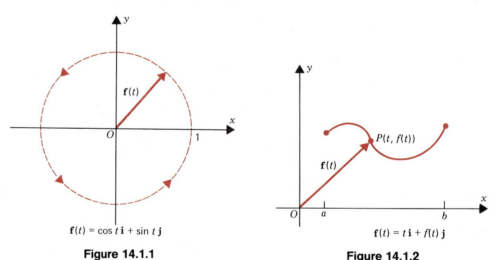

$$\mathbf{f}(t) = \cos t\, \mathbf{i} + \sin t\, \mathbf{j}$$

Figure 14.1.1

$$\mathbf{f}(t) = t\, \mathbf{i} + f(t)\, \mathbf{j}$$

Figure 14.1.2

Example 3. Each real-valued function f defined on some interval $[a, b]$ gives rise to a vector-valued function $\mathbf{f}$ in a natural way. Setting

$$f_1(t) = t, \quad f_2(t) = f(t), \quad f_3(t) = 0,$$

we obtain the vector function

$$\mathbf{f}(t) = t\mathbf{i} + f(t)\mathbf{j}.$$

As t ranges from a to b, the radius vector $\mathbf{f}(t)$ traces out the graph of f from left to right. (See Figure 14.1.2.)   ☐

The Limit Process

DEFINITION 14.1.1

$$\lim_{t \to t_0} \mathbf{f}(t) = \mathbf{L} \quad \text{iff} \quad \lim_{t \to t_0} \|\mathbf{f}(t) - \mathbf{L}\| = 0.$$

Note that $\|\mathbf{f}(t) - \mathbf{L}\|$ is real, and therefore the limit on the right is the limit of a real-valued function. Thus we are still in familiar territory.

The first thing we show is that

(14.1.2)

$$\text{if } \lim_{t \to t_0} \mathbf{f}(t) = \mathbf{L}, \qquad \text{then } \lim_{t \to t_0} \|\mathbf{f}(t)\| = \|\mathbf{L}\|.$$

Proof

$$0 \le |\,\|\mathbf{f}(t)\| - \|\mathbf{L}\|\,| \le \|\mathbf{f}(t) - \mathbf{L}\|.$$

Thus, by the pinching theorem,

$$\text{if } \lim_{t \to t_0} \|\mathbf{f}(t) - \mathbf{L}\| = 0, \qquad \text{then } \lim_{t \to t_0} |\,\|\mathbf{f}(t)\| - \|\mathbf{L}\|\,| = 0. \quad \square$$

Note that the converse of (14.1.2) is false: set $\mathbf{f}(t) = \mathbf{k}$ for all t and set $\mathbf{L} = -\mathbf{k}$.

We can indicate that $\lim_{t \to t_0} \mathbf{f}(t) = \mathbf{L}$ by writing

$$\text{as } t \to t_0, \quad \mathbf{f}(t) \to \mathbf{L}.$$

We will state the limit rules in this form. As you see below, there are no surprises.

THEOREM 14.1.3 LIMIT RULES

Suppose that, as $t \to t_0$,

$$\mathbf{f}(t) \to \mathbf{L}, \qquad \mathbf{g}(t) \to \mathbf{M}, \qquad u(t) \to A.$$

Then

$$\mathbf{f}(t) + \mathbf{g}(t) \to \mathbf{L} + \mathbf{M}, \qquad \alpha \mathbf{f}(t) \to \alpha \mathbf{L} \qquad (\alpha \text{ real})$$

$$u(t)\mathbf{f}(t) \to A\mathbf{L}, \qquad \mathbf{f}(t) \cdot \mathbf{g}(t) \to \mathbf{L} \cdot \mathbf{M}, \qquad \mathbf{f}(t) \times \mathbf{g}(t) \to \mathbf{L} \times \mathbf{M}.$$

Each of these limit rules is easy to verify. We will verify the last one. To do this, we have to show that

$$\text{as } t \to t_0, \quad \|[\mathbf{f}(t) \times \mathbf{g}(t)] - [\mathbf{L} \times \mathbf{M}]\| \to 0.$$

We do this as follows:

$$\|[\mathbf{f}(t) \times \mathbf{g}(t)] - [\mathbf{L} \times \mathbf{M}]\| = \|[\mathbf{f}(t) \times \mathbf{g}(t)] - [\mathbf{L} \times \mathbf{g}(t)] + [\mathbf{L} \times \mathbf{g}(t)] - [\mathbf{L} \times \mathbf{M}]\|$$

$$= \|[(\mathbf{f}(t) - \mathbf{L}) \times \mathbf{g}(t)] + [\mathbf{L} \times (\mathbf{g}(t) - \mathbf{M})]\|$$

$$\underset{\text{explain} \longrightarrow}{\le} \|(\mathbf{f}(t) - \mathbf{L}) \times \mathbf{g}(t)\| + \|\mathbf{L} \times (\mathbf{g}(t) - \mathbf{M})\|$$

$$\le \|\mathbf{f}(t) - \mathbf{L}\|\,\|\mathbf{g}(t)\| + \|\mathbf{L}\|\,\|\mathbf{g}(t) - \mathbf{M}\|.$$

As $t \to t_0$, $\|\mathbf{g}(t)\| \to \|\mathbf{M}\|$. [This follows from (14.1.2).] Therefore, as $t \to t_0$,

$$\|\mathbf{f}(t) - \mathbf{L}\|\,\|\mathbf{g}(t)\| + \|\mathbf{L}\|\,\|\mathbf{g}(t) - \mathbf{M}\| \to (0)\|\mathbf{M}\| + \|\mathbf{L}\|(0) = 0.$$

It follows from the pinching theorem that $\|[\mathbf{f}(t) \times \mathbf{g}(t)] - [\mathbf{L} \times \mathbf{M}]\| \to 0. \quad \square$

The limit process can be carried out component by component. As we show below,

(14.1.4) $\lim\limits_{t \to t_0} \mathbf{f}(t) = \mathbf{L}$ iff $\lim\limits_{t \to t_0} f_1(t) = L_1, \quad \lim\limits_{t \to t_0} f_2(t) = L_2, \quad \lim\limits_{t \to t_0} f_3(t) = L_3.$

Proof

$$\lim\limits_{t \to t_0} \mathbf{f}(t) = \mathbf{L} \quad \text{iff} \quad \lim\limits_{t \to t_0} \|\mathbf{f}(t) - \mathbf{L}\| = 0$$

$$\text{iff} \quad \lim\limits_{t \to t_0} \sqrt{[f_1(t) - L_1]^2 + [f_2(t) - L_2]^2 + [f_3(t) - L_3]^2} = 0$$

$$\text{iff} \quad \lim\limits_{t \to t_0} f_1(t) = L_1, \quad \lim\limits_{t \to t_0} f_2(t) = L_2, \quad \lim\limits_{t \to t_0} f_3(t) = L_3. \quad \square$$

Problem 4.   Find $\lim\limits_{t \to 0} \mathbf{f}(t)$ given that

$$\mathbf{f}(t) = \cos(t + \pi)\,\mathbf{i} + \sin(t + \pi)\,\mathbf{j} + e^{-t^2}\,\mathbf{k}.$$

Solution

$$\lim\limits_{t \to 0} \mathbf{f}(t) = \lim\limits_{t \to 0} \left[\cos(t + \pi)\,\mathbf{i} + \sin(t + \pi)\,\mathbf{j} + e^{-t^2}\,\mathbf{k} \right]$$

$$= \left[\lim\limits_{t \to 0} \cos(t + \pi) \right]\mathbf{i} + \left[\lim\limits_{t \to 0} \sin(t + \pi) \right]\mathbf{j} + \left[\lim\limits_{t \to 0} e^{-t^2} \right]\mathbf{k}$$

$$= (-1)\mathbf{i} + (0)\mathbf{j} + (1)\mathbf{k} = -\mathbf{i} + \mathbf{k}. \quad \square$$

Continuity and Differentiability

As you could expect, $\mathbf{f}$ is said to be *continuous* at t_0 iff

$$\lim\limits_{t \to t_0} \mathbf{f}(t) = \mathbf{f}(t_0).$$

Thus, by (14.1.4), $\mathbf{f}$ is continuous at t_0 iff each component of $\mathbf{f}$ is continuous at t_0.

The derivative of a vector function is defined as the limit of a *vector difference quotient*:

(14.1.5) $\mathbf{f}'(t) = \lim\limits_{h \to 0} \dfrac{\mathbf{f}(t + h) - \mathbf{f}(t)}{h}.$

Differentiation can be carried out component by component:

$$\mathbf{f}'(t) = f_1'(t)\mathbf{i} + f_2'(t)\mathbf{j} + f_3'(t)\mathbf{k}.$$

Proof

$$\mathbf{f}'(t) = \lim_{h \to 0} \frac{\mathbf{f}(t+h) - \mathbf{f}(t)}{h}$$

$$= \lim_{h \to 0} \left[\frac{f_1(t+h) - f_1(t)}{h} \mathbf{i} + \frac{f_2(t+h) - f_2(t)}{h} \mathbf{j} + \frac{f_3(t+h) - f_3(t)}{h} \mathbf{k} \right]$$

$$= \left[\lim_{h \to 0} \frac{f_1(t+h) - f_1(t)}{h} \right] \mathbf{i} + \left[\lim_{h \to 0} \frac{f_2(t+h) - f_2(t)}{h} \right] \mathbf{j} + \left[\lim_{h \to 0} \frac{f_3(t+h) - f_3(t)}{h} \right] \mathbf{k}$$

$$= f_1'(t)\mathbf{i} + f_2'(t)\mathbf{j} + f_3'(t)\mathbf{k}. \quad \square$$

As with real-valued functions, if $\mathbf{f}$ is differentiable at t, then $\mathbf{f}$ is continuous at t.
(Exercise 27)

Interpretations of the vector derivative and applications of vector differentiation are introduced later in the chapter. Here we limit ourselves to computation.

Problem 5. Given that

$$\mathbf{f}(t) = t\mathbf{i} + \sqrt{t+1}\,\mathbf{j} - e^t\mathbf{k},$$

find

(1) the domain of $\mathbf{f}$.   (2) $\mathbf{f}(0)$.   (3) $\mathbf{f}'(t)$.

(4) $\mathbf{f}'(0)$.   (5) $\|\mathbf{f}(t)\|$.   (6) $\mathbf{f}(t) \cdot \mathbf{f}'(t)$.

Solution

(1) For a number to be in the domain of $\mathbf{f}$ it is necessary only that it be in the domain of each of the components. The domain of $\mathbf{f}$ is $[-1, \infty)$.

(2) $\mathbf{f}(0) = 0\mathbf{i} + \sqrt{0+1}\,\mathbf{j} - e^0\mathbf{k} = \mathbf{j} - \mathbf{k}$.

(3) $\mathbf{f}'(t) = \mathbf{i} + \dfrac{1}{2\sqrt{t+1}}\mathbf{j} - e^t\mathbf{k}$.

(4) $\mathbf{f}'(0) = \mathbf{i} + \dfrac{1}{2\sqrt{0+1}}\mathbf{j} - e^0\mathbf{k} = \mathbf{i} + \tfrac{1}{2}\mathbf{j} - \mathbf{k}$.

(5) $\|\mathbf{f}(t)\| = \sqrt{t^2 + (\sqrt{t+1})^2 + (e^t)^2} = \sqrt{t^2 + t + 1 + e^{2t}}$.

(6) $\mathbf{f}(t) \cdot \mathbf{f}'(t) = (t\mathbf{i} + \sqrt{t+1}\,\mathbf{j} - e^t\mathbf{k}) \cdot \left(\mathbf{i} + \dfrac{1}{2\sqrt{t+1}}\mathbf{j} - e^t\mathbf{k} \right)$

$$= (t)(1) + (\sqrt{t+1})\left(\frac{1}{2\sqrt{t+1}} \right) + (-e^t)(-e^t) = t + \tfrac{1}{2} + e^{2t}. \quad \square$$

If $\mathbf{f}'$ is itself differentiable, we can form the second derivative $\mathbf{f}''$ and so on.

Problem 6. Find $\mathbf{f}''(t)$ for

$$\mathbf{f}(t) = t \sin t \, \mathbf{i} + e^{-t}\mathbf{j} + t\mathbf{k}.$$

Solution

$$\mathbf{f}'(t) = (t \cos t + \sin t)\mathbf{i} - e^{-t}\mathbf{j} + \mathbf{k},$$
$$\mathbf{f}''(t) = (-t \sin t + \cos t + \cos t)\mathbf{i} + e^{-t}\mathbf{j} = (2 \cos t - t \sin t)\mathbf{i} + e^{-t}\mathbf{j}. \quad \square$$

Integration

Since we can differentiate vector functions component by component, we can define integration component by component. For $\mathbf{f}$ continuous on $[a, b]$, we set

(14.1.6)
$$\int_a^b \mathbf{f}(t) \, dt = \left(\int_a^b f_1(t) \, dt \right) \mathbf{i} + \left(\int_a^b f_2(t) \, dt \right) \mathbf{j} + \left(\int_a^b f_3(t) \, dt \right) \mathbf{k}.$$

Problem 7. Find

$$\int_0^1 \mathbf{f}(t) \, dt \qquad \text{for} \qquad \mathbf{f}(t) = t\mathbf{i} + \sqrt{t + 1} \, \mathbf{j} - e^t\mathbf{k}.$$

Solution

$$\int_0^1 \mathbf{f}(t) \, dt = \left(\int_0^1 t \, dt \right) \mathbf{i} + \left(\int_0^1 \sqrt{t + 1} \, dt \right) \mathbf{j} + \left(\int_0^1 (-e^t) \, dt \right) \mathbf{k}$$
$$= \left[\tfrac{1}{2}t^2 \right]_0^1 \mathbf{i} + \left[\tfrac{2}{3}(t + 1)^{3/2} \right]_0^1 \mathbf{j} + \left[-e^t \right]_0^1 \mathbf{k}$$
$$= \tfrac{1}{2}\mathbf{i} + \tfrac{2}{3}(2\sqrt{2} - 1)\mathbf{j} + (1 - e)\mathbf{k}. \quad \square$$

We can calculate indefinite integrals.

Problem 8. Find $\mathbf{f}(t)$ given that

$$\mathbf{f}'(t) = 2 \cos t \, \mathbf{i} - t \sin t^2 \, \mathbf{j} + 2t\mathbf{k} \qquad \text{and} \qquad \mathbf{f}(0) = \mathbf{i} + 3\mathbf{k}.$$

Solution. By integrating $\mathbf{f}'(t)$ we find that

$$\mathbf{f}(t) = (2 \sin t + C_1)\mathbf{i} + (\tfrac{1}{2} \cos t^2 + C_2)\mathbf{j} + (t^2 + C_3)\mathbf{k},$$

where C_1, C_2, C_3 are constants to be determined. Since

$$\mathbf{i} + 3\mathbf{k} = \mathbf{f}(0) = C_1\mathbf{i} + (\tfrac{1}{2} + C_2)\mathbf{j} + C_3\mathbf{k},$$

you can see that

$$C_1 = 1, \quad C_2 = -\tfrac{1}{2}, \quad C_3 = 3.$$

Thus

$$\mathbf{f}(t) = (2\sin t + 1)\mathbf{i} + (\tfrac{1}{2}\cos t^2 - \tfrac{1}{2})\mathbf{j} + (t^2 + 3)\mathbf{k}. \quad \square$$

(Integration can also be carried out without direct reference to components. See Exercise 28.)

Properties of the Integral

It is easy to see that

(14.1.7)
$$\int_a^b [\mathbf{f}(t) + \mathbf{g}(t)]\, dt = \int_a^b \mathbf{f}(t)\, dt + \int_a^b \mathbf{g}(t)\, dt$$

and

(14.1.8)
$$\int_a^b [\alpha \mathbf{f}(t)]\, dt = \alpha \int_a^b \mathbf{f}(t)\, dt \qquad \text{for every scalar } \alpha.$$

It is also true that

(14.1.9)
$$\int_a^b [\mathbf{c} \cdot \mathbf{f}(t)]\, dt = \mathbf{c} \cdot \int_a^b \mathbf{f}(t)\, dt \qquad \text{for every vector } \mathbf{c}$$

and

(14.1.10)
$$\left\| \int_a^b \mathbf{f}(t)\, dt \right\| \le \int_a^b \|\mathbf{f}(t)\|\, dt.$$

The proof of (14.1.9) is left as an exercise. (See Exercise 30.) Here we prove (14.1.10). It is an important inequality.

Proof. Set

$$\mathbf{c} = \int_a^b \mathbf{f}(t)\, dt$$

and note that

$$\|\mathbf{c}\|^2 = \mathbf{c} \cdot \mathbf{c} = \mathbf{c} \cdot \int_a^b \mathbf{f}(t)\, dt = \int_a^b [\mathbf{c} \cdot \mathbf{f}(t)]\, dt \le \int_a^b \|\mathbf{c}\|\, \|\mathbf{f}(t)\|\, dt = \|\mathbf{c}\| \int_a^b \|\mathbf{f}(t)\|\, dt.$$

by (14.1.9) ———   ——— by Schwarz's inequality (13.4.17)

If $\mathbf{c} \ne \mathbf{0}$, we can divide by $\|\mathbf{c}\|$ and conclude that

$$\|\mathbf{c}\| \le \int_a^b \|\mathbf{f}(t)\|\, dt.$$

If $\mathbf{c} = \mathbf{0}$, the result is obvious in the first place. $\quad \square$

EXERCISES 14.1

Differentiate.

1. $\mathbf{f}(t) = (1 + 2t)\mathbf{i} + (3 - t)\mathbf{j} + (2 + 3t)\mathbf{k}$.   **2.** $\mathbf{f}(t) = 2\mathbf{i} - \cos t\,\mathbf{k}$.

3. $\mathbf{f}(t) = \sqrt{1 - t}\,\mathbf{i} + \sqrt{1 + t}\,\mathbf{j} + (1 - t)^{-1}\mathbf{k}$.   **4.** $\mathbf{f}(t) = e^t\mathbf{i} + \ln t\,\mathbf{j} + \tan^{-1} t\,\mathbf{k}$.

5. $\mathbf{f}(t) = \sin t\,\mathbf{i} + \cos t\,\mathbf{j} + \tan t\,\mathbf{k}$.   **6.** $\mathbf{f}(t) = e^t(\mathbf{i} + t\mathbf{j} + t^2\mathbf{k})$.

Calculate.

7. $\displaystyle\int_1^2 \mathbf{f}(t)\,dt$ for $\mathbf{f}(t) = \mathbf{i} + 2t\mathbf{j}$.   **8.** $\displaystyle\int_0^\pi \mathbf{r}(t)\,dt$ for $\mathbf{r}(t) = \sin t\,\mathbf{i} + \cos t\,\mathbf{j} + t\mathbf{k}$.

9. $\displaystyle\int_0^1 \mathbf{g}(t)\,dt$ for $\mathbf{g}(t) = e^t\mathbf{i} + e^{-t}\mathbf{k}$.   **10.** $\displaystyle\int_0^1 \mathbf{h}(t)\,dt$ for $\mathbf{h}(t) = e^{-t}[t^2\mathbf{i} + \sqrt{2}\,t\mathbf{j} + \mathbf{k}]$.

Find $\lim\limits_{t \to 0} \mathbf{f}(t)$ if it exists.

11. $\mathbf{f}(t) = \dfrac{\sin t}{2t}\mathbf{i} + e^{2t}\mathbf{j} + \dfrac{t^2}{e^t}\mathbf{k}$.   **12.** $\mathbf{f}(t) = 3(t^2 - 1)\mathbf{i} + \cos t\,\mathbf{j} + \dfrac{t}{|t|}\mathbf{k}$.

Find a vector-valued function $\mathbf{f}$ that traces out the given curve in the indicated direction.

13. $4x^2 + 9y^2 = 36$ (a) counterclockwise.   (b) clockwise.

14. $(x - 1)^2 + y^2 = 1$ (a) counterclockwise.   (b) clockwise.

15. $y = x^2$ (a) from left to right.   (b) from right to left.

16. $y = x^3$ (a) from left to right.   (b) from right to left.

17. The directed line segment from $(1, 4, -2)$ to $(3, 9, 6)$.

18. The directed line segment from $(3, 2, -5)$ to $(7, 2, 9)$.

19. Set $\mathbf{f}(t) = t\mathbf{i} + f(t)\mathbf{j}$ and calculate

$$\mathbf{f}'(t_0), \qquad \int_a^b \mathbf{f}(t)\,dt, \qquad \int_a^b \mathbf{f}'(t)\,dt$$

given that

$$f'(t_0) = m, \qquad f(a) = c, \qquad f(b) = d, \qquad \int_a^b f(t)\,dt = A.$$

Find $\mathbf{f}(t)$ from the following information.

20. $\mathbf{f}'(t) = t\mathbf{i} + t(1 + t^2)^{-1/2}\mathbf{j} + te^t\mathbf{k}$ and $\mathbf{f}(0) = \mathbf{i} + 2\mathbf{j} + 3\mathbf{k}$.

21. $\mathbf{f}'(t) = \mathbf{i} + t^2\mathbf{j}$ and $\mathbf{f}(0) = \mathbf{j} - \mathbf{k}$.   **22.** $\mathbf{f}'(t) = 2\mathbf{f}(t)$ and $\mathbf{f}(0) = \mathbf{i} - \mathbf{k}$.

23. $\mathbf{f}'(t) = \alpha\mathbf{f}(t)$ with α a real number   and $\mathbf{f}(0) = \mathbf{c}$.

24. No ϵ, δ's have surfaced so far, but they are still there at the heart of the limit process. Give an ϵ, δ characterization of

$$\lim_{t \to t_0} \mathbf{f}(t) = \mathbf{L}.$$

25. (a) Show that, if $\mathbf{f}'(t) = \mathbf{0}$ for all t in an interval I, then $\mathbf{f}$ is constant on I.
(b) Show that, if $\mathbf{f}'(t) = \mathbf{g}'(t)$ for all t in an interval I, then $\mathbf{f}$ and $\mathbf{g}$ differ by a constant on I.

26. Assume that, as $t \to t_0$, $\mathbf{f}(t) \to \mathbf{L}$ and $\mathbf{g}(t) \to \mathbf{M}$. Show that

$$\mathbf{f}(t) \cdot \mathbf{g}(t) \to \mathbf{L} \cdot \mathbf{M}.$$

27. Show that, if $\mathbf{f}$ is differentiable at t, then $\mathbf{f}$ is continuous at t.

28. A vector-valued function $\mathbf{G}$ is called an *antiderivative* for $\mathbf{f}$ on $[a, b]$ iff
    (i) $\mathbf{G}$ is continuous on $[a, b]$ and   (ii) $\mathbf{G}'(t) = \mathbf{f}(t)$ for all $t \in (a, b)$.
Show that, if $\mathbf{f}$ is continuous on $[a, b]$ and $\mathbf{G}$ is an antiderivative for $\mathbf{f}$ on $[a, b]$, then

$$\int_a^b \mathbf{f}(t)\, dt = \mathbf{G}(b) - \mathbf{G}(a).$$

(This is a vector version of the fundamental theorem of integral calculus.)

29. Is it always true that

$$\int_a^b [\mathbf{f}(t) \cdot \mathbf{g}(t)]\, dt = \left[\int_a^b \mathbf{f}(t)\, dt\right] \cdot \left[\int_a^b \mathbf{g}(t)\, dt\right]?$$

30. Prove that, if $\mathbf{f}$ is continuous on $[a, b]$, then for each vector $\mathbf{c}$

$$\int_a^b [\mathbf{c} \cdot \mathbf{f}(t)]\, dt = \mathbf{c} \cdot \int_a^b \mathbf{f}(t)\, dt \qquad \text{and} \qquad \int_a^b [\mathbf{c} \times \mathbf{f}(t)]\, dt = \mathbf{c} \times \int_a^b \mathbf{f}(t)\, dt.$$

14.2 DIFFERENTIATION FORMULAS

Vector functions with a common domain can be combined in many ways to form new functions. From $\mathbf{f}$ and $\mathbf{g}$ we can form the sum $\mathbf{f} + \mathbf{g}$:

$$(\mathbf{f} + \mathbf{g})(t) = \mathbf{f}(t) + \mathbf{g}(t).$$

We can form scalar multiples $\alpha \mathbf{f}$ and thus linear combinations $\alpha \mathbf{f} + \beta \mathbf{g}$:

$$(\alpha \mathbf{f})(t) = \alpha \mathbf{f}(t), \qquad (\alpha \mathbf{f} + \beta \mathbf{g})(t) = \alpha \mathbf{f}(t) + \beta \mathbf{g}(t).$$

We can form the dot product $\mathbf{f} \cdot \mathbf{g}$:

$$(\mathbf{f} \cdot \mathbf{g})(t) = \mathbf{f}(t) \cdot \mathbf{g}(t).$$

We can also form the cross product $\mathbf{f} \times \mathbf{g}$:

$$(\mathbf{f} \times \mathbf{g})(t) = \mathbf{f}(t) \times \mathbf{g}(t).$$

There are two ways of bringing *scalar functions* (real-valued functions) into play. If a scalar function u has the same domain as $\mathbf{f}$, we can form the scalar product $u\mathbf{f}$:

$$(u\mathbf{f})(t) = u(t)\mathbf{f}(t).$$

If $u(t)$ is in the domain of $\mathbf{f}$ for each t in some interval, then we can form the composition $\mathbf{f} \circ u$:

$$(\mathbf{f} \circ u)(t) = \mathbf{f}(u(t)).$$

For example, with $u(t) = \sin t$ and $\mathbf{f}(t) = e^t \mathbf{i} + e^{-t} \mathbf{j}$,

$$(u\mathbf{f})(t) = u(t)\mathbf{f}(t) = \sin t \, (e^t \mathbf{i} + e^{-t} \mathbf{j}) \quad \text{and} \quad (\mathbf{f} \circ u)(t) = \mathbf{f}(u(t)) = e^{\sin t} \mathbf{i} + e^{-\sin t} \mathbf{j}.$$

From Theorem 14.1.3 it is clear that, if $\mathbf{f}$, $\mathbf{g}$, and u are continuous on a common domain, then $\mathbf{f} + \mathbf{g}$, $\alpha \mathbf{f}$, $\mathbf{f} \cdot \mathbf{g}$, $\mathbf{f} \times \mathbf{g}$, and $u\mathbf{f}$ are all continuous on that same set. We have yet to show the continuity of $\mathbf{f} \circ u$. The verification of that is left to you. (Exercise 26) What interests us here is that, if $\mathbf{f}$, $\mathbf{g}$, and u are differentiable, then the newly constructed functions are also differentiable and their derivatives satisfy the following rules:

(14.2.1)

(1) $(\mathbf{f} + \mathbf{g})'(t) = \mathbf{f}'(t) + \mathbf{g}'(t).$

(2) $(\alpha\mathbf{f})'(t) = \alpha\mathbf{f}'(t).$

(3) $(u\mathbf{f})'(t) = u(t)\mathbf{f}'(t) + u'(t)\mathbf{f}(t).$

(4) $(\mathbf{f} \cdot \mathbf{g})'(t) = [\mathbf{f}(t) \cdot \mathbf{g}'(t)] + [\mathbf{f}'(t) \cdot \mathbf{g}(t)].$

(5) $(\mathbf{f} \times \mathbf{g})'(t) = [\mathbf{f}(t) \times \mathbf{g}'(t)] + [\mathbf{f}'(t) \times \mathbf{g}(t)].$

(6) $(\mathbf{f} \circ u)'(t) = \mathbf{f}'(u(t))u'(t) = u'(t)\mathbf{f}'(u(t)).$ (the chain rule)

Rules (3), (4), (5) are all "product" rules and should remind you of the rule for differentiating the product of ordinary functions. Keep in mind, however, that the cross product is not commutative and therefore the order in Rule (5) is important.

In Rule (6) we first wrote the scalar part $u'(t)$ on the right so that the formula would look like the chain rule for ordinary functions. In general, $\mathbf{a}\alpha$ has the same meaning as $\alpha\mathbf{a}$.

Example 1.   Taking

$$\mathbf{f}(t) = 2t^2\mathbf{i} - 3\mathbf{j}, \qquad \mathbf{g}(t) = \mathbf{i} + t\mathbf{j} + t^2\mathbf{k}, \qquad u(t) = \tfrac{1}{3}t^3$$

we have

$$\mathbf{f}'(t) = 4t\mathbf{i}, \qquad \mathbf{g}'(t) = \mathbf{j} + 2t\mathbf{k}, \qquad u'(t) = t^2.$$

Therefore

(a) $(\mathbf{f} + \mathbf{g})'(t) = \mathbf{f}'(t) + \mathbf{g}'(t) = 4t\mathbf{i} + (\mathbf{j} + 2t\mathbf{k}) = 4t\mathbf{i} + \mathbf{j} + 2t\mathbf{k};$

(b) $(2\mathbf{f})'(t) = 2\mathbf{f}'(t) = 2(4t\mathbf{i}) = 8t\mathbf{i};$

(c) $(u\mathbf{f})'(t) = u(t)\mathbf{f}'(t) + u'(t)\mathbf{f}(t) = \tfrac{1}{3}t^3(4t\mathbf{i}) + t^2(2t^2\mathbf{i} - 3\mathbf{j}) = \tfrac{10}{3}t^4\mathbf{i} - 3t^2\mathbf{j};$

(d) $(\mathbf{f} \cdot \mathbf{g})'(t) = [\mathbf{f}(t) \cdot \mathbf{g}'(t)] + [\mathbf{f}'(t) \cdot \mathbf{g}(t)]$

$\qquad = [(2t^2\mathbf{i} - 3\mathbf{j}) \cdot (\mathbf{j} + 2t\mathbf{k})] + [4t\mathbf{i} \cdot (\mathbf{i} + t\mathbf{j} + t^2\mathbf{k})] = -3 + 4t;$

(e) $(\mathbf{f} \times \mathbf{g})'(t) = [\mathbf{f}(t) \times \mathbf{g}'(t)] + [\mathbf{f}'(t) \times \mathbf{g}(t)]$

$\qquad = [(2t^2\mathbf{i} - 3\mathbf{j}) \times (\mathbf{j} + 2t\mathbf{k})] + [4t\mathbf{i} \times (\mathbf{i} + t\mathbf{j} + t^2\mathbf{k})]$

$\qquad = (2t^2\mathbf{k} - 4t^3\mathbf{j} - 6t\mathbf{i}) + (4t^2\mathbf{k} - 4t^3\mathbf{j}) = -6t\mathbf{i} - 8t^3\mathbf{j} + 6t^2\mathbf{k}$

while

$(\mathbf{g} \times \mathbf{f})'(t) = [\mathbf{g}(t) \times \mathbf{f}'(t)] + [\mathbf{g}'(t) \times \mathbf{f}(t)]$

$\qquad = [(\mathbf{i} + t\mathbf{j} + t^2\mathbf{k}) \times 4t\mathbf{i}] + [(\mathbf{j} + 2t\mathbf{k}) \times (2t^2\mathbf{i} - 3\mathbf{j})]$

$\qquad = (-4t^2\mathbf{k} + 4t^3\mathbf{j}) + (-2t^2\mathbf{k} + 4t^3\mathbf{j} + 6t\mathbf{i})$

$\qquad = 6t\mathbf{i} + 8t^3\mathbf{j} - 6t^2\mathbf{k} = -(\mathbf{f} \times \mathbf{g})'(t);$

(f) $(\mathbf{f} \circ u)'(t) = \mathbf{f}'(u(t))u'(t)$

$\qquad = [4u(t)\mathbf{i}]u'(t) = [4(\tfrac{1}{3}t^3)\mathbf{i}]\,t^2 = \tfrac{4}{3}t^5\mathbf{i}.$ □

The differentiation formulas that we have given can all be derived component by component, and they can all be derived in a component-free manner. Take for example formula (3):

$$(u\mathbf{f})'(t) = u(t)\mathbf{f}'(t) + u'(t)\mathbf{f}(t).$$

Component-by-Component Derivation. Set

$$\mathbf{f}(t) = f_1(t)\mathbf{i} + f_2(t)\mathbf{j} + f_3(t)\mathbf{k}.$$

Then

$$(u\mathbf{f})(t) = u(t)\mathbf{f}(t) = u(t)f_1(t)\mathbf{i} + u(t)f_2(t)\mathbf{j} + u(t)f_3(t)\mathbf{k}$$

and

$$(u\mathbf{f})'(t) = [u(t)f_1'(t) + u'(t)f_1(t)]\mathbf{i} + [u(t)f_2'(t) + u'(t)f_2(t)]\mathbf{j}$$
$$+ [u(t)f_3'(t) + u'(t)f_3(t)]\mathbf{k}$$
$$= u(t)[f_1'(t)\mathbf{i} + f_2'(t)\mathbf{j} + f_3'(t)\mathbf{k}] + u'(t)[f_1(t)\mathbf{i} + f_2(t)\mathbf{j} + f_3(t)\mathbf{k}]$$
$$= u(t)\mathbf{f}'(t) + u'(t)\mathbf{f}(t). \quad \square$$

Component-Free Derivation. We find $(u\mathbf{f})'(t)$ by taking the limit as $h \to 0$ of the difference quotient

$$\frac{u(t + h)\mathbf{f}(t + h) - u(t)\mathbf{f}(t)}{h}.$$

By adding and subtracting $u(t + h)\mathbf{f}(t)$ inside the brackets and then factoring, we can rewrite this quotient as

$$u(t + h)\frac{\mathbf{f}(t + h) - \mathbf{f}(t)}{h} + \frac{u(t + h) - u(t)}{h}\mathbf{f}(t).$$

As $h \to 0$,

$$u(t + h) \to u(t), \qquad \text{(differentiable functions are continuous)}$$

$$\frac{\mathbf{f}(t + h) - \mathbf{f}(t)}{h} \to \mathbf{f}'(t), \qquad \text{(definition of derivative for vector functions)}$$

$$\frac{u(t + h) - u(t)}{h} \to u'(t). \qquad \text{(definition of derivative for scalar functions)}$$

It follows from the limit rules (Theorem 14.1.3) that

$$u(t + h)\frac{\mathbf{f}(t + h) - \mathbf{f}(t)}{h} \to u(t)\mathbf{f}'(t), \qquad \frac{u(t + h) - u(t)}{h}\mathbf{f}(t) \to u'(t)\mathbf{f}(t)$$

and therefore

$$u(t + h)\frac{\mathbf{f}(t + h) - \mathbf{f}(t)}{h} + \frac{u(t + h) - u(t)}{h}\mathbf{f}(t) \to u(t)\mathbf{f}'(t) + u'(t)\mathbf{f}(t). \quad \square$$

In Leibniz's notation the formulas take the following form:

(14.2.2)

$$(1)\ \frac{d}{dt}(\mathbf{f}+\mathbf{g}) = \frac{d\mathbf{f}}{dt} + \frac{d\mathbf{g}}{dt}.$$

$$(2)\ \frac{d}{dt}(\alpha\mathbf{f}) = \alpha\,\frac{d\mathbf{f}}{dt}. \qquad (\alpha\ \text{constant})$$

$$(3)\ \frac{d}{dt}(u\mathbf{f}) = u\,\frac{d\mathbf{f}}{dt} + \frac{du}{dt}\,\mathbf{f}. \qquad (u=u(t))$$

$$(4)\ \frac{d}{dt}(\mathbf{f}\cdot\mathbf{g}) = \left(\mathbf{f}\cdot\frac{d\mathbf{g}}{dt}\right) + \left(\frac{d\mathbf{f}}{dt}\cdot\mathbf{g}\right).$$

$$(5)\ \frac{d}{dt}(\mathbf{f}\times\mathbf{g}) = \left(\mathbf{f}\times\frac{d\mathbf{g}}{dt}\right) + \left(\frac{d\mathbf{f}}{dt}\times\mathbf{g}\right).$$

$$(6)\ \frac{d\mathbf{f}}{dt} = \frac{d\mathbf{f}}{du}\frac{du}{dt}. \qquad (\text{chain rule})$$

We come now to a couple of results that will prove useful as we go on.

Problem 2. Set $r = \|\mathbf{r}\|$. Show that, if $\mathbf{r}$ is a differentiable function of t, then r is differentiable where not zero and

(14.2.3)

$$\mathbf{r}\cdot\frac{d\mathbf{r}}{dt} = r\,\frac{dr}{dt}.$$

Solution. If $\mathbf{r}$ is differentiable, then

$$\mathbf{r}\cdot\mathbf{r} = \|\mathbf{r}\|^2 = r^2$$

is differentiable. Let's assume now that $r \neq 0$. Since the square-root function is differentiable at all positive numbers and r^2 is positive, we can apply the square-root function to r^2 and conclude by the chain rule that r is itself differentiable.

To obtain the formula we differentiate the identity $\mathbf{r}\cdot\mathbf{r} = r^2$:

$$\mathbf{r}\cdot\frac{d\mathbf{r}}{dt} + \frac{d\mathbf{r}}{dt}\cdot\mathbf{r} = 2r\,\frac{dr}{dt}$$

$$2\mathbf{r}\cdot\frac{d\mathbf{r}}{dt} = 2r\,\frac{dr}{dt}$$

$$\mathbf{r}\cdot\frac{d\mathbf{r}}{dt} = r\,\frac{dr}{dt}. \qquad \square$$

Problem 3. Set $r = \|\mathbf{r}\|$. Show that, if $\mathbf{r}$ is a differentiable function of t, then where $r \neq 0$

(14.2.4)

$$\frac{d}{dt}\left(\frac{\mathbf{r}}{r}\right) = \frac{1}{r^3}\left[\left(\mathbf{r}\times\frac{d\mathbf{r}}{dt}\right)\times\mathbf{r}\right].$$

Solution. This is a little tricky:

$$\frac{d}{dt}\left(\frac{\mathbf{r}}{r}\right) = \frac{1}{r}\frac{d\mathbf{r}}{dt} - \frac{1}{r^2}\frac{dr}{dt}\mathbf{r}$$

$$= \frac{1}{r^3}\left[r^2\frac{d\mathbf{r}}{dt} - r\frac{dr}{dt}\mathbf{r}\right]$$

$$= \frac{1}{r^3}\left[(\mathbf{r}\cdot\mathbf{r})\frac{d\mathbf{r}}{dt} - \left(\mathbf{r}\cdot\frac{d\mathbf{r}}{dt}\right)\mathbf{r}\right] = \frac{1}{r^3}\left[\left(\mathbf{r}\times\frac{d\mathbf{r}}{dt}\right)\times\mathbf{r}\right]. \quad \square$$

$$(\mathbf{a}\times\mathbf{b})\times\mathbf{c} = (\mathbf{c}\cdot\mathbf{a})\mathbf{b} - (\mathbf{c}\cdot\mathbf{b})\mathbf{a}$$

EXERCISES 14.2

Find $\mathbf{f}'(t)$ and $\mathbf{f}''(t)$ for each of the following functions.

1. $\mathbf{f}(t) = \mathbf{a} + t\mathbf{b}.$

2. $\mathbf{f}(t) = \mathbf{a} + t\mathbf{b} + t^2\mathbf{c}.$

3. $\mathbf{f}(t) = e^{2t}\,\mathbf{i} - \sin t\,\mathbf{j}.$

4. $\mathbf{f}(t) = [(t^2\mathbf{i} - \mathbf{j})\cdot(\mathbf{i} - t^2\mathbf{j})]\,\mathbf{i}.$

5. $\mathbf{f}(t) = (e^t\mathbf{i} + t\mathbf{k})\times(t\mathbf{j} + e^{-t}\mathbf{k}).$

6. $\mathbf{f}(t) = t\mathbf{g}(t^2).$

7. $\mathbf{f}(t) = t\mathbf{g}(\sqrt{t}).$

8. $\mathbf{f}(t) = (\cos t\,\mathbf{i} - \mathbf{j})\times(t\mathbf{j} + \sin t\,\mathbf{k}).$

Find the indicated derivative.

9. $\dfrac{d}{dt}[e^{\cos t}\,\mathbf{i} + e^{\sin t}\,\mathbf{j}].$

10. $\dfrac{d^2}{dt^2}[e^t\cos t\,\mathbf{i} + e^t\sin t\,\mathbf{j}].$

11. $\dfrac{d}{dt}[(\mathbf{a} + t\mathbf{b})\times(\mathbf{c} + t\mathbf{d})].$

12. $\dfrac{d}{dt}[(\mathbf{a} + t\mathbf{b})\times(\mathbf{a} + t\mathbf{b} + t^2\mathbf{c})].$

13. $\dfrac{d}{dt}[(\mathbf{a} + t\mathbf{b})\cdot(\mathbf{c} + t\mathbf{d})].$

14. $\dfrac{d}{dt}[(\mathbf{a} + t\mathbf{b})\cdot(\mathbf{a} + t\mathbf{b} + t^2\mathbf{c})].$

Find $\mathbf{r}(t)$ given that:

15. $\mathbf{r}'(t) = \mathbf{b}$ for all real t, $\mathbf{r}(0) = \mathbf{a}.$

16. $\mathbf{r}''(t) = \mathbf{c}$ for all real t, $\mathbf{r}'(0) = \mathbf{b},$ $\mathbf{r}(0) = \mathbf{a}.$

17. $\mathbf{r}''(t) = \mathbf{a} + t\mathbf{b}$ for all real t, $\mathbf{r}'(0) = \mathbf{c},$ $\mathbf{r}(0) = \mathbf{d}.$

18. Show that, if $\mathbf{r}(t) = e^{kt}\mathbf{i} + e^{-kt}\mathbf{j}$, then $\mathbf{r}(t)$ and $\mathbf{r}''(t)$ are parallel.

19. Calculate $\mathbf{r}(t)\cdot\mathbf{r}'(t)$ and $\mathbf{r}(t)\times\mathbf{r}'(t)$ given that $\mathbf{r}(t) = \cos t\,\mathbf{i} + \sin t\,\mathbf{j}.$

Assume the rule for differentiating a cross product and show the following.

20. $(\mathbf{g}\times\mathbf{f})'(t) = -(\mathbf{f}\times\mathbf{g})'(t).$

21. $\dfrac{d}{dt}[\mathbf{f}(t)\times\mathbf{f}'(t)] = \mathbf{f}(t)\times\mathbf{f}''(t).$

22. $\dfrac{d}{dt}[u_1(t)\mathbf{r}_1(t)\times u_2(t)\mathbf{r}_2(t)] = u_1(t)u_2(t)\dfrac{d}{dt}[\mathbf{r}_1(t)\times\mathbf{r}_2(t)] + [\mathbf{r}_1(t)\times\mathbf{r}_2(t)]\dfrac{d}{dt}[u_1(t)u_2(t)].$

23. Assume the rule for differentiating a dot product and show that

$$\|\mathbf{r}(t)\| \text{ is constant} \quad\text{iff}\quad \mathbf{r}(t)\cdot\mathbf{r}'(t) = 0 \text{ identically.}$$

24. Derive the formula

$$(\mathbf{f}\cdot\mathbf{g})'(t) = [\mathbf{f}(t)\cdot\mathbf{g}'(t)] + [\mathbf{f}'(t)\cdot\mathbf{g}(t)]$$

(a) by appealing to components. (b) without appealing to components.

25. Derive the formula

$$(\mathbf{f} \times \mathbf{g})'(t) = [\mathbf{f}(t) \times \mathbf{g}'(t)] + [\mathbf{f}'(t) \times \mathbf{g}(t)]$$

without appealing to components.

26. (a) Show that, if u is continuous at t_0 and $\mathbf{f}$ is continuous at $u(t_0)$, then the composition $(\mathbf{f} \circ u)$ is continuous at t_0.
   (b) Derive the chain rule for vector functions:

$$\frac{d\mathbf{f}}{dt} = \frac{d\mathbf{f}}{du}\frac{du}{dt}.$$

14.3   CURVES

Introduction

Earlier we explained how every linear vector function

$$\mathbf{r}(t) = \mathbf{r}_0 + t\mathbf{d} \qquad \text{with } \mathbf{d} \neq \mathbf{0}$$

parametrizes a line. More generally, every differentiable vector function parametrizes a curve.

Suppose that

$$\mathbf{r}(t) = x(t)\mathbf{i} + y(t)\mathbf{j} + z(t)\mathbf{k}$$

is differentiable on some interval I. (At the endpoints, if any, we require only continuity.) For each number $t \in I$, the tip of the radius vector $\mathbf{r}(t)$ is the point $P(x(t), y(t), z(t))$. As t ranges over I, the point P traces out some path C. (Figure 14.3.1) We call C a *differentiable curve* and say that C is *parametrized* by $\mathbf{r}$.

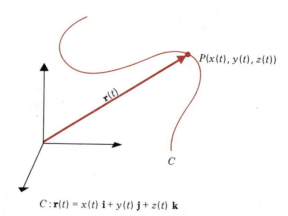

$$C : \mathbf{r}(t) = x(t)\,\mathbf{i} + y(t)\,\mathbf{j} + z(t)\,\mathbf{k}$$

Figure 14.3.1

Tangent Vector, Tangent Line

Let's try to interpret the derivative

$$\mathbf{r}'(t) = x'(t)\mathbf{i} + y'(t)\mathbf{j} + z'(t)\mathbf{k}$$

geometrically. First of all,

$$\mathbf{r}'(t) = \lim_{h \to 0} \frac{\mathbf{r}(t+h) - \mathbf{r}(t)}{h}.$$

If $\mathbf{r}'(t) \neq \mathbf{0}$, then we can be sure that for $t + h$ close enough to t, the vector

$$\mathbf{r}(t+h) - \mathbf{r}(t)$$

will not be **0**. (Explain.) Consequently, we can think of the vector $\mathbf{r}(t+h) - \mathbf{r}(t)$ as pictured in Figure 14.3.2.

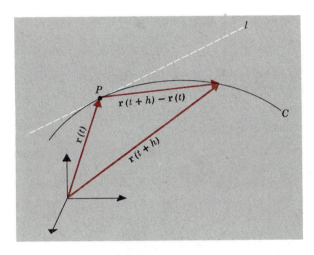

Figure 14.3.2

Let's agree that the line marked l in the figure corresponds to our intuitive notion of tangent line at the point P. As h tends to zero, the vector

$$\mathbf{r}(t+h) - \mathbf{r}(t)$$

comes increasingly closer to serving as a direction vector for that tangent line. It may be tempting therefore to take the limiting case

$$\lim_{h \to 0} [\mathbf{r}(t+h) - \mathbf{r}(t)]$$

and call that a direction vector for the tangent line. The trouble is that this limit vector is **0** and **0** has no direction.

We can circumvent this difficulty by replacing $\mathbf{r}(t+h) - \mathbf{r}(t)$ by a vector which, for small h, has greater length: the difference quotient

$$\frac{\mathbf{r}(t+h) - \mathbf{r}(t)}{h}.$$

For each $h \neq 0$, this vector is parallel to $\mathbf{r}(t+h) - \mathbf{r}(t)$, and therefore its limit,

$$\mathbf{r}'(t) = \lim_{h \to 0} \frac{\mathbf{r}(t+h) - \mathbf{r}(t)}{h},$$

which by assumption is not **0**, can be taken as a direction vector for the tangent line.

Hence the following definition:

DEFINITION 14.3.1

Let

$$C: \quad \mathbf{r}(t) = x(t)\mathbf{i} + y(t)\mathbf{j} + z(t)\mathbf{k}$$

be a differentiable curve. The vector $\mathbf{r}'(t)$, if not $\mathbf{0}$, is said to be *tangent* to the curve C at the point $P(x(t), y(t), z(t))$.

At each point of a line

$$l: \quad \mathbf{r}(t) = \mathbf{r}_0 + t\mathbf{d}$$

the tangent vector $\mathbf{r}'(t)$ is parallel to the line itself:

$$\mathbf{r}'(t) = \mathbf{d} \quad \text{and} \quad \mathbf{d} \text{ is parallel to } l.$$

In the case of a circle

$$C: \quad \mathbf{r}(t) = r(\cos t\, \mathbf{i} + \sin t\, \mathbf{j})$$

the tangent vector $\mathbf{r}'(t)$ is perpendicular to the radius vector $\mathbf{r}(t)$:

$$\mathbf{r}'(t) \cdot \mathbf{r}(t) = r(-\sin t\, \mathbf{i} + \cos t\, \mathbf{j}) \cdot r(\cos t\, \mathbf{i} + \sin t\, \mathbf{j})$$
$$= r^2(-\sin t \cos t + \cos t \sin t) = 0.$$

Problem 1. Find the point P on the curve

$$\mathbf{r}(t) = (1 - 2t)\,\mathbf{i} + t^2\mathbf{j} + 2e^{2(t-1)}\mathbf{k}$$

at which the tangent vector $\mathbf{r}'(t)$ is parallel to the radius vector $\mathbf{r}(t)$.

Solution

$$\mathbf{r}'(t) = -2\mathbf{i} + 2t\mathbf{j} + 4e^{2(t-1)}\mathbf{k}.$$

For $\mathbf{r}'(t)$ to be parallel to $\mathbf{r}(t)$ there must exist a scalar α such that

$$\mathbf{r}(t) = \alpha\mathbf{r}'(t).$$

This vector equation holds iff

$$1 - 2t = -2\alpha, \qquad t^2 = 2\alpha t, \qquad 2e^{2(t-1)} = 4\alpha e^{2(t-1)}.$$

The last scalar equation requires that $\alpha = \frac{1}{2}$. The only value of t that satisfies all three equations with $\alpha = \frac{1}{2}$ is $t = 1$. Therefore the only point at which $\mathbf{r}'(t)$ is parallel to $\mathbf{r}(t)$ is the tip of $\mathbf{r}(1)$. This is the point $P(-1, 1, 2)$. ☐

If $\mathbf{r}'(t_0) \neq \mathbf{0}$, then $\mathbf{r}'(t_0)$ is tangent to the curve at the tip of $\mathbf{r}(t_0)$. The *tangent line* at this point can be parametrized by setting

(14.3.2)

$$\mathbf{R}(u) = \mathbf{r}(t_0) + u\mathbf{r}'(t_0).$$

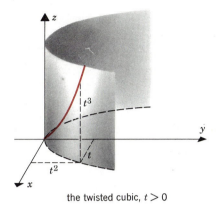

the twisted cubic, $t > 0$

Figure 14.3.3

Problem 2. Find a vector tangent to the *twisted cubic*

$$\mathbf{r}(t) = t\mathbf{i} + t^2\mathbf{j} + t^3\mathbf{k} \qquad \text{(Figure 14.3.3)}$$

at the point $P(2, 4, 8)$, and then parametrize the tangent line.

Solution. Here

$$\mathbf{r}'(t) = \mathbf{i} + 2t\mathbf{j} + 3t^2\mathbf{k}.$$

Since $P(2, 4, 8)$ is the tip of $\mathbf{r}(2)$, the vector

$$\mathbf{r}'(2) = \mathbf{i} + 4\mathbf{j} + 12\mathbf{k}$$

is tangent to the curve at the point $P(2, 4, 8)$.   The vector function

$$\mathbf{R}(u) = (2\mathbf{i} + 4\mathbf{j} + 8\mathbf{k}) + u(\mathbf{i} + 4\mathbf{j} + 12\mathbf{k})$$

parametrizes the tangent line.   ☐

Intersecting Curves

Two curves

$$C_1: \quad \mathbf{r}_1(t) = x_1(t)\mathbf{i} + y_1(t)\mathbf{j} + z_1(t)\mathbf{k}, \qquad C_2: \quad \mathbf{r}_2(u) = x_2(u)\mathbf{i} + y_2(u)\mathbf{j} + z_2(u)\mathbf{k}$$

intersect iff there are numbers t and u for which

$$\mathbf{r}_1(t) = \mathbf{r}_2(u).$$

The angle between two intersecting curves (which, by definition, is the angle between the corresponding tangent lines) can be obtained by examining the tangent vectors at the point of intersection.

Problem 3. Show that the circles

$$C_1: \quad \mathbf{r}_1(t) = \cos t \, \mathbf{i} + \sin t \, \mathbf{j}, \qquad C_2: \quad \mathbf{r}_2(u) = \cos u \, \mathbf{j} + \sin u \, \mathbf{k}$$

intersect at right angles at $P(0, 1, 0)$ and $Q(0, -1, 0)$.

Solution. Since $\mathbf{r}_1(\tfrac{1}{2}\pi) = \mathbf{j} = \mathbf{r}_2(0)$, the curves meet at the tip of $\mathbf{j}$, which is $P(0, 1, 0)$. Since $\mathbf{r}_1(\tfrac{3}{2}\pi) = -\mathbf{j} = \mathbf{r}_2(\pi)$, the curves meet at the tip of $-\mathbf{j}$, which is $Q(0, -1, 0)$. Differentiation gives

$$\mathbf{r}_1'(t) = -\sin t \, \mathbf{i} + \cos t \, \mathbf{j} \qquad \text{and} \qquad \mathbf{r}_2'(u) = -\sin u \, \mathbf{j} + \cos u \, \mathbf{k}.$$

Since $\mathbf{r}_1'(\tfrac{1}{2}\pi) = -\mathbf{i}$ and $\mathbf{r}_2'(0) = \mathbf{k}$, we have

$$\mathbf{r}_1'(\tfrac{1}{2}\pi) \cdot \mathbf{r}_2'(0) = 0.$$

This tells us that the curves are perpendicular at $P(0, 1, 0)$. Since $\mathbf{r}_1'(\tfrac{3}{2}\pi) = \mathbf{i}$ and $\mathbf{r}_2'(\pi) = -\mathbf{k}$, we have

$$\mathbf{r}_1'(\tfrac{3}{2}\pi) \cdot \mathbf{r}_2'(\pi) = 0.$$

This tells us that the curves are perpendicular at $Q(0, -1, 0)$. The curves appear in Figure 14.3.4.   ☐

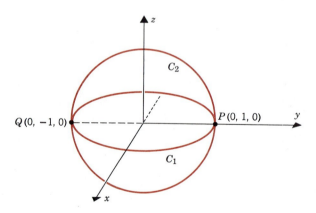

Figure 14.3.4

The Unit Tangent, the Principal Normal, the Osculating Plane

Suppose now that the curve

$$C: \quad \mathbf{r}(t) = x(t)\mathbf{i} + y(t)\mathbf{j} + z(t)\mathbf{k}$$

is twice differentiable and $\mathbf{r}'(t)$ is never zero. Then at each point $P(x(t), y(t), z(t))$ of the curve, there is a *unit tangent vector*:

(14.3.3)
$$T(t) = \frac{r'(t)}{\|r'(t)\|}.$$

Since $\|T(t)\| = 1$, we have $T(t) \cdot T(t) = 1$. Differentiation gives

$$T(t) \cdot T'(t) + T'(t) \cdot T(t) = 0.$$

Since the dot product is commutative, we have

$$2[T'(t) \cdot T(t)] = 0 \qquad \text{and thus} \qquad T'(t) \cdot T(t) = 0.$$

At each point of the curve *the vector $T'(t)$ is perpendicular to $T(t)$.*

The vector $T'(t)$ measures the rate of change of $T(t)$ with respect to t. Since the norm of $T(t)$ is constantly 1, $T(t)$ can change only in direction. The vector $T'(t)$ measures this change in direction.

If the unit tangent vector is not changing in direction (as in the case of a straight line), then $T'(t) = 0$. If $T'(t) \neq 0$, then we can form what is called the *principal normal vector*:

(14.3.4)
$$N(t) = \frac{T'(t)}{\|T'(t)\|}.$$

This is the unit vector in the direction of $T'(t)$. The *normal line* at P is the line through P parallel to the principal normal.

Figure 14.3.5 shows a curve on which we have marked several points. At each point we have drawn the unit tangent and the principal normal. The plane determined by these two vectors is called the *osculating plane* (literally, the "kissing plane"). This is the plane of greatest contact with the curve at the point in question.

If a curve is a plane curve but not a straight line, then the osculating plane is simply the plane of the curve. A straight line does not have an osculating plane. There is no principal normal vector $[T'(t) = 0]$ and there is no plane of greatest contact. Each straight line lies on an infinite number of planes.

Figure 14.3.5

circular helix

Figure 14.3.6

Example 4. In Figure 14.3.6 you can see a curve spiraling up a circular cylinder. If the rate of climb is constant, the curve is called a *circular helix*.

The simplest parametrization for a circular helix takes the form

$$\mathbf{r}(t) = a \cos t \, \mathbf{i} + a \sin t \, \mathbf{j} + bt\mathbf{k} \qquad \text{with } a > 0, \, b > 0.$$

The first two components produce the rotational effect; the third component gives the rate of climb.

We will find the unit tangent, the principal normal, and then an equation for the osculating plane. Since

$$\mathbf{r}'(t) = -a \sin t \, \mathbf{i} + a \cos t \, \mathbf{j} + b\mathbf{k},$$

we have

$$\|\mathbf{r}'(t)\| = \sqrt{a^2 \sin^2 t + a^2 \cos^2 t + b^2} = \sqrt{a^2 + b^2}$$

and therefore

$$\mathbf{T}(t) = \frac{\mathbf{r}'(t)}{\|\mathbf{r}'(t)\|} = \frac{1}{\sqrt{a^2 + b^2}} (-a \sin t \, \mathbf{i} + a \cos t \, \mathbf{j} + b\mathbf{k}).$$

This is the unit tangent vector.

The principal normal vector is the unit vector in the direction of $\mathbf{T}'(t)$. Since

$$\frac{d}{dt}(-a \sin t \, \mathbf{i} + a \cos t \, \mathbf{j} + b\mathbf{k}) = -a \cos t \, \mathbf{i} - a \sin t \, \mathbf{j} \qquad \text{and} \qquad a > 0,$$

you can see that

$$\mathbf{N}(t) = -\cos t \, \mathbf{i} - \sin t \, \mathbf{j}.$$

The principal normal is horizontal and points directly to the z-axis.

Now let's find an equation for the osculating plane p at an arbitrary point $P(a \cos t, a \sin t, bt)$. The cross product $\mathbf{T}(t) \times \mathbf{N}(t)$ is perpendicular to p. Therefore, as a normal for p, we can take any nonzero scalar multiple of $\mathbf{T}(t) \times \mathbf{N}(t)$. In particular, we can take

$$(a \sin t \, \mathbf{i} - a \cos t \, \mathbf{j} - b\mathbf{k}) \times (\cos t \, \mathbf{i} + \sin t \, \mathbf{j}).$$

As you can check, this simplifies to

$$b \sin t \, \mathbf{i} - b \cos t \, \mathbf{j} + a\mathbf{k}.$$

The equation for the osculating plane at the point $P(a \cos t, a \sin t, bt)$ thus takes the form

$$b \sin t \, (x - a \cos t) - b \cos t \, (y - a \sin t) + a \, (z - bt) = 0. \qquad (13.8.1)$$

This simplifies to

$$(b \sin t)x - (b \cos t)y + az = abt.$$

To visualize how this osculating plane changes from point to point, think of a playground spiral slide or the threaded surface on a bolt. ☐

Reversing the Sense of a Curve

A parametrized curve is an *oriented* curve: not just a set of points, but rather a succession of points, a set of points traversed in a certain order, a certain sense.
We make a distinction between the curve

$$\mathbf{r} = \mathbf{r}(t), \qquad t \in [a, b]$$

and the curve

$$\mathbf{R}(u) = \mathbf{r}(a + b - u), \qquad u \in [a, b].$$

Both vector functions trace out the same set of points (check that out), but the order has been reversed. Whereas the first curve starts at $\mathbf{r}(a)$ and ends at $\mathbf{r}(b)$, the second curve starts at $\mathbf{r}(b)$ and ends at $\mathbf{r}(a)$:

$$\mathbf{R}(a) = \mathbf{r}(a + b - a) = \mathbf{r}(b), \qquad \mathbf{R}(b) = \mathbf{r}(a + b - b) = \mathbf{r}(a).$$

Thus, for example,

$$\mathbf{r}(t) = \cos t \, \mathbf{i} + \sin t \, \mathbf{j}, \qquad t \in [0, 2\pi]$$

gives the unit circle traversed counterclockwise while

$$\mathbf{R}(u) = \cos (2\pi - u) \, \mathbf{i} + \sin (2\pi - u) \, \mathbf{j}, \qquad u \in [0, 2\pi]$$

gives the unit circle traversed clockwise.

What happens to the unit tangent $\mathbf{T}$, the principal normal $\mathbf{N}$, and the osculating plane when we reverse the sense of a curve? As you are asked to show in Exercise 29, $\mathbf{T}$ is replaced by $-\mathbf{T}$, but $\mathbf{N}$ remains the same. The osculating plane remains the same.

Other Changes of Parameter

Not all changes of parameter change the succession of points. Suppose that

$$\mathbf{r} = \mathbf{r}(t), \qquad t \in I$$

is a differentiable curve, and let ϕ be a function that maps some interval J onto the interval I. (domain J, range I) Now set

$$\mathbf{R}(u) = \mathbf{r}(\phi(u)) \qquad \text{for all } u \in J.$$

If $\phi'(u) > 0$ for all $u \in J$, then $\mathbf{r}$ and $\mathbf{R}$ are said to differ by a *sense-preserving change of parameter*. In this case, $\mathbf{r}$ and $\mathbf{R}$ take on exactly the same values in exactly the same order. In other words, they produce exactly the same oriented curve (Exercise 30).

If, on the other hand, $\phi'(u) < 0$ for all $u \in J$, then the change in parameter is said to be *sense-reversing*. In that case, $\mathbf{r}$ and $\mathbf{R}$ still take on exactly the same values but in opposite order. The paths are the same, but they are traversed in opposite directions. You have already seen one example of this:

$$\mathbf{R}(u) = \mathbf{r}(a + b - u).$$

EXERCISES 14.3

Find the tangent vector $\mathbf{r}'(t)$ and the tangent line at the indicated point.

1. $\mathbf{r}(t) = \cos \pi t \, \mathbf{i} + \sin \pi t \, \mathbf{j} + t \, \mathbf{k}$ at $t = 2$.
2. $\mathbf{r}(t) = e^t \mathbf{i} + e^{-t} \mathbf{j} - \ln t \, \mathbf{k}$ at $t = 1$.
3. $\mathbf{r}(t) = \mathbf{a} + t\mathbf{b} + t^2 \mathbf{c}$ at $t = -1$.
4. $\mathbf{r}(t) = (t + 1)\mathbf{i} + (t^2 + 1)\mathbf{j} + (t^3 + 1)\mathbf{k}$ at $P(1, 1, 1)$.
5. $\mathbf{r}(t) = 2t^2 \mathbf{i} + (1 - t)\mathbf{j} + (3 + 2t^2)\mathbf{k}$ at $P(2, 0, 5)$.
6. $\mathbf{r}(t) = 3t\mathbf{a} + \mathbf{b} - t^2 \mathbf{c}$ at $t = 2$.
7. Show that $\mathbf{r}(t) = at\mathbf{i} + bt^2 \mathbf{j}$ parametrizes a parabola. Find an equation in x and y for this parabola.
8. Show that $\mathbf{r}(t) = \frac{1}{2}a(e^{\omega t} + e^{-\omega t})\mathbf{i} + \frac{1}{2}a(e^{\omega t} - e^{-\omega t})\mathbf{j}$ parametrizes the right branch $(x > 0)$ of the hyperbola $x^2 - y^2 = a^2$.
9. Find (a) the points on the curve $\mathbf{r}(t) = t\mathbf{i} + (1 + t^2)\,\mathbf{j}$ at which $\mathbf{r}(t)$ and $\mathbf{r}'(t)$ are perpendicular; (b) the points at which they have the same direction; (c) the points at which they have opposite directions.
10. Find the curve given that $\mathbf{r}'(t) = \alpha \mathbf{r}(t)$ for all real t and $\mathbf{r}(0) = \mathbf{i} + 2\mathbf{j} + 3\mathbf{k}$.
11. Suppose that $\mathbf{r}'(t)$ and $\mathbf{r}(t)$ are parallel for all t. Show that, if $\mathbf{r}'(t)$ is never $\mathbf{0}$, then the tangent line at each point passes through the origin.
12. The curves

$$\mathbf{r}_1(t) = (e^t - 1)\mathbf{i} + 2 \sin t \, \mathbf{j} + \ln (t + 1) \, \mathbf{k},$$
$$\mathbf{r}_2(u) = (u + 1)\mathbf{i} + (u^2 - 1)\mathbf{j} + (u^3 + 1)\mathbf{k}$$

intersect at the origin. Find the angle of intersection.
13. Find the point at which the curves

$$\mathbf{r}_1(t) = e^t \mathbf{i} + 2 \sin (t + \tfrac{1}{2}\pi) \, \mathbf{j} + (t^2 - 2)\mathbf{k}, \qquad \mathbf{r}_2(u) = u\mathbf{i} + 2\mathbf{j} + (u^2 - 3)\mathbf{k}$$

intersect and find the angle of intersection.

14. Consider the vector function $\mathbf{f}(t) = t\mathbf{i} + f(t)\mathbf{j}$ formed from a differentiable real-valued function f. The vector function $\mathbf{f}$ parametrizes the graph of f.
(a) Parametrize the tangent line at $P(t_0, f(t_0))$.
(b) Show that the parametrization obtained in (a) reduces to the usual equation for the tangent line:

$$y - f(t_0) = f'(t_0)[x - t_0] \quad \text{if } f'(t_0) \neq 0; \qquad y = f(t_0) \quad \text{if } f'(t_0) = 0.$$

15. Define a vector function $\mathbf{r}$ on the interval $[0, 2\pi]$ that satisfies the initial condition $\mathbf{r}(0) = a\mathbf{i}$ and, as t increases to 2π, traces out the ellipse $b^2x^2 + a^2y^2 = a^2b^2$
(a) once in a counterclockwise manner.   (b) once in a clockwise manner.
(c) twice in a counterclockwise manner.   (d) three times in a clockwise manner.

16. Exercise 15 with $\mathbf{r}(0) = b\mathbf{j}$.

Find a vector parametrization for the curve.

17. $y^2 = x - 1$, $\quad y \geq 1$.
18. $r = 1 - \cos\theta$, $\quad \theta \in [0, 2\pi]$.   (polar coordinates)
19. $r = \sin 3\theta$, $\quad \theta \in [0, \pi]$.   (polar coordinates)
20. $y^4 = x^3$, $\quad y \leq 0$.

21. Find an equation in x and y for the curve $\mathbf{r}(t) = t^3\mathbf{i} + t^2\mathbf{j}$. Draw the curve. Does the curve have a tangent vector at the origin? If so, what is the unit tangent vector?

22. (a) Show that the curve

$$\mathbf{r}(t) = (t^2 - t + 1)\mathbf{i} + (t^3 - t + 2)\mathbf{j} + (\sin \pi t)\mathbf{k}$$

intersects itself at $P(1, 2, 0)$ by finding numbers $t_1 < t_2$ for which P is the tip of both $\mathbf{r}(t_1)$ and $\mathbf{r}(t_2)$.
(b) Find the unit tangents at $P(1, 2, 0)$ first taking $t = t_1$, then taking $t = t_2$.

23. Find the point(s) at which the twisted cubic

$$\mathbf{r}(t) = t\mathbf{i} + t^2\mathbf{j} + t^3\mathbf{k}$$

intersects the plane $4x + 2y + z = 24$. What is the angle of intersection between the curve and the plane?

24. (a) Find the unit tangent and the principal normal at an arbitrary point of the ellipse

$$\mathbf{r}(t) = a \cos t\, \mathbf{i} + b \sin t\, \mathbf{j}.$$

(b) Write vector equations for the tangent line and the normal line at the tip of $\mathbf{r}(\tfrac{1}{4}\pi)$.

At the point on the curve corresponding to the indicated value of t, find the unit tangent vector, the principal normal vector, and an equation in x, y, z for the osculating plane.

25. $\mathbf{r}(t) = \cos 2t\, \mathbf{i} + \sin 2t\, \mathbf{j} + t\mathbf{k}$ at $t = \tfrac{1}{4}\pi$.   **26.** $\mathbf{r}(t) = t\mathbf{i} + 2t\mathbf{j} + t^2\mathbf{k}$ at $t = 2$.
27. $\mathbf{r}(t) = t\mathbf{i} + t^2\mathbf{j} + t^3\mathbf{k}$ at $t = 1$.   **28.** $\mathbf{r}(t) = \cos 3t\, \mathbf{i} + t\mathbf{j} - \sin 3t\, \mathbf{k}$ at $t = \tfrac{1}{3}\pi$.

29. Let $\mathbf{r} = \mathbf{r}(t)$, $t \in [a, b]$ and set

$$\mathbf{R}(u) = \mathbf{r}(a + b - u), \qquad u \in [a, b].$$

Show that this change of parameter changes the sign of the unit tangent but does not alter the principal normal.   HINT: Let P be the tip of $\mathbf{R}(u) = \mathbf{r}(a + b - u)$. At that point, $\mathbf{R}$ produces a unit tangent $\mathbf{T}_1(u)$ and a principal normal $\mathbf{N}_1(u)$. At that same point, $\mathbf{r}$ produces a unit tangent $\mathbf{T}(a + b - u)$ and a principal normal $\mathbf{N}(a + b - u)$.

30. Show that two vector functions that differ by a sense-preserving change of parameter take on exactly the same values in exactly the same order. That is, set

$$\mathbf{r} = \mathbf{r}(t), \qquad t \in I.$$

Assume that ϕ maps an interval J onto the interval I and that $\phi'(u) > 0$ for all $u \in J$. Set

$$\mathbf{R}(u) = \mathbf{r}(\phi(u)),$$

and show that $\mathbf{R}$ and $\mathbf{r}$ take on exactly the same values in exactly the same order.

31. Show that the unit tangent vector, the principal normal vector, and the osculating plane are left invariant (left unchanged) by every sense-preserving change of parameter.

14.4   ARC LENGTH

Arc length for space curves is defined exactly as for plane curves. In Figure 14.4.1 we have sketched a space curve C and a polygonal approximation to it. The *length* of C is by definition the least upper bound of the lengths of such polygonal approximations. (Recall Definition 10.9.1.)

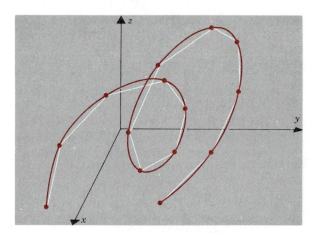

Figure 14.4.1

Earlier, working with plane curves, we used the arc length formula

$$L(C) = \int_a^b \sqrt{[x'(t)]^2 + [y'(t)]^2} \, dt.$$

For space curves the formula becomes

$$L(C) = \int_a^b \sqrt{[x'(t)]^2 + [y'(t)]^2 + [z'(t)]^2} \, dt.$$

In vector notation both formulas can be written

$$L(C) = \int_a^b \|\mathbf{r}'(t)\| \, dt.$$

We prove the result in this form.

THEOREM 14.4.1

The length of a continuously differentiable curve

$$C: \quad \mathbf{r} = \mathbf{r}(t), \quad t \in [a, b]$$

is given by the formula

$$L(C) = \int_a^b \|\mathbf{r}'(t)\| \, dt.$$

Proof. First we show that

$$L(C) \leq \int_a^b \|\mathbf{r}'(t)\| \, dt.$$

To do this, we begin with an arbitrary partition P of $[a, b]$:

$$P = \{a = t_0, \ \ldots, t_{i-1}, t_i, \ \ldots, t_n = b\}.$$

Such a partition gives rise to a finite number of points of C:

$$\mathbf{r}(a) = \mathbf{r}(t_0), \ \ldots, \mathbf{r}(t_{i-1}), \mathbf{r}(t_i), \ \ldots, \mathbf{r}(t_n) = \mathbf{r}(b)$$

and thus to an inscribed polygonal path of total length

$$L_P = \sum_{i=1}^n \|\mathbf{r}(t_i) - \mathbf{r}(t_{i-1})\|. \qquad \text{(Figure 14.4.2)}$$

For each i,

$$\mathbf{r}(t_i) - \mathbf{r}(t_{i-1}) = \int_{t_{i-1}}^{t_i} \mathbf{r}'(t) \, dt.$$

This gives

$$\|\mathbf{r}(t_i) - \mathbf{r}(t_{i-1})\| = \left\| \int_{t_{i-1}}^{t_i} \mathbf{r}'(t) \, dt \right\| \overset{\text{by (14.1.10)}}{\leq} \int_{t_{i-1}}^{t_i} \|\mathbf{r}'(t)\| \, dt$$

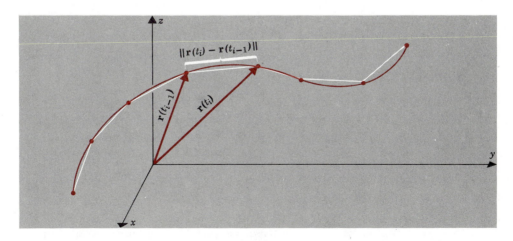

Figure 14.4.2

and thus

$$L_P = \sum_{i=1}^{n} \|\mathbf{r}(t_i) - \mathbf{r}(t_{i-1})\| \le \sum_{i=1}^{n} \int_{t_{i-1}}^{t_i} \|\mathbf{r}'(t)\| \, dt = \int_{a}^{b} \|\mathbf{r}'(t)\| \, dt.$$

From the arbitrariness of P, we know that the inequality

$$L_P \le \int_{a}^{b} \|\mathbf{r}'(t)\| \, dt$$

must hold for all the L_P. This makes the integral on the right an upper bound for all the L_P. Since $L(C)$ is the *least* upper bound of all the L_P, we can conclude right now that

$$L(C) \le \int_{a}^{b} \|\mathbf{r}'(t)\| \, dt.$$

To show equality, we need to know that arc length is additive: that, with P, Q, R as in Figure 14.4.3, the arc length from P to Q, plus the arc length from Q to R, equals the arc length from P to R.† A proof of the additivity of arc length is given in the supplement at the end of this section. From this point on we will assume that arc length is additive.

In Figure 14.4.4 we display the initial vector $\mathbf{r}(a)$, a general radius vector $\mathbf{r}(t)$, and a nearby vector $\mathbf{r}(t + h)$. Now set

$$s(t) = \text{length of the curve from } \mathbf{r}(a) \text{ to } \mathbf{r}(t),$$
$$s(t + h) = \text{length of the curve from } \mathbf{r}(a) \text{ to } \mathbf{r}(t + h).$$

Then, by the additivity of arc length (remember we are assuming that),

$$s(t + h) - s(t) = \text{length of the curve from } \mathbf{r}(t) \text{ to } \mathbf{r}(t + h).$$

From what we have shown already, you can see that

$$\|\mathbf{r}(t + h) - \mathbf{r}(t)\| \le s(t + h) - s(t) \le \int_{t}^{t+h} \|\mathbf{r}'(u)\| \, du.$$

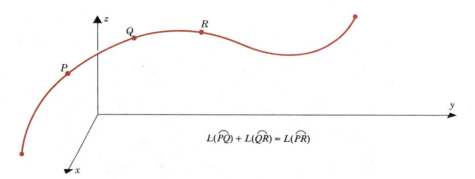

$$L(\widehat{PQ}) + L(\widehat{QR}) = L(\widehat{PR})$$

Figure 14.4.3

---

† It is obvious that we want arc length to have this property. What we have to verify is that arc length *as we have defined it* has this property.

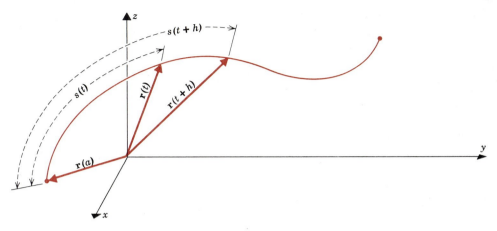

Figure 14.4.4

Dividing this inequality by h (which we are taking as positive), we get

$$\left\| \frac{\mathbf{r}(t+h) - \mathbf{r}(t)}{h} \right\| \leq \frac{s(t+h) - s(t)}{h} \leq \frac{1}{h} \int_t^{t+h} \|\mathbf{r}'(u)\| \, du.$$

As $h \to 0^+$, the left-hand side tends to $\|\mathbf{r}'(t)\|$ and, by the First Mean-Value Theorem For Integrals (Theorem 5.11.1), so does the right-hand side:

$$\frac{1}{h} \int_t^{t+h} \|\mathbf{r}'(u)\| \, du = \frac{1}{h} \|\mathbf{r}'(c_h)\|(t+h-t) = \|\mathbf{r}'(c_h)\| \to \|\mathbf{r}'(t)\|.$$
$$c_h \in (t, t+h) \underrightarrow{\qquad}$$

Therefore

$$\lim_{h \to 0^+} \frac{s(t+h) - s(t)}{h} = \|\mathbf{r}'(t)\|.$$

By taking $h < 0$ and proceeding in a similar manner, one can show that

$$\lim_{h \to 0^-} \frac{s(t+h) - s(t)}{h} = \|\mathbf{r}'(t)\|.$$

It follows that

$$s'(t) = \lim_{h \to 0} \frac{s(t+h) - s(t)}{h} = \|\mathbf{r}'(t)\|.$$

Recall now that $s(t) = $ length of the curve from $\mathbf{r}(a)$ to $\mathbf{r}(t)$. Obviously, then, $s(a) = 0$. This, together with the fact that $s'(t) = \|\mathbf{r}'(t)\|$, implies that

$$s(t) = \text{length of the curve from } \mathbf{r}(a) \text{ to } \mathbf{r}(t) = \int_a^t \|\mathbf{r}'(u)\| \, du.$$

The total length of C is therefore

$$s(b) = \int_a^b \|\mathbf{r}'(t)\| \, dt. \quad \square$$

Problem 1. Find the length of the curve

$$\mathbf{r}(t) = 2 \cos t \, \mathbf{i} + 2 \sin t \, \mathbf{j} + t^2 \mathbf{k} \quad \text{from } t = 0 \text{ to } t = 1$$

and compare it to the straight-line distance between the endpoints of the curve.

Solution

$$\mathbf{r}'(t) = -2 \sin t \, \mathbf{i} + 2 \cos t \, \mathbf{j} + 2t \mathbf{k}.$$

$$\|\mathbf{r}'(t)\| = \sqrt{4 \sin^2 t + 4 \cos^2 t + 4t^2} = 2\sqrt{\sin^2 t + \cos^2 t + t^2} = 2\sqrt{1 + t^2}.$$

$$L(C) = \int_0^1 \|\mathbf{r}'(t)\| \, dt = \int_0^1 2\sqrt{1 + t^2} \, dt = \left[t\sqrt{1 + t^2} + \ln\left(t + \sqrt{1 + t^2}\right) \right]_0^1$$

$$= \sqrt{2} + \ln\left(1 + \sqrt{2}\right) \cong 2.30.$$

The curve begins at $\mathbf{r}(0) = 2\mathbf{i}$ and ends at $\mathbf{r}(1) = (2 \cos 1)\mathbf{i} + (2 \sin 1)\mathbf{j} + \mathbf{k}$. The distance between these points is

$$\|\mathbf{r}(1) - \mathbf{r}(0)\| = \sqrt{[2(1 - \cos 1)]^2 + (2 \sin 1)^2 + 1} \cong 2.06.$$

The curve is about 12% longer than the straight-line distance between the endpoints of the curve. ☐

EXERCISES 14.4

Find the length of the curve.

1. $\mathbf{r}(t) = a \cos t \, \mathbf{i} + a \sin t \, \mathbf{j} + bt\mathbf{k}$ from $t = 0$ to $t = 2\pi$.
2. $\mathbf{r}(t) = t\mathbf{i} + \frac{2}{3}\sqrt{2} \, t^{3/2}\mathbf{j} + \frac{1}{2}t^2\mathbf{k}$ from $t = 0$ to $t = 2$.
3. $\mathbf{r}(t) = t\mathbf{i} + \ln(\sec t) \, \mathbf{j} + 3\mathbf{k}$ from $t = 0$ to $t = \frac{1}{4}\pi$.
4. $\mathbf{r}(t) = \tan^{-1} t \, \mathbf{i} + \frac{1}{2}\ln(1 + t^2) \, \mathbf{j}$ from $t = 0$ to $t = 1$.
5. $\mathbf{r}(t) = t^3\mathbf{i} + t^2\mathbf{j}$ from $t = 0$ to $t = 1$.
6. $\mathbf{r}(t) = t\mathbf{i} + \mathbf{j} + (\frac{1}{6}t^3 + \frac{1}{2}t^{-1})\mathbf{k}$ from $t = 1$ to $t = 3$.
7. $\mathbf{r}(t) = e^t[\cos t \, \mathbf{i} + \sin t \, \mathbf{j}]$ from $t = 0$ to $t = \pi$.
8. $\mathbf{r}(t) = 3t \cos t \, \mathbf{i} + 3t \sin t \, \mathbf{j} + 4t\mathbf{k}$ from $t = 0$ to $t = 4$.
9. $\mathbf{r}(t) = 2t\mathbf{i} + (t^2 - 2)\mathbf{j} + (1 - t^2)\mathbf{k}$ from $t = 0$ to $t = 2$.
10. $\mathbf{r}(t) = t^2\mathbf{i} + (t^2 - 2)\mathbf{j} + (1 - t^2)\mathbf{k}$ from $t = 0$ to $t = 2$.

11. (*Important*) Let $\mathbf{r}(t) = x(t)\mathbf{i} + y(t)\mathbf{j} + z(t)\mathbf{k}$, $t \in [a, b]$ be a continuously differentiable curve. Show that, if s is the length of the curve from the tip of $\mathbf{r}(a)$ to the tip of $\mathbf{r}(t)$, then

(14.4.2)
$$\frac{ds}{dt} = \sqrt{\left(\frac{dx}{dt}\right)^2 + \left(\frac{dy}{dt}\right)^2 + \left(\frac{dz}{dt}\right)^2}.$$

12. Use vector methods to show that, if $y = f(x)$ has a continuous first derivative, then the length of the graph from $x = a$ to $x = b$ is given by the integral

$$\int_a^b \sqrt{1 + [f'(x)]^2} \, dx.$$

13. (*Important*) Let $y = f(x)$, $x \in [a, b]$ be a continuously differentiable function. Show that, if s is the length of the graph from $(a, f(a))$ to $(x, f(x))$, then

(14.4.3)

$$\frac{ds}{dx} = \sqrt{1 + \left(\frac{dy}{dx}\right)^2}.$$

14. Let C_1 be the curve

$$\mathbf{r}(t) = (t - \ln t)\mathbf{i} + (t + \ln t)\mathbf{j}, \qquad 1 \le t \le e$$

and let C_2 be the graph of

$$y = e^x, \qquad 0 \le x \le 1.$$

Find a relation between the length of C_1 and the length of C_2.

15. Show that the length of a continuously differentiable curve is left invariant (left unchanged) by a sense-preserving (or a sense-reversing) change of parameter. That is, set

$$\mathbf{r} = \mathbf{r}(t), \qquad t \in [a, b].$$

Assume that ϕ maps $[c, d]$ onto $[a, b]$ and that ϕ' is positive (or negative) and continuous on $[a, b]$. Set $\mathbf{R}(u) = \mathbf{r}(\phi(u))$ and show that the length of the curve as computed from $\mathbf{R}$ is the length of the curve as computed from $\mathbf{r}$.

Estimate within 0.001 the length of the curve from $t = 0$ to $t = \frac{1}{2}$. Evaluate the integral by expanding the integrand in a power series.

16. $\mathbf{r}(t) = \frac{2}{5}t^{5/2}\mathbf{j} + t\mathbf{k}.$
17. $\mathbf{r}(t) = t\mathbf{i} + \frac{1}{3}t^3\mathbf{j}.$

*SUPPLEMENT TO SECTION 14.4

The Additivity of Arc Length We wish to show that with P, Q, R as in Figure 14.4.3,

$$L(\widehat{PQ}) + L(\widehat{QR}) = L(\widehat{PR}).$$

Let γ_1 be an arbitrary polygonal path inscribed in $\widehat{PQ}$ and γ_2 an arbitrary polygonal path inscribed in $\widehat{QR}$. Then $\gamma_1 \cup \gamma_2$ is a polygonal path inscribed in $\widehat{PR}$. Since

$$L(\gamma_1) + L(\gamma_2) = L(\gamma_1 \cup \gamma_2) \qquad \text{and} \qquad L(\gamma_1 \cup \gamma_2) \le L(\widehat{PR}),$$

we have

$$L(\gamma_1) + L(\gamma_2) \le L(\widehat{PR}) \qquad \text{and thus} \qquad L(\gamma_1) \le L(\widehat{PR}) - L(\gamma_2).$$

From the arbitrariness of γ_1 we can conclude that $L(\widehat{PR}) - L(\gamma_2)$ is an upper bound for the set of all lengths of polygonal paths inscribed in $\widehat{PQ}$. Since $L(\widehat{PQ})$ is the *least* upper bound of this set, we have

$$L(\widehat{PQ}) \le L(\widehat{PR}) - L(\gamma_2).$$

It follows that

$$L(\gamma_2) \le L(\widehat{PR}) - L(\widehat{PQ}).$$

Arguing as we did with γ_1, we can conclude that

$$L(\widehat{QR}) \le L(\widehat{PR}) - L(\widehat{PQ}).$$

This gives

$$L(\widehat{PQ}) + L(\widehat{QR}) \le L(\widehat{PR}).$$

We now set out to prove that $L(\widehat{PR}) \leq L(\widehat{PQ}) + L(\widehat{QR})$. To do this, we need only take $\gamma = \overline{T_0 T_1} \cup \cdots \cup \overline{T_{n-1} T_n}$ as an arbitrary polygonal path inscribed in $\widehat{PR}$ and show that

$$L(\gamma) \leq L(\widehat{PQ}) + L(\widehat{QR}).$$

If Q is one of the T_i, say $Q = T_k$, then

$$\gamma_1 = \overline{T_0 T_1} \cup \cdots \cup \overline{T_{k-1} T_k} \quad \text{is inscribed in } \widehat{PQ}$$

and

$$\gamma_2 = \overline{T_k T_{k+1}} \cup \cdots \cup \overline{T_{n-1} T_n} \quad \text{is inscribed in } \widehat{QR}.$$

Moreover, $L(\gamma) = L(\gamma_1) + L(\gamma_2)$, so that

$$L(\gamma) \leq L(\widehat{PQ}) + L(\widehat{QR}).$$

If Q is none of the T_i, then Q lies between two consecutive points T_k and T_{k+1}. Set

$$\gamma' = \overline{T_0 T_1} \cup \cdots \cup \overline{T_k Q} \cup \overline{Q T_{k+1}} \cup \cdots \cup \overline{T_{n-1} T_n}.$$

Since

$$d(T_k, T_{k+1}) \leq d(T_k, Q) + d(Q, T_{k+1}),$$

we have

$$L(\gamma) \leq L(\gamma').$$

Proceed as before and you will see that

$$L(\gamma') \leq L(\widehat{PQ}) + L(\widehat{QR}),$$

and once again

$$L(\gamma) \leq L(\widehat{PQ}) + L(\widehat{QR}). \quad \square$$

14.5 CURVILINEAR MOTION; VECTOR CALCULUS IN MECHANICS

Introduction

The tools we have just developed find their premier application in Newtonian mechanics, the study of bodies in motion subject to Newton's laws. The heart of Newton's mechanics is his Second Law of Motion:

$$\text{force} = \text{mass} \times \text{acceleration}.$$

We have worked with Newton's second law, but only in a very restricted context: motion along a coordinate line under the influence of a force directed along that same line. In that special setting Newton's law was written as a scalar equation: $F = ma$. In general, objects do not move along straight lines (they move along curved paths) and the forces on them vary in direction. What happens to Newton's second law then? It becomes the vector equation

$$\mathbf{F} = m\mathbf{a}.$$

This is Newton's Second Law of Motion in its full glory. To understand this law and be able to use it effectively, you must learn to look at curvilinear motion from a vector viewpoint.

Curvilinear Motion from a Vector Viewpoint

We can describe the position of a moving object at time t by a radius vector $\mathbf{r}(t)$. As t ranges over a time interval I, the object traces out some path

$$C: \quad \mathbf{r}(t) = x(t)\mathbf{i} + y(t)\mathbf{j} + z(t)\mathbf{k}, \quad t \in I.$$

If $\mathbf{r}$ is twice differentiable, we can form $\mathbf{r}'(t)$ and $\mathbf{r}''(t)$. In this context these vectors have special names and special significance: $\mathbf{r}'(t)$ is called the *velocity* of the object at time t, and $\mathbf{r}''(t)$ is called the *acceleration*. In symbols, we have

(14.5.1) $$\mathbf{r}'(t) = \mathbf{v}(t) \quad \text{and} \quad \mathbf{r}''(t) = \mathbf{v}'(t) = \mathbf{a}(t).$$

There should be nothing surprising about this. As before, velocity is the time rate of change of position and acceleration the time rate of change of velocity.

Since $\mathbf{v}(t) = \mathbf{r}'(t)$, the velocity vector, when not $\mathbf{0}$, is tangent to the path of the motion at the tip of $\mathbf{r}(t)$. (Section 14.3) The direction of the velocity vector at time t thus gives the direction of the motion at time t. (Figure 14.5.1)

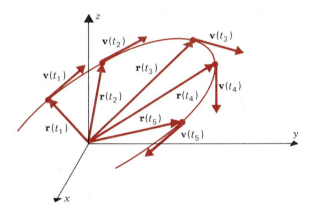

Figure 14.5.1

The magnitude of the velocity vector is called the *speed* of the object:

(14.5.2) $$\|\mathbf{v}(t)\| = \text{the speed at time } t.$$

The reasoning is as follows: during a time interval $[t_0, t]$ the object moves along its path from $\mathbf{r}(t_0)$ to $\mathbf{r}(t)$ for a total distance

$$s(t) = \int_{t_0}^{t} \|\mathbf{r}'(u)\| \, du. \qquad \text{(Section 14.4)}$$

Differentiating with respect to t, we have

$$s'(t) = \|\mathbf{r}'(t)\|.$$

The magnitude of the velocity vector is thus *the rate of change of arc distance with respect to time*. This is why we call it the speed of the object.

Speed is commonly denoted by the letter v. In symbols

(14.5.3) $$s'(t) = \|\mathbf{v}(t)\| = v(t).$$

Motion Along a Straight Line

The position at time t is given by a function of the form

$$\mathbf{r}(t) = \mathbf{r}_0 + f(t)\mathbf{d}, \qquad \mathbf{d} \neq \mathbf{0}.$$

For convenience we take $\mathbf{d}$ as a unit vector.

The velocity and acceleration vectors are both directed along the line of the motion:

$$\mathbf{v}(t) = f'(t)\mathbf{d} \qquad \text{and} \qquad \mathbf{a}(t) = f''(t)\mathbf{d}.$$

The speed is $|f'(t)|$:

$$v(t) = \|f'(t)\mathbf{d}\| = |f'(t)|\,\|\mathbf{d}\| = |f'(t)|,$$

and the magnitude of the acceleration is $|f''(t)|$:

$$\|\mathbf{a}(t)\| = \|f''(t)\mathbf{d}\| = |f''(t)|\,\|\mathbf{d}\| = |f''(t)|.$$

Circular Motion About the Origin

The position function can be written

$$\mathbf{r}(t) = r[\cos\theta(t)\,\mathbf{i} + \sin\theta(t)\,\mathbf{j}].$$

Here $\theta'(t)$ gives the time rate of change of the central angle θ. If $\theta'(t) > 0$, the motion is counterclockwise; if $\theta'(t) < 0$, the motion is clockwise. We call $\theta'(t)$ the *angular velocity* and $|\theta'(t)|$ the *angular speed*.

Uniform circular motion is circular motion with constant angular speed ω ($\omega > 0$). The position function for uniform circular motion in the counterclockwise direction can be written

$$\mathbf{r}(t) = r(\cos\omega t\,\mathbf{i} + \sin\omega t\,\mathbf{j}).$$

Differentiation gives

$$\mathbf{v}(t) = r\omega(-\sin\omega t\,\mathbf{i} + \cos\omega t\,\mathbf{j}),$$
$$\mathbf{a}(t) = -r\omega^2(\cos\omega t\,\mathbf{i} + \sin\omega t\,\mathbf{j}) = -r\omega^2\mathbf{r}(t).$$

The acceleration is directed along the line of the radius vector toward the center of the circle and is therefore perpendicular to the velocity vector, which, as always, is tangential. As you can verify, the speed is $r\omega$ and the magnitude of acceleration is $r\omega^2$. ◻

Motion along a circular helix is a combination of circular motion and rectilinear motion.

Problem 1. A particle moves along a circular helix (Figure 14.3.6) with position at time t given by the function

$$\mathbf{r}(t) = a \cos \omega t \, \mathbf{i} + a \sin \omega t \, \mathbf{j} + b\omega t \, \mathbf{k}. \qquad (a > 0, b > 0, \omega > 0)$$

For each time t, find

(a) the velocity of the particle.   (b) the speed.   (c) the acceleration.
(d) the magnitude of acceleration.
(e) the angle between the velocity vector and the acceleration vector.

Solution

(a) velocity: $\mathbf{v}(t) = \mathbf{r}'(t) = -a\omega \sin \omega t \, \mathbf{i} + a\omega \cos \omega t \, \mathbf{j} + b\omega \mathbf{k}.$

(b) speed: $v(t) = \|\mathbf{v}(t)\| = \sqrt{a^2\omega^2 \sin^2 \omega t + a^2\omega^2 \cos^2 \omega t + b^2\omega^2}$

$$= \sqrt{a^2\omega^2 + b^2\omega^2} = \omega\sqrt{a^2 + b^2};$$

(The speed is thus constant.)

(c) acceleration: $\mathbf{a}(t) = \mathbf{v}'(t) = -a\omega^2 \cos \omega t \, \mathbf{i} - a\omega^2 \sin \omega t \, \mathbf{j}$

$$= -a\omega^2(\cos \omega t \, \mathbf{i} + \sin \omega t \, \mathbf{j}).$$

(Since the speed is constant, the acceleration comes entirely from the change in direction.)

(d) magnitude of acceleration: $\|\mathbf{a}(t)\| = a\omega^2.$

(e) angle between $\mathbf{v}(t)$ and $\mathbf{a}(t)$:

$$\cos \theta = \frac{\mathbf{v}(t)}{\|\mathbf{v}(t)\|} \cdot \frac{\mathbf{a}(t)}{\|\mathbf{a}(t)\|}$$

$$= \left[\frac{-a\omega \sin \omega t \, \mathbf{i} + a\omega \cos \omega t \, \mathbf{j} + b\omega \mathbf{k}}{\omega\sqrt{a^2 + b^2}}\right] \cdot \left[\frac{-a\omega^2(\cos \omega t \, \mathbf{i} + \sin \omega t \, \mathbf{j})}{a\omega^2}\right]$$

$$= \frac{a(\sin \omega t \cos \omega t - \cos \omega t \sin \omega t)}{\sqrt{a^2 + b^2}} = 0;$$

Therefore $\theta = \tfrac{1}{2}\pi.$ (At each point the acceleration vector is perpendicular to the velocity vector.) ▨

An Introduction to Vector Mechanics

We are now ready to work with Newton's Second Law of Motion in its vector form: $\mathbf{F} = m\mathbf{a}$. Since at each time t we have $\mathbf{a}(t) = \mathbf{r}''(t)$, Newton's law can be written

(14.5.4) $\boxed{\mathbf{F}(t) = m\mathbf{a}(t) = m\mathbf{r}''(t).}$

This is a second-order differential equation in t. Much of the rest of this section will be devoted to searching for solutions of this equation in particular situations.

When objects are moving, certain quantities (positions, velocities, etc.) are continually changing. This can make a situation difficult to grasp. In these circumstances it is particularly satisfying to find quantities that do not change. Such quantities are said to be *conserved*. (These conserved quantities are called the *constants of the motion.*) Mathematically we can determine whether or not a quantity is conserved by looking at its time derivative: *the quantity is conserved (is constant) iff its time derivative remains zero.* A *conservation law* is the assertion that in such and such a context a certain quantity does not change.

Momentum

We start with the idea of momentum. The *momentum* **p** of an object is the mass of the object times the velocity of the object:

$$\mathbf{p} = m\mathbf{v}.$$

To indicate the time dependence we write

(14.5.5)
$$\mathbf{p}(t) = m\mathbf{v}(t) = m\mathbf{r}'(t).$$

Differentiation gives

$$\mathbf{p}'(t) = m\mathbf{r}''(t) = \mathbf{F}(t).$$

Thus, *the time derivative of the momentum of an object is the net force on the object.* If the net force on an object is continually zero, the momentum **p**(t) is constant. This is the law of *conservation of momentum*:

(14.5.6)
if the net force on an object is continually zero, then the momentum of the object is conserved.

Angular Momentum

The angular momentum of an object about any given point is a vector quantity that is intended to measure the extent to which the object is circling about that point. If the position of the object at time t is given by the radius vector **r**(t), then the object's *angular momentum about the origin* is defined by the formula

(14.5.7)
$$\mathbf{L}(t) = \mathbf{r}(t) \times \mathbf{p}(t) = \mathbf{r}(t) \times m\mathbf{v}(t).$$

At each time t of a motion, **L**(t) is perpendicular to **r**(t), perpendicular to **v**(t), and oriented so that **r**(t), **v**(t), **L**(t) form a right-handed triple. The magnitude of **L**(t) is given by the relation

$$\|\mathbf{L}(t)\| = \|\mathbf{r}(t)\|\,\|m\mathbf{v}(t)\| \sin \theta(t)$$

where $\theta(t)$ is the angle between **r**(t) and **v**(t). (All this, of course, comes simply from the definition of cross product.)

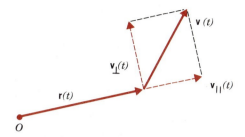

Figure 14.5.2

If $\mathbf{r}(t)$ and $\mathbf{v}(t)$ are not zero, then we can express $\mathbf{v}(t)$ as a vector parallel to $\mathbf{r}(t)$ plus a vector perpendicular to $\mathbf{r}(t)$:

$$\mathbf{v}(t) = \mathbf{v}_{\parallel}(t) + \mathbf{v}_{\perp}(t). \qquad \text{(See Figure 14.5.2.)}$$

Thus

$$\begin{aligned}
\mathbf{L}(t) &= \mathbf{r}(t) \times m\mathbf{v}(t) \\
&= \mathbf{r}(t) \times m[\mathbf{v}_{\parallel}(t) + \mathbf{v}_{\perp}(t)] \\
&= \underbrace{[\mathbf{r}(t) \times m\mathbf{v}_{\parallel}(t)]}_{\mathbf{0}} + [\mathbf{r}(t) \times m\mathbf{v}_{\perp}(t)] \\
&= \mathbf{r}(t) \times m\mathbf{v}_{\perp}(t).
\end{aligned}$$

The component of velocity that is parallel to the radius vector contributes nothing to angular momentum. *The angular momentum comes entirely from the component of velocity that is perpendicular to the radius vector.*

Example 2. In uniform circular motion about the origin,

$$\mathbf{r}(t) = r(\cos \omega t \, \mathbf{i} + \sin \omega t \, \mathbf{j}),$$

the velocity vector $\mathbf{v}(t)$ is always perpendicular to the radius vector $\mathbf{r}(t)$. In this case all of $\mathbf{v}(t)$ contributes to the angular momentum.

We can calculate $\mathbf{L}(t)$ as follows:

$$\begin{aligned}
\mathbf{L}(t) &= \mathbf{r}(t) \times m\mathbf{v}(t) \\
&= [r(\cos \omega t \, \mathbf{i} + \sin \omega t \, \mathbf{j})] \times [mr(-\omega \sin \omega t \, \mathbf{i} + \omega \cos \omega t \, \mathbf{j})] \\
&= mr^2\omega(\cos^2 \omega t + \sin^2 \omega t)\mathbf{k} = mr^2\omega\mathbf{k}.
\end{aligned}$$

The angular momentum is constant and is perpendicular to the xy-plane. If the motion is counterclockwise (if $\omega > 0$), then the angular momentum points up from the xy-plane. If the motion is clockwise (if $\omega < 0$), then the angular momentum points down from the xy-plane. (This is the right-handedness of the cross product coming in.) ◻

Example 3. In uniform straight-line motion with constant velocity **d**,

$$\mathbf{r}(t) = \mathbf{r}_0 + t\mathbf{d},$$

we have

$$\mathbf{L}(t) = \mathbf{r}(t) \times m\mathbf{v}(t) = (\mathbf{r}_0 + t\mathbf{d}) \times m\mathbf{d} = m(\mathbf{r}_0 \times \mathbf{d}).$$

Here again the angular momentum is constant, but all is not quite so simple as it looks. In Figure 14.5.3 you can see the radius vector $\mathbf{r}(t)$, the velocity vector $\mathbf{v}(t) = \mathbf{d}$, and the part of the velocity vector that gives rise to angular momentum, the part perpendicular to $\mathbf{r}(t)$. As before we have called this $\mathbf{v}_\perp(t)$. While $\mathbf{v}(t)$ is constant, $\mathbf{v}_\perp(t)$ is not constant. What happens here is that $\mathbf{r}(t)$ and $\mathbf{v}_\perp(t)$ vary in such a way that the cross product

$$\mathbf{L}(t) = \mathbf{r}(t) \times m\mathbf{v}_\perp(t)$$

remains constant. If, of course, the line of motion passes through the origin, then $\mathbf{v}(t)$ is parallel to $\mathbf{r}(t)$, $\mathbf{v}_\perp(t)$ is zero, and the angular momentum is zero. ■

Figure 14.5.3

Torque

How the angular momentum of an object changes in time depends on the force acting on the object and on the position of the object relative to the origin: since $\mathbf{L}(t) = \mathbf{r}(t) \times m\mathbf{r}'(t)$,

$$\mathbf{L}'(t) = [\mathbf{r}(t) \times m\mathbf{r}''(t)] + \underbrace{[\mathbf{r}'(t) \times m\mathbf{r}'(t)]}_{\mathbf{0}} = \mathbf{r}(t) \times \mathbf{F}(t).$$

The cross product

(14.5.8)
$$\boxed{\boldsymbol{\tau}(t) = \mathbf{r}(t) \times \mathbf{F}(t)}$$

is called the *torque* about the origin.†
    Since $\mathbf{L}'(t) = \boldsymbol{\tau}(t)$, we have the following conservation law:

(14.5.9)
> if the net torque on an object is continually zero,
> then the angular momentum of the object is conserved.

---

† The symbol τ is the Greek letter tau. The word *torque* comes from the Latin word *torquere*, "to twist".

A force $\mathbf{F}(t)$ is called a *central force* (or a *radial force*) iff $\mathbf{F}(t)$ is always parallel to $\mathbf{r}(t)$. (Gravitational force, for example, is a central force.) For a central force, the cross product $\mathbf{r}(t) \times \mathbf{F}(t)$ is always zero. Thus a central force produces no torque about the origin. As you will see, this places severe restrictions on the kind of motion possible under a central force.

THEOREM 14.5.10

If an object moves under a central force and the angular momentum $\mathbf{L}$ is different from zero, then

1. the object is confined to the plane that passes through the origin and is perpendicular to $\mathbf{L}$ and

2. the radius vector of the object sweeps out equal areas in equal times.

(This theorem plays an important role in astronomy and is embodied in Kepler's three celebrated laws of planetary motion. We will study Kepler's laws in Section *14.6.)

Proof of Theorem 14.5.10. The first assertion is easy to verify: all the radius vectors pass through the origin and (by the very definition of angular momentum) they are all perpendicular to the constant vector $\mathbf{L}$.

To verify the second assertion we introduce a right-handed coordinate system O-xyz setting the xy-plane as the plane of the motion, the positive z-axis pointing along $\mathbf{L}$. On the xy-plane we introduce polar coordinates r, θ. Thus, at time t, the object has some position $[r(t), \theta(t)]$.

Let's denote by $A(t)$ the area swept out by the radius vector from some fixed time t_0 up to time t. Our task is to show that $A'(t)$ is constant.

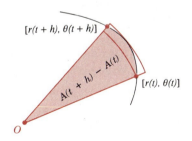

Figure 14.5.4

The area swept out during the time interval $[t, t + h]$ is simply

$$A(t + h) - A(t).$$

Assuming that the motion takes place in the direction of increasing polar angle θ (see Figure 14.5.4), we have, with obvious notation,

$$\tfrac{1}{2} \min [r(t)]^2 \cdot [\theta(t + h) - \theta(t)] \le A(t + h) - A(t) \le \tfrac{1}{2} \max [r(t)]^2 \cdot [\theta(t + h) - \theta(t)].$$

$$\underbrace{\phantom{\tfrac{1}{2} \min [r(t)]^2 \cdot [\theta(t + h) - \theta(t)]}}_{\text{area of inner sector}} \qquad\qquad \underbrace{\phantom{\tfrac{1}{2} \max [r(t)]^2 \cdot [\theta(t + h) - \theta(t)]}}_{\text{area of outer sector}}$$

Divide through by h, take the limit as h tends to 0, and you will see that

(*) $$A'(t) = \tfrac{1}{2}[r(t)]^2\theta'(t).$$

Now

$$\mathbf{r}(t) = r(t)[\cos\theta(t)\,\mathbf{i} + \sin\theta(t)\,\mathbf{j}],$$
$$\mathbf{v}(t) = r'(t)[\cos\theta(t)\,\mathbf{i} + \sin\theta(t)\,\mathbf{j}] + r(t)\theta'(t)[-\sin\theta(t)\,\mathbf{i} + \cos\theta(t)\,\mathbf{j}]$$
$$= [r'(t)\cos\theta(t) - r(t)\theta'(t)\sin\theta(t)]\mathbf{i} + [r'(t)\sin\theta(t) + r(t)\theta'(t)\cos\theta(t)]\mathbf{j}.$$

A calculation that you can carry out yourself shows that

$$\mathbf{L} = \mathbf{r}(t)\times m\mathbf{v}(t) = mr^2(t)\theta'(t)\mathbf{k}.$$

Since $\mathbf{L}$ is constant, $r^2(t)\theta'(t)$ is constant. Thus, by (*), $A'(t)$ is constant:

$$A'(t) = L/2m \qquad \text{where} \quad L = \|\mathbf{L}\|. \quad \blacksquare$$

Initial-Value Problems

In physics one tries to make predictions about the future on the basis of current information and a knowledge of the forces at work. In the case of an object in motion, the task can be to determine $\mathbf{r}(t)$ for all t given the law of force and some "initial conditions." Frequently the initial conditions give the position and velocity of the object at some time t_0. The problem then is to solve the differential equation $\mathbf{F} = m\mathbf{r}''$ subject to conditions of the form $\mathbf{r}(t_0) = \mathbf{r}_0$, $\mathbf{v}(t_0) = \mathbf{v}_0$.

By far the simplest problem of this sort concerns a *free particle*, an object on which there is no net force.

Problem 4. At time t_0 a free particle has position $\mathbf{r}(t_0) = \mathbf{r}_0$ and velocity $\mathbf{v}(t_0) = \mathbf{v}_0$. Find $\mathbf{r}(t)$ for all t.

Solution. Since there is no net force on the object, the acceleration is zero and the velocity is constant. Since $\mathbf{v}(t_0) = \mathbf{v}_0$,

$$\mathbf{v}(t) = \mathbf{v}_0 \qquad \text{for all } t.$$

Integration with respect to t gives

$$\mathbf{r}(t) = t\mathbf{v}_0 + \mathbf{c}$$

where $\mathbf{c}$, the constant of integration, is a vector that we can determine from the initial position. The initial position $\mathbf{r}(t_0) = \mathbf{r}_0$ gives

$$\mathbf{r}_0 = t_0\mathbf{v}_0 + \mathbf{c} \qquad \text{and therefore} \qquad \mathbf{c} = \mathbf{r}_0 - t_0\mathbf{v}_0.$$

Using this value for $\mathbf{c}$ in our equation for $\mathbf{r}(t)$, we have

$$\mathbf{r}(t) = t\mathbf{v}_0 + (\mathbf{r}_0 - t_0\mathbf{v}_0),$$

which we write as

$$\mathbf{r}(t) = \mathbf{r}_0 + (t - t_0)\mathbf{v}_0.$$

This is the equation of a straight line with direction vector $\mathbf{v}_0$. Free particles travel in straight lines with constant velocity. (We have tacitly assumed that $\mathbf{v}_0 \neq \mathbf{0}$. If $\mathbf{v}_0 = \mathbf{0}$, the particle remains at rest at $\mathbf{r}_0$.) ☐

Problem 5. An object of mass m is subject to a force of the form

$$\mathbf{F}(t) = -m\omega^2 \mathbf{r}(t) \qquad \text{with } \omega > 0.$$

Find $\mathbf{r}(t)$ for all t given that

$$\mathbf{r}(0) = a\mathbf{i} \quad \text{and} \quad \mathbf{v}(0) = \omega a\mathbf{j} \qquad \text{with } a > 0.$$

Solution. The force is a vector version of the restoring force exerted by a linear spring. (Hooke's law) Since the force is central, the angular momentum of the object, $\mathbf{L}(t) = \mathbf{r}(t) \times m\mathbf{v}(t)$, is conserved. So $\mathbf{L}(t)$ is constantly equal to the value it had at time $t = 0$: for all t

$$\mathbf{L}(t) = \mathbf{L}(0) = \mathbf{r}(0) \times m\mathbf{v}(0) = a\mathbf{i} \times m\omega a\mathbf{j} = ma^2\omega\,\mathbf{k}.$$

From our earlier discussion (Theorem 14.5.10) we can conclude that the motion takes place in the plane that passes through the origin and is perpendicular to $\mathbf{k}$. This is the xy-plane. Thus we can write

$$\mathbf{r}(t) = x(t)\mathbf{i} + y(t)\mathbf{j}.$$

Since $\mathbf{F}(t) = m\mathbf{r}''(t)$, the force equation can be written $\mathbf{r}''(t) = -\omega^2 \mathbf{r}(t)$. In terms of components we have

$$x''(t) = -\omega^2 x(t), \qquad y''(t) = -\omega^2 y(t).$$

These are the equations of simple harmonic motion. By (7.8.3) the solutions of these equations take the form

$$x(t) = A_1 \sin(\omega t + \phi_1), \qquad y(t) = A_2 \sin(\omega t + \phi_2).$$

To evaluate the constants we use the initial conditions. That $\mathbf{r}(0) = a\mathbf{i}$ means that $x(0) = a$ and $y(0) = 0$. So

(*) $$A_1 \sin \phi_1 = a, \qquad A_2 \sin \phi_2 = 0.$$

That $\mathbf{v}(0) = \omega a\mathbf{j}$ means that $x'(0) = 0$ and $y'(0) = \omega a$. Since

$$x'(t) = \omega A_1 \cos(\omega t + \phi_1) \quad \text{and} \quad y'(t) = \omega A_2 \cos(\omega t + \phi_2),$$

we have

(**) $$\omega A_1 \cos \phi_1 = 0, \qquad \omega A_2 \cos \phi_2 = \omega a.$$

Conditions (*) and (**) are met by setting $A_1 = a$, $A_2 = a$, $\phi_1 = \frac{1}{2}\pi$, $\phi_2 = 0$. Thus

$$x(t) = a \sin(\omega t + \tfrac{1}{2}\pi) = a \cos \omega t, \qquad y(t) = a \sin \omega t.$$

The vector equation reads

$$\mathbf{r}(t) = a \cos \omega t\,\mathbf{i} + a \sin \omega t\,\mathbf{j}.$$

The object moves in a circle of radius a about the origin with constant angular velocity ω. ☐

Problem 6.   A particle of charge q in a magnetic field $\mathbf{B}$ is subject to the force

$$\mathbf{F}(t) = \frac{q}{c}\,[\mathbf{v}(t) \times \mathbf{B}(t)]$$

where c is the speed of light and $\mathbf{v}$ is the velocity of the particle. Given that $\mathbf{r}(0) = \mathbf{r}_0$ and $\mathbf{v}(0) = \mathbf{v}_0$, find the path of the particle in the constant magnetic field $\mathbf{B}(t) = B_0\mathbf{k}$, $B_0 \neq 0$.

Solution.   There is no conservation law that we can conveniently appeal to here. Neither momentum nor angular momentum are conserved: the force is not zero and it is not central. We will start directly with Newton's $\mathbf{F} = m\mathbf{r}''$.

Since $\mathbf{r}'' = \mathbf{v}'$, we have

$$m\,\mathbf{v}'(t) = \frac{q}{c}\,[\mathbf{v}(t) \times B_0\mathbf{k}],$$

which we can write as

$$\mathbf{v}'(t) = \frac{qB_0}{mc}\,[\mathbf{v}(t) \times \mathbf{k}].$$

To simplify notation we set $qB_0/mc = \omega$. We then have

$$\mathbf{v}'(t) = \omega[\mathbf{v}(t) \times \mathbf{k}].$$

Placing $\mathbf{v}(t) = v_1(t)\mathbf{i} + v_2(t)\mathbf{j} + v_3(t)\mathbf{k}$ in this last equation and working out the cross product, we find that

$$v_1'(t)\mathbf{i} + v_2'(t)\mathbf{j} + v_3'(t)\mathbf{k} = \omega[v_2(t)\mathbf{i} - v_1(t)\mathbf{j}].$$

This gives the scalar equations

$$v_1'(t) = \omega v_2(t), \qquad v_2'(t) = -\omega v_1(t), \qquad v_3'(t) = 0.$$

The last equation is trivial. It says that v_3 is constant:

$$v_3(t) = C.$$

The equations for v_1 and v_2 are linked together. We can get an equation that involves only v_1 by differentiating the first equation:

$$v_1''(t) = \omega v_2'(t) = -\omega^2 v_1(t).$$

As we know from our earlier work, this gives

$$v_1(t) = A_1 \sin(\omega t + \phi_1).$$

Since $v_1'(t) = \omega v_2(t)$, we have

$$v_2(t) = \frac{v_1'(t)}{\omega} = \frac{A_1\omega}{\omega}\cos(\omega t + \phi_1) = A_1 \cos(\omega t + \phi_1).$$

Therefore

$$\mathbf{v}(t) = A_1 \sin(\omega t + \phi_1)\,\mathbf{i} + A_1 \cos(\omega t + \phi_1)\,\mathbf{j} + C\mathbf{k}.$$

A final integration with respect to t gives

$$\mathbf{r}(t) = \left[-\frac{A_1}{\omega} \cos (\omega t + \phi_1) + D_1 \right] \mathbf{i} + \left[\frac{A_1}{\omega} \sin (\omega t + \phi_1) + D_2 \right] \mathbf{j} + [Ct + D_3]\mathbf{k}$$

where D_1, D_2, D_3 are constants of integration. All six constants of integration — A_1, ϕ_1, C, D_1, D_2, D_3 — can be evaluated from the initial conditions. We will not pursue this. What is important here is that the path of the particle is a circular helix with axis parallel to **B**, in this case parallel to **k**. You should be able to see this from the equation for $\mathbf{r}(t)$: the z-component of $\mathbf{r}$ varies linearly with t from the value D_3, while the x and y components represent uniform motion with angular velocity ω in a circle of radius $|A_1/\omega|$ around the center $x = D_1$, $y = D_2$. ☐

(Physicists express the behavior just found by saying that charged particles *spiral around* the magnetic field lines. Qualitatively, this behavior still holds even if the magnetic field lines are "bent," as is the case with the earth's magnetic field. Many charged particles become trapped by the earth's magnetic field. They keep spiraling around the magnetic field lines that run from pole to pole.)

EXERCISES 14.5

1. A particle moves in a circle of radius r at constant speed v. Find the angular speed and the magnitude of acceleration.

2. A particle moves so that

$$\mathbf{r}(t) = (a \cos \pi t + bt^2)\mathbf{i} + (a \sin \pi t - bt^2) \mathbf{j}.$$

Find the velocity, speed, acceleration, and the magnitude of acceleration all at time $t = 1$.

3. A particle moves so that $\mathbf{r}(t) = at\mathbf{i} + b \sin at \mathbf{j}$. Show that the magnitude of the acceleration of the particle is proportional to the distance of the particle from the x-axis.

4. A particle moves so that $\mathbf{r}(t) = 2\mathbf{i} + t^2\mathbf{j} + (t - 1)^2\mathbf{k}$. At what time is the speed a minimum?

Sketch the curve. Then compute and sketch the acceleration vector at the indicated points.

5. $\mathbf{r}(t) = (t/\pi) \mathbf{i} + \cos t \mathbf{j}, \quad t \in [0, 2\pi];$ at $t = \frac{1}{4}\pi, \frac{1}{2}\pi, \pi$.

6. $\mathbf{r}(t) = t^3\mathbf{i} + t\mathbf{j}, \quad t$ real; at $t = -\frac{1}{2}, \frac{1}{2}, 1$.

7. $\mathbf{r}(t) = \sec t \mathbf{i} + \tan t \mathbf{j}, \quad t \in [-\frac{1}{4}\pi, \frac{1}{2}\pi);$ at $t = -\frac{1}{6}\pi, 0, \frac{1}{3}\pi$.

8. $\mathbf{r}(t) = \sin \pi t \mathbf{i} + t\mathbf{j}, \quad t \in [0, 2];$ at $t = \frac{1}{2}, 1, \frac{5}{4}$.

9. An object moves so that

$$\mathbf{r}(t) = x_0 \mathbf{i} + [y_0 + (\alpha \cos \theta)t] \mathbf{j} + [z_0 + (\alpha \sin \theta)t - 16t^2] \mathbf{k}, \quad t \geq 0.$$

Find (a) the initial position, (b) the initial velocity, (c) the initial speed, (d) the acceleration throughout the motion. Finally, (e) identify the curve.

10. A particle moves so that $\mathbf{r}(t) = 2 \cos 2t \mathbf{i} + 3 \cos t \mathbf{j}$. (a) Show that the particle oscillates on an arc of the parabola $4y^2 - 9x = 18$. (b) Draw the path. (c) What are the acceleration vectors at the points of zero velocity? (d) Draw these vectors at the points in question.

11. An object of mass m moves so that

$$\mathbf{r}(t) = \frac{1}{2}a(e^{\omega t} + e^{-\omega t})\mathbf{i} + \frac{1}{2}b(e^{\omega t} - e^{-\omega t})\mathbf{j}.$$

(a) What is the velocity at $t = 0$? (b) Show that the acceleration vector is a constant positive multiple of the radius vector. (This shows that the force is central and repelling.) (c) What does (b) imply about the angular momentum and the torque? Verify your answers by direct calculation.

12. (a) An object moves so that

$$\mathbf{r}(t) = a_1 e^{bt}\mathbf{i} + a_2 e^{bt}\mathbf{j} + a_3 e^{bt}\mathbf{k}.$$

Show that, if $b > 0$, the object experiences a repelling central force.
(b) An object moves so that

$$\mathbf{r}(t) = \sin t\,\mathbf{i} + \cos t\,\mathbf{j} + (\sin t + \cos t)\mathbf{k}.$$

Show that the object experiences an attracting central force.
(c) Compute the angular momentum $\mathbf{L}(t)$ for the motion in (b).

13. A constant force of magnitude α directed upward from the xy-plane is continually applied to an object of mass m. Given that the object starts at time 0 at the point $P(0, y_0, z_0)$ with initial velocity $2\mathbf{j}$, find: (a) the velocity of the object t seconds later. (b) the speed of the object t seconds later. (c) the momentum of the object t seconds later. (d) the path followed by the object, both in vector form and in Cartesian coordinates.

14. Show that, if the force on an object is always perpendicular to the velocity of the object, then the *speed* of the object is constant. (This tells us that the speed of a charged particle in a magnetic field is constant.)

15. Find the force required to propel a particle of mass m so that $\mathbf{r}(t) = t\mathbf{j} + t^2\mathbf{k}$.

16. Show that for an object of constant velocity the angular momentum is constant.

17. At each point $P(x(t), y(t), z(t))$ of its motion an object of mass m is subject to a force

$$\mathbf{F}(t) = m\pi^2[a \cos \pi t\, \mathbf{i} + b \sin \pi t\, \mathbf{j}]. \qquad (a > 0, b > 0)$$

Given that $\mathbf{v}(0) = -\pi b\mathbf{j} + \mathbf{k}$ and $\mathbf{r}(0) = b\mathbf{j}$ find the following at time $t = 1$:
(a) the velocity. (b) the speed. (c) the acceleration.
(d) the momentum. (e) the angular momentum. (f) the torque.

18. If an object of mass m moves with velocity $\mathbf{v}(t)$ subject to a force $\mathbf{F}(t)$, the scalar product

(14.5.11) $\qquad \mathbf{F}(t) \cdot \mathbf{v}(t)$

is called the *power* (expended by the force) and the number

(14.5.12) $\qquad \frac{1}{2}m[v(t)]^2$

is called the *kinetic energy* of the object. Show that the time rate of change of the kinetic energy of an object is the power expended on it:

(14.5.13) $\qquad \frac{d}{dt}(\frac{1}{2}m[v(t)]^2) = \mathbf{F}(t) \cdot \mathbf{v}(t).$

19. Two particles of equal mass m, one with constant velocity $\mathbf{v}$ and the other at rest, collide elastically (i.e., the kinetic energy of the system is preserved) and go off in different directions. Show that the two particles go off at right angles.

20. (*Elliptic harmonic motion*) Show that, if the force on a particle of mass m is of the form

$$\mathbf{F}(t) = -m\omega^2\mathbf{r}(t),$$

then the path of the particle may be written

$$\mathbf{r}(t) = \cos \omega t\,\mathbf{A} + \sin \omega t\,\mathbf{B}$$

with **A** and **B** constant vectors. Give the physical significance of **A** and **B** and specify conditions on **A** and **B** that restrict the particle to a circular path.   HINT: Recall that every solution of the differential equation

$$x''(t) = -\omega^2 x(t)$$

can be written in the form

$$x(t) = A \cos \omega t + B \sin \omega t. \qquad \text{(Exercise 6, Section 7.8)}$$

21. A particle moves with constant acceleration **a**. Show that the path of the particle lies entirely in some plane. Find a vector equation for this plane.

22. In Problem 6 we stated that the path

$$\mathbf{r}(t) = \left[-\frac{A_1}{\omega} \cos (\omega t + \phi_1) + D_1 \right] \mathbf{i} + \left[\frac{A_1}{\omega} \sin (\omega t + \phi_1) + D_2 \right] \mathbf{j} + [Ct + D_3]\mathbf{k}$$

was a circular helix. Set $\omega = -1$ and show that, if $\mathbf{r}(0) = a\mathbf{i}$ and $\mathbf{v}(0) = a\mathbf{j} + b\mathbf{k}$, then the path takes the form

$$\mathbf{r}(t) = a \cos t\, \mathbf{i} + a \sin t\, \mathbf{j} + bt\, \mathbf{k},$$

the circular helix described in Section 14.3.

23. A charged particle in a constant electric field **E** experiences the force $q\mathbf{E}$, where q is the charge of the particle. Assume that the field has the constant value $\mathbf{E} = E_0\mathbf{k}$ and find the path of the particle given that $\mathbf{r}(0) = \mathbf{i}$ and $\mathbf{v}(0) = \mathbf{j}$.

24. (*Important*) A wheel is rotating about an axle with angular speed ω. Let $\boldsymbol{\omega}$ be the *angular velocity vector*, the vector of length ω that points along the axis of the wheel in such a direction that, observed from the tip of $\boldsymbol{\omega}$, the wheel rotates counterclockwise. Take the origin as the center of the wheel and let **r** be the vector from the origin to a point P on the rim of the wheel. Express the velocity **v** of P in terms of $\boldsymbol{\omega}$ and **r**.

25. Solve the initial value problem

$$\mathbf{F}(t) = m\mathbf{r}''(t) = t\mathbf{i} + t^2\mathbf{j}, \qquad \mathbf{r}_0 = \mathbf{r}(0) = \mathbf{i}, \qquad \mathbf{v}_0 = \mathbf{v}(0) = \mathbf{k}.$$

26. Solve the initial value problem

$$\mathbf{F}(t) = m\mathbf{r}''(t) = -m\beta^2 z(t)\mathbf{k}, \qquad \mathbf{r}_0 = \mathbf{r}(0) = \mathbf{k}, \qquad \mathbf{v}_0 = \mathbf{v}(0) = \mathbf{0}.$$

[Here $z(t)$ is the z-component of $\mathbf{r}(t)$.]

27. An object of mass m moves subject to the force

$$\mathbf{F}(\mathbf{r}) = 4r^2\mathbf{r}$$

where **r** is the position of the object. Suppose $\mathbf{r}(0) = \mathbf{0}$ and $\mathbf{v}(0) = 2\mathbf{u}$, where **u** is a unit vector. Show that at each time t the speed of the object satisfies the relation

$$v = \sqrt{4 + \frac{2}{m} r^4}.$$

HINT: Examine the quantity $\frac{1}{2} mv^2 - r^4$. (This is the *energy* of the object, a notion we will take up in Chapter 18.)

28. A moving particle traces a path that at each point has a unit tangent vector $\mathbf{T}(t)$ and a principal normal $\mathbf{N}(t)$. In anticipation of Section 14.7, (a) write the velocity vector as a scalar multiple of $\mathbf{T}(t)$ and then (b) write the acceleration vector as a linear combination of $\mathbf{T}(t)$ and $\mathbf{N}(t)$. (Incidentally, this will show that the acceleration vector lies in the osculating plane.)

*14.6   PLANETARY MOTION

Tycho Brahe, Johannes Kepler

In the middle of the sixteenth century the arguments on planetary motion persisted. Was Copernicus right? Did the planets move in circles about the sun? Obviously not. Was not the earth the center of the universe?

In 1576, with the generous support of his king, the Danish astronomer Tycho Brahe built an elaborate astronomical observatory on the isle of Hveen and began his painstaking observations. For more than twenty years he looked through his telescopes and recorded what he saw. He was a meticulous observer, but he could draw no definite conclusions.

In 1599 the German astronomer-mathematician Johannes Kepler began his study of Brahe's voluminous tables. For a year and a half Brahe and Kepler worked together. Then Brahe died and Kepler went on wrestling with the data. His persistence paid off. By 1619 Kepler had made three stupendous discoveries, known today as *Kepler's Laws of Planetary Motion*:

  I. Each planet moves in a plane, not in a circle, but in an elliptic orbit with the sun at one focus.
 II. The radius vector from sun to planet sweeps out equal areas in equal times.
III. The square of the period of the motion varies directly as the cube of the major semiaxis, and the constant of proportionality is the same for all the planets.

What Kepler formulated, Newton was able to explain. Each of these laws, Newton showed, was deducible from his laws of motion and his law of gravitation.

Newton's Second Law of Motion for Extended Three-Dimensional Objects

Imagine an object that consists of n particles with masses $m_1, m_2, \ldots, m_n$ located at $\mathbf{r}_1, \mathbf{r}_2, \ldots, \mathbf{r}_n$.† The total mass M of the object is the sum of the masses of the constituent particles:

$$M = m_1 + \cdots + m_n.$$

The center of mass of the object is by definition the point $\mathbf{R}_M$ where

$$M\mathbf{R}_M = m_1\mathbf{r}_1 + \cdots + m_n\mathbf{r}_n.$$

The total force $\mathbf{F}_{TOT}$ on the object is by definition the sum of all the forces that act on the particles that constitute the object:

$$\mathbf{F}_{TOT} = \mathbf{F}_1 + \cdots + \mathbf{F}_n.$$

---

† The case of a continuously distributed mass is taken up in Chapter 17.

Since $\mathbf{F}_1 = m_1\mathbf{r}_1''$, . . . , $\mathbf{F}_n = m_n\mathbf{r}_n''$, we have

$$\mathbf{F}_{\text{TOT}} = m_1\mathbf{r}_1'' + \cdots + m_n\mathbf{r}_n'',$$

which we can write as

$$\mathbf{F}_{\text{TOT}} = M\mathbf{R}_M''.$$

The total force on an extended object is thus the total mass of the object times the acceleration of the center of mass.

We can simplify this still further. The forces that act between the constituent particles, the so-called "internal forces," cancel in pairs: if particle 23 tugs at particle 71 in a certain direction with a certain strength, then particle 71 tugs at particle 23 in the opposite direction with the same strength. (Newton's third law: to every action there is an equal reaction.) Therefore, in calculating the total force on our object, we can disregard the internal forces and simply add up the external forces. $\mathbf{F}_{\text{TOT}} = \mathbf{F}_{\text{TOT}}^{(\text{Ext})}$ and Newton's second law takes the form

(14.6.1)
$$\mathbf{F}_{\text{TOT}}^{(\text{Ext})} = M\mathbf{R}_M''.$$

The total external force on an extended three-dimensional object is thus the total mass of the object times the acceleration of the center of mass.

When an external force is applied to an extended object, we cannot predict the reaction of all the constituent particles, but we can predict the reaction of the center of mass. The center of mass will react to the force as if it were a particle with all the mass concentrated there. Suppose, for example, that a bomb is dropped from an airplane. The center of mass, "feeling" only the force of gravity (we are neglecting air resistance), falls in a parabolic arc toward the ground even if the bomb explodes at a thousand meters and individual pieces fly every which way. The forces of explosion are internal and do not affect the center of mass.

Some Preliminary Comments About the Planets

Roughly speaking, a planet is a massive object in the form of a ball with the center of mass at the center. In what follows, when we refer to the position of a planet, you are to understand that we really mean the position of the center of mass of the planet. Two other points require comment. First, we will write our equations as if the sun affected the motion of the planet but not vice versa: we will assume that the sun stays put and that the planet moves. Really each affects the other. Our viewpoint is justified by the immense difference in mass between the planets and the sun. In the case of the earth and the sun, for example, a reasonable analogy is to imagine a tug of war in space between someone who weighs three pounds and someone who weighs a million pounds: to a good approximation, the million-pound person does not move. The second point is that the planet is affected not only by the pull from the sun, but also by the gravitational pulls from the other planets and all the other celestial bodies. But these forces are much smaller, and they tend to cancel. We will ignore them. (Our results are only approximations, but they prove to be very good approximations.)

A Derivation of Kepler's Laws from Newton's Laws of Motion and His Law of Gravitation

The gravitational force exerted by the sun on a planet can be written in vector form as follows:

(*)
$$\mathbf{F}(\mathbf{r}) = -G\,\frac{mM}{r^3}\,\mathbf{r}.$$

Here m is the mass of the planet, M is the mass of the sun, G is the gravitational constant, $\mathbf{r}$ is the vector from the center of the sun to the center of the planet, and r is the magnitude of $\mathbf{r}$. (Thus we are placing the sun at the origin of our coordinate system; namely, we are using what is known as a *heliocentric* coordinate system, from *Hēlios,* the Greek word for sun.)

Let's make sure that equation (*) conforms to our earlier characterization of gravitational force. First of all, because of the minus sign, the direction is toward the origin, where the sun is located. Taking norms we have

$$\|\mathbf{F}(\mathbf{r})\| = \frac{GmM}{r^3}\,\|\mathbf{r}\| = \frac{GmM}{r^2}.$$

$\uparrow$ $\quad \|\mathbf{r}\| = r$

Thus the magnitude of the force is as expected: it does vary directly as the product of the masses and inversely as the square of the distance between them. So we are back in familiar territory.

We will derive Kepler's laws in a somewhat piecemeal manner. Since the force on the planet is a central force, $\mathbf{L}$, the angular momentum of the planet, is conserved. (Section 14.5) If $\mathbf{L}$ were zero, we would have

$$\|\mathbf{L}\| = m\|\mathbf{r}\|\,\|\mathbf{v}\|\sin\theta = 0.$$

This would mean that either $\mathbf{r} = \mathbf{0}$, $\mathbf{v} = \mathbf{0}$, or $\sin\theta = 0$. The first equation would place the planet at the center of the sun. The second equation could hold only if the planet stopped. The third equation could hold only if the planet moved directly toward the sun or directly away from the sun. We can be thankful that none of those things happen.

Since the planet moves under a central force and $\mathbf{L}$ is not zero, *the planet stays on the plane that passes through the center of the sun and is perpendicular to $\mathbf{L}$, and the radius vector does sweep out equal areas in equal times.* (We know all this from Theorem 14.5.10.)

Now we go on to show that the path is an ellipse with the sun at one focus.† The equation of motion of a planet of mass m can be written

$$m\mathbf{a}(t) = -GmM\,\frac{\mathbf{r}(t)}{[r(t)]^3}.$$

---

† If some of the steps seem unanticipated to you, you should realize that we are discussing one of the most celebrated physical problems in history and there have been three hundred years to think of ingenious ways to deal with it. The argument we give here follows the lines of the excellent discussion that appears in Harry Pollard's *Celestial Mechanics*, Mathematical Association of America (1976).

Clearly m drops out of the equation. Setting $GM = \rho$ and suppressing the explicit dependence on t, we have

$$\mathbf{a} = -\rho \, \frac{\mathbf{r}}{r^3},$$

which gives

(**)
$$\frac{\mathbf{r}}{r^3} = -\frac{1}{\rho} \, \mathbf{a} = -\frac{1}{\rho} \, \frac{d\mathbf{v}}{dt}.$$

From (14.2.4) we know that in general

$$\frac{d}{dt}\left(\frac{\mathbf{r}}{r}\right) = (\mathbf{r} \times \mathbf{v}) \times \frac{\mathbf{r}}{r^3}.$$

Since $\mathbf{L} = \mathbf{r} \times m\mathbf{v}$, we have

$$\frac{d}{dt}\left(\frac{\mathbf{r}}{r}\right) = \frac{\mathbf{L}}{m} \times \frac{\mathbf{r}}{r^3}.$$

Inserting (**) we see that

$$\frac{d}{dt}\left(\frac{\mathbf{r}}{r}\right) = \frac{\mathbf{L}}{m} \times \left(-\frac{1}{\rho} \, \frac{d\mathbf{v}}{dt}\right) = \frac{d\mathbf{v}}{dt} \times \frac{\mathbf{L}}{m\rho} = \frac{d}{dt}\left(\mathbf{v} \times \frac{\mathbf{L}}{m\rho}\right)$$

$\mathbf{L}, m, \rho$ are all constant

and therefore

$$\frac{d}{dt}\left[\left(\mathbf{v} \times \frac{\mathbf{L}}{m\rho}\right) - \frac{\mathbf{r}}{r}\right] = 0.$$

Integration with respect to t gives

(***)
$$\left(\mathbf{v} \times \frac{\mathbf{L}}{m\rho}\right) - \frac{\mathbf{r}}{r} = \mathbf{e}$$

where $\mathbf{e}$ is a constant vector that depends on the initial conditions. Dotting both sides with $\mathbf{r}$, we have

$$\mathbf{r} \cdot \left(\mathbf{v} \times \frac{\mathbf{L}}{m\rho}\right) - \frac{\mathbf{r} \cdot \mathbf{r}}{r} = \mathbf{r} \cdot \mathbf{e}.$$

Since

$$\mathbf{r} \cdot \left(\mathbf{v} \times \frac{\mathbf{L}}{m\rho}\right) = \frac{\mathbf{L}}{m\rho} \cdot (\mathbf{r} \times \mathbf{v}) = \frac{\mathbf{L}}{m\rho} \cdot \frac{\mathbf{L}}{m} = \frac{L^2}{m^2\rho} \quad \text{and} \quad \frac{\mathbf{r} \cdot \mathbf{r}}{r} = \frac{r^2}{r} = r,$$

(13.6.6)

we find that

$$\frac{L^2}{m^2\rho} - r = \mathbf{r} \cdot \mathbf{e},$$

which we write as

$$r + (\mathbf{r} \cdot \mathbf{e}) = \frac{L^2}{m^2 \rho}.$$ (orbit equation)

If $\mathbf{e} = \mathbf{0}$, the orbit is a circle:

$$r = \frac{L^2}{m^2 \rho}.$$

This is a possibility that requires very special initial conditions, conditions not met by any of the planets in our solar system. In our solar system at least $\mathbf{e} \neq \mathbf{0}$.

Given that $e \neq 0$, we can write the orbit equation as

(14.6.2)

$$r(1 + e \cos \theta) = \frac{L^2}{m^2 \rho}$$

where $e = \|\mathbf{e}\|$ and θ is the angle between $\mathbf{r}$ and $\mathbf{e}$.

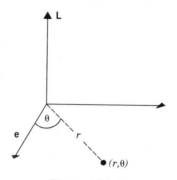

Figure 14.6.1

We are almost through. Since $\mathbf{v} \times \mathbf{L}$ and $\mathbf{r}$ are both perpendicular to $\mathbf{L}$, we know from (**∗∗**) that $\mathbf{e}$ is perpendicular to $\mathbf{L}$. Therefore, in the plane of $\mathbf{e}$ and $\mathbf{r}$ (see Figure 14.6.1) we can interpret r and θ as the usual polar coordinates. The orbit equation (14.6.2) is then a polar equation and, as such, represents a conic section with focus at the origin, which is to say, *focus at the sun*. According to (10.3.1) the conic section can be a parabola, a hyperbola, or an ellipse. The repetitiveness of planetary motion rules out the parabola and the hyperbola. *The orbit is therefore an ellipse.*

Finally we will verify Kepler's third law: that the square of the period of the motion varies directly as the cube of the major semiaxis and the constant of proportionality is the same for all planets.

The elliptic orbit has an equation of the form

$$r(1 + e \cos \theta) = ed \qquad \text{with} \quad 0 < e < 1, \ d > 0.$$

By Section 10.3 the major and minor semiaxes of the ellipse are given by

$$a = \frac{ed}{1 - e^2}, \qquad b = a\sqrt{1 - e^2}.$$

(See the footnote on page 512.) Denote the period of revolution by T. Since the radius vector sweeps out area at the constant rate of $L/2m$ (we know this from the proof of Theorem 14.5.10),

$$\left(\frac{L}{2m}\right) T = \text{area of the ellipse} = \pi ab = \pi a^2 \sqrt{1 - e^2}.$$

Thus

$$T = \frac{2\pi m a^2 \sqrt{1 - e^2}}{L} \qquad \text{and} \qquad T^2 = \frac{4\pi^2 m^2 a^4 (1 - e^2)}{L^2}.$$

From (14.6.2) we know that

$$ed = \frac{L^2}{m^2 \rho} = \frac{L^2}{m^2 GM} \qquad \text{and therefore} \qquad \frac{m^2}{L^2} = \frac{1}{edGM}.$$

It follows that

$$T^2 = \frac{4\pi^2 a^4 (1 - e^2)}{edGM} = \frac{4\pi^2 a^4 (1 - e^2)}{a(1 - e^2)GM} = \frac{4\pi^2}{GM} a^3.$$

$$\underset{\uparrow}{\rule{0pt}{0pt}} \quad ed = a(1 - e^2)$$

T^2 does vary directly with a^3, and the constant of proportionality $4\pi^2/GM$ is the same for all planets. ☐

EXERCISES *14.6

1. Kepler's third law can be stated as follows: for each planet the square of the period of revolution varies directly as the cube of the planet's average distance from the sun, and the constant of proportionality is the same for all planets. A *year* on a given planet is the time taken by the planet to make one circuit around the sun; thus a year on a planet is the period of revolution of that planet. Given that on average Venus is 0.72 times as far from the sun as the earth, how does the length of a "Venus year" compare with the length of an "earth year"?

2. Verify by differentiation with respect to time t that, if the acceleration of a planet is given by

$$\mathbf{a} = -\rho \frac{\mathbf{r}}{r^3},$$

then the energy $E = \frac{1}{2}mv^2 - m\rho r^{-1}$ is constant.

3. Given that a planet moves in a plane, its motion can be described by rectangular coordinates (x, y) or polar coordinates (r, θ), with origin at the sun. The kinetic energy of a planet is

$$\tfrac{1}{2}mv^2 = \tfrac{1}{2}m\left[\left(\frac{dx}{dt}\right)^2 + \left(\frac{dy}{dt}\right)^2\right].$$

Show that in polar coordinates

$$\tfrac{1}{2}mv^2 = \tfrac{1}{2}m\left[\left(\frac{dr}{dt}\right)^2 + r^2\left(\frac{d\theta}{dt}\right)^2\right].$$

4. We have seen that the energy of a planet

$$E = \tfrac{1}{2}mv^2 - \frac{m\rho}{r}$$

is constant. (Exercise 2) Setting $dr/dt = \dot{r}$, $d\theta/dt = \dot{\theta}$ and using Exercise 3, we have

$$E = \tfrac{1}{2}m(\dot{r}^2 + r^2\dot{\theta}^2) - \frac{m\rho}{r}.$$

We also know that the angular momentum **L** is constant and that $L = mr^2\dot{\theta}$. Use this fact to verify that

$$E = \frac{L^2}{2m}\left\{\frac{1}{r^2} + \frac{1}{r^4}\left(\frac{dr}{d\theta}\right)^2\right\} - \frac{m\rho}{r}.$$

Since E is a constant, this is a differential equation for r as a function of θ.

5. Show that the function

$$r = \frac{a}{1 + e\cos\theta} \quad \text{with} \quad a = \frac{L^2}{m^2\rho} \quad \text{and} \quad e^2 = \frac{2Ea}{m\rho} + 1$$

satisfies the equation derived in Exercise 4.

14.7   CURVATURE

The Curvature of a Plane Curve

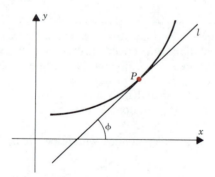

Figure 14.7.1

Figure 14.7.1 shows a plane curve which we assume to be twice differentiable. At the point P we have drawn the tangent line l. The angle that l makes with the x-axis has been labeled ϕ. As P moves along the curve, l changes and ϕ changes also. The magnitude k of the change in ϕ per unit of arc length is called the *curvature*:

(14.7.1)

$$k = \left|\frac{d\phi}{ds}\right|.$$

Calculating Curvature

If the curve is the graph of a twice differentiable function

$$y = y(x),$$

then the curvature can be calculated from the formula

(14.7.2)
$$k = \frac{|y''|}{[1 + (y')^2]^{3/2}}$$

where the primes indicate differentiation with respect to x.

Derivation of Formula 14.7.2. We know that $\tan \phi = y'$. Therefore,

$$\phi = \tan^{-1}(y').$$

Differentiating with respect to x, we have

$$\frac{d\phi}{dx} = \frac{1}{1 + (y')^2} \cdot \frac{d}{dx}(y') = \frac{y''}{1 + (y')^2}.$$

Since

$$\underset{\text{chain rule}}{\frac{d\phi}{dx}} = \frac{d\phi}{ds}\frac{ds}{dx} = \frac{d\phi}{ds}\underset{(14.4.3)}{\sqrt{1 + (y')^2}},$$

we have

$$\frac{d\phi}{ds}\sqrt{1 + (y')^2} = \frac{y''}{1 + (y')^2} \quad \text{and therefore} \quad \left|\frac{d\phi}{ds}\right| = \frac{|y''|}{[1 + (y')^2]^{3/2}}. \quad \blacksquare$$

If the curve is given parametrically by a twice differentiable vector function

$$\mathbf{r}(t) = x(t)\mathbf{i} + y(t)\mathbf{j},$$

then

(14.7.3)
$$k = \frac{|x'y'' - y'x''|}{[(x')^2 + (y')^2]^{3/2}}$$

where the primes now indicate differentiation with respect to t.

We will derive the formula under the assumption that $x' \neq 0$. Actually, the formula holds provided that $(x')^2 + (y')^2 \neq 0$.

Derivation of Formula 14.7.3

$$\frac{dy}{dx} = \frac{dy/dt}{dx/dt} = \frac{y'}{x'}.$$

Therefore, as you can verify,

$$\underset{(10.7.5)}{\frac{d^2y}{dx^2}} = \frac{(dx/dt)(d^2y/dt^2) - (dy/dt)(d^2x/dt^2)}{(dx/dt)^3} = \frac{x'y'' - y'x''}{(x')^3}.$$

Thus

$$k = \underset{(14.7.2)}{\frac{|d^2y/dx^2|}{[1 + (dy/dx)^2]^{3/2}}} = \left| \frac{x'y'' - y'x''}{(x')^3} \right| \frac{1}{[1 + (y'/x')^2]^{3/2}} = \frac{|x'y'' - y'x''|}{[(x')^2 + (y')^2]^{3/2}}. \quad \square$$

Example 1. Since a straight line has a constant angle of inclination, we have

$$\frac{d\phi}{ds} = 0 \quad \text{and thus} \quad k = 0.$$

(14.7.4) | Along a straight line the curvature is constantly zero. | $\square$

Example 2. For a circle of radius r,

$$\mathbf{r}(t) = r(\cos t \, \mathbf{i} + \sin t \, \mathbf{j}),$$

we have

$$x = r \cos t, \qquad y = r \sin t.$$

Differentiation with respect to t gives

$$x' = -r \sin t, \quad x'' = -r \cos t; \qquad y' = r \cos t, \quad y'' = -r \sin t.$$

Thus

$$k = \frac{|x'y'' - y'x''|}{[(x')^2 + (y')^2]^{3/2}} = \frac{|(-r \sin t)(-r \sin t) - (r \cos t)(-r \cos t)|}{[(-r \sin t)^2 + (r \cos t)^2]^{3/2}} = \frac{r^2}{r^3} = \frac{1}{r}.$$

(14.7.5) | Along a circle of radius r the curvature is constantly $1/r$.

Hardly surprising. It is geometrically evident that along a circular path the change in direction takes place at a constant rate. Since a complete revolution entails a change of direction of 2π radians and this change is effected on a path of length $2\pi r$, the change in direction per unit of arc length is $2\pi/2\pi r = 1/r$. $\square$

To say that a curve $y = f(x)$ has slope m at a point P is to say that at the point P the curve is rising or falling at the rate of a line of slope m. To say that a plane curve C has curvature $1/r$ at a point P is to say that at the point P the curve is turning at the rate of a circle of radius r. (Figure 14.7.2) The smaller the circle, the tighter the turn and, thus, the greater the curvature.

Figure 14.7.2

The reciprocal of the curvature,

$$\rho = \frac{1}{k},$$

is called the *radius of curvature*. The point at a distance ρ from the curve in the direction of the principal normal is called the *center of curvature*.

Problem 3. Take $a > 0$. Show that the point of maximal curvature on the catenary

$$y = \frac{a}{2}(e^{x/a} + e^{-x/a}) \qquad\qquad \text{(Figure 10.9.6)}$$

is the lowest point on the curve. What is the radius of curvature at that point?

Solution. We will use the formula

$$k = \frac{|y''|}{[1 + (y')^2]^{3/2}}.$$

Here

$$y' = \frac{1}{2}(e^{x/a} - e^{-x/a}) \qquad \text{and} \qquad y'' = \frac{1}{2a}(e^{x/a} + e^{-x/a}) = \frac{y}{a^2} > 0.$$

Observe that

$$[1 + (y')^2]^{3/2} = [1 + \tfrac{1}{4}(e^{x/a} - e^{-x/a})^2]^{3/2},$$

which, as you can check, reduces to give

$$[1 + (y')^2]^{3/2} = \frac{y^3}{a^3}.$$

Thus

$$k = \left(\frac{y}{a^2}\right)\left(\frac{a^3}{y^3}\right) = \frac{a}{y^2}.$$

The curvature is maximal where y is minimal; that is, at the lowest point of the curve.

The value of y at the lowest point of the curve is a. The curvature at that point is $1/a$; the radius of curvature ρ at that point is a. $\square$

Problem 4. Find the curvature at an arbitrary point (x, y) of the ellipse

$$\frac{x^2}{a^2} + \frac{y^2}{b^2} = 1. \qquad\qquad (a > b > 0)$$

Determine the points of maximal curvature and the points of minimal curvature. What is the radius of curvature at each of these points?

Solution. We parametrize the ellipse by setting

$$\mathbf{r}(t) = a \cos t \, \mathbf{i} + b \sin t \, \mathbf{j}$$

and use the fact that

$$k = \frac{|x'y'' - y'x''|}{[(x')^2 + (y')^2]^{3/2}}.$$

Here

$$x = a \cos t, \qquad y = b \sin t$$

and therefore

$$x' = -a \sin t, \quad x'' = -a \cos t; \qquad y' = b \cos t, \quad y'' = -b \sin t.$$

Thus

$$k = \frac{|(-a \sin t)(-b \sin t) - (b \cos t)(-a \cos t)|}{[(-a \sin t)^2 + (b \cos t)^2]^{3/2}} = \frac{ab}{[a^2 \sin^2 t + b^2 \cos^2 t]^{3/2}}.$$

As you can check, the curvature at the point (x, y) can be written

$$\frac{a^4 b^4}{(b^4 x^2 + a^4 y^2)^{3/2}}.$$

To find the points of maximal and minimal curvature, we go back to the parameter t. Observe that

$$a^2 \sin^2 t + b^2 \cos^2 t = (a^2 - b^2) \sin^2 t + b^2(\sin^2 t + \cos^2 t)$$
$$= (a^2 - b^2) \sin^2 t + b^2.$$

Thus we have

$$k = \frac{ab}{[(a^2 - b^2) \sin^2 t + b^2]^{3/2}}.$$

The curvature is maximal when $\sin^2 t = 0$; that is, when $t = 0$ and when $t = \pi$. Thus the points of maximal curvature are the points $P(\pm a, 0)$, the ends of the major axis. The curvature at these points is a/b^2 and the radius of curvature is b^2/a. The curvature is minimal when $\sin^2 t = 1$; that is, when $t = \frac{1}{2}\pi$ and when $t = \frac{3}{2}\pi$. The points of minimal curvature are the points $P(0, \pm b)$, the ends of the minor axis. The curvature at these points is b/a^2 and the radius of curvature is a^2/b. ∎

Figure 14.7.3

In Figure 14.7.3 you can see a plane curve. At the point P we have affixed the unit tangent vector **T**. As P moves along the curve, **T** changes, not in length, but in direction. The curvature of the curve is the magnitude of the change in **T** per unit of arc length:

(14.7.6)
$$k = \left\| \frac{d\mathbf{T}}{ds} \right\|.$$

Proof. Since **T** has length 1, we can write

$$\mathbf{T} = \cos \phi \, \mathbf{i} + \sin \phi \, \mathbf{j}.$$

Note that ϕ represents the angle between the tangent line and the x-axis. Differentiation with respect to s gives

$$\frac{d\mathbf{T}}{ds} = -\sin \phi \, \frac{d\phi}{ds} \, \mathbf{i} + \cos \phi \, \frac{d\phi}{ds} \, \mathbf{j} = \frac{d\phi}{ds} \, (-\sin \phi \, \mathbf{i} + \cos \phi \, \mathbf{j}).$$

Taking norms we have

$$\left\| \frac{d\mathbf{T}}{ds} \right\| = \left| \frac{d\phi}{ds} \right| \sqrt{\sin^2 \theta + \cos^2 \theta} = \left| \frac{d\phi}{ds} \right| = k$$

as asserted. ☐

This characterization of curvature generalizes to space curves.

The Curvature of a Space Curve

A space curve bends in two ways. It bends in the osculating plane (the plane of the unit tangent **T** and the principal normal **N**) and it bends away from that plane. The first form of bending is measured by the rate at which the unit tangent **T** changes direction. The second form of bending is measured by the rate at which the vector **T** × **N** changes direction. We will concentrate here on the first form of bending, the bending in the osculating plane. The measure of this is called *curvature*.

What was a theorem on curvature in the case of a plane curve becomes a definition of curvature in the case of a space curve; namely, in the case of a space curve, we *define* the curvature k by setting

$$k = \left\| \frac{d\mathbf{T}}{ds} \right\|.$$

If the space curve is given in terms of a parameter t, say

$$C: \quad \mathbf{r}(t) = x(t)\,\mathbf{i} + y(t)\,\mathbf{j} + z(t)\,\mathbf{k}, \quad t \in [a, b],$$

then the curvature can be calculated from the formula

(14.7.7)
$$k = \frac{\|d\mathbf{T}/dt\|}{ds/dt}.$$

We arrive at this by noting that

$$\frac{d\mathbf{T}}{ds}\frac{ds}{dt} = \frac{d\mathbf{T}}{dt},$$

dividing through by ds/dt and taking norms. The assumption here is that $ds/dt = \|\mathbf{r}'(t)\|$ remains nonzero.

Problem 5. Calculate the curvature of the circular helix

$$\mathbf{r}(t) = r \sin t\,\mathbf{i} + r \cos t\,\mathbf{j} + t\mathbf{k}. \qquad (r > 0)$$

Solution. We will use the Leibniz notation.

$$\frac{d\mathbf{r}}{dt} = r \cos t\,\mathbf{i} - r \sin t\,\mathbf{j} + \mathbf{k}, \qquad \frac{ds}{dt} = \left\| \frac{d\mathbf{r}}{dt} \right\| = \sqrt{r^2 + 1}$$

$$\mathbf{T} = \frac{d\mathbf{r}/dt}{\|d\mathbf{r}/dt\|} = \frac{r \cos t\,\mathbf{i} - r \sin t\,\mathbf{j} + \mathbf{k}}{\sqrt{r^2 + 1}}, \qquad \frac{d\mathbf{T}}{dt} = \frac{-r \sin t\,\mathbf{i} - r \cos t\,\mathbf{j}}{\sqrt{r^2 + 1}}$$

$$k = \frac{\|d\mathbf{T}/dt\|}{ds/dt} = \frac{1}{\sqrt{r^2 + 1}} \cdot \frac{r}{\sqrt{r^2 + 1}} = \frac{r}{r^2 + 1}. \quad \square$$

Components of Acceleration

In straight-line motion, acceleration is purely tangential; that is, the acceleration vector points in the direction of the motion. (Section 14.5) In uniform circular motion, the acceleration is normal; the acceleration vector is perpendicular to the tangent vector and points along the line of the normal vector toward the center of the circle. (Also Section 14.5) In general, acceleration has two components, a tangential component and a normal component. To see this, let's suppose that the position of an object at time t is given by the vector function

$$\mathbf{r}(t) = x(t)\mathbf{i} + y(t)\mathbf{j} + z(t)\mathbf{k}.$$

Since

$$T = \frac{d\mathbf{r}/dt}{\|d\mathbf{r}/dt\|} = \frac{\mathbf{v}}{ds/dt},$$

we have

$$\mathbf{v} = \frac{ds}{dt} \, T.$$

Differentiation gives

$$\mathbf{a} = \frac{d^2s}{dt^2} \, T + \frac{ds}{dt} \frac{d\mathbf{T}}{dt}.$$

Observe now that

$$\frac{d\mathbf{T}}{dt} = \left\| \frac{d\mathbf{T}}{dt} \right\| \mathbf{N} = k \frac{ds}{dt} \mathbf{N}.$$

(14.3.4) ——————⌐   ⌐—————— (14.7.7)

Substitution in the previous display gives

(14.7.8)
$$\mathbf{a} = \frac{d^2s}{dt^2} \, T + k \left(\frac{ds}{dt} \right)^2 \mathbf{N}.$$

The acceleration vector lies in the osculating plane, the plane of **T** and **N**. The tangential component of acceleration

$$a_T = \frac{d^2s}{dt^2}$$

depends only on the change of speed of the object; if the speed is constant, the tangential component of acceleration is zero and the acceleration is directed entirely toward the center of curvature of the path. On the other hand, the normal component of acceleration

$$a_N = k \left(\frac{ds}{dt} \right)^2$$

depends both on the speed of the object and the curvature of the path. At a point where the curvature is zero, the normal component of acceleration is zero and the acceleration is directed entirely along the path of the motion.

EXERCISES 14.7

Find the curvature of the graph.

1. $y = e^{-x}$.   **2.** $y = x^3$.   **3.** $y = \ln(\sec x)$

4. $y = x - x^2$.   **5.** $y = \sin x$.   **6.** $y = \tan x$.

Find the radius of curvature at the indicated point.

7. $6y = x^3$; $(2, \frac{4}{3})$ **8.** $2y = x^2$; $(0, 0)$.

9. $y^2 = 2x$; $(2, 2)$.   **10.** $y = 2 \sin 2x$; $(\frac{1}{4}\pi, 2)$

11. Find the point of maximal curvature on the curve $y = \ln x$.

12. Find the curvature of the graph of $y = 3x - x^3$ at the point where the function takes on its maximum value.

Express the curvature in terms of t.

13. $\mathbf{r}(t) = t\mathbf{i} + \frac{1}{2}t^2\mathbf{j}$.

14. $\mathbf{r}(t) = e^t\mathbf{i} + e^{-t}\mathbf{j}$.

15. $\mathbf{r}(t) = e^t(\cos t\,\mathbf{i} + \sin t\,\mathbf{j})$

16. $\mathbf{r}(t) = (\cos t + t \sin t)\mathbf{i} + (\sin t - t \cos t)\mathbf{j},\quad t > 0$.

17. Find the radius of curvature at the vertices of the hyperbola $xy = 1$.

18. Find the radius of curvature at the vertices of the hyperbola $x^2 - y^2 = 1$.

19. Find the curvature at each point (x, y) of the hyperbola $b^2x^2 - a^2y^2 = a^2b^2$.
HINT: Parametrize the hyperbola by setting $\mathbf{r}(t) = a \cosh t\,\mathbf{i} + b \sinh t\,\mathbf{j}$.

20. Find the curvature at the highest point of an arch of the cycloid

$$x(t) = r(t - \sin t), \quad y(t) = r(1 - \cos t).$$

21. Show that the curvature of the path of a moving object can be written

(14.7.9)
$$k = \frac{\|\mathbf{v} \times \mathbf{a}\|}{(ds/dt)^3}.$$

HINT: $\mathbf{v} = \dfrac{ds}{dt}\,\mathbf{T}$ and $\mathbf{a} = \dfrac{d^2s}{dt^2}\,\mathbf{T} + k\left(\dfrac{ds}{dt}\right)^2\mathbf{N}$.

Interpret $\mathbf{r}(t)$ as the position of a moving object at time t. Find the curvature of the path and determine the tangential and normal components of acceleration.

22. $\mathbf{r}(t) = (1 - 2t)\mathbf{i} + (3 + 4t)\mathbf{j} + (2 - 3t)\mathbf{k}$.

23. $\mathbf{r}(t) = e^t \cos t\,\mathbf{i} + e^t \sin t\,\mathbf{j} + e^t\mathbf{k}$.

24. $\mathbf{r}(t) = t\mathbf{i} + t^2\mathbf{j} + \frac{2}{3}t^3\mathbf{k}$.

25. $\mathbf{r}(t) = t\mathbf{i} + \frac{1}{2}t^2\mathbf{j} + \frac{1}{3}t^3\mathbf{k}$.

26. Show that the curvature of a polar curve $r = f(\theta)$ is given by

$$k = \frac{|[f(\theta)]^2 + 2[f'(\theta)]^2 - f(\theta)f''(\theta)|}{([f(\theta)]^2 + [f'(\theta)]^2)^{3/2}}.$$

27. Find the curvature of the logarithmic spiral $r = e^{a\theta}$.

28. Find the curvature of the spiral of Archimedes $r = a\theta$.

29. Find the curvature of the cardioid $r = a(1 - \cos \theta)$ in terms of r.

30. (*Transition Curves*) When engineers lay railroad track, they cannot allow abrupt changes in curvature. To join a straightaway that ends at a point P to a curved track that begins at a point Q, they need to lay some transitional track that has curvature zero at P and the curvature of the second piece at Q. Find an arc of the form

$$y = Cx^n, \qquad x \in [0, 1]$$

that joins the arcs C_1 and C_2 of Figure 14.7.4 without any discontinuities in curvature.

31. Let $s(\theta)$ be the arc distance from the highest point of the cycloidal arch

$$x(\theta) = R(\theta - \sin \theta), \quad y(\theta) = R(1 - \cos \theta) \qquad \theta \in [0, 2\pi]$$

to the point $(x(\theta), y(\theta))$ of that same arch. (The arch is pictured in Figure 10.11.1.) Let $\rho(\theta)$ be the radius of curvature at the point $(x(\theta), y(\theta))$. (a) Calculate $s(\theta)$. (b) Calculate $\rho(\theta)$. (c) Then find an equation in s and ρ for that arch. (Such an equation is called a *natural equation* for the curve.)

32. Let $s(\theta)$ be the arc distance from the origin to the point $(x(\theta), y(\theta))$ along the exponential spiral $r = ae^{c\theta}$. (Take $a > 0, c > 0$.) Let $\rho(\theta)$ be the radius of curvature at that same point. Find an equation in s and ρ for that curve.

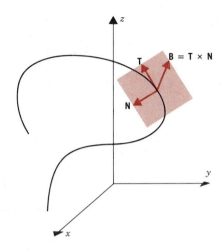

Figure 14.7.4   **Figure 14.7.5**

33. Let $s(\theta)$ be the arc distance from the point $(-2a, 0)$ to the point $(x(\theta), y(\theta))$ along the cardioid $r = a(1 - \cos \theta)$.   (Take $a > 0$.)   Let $\rho(\theta)$ be the radius of curvature at that same point.   Find an equation in s and ρ for that curve.

(*The Frenet Formulas*)   Figure 14.7.5 shows a space curve.   At a point of the curve we have drawn the unit tangent $\mathbf{T}$, the principal normal $\mathbf{N}$, and the vector $\mathbf{B} = \mathbf{T} \times \mathbf{N}$, which, being normal to both $\mathbf{T}$ and $\mathbf{N}$, is called the *binormal*.   At each point of the curve, the vectors $\mathbf{T}, \mathbf{N}, \mathbf{B}$ form what is called the *Frenet trihedral*, a set of mutually perpendicular unit vectors that, in the order given, form a local right-handed coordinate system.

34. (a) Show that $d\mathbf{B}/ds$ is parallel to $\mathbf{N}$ and therefore there is a scalar τ for which

$$\frac{d\mathbf{B}}{ds} = \tau \mathbf{N}.$$

HINT: Since $\mathbf{B}$ has constant length one, $d\mathbf{B}/ds \perp \mathbf{B}$.   Show that $d\mathbf{B}/ds \perp \mathbf{T}$ by carrying out the differentiation

$$\frac{d\mathbf{B}}{ds} = \frac{d}{ds}(\mathbf{T} \times \mathbf{N}).$$

(b) Now show that

$$\frac{d\mathbf{N}}{ds} = -k\mathbf{T} - \tau\mathbf{B}.$$

HINT: Since $\mathbf{T}, \mathbf{N}, \mathbf{B}$ form a right-handed system of mutually perpendicular unit vectors, we can show that

$$\mathbf{N} \times \mathbf{B} = \mathbf{T} \quad \text{and} \quad \mathbf{B} \times \mathbf{T} = \mathbf{N}.$$

You can assume these relations.

(c) The scalar τ is called the *torsion* of the curve.   Give a geometric interpretation to $|\tau|$.

The three formulas

(14.7.10) $$\frac{d\mathbf{T}}{ds} = k\mathbf{N}, \qquad \frac{d\mathbf{N}}{ds} = -k\mathbf{T} - \tau\mathbf{B}, \qquad \frac{d\mathbf{B}}{ds} = \tau\mathbf{N}$$

were developed in 1847 by the French geometer Frederic Frenet (1816–1900).   Another Frenchman, Alfred Serret (1819–1885), discovered these same relations independently in 1850.   Accordingly, the formulas are often called the Frenet–Serret formulas.

CHAPTER HIGHLIGHTS

14.1 Vector Functions

basic definition: $\lim\limits_{t \to t_0} \mathbf{f}(t) = \mathbf{L}$ iff $\lim\limits_{t \to t_0} \|\mathbf{f}(t) - \mathbf{L}\| = 0$

theorem: if $\lim\limits_{t \to t_0} \mathbf{f}(t) = \mathbf{L}$ then $\lim\limits_{t \to t_0} \|\mathbf{f}(t)\| = \|\mathbf{L}\|$

limit rules (p. 757)
limits can be taken component by component (p. 758)
continuity and differentiability (p. 758)   integration (p. 760)
properties of the integral (p. 761)

14.2 Differentiation Formulas

$$\frac{d}{dt}(\mathbf{f} + \mathbf{g}) = \frac{d\mathbf{f}}{dt} + \frac{d\mathbf{g}}{dt} \qquad\qquad \frac{d}{dt}(\alpha\mathbf{f}) = \alpha\,\frac{d\mathbf{f}}{dt} \quad (\alpha \text{ constant})$$

$$\frac{d}{dt}(u\mathbf{f}) = u\,\frac{d\mathbf{f}}{dt} + \frac{du}{dt}\mathbf{f} \quad (u = u(t)) \qquad \frac{d}{dt}(\mathbf{f} \cdot \mathbf{g}) = \left(\mathbf{f} \cdot \frac{d\mathbf{g}}{dt}\right) + \left(\frac{d\mathbf{f}}{dt} \cdot \mathbf{g}\right)$$

$$\frac{d}{dt}(\mathbf{f} \times \mathbf{g}) = \left(\mathbf{f} \times \frac{d\mathbf{g}}{dt}\right) + \left(\frac{d\mathbf{f}}{dt} \times \mathbf{g}\right) \qquad \frac{d\mathbf{f}}{dt} = \frac{d\mathbf{f}}{du}\frac{du}{dt} \quad \text{(chain rule)}$$

14.3 Curves

tangent vector (p. 770)   tangent line (p. 770)
unit tangent (p. 772)   principal normal (p. 773)
normal line (p. 773)   osculating plane (p. 773)
sense-preserving (sense-reversing) change of parameter (p. 776)

14.4 Arc Length

$$L(C) = \int_a^b \|\mathbf{r}'(t)\|\,dt \qquad \frac{ds}{dt} = \sqrt{\left(\frac{dx}{dt}\right)^2 + \left(\frac{dy}{dt}\right)^2 + \left(\frac{dz}{dt}\right)^2}$$

14.5 Curvilinear Motion; Vector Calculus in Mechanics

velocity, acceleration (p. 785)   speed (p. 785)
angular speed, uniform circular motion (p. 786)   conservation law (p. 788)
angular momentum (p. 788)   torque (p. 790)
central force (radial force) (p. 791)
effect of a central force on an object with nonzero angular momentum (p. 791)
initial value problems (p. 792)   power, kinetic energy (p. 796)

*14.6 Planetary Motion

Kepler's three laws of planetary motion (p. 798)

14.7 Curvature

curvature (pp. 804, 810)   radius of curvature (p. 807)
components of acceleration (p. 811)   Frenet formulas (p. 813)

CHAPTER 15

FUNCTIONS OF SEVERAL VARIABLES

15.1 ELEMENTARY EXAMPLES

First a remark on notation. Points $P(x, y)$ of the xy-plane will be written (x, y) and points $P(x, y, z)$ of three-space will be written (x, y, z).

Let D be a nonempty subset of the xy-plane. A rule f that assigns a real number $f(x, y)$ to each point (x, y) in D is called a *function of two variables*. The set D is called the *domain* of f and the set of all values $f(x, y)$ is called the *range*.

Example 1. Take D as the entire xy-plane and to each point (x, y) assign the number

$$f(x, y) = xy. \quad \square$$

Example 2. Take D as the set of all points (x, y) with $y \neq 0$. To each such point assign the number

$$f(x, y) = \tan^{-1}\left(\frac{x}{y}\right). \quad \square$$

Example 3. Take D as the *open unit disc*:

$$D = \{(x, y): x^2 + y^2 < 1\}.$$

This set consists of all points inside the *unit circle* $x^2 + y^2 = 1$; the circle itself is omitted.† To each point (x, y) in D assign the number

$$f(x, y) = \frac{1}{\sqrt{1 - (x^2 + y^2)}}. \quad \square$$

Now let D be a nonempty subset of three-space. A rule f that assigns a real number $f(x, y, z)$ to each point (x, y, z) in D is called a *function of three variables*. The set D is called the *domain* and the set of all values $f(x, y, z)$ is called the *range*.

Example 4. Take D as all of three-space and to each point (x, y, z) assign the number

$$g(x, y, z) = xyz. \quad \square$$

Example 5. Take D as the set of all points (x, y, z) with $z \neq x + y$. (Thus D consists of all points not on the plane $x + y - z = 0$.) To each point of D assign the number

$$h(x, y, z) = \cos\left(\frac{1}{x + y - z}\right). \quad \square$$

Example 6. Take D as the *open unit ball*:

$$D = \{(x, y, z): x^2 + y^2 + z^2 < 1\}.$$

This set consists of all points inside the *unit sphere* $x^2 + y^2 + z^2 = 1$; the sphere itself is omitted.†† To each point (x, y, z) in D assign the number

$$f(x, y, z) = \frac{1}{\sqrt{1 - (x^2 + y^2 + z^2)}}. \quad \square$$

Functions of several variables arise naturally in very elementary settings.

$f(x, y) = \sqrt{x^2 + y^2}$ gives the distance between $P(x, y)$ and the origin;
$f(x, y) = xy$ gives the area of a rectangle of dimensions x, y, and
$f(x, y) = 2(x + y)$ gives the perimeter.

$f(x, y, z) = \sqrt{x^2 + y^2 + z^2}$ gives the distance between $P(x, y, z)$ and the origin;
$f(x, y, z) = xyz$ gives the volume of a rectangular solid of dimensions x, y, z, and
$f(x, y, z) = 2(xy + xz + yz)$ gives the total surface area.

---

† The *closed unit disc* $D = \{(x, y): x^2 + y^2 \leq 1\}$ consists of all points on or inside the unit circle; the circle is included.

†† The *closed unit ball* $D = \{(x, y, z): x^2 + y^2 + z^2 \leq 1\}$ consists of all points on or inside the unit sphere; the sphere is included.

If the domain of a function of several variables is not explicitly given, it is to be understood that the domain is the set of all points for which the definition makes sense. Thus, in the case of

$$f(x, y) = \frac{1}{x - y},$$

the domain is understood to be all points (x, y) with $y \neq x$. That is, all points of the plane not on the line $y = x$. In the case of

$$g(x, y, z) = \sin^{-1}(x + y + z),$$

the domain is understood to be all points (x, y, z) with $-1 \leq x + y + z \leq 1$. This set is the slab bounded by the parallel planes

$$x + y + z = -1 \quad \text{and} \quad x + y + z = 1. \quad \square$$

To say that a function is *bounded* is to say that its range is bounded. Since the function

$$f(x, y) = \frac{1}{x - y}$$

takes on all values other than 0, its range is $(-\infty, 0) \cup (0, \infty)$. The function is unbounded. In the case of

$$g(x, y, z) = \sin^{-1}(x + y + z)$$

the range is the closed interval $[-\frac{1}{2}\pi, \frac{1}{2}\pi]$. This function is bounded, below by $-\frac{1}{2}\pi$ and above by $\frac{1}{2}\pi$.

Problem 7. Find the domain and range of

$$f(x, y) = \frac{1}{\sqrt{4x^2 - y^2}}.$$

Solution. A point (x, y) is in the domain of f iff

$$4x^2 - y^2 > 0.$$

This occurs iff

$$y^2 < 4x^2$$

and thus iff

$$-2|x| < y < 2|x|.$$

The domain of f is shaded in Figure 15.1.1. It consists of all the points of the xy-plane that lie between the graph of $y = -2|x|$ and the graph of $y = 2|x|$. On this set $\sqrt{4x^2 - y^2}$ takes on all positive values, and so does its reciprocal $f(x, y)$. The range of f is $(0, \infty)$. $\quad \square$

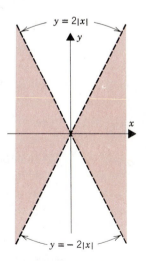

Figure 15.1.1

Problem 8. Find the domain and range of

$$f(x, y, z) = \frac{\sqrt{a^2 - z^2}}{b^2 + \sqrt{r^2 - (x^2 + y^2)}}.$$

$(a > 0, b > 0, r > 0)$

Solution. Here we must have

$$x^2 + y^2 \leq r^2 \qquad \text{and} \qquad -a \leq z \leq a.$$

(Explain.) The domain is the solid right circular cylinder shown in Figure 15.1.2.

Since the numerator $\sqrt{a^2 - z^2}$ takes on all values from 0 to a and the denominator independently (there are no z's in the denominator) takes on all values from b^2 to $b^2 + r$, the fraction takes on all values from 0 to a/b^2. The range is thus the closed interval $[0, a/b^2]$. ◻

Figure 15.1.2

EXERCISES 15.1

Find the domain and range of each of the following functions.

1. $f(x, y) = \sqrt{xy}.$ **2.** $f(x, y) = \sqrt{1 - xy}.$ **3.** $f(x, y) = \dfrac{1}{x + y}.$

4. $f(x, y) = \dfrac{1}{x^2 + y^2}.$ **5.** $f(x, y) = \dfrac{e^x - e^y}{e^x + e^y}.$ **6.** $f(x, y) = \dfrac{x^2}{x^2 + y^2}.$

7. $f(x, y) = \ln(xy).$ **8.** $f(x, y) = \ln(1 - xy).$ **9.** $f(x, y) = \dfrac{1}{\sqrt{y - x^2}}.$

10. $f(x, y) = \dfrac{\sqrt{9 - x^2}}{1 + \sqrt{1 - y^2}}.$ **11.** $f(x, y) = \sqrt{9 - x^2} - \sqrt{4 - y^2}.$

12. $f(x, y, z) = \cos x + \cos y + \cos z.$ **13.** $f(x, y, z) = \dfrac{x + y + z}{|x + y + z|}.$

14. $f(x, y, z) = \dfrac{z^2}{x^2 - y^2}.$ **15.** $f(x, y, z) = -\dfrac{z^2}{\sqrt{x^2 - y^2}}.$

16. $f(x, y, z) = \dfrac{z}{x - y}.$ **17.** $f(x, y) = \dfrac{2}{\sqrt{9 - (x^2 + y^2)}}.$

18. $f(x, y, z) = \ln(|x + 2y + 3z| + 1).$ **19.** $f(x, y, z) = \ln(x + 2y + 3z).$

20. $f(x, y, z) = e^{\sqrt{4 - (x^2 + y^2 + z^2)}}.$ **21.** $f(x, y, z) = e^{-(x^2 + y^2 + z^2)}.$

22. $f(x, y, z) = \dfrac{\sqrt{1 - x^2} + \sqrt{4 - y^2}}{1 + \sqrt{9 - z^2}}.$ **23.** $f(x, y, z) = \dfrac{\sqrt{9 - x^2} - \sqrt{4 - y^2}}{1 + \sqrt{1 - z^2}}.$

Compare the following functions.

24. $f(x) = 2, \quad g(x, y) = 2, \quad h(x, y, z) = 2.$

25. $f(x) = \sqrt{x}, \quad g(x, y) = \sqrt{x}, \quad h(x, y, z) = \sqrt{x}.$

26. $f(x, y) = \cos \pi x \sin \pi y, \quad g(x, y, z) = \cos \pi x \sin \pi y.$

15.2 A BRIEF CATALOG OF THE QUADRIC SURFACES; PROJECTIONS

We represented functions of one variable by curves in the xy-plane. We will represent functions of two variables by surfaces in three-space.

In this section we examine in a systematic manner the surfaces given by equations of the form

$$Ax^2 + By^2 + Cz^2 + Dxy + Exz + Fyz + Hx + Iy + Jz + K = 0.$$

Such surfaces are called *quadric surfaces.*

By suitable translations and rotations of the coordinate axes we can simplify such equations and thereby show that the nondegenerate† quadrics fall into nine distinct types:

1. the ellipsoid

2. the hyperboloid of one sheet   3. the hyperboloid of two sheets

4. the quadric cone

5. the elliptic paraboloid   6. the hyperbolic paraboloid

7. the parabolic cylinder   8. the elliptic cylinder   9. the hyperbolic cylinder.

As you go on with calculus, you will encounter these surfaces time and time again. Here we give you a picture of each one, together with its equation in standard form and some information about its special properties. These are some of the things to look for:

(a) the *intercepts* (the points at which the surface intersects the coordinate axes)

(b) the *traces* (the intersections with the coordinate planes)

(c) the *sections* (the intersections with planes in general)

(d) the *center* (some quadrics have a center; some do not)

(e) *symmetry*

(f) *boundedness, unboundedness.*

---

† We are excluding such degenerate quadrics as

$$1 + x^2 + y^2 = -z^2 \quad \text{and} \quad x^2 + y^2 = -z^2.$$

The first one has no points and the second consists of only one point, the origin.

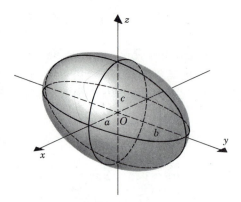

Figure 15.2.1

1. The Ellipsoid

$$\frac{x^2}{a^2} + \frac{y^2}{b^2} + \frac{z^2}{c^2} = 1.$$ (Figure 15.2.1)

The ellipsoid is centered at the origin and is symmetric about the three coordinate planes. It intersects the coordinate axes at six points: $(\pm a, 0, 0)$, $(0, \pm b, 0)$, $(0, 0, \pm c)$. These points are called the *vertices*. The surface is bounded, being contained in the rectangular solid: $|x| \le a$, $|y| \le b$, $|z| \le c$.† All three traces are ellipses; thus, for example, the trace in the xy-plane (set $z = 0$) is the ellipse

$$\frac{x^2}{a^2} + \frac{y^2}{b^2} = 1.$$

All sections parallel to the coordinate planes are also ellipses; for example, taking $y = y_0$ we have

$$\frac{x^2}{a^2} + \frac{z^2}{c^2} = 1 - \frac{y_0^2}{b^2}.$$

This ellipse is the intersection of the ellipsoid with the plane $y = y_0$. The numbers a, b, c are called the *semiaxes* of the ellipsoid. If two of the semiaxes are equal, then we have an *ellipsoid of revolution*. (If, for example, $a = c$, then all sections parallel to the xz-plane are circles and the surface can be obtained by revolving the trace in the xy-plane about the y-axis.) If all three semiaxes are equal, the surface is a *sphere*. ☐

2. The Hyperboloid of One Sheet

$$\frac{x^2}{a^2} + \frac{y^2}{b^2} - \frac{z^2}{c^2} = 1.$$ (Figure 15.2.2)

The surface is unbounded. It is centered at the origin and is symmetric about the three coordinate planes. The surface intersects the coordinate axes at four points: $(\pm a, 0, 0)$, $(0, \pm b, 0)$. The trace in the xy-plane (set $z = 0$) is the ellipse

---

† Throughout this section we take a, b, c as positive.

$$\frac{x^2}{a^2} + \frac{y^2}{b^2} = 1.$$

All sections parallel to the xy-plane are ellipses. The trace in the xz-plane (set $y = 0$) is the hyperbola

$$\frac{x^2}{a^2} - \frac{z^2}{c^2} = 1,$$

and the trace in the yz-plane (set $x = 0$) is the hyperbola

$$\frac{y^2}{b^2} - \frac{z^2}{c^2} = 1.$$

All sections parallel to the xz-plane or yz-plane are hyperbolas. If $a = b$, then all sections parallel to the xy-plane are circles and we have a *hyperboloid of revolution*. ☐

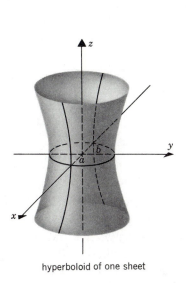

hyperboloid of one sheet

Figure 15.2.2

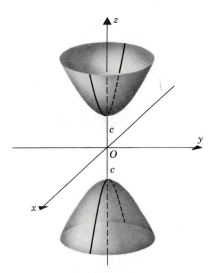

hyperboloid of two sheets

Figure 15.2.3

3. The Hyperboloid of Two Sheets

$$\frac{x^2}{a^2} + \frac{y^2}{b^2} - \frac{z^2}{c^2} = -1. \qquad \text{(Figure 15.2.3)}$$

The surface intersects the coordinate axes only at the two vertices $(0, 0, \pm c)$. The surface consists of two parts: one for which $z \geq c$, another for which $z \leq -c$. Each of these parts is unbounded. All sections parallel to the xy-plane are ellipses: for $z = z_0$ with $|z_0| \geq c$, we have

$$\frac{x^2}{a^2} + \frac{y^2}{b^2} = \frac{z_0^2}{c^2} - 1.$$

Sections parallel to the other coordinate planes are hyperbolas; for example, for $y = y_0$ we have

$$\frac{z^2}{c^2} - \frac{x^2}{a^2} = 1 + \frac{y_0^2}{b^2}.$$

The surface is symmetric about the three coordinate planes and is centered at the origin. ☐

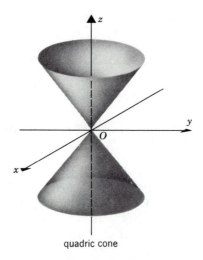

quadric cone

Figure 15.2.4

4. The Quadric Cone

$$\frac{x^2}{a^2} + \frac{y^2}{b^2} = z^2.$$ (Figure 15.2.4)

The surface intersects the coordinate axes only at the origin. The surface is un-bounded. Once again there is symmetry about the three coordinate planes. The trace in the xz-plane is a pair of intersecting lines: $z = \pm x/a$. The trace in the yz-plane is also a pair of intersecting lines: $z = \pm y/b$. The trace in the xy-plane is just the origin. Sections parallel to the xy-plane are ellipses. If $a = b$, these sections are circles and we have what is commonly called a *double circular cone* or simply a *cone*. The upper and lower portions of the cone are called *nappes*.   □

We come now to the *paraboloids*. The equations in standard form will involve x^2 and y^2, but then z instead of z^2.

5. The Elliptic Paraboloid

$$\frac{x^2}{a^2} + \frac{y^2}{b^2} = z.$$ (Figure 15.2.5)

The surface does not extend below the xy-plane. Otherwise it is unbounded. The origin is called the *vertex*. Sections parallel to the xy-plane are ellipses; sections parallel to the other coordinate planes are parabolas. Hence the term "elliptic paraboloid." The surface is symmetric about the xz-plane and about the yz-plane.

It is also symmetric about the z-axis. If $a = b$, then the surface is a *paraboloid of revolution.* ▫

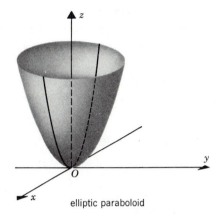

elliptic paraboloid

Figure 15.2.5

6. The Hyperbolic Paraboloid

$$\frac{x^2}{a^2} - \frac{y^2}{b^2} = z.$$ (Figure 15.2.6)

Here there is symmetry about the xz-plane and yz-plane. Sections parallel to the xy-plane are hyperbolas; sections parallel to the other coordinate planes are parabolas. Hence the term "hyperbolic paraboloid." The origin is a minimum point for the trace in the xz-plane but a maximum point for the trace in the yz-plane. The origin is called a *minimax* or *saddle point* of the surface. ▫

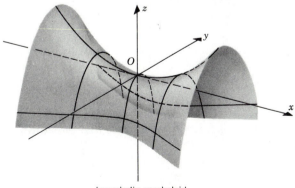

hyperbolic paraboloid

Figure 15.2.6

The rest of the quadric surfaces are *cylinders*. The term deserves definition. Take any plane curve C. All the lines through C that are perpendicular to the plane of C form a surface. Such a surface is called a *cylinder, the cylinder with base curve C*. The perpendicular lines are known as the *generators* of the cylinder.

If the base curve lies in the xy-plane (or in a plane parallel to the xy-plane), then the generators of the cylinder are parallel to the z-axis. In such a case the equation of the cylinder involves only x and y. The z-coordinate is left unrestricted; it can take on all values.

There are three types of quadric cylinders. We will give you their equations in standard form: base curve in the xy-plane, generators parallel to the z-axis.

7. The Parabolic Cylinder

$$x^2 = 4cy.$$
(Figure 15.2.7)

This surface is formed by all lines that pass through the parabola $x^2 = 4cy$ and are perpendicular to the xy-plane. ☐

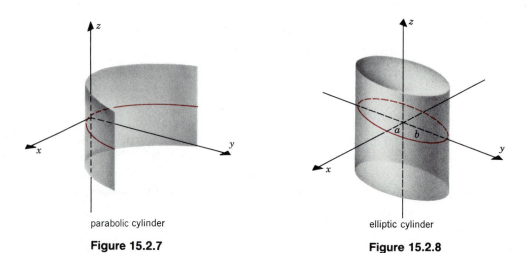

parabolic cylinder

Figure 15.2.7

elliptic cylinder

Figure 15.2.8

8. The Elliptic Cylinder

$$\frac{x^2}{a^2} + \frac{y^2}{b^2} = 1.$$
(Figure 15.2.8)

The surface is formed by all lines that pass through the ellipse

$$\frac{x^2}{a^2} + \frac{y^2}{b^2} = 1$$

and are perpendicular to the xy-plane. If $a = b$, we have the common *right circular cylinder*. ☐

9. The Hyperbolic Cylinder

$$\frac{x^2}{a^2} - \frac{y^2}{b^2} = 1.$$

(Figure 15.2.9) The surface has two parts, each generated by a branch of the hyperbola

$$\frac{x^2}{a^2} - \frac{y^2}{b^2} = 1. \quad \square$$

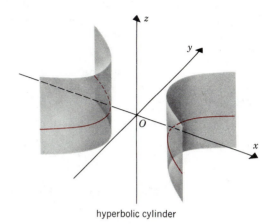

hyperbolic cylinder

Figure 15.2.9

Projections

Suppose that S_1: $z = f(x, y)$ and S_2: $z = g(x, y)$ are surfaces in three-space that intersect in a space curve C. See Figure 15.2.10.

The curve C is the set of all points (x, y, z) with

$$z = f(x, y) \qquad \text{and} \qquad z = g(x, y).$$

The set of all points (x, y, z) with

$$f(x, y) = g(x, y) \qquad \text{(here z is unrestricted)}$$

is the vertical cylinder that passes through C.

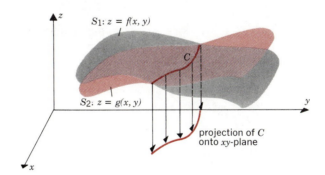

Figure 15.2.10

The set of all points $(x, y, 0)$ with

$$f(x, y) = g(x, y) \qquad \text{(here $z = 0$)}$$

is called the *projection of C onto the xy-plane.* In Figure 15.2.10 it appears as the curve in the xy-plane that lies directly below C.

Example 1. The paraboloid of revolution

$$z = x^2 + y^2$$

and the plane

$$z = 2y + 3$$

intersect in a curve C. See Figure 15.2.11. The projection of this curve onto the xy-plane is the set of all points $(x, y, 0)$ with

$$x^2 + y^2 = 2y + 3.$$

This equation can be written

$$x^2 + (y - 1)^2 = 4.$$

The projection of C onto the xy-plane is the circle of radius 2 centered at $(0, 1, 0)$. ☐

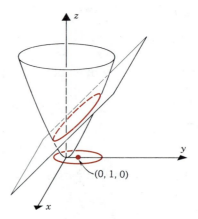

Figure 15.2.11

EXERCISES 15.2

Identify the surface.

1. $x^2 + 4y^2 - 16z^2 = 0$.

2. $x^2 + 4y^2 + 16z^2 - 12 = 0$.

3. $x - 4y^2 = 0$.

4. $x^2 - 4y^2 - 2z = 0$.

5. $5x^2 + 2y^2 - 6z^2 - 10 = 0$.

6. $2x^2 + 4y^2 - 1 = 0$.

7. $x^2 + y^2 + z^2 - 4 = 0$.

8. $5x^2 + 2y^2 - 6z^2 + 10 = 0$.

9. $x^2 + 2y^2 - 4z = 0$.

10. $2x^2 - 3y^2 - 6 = 0$.

11. $x - y^2 + 2z^2 = 0$.

12. $x - y^2 - 6z^2 = 0$.

Sketch the following cylinders.

13. $25y^2 + 4z^2 - 100 = 0$.　　**14.** $25x^2 + 4y^2 - 100 = 0$.　　**15.** $y^2 - z = 0$.

16. $x^2 - y + 1 = 0$.　　　　　**17.** $y^2 + z = 0$.　　　　　**18.** $25x^2 - 9y^2 - 225 = 0$.

Identify the surface and find the traces.　Then sketch the surface.

19. $9x^2 + 4y^2 - 36z = 0$.　　　　　　**20.** $9x^2 + 4y^2 + 36z^2 - 36 = 0$.

21. $9x^2 + 4y^2 - 36z^2 = 0$.　　　　　　**22.** $9x^2 + 4y^2 - 36z^2 - 36 = 0$.

23. $9x^2 + 4y^2 - 36z^2 + 36 = 0$.　　　　**24.** $9x^2 - 4y^2 - 36z = 0$.

25. Identify all possibilities for the surface

$$z = Ax^2 + By^2$$

taking (a) $AB > 0$.　　(b) $AB < 0$.　　(c) $AB = 0$.

26. Find the planes of symmetry for the cylinder $x - 4y^2 = 0$.

27. Write an equation for the surface obtained by revolving the parabola $4z - y^2 = 0$ about the z-axis.

28. The hyperbola $c^2y^2 - b^2z^2 - b^2c^2 = 0$ is revolved about the z-axis.　Find an equation for the resulting surface.

29. (a) The equation

$$\sqrt{x^2 + y^2} = kz \qquad \text{with } k > 0$$

represents the upper nappe of a cone, with vertex at the origin and the positive z-axis as the axis of symmetry.　Describe the section in the plane $z = z_0$, $z_0 > 0$.

   (b) Let S be one nappe of a cone, with vertex at the origin.　Write an equation for S given that

   　(i) the negative z-axis is the axis of symmetry and the section in the plane $z = -2$ is a circle of radius 6.

   　(ii) the positive y-axis is the axis of symmetry and the section in the plane $y = 3$ is a circle of radius 1.

30. Form the elliptic paraboloid

$$x^2 + \frac{y^2}{b^2} = z.$$

   (a) Describe the section in the plane $z = 1$.

   (b) What happens to this section as b tends to infinity?

   (c) What happens to the paraboloid as b tends to infinity?

The given surfaces intersect in a space curve C.　Determine the projection of C onto the xy-plane.

31. The planes $x + 2y + 3z = 6$ and $x + y - 2z = 6$.

32. The planes $x - 2y + z = 4$ and $3x + y - 2z = 1$.

33. The sphere $x^2 + y^2 + (z - 1)^2 = \frac{3}{2}$ and the hyperboloid $x^2 + y^2 - z^2 = 1$.

34. The sphere $x^2 + y^2 + (z - 2)^2 = 2$ and the cone $x^2 + y^2 = z^2$.

35. The paraboloids $x^2 + y^2 + z = 4$ and $x^2 + 3y^2 = z$.

36. The cylinder $y^2 + z - 4 = 0$ and the paraboloid $x^2 + 3y^2 = z$.

37. The cone $x^2 + y^2 = z^2$ and the plane $y + z = 2$.

38. The cone $x^2 + y^2 = z^2$ and the plane $y + 2z = 2$.

15.3   GRAPHS; LEVEL CURVES AND LEVEL SURFACES

We begin with a function f of two variables defined on a subset D of the xy-plane. By the *graph of f* we mean the graph of the equation

$$z = f(x, y), \qquad (x, y) \in D.$$

Example 1.   In the case of

$$f(x, y) = x^2 + y^2$$

the domain is the entire plane.   The graph of f is a paraboloid of revolution:

$$z = x^2 + y^2.$$

This surface can be generated by revolving the parabola

$$z = x^2$$

about the z-axis.   See Figure 15.3.1.   □

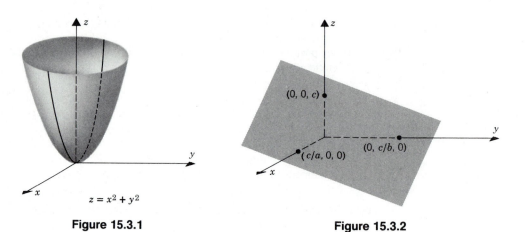

$$z = x^2 + y^2$$

Figure 15.3.1   **Figure 15.3.2**

Example 2.   The domain of the function

$$g(x, y) = c - ax - by$$

is also the entire xy-plane.   The graph of g is the plane

$$z = c - ax - by \qquad\qquad \text{(Figure 15.3.2)}$$

with intercepts $x = c/a$, $y = c/b$, $z = c$.   □

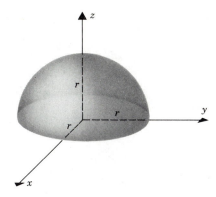

Figure 15.3.3

Example 3. The function

$$f(x, y) = \sqrt{r^2 - (x^2 + y^2)}$$

is defined only on the closed disc $x^2 + y^2 \leq r^2$. The graph is the surface

$$z = \sqrt{r^2 - (x^2 + y^2)},$$ (Figure 15.3.3)

the upper half of the sphere

$$x^2 + y^2 + z^2 = r^2. \quad \square$$

Example 4. The function

$$f(x, y) = xy$$

is simple enough, but its graph, the surface $z = xy$, is quite difficult to draw. It is a "saddle-shaped" surface, a hyperbolic paraboloid: rotate the x and y axes by $\frac{1}{4}\pi$ radians and the equation takes the form

$$\frac{X^2}{2} - \frac{Y^2}{2} = z.† \quad \square$$

Level Curves

In practice the graph of a function of two variables is usually too difficult to draw. Moreover, if drawn, it is often difficult to interpret. Fortunately, there is an easier way to convey the same information.

We take our cue from the art of the map maker. In mapping mountainous terrain it is a common practice to sketch curves joining points of constant elevation. A collection of such curves, properly labeled, gives a good idea of the altitude variations in a region. (See Figure 15.3.4.)

---

† To see this, apply Formula *9.6.2 setting $\alpha = \frac{1}{4}\pi$.

Figure 15.3.4

We can do the same thing to portray functions of two variables. Suppose that f is a nonconstant function defined on some portion of the xy-plane. If c is some value taken on by f, then we can sketch the curve $f(x, y) = c$. Such a curve is called a *level curve* for f. It can be obtained by intersecting the graph of f with the horizontal plane $z = c$ and then projecting this intersection onto the xy-plane. See Figure 15.3.5.

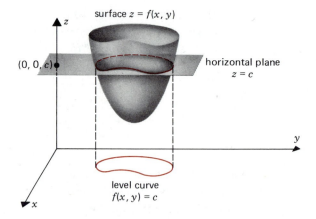

Figure 15.3.5

The level curve $f(x, y) = c$ lies entirely in the domain of f, and on this curve f is constantly c.

A collection of level curves, properly drawn and labeled, can lead to a good understanding of the overall behavior of a function.

Example 5. We begin with the function $f(x, y) = x^2 + y^2$. The graph appeared in Figure 15.3.1. The level curves are circles centered at the origin:

$$x^2 + y^2 = c, \qquad c \geq 0. \qquad \text{(Figure 15.3.6)}$$

The function takes on the value c on the circle of radius $\sqrt{c}$. At the origin the function is zero. ☐

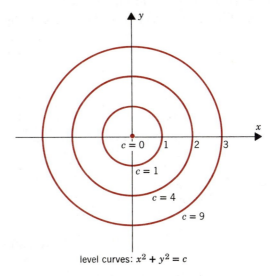

level curves: $x^2 + y^2 = c$

Figure 15.3.6

Example 6. The graph of the function $g(x, y) = x + y + 1$ is a plane. The level curves are parallel lines of the form

$$x + y + 1 = c.$$

Some are drawn in Figure 15.3.7. ☐

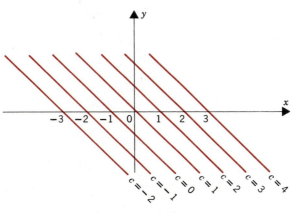

level curves: $x + y + 1 = c$

Figure 15.3.7

Example 7. Now we consider the function

$$h(x, y) = \begin{cases} \sqrt{x^2 + y^2}, & x \geq 0 \\ |y|, & x < 0 \end{cases}.$$

For $x \geq 0$, $h(x, y)$ is the distance from (x, y) to the origin. For $x < 0$, $h(x, y)$ is the distance from (x, y) to the x-axis. The level curves are pictured in Figure 15.3.8. The 0-level curve is the nonpositive x-axis. The other level curves are horseshoe-shaped: pairs of horizontal rays capped on the right by semicircles. ☐

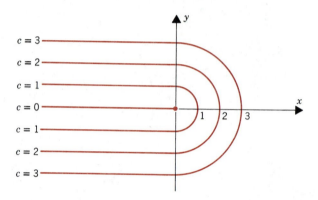

Figure 15.3.8

Example 8. Let's return to the function $f(x, y) = xy$. Earlier we noted that the graph is a "saddle-shaped" surface. You can visualize the surface from the few level curves sketched in Figure 15.3.9. The 0-level curve, $xy = 0$, consists of the two coordinate axes. The other level curves, $xy = c$ with $c \neq 0$, are hyperbolas. ☐

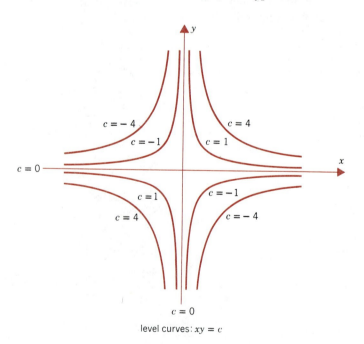

level curves: $xy = c$

Figure 15.3.9

Level Surfaces

Drawing graphs for functions of two variables is usually quite difficult. Drawing graphs for functions of three variables is actually impossible. To draw such figures we would need four dimensions at our disposal; the domain itself is a portion of three-space.

One can try to visualize the behavior of a function of three variables by examining the *level surfaces*. These are the subsets of the domain with equations of the form

$$f(x, y, z) = c.$$

Level surfaces are usually difficult to draw. Nevertheless, a knowledge of what they are can be helpful. Here we restrict ourselves to a few simple examples.

Example 9. For the function $f(x, y, z) = Ax + By + Cz$ the level surfaces are parallel planes

$$Ax + By + Cz = c. \quad \square$$

Example 10. For the function $g(x, y, z) = \sqrt{x^2 + y^2 + z^2}$ the level surfaces are concentric spheres

$$x^2 + y^2 + z^2 = c^2. \quad \square$$

Example 11. As our final example we take the function

$$f(x, y, z) = \frac{|z|}{x^2 + y^2}.$$

We extend this function to the origin by defining it to be zero there. At other points of the z-axis we leave f undefined.

In the first place note that f takes on only nonnegative values. Since f is zero only when $z = 0$, the 0-level surface is the xy-plane. To find the other level surfaces, we take $c > 0$ and set $f(x, y, z) = c$. This gives

$$\frac{|z|}{x^2 + y^2} = c \quad \text{and thus} \quad |z| = c(x^2 + y^2).$$

(Figure 15.3.10) Each of these surfaces is a double-paraboloid of revolution.† $\square$

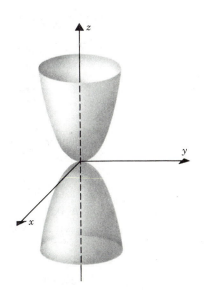

level surface: $|z| = c(x^2 + y^2)$, $(c > 0)$

Figure 15.3.10

---

† Surface 5 of the last section together with its mirror image below the xy-plane.

EXERCISES 15.3

Identify the level curves and sketch some of them.

1. $f(x, y) = x - y.$ **2.** $f(x, y) = 2x - y.$ **3.** $f(x, y) = x^2 - y.$

4. $f(x, y) = \dfrac{1}{x - y^2}.$ **5.** $f(x, y) = \dfrac{x}{x + y}.$ **6.** $f(x, y) = \dfrac{y}{x^2}.$

7. $f(x, y) = x^3 - y.$ **8.** $f(x, y) = e^{xy}.$ **9.** $f(x, y) = x^2 - y^2.$
10. $f(x, y) = x^2.$ **11.** $f(x, y) = y^2.$ **12.** $f(x, y) = x(y - 1).$

13. $f(x, y) = \ln(x^2 + y^2).$ **14.** $f(x, y) = \ln\left(\dfrac{y}{x^2}\right).$ **15.** $f(x, y) = \dfrac{\ln y}{x^2}.$

16. $f(x, y) = x^2 y^2.$ **17.** $f(x, y) = \dfrac{x^2}{x^2 + y^2}.$ **18.** $f(x, y) = \dfrac{\ln y}{x}.$

Identify the c-level surface and sketch it.

19. $f(x, y, z) = x + 2y + 3z, \quad c = 0.$ **20.** $f(x, y, z) = x^2 + y^2, \quad c = 4.$
21. $f(x, y, z) = z(x^2 + y^2)^{-1/2}, \quad c = 1.$ **22.** $f(x, y, z) = x^2/4 + y^2/6 + z^2/9, \quad c = 1.$
23. $f(x, y, z) = 4x^2 + 9y^2 - 72z, \quad c = 0.$ **24.** $f(x, y, z) = z^2 - 36x^2 - 9y^2, \quad c = 1.$

25. Identify the c-level surfaces of

$$f(x, y, z) = x^2 + y^2 - z^2$$

taking (i) $c < 0.$ (ii) $c = 0.$ (iii) $c > 0.$

15.4   PARTIAL DERIVATIVES

Two Variables

Let f be a function of x and y; for example

$$f(x, y) = 3x^2 y - 5x \cos \pi y.$$

The *partial derivative of f with respect to x* is the function f_x obtained by differentiating f with respect to x, treating y as a constant; in this case

$$f_x(x, y) = 6xy - 5 \cos \pi y.$$

The *partial derivative of f with respect to y* is the function f_y obtained by differentiating f with respect to y, treating x as a constant; in this case

$$f_y(x, y) = 3x^2 + 5\pi x \sin \pi y.$$

These partial derivatives are formally defined as limits:

(15.4.1)

$$f_x(x, y) = \lim_{h \to 0} \frac{f(x + h, y) - f(x, y)}{h},$$

$$f_y(x, y) = \lim_{h \to 0} \frac{f(x, y + h) - f(x, y)}{h}.$$

Example 1. For

$$f(x, y) = x \tan^{-1} xy$$

we have

$$f_x(x, y) = x \frac{y}{1 + (xy)^2} + \tan^{-1} xy = \frac{xy}{1 + x^2y^2} + \tan^{-1} xy$$

and

$$f_y(x, y) = x \frac{x}{1 + (xy)^2} = \frac{x^2}{1 + x^2y^2}. \quad \square$$

In the one-variable case, $f'(x_0)$ gives the rate of change with respect to x of $f(x)$ at $x = x_0$. In the two-variable case, $f_x(x_0, y_0)$ *gives the rate of change with respect to x of $f(x, y_0)$ at $x = x_0$, and $f_y(x_0, y_0)$ gives the rate of change with respect to y of $f(x_0, y)$ at $y = y_0$.*

Example 2. For the function

$$f(x, y) = e^{xy} + \ln (x^2 + y)$$

we have

$$f_x(x, y) = ye^{xy} + \frac{2x}{x^2 + y} \qquad \text{and} \qquad f_y(x, y) = xe^{xy} + \frac{1}{x^2 + y}.$$

The number

$$f_x(2, 1) = e^2 + \frac{4}{4 + 1} = e^2 + \frac{4}{5}$$

gives the rate of change with respect to x of the function

$$f(x, 1) = e^x + \ln (x^2 + 1) \qquad \text{at } x = 2;$$

the number

$$f_y(2, 1) = 2e^2 + \frac{1}{4 + 1} = 2e^2 + \frac{1}{5}$$

gives the rate of change with respect to y of the function

$$f(2, y) = e^{2y} + \ln (4 + y) \qquad \text{at } y = 1. \quad \square$$

A Geometric Interpretation

In Figure 15.4.1 we have sketched a surface $z = f(x, y)$, which you can take as everywhere defined. Through the surface we have passed a plane $y = y_0$ parallel to the xz-plane. The plane $y = y_0$ intersects the surface in a curve, the y_0-section of the surface.

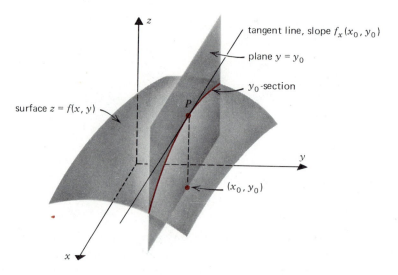

Figure 15.4.1

The y_0-section of the surface is the graph of the function

$$g(x) = f(x, y_0).$$

Differentiating with respect to x, we have

$$g'(x) = f_x(x, y_0)$$

and in particular

$$g'(x_0) = f_x(x_0, y_0).$$

The number $f_x(x_0, y_0)$ is thus the slope of the y_0-section of the surface $z = f(x, y)$ at the point $P = P(x_0, y_0, f(x_0, y_0))$.

   The other partial derivative f_y can be given a similar interpretation. In Figure 15.4.2 you can see the same surface $z = f(x, y)$, this time sliced by a plane $x = x_0$ parallel to the yz-plane. The plane $x = x_0$ intersects the surface in a curve, the x_0-section of the surface.

   The x_0-section of the surface is the graph of the function

$$h(y) = f(x_0, y).$$

Differentiating, this time with respect to y, we have

$$h'(y) = f_y(x_0, y)$$

and thus

$$h'(y_0) = f_y(x_0, y_0).$$

The number $f_y(x_0, y_0)$ is the slope of the x_0-section of the surface $z = f(x, y)$ at the point $P = P(x_0, y_0, f(x_0, y_0))$.

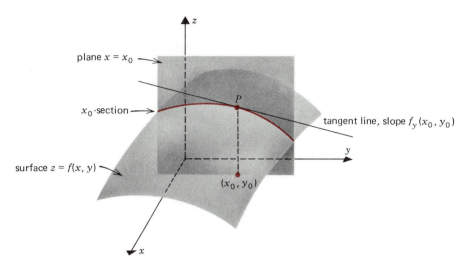

Figure 15.4.2

Three Variables

In the case of a function of three variables, you can look for three partial derivatives: the partial with respect to x, the partial with respect to y, and also the partial with respect to z. These partials

$$f_x(x, y, z), \qquad f_y(x, y, z), \qquad f_z(x, y, z)$$

are defined as follows:

(15.4.2)

$$f_x(x, y, z) = \lim_{h \to 0} \frac{f(x + h, y, z) - f(x, y, z)}{h},$$

$$f_y(x, y, z) = \lim_{h \to 0} \frac{f(x, y + h, z) - f(x, y, z)}{h},$$

$$f_z(x, y, z) = \lim_{h \to 0} \frac{f(x, y, z + h) - f(x, y, z)}{h}.$$

Each partial can be found by differentiating with respect to the subscript variable, treating the other two variables as constants.

Example 3. For the function

$$f(x, y, z) = xy^2z^3$$

the partial derivatives are

$$f_x(x, y, z) = y^2z^3, \qquad f_y(x, y, z) = 2xyz^3, \qquad f_z(x, y, z) = 3xy^2z^2.$$

In particular

$$f_x(1, -2, -1) = -4, \qquad f_y(1, -2, -1) = 4, \qquad f_z(1, -2, -1) = 12. \quad \square$$

Example 4. For

$$g(x, y, z) = x^2 e^{y/z}$$

we have

$$g_x(x, y, z) = 2xe^{y/z}, \qquad g_y(x, y, z) = \frac{x^2}{z} e^{y/z}, \qquad g_z(x, y, z) = -\frac{x^2 y}{z^2} e^{y/z}. \quad \square$$

Example 5. For a function of the form

$$f(x, y, z) = F(x, y)G(y, z)$$

we can write

$$f_x(x, y, z) = F_x(x, y)G(y, z),$$
$$f_y(x, y, z) = F(x, y)G_y(y, z) + F_y(x, y)G(y, z),$$
$$f_z(x, y, z) = F(x, y)G_z(y, z). \quad \square$$

The number $f_x(x_0, y_0, z_0)$ gives the rate of change with respect to x of $f(x, y_0, z_0)$ at $x = x_0$; $f_y(x_0, y_0, z_0)$ gives the rate of change with respect to y of $f(x_0, y, z_0)$ at $y = y_0$; and $f_z(x_0, y_0, z_0)$ gives the rate of change with respect to z of $f(x_0, y_0, z)$ at $z = z_0$.

Example 6. The function

$$f(x, y, z) = xy^2 - yz^2$$

has partial derivatives

$$f_x(x, y, z) = y^2, \qquad f_y(x, y, z) = 2xy - z^2, \qquad f_z(x, y, z) = -2yz.$$

The number $f_x(1, 2, 3) = 4$ gives the rate of change with respect to x of the function

$$f(x, 2, 3) = 4x - 18 \qquad \text{at } x = 1;$$

$f_y(1, 2, 3) = -5$ gives the rate of change with respect to y of the function

$$f(1, y, 3) = y^2 - 9y \qquad \text{at } y = 2;$$

$f_z(1, 2, 3) = -12$ gives the rate of change with respect to z of the function

$$f(1, 2, z) = 4 - 2z^2 \qquad \text{at } z = 3. \quad \square$$

The geometric significance that we attached to the partials of a function of two variables has no analog in the case of three variables. Figures 15.4.1 and 15.4.2 do not generalize to three variables.

Other Notations

There is obviously no need to restrict ourselves to the variables x, y, z. Where more convenient we can use other letters.

Problem 7.   The volume of the frustum of a cone
(Figure 15.4.3) is given by the function

$$V(R, r, h) = \tfrac{1}{3}\pi h(R^2 + Rr + r^2).$$

At time t_0, $R = 8$, $r = 4$, $h = 6$.   Find the rate of
change of the volume with respect to each of its
dimensions at time t_0 if the other dimensions are
held constant.

frustum of a cone

Figure 15.4.3

Solution.   The partial derivatives of V are as follows:

$$V_R(R, r, h) = \tfrac{1}{3}\pi h(2R + r),$$
$$V_r(R, r, h) = \tfrac{1}{3}\pi h(R + 2r),$$
$$V_h(R, r, h) = \tfrac{1}{3}\pi(R^2 + Rr + r^2).$$

At time t_0,

the rate of change of V with respect to R is $V_R(8, 4, 6) = 40\pi$,
the rate of change of V with respect to r is $V_r(8, 4, 6) = 32\pi$,
the rate of change of V with respect to h is $V_h(8, 4, 6) = \frac{112}{3}\pi$.   ◻

The subscript notation is not the only one used for partial differentiation.   A
variant of Leibniz's double-d notation is also commonly used.   In this notation the
partials f_x, f_y, f_z are denoted by

$$\frac{\partial f}{\partial x}, \quad \frac{\partial f}{\partial y}, \quad \frac{\partial f}{\partial z}.$$

Thus, for

$$f(x, y, z) = x^3 y^2 z + \sin xy$$

we have

$$\frac{\partial f}{\partial x}(x, y, z) = 3x^2 y^2 z + y\cos xy,$$

$$\frac{\partial f}{\partial y}(x, y, z) = 2x^3 yz + x\cos xy,$$

$$\frac{\partial f}{\partial z}(x, y, z) = x^3 y^2,$$

or more simply,

$$\frac{\partial f}{\partial x} = 3x^2 y^2 z + y\cos xy,$$

$$\frac{\partial f}{\partial y} = 2x^3 yz + x\cos xy,$$

$$\frac{\partial f}{\partial z} = x^3 y^2.$$

We can also write

$$\frac{\partial}{\partial x}(x^3y^2z + \sin xy) = 3x^2y^2z + y\cos xy,$$

$$\frac{\partial}{\partial y}(x^3y^2z + \sin xy) = 2x^3yz + x\cos xy,$$

$$\frac{\partial}{\partial z}(x^3y^2z + \sin xy) = x^3y^2.$$

The "double-decker" notation is not restricted to the letters x, y, z.   For instance, we can write

$$\frac{\partial}{\partial r}(r^2\cos\theta + e^{\theta r}) = 2r\cos\theta + \theta e^{\theta r},$$

$$\frac{\partial}{\partial\theta}(r^2\cos\theta + e^{\theta r}) = -r^2\sin\theta + re^{\theta r}.$$

For the function

$$\rho = \sin 2\theta \cos 3\phi$$

we have

$$\frac{\partial\rho}{\partial\theta} = 2\cos 2\theta\cos 3\phi \quad\text{and}\quad \frac{\partial\rho}{\partial\phi} = -3\sin 2\theta\sin 3\phi.$$

EXERCISES 15.4

Calculate the partial derivatives.

1. $f(x, y) = 3x^2 - xy + y.$

2. $g(x, y) = x^2e^{-y}.$

3. $\rho = \sin\phi\cos\theta.$

4. $\rho = \sin^2(\theta - \phi).$

5. $f(x, y) = e^{x-y} - e^{y-x}.$

6. $z = \sqrt{x^2 - 3y}.$

7. $g(x, y) = \dfrac{Ax + By}{Cx + Dy}.$

8. $u = \dfrac{e^z}{xy^2}.$

9. $u = xy + yz + zx.$

10. $z = Ax^2 + Bxy + Cy^2.$

11. $f(x, y, z) = z\sin(x - y).$

12. $g(u, v, w) = \ln(u^2 + vw - w^2).$

13. $\rho = e^{\theta+\phi}\cos(\theta - \phi).$

14. $f(x, y) = (x + y)\sin(x - y).$

15. $h(x, y) = [f(x)]^2g(y).$

16. $h(x, y) = e^{f(x)g(y)}.$

17. $f(x, y, z) = z^{xy^2}.$

18. $h(x, y, z) = [f(x, y)]^3[g(x, z)]^2.$

19. Find $f_x(0, e)$ and $f_y(0, e)$ given that $f(x, y) = e^x\ln y.$

20. Find $g_x(0, \frac{1}{4}\pi)$ and $g_y(0, \frac{1}{4}\pi)$ given that $g(x, y) = e^{-x}\sin(x + 2y).$

21. Find $f_x(1, 2)$ and $f_y(1, 2)$ given that $f(x, y) = \dfrac{x}{x + y}.$

22. Find $g_x(1, 2)$ and $g_y(1, 2)$ given that $g(x, y) = \dfrac{y}{x + y^2}.$

Find $f_x(x, y)$ and $f_y(x, y)$ by forming the appropriate difference quotient and taking the limit as h tends to zero (Definition 15.4.1).

23. $f(x, y) = x^2y$.   **24.** $f(x, y) = y^2$.   **25.** $f(x, y) = \ln(x^2y)$.

26. $f(x, y) = \dfrac{1}{x + 4y}$.   **27.** $f(x, y) = \dfrac{1}{x - y}$.   **28.** $f(x, y) = e^{2x+3y}$.

Find $f_x(x, y, z)$, $f_y(x, y, z)$, and $f_z(x, y, z)$ by forming the appropriate difference quotient and taking the limit as h tends to zero (Definition 15.4.2).

29. $f(x, y, z) = xy^2z$.   **30.** $f(x, y, z) = \dfrac{x^2y}{z}$.

31. Show that

$$u = Ax^4 + 2Bx^2y^2 + Cy^4 \quad \text{satisfies the equation} \quad x\frac{\partial u}{\partial x} + y\frac{\partial u}{\partial y} = 4u.\dagger$$

32. Show that

$$u = \frac{x^2y^2}{x + y} \quad \text{satisfies the equation} \quad x\frac{\partial u}{\partial x} + y\frac{\partial u}{\partial y} = 3u.$$

33. Given that $x = r\cos\theta$ and $y = r\sin\theta$, find

$$\frac{\partial x}{\partial r}\frac{\partial y}{\partial \theta} - \frac{\partial x}{\partial \theta}\frac{\partial y}{\partial r}.$$

34. Show that

$$u = x^2y + y^2z + z^2x \quad \text{satisfies the equation} \quad \frac{\partial u}{\partial x} + \frac{\partial u}{\partial y} + \frac{\partial u}{\partial z} = (x + y + z)^2.$$

35. For a gas confined in a container, the ideal gas law states that the pressure P is related to the volume V and the temperature T by an equation of the form

$$P = k\frac{T}{V}$$

where k is a positive constant. Show that

$$V\frac{\partial P}{\partial V} = -P \quad \text{and} \quad V\frac{\partial P}{\partial V} + T\frac{\partial P}{\partial T} = 0.$$

36. The law of cosines for a triangle can be written

$$a^2 = b^2 + c^2 - 2bc\cos\theta.$$

At time t_0 we have $b_0 = 10$ inches, $c_0 = 15$ inches, $\theta_0 = \frac{1}{3}\pi$ radians.
(a) Find a_0.
(b) Find the rate of change of a with respect to b at time t_0 if c and θ remain constant.
(c) Using the rate found in (b), calculate (by differentials) the approximate change in a if b is decreased by 1 inch.
(d) Find the rate of change of a with respect to θ at time t_0 if b and c remain constant.
(e) Find the rate of change of c with respect to θ at time t_0 if a and b remain constant.

---

† Equations that involve partial derivatives are called *partial differential equations*. Such equations play a prominent role in mathematical physics.

37. The area of a triangle is given by the formula

$$A = \tfrac{1}{2}bc \sin \theta.$$

At time t_0 we have $b_0 = 10$ inches, $c_0 = 20$ inches, $\theta_0 = \tfrac{1}{3}\pi$ radians.
(a) Find the area of the triangle at time t_0.
(b) Find the rate of change of the area with respect to b at time t_0 if c and θ remain constant.
(c) Find the rate of change of the area with respect to θ at time t_0 if b and c remain constant.
(d) Using the rate found in (c), calculate (by differentials) the approximate change in area if the angle is increased by one degree.
(e) Find the rate of change of c with respect to b at time t_0 if the area and the angle are to remain constant.

38. Let f be the function of x and y that satisfies a relation of the form

$$\frac{\partial f}{\partial x} = kf, \qquad k \text{ a constant}.$$

Show that

$$f(x, y) = g(y)e^{kx}$$

where g is some function of y.

39. Let $z = f(x, y)$ be a surface everywhere defined.
(a) Find a vector function that parametrizes the y_0-section of the surface. (See Figure 15.4.1.) Find a vector function that parametrizes the line tangent to this section at the point $P(x_0, y_0, f(x_0, y_0))$.
(b) Find a vector function that parametrizes the x_0-section of the surface. (See Figure 15.4.2.) Find a vector function that parametrizes the line tangent to this section at the point $P(x_0, y_0, f(x_0, y_0))$.
(c) Show that the equation of the plane determined by the tangent lines found in (a) and (b) can be written

$$z - f(x_0, y_0) = (x - x_0)\frac{\partial f}{\partial x}(x_0, y_0) + (y - y_0)\frac{\partial f}{\partial y}(x_0, y_0).$$

(We will show in Chapter 16 that, for certain functions of two variables, this plane plays a role similar to the role played by the tangent line for differentiable functions of a single variable.)

40. (*A chain rule*) Let f be a function of x and y, and g be a function of a single variable. Form the composition $h(x, y) = g(f(x, y))$ and show that

$$h_x(x, y) = g'(f(x, y))f_x(x, y) \qquad \text{and} \qquad h_y(x, y) = g'(f(x, y))f_y(x, y).$$

In Leibniz's notation, setting $u = f(x, y)$, we have

$$\frac{\partial h}{\partial x} = \frac{dg}{du}\frac{\partial u}{\partial x} \qquad \text{and} \qquad \frac{\partial h}{\partial y} = \frac{dg}{du}\frac{\partial u}{\partial y}.$$

41. Let g be a differentiable function of a single variable. Use Exercise 40 to verify the following results.
(a) If $w = g(ax + by)$, then

$$b\frac{\partial w}{\partial x} = a\frac{\partial w}{\partial y}.$$

(b) If m and n are nonzero integers and $w = g(x^m y^n)$, then

$$nx\frac{\partial w}{\partial x} = my\frac{\partial w}{\partial y}.$$

15.5 OPEN SETS AND CLOSED SETS

Points in the domain of a function of several variables can be written in vector notation. In the two-variable case, set

$$\mathbf{x} = (x, y),$$

and, in the three-variable case, set

$$\mathbf{x} = (x, y, z).$$

The vector notation enables us to treat the two cases together.

In this section we introduce five important notions:

        (1) neighborhood of a point
        (2) interior of a set
        (3) boundary of a set
        (4) open set
        (5) closed set.

For our purposes, the fundamental notion here is the notion of neighborhood. The other four notions can be derived from it.

DEFINITION 15.5.1 A NEIGHBORHOOD OF A POINT

A *neighborhood* of a point $\mathbf{x}_0$ is a set of the form

$$\{\mathbf{x}: \|\mathbf{x} - \mathbf{x}_0\| < \delta\}$$

where δ is some number greater than zero.

In the plane, a neighborhood of $\mathbf{x}_0 = (x_0, y_0)$ consists of all the points inside of a disc centered at (x_0, y_0). In three-space, a neighborhood of $\mathbf{x}_0 = (x_0, y_0, z_0)$ consists of all the points inside of a ball centered at (x_0, y_0, z_0). See Figure 15.5.1.

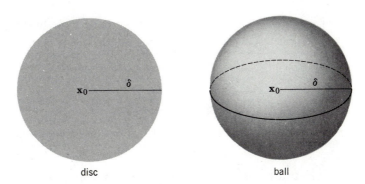

disc   ball

Figure 15.5.1

DEFINITION 15.5.2 THE INTERIOR OF A SET

A point x_0 is said to be an *interior point* of the set S iff the set S contains some neighborhood of x_0. The set of all interior points of S is called the *interior* of S.

---

a neighborhood of x_1

Ω

a neighborhood of x_2

Figure 15.5.2

Example 1. Let Ω be the plane set shown in Figure 15.5.2. The point marked x_1 is an interior point of Ω because Ω contains a neighborhood of x_1. The point x_2 is not an interior point of Ω because *no* neighborhood of x_2 is completely contained in Ω. (Every neighborhood of x_2 has points that lie outside of Ω.) ☐

DEFINITION 15.5.3 THE BOUNDARY OF A SET

A point x_0 is said to be a *boundary point* of the set S iff every neighborhood of x_0 contains points that are in S and points that are not in S. The set of all boundary points of S is called the *boundary* of S.

Example 2. The point marked x_2 in Figure 15.5.2 is a boundary point of Ω: each neighborhood of x_2 contains points in Ω and points not in Ω. ☐

DEFINITION 15.5.4 OPEN SET

A set S is said to be *open* iff it contains a neighborhood of each of its points (iff each of its points is an interior point) (iff it contains no boundary points).

DEFINITION 15.5.5 CLOSED SET

A set S is said to be *closed* iff it contains its boundary.

Here are some examples of sets that are open, sets that are closed, and sets that are neither open nor closed:

Two-Dimensional Examples

The sets

$$S_1 = \{(x, y): 1 < x < 2,\ 1 < y < 2\},$$
$$S_2 = \{(x, y): 3 \le x \le 4,\ 1 \le y \le 2\},$$
$$S_3 = \{(x, y): 5 \le x \le 6,\ 1 < y < 2\}$$

are displayed in Figure 15.5.3. S_1 is the inside of the first square. S_1 is open because it contains a neighborhood of each of its points. S_2 is the inside of the second square together with the four bounding line segments. S_2 is closed because it contains its entire boundary. S_3 is the inside of the last square together with the two vertical bounding line segments. S_3 is not open because it contains part of its boundary, and it is not closed because it does not contain all of its boundary. ☐

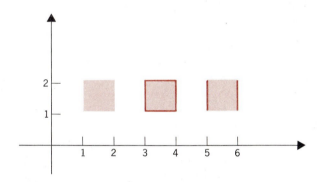

Figure 15.5.3

Three-Dimensional Examples

We now examine some three-dimensional sets:

$$S_1 = \{(x, y, z): z > x^2 + y^2\},$$
$$S_2 = \{(x, y, z): z \ge x^2 + y^2\},$$
$$S_3 = \left\{(x, y, z): 1 \ge \frac{x^2 + y^2}{z}\right\}.$$

The boundary of each of these sets is the paraboloid of revolution

$$z = x^2 + y^2. \qquad\qquad \text{(Figure 15.3.1)}$$

The first set consists of all points above this surface. This set is open because, if a point is above this surface, then all points sufficiently close to it are also above this surface. Thus the set contains a neighborhood of each of its points. The second set is closed because it contains all of its boundary. The third set is neither open nor closed. It is not open because it contains some boundary points; for example, it contains the point $(1, 1, 2)$. It is not closed because it fails to contain the boundary point $(0, 0, 0)$. ☐

A Final Remark   A neighborhood of $\mathbf{x}_0$ is a set of the form

$$\{\mathbf{x}\colon \|\mathbf{x} - \mathbf{x}_0\| < \delta\}.$$

If we remove $\mathbf{x}_0$ from the set, we have the set

$$\{\mathbf{x}\colon 0 < \|\mathbf{x} - \mathbf{x}_0\| < \delta\}.$$

Such a set is called a *deleted neighborhood* of $\mathbf{x}_0$.   (We will use deleted neighborhoods in the next section.)

EXERCISES 15.5

Specify the interior and the boundary of the set.   State whether the set is open, closed, or neither.   Then sketch the set.

1. $\{(x, y)\colon 2 \le x \le 4, 1 \le y \le 3\}$.
2. $\{(x, y)\colon 2 < x < 4, 1 < y < 3\}$.
3. $\{(x, y)\colon 1 < x^2 + y^2 < 4\}$.
4. $\{(x, y)\colon 1 \le x^2 \le 4\}$.
5. $\{(x, y)\colon 1 < x^2 \le 4\}$.
6. $\{(x, y)\colon y < x^2\}$.
7. $\{(x, y)\colon y \le x^2\}$.
8. $\{(x, y, z)\colon 1 \le x \le 2, 1 \le y \le 2, 1 \le z < 2\}$.
9. $\{(x, y, z)\colon x^2 + y^2 \le 1, 0 \le z \le 4\}$.
10. $\{(x, y, z)\colon (x - 1)^2 + (y - 1)^2 + (z - 1)^2 < \frac{1}{4}\}$.

11. Let $S = \{\mathbf{x}_1, \mathbf{x}_2, \ldots, \mathbf{x}_n\}$ be a finite set of points.   (a) What is the interior of S?   (b) What is the boundary of S?   (c) Is S open, closed, or neither?

All the notions introduced in this section can be applied to sets of real numbers: write x for $\mathbf{x}$ and $|x|$ for $\|\mathbf{x}\|$.   A *neighborhood* of a number x_0 is then a set of the form

$$\{x\colon |x - x_0| < \delta\} \qquad \text{with } \delta > 0.$$

This is just an open interval $(x_0 - \delta, x_0 + \delta)$.   In Exercises 12–19 specify the interior and boundary of the given set of real numbers.   State whether the set is open, closed, or neither.

12. $\{x\colon 1 < x < 3\}$.
13. $\{x\colon 1 \le x \le 3\}$.
14. $\{x\colon 1 \le x < 3\}$.
15. $\{x\colon x > 1\}$.
16. $\{x\colon x \le -1\}$.
17. $\{x\colon x < -1 \text{ or } x \ge 1\}$.
18. The set of positive integers: $\{1, 2, 3, \cdots, n, \cdots\}$.
19. The set of reciprocals: $(1, 1/2, 1/3, \cdots, 1/n, \cdots)$.

20. Let ϕ be the empty set.   Let X be the real line, the entire plane, or, in the three-dimensional case, all of three-space.   For each subset A of X, let $X - A$ be the set of all points $\mathbf{x} \in X$ such that $\mathbf{x} \notin A$.
(a) Show that ϕ is both open and closed.
(b) Show that X is both open and closed.   (It can be shown that ϕ and X are the only subsets of X that are both open and closed.)
(c) Let U be a subset of X.   Show that U is open iff $X - U$ is closed.
(d) Let F be a subset of X.   Show that F is closed iff $X - F$ is open.

15.6 LIMITS AND CONTINUITY; EQUALITY OF MIXED PARTIALS

The Basic Notions

The limit process used in taking partial derivatives involved nothing new because in each instance all but one of the variables remained fixed. In this section we take up limits of the form

$$\lim_{(x, y) \to (x_0, y_0)} f(x, y) \quad \text{and} \quad \lim_{(x, y, z) \to (x_0, y_0, z_0)} f(x, y, z).$$

To avoid having to treat the two- and three-variable cases separately, we will write instead

$$\lim_{\mathbf{x} \to \mathbf{x}_0} f(\mathbf{x}).$$

This gives us both the two-variable case [set $\mathbf{x} = (x, y)$ and $\mathbf{x}_0 = (x_0, y_0)$] and the three-variable case [set $\mathbf{x} = (x, y, z)$ and $\mathbf{x}_0 = (x_0, y_0, z_0)$].

To take the limit of $f(\mathbf{x})$ as $\mathbf{x}$ tends to $\mathbf{x}_0$ we do not need that f be defined at $\mathbf{x}_0$, but we do need that f be defined at points $\mathbf{x}$ close to $\mathbf{x}_0$. At this stage we will assume that f is defined at all points $\mathbf{x}$ in some deleted neighborhood of $\mathbf{x}_0$. (This will guarantee that we can form $f(\mathbf{x})$ for all $\mathbf{x} \neq \mathbf{x}_0$ that are "sufficiently close" to $\mathbf{x}_0$.)

To say that

$$\lim_{\mathbf{x} \to \mathbf{x}_0} f(\mathbf{x}) = l$$

is to say that for $\mathbf{x}$ sufficiently close to $\mathbf{x}_0$ but different from $\mathbf{x}_0$, the number $f(\mathbf{x})$ is close to l; or, to put it another way, as $\|\mathbf{x} - \mathbf{x}_0\|$ tends to zero but remains different from zero, $|f(\mathbf{x}) - l|$ tends to zero. The ϵ-δ definition is a direct generalization of the ϵ-δ definition in the single-variable case.

DEFINITION 15.6.1 THE LIMIT OF A FUNCTION OF SEVERAL VARIABLES

Let f be a function defined at least on some deleted neighborhood of $\mathbf{x}_0$.

$$\lim_{\mathbf{x} \to \mathbf{x}_0} f(\mathbf{x}) = l \quad \text{iff} \quad \begin{cases} \text{for each } \epsilon > 0 \text{ there exists } \delta > 0 \text{ such that,} \\ \text{if } \ 0 < \|\mathbf{x} - \mathbf{x}_0\| < \delta, \quad \text{then } \ |f(\mathbf{x}) - l| < \epsilon. \end{cases}$$

Observe that, if the limit l exists, then the number l must be the limiting value of $f(\mathbf{x})$ along *all* paths to $\mathbf{x}_0$. If all paths to $\mathbf{x}_0$ do not yield the same limiting value, then the limit does not exist.

Example 1. We will show that the function

$$f(x, y) = \frac{xy + y^3}{x^2 + y^2}$$

does not have a limit at $(0, 0)$.

Along the obvious paths to $(0, 0)$, the coordinate axes, the limiting value is 0:

along the x-axis, $y = 0$ and thus $f(x, y) = f(x, 0) = 0$ tends to 0;

along the y-axis, $x = 0$ and thus $f(x, y) = f(0, y) = y$ tends to 0.

However, along the line $y = 2x$, the limiting value is $\frac{2}{5}$:

$$f(x, y) = f(x, 2x) = \frac{2x^2 + 8x^3}{x^2 + 4x^2} = \tfrac{2}{5} + \tfrac{8}{5}x \quad \text{tends to } \tfrac{2}{5}.$$

We have shown that not all paths to $(0, 0)$ yield the same limiting value. It follows that f does not have a limit at $(0, 0)$. ☐

As in the one-variable case, the limit (if it exists) is unique. Moreover, if

$$\lim_{\mathbf{x} \to \mathbf{x}_0} f(\mathbf{x}) = l \quad \text{and} \quad \lim_{\mathbf{x} \to \mathbf{x}_0} g(\mathbf{x}) = m,$$

then

$$\lim_{\mathbf{x} \to \mathbf{x}_0} [f(\mathbf{x}) + g(\mathbf{x})] = l + m, \quad \lim_{\mathbf{x} \to \mathbf{x}_0} [\alpha f(\mathbf{x})] = \alpha l, \quad \lim_{\mathbf{x} \to \mathbf{x}_0} [f(\mathbf{x})g(\mathbf{x})] = lm,$$

and

$$\lim_{\mathbf{x} \to \mathbf{x}_0} [f(\mathbf{x})/g(\mathbf{x})] = l/m \quad \text{provided } m \neq 0.$$

These results are not hard to derive. You can do it simply by imitating the comparable arguments in the one-variable case.

Suppose now that $\mathbf{x}_0$ is an interior point of the domain of f. To say that f is *continuous* at $\mathbf{x}_0$ is to say that

(15.6.2)
$$\lim_{\mathbf{x} \to \mathbf{x}_0} f(\mathbf{x}) = f(\mathbf{x}_0)$$

or, equivalently, that

(15.6.3)
$$\lim_{\mathbf{h} \to 0} f(\mathbf{x}_0 + \mathbf{h}) = f(\mathbf{x}_0).$$

For two variables we can write

$$\lim_{(x, y) \to (x_0, y_0)} f(x, y) = f(x_0, y_0)$$

and for three variables

$$\lim_{(x, y, z) \to (x_0, y_0, z_0)} f(x, y, z) = f(x_0, y_0, z_0).$$

To say that f is *continuous on an open set S* is to say that f is continuous at all points of S.

Some Examples of Continuous Functions

Polynomials in several variables, for example,

$$P(x, y) = x^2y + 3x^3y^4 - x + 2y \qquad \text{and} \qquad Q(x, y, z) = 6x^3z - z^3y + 2xyz,$$

are everywhere continuous.   In the two-variable case, that means continuity at each point of the xy-plane; in the three-variable case, continuity at each point of three-space.

Rational functions (quotients of polynomials) are continuous everywhere except where the denominator is zero.   Thus

$$f(x, y) = \frac{2x - y}{x^2 + y^2}$$

is continuous at each point of the xy-plane other than the origin $(0, 0)$;

$$g(x, y) = \frac{x^4}{x - y}$$

is continuous except on the line $y = x$;

$$h(x, y) = \frac{1}{x^2 - y}$$

is continuous except on the parabola $y = x^2$;

$$F(x, y, z) = \frac{2x}{x^2 + y^2 + z^2}$$

is continuous at each point of three-space other than the origin $(0, 0, 0)$;

$$G(x, y, z) = \frac{x^5 - y}{ax + by + cz}$$

is continuous except on the plane $ax + by + cz = 0$.
   You can construct more elaborate continuous functions by forming composites: take, for example,

$$\tan^{-1}\left(\frac{xz^2}{x + y}\right), \qquad \sqrt{x^2 + y^4 + z^6}, \qquad \sin xyz.$$

The first function is continuous except along the vertical plane $x + y = 0$.   The other two functions are continuous at each point of space.   The continuity of such composites follows from a simple theorem that we state and prove below.   In the theorem, g is a function of several variables, but f is a function of a single variable.

THEOREM 15.6.4   THE CONTINUITY OF COMPOSITE FUNCTIONS

If g is continuous at the point $\mathbf{x}_0$ and f is continuous at the number $g(\mathbf{x}_0)$, then the composition $f \circ g$ is continuous at the point $\mathbf{x}_0$.

Proof. We begin with $\epsilon > 0$. We must show that there exists $\delta > 0$ such that

$$\text{if} \quad \|\mathbf{x} - \mathbf{x}_0\| < \delta \quad \text{then} \quad |f(g(\mathbf{x})) - f(g(\mathbf{x}_0))| < \epsilon.$$

From the continuity of f at $g(\mathbf{x}_0)$, we know that there exists $\delta_1 > 0$ such that

$$\text{if} \quad |y - g(\mathbf{x}_0)| < \delta_1 \quad \text{then} \quad |f(y) - f(g(\mathbf{x}_0))| < \epsilon.$$

From the continuity of g at $\mathbf{x}_0$, we know that there exists $\delta > 0$ such that

$$\text{if} \quad \|\mathbf{x} - \mathbf{x}_0\| < \delta \quad \text{then} \quad |g(\mathbf{x}) - g(\mathbf{x}_0)| < \delta_1.$$

This last δ obviously works; namely,

$$\text{if} \quad \|\mathbf{x} - \mathbf{x}_0\| < \delta \quad \text{then} \quad |g(\mathbf{x}) - g(\mathbf{x}_0)| < \delta_1 \quad \text{and} \quad |f(g(\mathbf{x})) - f(g(\mathbf{x}_0))| < \epsilon. \quad \square$$

Continuity in Each Variable Separately

A *continuous function of several variables is continuous in each of its variables separately.* In the two-variable case, this means that, if

$$\lim_{(x, y) \to (x_0, y_0)} f(x, y) = f(x_0, y_0),$$

then

$$\lim_{x \to x_0} f(x, y_0) = f(x_0, y_0) \quad \text{and} \quad \lim_{y \to y_0} f(x_0, y) = f(x_0, y_0).$$

(This is not hard to prove.) The converse is false. *It is possible for a function to be continuous in each variable separately and yet fail to be continuous as a function of several variables.* You can see this in the next example.

Example 2. We set

$$f(x, y) = \left\{ \begin{array}{ll} \dfrac{2xy}{x^2 + y^2}, & (x, y) \neq (0, 0) \\ 0, & (x, y) = (0, 0) \end{array} \right].$$

Since

$$f(x, 0) = 0 \quad \text{for all } x \quad \text{and} \quad f(0, y) = 0 \quad \text{for all } y,$$

we have

$$\lim_{x \to 0} f(x, 0) = f(0, 0) \quad \text{and} \quad \lim_{y \to 0} f(0, y) = f(0, 0).$$

Thus, at the point $(0, 0)$, f is continuous in x and continuous in y. However, as a function of two variables, f is not continuous at $(0, 0)$. One way to see this is to note that we can approach $(0, 0)$ as closely as we wish by points of the form (t, t) with $t \neq 0$. At such points f takes on the value 1:

$$f(t, t) = \frac{2tt}{t^2 + t^2} = 1.$$

Hence, f cannot tend to $f(0, 0) = 0$ as required. $\square$

Continuity and Partial Differentiability

For functions of a single variable the existence of the derivative guarantees continuity (3.1.4). *For functions of several variables the existence of partial derivatives fails to guarantee continuity.*†

To show this, we can use the same function

$$f(x, y) = \begin{cases} \dfrac{2xy}{x^2 + y^2}, & (x, y) \neq (0, 0) \\ 0, & (x, y) = (0, 0) \end{cases}.$$

Since both $f(x, 0)$ and $f(0, y)$ are constantly zero, both partials exist (and are zero) at $(0, 0)$, and yet, as you saw, the function is discontinuous at $(0, 0)$.

It is not hard to understand how a function can have partial derivatives and yet fail to be continuous. The existence of $\partial f/\partial x$ at (x_0, y_0) depends on the behavior of f only at points of the form $(x_0 + h, y_0)$. Similarly, the existence of $\partial f/\partial y$ at (x_0, y_0) depends on the behavior of f only at points of the form $(x_0, y_0 + k)$. On the other hand, continuity at (x_0, y_0) depends on the behavior of f at points of the more general form $(x_0 + h, y_0 + k)$. More briefly, we can put it this way: *the existence of a partial derivative depends on the behavior of the function along a line segment* (two directions), *whereas continuity depends on the behavior of the function in all directions.*

Equality of Mixed Partials

Suppose that f is a function of x and y with first partials

$$\frac{\partial f}{\partial x} \quad \text{and} \quad \frac{\partial f}{\partial y}.$$

These are again functions of x and y and may themselves possess partial derivatives:

$$\frac{\partial}{\partial x}\left(\frac{\partial f}{\partial x}\right) = \frac{\partial^2 f}{\partial x^2}, \qquad \frac{\partial}{\partial y}\left(\frac{\partial f}{\partial x}\right) = \frac{\partial^2 f}{\partial y \partial x},$$

$$\frac{\partial}{\partial x}\left(\frac{\partial f}{\partial y}\right) = \frac{\partial^2 f}{\partial x \partial y}, \qquad \frac{\partial}{\partial y}\left(\frac{\partial f}{\partial y}\right) = \frac{\partial^2 f}{\partial y^2}.$$

These last functions are called the *second-order partials*. Note that there are two "mixed" partials.

$$\frac{\partial^2 f}{\partial y \partial x} \quad \text{and} \quad \frac{\partial^2 f}{\partial x \partial y}.$$

The first of these is obtained by differentiating first with respect to x and then with respect to y. The second is obtained by differentiating first with respect to y and then with respect to x.

---

† See, however, Exercise 18.

Example 3. The function $f(x, y) = \sin x^2 y$ has first partials

$$\frac{\partial f}{\partial x} = 2xy \cos x^2 y \qquad \text{and} \qquad \frac{\partial f}{\partial y} = x^2 \cos x^2 y.$$

The second-order partials are

$$\frac{\partial^2 f}{\partial x^2} = -4x^2 y^2 \sin x^2 y + 2y \cos x^2 y, \qquad \frac{\partial^2 f}{\partial y \partial x} = -2x^3 y \sin x^2 y + 2x \cos x^2 y,$$

$$\frac{\partial^2 f}{\partial x \partial y} = -2x^3 y \sin x^2 y + 2x \cos x^2 y, \qquad \frac{\partial^2 f}{\partial y^2} = -x^4 \sin x^2 y. \quad \square$$

Example 4. Setting $f(x, y) = \ln (x^2 + y^3)$, we have

$$\frac{\partial f}{\partial x} = \frac{2x}{x^2 + y^3} \qquad \text{and} \qquad \frac{\partial f}{\partial y} = \frac{3y^2}{x^2 + y^3}.$$

The second-order partials are

$$\frac{\partial^2 f}{\partial x^2} = \frac{(x^2 + y^3)2 - 2x(2x)}{(x^2 + y^3)^2} = \frac{2(y^3 - x^2)}{(x^2 + y^3)^2},$$

$$\frac{\partial^2 f}{\partial y \partial x} = \frac{-2x(3y^2)}{(x^2 + y^3)^2} = -\frac{6xy^2}{(x^2 + y^3)^2},$$

$$\frac{\partial^2 f}{\partial x \partial y} = \frac{-3y^2(2x)}{(x^2 + y^3)^2} = -\frac{6xy^2}{(x^2 + y^3)^2},$$

$$\frac{\partial^2 f}{\partial y^2} = \frac{(x^2 + y^3)6y - 3y^2(3y^2)}{(x^2 + y^3)^2} = \frac{3y(2x^2 - y^3)}{(x^2 + y^3)^2}. \quad \square$$

Perhaps you noticed that in both examples we had

$$\frac{\partial^2 f}{\partial y \partial x} = \frac{\partial^2 f}{\partial x \partial y}.$$

Since in neither case was f symmetric in x and y, this equality of the mixed partials was not due to symmetry. Actually it was due to continuity. It can be proved that

(15.6.5)

$$\frac{\partial^2 f}{\partial y \partial x} = \frac{\partial^2 f}{\partial x \partial y}$$

on every open set U on which f and its partials

$$\frac{\partial f}{\partial x}, \quad \frac{\partial f}{\partial y}, \quad \frac{\partial^2 f}{\partial y \partial x}, \quad \frac{\partial^2 f}{\partial x \partial y}$$

are all continuous.†

---

† For a proof consult a text on advanced calculus.

In the case of a function of three variables you can look for three first partials

$$\frac{\partial f}{\partial x}, \quad \frac{\partial f}{\partial y}, \quad \frac{\partial f}{\partial z}$$

and nine second partials

$$\frac{\partial^2 f}{\partial x^2}, \quad \frac{\partial^2 f}{\partial x \partial y}, \quad \frac{\partial^2 f}{\partial x \partial z} \quad \frac{\partial^2 f}{\partial y \partial x}, \quad \frac{\partial^2 f}{\partial y^2}, \quad \frac{\partial^2 f}{\partial y \partial z} \quad \frac{\partial^2 f}{\partial z \partial x}, \quad \frac{\partial^2 f}{\partial z \partial y}, \quad \frac{\partial^2 f}{\partial z^2}.$$

Here again, there is equality of the mixed partials

$$\frac{\partial^2 f}{\partial y \partial x} = \frac{\partial^2 f}{\partial x \partial y}, \qquad \frac{\partial^2 f}{\partial z \partial x} = \frac{\partial^2 f}{\partial x \partial z}, \qquad \frac{\partial^2 f}{\partial y \partial z} = \frac{\partial^2 f}{\partial z \partial y}$$

provided that f and its first and second partials are continuous.

Example 5. For

$$f(x, y, z) = xe^y \sin \pi z$$

we have

$$\frac{\partial f}{\partial x} = e^y \sin \pi z, \qquad \frac{\partial f}{\partial y} = xe^y \sin \pi z, \qquad \frac{\partial f}{\partial z} = \pi xe^y \cos \pi z,$$

$$\frac{\partial^2 f}{\partial x^2} = 0, \qquad \frac{\partial^2 f}{\partial y^2} = xe^y \sin \pi z, \qquad \frac{\partial^2 f}{\partial z^2} = -\pi^2 xe^y \sin \pi z,$$

$$\frac{\partial^2 f}{\partial y \partial x} = e^y \sin \pi z = \frac{\partial^2 f}{\partial x \partial y},$$

$$\frac{\partial^2 f}{\partial z \partial x} = \pi e^y \cos \pi z = \frac{\partial^2 f}{\partial x \partial z},$$

$$\frac{\partial^2 f}{\partial y \partial z} = \pi xe^y \cos \pi z = \frac{\partial^2 f}{\partial z \partial y}. \quad \square$$

EXERCISES 15.6

Find the second partials.

1. $f(x, y) = Ax^2 + 2Bxy + Cy^2$.

2. $f(x, y) = Ax^3 + Bx^2y + Cxy^2$.

3. $f(x, y) = Ax + By + Ce^{xy}$

4. $f(x, y) = x^2 \cos y + y^2 \sin x$.

5. $f(x, y, z) = (x + y^2 + z^3)^2$.

6. $f(x, y) = \sqrt{x + y^2}$.

7. $f(x, y) = \ln \left(\dfrac{x}{x + y} \right)$.

8. $f(x, y) = \dfrac{Ax + By}{Cx + Dy}$.

9. $f(x, y, z) = (x + y)(y + z)(z + x)$.

10. $f(x, y, z) = \tan^{-1} xyz$.

11. $f(x, y) = x^y$.

12. $f(x, y, z) = \sin (x + z^y)$.

13. Show that

$$\text{if} \quad u = \frac{xy}{x+y} \quad \text{then} \quad x^2\frac{\partial^2 u}{\partial x^2} + 2xy\frac{\partial^2 u}{\partial x\partial y} + y^2\frac{\partial^2 u}{\partial y^2} = 0.$$

14. Show that the following functions are *harmonic* on their domains; that is, they satisfy *Laplace's equation*†

$$\frac{\partial^2 f}{\partial x^2} + \frac{\partial^2 f}{\partial y^2} = 0.$$

(a) $f(x, y) = x^3 - 3xy^2$.
(c) $f(x, y) = e^{-y}\cos x$.

(b) $f(x, y) = e^x \sin y$.
(d) $f(x, y) = \ln\sqrt{x^2 + y^2}$.

15. Show that the following functions are solutions of the wave equation

$$\frac{\partial^2 f}{\partial t^2} - c^2\frac{\partial^2 f}{\partial x^2} = 0.\dagger\dagger$$

(a) $f(x, t) = (Ax + B)(Ct + D)$.
(b) $f(x, t) = (Ae^{kx} + Be^{-kx})(Ce^{ckt} + De^{-ckt})$, k constant.
(c) $f(x, t) = g(x - ct)$, where g is any twice differentiable function.

16. Show that

$$\text{if} \quad u = \frac{1}{\sqrt{x^2 + y^2 + z^2}} \quad \text{then} \quad \frac{\partial^2 u}{\partial x^2} + \frac{\partial^2 u}{\partial y^2} + \frac{\partial^2 u}{\partial z^2} = 0.$$

17. Verify that

$$\frac{\partial^2 f}{\partial y\partial x} = \frac{\partial^2 f}{\partial x\partial y}$$

given that
(a) $f(x, y) = g(x) + h(y)$ with g and h differentiable.
(b) $f(x, y) = g(x)h(y)$ with g and h differentiable.
(c) $f(x, y)$ is a polynomial in x and y.   HINT: Check each term $x^m y^n$ separately.

18. If a function of several variables has all first partials at a point, then it is continuous in each variable separately at that point. Show, for example, that

$$\text{if} \frac{\partial f}{\partial x} \text{ exists at } (x_0, y_0), \text{ then } f \text{ is continuous in } x \text{ at } (x_0, y_0).$$

19. Let f be a function of x and y with everywhere continuous second partials. Is it possible that

(a) $\dfrac{\partial f}{\partial x} = x + y$ and $\dfrac{\partial f}{\partial y} = y - x$?   (b) $\dfrac{\partial f}{\partial x} = xy$ and $\dfrac{\partial f}{\partial y} = xy$?

---

† Named after the French mathematician Pierre-Simon Laplace (1749–1827). Laplace wrote two monumental works: one on celestial mechanics, the other on probability theory. He also made major contributions to the theory of differential equations.

†† Such equations play an important role in electromagnetism.

20. Let g be a twice-differentiable function of one variable and set

$$h(x, y) = g(x + y) + g(x - y).$$

Show that

$$\frac{\partial^2 h}{\partial x^2} = \frac{\partial^2 h}{\partial y^2}.$$

HINT: The chain rule (Exercise 40, Section 15.4).

21. Let f be a function of x and y with third-order partials

$$\frac{\partial^3 f}{\partial x^2 \partial y} = \frac{\partial}{\partial x}\left(\frac{\partial^2 f}{\partial x \partial y}\right) \quad \text{and} \quad \frac{\partial^3 f}{\partial y \partial x^2} = \frac{\partial}{\partial y}\left(\frac{\partial^2 f}{\partial x^2}\right).$$

Show that, if all the partials are continuous, then

$$\frac{\partial^3 f}{\partial x^2 \partial y} = \frac{\partial^3 f}{\partial y \partial x^2}.$$

22. Show that the following functions do not have a limit at $(0, 0)$:

(a) $f(x, y) = \dfrac{x^2 - y^2}{x^2 + y^2}.$ (b) $f(x, y) = \dfrac{y^2}{x^2 + y^2}.$

Evaluate the limit as (x, y) approaches the origin along
(a) the x-axis.   (b) the y-axis.   (c) the line $y = mx$.   (d) the spiral $r = \theta, \theta > 0$.
(e) the differentiable curve $y = f(x)$, given that $f(0) = 0$.

(f) the arc $r = \sin 3\theta$, $\frac{1}{6}\pi < \theta < \frac{1}{3}\pi$.   (g) the path $\mathbf{r}(t) = \dfrac{1}{t}\mathbf{i} + \dfrac{\sin t}{t}\mathbf{j}$, $t > 0$.

23. $\displaystyle\lim_{(x, y)\to(0, 0)} \frac{xy}{x^2 + y^2}.$ **24.** $\displaystyle\lim_{(x, y)\to(0, 0)} \frac{xy^2}{(x^2 + y^2)^{3/2}}.$

25. Set

$$g(x, y) = \begin{cases} \dfrac{x^2 y^2}{x^4 + y^4}, & (x, y) \neq (0, 0) \\ 0, & (x, y) = (0, 0). \end{cases}$$

(a) Show that $\dfrac{\partial g}{\partial x}$ and $\dfrac{\partial g}{\partial y}$ both exist at $(0, 0)$. What are their values at $(0, 0)$?
(b) Show that $\displaystyle\lim_{(x, y)\to(0, 0)} g(x, y)$ does not exist.

26. Set

$$f(x, y) = \frac{x - y^4}{x^3 - y^4}.$$

Determine whether or not f has a limit at $(1, 1)$.   HINT: Let (x, y) tend to $(1, 1)$ along
the lines $x = 1$ and $y = 1$.

27. Set

$$f(x, y) = \begin{cases} \dfrac{xy(y^2 - x^2)}{x^2 + y^2}, & (x, y) \neq (0, 0) \\ 0, & (x, y) = (0, 0). \end{cases}$$

It can be shown that some of the second partials are discontinuous at $(0, 0)$. Show that

$$\frac{\partial^2 f}{\partial y \partial x}(0, 0) \neq \frac{\partial^2 f}{\partial x \partial y}(0, 0).$$

CHAPTER HIGHLIGHTS

15.1 Elementary Examples

domain, range (pp. 815, 816) open (closed) unit disc (p. 816)
open (closed) unit ball (p. 816)

15.2 A Brief Catalog of the Quadric Surfaces; Projections

traces, sections (p. 819) ellipsoid (p. 820)
hyperboloid of one sheet (p. 820) hyperboloid of two sheets (p. 821)
quadric cone, nappe (p. 822) elliptic paraboloid (p. 822)
hyperbolic paraboloid (p. 823) cylinder (p. 824)
parabolic cylinder (p. 824) elliptic cylinder (p. 824)
hyperbolic cylinder (p. 825) projection (p. 825)

15.3 Graphs; Level Curves and Level Surfaces

level curve (p. 830) level surface (p. 833)

15.4 Partial Derivatives

for a function of two variables:
    limit definition (p. 834)
    geometric interpretation (Figs. 15.4.1 and 15.4.2)
for a function of three variables:
    limit definition (p. 837) interpretation (p. 838)

15.5 Open Sets and Closed Sets

neighborhood (p. 843) interior (p. 844) boundary (p. 844)
open set (p. 844) closed set (p. 844) deleted neighborhood (p. 846)

Some sets are neither open nor closed.

15.6 Limits and Continuity; Equality of Mixed Partials

limit of a function (p. 847) continuity at $\mathbf{x}_0$ (p. 848)

Points in the domain of a function of several variables can be written in vector notation. In the two-variable case, set $\mathbf{x} = (x, y)$, and, in the three-variable case, set $\mathbf{x} = (x, y, z)$. The vector notation enables us to treat the two cases together simply by writing $f(\mathbf{x})$.

A continuous function of several variables is continuous in each of its variables. The converse is false: it is possible for a function to be continuous in each variable separately and yet fail to be continuous as a function of several variables. (p. 850)

For functions of several variables the existence of partial derivatives fails to guarantee continuity. (p. 851)

The existence of a partial derivative depends on the behavior of the function along a line segment (two directions), whereas continuity depends on the behavior of the function in all directions. (p. 851)

equality of mixed partials (p. 852)

16

GRADIENTS; EXTREME VALUES; DIFFERENTIALS

16.1 DIFFERENTIABILITY AND GRADIENT

The Notion of Differentiability

Our object here is to extend the notion of differentiability from functions of one variable to functions of several variables. Partial derivatives alone do not fulfill this role because they reflect behavior only in particular directions.

In the one-variable case we formed the difference quotient

$$\frac{f(x + h) - f(x)}{h}$$

and called f differentiable at x provided that this quotient had a limit as h tended to zero. In the multivariable case we can still form the difference

$$f(\mathbf{x} + \mathbf{h}) - f(\mathbf{x}),$$

but the "quotient"

$$\frac{f(\mathbf{x} + \mathbf{h}) - f(\mathbf{x})}{\mathbf{h}}$$

makes no sense because to divide by a vector makes no sense.

We can get around this difficulty by going back to an idea introduced in the exercises to Section 3.12. We will review the idea here.

If $g(h)$ is an expression in h, we say that $g(h)$ is *little-o(h)* and write $g(h) = o(h)$ iff

$$\lim_{h \to 0} \frac{g(h)}{|h|} = 0.$$

For a function of one variable the following statements are equivalent:

$$\lim_{h \to 0} \frac{f(x + h) - f(x)}{h} = f'(x)$$

$$\lim_{h \to 0} \frac{[f(x + h) - f(x)] - f'(x)h}{h} = 0$$

$$\lim_{h \to 0} \frac{[f(x + h) - f(x)] - f'(x)h}{|h|} = 0$$

$$[f(x + h) - f(x)] - f'(x)h = o(h)$$

$$[f(x + h) - f(x)] = f'(x)h + o(h).$$

Thus, for a function of one variable, the derivative of f at x is the unique number $f'(x)$ such that

$$f(x + h) - f(x) = f'(x)h + o(h).$$

It is this view of the derivative that inspires the notion of differentiability in the multivariable case.

Let us agree to call $o(\mathbf{h})$ any expression $g(\mathbf{h})$ for which

$$\lim_{h \to 0} \frac{g(\mathbf{h})}{\|\mathbf{h}\|} = 0.$$

Now let f be a function of several variables *defined at least in some neighborhood of* $\mathbf{x}$. [In the three-variable case, $\mathbf{x} = (x, y, z)$; in the two-variable case, $\mathbf{x} = (x, y)$.]

DEFINITION 16.1.1   DIFFERENTIABILITY

We say that f is *differentiable* at $\mathbf{x}$ iff there exists a vector $\mathbf{y}$ such that

$$f(\mathbf{x} + \mathbf{h}) - f(\mathbf{x}) = \mathbf{y} \cdot \mathbf{h} + o(\mathbf{h}).$$

It is not hard to show that, if such a vector $\mathbf{y}$ exists, it is unique (Exercise 23). We call this unique vector *the gradient of f at* $\mathbf{x}$ and denote it by $\nabla f(\mathbf{x})$:

(16.1.2)

if f is differentiable at $\mathbf{x}$, the gradient of f at $\mathbf{x}$ is the unique vector $\nabla f(\mathbf{x})$ such that

$$f(\mathbf{x} + \mathbf{h}) - f(\mathbf{x}) = \nabla f(\mathbf{x}) \cdot \mathbf{h} + o(\mathbf{h}).$$

The similarities between the one-variable case,

$$f(x + h) - f(x) = f'(x)h + o(h),$$

and the multivariable case,

$$f(\mathbf{x} + \mathbf{h}) - f(\mathbf{x}) = \nabla f(\mathbf{x}) \cdot \mathbf{h} + o(\mathbf{h}),$$

are obvious. We point to the differences. There are essentially two of them:

(1) While the derivative $f'(x)$ is a number, the gradient $\nabla f(\mathbf{x})$ is a vector.

(2) While $f'(x)h$ is the ordinary product of two real numbers, $\nabla f(\mathbf{x}) \cdot \mathbf{h}$ is the dot product of two vectors.

Calculating Gradients

First we calculate some gradients by applying the definition directly. Then we give a theorem that makes such calculations much easier. Finally, we calculate some gradients with the aid of the theorem. As for notation, in the two-variable case we will write

$$\nabla f(\mathbf{x}) = \nabla f(x, y) \qquad \text{and} \qquad \mathbf{h} = (h_1, h_2),$$

and in the three-variable case,

$$\nabla f(\mathbf{x}) = \nabla f(x, y, z) \qquad \text{and} \qquad \mathbf{h} = (h_1, h_2, h_3).$$

Example 1. For the function

$$f(x, y) = x^2 + y^2$$

we have

$$\begin{aligned}
f(\mathbf{x} + \mathbf{h}) - f(\mathbf{x}) &= f(x + h_1, y + h_2) - f(x, y) \\
&= [(x + h_1)^2 + (y + h_2)^2] - [x^2 + y^2] \\
&= [2xh_1 + 2yh_2] + [h_1^2 + h_2^2] \\
&= [2x\mathbf{i} + 2y\mathbf{j}] \cdot \mathbf{h} + \|\mathbf{h}\|^2.
\end{aligned}$$

The remainder $\|\mathbf{h}\|^2$ is $o(\mathbf{h})$:

$$\frac{\|\mathbf{h}\|^2}{\|\mathbf{h}\|} = \|\mathbf{h}\| \to 0 \quad \text{as} \quad \mathbf{h} \to \mathbf{0}.$$

Thus

$$\nabla f(\mathbf{x}) = \nabla f(x, y) = 2x\mathbf{i} + 2y\mathbf{j}. \quad \square$$

Example 2. For the function

$$f(x, y, z) = ax + by + cz$$

we have

$$\begin{aligned}
f(\mathbf{x} + \mathbf{h}) - f(\mathbf{x}) &= f(x + h_1, y + h_2, z + h_3) - f(x, y, z) \\
&= [a(x + h_1) + b(y + h_2) + c(z + h_3)] - [ax + by + cz]
\end{aligned}$$

$$= ah_1 + bh_2 + ch_3$$
$$= [a\mathbf{i} + b\mathbf{j} + c\mathbf{k}] \cdot \mathbf{h}.$$

Here the remainder is identically zero and is thus certainly $o(\mathbf{h})$. Therefore

$$\nabla f(\mathbf{x}) = a\mathbf{i} + b\mathbf{j} + c\mathbf{k}. \quad \square$$

THEOREM 16.1.3

If f has continuous first partials in a neighborhood of $\mathbf{x}$, then f is differentiable at $\mathbf{x}$ and

$$\nabla f(\mathbf{x}) = \frac{\partial f}{\partial x}(\mathbf{x})\,\mathbf{i} + \frac{\partial f}{\partial y}(\mathbf{x})\,\mathbf{j} + \frac{\partial f}{\partial z}(\mathbf{x})\,\mathbf{k}.$$

In two variables,

$$\nabla f(\mathbf{x}) = \frac{\partial f}{\partial x}(\mathbf{x})\,\mathbf{i} + \frac{\partial f}{\partial y}(\mathbf{x})\,\mathbf{j}.$$

The proof is somewhat difficult. We have placed it in a supplement at the end of this section. $\square$

Example 3. For

$$f(x, y) = xe^y - ye^x$$

we have

$$\frac{\partial f}{\partial x}(x, y) = e^y - ye^x, \qquad \frac{\partial f}{\partial y}(x, y) = xe^y - e^x$$

and therefore

$$\nabla f(x, y) = (e^y - ye^x)\mathbf{i} + (xe^y - e^x)\mathbf{j}. \quad \square$$

When there is no reason to emphasize the point of evaluation, we don't write

$$\nabla f(\mathbf{x}) \quad \text{or} \quad \nabla f(x, y) \quad \text{or} \quad \nabla f(x, y, z)$$

but simply ∇f. Thus for the function

$$f(x, y) = xe^y - ye^x$$

we write

$$\frac{\partial f}{\partial x} = e^y - ye^x, \qquad \frac{\partial f}{\partial y} = xe^y - e^x$$

and

$$\nabla f = (e^y - ye^x)\mathbf{i} + (xe^y - e^x)\mathbf{j}.$$

Example 4. For

$$f(x, y, z) = \sin xy^2z^3$$

we have

$$\frac{\partial f}{\partial x} = y^2z^3 \cos xy^2z^3, \qquad \frac{\partial f}{\partial y} = 2xyz^3 \cos xy^2z^3, \qquad \frac{\partial f}{\partial z} = 3xy^2z^2 \cos xy^2z^3$$

and

$$\nabla f = (\cos xy^2z^3)[y^2z^3\mathbf{i} + 2xyz^3\mathbf{j} + 3xy^2z^2\mathbf{k}]. \quad \square$$

Example 5. We take the function

$$f(x, y, z) = x \sin \pi y + y \cos \pi z$$

and evaluate ∇f at $(0, 1, 2)$.

Here

$$\frac{\partial f}{\partial x} = \sin \pi y, \qquad \frac{\partial f}{\partial y} = \pi x \cos \pi y + \cos \pi z, \qquad \frac{\partial f}{\partial z} = -\pi y \sin \pi z.$$

At $(0, 1, 2)$

$$\frac{\partial f}{\partial x} = 0, \quad \frac{\partial f}{\partial y} = 1, \quad \frac{\partial f}{\partial z} = 0 \qquad \text{and thus} \qquad \nabla f = \mathbf{j}. \quad \square$$

Of special interest for later work are the powers of r where, as usual,

$$r = \|\mathbf{r}\| \qquad \text{and} \qquad \mathbf{r} = x\mathbf{i} + y\mathbf{j} + z\mathbf{k}.$$

We begin by showing that

(16.1.4)
$$\nabla r = \frac{\mathbf{r}}{r} \qquad \text{and} \qquad \nabla \left(\frac{1}{r}\right) = -\frac{\mathbf{r}}{r^3}.$$

Proof

$$\nabla r = \nabla(x^2 + y^2 + z^2)^{1/2}$$

$$= \frac{\partial}{\partial x}(x^2 + y^2 + z^2)^{1/2}\,\mathbf{i} + \frac{\partial}{\partial y}(x^2 + y^2 + z^2)^{1/2}\,\mathbf{j} + \frac{\partial}{\partial z}(x^2 + y^2 + z^2)^{1/2}\,\mathbf{k}$$

$$= \frac{x}{(x^2 + y^2 + z^2)^{1/2}}\,\mathbf{i} + \frac{y}{(x^2 + y^2 + z^2)^{1/2}}\,\mathbf{j} + \frac{z}{(x^2 + y^2 + z^2)^{1/2}}\,\mathbf{k}$$

$$= \frac{1}{(x^2 + y^2 + z^2)^{1/2}}(x\mathbf{i} + y\mathbf{j} + z\mathbf{k}) = \frac{\mathbf{r}}{r}.$$

$$\nabla\left(\frac{1}{r}\right) = \nabla(x^2 + y^2 + z^2)^{-1/2}$$

$$= \frac{\partial}{\partial x}(x^2 + y^2 + z^2)^{-1/2}\,\mathbf{i} + \frac{\partial}{\partial y}(x^2 + y^2 + z^2)^{-1/2}\,\mathbf{j} + \frac{\partial}{\partial z}(x^2 + y^2 + z^2)^{-1/2}\,\mathbf{k}$$

$$= -\frac{x}{(x^2 + y^2 + z^2)^{3/2}}\,\mathbf{i} - \frac{y}{(x^2 + y^2 + z^2)^{3/2}}\,\mathbf{j} - \frac{z}{(x^2 + y^2 + z^2)^{3/2}}\,\mathbf{k}$$

$$= -\frac{1}{(x^2 + y^2 + z^2)^{3/2}}(x\mathbf{i} + y\mathbf{j} + z\mathbf{k}) = -\frac{\mathbf{r}}{r^3}. \quad \square$$

The formulas we just derived can be generalized. As you are asked to show in the exercises, for each integer n and all $\mathbf{r} \neq \mathbf{0}$,

(16.1.5)
$$\nabla r^n = nr^{n-2}\mathbf{r}.$$

(If n is positive and even, the result also holds at $\mathbf{r} = \mathbf{0}$.)

Differentiability Implies Continuity

As in the one-variable case, differentiability implies continuity; namely,

(16.1.6)
$$\text{if } f \text{ is differentiable at } \mathbf{x}, \text{ then } f \text{ is continuous at } \mathbf{x}.$$

To see this, write

$$f(\mathbf{x} + \mathbf{h}) - f(\mathbf{x}) = \nabla f(\mathbf{x}) \cdot \mathbf{h} + o(\mathbf{h})$$

and note that

$$|f(\mathbf{x} + \mathbf{h}) - f(\mathbf{x})| = |\nabla f(\mathbf{x}) \cdot \mathbf{h} + o(\mathbf{h})| \le |\nabla f(\mathbf{x}) \cdot \mathbf{h}| + |o(\mathbf{h})|.$$

As $\mathbf{h} \to \mathbf{0}$,

$$|\nabla f(\mathbf{x}) \cdot \mathbf{h}| \le \|\nabla f(\mathbf{x})\|\,\|\mathbf{h}\| \to 0 \quad \text{and} \quad |o(\mathbf{h})| \to 0.$$
$$\text{Exercise 24}$$

It follows that

$$[f(\mathbf{x} + \mathbf{h}) - f(\mathbf{x})] \to 0 \quad \text{and therefore} \quad f(\mathbf{x} + \mathbf{h}) \to f(\mathbf{x}). \quad \square$$

EXERCISES 16.1

Find the gradient.

1. $f(x, y) = xe^{xy}$.

2. $f(x, y) = x^3 + y^2$.

3. $f(x, y) = 3x^2 - xy + y$.

4. $f(x, y) = Ax^2 + Bxy + Cy^2$.

5. $f(x, y) = x^2y^{-2}$.

6. $f(x, y) = e^{x-y} - e^{y-x}$.

7. $f(x, y, z) = z \sin(x - y)$.

8. $f(x, y, z) = xy^2e^{-z}$.

9. $f(x, y, z) = xy + yz + zx$.

10. $f(x, y, z) = \sqrt{x + y + z}$.

11. $f(x, y) = (x + y)e^{x-y}$.

12. $f(x, y) = (x + y) \sin (x - y)$.

13. $f(x, y) = e^x \ln y$.

14. $f(x, y, z) = x^2y + y^2z + z^2x$.

15. $f(x, y) = \dfrac{Ax + By}{Cx + Dy}$.

16. $f(x, y) = \dfrac{x - y}{x^2 + y^2}$.

Find the gradient vector at the indicated point.

17. $f(x, y) = 2x^2 - 3xy + 4y^2$ at $(2, 3)$.   18. $f(x, y) = 2x(x - y)^{-1}$ at $(3, 1)$.

19. $f(x, y, z) = e^{-x} \sin (z + 2y)$ at $(0, \frac{1}{4}\pi, \frac{1}{4}\pi)$.

20. $f(x, y, z) = (x - y) \cos \pi z$ at $(1, 0, \frac{1}{2})$.

21. Calculate: (a) $\nabla(\ln r)$. (b) $\nabla(\sin r)$. (c) $\nabla(e^r)$.

22. Derive (16.1.5).

23. (a) Show that, if $\mathbf{c} \cdot \mathbf{h}$ is $o(\mathbf{h})$, then $\mathbf{c} = \mathbf{0}$.   HINT: First set $\mathbf{h} = h\mathbf{i}$, then set $\mathbf{h} = h\mathbf{j}$, then $\mathbf{h} = h\mathbf{k}$.

   (b) Show that, if

$$f(\mathbf{x} + \mathbf{h}) - f(\mathbf{x}) = \mathbf{y} \cdot \mathbf{h} + o(\mathbf{h}) \qquad \text{and} \qquad f(\mathbf{x} + \mathbf{h}) - f(\mathbf{x}) = \mathbf{z} \cdot \mathbf{h} + o(\mathbf{h}),$$

   then

$$\mathbf{y} = \mathbf{z}.$$

   HINT: Verify that $\mathbf{y} \cdot \mathbf{h} - \mathbf{z} \cdot \mathbf{h} = (\mathbf{y} - \mathbf{z}) \cdot \mathbf{h}$ is $o(\mathbf{h})$.

24. Show that, if g is $o(\mathbf{h})$, then

$$\lim_{\mathbf{h} \to \mathbf{0}} g(\mathbf{h}) = 0.$$

25. Set

$$f(x, y) = \left\{ \begin{array}{ll} \dfrac{2xy}{x^2 + y^2}, & (x, y) \neq (0, 0) \\[2mm] 0, & (x, y) = (0, 0) \end{array} \right].$$

   (a) Show that f is not differentiable at $(0, 0)$.

   (b) In Section 15.6 you saw that the first partials $\partial f/\partial x$ and $\partial f/\partial y$ exist at $(0, 0)$. Since these partials obviously exist at every other point of the plane, we can conclude from Theorem 16.1.3 that at least one of these partials is not continuous in a neighborhood of $(0, 0)$. Show that $\partial f/\partial x$ is discontinuous at $(0, 0)$.

*SUPPLEMENT TO SECTION 16.1

Proof of Theorem 16.1.3. We prove the theorem in the two-variable case. A similar argument yields a proof in the three-variable case, but there the details are more burdensome.
   In the first place,

$$f(\mathbf{x} + \mathbf{h}) - f(\mathbf{x}) = f(x + h_1, y + h_2) - f(x, y).$$

Adding and subtracting $f(x, y + h_2)$, we have

(1) $f(\mathbf{x} + \mathbf{h}) - f(\mathbf{x}) = [f(x + h_1, y + h_2) - f(x, y + h_2)] + [f(x, y + h_2) - f(x, y)]$.

By the mean-value theorem for functions of one variable, we know that there are numbers

$$0 < \theta_1 < 1 \qquad \text{and} \qquad 0 < \theta_2 < 1$$

such that

$$f(x + h_1, y + h_2) - f(x, y + h_2) = \frac{\partial f}{\partial x}(x + \theta_1 h_1, y + h_2)\, h_1$$

and

$$f(x, y + h_2) - f(x, y) = \frac{\partial f}{\partial y}(x, y + \theta_2 h_2)\, h_2.$$

(Exercise 24, Section 4.1) By the continuity of $\partial f/\partial x$,

$$\frac{\partial f}{\partial x}(x + \theta_1 h_1, y + h_2) = \frac{\partial f}{\partial x}(x, y) + \epsilon_1(\mathbf{h})$$

where

$$\epsilon_1(\mathbf{h}) \to 0 \quad \text{as} \quad \mathbf{h} \to \mathbf{0}.\dagger$$

By the continuity of $\partial f/\partial y$,

$$\frac{\partial f}{\partial y}(x, y + \theta_2 h_2) = \frac{\partial f}{\partial y}(x, y) + \epsilon_2(\mathbf{h})$$

where

$$\epsilon_2(\mathbf{h}) \to 0 \quad \text{as} \quad \mathbf{h} \to \mathbf{0}.$$

Substituting these expressions in equation (1), we find that

$$\begin{aligned}
f(\mathbf{x} + \mathbf{h}) - f(\mathbf{x}) &= \left[\frac{\partial f}{\partial x}(x, y) + \epsilon_1(\mathbf{h})\right] h_1 + \left[\frac{\partial f}{\partial y}(x, y) + \epsilon_2(\mathbf{h})\right] h_2 \\
&= \left[\frac{\partial f}{\partial x}(x, y) + \epsilon_1(\mathbf{h})\right](\mathbf{i} \cdot \mathbf{h}) + \left[\frac{\partial f}{\partial y}(x, y) + \epsilon_2(\mathbf{h})\right](\mathbf{j} \cdot \mathbf{h}) \\
&= \left[\frac{\partial f}{\partial x}(x, y)\mathbf{i} + \epsilon_1(\mathbf{h})\mathbf{i}\right] \cdot \mathbf{h} + \left[\frac{\partial f}{\partial y}(x, y)\mathbf{j} + \epsilon_2(\mathbf{h})\mathbf{j}\right] \cdot \mathbf{h} \\
&= \left[\frac{\partial f}{\partial x}(x, y)\mathbf{i} + \frac{\partial f}{\partial y}(x, y)\mathbf{j}\right] \cdot \mathbf{h} + \left[\epsilon_1(\mathbf{h})\mathbf{i} + \epsilon_2(\mathbf{h})\mathbf{j}\right] \cdot \mathbf{h}.
\end{aligned}$$

To complete the proof of the theorem we need only show that

(2) $$[\epsilon_1(\mathbf{h})\mathbf{i} + \epsilon_2(\mathbf{h})\mathbf{j}] \cdot \mathbf{h} = o(\mathbf{h}).$$

From Schwarz's inequality, $|\mathbf{a} \cdot \mathbf{b}| \leq \|\mathbf{a}\|\|\mathbf{b}\|$, we know that

$$|[\epsilon_1(\mathbf{h})\mathbf{i} + \epsilon_2(\mathbf{h})\mathbf{j}] \cdot \mathbf{h}| \leq \|\epsilon_1(\mathbf{h})\mathbf{i} + \epsilon_2(\mathbf{h})\mathbf{j}\|\,\|\mathbf{h}\|.$$

It follows that

$$\frac{|[\epsilon_1(\mathbf{h})\mathbf{i} + \epsilon_2(\mathbf{h})\mathbf{j}] \cdot \mathbf{h}|}{\|\mathbf{h}\|} \leq \|\epsilon_1(\mathbf{h})\mathbf{i} + \epsilon_2(\mathbf{h})\mathbf{j}\| \leq \|\epsilon_1(\mathbf{h})\mathbf{i}\| + \|\epsilon_2(\mathbf{h})\mathbf{j}\| = |\epsilon_1(\mathbf{h})| + |\epsilon_2(\mathbf{h})|.$$

⌐———— by the triangle inequality

As $\mathbf{h} \to \mathbf{0}$, the expression on the right tends to 0. This shows that (2) holds and completes the proof of the theorem. ▢

———

$\dagger$ $$\epsilon(\mathbf{h}) = \frac{\partial f}{\partial x}(x + \theta_1 h_1, y + h_2) - \frac{\partial f}{\partial x}(x, y) \to 0$$

since, by the continuity of $\partial f/\partial x$,

$$\frac{\partial f}{\partial x}(x + \theta_1 h_1, y + h_2) \to \frac{\partial f}{\partial x}(x, y).$$

16.2 GRADIENTS AND DIRECTIONAL DERIVATIVES

Some Elementary Formulas

In many respects gradients behave just as derivatives do in the one-variable case. In particular, if $\nabla f(\mathbf{x})$ and $\nabla g(\mathbf{x})$ exist, then $\nabla[f(\mathbf{x}) + g(\mathbf{x})]$, $\nabla[\alpha f(\mathbf{x})]$, and $\nabla[f(\mathbf{x})g(\mathbf{x})]$ all exist, and

(16.2.1)

$$\nabla[f(\mathbf{x}) + g(\mathbf{x})] = \nabla f(\mathbf{x}) + \nabla g(\mathbf{x}),$$
$$\nabla[\alpha f(\mathbf{x})] = \alpha \, \nabla f(\mathbf{x}),$$
$$\nabla[f(\mathbf{x})g(\mathbf{x})] = f(\mathbf{x}) \, \nabla g(\mathbf{x}) + g(\mathbf{x}) \, \nabla f(\mathbf{x}).$$

The first two formulas are easy to derive. To derive the third formula, let's assume that $\nabla f(\mathbf{x})$ and $\nabla g(\mathbf{x})$ both exist. Our task is to show that

$$f(\mathbf{x} + \mathbf{h})g(\mathbf{x} + \mathbf{h}) - f(\mathbf{x})g(\mathbf{x}) = [f(\mathbf{x}) \, \nabla g(\mathbf{x}) + g(\mathbf{x}) \, \nabla f(\mathbf{x})] \cdot \mathbf{h} + o(\mathbf{h}).$$

A sketch of how this can be done is given below. We leave it to you to justify each step. To have enough room on the page we set

$$D(\mathbf{x}, \mathbf{h}) = f(\mathbf{x} + \mathbf{h})g(\mathbf{x} + \mathbf{h}) - f(\mathbf{x})g(\mathbf{x}).$$

Then

$$
\begin{aligned}
D(\mathbf{x}, \mathbf{h}) &= [f(\mathbf{x} + \mathbf{h})g(\mathbf{x} + \mathbf{h}) - f(\mathbf{x})g(\mathbf{x} + \mathbf{h})] + [f(\mathbf{x})g(\mathbf{x} + \mathbf{h}) - f(\mathbf{x})g(\mathbf{x})] \\
&= [f(\mathbf{x} + \mathbf{h}) - f(\mathbf{x})]g(\mathbf{x} + \mathbf{h}) + f(\mathbf{x})[g(\mathbf{x} + \mathbf{h}) - g(\mathbf{x})] \\
&= [\nabla f(\mathbf{x}) \cdot \mathbf{h} + o(\mathbf{h})]g(\mathbf{x} + \mathbf{h}) + f(\mathbf{x})[\nabla g(\mathbf{x}) \cdot \mathbf{h} + o(\mathbf{h})] \\
&= g(\mathbf{x} + \mathbf{h}) \, \nabla f(\mathbf{x}) \cdot \mathbf{h} + f(\mathbf{x}) \, \nabla g(\mathbf{x}) \cdot \mathbf{h} + o(\mathbf{h}) \qquad \text{(Exercise 20)} \\
&= g(\mathbf{x}) \, \nabla f(\mathbf{x}) \cdot \mathbf{h} + f(\mathbf{x}) \, \nabla g(\mathbf{x}) \cdot \mathbf{h} + [g(\mathbf{x} + \mathbf{h}) - g(\mathbf{x})] \, \nabla f(\mathbf{x}) \cdot \mathbf{h} + o(\mathbf{h}) \\
&= [g(\mathbf{x}) \, \nabla f(\mathbf{x}) + f(\mathbf{x}) \, \nabla g(\mathbf{x})] \cdot \mathbf{h} + o(\mathbf{h}). \quad \square \qquad \text{(Exercise 20)}
\end{aligned}
$$

Directional Derivatives

Here we take up an idea that generalizes the notion of partial derivative. Its connection to gradients will be made clear as we go on.

The partial derivatives

$$\frac{\partial f}{\partial x}(x, y, z) = \lim_{h \to 0} \frac{f(x + h, y, z) - f(x, y, z)}{h},$$

$$\frac{\partial f}{\partial y}(x, y, z) = \lim_{h \to 0} \frac{f(x, y + h, z) - f(x, y, z)}{h},$$

$$\frac{\partial f}{\partial z}(x, y, z) = \lim_{h \to 0} \frac{f(x, y, z + h) - f(x, y, z)}{h}$$

expressed in vector notation take the form

$$\frac{\partial f}{\partial x}(\mathbf{x}) = \lim_{h \to 0} \frac{f(\mathbf{x} + h\mathbf{i}) - f(\mathbf{x})}{h},$$

$$\frac{\partial f}{\partial y}(\mathbf{x}) = \lim_{h \to 0} \frac{f(\mathbf{x} + h\mathbf{j}) - f(\mathbf{x})}{h},$$

$$\frac{\partial f}{\partial z}(\mathbf{x}) = \lim_{h \to 0} \frac{f(\mathbf{x} + h\mathbf{k}) - f(\mathbf{x})}{h}.$$

Each partial is thus the limit of a quotient

$$\frac{f(\mathbf{x} + h\mathbf{u}) - f(\mathbf{x})}{h}$$

where $\mathbf{u}$ is one of the unit coordinate vectors $\mathbf{i}$, $\mathbf{j}$, or $\mathbf{k}$. There is no reason to be so restrictive on $\mathbf{u}$. If f is defined in a neighborhood of $\mathbf{x}$, then, for small h, the difference quotient

$$\frac{f(\mathbf{x} + h\mathbf{u}) - f(\mathbf{x})}{h}$$

makes sense for any unit vector $\mathbf{u}$.

DEFINITION 16.2.2 DIRECTIONAL DERIVATIVE

For each unit vector $\mathbf{u}$, the limit

$$f_{\mathbf{u}}'(\mathbf{x}) = \lim_{h \to 0} \frac{f(\mathbf{x} + h\mathbf{u}) - f(\mathbf{x})}{h},$$

if it exists, is called the *directional derivative of f at $\mathbf{x}$ in the direction $\mathbf{u}$.*

The partials are of course themselves directional derivatives:

(16.2.3) $$\frac{\partial f}{\partial x}(\mathbf{x}) = f_{\mathbf{i}}'(\mathbf{x}), \qquad \frac{\partial f}{\partial y}(\mathbf{x}) = f_{\mathbf{j}}'(\mathbf{x}), \qquad \frac{\partial f}{\partial z}(\mathbf{x}) = f_{\mathbf{k}}'(\mathbf{x}).$$

As you know, the partials of f give the rates of change of f in the $\mathbf{i}$, $\mathbf{j}$, $\mathbf{k}$ directions. The directional derivative $f_{\mathbf{u}}'$ gives *the rate of change of f in the direction $\mathbf{u}$.*

There is an important connection between the gradient at $\mathbf{x}$ and the directional derivatives at $\mathbf{x}$.

THEOREM 16.2.4

If f is differentiable at $\mathbf{x}$, then f has a directional derivative at $\mathbf{x}$ in every direction $\mathbf{u}$ and

$$f_{\mathbf{u}}'(\mathbf{x}) = \nabla f(\mathbf{x}) \cdot \mathbf{u}.$$

Proof. We take **u** as a unit vector and assume that f is differentiable at **x**. The differentiability at **x** tells us that $\nabla f(\mathbf{x})$ exists and

$$f(\mathbf{x} + h\mathbf{u}) - f(\mathbf{x}) = \nabla f(\mathbf{x}) \cdot h\mathbf{u} + o(h\mathbf{u}).$$

Division by h gives

$$\frac{f(\mathbf{x} + h\mathbf{u}) - f(\mathbf{x})}{h} = \nabla f(\mathbf{x}) \cdot \mathbf{u} + \frac{o(h\mathbf{u})}{h}.$$

Since

$$\left| \frac{o(h\mathbf{u})}{h} \right| = \frac{|o(h\mathbf{u})|}{|h|} = \frac{|o(h\mathbf{u})|}{\|h\mathbf{u}\|} \to 0,$$

you can see that

$$\frac{o(h\mathbf{u})}{h} \to 0 \qquad \text{and thus} \qquad \frac{f(\mathbf{x} + h\mathbf{u}) - f(\mathbf{x})}{h} \to \nabla f(\mathbf{x}) \cdot \mathbf{u}. \quad \square$$

Earlier (Theorem 16.1.3) you saw that, *if f has continuous first partials in a neighborhood of* **x**, *then f is differentiable at* **x** *and*

$$\nabla f(\mathbf{x}) = \frac{\partial f}{\partial x}(\mathbf{x})\,\mathbf{i} + \frac{\partial f}{\partial y}(\mathbf{x})\,\mathbf{j} + \frac{\partial f}{\partial z}(\mathbf{x})\,\mathbf{k}.$$

The next theorem shows that this formula for $\nabla f(\mathbf{x})$ holds wherever f is differentiable.

THEOREM 16.2.5

If f is differentiable at **x**, then all the first partials exist at **x** and

$$\nabla f(\mathbf{x}) = \frac{\partial f}{\partial x}(\mathbf{x})\,\mathbf{i} + \frac{\partial f}{\partial y}(\mathbf{x})\,\mathbf{j} + \frac{\partial f}{\partial z}(\mathbf{x})\,\mathbf{k}.$$

Proof. Assume that f is differentiable at **x**. Then $\nabla f(\mathbf{x})$ exists and we can write

$$\nabla f(\mathbf{x}) = [\nabla f(\mathbf{x}) \cdot \mathbf{i}]\,\mathbf{i} + [\nabla f(\mathbf{x}) \cdot \mathbf{j}]\,\mathbf{j} + [\nabla f(\mathbf{x}) \cdot \mathbf{k}]\,\mathbf{k}. \qquad (13.4.13)$$

The result follows from observing that

$$\begin{array}{cc} (16.2.4) & (16.2.3) \\ \downarrow & \downarrow \end{array}$$

$$\nabla f(\mathbf{x}) \cdot \mathbf{i} = f'_{\mathbf{i}}(\mathbf{x}) = \frac{\partial f}{\partial x}(\mathbf{x}),$$

$$\nabla f(\mathbf{x}) \cdot \mathbf{j} = f'_{\mathbf{j}}(\mathbf{x}) = \frac{\partial f}{\partial y}(\mathbf{x}),$$

$$\nabla f(\mathbf{x}) \cdot \mathbf{k} = f'_{\mathbf{k}}(\mathbf{x}) = \frac{\partial f}{\partial z}(\mathbf{x}). \quad \square$$

Problem 1. Find the directional derivative of the function

$$f(x, y) = x^2 + y^2$$

at the point $(1, 2)$ in the direction of the vector $2\mathbf{i} - 3\mathbf{j}$.

Solution. In the first place, $2\mathbf{i} - 3\mathbf{j}$ is not a unit vector; its norm is $\sqrt{13}$. The unit vector in the direction of $2\mathbf{i} - 3\mathbf{j}$ is the vector

$$\mathbf{u} = \tfrac{1}{13}\sqrt{13}[2\mathbf{i} - 3\mathbf{j}].$$

Next

$$\nabla f = 2x\mathbf{i} + 2y\mathbf{j},$$

and therefore

$$\nabla f(1, 2) = 2\mathbf{i} + 4\mathbf{j}.$$

By (16.2.4) we have

$$f'_{\mathbf{u}}(1, 2) = \nabla f(1, 2) \cdot \mathbf{u}$$
$$= (2\mathbf{i} + 4\mathbf{j}) \cdot \tfrac{1}{13}\sqrt{13}[2\mathbf{i} - 3\mathbf{j}] = -\tfrac{8}{13}\sqrt{13} \cong -2.22. \quad \square$$

Problem 2. Find the directional derivative of the function

$$f(x, y, z) = x \cos y \sin z$$

at the point $(1, \pi, \tfrac{1}{4}\pi)$ in the direction of the vector $2\mathbf{i} - \mathbf{j} + 4\mathbf{k}$.

Solution. The unit vector in the direction of $2\mathbf{i} - \mathbf{j} + 4\mathbf{k}$ is the vector

$$\mathbf{u} = \tfrac{1}{21}\sqrt{21}[2\mathbf{i} - \mathbf{j} + 4\mathbf{k}].$$

Here

$$\frac{\partial f}{\partial x}(x, y, z) = \cos y \sin z, \qquad \frac{\partial f}{\partial y}(x, y, z) = -x \sin y \sin z,$$

$$\frac{\partial f}{\partial z}(x, y, z) = x \cos y \cos z$$

so that

$$\frac{\partial f}{\partial x}(1, \pi, \tfrac{1}{4}\pi) = -\tfrac{1}{2}\sqrt{2}, \qquad \frac{\partial f}{\partial y}(1, \pi, \tfrac{1}{4}\pi) = 0, \qquad \frac{\partial f}{\partial z}(1, \pi, \tfrac{1}{4}\pi) = -\tfrac{1}{2}\sqrt{2}.$$

Therefore

$$\nabla f(1, \pi, \tfrac{1}{4}\pi) = -\tfrac{1}{2}\sqrt{2}\,\mathbf{i} - \tfrac{1}{2}\sqrt{2}\,\mathbf{k}$$

and

$$f'_{\mathbf{u}}(1, \pi, \tfrac{1}{4}\pi) = \nabla f(1, \pi, \tfrac{1}{4}\pi) \cdot \mathbf{u}$$
$$= -\tfrac{1}{2}\sqrt{2}[\mathbf{i} + \mathbf{k}] \cdot \tfrac{1}{21}\sqrt{21}[2\mathbf{i} - \mathbf{j} + 4\mathbf{k}] = -\tfrac{1}{7}\sqrt{42} \cong -0.93. \quad \square$$

You know that for each unit vector **u**

$$f'_{\mathbf{u}}(\mathbf{x}) = \nabla f(\mathbf{x}) \cdot \mathbf{u}.$$

Since

$$\nabla f(\mathbf{x}) \cdot \mathbf{u} = \text{comp}_{\mathbf{u}} \, \nabla f(\mathbf{x}), \qquad (13.4.11)$$

we have

(16.2.6)
$$\boxed{f'_{\mathbf{u}}(\mathbf{x}) = \text{comp}_{\mathbf{u}} \, \nabla f(\mathbf{x}).}$$

Namely, the directional derivative in a direction **u** *is the component of the gradient vector in that direction.*   (Figure 16.2.1)

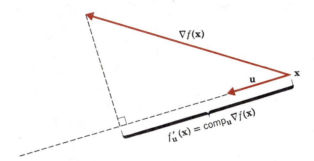

Figure 16.2.1

If $\nabla f(\mathbf{x}) \neq \mathbf{0}$, then

$$f'_{\mathbf{u}}(\mathbf{x}) = \nabla f(\mathbf{x}) \cdot \mathbf{u} = \|\nabla f(\mathbf{x})\| \, \|\mathbf{u}\| \cos \theta = \|\nabla f(\mathbf{x})\| \cos \theta$$
$$\underset{\underline{\qquad\qquad (13.4.7)}}{\uparrow}$$

where θ is the angle between $\nabla f(\mathbf{x})$ and **u**. Since $-1 \leq \cos \theta \leq 1$, we have

$$-\|\nabla f(\mathbf{x})\| \leq f'_{\mathbf{u}}(\mathbf{x}) \leq \|\nabla f(\mathbf{x})\| \qquad \text{for all directions } \mathbf{u}.$$

If **u** points in the direction of $\nabla f(\mathbf{x})$, then

$$f'_{\mathbf{u}}(\mathbf{x}) = \|\nabla f(\mathbf{x})\|, \qquad (\theta = 0, \cos \theta = 1)$$

and, if **u** points in the direction of $-\nabla f(\mathbf{x})$, then

$$f'_{\mathbf{u}}(\mathbf{x}) = -\|\nabla f(\mathbf{x})\|. \qquad (\theta = \pi, \cos \theta = -1)$$

Since the directional derivative gives the rate of change of the function in that direction, it is clear that

(16.2.7) a differentiable function f increases most rapidly in the direction of the gradient (the rate of change is then $\|\nabla f(\mathbf{x})\|$) and it decreases most rapidly in the opposite direction (the rate of change is then $-\|\nabla f(\mathbf{x})\|$).

Example 3.   The graph of the function

$$f(x, y) = \sqrt{1 - (x^2 + y^2)}$$

is the upper half of the unit sphere $x^2 + y^2 + z^2 = 1$.   The function is defined on the closed unit disc but differentiable only on the open unit disc.

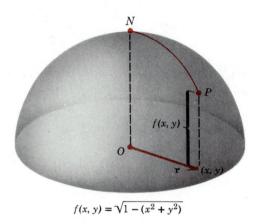

$$f(x, y) = \sqrt{1 - (x^2 + y^2)}$$

Figure 16.2.2

In Figure 16.2.2 we have marked a point (x, y) and drawn the corresponding radius vector $\mathbf{r} = x\mathbf{i} + y\mathbf{j}$.   The gradient

$$\nabla f(x, y) = \frac{-x}{\sqrt{1 - (x^2 + y^2)}} \mathbf{i} + \frac{-y}{\sqrt{1 - (x^2 + y^2)}} \mathbf{j}$$

is a negative multiple of $\mathbf{r}$:

$$\nabla f(x, y) = -\frac{1}{\sqrt{1 - (x^2 + y^2)}} (x\mathbf{i} + y\mathbf{j}) = -\frac{1}{\sqrt{1 - (x^2 + y^2)}} \mathbf{r}.$$

Since $\mathbf{r}$ points from the origin to (x, y), the gradient points from (x, y) to the origin. This means that f increases most rapidly toward the origin.   This is borne out by the observation that along the hemispherical surface the path of steepest ascent from the point $P(x, y, f(x, y))$ is the "great circle route to the north pole."   ☐

Problem 4.   The temperature at each point of a metal plate is given by the function

$$T(x, y) = e^x \cos y + e^y \cos x.$$

(a) In what direction does the temperature increase most rapidly at the point $(0, 0)$? What is this rate of increase?

(b) In what direction does the temperature decrease most rapidly at $(0, 0)$?

Solution

$$\nabla T(x, y) = \frac{\partial T}{\partial x}(x, y)\,\mathbf{i} + \frac{\partial T}{\partial y}(x, y)\,\mathbf{j}$$
$$= (e^x \cos y - e^y \sin x)\,\mathbf{i} + (e^y \cos x - e^x \sin y)\,\mathbf{j}.$$

(a) At $(0, 0)$ the temperature increases most rapidly in the direction of the gradient

$$\nabla T(0, 0) = \mathbf{i} + \mathbf{j}.$$

This rate of increase is

$$\|\nabla T(0, 0)\| = \|\mathbf{i} + \mathbf{j}\| = \sqrt{2}.$$

(b) The temperature decreases most rapidly in the direction of

$$-\nabla T(0, 0) = -\mathbf{i} - \mathbf{j}. \quad \square$$

Problem 5. The mass density (mass per unit volume) of a metal ball centered at the origin is given by the function

$$\lambda(x, y, z) = ke^{-(x^2+y^2+z^2)}, \qquad k \text{ a positive constant.}$$

(a) In what direction does the density increase most rapidly at the point (x, y, z)? What is this rate of density increase?
(b) In what direction does the density decrease most rapidly?
(c) What are the rates of density change at (x, y, z) in the $\mathbf{i}, \mathbf{j}, \mathbf{k}$ directions?

Solution. The gradient

$$\nabla \lambda(x, y, z) = \frac{\partial \lambda}{\partial x}(x, y, z)\,\mathbf{i} + \frac{\partial \lambda}{\partial y}(x, y, z)\,\mathbf{j} + \frac{\partial \lambda}{\partial z}(x, y, z)\,\mathbf{k}$$
$$= -2ke^{-(x^2+y^2+z^2)}[x\mathbf{i} + y\mathbf{j} + z\mathbf{k}]$$
$$= -2\lambda(x, y, z)\mathbf{r}$$

points from (x, y, z) in the direction opposite to that of the radius vector.

(a) The density increases most rapidly toward the origin. The rate of increase is

$$\|\nabla \lambda(x, y, z)\| = 2\lambda(x, y, z)\|\mathbf{r}\| = 2\lambda(x, y, z)\sqrt{x^2 + y^2 + z^2}.$$

(b) The density decreases most rapidly directly away from the origin.

(c) The rates of density change in the $\mathbf{i}, \mathbf{j}, \mathbf{k}$ directions are given by the directional derivatives

$$\lambda_i'(x, y, z) = \nabla \lambda(x, y, z) \cdot \mathbf{i} = -2x\,\lambda(x, y, z),$$
$$\lambda_j'(x, y, z) = \nabla \lambda(x, y, z) \cdot \mathbf{j} = -2y\,\lambda(x, y, z),$$
$$\lambda_k'(x, y, z) = \nabla \lambda(x, y, z) \cdot \mathbf{k} = -2z\,\lambda(x, y, z).$$

These are just the first partials of λ. $\quad \square$

Problem 6.   The temperature at each point of a metal plate is given by the function

$$T(x, y) = A + x^2 - y^2.$$

Find the path followed by a heat-seeking particle that originates at $(-2, 1)$.

Solution.   The particle moves in the direction of the gradient vector

$$\nabla T = 2x\mathbf{i} - 2y\mathbf{j}.$$

We want the curve

$$C: \mathbf{r}(t) = x(t)\mathbf{i} + y(t)\mathbf{j}$$

which begins at $(-2, 1)$ and at each point has its tangent vector in the direction of ∇T.   We can satisfy the first condition by setting

$$x(0) = -2, \quad y(0) = 1.$$

We can satisfy the second condition by setting

$$x'(t) = 2x(t), \quad y'(t) = -2y(t). \qquad \text{(explain)}$$

These differential equations, together with initial conditions at $t = 0$, imply that

$$x(t) = -2e^{2t}, \quad y(t) = e^{-2t}. \qquad \text{(Section 7.6)}$$

We can eliminate the parameter t by noting that

$$x(t)y(t) = (-2e^{2t})(e^{-2t}) = -2.$$

In terms of just x and y we have

$$xy = -2.$$

The particle moves from the point $(-2, 1)$ along the left branch of the hyperbola $xy = -2$ in the direction of decreasing x.   ◻

Remark.   The pair of differential equations

$$x'(t) = 2x(t), \quad y'(t) = -2y(t)$$

can be set as a single differential equation in x and y: the relation

$$\frac{y'(t)}{x'(t)} = -\frac{y(t)}{x(t)}$$

gives

$$\frac{dy}{dx} = -\frac{y}{x}. \qquad (10.7.2)$$

This equation is readily solved directly:

$$\frac{1}{y}\frac{dy}{dx} = -\frac{1}{x}$$

$$\ln|y| = -\ln|x| + C$$

$$\ln|x| + \ln|y| = C$$

$$\ln|xy| = C.$$

Thus xy is constant:

$$xy = k.$$

Since the curve passes through the point $(-2, 1)$, $k = -2$ and once again we have the curve

$$xy = -2.$$

You will be called upon to use this method of solution in some of the exercises. ☐

EXERCISES 16.2

Find the directional derivative at the given point in the direction indicated.

1. $f(x, y) = x^2 + 3y^2$ at $(1, 1)$ in the direction of $\mathbf{i} - \mathbf{j}$.

2. $f(x, y) = x + \sin(x + y)$ at $(0, 0)$ in the direction of $2\mathbf{i} + \mathbf{j}$.

3. $f(x, y) = xe^y - ye^x$ at $(1, 0)$ in the direction of $3\mathbf{i} + 4\mathbf{j}$.

4. $f(x, y) = \dfrac{2x}{x - y}$ at $(1, 0)$ in the direction of $\mathbf{i} - \sqrt{3}\mathbf{j}$.

5. $f(x, y) = \dfrac{ax + by}{x + y}$ at $(1, 1)$ in the direction of $\mathbf{i} - \mathbf{j}$.

6. $f(x, y) = \dfrac{x + y}{cx + dy}$ at $(1, 1)$ in the direction of $c\mathbf{i} - d\mathbf{j}$.

7. $f(x, y, z) = xy + yz + zx$ at $(1, -1, 1)$ in the direction of $\mathbf{i} + 2\mathbf{j} + \mathbf{k}$.

8. $f(x, y, z) = x^2y + y^2z + z^2x$ at $(1, 0, 1)$ in the direction of $3\mathbf{j} - \mathbf{k}$.

9. $f(x, y, z) = (x + y^2 + z^3)^2$ at $(1, -1, 1)$ in the direction of $\mathbf{i} + \mathbf{j}$.

10. $f(x, y, z) = Ax^2 + Bxyz + Cy^2$ at $(1, 2, 1)$ in the direction of $A\mathbf{i} + B\mathbf{j} + C\mathbf{k}$.

11. Find the directional derivative of $f(x, y) = \ln\sqrt{x^2 + y^2}$ at $(x, y) \neq (0, 0)$ toward the origin.

12. Find the directional derivative of $f(x, y) = (x - 1)y^2e^{xy}$ at $(0, 1)$ toward the point $(-1, 3)$.

13. Find the directional derivative of $f(x, y) = Ax^2 + 2Bxy + Cy^2$ at (a, b) toward (b, a) (a) if $a > b$.   (b) if $a < b$.

14. Find the directional derivative of $f(x, y, z) = xy^2 + y^2z^3 + z^3x$ at $(2, -1, 1)$ in the direction of the most rapid increase of f.

15. Find the directional derivative of $f(x, y, z) = xe^{y^2 - z^2}$ at $(1, 2, -2)$ in the direction of the path $\mathbf{r}(t) = t\mathbf{i} + 2\cos(t - 1)\mathbf{j} - 2e^{t-1}\mathbf{k}$.

16. Find the directional derivative of $f(x, y, z) = x^2 + yz$ at $(1, -3, 2)$ in the direction of the path $\mathbf{r}(t) = t^2\mathbf{i} + 3t\mathbf{j} + (1 - t^3)\mathbf{k}$.

17. Let $f = f(x)$ be a differentiable function of one variable. What is the gradient of f at x_0? What is the geometric significance of the direction of this gradient?

18. Suppose that f is differentiable at (x_0, y_0) and $\nabla f(x_0, y_0) \neq 0$. Compute the rate of change of f in the direction of the vector

$$\frac{\partial f}{\partial y}(x_0, y_0)\, \mathbf{i} - \frac{\partial f}{\partial x}(x_0, y_0)\, \mathbf{j}.$$

Give a geometric interpretation to your answer.

19. Let

$$f(x, y) = \sqrt{x^2 + y^2}.$$

(a) Show that $\partial f/\partial x$ is not defined at $(0, 0)$.
(b) Is f differentiable at $(0, 0)$?

20. Verify that, if g is continuous at $\mathbf{x}$, then
(a) $g(\mathbf{x} + \mathbf{h})o(\mathbf{h}) = o(\mathbf{h})$ and   (b) $[g(\mathbf{x} + \mathbf{h}) - g(\mathbf{x})]\, \nabla f(\mathbf{x}) \cdot \mathbf{h} = o(\mathbf{h})$.

21. Given the density function $\lambda(x, y) = 48 - \frac{4}{3}x^2 - 3y^2$, find the rate of density change (a) at $(1, -1)$ in the direction of the most rapid density decrease. (b) at $(1, 2)$ in the $\mathbf{i}$ direction. (c) at $(2, 2)$ away from the origin.

22. The intensity of light in a neighborhood of the point $(-2, 1)$ is given by a function of the form $I(x, y) = A - 2x^2 - y^2$. Find the path followed by a light-seeking particle that originates at the center of the neighborhood.

23. Determine the path of steepest descent along the surface $z = x^2 + 3y^2$ from each of the following points: (a) $(1, 1, 4)$. (b) $(1, -2, 13)$.

24. Determine the path of steepest ascent along the hyperbolic paraboloid $z = \frac{1}{2}x^2 - y^2$ from each of the following points: (a) $(-1, 1, -\frac{1}{2})$. (b) $(1, 0, \frac{1}{2})$.

25. Determine the path of steepest descent along the surface $z = a^2x^2 + b^2y^2$ from the point $(a^2, b^2, a^4 + b^4)$.

26. The temperature in a neighborhood of the origin is given by a function of the form

$$T(x, y) = T_0 + e^y \sin x.$$

Find the path followed by a heat-fleeing particle that originates at the origin.

27. The temperature in a neighborhood of the point $(\frac{1}{4}\pi, 0)$ is given by the function

$$T(x, y) = \sqrt{2}\, e^{-y} \cos x.$$

Find the path followed by a heat-seeking particle that originates at the center of the neighborhood.

28. Determine the path of steepest descent along the surface $z = A + x + 2y - x^2 - 3y^2$ from the point $(0, 0, A)$.

29. Set $f(x, y) = 3x^2 + y$.
(a) Find

$$\lim_{h \to 0} \frac{f(x(2 + h), y(2 + h)) - f(2, 4)}{h}$$

given that $x(t) = t$ and $y(t) = t^2$. (These functions parametrize the parabola $y = x^2$.)
(b) Find

$$\lim_{h \to 0} \frac{f(x(4 + h), y(4 + h)) - f(2, 4)}{h}$$

given that $x(t) = \frac{1}{4}(t + 4)$ and $y(t) = t$. (These functions parametrize the line $y = 4x - 4$.)
(c) Compute the directional derivative of f at $(2, 4)$ in the direction of $\mathbf{i} + 4\mathbf{j}$.
(d) Notice that $\mathbf{i} + 4\mathbf{j}$ is a direction vector for the line $y = 4x - 4$ and this line is tangent to the parabola $y = x^2$ at $(2, 4)$. Explain then why the computations in (a), (b), and (c) yield different values.

16.3   THE MEAN-VALUE THEOREM; TWO INTERMEDIATE-VALUE THEOREMS

You have seen the large role played by the mean-value theorem in the calculus of functions of one variable. Here we take up the analogous result for functions of several variables. By **ab** we mean the line segment that joins point **a** to point **b**.

THEOREM 16.3.1   THE MEAN-VALUE THEOREM (SEVERAL VARIABLES)

If f is differentiable at each point of the line segment **ab**, then there exists on that line segment a point **c** between **a** and **b** such that

$$f(\mathbf{b}) - f(\mathbf{a}) = \nabla f(\mathbf{c}) \cdot (\mathbf{b} - \mathbf{a}).$$

Proof. As t ranges from 0 to 1, $\mathbf{a} + t(\mathbf{b} - \mathbf{a})$ traces out the line segment **ab**. The idea of the proof is to apply the one-variable mean-value theorem to the function

$$g(t) = f(\mathbf{a} + t[\mathbf{b} - \mathbf{a}]), \qquad t \in [0, 1].$$

To show that g is differentiable on the open interval $(0, 1)$, we take $t \in (0, 1)$ and form

$$
\begin{aligned}
g(t + h) - g(t) &= f(\mathbf{a} + (t + h)[\mathbf{b} - \mathbf{a}]) - f(\mathbf{a} + t[\mathbf{b} - \mathbf{a}]) \\
&= f(\mathbf{a} + t[\mathbf{b} - \mathbf{a}] + h[\mathbf{b} - \mathbf{a}]) - f(\mathbf{a} + t[\mathbf{b} - \mathbf{a}]) \\
&= \nabla f(\mathbf{a} + t[\mathbf{b} - \mathbf{a}]) \cdot h[\mathbf{b} - \mathbf{a}] + o(h[\mathbf{b} - \mathbf{a}]).
\end{aligned}
$$

Since

$$\nabla f(\mathbf{a} + t[\mathbf{b} - \mathbf{a}]) \cdot h(\mathbf{b} - \mathbf{a}) = [\nabla f(\mathbf{a} + t[\mathbf{b} - \mathbf{a}]) \cdot (\mathbf{b} - \mathbf{a})]h$$

and the $o(h(\mathbf{b} - \mathbf{a}))$ term is obviously $o(h)$, we can write

$$g(t + h) - g(t) = [\nabla f(\mathbf{a} + t[\mathbf{b} - \mathbf{a}]) \cdot (\mathbf{b} - \mathbf{a})]h + o(h).$$

Dividing both sides by h, we see that g is differentiable and

$$g'(t) = \nabla f(\mathbf{a} + t[\mathbf{b} - \mathbf{a}]) \cdot (\mathbf{b} - \mathbf{a}).$$

The function g is clearly continuous at 0 and at 1. Applying the one-variable mean-value theorem to g, we can conclude that there exists a number t_0 between 0 and 1 such that

$$(1) \qquad\qquad g(1) - g(0) = g'(t_0)(1 - 0).$$

Since $g(1) = f(\mathbf{b})$, $g(0) = f(\mathbf{a})$, and $g'(t_0) = \nabla f(\mathbf{a} + t_0[\mathbf{b} - \mathbf{a}]) \cdot (\mathbf{b} - \mathbf{a})$, condition (1) gives

$$f(\mathbf{b}) - f(\mathbf{a}) = \nabla f(\mathbf{a} + t_0[\mathbf{b} - \mathbf{a}]) \cdot (\mathbf{b} - \mathbf{a}).$$

Setting $\mathbf{c} = \mathbf{a} + t_0[\mathbf{b} - \mathbf{a}]$, we have

$$f(\mathbf{b}) - f(\mathbf{a}) = \nabla f(\mathbf{c}) \cdot (\mathbf{b} - \mathbf{a}). \qquad \square$$

A nonempty open set U (in the plane or in three-space) is said to be *connected* iff any two points of U can be joined by a polygonal path that lies entirely in U. You can see such a set in Figure 16.3.1.

The set shown in Figure 16.3.2 is the union of two disjoint open sets. The set is open but not connected: it is impossible to join **a** and **b** by a polygonal path that lies within the set.

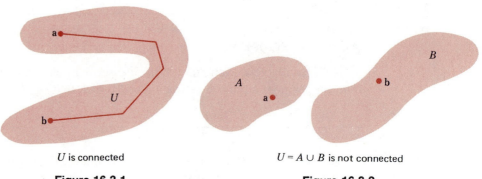

U is connected $U = A \cup B$ is not connected

Figure 16.3.1   **Figure 16.3.2**

In Chapter 4 you saw that, if $f'(x) = 0$ for all x in an open interval I, then f is constant on I. We have a similar result for functions of several variables.

THEOREM 16.3.2

Let U be an open connected set:

$$\text{if } \nabla f(\mathbf{x}) = \mathbf{0} \text{ for all } \mathbf{x} \text{ in } U, \text{ then } f \text{ is constant on } U.$$

Proof. Let **a** and **b** be any two points in U. Since U is open and connected, we can join these points by a polygonal path with vertices $\mathbf{a} = \mathbf{c}_0, \mathbf{c}_1, \ldots, \mathbf{c}_{n-1}, \mathbf{c}_n = \mathbf{b}$. By the mean-value theorem there exist points

$$\mathbf{c}_1^* \text{ on } \overline{\mathbf{c}_0 \mathbf{c}_1} \quad \text{such that} \quad f(\mathbf{c}_1) - f(\mathbf{c}_0) = \nabla f(\mathbf{c}_1^*) \cdot (\mathbf{c}_1 - \mathbf{c}_0),$$
$$\mathbf{c}_2^* \text{ on } \overline{\mathbf{c}_1 \mathbf{c}_2} \quad \text{such that} \quad f(\mathbf{c}_2) - f(\mathbf{c}_1) = \nabla f(\mathbf{c}_2^*) \cdot (\mathbf{c}_2 - \mathbf{c}_1),$$
$$\vdots \qquad\qquad\qquad\qquad \vdots$$
$$\mathbf{c}_n^* \text{ on } \overline{\mathbf{c}_{n-1} \mathbf{c}_n} \quad \text{such that} \quad f(\mathbf{c}_n) - f(\mathbf{c}_{n-1}) = \nabla f(\mathbf{c}_n^*) \cdot (\mathbf{c}_n - \mathbf{c}_{n-1}).$$

If $\nabla f(\mathbf{x}) = \mathbf{0}$ for all $\mathbf{x}$ in U, then

$$f(\mathbf{c}_1) - f(\mathbf{c}_0) = 0, \quad f(\mathbf{c}_2) - f(\mathbf{c}_1) = 0, \quad \ldots, f(\mathbf{c}_n) - f(\mathbf{c}_{n-1}) = 0.$$

This shows that

$$f(\mathbf{a}) = f(\mathbf{c}_0) = f(\mathbf{c}_1) = f(\mathbf{c}_2) = \cdots = f(\mathbf{c}_{n-1}) = f(\mathbf{c}_n) = f(\mathbf{b}).$$

Since **a** and **b** are arbitrary points of U, f must be constant on U. ☐

THEOREM 16.3.3

Let U be an open connected set:

   if $\nabla f(\mathbf{x}) = \nabla g(\mathbf{x})$ for all $\mathbf{x}$ in U, then f and g differ by a constant on U.

Proof. If $\nabla f(\mathbf{x}) = \nabla g(\mathbf{x})$ for all $\mathbf{x}$ in U, then

$$\nabla[f(\mathbf{x}) - g(\mathbf{x})] = \nabla f(\mathbf{x}) - \nabla g(\mathbf{x}) = \mathbf{0} \qquad \text{for all } \mathbf{x} \text{ in } U.$$

By Theorem 16.3.2, $f - g$ must be constant on U. $\quad\square$

You have seen that a function continuous on an interval skips no values. (Theorem 2.6.1) There are analogous results for functions of several variables. Here is one of them.

THEOREM 16.3.4   AN INTERMEDIATE-VALUE THEOREM (SEVERAL VARIABLES)

Suppose that f is continuous on an open connected set U and $A < C < B$. If, somewhere on U, f takes on the value A and, somewhere on U, f takes on the value B, then, somewhere on U, f takes on the value C.

Proof. Let $\mathbf{a}$ and $\mathbf{b}$ be points of U for which

$$f(\mathbf{a}) = A \qquad \text{and} \qquad f(\mathbf{b}) = B.$$

We must show that there exists a point $\mathbf{c}$ in U for which $f(\mathbf{c}) = C$.

Since U is polygonally connected, there is a polygonal path in U

$$\mathbf{r} = \mathbf{r}(t), \qquad t \in [a, b]$$

that joins $\mathbf{a}$ to $\mathbf{b}$. Since $\mathbf{r}$ is continuous on $[a, b]$, the composition

$$g(t) = f(\mathbf{r}(t))$$

is also continuous on $[a, b]$. Since

$$g(a) = f(\mathbf{r}(a)) = f(\mathbf{a}) = A \qquad \text{and} \qquad g(b) = f(\mathbf{r}(b)) = f(\mathbf{b}) = B,$$

we know from Theorem 2.6.1 that there is a number c in $[a, b]$ for which $g(c) = C$. Setting $\mathbf{c} = \mathbf{r}(c)$ we have $f(\mathbf{c}) = C$. $\quad\square$

Continuity on a Closed Region

An open connected set is called an *open region*. If we start with an open region and adjoin to it the boundary, then we have what is called a *closed region*. (A closed region is therefore a closed set, the interior of which is an open region.)

Continuity on a closed region Ω requires continuity at the boundary points of Ω as well as the interior points.

If $\mathbf{x}_0$ is an interior point of Ω, then all points $\mathbf{x}$ sufficiently close to $\mathbf{x}_0$ are in Ω and, by definition, f is continuous at $\mathbf{x}_0$ iff

$$\text{as } \mathbf{x} \text{ approaches } \mathbf{x}_0, \quad f(\mathbf{x}) \text{ approaches } f(\mathbf{x}_0).$$

If $\mathbf{x}_0$ is a boundary point of Ω, then we have to modify the definition and say: f is continuous at $\mathbf{x}_0$ iff

$$\text{as } \mathbf{x} \text{ approaches } \mathbf{x}_0 \text{ } within \text{ } \Omega, \quad f(\mathbf{x}) \text{ approaches } f(\mathbf{x}_0).$$

In terms of ϵ-δ, f is continuous at a boundary point $\mathbf{x}_0$ iff for each $\epsilon > 0$ there exists $\delta > 0$ such that

$$\text{if } \|\mathbf{x} - \mathbf{x}_0\| < \delta \text{ and } \mathbf{x} \in \Omega, \quad \text{then } |f(\mathbf{x}) - f(\mathbf{x}_0)| < \epsilon.$$

(This is completely analogous to one-sided continuity at an endpoint of a closed interval $[a, b]$ as defined in Section 2.4.)

The intermediate-value result that we just proved for open connected sets can be extended to closed regions.

THEOREM 16.3.5 A SECOND INTERMEDIATE-VALUE THEOREM [SEVERAL VARIABLES]

Suppose that f is continuous on a closed region Ω and $A < C < B$. If, somewhere on Ω, f takes on the value A and, somewhere on Ω, f takes on the value B, then, somewhere on Ω, f takes on the value C.

Proof. Let $\mathbf{a}$ and $\mathbf{b}$ be points of Ω for which

$$f(\mathbf{a}) = A \quad \text{and} \quad f(\mathbf{b}) = B.$$

If $\mathbf{a}$ and $\mathbf{b}$ are both in the interior of Ω, then the result follows from the previous theorem. But one or both of these points could lie on the boundary of Ω. To take care of that possibility, we can proceed as follows. Take ϵ small enough that

$$A + \epsilon < C < B - \epsilon.$$

By continuity there exist points $\mathbf{x}_1$, $\mathbf{x}_2$ in the interior of Ω for which

$$f(\mathbf{x}_1) < A + \epsilon \quad \text{and} \quad B - \epsilon < f(\mathbf{x}_2). \qquad \text{(Exercise 5)}$$

Then

$$f(\mathbf{x}_1) < C < f(\mathbf{x}_2)$$

and the result follows from the previous theorem. $\square$

EXERCISES 16.3

1. (a) Find f if $\nabla f(x, y, z) = a_1\mathbf{i} + a_2\mathbf{j} + a_3\mathbf{k}$ for all (x, y, z).
   (b) What can you conclude about f and g if $\nabla f(x, y, z) - \nabla g(x, y, z) = a_1\mathbf{i} + a_2\mathbf{j} + a_3\mathbf{k}$ for all (x, y, z)?

2. (*Rolle's theorem for functions of several variables*) Show that, if f is differentiable at each point of the line segment **ab** and $f(\mathbf{a}) = f(\mathbf{b})$, then there exists a point **c** between **a** and **b** for which $\nabla f(\mathbf{c}) \perp \mathbf{b} - \mathbf{a}$.

3. Let $U = \{\mathbf{x}: \|\mathbf{x}\| \neq 1\}$. Define f on U by setting

$$f(\mathbf{x}) = \begin{bmatrix} 0, & \|\mathbf{x}\| < 1 \\ 1, & \|\mathbf{x}\| > 1 \end{bmatrix}.$$

(a) Note that $\nabla f(\mathbf{x}) = \mathbf{0}$ for all $\mathbf{x}$ in U, but f is not constant on U. Explain how this does not contradict Theorem 16.3.2.
(b) Define a function g on U different from f such that $\nabla f(\mathbf{x}) = \nabla g(\mathbf{x})$ for all $\mathbf{x}$ in U and $f - g$ is (i) constant on U, (ii) not constant on U.
(c) The function f takes on the value 0, takes on the value 1, but takes on no value in between. Explain how this does not contradict Theorem 16.3.4.

4. A set of points is said to be *convex* provided that every pair of points in the set can be joined by a line segment that lies entirely within the set. Show that, if $\|\nabla f(\mathbf{x})\| \leq M$ for all $\mathbf{x}$ in some convex set Ω, then

$$|f(\mathbf{x}_1) - f(\mathbf{x}_2)| \leq M \|\mathbf{x}_1 - \mathbf{x}_2\| \qquad \text{for all } \mathbf{x}_1 \text{ and } \mathbf{x}_2 \text{ in } \Omega.$$

5. Justify the assertion made in the proof of Theorem 16.3.5 that there exist points $\mathbf{x}_1, \mathbf{x}_2$ in Ω for which

$$f(\mathbf{x}_1) < A + \epsilon \qquad \text{and} \qquad B - \epsilon < f(\mathbf{x}_2).$$

16.4 CHAIN RULES

For functions of a single variable there is basically only one chain rule: Theorem 3.6.7. For functions of several variables there are many chain rules.

A vector-valued function is said to be *continuous* provided that its components are continuous. If $f = f(x, y, z)$ is a scalar-valued function (a real-valued function), then its gradient ∇f is a vector-valued function. We say that f is *continuously differentiable* on an *open set* U iff f is differentiable on U and ∇f is continuous on U.

If a curve $\mathbf{r}$ lies in the domain of f, then we can form the composition

$$(f \circ \mathbf{r})(t) = f(\mathbf{r}(t)).$$

The composition $f \circ \mathbf{r}$ is a real-valued function of a real variable t. The numbers $f(\mathbf{r}(t))$ are the values taken on by f *along the curve* $\mathbf{r}$.

THEOREM 16.4.1 CHAIN RULE (ALONG A CURVE)

If f is continuously differentiable on an open set U and $\mathbf{r} = \mathbf{r}(t)$ is a differentiable curve that lies entirely in U, then the composition $f \circ \mathbf{r}$ is differentiable and

$$\frac{d}{dt}[f(\mathbf{r}(t))] = \nabla f(\mathbf{r}(t)) \cdot \mathbf{r}'(t).$$

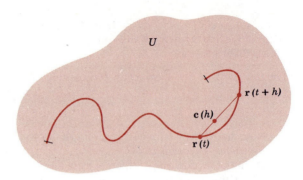

Figure 16.4.1

Proof.   We will show that

$$\lim_{h \to 0} \frac{f(\mathbf{r}(t + h)) - f(\mathbf{r}(t))}{h} = \nabla f(\mathbf{r}(t)) \cdot \mathbf{r}'(t).$$

For $h \neq 0$ and sufficiently small, the line segment that joins $\mathbf{r}(t)$ to $\mathbf{r}(t + h)$ lies entirely in U. This we know because U is open and $\mathbf{r}$ is continuous. (See Figure 16.4.1.) For such h, the mean-value theorem we proved in Section 16.3 assures us that there exists a point $\mathbf{c}(h)$ between $\mathbf{r}(t)$ and $\mathbf{r}(t + h)$ such that

$$f(\mathbf{r}(t + h)) - f(\mathbf{r}(t)) = \nabla f(\mathbf{c}(h)) \cdot [\mathbf{r}(t + h) - \mathbf{r}(t)].$$

Dividing both sides by h, we have

$$\frac{f(\mathbf{r}(t + h)) - f(\mathbf{r}(t))}{h} = \nabla f(\mathbf{c}(h)) \cdot \left[\frac{\mathbf{r}(t + h) - \mathbf{r}(t)}{h} \right].$$

As h tends to zero, $\mathbf{c}(h)$ tends to $\mathbf{r}(t)$ and by the continuity of ∇f,

$$\nabla f(\mathbf{c}(h)) \to \nabla f(\mathbf{r}(t)).$$

Since

$$\frac{\mathbf{r}(t + h) - \mathbf{r}(t)}{h} \to \mathbf{r}'(t),$$

the result follows.   ☐

Problem 1.   Use the chain rule to find the rate of change of

$$f(x, y) = \tfrac{1}{3}(x^3 + y^3)$$

with respect to t along the curve $\mathbf{r}(t) = a \cos t \, \mathbf{i} + b \sin t \, \mathbf{j}$.

Solution.   The rate of change of f with respect to t along the curve $\mathbf{r}$ is the derivative

$$\frac{d}{dt} [f(\mathbf{r}(t))].$$

By the chain rule

$$\frac{d}{dt}[f(\mathbf{r}(t))] = \nabla f(\mathbf{r}(t)) \cdot \mathbf{r}'(t).$$

Here

$$\nabla f = x^2\mathbf{i} + y^2\mathbf{j}.$$

With $x(t) = a \cos t$ and $y(t) = b \sin t$, we have

$$\nabla f(\mathbf{r}(t)) = a^2 \cos^2 t \, \mathbf{i} + b^2 \sin^2 t \, \mathbf{j}.$$

Since $\mathbf{r}'(t) = -a \sin t \, \mathbf{i} + b \cos t \, \mathbf{j}$, you can see that

$$\frac{d}{dt}[f(\mathbf{r}(t))] = \nabla f(\mathbf{r}(t)) \cdot \mathbf{r}'(t)$$

$$= (a^2 \cos^2 t \, \mathbf{i} + b^2 \sin^2 t \, \mathbf{j}) \cdot (-a \sin t \, \mathbf{i} + b \cos t \, \mathbf{j})$$

$$= -a^3 \sin t \cos^2 t + b^3 \sin^2 t \cos t$$

$$= \sin t \cos t \, (b^3 \sin t - a^3 \cos t). \quad \square$$

Remark. Note that we could have obtained the same result without invoking Theorem 16.4.1 by first forming $f(\mathbf{r}(t))$ and then differentiating:

$$f(\mathbf{r}(t)) = f(x(t), y(t)) = \tfrac{1}{3}([x(t)]^3 + [y(t)]^3) = \tfrac{1}{3}(a^3 \cos^3 t + b^3 \sin^3 t)$$

so that

$$\frac{d}{dt}[f(\mathbf{r}(t))] = \tfrac{1}{3}[3a^3 (\cos^2 t)(-\sin t) + 3b^3 \sin^2 t \cos t]$$

$$= \sin t \cos t \, (b^3 \sin t - a^3 \cos t). \quad \square$$

Problem 2. Use the chain rule to find the rate of change of

$$f(x, y, z) = x^2y + z \cos x$$

with respect to t along the twisted cubic $\mathbf{r}(t) = t\mathbf{i} + t^2\mathbf{j} + t^3\mathbf{k}$.

Solution. Once again we use the relation

$$\frac{d}{dt}[f(\mathbf{r}(t))] = \nabla f(\mathbf{r}(t)) \cdot \mathbf{r}'(t).$$

This time

$$\nabla f = (2xy - z \sin x)\mathbf{i} + x^2\mathbf{j} + \cos x \, \mathbf{k}.$$

With $x(t) = t, \ y(t) = t^2, \ z(t) = t^3$, we have

$$\nabla f(\mathbf{r}(t)) = (2t^3 - t^3 \sin t)\mathbf{i} + t^2\mathbf{j} + \cos t \, \mathbf{k}.$$

Since $\mathbf{r}'(t) = \mathbf{i} + 2t\mathbf{j} + 3t^2\mathbf{k}$, we have

$$\frac{d}{dt}[f(\mathbf{r}(t))] = \nabla f(\mathbf{r}(t)) \cdot \mathbf{r}'(t)$$

$$= [(2t^3 - t^3 \sin t)\mathbf{i} + t^2\mathbf{j} + \cos t \, \mathbf{k}] \cdot [\mathbf{i} + 2t\mathbf{j} + 3t^2\mathbf{k}]$$

$$= 2t^3 - t^3 \sin t + 2t^3 + 3t^2 \cos t$$

$$= 4t^3 - t^3 \sin t + 3t^2 \cos t.$$

You can check this answer by first forming $f(\mathbf{r}(t))$ and then differentiating.   □

Another Formulation of (16.4.1)

The chain rule for functions of one variable,

$$\frac{d}{dt}[u(x(t))] = u'(x(t))x'(t),$$

can be written

$$\frac{du}{dt} = \frac{du}{dx}\frac{dx}{dt}.$$

In a similar manner, the relation

$$\frac{d}{dt}[u(\mathbf{r}(t))] = \nabla u(\mathbf{r}(t)) \cdot \mathbf{r}'(t)$$

can be written

(1)
$$\frac{du}{dt} = \nabla u \cdot \frac{d\mathbf{r}}{dt}.$$

With

$$\nabla u = \frac{\partial u}{\partial x}\mathbf{i} + \frac{\partial u}{\partial y}\mathbf{j} + \frac{\partial u}{\partial z}\mathbf{k} \qquad \text{and} \qquad \frac{d\mathbf{r}}{dt} = \frac{dx}{dt}\mathbf{i} + \frac{dy}{dt}\mathbf{j} + \frac{dz}{dt}\mathbf{k},$$

equation (1) takes the form

(16.4.2)
$$\frac{du}{dt} = \frac{\partial u}{\partial x}\frac{dx}{dt} + \frac{\partial u}{\partial y}\frac{dy}{dt} + \frac{\partial u}{\partial z}\frac{dz}{dt}.$$

In the two-variable case, the z-term drops out and we have

(16.4.3)
$$\frac{du}{dt} = \frac{\partial u}{\partial x}\frac{dx}{dt} + \frac{\partial u}{\partial y}\frac{dy}{dt}.$$

Problem 3.   Find du/dt if

$$u = x^2 - y^2 \qquad \text{and} \qquad x = t^2 - 1, \quad y = 3 \sin \pi t.$$

Solution. Here we are in the two-variable case

$$\frac{du}{dt} = \frac{\partial u}{\partial x}\frac{dx}{dt} + \frac{\partial u}{\partial y}\frac{dy}{dt}.$$

Since

$$\frac{\partial u}{\partial x} = 2x, \quad \frac{\partial u}{\partial y} = -2y \quad \text{and} \quad \frac{dx}{dt} = 2t, \quad \frac{dy}{dt} = 3\pi \cos \pi t,$$

we have

$$\frac{du}{dt} = (2x)(2t) + (-2y)(3\pi \cos \pi t)$$

$$= 2(t^2 - 1)(2t) + (-2)(3 \sin \pi t)(3\pi \cos \pi t)$$

$$= 4t^3 - 4t - 18\pi \sin \pi t \cos \pi t.$$

You can obtain this same result by first writing u directly as a function of t and then differentiating:

$$u = x^2 - y^2 = (t^2 - 1)^2 - (3 \sin \pi t)^2$$

so that

$$\frac{du}{dt} = 2(t^2 - 1)2t - 2(3 \sin \pi t)3\pi \cos \pi t = 4t^3 - 4t - 18\pi \sin \pi t \cos \pi t. \quad \square$$

Problem 4. The upper radius of the frustum of a cone is 10 inches, the lower radius is 12 inches, and the height is 18 inches. At what rate is the volume changing if the upper radius decreases at the rate of 2 inches per minute, the lower radius increases at the rate of 3 inches per minute, and the height decreases at the rate of 4 inches per minute?

Solution. Let x be the upper radius, y the lower radius, and z the height. Then

$$V = \frac{1}{3}\pi z(x^2 + xy + y^2)$$

so that

$$\frac{\partial V}{\partial x} = \frac{1}{3}\pi z(2x + y), \qquad \frac{\partial V}{\partial y} = \frac{1}{3}\pi z(x + 2y), \qquad \frac{\partial V}{\partial z} = \frac{1}{3}\pi(x^2 + xy + y^2).$$

Since

$$\frac{dV}{dt} = \frac{\partial V}{\partial x}\frac{dx}{dt} + \frac{\partial V}{\partial y}\frac{dy}{dt} + \frac{\partial V}{\partial z}\frac{dz}{dt},$$

we have here

$$\frac{dV}{dt} = \frac{1}{3}\pi z(2x + y)\frac{dx}{dt} + \frac{1}{3}\pi z(x + 2y)\frac{dy}{dt} + \frac{1}{3}\pi(x^2 + xy + y^2)\frac{dz}{dt}.$$

Set

$$x = 10, \quad y = 12, \quad z = 18, \quad \frac{dx}{dt} = -2, \quad \frac{dy}{dt} = 3, \quad \frac{dz}{dt} = -4,$$

and you will find that

$$\frac{dV}{dt} = -\frac{772}{3}\, \pi.$$

The volume decreases at the rate of $\frac{772}{3}\pi$ cubic inches per minute. □

Other Chain Rules

In the setting of functions of several variables there are numerous chain rules. Some will be stated here, others in the exercises. They can all be deduced from Theorem 16.4.1 and its corollaries, (16.4.2) and (16.4.3).

If, for example,

$$u = u(x, y) \qquad \text{where} \quad x = x(s, t) \quad \text{and} \quad y = y(s, t),$$

then

(16.4.4)
$$\frac{\partial u}{\partial s} = \frac{\partial u}{\partial x}\frac{\partial x}{\partial s} + \frac{\partial u}{\partial y}\frac{\partial y}{\partial s} \quad \text{and} \quad \frac{\partial u}{\partial t} = \frac{\partial u}{\partial x}\frac{\partial x}{\partial t} + \frac{\partial u}{\partial y}\frac{\partial y}{\partial t}.$$

To obtain the first equation, keep t fixed and differentiate u with respect to s according to Formula 16.4.3; to obtain the second equation, keep s fixed and differentiate u with respect to t. □

In Figure 16.4.2 we have drawn a *tree diagram* for Formula 16.4.4. We construct such a tree by branching at each stage from a function to all the variables that directly determine it. Each path starting at u and ending at a variable determines a product of (partial) derivatives. The partial derivative of u with respect to each variable is the sum of the products generated by all the direct paths to that variable.

Figure 16.4.2

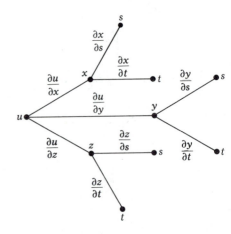

Figure 16.4.3

Suppose, for example, that

$$u = u(x, y, z) \quad \text{where} \quad x = x(s, t), \quad y = y(s, t), \quad z = z(s, t).$$

A tree diagram for the partials of u appears in Figure 16.4.3. The partials of u with respect to s and t can be read from the diagram:

(16.4.5)
$$\frac{\partial u}{\partial s} = \frac{\partial u}{\partial x}\frac{\partial x}{\partial s} + \frac{\partial u}{\partial y}\frac{\partial y}{\partial s} + \frac{\partial u}{\partial z}\frac{\partial z}{\partial s}, \qquad \frac{\partial u}{\partial t} = \frac{\partial u}{\partial x}\frac{\partial x}{\partial t} + \frac{\partial u}{\partial y}\frac{\partial y}{\partial t} + \frac{\partial u}{\partial z}\frac{\partial z}{\partial t}.$$

Problem 5. (*Implicit differentiation*) Let $u = u(x, y, z)$ be a continuously differentiable function, and suppose that the equation

$$u(x, y, z) = 0$$

defines z implicitly as a differentiable function of x and y. Show that

$$\text{if} \quad \frac{\partial u}{\partial z} \neq 0, \quad \text{then} \quad \frac{\partial z}{\partial x} = -\frac{\partial u/\partial x}{\partial u/\partial z} \quad \text{and} \quad \frac{\partial z}{\partial y} = -\frac{\partial u/\partial y}{\partial u/\partial z}.$$

Solution. To be able to apply (16.4.5), we write

$$u = u(x, y, z) \quad \text{with} \quad x = s, \quad y = t, \quad z = z(s, t).$$

Then

$$\frac{\partial u}{\partial s} = \frac{\partial u}{\partial x}\frac{\partial x}{\partial s} + \frac{\partial u}{\partial y}\frac{\partial y}{\partial s} + \frac{\partial u}{\partial z}\frac{\partial z}{\partial s}.$$

Since $u(s, t, z(s, t)) = 0$ for all s and t,

$$\frac{\partial u}{\partial s} = 0.$$

Since

$$\frac{\partial x}{\partial s} = 1 \quad \text{and} \quad \frac{\partial y}{\partial s} = 0,$$

we have

$$0 = \frac{\partial u}{\partial x} \cdot 1 + \frac{\partial u}{\partial y} \cdot 0 + \frac{\partial u}{\partial z}\frac{\partial z}{\partial s} = \frac{\partial u}{\partial x} + \frac{\partial u}{\partial z}\frac{\partial z}{\partial s} = \frac{\partial u}{\partial x} + \frac{\partial u}{\partial z}\frac{\partial z}{\partial x}.$$

$$\underset{x = s}{}$$

If $\partial u/\partial z \neq 0$, then

$$\frac{\partial z}{\partial x} = -\frac{\partial u/\partial x}{\partial u/\partial z}.$$

The formula for $\partial z/\partial y$ can be obtained in a similar manner. ☐

EXERCISES 16.4

Find the rate of change of f with respect to t along the given curve.

1. $f(x, y) = x^2 y$, $\quad \mathbf{r}(t) = e^t \mathbf{i} + e^{-t} \mathbf{j}$.
2. $f(x, y) = x - y$, $\quad \mathbf{r}(t) = at \mathbf{i} + b \cos at \, \mathbf{j}$.
3. $f(x, y, z) = xy - yz$, $\quad \mathbf{r}(t) = t \mathbf{i} + t^2 \mathbf{j} + t^3 \mathbf{k}$.
4. $f(x, y, z) = x^2 + y^2$, $\quad \mathbf{r}(t) = a \cos \omega t \, \mathbf{i} + b \sin \omega t \, \mathbf{j} + b \omega t \mathbf{k}$.
5. $f(x, y, z) = x^2 + y^2 + z$, $\quad \mathbf{r}(t) = a \cos \omega t \, \mathbf{i} + b \sin \omega t \, \mathbf{j} + b \omega t \mathbf{k}$.
6. $f(x, y, z) = y^2 \sin (x + z)$, $\quad \mathbf{r}(t) = 2t \mathbf{i} + \cos t \, \mathbf{j} + t^3 \mathbf{k}$.

Find du/dt by applying (16.4.2) or (16.4.3).

7. $u = x^2 - 3xy + 2y^2$; $\quad x = \cos t$, $\quad y = \sin t$.
8. $u = x + 4\sqrt{xy} - 3y$; $\quad x = t^3$, $\quad y = t^{-1}$. $\quad (t > 0)$
9. $u = e^x \sin y + e^y \sin x$; $\quad x = \frac{1}{2} t$, $\quad y = 2t$.
10. $u = 2x^2 - xy + y^2$; $\quad x = \cos 2t$, $\quad y = \sin t$.
11. $u = xy + yz + zx$; $\quad x = t^2$, $\quad y = t(1 - t)$, $\quad z = (1 - t)^2$.
12. $u = x \sin \pi y - z \cos \pi x$; $\quad x = t^2$, $\quad y = 1 - t$, $\quad z = 1 - t^2$.

13. The height of a cone is 20 inches and the radius is 14 inches. Use the methods of this section to find the rate at which the volume is changing if the height decreases at the rate of 2 inches per second and the radius increases at the rate of 3 inches per second.

14. The height of a cylinder is 18 inches and the radius is 13 inches. Use the methods of this section to find the rate at which the surface area is changing if the height increases at the rate of 3 inches per second and the radius decreases at the rate of 2 inches per second.

Find $\partial u/\partial s$ and $\partial u/\partial t$.

15. $u = x^2 - xy$ $\quad$ where $\quad x = s \cos t$, $\quad y = t \sin s$.
16. $u = \sin (x - y) + \cos (x + y)$ $\quad$ where $\quad x = st$, $\quad y = s^2 - t^2$.
17. $u = x^2 - xy + z^2$ $\quad$ where $\quad x = s \cos t$, $\quad y = \sin (t - s)$, $\quad z = t \sin s$.
18. $u = xe^{yz^2}$ $\quad$ where $\quad x = \ln st$, $\quad y = t^3$, $\quad z = s^2 + t^2$.

19. An object moves so that at time t it has position $\mathbf{r}(t) = x(t)\mathbf{i} + y(t)\mathbf{j} + z(t)\mathbf{k}$. Show that

$$\frac{d}{dt} [f(\mathbf{r}(t))]$$

is the directional derivative of f in the direction of the motion times the speed of the object.

20. *(Important)* Set $r = \|\mathbf{r}\|$ where $\mathbf{r} = x\mathbf{i} + y\mathbf{j} + z\mathbf{k}$. If f is a continuously differentiable function of r, then

(16.4.6) $\qquad \boxed{\nabla [f(r)] = f'(r) \dfrac{\mathbf{r}}{r}.}$

Derive this formula.

Calculate the following gradients.

21. (a) $\nabla(\sin r)$. (b) $\nabla(r \sin r)$. $\qquad$ **22.** (a) $\nabla(r \ln r)$. (b) $\nabla(e^{1-r^2})$.

23. (a) $\nabla \left(\dfrac{\sin r}{r} \right)$. (b) $\nabla \left(\dfrac{r}{\sin r} \right)$.

24. (a) Draw a tree diagram for du/dt given that

$$u = u(x, y) \quad \text{where} \quad x = x(s), \quad y = y(s) \quad \text{and} \quad s = s(t).$$

(b) Calculate du/dt.

25. (a) Let $u = u(x, y)$. Show that the change of variables to polar coordinates

$$x = r \cos \theta, \qquad y = r \sin \theta$$

gives

$$\frac{\partial u}{\partial r} = \frac{\partial u}{\partial x} \cos \theta + \frac{\partial u}{\partial y} \sin \theta, \qquad \frac{\partial u}{\partial \theta} = -\frac{\partial u}{\partial x} r \sin \theta + \frac{\partial u}{\partial y} r \cos \theta.$$

(b) Express

$$\left(\frac{\partial u}{\partial r}\right)^2 + \frac{1}{r^2} \left(\frac{\partial u}{\partial \theta}\right)^2$$

entirely in terms of $\partial u/\partial x$ and $\partial u/\partial y$.

26. Let w be a function of polar coordinates r and θ. Then w is also a function of rectangular coordinates

$$x = r \cos \theta \quad \text{and} \quad y = r \sin \theta.$$

(a) Using the first part of Exercise 25, verify that

$$\frac{\partial w}{\partial x} = \frac{\partial w}{\partial r} \cos \theta - \frac{1}{r} \frac{\partial w}{\partial \theta} \sin \theta, \qquad \frac{\partial w}{\partial y} = \frac{\partial w}{\partial r} \sin \theta + \frac{1}{r} \frac{\partial w}{\partial \theta} \cos \theta.$$

(b) Deduce from (a) that

$$\frac{\partial r}{\partial x} = \cos \theta, \qquad \frac{\partial r}{\partial y} = \sin \theta; \qquad \frac{\partial \theta}{\partial x} = -\frac{1}{r} \sin \theta, \qquad \frac{\partial \theta}{\partial y} = \frac{1}{r} \cos \theta.$$

(c) Find the fallacy:

$$x = r \cos \theta, \qquad r = \frac{x}{\cos \theta}, \qquad \frac{\partial r}{\partial x} = \frac{1}{\cos \theta}.$$

Find $\partial z/\partial x$ and $\partial z/\partial y$ given that z is defined implicitly as a function of x and y by the equation.

27. $z^4 + x^2 z^3 + y^2 + xy = 2.$ **28.** $\cos xyz + \ln (x^2 + y^2 + z^2) = 0.$

29. Set $u = u(x, y, z)$ where

$$x = x(w, t), \quad y = y(w, t), \quad z = z(w, t) \quad \text{and} \quad w = w(r, s), \quad t = t(r, s).$$

(a) Draw a tree diagram for the partials of u.   (b) Compute $\partial u/\partial r$ and $\partial u/\partial s$.

30. Set $u = u(x, y, z, w)$ where

$$x = x(r, s, t), \quad y = y(s, t, v), \quad z = z(r, t) \quad \text{and} \quad w = w(r, s, t, v).$$

(a) Draw a tree diagram for the partials of u.   (b) Compute $\partial u/\partial r$ and $\partial u/\partial v$.

31. (a) Let $u = u(x, y)$ be a differentiable function and suppose that the equation $u(x, y) = 0$ gives y implicitly as a function of x. Show that

$$\text{if} \quad \frac{\partial u}{\partial y} \neq 0 \quad \text{then} \quad \frac{dy}{dx} = -\frac{\partial u/\partial x}{\partial u/\partial y}.$$

(b) Use part (a) to find dy/dx given that $x^2 y^4 + \sin (x + y) = 0.$

32. (*A chain rule for vector-valued functions*) Suppose that

$$\mathbf{u}(x, y) = u_1(x, y)\mathbf{i} + u_2(x, y)\mathbf{j} \qquad \text{where} \quad x = x(t), \quad y = y(t).$$

(a) Show that

(16.4.7)

$$\frac{d\mathbf{u}}{dt} = \frac{\partial \mathbf{u}}{\partial x}\frac{dx}{dt} + \frac{\partial \mathbf{u}}{\partial y}\frac{dy}{dt}$$

where

$$\frac{\partial \mathbf{u}}{\partial x} = \frac{\partial u_1}{\partial x}\mathbf{i} + \frac{\partial u_2}{\partial x}\mathbf{j} \quad \text{and} \quad \frac{\partial \mathbf{u}}{\partial y} = \frac{\partial u_1}{\partial y}\mathbf{i} + \frac{\partial u_2}{\partial y}\mathbf{j}.$$

(b) Let

$$\mathbf{u} = e^x \cos y\,\mathbf{i} + e^x \sin y\,\mathbf{j} \qquad \text{where} \quad x = \tfrac{1}{2}t^2, \quad y = \pi t.$$

Compute $d\mathbf{u}/dt$ (i) by applying (16.4.7), (ii) by forming $\mathbf{u}(t)$ directly.

33. Set

$$\mathbf{u}(x, y) = u_1(x, y)\mathbf{i} + u_2(x, y)\mathbf{j} \qquad \text{where} \quad x = x(s, t), \quad y = y(s, t).$$

Find $\partial \mathbf{u}/\partial s$ and $\partial \mathbf{u}/\partial t$.

34. Set

$$\mathbf{u}(x, y, z) = u_1(x, y, z)\mathbf{i} + u_2(x, y, z)\mathbf{j} + u_3(x, y, z)\mathbf{k}$$

where

$$x = x(t), \quad y = y(t), \quad z = z(t).$$

Derive a formula for $d\mathbf{u}/dt$ analogous to (16.4.7).

16.5   THE GRADIENT AS A NORMAL; TANGENT LINES AND TANGENT PLANES

Functions of Two Variables

We begin with a nonconstant function $f = f(x, y)$ that is continuously differentiable. (*Remember*: that means f is differentiable and its gradient ∇f is continuous.)   You have seen that at each point of the domain the gradient vector, if not $\mathbf{0}$, points in the direction of the most rapid increase of f.   Here we show that

(16.5.1)

at each point of the domain, the gradient vector, if not $\mathbf{0}$, is perpendicular to the level curve that passes through that point.

Proof.   We choose a point (x_0, y_0) in the domain and assume that $\nabla f(x_0, y_0) \neq \mathbf{0}$. The level curve through this point has equation

$$f(x, y) = c \qquad \text{where } c = f(x_0, y_0).$$

Under our assumptions on f, this curve can be parametrized in a neighborhood of (x_0, y_0) by a continuously differentiable vector function

$$\mathbf{r}(t) = x(t)\mathbf{i} + y(t)\mathbf{j}, \qquad t \in I$$

with nonzero tangent vector $\mathbf{r}'(t)$.†

Now take t_0 such that

$$\mathbf{r}(t_0) = x_0\mathbf{i} + y_0\mathbf{j} = (x_0, y_0).$$

We will show that

$$\nabla f(\mathbf{r}(t_0)) \perp \mathbf{r}'(t_0).$$

Since f is constantly c on the curve, we have

$$f(\mathbf{r}(t)) = c \qquad \text{for all } t \in I.$$

For such t

$$\frac{d}{dt}[f(\mathbf{r}(t))] = \nabla f(\mathbf{r}(t)) \cdot \mathbf{r}'(t) = 0. \qquad \text{(Theorem 16.4.1)}$$

In particular

$$\nabla f(\mathbf{r}(t_0)) \cdot \mathbf{r}'(t_0) = 0,$$

and thus

$$\nabla f(\mathbf{r}(t_0)) \perp \mathbf{r}'(t_0). \quad \square$$

Figure 16.5.1 illustrates the result.

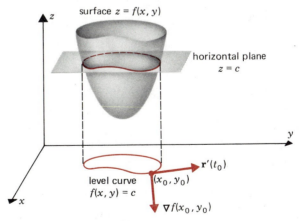

the gradient vector $\nabla f(x_0, y_0)$ is perpendicular to the level curve at (x_0, y_0)

Figure 16.5.1

---

† This follows from a result of advanced calculus known as the *implicit function theorem*.

Example 1. For the function

$$f(x, y) = x^2 + y^2$$

the level curves are concentric circles:

$$x^2 + y^2 = c.$$

At each point $(x, y) \neq (0, 0)$ the gradient vector

$$\nabla f(x, y) = 2x\mathbf{i} + 2y\mathbf{j} = 2\mathbf{r}$$

points away from the origin along the line of the radius vector and is thus perpendicular to the circle in question. At the origin the level curve is reduced to a point and the gradient is simply $\mathbf{0}$. See Figure 16.5.2. ☐

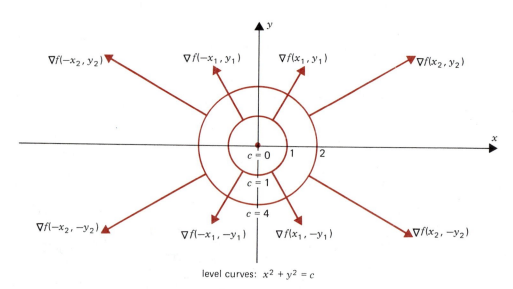

level curves: $x^2 + y^2 = c$

Figure 16.5.2

Consider now a curve in the xy-plane

$$C: \quad f(x, y) = c.$$

As before we assume that f is nonconstant and continuously differentiable. Let's suppose that (x_0, y_0) lies on the curve and $\nabla f(x_0, y_0) \neq \mathbf{0}$.

We can view C as the c-level curve of f and conclude from (16.5.1) that the gradient

(16.5.2)
$$\nabla f(x_0, y_0) = \frac{\partial f}{\partial x}(x_0, y_0)\, \mathbf{i} + \frac{\partial f}{\partial y}(x_0, y_0)\, \mathbf{j}$$

is perpendicular to C at (x_0, y_0). We call it a *normal vector*.

The vector

(16.5.3)
$$\mathbf{t}(x_0, y_0) = \frac{\partial f}{\partial y}(x_0, y_0)\,\mathbf{i} - \frac{\partial f}{\partial x}(x_0, y_0)\,\mathbf{j}$$

is perpendicular to the gradient:

$$\nabla f(x_0, y_0) \cdot \mathbf{t}(x_0, y_0) = \frac{\partial f}{\partial x}(x_0, y_0)\frac{\partial f}{\partial y}(x_0, y_0) - \frac{\partial f}{\partial y}(x_0, y_0)\frac{\partial f}{\partial x}(x_0, y_0) = 0.$$

It is therefore a *tangent vector*.

The line through (x_0, y_0) perpendicular to the gradient is the tangent line. To obtain an equation for the tangent line we refer to Figure 16.5.3. A point (x, y) will lie on the tangent line iff

$$[(x - x_0)\mathbf{i} + (y - y_0)\mathbf{j}] \cdot \nabla f(x_0, y_0) = 0,$$

that is, iff

(16.5.4)
$$(x - x_0)\frac{\partial f}{\partial x}(x_0, y_0) + (y - y_0)\frac{\partial f}{\partial y}(x_0, y_0) = 0.$$

This is an equation for the *tangent line*.

Figure 16.5.3

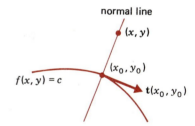

Figure 16.5.4

The line through (x_0, y_0) perpendicular to the tangent vector $\mathbf{t}(x_0, y_0)$ is the normal line (Figure 16.5.4). A point (x, y) will lie on the normal line iff

$$[(x - x_0)\mathbf{i} + (y - y_0)\mathbf{j}] \cdot \mathbf{t}(x_0, y_0) = 0,$$

that is, iff

(16.5.5)
$$(x - x_0)\frac{\partial f}{\partial y}(x_0, y_0) - (y - y_0)\frac{\partial f}{\partial x}(x_0, y_0) = 0.$$

This is an equation for the *normal line*.

Example 2. Choose a point (x_0, y_0) on the hyperbola

$$\frac{x^2}{a^2} - \frac{y^2}{b^2} = 1.$$

To avoid denominators, we write

$$b^2x^2 - a^2y^2 = a^2b^2.$$

This equation is of the form

$$f(x, y) = c \quad \text{with} \quad f(x, y) = b^2x^2 - a^2y^2 \quad \text{and} \quad c = a^2b^2.$$

Partial differentiation gives

$$\frac{\partial f}{\partial x}(x, y) = 2b^2x, \qquad \frac{\partial f}{\partial y}(x, y) = -2a^2y.$$

At (x_0, y_0) the gradient

$$\frac{\partial f}{\partial x}(x_0, y_0)\,\mathbf{i} + \frac{\partial f}{\partial y}(x_0, y_0)\,\mathbf{j} = 2b^2x_0\mathbf{i} - 2a^2y_0\mathbf{j}$$

is normal to the curve and the vector

$$\frac{\partial f}{\partial y}(x_0, y_0)\,\mathbf{i} - \frac{\partial f}{\partial x}(x_0, y_0)\,\mathbf{j} = -2a^2y_0\mathbf{i} - 2b^2x_0\mathbf{j}$$

is tangent to the curve.

The equation of the tangent line can be written

$$(x - x_0)2b^2x_0 + (y - y_0)(-2a^2y_0) = 0.$$

Dividing by 2 and noting that $b^2x_0^2 - a^2y_0^2 = a^2b^2,$† we can simplify the equation to

$$(b^2x_0)x - (a^2y_0)y = a^2b^2.$$

The equation of the normal line takes the form

$$(x - x_0)(-2a^2y_0) + (y - y_0)(-2b^2x_0) = 0.$$

This simplifies to

$$(a^2y_0)x + (b^2x_0)y = (a^2 + b^2)x_0y_0. \quad \square$$

Functions of Three Variables

Here, instead of level curves, we have level surfaces, but the results are similar. If $f = f(x, y, z)$ is nonconstant and continuously differentiable, then

(16.5.6)   at each point of the domain, the gradient vector, if not **0**, is perpendicular to the level surface that passes through that point.

---

† The point (x_0, y_0) lies on the hyperbola.

Proof. We choose a point $\mathbf{x}_0 = (x_0, y_0, z_0)$ in the domain and assume that $\nabla f(x_0, y_0, z_0) \neq \mathbf{0}$. The level surface through this point has equation

$$f(x, y, z) = c \qquad \text{where } c = f(x_0, y_0, z_0).$$

We suppose now that

$$\mathbf{r}(t) = x(t)\mathbf{i} + y(t)\mathbf{j} + z(t)\mathbf{k}, \qquad t \in I$$

is a differentiable curve that lies on this surface and passes through the point $\mathbf{x}_0 = (x_0, y_0, z_0)$. We choose t_0 so that

$$\mathbf{r}(t_0) = \mathbf{x}_0 = (x_0, y_0, z_0)$$

and suppose that $\mathbf{r}'(t_0) \neq \mathbf{0}$.

Since the curve lies on the given surface, we have

$$f(\mathbf{r}(t)) = c \qquad \text{for all } t \in I.$$

For such t

$$\frac{d}{dt}[f(\mathbf{r}(t))] = \nabla f(\mathbf{r}(t)) \cdot \mathbf{r}'(t) = 0.$$

In particular,

$$\nabla f(\mathbf{r}(t_0)) \cdot \mathbf{r}'(t_0) = 0.$$

The gradient vector

$$\nabla f(\mathbf{r}(t_0)) = \nabla f(\mathbf{x}_0) = \nabla f(x_0, y_0, z_0)$$

is thus perpendicular to the curve in question.

This same argument applies to *every* differentiable curve that lies on this level surface and passes through the point $\mathbf{x}_0 = (x_0, y_0, z_0)$ with nonzero tangent vector. (See Figure 16.5.5.) Consequently, we view $\nabla f(\mathbf{x}_0)$ as perpendicular to the surface itself. $\square$

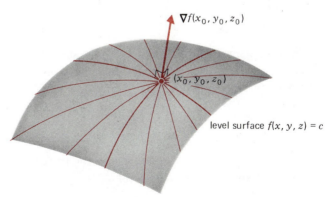

the gradient vector $\nabla f(x_0, y_0, z_0)$ is perpendicular
to the level surface at (x_0, y_0, z_0)

Figure 16.5.5

Example 3.   For the function

$$f(x, y, z) = x^2 + y^2 + z^2$$

the level surfaces are concentric spheres:

$$x^2 + y^2 + z^2 = c.$$

At each point $(x, y, z) \neq (0, 0, 0)$ the gradient vector

$$\nabla f(x, y, z) = 2x\mathbf{i} + 2y\mathbf{j} + 2z\mathbf{k} = 2\mathbf{r}$$

points away from the origin along the line of the radius vector and is thus perpendicular to the sphere in question.   At the origin the level surface is reduced to a point and the gradient is $\mathbf{0}$.   ☐

The *tangent plane* to a surface

$$f(x, y, z) = c$$

at a point $\mathbf{x}_0 = (x_0, y_0, z_0)$ is the plane through $\mathbf{x}_0$ with normal $\nabla f(\mathbf{x}_0)$.   See Figure 16.5.6.

tangent plane to a surface $f(x, y, z) = c$

Figure 16.5.6

The tangent plane at a point $\mathbf{x}_0$ is the plane through $\mathbf{x}_0$ that best approximates the surface in a neighborhood of $\mathbf{x}_0$.   (We will return to this later.)

A point $\mathbf{x}$ lies on the tangent plane through $\mathbf{x}_0$ iff

(16.5.7) $$\boxed{(\mathbf{x} - \mathbf{x}_0) \cdot \nabla f(\mathbf{x}_0) = 0.}$$ (Figure 16.5.6)

This is an equation for the tangent plane in vector notation.   In Cartesian coordinates the equation takes the form

$$(x - x_0) \frac{\partial f}{\partial x}(x_0, y_0, z_0) + (y - y_0) \frac{\partial f}{\partial y}(x_0, y_0, z_0) + (z - z_0) \frac{\partial f}{\partial z}(x_0, y_0, z_0) = 0.$$

Problem 4. Find an equation in x, y, z for the plane tangent to the surface

$$xy + yz + zx = 11 \qquad \text{at the point } (1, 2, 3).$$

Solution. The surface is of the form

$$f(x, y, z) = c \qquad \text{with} \quad f(x, y, z) = xy + yz + zx \quad \text{and} \quad c = 11.$$

Observe that

$$\frac{\partial f}{\partial x} = y + z, \qquad \frac{\partial f}{\partial y} = x + z, \qquad \frac{\partial f}{\partial z} = x + y.$$

At the point $(1, 2, 3)$

$$\frac{\partial f}{\partial x} = 5, \qquad \frac{\partial f}{\partial y} = 4, \qquad \frac{\partial f}{\partial z} = 3.$$

The equation for the tangent plane can therefore be written

$$5(x - 1) + 4(y - 2) + 3(z - 3) = 0.$$

This simplifies to

$$5x + 4y + 3z - 22 = 0. \quad \square$$

Problem 5. The curve

$$\mathbf{r}(t) = \tfrac{1}{2}t^2\mathbf{i} + 4t^{-1}\mathbf{j} + (\tfrac{1}{2}t - t^2)\mathbf{k}$$

intersects the surface $x^2 - 4y^2 - 4z = 0$ at the point $(2, 2, -3)$. What is the angle of intersection?

Solution. We want the angle between the tangent vector of the curve and the tangent plane of the surface at the point of intersection.

A simple calculation shows that the curve passes through the point $(2, 2, -3)$ at $t = 2$. Since

$$\mathbf{r}'(t) = t\mathbf{i} - 4t^{-2}\mathbf{j} + (\tfrac{1}{2} - 2t)\mathbf{k},$$

we have

$$\mathbf{r}'(2) = 2\mathbf{i} - \mathbf{j} - \tfrac{7}{2}\mathbf{k}.$$

Now set

$$f(x, y, z) = x^2 - 4y^2 - 4z.$$

This function has gradient $2x\mathbf{i} - 8y\mathbf{j} - 4\mathbf{k}$. At the point $(2, 2, -3)$

$$\nabla f = 4\mathbf{i} - 16\mathbf{j} - 4\mathbf{k}.$$

Now let θ be the angle between $\mathbf{r}'(2)$ and this gradient. By (13.4.9)

$$\cos \theta = \frac{\mathbf{r}'(2)}{\|\mathbf{r}'(2)\|} \cdot \frac{\nabla f}{\|\nabla f\|} = \frac{19}{414} \sqrt{138} \approx 0.539$$

so that $\theta \cong 1.00$ radian. Since the gradient is normal to the tangent plane, the angle we want is

$$\tfrac{1}{2}\pi - \theta \cong 1.57 - 1.00 = 0.57 \text{ radians.} \quad \square$$

A surface of the form

$$z = g(x, y)$$

can be written in the form

$$f(x, y, z) = 0$$

by setting

$$f(x, y, z) = g(x, y) - z.$$

If g is differentiable, so is f. Moreover,

$$\frac{\partial f}{\partial x}(x, y, z) = \frac{\partial g}{\partial x}(x, y), \qquad \frac{\partial f}{\partial y}(x, y, z) = \frac{\partial g}{\partial y}(x, y), \qquad \frac{\partial f}{\partial z}(x, y, z) = -1.$$

By (16.5.7) the tangent plane at (x_0, y_0, z_0) has equation

$$(x - x_0)\frac{\partial g}{\partial x}(x_0, y_0) + (y - y_0)\frac{\partial g}{\partial y}(x_0, y_0) + (z - z_0)(-1) = 0,$$

which we can rewrite as

(16.5.8)
$$z - z_0 = (x - x_0)\frac{\partial g}{\partial x}(x_0, y_0) + (y - y_0)\frac{\partial g}{\partial y}(x_0, y_0).$$

If $\nabla g(x_0, y_0) = \mathbf{0}$, then both partials of g are zero at (x_0, y_0) and the equation reduces to

(16.5.9)
$$z = z_0.$$

In this case the tangent plane is *horizontal*.

Problem 6. Find an equation for the plane tangent to the surface

$$z = \frac{x^2}{2p} + \frac{y^2}{2q} \qquad \text{at the point } (2p^2, 2q^2, 2p^3 + 2q^3).$$

Solution. Set

$$g(x, y) = \frac{x^2}{2p} + \frac{y^2}{2q}.$$

Partial differentiation gives

$$\frac{\partial g}{\partial x}(x, y) = \frac{x}{p}, \qquad \frac{\partial g}{\partial y}(x, y) = \frac{y}{q}.$$

When $x = 2p^2$ and $y = 2q^2$,

$$\frac{\partial g}{\partial x} = 2p \qquad \text{and} \qquad \frac{\partial g}{\partial y} = 2q.$$

At the point $(2p^2, 2q^2, 2p^3 + 2q^3)$ the tangent plane has equation

$$z - (2p^3 + 2q^3) = (x - 2p^2)2p + (y - 2q^2)2q.$$

This simplifies to

$$z = 2px + 2qy - 2p^3 - 2q^3. \quad \square$$

Problem 7. At what points of the surface

$$z = 3xy - x^3 - y^3$$

is the tangent plane horizontal?

Solution. The function

$$g(x, y) = 3xy - x^3 - y^3$$

has first partials

$$\frac{\partial g}{\partial x}(x, y) = 3y - 3x^2, \qquad \frac{\partial g}{\partial y}(x, y) = 3x - 3y^2.$$

These partials are both zero only at $(0, 0)$ and $(1, 1)$. The surface has a horizontal tangent plane only at $(0, 0, 0)$ and $(1, 1, 1)$. $\quad \square$

EXERCISES 16.5

Find a normal vector and a tangent vector at the point indicated. Write an equation for the tangent line and an equation for the normal line.

1. $x^2 + xy + y^2 = 3$; $(-1, -1)$.
2. $(y - x)^2 = 2x$; $(2, 4)$.
3. $(x^2 + y^2)^2 = 9(x^2 - y^2)$; $(\sqrt{2}, 1)$.
4. $x^3 + y^3 = 9$; $(1, 2)$.
5. $xy^2 - 2x^2 + y + 5x = 6$; $(4, 2)$.
6. $x^5 + y^5 = 2x^3$; $(1, 1)$.
7. $2x^3 - x^2y^2 = 3x - y - 7$; $(1, -2)$.
8. $x^3 + y^2 + 2x = 6$; $(-1, 3)$.

9. Find the slope of the curve $x^2y = a^2(a - y)$ at the point $(0, a)$.

Find an equation for the tangent plane at the point indicated.

10. $z = (x^2 + y^2)^2$;   (1, 1, 4).

11. $x^3 + y^3 = 3xyz$;   (1, 2, $\frac{3}{2}$).

12. $xy^2 + 2z^2 = 12$;   (1, 2, 2).

13. $z = axy$;   (1, 1/a, 1).

14. $\sqrt{x} + \sqrt{y} + \sqrt{z} = 4$;   (1, 4, 1).

15. $z = \sin x + \sin y + \sin (x + y)$;   (0, 0, 0).

16. $z = x^2 + xy + y^2 - 6x + 2$;   (4, -2, -10).

17. $b^2c^2x^2 - a^2c^2y^2 - a^2b^2z^2 = a^2b^2c^2$; (x_0, y_0, z_0).

18. $z = \sin (x \cos y)$;   (0, $\frac{1}{2}\pi$, 0).

Find the point(s) on the surface at which the tangent plane is horizontal.

19. $xy + a^3x^{-1} + b^3y^{-1} - z = 0$.

20. $z = 4x + 2y - x^2 + xy - y^2$.

21. $z = xy$.

22. $x + y + z + xy - x^2 - y^2 = 0$.

23. $z - 2x^2 - 2xy + y^2 + 5x - 3y + 2 = 0$.

24. (a) Find the *upper unit normal* (the unit normal with positive **k** component) for the surface $z = xy$ at the point (1, 1, 1).

(b) Find the *lower unit normal* (the unit normal with negative **k** component) for the surface $z = 1/x - 1/y$ at the point (1, 1, 0).

25. Write equations in symmetric form for the line normal to the surface $f(x, y, z) = c$ at the point (x_0, y_0, z_0).

26. Show that in the case of a surface of the form $z = xf(x/y)$ with f continuously differentiable, all the tangent planes share a point in common.

27. Given that the surfaces $F(x, y, z) = 0$ and $G(x, y, z) = 0$ intersect at right angles in a curve γ, what condition must be satisfied by the partial derivatives of F and G on γ?

28. Show that, for all planes tangent to the surface $\sqrt{x} + \sqrt{y} + \sqrt{z} = \sqrt{a}$, the sum of the intercepts is the same.

29. Show that all pyramids formed by the coordinate planes and a plane tangent to the surface $xyz = a^3$ have the same volume.   What is this volume?

30. Show that, for all planes tangent to the surface $x^{2/3} + y^{2/3} + z^{2/3} = a^{2/3}$, the sum of the squares of the intercepts is the same.

31. The curve $\mathbf{r}(t) = 2t\mathbf{i} + 3t^{-1}\mathbf{j} - 2t^2\mathbf{k}$ and the ellipsoid $x^2 + y^2 + 3z^2 = 25$ intersect at (2, 3, -2).   What is the angle of intersection?

32. Show that the curve $\mathbf{r}(t) = \frac{3}{2}(t^2 + 1)\mathbf{i} + (t^4 + 1)\mathbf{j} + t^3\mathbf{k}$ is perpendicular to the ellipsoid $x^2 + 2y^2 + 3z^2 = 20$ at the point (3, 2, 1).

33. The surfaces $x^2y^2 + 2x + z^3 = 16$ and $3x^2 + y^2 - 2z = 9$ intersect in a curve that passes through the point (2, 1, 2).   What are the equations of the respective tangent planes to the two surfaces at this point?

34. Show that the sphere $x^2 + y^2 + z^2 - 8x - 8y - 6z + 24 = 0$ is tangent to the ellipsoid $x^2 + 3y^2 + 2z^2 = 9$ at the point (2, 1, 1).

35. Show that the sphere $x^2 + y^2 + z^2 - 4y - 2z + 2 = 0$ is perpendicular to the paraboloid $3x^2 + 2y^2 - 2z = 1$ at the point (1, 1, 2).

36. Show that the following surfaces are mutually perpendicular: $xy = az^2, x^2 + y^2 + z^2 = b$, $z^2 + 2x^2 = c(z^2 + 2y^2)$.

37. The surface $S: z = x^2 + 3y^2 + 2$ intersects the vertical plane $p: 3x + 4y + 6 = 0$ in a space curve C.

(a) Let C_1 be the projection of C in the xy-plane.   Find an equation for C_1.

(b) Find a parametrization $\mathbf{r}(t) = x(t)\mathbf{i} + y(t)\mathbf{j} + z(t)\mathbf{k}$ for C setting $x(t) = 4t - 2$.

(c) Find a parametrization $\mathbf{R}(s) = \mathbf{R}_0 + s\mathbf{d}$ for the line l tangent to C at (2, -3, 33).

(d) Find an equation for the plane p_1 tangent to S at (2, -3, 33).

(e) Find a parametrization $\mathbf{r}(t) = x(t)\mathbf{i} + y(t)\mathbf{j} + z(t)\mathbf{k}$ for the line l' formed by the intersection of p with p_1 taking $x(t) = t$.   What is the relation between l and l'?

16.6 MAXIMUM AND MINIMUM VALUES

In Chapter 4 we discussed the local extreme values of a function of one variable. Here we take up the same subject for functions of several variables. The ideas are very similar.

DEFINITION 16.6.1 LOCAL MAXIMUM AND LOCAL MINIMUM

Let f be a function of several variables and let $\mathbf{x}_0$ be an interior point of the domain:

f is said to have a *local maximum* at $\mathbf{x}_0$ iff

$$f(\mathbf{x}_0) \geq f(\mathbf{x}) \qquad \text{for all } \mathbf{x} \text{ in some neighborhood of } \mathbf{x}_0;$$

f is said to have a *local minimum* at $\mathbf{x}_0$ iff

$$f(\mathbf{x}_0) \leq f(\mathbf{x}) \qquad \text{for all } \mathbf{x} \text{ in some neighborhood of } \mathbf{x}_0.$$

As in the one-variable case, the local maxima and local minima together comprise the *local extreme values*.

In the one-variable case we know that, if f has a local extreme value at x_0, then

$$\text{either} \quad f'(x_0) = 0 \quad \text{or} \quad f'(x_0) \text{ does not exist.}$$

We have a similar result for functions of several variables.

THEOREM 16.6.2

If f has a local extreme value at $\mathbf{x}_0$, then

$$\text{either} \quad \nabla f(\mathbf{x}_0) = \mathbf{0} \quad \text{or} \quad \nabla f(\mathbf{x}_0) \text{ does not exist.}$$

Proof. We assume that f has a local extreme value at $\mathbf{x}_0$ and that f is differentiable at $\mathbf{x}_0$ [that $\nabla f(\mathbf{x}_0)$ exists]. We need to show that $\nabla f(\mathbf{x}_0) = \mathbf{0}$. For simplicity we set $\mathbf{x}_0 = (x_0, y_0)$. The three-variable case is similar.

Since f has a local extreme value at (x_0, y_0), the function $g(x) = f(x, y_0)$ has a local extreme value at x_0. Since f is differentiable at (x_0, y_0), g is differentiable at x_0 and therefore

$$g'(x_0) = \frac{\partial f}{\partial x}(x_0, y_0) = 0.$$

Similarly, the function $h(y) = f(x_0, y)$ has a local extreme value at y_0 and, being differentiable there, satisfies the relation

$$h'(y_0) = \frac{\partial f}{\partial y}(x_0, y_0) = 0.$$

The gradient is $\mathbf{0}$ since both partials are 0. ∎

Interior points of the domain at which the gradient is zero or the gradient does not exist are called *critical points*. By Theorem 16.6.2 these are the only points at which a local extreme value can occur.

Although the ideas introduced so far are completely general, their application to functions of more than two variables is generally laborious. We restrict ourselves mostly to functions of two variables. Not only are the computations less formidable, but also we can make use of our geometric intuition.

Two Variables

We suppose for the moment that $f = f(x, y)$ is defined on an open set and is continuously differentiable there. The graph of f is a surface

$$z = f(x, y).$$

Where f has a local maximum, the surface has a local high point. Where f has a local minimum, the surface has a local low point. Where f has either a local maximum or a local minimum, the gradient is $\mathbf{0}$ and therefore the tangent plane is horizontal. See Figure 16.6.1.

Figure 16.6.1

A zero gradient signals the possibility of a local extreme value; it does not guarantee it. For example, in the case of the saddle-shaped surface of Figure 16.6.2, there is a horizontal tangent plane at the origin and therefore the gradient is zero there, yet the origin gives neither a local maximum nor a local minimum.

saddle-shaped surface

Figure 16.6.2

Critical points at which the gradient is zero are called *stationary points*. The stationary points that do not give rise to extreme values are called *saddle points*.

Below we test some differentiable functions for extreme values. In each case, our first step is to seek out the stationary points.

Example 1. For the function

$$f(x, y) = 2x^2 + y^2 - xy - 7y,$$

we have

$$\nabla f(x, y) = (4x - y)\mathbf{i} + (2y - x - 7)\mathbf{j}.$$

To find the stationary points, we set $\nabla f(x, y) = \mathbf{0}$. This gives

$$4x - y = 0 \quad \text{and} \quad 2y - x - 7 = 0.$$

The only simultaneous solution to these equations is $x = 1, y = 4$. The point $(1, 4)$ is therefore the only stationary point.

We now compare the value of f at $(1, 4)$ with the values of f at nearby points $(1 + h, 4 + k)$:

$$f(1, 4) = 2 + 16 - 4 - 28 = -14,$$

$$
\begin{aligned}
f(1 + h, 4 + k) &= 2(1 + h)^2 + (4 + k)^2 - (1 + h)(4 + k) - 7(4 + k) \\
&= 2 + 4h + 2h^2 + 16 + 8k + k^2 - 4 - 4h - k - hk - 28 - 7k \\
&= 2h^2 + k^2 - hk - 14.
\end{aligned}
$$

The difference

$$
\begin{aligned}
f(1 + h, 4 + k) - f(1, 4) &= 2h^2 + k^2 - hk \\
&= h^2 + (h^2 - hk + k^2) \\
&\geq h^2 + (h^2 - 2|h||k| + k^2) \\
&= h^2 + (|h| - |k|)^2
\end{aligned}
$$

is nonnegative for all small h and k (in fact for all real h and k).† It follows that f has a local minimum at $(1, 4)$. This local minimum is -14. ◻

---

† Another way to see that $2h^2 + k^2 - hk$ is nonnegative is to complete the square:

$$2h^2 + k^2 - hk = \tfrac{1}{4}h^2 - hk + k^2 + \tfrac{7}{4}h^2 = (\tfrac{1}{2}h - k)^2 + \tfrac{7}{4}h^2 \geq 0.$$

Example 2.   In the case of

$$f(x, y) = y^2 - xy + 2x + y + 1$$

we have

$$\nabla f(x, y) = (2 - y)\mathbf{i} + (2y - x + 1)\mathbf{j}.$$

The gradient is **0** where

$$2 - y = 0 \quad \text{and} \quad 2y - x + 1 = 0.$$

The only simultaneous solution to these equations is $x = 5$, $y = 2$.   The point $(5, 2)$ is the only stationary point.

We now compare the value of f at $(5, 2)$ with the values of f at nearby points $(5 + h, 2 + k)$:

$$f(5, 2) = 4 - 10 + 10 + 2 + 1 = 7,$$

$$\begin{aligned} f(5 + h, 2 + k) &= (2 + k)^2 - (5 + h)(2 + k) + 2(5 + h) + (2 + k) + 1 \\ &= 4 + 4k + k^2 - 10 - 2h - 5k - hk + 10 + 2h + 2 + k + 1 \\ &= k^2 - hk + 7. \end{aligned}$$

The difference

$$f(5 + h, 2 + k) - f(5, 2) = k^2 - hk = k(k - h)$$

does not keep a constant sign for small h and k.†   It follows that $(5, 2)$ is a saddle point.   ☐

The function in the next example has an infinite number of stationary points.

———
† See Figure 16.6.3 below.

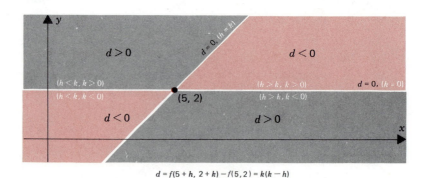

$$d = f(5 + h, 2 + k) - f(5, 2) = k(k - h)$$

Figure 16.6.3

Example 3. For all points (x, y) with $x + y \neq 0$, set

$$f(x, y) = -\frac{x^2 + 2y^2}{(x + y)^2}.$$

Then

$$\nabla f(x, y) = 2y \frac{2y - x}{(x + y)^3} \mathbf{i} + 2x \frac{x - 2y}{(x + y)^3} \mathbf{j}.$$

The gradient is **0** where

$$2y \frac{2y - x}{(x + y)^3} = 0 \quad \text{and} \quad 2x \frac{x - 2y}{(x + y)^3} = 0.$$

Solving these two equations simultaneously, we get

$$x = 2y.$$

The stationary points are all the points of the line $x = 2y$ other than the point $(0, 0)$, which is not in the domain of f. At each stationary point (x_0, y_0) the function takes on the value $-\frac{2}{3}$:

$$f(x_0, y_0) = f(2y_0, y_0) = -\frac{(2y_0)^2 + 2y_0^2}{(2y_0 + y_0)^2} = -\frac{6y_0^2}{9y_0^2} = -\frac{2}{3}.$$

In a moment we will show that for all (x, y) in the domain,

$$f(x, y) = -\frac{x^2 + 2y^2}{(x + y)^2} \leq -\frac{2}{3}.$$

It will then follow that the number $-\frac{2}{3}$ is a local maximum value.

The inequality

$$-\frac{x^2 + 2y^2}{(x + y)^2} \leq -\frac{2}{3}$$

can be justified by the following sequence of equivalent inequalities, the last of which is obvious:

$$-3(x^2 + 2y^2) \leq -2(x + y)^2,$$
$$3(x^2 + 2y^2) \geq 2(x + y)^2,$$
$$3x^2 + 6y^2 \geq 2x^2 + 4xy + 2y^2,$$
$$x^2 - 4xy + 4y^2 \geq 0,$$
$$(x - 2y)^2 \geq 0. \quad \blacksquare$$

Each function we have considered so far was differentiable on its entire domain. The only critical points were thus the stationary points. Our next example exhibits a function that is everywhere defined but is not differentiable at the origin. The origin is therefore a critical point, though not a stationary point.

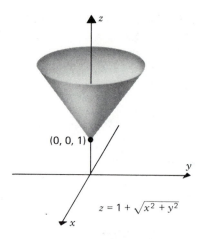

Figure 16.6.4

Example 4. The function

$$f(x, y) = 1 + \sqrt{x^2 + y^2}$$

is everywhere defined and everywhere continuous. The graph is the upper nappe of a cone. (See Figure 16.6.4.) The number $f(0, 0) = 1$ is obviously a local minimum.

Since the partials

$$\frac{\partial f}{\partial x} = \frac{x}{\sqrt{x^2 + y^2}}, \qquad \frac{\partial f}{\partial y} = \frac{y}{\sqrt{x^2 + y^2}}$$

are not defined at $(0, 0)$, the gradient is not defined at $(0, 0)$. (Theorem 16.2.5) The point $(0, 0)$ is thus a critical point, but not a stationary point. At $(0, 0, 1)$ the surface comes to a sharp point and there is no tangent plane. ☐

Absolute Extreme Values

In Chapter 2 you saw that a function of one variable that is continuous on a bounded closed interval takes on both an absolute maximum and an absolute minimum on that interval. (Theorem 2.6.2) More generally, it can be proved that, if f, now a function of one or several variables, is continuous on a *bounded closed set* (a closed set that can be contained in a ball of finite radius), then f takes on both an absolute maximum and an absolute minimum on that set.

In the search for local extreme values the critical points are interior points of the domain: the stationary points and the interior points at which the gradient does not exist. In the search for absolute extreme values we must also test the boundary points. This usually requires special methods. One approach is to try to parametrize the boundary by some vector function $\mathbf{r} = \mathbf{r}(t)$ and then work with $f(\mathbf{r}(t))$. This is the approach we take in the example below. Another, altogether different approach is outlined in Section 16.8.

Example 5. The triangular region

$$\Omega = \{(x, y): 0 \le x \le 4, 0 \le y \le 2x\} \qquad \text{(Figure 16.6.5)}$$

is a closed, bounded set. The function

$$f(x, y) = xy - 2x - 3y,$$

being continuous everywhere, is continuous on Ω. Therefore, we know that f takes on an absolute maximum on Ω and an absolute minimum.

To see whether either of these values is taken on in the interior of the set, we form the gradient

$$\nabla f(x, y) = (y - 2)\mathbf{i} + (x - 3)\mathbf{j}.$$

The gradient, defined everywhere, is $\mathbf{0}$ only at the point $(3, 2)$. As you can check,

$$f(3 + h, 2 + k) - f(3, 2) = hk.$$

Since this difference does not keep a constant sign for small h and k, the point $(3, 2)$ does not give rise to an extreme value. It is a saddle point.

Now we will look for extreme values on the boundary by writing each side of the triangle in the form $\mathbf{r} = \mathbf{r}(t)$ and then analyzing $f(\mathbf{r}(t))$. With C_1, C_2, C_3 as in the figure, we have

$$C_1: \quad \mathbf{r}_1(t) = t\mathbf{i} + 2t\mathbf{j}, \quad t \in [0, 4]$$
$$C_2: \quad \mathbf{r}_2(t) = 4\mathbf{i} + t\mathbf{j}, \quad t \in [0, 8]$$
$$C_3: \quad \mathbf{r}_3(t) = t\mathbf{i}, \quad\quad\ t \in [0, 4].$$

The values of f on these line segments are given by the functions

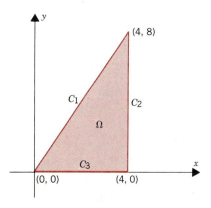

Figure 16.6.5

$$f_1(t) = f(\mathbf{r}_1(t)) = 2t^2 - 8t, \quad t \in [0, 4]$$
$$f_2(t) = f(\mathbf{r}_2(t)) = t - 8, \quad\quad t \in [0, 8]$$
$$f_3(t) = f(\mathbf{r}_3(t)) = -2t, \quad\quad t \in [0, 4].$$

The maximum value of f is the maximum of the maxima of f_1, f_2, f_3. And, the minimum value of f is the minimum of the minima of f_1, f_2, f_3. In this case, it happens that each of the f_i has a maximum of 0 and a minimum of -8. (Check this out.) It follows that the maximum value of f is 0 and the minimum value of f is -8. ☐

In some cases, physical or geometric considerations may allow us to conclude that an absolute maximum or an absolute minimum exists even if the function is not continuous, or the domain is not bounded or not closed.

Example 6. The rectangle

$$\{(x, y): 0 \le x \le a, -b \le y \le b\}$$

is a bounded closed subset of the plane. The function

$$f(x, y) = 1 + \sqrt{x^2 + y^2},$$

being everywhere continuous, is continuous on this rectangle. Thus we can be sure that f takes on both an absolute maximum and an absolute minimum on this set. The absolute maximum is taken on at the points $(a, -b)$ and (a, b), the points of the rectangle furthest away from the origin. (This should be clear from Figure 16.6.4.) The value at these points is $1 + \sqrt{a^2 + b^2}$. The absolute minimum is taken on at the origin $(0, 0)$. The value there is 1.

Now let's continue with the same function but apply it instead to the rectangle

$$\{(x, y): 0 < x \le a, -b \le y \le b\}.$$

This rectangle is bounded but not closed. On this set f takes on an absolute maximum (the same maximum as before and at the same points), but it takes on no absolute minimum (the origin is not in the set).

Finally, on the entire plane (which is closed but not bounded), f takes on an absolute minimum (1 at the origin) but no maximum. ☐

EXERCISES 16.6

Find the stationary points and determine the local extreme values.

1. $2x - x^2 - y^2$.
2. $x^2 + 2y^2 - 4y$.
3. $2x + 2y - x^2 + y^2 + 5$.
4. $x^2 - xy + y^2 + 1$.
5. $x^2 + xy + y^2 + 3x + 1$.
6. $x^3 - y^2 + 2y$.
7. $x^3 - 3x + y$.
8. $x^5 - 3x^2 - 2y$.
9. $x^2 + xy + y^2 - 3x - 3y$.
10. $x^2 - xy + y^2 + 2x + 2y - 6$.
11. $x^2 + xy + 2x + 2y + 1$.
12. $5xy - 7x^2 - y^2 + 3x - 6y$.
13. $4x^3y - 4xy^3$.
14. $x \sin y, -\pi < y < \pi$.

Find the absolute extreme values taken on by the function on the set D.

15. $f(x, y) = \dfrac{1}{\sqrt{x^2 + y^2}}, \quad D = \{(x, y): 1 \le x \le 3, 1 \le y \le 4\}$.

16. $f(x, y) = |x| + |y|, \quad D = \{(x, y): -1 \le x \le 1, -1 \le y \le 1\}$.

17. $f(x, y) = 4x^2 - 9y^2, \quad D = \{(x, y): -1 \le x \le 1\}$.

18. $f(x, y) = (x^2 + y^2)e^{-(x^2+y^2)}, \quad D = \{(x, y): x^2 + y^2 \le 25\}$.   HINT: Consider te^{-t}.

19. $f(x, y) = (x - 1)^2 + (y - 1)^2, \quad D = \{(x, y): x^2 + y^2 \le 4\}$.

20. $f(x, y) = y(x - 3), \quad D = \{(x, y): x^2 + y^2 \le 9\}$.

21. $f(x, y) = (x - y)^2, \quad D = \{(x, y): 0 \le x \le 6, 0 \le y \le 12 - 2x\}$.

22. $f(x, y) = (x - 3)^2 + y^2, \quad D = \{(x, y): 0 \le x \le 4, x^2 \le y \le 4x\}$.

23. $f(x, y) = (x - 4)^2 + y^2, \quad D = \{(x, y): 0 \le x \le 2, x^3 \le y \le 4x\}$.

24. $f(x, y) = (x + y - 2)^2, \quad D = \{(x, y): 0 \le x \le 3, x \le y \le 3\}$.

25. Define $f(x, y) = \frac{1}{4}x^2 - \frac{1}{9}y^2$ on the closed-unit disc. Find (a) the stationary points, (b) the local extreme values, (c) the absolute extreme values.

26. Find the dimensions of the rectangular solid of maximum volume given that the sum of these dimensions is b.

27. Find the point with the property that the sum of the squares of its distances from $P_1(x_1, y_1)$, $P_2(x_2, y_2)$, $P_3(x_3, y_3)$ is an absolute minimum.

28. A manufacturer finds that it can sell in two distinct markets. If it sells Q_1 units in the first market and Q_2 units in the second market, the revenue functions are

$$R_1 = A_1Q_1 - B_1Q_1^2, \qquad R_2 = A_2Q_2 - B_2Q_2^2$$

and the total cost function is

$$C = A_3 + B_3(Q_1 + Q_2).$$

Given that the A's and B's are known positive constants, find the quantities Q_1 and Q_2 that will maximize profit.

29. Find the distance from the point $(\frac{1}{2}, \frac{1}{2}, \frac{1}{2})$ to the sphere $x^2 + y^2 + z^2 = 1$.

30. Given that $0 < a < b$, find the absolute maximum value taken on by the function

$$f(x, y) = \frac{xy}{(a + x)(x + y)(b + y)}$$

on the open square $\{(x, y): a < x < b, a < y < b\}$.

31. A manufacturer produces razors and blades at a constant average cost of 80 cents per razor and 60 cents per dozen blades. If the razors are sold at x cents each and the blades at y cents per dozen, the demand of the market each week is

$$1000(320 - x) \text{ razors } \quad \text{and} \quad 1000(140 - y) \text{ dozen blades.}$$

Find the selling prices for maximum profit.

16.7 SECOND-PARTIALS TEST

We start with a function g of one variable and suppose that $g'(x_0) = 0$. According to the second-derivative test, g has

$$\text{a local minimum at } x_0 \quad \text{if} \quad g''(x_0) > 0,$$
$$\text{a local maximum at } x_0 \quad \text{if} \quad g''(x_0) < 0.$$

We have a similar test for functions of two variables. As one might expect, the test is somewhat more complicated to state and definitely more difficult to prove. We will omit the proof.†

———

† You can find a proof in most texts on advanced calculus. See, for example, Section 2-19 in Wilfred Kaplan's *Advanced Calculus* (Addison-Wesley, Reading, Mass., 1984).

THEOREM 16.7.1   THE SECOND-PARTIALS TEST

Suppose that f has continuous second-order partials in a neighborhood of (x_0, y_0) and that $\nabla f(x_0, y_0) = \mathbf{0}$. Set

$$A = \frac{\partial^2 f}{\partial x^2}(x_0, y_0), \qquad B = \frac{\partial^2 f}{\partial y \partial x}(x_0, y_0), \qquad C = \frac{\partial^2 f}{\partial y^2}(x_0, y_0)$$

and form the *discriminant* $D = B^2 - AC$.

1. If $D > 0$, then (x_0, y_0) is a saddle point.

2. If $D < 0$, then f has

$$\text{a local minimum at } (x_0, y_0) \quad \text{if} \quad A > 0,$$
$$\text{a local maximum at } (x_0, y_0) \quad \text{if} \quad A < 0.$$

The test is geometrically evident for functions of the form

$$f(x, y) = \tfrac{1}{2}Ax^2 + \tfrac{1}{2}Cy^2. \qquad (A \neq 0, C \neq 0)$$

The graph of such a function is a paraboloid:

$$z = \tfrac{1}{2}Ax^2 + \tfrac{1}{2}Cy^2.$$

The gradient is $\mathbf{0}$ at the origin $(0, 0)$. Moreover

$$A = \frac{\partial^2 f}{\partial x^2}(0, 0), \qquad B = \frac{\partial^2 f}{\partial y \partial x}(0, 0) = 0, \qquad C = \frac{\partial^2 f}{\partial y^2}(0, 0)$$

and $D = B^2 - AC = -AC$. If $D > 0$, then A and C have opposite signs and the surface has a saddle point. (Figure 16.7.1)

Suppose now that $D < 0$. If $A > 0$, then $C > 0$ and the surface has a minimum point; if $A < 0$, then $C < 0$ and the surface has a maximum point.

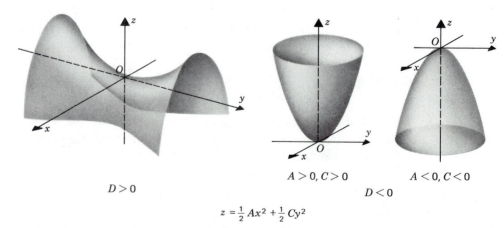

$A > 0, C > 0$ $A < 0, C < 0$

$D > 0$ $D < 0$

$$z = \tfrac{1}{2}Ax^2 + \tfrac{1}{2}Cy^2$$

Figure 16.7.1

In the examples that follow we apply the second-partials test to a variety of functions.

Example 1. For the function

$$f(x, y) = 2x^2 + y^2 - xy - 7y$$

we have

$$\frac{\partial f}{\partial x} = 4x - y, \qquad \frac{\partial f}{\partial y} = 2y - x - 7.$$

Setting both partials equal to zero, we have

$$4x - y = 0, \qquad 2y - x - 7 = 0.$$

The only simultaneous solution to these equations is $x = 1, y = 4$. The point $(1, 4)$ is thus the only stationary point.

The second partials are constant:

$$\frac{\partial^2 f}{\partial x^2} = 4, \qquad \frac{\partial^2 f}{\partial y \partial x} = -1, \qquad \frac{\partial^2 f}{\partial y^2} = 2.$$

Thus $A = 4, B = -1, C = 2$, and

$$D = B^2 - AC = -7 < 0.$$

Since $A > 0$, it follows from the second-partials test that

$$f(1, 4) = 2 + 16 - 4 - 28 = -14$$

is a local **minimum.** (We obtained this result in Section 16.6 without benefit of the second-partials test. See page 901.) ☐

Example 2. The function

$$f(x, y) = \frac{x}{y^2} + xy$$

has partial derivatives

$$\frac{\partial f}{\partial x} = \frac{1}{y^2} + y, \qquad \frac{\partial f}{\partial y} = -\frac{2x}{y^3} + x.$$

Setting both of these partials equal to zero, we have

$$\frac{1}{y^2} + y = 0, \qquad x\left(-\frac{2}{y^3} + 1\right) = 0.$$

As you can check, the only simultaneous solution to these equations is $x = 0$, $y = -1$. The point $(0, -1)$ is thus the only stationary point.

The second partials are

$$\frac{\partial^2 f}{\partial x^2} = 0, \qquad \frac{\partial^2 f}{\partial y \partial x} = -\frac{2}{y^3} + 1, \qquad \frac{\partial^2 f}{\partial y^2} = \frac{6x}{y^4}.$$

Evaluating these partials at the point $(0, -1)$, we find that $A = 0$, $B = 3$, $C = 0$, and thus

$$D = B^2 - AC = 9 > 0.$$

By the second-partials test, $(0, -1)$ is a saddle point.

We could have determined that $(0, -1)$ is a saddle point by comparing the value of f at $(0, -1)$ with its value at nearby points $(h, -1 + k)$:

$$f(h, -1 + k) - f(0, -1) = \frac{h}{(k-1)^2} + h(k-1) - 0 = h\left[\frac{1 + (k-1)^3}{(k-1)^2}\right].$$

For small $k > 0$, this expression is positive if $h > 0$ and negative if $h < 0$.   ☐

Example 3.   The function

$$f(x, y) = x^3 + y^3 - 3xy$$

has partial derivatives

$$\frac{\partial f}{\partial x} = 3x^2 - 3y, \qquad \frac{\partial f}{\partial y} = 3y^2 - 3x.$$

Setting both partials equal to zero, we have

$$x^2 - y = 0, \qquad y^2 - x = 0.$$

The only simultaneous solutions are $x = 1$, $y = 1$ and $x = 0$, $y = 0$. The points $(1, 1)$ and $(0, 0)$ are the only stationary points.

The second partials are

$$\frac{\partial^2 f}{\partial x^2} = 6x, \qquad \frac{\partial^2 f}{\partial y \partial x} = -3, \qquad \frac{\partial^2 f}{\partial y^2} = 6y.$$

At $(1, 1)$, we have $A = 6$, $B = -3$, $C = 6$, and thus

$$D = B^2 - AC = 9 - 36 = -27 < 0.$$

Since $A > 0$, we know that

$$f(1, 1) = 1 + 1 - 3 = -1$$

is a local minimum. At $(0, 0)$, we have $A = 0$, $B = -3$, $C = 0$. Thus

$$D = B^2 - AC = 9 > 0.$$

The origin is a saddle point.   ☐

Example 4. Here we test the function

$$f(x, y) = \sin x + \sin y + \sin (x + y)$$

on the open square $0 < x < \pi$, $0 < y < \pi$.

In the first place

$$\frac{\partial f}{\partial x} = \cos x + \cos (x + y), \qquad \frac{\partial f}{\partial y} = \cos y + \cos (x + y).$$

Setting both of these partials equal to zero, we have

$$\cos x + \cos (x + y) = 0, \qquad \cos y + \cos (x + y) = 0.$$

Together these equations imply that $\cos x = \cos y$. Since both x and y lie between 0 and π, we can conclude that $x = y$. The condition

$$\cos x + \cos (x + y) = 0$$

now gives

$$\cos x + \cos 2x = 0.$$

Into this last equation we substitute the identity $\cos 2x = 2 \cos^2 x - 1$ and thereby obtain the relation

$$2 \cos^2 x + \cos x - 1 = 0.$$

This is a quadratic in $\cos x$ with solutions

$$\cos x = \tfrac{1}{4}(-1 \pm \sqrt{1 + 8}) = -1 \text{ or } \tfrac{1}{2}.$$

We throw out the solution -1 because x has to lie strictly between 0 and π. The other possibility, $\cos x = \tfrac{1}{2}$, gives $x = \tfrac{1}{3}\pi$. Since $x = y$, we also have $y = \tfrac{1}{3}\pi$. This shows that $(\tfrac{1}{3}\pi, \tfrac{1}{3}\pi)$ is the only stationary point.

The second partials are

$$\frac{\partial^2 f}{\partial x^2} = -\sin x - \sin (x + y), \qquad \frac{\partial^2 f}{\partial y \partial x} = -\sin (x + y),$$

$$\frac{\partial^2 f}{\partial y^2} = -\sin y - \sin (x + y).$$

At $(\tfrac{1}{3}\pi, \tfrac{1}{3}\pi)$ we have

$$A = -\sin \tfrac{1}{3}\pi - \sin \tfrac{2}{3}\pi = -\tfrac{1}{2}\sqrt{3} - \tfrac{1}{2}\sqrt{3} = -\sqrt{3}, \qquad B = -\sin \tfrac{2}{3}\pi = -\tfrac{1}{2}\sqrt{3},$$
$$C = -\sin \tfrac{1}{3}\pi - \sin \tfrac{2}{3}\pi = -\tfrac{1}{2}\sqrt{3} - \tfrac{1}{2}\sqrt{3} = -\sqrt{3},$$

and thus

$$D = B^2 - AC = \tfrac{3}{4} - 3 < 0.$$

Since $A < 0$, we can conclude that

$$f(\tfrac{1}{3}\pi, \tfrac{1}{3}\pi) = \sin \tfrac{1}{3}\pi + \sin \tfrac{1}{3}\pi + \sin \tfrac{2}{3}\pi = \tfrac{1}{2}\sqrt{3} + \tfrac{1}{2}\sqrt{3} + \tfrac{1}{2}\sqrt{3} = \tfrac{3}{2}\sqrt{3}$$

is a local maximum. ◻

The second-derivative test applies to points x_0 where $g'(x_0) = 0$ but $g''(x_0) \neq 0$. If $g''(x_0) = 0$, the second-derivative test provides no information. The second-partials test suffers from a similar limitation. It applies to points (x_0, y_0) where $\nabla f(x_0, y_0) = \mathbf{0}$ but $D \neq 0$. If $D = 0$, the second-partials test provides no information.

Consider, for example, the functions

$$F(x, y) = x^4 + y^4, \qquad G(x, y) = -(x^4 + y^4), \qquad H(x, y) = x^4 - y^4.$$

Each of these functions has zero gradient at the origin, and, as you can check, in each case $D = 0$. Yet,

(1) for F, $(0, 0)$ gives a local minimum;

(2) for G, $(0, 0)$ gives a local maximum;

(3) for H, $(0, 0)$ is a saddle point.

Statements (1) and (2) are obvious. To confirm (3), note that $H(0, 0) = 0$, but in every neighborhood of $(0, 0)$ the function H takes on both positive and negative values:

$$H(x, 0) > 0 \quad \text{for } x \neq 0 \qquad \text{while} \qquad H(0, y) < 0 \quad \text{for } y \neq 0. \quad \square$$

EXERCISES 16.7

Find the stationary points and the local extreme values.

1. $x^2 + xy + y^2 - 6x + 2.$
2. $x^2 + 2xy + 3y^2 + 2x + 10y + 1.$
3. $4x + 2y - x^2 + xy - y^2.$
4. $x^2 - xy + y^2 + 2x + 2y - 1.$
5. $x^3 - 6xy + y^3.$
6. $3x^2 + xy - y^2 + 5x - 5y + 4.$
7. $x^3 + y^2 - 6xy + 6x + 3y - 2.$
8. $x^2 + xy + y^2 - 6y + 5.$
9. $e^x \cos y.$
10. $x^2 - 2xy + 2y^2 - 3x + 5y.$
11. $x \sin y.$
12. $y + x \sin y.$
13. $(x + y)(xy + 1).$
14. $xy^{-1} - yx^{-1}.$
15. $xy + x^{-1} + 8y^{-1}.$
16. $x^2 - 2xy - y^2 + 1.$
17. $xy + x^{-1} + y^{-1}.$
18. $(x - y)(xy - 1).$

19. Find the maximum volume for a rectangular solid in the first octant $(x \geq 0, y \geq 0, z \geq 0)$ with one vertex at the origin and opposite vertex on the plane $x + y + z = 1.$

20. Let n be an integer greater than 2 and set $f(x, y) = ax^n + cy^n$ taking $ac \neq 0.$
    (a) Find the stationary points.   (b) Find the discriminant at each stationary point.
    (c) Find the local and absolute extreme values given that
        (i) $a > 0, c > 0.$ (ii) $a < 0, c < 0.$ (iii) $a > 0, c < 0.$

21. Describe the behavior of the function at the origin.
    (a) $f(x, y) = x^2 - y^3.$ (b) $f(x, y) = 2 \cos (x + y) + e^{xy}.$

22. Find the maximum volume for a rectangular solid inscribed in the ellipsoid

$$\frac{x^2}{a^2} + \frac{y^2}{b^2} + \frac{z^2}{c^2} = 1.$$

23. A pentagon is composed of a rectangle surmounted by an isosceles triangle (Figure 16.7.2). Given that the perimeter of the pentagon has a fixed value P, find the dimensions for maximum area.

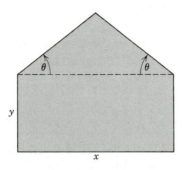

Figure 16.7.2

24. A bakery produces two types of bread, one at a cost of 50 cents per loaf, the other at a cost of 60 cents per loaf. Assume that, if the first bread is sold at x cents a loaf and the second at y cents a loaf, then the number of loaves that can be sold each week is given by the formulas

$$N_1 = 250(y - x), \qquad N_2 = 32{,}000 + 250(x - 2y).$$

Determine x and y for maximum profit.

25. Find the distance between the lines $x = \frac{1}{2}y = \frac{1}{3}z$ and $x = y - 2 = z$.

26. Find the absolute maximum value of

$$f(x, y) = \frac{(ax + by + c)^2}{x^2 + y^2 + 1}.$$

27. Find the dimensions of the most economical open-top rectangular crate 96 cubic meters in volume given that the base costs 30 cents per square meter and the sides cost 10 cents per square meter.

28. Let $f(x, y) = ax^2 + bxy + cy^2$ taking $abc \neq 0$.
(a) Find the discriminant D.
(b) Find the stationary points and local extreme values if $D \neq 0$.
(c) Suppose that $D = 0$. Find the stationary points and the local and absolute extreme values given that (i) $a > 0, c > 0$. (ii) $a < 0, c < 0$.

29. (*The method of least squares*) In this exercise we illustrate an important method of fitting a curve to a collection of points. Consider three points

$$(x_1, y_1) = (0, 2), \qquad (x_2, y_2) = (1, -5), \qquad (x_3, y_3) = (2, 4).$$

(a) Find the line $y = mx + b$ that minimizes the sum of the squares of the vertical distances $d_i = |y_i - (mx_i + b)|$ from these points to the line.
(b) Find the parabola $y = \alpha x^2 + \beta$ that minimizes the sum of the squares of the vertical distances $d_i = |y_i - (\alpha x_i^2 + \beta)|$ from the points to the parabola.

30. Repeat Exercise 29 taking $(x_1, y_1) = (-1, 2)$, $(x_2, y_2) = (0, -1)$, $(x_3, y_3) = (1, 1)$.

16.8   MAXIMA AND MINIMA WITH SIDE CONDITIONS

When we ask for the distance from a point $P(x_0, y_0)$ to a line l: $Ax + By + C = 0$, we are asking for the minimum value of

$$f(x, y) = \sqrt{(x - x_0)^2 + (y - y_0)^2}$$

with (x, y) subject to the side condition $Ax + By + C = 0$.   When we ask for the distance from a point $P(x_0, y_0, z_0)$ to a plane p: $Ax + By + Cz + D = 0$, we are asking for the minimum value of

$$f(x, y, z) = \sqrt{(x - x_0)^2 + (y - y_0)^2 + (z - z_0)^2}$$

with (x, y, z) subject to the side condition $Ax + By + Cz + D = 0$.

   We have already treated these particular problems by special techniques.   Our interest here is to present techniques for handling problems of this sort in general.   In the two-variable case, the problems will take the form of maximizing (or minimizing) some expression $f(x, y)$ subject to a side condition $g(x, y) = 0$.   In the three-variable case, we will seek to maximize (or minimize) some expression $f(x, y, z)$ subject to a side condition $g(x, y, z) = 0$.   We begin with two simple problems.

Problem 1.   Maximize the product xy subject to the side condition $x + y - 1 = 0$.

Solution.   The condition $x + y - 1 = 0$ gives $y = 1 - x$.   Our initial problem can therefore be solved simply by maximizing the product $h(x) = x(1 - x)$.   The derivative $h'(x) = 1 - 2x$ is 0 only at $x = \frac{1}{2}$.   Since $h''(x) = -2 < 0$, we know from the second-derivative test that $h(\frac{1}{2}) = \frac{1}{2}(1 - \frac{1}{2}) = \frac{1}{4}$ is the desired maximum.   ☐

Problem 2.   Find the maximum volume of a rectangular solid given that the sum of the lengths of its edges is $12a$.

Solution.   We denote the dimensions of the solid by x, y, z.   The volume is given by

$$V = xyz.$$

The stipulation on the edges requires that

$$4(x + y + z) = 12a.$$

Solving this last equation for z, we find that

$$z = 3a - (x + y).$$

Substituting this expression for z in the volume formula, we have

$$V = xy[3a - (x + y)].$$

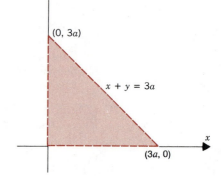

Figure 16.8.1

Since x, y, and $z = 3a - (x + y)$ must all remain positive, our problem is to find the maximum value of V on the interior of the triangle shown in Figure 16.8.1.

The first partials are

$$\frac{\partial V}{\partial x} = xy(-1) + y[3a - (x + y)] = 3ay - 2xy - y^2,$$

$$\frac{\partial V}{\partial y} = xy(-1) + x[3a - (x + y)] = 3ax - x^2 - 2xy.$$

Setting both partials equal to zero, we have

$$(3a - 2x - y)y = 0 \qquad \text{and} \qquad (3a - x - 2y)x = 0.$$

Since x and y are assumed positive, we can divide by x and y and get

$$3a - 2x - y = 0 \qquad \text{and} \qquad 3a - x - 2y = 0.$$

Solving these equations simultaneously, we find that $x = y = a$. The point (a, a), which does lie within the triangle, is the only stationary point. The value of V at that point is a^3. The conditions of the problem make it clear that this is a maximum. (If you are skeptical, you can confirm this by appealing to the second-partials test.) ☐

The last two problems were easy. They were easy in part because the side conditions were such that we could solve for one of the variables in terms of the other(s). In general this is not possible and a more sophisticated approach is required.

The Method of Lagrange†

We begin with what looks like a detour. To avoid having to make separate statements for the two- and three-variable cases, we will use vector notation.

Throughout the discussion f will be a function of two or three variables continuously differentiable on some open set U. We take

$$C: \quad \mathbf{r} = \mathbf{r}(t), \quad t \in I$$

as a curve that lies entirely in U and has at each point a nonzero tangent vector $\mathbf{r}'(t)$. The basic result is this:

(16.8.1)
> if $\mathbf{x}_0$ maximizes (or minimizes) $f(\mathbf{x})$ on C, then $\nabla f(\mathbf{x}_0)$ is perpendicular to C at $\mathbf{x}_0$.

Proof. Choose t_0 so that

$$\mathbf{r}(t_0) = \mathbf{x}_0.$$

The composition $f(\mathbf{r}(t))$ has a maximum (or minimum) at t_0. Consequently, its derivative,

---

† Another contribution of the French mathematician Joseph Louis Lagrange.

$$\frac{d}{dt}[f(\mathbf{r}(t))] = \nabla f(\mathbf{r}(t)) \cdot \mathbf{r}'(t),$$

must be zero at t_0:

$$0 = \nabla f(\mathbf{r}(t_0)) \cdot \mathbf{r}'(t_0) = \nabla f(\mathbf{x}_0) \cdot \mathbf{r}'(t_0).$$

This shows that

$$\nabla f(\mathbf{x}_0) \perp \mathbf{r}'(t_0).$$

Since $\mathbf{r}'(t_0)$ is tangent to C at $\mathbf{x}_0$, $\nabla f(\mathbf{x}_0)$ is perpendicular to C at $\mathbf{x}_0$. □

We are now ready for the side condition problems. Suppose that g is a continuously differentiable function of two or three variables defined on a subset of the domain of f. Lagrange made the following observation:

(16.8.2)
> if $\mathbf{x}_0$ maximizes (or minimizes) $f(\mathbf{x})$ subject to the side condition $g(\mathbf{x}) = 0$, then $\nabla f(\mathbf{x}_0)$ and $\nabla g(\mathbf{x}_0)$ are parallel. Thus, if $\nabla g(\mathbf{x}_0) \neq \mathbf{0}$, then there exists a scalar λ such that
>
> $$\nabla f(\mathbf{x}_0) = \lambda \, \nabla g(\mathbf{x}_0).$$

Such a scalar λ has come to be called a *Lagrange multiplier*.

Proof of (16.8.2). Let's suppose that $\mathbf{x}_0$ maximizes (or minimizes) $f(\mathbf{x})$ subject to the side condition $g(\mathbf{x}) = 0$. If $\nabla g(\mathbf{x}_0) = \mathbf{0}$, the result is trivially true: every vector is parallel to the zero vector. We suppose therefore that $\nabla g(\mathbf{x}_0) \neq \mathbf{0}$.

In the two-variable case we have

$$\mathbf{x}_0 = (x_0, y_0) \qquad \text{and} \qquad \text{the side condition} \quad g(x, y) = 0.$$

The side condition defines a curve C that has a nonzero tangent vector at (x_0, y_0).†
Since (x_0, y_0) maximizes (or minimizes) $f(x, y)$ on C, we know from (16.8.1) that $\nabla f(x_0, y_0)$ is perpendicular to C at (x_0, y_0). By (16.5.2), $\nabla g(x_0, y_0)$ is also perpendicular to C at (x_0, y_0). The two gradients are therefore parallel.

In the three-variable case we have

$$\mathbf{x}_0 = (x_0, y_0, z_0) \qquad \text{and} \qquad \text{the side condition} \quad g(x, y, z) = 0.$$

The side condition defines a surface Γ that lies in the domain of f. Now let C be a curve that lies on Γ and passes through (x_0, y_0, z_0) with nonzero tangent vector. We know that (x_0, y_0, z_0) maximizes (or minimizes) $f(x, y, z)$ on C. Consequently, $\nabla f(x_0, y_0, z_0)$ is perpendicular to C at (x_0, y_0, z_0). Since this is true for each such curve C, $\nabla f(x_0, y_0, z_0)$ must be perpendicular to Γ itself. But $\nabla g(x_0, y_0, z_0)$ is also perpendicular to Γ at (x_0, y_0, z_0). (Section 16.5) It follows that $\nabla f(x_0, y_0, z_0)$ and $\nabla g(x_0, y_0, z_0)$ are parallel. □

---

† $\mathbf{t}(x_0, y_0) = \dfrac{\partial g}{\partial y}(x_0, y_0)\mathbf{i} - \dfrac{\partial g}{\partial x}(x_0, y_0)\mathbf{j} \neq \mathbf{0}.$

We come now to some problems that are susceptible to Lagrange's method. In each case ∇g is not $\mathbf{0}$ where g is 0 and therefore we can focus entirely on those points $\mathbf{x}$ that satisfy a Lagrange condition

$$\nabla f(\mathbf{x}) = \lambda \, \nabla g(\mathbf{x}).$$

Problem 3. Maximize and minimize

$$f(x, y) = xy \quad \text{on the unit circle} \quad x^2 + y^2 = 1.$$

Solution. Since f is continuous and the unit circle is closed and bounded, it is clear that both a maximum and a minimum exist.

To apply Lagrange's method we set

$$g(x, y) = x^2 + y^2 - 1.$$

We want to maximize and minimize

$$f(x, y) = xy \quad \text{subject to the side condition} \quad g(x, y) = 0.$$

The gradients are

$$\nabla f(x, y) = y\mathbf{i} + x\mathbf{j}, \qquad \nabla g(x, y) = 2x\mathbf{i} + 2y\mathbf{j}.$$

Setting

$$\nabla f(x, y) = \lambda \, \nabla g(x, y),$$

we obtain

$$y = 2\lambda x, \qquad x = 2\lambda y.$$

Multiplying the first equation by y and the second equation by x, we find that

$$y^2 = 2\lambda xy, \qquad x^2 = 2\lambda xy$$

and thus

$$y^2 = x^2.$$

The side condition $x^2 + y^2 = 1$ now implies that $2x^2 = 1$ and therefore that $x = \pm\frac{1}{2}\sqrt{2}$. The only points that can give rise to an extreme value are

$$(\tfrac{1}{2}\sqrt{2}, \tfrac{1}{2}\sqrt{2}), \quad (\tfrac{1}{2}\sqrt{2}, -\tfrac{1}{2}\sqrt{2}), \quad (-\tfrac{1}{2}\sqrt{2}, \tfrac{1}{2}\sqrt{2}), \quad (-\tfrac{1}{2}\sqrt{2}, -\tfrac{1}{2}\sqrt{2}).$$

At the first and fourth points f takes on the value $\frac{1}{2}$. At the second and third points f takes on the value $-\frac{1}{2}$. Clearly then $\frac{1}{2}$ is the maximum value and $-\frac{1}{2}$ the minimum value. $\square$

Problem 4. Find the minimum value taken on by the function

$$f(x, y) = x^2 + (y - 2)^2 \quad \text{on the hyperbola} \quad x^2 - y^2 = 1.$$

Solution. This minimum is simply the square of the distance from the point (0, 2) to the hyperbola and thus clearly exists.

Now set

$$g(x, y) = x^2 - y^2 - 1.$$

We want to minimize

$$f(x, y) = x^2 + (y - 2)^2 \quad \text{subject to the side condition} \quad g(x, y) = 0.$$

Here

$$\nabla f(x, y) = 2x\mathbf{i} + 2(y - 2)\mathbf{j}, \qquad \nabla g(x, y) = 2x\mathbf{i} - 2y\mathbf{j}.$$

The Lagrange condition $\nabla f(x, y) = \lambda \nabla g(x, y)$ gives

$$2x = 2\lambda x, \qquad 2(y - 2) = -2\lambda y,$$

which we can simplify to

$$x = \lambda x, \qquad y - 2 = -\lambda y.$$

The side condition $x^2 - y^2 = 1$ shows that x cannot be zero. Dividing $x = \lambda x$ by x, we get $\lambda = 1$. This means that $y - 2 = -y$ and therefore $y = 1$. With $y = 1$, the side condition gives $x = \pm\sqrt{2}$. The points to be checked are therefore $(-\sqrt{2}, 1)$ and $(\sqrt{2}, 1)$. At each of these points f takes on the value 3. This is the desired minimum. ☐

Remark. The last problem could have been solved more simply by rewriting the side condition as $x^2 = 1 + y^2$ and eliminating x from $f(x, y)$ by substitution. It would then have been only a matter of minimizing $1 + y^2 + (y - 2)^2 = 2y^2 - 4y + 5$. ☐

Problem 5. Maximize

$$f(x, y, z) = xyz \quad \text{subject to the side condition} \quad x^3 + y^3 + z^3 = 1$$

with $x \geq 0, y \geq 0, z \geq 0$.

Solution. The set of all points (x, y, z) that satisfy the side condition can be shown to be closed and bounded. Since f is continuous, we can be sure that the desired maximum exists.

We begin by setting

$$g(x, y, z) = x^3 + y^3 + z^3 - 1$$

so that the side condition becomes $g(x, y, z) = 0$. We seek those triples (x, y, z) that simultaneously satisfy a Lagrange condition

$$\nabla f(x, y, z) = \lambda \nabla g(x, y, z) \quad \text{and the side condition} \quad g(x, y, z) = 0.$$

The gradients are

$$\nabla f(x, y, z) = yz\mathbf{i} + xz\mathbf{j} + xy\mathbf{k}, \qquad \nabla g(x, y, z) = 3x^2\mathbf{i} + 3y^2\mathbf{j} + 3z^2\mathbf{k}.$$

The Lagrange condition $\nabla f(x, y, z) = \lambda \nabla g(x, y, z)$ gives

$$yz = \lambda 3x^2, \qquad xz = \lambda 3y^2, \qquad xy = \lambda 3z^2.$$

Multiplying the first equation by x, the second by y, and the third by z, we get

$$xyz = \lambda 3x^3, \qquad xyz = \lambda 3y^3, \qquad xyz = \lambda 3z^3$$

and consequently

$$\lambda x^3 = \lambda y^3 = \lambda z^3.$$

We can exclude $\lambda = 0$ because, if $\lambda = 0$, then x, y, or z would have to be zero. That would force xyz to be 0, and 0 is obviously not a maximum. Having excluded $\lambda = 0$, we can divide by λ and get $x^3 = y^3 = z^3$ and thus $x = y = z$. The side condition $x^3 + y^3 + z^3 = 1$ now gives

$$x = (\tfrac{1}{3})^{1/3}, \qquad y = (\tfrac{1}{3})^{1/3}, \qquad z = (\tfrac{1}{3})^{1/3}.$$

The desired maximum is $\tfrac{1}{3}$. $\square$

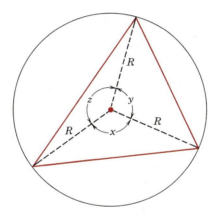

Figure 16.8.2

Problem 6. Show that, of all the triangles inscribed in a fixed circle, the equilateral triangle has the largest perimeter.

Solution. It is intuitively clear that this maximum exists and geometrically clear that the triangle that offers this maximum contains the center of the circle in its interior or on its boundary. As in Figure 16.8.2 we denote by x, y, z the central angles that subtend the three sides. As you can verify by trigonometry, the perimeter of the triangle is given by the function

$$f(x, y, z) = 2R(\sin \tfrac{1}{2}x + \sin \tfrac{1}{2}y + \sin \tfrac{1}{2}z).$$

As a side condition we have

$$g(x, y, z) = x + y + z - 2\pi = 0.$$

To maximize the perimeter we form the gradients

$$\nabla f(x, y, z) = R[\cos \tfrac{1}{2}x \, \mathbf{i} + \cos \tfrac{1}{2}y \, \mathbf{j} + \cos \tfrac{1}{2}z \, \mathbf{k}], \qquad \nabla g(x, y, z) = \mathbf{i} + \mathbf{j} + \mathbf{k}.$$

The Lagrange condition $\nabla f(x, y, z) = \lambda \, \nabla g(x, y, z)$ gives

$$\lambda = R \cos \tfrac{1}{2}x, \qquad \lambda = R \cos \tfrac{1}{2}y, \qquad \lambda = R \cos \tfrac{1}{2}z$$

and therefore

$$\cos \tfrac{1}{2}x = \cos \tfrac{1}{2}y = \cos \tfrac{1}{2}z.$$

With x, y, z all in $(0, \pi]$, we can conclude that $x = y = z$. Since the central angles are equal, the sides are equal. The triangle is therefore equilateral. ☐

The Lagrange equation can be replaced by a cross-product equation: points that satisfy $\nabla f = \lambda \, \nabla g$ satisfy

(16.8.3)
$$\boxed{\nabla f \times \nabla g = \mathbf{0}.}$$

In two variables (16.8.3) reduces to

(16.8.4)
$$\boxed{\frac{\partial f}{\partial x} \frac{\partial g}{\partial y} - \frac{\partial f}{\partial y} \frac{\partial g}{\partial x} = 0.}$$

Problem 7. Maximize and minimize $f(x, y) = xy$ on the unit circle $x^2 + y^2 = 1$.

Solution. This problem was solved earlier by means of the Lagrange equation. This time we will use (16.8.4) instead.

As before we set

$$g(x, y) = x^2 + y^2 - 1$$

so that the side condition takes the form $g(x, y) = 0$. Since

$$\frac{\partial f}{\partial x} = y, \quad \frac{\partial f}{\partial y} = x \qquad \text{and} \qquad \frac{\partial g}{\partial x} = 2x, \quad \frac{\partial g}{\partial y} = 2y,$$

(16.8.4) takes the form

$$y(2y) - x(2x) = 0.$$

This gives $x^2 = y^2$.

As before, the side condition $x^2 + y^2 = 1$ now implies that $2x^2 = 1$ and therefore that $x = \pm\tfrac{1}{2}\sqrt{2}$. The points under consideration are

$$(\tfrac{1}{2}\sqrt{2}, \tfrac{1}{2}\sqrt{2}), \quad (\tfrac{1}{2}\sqrt{2}, -\tfrac{1}{2}\sqrt{2}), \quad (-\tfrac{1}{2}\sqrt{2}, \tfrac{1}{2}\sqrt{2}), \quad (-\tfrac{1}{2}\sqrt{2}, -\tfrac{1}{2}\sqrt{2}).$$

At the first and fourth points f takes on the value $\tfrac{1}{2}$. At the second and third points f takes on the value $-\tfrac{1}{2}$. The first is a maximum and the second a minimum. ☐

In three variables the computations demanded by the cross-product equation are often quite complicated, and it is usually easier to follow the method of Lagrange.

EXERCISES 16.8

1. Minimize $x^2 + y^2$ on the hyperbola $xy = 1$.
2. Maximize xy on the ellipse $b^2x^2 + a^2y^2 = a^2b^2$.
3. Minimize xy on the ellipse $b^2x^2 + a^2y^2 = a^2b^2$.
4. Minimize xy^2 on the unit circle $x^2 + y^2 = 1$.
5. Maximize xy^2 on the ellipse $b^2x^2 + a^2y^2 = a^2b^2$.
6. Maximize $x + y$ on the curve $x^4 + y^4 = 1$.
7. Maximize $x^2 + y^2$ on the curve $x^4 + 7x^2y^2 + y^4 = 1$.
8. Minimize xyz on the unit sphere $x^2 + y^2 + z^2 = 1$.
9. Maximize xyz on the ellipsoid $x^2/a^2 + y^2/b^2 + z^2/c^2 = 1$.
10. Minimize $x + 2y + 4z$ on the sphere $x^2 + y^2 + z^2 = 7$.
11. Maximize $2x + 3y + 5z$ on the sphere $x^2 + y^2 + z^2 = 19$.
12. Minimize $x^4 + y^4 + z^4$ on the plane $x + y + z = 1$.
13. Maximize the volume of a rectangular solid in the first octant with one vertex at the origin and opposite vertex on the plane $x/a + y/b + z/c = 1$. (Take $a > 0, b > 0, c > 0$.)
14. Show that the square has the largest area of all the rectangles with a given perimeter.
15. Find the distance from the point $(0, 1)$ to the parabola $x^2 = 4y$.
16. Find the distance from the point $(p, 4p)$ to the parabola $y^2 = 2px$.
17. Use the method of Lagrange to find the distance from the origin to the plane with equation $Ax + By + Cz + D = 0$.
18. Maximize the volume of a rectangular solid given that the sum of the areas of the six faces is $6a^2$.
19. Within a triangle there is a point P such that the sum of the squares of its distances from the sides of the triangle is a minimum.   Find this minimum.
20. Show that of all the triangles inscribed in a fixed circle the equilateral one has the largest (a) product of the lengths of the sides, (b) sum of the squares of the lengths of the sides.
21. The curve $x^3 - y^3 = 1$ is asymptotic to the line $y = x$. Find the point(s) on the curve $x^3 - y^3 = 1$ furthest from the line $y = x$.
22. A plane passes through the point (a, b, c). Find its intercepts with the coordinate axes if the volume of the solid bounded by the plane and the coordinate planes is to be a minimum.
23. Show that, of all the triangles with a given perimeter, the equilateral triangle has the largest area.   HINT: area $= \sqrt{s(s - a)(s - b)(s - c)}$, where s represents the semiperimeter $s = \frac{1}{2}(a + b + c)$.
24. (a) Maximize $(xyz)^{1/3}$ given that x, y, z are nonnegative and $x + y + z = a$.
    (b) Use (a) to show that, if x, y, z are nonnegative, then $(xyz)^{1/3} \leq \frac{1}{3}(x + y + z)$.
25. A manufacturer can produce three distinct products in quantities Q_1, Q_2, Q_3, respectively, and thereby derive a profit $P(Q_1, Q_2, Q_3) = 2Q_1 + 8Q_2 + 24Q_3$. Find the values of Q_1, Q_2, Q_3 that maximize profit if production is subject to the constraint $Q_1^2 + 2Q_2^2 + 4Q_3^2 = 4.5 \times 10^9$.
26. (*Refraction and Fermat's Principle of Least Time*) Snell's Law of Refraction

$$n_i \sin \theta_i = n_r \sin \theta_r$$

was explained in detail in the exercises to Section 7.13. Derive Snell's law from Fermat's Principle of Least Time: that of all (neighboring) paths light chooses the one that demands the least time.

16.9 DIFFERENTIALS

We begin by reviewing the one-variable case. If f is differentiable at x, then, for small h, the increment

$$\Delta f = f(x + h) - f(x)$$

can be approximated by the differential

$$df = f'(x)h.$$

We write

$$\Delta f \cong df.$$

How good is this approximation? Good enough that the ratio

$$\frac{\Delta f - df}{|h|}$$

tends to 0 as h tends to 0.

The differential of a function of several variables, defined in an analogous manner, plays a similar approximating role. Let's suppose that f, now a function of several variables, is differentiable at $\mathbf{x}$. The difference

(16.9.1) $$\boxed{\Delta f = f(\mathbf{x} + \mathbf{h}) - f(\mathbf{x})}$$

is called the *increment* of f, and the dot product

(16.9.2) $$\boxed{df = \nabla f(\mathbf{x}) \cdot \mathbf{h}}$$

is called the *differential* (more formally, the *total differential*). As in the one-variable case, for small $\mathbf{h}$, the differential and the increment are approximately equal:

(16.9.3) $$\boxed{\Delta f \cong df.}$$

How approximately equal are they? Enough so that the ratio

$$\frac{\Delta f - df}{\|\mathbf{h}\|}$$

tends to 0 as $\mathbf{h}$ tends to $\mathbf{0}$. How do we know this? We know that

$$f(\mathbf{x} + \mathbf{h}) - f(\mathbf{x}) = \nabla f(\mathbf{x}) \cdot \mathbf{h} + o(\mathbf{h}).$$

It follows that

$$[f(\mathbf{x} + \mathbf{h}) - f(\mathbf{x})] - \nabla f(\mathbf{x}) \cdot \mathbf{h} = o(\mathbf{h});$$

that is, as $\mathbf{h} \to \mathbf{0}$,

$$\frac{\overbrace{[f(\mathbf{x} + \mathbf{h}) - f(\mathbf{x})]}^{\Delta f} - \overbrace{\nabla f(\mathbf{x}) \cdot \mathbf{h}}^{df}}{\|\mathbf{h}\|} \to 0.$$

In the two-variable case we set $\mathbf{x} = (x, y)$ and $\mathbf{h} = (\Delta x, \Delta y)$.   The increment $\Delta f = f(\mathbf{x} + \mathbf{h}) - f(\mathbf{x})$ then takes the form

$$\Delta f = f(x + \Delta x, y + \Delta y) - f(x, y),$$

and the differential $df = \nabla f(\mathbf{x}) \cdot \mathbf{h}$ takes the form

$$df = \frac{\partial f}{\partial x}(x, y)\, \Delta x + \frac{\partial f}{\partial y}(x, y)\, \Delta y.$$

By suppressing the point of evaluation, we can write

(16.9.4)
$$df = \frac{\partial f}{\partial x}\, \Delta x + \frac{\partial f}{\partial y}\, \Delta y.$$

The approximation $\Delta f \cong df$ is illustrated in Figure 16.9.1.   There we have represented f as a surface $z = f(x, y)$.   Through a point $P(x_0, y_0, f(x_0, y_0))$ we have drawn the tangent plane.   *The difference $df - \Delta f$ is the vertical separation between this tangent plane and the surface as measured at the point $(x_0 + \Delta x, y_0 + \Delta y)$.*

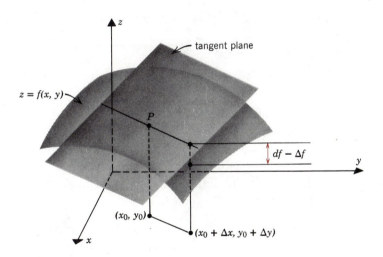

Figure 16.9.1

Proof.   The tangent plane at P has equation

$$z - f(x_0, y_0) = (x - x_0)\frac{\partial f}{\partial x}(x_0, y_0) + (y - y_0)\frac{\partial f}{\partial y}(x_0, y_0).$$

The z-coordinate of this plane at the point $(x_0 + \Delta x, y_0 + \Delta y)$ is

$$f(x_0, y_0) + \frac{\partial f}{\partial x}(x_0, y_0)\, \Delta x + \frac{\partial f}{\partial y}(x_0, y_0)\, \Delta y. \qquad \text{(check this)}$$

The z-coordinate of the surface at this same point is

$$f(x_0 + \Delta x, y_0 + \Delta y).$$

The difference

$$\left[f(x_0, y_0) + \frac{\partial f}{\partial x} (x_0, y_0) \Delta x + \frac{\partial f}{\partial y} (x_0, y_0) \Delta y \right] - \left[f(x_0 + \Delta x, \ y_0 + \Delta y) \right]$$

can be written as

$$\left[\frac{\partial f}{\partial x} (x_0, y_0) \Delta x + \frac{\partial f}{\partial y} (x_0, y_0) \Delta y \right] - \left[f(x_0 + \Delta x, y_0 + \Delta y) - f(x_0, y_0) \right].$$

This is just $df - \Delta f$.   □ †

For the three-variable case we set $\mathbf{x} = (x, y, z)$ and $\mathbf{h} = (\Delta x, \Delta y, \Delta z)$. The increment becomes

$$\Delta f = f(x + \Delta x, y + \Delta y, z + \Delta z) - f(x, y, z),$$

and the approximating differential becomes

$$df = \frac{\partial f}{\partial x} (x, y, z) \Delta x + \frac{\partial f}{\partial y} (x, y, z) \Delta y + \frac{\partial f}{\partial z} (x, y, z) \Delta z.$$

Suppressing the point of evaluation we have

(16.9.5)
$$df = \frac{\partial f}{\partial x} \Delta x + \frac{\partial f}{\partial y} \Delta y + \frac{\partial f}{\partial z} \Delta z.$$

To illustrate the use of differentials we begin with a rectangle of sides x and y. The area is given by

$$A(x, y) = xy.$$

An increase in the dimensions of the rectangle to $x + \Delta x$ and $y + \Delta y$ produces a change in area

$$\begin{aligned}
\Delta A &= (x + \Delta x)(y + \Delta y) - xy \\
&= (xy + x \Delta y + y \Delta x + \Delta x \Delta y) - xy \\
&= x \Delta y + y \Delta x + \Delta x \Delta y.
\end{aligned}$$

The differential estimate for this change in area is

$$dA = \frac{\partial A}{\partial x} \Delta x + \frac{\partial A}{\partial y} \Delta y = y \Delta x + x \Delta y. \qquad \text{(Figure 16.9.2)}$$

The error of our estimate, the difference between the actual change and the estimated change, is the difference $\Delta A - dA = \Delta x \Delta y$.   □

---

† As in the figure, we have been assuming that the tangent plane lies above the surface. If the tangent plane lies below the surface, then $df - \Delta f$ is negative. The vertical separation between the tangent plane and the surface is then $\Delta f - df$.

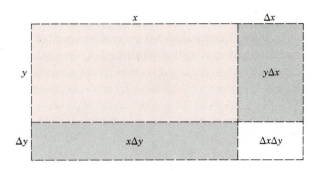

Figure 16.9.2

Problem 1. Given that

$$f(x, y) = yx^{2/5} + x\sqrt{y},$$

estimate by a differential the change in f from (32, 16) to (35, 18).

Solution. Since

$$\frac{\partial f}{\partial x} = \frac{2}{5}\left(\frac{1}{x}\right)^{3/5} y + \sqrt{y} \quad \text{and} \quad \frac{\partial f}{\partial y} = x^{2/5} + \frac{x}{2\sqrt{y}},$$

we have

$$df = \left[\frac{2}{5}\left(\frac{1}{x}\right)^{3/5} y + \sqrt{y}\right]\Delta x + \left[x^{2/5} + \frac{x}{2\sqrt{y}}\right]\Delta y.$$

At $x = 32$, $y = 16$, $\Delta x = 3$, $\Delta y = 2$,

$$df = \left[\frac{2}{5}\left(\frac{1}{32}\right)^{3/5} 16 + \sqrt{16}\right] 3 + \left[32^{2/5} + \frac{32}{2\sqrt{16}}\right] 2 = 30.4.$$

The change increases the value of f by approximately 30.4. □

Problem 2. Use differentials to estimate $\sqrt{27}\sqrt[3]{1021}$.

Solution. We know $\sqrt{25}$ and $\sqrt[3]{1000}$. What we need is an estimate for the increase of

$$f(x, y) = \sqrt{x}\sqrt[3]{y} = x^{1/2}y^{1/3}$$

from $x = 25$, $y = 1000$ to $x = 27$, $y = 1021$. The differential is

$$df = \tfrac{1}{2}x^{-1/2}y^{1/3}\,\Delta x + \tfrac{1}{3}x^{1/2}y^{-2/3}\,\Delta y.$$

With $x = 25$, $y = 1000$, $\Delta x = 2$, $\Delta y = 21$, df becomes

$$(\tfrac{1}{2} \cdot 25^{-1/2} \cdot 1000^{1/3})2 + (\tfrac{1}{3} \cdot 25^{1/2} \cdot 1000^{-2/3})21 = 2.35.$$

The change increases the value of the function by about 2.35. It follows that

$$\sqrt{27}\sqrt[3]{1021} \cong \sqrt{25}\sqrt[3]{1000} + 2.35 = 52.35.$$

(Our calculator gives $\sqrt{27}\sqrt[3]{1021} \cong 52.32$.) □

Problem 3. Estimate by a differential the change in the volume of the frustum of a cone if the upper radius r is decreased from 3 to 2.7 centimeters, the base radius R is increased from 8 to 8.1 centimeters, and the height h is increased from 6 to 6.3 centimeters.

Solution.   Since

$$V(r, R, h) = \tfrac{1}{3}\pi h(R^2 + Rr + r^2),$$

we have

$$dV = \tfrac{1}{3}\pi h(R + 2r)\,\Delta r + \tfrac{1}{3}\pi h(2R + r)\,\Delta R + \tfrac{1}{3}\pi(R^2 + Rr + r^2)\,\Delta h.$$

At $r = 3$, $R = 8$, $h = 6$, $\Delta r = -0.3$, $\Delta R = 0.1$, and $\Delta h = 0.3$,

$$dV = (28\pi)(-0.3) + (38\pi)(0.1) + \tfrac{1}{3}(97\pi)(0.3) = 5.1\pi.$$

According to our differential estimate, the volume increases by about 5.1π cubic centimeters.   (As you can check, the exact change in volume is 5.015π cubic centimeters.)   ☐

EXERCISES 16.9

Find the differential df.

1. $f(x, y) = x^3 y - x^2 y^2$.

2. $f(x, y) = xy + yz + xz$.

3. $f(x, y) = x \cos y - y \cos x$.

4. $f(x, y, z) = x^2 y e^{2z}$.

5. $f(x, y, z) = x - y \tan z$.

6. $f(x, y) = (x - y) \ln(x + y)$.

7. Compute Δu and du for $u = x^2 - 3xy + 2y^2$ at $x = 2$, $y = -3$, $\Delta x = -0.3$, $\Delta y = 0.2$.

8. Compute du for $u = (x + y)\sqrt{x - y}$ at $x = 6$, $y = 2$, $\Delta x = \tfrac{1}{4}$, $\Delta y = -\tfrac{1}{2}$.

Use differentials to find the approximate value.

9. $\sqrt{125}\,\sqrt[4]{17}$.

10. $(1 - \sqrt{10})(1 + \sqrt{24})$.

11. $\sin \tfrac{6}{7}\pi \cos \tfrac{1}{5}\pi$.

12. $\sqrt{8} \tan \tfrac{5}{16}\pi$.

13. Given that $z = (x - y)(x + y)^{-1}$, use dz to find the approximate change in z if x is increased from 4 to $4\tfrac{1}{10}$ and y is increased from 2 to $2\tfrac{1}{10}$.   What is the exact change?

14. Estimate by a differential the change in the volume of a right circular cylinder if the height is increased from 12 to 12.2 inches and the radius is decreased from 8 to 7.7 inches.

15. Estimate the change in total surface area for the cylinder of Exercise 14.

16. Use a differential to estimate the change in $T = x^2 \cos \pi z - y^2 \sin \pi z$ from $x = 2$, $y = 2$, $z = 2$ to $x = 2.1$, $y = 1.9$, $z = 2.2$.

17. The radius of a circular cylinder of height h is increased from r to $r + \Delta r$.
   (a) Determine the exact change in h that will keep the volume constant.   Then estimate this change in h by using a differential.
   (b) Determine the exact change in h that will keep the total surface area constant.   Then estimate this change in h by using a differential.

18. Give bounding estimates for the volume and the lateral area of a right circular cone given that the radius is between 4.8 and 5.2 inches and the height is between 11.7 and 12.3 inches.

19. The specific gravity of a solid is given by the formula $s = A(A - W)^{-1}$ where A is the weight in air and W is the weight in water.   Estimate s if A is measured to be 9 pounds (within a tolerance of 0.01 pounds) and W is measured to be 5 pounds (within a tolerance of 0.02 pounds).

16.10   RECONSTRUCTING A FUNCTION FROM ITS GRADIENT

This section has three parts.   In Part 1 we show how to find $f(x, y)$ given its gradient

$$\nabla f(x, y) = \frac{\partial f}{\partial x}(x, y)\, \mathbf{i} + \frac{\partial f}{\partial y}(x, y)\, \mathbf{j}.$$

In Part 2 we show that, although all gradients $\nabla f(x, y)$ are expressions of the form

$$P(x, y)\mathbf{i} + Q(x, y)\mathbf{j}$$

(set $P = \partial f/\partial x$ and $Q = \partial f/\partial y$), not all such expressions are gradients.   In Part 3 we tackle the problem of recognizing which expressions $P(x, y)\mathbf{i} + Q(x, y)\mathbf{j}$ are actually gradients.

Part 1

Problem 1.   Find f given that

$$\nabla f(x, y) = y^2\mathbf{i} + (2xy - 1)\mathbf{j}.$$

Solution.   Since

$$\nabla f(x, y) = \frac{\partial f}{\partial x}(x, y)\, \mathbf{i} + \frac{\partial f}{\partial y}(x, y)\, \mathbf{j},$$

we have

$$\frac{\partial f}{\partial x}(x, y) = y^2, \qquad \frac{\partial f}{\partial y}(x, y) = 2xy - 1.$$

Integrating $\partial f/\partial x$ with respect to x, we have

$$f(x, y) = xy^2 + \phi(y)$$

where $\phi(y)$ is independent of x but may depend on y.   Differentiation with respect to y gives

$$\frac{\partial f}{\partial y}(x, y) = 2xy + \phi'(y).$$

The two equations for $\partial f/\partial y$ can be reconciled only by having

$$\phi'(y) = -1 \quad \text{and thus} \quad \phi(y) = -y + C.$$

This means that

$$f(x, y) = x^2y - y + C. \quad \square$$

Remark.   We could have started by integrating $\partial f/\partial y$ with respect to y and then differentiating with respect to x.   This procedure would have yielded the same result. $\square$

Problem 2. Find f given that

$$\nabla f(x, y) = \left(\sqrt{y} - \frac{y}{2\sqrt{x}} + 2x\right)\mathbf{i} + \left(\frac{x}{2\sqrt{y}} - \sqrt{x} + 1\right)\mathbf{j}.$$

Solution. Here we have

$$\frac{\partial f}{\partial x}(x, y) = \sqrt{y} - \frac{y}{2\sqrt{x}} + 2x, \qquad \frac{\partial f}{\partial y}(x, y) = \frac{x}{2\sqrt{y}} - \sqrt{x} + 1.$$

We will proceed as in the first problem. Integrating $\partial f/\partial x$ with respect to x, we have

$$f(x, y) = x\sqrt{y} - \sqrt{x}\,y + x^2 + \phi(y)$$

with $\phi(y)$ independent of x. Differentiation with respect to y gives

$$\frac{\partial f}{\partial y}(x, y) = \frac{x}{2\sqrt{y}} - \sqrt{x} + \phi'(y).$$

The two equations for $\partial f/\partial y$ can be reconciled only by having

$$\phi'(y) = 1 \quad \text{and thus} \quad \phi(y) = y + C.$$

This means that

$$f(x, y) = x\sqrt{y} - \sqrt{x}\,y + x^2 + y + C. \quad \square$$

The function

$$f(x, y) = x\sqrt{y} - \sqrt{x}\,y + x^2 + y + C$$

is the *general solution* of the vector differential equation

$$\nabla f(x, y) = \left(\sqrt{y} - \frac{y}{2\sqrt{x}} + 2x\right)\mathbf{i} + \left(\frac{x}{2\sqrt{y}} - \sqrt{x} + 1\right)\mathbf{j}.$$

Each *particular* solution can be obtained by assigning a particular value to the constant of integration C.

Part 2

The next problem shows that not all linear combinations $P(x, y)\mathbf{i} + Q(x, y)\mathbf{j}$ are gradients.

Problem 3. Show that $y\mathbf{i} - x\mathbf{j}$ is not a gradient.

Solution. Suppose on the contrary that it is a gradient. Then there exists a function f such that

$$\nabla f(x, y) = y\mathbf{i} - x\mathbf{j}.$$

Obviously

$$\frac{\partial f}{\partial x}(x,\, y) = y, \qquad \frac{\partial f}{\partial y}(x,\, y) = -x$$

$$\frac{\partial^2 f}{\partial y \partial x}(x,\, y) = 1, \qquad \frac{\partial^2 f}{\partial x \partial y}(x,\, y) = -1$$

and thus

$$\frac{\partial^2 f}{\partial y \partial x}(x,\, y) \neq \frac{\partial^2 f}{\partial x \partial y}(x,\, y).$$

This contradicts (15.6.5): the four partial derivatives under consideration are everywhere continuous and thus, according to (15.6.5), we must have

$$\frac{\partial^2 f}{\partial y \partial x}(x,\, y) = \frac{\partial^2 f}{\partial x \partial y}(x,\, y).$$

This contradiction shows that $y\mathbf{i} - x\mathbf{j}$ is not a gradient. ☐

Part 3

We come now to the problem of recognizing which linear combinations

$$P(x,\, y)\mathbf{i} + Q(x,\, y)\mathbf{j}$$

are actually gradients. But first we need to review some ideas and establish some new terminology.

As indicated earlier (Section 16.3), an open set (in the plane or in three-space) is said to be *connected* iff any two points of the set can be joined by a polygonal path that lies entirely within the set. An open connected set is called an *open region.* A curve

$$C: \quad \mathbf{r} = \mathbf{r}(t), \quad t \in [a,\, b]$$

is said to be *closed* iff it begins and ends at the same point:

$$\mathbf{r}(a) = \mathbf{r}(b).$$

It is said to be *simple* iff it does not intersect itself:

$$a < t_1 < t_2 < b \qquad \text{implies} \qquad \mathbf{r}(t_1) \neq \mathbf{r}(t_2).$$

These notions are illustrated in Figure 16.10.1.

a simple,
closed curve

a simple curve
that is not closed

a closed curve
that is not simple

Figure 16.10.1

As is intuitively clear (Figure 16.10.2), a simple closed curve in the plane separates the plane into two disjoint open connected sets: a bounded inner region consisting of all points surrounded by the curve and an unbounded outer region consisting of all points not surrounded by the curve.†

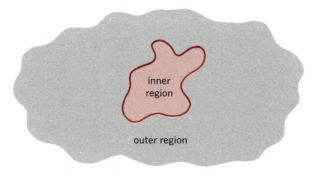

Figure 16.10.2

Finally, we come to the notion we need in our work with gradients:

(16.10.1)
> Let Ω be an open region of the plane, and let C be an arbitrary simple closed curve. Ω is said to be *simply connected* iff
>
> C is in Ω implies   the inner region of C is in Ω.

The first two regions in Figure 16.10.3 are simply connected. The annular region is not; the annular region contains the simple closed curve drawn there, but it does not contain all of the inner region of that curve.

---

† Add to this the assertion that the curve in question constitutes the total boundary of both regions and you have what is called the Jordan curve theorem, named after the French mathematician Camille Jordan (1838–1922). Jordan was the first to point out that, although all this is apparently obvious, it nevertheless requires proof. In recognition of Jordan, plane simple closed curves are now commonly known as *Jordan curves.*

open rectangle   open disc   open annular region

Figure 16.10.3

THEOREM 16.10.2

Let P and Q be functions of two variables, each continuously differentiable on a simply connected open region Ω. The linear combination $P(x, y)\mathbf{i} + Q(x, y)\mathbf{j}$ is a gradient on Ω iff

$$\frac{\partial P}{\partial y}(x, y) = \frac{\partial Q}{\partial x}(x, y) \qquad \text{for all } (x, y) \in \Omega.$$

Proof. A complete proof of this theorem is beyond the scope of this text. We will prove the result under the additional assumption that Ω has the form of an open rectangle with sides parallel to the coordinate axes.

Suppose that $P(x, y)\mathbf{i} + Q(x, y)\mathbf{j}$ is a gradient on this open rectangle Ω, say

$$\nabla f(x, y) = P(x, y)\mathbf{i} + Q(x, y)\mathbf{j}.$$

Since

$$\nabla f(x, y) = \frac{\partial f}{\partial x}(x, y)\,\mathbf{i} + \frac{\partial f}{\partial y}(x, y)\,\mathbf{j},$$

we have

$$P = \frac{\partial f}{\partial x} \qquad \text{and} \qquad Q = \frac{\partial f}{\partial y}.$$

Since P and Q have continuous first partials, f has continuous second partials. Thus, according to (15.6.5), the mixed partials are equal and we have

$$\frac{\partial P}{\partial y} = \frac{\partial^2 f}{\partial y \partial x} = \frac{\partial^2 f}{\partial x \partial y} = \frac{\partial Q}{\partial x}.$$

Conversely, suppose that

$$\frac{\partial P}{\partial y}(x, y) = \frac{\partial Q}{\partial x}(x, y) \qquad \text{for all } (x, y) \in \Omega.$$

We must show that $P(x, y)\mathbf{i} + Q(x, y)\mathbf{j}$ is a gradient on Ω. To do this, we choose a point (x_0, y_0) from Ω and form the function

$$f(x, y) = \int_{x_0}^{x} P(u, y_0)\, du + \int_{y_0}^{y} Q(x, v)\, dv, \qquad (x, y) \in \Omega.$$

[If you want to visualize f, you can refer to Figure 16.10.4. The function P is being integrated along the horizontal line segment joining (x_0, y_0) to (x, y_0), and Q is being integrated along the vertical line segment joining (x, y_0) to (x, y). Our assumptions on Ω guarantee that these line segments remain in Ω.]

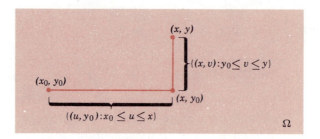

Figure 16.10.4

The first integral is independent of y. Hence

$$\frac{\partial f}{\partial y}(x, y) = \frac{\partial}{\partial y}\left(\int_{y_0}^{y} Q(x, v)\, dv \right) = Q(x, y).$$

The last equality holds because we are differentiating an integral with respect to its upper limit.† Differentiating f with respect to x we have

$$\frac{\partial f}{\partial x}(x, y) = \frac{\partial}{\partial x}\left(\int_{x_0}^{x} P(u, y_0)\, du \right) + \frac{\partial}{\partial x}\left(\int_{y_0}^{y} Q(x, v)\, dv \right).$$

The first term is $P(x, y_0)$ since once again we are differentiating with respect to the upper limit. In the second term the variable x appears under the sign of integration. It can be shown that, since Q and $\partial Q/\partial x$ are continuous,

$$\frac{\partial}{\partial x}\left(\int_{y_0}^{y} Q(x, v)\, dv \right) = \int_{y_0}^{y} \frac{\partial Q}{\partial x}(x, v)\, dv.\dagger\dagger$$

Anticipating this result, we have

$$\frac{\partial f}{\partial x}(x, y) = P(x, y_0) + \int_{y_0}^{y} \frac{\partial Q}{\partial x}(x, v)\, dv = P(x, y_0) + \int_{y_0}^{y} \frac{\partial P}{\partial y}(x, v)\, dv$$

$$\text{explain} \longrightarrow$$

$$= P(x, y_0) + P(x, y) - P(x, y_0) = P(x, y).$$

---

† This is a fundamental notion. If you are unsure about it, reread Theorem 5.3.5.

†† See Exercise 50, Section 17.4.

We have now shown that

$$P(x, y) = \frac{\partial f}{\partial x}(x, y) \quad \text{and} \quad Q(x, y) = \frac{\partial f}{\partial y}(x, y) \quad \text{for all } (x, y) \in \Omega.$$

It follows that $P(x, y)\mathbf{i} + Q(x, y)\mathbf{j}$ is the gradient of f on Ω. □

Example 4. The vector functions

$$\mathbf{F}(x, y) = 2x \sin y \, \mathbf{i} + x^2 \cos y \, \mathbf{j} \quad \text{and} \quad \mathbf{F}(x, y) = xy\mathbf{i} + \tfrac{1}{2}(x + 1)^2 y^2 \mathbf{j}$$

are both everywhere defined.   The first vector function is the gradient of a function everywhere defined:

$$\frac{\partial P}{\partial y}(x, y) = 2x \cos y = \frac{\partial Q}{\partial x}(x, y) \quad \text{for all } (x, y).$$

The second is not a gradient:

$$\frac{\partial P}{\partial y}(x, y) = x, \quad \frac{\partial Q}{\partial x}(x, y) = (x + 1)y^2$$

and so

$$\frac{\partial P}{\partial y}(x, y) \neq \frac{\partial Q}{\partial x}(x, y). □$$

Example 5. The vector function defined on *the punctured disc* $0 < x^2 + y^2 < 1$ by setting

$$\mathbf{F}(x, y) = \frac{y}{x^2 + y^2}\mathbf{i} - \frac{x}{x^2 + y^2}\mathbf{j}$$

satisfies

$$\frac{\partial P}{\partial y}(x, y) = \frac{\partial Q}{\partial x}(x, y) \quad \text{on that punctured disc.}$$

(Check this out.) Nevertheless, as you will see in Chapter 18, $\mathbf{F}$ is not a gradient on that set.   That set is not simply connected and therefore Theorem 16.10.2 does not apply. □

EXERCISES 16.10

Determine whether the vector function is the gradient $\nabla f(x, y)$ of a function everywhere defined. If so, find f.

1. $xy^2\mathbf{i} + x^2y\mathbf{j}$.

2. $x\mathbf{i} + y\mathbf{j}$.

3. $y\mathbf{i} + x\mathbf{j}$.

4. $(x^2 + y)\mathbf{i} + (y^3 + x)\mathbf{j}$.

5. $(y^3 + x)\mathbf{i} + (x^2 + y)\mathbf{j}$.

6. $(y^2e^x - y)\mathbf{i} + (2ye^x - x)\mathbf{j}$.

7. $(\cos x - y \sin x)\mathbf{i} + \cos x\,\mathbf{j}$.

8. $(1 + e^y)\mathbf{i} + (xe^y + y^2)\mathbf{j}$.

9. $e^x \cos y^2\,\mathbf{i} - 2ye^x \sin y^2\,\mathbf{j}$.

10. $e^x \cos y\,\mathbf{i} + e^x \sin y\,\mathbf{j}$.

11. $ye^x (1 + x)\mathbf{i} + (xe^x - e^{-y})\mathbf{j}$.

12. $(e^x + 2xy)\mathbf{i} + (x^2 + \sin y)\mathbf{j}$.

Find the most general function with the given gradient.

13. $\dfrac{x}{\sqrt{x^2 + y^2}}\,\mathbf{i} + \dfrac{y}{\sqrt{x^2 + y^2}}\,\mathbf{j}$.

14. $(x \tan y + \sec^2 x)\mathbf{i} + (\tfrac{1}{2}x^2 \sec^2 y + \pi y)\mathbf{j}$.

15. $(x^2 \sin^{-1} y)\mathbf{i} + \left(\dfrac{x^3}{3\sqrt{1 - y^2}} - \ln y\right)\mathbf{j}$.

16. $\left(\dfrac{\tan^{-1} y}{\sqrt{1 - x^2}} + \dfrac{x}{y}\right)\mathbf{i} + \left(\dfrac{\sin^{-1} x}{1 + y^2} - \dfrac{x^2}{2y^2} + 1\right)\mathbf{j}$.

17. Find the general solution of the differential equation $\nabla f(x, y) = f(x, y)\mathbf{i} + f(x, y)\mathbf{j}$.

18. Given that g and its first and second partials are everywhere continuous, find the general solution of the differential equation $\nabla f(x, y) = e^{g(x, y)}[g_x(x, y)\mathbf{i} + g_y(x, y)\mathbf{j}]$.

Theorem 16.10.2 has a three-dimensional analog. In particular we can show that, if P, Q, R are continuously differentiable on an open rectangular box S, then the vector function

$$P(x, y, z)\mathbf{i} + Q(x, y, z)\mathbf{j} + R(x, y, z)\mathbf{k}$$

is a gradient on S iff

$$\frac{\partial P}{\partial y} = \frac{\partial Q}{\partial x}, \qquad \frac{\partial P}{\partial z} = \frac{\partial R}{\partial x}, \qquad \frac{\partial Q}{\partial z} = \frac{\partial R}{\partial y} \qquad \text{throughout } S.$$

19. (a) Verify that $2x\mathbf{i} + z\mathbf{j} + y\mathbf{k}$ is the gradient of a function f that is everywhere defined.

(b) Deduce from the relation $\partial f/\partial x = 2x$ that $f(x, y, z) = x^2 + g(y, z)$.   (c) Verify then that $\partial f/\partial y = z$ gives $g(y, z) = zy + h(z)$ and finally that $\partial f/\partial z = y$ gives $h(z) = C$. (d) What is $f(x, y, z)$?

Determine whether the function is a gradient $\nabla f(x, y, z)$ and, if so, find $f(x, y, z)$.

20. $yz\mathbf{i} + xz\mathbf{j} + xy\mathbf{k}$.

21. $(2x + y)\mathbf{i} + (2y + x + z)\mathbf{j} + (y - 2z)\mathbf{k}$.

22. $(2x \sin 2y \cos z)\mathbf{i} + (2x^2 \cos 2y \cos z)\mathbf{j} - (x^2 \sin 2y \sin z)\mathbf{k}$.

23. Verify that the gravitational force function

$$\mathbf{F}(\mathbf{r}) = -G\,\frac{mM}{r^3}\,\mathbf{r}$$

is a gradient.

24. Verify that every vector function of the form

$$\mathbf{h}(\mathbf{r}) = kr^n\mathbf{r} \qquad\qquad (k \text{ constant, } n \text{ an integer})$$

is a gradient.

CHAPTER HIGHLIGHTS

16.1 Differentiability and Gradient

gradient of f at $\mathbf{x}$: $\nabla f(\mathbf{x})$ (p. 858)

$$\nabla f(x, y) = \frac{\partial f}{\partial x}\mathbf{i} + \frac{\partial f}{\partial y}\mathbf{j} \qquad \nabla f(x, y, z) = \frac{\partial f}{\partial x}\mathbf{i} + \frac{\partial f}{\partial y}\mathbf{j} + \frac{\partial f}{\partial z}\mathbf{k}$$

If f is differentiable at $\mathbf{x}$, then f is continuous at $\mathbf{x}$.

$$\nabla r^n = nr^{n-2}\mathbf{r}$$

16.2 Gradients and Directional Derivatives

directional derivative: $f'_{\mathbf{u}}(\mathbf{x}) = \nabla f(\mathbf{x}) \cdot \mathbf{u}$ (p. 866)

The directional derivative $f'_{\mathbf{u}}$ gives the rate of change of f in the direction of the unit vector $\mathbf{u}$.

The directional derivative in a direction $\mathbf{u}$ is the component of the gradient vector in that direction. (p. 869)

A differentiable function f increases most rapidly in the direction of the gradient (the rate of change is then $\|\nabla f(\mathbf{x})\|$) and it decreases most rapidly in the opposite direction (the rate of change is then $-\|\nabla f(\mathbf{x})\|$).

16.3 The Mean-Value Theorem; Two Intermediate-Value Theorems

The Mean-Value Theorem (p. 875)   open, connected set (p. 876)
intermediate-value theorems (pp. 877, 878)   open, closed regions (p. 877)

16.4 Chain Rules

chain rule along a curve (p. 879)   tree diagram (p. 884)

In the setting of functions of several variables, there are numerous chain rules. They can all be deduced from the chain rule along a curve. If, for example, $u = u(x, y)$ where $x = x(s, t)$ and $y = y(s, t)$, then

$$\frac{\partial u}{\partial s} = \frac{\partial u}{\partial x}\frac{\partial x}{\partial s} + \frac{\partial u}{\partial y}\frac{\partial y}{\partial s} \qquad \text{and} \qquad \frac{\partial u}{\partial t} = \frac{\partial u}{\partial x}\frac{\partial x}{\partial t} + \frac{\partial u}{\partial y}\frac{\partial y}{\partial t}.$$

If f is a continuously differentiable function of $r = \|\mathbf{r}\|$, then

$$\nabla[f(r)] = f'(r)\frac{\mathbf{r}}{r}.$$

16.5 The Gradient as a Normal; Tangent Lines and Tangent Planes

tangent and normal lines to a curve $f(x, y) = c$ (p. 891)
tangent plane to a surface $f(x, y, z) = c$ (p. 894)
upper and lower unit normals (p. 898)

At each point of the domain of a function, the gradient vector, if not $\mathbf{0}$, is perpendicular to the level curve (level surface) that passes through that point.

16.6 Maximum and Minimum Values

If f has a local extreme value at $\mathbf{x}_0$, then either $\nabla f(\mathbf{x}_0) = \mathbf{0}$ or $\nabla f(\mathbf{x}_0)$ does not exist.

16.7 Second-Partials Test

16.8 Maxima and Minima with Side Conditions

If $\mathbf{x}_0$ maximizes (or minimizes) $f(\mathbf{x})$ subject to the side condition $g(\mathbf{x}) = 0$, then $\nabla f(\mathbf{x}_0)$ and $\nabla g(\mathbf{x}_0)$ are parallel. Thus, if $\nabla g(\mathbf{x}_0) \neq \mathbf{0}$, then there exists a scalar λ, called a Lagrange multiplier, such that $\nabla f(\mathbf{x}_0) = \lambda \, \nabla g(\mathbf{x}_0)$.

16.9 Differentials

increment: $\Delta f = f(\mathbf{x} + \mathbf{h}) - f(\mathbf{x})$ differential: $df = \nabla f(\mathbf{x}) \cdot \mathbf{h}$

$\Delta f \cong df$ in the sense that $\dfrac{\Delta f - df}{\|\mathbf{h}\|} \to 0$ as $\mathbf{h} \to \mathbf{0}$

two variables: $df = \dfrac{\partial f}{\partial x} \Delta x + \dfrac{\partial f}{\partial y} \Delta y$

three variables: $df = \dfrac{\partial f}{\partial x} \Delta x + \dfrac{\partial f}{\partial y} \Delta y + \dfrac{\partial f}{\partial z} \Delta z$

16.10 Reconstructing a Function from its Gradient

necessary and sufficient conditions for a vector-valued function to be a gradient:

DOUBLE
AND
TRIPLE INTEGRALS

We began with ordinary integrals

$$\int_a^b f(x)\, dx.$$

Here we will study double integrals

$$\iint_\Omega f(x, y)\, dxdy$$

and a little later triple integrals

$$\iiint_T f(x, y, z)\, dxdydz.$$

Our first step is to introduce some new notation.

17.1 MULTIPLE-SIGMA NOTATION

In an ordinary sequence each term a_i is indexed by a single integer i. The sum of all the a_i from $i = 1$ to $i = m$ is then denoted by

$$\sum_{i=1}^m a_i.$$

When two indices are involved, say

$$a_{ij} = 2^i 5^j, \qquad a_{ij} = i/j, \qquad \text{or} \qquad a_{ij} = (1 + i)^j,$$

then we use double-sigma notation.   By

(17.1.1)

$$\sum_{i=1}^{m} \sum_{j=1}^{n} a_{ij}$$

we mean *the sum of all the a_{ij} where i ranges from 1 to m and j ranges from 1 to n.*

Since addition is associative and commutative, we can add the terms of (17.1.1) in any order we choose.   Usually we set

(17.1.2)

$$\sum_{i=1}^{m} \sum_{j=1}^{n} a_{ij} = \sum_{i=1}^{m} \left(\sum_{j=1}^{n} a_{ij} \right).$$

We can expand the expression on the right by expanding first with respect to i and then with respect to j:

$$\sum_{i=1}^{m} \left(\sum_{j=1}^{n} a_{ij} \right) = \sum_{j=1}^{n} a_{1j} + \sum_{j=1}^{n} a_{2j} + \cdots + \sum_{j=1}^{n} a_{mj}$$
$$= (a_{11} + a_{12} + \cdots + a_{1n}) + (a_{21} + a_{22} + \cdots + a_{2n})$$
$$+ \cdots + (a_{m1} + a_{m2} + \cdots + a_{mn})$$

or we can expand first with respect to j and then with respect to i:

$$\sum_{i=1}^{m} \left(\sum_{j=1}^{n} a_{ij} \right) = \sum_{i=1}^{m} (a_{i1} + a_{i2} + \cdots + a_{in})$$
$$= (a_{11} + a_{12} + \cdots + a_{1n}) + (a_{21} + a_{22} + \cdots + a_{2n})$$
$$+ \cdots + (a_{m1} + a_{m2} + \cdots + a_{mn}).$$

The results are the same.

For example, we can write

$$\sum_{i=1}^{3} \left(\sum_{j=1}^{2} a_{ij} \right) = \sum_{j=1}^{2} a_{1j} + \sum_{j=1}^{2} a_{2j} + \sum_{j=1}^{2} a_{3j} = (a_{11} + a_{12}) + (a_{21} + a_{22}) + (a_{31} + a_{32}),$$

or we can write

$$\sum_{i=1}^{3} \left(\sum_{j=1}^{2} a_{ij} \right) = \sum_{i=1}^{3} (a_{i1} + a_{i2}) = (a_{11} + a_{12}) + (a_{21} + a_{22}) + (a_{31} + a_{32}).$$

Since constants can be factored through single sums, they can also be factored through double sums; namely,

(17.1.3)

$$\sum_{i=1}^{m} \sum_{j=1}^{n} \alpha a_{ij} = \alpha \sum_{i=1}^{m} \sum_{j=1}^{n} a_{ij}.$$

Moreover

(17.1.4)
$$\sum_{i=1}^{m} \sum_{j=1}^{n} (a_{ij} + b_{ij}) = \sum_{i=1}^{m} \sum_{j=1}^{n} a_{ij} + \sum_{i=1}^{m} \sum_{j=1}^{n} b_{ij}.$$

The easiest double sums to handle are those where each term a_{ij} appears as a product $b_i c_j$ in which each of the factors bears only one index. In that case, we can express the double sum as the product of two single sums:

(17.1.5)
$$\sum_{i=1}^{m} \sum_{j=1}^{n} b_i c_j = \left(\sum_{i=1}^{m} b_i\right)\left(\sum_{j=1}^{n} c_j\right).$$

Proof. Set

$$B = \sum_{i=1}^{m} b_i, \qquad C = \sum_{j=1}^{n} c_j.$$

Then

$$\sum_{i=1}^{m} \sum_{j=1}^{n} b_i c_j = \sum_{i=1}^{m} \left(\sum_{j=1}^{n} b_i c_j\right) \overset{\dagger}{=} \sum_{i=1}^{m} b_i \left(\sum_{j=1}^{n} c_j\right) = \sum_{i=1}^{m} b_i C = C \sum_{i=1}^{m} b_i = CB = BC. \quad \square$$

For example,

$$\sum_{i=1}^{4} \sum_{j=1}^{2} 2^i 5^j = \left(\sum_{i=1}^{4} 2^i\right)\left(\sum_{j=1}^{2} 5^j\right) = (2 + 2^2 + 2^3 + 2^4)(5 + 5^2) = (30)(30) = 900.$$

Triple-sigma notation is used when three indices are involved. The sum of all the a_{ijk} where i ranges from 1 to m, j from 1 to n, and k from 1 to q can be written

(17.1.6)
$$\sum_{i=1}^{m} \sum_{j=1}^{n} \sum_{k=1}^{q} a_{ijk}.$$

Multiple sums will appear in the following sections. We introduced them here so as to avoid lengthy asides later.

EXERCISES 17.1

Let

$$P_1 = \{x_0, x_1, \ldots, x_m\} \text{ be a partition of } [a_1, a_2],$$
$$P_2 = \{y_0, y_1, \ldots, y_n\} \text{ be a partition of } [b_1, b_2],$$
$$P_3 = \{z_0, z_1, \ldots, z_q\} \text{ be a partition of } [c_1, c_2].$$

---

† Since b_i is independent of j, it can be factored through the j-summation.

As usual take

$$\Delta x_i = x_i - x_{i-1}, \qquad \Delta y_j = y_j - y_{j-1}, \qquad \Delta z_k = z_k - z_{k-1}.$$

Now evaluate the following sums.

1. $\displaystyle\sum_{i=1}^{m} \Delta x_i.$

2. $\displaystyle\sum_{j=1}^{n} \Delta y_j.$

3. $\displaystyle\sum_{i=1}^{m} \sum_{j=1}^{n} \Delta x_i \, \Delta y_j.$

4. $\displaystyle\sum_{j=1}^{n} \sum_{k=1}^{q} \Delta y_j \, \Delta z_k.$

5. $\displaystyle\sum_{i=1}^{m} (x_i + x_{i-1}) \Delta x_i.$

6. $\displaystyle\sum_{j=1}^{n} \tfrac{1}{2}(y_j^2 + y_j y_{j-1} + y_{j-1}^2) \Delta y_j.$

7. $\displaystyle\sum_{i=1}^{m} \sum_{j=1}^{n} (x_i + x_{i-1}) \Delta x_i \, \Delta y_j.$

8. $\displaystyle\sum_{i=1}^{m} \sum_{j=1}^{n} (y_j + y_{j-1}) \Delta x_i \, \Delta y_j.$

9. $\displaystyle\sum_{i=1}^{m} \sum_{j=1}^{n} (2\Delta x_i - 3\Delta y_j).$

10. $\displaystyle\sum_{i=1}^{m} \sum_{j=1}^{n} (3\Delta x_i - 2\Delta y_j).$

11. $\displaystyle\sum_{i=1}^{m} \sum_{j=1}^{n} \sum_{k=1}^{q} \Delta x_i \, \Delta y_j \, \Delta z_k.$

12. $\displaystyle\sum_{i=1}^{m} \sum_{j=1}^{n} \sum_{k=1}^{q} (x_i + x_{i-1}) \, \Delta x_i \, \Delta y_j \, \Delta z_k.$

13. Evaluate

$$\sum_{i=1}^{n} \sum_{j=1}^{n} \sum_{k=1}^{n} \delta_{ijk} a_{ijk} \qquad \text{where} \qquad \delta_{ijk} = \begin{cases} 1, & \text{if } i = j = k \\ 0, & \text{otherwise} \end{cases}.$$

14. Show that every sum written in multiple-sigma notation can be expressed in ordinary sigma notation.

17.2   THE DOUBLE INTEGRAL OVER A RECTANGLE

We start with a function f continuous on a rectangle

$$R: \quad a \le x \le b, \quad c \le y \le d.$$

See Figure 17.2.1.

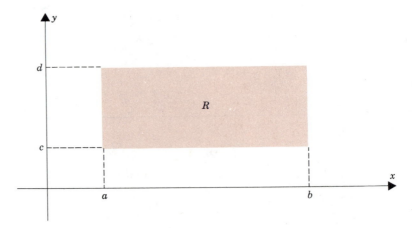

Figure 17.2.1

Our object is to define the double integral

$$\iint\limits_{R} f(x, y)\ dxdy.$$

To define the integral

$$\int_{a}^{b} f(x)\ dx$$

we first introduced some auxiliary notions: partition P, upper sum $U_f(P)$, and lower sum $L_f(P)$. We were then able to define

$$\int_{a}^{b} f(x)\ dx$$

as the unique number I that satisfies the inequality

$$L_f(P) \le I \le U_f(P) \qquad \text{for all partitions } P \text{ of } [a, b].$$

We will follow exactly the same procedure to define the double integral

$$\iint\limits_{R} f(x, y)\ dxdy.$$

First we explain what we mean by a partition of the rectangle R. To do this, we begin with a partition

$$P_1 = \{x_0, x_1, \ldots, x_m\} \quad \text{of } [a, b]$$

and a partition

$$P_2 = \{y_0, y_1, \ldots, y_n\} \quad \text{of } [c, d].$$

The set

$$P = P_1 \times P_2 = \{(x_i, y_j): x_i \in P_1, y_j \in P_2\}\dagger$$

is called a *partition of R*. (See Figure 17.2.2.) P consists of all the grid points (x_i, y_j).

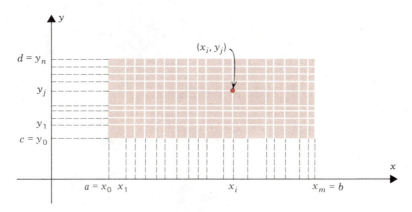

Figure 17.2.2

$\dagger$ $P_1 \times P_2$ is a Cartesian product. Cartesian products are discussed briefly in Appendix A.1.

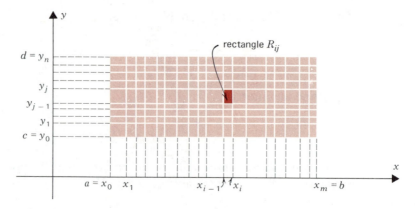

Figure 17.2.3

The partition P breaks up R into $m \times n$ nonoverlapping rectangles

$$R_{ij}: \quad x_{i-1} \leq x \leq x_i, \quad y_{j-1} \leq y \leq y_j. \qquad \text{(Figure 17.2.3)}$$

On each rectangle R_{ij}, the function f takes on a maximum value M_{ij} and a minimum value m_{ij}.† The sum of all the products

$$M_{ij}(\text{area of } R_{ij}) = M_{ij}(x_i - x_{i-1})(y_j - y_{j-1}) = M_{ij}\,\Delta x_i\,\Delta y_j$$

is called the *P upper sum* for f:

(17.2.1)
$$U_f(P) = \sum_{i=1}^{m}\sum_{j=1}^{n} M_{ij}(\text{area of } R_{ij}) = \sum_{i=1}^{m}\sum_{j=1}^{n} M_{ij}\,\Delta x_i\,\Delta y_j.$$

The sum of all the products

$$m_{ij}(\text{area of } R_{ij}) = m_{ij}(x_i - x_{i-1})(y_j - y_{j-1}) = m_{ij}\,\Delta x_i\,\Delta y_j$$

is called the *P lower sum* for f:

(17.2.2)
$$L_f(P) = \sum_{i=1}^{m}\sum_{j=1}^{n} m_{ij}(\text{area of } R_{ij}) = \sum_{i=1}^{m}\sum_{j=1}^{n} m_{ij}\,\Delta x_i\,\Delta y_j.$$

Example 1. Consider the function

$$f(x, y) = x + y - 2$$

on the rectangle

$$R: \quad 1 \leq x \leq 4, \quad 1 \leq y \leq 3.$$

As a partition of $[1, 4]$ take

$$P_1 = \{1, 2, 3, 4\},$$

---

† We know this because f is continuous and R_{ij} is closed and bounded (Section 16.6).

and as a partition of $[1, 3]$ take

$$P_2 = \{1, \tfrac{3}{2}, 3\}.$$

The partition $P = P_1 \times P_2$ then breaks up the initial rectangle into the six rectangles marked in Figure 17.2.4.

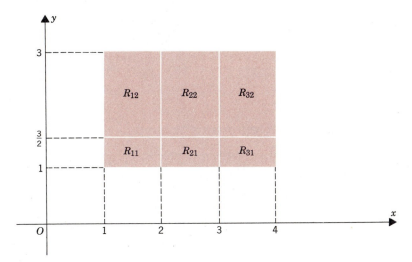

Figure 17.2.4

On each rectangle R_{ij}, the function f takes on its maximum value M_{ij} at the point (x_i, y_j), the corner farthest from the origin:

$$M_{ij} = f(x_i, y_j) = x_i + y_j - 2.$$

Thus

$$U_f(P) = M_{11}(\text{area of } R_{11}) + M_{12}(\text{area of } R_{12}) + M_{21}(\text{area of } R_{21})$$
$$+ M_{22}(\text{area of } R_{22}) + M_{31}(\text{area of } R_{31}) + M_{32}(\text{area of } R_{32})$$
$$= \tfrac{3}{2}(\tfrac{1}{2}) + 3(\tfrac{3}{2}) + \tfrac{5}{2}(\tfrac{1}{2}) + 4(\tfrac{3}{2}) + \tfrac{7}{2}(\tfrac{1}{2}) + 5(\tfrac{3}{2}) = \tfrac{87}{4}.$$

On each rectangle R_{ij}, f takes on its minimum value m_{ij} at the point (x_{i-1}, y_{j-1}), the corner closest to the origin:

$$m_{ij} = f(x_{i-1}, y_{j-1}) = x_{i-1} + y_{j-1} - 2.$$

Thus

$$L_f(P) = m_{11}(\text{area of } R_{11}) + m_{12}(\text{area of } R_{12}) + m_{21}(\text{area of } R_{21})$$
$$+ m_{22}(\text{area of } R_{22}) + m_{31}(\text{area of } R_{31}) + m_{32}(\text{area of } R_{32})$$
$$= 0(\tfrac{1}{2}) + \tfrac{1}{2}(\tfrac{3}{2}) + 1(\tfrac{1}{2}) + \tfrac{3}{2}(\tfrac{3}{2}) + 2(\tfrac{1}{2}) + \tfrac{5}{2}(\tfrac{3}{2}) = \tfrac{33}{4}. \quad \square$$

We return now to the general situation. As in the one-variable case, it can be shown that, if f is continuous, then there exists one and only one number I that satisfies the inequality

$$L_f(P) \leq I \leq U_f(P) \qquad \text{for all partitions } P \text{ of } R.$$

DEFINITION 17.2.3 THE DOUBLE INTEGRAL OVER A RECTANGLE R

The unique number I that satisfies the inequality

$$L_f(P) \le I \le U_f(P) \qquad \text{for all partitions } P \text{ of } R$$

is called the *double integral* of f over R and is denoted by

$$\iint_R f(x, y)\, dxdy.$$

The Double Integral as a Volume

If f is nonnegative on the rectangle R, the equation

$$z = f(x, y)$$

represents a surface that lies above R. (See Figure 17.2.5.) In this case the double integral

$$\iint_R f(x, y)\, dxdy$$

gives the volume of the solid that is bounded below by R and bounded above by the surface $z = f(x, y)$.

Figure 17.2.5

To see this, consider a partition P of R. P breaks up R into subrectangles R_{ij} and thus the entire solid T into parts T_{ij}. Since T_{ij} contains a rectangular solid with base R_{ij} and height m_{ij} (Figure 17.2.6), we must have

$$m_{ij}(\text{area of } R_{ij}) \le \text{volume of } T_{ij}.$$

Since T_{ij} is contained in a rectangular solid with base R_{ij} and height M_{ij} (Figure 17.2.7), we must have

$$\text{volume of } T_{ij} \le M_{ij}(\text{area of } R_{ij}).$$

Figure 17.2.6

Figure 17.2.7

In short, for each pair of indices i and j, we must have

$$m_{ij}(\text{area of } R_{ij}) \leq \text{volume of } T_{ij} \leq M_{ij}(\text{area of } R_{ij}).$$

Adding up these inequalities, we are forced to conclude that

$$L_f(P) \leq \text{volume of } T \leq U_f(P).$$

Since P is arbitrary, the volume of T must be the double integral:

(17.2.4)

$$\text{volume of } T = \iint_R f(x, y)\, dxdy.$$

The double integral

$$\iint_R 1\, dxdy = \iint_R dxdy$$

gives the volume of a solid of constant height 1 erected over R. In square units this is just the area of R:

(17.2.5)

$$\text{area of } R = \iint_R dxdy.$$

Some Computations

Double integrals are generally computed by techniques that we will take up later. It is possible, however, to evaluate simple double integrals directly from the definition.

Problem 2. Evaluate

$$\iint_R \alpha\, dxdy$$

where R is the rectangle

$$a \leq x \leq b, \quad c \leq y \leq d.$$

Solution. Here

$$f(x, y) = \alpha \qquad \text{for all } (x, y) \in R.$$

We begin with $P_1 = \{x_0, x_1, \ldots, x_m\}$ as an arbitrary partition of $[a, b]$ and $P_2 = \{y_0, y_1, \ldots, y_n\}$ as an arbitrary partition of $[c, d]$. This gives

$$P = P_1 \times P_2 = \{(x_i, y_j): x_i \in P_1, y_j \in P_2\}$$

as an arbitrary partition of R. On each rectangle R_{ij}, f has constant value α.

This gives $M_{ij} = \alpha$ and $m_{ij} = \alpha$ throughout. Thus

$$U_f(P) = \sum_{i=1}^{m} \sum_{j=1}^{n} \alpha \, \Delta x_i \, \Delta y_j = \alpha \left(\sum_{i=1}^{m} \Delta x_i \right) \left(\sum_{j=1}^{n} \Delta y_j \right) = \alpha(b - a)(d - c).$$

Similarly

$$L_f(P) = \alpha(b - a)(d - c).$$

Since

$$L_f(P) \leq \alpha(b - a)(d - c) \leq U_f(P)$$

and P was chosen arbitrarily, the inequality must hold for all partitions P of R. This means that

$$\iint_R f(x, y) \, dxdy = \alpha(b - a)(d - c). \quad \square$$

Remark. If $\alpha > 0$,

$$\iint_R \alpha \, dxdy = \alpha(b - a)(d - c)$$

gives the volume of the rectangular solid of constant height α erected over the rectangle R. $\quad \square$

Problem 3. Evaluate

$$\iint_R (y - 2x) \, dxdy$$

where

$$R: \quad 1 \leq x \leq 2, \quad 3 \leq y \leq 5.$$

Solution. With

$$P_1 = \{x_0, x_1, \ldots, x_m\} \text{ as an arbitrary partition of } [1, 2]$$

and

$$P_2 = \{y_0, y_1, \ldots, y_n\} \text{ as an arbitrary partition of } [3, 5]$$

we have

$$P = P_1 \times P_2 = \{(x_i, y_j): x_i \in P_1, y_j \in P_2\}$$

as an arbitrary partition of R. On each rectangle

$$R_{ij}: \quad x_{i-1} \leq x \leq x_i, \quad y_{j-1} \leq y \leq y_j$$

the function

$$f(x, y) = y - 2x$$

has a maximum

$$M_{ij} = y_j - 2x_{i-1} \qquad (y \text{ is maximized and } x \text{ is minimized})$$

and a minimum

$$m_{ij} = y_{j-1} - 2x_i. \qquad (y \text{ is minimized and } x \text{ is maximized})$$

Thus

$$U_f(P) = \sum_{i=1}^{m} \sum_{j=1}^{n} (y_j - 2x_{i-1}) \, \Delta x_i \, \Delta y_j \quad \text{and} \quad L_f(P) = \sum_{i=1}^{m} \sum_{j=1}^{n} (y_{j-1} - 2x_i) \, \Delta x_i \, \Delta y_j.$$

For each pair of indices i and j,

$$y_{j-1} - 2x_i \leq \tfrac{1}{2}(y_j + y_{j-1}) - (x_i + x_{i-1}) \leq y_j - 2x_{i-1}. \qquad \text{(explain)}$$

This means that for arbitrary P we have

$$L_f(P) \leq \sum_{i=1}^{m} \sum_{j=1}^{n} [\tfrac{1}{2}(y_j + y_{j-1}) - (x_i + x_{i-1})] \, \Delta x_i \, \Delta y_j \leq U_f(P).$$

The middle term can be written

$$\sum_{i=1}^{m} \sum_{j=1}^{n} \tfrac{1}{2}(y_j + y_{j-1}) \, \Delta x_i \, \Delta y_j + \sum_{i=1}^{m} \sum_{j=1}^{n} - (x_i + x_{i-1}) \, \Delta x_i \, \Delta y_j.$$

The first double sum reduces to

$$\sum_{i=1}^{m} \sum_{j=1}^{n} \tfrac{1}{2} \Delta x_i \, (y_j^2 - y_{j-1}^2) = \tfrac{1}{2} \left(\sum_{i=1}^{m} \Delta x_i \right) \left(\sum_{j=1}^{n} (y_j^2 - y_{j-1}^2) \right) = \tfrac{1}{2}(2 - 1)(25 - 9) = 8.$$

The second double sum reduces to

$$-\sum_{i=1}^{m} \sum_{j=1}^{n} (x_i^2 - x_{i-1}^2) \, \Delta y_j = -\left(\sum_{i=1}^{m} (x_i^2 - x_{i-1}^2) \right) \left(\sum_{j=1}^{n} \Delta y_j \right) = -(4 - 1)(5 - 3) = -6.$$

The sum of these two numbers

$$I = 2$$

satisfies the inequality

$$L_f(P) \leq 2 \leq U_f(P) \qquad \text{for arbitrary } P.$$

The integral is therefore 2:

$$\iint_R (y - 2x) \, dx\,dy = 2. \quad \square$$

Remark. This last integral should not be interpreted as a volume. The expression $y - 2x$ does not keep a constant sign on

$$R: \quad 1 \leq x \leq 2, \quad 3 \leq y \leq 5. \quad \square$$

EXERCISES 17.2

For Exercises 1–3 take

$$f(x, y) = x + 2y \qquad \text{on} \qquad R: \quad 0 \le x \le 2, \quad 0 \le y \le 1$$

and P as the partition $P = P_1 \times P_2$.

1. Find $L_f(P)$ and $U_f(P)$ if $P_1 = \{0, 1, \frac{3}{2}, 2\}$ and $P_2 = \{0, \frac{1}{2}, 1\}$.

2. Find $L_f(P)$ and $U_f(P)$ if $P_1 = \{0, \frac{1}{2}, 1, \frac{3}{2}, 2\}$ and $P_2 = \{0, \frac{1}{4}, \frac{1}{2}, \frac{3}{4}, 1\}$.

3. (a) Find $L_f(P)$ and $U_f(P)$ if

$$P_1 = \{x_0, x_1, \ldots, x_m\} \text{ is an arbitrary partition of } [0, 2] \quad \text{and}$$
$$P_2 = \{y_0, y_1, \ldots, y_n\} \text{ is an arbitrary partition of } [0, 1].$$

   (b) Use (a) to evaluate the double integral

$$\iint_R (x + 2y) \, dx dy$$

and give a geometric interpretation to your answer.

For Exercises 4–6 take

$$f(x, y) = x - y \qquad \text{on} \qquad R: \quad 0 \le x \le 1, \quad 0 \le y \le 1$$

and P as the partition $P = P_1 \times P_2$.

4. Find $L_f(P)$ and $U_f(P)$ if $P_1 = \{0, \frac{1}{2}, \frac{3}{4}, 1\}$ and $P_2 = \{0, \frac{1}{2}, 1\}$.

5. Find $L_f(P)$ and $U_f(P)$ if $P_1 = \{0, \frac{1}{4}, \frac{1}{2}, \frac{3}{4}, 1\}$ and $P_2 = \{0, \frac{1}{3}, \frac{2}{3}, 1\}$.

6. (a) Find $L_f(P)$ and $U_f(P)$ if

$$P_1 = \{x_0, x_1, \ldots, x_m\} \text{ and } P_2 = \{y_0, y_1, \ldots, y_n\} \text{ are arbitrary partitions of } [0, 1].$$

   (b) Use (a) to evaluate the double integral

$$\iint_R (x - y) \, dx dy.$$

For Exercises 7–9 take $R: 0 \le x \le b, 0 \le y \le d$ and let $P = P_1 \times P_2$ where

$$P_1 = \{x_0, x_1, \ldots, x_m\} \text{ is an arbitrary partition of } [0, b],$$
$$P_2 = \{y_0, y_1, \ldots, y_n\} \text{ is an arbitrary partition of } [0, d].$$

7. (a) Find $L_f(P)$ and $U_f(P)$ for $f(x, y) = 4xy$.

   (b) Calculate

$$\iint_R 4xy \, dx dy.$$

HINT: $4x_{i-1}y_{j-1} \le (x_i + x_{i-1})(y_j + y_{j-1}) \le 4x_i y_j$.

8. (a) Find $L_f(P)$ and $U_f(P)$ for $f(x, y) = 3(x^2 + y^2)$.

   (b) Calculate

$$\iint_R 3(x^2 + y^2) \, dx dy.$$

HINT: If $0 \le s \le t$, then $3s^2 \le t^2 + ts + s^2 \le 3t^2$.

9. (a) Find $L_f(P)$ and $U_f(P)$ for $f(x, y) = 3(x^2 - y^2)$.
(b) Calculate

$$\iint\limits_{R} 3(x^2 - y^2) \, dx\,dy.$$

17.3   THE DOUBLE INTEGRAL OVER A REGION

We start with a closed bounded region Ω in the xy-plane that we depict in Figure 17.3.1. We assume that Ω is a *basic region*; that is, we assume that the boundary of Ω consists of a finite number of continuous arcs $y = \phi(x)$, $x = \psi(y)$. Now let's suppose that f is some function continuous on Ω. We want to define the double integral

$$\iint\limits_{\Omega} f(x, y) \, dx\,dy.$$

Figure 17.3.1

Figure 17.3.2

To do this, we surround Ω by a rectangle R as in Figure 17.3.2. We now extend f to all of R by setting f equal to 0 outside of Ω. This extended function f is bounded on R, and it is continuous on all of R except possibly at the boundary of Ω. In spite of these possible discontinuities, f is still integrable on R; that is, there still exists a unique number I such that

$$L_f(P) \le I \le U_f(P) \qquad \text{for all partitions } P \text{ of } R.$$

(We will not attempt to prove this.) This number I is by definition the double integral

$$\iint\limits_{R} f(x, y) \, dx\,dy.$$

As you probably guessed by now, we define the double integral over Ω by setting

(17.3.1)
$$\iint\limits_{\Omega} f(x, y) \, dx\,dy = \iint\limits_{R} f(x, y) \, dx\,dy.$$

If f is nonnegative over Ω, the extended f is nonnegative on all of R. (Figure 17.3.3) The double integral gives the volume of the solid trapped between the surface $z = f(x, y)$ and the rectangle R. But since the surface has height 0 outside of Ω, the volume outside of Ω is 0. It follows then that

$$\iint\limits_{\Omega} f(x, y) \, dxdy$$

gives the volume of the solid T bounded above by $z = f(x, y)$ and below by Ω:

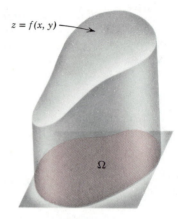

Figure 17.3.3

(17.3.2)
$$\text{volume of } T = \iint\limits_{\Omega} f(x, y) \, dxdy.$$

The double integral

$$\iint\limits_{\Omega} 1 \, dxdy = \iint\limits_{\Omega} dxdy$$

gives the volume of a solid of constant height 1 over Ω. In square units this is the area of Ω:

(17.3.3)
$$\text{area of } \Omega = \iint\limits_{\Omega} dxdy.$$

Below we list four elementary properties of the double integral. They are all analogous to what you saw in the one-variable case. The Ω referred to is a basic region. The functions f and g are assumed to be continuous on Ω.

I. The double integral is linear:

$$\iint\limits_{\Omega} [\alpha f(x, y) + \beta g(x, y)] \, dxdy = \alpha \iint\limits_{\Omega} f(x, y) \, dxdy + \beta \iint\limits_{\Omega} g(x, y) \, dxdy.$$

II. The double integral preserves order:

$$\text{if } f \geq 0 \text{ on } \Omega, \text{ then } \iint\limits_{\Omega} f(x, y) \, dxdy \geq 0;$$

$$\text{if } f \leq g \text{ on } \Omega, \text{ then } \iint\limits_{\Omega} f(x, y) \, dxdy \leq \iint\limits_{\Omega} g(x, y) \, dxdy.$$

III. The double integral is additive:

if Ω is broken up into a finite number of basic regions $\Omega_1, \ldots, \Omega_n$, then

$$\iint\limits_{\Omega} f(x, y)\, dxdy = \iint\limits_{\Omega_1} f(x, y)\, dxdy + \cdots + \iint\limits_{\Omega_n} f(x, y)\, dxdy.$$

IV. The double integral satisfies a mean-value condition; namely, there is a point (x_0, y_0) in Ω for which

$$\iint\limits_{\Omega} f(x, y)\, dxdy = f(x_0, y_0)(\text{the area of } \Omega).$$

We call $f(x_0, y_0)$ *the average value of f on* Ω.

The notion of average given in (IV) enables us to write

(17.3.4) $$\iint\limits_{\Omega} f(x, y)\, dxdy = \begin{pmatrix} \text{the average value} \\ \text{of } f \text{ on } \Omega \end{pmatrix} (\text{the area of } \Omega).$$

This is a powerful intuitive way of viewing the double integral. We will capitalize on it as we go on.

> **THEOREM 17.3.5 MEAN-VALUE THEOREM FOR DOUBLE INTEGRALS**
>
> Let f and g be functions continuous on a basic region Ω. If g is nonnegative on Ω, then there exists a point (x_0, y_0) in Ω for which
>
> $$\iint\limits_{\Omega} f(x, y)g(x, y)\, dxdy = f(x_0, y_0) \iint\limits_{\Omega} g(x, y)\, dxdy.\dagger$$
>
> We call $f(x_0, y_0)$ *the g-weighted average of f on* Ω.

Proof. Since f is continuous on Ω and Ω is closed and bounded, we know that f takes on a minimum value m and a maximum value M. Since g is nonnegative on Ω,

$$mg(x, y) \leq f(x, y)g(x, y) \leq Mg(x, y) \qquad \text{for all } (x, y) \text{ in } \Omega.$$

Therefore [by property II]

$$\iint\limits_{\Omega} mg(x, y)\, dxdy \leq \iint\limits_{\Omega} f(x, y)g(x, y)\, dxdy \leq \iint\limits_{\Omega} Mg(x, y)\, dxdy$$

$\dagger$ Property IV is this equation with g constantly 1.

and [by property I]

(*) $$ m \iint_\Omega g(x, y)\, dxdy \le \iint_\Omega f(x, y)g(x, y)\, dxdy \le M \iint_\Omega g(x, y)\, dxdy. $$

We know that $\iint_\Omega g(x, y)\, dxdy \ge 0$. If $\iint_\Omega g(x, y)\, dxdy = 0$, then by (*) we have $\iint_\Omega f(x, y)g(x, y)\, dxdy = 0$ and the theorem holds for all choices of (x_0, y_0) in Ω. If $\iint_\Omega g(x, y)\, dxdy > 0$, then

$$ m \le \frac{\iint_\Omega f(x, y)g(x, y)\, dxdy}{\iint_\Omega g(x, y)\, dxdy} \le M $$

and by the intermediate-value theorem (Theorem 16.3.5) there exists (x_0, y_0) in Ω for which

$$ f(x_0, y_0) = \frac{\iint_\Omega f(x, y)g(x, y)\, dxdy}{\iint_\Omega g(x, y)\, dxdy}. $$

Obviously then

$$ f(x_0, y_0) \iint_\Omega g(x, y)\, dxdy = \iint_\Omega f(x, y)g(x, y)\, dxdy. \quad \square $$

EXERCISES 17.3

1. Let ϕ be a function continuous and nonnegative on $[a, b]$, and set

$$ \Omega = \{(x, y): \quad a \le x \le b, \quad 0 \le y \le \phi(x)\}. $$

Compare

$$ \iint_\Omega dxdy \quad \text{to} \quad \int_a^b \phi(x)\, dx. $$

2. Begin with a function f continuous on a closed bounded region Ω. Now surround Ω by a rectangle R as in Figure 17.3.2 and extend f to all of R by defining f to be 0 outside of Ω. Explain how the extended f can fail to be continuous on the boundary of Ω although the original function f, being continuous on all of Ω, was continuous on the boundary of Ω.

17.4   THE EVALUATION OF DOUBLE INTEGRALS BY REPEATED INTEGRALS

The Reduction Formulas

If an integral

$$ \int_a^b f(x)\, dx $$

proves difficult to evaluate, it is not because of the interval $[a, b]$ but because of the integrand f. Difficulty in evaluating a double integral

$$\iint_{\Omega} f(x, y) \, dxdy$$

can come from two sources: from the integrand f and from the base region Ω. Even such a simple looking integral as $\iint_{\Omega} 1 \, dxdy$ is difficult to evaluate if Ω is complicated.

In this section we introduce a technique for evaluating double integrals of continuous functions over Ω's that have a simple structure, the structure shown in Figure 17.4.1. Ω is then a basic region and we know that the double integral exists. The fundamental idea of this section is that double integrals over sets of this structure can be reduced to a pair of ordinary integrals.

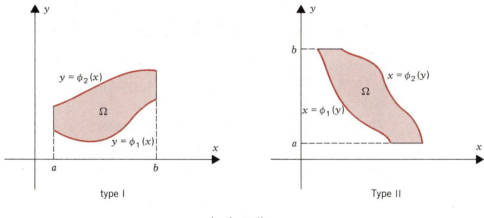

ϕ_1, ϕ_2 continuous

Figure 17.4.1

Type I. The *projection* of Ω onto the x-axis is a closed interval $[a, b]$ and Ω consists of all points (x, y) with

$$a \le x \le b \quad \text{and} \quad \phi_1(x) \le y \le \phi_2(x).$$

Then

(17.4.1)
$$\iint_{\Omega} f(x, y) \, dxdy = \int_a^b \left(\int_{\phi_1(x)}^{\phi_2(x)} f(x, y) \, dy \right) dx.$$

Here we first calculate

$$\int_{\phi_1(x)}^{\phi_2(x)} f(x, y) \, dy$$

by integrating $f(x, y)$ with respect to y from $y = \phi_1(x)$ to $y = \phi_2(x)$. The resulting expression is a function of x alone, which we then integrate with respect to x from $x = a$ to $x = b$.

Type II. The *projection* of Ω onto the y-axis is a closed interval $[a, b]$ and Ω consists of all points (x, y) with

$$a \le y \le b \quad \text{and} \quad \phi_1(y) \le x \le \phi_2(y).$$

Then

(17.4.2)
$$\iint_\Omega f(x, y)\, dxdy = \int_a^b \left(\int_{\phi_1(y)}^{\phi_2(y)} f(x, y)\, dx \right) dy.$$

This time we first calculate

$$\int_{\phi_1(y)}^{\phi_2(y)} f(x, y)\, dx$$

by integrating $f(x, y)$ with respect to x from $x = \phi_1(y)$ to $x = \phi_2(y)$. The resulting expression is a function of y alone, which we then integrate with respect to y from $y = a$ to $y = b$.

These formulas are easy to understand geometrically.

The Reduction Formulas Viewed Geometrically

Let's take f as nonnegative and Ω of type I. The double integral over Ω gives the volume of the solid T that is trapped between Ω and the surface $z = f(x, y)$:

(1)
$$\iint_\Omega f(x, y)\, dxdy = \text{volume of } T. \qquad \text{(Figure 17.4.2)}$$

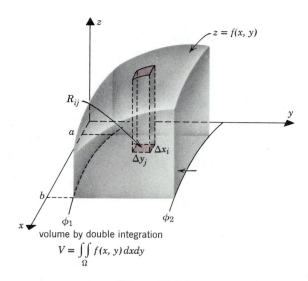

volume by double integration
$$V = \iint_\Omega f(x, y)\, dxdy$$

Figure 17.4.2

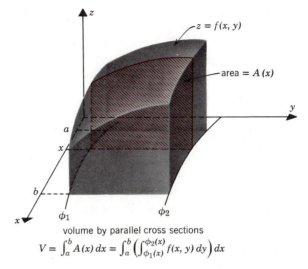

volume by parallel cross sections

$$V = \int_a^b A(x)\, dx = \int_a^b \left(\int_{\phi_1(x)}^{\phi_2(x)} f(x,\, y)\, dy \right) dx$$

Figure 17.4.3

We can also compute the volume of T by the method of parallel cross sections. As in Figure 17.4.3, let $A(x)$ be the area of that cross section of T that has first coordinate x. Then by (6.2.1)

$$\int_a^b A(x)\, dx = \text{volume of } T.$$

Since

$$A(x) = \int_{\phi_1(x)}^{\phi_2(x)} f(x,\, y)\, dy,$$

we have

(2) $$\int_a^b \left(\int_{\phi_1(x)}^{\phi_2(x)} f(x,\, y)\, dy \right) dx = \text{volume of } T.$$

Combining (1) with (2), we have the first reduction formula

$$\iint_\Omega f(x,\, y)\, dxdy = \int_a^b \left(\int_{\phi_1(x)}^{\phi_2(x)} f(x,\, y)\, dy \right) dx.$$

The other reduction formula can be obtained in a similar manner.   □

Remark. Note that our argument was a very loose one and certainly not a proof. How do we know, for example, that the "volume" obtained by double integration is the same as the "volume" obtained by the method of parallel cross sections? Intuitively it seems evident, but actually it is quite difficult to prove.   □

Computations

Problem 1. Evaluate

$$\iint_{\Omega} (x^2 - y)\, dxdy$$

with Ω as in Figure 17.4.4.

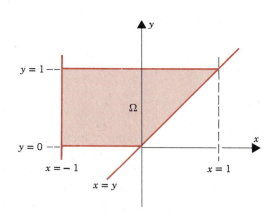

Solution. By projecting Ω onto the x-axis we obtain the interval $[-1, 1]$. The region Ω consists of all points (x, y) with

$$-1 \leq x \leq 1 \quad \text{and} \quad -x^2 \leq y \leq x^2.$$

This is a region of type I. By (17.4.1)

$$\iint_{\Omega} (x^2 - y)\, dxdy = \int_{-1}^{1} \left(\int_{-x^2}^{x^2} (x^2 - y)\, dy \right) dx$$

$$= \int_{-1}^{1} \left[x^2 y - \tfrac{1}{2} y^2 \right]_{-x^2}^{x^2} dx$$

$$= \int_{-1}^{1} [(x^4 - \tfrac{1}{2} x^4) - (-x^4 - \tfrac{1}{2} x^4)]\, dx$$

$$= \int_{-1}^{1} 2x^4\, dx = \left[\tfrac{2}{5} x^5 \right]_{-1}^{1} = \tfrac{4}{5}. \quad \square$$

Figure 17.4.4

Problem 2. Evaluate

$$\iint_{\Omega} (xy - y^3)\, dxdy$$

with Ω as in Figure 17.4.5.

Solution. By projecting Ω onto the y-axis we obtain the interval $[0, 1]$. The region Ω consists of all points (x, y) with

$$0 \leq y \leq 1 \quad \text{and} \quad -1 \leq x \leq y.$$

Figure 17.4.5

This is a region of type II.   By (17.4.2)

$$\iint_{\Omega} (xy - y^3)\, dxdy = \int_0^1 \left(\int_{-1}^{y} (xy - y^3)\, dx \right) dy$$

$$= \int_0^1 \left[\tfrac{1}{2}x^2 y - xy^3 \right]_{-1}^{y} dy$$

$$= \int_0^1 (-\tfrac{1}{2}y - \tfrac{1}{2}y^3 - y^4)\, dy$$

$$= \left[-\tfrac{1}{4}y^2 - \tfrac{1}{8}y^4 - \tfrac{1}{5}y^5 \right]_0^1 = -\tfrac{23}{40}.$$

We can also project Ω onto the x-axis and express Ω as a region of type I, but then the lower boundary is defined piecewise and the calculations are somewhat more complicated: setting

$$\phi(x) = \begin{cases} 0, & -1 \leq x \leq 0 \\ x, & 0 \leq x \leq 1 \end{cases},$$

we have Ω as the set of all points (x, y) with

$$-1 \leq x \leq 1 \quad \text{and} \quad \phi(x) \leq y \leq 1;$$

thus

$$\iint_{\Omega} (xy - y^3)\, dxdy = \int_{-1}^{1} \left(\int_{\phi(x)}^{1} (xy - y^3)\, dy \right) dx$$

$$= \int_{-1}^{0} \left(\int_{\phi(x)}^{1} (xy - y^3)\, dy \right) dx + \int_{0}^{1} \left(\int_{\phi(x)}^{1} (xy - y^3)\, dy \right) dx$$

$$= \int_{-1}^{0} \left(\int_{0}^{1} (xy - y^3)\, dy \right) dx + \int_{0}^{1} \left(\int_{x}^{1} (xy - y^3)\, dy \right) dx$$

as you can check ⟶

$$= (-\tfrac{1}{2}) + (-\tfrac{3}{40}) = -\tfrac{23}{40}. \quad \square$$

Repeated integrals

$$\int_a^b \left(\int_{\phi_1(x)}^{\phi_2(x)} f(x, y)\, dy \right) dx \quad \text{and} \quad \int_a^b \left(\int_{\phi_1(y)}^{\phi_2(y)} f(x, y)\, dx \right) dy$$

can be written in more compact form by omitting the large parentheses.   From now on we will simply write

$$\int_a^b \int_{\phi_1(x)}^{\phi_2(x)} f(x, y)\, dy\, dx \quad \text{and} \quad \int_a^b \int_{\phi_1(y)}^{\phi_2(y)} f(x, y)\, dx\, dy.$$

Problem 3. Evaluate

$$\iint_{\Omega} (x^{1/2} - y^2)\, dxdy$$

with Ω as in Figure 17.4.6.

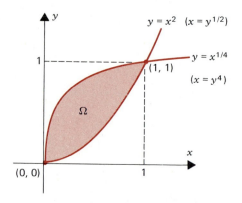

Figure 17.4.6

Solution. The projection of Ω onto the x-axis is the closed interval $[0, 1]$, and Ω can be characterized as the set of all (x, y) with

$$0 \le x \le 1 \quad \text{and} \quad x^2 \le y \le x^{1/4}.$$

Thus

$$\iint_{\Omega} (x^{1/2} - y^2)\, dxdy = \int_0^1 \int_{x^2}^{x^{1/4}} (x^{1/2} - y^2)\, dy\, dx$$

$$= \int_0^1 \left[x^{1/2}y - \tfrac{1}{3}y^3 \right]_{x^2}^{x^{1/4}} dx$$

$$= \int_0^1 (\tfrac{2}{3}x^{3/4} - x^{5/2} + \tfrac{1}{3}x^6)\, dx$$

$$= \left[\tfrac{8}{21}x^{7/4} - \tfrac{2}{7}x^{7/2} + \tfrac{1}{21}x^7 \right]_0^1 = \tfrac{8}{21} - \tfrac{2}{7} + \tfrac{1}{21} = \tfrac{1}{7}.$$

We can also integrate in the other order. The projection of Ω onto the y-axis is the closed interval $[0, 1]$, and Ω can be characterized as the set of all (x, y) with

$$0 \le y \le 1 \quad \text{and} \quad y^4 \le x \le y^{1/2}.$$

This gives the same result:

$$\iint_{\Omega} (x^{1/2} - y^2)\, dxdy = \int_0^1 \int_{y^4}^{y^{1/2}} (x^{1/2} - y^2)\, dx\, dy$$

$$= \int_0^1 \left[\tfrac{2}{3}x^{3/2} - y^2x \right]_{y^4}^{y^{1/2}} dy$$

$$= \int_0^1 (\tfrac{2}{3}y^{3/4} - y^{5/2} + \tfrac{1}{3}y^6)\, dy$$

$$= \left[\tfrac{8}{21}y^{7/4} - \tfrac{2}{7}y^{7/2} + \tfrac{1}{21}y^7 \right]_0^1 = \tfrac{8}{21} - \tfrac{2}{7} + \tfrac{1}{21} = \tfrac{1}{7}. \quad \square$$

Problem 4. Calculate by double integration the area of the region Ω that lies between

$$\sqrt{x} + \sqrt{y} = \sqrt{a} \quad \text{and} \quad x + y = a.$$

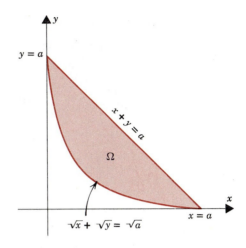

Figure 17.4.7

Solution. The region Ω is pictured in Figure 17.4.7. Its area is given by the double integral

$$\iint\limits_{\Omega} dx\,dy. \qquad (17.3.3)$$

Here again we can integrate in either order. We can project Ω onto the x-axis and write the boundaries as functions of x:

$$y = (\sqrt{a} - \sqrt{x})^2 \quad \text{and} \quad y = a - x;$$

Ω is then the set of all (x, y) with

$$0 \le x \le a \quad \text{and} \quad (\sqrt{a} - \sqrt{x})^2 \le y \le a - x.$$

This gives

$$\iint\limits_{\Omega} dx\,dy = \int_0^a \int_{(\sqrt{a}-\sqrt{x})^2}^{a-x} dy\,dx$$

$$= \int_0^a [(a - x) - (\sqrt{a} - \sqrt{x})^2]\,dx$$

$$= \int_0^a (-2x + 2\sqrt{a}\,\sqrt{x})\,dx = \left[-x^2 + \tfrac{4}{3}\sqrt{a}x\sqrt{x}\right]_0^a = -a^2 + \tfrac{4}{3}a^2 = \tfrac{1}{3}a^2.$$

We can also project Ω onto the y-axis and write the boundaries as functions of y:

$$x = (\sqrt{a} - \sqrt{y})^2 \quad \text{and} \quad x = a - y;$$

Ω is then the set of all (x, y) with

$$0 \le y \le a \quad \text{and} \quad (\sqrt{a} - \sqrt{y})^2 \le x \le a - y.$$

This gives

$$\iint\limits_{\Omega} dx\,dy = \int_0^a \int_{(\sqrt{a}-\sqrt{y})^2}^{a-y} dx\,dy,$$

which is also $\tfrac{1}{3}a^2$. ☐

Symmetry in Double Integration

First we go back to the one-variable case. Let's suppose that g is continuous on an interval that is symmetric about the origin, say $[-a, a]$.

$$\text{If } g \text{ is odd, then } \int_{-a}^{a} g(x)\, dx = 0.$$

$$\text{If } g \text{ is even, then } \int_{-a}^{a} g(x)\, dx = 2 \int_{0}^{a} g(x)\, dx.$$

We have similar results for double integrals.

Suppose that Ω is symmetric about the y-axis.

If f is odd in x [if $f(-x, y) = -f(x, y)$], then $\displaystyle\iint_{\Omega} f(x, y)\, dxdy = 0.$

If f is even in x [if $f(-x, y) = f(x, y)$], then $\displaystyle\iint_{\Omega} f(x, y)\, dxdy = 2 \iint_{\substack{\text{right half} \\ \text{of } \Omega}} f(x, y)\, dxdy.$

Suppose that Ω is symmetric about the x-axis.

If f is odd in y [if $f(x, -y) = -f(x, y)$], then $\displaystyle\iint_{\Omega} f(x, y)\, dxdy = 0.$

If f is even in y [if $f(x, -y) = f(x, y)$], then $\displaystyle\iint_{\Omega} f(x, y)\, dxdy = 2 \iint_{\substack{\text{upper half} \\ \text{of } \Omega}} f(x, y)\, dxdy.$

As an example take the region Ω of Figure 17.4.8. Suppose we wanted to calculate

$$\iint_{\Omega} (2x - \sin x^2 y)\, dxdy.$$

Symmetry about the y-axis gives

$$\iint_{\Omega} 2x\, dxdy = 0. \qquad \text{(integrand is odd in } x\text{)}$$

Symmetry about the x-axis gives

$$\iint_{\Omega} \sin x^2 y\, dxdy = 0. \qquad \text{(integrand is odd in } y\text{)}$$

Therefore

$$\iint_{\Omega} (2x - \sin x^2 y)\, dxdy = 0. \quad \square$$

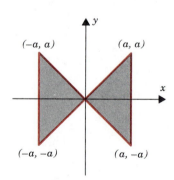

Figure 17.4.8

As another example we go to Problem 1 and reevaluate

$$\iint_{\Omega} (x^2 - y)\, dxdy,$$

this time capitalizing on the symmetry of Ω.   Note that

symmetry about x-axis   symmetry about y-axis   symmetry about x-axis

$$\iint_{\Omega} y\, dxdy \overset{\downarrow}{=} 0 \quad \text{and} \quad \iint_{\Omega} x^2\, dxdy \overset{\downarrow}{=} 2 \underset{\substack{\text{right half}\\ \text{of } \Omega}}{\iint} x^2\, dxdy \overset{\downarrow}{=} 4 \underset{\substack{\text{upper part}\\ \text{of right half}\\ \text{of } \Omega}}{\iint} x^2\, dxdy.$$

Therefore

$$\iint_{\Omega} (x^2 - y)\, dxdy = 4 \int_0^1 \int_0^{x^2} x^2\, dy\, dx = 4 \int_0^1 x^4\, dx = \tfrac{4}{5}. \quad \square$$

Problem 5.   Calculate the volume within the cylinder $x^2 + y^2 = b^2$ between the planes $y + z = a^2$ and $z = 0$ given that $a^2 \geq b > 0$.

Solution.   See Figure 17.4.9. The solid in question is bounded below by the disc

Ω: $0 \leq x^2 + y^2 \leq b^2$

and above by the plane

$z = a^2 - y$.

The volume is given by the double integral

$$\iint_{\Omega} (a^2 - y)\, dxdy.$$

Since Ω is symmetric about the x-axis,

$$\iint_{\Omega} y\, dxdy = 0.$$

Figure 17.4.9

Thus

$$\iint_{\Omega} (a^2 - y)\, dxdy = \iint_{\Omega} a^2\, dxdy = a^2 \iint_{\Omega} dxdy = a^2(\text{area of } \Omega) = \pi a^2 b^2. \quad \square$$

Concluding Remarks

When two orders of integration are possible, one order may be easy to carry out, while the other may present serious difficulties. Take as an example the double integral

$$\iint_{\Omega} \cos \tfrac{1}{2}\pi x^2 \, dx dy$$

with Ω as in Figure 17.4.10. Projection onto the x-axis leads to

$$\int_0^1 \int_0^x \cos \tfrac{1}{2}\pi x^2 \, dy \, dx.$$

Projection onto the y-axis leads to

$$\int_0^1 \int_y^1 \cos \tfrac{1}{2}\pi x^2 \, dx \, dy.$$

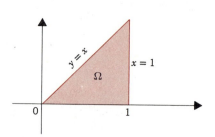

Figure 17.4.10

The first expression is easy to calculate, but the second is not. The only practical way to handle the second expression is to expand $\cos \tfrac{1}{2}\pi x^2$ as a series in x and then integrate term by term.

Finally, if Ω, the region of integration, is neither of type I nor of type II, it is usually possible to break it up into a finite number of regions $\Omega_1, \ldots, \Omega_n$, each of type I or type II. (See Figure 17.4.11.) Since the double integral is additive,

$$\iint_{\Omega_1} f(x, y) \, dx dy + \cdots + \iint_{\Omega_n} f(x, y) \, dx dy = \iint_{\Omega} f(x, y) \, dx dy.$$

Each of the integrals on the left can be evaluated by the methods of this section.

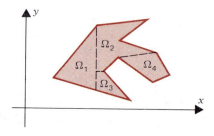

Figure 17.4.11

EXERCISES 17.4

Evaluate taking Ω: $0 \le x \le 1$, $0 \le y \le 3$.

1. $\displaystyle\iint_{\Omega} x^2 \, dx dy.$ **2.** $\displaystyle\iint_{\Omega} e^{x+y} \, dx dy.$ **3.** $\displaystyle\iint_{\Omega} xy^2 \, dx dy.$

Evaluate taking Ω: $0 \leq x \leq 1$, $0 \leq y \leq x$.

4. $\displaystyle\iint_{\Omega} x^3 y \, dxdy.$ **5.** $\displaystyle\iint_{\Omega} xy^3 \, dxdy.$ **6.** $\displaystyle\iint_{\Omega} x^2 y^2 \, dxdy.$

Evaluate taking Ω: $0 \leq x \leq \tfrac{1}{2}\pi$, $0 \leq y \leq \tfrac{1}{2}\pi$.

7. $\displaystyle\iint_{\Omega} \sin(x+y) \, dxdy.$ **8.** $\displaystyle\iint_{\Omega} \cos(x+y) \, dxdy.$ **9.** $\displaystyle\iint_{\Omega} (1 + xy) \, dxdy.$

Evaluate the following double integrals.

10. $\displaystyle\iint_{\Omega} (x^2 + 3y^3) \, dxdy,$ Ω: $0 \leq x^2 + y^2 \leq 1.$

11. $\displaystyle\iint_{\Omega} \sqrt{xy} \, dxdy,$ Ω: $0 \leq y \leq 1, \quad y^2 \leq x \leq y.$

12. $\displaystyle\iint_{\Omega} y e^x \, dxdy,$ Ω: $0 \leq y \leq 1, \quad 0 \leq x \leq y^2.$

13. $\displaystyle\iint_{\Omega} (4 - y^2) \, dxdy,$ Ω the bounded region between $y^2 = 2x$ and $y^2 = 8 - 2x.$

14. $\displaystyle\iint_{\Omega} (x^4 + y^2) \, dxdy,$ Ω the bounded region between $y = x^3$ and $y = x^2.$

15. $\displaystyle\iint_{\Omega} (3xy^2 - y) \, dxdy,$ Ω the region between $y = |x|$ and $y = -|x|, \quad x \in [-1, 1].$

16. $\displaystyle\iint_{\Omega} e^{-y^2/2} \, dxdy,$ Ω the triangle formed by the y-axis, $2y = x$, $y = 1.$

17. $\displaystyle\iint_{\Omega} e^{x^2} \, dxdy,$ Ω the triangle formed by the x-axis, $2y = x$, $x = 2.$

18. $\displaystyle\iint_{\Omega} (x + y) \, dxdy,$ Ω the region between $y = x^3$ and $y = x^4, \quad x \in [-1, 1].$

Sketch the region Ω that gives rise to the repeated integral and change the order of integration.

19. $\displaystyle\int_{0}^{1} \int_{x^4}^{x^2} f(x, y) \, dy \, dx.$ **20.** $\displaystyle\int_{0}^{1} \int_{0}^{y^2} f(x, y) \, dx \, dy.$ **21.** $\displaystyle\int_{0}^{1} \int_{-y}^{y} f(x, y) \, dx \, dy.$

22. $\displaystyle\int_{1/2}^{1} \int_{x^3}^{x} f(x, y) \, dy \, dx.$ **23.** $\displaystyle\int_{1}^{4} \int_{x}^{2x} f(x, y) \, dy \, dx.$ **24.** $\displaystyle\int_{1}^{3} \int_{-x}^{x^2} f(x, y) \, dy \, dx.$

Calculate by double integration the area of the bounded region determined by the following pairs of curves.

25. $x^2 = 4y, \quad 2y - x - 4 = 0.$ **26.** $y = x, \quad x = 4y - y^2.$

27. $y = x, \quad 4y^3 = x^2.$ **28.** $x + y = 5, \quad xy = 6.$

Sketch the region Ω that gives rise to the repeated integral, change the order of integration, and then evaluate.

29. $\displaystyle\int_0^1 \int_{\sqrt{x}}^1 \sin\left(\frac{y^3 + 1}{2}\right) dy\, dx.$

30. $\displaystyle\int_{-1}^0 \int_{-\sqrt{y+1}}^{\sqrt{y+1}} x^2\, dx\, dy.$

31. $\displaystyle\int_1^2 \int_0^{\ln y} e^{-x}\, dx\, dy.$

32. $\displaystyle\int_0^1 \int_{x^2}^1 \frac{x^3}{\sqrt{x^4 + y^2}}\, dy\, dx.$

33. Find the area of the first quadrant region bounded by $xy = 2$, $y = 1$, $y = x + 1$.

34. Find the volume of the solid bounded above by $z = x + y$ and below by the triangle with vertices $(0, 0)$, $(0, 1)$, $(1, 0)$.

35. Find the volume of the solid bounded by $\frac{1}{2}x + \frac{1}{3}y + \frac{1}{4}z = 1$ and the coordinate planes.

36. Find the volume of the solid bounded above by the plane $z = 2x + 3y$ and below by the unit square $0 \le x \le 1, 0 \le y \le 1$.

37. Find the volume of the solid bounded above by $z = x^3 y$ and below by the triangle with vertices $(0, 0)$, $(2, 0)$, $(0, 1)$.

38. Find the volume under the paraboloid $z = x^2 + y^2$ within the cylinder $x^2 + y^2 \le 1$.

39. Find the volume of the solid bounded above by the plane $z = 2x + 1$ and below by the disc $(x - 1)^2 + y^2 \le 1$.

40. Find the volume of the solid bounded above by $z = 4 - y^2 - \frac{1}{4}x^2$ and below by the disc $(y - 1)^2 + x^2 \le 1$.

Evaluate.

41. $\displaystyle\int_0^1 \int_y^1 e^{y/x}\, dx\, dy.$

42. $\displaystyle\int_0^1 \int_0^{\cos^{-1} y} e^{\sin x}\, dx\, dy.$

43. $\displaystyle\int_0^1 \int_x^1 x^2\, e^{y^4}\, dy\, dx.$

44. $\displaystyle\int_0^1 \int_x^1 e^{y^2}\, dy\, dx.$

45. (*Separated variables over a rectangle*) Let R be the rectangle $a \le x \le b, c \le y \le d$. Show that, if f is continuous on $[a, b]$ and g is continuous on $[c, d]$, then

(17.4.3)
$$\iint_R f(x)g(y)\, dx\, dy = \left[\int_a^b f(x)\, dx\right] \cdot \left[\int_c^d g(y)\, dy\right].$$

46. Let R be a rectangle symmetric about the x-axis, sides parallel to the coordinate axes. Show that, if f is odd with respect to y, then the double integral of f over R is 0.

47. Let R be a rectangle symmetric about the y-axis, sides parallel to the coordinate axes. Show that, if f is odd with respect to x, then the double integral of f over R is 0.

48. Given that $f(-x, -y) = -f(x, y)$ for all (x, y) in Ω, what form of symmetry in Ω will ensure that the double integral of f over Ω is zero?

49. Let Ω be the triangle with vertices $(0, 0)$, $(0, 1)$, $(1, 1)$. Show that

$$\text{if } \int_0^1 f(x)\, dx = 0 \quad \text{then} \quad \iint_\Omega f(x)f(y)\, dx\, dy = 0.$$

50. (*Differentiation under the integral sign*) If f and $\partial f/\partial x$ are continuous, then the function

$$H(t) = \int_a^b \frac{\partial f}{\partial x}(t, y)\, dy$$

can be shown to be continuous. Use the identity

$$\int_0^x \int_a^b \frac{\partial f}{\partial x}(t, y)\, dy\, dt = \int_a^b \int_0^x \frac{\partial f}{\partial x}(t, y)\, dt\, dy$$

to verify that

$$\frac{d}{dx}\left[\int_a^b f(x, y) \, dy\right] = \int_a^b \frac{\partial f}{\partial x}(x, y) \, dy.$$

51. We integrate over regions of type I by means of the formula

$$\iint_\Omega f(x, y) \, dxdy = \int_a^b \int_{\phi_1(x)}^{\phi_2(x)} f(x, y) \, dy \, dx.$$

Here f is assumed to be continuous on Ω and ϕ_1, ϕ_2 are assumed to be continuous on $[a, b]$. Show that the function

$$F(x) = \int_{\phi_1(x)}^{\phi_2(x)} f(x, y) \, dy$$

is continuous on $[a, b]$.

17.5   THE DOUBLE INTEGRAL AS THE LIMIT OF RIEMANN SUMS; POLAR COORDINATES

The Double Integral as the Limit of Riemann Sums

In the one-variable case we can write the integral as the limit of Riemann sums:

$$\int_a^b f(x) \, dx = \lim_{\max \Delta x_i \to 0} \sum_{i=1}^n f(x_i^*) \, \Delta x_i.$$

The same approach works with double integrals. To explain it we need to explain what we mean by the *diameter of a set*.

Suppose that S is a bounded closed set (on the line, in the plane, or in three-space). For any two points P and Q of S we can measure their separation, $d(P, Q)$. The greatest separation between points of S is called *the diameter of S*:

$$\text{diam } S = \max_{P, Q \in S} d(P, Q).$$

For a circle, a circular disc, a sphere, or a ball, this sense of diameter agrees with the usual one. Figure 17.5.1 gives some other examples.

diam = Δx_i

(Length of the interval)

diam = $\sqrt{(\Delta x_i)^2 + (\Delta y_j)^2}$

(Length of the diagonal)

diam = $\sqrt{(\Delta x_i)^2 + (\Delta y_j)^2 + (\Delta z_k)^2}$

(Length of the diagonal)

Figure 17.5.1

Now let's start with a basic region Ω and decompose it into a finite number of basic subregions $\Omega_1, \ldots, \Omega_N$.   (See Figure 17.5.2.)

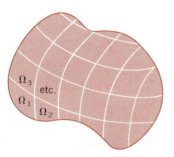

Figure 17.5.2

If f is continuous on Ω, then f is continuous on each Ω_i.   Now from each Ω_i we pick an arbitrary point (x_i^*, y_i^*) and form the *Riemann sum*

$$\sum_{i=1}^{N} f(x_i^*, y_i^*)(\text{area of } \Omega_i).$$

As you would expect, the double integral over Ω can be obtained as the limit of such sums; namely, given any $\epsilon > 0$, there exists $\delta > 0$ such that, if the diameters of the Ω_i are all less than δ, then

$$\left| \sum_{i=1}^{N} f(x_i^*, y_i^*)(\text{area of } \Omega_i) - \iint_{\Omega} f(x, y)\, dxdy \right| < \epsilon$$

no matter how the (x_i^*, y_i^*) are chosen within the Ω_i.   We express this by writing

(17.5.1)
$$\iint_{\Omega} f(x, y)\, dxdy = \lim_{\text{diam } \Omega_i \to 0} \sum_{i=1}^{N} f(x_i^*, y_i^*)(\text{area of } \Omega_i).$$

Evaluating Double Integrals Using Polar Coordinates

Here we explain how to calculate double integrals

$$\iint_{\Omega} f(x, y)\, dxdy$$

using polar coordinates (r, θ).   Throughout we take $r \geq 0$ and restrict θ to $[0, 2\pi]$.

We will work with the type of region shown in Figure 17.5.3. Ω is then the set of all points (x, y) that have polar coordinates (r, θ) in the set

$$\Gamma: \quad \alpha \leq \theta \leq \beta, \quad \rho(\theta) \leq r \leq P(\theta).$$

Figure 17.5.3

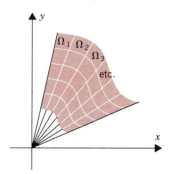

Figure 17.5.4

You already know how to calculate the area of Ω. By (10.5.3)

$$\text{area of } \Omega = \int_\alpha^\beta \tfrac{1}{2}[P^2(\theta) - \rho^2(\theta)]\, d\theta.$$

We can write this as a double integral over Γ:

(17.5.2)

$$\text{area of } \Omega = \iint_\Gamma r\, dr d\theta.$$

Proof. Simply note that

$$\tfrac{1}{2}[P^2(\theta) - \rho^2(\theta)] = \int_{\rho(\theta)}^{P(\theta)} r\, dr$$

and therefore

$$\text{area of } \Omega = \int_\alpha^\beta \int_{\rho(\theta)}^{P(\theta)} r\, dr\, d\theta = \iint_\Gamma r\, dr d\theta. \quad \square$$

Now let's suppose that f is some function continuous at each point (x, y) of Ω. Then the composition

$$F(r, \theta) = f(r \cos \theta, r \sin \theta)$$

is continuous at each point (r, θ) of Γ. We will show that

(17.5.3)

$$\iint_\Omega f(x, y)\, dxdy = \iint_\Gamma f(r \cos \theta, r \sin \theta)\, r\, dr d\theta.$$

(note the extra r)

Proof. Our first step is to place a grid on Ω by using a finite number of rays $\theta = \theta_j$ and a finite number of continuous curves $r = \rho_k(\theta)$ in the manner of Figure 17.5.4. This grid decomposes Ω into a finite number of little regions

$$\Omega_1, \ldots, \Omega_N$$

with polar coordinates in sets $\Gamma_1, \ldots, \Gamma_N$. Note that by (17.5.2)

$$\text{the area of each } \Omega_i = \iint_{\Gamma_i} r\, dr d\theta.$$

Writing $F(r, \theta)$ for $f(r \cos \theta, r \sin \theta)$, we have

$$\iint_{\Gamma} F(r, \theta)\, r\, dr d\theta = \sum_{i=1}^{N} \iint_{\Gamma_i} F(r, \theta)\, r\, dr d\theta$$

additivity ————————————⌐

$$= \sum_{i=1}^{N} F(r_i^*, \theta_i^*) \iint_{\Gamma_i} r\, dr d\theta$$

with $(r_i^*, \theta_i^*) \in \Gamma_i$ ————————⌐
(Theorem 17.3.5)

$$= \sum_{i=1}^{N} F(r_i^*, \theta_i^*)(\text{area of } \Omega_i)$$

(with $x_i^* = r_i^* \cos \theta_i^*$, $y_i^* = r_i^* \sin \theta_i^*$) ——↓

$$= \sum_{i=1}^{N} f(x_i^*, y_i^*)(\text{area of } \Omega_i).$$

This last expression is a Riemann sum for the double integral

$$\iint_{\Omega} f(x, y)\, dxdy$$

and, as such, by (17.5.1), differs from that integral by less than any preassigned positive ϵ provided only that the diameters of all the Ω_i are sufficiently small. This we can guarantee by making our grid sufficiently fine. ☐

Problem 1. Use polar coordinates to evaluate

$$\iint_{\Omega} xy\, dxdy$$

where Ω is the portion of the unit disc that lies in the first quadrant.

Solution. Here

$$\Gamma: \quad 0 \le \theta \le \tfrac{1}{2}\pi, \quad 0 \le r \le 1.$$

Therefore

$$\iint_{\Omega} xy\, dxdy = \iint_{\Gamma} (r \cos \theta)(r \sin \theta)\, r\, dr d\theta$$

$$= \int_0^{\pi/2} \int_0^1 r^3 \cos \theta \sin \theta\, dr\, d\theta = \tfrac{1}{8}. ☐$$

check this ——————⌐

Problem 2. Use polar coordinates to calculate the volume of a sphere of radius R.

Solution. In rectangular coordinates

$$V = 2 \iint_{\Omega} \sqrt{R^2 - (x^2 + y^2)} \, dxdy$$

where Ω is the disc of radius R centered at the origin. (Make sure of this.) Here

$$\Gamma: \quad 0 \le \theta \le 2\pi, \quad 0 \le r \le R.$$

Therefore

$$V = 2 \iint_{\Gamma} \sqrt{R^2 - r^2} \, r \, drd\theta = 2 \int_0^{2\pi} \int_0^R \sqrt{R^2 - r^2} \, r \, dr \, d\theta = \tfrac{4}{3}\pi r^3. \quad \square$$

check this ⟶

Problem 3. Calculate the volume of the solid bounded above by the lower nappe of the cone $z = 2 - \sqrt{x^2 + y^2}$ and bounded below by the disc $\Omega: (x - 1)^2 + y^2 \le 1$. (See Figure 17.5.5.)

Solution

$$V = \iint_{\Omega} (2 - \sqrt{x^2 + y^2}) \, dxdy$$

$$= 2 \iint_{\Omega} dxdy - \iint_{\Omega} \sqrt{x^2 + y^2} \, dxdy.$$

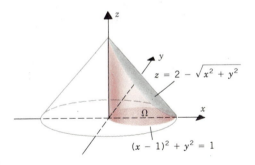

$$z = 2 - \sqrt{x^2 + y^2}$$

$$(x - 1)^2 + y^2 = 1$$

Figure 17.5.5

The first integral is 2(area of Ω) = 2π. We evaluate the second integral by changing to polar coordinates.

The equation $(x - 1)^2 + y^2 = 1$ simplifies to $x^2 + y^2 = 2x$. In polar coordinates this becomes $r^2 = 2r \cos \theta$, which simplifies to $r = 2 \cos \theta$. The disc Ω is the set of all points with polar coordinates in the set

$$\Gamma: \quad -\tfrac{1}{2}\pi \le \theta \le \tfrac{1}{2}\pi, \quad 0 \le r \le 2 \cos \theta.$$

Therefore

$$\iint_{\Omega} \sqrt{x^2 + y^2} \, dxdy = \iint_{\Gamma} r^2 \, drd\theta = \int_{-\pi/2}^{\pi/2} \int_0^{2 \cos \theta} r^2 \, dr \, d\theta = \tfrac{32}{9}.$$

check this ⟶

We then have $V = 2\pi - \tfrac{32}{9} \cong 2.73$. $\square$

Problem 4. Evaluate

$$\iint\limits_{\Omega} \frac{1}{(1 + x^2 + y^2)^{3/2}} \, dxdy$$

where Ω is the triangle of Figure 17.5.6.

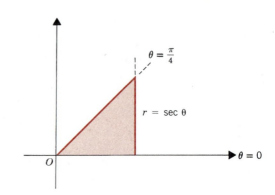

Figure 17.5.6   **Figure 17.5.7**

Solution. The vertical side of the triangle is part of the line $x = 1$. In polar coordinates this is $r \cos \theta = 1$, which can be written $r = \sec \theta$. Therefore

$$\iint\limits_{\Omega} \frac{1}{(1 + x^2 + y^2)^{3/2}} \, dxdy = \iint\limits_{\Gamma} \frac{r}{(1 + r^2)^{3/2}} \, drd\theta$$

where

$$\Gamma: \quad 0 \leq \theta \leq \pi/4, \quad 0 \leq r \leq \sec \theta. \qquad \text{(Figure 17.5.7)}$$

The double integral over Γ reduces to

$$\int_0^{\pi/4} \int_0^{\sec \theta} \frac{r}{(1 + r^2)^{3/2}} \, dr \, d\theta = \int_0^{\pi/4} \left[\frac{1}{\sqrt{1 + r^2}} \right]_{\sec \theta}^0 d\theta = \int_0^{\pi/4} \left(1 - \frac{1}{\sqrt{1 + \sec^2 \theta}} \right) d\theta.$$

For $\theta \in [0, \pi/4]$

$$\frac{1}{\sqrt{1 + \sec^2 \theta}} = \frac{\cos \theta}{\sqrt{\cos^2 \theta + 1}} = \frac{\cos \theta}{\sqrt{2 - \sin^2 \theta}}.$$

Therefore the integral can be written

$$\int_0^{\pi/4} \left(1 - \frac{\cos \theta}{\sqrt{2 - \sin^2 \theta}} \right) d\theta = \left[\theta - \sin^{-1} \left(\frac{\sin \theta}{\sqrt{2}} \right) \right]_0^{\pi/4} = \frac{\pi}{4} - \frac{\pi}{6} = \frac{\pi}{12}. \quad \square$$

$$\text{(7.10.4)} \longrightarrow$$

The function $f(x) = e^{-x^2}$ has no elementary antiderivative. Nevertheless, by taking a circuitous route and then using polar coordinates, we can show that

(17.5.4)
$$\int_{-\infty}^{\infty} e^{-x^2}\, dx = \sqrt{\pi}. \quad \dagger$$

Proof. The circular disc D_b: $x^2 + y^2 \leq b^2$ is the set of all (x, y) with polar coordinates (r, θ) in the set Γ: $0 \leq \theta \leq 2\pi$, $0 \leq r \leq b$. Therefore

$$\iint_{D_b} e^{-(x^2+y^2)}\, dxdy = \iint_{\Gamma} e^{-r^2}\, r\, drd\theta = \int_0^{2\pi} \int_0^b e^{-r^2}\, r\, dr\, d\theta$$

$$= \int_0^{2\pi} \tfrac{1}{2}(1 - e^{-b^2})\, d\theta = \pi(1 - e^{-b^2}).$$

Let S_a be the square $-a \leq x \leq a$, $-a \leq y \leq a$. Since $D_a \subseteq S_a \subseteq D_{2a}$ and $e^{-(x^2+y^2)}$ is positive,

$$\iint_{D_a} e^{-(x^2+y^2)}\, dxdy \leq \iint_{S_a} e^{-(x^2+y^2)}\, dxdy \leq \iint_{D_{2a}} e^{-(x^2+y^2)}\, dxdy.$$

It follows that

$$\pi(1 - e^{-a^2}) \leq \iint_{S_a} e^{-(x^2+y^2)}\, dxdy \leq \pi(1 - e^{-4a^2}).$$

As $a \to \infty$, $\pi(1 - e^{-a^2}) \to \pi$ and $\pi(1 - e^{-4a^2}) \to \pi$. Therefore

$$\lim_{a \to \infty} \iint_{S_a} e^{-(x^2+y^2)}\, dxdy = \pi.$$

But

$$\iint_{S_a} e^{-(x^2+y^2)}\, dxdy = \int_{-a}^a \int_{-a}^a e^{-(x^2+y^2)}\, dx\, dy$$

$$= \int_{-a}^a \int_{-a}^a e^{-x^2} \cdot e^{-y^2}\, dx\, dy$$

$$= \left(\int_{-a}^a e^{-x^2}\, dx\right)\left(\int_{-a}^a e^{-y^2}\, dy\right) = \left(\int_{-a}^a e^{-x^2}\, dx\right)^2.$$

Therefore

$$\lim_{a \to \infty} \int_{-a}^a e^{-x^2}\, dx = \lim_{a \to \infty} \left(\iint_{S_a} e^{-(x^2+y^2)}\, dxdy\right)^{1/2} = \sqrt{\pi}. \quad \square$$

---

† This integral comes up frequently in probability theory and plays an important role in what is called "statistical mechanics." You ran across a related integral in Exercise 34, Section 11.7, in conjunction with a problem on the speed distribution in a sample of gas molecules.

EXERCISES 17.5

1. Integrate $f(x, y) = \cos (x^2 + y^2)$ over
   (a) the closed unit disc.   (b) the annular region $1 \le x^2 + y^2 \le 4$.

2. Integrate $f(x, y) = \sin (\sqrt{x^2 + y^2})$ over
   (a) the closed unit disc.   (b) the annular region $1 \le x^2 + y^2 \le 4$.

3. Integrate $f(x, y) = x + y$ over
   (a) $0 \le x^2 + y^2 \le 1$, $x \ge 0$, $y \ge 0$.   (b) $1 \le x^2 + y^2 \le 4$, $x \ge 0$, $y \ge 0$.

4. Integrate $f(x, y) = \sqrt{x^2 + y^2}$ over the triangle with vertices $(0, 0)$, $(1, 0)$, $(1, \sqrt{3})$.

Calculate by changing to polar coordinates.

5. $\displaystyle\int_{-1}^{1} \int_{0}^{\sqrt{1-y^2}} \sqrt{x^2 + y^2}\ dx\ dy$.

6. $\displaystyle\int_{0}^{2} \int_{0}^{\sqrt{4-x^2}} \sqrt{x^2 + y^2}\ dy\ dx$.

7. $\displaystyle\int_{1/2}^{1} \int_{0}^{\sqrt{1-x^2}} (x^2 + y^2)^{3/2}\ dy\ dx$.

8. $\displaystyle\int_{0}^{1/2} \int_{0}^{\sqrt{1-x^2}} xy\sqrt{x^2 + y^2}\ dy\ dx$.

9. Find the volume of the solid bounded above by the plane $z = y + b$, below by the xy-plane, and on the sides by the circular cylinder $x^2 + y^2 = b^2$.

10. Find the volume of the solid bounded below by the xy-plane and above by the paraboloid $z = 1 - (x^2 + y^2)$.

11. Find the volume of the ellipsoid $x^2/4 + y^2/4 + z^2/3 = 1$.

12. Find the volume of the solid bounded below by the xy-plane and above by the surface $x^2 + y^2 + z^6 = 5$.

13. Find the volume of the solid bounded below by the xy-plane, above by the spherical surface $x^2 + y^2 + z^2 = 4$, and on the sides by the cylinder $x^2 + y^2 = 1$.

14. Find the volume of the solid bounded above by $z = 1 - (x^2 + y^2)$, below by the xy-plane, and on the sides by the cylinder $x^2 + y^2 - x = 0$.

15. Find the volume of the solid bounded above by the plane $z = 2x$ and below by the disc $(x - 1)^2 + y^2 \le 1$.

16. Find the volume of the solid bounded above by the cone $z^2 = x^2 + y^2$ and below by the region Ω which lies inside the curve $x^2 + y^2 = 2ax$.

17. Find the volume of the solid bounded below by the xy-plane, above by the ellipsoid of revolution $b^2x^2 + b^2y^2 + a^2z^2 = a^2b^2$, and on the sides by the cylinder $x^2 + y^2 - ay = 0$.

17.6   SOME APPLICATIONS OF DOUBLE INTEGRATION

The Mass of a Plate

Suppose that a thin distribution of matter, what we call a *plate*, is laid out in the xy-plane in the form of a basic region Ω. If the mass density of the plate (the mass per unit area) is constant, then the total mass M of the plate is simply the density λ times the area of the plate:

$$M = \lambda \times \text{the area of } \Omega.$$

If the density varies continuously from point to point, say $\lambda = \lambda(x, y)$, then the mass of the plate is the average density of the plate times the area of the plate:

$$M = \text{average density} \times \text{the area of } \Omega.$$

This is an integral:

(17.6.1)
$$M = \iint_\Omega \lambda(x, y)\, dxdy.$$

The Center of Mass of a Plate

The center of mass of a rod x_M is a density-weighted average taken over the interval occupied by the rod:

$$x_M M = \int_a^b x\lambda(x)\, dx. \qquad \text{[This you have seen: (5.11.5).]}$$

The center of mass of a plate (x_M, y_M) is determined by two density-weighted averages, each taken over the region occupied by the plate:

(17.6.2)
$$x_M M = \iint_\Omega x\lambda(x, y)\, dxdy, \qquad y_M M = \iint_\Omega y\lambda(x, y)\, dxdy.$$

Problem 1. A plate is in the form of a half-disc of radius a. Find the mass of the plate and the center of mass given that the mass density of the plate varies directly as the distance from the center of the straight edge of the plate.

Solution. Place the plate over the region Ω: $-a \le x \le a$, $0 \le y \le \sqrt{a^2 - x^2}$. The mass density can then be written $\lambda(x, y) = k\sqrt{x^2 + y^2}$. Here

$$M = \iint_\Omega k\sqrt{x^2 + y^2}\, dxdy = \int_0^\pi \int_0^a (kr)r\, dr\, d\theta = k\left(\int_0^\pi 1\, d\theta\right)\left(\int_0^a r^2\, dr\right)$$

$$= k(\pi)(\tfrac{1}{3}a^3) = \tfrac{1}{3}ka^3\pi.$$

$$x_M M = \iint_\Omega x(k\sqrt{x^2 + y^2})\, dxdy = 0. \qquad \text{(the integrand is odd with respect to } x\text{)}$$

Thus $x_M = 0$.

$$y_M M = \iint_\Omega y(k\sqrt{x^2 + y^2})\, dxdy = \int_0^\pi \int_0^a (r\sin\theta)(kr)r\, dr\, d\theta$$

$$= k\left(\int_0^\pi \sin\theta\, d\theta\right)\left(\int_0^a r^3\, dr\right) = k(2)(\tfrac{1}{4}a^4) = \tfrac{1}{2}ka^4.$$

Since $M = \tfrac{1}{3}ka^3\pi$, we have $y_M = (\tfrac{1}{2}ka^4)/(\tfrac{1}{3}ka^3\pi) = 3a/2\pi.$ □

Centroids

If the plate is homogeneous, then the mass density λ is constantly M/A where A is the area of the base region Ω. In this case the center of mass of the plate falls on the *centroid* of the base region (a notion with which you are already familiar). The centroid $(\bar{x}, \bar{y})$ depends only on the geometry of Ω:

$$\bar{x}\,M = \iint_{\Omega} x(M/A)\,dxdy = (M/A)\iint_{\Omega} x\,dxdy,$$

$$\bar{y}\,M = \iint_{\Omega} y(M/A)\,dxdy = (M/A)\iint_{\Omega} y\,dxdy.$$

Dividing by M and multiplying through by A we have

(17.6.3)
$$\bar{x}\,A = \iint_{\Omega} x\,dxdy, \qquad \bar{y}\,A = \iint_{\Omega} y\,dxdy.$$

Thus $\bar{x}$ is the average x-coordinate on Ω and $\bar{y}$ is the average y-coordinate on Ω. The mass of the plate does not enter into this at all.

Problem 2. Find the centroid of the region

$$\Omega: \quad a \le x \le b, \quad \phi_1(x) \le y \le \phi_2(x).$$

Solution

$$\bar{x}\,A = \iint_{\Omega} x\,dxdy = \int_a^b \int_{\phi_1(x)}^{\phi_2(x)} x\,dy\,dx = \int_a^b x[\phi_2(x) - \phi_1(x)]\,dx;$$

$$\bar{y}\,A = \iint_{\Omega} y\,dxdy = \int_a^b \int_{\phi_1(x)}^{\phi_2(x)} y\,dy\,dx = \int_a^b \tfrac{1}{2}([\phi_2(x)]^2 - [\phi_1(x)]^2)\,dx.$$

These are the formulas for the centroid that we developed in Section 6.4. Having calculated many centroids at that time, we won't do so here. ▢

Kinetic Energy and Moment of Inertia

A particle of mass m at a distance r from a given line rotates about that line with angular speed ω. The speed v of the particle is then $r\omega$ and the kinetic energy is given by the formula

$$\mathrm{KE} = \tfrac{1}{2}mv^2 = \tfrac{1}{2}mr^2\omega^2.$$

Imagine now a rigid body composed of a finite number of point masses m_i located at distances r_i from some fixed line. If the rigid body rotates about that line with angular speed ω, then all the point masses rotate about that same line with that same angular speed ω. The kinetic energy of the body can be obtained by adding up the kinetic energies of all the individual particles:

$$KE = \sum_i \tfrac{1}{2} m_i r_i^2 \omega^2 = \tfrac{1}{2} \left(\sum_i m_i r_i^2 \right) \omega^2.$$

The expression in parentheses is called the *moment of inertia* (or *rotational inertia*) of the body and is denoted by the letter I:

(17.6.4)

$$I = \sum_i m_i r_i^2 \, .$$

For a rigid body in straight-line motion

$$KE = \tfrac{1}{2} M v^2 \qquad \text{where } v \text{ is the speed of the body.}$$

For a rigid body in rotational motion

$$KE = \tfrac{1}{2} I \omega^2 \qquad \text{where } \omega \text{ is the angular speed of the body.}$$

The Moment of Inertia of a Plate

Suppose that a plate in the shape of a basic region Ω rotates about an axis. The moment of inertia of the plate about that axis is given by the formula

(17.6.5)

$$I = \iint_\Omega \lambda(x, y) [r(x, y)]^2 \, dx dy$$

where $\lambda = \lambda(x, y)$ is the mass density function and $r(x, y)$ is the distance from the axis to the point (x, y).

Derivation of (17.6.5). Decompose the plate into N pieces in the form of basic regions $\Omega_1, \ldots, \Omega_N$. From each Ω_i choose a point (x_i^*, y_i^*) and view all the mass of the ith piece as concentrated there. The moment of inertia of this piece is then approximately

$$\underbrace{[\lambda(x_i^*, y_i^*)(\text{area of } \Omega_i)]}_{\text{approx. mass of piece}} \underbrace{[r(x_i^*, y_i^*)]^2}_{(\text{approx. distance})^2} = \lambda(x_i^*, y_i^*)[r(x_i^*, y_i^*)]^2 (\text{area of } \Omega_i).$$

The sum of these approximations,

$$\sum_{i=1}^N \lambda(x_i^*, y_i^*)[r(x_i^*, y_i^*)]^2 (\text{area of } \Omega_i),$$

is a Riemann sum for the double integral

$$\iint_\Omega \lambda(x, y)[r(x, y)]^2 \, dx dy.$$

As the maximum diameter of the Ω_i tends to zero, the Riemann sum tends to this integral. ☐

Problem 3.   A rectangular plate of mass M, length L, and width W rotates about the axis shown in Figure 17.6.1. Find the moment of inertia of the plate about that axis (a) given that the plate has uniform mass density. (b) given that the mass density of the plate varies directly as the square of the distance from the rightmost side.

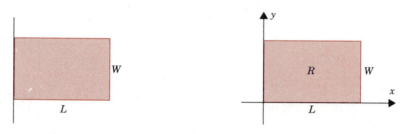

Figure 17.6.1   Figure 17.6.2

Solution.   Coordinatize the plate as in Figure 17.6.2 and call the base region R.
(a) Here $\lambda(x, y) = M/LW$ and $r(x, y) = x$. Thus

$$I = \iint_R \frac{M}{LW} x^2 \, dxdy = \frac{M}{LW} \int_0^W \int_0^L x^2 \, dx \, dy$$

$$= \frac{M}{LW} W \int_0^L x^2 \, dx = \frac{M}{L} (\tfrac{1}{3} L^3) = \tfrac{1}{3} ML^2.$$

(b) In this case $\lambda(x, y) = k(L - x)^2$ but we still have $r(x, y) = x$. Therefore

$$I = \iint_R k(L - x)^2 x^2 \, dxdy = k \int_0^W \int_0^L (L - x)^2 x^2 \, dx \, dy$$

$$= kW \int_0^L (L^2 x^2 - 2Lx^3 + x^4) \, dx = \tfrac{1}{30} kL^5 W.$$

We can eliminate the constant of proportionality k by noting that

$$M = \iint_R k(L - x)^2 \, dxdy = k \int_0^W \int_0^L (L^2 - 2Lx + x^2) \, dx \, dy$$

$$= kW \int_0^L (L^2 - 2Lx + x^2) \, dx = \tfrac{1}{3} kL^3 W.$$

Therefore $k = 3M/L^3 W$ and

$$I = \tfrac{1}{30} \left(\frac{3M}{L^3 W} \right) L^5 W = \tfrac{1}{10} ML^2. \quad \square$$

Radius of Gyration

If the mass M of an object is all concentrated at a distance r from a given axis, then the moment of inertia about that axis is given by the product Mr^2.

Suppose now that we have a plate of mass M (actually any object of mass M will do here), and suppose that l is some line. The object has some moment of inertia I about l. Its *radius of gyration* about l is the distance K for which

$$I = MK^2.$$

Namely, the radius of gyration about l is the distance from l at which all the mass of the object would have to be concentrated to effect the same moment of inertia. The formula for radius of gyration is usually written

(17.6.6)
$$K = \sqrt{I/M}.$$

Problem 4. A homogeneous circular plate of mass M and radius R rotates about an axle that passes through the center of the plate and is perpendicular to the plate. Calculate the moment of inertia and the radius of gyration.

Solution. Take the axle as the z-axis and let the plate rest on the circular region Ω: $x^2 + y^2 \le R^2$. The density of the plate is $M/A = M/\pi R^2$ and $r(x, y) = \sqrt{x^2 + y^2}$. Hence

$$I = \iint_\Omega \frac{M}{\pi R^2} (x^2 + y^2)\, dxdy = \frac{M}{\pi R^2} \int_0^{2\pi} \int_0^R r^3\, dr\, d\theta = \tfrac{1}{2} MR^2.$$

The radius of gyration K is $\sqrt{I/M} = R/\sqrt{2}$.

The circular plate of radius R has the same moment of inertia about the central axle as a circular wire of the same mass with radius $R/\sqrt{2}$. The circular wire is a more efficient carrier of moment of inertia than the circular plate. ☐

The Parallel Axis Theorem

Suppose we have an object of mass M and a line l_M that passes through the center of mass of the object. The object has some moment of inertia about that line; call it I_M. If l is any line parallel to l_M, then the object has a certain moment of inertia about l; call that I. The parallel axis theorem states that

(17.6.7)
$$I = I_M + d^2 M$$

where d is the distance between the axes.

We will prove the theorem under somewhat restrictive assumptions. Assume that the object is a plate of mass M in the shape of a basic region Ω, and assume that l_M is perpendicular to the plate. Call l the z-axis. Call the plane of the plate the xy-plane. (See Figure 17.6.3.)

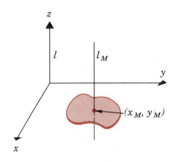

Figure 17.6.3

Denoting the points of Ω by (x, y) we have

$$I - I_M = \iint\limits_\Omega \lambda(x, y)(x^2 + y^2)\, dxdy - \iint\limits_\Omega \lambda(x, y)[(x - x_M)^2 + (y - y_M)^2]\, dxdy$$

$$= \iint\limits_\Omega \lambda(x, y)[2x_M x + 2y_M y - (x_M^2 + y_M^2)]\, dxdy$$

$$= 2x_M \iint\limits_\Omega x\lambda(x, y)\, dxdy + 2y_M \iint\limits_\Omega y\lambda(x, y)\, dxdy$$

$$- (x_M^2 + y_M^2) \iint\limits_\Omega \lambda(x, y)\, dxdy$$

$$= 2x_M^2 M + 2y_M^2 M - (x_M^2 + y_M^2)M = (x_M^2 + y_M^2)M = d^2 M. \quad \square$$

An obvious consequence of the parallel axis theorem is that $I_M \le I$ for all lines l parallel to l_M. To minimize moment of inertia we must pass our axis through the center of mass.

EXERCISES 17.6

(By I_x, I_y, I_z we mean the moments of inertia about the x, y, z axes. This is common notation.)

1. A rectangular plate of mass M, length L, and width W is placed on the xy-plane with center at the origin, long sides parallel to the x-axis. (We assume here that $L \ge W$.) Find I_x, I_y, I_z if the plate is homogeneous. Determine the corresponding radii of gyration K_x, K_y, K_z.

2. Verify that I_x, I_y, I_z are unchanged if the mass density of the plate of Exercise 1 varies directly as the distance from the leftmost side.

3. Determine the center of mass of the plate of Exercise 1 if the mass density varies as in Exercise 2.

4. Show that for any plate in the xy-plane

$$I_z = I_x + I_y.$$

How are the corresponding radii of gyration K_x, K_y, K_z related?

5. A homogeneous plate of mass M in the form of a quarter-disc of radius R is placed in the xy-plane as in Figure 17.6.4. Find I_x, I_y, I_z and the corresponding radii of gyration.

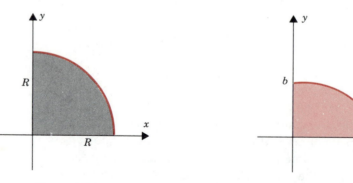

Figure 17.6.4   **Figure 17.6.5**

6. A plate in the xy-plane undergoes a rotation in that plane about its center of mass. Show that I_z remains unchanged.

7. A homogeneous disc of mass M and radius R is to be placed on the xy-plane so that it has moment of inertia I_0 about the z-axis. Where should the disc be placed?

8. A homogeneous plate of mass density λ occupies the region under the curve $y = f(x)$ from $x = a$ to $x = b$. Show that

$$I_x = \tfrac{1}{3}\lambda \int_a^b [f(x)]^3 \, dx \qquad \text{and} \qquad I_y = \lambda \int_a^b x^2 f(x) \, dx.$$

9. A homogeneous plate of mass M in the form of an elliptical quadrant is placed on the xy-plane. (Figure 17.6.5) Find I_x, I_y, I_z.

10. A plate of varying density occupies the region $\Omega = \Omega_1 \cup \Omega_2$ shown in Figure 17.6.6. Find the center of mass of the plate given that the Ω_1 piece has mass M_1 and center of mass (x_1, y_1), and the Ω_2 piece has mass M_2 and center of mass (x_2, y_2).

Figure 17.6.6

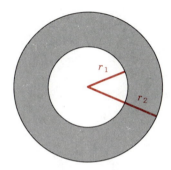

Figure 17.6.7

11. A homogeneous plate of mass M is in the form of a ring. (Figure 17.6.7) Calculate the moment of inertia of the plate
(a) about a diameter.   (b) about a tangent to the inner circle.
(c) about a tangent to the outer circle.

12. Find the moment of inertia of a homogeneous circular wire of mass M and radius r
(a) about a diameter.   (b) about a tangent.   HINT: Use the previous exercise.

13. The plate of Exercise 11 rotates about the axis that is perpendicular to the plate and passes through the center. Find the moment of inertia.

14. Prove the parallel axis theorem for the case where the line through the center of mass lies in the plane of the plate.

15. A plate of mass M has the form of a half disc Ω: $-R \leq x \leq R$, $0 \leq y \leq \sqrt{R^2 - x^2}$. Find the center of mass given that the mass density varies directly as the distance from the curved boundary.

16. Find I_x, I_y, I_z for the plate of Exercise 15.

17. A plate of mass M is in the form of a disc of radius R. Given that the mass density of the plate varies directly as the distance from a point P on the boundary of the plate, locate the center of mass.

18. A plate of mass M is in the form of a right triangle of base b and height h. Given that the mass density of the plate varies directly as the square of the distance from the vertex of the right angle, locate the center of mass of the plate.

19. Use double integrals to justify an assumption we made about centroids in Chapter 6: Formula 6.4.1.

17.7 TRIPLE INTEGRALS

Now that you are familiar with double integrals

$$\iint\limits_{\Omega} f(x, y)\,dxdy,$$

you will find it easy to understand triple integrals

$$\iiint\limits_{T} f(x, y, z)\,dxdydz.$$

Basically the only difference is this: instead of working with functions of two variables continuous on a plane region Ω, we will be working with functions of three variables continuous on some portion T of three-space.

The Triple Integral over a Box

Figure 17.7.1

For double integration we began with a rectangle

$$R: \quad a_1 \leq x \leq a_2, \quad b_1 \leq y \leq b_2.$$

For triple integration we begin with a *box* (a rectangular solid)

$$\Pi: \quad a_1 \leq x \leq a_2, \quad b_1 \leq y \leq b_2, \quad c_1 \leq z \leq c_2. \qquad \text{(Figure 17.7.1)}$$

To partition this box, we first partition the edges. Taking

$$P_1 = \{x_0, \dots, x_m\} \quad \text{as a partition of } [a_1, a_2],$$
$$P_2 = \{y_0, \dots, y_n\} \quad \text{as a partition of } [b_1, b_2],$$
$$P_3 = \{z_0, \dots, z_q\} \quad \text{as a partition of } [c_1, c_2],$$

we form the set

$$P = P_1 \times P_2 \times P_3 = \{(x_i, y_j, z_k): x_i \in P_1, \; y_j \in P_2, \; z_k \in P_3\}\dagger$$

---

† $P_1 \times P_2 \times P_3$ is a Cartesian product. Cartesian products are discussed briefly in Appendix A.1.

and call this a *partition of* Π. P breaks up Π into $m \times n \times q$ nonoverlapping boxes

$$\Pi_{ijk}: \quad x_{i-1} \le x \le x_i, \quad y_{j-1} \le y \le y_j, \quad z_{k-1} \le z \le z_k.$$

A typical such box is pictured in Figure 17.7.2.

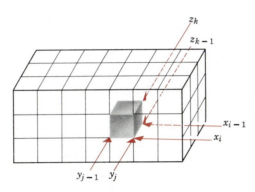

Figure 17.7.2

Taking

$$M_{ijk} \text{ as the maximum value of } f \text{ on } \Pi_{ijk}$$

and

$$m_{ijk} \text{ as the minimum value of } f \text{ on } \Pi_{ijk},$$

we form the *upper sum*

$$U_f(P) = \sum_{i=1}^{m} \sum_{j=1}^{n} \sum_{k=1}^{q} M_{ijk}(\text{volume of } \Pi_{ijk}) = \sum_{i=1}^{m} \sum_{j=1}^{n} \sum_{k=1}^{q} M_{ijk} \, \Delta x_i \, \Delta y_j \, \Delta z_k$$

and the *lower sum*

$$L_f(P) = \sum_{i=1}^{m} \sum_{j=1}^{n} \sum_{k=1}^{q} m_{ijk}(\text{volume of } \Pi_{ijk}) = \sum_{i=1}^{m} \sum_{j=1}^{n} \sum_{k=1}^{q} m_{ijk} \, \Delta x_i \, \Delta y_j \, \Delta z_k.$$

As in the case of functions of one and two variables, it turns out that, with f continuous on Π, there is one and only one number I that satisfies the inequality

$$L_f(P) \le I \le U_f(P) \qquad \text{for all partitions } P \text{ of } \Pi.$$

DEFINITION 17.7.1   THE TRIPLE INTEGRAL OVER A BOX Π

The unique number I that satisfies the inequality

$$L_f(P) \le I \le U_f(P) \qquad \text{for all partitions } P \text{ of } \Pi$$

is called the *triple integral* of f over Π and is denoted by

$$\iiint_{\Pi} f(x, y, z) \, dxdydz.$$

The Triple Integral over a More General Solid

We start with a three-dimensional, bounded, open, connected set and adjoin to it the boundary. We now have a three-dimensional, bounded, closed, connected set T. We assume that T is a *basic solid*; that is, we assume that the boundary of T consists of a finite number of continuous surfaces $z = \alpha(x, y)$, $y = \beta(x, z)$, $x = \gamma(y, z)$.

Now let's suppose that f is some function continuous on T. To define the triple integral of f over T we first encase T in a rectangular box Π. We then extend f to all of Π by defining f to be zero outside of T. This extended function f is bounded on Π, and it is continuous on all of Π except possibly at the boundary of T. In spite of these possible discontinuities, f is still integrable over Π; that is, there still exists a unique number I such that

$$L_f(P) \leq I \leq U_f(P) \qquad \text{for all partitions } P \text{ of } \Pi.$$

(We will not attempt to prove this.) The number I is by definition the triple integral

$$\iiint_\Pi f(x, y, z)\, dxdydz.$$

We define the triple integral over T by setting

(17.7.2)
$$\iiint_T f(x, y, z)\, dxdydz = \iiint_\Pi f(x, y, z)\, dxdydz.$$

Volume as a Triple Integral

The simplest triple integral of interest is the triple integral of the function that is constantly one on T. This gives the volume of T:

(17.7.3)
$$\text{volume of } T = \iiint_T dxdydz.$$

Proof. Set $f(x, y, z) = 1$ for all (x, y, z) in T. Encase T in a box Π. Define f to be zero outside of T. An arbitrary partition P of Π breaks up T into little boxes Π_{ijk}. Note that

$L_f(P) = $ the sum of the volumes of all the Π_{ijk} that are contained in T

$U_f(P) = $ the sum of the volumes of all the Π_{ijk} that intersect T.

It follows that

$$L_f(P) \leq \text{the volume of } T \leq U_f(P).$$

The arbitrariness of P gives the formula. $\square$

Some Properties of the Triple Integral

Below we give without proof the salient elementary properties of the triple integral. They are all analogous to what you saw in the one- and two-variable cases. The T referred to is a basic solid. The functions f and g are assumed to be continuous on T.

I. The triple integral is linear:

$$\iiint_T [\alpha f(x, y, z) + \beta g(x, y, z)]\, dxdydz = \alpha \iiint_T f(x, y, z)\, dxdydz + \beta \iiint_T g(x, y, z)\, dxdydz.$$

II. It preserves order:

if $f \geq 0$ on T, then $\iiint_T f(x, y, z)\, dxdydz \geq 0$;

if $f \leq g$ on T, then $\iiint_T f(x, y, z)\, dxdydz \leq \iiint_T g(x, y, z)\, dxdydz.$

III. It is additive:

if T is broken up into a finite number of basic solids $T_1, \ldots, T_n$, then

$$\iiint_T f(x, y, z)\, dxdydz = \iiint_{T_1} f(x, y, z)\, dxdydz + \cdots + \iiint_{T_n} f(x, y, z)\, dxdydz.$$

IV. It satisfies a mean-value condition: namely, there is a point (x_0, y_0, z_0) in T for which

$$\iiint_T f(x, y, z)\, dxdydz = f(x_0, y_0, z_0)(\text{the volume of } T).$$

We call $f(x_0, y_0, z_0)$ the average value of f on T.

The notion of average given above enables us to write

(17.7.4)
$$\iiint_T f(x, y, z)\, dxdydz = \left(\begin{array}{c}\text{the average value} \\ \text{of } f \text{ on } T\end{array}\right)(\text{the volume of } T).$$

We can also take weighted averages: if f and g are continuous and g is nonnegative on T, then there is a point (x_0, y_0, z_0) in T for which

(17.7.5)
$$\iiint_T f(x, y, z)g(x, y, z)\, dxdydz = f(x_0, y_0, z_0)\iiint_T g(x, y, z)\, dxdydz.$$

As you would expect, we call $f(x_0, y_0, z_0)$ the g-weighted average of f on T.

The formulas for mass, center of mass, and moments of inertia derived in the previous section for two-dimensional plates are easily extended to three-dimensional objects.

Suppose that T is an object in the form of a basic solid. If T has constant mass density λ (here density is mass per unit volume), then the mass of T is the density λ times the volume of T:

$$M = \lambda V.$$

If the mass density varies continuously over T, say $\lambda = \lambda(x, y, z)$, then *the mass of T is the average density of T times the volume of T*:

(17.7.6)
$$M = \iiint_T \lambda(x, y, z)\, dxdydz.$$

The coordinates of the center of mass (x_M, y_M, z_M) are density-weighted averages:

(17.7.7)
$$x_M M = \iiint_T x\lambda(x, y, z)\, dxdydz, \qquad \text{etc.}$$

If the object T is homogeneous (constant mass density M/V), then the center of mass of T depends only on the geometry of T and falls on the centroid $(\bar{x}, \bar{y}, \bar{z})$ of the space occupied by T. The density is irrelevant. The coordinates of the centroid are simple averages over T:

(17.7.8)
$$\bar{x} V = \iiint_T x\, dxdydz, \qquad \text{etc.}$$

The moment of inertia of T about a line is given by the formula

(17.7.9)
$$I = \iiint_T \lambda(x, y, z)[r(x, y, z)]^2\, dxdydz.$$

Here $\lambda(x, y, z)$ is the mass density of T at (x, y, z) and $r(x, y, z)$ is the distance of (x, y, z) from the line in question. The moments of inertia about the x, y, z axes are again denoted by I_x, I_y, I_z.

All of this should be readily understandable. Techniques for evaluating triple integrals take up the next three sections.

EXERCISES 17.7

1. Let $f = f(x, y)$ be a function continuous and nonnegative on a basic region Ω and set

$$T = \{(x, y, z): \ (x, y) \in \Omega, \ 0 \le z \le f(x, y)\}.$$

Compare

$$\iiint_T dx\,dy\,dz \quad \text{to} \quad \iint_\Omega f(x, y) \, dx\,dy.$$

2. Set $f(x, y, z) = xyz$ on Π: $0 \le x \le 1, \ 0 \le y \le 1, \ 0 \le z \le 1$ and take P as the partition $P_1 \times P_2 \times P_3$.
   (a) Find $L_f(P)$ and $U_f(P)$ given that

$$P_1 = \{x_0, \ \ldots \ , x_m\}, \qquad P_2 = \{y_0, \ \ldots \ , y_n\}, \qquad P_3 = \{z_0, \ \ldots \ , z_q\}$$

   are all arbitrary partitions of $[0, 1]$.
   (b) Use your answer to (a) to calculate

$$\iiint_\Pi xyz \, dx\,dy\,dz.$$

3. Calculate

$$\iiint_\Pi xy \, dx\,dy\,dz \qquad \text{where} \quad \Pi: \ 0 \le x \le a, \ 0 \le y \le b, \ 0 \le z \le c.$$

4. Let T be a basic solid of varying mass density $\lambda = \lambda(x, y, z)$. The moment of inertia of T about the xy-plane is defined by setting

$$I_{xy} = \iiint_T \lambda(x, y, z)z^2 \, dx\,dy\,dz.$$

   The other plane moments of inertia, I_{xz} and I_{yz}, have comparable definitions. Express I_x, I_y, I_z in terms of the plane moments of inertia.

5. A box Π_1: $0 \le x \le 2a, \ 0 \le y \le 2b, \ 0 \le z \le 2c$ is cut away from a larger box Π_0: $0 \le x \le 2A, \ 0 \le y \le 2B, \ 0 \le z \le 2C$. Locate the centroid of the remaining solid.

6. Show that, if f is continuous and nonnegative on a basic solid T, then the triple integral of f over T is nonnegative.

7. Calculate the mass M of the cube Π: $0 \le x \le a, \ 0 \le y \le a, \ 0 \le z \le a$ given that the mass density varies directly with the distance from the face on the xy-plane.

8. Locate the center of mass of the cube of Exercise 7.

9. Find the moment of inertia I_z of the cube of Exercise 7.

17.8 REDUCTION TO REPEATED INTEGRALS

(In this section we give no proofs. You can assume that all the solids that appear are basic solids and all the functions that you encounter are continuous.)

In Figure 17.8.1 we have sketched a solid T. The projection of T onto the xy-plane

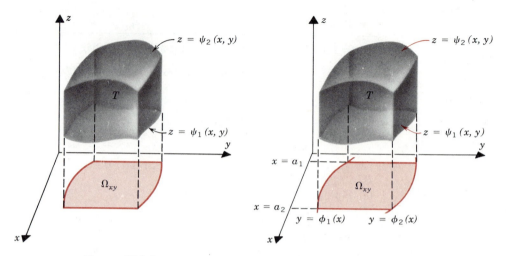

Figure 17.8.1   **Figure 17.8.2**

has been labeled Ω_{xy}. The solid T is then the set of all (x, y, z) with

$$(x, y) \text{ in } \Omega_{xy} \quad \text{and} \quad \psi_1(x, y) \leq z \leq \psi_2(x, y).$$

The triple integral over T can be evaluated by setting

(∗) $$\iiint_T f(x, y, z) \, dxdydz = \iint_{\Omega_{xy}} \left(\int_{\psi_1(x, y)}^{\psi_2(x, y)} f(x, y, z) \, dz \right) dxdy.$$

Moving over to Figure 17.8.2 we see that in this case Ω_{xy} is the region

$$a_1 \leq x \leq a_2, \quad \phi_1(x) \leq y \leq \phi_2(x).$$

T itself is then the set of all (x, y, z) with

$$a_1 \leq x \leq a_2, \quad \phi_1(x) \leq y \leq \phi_2(x), \quad \psi_1(x, y) \leq z \leq \psi_2(x, y).$$

The triple integral over T can then be expressed by three ordinary integrals:

$$\iiint_T f(x, y, z) \, dxdydz = \int_{a_1}^{a_2} \left[\int_{\phi_1(x)}^{\phi_2(x)} \left(\int_{\psi_1(x,y)}^{\psi_2(x,y)} f(x, y, z) \, dz \right) dy \right] dx.$$

It is customary to omit the brackets and parentheses and write

(17.8.1) $$\iiint_T f(x, y, z) \, dxdydz = \int_{a_1}^{a_2} \int_{\phi_1(x)}^{\phi_2(x)} \int_{\psi_1(x,y)}^{\psi_2(x,y)} f(x, y, z) \, dz \, dy \, dx. \quad †$$

Here we first integrate with respect to z [from $z = \psi_1(x, y)$ to $z = \psi_2(x, y)$], then with respect to y [from $y = \phi_1(x)$ to $y = \phi_2(x)$], and finally with respect to x [from $x = a_1$ to $x = a_2$].

---

† This formula is formula (∗) taken one step further. Usually we skip the double integral stage and go directly to three integrals.

There is nothing sacred about this order of integration. Other orders of integration are possible and in some cases more convenient. Suppose, for example, that the projection of T onto the xz-plane is a region of the form

$$\Omega_{xz}: \quad a_1 \leq z \leq a_2, \quad \phi_1(z) \leq x \leq \phi_2(z).$$

If T is the set of all (x, y, z) with

$$a_1 \leq z \leq a_2, \quad \phi_1(z) \leq x \leq \phi_2(z), \quad \psi_1(x, z) \leq y \leq \psi_2(x, z)$$

then

$$\iiint_T f(x, y, z)\, dxdydz = \int_{a_1}^{a_2} \int_{\phi_1(z)}^{\phi_2(z)} \int_{\psi_1(x,z)}^{\psi_2(x,z)} f(x, y, z)\, dy\, dx\, dz.$$

In this case we integrate first with respect to y, then with respect to x, and finally with respect to z. Still four other orders of integration are possible.

Problem 1. Use triple integration to find the volume of the tetrahedron T shown in Figure 17.8.3. Where is the centroid?

Figure 17.8.3

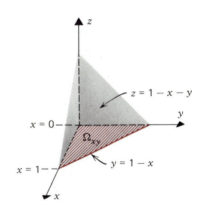

Figure 17.8.4

Solution. The volume of T is given by the triple integral

$$V = \iiint_T dxdydz.$$

To evaluate this triple integral we can project T onto any one of the three coordinate planes. We will project onto the xy-plane. The base region is then the triangle

$$\Omega_{xy}: \quad 0 \leq x \leq 1, \quad 0 \leq y \leq 1 - x. \qquad \text{(Figure 17.8.4)}$$

Since the inclined face is part of the plane $z = 1 - x - y$, we have T as the set of all (x, y, z) with

$$0 \leq x \leq 1, \quad 0 \leq y \leq 1 - x, \quad 0 \leq z \leq 1 - x - y.$$

It follows that

$$V = \iiint_T dxdydz = \int_0^1 \int_0^{1-x} \int_0^{1-x-y} dz\,dy\,dx$$

$$= \int_0^1 \int_0^{1-x} (1-x-y)\,dy\,dx$$

$$= \int_0^1 \left[(1-x)y - \tfrac{1}{2}y^2 \right]_0^{1-x} dx$$

$$= \int_0^1 \tfrac{1}{2}(1-x)^2\,dx = \left[-\tfrac{1}{6}(1-x)^3 \right]_0^1 = \tfrac{1}{6}.$$

By symmetry $\bar{x} = \bar{y} = \bar{z}$. We can calculate $\bar{x}$ as follows:

$$\bar{x}\,V = \iiint_T x\,dxdydz = \int_0^1 \int_0^{1-x} \int_0^{1-x-y} x\,dz\,dy\,dx = \tfrac{1}{24}.$$

check this

Since $V = \tfrac{1}{6}$, we have $\bar{x} = \tfrac{1}{4}$. The centroid is the point $(\tfrac{1}{4}, \tfrac{1}{4}, \tfrac{1}{4})$. □

Problem 2. Find the mass of a solid right circular cylinder of radius r and height h given that the mass density varies directly with the distance from one of the bases.

Solution. Call the solid T. In the setup of Figure 17.8.5 we can characterize T by the following inequalities:

$$-r \le x \le r, \quad -\sqrt{r^2 - x^2} \le y \le \sqrt{r^2 - x^2}, \quad 0 \le z \le h.$$

The first two inequalities define the base region Ω_{xy}. Supposing that the density varies directly with the distance from the lower base, we have $\lambda(x, y, z) = kz$. Then

$$M = \iiint_T kz\,dxdydz$$

$$= \int_{-r}^r \int_{-\sqrt{r^2-x^2}}^{\sqrt{r^2-x^2}} \int_0^h kz\,dz\,dy\,dx$$

$$= \int_{-r}^r \int_{-\sqrt{r^2-x^2}}^{\sqrt{r^2-x^2}} \tfrac{1}{2}kh^2\,dy\,dx$$

$$= 4 \int_0^r \int_0^{\sqrt{r^2-x^2}} \tfrac{1}{2}kh^2\,dy\,dx$$

$$= 2kh^2 \int_0^r \sqrt{r^2 - x^2}\,dx$$

$$= 2kh^2(\tfrac{1}{4}\pi r^2) = \tfrac{1}{2}kh^2r^2\pi. □$$

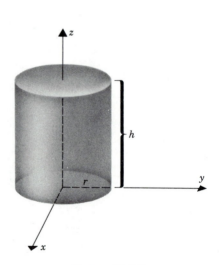

Figure 17.8.5

Remark. In Problem 2 we would have profited by not skipping the double integral stage; namely, we could have written

$$M = \iint_{\Omega_{xy}} \left(\int_0^h kz\ dz \right) dxdy = \iint_{\Omega_{xy}} \tfrac{1}{2}kh^2\ dxdy = \tfrac{1}{2}kh^2(\text{area of } \Omega_{xy}) = \tfrac{1}{2}kh^2 r^2 \pi. \quad \square$$

Problem 3. Integrate $f(x, y, z) = yz$ over that part of the first octant $x \geq 0$, $y \geq 0$, $z \geq 0$ that is cut off by the ellipsoid

$$\frac{x^2}{a^2} + \frac{y^2}{b^2} + \frac{z^2}{c^2} = 1.$$

Solution. Call the solid T. The upper boundary of T has equation

$$z = \psi(x, y) = \frac{c}{ab}\sqrt{a^2b^2 - b^2x^2 - a^2y^2}.$$

This surface intersects the xy-plane in the curve

$$y = \phi(x) = \frac{b}{a}\sqrt{a^2 - x^2}.$$

We can take

$$\Omega_{xy}: \quad 0 \leq x \leq a, \quad 0 \leq y \leq \phi(x)$$

as the base region (see Figure 17.8.6) and characterize T as the set of all (x, y, z) with

$$0 \leq x \leq a, \quad 0 \leq y \leq \phi(x), \quad 0 \leq z \leq \psi(x, y).$$

We can therefore calculate the triple integral by evaluating

$$\int_0^a \int_0^{\phi(x)} \int_0^{\psi(x, y)} yz\ dz\ dy\ dx.$$

A straightforward (but somewhat lengthy) computation that you can carry out yourself gives an answer of $\frac{1}{15}ab^2c^2$. $\quad \square$

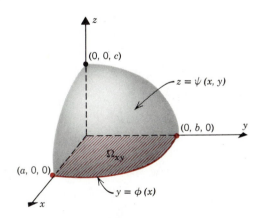

Figure 17.8.6

Another Solution. We return to Problem 3 but this time carry out the integration in a different order. In Figure 17.8.7 you can see the same solid projected this time onto the yz-plane. In terms of y and z the curved surface has equation

$$x = \Psi(y, z) = \frac{a}{bc} \sqrt{b^2 c^2 - c^2 y^2 - b^2 z^2}.$$

This surface intersects the yz-plane in the curve

$$y = \Phi(z) = \frac{b}{c} \sqrt{c^2 - z^2}.$$

We can take

$$\Omega_{yz}: \quad 0 \le z \le c, \quad 0 \le y \le \Phi(z)$$

as the base region and characterize T as the set of all (x, y, z) with

$$0 \le z \le c, \quad 0 \le y \le \Phi(z), \quad 0 \le x \le \Psi(y, z).$$

This leads to the repeated integral

$$\int_0^c \int_0^{\Phi(z)} \int_0^{\Psi(y,z)} yz \, dx \, dy \, dz$$

which, as you can check, also gives $\frac{1}{15} ab^2 c^2$. $\square$

Figure 17.8.7

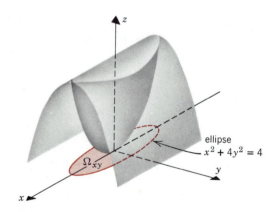

Figure 17.8.8

Problem 4. Use triple integration to find the volume of the solid T bounded above by the parabolic cylinder $z = 4 - y^2$ and bounded below by the elliptic paraboloid $z = x^2 + 3y^2$.

Solution. Solving the two equations simultaneously, we have

$$4 - y^2 = x^2 + 3y^2 \quad \text{and thus} \quad x^2 + 4y^2 = 4.$$

This tells us that the two surfaces intersect in a space curve that lies along the elliptic cylinder $x^2 + 4y^2 = 4$. The projection of this intersection onto the xy-plane is the ellipse $x^2 + 4y^2 = 4$. (See Figure 17.8.8.)

The projection of T onto the xy-plane is the region

$$\Omega_{xy}: \quad -2 \leq x \leq 2, \quad -\tfrac{1}{2}\sqrt{4 - x^2} \leq y \leq \tfrac{1}{2}\sqrt{4 - x^2}.$$

The solid T is then the set of all (x, y, z) with

$$-2 \leq x \leq 2, \quad -\tfrac{1}{2}\sqrt{4 - x^2} \leq y \leq \tfrac{1}{2}\sqrt{4 - x^2}, \quad x^2 + 3y^2 \leq z \leq 4 - y^2.$$

Its volume is given by

$$V = \int_{-2}^{2} \int_{-\sqrt{4-x^2}/2}^{\sqrt{4-x^2}/2} \int_{x^2+3y^2}^{4-y^2} dz\, dy\, dx = 4 \int_{0}^{2} \int_{0}^{\sqrt{4-x^2}/2} \int_{x^2+3y^2}^{4-y^2} dz\, dy\, dx = 4\pi. \quad \square$$

$\llcorner$ check this

EXERCISES 17.8

Evaluate.

1. $\displaystyle\int_0^a \int_0^b \int_0^c dx\, dy\, dz.$

2. $\displaystyle\int_0^1 \int_0^x \int_0^y y\, dz\, dy\, dx.$

3. $\displaystyle\int_0^1 \int_1^{2y} \int_0^x (x + 2z)\, dz\, dx\, dy.$

4. $\displaystyle\int_0^1 \int_{1-x}^{1+x} \int_0^{xy} 4z\, dz\, dy\, dx.$

5. (*Separated variables over a box*) Set Π: $a_1 \leq x \leq a_2,\ b_1 \leq y \leq b_2,\ c_1 \leq z \leq c_2$. Show that, if f is continuous on $[a_1, a_2]$, g is continuous on $[b_1, b_2]$ and h is continuous on $[c_1, c_2]$, then

$$\iiint_\Pi f(x)g(y)h(z)\, dxdydz = \left(\int_{a_1}^{a_2} f(x)\, dx \right)\left(\int_{b_1}^{b_2} g(y)\, dy \right)\left(\int_{c_1}^{c_2} h(z)\, dz \right).$$

Evaluate taking Π: $0 \leq x \leq 1,\ 0 \leq y \leq 2,\ 0 \leq z \leq 3$.

6. $\displaystyle\iiint_\Pi x^3 y^2 z\, dxdydz.$

7. $\displaystyle\iiint_\Pi x^2 y^2 z^2\, dxdydz.$

The mass density of a box Π: $0 \leq x \leq a,\ 0 \leq y \leq b,\ 0 \leq z \leq c$ varies directly with the product xyz.

8. Calculate the mass of Π.   **9.** Locate the center of mass.

10. Determine the moment of inertia of Π (a) about the vertical line that passes through the point (a, b, c). (b) about the vertical line that passes through the center of mass.

A homogeneous solid T of mass M consists of all points (x, y, z) with $0 \leq x \leq 1,\ 0 \leq y \leq 1$, $0 \leq z \leq 1 - y$.

11. Sketch T.   **12.** Find the volume of T.   **13.** Locate the center of mass.

14. Find the moments of inertia of T about the coordinate axes.

Express 15–20 by repeated integrals. Do not evaluate.

15. The mass of a ball $x^2 + y^2 + z^2 \leq r^2$ given that the density varies directly with the distance from the outer shell.

16. The mass of the solid bounded above by $z = 1$ and bounded below by $z = \sqrt{x^2 + y^2}$ given that the density varies directly with the distance from the origin. Identify the solid.

17. The volume of the solid bounded above by the parabolic cylinder $z = 1 - y^2$, below by the plane $2x + 3y + z + 10 = 0$, and on the sides by the circular cylinder $x^2 + y^2 - x = 0$.

18. The volume of the solid bounded above by the paraboloid $z = 4 - x^2 - y^2$ and bounded below by the parabolic cylinder $z = 2 + y^2$.

19. The mass of the solid bounded by the elliptic paraboloids $z = 4 - x^2 - \frac{1}{4}y^2$ and $z = 3x^2 + \frac{1}{4}y^2$ given that the density varies directly with the vertical distance from the lower surface.

20. The mass of the solid bounded by the paraboloid $x = z^2 + 2y^2$ and the parabolic cylinder $x = 4 - z^2$ given that the density varies directly with the distance from the z-axis.

21. Find the volume of the portion of the first octant bounded by the planes $z = x$, $y - x = 2$ and the cylinder $y = x^2$. Where is the centroid?

22. Find the mass of a block in the shape of a unit cube given that the density varies directly (a) with the distance from one of the faces. (b) with the square of the distance from one of the vertices.

23. Find the volume and the centroid of the solid bounded above by the cylindrical surface $x^2 + z = 4$, below by the plane $x + z = 2$, and on the sides by the planes $y = 0$ and $y = 3$.

24. Show that, if $(\bar{x}, \bar{y}, \bar{z})$ is the centroid of a solid T, then

$$\iiint_T (x - \bar{x})\, dxdydz = 0, \quad \iiint_T (y - \bar{y})\, dxdydz = 0, \quad \iiint_T (z - \bar{z})\, dxdydz = 0.$$

25. Taking a, b, c as positive, find the volume of the tetrahedron with vertices $(0, 0, 0)$, $(a, 0, 0)$, $(0, b, 0)$, and $(0, 0, c)$. Where is the centroid?

26. A homogeneous solid of mass M in the form and position of the tetrahedron of Figure 17.8.3 rotates about the z-axis with moment of inertia I_z. Find I_z.

27. A homogeneous box of mass M has edges a, b, c. Calculate the moment of inertia
    (a) about the edge c.
    (b) about the line that passes through the center of the box and is parallel to the edge c.
    (c) about the line that passes through the center of the face bc and is parallel to the edge c.

28. Where is the centroid of the solid bounded above by the plane $z = 1 + x + y$, below by the plane $z = -2$, and on the sides by the planes $x = 1$, $x = 2$, $y = 1$, $y = 2$?

29. Let T be the solid bounded above by the plane $z = y$, below by the xy-plane, and on the sides by the planes $x = 0$, $x = 1$, $y = 0$, $y = 1$. Find the mass of T given that the density varies directly with the square of the distance from the origin. Where is the center of mass?

30. What can you conclude about T given that

$$\iiint_T f(x, y, z)\, dxdydz = 0$$

(a) for every continuous function f that is odd in x? (b) for every continuous function f that is odd in y? (c) for every continuous function f that is odd in z? (d) for every continuous function f that satisfies the relation $f(-x, -y, -z) = -f(x, y, z)$?

31. (a) Integrate $f(x, y, z) = x + y^3 + z$ over the unit ball centered at the origin.
(b) Integrate $f(x, y, z) = a_1x + a_2y + a_3z + a_4$ over the unit ball centered at the origin.

32. Integrate $f(x, y, z) = x^2y^2$ over the solid bounded above by the cylinder $y^2 + z = 4$, below by the plane $y + z = 2$, and on the sides by the planes $x = 0$ and $x = 2$.

33. Find the mass of the solid of Problem 4 given that the density varies directly with $|x|$.

34. Find the volume of the solid bounded by the paraboloids $z = 2 - x^2 - y^2$ and $z = x^2 + y^2$.

35. Find the mass and the center of mass of the solid of Exercise 23 given that the density varies directly with $1 + y$.

36. Let T be a solid with volume

$$V = \iiint_T dx\,dy\,dz = \int_0^2 \int_0^{9-x^2} \int_0^{2-x} dz\,dy\,dx.$$

Sketch T and fill in the blanks.

(a) $V = \int_\square^\square \int_\square^\square \int_\square^\square dy\,dx\,dz.$ (b) $V = \int_\square^\square \int_\square^\square \int_\square^\square dy\,dz\,dx.$

(c) $V = \int_0^5 \int_\square^\square \int_\square^\square dz\,dx\,dy + \int_5^9 \int_\square^\square \int_\square^\square dz\,dx\,dy.$

37. Let T be a solid with volume

$$V = \iiint_T dx\,dy\,dz = \int_0^3 \int_x^{6-x} \int_0^{2x} dz\,dy\,dx.$$

Sketch T and fill in the blanks.

(a) $V = \int_\square^\square \int_\square^\square \int_\square^\square dy\,dx\,dz.$ (b) $V = \int_\square^\square \int_\square^\square \int_\square^\square dy\,dz\,dx.$

(c) $V = \int_\square^\square \int_\square^\square \int_\square^\square dx\,dy\,dz + \int_\square^\square \int_\square^\square \int_\square^\square dx\,dy\,dz.$

For the remaining exercises, let V be the volume of the solid T enclosed by the parabolic cylinder $y = 4 - z^2$ and the V-shaped cylinder $y = |x|$. Let $\Omega_{xy}, \Omega_{yz}, \Omega_{xz}$ be the projections of T onto the xy, yz, xz planes, respectively. Fill in the blanks.

38. (a) $V = \iint_{\Omega_{xy}} \boxed{}\, dx\,dy.$ (b) $V = \iint_{\Omega_{xy}} \left(\int_\square^\square dz \right) dx\,dy.$

(c) $V = \int_\square^\square \int_\square^\square \int_\square^\square dz\,dy\,dx.$ (d) $V = \int_\square^\square \int_\square^\square \int_\square^\square dz\,dx\,dy.$

39. (a) $V = \iint_{\Omega_{yz}} \boxed{}\, dy\,dz.$ (b) $V = \iint_{\Omega_{yz}} \left(\int_\square^\square dx \right) dy\,dz.$

(c) $V = \int_\square^\square \int_\square^\square \int_\square^\square dx\,dz\,dy.$ (d) $V = \int_\square^\square \int_\square^\square \int_\square^\square dx\,dy\,dz.$

40. (a) $V = \iint_{\Omega_{xz}} \boxed{}\, dx\,dz.$ (b) $V = \iint_{\Omega_{xz}} \left(\int_\square^\square dy \right) dx\,dz.$

(c) $V = \int_\square^\square \int_\square^\square \int_\square^\square dy\,dx\,dz.$ (d) $V = \int_{-2}^0 \int_\square^\square \int_\square^\square dy\,dz\,dx + \int_0^2 \int_\square^\square \int_\square^\square dy\,dz\,dx.$

17.9   CYLINDRICAL COORDINATES

Introduction to Cylindrical Coordinates

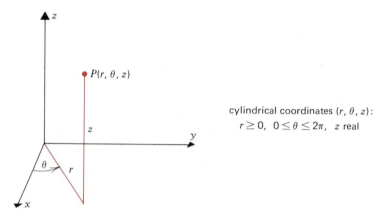

cylindrical coordinates (r, θ, z):
$r \geq 0, \ 0 \leq \theta \leq 2\pi, \ z$ real

Figure 17.9.1

The cylindrical coordinates (r, θ, z) of a point P in xyz-space are shown geometrically in Figure 17.9.1.   The first two coordinates, r and θ, are the usual plane polar coordinates, except that r is taken as nonnegative and θ is restricted to the interval $[0, 2\pi]$.   The third coordinate is the third rectangular coordinate z.

In rectangular coordinates the coordinate surfaces

$$x = x_0, \quad y = y_0, \quad z = z_0$$

are three mutually perpendicular planes.   In cylindrical coordinates the coordinate surfaces take the form

$$r = r_0, \quad \theta = \theta_0, \quad z = z_0.$$

These surfaces are drawn in Figure 17.9.2.

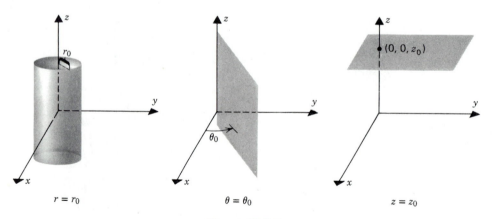

Figure 17.9.2

The first surface is a circular cylinder of radius r_0. The central axis of the cylinder is the z-axis. The surface $\theta = \theta_0$ is a vertical half-plane hinged at the z-axis. The plane stands at an angle of θ_0 radians from the positive x-axis. The last coordinate surface is the plane $z = z_0$ of rectangular coordinates.

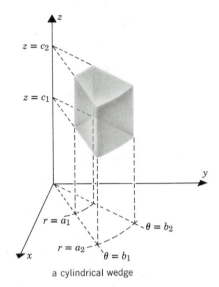

a cylindrical wedge

Figure 17.9.3

The xyz-solids easiest to describe in cylindrical coordinates are the *cylindrical wedges*. One is pictured in Figure 17.9.3. That wedge consists of all points (x, y, z) that have cylindrical coordinates (r, θ, z) in the box

$$\Pi: \quad a_1 \leq r \leq a_2, \quad b_1 \leq \theta \leq b_2, \quad c_1 \leq z \leq c_2.$$

Evaluating Triple Integrals Using Cylindrical Coordinates

Suppose that T is some basic solid in xyz-space, not necessarily a wedge. If T is the set of all (x, y, z) with cylindrical coordinates in some basic solid S in $r\theta z$-space, then

(17.9.1)
$$\iiint_T f(x, y, z)\, dxdydz = \iiint_S f(r \cos \theta, r \sin \theta, z)\, r\, drd\theta dz.$$

Derivation of (17.9.1). We will carry out the argument on the assumption that T is projectable onto some basic region Ω_{xy} of the xy-plane. (It is for such solids that the formula is most useful.) T has some lower boundary $z = \psi_1(x, y)$ and some upper boundary

$z = \psi_2(x, y)$. T is then the set of all (x, y, z) with

$$(x, y) \in \Omega_{xy} \quad \text{and} \quad \psi_1(x, y) \leq z \leq \psi_2(x, y).$$

The region Ω_{xy} has polar coordinates in some set $\Omega_{r\theta}$ (which we assume is a basic region). Then S is the set of all (r, θ, z) with

$$(r, \theta) \in \Omega_{r\theta} \quad \text{and} \quad \psi_1(r \cos \theta, r \sin \theta) \leq z \leq \psi_2(r \cos \theta, r \sin \theta).$$

Therefore

$$\iiint_T f(x, y, z)\, dxdydz = \iint_{\Omega_{xy}} \left(\int_{\psi_1(x,\, y)}^{\psi_2(x,\, y)} f(x, y, z)\, dz \right) dxdy$$

$$(17.5.3) \quad = \iint_{\Omega_{r\theta}} \left(\int_{\psi_1(r \cos \theta,\, r \sin \theta)}^{\psi_2(r \cos \theta,\, r \sin \theta)} f(r \cos \theta, r \sin \theta, z)\, dz \right) r\, drd\theta$$

$$= \iiint_S f(r \cos \theta, r \sin \theta, z)\, r\, drd\theta dz. \quad \square$$

Volume Formula

If $f(x, y, z) = 1$ for all (x, y, z) in T, then the formula reduces to

$$\iiint_T dxdydz = \iiint_S r\, drd\theta dz.$$

The triple integral on the left is the volume of T. In summary, if T is a basic solid in xyz-space and the cylindrical coordinates of T constitute a basic solid S in $r\theta z$-space, then the volume of T is given by the formula

(17.9.2)

$$V = \iiint_S r\, drd\theta dz.$$

Calculations

Cylindrical coordinates are particularly useful when an axis of symmetry is present. The axis of symmetry is then taken as the z-axis.

Problem 1. Find the mass of a solid cylinder T of radius R and height h given that the density varies directly with the distance from the axis of the cylinder.

Solution. Place the cylinder T on the xy-plane so that the axis of T coincides with the z-axis. The density function then takes the form $\lambda(x, y, z) = k\sqrt{x^2 + y^2}$ and T consists of all points (x, y, z) with cylindrical coordinates (r, θ, z) in the set

$$S: \quad 0 \le r \le R, \quad 0 \le \theta \le 2\pi, \quad 0 \le z \le h.$$

Therefore

$$M = \iiint_T k\sqrt{x^2 + y^2} \, dxdydz = \iiint_S (kr) r \, drd\theta dz$$

$$= k \int_0^R \int_0^{2\pi} \int_0^h r^2 \, dz \, d\theta \, dr = \tfrac{2}{3} k\pi R^3 h. \quad \square$$

└── check this

Problem 2. Use cylindrical coordinates to find the volume of the solid T bounded above by the plane $z = y$ and below by the paraboloid $z = x^2 + y^2$.

Solution. In cylindrical coordinates the plane has equation $z = r \sin \theta$ and the paraboloid has equation $z = r^2$. Solving these two equations simultaneously, we have $r = \sin \theta$. This tells us that the two surfaces intersect in a space curve that lies along the circular cylinder $r = \sin \theta$. The projection of this intersection onto the xy-plane is the circle with polar equation $r = \sin \theta$. (See Figure 17.9.4.)

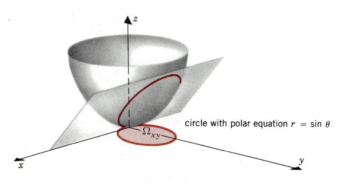

circle with polar equation $r = \sin \theta$

Figure 17.9.4

The base region Ω_{xy} is thus the set of all (x, y) with polar coordinates in the set

$$0 \le \theta \le \pi, \quad 0 \le r \le \sin \theta.$$

T itself is the set of all (x, y, z) with cylindrical coordinates in the set

$$S: \quad 0 \le \theta \le \pi, \quad 0 \le r \le \sin \theta, \quad r^2 \le z \le r \sin \theta. \qquad \text{(check this)}$$

Therefore

$$V = \iiint_T dxdydz = \iiint_S r \, drd\theta dz$$

$$= \int_0^\pi \int_0^{\sin\theta} \int_{r^2}^{r\sin\theta} r \, dz \, dr \, d\theta$$

$$= \int_0^\pi \int_0^{\sin\theta} (r^2 \sin\theta - r^3) \, dr \, d\theta$$

$$= \int_0^\pi \left[\tfrac{1}{3} r^3 \sin\theta - \tfrac{1}{4} r^4 \right]_0^{\sin\theta} d\theta$$

$$= \tfrac{1}{12} \int_0^\pi \sin^4\theta \, d\theta = \tfrac{1}{12} (\tfrac{3}{8}\pi) = \tfrac{1}{32}\pi. \quad \square$$

Problem 3. Locate the centroid of the solid in Problem 2.

Solution. Since T is symmetric about the yz-plane, we see that $\bar{x} = 0$. To get $\bar{y}$ we begin as usual:

$$\bar{y} V = \iiint_T y \, dxdydz = \iiint_S (r\sin\theta) \, r \, drd\theta dz$$

$$= \int_0^\pi \int_0^{\sin\theta} \int_{r^2}^{r\sin\theta} r^2 \sin\theta \, dz \, dr \, d\theta$$

$$= \int_0^\pi \int_0^{\sin\theta} (r^3 \sin^2\theta - r^4 \sin\theta) \, dr \, d\theta$$

$$= \int_0^\pi \left[\tfrac{1}{4} r^4 \sin^2\theta - \tfrac{1}{5} r^5 \sin\theta \right]_0^{\sin\theta} dr \, d\theta$$

$$= \int_0^\pi \tfrac{1}{20} \sin^6\theta \, d\theta = \tfrac{1}{20}(\tfrac{5}{16}\pi) = \tfrac{1}{64}\pi.$$

Since $V = \tfrac{1}{32}\pi$, we have $\bar{y} = \tfrac{1}{2}$. Now for $\bar{z}$:

$$\bar{z} V = \iiint_T z \, dxdydz$$

$$= \iiint_S zr \, drd\theta dz$$

$$= \int_0^\pi \int_0^{\sin\theta} \int_{r^2}^{r\sin\theta} zr \, dz \, dr \, d\theta = \cdots = \int_0^\pi \tfrac{1}{24} \sin^6\theta \, d\theta = \tfrac{1}{24}(\tfrac{5}{16}\pi) = \tfrac{5}{384}\pi.$$

details are left to you ⟶

Division by $V = \tfrac{1}{32}\pi$ gives $\bar{z} = \tfrac{5}{12}$. The centroid is thus the point $(0, \tfrac{1}{2}, \tfrac{5}{12})$. $\quad \square$

EXERCISES 17.9

Let T be a solid cylinder of base radius R and height h. Assume that the mass density varies directly with the distance from one of the bases.

1. Use cylindrical coordinates to find the mass M of T.
2. Locate the center of mass of T.
3. Find the moment of inertia of T about the axis of the cylinder.
4. Let T be a homogeneous cylinder of mass M, base radius R, and height h. Find the moment of inertia of the cylinder (a) about the central axis. (b) about a line that lies in the plane of one of the bases and passes through the center of that base. (c) about a line that passes through the center of the cylinder and is parallel to the bases.

Let T be a homogeneous solid circular cone of mass M, base radius R, and height h.

5. Use cylindrical coordinates to verify that the volume of the cone is given by the formula $V = \frac{1}{3}\pi R^2 h$.
6. Locate the center of mass.
7. Find the moment of inertia about the axis of the cone.
8. Find the moment of inertia about a line that passes through the vertex and is parallel to the base.

Let T be the solid bounded above by the paraboloid $z = 1 - (x^2 + y^2)$ and bounded below by the xy-plane.

9. Use cylindrical coordinates to find the volume of T.
10. Find the mass of T if the density varies directly with the distance to the xy-plane.
11. Find the mass of T if the density varies directly with the square of the distance from the origin.

Evaluate using cylindrical coordinates.

12. $\displaystyle\int_{-1}^{1} \int_{0}^{\sqrt{1-x^2}} \int_{\sqrt{x^2+y^2}}^{1} z^3 \, dz \, dy \, dx.$ 13. $\displaystyle\int_{0}^{1} \int_{0}^{\sqrt{1-x^2}} \int_{0}^{\sqrt{4-(x^2+y^2)}} dz \, dy \, dx.$

14. Find the volume of the solid bounded by the paraboloid of revolution $x^2 + y^2 = az$, the xy-plane, and the cylinder $x^2 + y^2 = 2ax$.
15. Find the volume of the solid bounded above by the cone $z^2 = x^2 + y^2$, below by the xy-plane, and on the sides by the cylinder $x^2 + y^2 = 2ax$.
16. Find the volume of the solid bounded above by the plane $2z = 4 + x$, below by the xy-plane, and on the sides by the cylinder $x^2 + y^2 = 2x$.
17. Find the volume of the solid bounded above by $z = a - \sqrt{x^2 + y^2}$, below by the xy-plane, and on the sides by the cylinder $x^2 + y^2 = ax$.
18. Find the volume of the solid that is bounded above by $x^2 + y^2 + z^2 = 25$ and below by $z = \sqrt{x^2 + y^2} + 1$.
19. Find the volume of the solid bounded by the paraboloid $z = x^2 + y^2$ and the plane $z = x$.
20. Find the volume of the solid bounded by the hyperboloid $z^2 = a^2 + x^2 + y^2$ and the upper nappe of the cone $z^2 = 2(x^2 + y^2)$.
21. Find the volume of the "ice cream cone" bounded below by the half-cone $z = \sqrt{3(x^2 + y^2)}$ and above by the unit sphere $x^2 + y^2 + z^2 = 1$.
22. Find the volume of the solid that lies between the cylinders $x^2 + y^2 = 1$ and $x^2 + y^2 = 4$, and is bounded above by the ellipsoid $x^2 + y^2 + 4z^2 = 36$ and below by the xy-plane.

17.10   THE TRIPLE INTEGRAL AS THE LIMIT OF RIEMANN SUMS; SPHERICAL COORDINATES

The Triple Integral as the Limit of Riemann Sums

You have seen how single integrals and double integrals can be obtained as limits of Riemann sums. The same holds true for triple integrals.

Start with a basic solid T in xyz-space and decompose it into a finite number of basic solids $T_1, \ldots, T_N$. If f is continuous on T, then f is continuous on each T_i. From each T_i pick an arbitrary point (x_i^*, y_i^*, z_i^*) and form the *Riemann sum*

$$\sum_{i=1}^{N} f(x_i^*, y_i^*, z_i^*)(\text{volume of } T_i).$$

As you would expect, the triple integral over T is the limit of such sums; namely, given any $\epsilon > 0$, there exists $\delta > 0$ such that, if the diameters of the T_i are all less than δ, then

$$\left| \sum_{i=1}^{N} f(x_i^*, y_i^*, z_i^*)(\text{volume of } T_i) - \iiint_T f(x, y, z) \, dxdydz \right| < \epsilon$$

no matter how the (x_i^*, y_i^*, z_i^*) are chosen within the T_i. We express this by writing

(17.10.1)
$$\iiint_T f(x, y, z) \, dxdydz = \lim_{\text{diam } T_i \to 0} \sum_{i=1}^{N} f(x_i^*, y_i^*, z_i^*)(\text{volume of } T_i).$$

Introduction to Spherical Coordinates

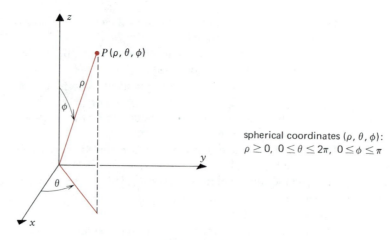

spherical coordinates (ρ, θ, ϕ):
$\rho \geq 0, \ 0 \leq \theta \leq 2\pi, \ 0 \leq \phi \leq \pi$

Figure 17.10.1

The spherical coordinates (ρ, θ, ϕ) of a point P in xyz-space are shown geometrically in Figure 17.10.1. The first coordinate, ρ, is the distance from the origin; thus $\rho \geq 0$. The second coordinate, the angle marked θ, is the second coordinate of cylindrical coordinates; θ ranges from 0 to 2π. We call θ the *longitude*. The third coordinate, the angle marked ϕ, ranges only from 0 to π. We call ϕ the *colatitude*, or more simply the *polar angle*. (The complement of ϕ would be the *latitude* on a globe.)

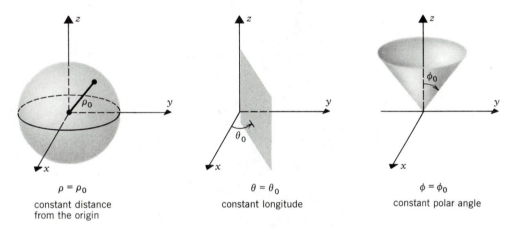

| $\rho = \rho_0$ | $\theta = \theta_0$ | $\phi = \phi_0$ |
| constant distance from the origin | constant longitude | constant polar angle |

Figure 17.10.2

The coordinate surfaces

$$\rho = \rho_0, \quad \theta = \theta_0, \quad \phi = \phi_0$$

are shown in Figure 17.10.2. The surface $\rho = \rho_0$ is a sphere; the radius is ρ_0 and the center is the origin. The second surface, $\theta = \theta_0$, is the same as in cylindrical coordinates: the vertical half-plane hinged at the z-axis and standing at an angle of θ_0 radians from the positive x-axis. The surface $\phi = \phi_0$ requires detailed explanation. If $0 < \phi_0 < \frac{1}{2}\pi$ or $\frac{1}{2}\pi < \phi_0 < \pi$, the surface is a nappe of a cone; it is generated by revolving about the z-axis any ray that emerges from the origin at an angle of ϕ_0 radians from the positive z-axis. The surface $\phi = \frac{1}{2}\pi$ is the xy-plane. (The nappe of the cone has opened up completely.) The equation $\phi = 0$ gives the positive z-axis, and the equation $\phi = \pi$ gives the negative z-axis. (When $\phi = 0$ or $\phi = \pi$, the nappe of the cone has closed up completely.)

Rectangular coordinates (x, y, z) are related to spherical coordinates (ρ, θ, ϕ) by the following equations:

$$x = \rho \sin \phi \cos \theta, \quad y = \rho \sin \phi \sin \theta, \quad z = \rho \cos \phi.$$

You can verify these relations by referring to Figure 17.10.1. (Note that the factor $\rho \sin \phi$ appearing in the first two equations is the r of cylindrical coordinates: $r = \rho \sin \phi$.) Conversely, with obvious exclusions, we have

$$\rho = \sqrt{x^2 + y^2 + z^2}, \quad \tan \theta = \frac{y}{x}, \quad \cos \phi = \frac{z}{\sqrt{x^2 + y^2 + z^2}}.$$

The Volume of a Spherical Wedge

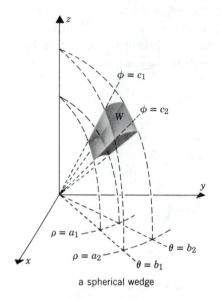

a spherical wedge

Figure 17.10.3

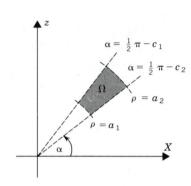

Figure 17.10.4

Figure 17.10.3 shows a *spherical wedge* W in xyz-space. W consists of all points (x, y, z) that have spherical coordinates in the box

$$\Pi: \quad a_1 \le \rho \le a_2, \quad b_1 \le \theta \le b_2, \quad c_1 \le \phi \le c_2.$$

The volume of this wedge is given by the formula

(17.10.2)
$$V = \int\!\!\int\!\!\int_{\Pi} \rho^2 \sin \phi \, d\rho d\theta d\phi.$$

Proof. Note first that W is a solid of revolution. One way to obtain W is to rotate the $\theta = b_1$ face of W, call it Ω, about the z-axis for $b_2 - b_1$ radians. (See Figure 17.10.3.) On that face ρ and $\alpha = \frac{1}{2}\pi - \phi$ play the role of polar coordinates. (See Figure 17.10.4.) In the setup of Figure 17.10.4 the face Ω is the set of all (z, X) with polar coordinates (ρ, α) in the set Γ: $a_1 \le \rho \le a_2$, $\frac{1}{2}\pi - c_2 \le \alpha \le \frac{1}{2}\pi - c_1$. The centroid of Ω is at a distance $\bar{X}$ from the z-axis where

$$\bar{X}(\text{area of } \Omega) = \int\!\!\int_{\Omega} X \, dXdz = \int\!\!\int_{\Gamma} \rho^2 \cos \alpha \, d\rho d\alpha$$

this follows from (17.5.3) together with the fact that here (ρ, α) play the role of polar coordinates

$$= \left(\int_{a_1}^{a_2} \rho^2 \, d\rho \right) \left(\int_{\frac{1}{2}\pi - c_2}^{\frac{1}{2}\pi - c_1} \cos \alpha \, d\alpha \right)$$

$$= \left(\int_{a_1}^{a_2} \rho^2 \, d\rho \right) \left(\int_{c_1}^{c_2} \sin \phi \, d\phi \right).$$

$\phi = \frac{1}{2}\pi - \alpha$

As the face Ω is rotated from $\theta = b_1$ to $\theta = b_2$, the centroid travels through a circular arc of length

$$s = (b_2 - b_1)\overline{X} = (b_2 - b_1)\frac{1}{\text{area of }\Omega}\left(\int_{a_1}^{a_2}\rho^2\,d\rho\right)\left(\int_{c_1}^{c_2}\sin\phi\,d\phi\right).$$

From Pappus (see the remark on p. 322) we know that

$$\text{the volume of } W = s(\text{area of }\Omega) = (b_2 - b_1)\left(\int_{a_1}^{a_2}\rho^2\,d\rho\right)\left(\int_{c_1}^{c_2}\sin\phi\,d\phi\right)$$

$$= \left(\int_{b_1}^{b_2}d\theta\right)\left(\int_{a_1}^{a_2}\rho^2\,d\rho\right)\left(\int_{c_1}^{c_2}\sin\phi\,d\phi\right)$$

$$= \int_{a_1}^{a_2}\int_{b_1}^{b_2}\int_{c_1}^{c_2}\rho^2\sin\phi\,d\phi\,d\theta\,d\rho$$

$$= \iiint_\Pi \rho^2\sin\phi\,d\rho d\theta d\phi. \quad \square$$

Evaluating Triple Integrals Using Spherical Coordinates

Suppose that T is a basic solid in xyz-space with spherical coordinates in some basic solid S of $\rho\theta\phi$-space.   Then

(17.10.3)
$$\iiint_T f(x, y, z)\,dxdydz = \iiint_S f(\rho\sin\phi\cos\theta, \rho\sin\phi\sin\theta, \rho\cos\phi)\,\rho^2\sin\phi\,d\rho d\theta d\phi.$$

Derivation of (17.10.3).   Assume first that T is a spherical wedge W. S is then a box Π.   Now decompose Π into N boxes $\Pi_1, \ldots, \Pi_N$.   This induces a subdivision of W into N spherical wedges $W_1, \ldots, W_N$.

Writing $F(\rho, \theta, \phi)$ for $f(\rho\sin\phi\cos\theta, \rho\sin\phi\sin\theta, \rho\cos\phi)$ to save space, we have

$$\iiint_\Pi F(\rho, \theta, \phi)\,\rho^2\sin\phi\,d\rho d\theta d\phi = \sum_{i=1}^N \iiint_{\Pi_i} F(\rho, \theta, \phi)\,\rho^2\sin\phi\,d\rho d\theta d\phi$$

additivity

$$= \sum_{i=1}^N F(\rho_i^*, \theta_i^*, \phi_i^*)\iiint_{\Pi_i}\rho^2\sin\phi\,d\rho d\theta d\phi$$

with $(\rho_i^*, \theta_i^*, \phi_i^*) \in \Pi_i$
(Theorem 17.7.5)

$$= \sum_{i=1}^N F(\rho_i^*, \theta_i^*, \phi_i^*)(\text{volume of } W_i)$$

(17.10.2) applied to Π_i

$x_i^* = \rho_i^*\sin\theta_i^*\cos\phi_i^*,$
$y_i^* = \rho_i^*\sin\theta_i^*\sin\phi_i^*,$
$z_i^* = \rho_i^*\cos\phi_i^*$

$$= \sum_{i=1}^N f(x_i^*, y_i^*, z_i^*)(\text{volume of } W_i).$$

This last expression is a Riemann sum for

$$\iiint_W f(x, y, z)\, dxdydz$$

and, as such, by (17.10.1), will differ from that integral by less than any preassigned positive number ϵ provided only that the diameters of all the W_i are sufficiently small. This we can guarantee by making the diameters of all the Π_i sufficiently small.

This verifies the formula for the case where T is a spherical wedge. The more general case is left to you.   HINT: Encase T in a wedge W and define f to be zero outside of T.   ☐

Volume Formula

If $f(x, y, z) = 1$ for all (x, y, z) in T, then the change of variables formula reduces to

$$\iiint_T dxdydz = \iiint_S \rho^2 \sin \phi \, d\rho d\theta d\phi.$$

The integral on the left is the volume of T. It follows that the volume of T is given by the formula

(17.10.4)
$$V = \iiint_S \rho^2 \sin \phi \, d\rho d\theta d\phi.$$

Spherical coordinates are commonly used in applications where there is a center of symmetry. The center of symmetry is then taken as the origin.

Calculations

Problem 1. Calculate the mass M of a solid ball of radius 1 given that the density varies directly with the square of the distance from the center of the ball.

Solution. Center the ball at the origin. The ball, call it T, is now the set of all (x, y, z) with spherical coordinates (ρ, θ, ϕ) in the box

$$S: \quad 0 \le \rho \le 1, \quad 0 \le \theta \le 2\pi, \quad 0 \le \phi \le \pi.$$

Therefore

$$M = \iiint_T k(x^2 + y^2 + z^2)\, dxdydz$$

$$= \iiint_S (k\rho^2)\rho^2 \sin\phi\, d\rho d\theta d\phi$$

$$= k \int_0^\pi \int_0^{2\pi} \int_0^1 \rho^4 \sin\phi\, d\rho\, d\theta\, d\phi$$

$$= k \left(\int_0^\pi \sin\phi\, d\phi \right)\left(\int_0^{2\pi} d\theta \right)\left(\int_0^1 \rho^4\, d\rho \right)$$

$$= k(2)(2\pi)(\tfrac{1}{5}) = \tfrac{4}{5}k\pi. \quad \square$$

Problem 2. Find the volume of the solid T enclosed by the surface

$$(x^2 + y^2 + z^2)^2 = 2z(x^2 + y^2).$$

Solution. In spherical coordinates the bounding surface takes the form

$$\rho = 2 \sin^2 \phi \cos \phi. \qquad \text{(check this out)}$$

This equation places no restriction on θ; thus θ can range from 0 to 2π. Since ρ remains nonnegative, ϕ can range only from 0 to $\frac{1}{2}\pi$. Thus the solid T is the set of all (x, y, z) with spherical coordinates (ρ, θ, ϕ) in the set

$$S: \quad 0 \le \theta \le 2\pi, \quad 0 \le \phi \le \tfrac{1}{2}\pi, \quad 0 \le \rho \le 2 \sin^2 \phi \cos \phi.$$

The rest is straightforward:

$$V = \iiint_T dxdydz$$

$$= \iiint_S \rho^2 \sin\phi\, d\rho d\theta d\phi$$

$$= \int_0^{2\pi} \int_0^{\pi/2} \int_0^{2 \sin^2 \phi \cos \phi} \rho^2 \sin\phi\, d\rho\, d\phi\, d\theta$$

$$= \int_0^{2\pi} \int_0^{\pi/2} \tfrac{8}{3} \sin^7 \phi \cos^3 \phi\, d\phi\, d\theta$$

$$= \tfrac{8}{3}\left(\int_0^{2\pi} d\theta \right)\left(\int_0^{\pi/2} (\sin^7 \phi \cos \phi - \sin^9 \phi \cos \phi)\, d\phi \right)$$

$$= \tfrac{8}{3}(2\pi)(\tfrac{1}{40}) = \tfrac{2}{15}\pi. \quad \square$$

EXERCISES 17.10

1. Find the spherical coordinates (ρ, θ, ϕ) of the point with rectangular coordinates $(1, 1, 1)$.

2. Find the rectangular coordinates of the point with spherical coordinates $(2, \frac{1}{6}\pi, \frac{1}{4}\pi)$.

3. Find the rectangular coordinates of the point with spherical coordinates $(3, \frac{1}{3}\pi, \frac{1}{6}\pi)$.

4. Find the spherical coordinates of the point with cylindrical coordinates $(2, \frac{2}{3}\pi, 6)$.

The following equations are given in spherical coordinates. Interpret each one geometrically.

5. $\rho \sin \phi = 1$.   **6.** $\sin \phi = 1$.   **7.** $\cos \phi = -\frac{1}{2}\sqrt{2}$.

8. $\tan \theta = 1$.   **9.** $\rho \cos \phi = 1$.   **10.** $\rho = \cos \phi$.

11. Derive the formula for the volume of a sphere of radius R using spherical coordinates.

12. Express cylindrical coordinates in terms of spherical coordinates.

13. A wedge is cut from a ball of radius R by two planes that meet in a diameter. Find the volume of the wedge if the angle between the planes is α radians.

14. Find the mass of a ball of radius R given that the density varies directly with the distance from the boundary.

15. Find the mass of a circular cone of base radius r and height h given that the density varies directly with the distance from the vertex.

16. Use spherical coordinates to derive the formula for the volume of a circular cone of base radius r and height h.

Let T be a homogeneous ball of mass M, radius R.

17. Calculate the moment of inertia (a) about a diameter. (b) about a tangent line.

18. Locate the center of mass of the upper half given that the center of the ball is at the origin.

Let T be a homogeneous solid bounded by two concentric spherical shells, an outer shell of radius R_2 and an inner shell of radius R_1.

19. (a) Calculate the moment of inertia about a diameter.
    (b) Use your result in (a) to determine the moment of inertia of a spherical shell of radius R and mass M about a diameter.
    (c) What is the moment of inertia of that same shell about a tangent line?

20. (a) Locate the center of mass of the upper half of T given that the center of T is at the origin.
    (b) Use your result in (a) to locate the center of mass of a homogeneous hemispherical shell of radius R.

21. Find the volume of the solid common to the sphere $\rho = a$ and the cone $\phi = \alpha$. Take $\alpha \in (0, \frac{1}{2}\pi)$.

22. Let T be the solid bounded below by the half-cone $z = \sqrt{x^2 + y^2}$ and above by the spherical surface $x^2 + y^2 + z^2 = 1$. Use spherical coordinates to evaluate

$$\iiint_T e^{(x^2+y^2+z^2)^{3/2}} \, dxdydz.$$

23. (a) Find an equation for the sphere $x^2 + y^2 + (z - R)^2 = R^2$ in spherical coordinates.
    (b) Express the upper half of the ball $x^2 + y^2 + (z - R)^2 \le R^2$ by inequalities in spherical coordinates.

24. Find the mass of the ball $\rho \le 2R \cos \phi$ given that the density varies directly
    (a) with ρ.   (b) with $\rho \sin \phi$.   (c) with $\rho \cos^2 \theta \sin \phi$.

25. Find the volume of the solid common to the spheres $\rho = 2\sqrt{2} \cos \phi$ and $\rho = 2$.

26. Find the volume of the solid enclosed by the surface $\rho = 1 - \cos \phi$.

27. Finish the argument in (17.10.3).

28. (*Gravitational Attraction*) Let T be a basic solid and let (a, b, c) be a point not in T. Show that, if T has continuously varying mass density $\lambda = \lambda(x, y, z)$, then T attracts a point mass m at (a, b, c) with a force

$$\mathbf{F} = \int \int \int_{T} \frac{Gm\lambda(x, y, z)[(x - a)\mathbf{i} + (y - b)\mathbf{j} + (z - c)\mathbf{k}]}{[(x - a)^2 + (y - b)^2 + (z - c)^2]^{3/2}} \, dxdydz.$$

(Assume that a point mass m_1 at P_1 attracts a point mass m_2 at P_2 with a force $\mathbf{F} = -(Gm_1m_2/r^3)\mathbf{r}$ where $\mathbf{r}$ is the vector $\overrightarrow{P_1P_2}$. Interpret the triple integral component by component.)

29. Let T be the upper half of the ball $x^2 + y^2 + (z - R)^2 \le R^2$. Given that T is homogeneous and has mass M, find the gravitational force exerted by T on a point mass m located at the origin. (Note Exercise 28.)

30. A point mass m is placed on the axis of a homogeneous solid cylinder at a distance α from the nearest base of the cylinder. Find the gravitational force exerted by the cylinder on the point mass given that the cylinder has base radius R, height h, and mass M. (Note Exercise 28.)

17.11 JACOBIANS; CHANGING VARIABLES IN MULTIPLE INTEGRATION

During the course of the last few sections you have met several formulas for changing variables in multiple integration: to polar coordinates, to cylindrical coordinates, to spherical coordinates. The purpose of this section is to bring some unity into that material and provide a general prescription for other changes of variable.

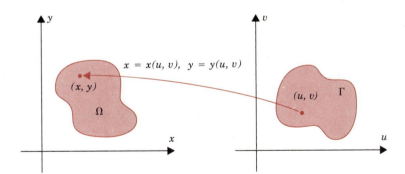

Figure 17.11.1

We begin with a consideration of area. Figure 17.11.1 shows a basic region Γ in a plane that we are calling the uv-plane. (In this plane we denote the abscissa of a point by u and the ordinate by v.) Suppose that

$$x = x(u, v), \qquad y = y(u, v)$$

are functions continuously differentiable on Γ. As (u, v) ranges over Γ, the point $(x, y) = (x(u, v), y(u, v))$ generates a region Ω in the xy-plane. If the mapping

$$(u, v) \rightarrow (x, y)$$

is one-to-one on the interior of Γ and the *Jacobian*

$$J = \begin{vmatrix} \dfrac{\partial x}{\partial u} & \dfrac{\partial y}{\partial u} \\ \dfrac{\partial x}{\partial v} & \dfrac{\partial y}{\partial v} \end{vmatrix} = \frac{\partial x}{\partial u}\frac{\partial y}{\partial v} - \frac{\partial x}{\partial v}\frac{\partial y}{\partial u}$$

is never zero on the interior of Γ, then

(17.11.1)

$$\text{the area of } \Omega = \int\!\!\int_{\Gamma} |J(u, v)|\, du\, dv.$$

It is very difficult to prove this assertion without making additional assumptions. A proof valid for all cases of practical interest is given in the supplement to Section 18.5. At this point we will simply assume this area formula and go on from there.

Suppose now that we want to integrate some continuous function $f = f(x, y)$ over Ω. If this proves difficult to do directly, then we can change variables to u, v and try to integrate over Γ instead. Thanks to (17.11.1),

(17.11.2)

$$\int\!\!\int_{\Omega} f(x, y)\, dx\, dy = \int\!\!\int_{\Gamma} f(x(u, v), y(u, v))\, |J(u, v)|\, du\, dv.$$

The derivation of this formula from (17.11.1) follows the usual lines. Break up Γ into N little basic regions $\Gamma_1, \ldots, \Gamma_N$. These induce a decomposition of Ω into N little basic regions $\Omega_1, \ldots, \Omega_N$. We can then write

$$\int\!\!\int_{\Gamma} f(x(u, v), y(u, v))\, |J(u, v)|\, du\, dv = \sum_{i=1}^{N} \int\!\!\int_{\Gamma_i} f(x(u, v), y(u, v))\, |J(u, v)|\, du\, dv$$

$$\text{additivity} \longrightarrow$$

$$= \sum_{i=1}^{N} f(x(u_i^*, v_i^*), y(u_i^*, v_i^*)) \int\!\!\int_{\Gamma_i} |J(u, v)|\, du\, dv$$

$$\text{Theorem 17.3.5 applied to } \Gamma_i \longrightarrow$$

$$= \sum_{i=1}^{N} f(x_i^*, y_i^*) \int\!\!\int_{\Gamma_i} |J(u, v)|\, du\, dv$$

$$\text{set } x_i^* = x(u_i^*, v_i^*), \quad y_i^* = y(u_i^*, v_i^*) \longrightarrow$$

$$= \sum_{i=1}^{N} f(x_i^*, y_i^*)(\text{area of } \Omega_i).$$

$$\text{(17.11.1) applied to } \Gamma_i \longrightarrow$$

This last expression is a Riemann sum for

$$\int\!\!\int_{\Omega} f(x, y)\, dx\, dy$$

and tends to that integral as the maximum diameter of the Ω_i tends to zero. This we can ensure by letting the maximum diameter of the Γ_i tend to zero. $\square$

Problem 1. Evaluate

$$\iint_\Omega (x + y)^2 \, dxdy$$

where Ω is the parallelogram bounded by the lines

$$x + y = 0, \qquad x + y = 1, \qquad 2x - y = 0, \qquad 2x - y = 3.$$

Figure 17.11.2

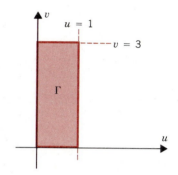

Figure 17.11.3

Solution. The parallelogram is shown in Figure 17.11.2. The boundaries suggest that we set

$$u = x + y, \qquad v = 2x - y.$$

We want x and y in terms of u and v. Since

$$u + v = (x + y) + (2x - y) = 3x \qquad \text{and} \qquad 2u - v = (2x + 2y) - (2x - y) = 3y,$$

we have

$$x = \frac{u + v}{3}, \qquad y = \frac{2u - v}{3}.$$

This transformation maps the rectangle Γ of Figure 17.11.3 onto Ω. The Jacobian is given by

$$J(u, v) = \begin{vmatrix} \dfrac{\partial}{\partial u}\left(\dfrac{u + v}{3}\right) & \dfrac{\partial}{\partial u}\left(\dfrac{2u - v}{3}\right) \\ \dfrac{\partial}{\partial v}\left(\dfrac{u + v}{3}\right) & \dfrac{\partial}{\partial v}\left(\dfrac{2u - v}{3}\right) \end{vmatrix} = \begin{vmatrix} \frac{1}{3} & \frac{2}{3} \\ \frac{1}{3} & -\frac{1}{3} \end{vmatrix} = -\frac{1}{3}.$$

Therefore

$$\iint_\Omega (x + y)^2 \, dxdy = \iint_\Gamma u^2 \, |J(u, v)| \, dudv$$

$$= \frac{1}{3} \int_0^3 \int_0^1 u^2 \, du \, dv$$

$$= \frac{1}{3}\left(\int_0^3 dv\right)\left(\int_0^1 u^2 \, du\right) = \frac{1}{3}(3)\frac{1}{3} = \frac{1}{3}. \quad \square$$

Problem 2. Evaluate

$$\iint_{\Omega} xy \, dxdy$$

where Ω is the first-quadrant region bounded by the curves

$$x^2 + y^2 = 4, \qquad x^2 + y^2 = 9, \qquad x^2 - y^2 = 1, \qquad x^2 - y^2 = 4.$$

Figure 17.11.4

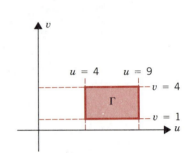

Figure 17.11.5

Solution. The region is shown in Figure 17.11.4. The boundaries suggest that we set

$$u = x^2 + y^2, \qquad v = x^2 - y^2.$$

We want x and y in terms of u and v. Since

$$u + v = 2x^2 \qquad \text{and} \qquad u - v = 2y^2,$$

we have

$$x = \sqrt{\frac{u + v}{2}}, \qquad y = \sqrt{\frac{u - v}{2}}.$$

The transformation maps the rectangle Γ of Figure 17.11.5 onto Ω with Jacobian

$$J(u, v) = \begin{vmatrix} \dfrac{\partial}{\partial u}\left(\sqrt{\dfrac{u + v}{2}}\right) & \dfrac{\partial}{\partial u}\left(\sqrt{\dfrac{u - v}{2}}\right) \\[2mm] \dfrac{\partial}{\partial v}\left(\sqrt{\dfrac{u + v}{2}}\right) & \dfrac{\partial}{\partial v}\left(\sqrt{\dfrac{u - v}{2}}\right) \end{vmatrix} = -\frac{1}{4\sqrt{u^2 - v^2}}.$$

— check this

Therefore

$$\iint_{\Omega} xy \, dxdy = \iint_{\Gamma} \left(\sqrt{\frac{u + v}{2}}\right)\left(\sqrt{\frac{u - v}{2}}\right)\left(\frac{1}{4\sqrt{u^2 - v^2}}\right) dudv$$

$$= \iint_{\Gamma} \tfrac{1}{8} \, dudv = \tfrac{1}{8} \,(\text{area of } \Gamma) = \tfrac{15}{8}. \quad \square$$

In Section 17.5 you saw the formula for changing variables from rectangular coordinates (x, y) to polar coordinates (r, θ). The formula reads

$$\iint_\Omega f(x, y)\, dxdy = \iint_\Gamma f(r \cos \theta, r \sin \theta)\, r\, drd\theta. \qquad (17.5.3)$$

The factor r in the double integral over Γ is the Jacobian of the transformation $x = r \cos \theta,\ y = r \sin \theta$:

$$J(r, \theta) = \begin{vmatrix} \dfrac{\partial}{\partial r}(r \cos \theta) & \dfrac{\partial}{\partial r}(r \sin \theta) \\[2mm] \dfrac{\partial}{\partial \theta}(r \cos \theta) & \dfrac{\partial}{\partial \theta}(r \sin \theta) \end{vmatrix} = \begin{vmatrix} \cos \theta & \sin \theta \\ -r \sin \theta & r \cos \theta \end{vmatrix} = r(\cos^2 \theta + \sin^2 \theta) = r.$$

As you can see, (17.5.3) is a special case of (17.11.2).

When changing variables in a triple integral we make three coordinate changes:

$$x = x(u, v, w), \qquad y = y(u, v, w), \qquad z = z(u, v, w).$$

If these functions carry a basic solid Γ onto a solid T, then, under conditions analogous to the two-dimensional case,

$$\text{volume of } T = \iiint_\Gamma |J(u, v, w)|\, dudvdw$$

where now the Jacobian is a three-by-three determinant:

$$J(u, v, w) = \begin{vmatrix} \dfrac{\partial x}{\partial u} & \dfrac{\partial y}{\partial u} & \dfrac{\partial z}{\partial u} \\[2mm] \dfrac{\partial x}{\partial v} & \dfrac{\partial y}{\partial v} & \dfrac{\partial z}{\partial v} \\[2mm] \dfrac{\partial x}{\partial w} & \dfrac{\partial y}{\partial w} & \dfrac{\partial z}{\partial w} \end{vmatrix}.$$

In this case the change of variables formula reads

$$\iiint_T f(x, y, z)\, dxdydz = \iiint_\Gamma f(x(u, v, w), y(u, v, w), z(u, v, w))\, |J(u, v, w)|\, dudvdw.$$

(A historical note on Jacobians: the study of these functional determinants goes back to a memoir by the German mathematician C. G. Jacobi, 1804–1851.)

EXERCISES 17.11

Find the Jacobian of the transformation.

1. $x = au + bv,\quad y = cu + dv.$ (*linear transformation*)
2. $x = u \cos \theta - v \sin \theta,\quad y = u \sin \theta + v \cos \theta.$ (*rotation by θ*)
3. $x = uv,\quad y = u^2 + v^2.$
4. $x = u \ln v,\quad y = uv.$
5. $x = uv^2,\quad y = u^2 v.$
6. $x = u - \ln v,\quad y = \ln u + v.$
7. $x = au,\quad y = bv,\quad z = cw.$
8. $x = v + w,\quad y = u + w,\quad z = u + v.$

9. $x = r \cos \theta$, $y = r \sin \theta$, $z = z$.   *(cylindrical coordinates)*

10. $x = \rho \sin \phi \cos \theta$, $y = \rho \sin \phi \sin \theta$, $z = \rho \cos \phi$.   *(spherical coordinates)*

11. $x = (1 + w \cos v) \cos u$, $y = (1 + w \cos v) \sin u$, $z = w \sin v$.

12. Every linear transformation

$$x = au + bv, \quad y = cu + dv \qquad \text{with } ad - bc \neq 0$$

maps lines of the uv-plane onto lines of the xy-plane. Find the image
(a) of a vertical line $u = u_0$.   (b) of a horizontal line $v = v_0$.

For Exercises 13–15 take Ω as the parallelogram bounded by

$$x + y = 0, \qquad x + y = 1, \qquad x - y = 0, \qquad x - y = 2.$$

Evaluate.

13. $\displaystyle\iint_\Omega (x^2 - y^2)\, dxdy.$ **14.** $\displaystyle\iint_\Omega 4xy\, dxdy.$ **15.** $\displaystyle\iint_\Omega (x + y) \cos\,[\pi(x - y)]\, dxdy.$

For Exercises 16–18 take Ω as the parallelogram bounded by

$$x - y = 0, \qquad x - y = \pi, \qquad x + 2y = 0, \qquad x + 2y = \tfrac{1}{2}\pi.$$

Evaluate.

16. $\displaystyle\iint_\Omega (x + y)\, dxdy.$ **17.** $\displaystyle\iint_\Omega \sin\,(x - y) \cos\,(x + 2y)\, dxdy.$ **18.** $\displaystyle\iint_\Omega \sin 3x\, dxdy.$

19. Let Ω be the first-quadrant region bounded by the curves $xy = 1$, $xy = 4$, $y = x$, $y = 4x$. (a) Determine the area of Ω and (b) locate the centroid.

20. Show that the ellipse $b^2x^2 + a^2y^2 = a^2b^2$ has area πab setting $x = ar \cos \theta$, $y = br \cos \theta$.

21. A homogeneous plate in the xy-plane is in the form of a parallelogram. The parallelogram is bounded by the lines $x + y = 0$, $x + y = 1$, $3x - 2y = 0$, $3x - 2y = 2$. Calculate the moments of inertia of the plate about the three coordinate axes. Express your answers in terms of the mass of the plate.

22. Calculate the area of the region Ω bounded by the curves

$$x^2 - 2xy + y^2 + x + y = 0, \qquad x + y + 4 = 0.$$

HINT: Set $u = x - y$, $v = x + y$.

23. Calculate the area of the region Ω bounded by the curves

$$x^2 - 4xy + 4y^2 - 2x - y - 1 = 0, \qquad y = \tfrac{2}{5}.$$

24. Locate the centroid of the region in Exercise 22.

25. Calculate the area of the region Ω enclosed by the curve

$$11x^2 + 4\sqrt{3}\,xy + 7y^2 - 1 = 0.$$

HINT: Use a rotation $x = u \cos \theta - v \sin \theta$, $y = u \sin \theta + v \cos \theta$ such that the resulting uv-equation has no uv-term.

26. (*Generalized Polar Coordinates*)   For a, b, α fixed positive numbers, define

$$x = ar(\cos \theta)^\alpha, \qquad y = br(\sin \theta)^\alpha.$$

(a) Show that the mapping carries the strip

$$0 \le r < \infty, \quad 0 \le \theta \le \tfrac{1}{2}\pi$$

onto the first quadrant of the xy-plane.   [HINT: Find a point (r, θ) in the strip that maps onto (x, y) given that $x \ge 0$, $y \ge 0$.]

(b) Show that the mapping is one-to-one on the interior of the strip.

27. Determine the Jacobian of the mapping of Exercise 26.

28. Calculate the area enclosed by the curve $(x/a)^{1/4} + (y/b)^{1/4} = 1$ in the first quadrant by setting $x = ar \cos^8 \theta$, $y = br \sin^8 \theta$.

For Exercises 29–32 let T be the solid ellipsoid $x^2/a^2 + y^2/b^2 + z^2/c^2 \le 1$.

29. Calculate the volume of T by setting

$$x = a\rho \sin \phi \cos \theta, \qquad y = b\rho \sin \phi \sin \theta, \qquad z = c\rho \cos \phi.$$

30. Locate the centroid of the upper half of T.

31. View the upper half of T as a homogeneous solid of mass M.   Find the moments of inertia of this solid about the coordinate axes.

32. Evaluate

$$\iiint_T \left(\frac{x^2}{a^2} + \frac{y^2}{b^2} + \frac{z^2}{c^2} \right) dx\,dy\,dz.$$

33. Evaluate

$$\int_{-\infty}^{\infty} \int_{-\infty}^{\infty} \frac{e^{-(x-y)^2}}{1 + (x+y)^2}\, dx\, dy$$

by integrating over the square S_a: $-a \le x \le a$, $-a \le y \le a$ and taking the limit as $a \to \infty$.   HINT: Set $u = x - y$, $v = x + y$ and see (17.5.4).

CHAPTER HIGHLIGHTS

17.1 Multiple-Sigma Notation

17.2 The Double Integral Over a Rectangle

partition (p. 941)   upper sum, lower sum (p. 942)
double integral over a rectangle (p. 944)

17.3 The Double Integral Over a Region

basic region (p. 949)   double integral over a region (p. 949)
double integral as a volume (p. 950)   area as a double integral (p. 950)
properties of the double integral (pp. 950–951)
average value over a region (p. 951)   weighted average over a region (p. 951)

LINE INTEGRALS
AND
SURFACE INTEGALS

In this chapter we will study integration along curves and integration over surfaces. At the heart of this subject lie three great integration theorems: *Green's Theorem, Gauss's Theorem* (commonly known as *The Divergence Theorem*), and *Stokes's Theorem.*

All three theorems are ultimately based on *The Fundamental Theorem of Integral Calculus,* and all can be cast in the same general form:

an integral over a set S = a related integral over the boundary of S.

A word about terminology. Suppose that S is some subset of the plane or of three-dimensional space. A function that assigns a scalar to each point of S (say, the temperature at that point or the mass density at that point) is known in science as a *scalar field.* A function that assigns a vector to each point of S (say, the wind velocity at that point) is called a *vector field.* We will be using this "field" language throughout.

18.1 LINE INTEGRALS

We will be led to the definition of *line integral* by the notion of work.

The Work Done by a Varying Force over a Curved Path

The work done by a constant force **F** on an object that moves along a straight line is, by definition, the component of **F** in the direction of the displacement multiplied by the length of the displacement vector **d**:

$$W = (\text{comp}_\mathbf{d}\ \mathbf{F}) \times \|\mathbf{d}\|.$$

We can write this more briefly as a dot product:

(18.1.1)
$$\boxed{W = \mathbf{F} \cdot \mathbf{d}.}$$

This elementary notion of work is useful, but it is not sufficient. Consider, for example, an object that moves through a magnetic field or a gravitational field. The path of the motion is then usually not a straight line but a curve, and the force, rather than remain constant, tends to vary from point to point. What we want now is a notion of work that applies to this more general situation.

Let's suppose that an object moves along a curve

$$C: \qquad \mathbf{r}(u) = x(u)\mathbf{i} + y(u)\mathbf{j} + z(u)\mathbf{k}, \quad u \in [a, b]$$

subject to a continuous force **F**. (The vector field **F** may vary from point to point, not only in magnitude but also in direction.) We will suppose that the curve is *smooth*; namely, we will suppose that the tangent vector **r**′ is continuous and never zero. What we want to do here is define the total work done by **F** along the curve C.

To decide how to do this, we begin by focusing our attention on what happens over a short parameter interval $[u, u + h]$. As an estimate for the work done over this interval we can use the dot product

$$\mathbf{F}(\mathbf{r}(u)) \cdot [\mathbf{r}(u + h) - \mathbf{r}(u)].$$

In making this estimate we are using the force vector at **r**(u) and we are replacing the curved path from **r**(u) to **r**(u + h) by the line segment from **r**(u) to **r**(u + h). (See Figure 18.1.1.)

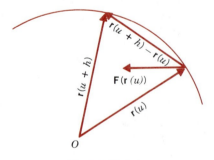

Figure 18.1.1

If we set

$$W(u) = \text{total work done by } \mathbf{F} \text{ from } \mathbf{r}(a) \text{ to } \mathbf{r}(u) \quad \text{and}$$
$$W(u + h) = \text{total work done by } \mathbf{F} \text{ from } \mathbf{r}(a) \text{ to } \mathbf{r}(u + h),$$

then the work done by **F** from **r**(u) to **r**(u + h) must be the difference

$$W(u + h) - W(u).$$

Bringing our estimate into play, we are led to the approximate equation

$$W(u + h) - W(u) \cong \mathbf{F}(\mathbf{r}(u)) \cdot [\mathbf{r}(u + h) - \mathbf{r}(u)],$$

which, upon division by h, becomes

$$\frac{W(u + h) - W(u)}{h} \cong \mathbf{F}(\mathbf{r}(u)) \cdot \frac{[\mathbf{r}(u + h) - \mathbf{r}(u)]}{h}.$$

The quotients here are average rates of change, and the equation is only an approximate one. The notion of work is made precise by requiring that both sides have exactly the same limit as h tends to zero; in other words, by requiring that

$$W'(u) = \mathbf{F}(\mathbf{r}(u)) \cdot \mathbf{r}'(u).$$

The rest is now determined. Since

$$W(a) = 0 \quad \text{and} \quad W(b) = \text{total work done on } C,$$

we have

$$\text{total work done on } C = W(b) - W(a) = \int_a^b W'(u) \, du = \int_a^b [\mathbf{F}(\mathbf{r}(u)) \cdot \mathbf{r}'(u)] \, du.$$

In short, we have arrived at the following definition of work:

(18.1.2)

$$W = \int_a^b [\mathbf{F}(\mathbf{r}(u)) \cdot \mathbf{r}'(u)] \, du.$$

Line Integrals

The integral on the right of (18.1.2) can be calculated not only for a force function $\mathbf{F}$ but for any vector field $\mathbf{h}$ continuous on C.

DEFINITION 18.1.3 LINE INTEGRAL

Let $\mathbf{h}$ be a vector field continuous on a smooth curve

$$C: \quad \mathbf{r}(u) = x(u)\mathbf{i} + y(u)\mathbf{j} + z(u)\mathbf{k}, \quad u \in [a, b].$$

The line integral of $\mathbf{h}$ over C is the number

$$\int_C \mathbf{h}(\mathbf{r}) \cdot d\mathbf{r} = \int_a^b [\mathbf{h}(\mathbf{r}(u)) \cdot \mathbf{r}'(u)] \, du.$$

Note that, although we speak of integrating over C, we actually carry out the calculations over the parameter set $[a, b]$. If our definition of line integral is to make sense, the line integral as defined must be independent of the particular parametrization chosen for C. Within the limitations spelled out below this is indeed the case:

(18.1.4)

the line integral

$$\int_C \mathbf{h}(\mathbf{r}) \cdot d\mathbf{r} = \int_a^b [\mathbf{h}(\mathbf{r}(u)) \cdot \mathbf{r}'(u)] \, du$$

is left invariant by every *sense-preserving* change of parameter.†

Proof. Suppose that ϕ maps $[c, d]$ onto $[a, b]$ and that ϕ' is positive and continuous on $[c, d]$. We must show that the line integral over C as parametrized by

$$\mathbf{R}(w) = \mathbf{r}(\phi(w)), \qquad w \in [c, d]$$

equals the line integral over C as parametrized by $\mathbf{r}$. The argument is straightforward:

$$\int_C \mathbf{h}(\mathbf{R}) \cdot d\mathbf{R} = \int_c^d [\mathbf{h}(\mathbf{R}(w)) \cdot \mathbf{R}'(w)] \, dw$$

$$= \int_c^d [\mathbf{h}(\mathbf{r}(\phi(w))) \cdot \mathbf{r}'(\phi(w)) \, \phi'(w)] \, dw$$

Set $u = \phi(w)$, $du = \phi'(w) \, du$. $\left.\right\}$ $= \int_c^d [\mathbf{h}(\mathbf{r}(\phi(w))) \cdot \mathbf{r}'(\phi(w))] \, \phi'(w) \, dw$
At $w = c$, $u = a$; at $w = d$, $u = b$.

$$= \int_a^b [\mathbf{h}(\mathbf{r}(u)) \cdot \mathbf{r}'(u)] \, du = \int_C \mathbf{h}(\mathbf{r}) \cdot d\mathbf{r}. \quad \square$$

Problem 1. Calculate

$$\int_C \mathbf{h}(\mathbf{r}) \cdot d\mathbf{r}$$

given that

$$\mathbf{h}(x, y) = xy\mathbf{i} + y^2\mathbf{j} \qquad \text{and} \qquad C: \quad \mathbf{r}(u) = u\mathbf{i} + u^2\mathbf{j}, \quad u \in [0, 1].$$

Solution. Here

$$x(u) = u, \qquad y(u) = u^2$$

and

$$\mathbf{h}(\mathbf{r}(u)) \cdot \mathbf{r}'(u) = [x(u)y(u)\mathbf{i} + [y(u)]^2\mathbf{j}] \cdot [x'(u)\mathbf{i} + y'(u)\mathbf{j}]$$

$$= x(u)y(u)x'(u) + [y(u)]^2 y'(u)$$

$$= u(u^2)1 + u^4(2u) = u^3 + 2u^5.$$

It follows that

$$\int_C \mathbf{h}(\mathbf{r}) \cdot d\mathbf{r} = \int_0^1 (u^3 + 2u^5) \, du = \tfrac{7}{12}. \quad \square$$

---

† Changes of parameter were explained in Section 14.3.

Problem 2. Integrate the vector field $h(x, y, z) = xy\mathbf{i} + yz\mathbf{j} + xz\mathbf{k}$ over the twisted cubic

$$\mathbf{r}(u) = u\mathbf{i} + u^2\mathbf{j} + u^3\mathbf{k}$$

from $(-1, 1, -1)$ to $(1, 1, 1)$.

Solution. The path of integration begins at $u = -1$ and ends at $u = 1$. In this case

$$x(u) = u, \qquad y(u) = u^2, \qquad z(u) = u^3.$$

Therefore

$$\begin{aligned}
\mathbf{h}(\mathbf{r}(u)) \cdot \mathbf{r}'(u) &= [x(u)y(u)\mathbf{i} + y(u)z(u)\mathbf{j} + x(u)z(u)\mathbf{k}] \cdot [x'(u)\mathbf{i} + y'(u)\mathbf{j} + z'(u)\mathbf{k}] \\
&= x(u)y(u)x'(u) + y(u)z(u)y'(u) + x(u)z(u)z'(u) \\
&= u(u^2)1 + u^2(u^3)2u + u(u^3)3u^2 \\
&= u^3 + 5u^6
\end{aligned}$$

and

$$\int_C \mathbf{h}(\mathbf{r}) \cdot d\mathbf{r} = \int_{-1}^{1} (u^3 + 5u^6) \, du = 2 \int_0^1 5u^6 \, du = \tfrac{10}{7}. \quad \square$$

If a curve C is not smooth but is made up of a finite number of adjoining smooth pieces $C_1, C_2, \ldots, C_n$, then we define the integral over C as the sum of the integrals over the C_i:

(18.1.5)
$$\int_C = \int_{C_1} + \int_{C_2} + \cdots + \int_{C_n}.$$

A curve of this type is said to be *piecewise smooth.* See Figure 18.1.2.

Figure 18.1.2

All polygons are piecewise-smooth curves. In the next problem we integrate over a triangle. We do this by integrating over each of the sides and then adding up the results. Observe that the directed line segment that begins at $\mathbf{a} = (a_1, a_2)$ and ends at $\mathbf{b} = (b_1, b_2)$ can be parametrized by setting

$$\mathbf{r}(u) = (1 - u)\mathbf{a} + u\mathbf{b}, \qquad u \in [0, 1].$$

Problem 3. Evaluate the line integral

$$\int_C \mathbf{h}(\mathbf{r}) \cdot d\mathbf{r}$$

if $\mathbf{h}(x, y) = e^y \mathbf{i} - \sin \pi x \, \mathbf{j}$ and C is the triangle with vertices $(1, 0)$, $(0, 1)$, $(-1, 0)$ traversed counterclockwise.

Solution. The path C is made up of three line segments:

$$C_1: \quad \mathbf{r}(u) = (1 - u)\mathbf{i} + u\mathbf{j}, \quad u \in [0, 1],$$
$$C_2: \quad \mathbf{r}(u) = (1 - u)\mathbf{j} + u(-\mathbf{i}) = -u\mathbf{i} + (1 - u)\mathbf{j}, \quad u \in [0, 1],$$
$$C_3: \quad \mathbf{r}(u) = (1 - u)(-\mathbf{i}) + u\mathbf{i} = (2u - 1)\mathbf{i}, \quad u \in [0, 1].$$

See Figure 18.1.3.

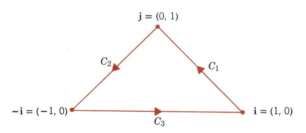

Figure 18.1.3

As you can verify,

$$\int_{C_1} \mathbf{h}(\mathbf{r}) \cdot d\mathbf{r} = \int_0^1 \left[e^{y(u)} x'(u) - \sin \left[\pi x(u) \right] y'(u) \right] du$$

$$= \int_0^1 \left[-e^u - \sin \left[\pi (1 - u) \right] \right] du = 1 - e - \frac{2}{\pi};$$

$$\int_{C_2} \mathbf{h}(\mathbf{r}) \cdot d\mathbf{r} = \int_0^1 \left[e^{y(u)} x'(u) - \sin \left[\pi x(u) \right] y'(u) \right] du$$

$$= \int_0^1 \left[-e^{1-u} + \sin \left(-\pi u \right) \right] du = 1 - e - \frac{2}{\pi};$$

$$\int_{C_3} \mathbf{h}(\mathbf{r}) \cdot d\mathbf{r} = \int_0^1 \left[e^{y(u)} x'(u) - \sin \left[\pi x(u) \right] y'(u) \right] du = \int_0^1 2 \, du = 2.$$

The integral over the entire triangle is the sum of these integrals:

$$\int_C \mathbf{h}(\mathbf{r}) \cdot d\mathbf{r} = \left(1 - e - \frac{2}{\pi}\right) + \left(1 - e - \frac{2}{\pi}\right) + 2 = 4 - 2e - \frac{4}{\pi} \cong -2.71. \quad \square$$

Figure 18.1.4

When we integrate over a parametrized curve, we integrate in the direction determined by the parametrization. If we integrate in the opposite direction, our answer is altered by a factor of -1. To be precise, let C be a piecewise-smooth curve and let $-C$ denote the same path traversed in the *opposite direction*. (See Figure 18.1.4.) If C is parametrized by a vector function $\mathbf{r}$ defined on $[a, b]$, then $-C$ can be parametrized by setting

$$\mathbf{R}(w) = \mathbf{r}(a + b - w), \qquad w \in [a, b]. \qquad \text{(Section 14.3)}$$

Our assertion is that

(18.1.6)
$$\int_{-C} \mathbf{h}(\mathbf{R}) \cdot d\mathbf{R} = -\int_C \mathbf{h}(\mathbf{r}) \cdot d\mathbf{r}$$

or more briefly that

(18.1.7)
$$\int_{-C} = -\int_C .$$

We leave the proof of this to you. $\square$

We were led to the definition of line integral by the notion of work. From (18.1.2) it is clear that, if a force $\mathbf{F}$ is continually applied to an object that moves over a piecewise-smooth curve C, then the work done by $\mathbf{F}$ is the line integral of $\mathbf{F}$ over C:

(18.1.8)
$$W = \int_C \mathbf{F}(\mathbf{r}) \cdot d\mathbf{r}.$$

Problem 4. An object, acted upon by various forces, moves along the parabola $y = 3x^2$ from the origin to the point $(1, 3)$. One of the forces acting on the object is $\mathbf{F}(x, y) = x^3\mathbf{i} + y\mathbf{j}$. Calculate the work done by $\mathbf{F}$.

Solution. We can parametrize the path by setting

$$C: \qquad \mathbf{r}(u) = u\mathbf{i} + 3u^2\mathbf{j}, \quad u \in [0, 1].$$

Here

$$x(u) = u, \qquad y(u) = 3u^2$$

and

$$\mathbf{F}(\mathbf{r}(u)) \cdot \mathbf{r}'(u) = [x(u)]^3 x'(u) + y(u) y'(u) = u^3(1) + 3u^2(6u) = 19u^3.$$

It follows that

$$W = \int_C \mathbf{F}(\mathbf{r}) \cdot d\mathbf{r} = \int_0^1 19u^3 \, du = \tfrac{19}{4}. \quad \square$$

If an object of mass m moves so that at time t it has position $\mathbf{r}(t)$, then, from Newton's second law, the total force acting on the object at time t must be $m\mathbf{r}''(t)$.

Problem 5. An object of mass m moves from time $t = 0$ to $t = 1$ so that its position at time t is given by the vector function

$$\mathbf{r}(t) = \alpha t^2\mathbf{i} + \sin \beta t \, \mathbf{j} + \cos \beta t \, \mathbf{k}.$$

Find the total force acting on the object at time t and calculate the total work done by this force.

Solution. Differentiation gives

$$\mathbf{r}'(t) = 2\alpha t\mathbf{i} + \beta \cos \beta t \, \mathbf{j} - \beta \sin \beta t \, \mathbf{k}, \qquad \mathbf{r}''(t) = 2\alpha\mathbf{i} - \beta^2 \sin \beta t \, \mathbf{j} - \beta^2 \cos \beta t \, \mathbf{k}.$$

The total force on the object at time t is therefore

$$m\mathbf{r}''(t) = m(2\alpha\mathbf{i} - \beta^2 \sin \beta t \, \mathbf{j} - \beta^2 \cos \beta t \, \mathbf{k}).$$

We can calculate the total work done by this force by integrating the force over the curve

$$C: \qquad \mathbf{r}(t) = \alpha t^2\mathbf{i} + \sin \beta t \, \mathbf{j} + \cos \beta t \, \mathbf{k}, \quad t \in [0, 1].$$

We leave it to you to verify that

$$W = \int_0^1 [m\mathbf{r}''(t) \cdot \mathbf{r}'(t)] \, dt = m \int_0^1 4\alpha^2 t \, dt = 2\alpha^2 m. \quad \square$$

EXERCISES 18.1

1. Integrate $\mathbf{h}(x, y) = y\mathbf{i} + x\mathbf{j}$ over the indicated path:
   (a) $\mathbf{r}(u) = u\mathbf{i} + u^2\mathbf{j}, \quad u \in [0, 1]$.   (b) $\mathbf{r}(u) = u^3\mathbf{i} - 2u\mathbf{j}, \quad u \in [0, 1]$.

2. Integrate $\mathbf{h}(x, y) = x\mathbf{i} + y\mathbf{j}$ over the paths of Exercise 1.

3. Integrate $\mathbf{h}(x, y) = y\mathbf{i} + x\mathbf{j}$ over the unit circle traversed clockwise.

4. Integrate $\mathbf{h}(x, y) = y^2 x\mathbf{i} + 2\mathbf{j}$ over the indicated path:
   (a) $\mathbf{r}(u) = e^u\mathbf{i} + e^{-u}\mathbf{j}, \quad u \in [0, 1]$.   (b) $\mathbf{r}(u) = (1 - u)\mathbf{i}, \quad u \in [0, 2]$.

5. Integrate $\mathbf{h}(x, y) = (x - y)\mathbf{i} + xy\mathbf{j}$ over the indicated path:
   (a) the line segment from $(2, 3)$ to $(1, 2)$.   (b) the line segment from $(1, 2)$ to $(2, 3)$.

6. Integrate $\mathbf{h}(x, y) = x^{-1}y^{-2}\mathbf{i} + x^{-2}y^{-1}\mathbf{j}$ over the indicated path:
   (a) $\mathbf{r}(u) = \sqrt{u}\mathbf{i} + \sqrt{1 + u}\mathbf{j}, \quad u \in [1, 4]$.   (b) the line segment from $(1, 1)$ to $(2, 2)$.

7. Integrate $\mathbf{h}(x, y) = y\mathbf{i} - x\mathbf{j}$ over the triangle with vertices $(-2, 0), (2, 0), (0, 2)$ traversed counterclockwise.

8. Integrate $\mathbf{h}(x, y) = e^{x-y}\mathbf{i} + e^{x+y}\mathbf{j}$ over the line segment from $(-1, 1)$ to $(1, 2)$.

9. Integrate $\mathbf{h}(x, y) = (x + y)\mathbf{i} + (y^2 - x)\mathbf{j}$ over the closed curve that begins at $(-1, 0)$, goes along the x-axis to $(1, 0)$, and returns to $(-1, 0)$ by the upper part of the unit circle.

10. Integrate $\mathbf{h}(x, y) = 3x^2 y\mathbf{i} + (x^3 + 2y)\mathbf{j}$ over the square with vertices $(0, 0), (1, 0), (1, 1), (0, 1)$ traversed counterclockwise.

11. Integrate $\mathbf{h}(x, y, z) = yz\mathbf{i} + x^2\mathbf{j} + xz\mathbf{k}$ over the indicated path:
    (a) the line segment from $(0, 0, 0)$ to $(1, 1, 1)$.   (b) $\mathbf{r}(u) = u\mathbf{i} + u^2\mathbf{j} + u^3\mathbf{k}, \quad u \in [0, 1]$.

12. Integrate $\mathbf{h}(x, y, z) = e^x\mathbf{i} + e^y\mathbf{j} + e^z\mathbf{k}$ over the paths of Exercise 11.

13. Calculate the work done by the force $\mathbf{F}(x, y, z) = x\mathbf{i} + xy\mathbf{j} + xyz\mathbf{k}$ applied to an object that moves in a straight line from $(0, 1, 4)$ to $(1, 0, -4)$.

14. A mass m, moving in a force field, traces out a circular arc at constant speed. Show that the force field does no work. Give a physical explanation for this.

15. Let $C: \mathbf{r} = \mathbf{r}(u), \ u \in [a, b]$ be a smooth curve and $\mathbf{q}$ a fixed vector. Show that
$$\int_C \mathbf{q} \cdot d\mathbf{r} = \mathbf{q} \cdot [\mathbf{r}(b) - \mathbf{r}(a)] \quad \text{and} \quad \int_C \mathbf{r} \cdot d\mathbf{r} = \frac{\|\mathbf{r}(b)\|^2 - \|\mathbf{r}(a)\|^2}{2}.$$

C_1: straight-line path C_2: rectangular path C_3: semicircular path

Figure 18.1.5

16. Figure 18.1.5 shows three paths from $(1, 0)$ to $(-1, 0)$. Calculate the line integral of
$$\mathbf{h}(x, y) = x^2\mathbf{i} + y\mathbf{j}$$
over (a) the straight-line path. (b) the rectangular path. (c) the semicircular path.

17. Let f be a continuous real-valued function of a real variable. Show that, if

$$\mathbf{f}(x, y, z) = f(x)\mathbf{i} \qquad \text{and} \qquad C: \ \mathbf{r}(u) = u\mathbf{i}, \quad u \in [a, b],$$

then

$$\int_C \mathbf{f}(\mathbf{r}) \cdot d\mathbf{r} = \int_a^b f(u) \, du.$$

18. (*Linearity*) Show that, if $\mathbf{f}$ and $\mathbf{g}$ are continuous vector fields and C is piecewise smooth, then

(18.1.9)
$$\int_C [\alpha \mathbf{f}(\mathbf{r}) + \beta \mathbf{g}(\mathbf{r})] \cdot d\mathbf{r} = \alpha \int_C \mathbf{f}(\mathbf{r}) \cdot d\mathbf{r} + \beta \int_C \mathbf{g}(\mathbf{r}) \cdot d\mathbf{r}$$

for all real α, β.

19. The force $\mathbf{F}(x, y) = -\frac{1}{2}[y\mathbf{i} - x\mathbf{j}]$ is continually applied to an object that orbits an ellipse in standard position. Find a relation between the work done and the area of the ellipse.

20. An object of mass m moves from time $t = 0$ to $t = 1$ so that its position at time t is given by the vector function

$$\mathbf{r}(t) = \alpha t\mathbf{i} + \beta t^2 \mathbf{j}.$$

Find the total force acting on the object at time t and calculate the work done by that force during the time interval $[0, 1]$.

21. Exercise 20 with $\mathbf{r}(t) = \alpha t\mathbf{i} + \beta t^2 \mathbf{j} + \gamma t^3 \mathbf{k}$.

22. (*Important*) The *circulation* of a vector field $\mathbf{v}$ around an oriented closed curve C is by definition the line integral

$$\int_C \mathbf{v}(\mathbf{r}) \cdot d\mathbf{r}.$$

Let $\mathbf{v}$ be the velocity field of a fluid in counterclockwise circular motion about the z-axis with constant angular speed ω.

(a) Verify that $\mathbf{v}(\mathbf{r}) = \omega \mathbf{k} \times \mathbf{r}$.

(b) Show that the circulation of $\mathbf{v}$ around any circle C in the xy-plane with center at the origin is $\pm 2\omega$ times the area of the circle.

23. Let $\mathbf{v}$ be the velocity field of a fluid that moves radially from the origin: $\mathbf{v} = f(x, y)\mathbf{r}$. What is the circulation of $\mathbf{v}$ around a circle C centered at the origin?

18.2 THE FUNDAMENTAL THEOREM FOR LINE INTEGRALS

In general, if we integrate a vector field $\mathbf{h}$ from one point to another, the value of the line integral depends upon the path chosen. There is, however, an important exception. If the vector field is a *gradient field,*

$$\mathbf{h} = \nabla f,$$

then the value of the line integral depends only on the endpoints of the path and not on the path itself. The details are spelled out in the following theorem.

> **THEOREM 18.2.1 THE FUNDAMENTAL THEOREM FOR LINE INTEGRALS**
>
> Let C: $\mathbf{r} = \mathbf{r}(u)$, $u \in [a, b]$ be a piecewise-smooth curve that begins at $\mathbf{a}$ and ends at $\mathbf{b}$. If the scalar field f is continuously differentiable on an open set that contains the curve C, then
>
> $$\int_C \nabla f(\mathbf{r}) \cdot d\mathbf{r} = f(\mathbf{b}) - f(\mathbf{a}).$$

Proof. If C is smooth,

$$\int_C \nabla f(\mathbf{r}) \cdot d\mathbf{r} = \int_a^b [\nabla f(\mathbf{r}(u)) \cdot \mathbf{r}'(u)] \, du$$

$$\underset{\text{chain rule (16.4.1)}}{=} \int_a^b \frac{d}{du} [f(\mathbf{r}(u))] \, du$$

$$= f(\mathbf{r}(b)) - f(\mathbf{r}(a))$$

$$= f(\mathbf{b}) - f(\mathbf{a}).$$

If C is not smooth but only piecewise smooth, then we break up C into smooth pieces

$$C = C_1 \cup C_2 \cup \cdots \cup C_n.$$

With obvious notation,

$$\int_C \nabla f(\mathbf{r}) \cdot d\mathbf{r} = \int_{C_1} \nabla f(\mathbf{r}) \cdot d\mathbf{r} + \int_{C_2} \nabla f(\mathbf{r}) \cdot d\mathbf{r} + \cdots + \int_{C_n} \nabla f(\mathbf{r}) \cdot d\mathbf{r}$$

$$= [f(\mathbf{a}_1) - f(\mathbf{a}_0)] + [f(\mathbf{a}_2) - f(\mathbf{a}_1)] + \cdots + [f(\mathbf{a}_n) - f(\mathbf{a}_{n-1})]$$

$$= f(\mathbf{a}_n) - f(\mathbf{a}_0)$$

$$= f(\mathbf{b}) - f(\mathbf{a}). \quad \square$$

The theorem we just proved has an important corollary:

(18.2.2)

> if the curve C is *closed* (that is, if $\mathbf{a} = \mathbf{b}$), then of course $f(\mathbf{b}) = f(\mathbf{a})$ and
>
> $$\int_C \nabla f(\mathbf{r}) \cdot d\mathbf{r} = 0.$$

Problem 1.   Integrate the vector field $\mathbf{h}(x, y) = y^2\mathbf{i} + (2xy - e^y)\mathbf{j}$ over the circular arc

$$C: \quad \mathbf{r}(u) = \cos u\,\mathbf{i} + \sin u\,\mathbf{j}, \quad u \in [0, \tfrac{1}{2}\pi].$$

Solution.   First we try to determine whether $\mathbf{h}$ is a gradient.   We do this by applying Theorem 16.10.2.

Note that $\mathbf{h}(x, y)$ has the form $P(x, y)\mathbf{i} + Q(x, y)\mathbf{j}$ with

$$P(x, y) = y^2 \quad \text{and} \quad Q(x, y) = 2xy - e^y.$$

Since P and Q are continuously differentiable everywhere and

$$\frac{\partial P}{\partial y}(x, y) = 2y = \frac{\partial Q}{\partial x}(x, y),$$

we can conclude that $\mathbf{h}$ is a gradient.   Since the integral depends then only on the endpoints of C and not on C itself, we can simplify the computations by integrating over the line segment C' that joins these same endpoints.   (See Figure 18.2.1.)

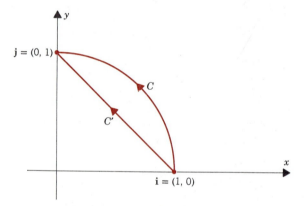

Figure 18.2.1

We parametrize C' by setting

$$\mathbf{r}(u) = (1 - u)\mathbf{i} + u\mathbf{j}, \quad u \in [0, 1].$$

We then have

$$\int_C \mathbf{h}(\mathbf{r}) \cdot d\mathbf{r} = \int_{C'} \mathbf{h}(\mathbf{r}) \cdot d\mathbf{r}$$

$$= \int_0^1 [\mathbf{h}(\mathbf{r}(u)) \cdot \mathbf{r}'(u)]\, du$$

$$= \int_0^1 \left[[y(u)]^2 x'(u) + [2x(u)y(u) - e^{y(u)}]y'(u) \right] du$$

$$= \int_0^1 [2u - 3u^2 - e^u]\, du = \left[u^2 - u^3 - e^u \right]_0^1 = 1 - e. \quad \square$$

Alternative Solution. Once we have recognized that $\mathbf{h}(x, y) = y^2\mathbf{i} + (2xy - e^y)\mathbf{j}$ is a gradient $\nabla f(x, y)$, we can determine $f(x, y)$ by the methods of Section 16.10. Since

$$\frac{\partial f}{\partial x}(x, y) = y^2 \quad \text{and} \quad \frac{\partial f}{\partial y}(x, y) = 2xy - e^y,$$

we have

$$f(x, y) = xy^2 + \phi(y) \quad \text{and therefore} \quad \frac{\partial f}{\partial y}(x, y) = 2xy + \phi'(y).$$

The two expressions for $\partial f/\partial y$ can be reconciled only if

$$\phi'(y) = -e^y \quad \text{and thus} \quad \phi(y) = -e^y + c.$$

This means that

$$f(x, y) = xy^2 - e^y + c.$$

Since the curve C begins at $(1, 0)$ and ends at $(0, 1)$, we see that

$$\int_C \mathbf{h}(\mathbf{r}) \cdot d\mathbf{r} = f(0, 1) - f(1, 0) = (-e + c) - (-1 + c) = 1 - e. \quad \square$$

Problem 2. Evaluate the line integral

$$\int_C \mathbf{h}(\mathbf{r}) \cdot d\mathbf{r}$$

where C is the unit circle

$$\mathbf{r}(u) = \cos u \, \mathbf{i} + \sin u \, \mathbf{j}, \qquad u \in [0, 2\pi]$$

and

$$\mathbf{h}(x, y) = (y^2 + y)\mathbf{i} + (2xy - e^y)\mathbf{j}.$$

Solution. Although $(y^2 + y)\mathbf{i} + (2xy - e^y)\mathbf{j}$ is not a gradient, part of it,

$$y^2\mathbf{i} + (2xy - e^y)\mathbf{j},$$

is a gradient. (See Problem 1.) Since we are integrating over a closed curve, the contribution of the gradient part is 0. The contribution of the remaining part is

$$\int_C \mathbf{g}(\mathbf{r}) \cdot d\mathbf{r} \quad \text{with} \quad \mathbf{g}(x, y) = y\mathbf{i}.$$

This last integral is easy to evaluate:

$$\int_C \mathbf{g}(\mathbf{r}) \cdot d\mathbf{r} = \int_0^{2\pi} [\mathbf{g}(\mathbf{r}(u)) \cdot \mathbf{r}'(u)] \, du = \int_0^{2\pi} y(u)x'(u) \, du = \int_0^{2\pi} -\sin^2 u \, du = -\pi. \quad \square$$

EXERCISES 18.2

Calculate the line integral of **h** over the indicated curve.

1. $\mathbf{h}(x, y) = x\mathbf{i} + y\mathbf{j}$; $\mathbf{r}(u) = a \cos u\, \mathbf{i} + b \sin u\, \mathbf{j}$, $u \in [0, 2\pi]$.
2. $\mathbf{h}(x, y) = (x + y)\mathbf{i} + y\mathbf{j}$; the curve of Exercise 1.
3. $\mathbf{h}(x, y) = \cos \pi y\, \mathbf{i} - \pi x \sin \pi y\, \mathbf{j}$; $\mathbf{r}(u) = u^2\mathbf{i} - u^3\mathbf{j}$, $u \in [0, 1]$.
4. $\mathbf{h}(x, y) = (x^2 - y)\mathbf{i} + (y^2 - x)\mathbf{j}$; the curve of Exercise 1.
5. $\mathbf{h}(x, y) = xy^2\mathbf{i} + x^2y\mathbf{j}$; $\mathbf{r}(u) = u \sin \pi u\, \mathbf{i} + \cos \pi u^2\, \mathbf{j}$, $u \in [0, 1]$.
6. $\mathbf{h}(x, y) = (1 + e^y)\mathbf{i} + (xe^y - x)\mathbf{j}$; the square with vertices $(-1, -1), (1, -1), (1, 1), (-1, 1)$ traversed counterclockwise.
7. $\mathbf{h}(x, y) = (2xy - y^2)\mathbf{i} + (x^2 - 2xy)\mathbf{j}$; $\mathbf{r}(u) = \cos u\, \mathbf{i} + \sin u\, \mathbf{j}$, $u \in [0, \pi]$.
8. $\mathbf{h}(x, y) = 3x(x^2 + y^4)^{1/2}\mathbf{i} + 6y^3(x^2 + y^4)^{1/2}\mathbf{j}$; the circular arc $y = (1 - x^2)^{1/2}$ from $(1, 0)$ to $(-1, 0)$.
9. $\mathbf{h}(x, y) = 3x(x^2 + y^4)^{1/2}\mathbf{i} + 6y^3(x^2 + y^4)^{1/2}\mathbf{j}$; the arc $y = -(1 - x^2)^{1/2}$ from $(-1, 0)$ to $(1, 0)$.
10. $\mathbf{h}(x, y) = 2xy \sinh x^2y\, \mathbf{i} + x^2 \sinh x^2y\, \mathbf{j}$; the curve of Exercise 1.
11. $\mathbf{h}(x, y) = (2x \cosh y - y)\mathbf{i} + (x^2 \sinh y - y)\mathbf{j}$; the square of Exercise 6.
12. $\mathbf{h}(x, y, z) = y^2z^3\mathbf{i} + 2xyz^3\mathbf{j} + 3xy^2z^2\mathbf{k}$; $\mathbf{r}(u) = u^2\mathbf{i} + u^4\mathbf{j} + u^6\mathbf{k}$, $u \in [0, 1]$.
    HINT: Note the three-dimensional analog of Theorem 10.6.2 given on page 934.
13. If g is a continuously differentiable real-valued function defined on $[a, b]$, then by the fundamental theorem of integral calculus

$$\int_a^b g'(u)\, du = g(b) - g(a).$$

Show how this result is included in Theorem 18.2.1.

14. Let $\mathbf{r} = x\mathbf{i} + y\mathbf{j} + z\mathbf{k}$ and set $r = \|\mathbf{r}\|$. The central force field

$$\mathbf{F}(\mathbf{r}) = \frac{K}{r^n}\, \mathbf{r}, \qquad n \text{ a positive integer}$$

is a gradient field. Find f such that $\nabla f(\mathbf{r}) = \mathbf{F}(\mathbf{r})$ if (a) $n = 2$, (b) $n \neq 2$.

15. Let $\mathbf{r} = x\mathbf{i} + y\mathbf{j} + z\mathbf{k}$ and set $r = \|\mathbf{r}\|$. The function

$$\mathbf{F}(\mathbf{r}) = -\frac{mG}{r^3}\, \mathbf{r} \qquad\qquad (G \text{ is a gravitational constant})$$

gives the gravitational force exerted by a unit mass at the origin on a mass m located at $\mathbf{r}$. What is the work done by $\mathbf{F}$ if m moves from $\mathbf{r}_1$ to $\mathbf{r}_2$?

16. Set

$$P(x, y) = \frac{y}{x^2 + y^2} \qquad \text{and} \qquad Q(x, y) = -\frac{x}{x^2 + y^2}$$

on *the punctured unit disc* Ω: $0 < x^2 + y^2 < 1$.

(a) Verify that P and Q are continuously differentiable on Ω and that

$$\frac{\partial P}{\partial y}(x, y) = \frac{\partial Q}{\partial x}(x, y) \qquad \text{for all } (x, y) \in \Omega.$$

(b) Verify that, in spite of (a), the vector field $\mathbf{h}(x, y) = P(x, y)\mathbf{i} + Q(x, y)\mathbf{j}$ is not a gradient on Ω.   HINT: Integrate **h**, over a circle of radius less than one centered at the origin.

(c) Explain how (b) does not contradict Theorem 16.10.2.

*18.3 WORK–ENERGY FORMULA; CONSERVATION OF MECHANICAL ENERGY

Suppose that a continuous force field $\mathbf{F} = \mathbf{F}(\mathbf{r})$ accelerates a mass m from $\mathbf{r} = \mathbf{a}$ to $\mathbf{r} = \mathbf{b}$ along some smooth curve C. The object undergoes a change in kinetic energy:

$$\tfrac{1}{2}m[v(b)]^2 - \tfrac{1}{2}m[v(a)]^2.$$

The force does a certain amount of work W. How are these quantities related? They are equal:

(18.3.1)
$$\boxed{W = \tfrac{1}{2}m[v(b)]^2 - \tfrac{1}{2}m[v(a)]^2.}$$

This relation is called the *Work–Energy Formula*.

Derivation of the Work–Energy Formula

We parametrize the path of the motion by the time parameter t:

$$C: \quad \mathbf{r} = \mathbf{r}(t), \quad t \in [a, b].$$

Note that $\mathbf{r}(a) = \mathbf{a}$ and $\mathbf{r}(b) = \mathbf{b}$. The work done by $\mathbf{F}$ is given by the formula

$$W = \int_C \mathbf{F}(\mathbf{r}) \cdot d\mathbf{r} = \int_a^b [\mathbf{F}(\mathbf{r}(t)) \cdot \mathbf{r}'(t)]\, dt.$$

From Newton's second law of motion, we know that at time t,

$$\mathbf{F}(\mathbf{r}(t)) = m\,\mathbf{a}(t) = m\,\mathbf{r}''(t).$$

It follows that

$$\mathbf{F}(\mathbf{r}(t)) \cdot \mathbf{r}'(t) = m\,\mathbf{r}''(t) \cdot \mathbf{r}'(t)$$
$$= \frac{d}{dt}\left[\frac{1}{2}m(\mathbf{r}'(t) \cdot \mathbf{r}'(t))\right] = \frac{d}{dt}\left[\frac{1}{2}m\|\mathbf{r}'(t)\|^2\right] = \frac{d}{dt}\left[\frac{1}{2}m[v(t)]^2\right].$$

Substituting this last expression into the work integral, we see that

$$W = \int_a^b \frac{d}{dt}\left(\frac{1}{2}m[v(t)]^2\right) dt = \frac{1}{2}m[v(b)]^2 - \frac{1}{2}m[v(a)]^2$$

as asserted. □

Conservative Force Fields

In general, if an object moves from one point to another, the work done (and hence the change in kinetic energy) depends on the path of the motion. There is, however, an important exception: if the force field is a gradient field,

$$\mathbf{F} = \nabla f,$$

then the work done (and hence the change in kinetic energy) depends only on the endpoints of the path and not on the path itself. (This follows directly from The Fundamental Theorem for Line Integrals.) A force field that is a gradient field is called a *conservative field*.

Since the line integral over a closed path is zero, *the work done by a conservative field over a closed path is always zero. An object that passes through a given point with a certain kinetic energy returns to that same point with exactly the same kinetic energy.*

Potential Energy Functions

Suppose that **F** is a conservative force field. It is then a gradient field. Then $-$**F** is a gradient field. The functions U for which $\nabla U = -$**F** are called *potential energy functions* for **F**.

The Conservation of Mechanical Energy

Suppose that **F** is a conservative force field: $\mathbf{F} = -\nabla U$. In our derivation of the work–energy formula we showed that

$$\frac{d}{dt}\left(\tfrac{1}{2}m[v(t)]^2\right) = \mathbf{F}(\mathbf{r}(t)) \cdot \mathbf{r}'(t).$$

Since

$$\frac{d}{dt}[U(\mathbf{r}(t))] = \nabla U(\mathbf{r}(t)) \cdot \mathbf{r}'(t) = -\mathbf{F}(\mathbf{r}(t)) \cdot \mathbf{r}'(t),$$

we have

$$\frac{d}{dt}[\tfrac{1}{2}m[v(t)]^2 + U(\mathbf{r}(t))] = 0$$

and therefore

$$\underbrace{\tfrac{1}{2}m[v(t)]^2}_{\text{KE}} + \underbrace{U(\mathbf{r}(t))}_{\text{PE}} = \text{a constant.}$$

An an object moves in a conservative force field, its kinetic energy can vary and its potential energy can vary, but the sum of these two quantities remains constant. We call this constant *the total mechanical energy.*

The total mechanical energy is usually denoted by the letter E. The law of conservation of mechanical energy can then be written

(18.3.2)
$$\tfrac{1}{2}mv^2 + U = E.$$

The conservation of energy is one of the cornerstones of physics. Here we have been talking about mechanical energy. There are other forms of energy and other energy conservation laws.

Differences in Potential Energy

Potential energy at a particular point has no physical significance. Only differences in potential energy are significant:

$$U(\mathbf{b}) - U(\mathbf{a}) = \int_C -\mathbf{F}(\mathbf{r}) \cdot d\mathbf{r}$$

is the work required to move from $\mathbf{r} = \mathbf{a}$ to $\mathbf{r} = \mathbf{b}$ *against* the force field $\mathbf{F}$.

Problem 1. A planet moves in the gravitational field of the sun:

$$\mathbf{F}(\mathbf{r}) = -m\rho\,\frac{\mathbf{r}}{r^3}. \qquad (m \text{ is the mass of the planet})$$

Show that the force field is conservative, find a potential energy function, and determine the total energy of the planet. How does the planet's speed vary with the planet's distance from the sun?

Solution. The field is conservative since

$$\mathbf{F}(\mathbf{r}) = -m\rho\,\frac{\mathbf{r}}{r^3} = \nabla\left(\frac{\rho m}{r}\right). \qquad (\text{check this out})$$

As a potential energy function we can use

$$U(\mathbf{r}) = -\frac{\rho m}{r}.$$

The total energy of the planet is the constant

$$E = \tfrac{1}{2}mv^2 - \frac{\rho m}{r}.$$

(You met this quantity before: Exercises 2 and 4 of Section *14.6.)

Solving the energy equation for v we have

$$v = \sqrt{\frac{2E}{m} + \frac{2\rho}{r}}.$$

As r decreases, $2\rho/r$ increases, and v increases; as r increases, $2\rho/r$ decreases, and v decreases. Thus every planet speeds up as it comes near the sun and slows down as it moves away. (The same holds true for Halley's comet. The fact that it slows down as it gets farther away helps explain why it comes back. The simplicity of all this is a testimony to the power of the principle of energy conservation.) ☐

EXERCISES *18.3

1. Let f be a real-valued function of a real variable continuous in x. Show that the force field $\mathbf{F}(x, y, z) = f(x)\mathbf{i}$ is conservative and the potential functions for $\mathbf{F}$ are (except for notation) the antiderivatives of $-f$.

2. A particle with electric charge e and velocity $\mathbf{v}$ moves in a magnetic field $\mathbf{B}$, experiencing the force

$$\mathbf{F} = \frac{e}{c}[\mathbf{v} \times \mathbf{B}].\qquad\qquad \text{(c is the velocity of light)}$$

   $\mathbf{F}$ is not a gradient—it can't be, depending as it does on the *velocity* of the particle. Still, we can find a conserved quantity: the *kinetic energy* $\frac{1}{2}mv^2$. Show by differentiation with respect to t that this quantity is constant. (Assume Newton's second law.)

3. An object is subject to a constant force in the direction of $-\mathbf{k}$: $\mathbf{F} = -c\mathbf{k}$ with $c > 0$. Find a potential energy function for $\mathbf{F}$, and use energy conservation to show that the speed of the object at time t_2 is related to that at time t_1 by the equation

$$v(t_2) = \sqrt{[v(t_1)]^2 + \frac{2c}{m}[z(t_1) - z(t_2)]}$$

   where $z(t_1)$ and $z(t_2)$ are the z-coordinates of the object at times t_1 and t_2. (This analysis is sometimes used to model the behavior of an object in the gravitational field near the surface of the earth.)

4. (*Escape velocity*) An object is to be fired straight up from the surface of the earth. Assume that the only force acting on the object is the gravitational pull of the earth and determine the initial speed v_0 necessary to send the object off to infinity.   HINT: Appeal to conservation of energy and use the idea that the object is to arrive at infinity with zero speed.

5. (a) Justify the statement that a conservative force field $\mathbf{F}$ always acts so as to encourage motion toward regions of lower potential energy U.

   (b) Evaluate $\mathbf{F}$ at a point where U has a minimum.

6. A harmonic oscillator has a restoring force $\mathbf{F} = -\lambda x\mathbf{i}$. The associated potential is $U(x, y, z) = -\frac{1}{2}\lambda x^2$, and the constant total energy is

$$E = \tfrac{1}{2}mv^2 + U(x, y, z) = \tfrac{1}{2}mv^2 + \tfrac{1}{2}\lambda x^2.$$

   Given that $x(0) = 2$ and $x'(0) = 1$, calculate the maximum speed of the oscillator and the maximum value of x.

7. The *equipotential surfaces* of a conservative field $\mathbf{F}$ are the surfaces where the potential energy is constant. Show that (a) the speed of an object in such a field is constant on every equipotential surface and (b) at each point of such a surface the force field is perpendicular to the surface.

18.4   ANOTHER NOTATION FOR LINE INTEGRALS; LINE INTEGRALS WITH RESPECT TO ARC LENGTH

If $\mathbf{h}(x, y, z) = P(x, y, z)\mathbf{i} + Q(x, y, z)\mathbf{j} + R(x, y, z)\mathbf{k}$, the line integral

$$\int_C \mathbf{h}(\mathbf{r}) \cdot d\mathbf{r} \quad \text{can be written} \quad \int_C P(x, y, z)\, dx + Q(x, y, z)\, dy + R(x, y, z)\, dz.$$

The notation arises as follows. With

$$C: \quad \mathbf{r}(u) = x(u)\mathbf{i} + y(u)\mathbf{j} + z(u)\mathbf{k}, \quad u \in [a, b]$$

the line integral

$$\int_C \mathbf{h}(\mathbf{r}) \cdot d\mathbf{r} = \int_a^b [\mathbf{h}(\mathbf{r}(t)) \cdot \mathbf{r}'(t)] \, dt$$

expands to

$$\int_a^b \{P[x(u), y(u), z(u)]x'(u) + Q[x(u), y(u), z(u)]y'(u) + R[x(u), y(u), z(u)]z'(u)\} \, du.$$

Now set

$$\int_C P(x, y, z) \, dx = \int_a^b P[x(u), y(u), z(u)]x'(u) \, du,$$

$$\int_C Q(x, y, z) \, dy = \int_a^b Q[x(u), y(u), z(u)]y'(u) \, du,$$

$$\int_C R(x, y, z) \, dz = \int_a^b R[x(u), y(u), z(u)]z'(u) \, du.$$

Writing the sum of these integrals as

$$\int_C P(x, y, z) \, dx + Q(x, y, z) \, dy + R(x, y, z) \, dz,$$

we have

$$\int_C P(x, y, z) \, dx + Q(x, y, z) \, dy + R(x, y, z) \, dz = \int_C \mathbf{h}(\mathbf{r}) \cdot d\mathbf{r}. \quad \square$$

If C lies in the xy-plane and $\mathbf{h}(x, y) = P(x, y)\mathbf{i} + Q(x, y)\mathbf{j}$, then the z's drop out and the line integral reduces to

$$\int_C P(x, y) \, dx + Q(x, y) \, dy.$$

The notation is easy to use; for example, with C as above,

$$\int_C y \, dx + 2z \, dy - x \, dz = \int_a^b [y(u)x'(u) + 2z(u)y'(u) - x(u)z'(u)] \, du,$$

$$\int_C \sin(xy) \, dx - \cos(xy) \, dy = \int_a^b \{\sin[x(u)y(u)] \, x'(u) - \cos[x(u)y(u)] \, y'(u)\} \, du.$$

Line Integrals with Respect to Arc Length

Suppose that f is a scalar field continuous on a piecewise-smooth curve

$$C: \quad \mathbf{r}(u) = x(u)\mathbf{i} + y(u)\mathbf{j} + z(u)\mathbf{k}, \quad u \in [a, b].$$

If $s(u)$ is the length of the curve from the tip of $\mathbf{r}(a)$ to the tip of $\mathbf{r}(u)$, then, as you have seen,

$$s'(u) = \|\mathbf{r}'(u)\| = \sqrt{[x'(u)]^2 + [y'(u)]^2 + [z'(u)]^2}.$$

The line integral of f over C *with respect to arc length s* is defined by setting

(18.4.1)
$$\int_C f(\mathbf{r}) \, ds = \int_a^b f(\mathbf{r}(u)) \, s'(u) \, du.$$

In the *xyz*-notation we have

$$\int_C f(x, y, z) \, ds = \int_a^b f(x(u), y(u), z(u)) \, s'(u) \, du,$$

which, in the two-dimensional case, becomes

$$\int_C f(x, y) \, ds = \int_a^b f(x(u), y(u)) \, s'(u) \, du.$$

Suppose now that C represents a thin wire (a material curve) of varying mass density $\lambda = \lambda(\mathbf{r})$. [Here mass density is mass per unit length.] The *length* of the wire can be written

(18.4.2)
$$L = \int_C ds.$$

The *mass* of the wire is given by

(18.4.3)
$$M = \int_C \lambda(\mathbf{r}) \, ds$$

and the *center of mass* $\mathbf{r}_M$ can be obtained from the vector equation

(18.4.4)
$$\mathbf{r}_M M = \int_C \mathbf{r} \lambda(\mathbf{r}) \, ds.$$

The equivalent scalar equations read

$$x_M M = \int_C x \lambda(\mathbf{r}) \, ds, \qquad y_M M = \int_C y \lambda(\mathbf{r}) \, ds, \qquad z_M M = \int_C z \lambda(\mathbf{r}) \, ds.$$

Finally, the *moment of inertia* about an axis is given by the formula

(18.4.5)
$$I = \int_C \lambda(\mathbf{r})[R(\mathbf{r})]^2 \, ds$$

where $R(\mathbf{r})$ is the distance from the axis to the tip of $\mathbf{r}$.

Problem 1. The mass density of a semicircular wire of radius a varies directly as the distance from the diameter that joins the two endpoints of the wire. (a) Find the mass of the wire. (b) Locate the center of mass. (c) Determine the moment of inertia of the wire about the diameter.

Solution. Placed as in Figure 18.4.1 the wire can be parametrized by

$$\mathbf{r}(u) = a \cos u \, \mathbf{i} + a \sin u \, \mathbf{j}, \qquad u \in [0, \pi]$$

and the mass density function can be written $\lambda(x, y) = ky$. Since $\mathbf{r}'(u) = -a \sin u \, \mathbf{i} + a \cos u \, \mathbf{j}$, we have

$$s'(u) = \|\mathbf{r}'(u)\| = a.$$

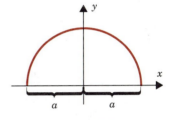

Figure 18.4.1

Therefore

$$M = \int_C \lambda(x, y) \, ds = \int_C ky \, ds$$
$$= \int_0^\pi k y(u) s'(u) \, du$$
$$= \int_0^\pi k(a \sin u) a \, du = ka^2 \int_0^\pi \sin u \, du = 2ka^2.$$

By the symmetry of the configuration, $x_M = 0$. To find y_M we have to integrate:

$$y_M M = \int_C y\lambda(x, y) \, ds = \int_C ky^2 \, ds$$
$$= \int_0^\pi k[y(u)]^2 s'(u) \, du$$
$$= \int_0^\pi k(a \sin u)^2 a \, du = ka^3 \int_0^\pi \sin^2 u \, du = \tfrac{1}{2}ka^3\pi.$$

Since $M = 2ka^2$, we have $y_M = (\tfrac{1}{2}ka^3\pi)/(2ka^2) = \tfrac{1}{4}a\pi$. The center of mass lies on the perpendicular bisector of the wire at a distance $\tfrac{1}{4}a\pi$ from the diameter.

Now let's find the moment of inertia about the diameter:

$$I = \int_C \lambda(x, y)[R(x, y)]^2 \, ds = \int_C (ky)y^2 \, ds$$

$$= \int_C k[y(u)]^3 s'(u) \, du$$

$$= \int_0^\pi k(a \sin u)^3 a \, du$$

$$= ka^4 \int_0^\pi \sin^3 u \, du = \tfrac{4}{3} ka^4.$$

It is customary to express I in terms of M. With $M = 2ka^2$, we have

$$I = \tfrac{2}{3} (2ka^2)a^2 = \tfrac{2}{3} Ma^2. \quad \square$$

EXERCISES 18.4

Evaluate

$$\int_C (x - 2y) \, dx + 2x \, dy \qquad \text{along the given path } C \text{ from } (0, 0) \text{ to } (1, 2).$$

1. The straight-line path.   **2.** The parabolic path $y = 2x^2$.
3. The polygonal path $(0, 0)$, $(1, 0)$, $(1, 2)$.   **4.** The polygonal path $(0, 0)$, $(0, 2)$, $(1, 2)$.

Evaluate

$$\int_C y \, dx + xy \, dy \qquad \text{along the given path } C \text{ from } (0, 0) \text{ to } (2, 1).$$

5. The parabolic path $x = 2y^2$.   **6.** The straight-line path.
7. The polygonal path $(0, 0)$, $(0, 1)$, $(2, 1)$.   **8.** The cubic path $x = 2y^3$.

Evaluate

$$\int_C y \, dx + 2z \, dy + x \, dz \qquad \text{along the given path } C \text{ from } (0, 0, 0) \text{ to } (1, 1, 1).$$

9. The straight-line path.   **10.** $\mathbf{r}(u) = u\mathbf{i} + u^2\mathbf{j} + u^3\mathbf{k}$.
11. The polygonal path $(0, 0, 0)$, $(0, 0, 1)$, $(0, 1, 1)$, $(1, 1, 1)$.
12. The polygonal path $(0, 0, 0)$, $(1, 0, 0)$, $(1, 1, 0)$, $(1, 1, 1)$.

13. A wire in the shape of the quarter-circle

$$C: \quad \mathbf{r}(u) = a(\cos u \, \mathbf{i} + \sin u \, \mathbf{j}), \quad u \in [0, \tfrac{1}{2}\pi]$$

has varying mass density $\lambda(x, y) = k(x + y)$ where k is a positive constant.
(a) Find the total mass of the wire and locate the center of mass.
(b) What is the moment of inertia of the wire about the x-axis?

14. Find the moment of inertia of a homogeneous circular wire of radius a and mass M (a) about a diameter. (b) about the axis through the center that is perpendicular to the plane of the wire.

15. Find the moment of inertia of the wire of Exercise 13
    (a) about the z-axis.   (b) about the line $y = x$.

16. A wire of constant mass density k has the form

$$\mathbf{r}(u) = (1 - \cos u)\mathbf{i} + (u - \sin u)\mathbf{j}, \qquad u \in [0, 2\pi].$$

   (a) Determine the mass of the wire.   (b) Locate the center of mass.

17. A homogeneous wire of mass M winds around the z-axis as

$$C: \quad \mathbf{r}(u) = a \cos u\, \mathbf{i} + a \sin u\, \mathbf{j} + bu\mathbf{k}, \quad u \in [0, 2\pi].$$

   (a) Find the length of the wire.   (b) Locate the center of mass.
   (c) Determine the moments of inertia of the wire about the coordinate axes.

18. A homogeneous wire of mass M is of the form

$$C: \quad \mathbf{r}(u) = u\mathbf{i} + u^2\mathbf{j} + \tfrac{2}{3}u^3\mathbf{k}, \quad u \in [0, a].$$

   (a) Find the length of the wire.   (b) Locate the center of mass.
   (c) Determine the moment of inertia of the wire about the z-axis.

19. Calculate the mass of the wire of Exercise 17 if the mass density varies directly as the square of the distance from the origin.

20. Show that

(18.4.6)
$$\int_C \mathbf{h}(\mathbf{r}) \cdot d\mathbf{r} = \int_C [\mathbf{h}(\mathbf{r}) \cdot \mathbf{T}(\mathbf{r})]\, ds$$

where $\mathbf{T}$ is the unit tangent vector.

18.5 GREEN'S THEOREM

(The first of the three integration theorems heralded at the beginning of this chapter.)

A Jordan curve, you will recall, is a plane curve that is both closed and simple. Thus circles, ellipses, and triangles are Jordan curves; figure eights are not.

Figure 18.5.1 depicts a closed region Ω, the total boundary of which is a Jordan curve C. Such a region is called a *Jordan region.*

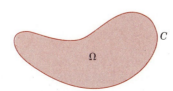

Figure 18.5.1

We know how to integrate over Ω, and, if the boundary C is piecewise smooth, we know how to integrate over C. Green's theorem expresses a double integral over Ω as a line integral over C.

THEOREM 18.5.1   GREEN'S THEOREM†

Let Ω be a Jordan region with a piecewise-smooth boundary C. If P and Q are scalar fields continuously differentiable on an open set that contains Ω, then

$$\iint_{\Omega} \left[\frac{\partial Q}{\partial x}(x, y) - \frac{\partial P}{\partial y}(x, y) \right] dx\,dy = \oint_C P(x, y)\,dx + Q(x, y)\,dy$$

where the integral on the right is the line integral over C taken in the counterclockwise direction.††

We will prove the theorem only for special cases. First of all let's assume that Ω is an *elementary region,* a region that is both of type I and type II as defined in Section 17.4. For simplicity we take Ω as in Figure 18.5.2.

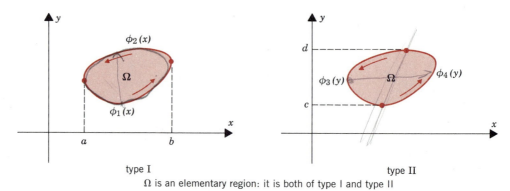

type I   type II

Ω is an elementary region: it is both of type I and type II

Figure 18.5.2

Ω being of type I, we can show that

(1)
$$\oint_C P(x, y)\,dx = \iint_{\Omega} -\frac{\partial P}{\partial y}(x, y)\,dx\,dy.$$

In the first place

---

† The result was established in 1828 by the English mathematician George Green (1793–1841).

†† The integral over C taken in the clockwise direction would be written $\oint_C$.

$$\iint\limits_{\Omega} -\frac{\partial P}{\partial y}(x, y)\, dxdy = -\int_a^b \int_{\phi_1(x)}^{\phi_2(x)} \frac{\partial P}{\partial y}(x, y)\, dy\, dx.$$

$$= -\int_a^b \{P[x, \phi_2(x)] - P[x, \phi_1(x)]\}\, dx$$

by The Fundamental Theorem of Calculus ⟶

$$(*) \qquad = \int_a^b P[x, \phi_1(x)]\, dx - \int_a^b P[x, \phi_2(x)]\, dx.$$

The graph of ϕ_1 parametrized from left to right is the curve

$$C_1: \qquad \mathbf{r}_1(u) = u\mathbf{i} + \phi_1(u)\mathbf{j}, \quad u \in [a, b];$$

the graph of ϕ_2, also parametrized from left to right, is the curve

$$C_2: \qquad \mathbf{r}_2(u) = u\mathbf{i} + \phi_2(u)\mathbf{j}, \quad u \in [a, b].$$

Since C traversed counterclockwise consists of C_1 followed by $-C_2$ (C_2 traversed from right to left), you can see that

$$\oint_C P(x, y)\, dx = \int_{C_1} P(x, y)\, dx - \int_{C_2} P(x, y)\, dx$$

$$= \int_a^b P[u, \phi_1(u)]\, du - \int_a^b P[u, \phi_2(u)]\, du.$$

Since u is a dummy variable, it can be replaced by x. Comparison with (∗) proves (1).

We leave it to you to verify that

$$(2) \qquad \oint_C Q(x, y)\, dy = \iint\limits_{\Omega} \frac{\partial Q}{\partial x}(x, y)\, dxdy$$

by using the fact that Ω is of type II. This will complete the proof of the theorem for Ω as in Figure 18.5.2.

A slight modification of this argument applies to elementary regions which are bordered entirely or in part by line segments parallel to the coordinate axes.

Figure 18.5.3 shows a Jordan region that is not elementary but can be broken up into two elementary regions. (Figure 18.5.4)

Figure 18.5.3

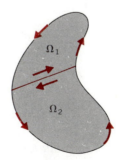

Figure 18.5.4

Green's theorem applied to the elementary parts tells us that

$$\iint\limits_{\Omega_1} \left(\frac{\partial Q}{\partial x} - \frac{\partial P}{\partial y}\right) dxdy = \oint\limits_{\text{bdry of } \Omega_1} P\,dx + Q\,dy,$$

$$\iint\limits_{\Omega_2} \left(\frac{\partial Q}{\partial x} - \frac{\partial P}{\partial y}\right) dxdy = \oint\limits_{\text{bdry of } \Omega_2} P\,dx + Q\,dy.$$

We now add these equations. The sum of the double integrals is, by additivity, the double integral over Ω. The sum of the line integrals is the integral over C (see the figure) plus the integrals over the crosscut. Since the crosscut is traversed twice and in opposite directions, the total contribution of the crosscut is zero and therefore Green's theorem holds:

$$\iint\limits_{\Omega} \left(\frac{\partial Q}{\partial x} - \frac{\partial P}{\partial y}\right) dxdy = \oint\limits_{C} P\,dx + Q\,dy.$$

This same argument can be extended to a Jordan region Ω that breaks up into n elementary regions $\Omega_1, \ldots, \Omega_n$. (Figure 18.5.5 gives an example with $n = 4$.) The double integrals over the Ω_i add up to the double integral over Ω, and, since the line integrals over the crosscuts cancel, the line integrals over the boundaries of the Ω_i add up to the line integral over C. (This is as far as we will carry the proof of Green's theorem. It is far enough to cover all the Jordan regions encountered in practice.)   ☐

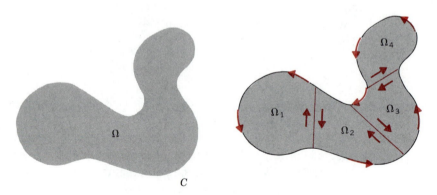

Figure 18.5.5

Problem 1. Use Green's theorem to evaluate

$$\oint\limits_{C} xy^2\,dx + x^2y\,dy$$

where C is the circle $x^2 + y^2 = a^2$.

Solution. Let Ω be the closed disc $0 \le x^2 + y^2 \le a$. With

$$P(x, y) = xy^2 \qquad \text{and} \qquad Q(x, y) = x^2y,$$

we have

$$\frac{\partial P}{\partial y} = 2xy, \qquad \frac{\partial Q}{\partial x} = 2xy, \qquad \text{and} \qquad \frac{\partial Q}{\partial x} - \frac{\partial P}{\partial y} = 0.$$

By Green's theorem

$$\oint_C xy^2 \, dx + x^2y \, dy = \iint_\Omega 0 \, dxdy = 0. \quad \square$$

Problem 2. Use Green's theorem to evaluate

$$\oint_C (1 + 10xy + y^2) \, dx + (6xy + 5x^2) \, dy$$

where C is the square with vertices $(0, 0)$, $(a, 0)$, (a, a), $(0, a)$.

Solution. Let Ω be the square region enclosed by C. With

$$P(x, y) = 1 + 10xy + y^2 \qquad \text{and} \qquad Q(x, y) = 6xy + 5x^2,$$

we have

$$\frac{\partial P}{\partial y} = 10x + 2y, \qquad \frac{\partial Q}{\partial x} = 6y + 10x, \qquad \text{and} \qquad \frac{\partial Q}{\partial x} - \frac{\partial P}{\partial y} = 4y.$$

By Green's theorem the line integral equals

$$\iint_\Omega 4y \, dxdy = 4 \iint_\Omega y \, dxdy = 4\bar{y} \,(\text{area of } \Omega) = 4a^2\bar{y}$$

where $\bar{y}$ is the y-coordinate of the centroid of the square region. Obviously $\bar{y} = \frac{1}{2}a$ and the integral equals $2a^3$. $\square$

Problem 3. Use Green's theorem to evaluate

$$\oint_C (4xy + y^2) \, dx - (xy + 3x^2) \, dy$$

where C is the circle $(x + 1)^2 + (y - 2)^2 = 9$.

Solution. The curve bounds the closed disc Ω of radius 3 centered at $(-1, 2)$. Here

$$\frac{\partial P}{\partial y} = 4x + 2y, \qquad \frac{\partial Q}{\partial x} = -y - 6x, \qquad \text{and} \qquad \frac{\partial Q}{\partial x} - \frac{\partial P}{\partial y} = -10x - 3y.$$

Therefore the line integral over C equals

$$\iint_\Omega (-10x - 3y) \, dxdy = (-10\bar{x} - 3\bar{y})(\text{area of } \Omega) = (10 - 6)9\pi = 36\pi. \quad \square$$

With Green's theorem we can show that, if C is a piecewise-smooth Jordan curve, then

(18.5.2)
$$\text{the area enclosed by } C = \oint_C - y \, dx = \oint_C x \, dy.$$

Proof. Let Ω be the region enclosed by C. In the first integral

$$P(x, y) = -y, \qquad Q(x, y) = 0.$$

Therefore

$$\frac{\partial P}{\partial y} = -1, \qquad \frac{\partial Q}{\partial x} = 0, \qquad \text{and} \qquad \frac{\partial Q}{\partial x} - \frac{\partial P}{\partial y} = 1.$$

Thus by Green's theorem

$$\oint_C - y \, dx = \iint_\Omega 1 \, dx dy = \text{area of } \Omega.$$

That the second integral also gives the area of Ω can be verified in a similar manner. $\square$

Problem 4. Let Ω be a Jordan region of area A with a piecewise-smooth boundary C. Show that the coordinates of the centroid of Ω are given by

$$\bar{x} A = \tfrac{1}{2} \oint_C x^2 \, dy, \qquad \bar{y} A = -\tfrac{1}{2} \oint_C y^2 \, dx.$$

Solution

$$\tfrac{1}{2} \oint_C x^2 \, dy = \tfrac{1}{2} \iint_\Omega 2x \, dx dy = \iint_\Omega x \, dx dy = \bar{x} A,$$

$$\bigg\downarrow \!\!\!- \text{ by Green's theorem}$$

$$-\tfrac{1}{2} \oint_C y^2 \, dx = -\tfrac{1}{2} \iint_\Omega (-2y) \, dx dy = \iint_\Omega y \, dx dy = \bar{y} A. \quad \square$$

Regions Bounded by Two or More Jordan Curves

(All the curves appearing below are assumed to be piecewise smooth.)

Figure 18.5.6 shows an annular region Ω. The region is not a Jordan region: the boundary consists of two Jordan curves C_1 and C_2. We cannot apply Green's theorem to Ω directly, but we can break up Ω into two Jordan regions as in Figure

Figure 18.5.6

Figure 18.5.7

18.5.7 and then apply Green's theorem to each piece. With Ω_1 and Ω_2 as in Figure 18.5.7,

$$\iint_{\Omega_1} \left(\frac{\partial Q}{\partial x} - \frac{\partial P}{\partial y} \right) dxdy = \oint_{\text{bdry of } \Omega_1} P\, dx + Q\, dy,$$

$$\iint_{\Omega_2} \left(\frac{\partial Q}{\partial x} - \frac{\partial P}{\partial y} \right) dxdy = \oint_{\text{bdry of } \Omega_2} P\, dx + Q\, dy.$$

When we add the double integrals, we get the double integral over Ω. When we add the line integrals, the integrals over the crosscuts cancel and we are left with the *counterclockwise integral* over C_1 and the *clockwise integral* over C_2. (See the figure.) Thus, for the annular region,

(18.5.3)
$$\iint_{\Omega} \left(\frac{\partial Q}{\partial x} - \frac{\partial P}{\partial y} \right) dxdy = \oint_{C_1} P\, dx + Q\, dy + \oint_{C_2} P\, dx + Q\, dy.$$

As a corollary to this we see that, if $\partial Q/\partial x = \partial P/\partial y$ throughout Ω, then the double integral on the left is 0, the sum of the integrals on the right is 0, and therefore

(18.5.4)
$$\oint_{C_1} P\, dx + Q\, dy = \oint_{C_2} P\, dx + Q\, dy.$$

Problem 5. Let C_1 be a Jordan curve that does not pass through the origin $(0, 0)$. Show that

$$\oint_{C_1} -\frac{y}{x^2 + y^2}\, dx + \frac{x}{x^2 + y^2}\, dy = \left. \begin{cases} 0, & \text{if } C_1 \text{ does not enclose the origin} \\ 2\pi, & \text{if } C_1 \text{ does enclose the origin} \end{cases} \right].$$

Solution. In this case

$$\frac{\partial P}{\partial y} = \frac{\partial}{\partial y}\left(-\frac{y}{x^2 + y^2} \right) = -\left[\frac{(x^2 + y^2)1 - 2y^2}{(x^2 + y^2)^2} \right] = \frac{y^2 - x^2}{(x^2 + y^2)^2},$$

$$\frac{\partial Q}{\partial x} = \frac{\partial}{\partial x}\left(\frac{x}{x^2 + y^2} \right) = \frac{(x^2 + y^2)1 - 2x^2}{(x^2 + y^2)^2} = \frac{y^2 - x^2}{(x^2 + y^2)^2}.$$

Thus

$$\frac{\partial Q}{\partial x} = \frac{\partial P}{\partial y} \qquad \text{except at the origin.}$$

If C_1 does not enclose the origin, then $\partial Q/\partial x - \partial P/\partial y = 0$ throughout the region enclosed by C_1, and, by Green's theorem, the line integral is 0.

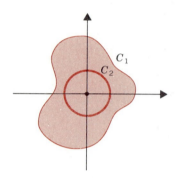

Figure 18.5.8

If C_1 does enclose the origin, we draw within the inner region of C_1 a small circle centered at the origin

$$C_2: \quad x^2 + y^2 = a^2. \qquad \text{(See Figure 18.5.8.)}$$

Since $\partial Q/\partial x - \partial P/\partial y = 0$ on the annular region bounded by C_1 and C_2, we know from (18.5.4) that the line integral over C_1 equals the line integral over C_2. All we have to show now is that the line integral over C_2 is 2π. This is straightforward. Parametrizing the circle by

$$\mathbf{r}(u) = a\cos u\,\mathbf{i} + a\sin u\,\mathbf{j} \qquad \text{with} \quad u \in [0, 2\pi],$$

we have

$$\oint_{C_2} -\frac{y}{x^2 + y^2}\,dx + \frac{x}{x^2 + y^2}\,dy = \int_0^{2\pi} (\sin^2 u \cos^2 u)\,du = 2\pi. \quad \square$$

$$\underset{\text{check this}}{\longrightarrow}$$

Figure 18.5.9 shows a region bounded by three Jordan curves: C_2 and C_3 both exterior to one another, both within C_1. For such a region Green's theorem gives

$$\iint_\Omega \left(\frac{\partial Q}{\partial x} - \frac{\partial P}{\partial y}\right) dxdy = \oint_{C_1} P\,dx + Q\,dy + \oint_{C_2} P\,dx + Q\,dy + \oint_{C_3} P\,dx + Q\,dy.$$

To see this, break up Ω into two regions by making the crosscuts shown in Figure 18.5.10.

The general formula for configurations of this type reads

$$\iint_\Omega \left(\frac{\partial Q}{\partial x} - \frac{\partial P}{\partial y}\right) dxdy = \oint_{C_1} P\,dx + Q\,dy + \sum_{i=2}^n \oint_{C_i} P\,dx + Q\,dy.$$

Figure 18.5.9

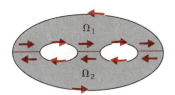

Figure 18.5.10

EXERCISES 18.5

Evaluate by Green's theorem.

1. $\oint_C 3y\, dx + 5x\, dy$; C: $x^2 + y^2 = 1$.

2. $\oint_C 5x\, dx + 3y\, dy$; C: $(x - 1)^2 + (y + 1)^2 = 1$.

3. $\oint_C x^2\, dy$ where C is the rectangle with vertices $(0, 0)$, $(a, 0)$, (a, b), $(0, b)$.

4. $\oint_C y^2\, dx$ where C is the rectangle of Exercise 3.

5. $\oint_C (3xy + y^2)\, dx + (2xy + 5x^2)\, dy$; C: $(x - 1)^2 + (y + 2)^2 = 1$.

6. $\oint_C (xy + 3y^2)\, dx + (5xy + 2x^2)\, dy$; C: $(x - 1)^2 + (y + 2)^2 = 1$.

7. $\oint_C (2x^2 + xy - y^2)\, dx + (3x^2 - xy + 2y^2)\, dy$; C: $(x - a)^2 + y^2 = r^2$.

8. $\oint_C (x^2 - 2xy + 3y^2)\, dx + (5x + 1)\, dy$; C: $x^2 + (y - b)^2 = r^2$.

9. $\oint_C e^x \sin y\, dx + e^x \cos y\, dy$; C: $(x - a)^2 + (y - b)^2 = r^2$.

10. $\oint_C e^x \cos y\, dx + e^x \sin y\, dy$; C the rectangle with vertices $(0, 0)$, $(1, 0)$, $(1, \pi)$, $(0, \pi)$.

Find the area enclosed by the curve by integrating over the curve.

11. The ellipse $b^2x^2 + a^2y^2 = a^2b^2$.   12. The astroid $x^{2/3} + y^{2/3} = a^{2/3}$.

13. Let C be a piecewise-smooth Jordan curve. Calculate

$$\oint_C (ay + b)\, dx + (cx + d)\, dy$$

given that C encloses a region of area A.

14. Calculate

$$\oint_C \mathbf{F(r)} \cdot \mathbf{dr}$$

given that $\mathbf{F}(x, y) = 2y\mathbf{i} - 3x\mathbf{j}$ and C is the astroid $x^{2/3} + y^{2/3} = a^{2/3}$.

15. Use Green's theorem to find the area under one arch of the cycloid

$$x(\theta) = R(\theta - \sin\theta), \quad y(\theta) = R(1 - \cos\theta).$$

16. Find the Jordan curve C that maximizes the line integral

$$\oint_C y^3\, dx + (3x - x^3)\, dy.$$

17. Complete the proof of Green's theorem for the elementary region of Figure 18.5.2 by showing that

$$\oint_C Q(x, y)\, dy = \int\int_\Omega \frac{\partial Q}{\partial x}(x, y)\, dxdy.$$

18. Let Ω be a plate of constant mass density λ in the form of a Jordan region with a piecewise-smooth boundary C. Show that the moments of inertia of the plate about the coordinate axes are given by the formulas

(18.5.5) $$I_x = -\frac{\lambda}{3}\oint_C y^3\, dx, \qquad I_y = \frac{\lambda}{3}\oint_C x^3\, dy.$$

19. Let P and Q be functions continuously differentiable on the region Ω of Figure 18.5.9. Given that $\partial P/\partial y = \partial Q/\partial x$ on Ω, find a relation between the line integrals

$$\oint_{C_1} P\, dx + Q\, dy, \qquad \oint_{C_2} P\, dx + Q\, dy, \qquad \oint_{C_3} P\, dx + Q\, dy.$$

20. Show that, if $f = f(x)$ and $g = g(y)$ are continuously differentiable, then

$$\int_C f(x)\, dx + g(y)\, dy = 0$$

for all piecewise-smooth Jordan curves C.

21. Let C be a piecewise-smooth Jordan curve that does not pass through the origin. Evaluate

$$\oint_C \frac{x}{x^2 + y^2}\, dx + \frac{y}{x^2 + y^2}\, dy$$

(a) if C does not enclose the origin.   (b) if C does enclose the origin.

22. Let C be a piecewise-smooth Jordan curve that does not pass through the origin. Evaluate

$$\oint_C -\frac{y^3}{(x^2 + y^2)^2}\, dx + \frac{xy^2}{(x^2 + y^2)^2}\, dy$$

(a) if C does not enclose the origin.   (b) if C does enclose the origin.

23. Let $\mathbf{v}$ be a vector field continuously differentiable on the entire plane. Use Green's theorem to verify that, if $\mathbf{v}$ is a gradient field [$\mathbf{v} = \nabla\phi$], then

$$\oint_C \mathbf{v} \cdot \mathbf{dr} = 0$$

for every piecewise-smooth Jordan curve C.

*SUPPLEMENT TO SECTION 18.5   A Justification of the Jacobian Area Formula

We based the change of variables for double integrals on the Jacobian area formula (Formula 17.11.1). Green's theorem enables us to derive this formula under the conditions spelled out below.

Let Γ be a Jordan region in the uv-plane with a piecewise-smooth boundary C_Γ. A vector function $\mathbf{r}(u, v) = x(u, v)\mathbf{i} + y(u, v)\mathbf{j}$ with continuous second partials maps Γ onto a region Ω of the xy-plane. If $\mathbf{r}$ is one-to-one on the interior of Γ and the Jacobian J of the components of $\mathbf{r}$ is different from zero on the interior of Γ, then

$$\text{area of } \Omega = \iint_\Gamma |J(u, v)|\, du\, dv.$$

Proof. Suppose that C_Γ is parametrized $u = u(t),\ v = v(t)$ with $t \in [a, b]$. Then the boundary of Ω is a piecewise-smooth curve C given by

$$\mathbf{r}[u(t), v(t)] = x[u(t), v(t)]\mathbf{i} + y[u(t), v(t)]\mathbf{j}, \qquad t \in [a, b].$$

By Green's theorem, the area of Ω is

$$\oint_C x\, dy = \left| \int_a^b x[u(t), v(t)]\, \frac{d}{dt}\, (y[u(t), v(t)])\, dt \right|$$

$$= \left| \int_a^b x[u(t), v(t)] \left(\frac{\partial y}{\partial u} [u(t), v(t)]\, u'(t) + \frac{\partial y}{\partial v} [u(t), v(t)]\, v'(t) \right) dt \right|$$

$$= \left| \int_a^b \left(x[u(t), v(t)]\, \frac{\partial y}{\partial u} [u(t), v(t)]\, u'(t) + x[u(t), v(t)]\, \frac{\partial y}{\partial v} [u(t), v(t)]\, v'(t) \right) dt \right|$$

$$= \left| \int_{C_\Gamma} x\, \frac{\partial y}{\partial u}\, du + x\, \frac{\partial y}{\partial v}\, dv \right|$$

─ again by Green's theorem

$$= \left| \iint_\Gamma \left[\frac{\partial}{\partial u} \left(x\, \frac{\partial y}{\partial v} \right) - \frac{\partial}{\partial v} \left(x\, \frac{\partial y}{\partial u} \right) \right] du\, dv \right|.$$

Now

$$\frac{\partial}{\partial u} \left(x\, \frac{\partial y}{\partial v} \right) - \frac{\partial}{\partial v} \left(x\, \frac{\partial y}{\partial u} \right) = \frac{\partial x}{\partial u}\, \frac{\partial y}{\partial v} + x\, \frac{\partial^2 y}{\partial u \partial v} - \frac{\partial x}{\partial v}\, \frac{\partial y}{\partial u} - x\, \frac{\partial^2 y}{\partial v \partial u}$$

$$= \frac{\partial x}{\partial u}\, \frac{\partial y}{\partial v} - \frac{\partial x}{\partial v}\, \frac{\partial y}{\partial u} = J(u, v).$$

Therefore

$$\text{area of } \Omega = \left| \iint_\Gamma J(u, v)\, du\, dv \right| = \iint_\Gamma |J(u, v)|\, du\, dv,$$

the final equality holding because $J(u, v)$ cannot change sign on Γ. ☐

18.6   PARAMETRIZED SURFACES; SURFACE AREA

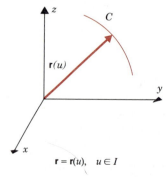

$$\mathbf{r} = \mathbf{r}(u), \quad u \in I$$

Figure 18.6.1

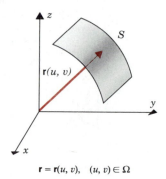

$$\mathbf{r} = \mathbf{r}(u, v), \quad (u, v) \in \Omega$$

Figure 18.6.2

You have seen that a space curve C can be parametrized by a vector function $\mathbf{r} = \mathbf{r}(u)$ where u ranges over some interval I of the u-axis. (Figure 18.6.1) In an analogous manner we can parametrize a surface S in space by a vector function $\mathbf{r} = \mathbf{r}(u, v)$ where (u, v) ranges over some region Ω of the uv-plane. (Figure 18.6.2)

Example 1. (*The graph of a function*) As you have seen, the graph of a function

$$y = f(x), \qquad x \in [a, b]$$

can be parametrized by setting

$$\mathbf{r}(u) = u\mathbf{i} + f(u)\mathbf{j}, \qquad u \in [a, b].$$

In the same vein the graph of a function

$$y = f(x, y), \qquad (x, y) \in \Omega$$

can be parametrized by setting

$$\mathbf{r}(u, v) = u\mathbf{i} + v\mathbf{j} + f(u, v)\mathbf{k}, \qquad (u, v) \in \Omega.$$

As (u, v) ranges over Ω, the tip of $\mathbf{r}(u, v)$ traces out the graph of f. ☐

Example 2. (*A plane*) If two vectors $\mathbf{a}$ and $\mathbf{b}$ are not parallel, then the set of all linear combinations $u\mathbf{a} + v\mathbf{b}$ generate a plane p_0 that passes through the origin. We can parametrize this plane by setting

$$\mathbf{r}(u, v) = u\mathbf{a} + v\mathbf{b}, \qquad (u, v) \in R \times R.$$

The plane p that is parallel to p_0 and passes through the tip of $\mathbf{c}$ can be parametrized by setting

$$\mathbf{r}(u, v) = u\mathbf{a} + v\mathbf{b} + \mathbf{c}, \qquad (u, v) \in R \times R.$$

Note that the plane contains the lines

$$l_1: \quad \mathbf{r}(u, 0) = u\mathbf{a} + \mathbf{c} \qquad \text{and} \qquad l_2: \quad \mathbf{r}(0, v) = v\mathbf{b} + \mathbf{c}. \quad ☐$$

Example 3. (*A sphere*) The sphere of radius a centered at the origin can be parametrized by

$$\mathbf{r}(u, v) = a \cos u \cos v \, \mathbf{i} + a \sin u \cos v \, \mathbf{j} + a \sin v \, \mathbf{k}$$

with (u, v) ranging over the rectangle R: $0 \le u \le 2\pi$, $-\frac{1}{2}\pi \le v \le \frac{1}{2}\pi$.

To derive this parametrization we refer to Figure 18.6.3. The points of latitude v (see the figure) form a circle of radius $a \cos v$ on the horizontal plane $z = a \sin v$. This circle can be parametrized by

$$\mathbf{R}(u) = a \cos v \, (\cos u \, \mathbf{i} + \sin u \, \mathbf{j}) + a \sin v \, \mathbf{k}, \qquad u \in [0, 2\pi].$$

This expands to give

$$\mathbf{R}(u) = a \cos u \cos v \, \mathbf{i} + a \sin u \cos v \, \mathbf{j} + a \sin v \, \mathbf{k}, \qquad u \in [0, 2\pi].$$

Letting $\mathbf{v}$ range from $-\frac{1}{2}\pi$ to $\frac{1}{2}\pi$, we obtain the entire sphere.

The xyz-equation for this same sphere is $x^2 + y^2 + z^2 = a^2$. It is easy to verify that the parametrization satisfies this equation:

$$\begin{aligned} x^2 + y^2 + z^2 &= a^2 \cos^2 u \cos^2 v + a^2 \sin^2 u \cos^2 v + a^2 \sin^2 v \\ &= a^2(\cos^2 u + \sin^2 u) \cos^2 v + a^2 \sin^2 v \\ &= a^2(\cos^2 v + \sin^2 v) = a^2. \quad \square \end{aligned}$$

Figure 18.6.3

Figure 18.6.4

Example 4. (*A cone*) Figure 18.6.4 shows a cone with apex semiangle α and slant height s. The points of slant height v (see the figure) form a circle of radius $v \sin \alpha$ on the horizontal plane $z = v \cos \alpha$. This circle can be parametrized by

$$\begin{aligned} \mathbf{R}(u) &= v \sin \alpha \, (\cos u \, \mathbf{i} + \sin u \, \mathbf{j}) + v \cos \alpha \, \mathbf{k} \\ &= v \cos u \sin \alpha \, \mathbf{i} + v \sin u \sin \alpha \, \mathbf{j} + v \cos \alpha \, \mathbf{k}, \qquad u \in [0, 2\pi]. \end{aligned}$$

Since we can obtain the entire cone by letting v range from 0 to s, the cone is parametrized by

$$\mathbf{r}(u, v) = v \cos u \sin \alpha \, \mathbf{i} + v \sin u \sin \alpha \, \mathbf{j} + v \cos \alpha \, \mathbf{k},$$

with $0 \le u \le 2\pi$, $0 \le v \le s$. $\quad \square$

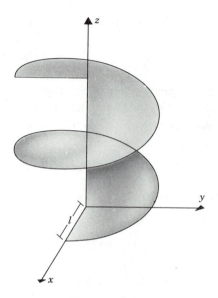

Figure 18.6.5

Example 5. (*A spiral ramp*) A rod of length l initially resting on the x-axis and attached at one end to the z-axis sweeps out a surface by rotating about the z-axis at constant rate ω while climbing at a constant rate b. The surface is pictured in Figure 18.6.5.

To parametrize this surface we mark the point of the rod at a distance u from the z-axis ($0 \le u \le l$) and ask for the position of this point at time v. At time v the rod will have climbed a distance bv and rotated through an angle ωv. Thus the point will be found at the tip of the vector

$$u(\cos \omega v \, \mathbf{i} + \sin \omega v \, \mathbf{j}) + bv\mathbf{k} = u \cos \omega v \, \mathbf{i} + u \sin \omega v \, \mathbf{j} + bv\mathbf{k}.$$

The entire surface can be parametrized by

$$\mathbf{r}(u, v) = u \cos \omega v \, \mathbf{i} + u \sin \omega v \, \mathbf{j} + bv\mathbf{k} \qquad \text{with } 0 \le u \le l, \ 0 \le v. \quad \square$$

The Fundamental Vector Product

Let S be a surface parametrized by a differentiable vector function $\mathbf{r} = \mathbf{r}(u, v)$. For simplicity let us suppose that (u, v) varies over an open rectangle

$$R: \quad a < u < b, \quad c < v < d.$$

Since $\mathbf{r}$ is a function of two variables, we can take two partial derivatives: $\mathbf{r}'_u$, the partial of $\mathbf{r}$ with respect to u, and $\mathbf{r}'_v$, the partial with respect to v. Now let (u_0, v_0) be a point of R for which

$$\mathbf{r}'_u(u_0, v_0) \times \mathbf{r}'_v(u_0, v_0) \ne \mathbf{0}.$$

(The reason for this condition will be apparent as we go on.) The vector function

$$\mathbf{r}_1(u) = \mathbf{r}(u, v_0), \qquad u \in (a, b)$$

(here we are keeping v fixed at v_0) traces a differentiable curve C_1 that lies on S. (Figure 18.6.6) The vector function

$$\mathbf{r}_2(v) = \mathbf{r}(u_0, v), \qquad v \in (c, d)$$

(this time we are keeping u fixed at u_0) traces a differentiable curve C_2 that also lies on S. Both curves pass through the tip of $\mathbf{r}(u_0, v_0)$:

$$C_1 \text{ with tangent vector } \mathbf{r}_1'(u_0) = \mathbf{r}_u'(u_0, v_0)$$
$$C_2 \text{ with tangent vector } \mathbf{r}_2'(v_0) = \mathbf{r}_v'(u_0, v_0).$$

The cross product $\mathbf{N}(u_0, v_0) = \mathbf{r}_u'(u_0, v_0) \times \mathbf{r}_v'(u_0, v_0)$, which we have assumed to be different from zero, is thus perpendicular to both curves at the tip of $\mathbf{r}(u_0, v_0)$ and can be taken as a normal to the surface at that point.

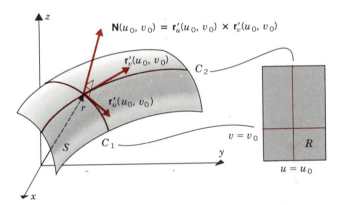

Figure 18.6.6

We record the result as follows:

> if S is a surface given by a differentiable function $\mathbf{r} = \mathbf{r}(u, v)$, then the vector $\mathbf{N}(u, v) = \mathbf{r}_u'(u, v) \times \mathbf{r}_v'(u, v)$ is perpendicular to the surface at the tip of $\mathbf{r}(u, v)$ and, if different from zero, can be taken as a normal to the surface at that point.

The cross product $\mathbf{N} = \mathbf{r}_u' \times \mathbf{r}_v'$ is called the *fundamental vector product* of the surface.

Example 6. For the plane $\mathbf{r}(u, v) = u\mathbf{a} + v\mathbf{b} + \mathbf{c}$ we have

$$\mathbf{r}_u'(u, v) = \mathbf{a}, \qquad \mathbf{r}_v'(u, v) = \mathbf{b} \qquad \text{and therefore} \qquad \mathbf{N}(u, v) = \mathbf{a} \times \mathbf{b}.$$

The vector $\mathbf{a} \times \mathbf{b}$ is normal to the plane. ☐

Example 7. We parametrized the sphere $x^2 + y^2 + z^2 = a^2$ by setting

$$\mathbf{r}(u, v) = a \cos u \cos v \, \mathbf{i} + a \sin u \cos v \, \mathbf{j} + a \sin v \, \mathbf{k}$$

with $0 \le u \le 2\pi$, $-\frac{1}{2}\pi \le v \le \frac{1}{2}\pi$. In this case

$$\mathbf{r}'_u(u, v) = -a \sin u \cos v \, \mathbf{i} + a \cos u \cos v \, \mathbf{j}$$

and

$$\mathbf{r}'_v(u, v) = -a \cos u \sin v \, \mathbf{i} - a \sin u \sin v \, \mathbf{j} + a \cos v \, \mathbf{k}.$$

Thus

$$\mathbf{N}(u, v) = \begin{vmatrix} \mathbf{i} & \mathbf{j} & \mathbf{k} \\ -a \sin u \cos v & a \cos u \cos v & 0 \\ -a \cos u \sin v & -a \sin u \sin v & a \cos v \end{vmatrix}$$

check this $\longrightarrow$

$$= a \cos v \, (a \cos u \cos v \, \mathbf{i} + a \sin u \cos v \, \mathbf{j} + a \sin v \, \mathbf{k})$$

$$= a \cos v \, \mathbf{r}(u, v).$$

As was to be expected, the fundamental vector product of a sphere is parallel to the radius vector $\mathbf{r}(u, v)$. ☐

The Area of a Parametrized Surface

A linear function

$$\mathbf{r}(u, v) = u\mathbf{a} + v\mathbf{b} + \mathbf{c} \qquad \text{(a and b not parallel)}$$

parametrizes a plane p. Horizontal lines from the uv-plane, lines with equations of the form $v = v_0$, are carried onto lines parallel to $\mathbf{a}$, and vertical lines, $u = u_0$, are carried onto lines parallel to $\mathbf{b}$:

$$\mathbf{r}(u, v_0) = u\mathbf{a} + v_0\mathbf{b} + \mathbf{c}, \qquad \mathbf{r}(u_0, v) = v\mathbf{b} + u_0\mathbf{a} + \mathbf{c}.$$

direction vector ↗   constant   direction vector ↑   constant

Thus a rectangle R in the uv-plane with sides parallel to the u and v axes,

$$R: \qquad u_1 \le u \le u_2, \quad v_1 \le v \le v_2,$$

is carried onto a parallelogram on p with sides parallel to $\mathbf{a}$ and $\mathbf{b}$. What is important to us here is that

the area of the parallelogram $= \|\mathbf{a} \times \mathbf{b}\|$ (the area of R).

To show this, we refer to Figure 18.6.7. The parallelogram is generated by the vectors

$$\mathbf{r}(u_2, v_1) - \mathbf{r}(u_1, v_1) = (u_2\mathbf{a} + v_1\mathbf{b} + \mathbf{c}) - (u_1\mathbf{a} + v_1\mathbf{b} + \mathbf{c}) = (u_2 - u_1)\mathbf{a},$$

$$\mathbf{r}(u_1, v_2) - \mathbf{r}(u_1, v_1) = (u_1\mathbf{a} + v_2\mathbf{b} + \mathbf{c}) - (u_1\mathbf{a} + v_1\mathbf{b} + \mathbf{c}) = (v_2 - v_1)\mathbf{b}.$$

The area of the parallelogram is thus

$$\|(u_2 - u_1)\mathbf{a} \times (v_2 - v_1)\mathbf{b}\| = \|\mathbf{a} \times \mathbf{b}\|(u_2 - u_1)(v_2 - v_1) = \|\mathbf{a} \times \mathbf{b}\|(\text{area of } R). \quad ☐$$

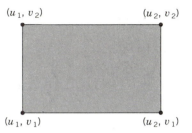

Figure 18.6.7

We can summarize as follows:

(18.6.2)

> Let R be a rectangle in the uv-plane with sides parallel to the coordinate axes. If $\mathbf{a}$ and $\mathbf{b}$ are not parallel, the linear function
>
> $$\mathbf{r}(u, v) = u\mathbf{a} + v\mathbf{b} + \mathbf{c}, \qquad (u, v) \in R$$
>
> parametrizes a parallelogram with sides parallel to $\mathbf{a}$ and $\mathbf{b}$ and
>
> the area of the parallelogram $= \|\mathbf{a} \times \mathbf{b}\|$ (the area of R).

More generally, let's suppose that we have a surface S parametrized by a continuously differentiable function

$$\mathbf{r} = \mathbf{r}(u, v), \qquad (u, v) \in \Omega.$$

We will assume that Ω is a basic region in the uv-plane and that $\mathbf{r}$ is one-to-one on the interior of Ω. (We don't want $\mathbf{r}$ to give chunks of S more than once.) Also we will assume that the fundamental vector product $\mathbf{N} = \mathbf{r}'_u \times \mathbf{r}'_v$ is never zero on the interior of Ω. (We can then use it as a normal.) Under these conditions we call S a *continuously differentiable surface* and define

(18.6.3)

$$\text{area of } S = \iint_{\Omega} \|\mathbf{N}(u, v)\| \, du dv.$$

We will try to show the reasoning behind this definition in the case that Ω is a rectangle R with sides parallel to the coordinate axes. We begin by breaking up R into N little rectangles $R_1, \ldots, R_N$. This induces a decomposition of S into little pieces $S_1, \ldots, S_N$. Taking (u_i^*, v_i^*) as the center of R_i, we have the tip of $\mathbf{r}(u_i^*, v_i^*)$ in S_i. Since the vector $\mathbf{r}'_u(u_i^*, v_i^*) \times \mathbf{r}'_v(u_i^*, v_i^*)$ is normal to the surface at the tip of $\mathbf{r}(u_i^*, v_i^*)$, we can parametrize the tangent plane at this point by the linear function

$$\mathbf{f}(u, v) = u\mathbf{r}'_u(u_i^*, v_i^*) + v\mathbf{r}'_v(u_i^*, v_i^*) + [\mathbf{r}(u_i^*, v_i^*) - u_i^*\mathbf{r}'_u(u_i^*, v_i^*) - v_i^*\mathbf{r}'_v(u_i^*, v_i^*)].$$

(Check that this linear function gives the right plane.) S_i is the portion of S that corresponds to R_i. The portion of the tangent plane that corresponds to this same R_i is a parallelogram with area

$$\|\mathbf{r}'_u(u_i^*, v_i^*) \times \mathbf{r}'_v(u_i^*, v_i^*)\|(\text{area of } R_i) = \|\mathbf{N}(u_i^*, v_i^*)\|(\text{area of } R_i). \qquad (18.6.2)$$

Taking this as our estimate for the area of S_i, we have

$$\text{area of } S = \sum_{i=1}^{N} \text{area of } S_i \cong \sum_{i=1}^{N} \|\mathbf{N}(u_i^*, v_i^*)\|(\text{area of } R_i).$$

This is a Riemann sum for

$$\iint_R \|\mathbf{N}(u, v)\| \, du dv$$

and tends to this integral as the maximum diameter of the R_i tends to zero.   ☐

To make sure that Formula 18.6.3 does not violate our previously established notion of area, we must verify that it gives the expected result both for plane regions and surfaces of revolution. This is done in Examples 9 and 10. By way of introduction we begin with the sphere.

Example 8.   (*The surface area of a sphere*)   The function

$$\mathbf{r}(u, v) = a \cos u \cos v \, \mathbf{i} + a \sin u \cos v \, \mathbf{j} + a \sin v \, \mathbf{k}$$

with (u, v) ranging over the set Ω: $0 \le u \le 2\pi$, $-\frac{1}{2}\pi \le v \le \frac{1}{2}\pi$ parametrizes a sphere of radius a. For this parametrization

$$\mathbf{N}(u, v) = a \cos v \, \mathbf{r}(u, v) \qquad \text{and} \qquad \|\mathbf{N}(u, v)\| = a^2 |\cos v| = a^2 \cos v.$$

$\text{Example 7} \longrightarrow$ $-\frac{1}{2}\pi \le v \le \frac{1}{2}\pi \longrightarrow$

According to the new formula

$$\text{area of the sphere} = \iint_\Omega a^2 \cos v \, du dv$$

$$= \int_0^{2\pi} \left(\int_{-\frac{1}{2}\pi}^{\frac{1}{2}\pi} a^2 \cos v \, dv \right) du = 2\pi a^2 \int_{-\frac{1}{2}\pi}^{\frac{1}{2}\pi} \cos v \, dv = 4\pi a^2,$$

which, as you know, is correct.   ☐

Example 9.   (*The area of a region*)   If S is a plane region Ω, then S can be parametrized by setting

$$\mathbf{r}(u, v) = u\mathbf{i} + v\mathbf{j}, \qquad (u, v) \in \Omega.$$

Here $\mathbf{N}(u, v) = \mathbf{r}_u'(u, v) \times \mathbf{r}_v'(u, v) = \mathbf{i} \times \mathbf{j} = \mathbf{k}$ and $\|\mathbf{N}(u, v)\| = 1$. In this case (18.6.3) reduces to the familiar formula

$$A = \iint_\Omega du dv.$$ ☐

Example 10. (*The area of a surface of revolution*) Let S be the surface generated by revolving the graph of a function

$$y = f(x), \qquad x \in [a, b]$$

about the x-axis. We will assume that f is positive and continuously differentiable.
We can parametrize S by setting

$$\mathbf{r}(u, v) = v\mathbf{i} + f(v) \cos u \, \mathbf{j} + f(v) \sin u \, \mathbf{k}$$

with (u, v) ranging over the set Ω: $0 \le u \le 2\pi$, $a \le v \le b$. (We leave it to you to verify that this is right.) In this case

$$\mathbf{N}(u, v) = \mathbf{r}'_u(u, v) \times \mathbf{r}'_v(u, v) = \begin{vmatrix} \mathbf{i} & \mathbf{j} & \mathbf{k} \\ 0 & -f(v) \sin u & f(v) \cos u \\ 1 & f'(v) \cos u & f'(v) \sin u \end{vmatrix}$$

$$= -f(v) \, f'(v)\mathbf{i} + f(v) \cos u \, \mathbf{j} + f(v) \sin u \, \mathbf{k}.$$

Therefore $\|\mathbf{N}(u, v)\| = f(v) \sqrt{[f'(v)]^2 + 1}$ and

$$\text{area of } S = \iint_\Omega f(v) \sqrt{[f'(v)]^2 + 1} \, du dv$$

$$= \int_0^{2\pi} \left(\int_a^b f(v) \sqrt{[f'(v)]^2 + 1} \, dv \right) du = \int_a^b 2\pi f(v) \sqrt{[f'(v)]^2 + 1} \, dv.$$

This is in agreement with Formula 10.10.3. ☐

Example 11. (*Spiral ramp*) One turn of the spiral ramp of Example 5 is the surface

$$S: \quad \mathbf{r}(u, v) = u \cos \omega v \, \mathbf{i} + u \sin \omega v \, \mathbf{j} + bv\mathbf{k}$$

with (u, v) ranging over the set Ω: $0 \le u \le l$, $0 \le v \le 2\pi/\omega$. In this case

$$\mathbf{r}'_u(u, v) = \cos \omega v \, \mathbf{i} + \sin \omega v \, \mathbf{j}, \qquad \mathbf{r}'_v(u, v) = -\omega u \sin \omega v \, \mathbf{i} + \omega u \cos \omega v \, \mathbf{j} + b\mathbf{k}.$$

Therefore

$$\mathbf{N}(u, v) = \begin{vmatrix} \mathbf{i} & \mathbf{j} & \mathbf{k} \\ \cos \omega v & \sin \omega v & 0 \\ -\omega u \sin \omega v & \omega u \cos \omega v & b \end{vmatrix} = b \sin \omega v \, \mathbf{i} - b \cos \omega v \, \mathbf{j} + \omega u \mathbf{k}$$

and

$$\|\mathbf{N}(u, v)\| = \sqrt{b^2 + \omega^2 u^2}.$$

Thus

$$\text{area of } S = \iint_\Omega \sqrt{b^2 + \omega^2 u^2} \, du dv$$

$$= \int_0^{2\pi/\omega} \left(\int_0^l \sqrt{b^2 + \omega^2 u^2} \, du \right) dv = \frac{2\pi}{\omega} \int_0^l \sqrt{b^2 + \omega^2 u^2} \, du.$$

The integral can be evaluated by setting $u = (b/\omega) \tan x$. ☐

The Area of a Surface z = f (x, y)

Figure 18.6.8

Figure 18.6.8 shows a surface that projects onto a basic region Ω of the xy-plane. Above each point (x, y) of Ω there is one and only one point of S. The surface S is then the graph of a function

$$z = f(x, y), \qquad (x, y) \in \Omega.$$

As we show below, if f is continuously differentiable, then

(18.6.4)
$$\text{area of } S = \iint_{\Omega} \sqrt{[f_x'(x, y)]^2 + [f_y'(x, y)]^2 + 1} \; dxdy.$$

Derivation of Formula 18.6.4. We can parametrize S by setting

$$\mathbf{r}(u, v) = u\mathbf{i} + v\mathbf{j} + f(u, v)\mathbf{k}, \qquad (u, v) \in \Omega.$$

We may just as well use x and y and write

$$\mathbf{r}(x, y) = x\mathbf{i} + y\mathbf{j} + f(x, y)\mathbf{k}, \qquad (x, y) \in \Omega.$$

Clearly

$$\mathbf{r}_x'(x, y) = \mathbf{i} + f_x'(x, y)\mathbf{k} \qquad \text{and} \qquad \mathbf{r}_y'(x, y) = \mathbf{j} + f_y'(x, y)\mathbf{k}.$$

Thus

$$\mathbf{N}(x, y) = \begin{vmatrix} \mathbf{i} & \mathbf{j} & \mathbf{k} \\ 1 & 0 & f_x'(x, y) \\ 0 & 1 & f_y'(x, y) \end{vmatrix} = -f_x'(x, y)\mathbf{i} - f_y'(x, y)\mathbf{j} + \mathbf{k}.$$

Therefore $\|\mathbf{N}(x, y)\| = \sqrt{[f_x'(x, y)]^2 + [f_y'(x, y)]^2 + 1}$ and the formula is verified. ☐

Problem 12. Find the surface area of that part of the parabolic cylinder $z = y^2$ that lies over the triangle with vertices $(0, 0)$, $(0, 1)$, $(1, 1)$ in the xy-plane.

Solution. Here $f(x, y) = y^2$ so that

$$f_x'(x, y) = 0, \qquad f_y'(x, y) = 2y.$$

The base triangle can be expressed by writing

$$\Omega: \quad 0 \le y \le 1, \quad 0 \le x \le y.$$

The surface has area

$$A = \iint_{\Omega} \sqrt{[f_x'(x, y)]^2 + [f_y'(x, y)]^2 + 1} \, dxdy$$

$$= \int_0^1 \int_0^y \sqrt{4y^2 + 1} \, dx \, dy$$

$$= \int_0^1 y\sqrt{4y^2 + 1} \, dy = \left[\tfrac{1}{12}(4y^2 + 1)^{3/2} \right]_0^1 = \tfrac{1}{12}(5\sqrt{5} - 1) \cong 0.85. \quad \square$$

Problem 13. Find the surface area of that part of the hyperbolic paraboloid $z = xy$ that lies inside the cylinder $x^2 + y^2 = a^2$.

Solution. Here $f(x, y) = xy$ so that

$$f_x'(x, y) = y, \qquad f_y'(x, y) = x.$$

The formula gives

$$A = \iint_{\Omega} \sqrt{y^2 + x^2 + 1} \, dxdy.$$

In polar coordinates the base region takes the form

$$0 \le r \le a, \quad 0 \le \theta \le 2\pi.$$

Thus we have

$$A = \iint_{\Omega} \sqrt{r^2 + 1} \, r \, drd\theta = \int_0^{2\pi} \int_0^a \sqrt{r^2 + 1} \, r \, dr \, d\theta = \tfrac{2}{3}\pi[(a^2 + 1)^{3/2} - 1]. \quad \square$$

There is an elegant version of this last area formula (Formula 18.6.4) that is geometrically vivid. We know that the vector

$$\mathbf{r}_x'(x, y) \times \mathbf{r}_y'(x, y) = -f_x'(x, y)\mathbf{i} - f_y'(x, y)\mathbf{j} + \mathbf{k}$$

is normal to the surface at the point $(x, y, f(x, y))$. The unit vector in that direction, the vector

$$\mathbf{n}(x, y) = \frac{-f_x'(x, y)\mathbf{i} - f_y'(x, y)\mathbf{j} + \mathbf{k}}{\sqrt{[f_x'(x, y)]^2 + [f_y'(x, y)]^2 + 1}},$$

is called the *upper unit normal.*   (It is the unit normal with a nonnegative **k**-component.)

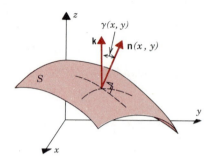

Figure 18.6.9

Now let $\gamma(x, y)$ be the angle between $\mathbf{n}(x, y)$ and $\mathbf{k}$. (Figure 18.6.9) Since $\mathbf{n}(x, y)$ and $\mathbf{k}$ are both unit vectors,

$$\cos [\gamma(x, y)] = \mathbf{n}(x, y) \cdot \mathbf{k} = \frac{1}{\sqrt{[f_x'(x, y)]^2 + [f_y'(x, y)]^2 + 1}}.$$

Taking reciprocals we have

$$\sec [\gamma(x, y)] = \sqrt{[f_x'(x, y)]^2 + [f_y'(x, y)]^2 + 1}.$$

The area formula can therefore be written

(18.6.5)
$$A = \iint_\Omega \sec [\gamma(x, y)] \, dxdy.$$

EXERCISES 18.6

Calculate the fundamental vector product.

1. $\mathbf{r}(u, v) = (u^2 - v^2)\mathbf{i} + (u^2 + v^2)\mathbf{j} + 2uv\mathbf{k}.$
2. $\mathbf{r}(u, v) = u \cos v \, \mathbf{i} + u \sin v \, \mathbf{j} + \mathbf{k}.$
3. $\mathbf{r}(u, v) = (u + v)\mathbf{i} + (u + v)\mathbf{j} + (u - v)\mathbf{k}.$
4. $\mathbf{r}(u, v) = \cos u \sin v \, \mathbf{i} + \sin u \cos v \, \mathbf{j} + u\mathbf{k}.$

Parametrize the given surface.

5. $S: y = g(x, z), (x, z) \in \Omega.$
6. $S: x = h(y, z), (y, z) \in \Gamma.$

Find an equation in x, y, z for the given surface and identify the surface.

7. $\mathbf{r}(u, v) = a \cos u \cos v \, \mathbf{i} + b \sin u \cos v \, \mathbf{j} + c \sin v \, \mathbf{k}; \quad 0 \le u \le 2\pi, -\frac{1}{2}\pi \le v \le \frac{1}{2}\pi.$
8. $\mathbf{r}(u, v) = au \cos v \, \mathbf{i} + bu \sin v \, \mathbf{j} + u^2\mathbf{k}; \quad 0 \le u, 0 \le v \le 2\pi.$
9. $\mathbf{r}(u, v) = au \cosh v \, \mathbf{i} + bu \sinh v \, \mathbf{j} + u^2\mathbf{k}; \quad u \text{ real}, v \text{ real}.$

10. Show that the hyperboloid of one sheet

$$\frac{x^2}{a^2} + \frac{y^2}{b^2} - \frac{z^2}{c^2} = 1 \qquad \text{(Figure 15.2.2)}$$

can be parametrized by the vector function

$$\mathbf{r}(u, v) = a \cos u \cosh v \, \mathbf{i} + b \sin u \cosh v \, \mathbf{j} + c \sinh v \, \mathbf{k}; \quad 0 \le u \le 2\pi, v \text{ real}.$$

11. The graph of a continuously differentiable function $y = f(x), x \in [a, b]$ is revolved about the y-axis. Parametrize the surface given that $a \ge 0.$

12. Show that the area of the surface of Exercise 11 is given by the formula

$$A = \int_a^b 2\pi x \sqrt{1 + [f'(x)]^2}\, dx.$$

13. A plane p intersects the xy-plane at an angle γ. (Draw a figure.) Find the area of a region Γ on p given that the projection of Γ onto the xy-plane is a region Ω of area A_Ω.

14. Determine the area of the portion of the plane $x + y + z = a$ that lies within the cylinder $x^2 + y^2 = b^2$.

15. Find the area of the part of the plane $bcx + acy + abz = abc$ that lies within the first octant: $x \geq 0$, $y \geq 0$, $z \geq 0$.

16. Find the area of the surface $z^2 = x^2 + y^2$ from $z = 0$ to $z = 1$.

17. Find the area of the surface $z = x^2 + y^2$ from $z = 0$ to $z = 4$.

Calculate the area of the surface.

18. $z^2 = 2xy$ with $0 \leq x \leq a$, $0 \leq y \leq b$, $z \geq 0$.

19. $z = a^2 - (x^2 + y^2)$ with $\frac{1}{4}a^2 \leq x^2 + y^2 \leq a^2$.

20. $3z^2 = (x + y)^3$ with $x + y \leq 2$, $x \geq 0$, $y \geq 0$.

21. $3z = x^{3/2} + y^{3/2}$ with $0 \leq x \leq 1$, $0 \leq y \leq x$.

22. $z = y^2$ with $0 \leq x \leq 1$, $0 \leq y \leq 1$.

23. $x^2 + y^2 + z^2 - 4z = 0$ with $0 \leq 3(x^2 + y^2) \leq z^2$, $z \geq 2$.

24. $x^2 + y^2 + z^2 - 2az = 0$ with $0 \leq x^2 + y^2 \leq bz$. Assume $a > b > 0$.

25. (a) Find a formula for the area of a surface that is projectable onto a region Ω of the yz-plane; say,

$$S: \quad x = g(y, z), \quad (y, z) \in \Omega.$$

Assume that g is continuously differentiable.

(b) Find a formula for the area of a surface that is projectable onto a region Ω of the xz-plane; say,

$$S: \quad y = h(x, z), \quad (x, z) \in \Omega.$$

Assume that h is continuously differentiable.

26. (a) Determine the fundamental vector product for the cylindrical surface

$$\mathbf{r}(u, v) = a \cos u\, \mathbf{i} + a \sin u\, \mathbf{j} + v\mathbf{k}; \quad 0 \leq u \leq 2\pi, \quad 0 \leq v \leq l.$$

(b) Use your answer to (a) to find the area of the surface.

27. (a) Determine the fundamental vector product for the cone of Example 4:

$$\mathbf{r}(u, v) = v \cos u \sin \alpha\, \mathbf{i} + v \sin u \sin \alpha\, \mathbf{j} + v \cos \alpha\, \mathbf{k}; \quad 0 \leq u \leq 2\pi, \quad 0 \leq v \leq s.$$

(b) Use your answer to (a) to calculate the area of the cone.

28. Verify that the ellipsoid of revolution

$$\frac{x^2}{a^2} + \frac{y^2}{a^2} + \frac{z^2}{b^2} = 1$$

can be parametrized by

$$\mathbf{r}(u, v) = a \cos u \sin v\, \mathbf{i} + a \sin u \sin v\, \mathbf{j} + b \cos v\, \mathbf{k}; \quad 0 \leq u \leq 2\pi, \quad 0 \leq v \leq \pi.$$

Then show that the surface area of the ellipsoid is given by the formula

$$A = 2\pi a \int_0^\pi \sin v \sqrt{b^2 \sin^2 v + a^2 \cos^2 v}\, dv.$$

29. Let Ω be a plane region in space and let A_1, A_2, A_3 be the areas of the projections of Ω onto the three coordinate planes. Express the area of Ω in terms of A_1, A_2, A_3.

30. Let S be a surface given in cylindrical coordinates by $z = f(r, \theta)$, $(r, \theta) \in \Omega$. Show that, if f is continuously differentiable, then

$$\text{area of } S = \iint\limits_{\Omega} \sqrt{r^2[f_r'(r, \theta)]^2 + [f_\theta'(r, \theta)]^2 + r^2} \; drd\theta$$

provided the integrand is never zero on the interior of Ω.

31. The following surfaces are given in cylindrical coordinates. Find the surface area.
(a) $z = r + \theta$; $0 \le r \le 1$, $0 \le \theta \le \pi$.   (b) $z = re^\theta$; $0 \le r \le a$, $0 \le \theta \le 2\pi$.

32. Show that, for a flat surface S that is part of the xy-plane, (18.6.3) gives

$$\text{area of } S = \iint\limits_{\Omega} |J(u, v)| \; dudv$$

where J is the Jacobian of the components of the vector function that parametrizes S. Except for notation this is Formula 17.10.1.

18.7   SURFACE INTEGRALS

The Mass of a Material Surface

Imagine a thin distribution of matter spread out over a surface S. We call this a *material surface*.

If the mass density (the mass per unit area) is constant throughout, then the total mass of the material surface is the density λ times the area of S:

$$M = \lambda(\text{area of } S).$$

If, however, the mass density varies continuously from point to point, $\lambda = \lambda(x, y, z)$, then the total mass must be calculated by integration.

To develop the appropriate integral we suppose that

$$S: \quad \mathbf{r} = \mathbf{r}(u, v) = x(u, v)\mathbf{i} + y(u, v)\mathbf{j} + z(u, v)\mathbf{k}, \quad (u, v) \in \Omega$$

is a continuously differentiable surface, a surface that meets the conditions for area formula (18.6.3).† Our first step is to break up Ω into N little basic regions $\Omega_1, \ldots, \Omega_N$. This decomposes the surface into N little pieces $S_1, \ldots, S_N$. The area of S_i is given by the integral

$$\iint\limits_{\Omega_i} \|\mathbf{N}(u, v)\| \; dudv. \qquad \text{(Formula 18.6.3)}$$

By the mean-value theorem for double integrals, there exists a point (u_i^*, v_i^*) in Ω_i for which

$$\iint\limits_{\Omega_i} \|\mathbf{N}(u, v)\| \; dudv = \|\mathbf{N}(u_i^*, v_i^*)\|(\text{area of } \Omega_i).$$

---

† We repeat the conditions here: $\mathbf{r}$ is continuously differentiable, Ω is a basic region in the uv-plane, $\mathbf{r}$ is one-to-one on the interior of Ω, $\mathbf{N} = \mathbf{r}_u' \times \mathbf{r}_v'$ is never zero on the interior of Ω.

It follows that

$$\text{area of } S_i = \|\mathbf{N}(u_i^*, v_i^*)\|(\text{area of } \Omega_i).$$

Since the point (u_i^*, v_i^*) is in Ω_i, the tip of $\mathbf{r}(u_i^*, v_i^*)$ is on S_i. The mass density at this point is $\lambda[\mathbf{r}(u_i^*, v_i^*)]$. If S_i is small (which we can guarantee by choosing Ω_i small), then the mass density on S_i is approximately the same throughout. Thus we can estimate M_i, the mass contribution of S_i, by writing

$$M_i \cong \lambda[\mathbf{r}(u_i^*, v_i^*)](\text{area of } S_i) = \lambda[\mathbf{r}(u_i^*, v_i^*)]\|\mathbf{N}(u_i^*, v_i^*)\|(\text{area of } \Omega_i).$$

Adding up these estimates we have an estimate for the total mass of the surface:

$$M \cong \sum_{i=1}^{N} \lambda[\mathbf{r}(u_i^*, v_i^*)]\|\mathbf{N}(u_i^*, v_i^*)\|(\text{area of } \Omega_i)$$

$$= \sum_{i=1}^{N} \lambda[x(u_i^*, v_i^*), y(u_i^*, v_i^*), z(u_i^*, v_i^*)]\|\mathbf{N}(u_i^*, v_i^*)\|(\text{area of } \Omega_i).$$

This last expression is a Riemann sum for

$$\iint_{\Omega} \lambda[x(u, v), y(u, v), z(u, v)]\|\mathbf{N}(u, v)\| \, du dv$$

and tends to this integral as the maximum diameter of the Ω_i tends to zero. We can therefore conclude that

(18.7.1) $$M = \iint_{\Omega} \lambda[x(u, v), y(u, v), z(u, v)]\|\mathbf{N}(u, v)\| \, du dv.$$

Surface Integrals

The double integral in (18.7.1) can be calculated not only for a mass density function λ but for any scalar field H continuous over S. We call this integral *the surface integral of H over S* and write

(18.7.2) $$\iint_{S} H(x, y, z) \, d\sigma = \iint_{\Omega} H[x(u, v), y(u, v), z(u, v)]\|\mathbf{N}(u, v)\| \, du dv.$$

Note that, if $H(x, y, z)$ is identically 1, then the right-hand side of (18.7.2) gives the area of S. Thus

(18.7.3) $$\iint_{S} d\sigma = \text{area of } S.$$

Problem 1. Calculate

$$\iint\limits_{S} xy \, d\sigma \qquad \text{where} \qquad S: \quad \mathbf{r}(u, v) = u\mathbf{a} + v\mathbf{b}; \quad 0 \le u \le 1, \quad 0 \le v \le 1.$$

Solution. Call the parameter set Ω. Then

$$\iint\limits_{S} xy \, d\sigma = \iint\limits_{\Omega} x(u, v) \, y(u, v) \, \|\mathbf{N}(u, v)\| \, dudv.$$

By a previous calculation $\|\mathbf{N}(u, v)\| = \|\mathbf{a} \times \mathbf{b}\|$. Thus

$$\iint\limits_{S} xy \, d\sigma = \|\mathbf{a} \times \mathbf{b}\| \iint\limits_{\Omega} x(u, v) \, y(u, v) \, dudv.$$

To find $x(u, v)$ and $y(u, v)$ we need the $\mathbf{i}$ and $\mathbf{j}$ components of $\mathbf{r}(u, v)$. We can get these as follows:

$$\mathbf{r}(u, v) = u\mathbf{a} + v\mathbf{b} = u(a_1\mathbf{i} + a_2\mathbf{j} + a_3\mathbf{k}) + v(b_1\mathbf{i} + b_2\mathbf{j} + b_3\mathbf{k})$$

$$= (a_1 u + b_1 v)\mathbf{i} + (a_2 u + b_2 v)\mathbf{j} + (a_3 u + b_3 v)\mathbf{k}.$$

Therefore $x(u, v) = a_1 u + b_1 v$ and $y(u, v) = a_2 u + b_2 v$. We can now write

$$\iint\limits_{S} xy \, d\sigma = \|\mathbf{a} \times \mathbf{b}\| \iint\limits_{\Omega} (a_1 u + b_1 v)(a_2 u + b_2 v) \, dudv$$

$$= \|\mathbf{a} \times \mathbf{b}\| \int_{0}^{1} \left(\int_{0}^{1} [a_1 a_2 u^2 + (a_1 b_2 + b_1 a_2)uv + b_1 b_2 v^2] \, du \right) dv$$

$$= \|\mathbf{a} \times \mathbf{b}\| [\tfrac{1}{3} a_1 a_2 + \tfrac{1}{4}(a_1 b_2 + b_1 a_2) + \tfrac{1}{3} b_1 b_2]. \quad \square$$

check this ⟶

Problem 2. Calculate

$$\iint\limits_{S} \sqrt{x^2 + y^2} \, d\sigma$$

where S is the ramp of Example 11, Section 18.6:

$$S: \quad \mathbf{r}(u, v) = u \cos \omega v \, \mathbf{i} + u \sin \omega v \, \mathbf{j} + bv\mathbf{k}; \quad 0 \le u \le l, \quad 0 \le v \le 2\pi/\omega.$$

Solution. Call the parameter set Ω. By a previous calculation

$$\|\mathbf{N}(u, v)\| = \sqrt{b^2 + \omega^2 u^2}.$$

Therefore

$$\iint_S \sqrt{x^2 + y^2}\, d\sigma = \iint_\Omega \sqrt{[x(u,\, v)]^2 + [y(u,\, v)]^2}\, \|\mathbf{N}(u,\, v)\|\, du dv$$

$$= \iint_\Omega \sqrt{u^2 \cos^2 \omega v + u^2 \sin^2 \omega v}\, \sqrt{b^2 + \omega^2 u^2}\, du dv$$

$$u \geq 0 \text{ on } \Omega \longrightarrow \quad = \iint_\Omega u \sqrt{b^2 + \omega^2 u^2}\, du dv$$

$$= \int_0^{2\pi/\omega} \left(\int_0^l u \sqrt{b^2 + \omega^2 u^2}\, du \right) dv$$

$$= \frac{2\pi}{\omega} \int_0^l u \sqrt{b^2 + \omega^2 u^2}\, du = \frac{2\pi}{3\omega^3} [(b^2 + \omega^2 l^2)^{3/2} - b^3]. \quad \square$$

Like the other integrals you have studied, the surface integral satisfies a mean-value condition; namely, if H is continuous, then there is a point (x_0, y_0, z_0) on S for which

$$\iint_S H(x, y, z)\, d\sigma = H(x_0, y_0, z_0)(\text{area of } S).$$

We call $H(x_0, y_0, z_0)$ *the average value of H on S.* We can thus write

(18.7.4)
$$\iint_S H(x, y, z)\, d\sigma = \left(\begin{array}{c} \text{average value} \\ \text{of } H \text{ on } S \end{array} \right) (\text{area of } S).$$

We can also take weighted averages: if H and G are continuous on S, then there is a point (x_0, y_0, z_0) on S for which

(18.7.5)
$$\iint_S H(x, y, z) G(x, y, z)\, d\sigma = H(x_0, y_0, z_0) \iint_S G(x, y, z)\, d\sigma.$$

As you would expect, we call $H(x_0, y_0, z_0)$ *the G-weighted average of H on S.*

The coordinates of the centroid $(\bar{x}, \bar{y}, \bar{z})$ of a surface are simple averages over the surface: for a surface S of area A

$$\bar{x}A = \iint_S x\, d\sigma, \qquad \bar{y}A = \iint_S y\, d\sigma, \qquad \bar{z}A = \iint_S z\, d\sigma.$$

In the case of a material surface of mass density $\lambda = \lambda(x, y, z)$, the coordinates of the center of mass (x_M, y_M, z_M) are density-weighted averages: for a surface S of total mass M

$$x_M M = \iint_S x\lambda(x, y, z)\, d\sigma, \qquad y_M M = \iint_S y\lambda(x, y, z)\, d\sigma, \qquad z_M M = \iint_S z\lambda(x, y, z)\, d\sigma.$$

Problem 3. Locate the center of mass of a material surface in the form of a hemisphere $x^2 + y^2 + z^2 = a^2$ with $z \geq 0$ given that the mass density varies directly as the distance from the xy-plane.

Solution. The surface S can be parametrized by

$$\mathbf{r}(u, v) = a \cos u \cos v \, \mathbf{i} + a \sin u \cos v \, \mathbf{j} + a \sin v \, \mathbf{k}; \quad 0 \leq u \leq 2\pi, 0 \leq v \leq \tfrac{1}{2}\pi.$$

Call the parameter set Ω and recall that $\mathbf{N}(u, v) = a^2 \cos v$. The density function can be written $\lambda(x, y, z) = kz$. We can calculate the mass as follows:

$$M = \iint\limits_{S} \lambda(x, y, z) \, d\sigma = k \iint\limits_{S} z \, d\sigma = k \iint\limits_{\Omega} z(u, v)\|\mathbf{N}(u, v)\| \, dudv$$

$$= k \int_{0}^{2\pi} \left(\int_{0}^{\pi/2} (a \sin v)(a^2 \cos v) \, dv \right) du$$

$$= 2\pi ka^3 \int_{0}^{\pi/2} \sin v \cos v \, du = \pi ka^3.$$

By symmetry $x_M = 0$ and $y_M = 0$. To find z_M we write

$$z_M M = \iint\limits_{S} z\lambda(x, y, z) \, d\sigma = k \iint\limits_{S} z^2 \, d\sigma$$

$$= k \iint\limits_{\Omega} [z(u, v)]^2 \|\mathbf{N}(u, v)\| \, dudv$$

$$= k \int_{0}^{2\pi} \left(\int_{0}^{\pi/2} (a^2 \sin^2 v)(a^2 \cos v) \, dv \right) du$$

$$= 2\pi ka^4 \int_{0}^{\pi/2} \sin^2 v \cos v \, dv = \tfrac{2}{3}\pi ka^4.$$

Since $M = \pi ka^3$, we see that $z_M = \tfrac{2}{3}\pi ka^4/M = \tfrac{2}{3}a$. The center of mass is the point $(0, 0, \tfrac{2}{3}a)$. $\square$

Suppose that a material surface S rotates about an axis. The moment of inertia of the surface about that axis is given by the formula

(18.7.6)

$$I = \iint\limits_{S} \lambda(x, y, z)[R(x, y, z)]^2 \, d\sigma$$

where $\lambda = \lambda(x, y, z)$ is the mass density function and $R(x, y, z)$ is the distance from the axis to the point (x, y, z). (As with other configurations the moments of inertia about the x, y, z axes are denoted by I_x, I_y, I_z.)

Problem 4. Calculate the moment of inertia about the z-axis of a material sphere

$$S: \quad x^2 + y^2 + z^2 = a^2.$$

Assume that the surface is homogeneous and has constant mass density 1.

Solution. We parametrize S by setting

$$\mathbf{r}(u, v) = a \cos u \cos v \, \mathbf{i} + a \sin u \cos v \, \mathbf{j} + a \sin v \, \mathbf{k}; \quad 0 \le u \le 2\pi, \ -\tfrac{1}{2}\pi \le v \le \tfrac{1}{2}\pi.$$

Call the parameter set Ω and recall that $\|\mathbf{N}(u, v)\| = a^2 \cos v$. We can calculate the moment of inertia as follows:

$$I_z = \iint_S 1(x^2 + y^2) \, d\sigma = \iint_\Omega ([x(u, v)]^2 + [y(u, v)]^2)\|\mathbf{N}(u, v)\| \, d\sigma$$

$$= \iint_\Omega (a^2 \cos^2 v)(a^2 \cos v) \, dudv$$

$$= a^4 \int_0^{2\pi} \left(\int_{-\pi/2}^{\pi/2} \cos^3 v \, dv \right) du$$

$$= 2\pi a^4 \int_{-\pi/2}^{\pi/2} \cos^3 v \, dv = \tfrac{8}{3}\pi a^4.$$
$$\underset{\text{check this}}{\Big\uparrow\underline{\qquad}}$$

Since the surface has mass $M = V = 4\pi a^2$, we can write $I_z = \tfrac{2}{3}Ma^2$. $\quad\square$

A surface

$$S: \quad z = f(x, y), \quad (x, y) \in \Omega$$

can be parametrized by

$$\mathbf{r}(x, y) = x\mathbf{i} + y\mathbf{j} + f(x, y)\mathbf{k}, \quad (x, y) \in \Omega.$$

As you saw in Section 18.6, $\|\mathbf{N}(x, y)\| = \sec[\gamma(x, y)]$ where $\gamma(x, y)$ is the angle between $\mathbf{k}$ and the upper unit normal. Therefore

(18.7.7)
$$\iint_S H(x, y, z) \, d\sigma = \iint_\Omega H(x, y, z) \sec[\gamma(x, y)] \, dxdy.$$

In evaluating this last integral we use the fact that

$$\sec[\gamma(x, y)] = \sqrt{[f_x'(x, y)]^2 + [f_y'(x, y)]^2 + 1}. \qquad \text{(Section 18.6)}$$

Problem 5.   Calculate

$$\iint\limits_{S} \sqrt{x^2 + y^2}\, d\sigma \qquad \text{with} \qquad S: \quad z = xy, \quad 0 \le x^2 + y^2 \le 1.$$

Solution.   The base region Ω is the unit disc.   The function $z = f(x, y) = xy$ has partial derivatives $f'_x(x, y) = y$, $f'_y(x, y) = x$. Therefore

$$\sec\left[\gamma(x, y)\right] = \sqrt{y^2 + x^2 + 1} = \sqrt{x^2 + y^2 + 1}$$

and

$$\iint\limits_{S} \sqrt{x^2 + y^2}\, d\sigma = \iint\limits_{\Omega} \sqrt{x^2 + y^2}\, \sqrt{x^2 + y^2 + 1}\, dxdy.$$

We evaluate this last integral by changing to polar coordinates. Ω is the set of all (x, y) with polar coordinates (r, θ) in the set

$$\Gamma: \quad 0 \le \theta \le 2\pi, \quad 0 \le r \le 1.$$

Therefore

$$\iint\limits_{S} \sqrt{x^2 + y^2}\, d\sigma = \iint\limits_{\Gamma} r\sqrt{r^2 + 1}\; r\, drd\theta = \int_0^{2\pi} \left(\int_0^1 r^2\sqrt{r^2 + 1}\, dr \right) d\theta$$

$$= 2\pi \int_0^1 r^2 \sqrt{r^2 + 1}\, dr$$

$$\text{set } r = \tan\phi \quad\longrightarrow\qquad = 2\pi \int_0^{\pi/4} \tan^2\phi \, \sec^3\phi \, d\phi$$

$$= 2\pi \int_0^{\pi/4} [\sec^5\phi - \sec^3\phi]\, d\phi$$

$$= \tfrac{1}{4}\pi\, [3\sqrt{2} - \ln(\sqrt{2} + 1)] \cong 2.64. \quad \square$$

The Flux of a Vector Field

Suppose that

$$S: \quad \mathbf{r} = \mathbf{r}(u, v), \quad (u, v) \in \Omega$$

is a continuously differentiable surface with a unit normal $\mathbf{n} = \mathbf{n}(x, y, z)$ that is continuous on all of S.   Such a surface is called a *smooth surface*.   If $\mathbf{v} = \mathbf{v}(x, y, z)$ is a vector field continuous on S, then we can form the surface integral

$$\iint\limits_{S} (\mathbf{v} \cdot \mathbf{n})\, d\sigma = \iint\limits_{S} [\mathbf{v}(x, y, z) \cdot \mathbf{n}(x, y, z)]\, d\sigma.$$

This surface integral is called *the flux of* $\mathbf{v}$ *across S in the direction of* $\mathbf{n}$.

Note that the flux across a surface depends on the choice of unit normal. If $-\mathbf{n}$ is chosen instead of $\mathbf{n}$, the sign of the flux is reversed:

$$\iint_S (\mathbf{v} \cdot -\mathbf{n})\, d\sigma = \iint_S -(\mathbf{v} \cdot \mathbf{n})\, d\sigma = -\iint_S (\mathbf{v} \cdot \mathbf{n})\, d\sigma.$$

Problem 6. Calculate the flux of the field $\mathbf{v}(x, y, z) = x\mathbf{i} + y\mathbf{j}$ out of the sphere $S\colon x^2 + y^2 + z^2 = a^2$.

Solution. The outward unit normal is the vector

$$\mathbf{n}(x, y, z) = \frac{1}{a}(x\mathbf{i} + y\mathbf{j} + z\mathbf{k}).$$

Here

$$\mathbf{v} \cdot \mathbf{n} = (x\mathbf{i} + y\mathbf{j}) \cdot \frac{1}{a}(x\mathbf{i} + y\mathbf{j} + z\mathbf{k}) = \frac{1}{a}(x^2 + y^2).$$

Therefore

$$\text{flux out of } S = \frac{1}{a} \iint_S (x^2 + y^2)\, d\sigma.$$

To evaluate this integral we use the usual parametrization

$$\mathbf{r}(u, v) = a\cos u \cos v\,\mathbf{i} + a\sin u\cos v\,\mathbf{j} + a\sin v\,\mathbf{k}; \quad 0 \le u \le 2\pi,\ -\tfrac{1}{2}\pi \le v \le \tfrac{1}{2}\pi$$

recalling that $\|\mathbf{N}(u, v)\| = a^2 \cos v$. Thus

$$\text{flux out of } S = \frac{1}{a} \iint_\Omega \left([x(u, v)]^2 + [y(u, v)]^2\right)\|\mathbf{N}(u, v)\|\, du\,dv$$

$$= \frac{1}{a} \iint_\Omega (a^2 \cos^2 u \cos^2 v + a^2 \sin^2 u \cos^2 v)(a^2 \cos v)\, du\,dv$$

$$= a^3 \iint_\Omega \cos^3 v\, du\,dv$$

$$= a^3 \int_0^{2\pi} \left(\int_{-\pi/2}^{\pi/2} \cos^3 v\, dv\right) du = 2\pi a^3 \int_{-\pi/2}^{\pi/2} \cos^3 v\, dv = \tfrac{8}{3}\pi a^3. \quad \square$$

If S is the graph of a function $z = f(x, y)$, $(x, y) \in \Omega$ and $\mathbf{n}$ is the upper unit normal, then

(18.7.8)

$$\iint_S (\mathbf{v} \cdot \mathbf{n})\, d\sigma = \iint_\Omega (-v_1 f'_x - v_2 f'_y + v_3)\, dx\,dy.$$

Proof. From Section 18.6 we know that

$$\mathbf{n} = \frac{-f'_x\mathbf{i} - f'_y\mathbf{j} + \mathbf{k}}{\sqrt{(f'_x)^2 + (f'_y)^2 + 1}} = (-f'_x\mathbf{i} - f'_y\mathbf{j} + \mathbf{k}) \cos \gamma.$$

Therefore $\mathbf{v} \cdot \mathbf{n} = (-v_1 f'_x - v_2 f'_y + v_3) \cos \gamma$ and

$$\iint_S (\mathbf{v} \cdot \mathbf{n}) \, d\sigma = \iint_\Omega (\mathbf{v} \cdot \mathbf{n}) \sec \gamma \, dxdy = \iint_\Omega (-v_1 f'_x - v_2 f'_y + v_3) \, dxdy. \quad \square$$

Problem 7. Let S be the portion of the paraboloid $z = 1 - (x^2 + y^2)$ that lies above the unit disc Ω. Calculate the flux of $\mathbf{v} = x\mathbf{i} + y\mathbf{j} + z\mathbf{k}$ across this surface in the direction of the upper unit normal.

Solution

$$f = 1 - (x^2 + y^2); \quad f'_x = -2x, \quad f'_y = -2y.$$

$$\text{Flux} = \iint_\Omega (-v_1 f'_x - v_2 f'_y + v_3) \, dxdy$$

$$= \iint_\Omega [(-x)(-2x) + y(-2y) + 1 - (x^2 + y^2)] \, dxdy$$

$$= \iint_\Omega (1 + x^2 + y^2) \, dxdy = \int_0^{2\pi} \left(\int_0^1 (1 + r^2)r \, dr \right) d\theta = \tfrac{3}{2}\pi. \quad \square$$

in polar coordinates

The flux through a closed *piecewise-smooth surface* (a closed surface that consists of a finite number of smooth pieces joined together at the boundaries) can be evaluated by integrating over each smooth piece and adding up the results.

Problem 8. Calculate the total flux of $\mathbf{v}(x, y, z) = xy\mathbf{i} + 4yz^2\mathbf{j} + yz\mathbf{k}$ out of the unit cube: $0 \le x \le 1, \ 0 \le y \le 1, \ 0 \le z \le 1$.

Solution. The total flux is the sum of the fluxes across the faces of the cube:

face	$\mathbf{n}$	$\mathbf{v} \cdot \mathbf{n}$	flux
$x = 0$	$-\mathbf{i}$	$-xy = 0$	0
$x = 1$	$\mathbf{i}$	$xy = y$	$\tfrac{1}{2}$
$y = 0$	$-\mathbf{j}$	$-4yz^2 = 0$	0
$y = 1$	$\mathbf{j}$	$4yz^2 = 4z^2$	$\tfrac{4}{3}$
$z = 0$	$-\mathbf{k}$	$-yz = 0$	0
$z = 1$	$\mathbf{k}$	$yz = y$	$\tfrac{1}{2}$

The total flux is $\tfrac{1}{2} + \tfrac{4}{3} + \tfrac{1}{2} = \tfrac{7}{3}$. $\square$

Flux plays an important role in the study of fluid motion. Imagine a surface S within a fluid and choose a unit normal $\mathbf{n}$. Take $\mathbf{v}$ as the velocity of the fluid that passes through S. You can expect $\mathbf{v}$ to vary from point to point, but, for simplicity, let's assume that $\mathbf{v}$ does not change with time. (Such a time-independent flow is said to be *steady-state*.) The flux of $\mathbf{v}$ across S is the average component of $\mathbf{v}$ in the direction of $\mathbf{n}$ times the area of the surface S. This is just the volume of fluid that passes through S in unit time, from the $-\mathbf{n}$ side of S to the $\mathbf{n}$ side of S. If S is a closed surface (such as a cube, a sphere, or an ellipsoid) and $\mathbf{n}$ is chosen as the *outer unit normal,* then the flux across S gives the volume of fluid that flows *out* through S in unit time; if $\mathbf{n}$ is chosen as the *inner unit normal,* then the flux gives the volume of liquid that flows *in* through S in unit time.

EXERCISES 18.7

Evaluate these integrals over the surface S: $z = \frac{1}{2}y^2,\ 0 \le x \le 1,\ 0 \le y \le 1$.

1. $\displaystyle\iint_S d\sigma$.

2. $\displaystyle\iint_S x^2\, d\sigma$.

3. $\displaystyle\iint_S 3y\, d\sigma$.

4. $\displaystyle\iint_S (x-y)\, d\sigma$.

5. $\displaystyle\iint_S \sqrt{2z}\, d\sigma$.

6. $\displaystyle\iint_S \sqrt{1+y^2}\, d\sigma$.

Find the mass of a material surface in the shape of a triangle $(a, 0, 0), (0, a, 0), (0, 0, a)$ with the given density function.

7. $\lambda(x, y, z) = k$.

8. $\lambda(x, y, z) = k(x + y)$.

9. $\lambda(x, y, z) = kx^2$.

10. Locate the centroid of the triangle $(a, 0, 0), (0, a, 0), (0, 0, a)$.
11. Locate the centroid of the hemisphere $x^2 + y^2 + z^2 = a^2,\ z \ge 0$.

Let S be the parallelogram given by $\mathbf{r}(u, v) = (u + v)\mathbf{i} + (u - v)\mathbf{j} + 2u\mathbf{k};\ \ 0 \le u \le 1, 0 \le v \le 1$.

12. Find the area of S.
13. Determine the flux of $\mathbf{v} = x\mathbf{i} - y\mathbf{j}$ across S in the direction of the fundamental vector product.
14. Find the mass of the material surface S: $z = 1 - \frac{1}{2}(x^2 + y^2)$ with $0 \le x \le 1,\ 0 \le y \le 1$ if the density at each point (x, y, z) is proportional to xy.

Calculate the flux out of the sphere $x^2 + y^2 + z^2 = a^2$.

15. $\mathbf{v} = z\mathbf{k}$.

16. $\mathbf{v} = x\mathbf{i} + y\mathbf{j} + z\mathbf{k}$.

17. $\mathbf{v} = y\mathbf{i} - x\mathbf{j}$.

18. A homogeneous plate of mass density 1 is in the form of the parallelogram of Exercises 12–13. Determine the moments of inertia about the coordinate axes: (a) I_x. (b) I_y. (c) I_z.

Determine the flux across the triangle $(a, 0, 0), (0, a, 0), (0, 0, a)$ in the direction of the upper unit normal.

19. $\mathbf{v} = x\mathbf{i} + y\mathbf{j} + z\mathbf{k}$.

20. $\mathbf{v} = (x + z)\mathbf{k}$.

21. $\mathbf{v} = x^2\mathbf{i} - y^2\mathbf{j}$.

Determine the flux across S: $z = xy$ with $0 \le x \le 1,\ 0 \le y \le 2$ in the direction of the upper unit normal.

22. $\mathbf{v} = -xy^2\mathbf{i} + z\mathbf{j}$.

23. $\mathbf{v} = xz\mathbf{j} - xy\mathbf{k}$.

24. $\mathbf{v} = x^2y\mathbf{i} + z^2\mathbf{k}$.

25. Calculate the flux of $\mathbf{v} = x\mathbf{i} + y\mathbf{j} + z\mathbf{k}$ out of the cylindrical surface

$$S: \quad \mathbf{r}(u, v) = a \cos u \, \mathbf{i} + a \sin u \, \mathbf{j} + v\mathbf{k}; \quad 0 \le u \le 2\pi, \ 0 \le v \le l.$$

26. Calculate the flux of the gravitational force function

$$\mathbf{F}(\mathbf{r}) = G \frac{mM}{r^3} \mathbf{r} \quad \text{out of the sphere } x^2 + y^2 + z^2 = a^2.$$

Find the flux across S: $z = \frac{2}{3}(x^{3/2} + y^{3/2})$ with $0 \le x \le 1$, $0 \le y \le 1 - x$ in the direction of the upper unit normal.

27. $\mathbf{v} = x\mathbf{i} - y\mathbf{j} + \frac{3}{2}z\mathbf{k}$.   28. $\mathbf{v} = x^2\mathbf{i}$.   29. $\mathbf{v} = y^2\mathbf{j}$.   30. $\mathbf{v} = y\mathbf{i} - \sqrt{xy}\,\mathbf{j}$.

31. The cone

$$\mathbf{r}(u, v) = v \cos u \sin \alpha \, \mathbf{i} + v \sin u \sin \alpha \, \mathbf{j} + v \cos \alpha \, \mathbf{k}; \quad 0 \le u \le 2\pi, \ 0 \le v \le s$$

has area $A = \pi s^2 \sin \alpha$. Locate the centroid.

The mass density of a material cone $z = \sqrt{x^2 + y^2}$ with $0 \le z \le 1$ varies directly as the distance from the z-axis.

32. Find the mass of the cone.   33. Locate the center of mass.

34. Determine the moments of inertia about the coordinate axes: (a) I_x.   (b) I_y.   (c) I_z.

A   *D*

B   *C*

Figure 18.7.1

35. You have seen that, if S is a smooth surface immersed in a fluid of velocity $\mathbf{v}$, then the flux

$$\iint_S (\mathbf{v} \cdot \mathbf{n}) \, d\sigma$$

is the volume of fluid that passes through S in unit time from the $-\mathbf{n}$ side of S to the $\mathbf{n}$ side of S. This requires that S be a two-sided surface. There are, however, one-sided surfaces: for example, the Möbius band. To construct a material Möbius band start with a piece of paper in the form of the rectangle of Figure 18.7.1. Now give the piece of paper a single twist and join the two far edges together so that C coincides with A and D coincides with B. (a) Convince yourself that this surface is one-sided and therefore the notion of flux cannot be applied to it. (b) The surface is not smooth because it is impossible to erect a unit normal $\mathbf{n}$ that varies continuously over the entire surface. Convince yourself of this as follows: erect a unit normal $\mathbf{n}$ at some point P and make a circuit of the surface with $\mathbf{n}$. Note that, as $\mathbf{n}$ returns to P, the direction of $\mathbf{n}$ has been reversed.

Assume that the parallelogram of Exercises 12 and 13 is a material surface with a mass density that varies directly as the square of the distance from the x-axis.

36. Determine the mass.   37. Find the x-coordinate of the center of mass.

38. Find the moment of inertia about the z-axis.

39. Calculate the total flux of $\mathbf{v}(x, y, z) = y\mathbf{i} - x\mathbf{j}$ out of the solid bounded on the sides by the cylinder $x^2 + y^2 = 1$ and above and below by the planes $z = 1$ and $z = 0$.   HINT: Draw a figure.

40. Calculate the total flux of $\mathbf{v}(x, y, z) = y\mathbf{i} - x\mathbf{j}$ out of the solid bounded above by $z = 4$ and below by $z = x^2 + y^2$.

41. Calculate the total flux of $\mathbf{v}(x, y, z) = x\mathbf{i} + y\mathbf{j} + z\mathbf{k}$ out of the solid bounded above by $z = \sqrt{2 - (x^2 + y^2)}$ and below by $z = x^2 + y^2$.

42. Calculate the total flux of $\mathbf{v}(x, y, z) = xz\mathbf{i} + 4xyz^2\mathbf{j} + 2z\mathbf{k}$ out of the unit cube: $0 \leq x \leq 1$, $0 \leq y \leq 1$, $0 \leq z \leq 1$.

18.8 THE VECTOR DIFFERENTIAL OPERATOR ∇

Divergence ∇ · v, Curl ∇ × v

The *vector differential operator* ∇ (this is an inverted delta and it's read *del*) is defined by setting

(18.8.1)
$$\nabla = \frac{\partial}{\partial x}\mathbf{i} + \frac{\partial}{\partial y}\mathbf{j} + \frac{\partial}{\partial z}\mathbf{k}.$$

This is no ordinary vector. Its "components" are differentiation symbols. As the term "operator" suggests, ∇ is to be thought of as something that "operates" on things. What sort of things? Scalar fields and vector fields.

Suppose that f is a differentiable scalar field. Then ∇ operates on f as follows:

$$\nabla f = \left(\frac{\partial}{\partial x}\mathbf{i} + \frac{\partial}{\partial y}\mathbf{j} + \frac{\partial}{\partial z}\mathbf{k}\right) f = \frac{\partial f}{\partial x}\mathbf{i} + \frac{\partial f}{\partial y}\mathbf{j} + \frac{\partial f}{\partial z}\mathbf{k}.$$

This is just the *gradient* of f, with which you are already familiar.

How does ∇ operate on vector fields? In two ways. If $\mathbf{v} = v_1\mathbf{i} + v_2\mathbf{j} + v_3\mathbf{k}$ is a differentiable vector field, then by definition

(18.8.2)
$$\nabla \cdot \mathbf{v} = \frac{\partial v_1}{\partial x} + \frac{\partial v_2}{\partial y} + \frac{\partial v_3}{\partial z}$$

and

(18.8.3)
$$\nabla \times \mathbf{v} = \begin{vmatrix} \mathbf{i} & \mathbf{j} & \mathbf{k} \\ \dfrac{\partial}{\partial x} & \dfrac{\partial}{\partial y} & \dfrac{\partial}{\partial z} \\ v_1 & v_2 & v_3 \end{vmatrix} = \left(\frac{\partial v_3}{\partial y} - \frac{\partial v_2}{\partial z}\right)\mathbf{i} + \left(\frac{\partial v_1}{\partial z} - \frac{\partial v_3}{\partial x}\right)\mathbf{j} + \left(\frac{\partial v_2}{\partial x} - \frac{\partial v_1}{\partial y}\right)\mathbf{k}.$$

The first "product," ∇ · v, defined in imitation of the ordinary dot product, is called the *divergence* of **v**:

$$\nabla \cdot \mathbf{v} = \text{div } \mathbf{v}.$$

The second "product," ∇ × v, defined in imitation of the ordinary cross product, is called the *curl* of **v**:

$$\nabla \times \mathbf{v} = \text{curl } \mathbf{v}.$$

Interpretation of Divergence and Curl

Suppose we know the divergence of a field and also the curl. What does that tell us? For definitive answers we must wait for the divergence theorem and Stokes's theorem, but, in a preliminary way, we can give you some rough answers right now. View $\mathbf{v}$ as the velocity field of some fluid. The divergence of $\mathbf{v}$ at a point P gives us an indication of whether the fluid tends to accumulate near P (negative divergence) or tends to move away from P (positive divergence). The curl at P measures the rotational tendency of the fluid.

Example 1. Set

$$\mathbf{v}(x, y, z) = \alpha x \mathbf{i} + \alpha y \mathbf{j} + \alpha z \mathbf{k}. \qquad (\alpha \text{ a constant})$$

Divergence:

$$\nabla \cdot \mathbf{v} = \alpha \frac{\partial x}{\partial x} + \alpha \frac{\partial y}{\partial y} + \alpha \frac{\partial z}{\partial z} = 3\alpha.$$

Curl:

$$\nabla \times \mathbf{v} = \begin{vmatrix} \mathbf{i} & \mathbf{j} & \mathbf{k} \\ \dfrac{\partial}{\partial x} & \dfrac{\partial}{\partial y} & \dfrac{\partial}{\partial z} \\ \alpha x & \alpha y & \alpha z \end{vmatrix} = \alpha \begin{vmatrix} \mathbf{i} & \mathbf{j} & \mathbf{k} \\ \dfrac{\partial}{\partial x} & \dfrac{\partial}{\partial y} & \dfrac{\partial}{\partial z} \\ x & y & z \end{vmatrix} = \mathbf{0}$$

because the partial derivatives that appear in the expanded determinant

$$\frac{\partial y}{\partial x}, \qquad \frac{\partial x}{\partial y}, \qquad \text{etc.}$$

are all zero. ☐

The field of Example 1

$$\mathbf{v}(x, y, z) = \alpha(x\mathbf{i} + y\mathbf{j} + z\mathbf{k}) = \alpha \mathbf{r}$$

can be viewed as the velocity field of a fluid in radial motion—toward the origin if $\alpha < 0$, away from the origin if $\alpha > 0$. Figure 18.8.1 shows a point (x, y, z), a spherical neighborhood of that point, and a cone emanating from the origin that is tangent to the boundary of the neighborhood.

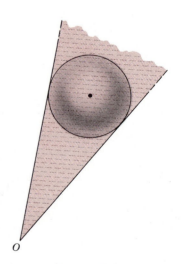

Figure 18.8.1

Note two things: all the fluid in the cone stays in the cone, and the speed of the fluid is proportional to its distance from the origin. Therefore, if the divergence 3α is negative, then α is negative, the motion is toward the origin, and the neighborhood *gains fluid* because the fluid coming in is moving more quickly than the fluid going out. (Also the entry area is greater than the exit area.) If, however, the divergence 3α is positive, then α is positive, the motion is away from the origin, and the neighborhood *loses fluid* because the fluid coming in is moving more slowly than the fluid going out. (Also the entry area is smaller than the exit area.)

Since the motion is radial, the fluid has no rotational tendency whatsoever, and we would expect the curl to be identically zero. It is. ☐

Example 2. Set

$$\mathbf{v}(x, y, z) = -\omega y\mathbf{i} + \omega x\mathbf{j}. \qquad \text{(ω a positive constant)}$$

Divergence:

$$\nabla \cdot \mathbf{v} = -\omega\,\frac{\partial y}{\partial x} + \omega\,\frac{\partial x}{\partial y} = 0 + 0 = 0.$$

Curl:

$$\nabla \times \mathbf{v} = \begin{vmatrix} \mathbf{i} & \mathbf{j} & \mathbf{k} \\ \dfrac{\partial}{\partial x} & \dfrac{\partial}{\partial y} & \dfrac{\partial}{\partial z} \\ -\omega y & \omega x & 0 \end{vmatrix} = \left\{\omega\,\frac{\partial x}{\partial x} + \omega\,\frac{\partial y}{\partial y}\right\}\mathbf{k} = 2\omega\mathbf{k}. \quad ☐$$

The field of Example 2

$$\mathbf{v}(x, y, z) = -\omega y\mathbf{i} + \omega x\mathbf{j} \qquad \text{(ω a positive constant)}$$

is the velocity field of uniform counterclockwise rotation about the z-axis with angular speed ω. You can see this by noting that $\mathbf{v}$ is perpendicular to $\mathbf{r}$:

$$\mathbf{v} \cdot \mathbf{r} = (-\omega y\mathbf{i} + \omega x\mathbf{j}) \cdot (x\mathbf{i} + y\mathbf{j} + z\mathbf{k}) = -\omega yx + \omega xy = 0$$

and the speed at each point is ωR where R is the radius of rotation:

$$v = \sqrt{\omega^2 y^2 + \omega^2 x^2} = \omega\sqrt{x^2 + y^2} = \omega R.$$

How is the curl, $2\omega\mathbf{k}$, related to the rotation? The angular velocity vector (see Exercise 24, Section 14.5) is the vector $\boldsymbol{\omega} = \omega\mathbf{k}$. In this case then the curl of $\mathbf{v}$ is twice the angular velocity vector.

With this rotation no neighborhood gains any fluid and no neighborhood loses any fluid. As we saw, the divergence is identically zero. ☐

Basic Identities

For vectors we have $\mathbf{a} \times \mathbf{a} = \mathbf{0}$. Is it true that $\nabla \times \nabla = \mathbf{0}$? We make it true by defining

$$(\nabla \times \nabla)f = \nabla \times (\nabla f).$$

THEOREM 18.8.4   (THE CURL OF A GRADIENT IS ZERO)

If f is a scalar field with continuous second partials, then

$$\nabla \times (\nabla f) = \mathbf{0}.$$

Proof

$$\nabla \times (\nabla f) = \begin{vmatrix} \mathbf{i} & \mathbf{j} & \mathbf{k} \\ \dfrac{\partial}{\partial x} & \dfrac{\partial}{\partial y} & \dfrac{\partial}{\partial z} \\ \dfrac{\partial f}{\partial x} & \dfrac{\partial f}{\partial y} & \dfrac{\partial f}{\partial z} \end{vmatrix} = \mathbf{0}.$$

by equality of the mixed partials $\square$

For vectors we have $\mathbf{a} \cdot (\mathbf{a} \times \mathbf{c}) = 0$. The analogous operator formula, $\nabla \cdot (\nabla \times \mathbf{v}) = 0$, is also valid.

THEOREM 18.8.5   (THE DIVERGENCE OF A CURL IS ZERO)

If the components of the vector field $\mathbf{v} = v_1\mathbf{i} + v_2\mathbf{j} + v_3\mathbf{k}$ have continuous second partials, then

$$\nabla \cdot (\nabla \times \mathbf{v}) = 0.$$

Proof. Again the key is the equality of the mixed partials:

$$\nabla \cdot (\nabla \times \mathbf{v}) = \frac{\partial}{\partial x}\left(\frac{\partial v_3}{\partial y} - \frac{\partial v_2}{\partial z}\right) + \frac{\partial}{\partial y}\left(\frac{\partial v_1}{\partial z} - \frac{\partial v_3}{\partial x}\right) + \frac{\partial}{\partial z}\left(\frac{\partial v_2}{\partial x} - \frac{\partial v_1}{\partial y}\right) = 0$$

since for each component the mixed partials cancel. Try it for v_1. $\square$

The next two identities are product rules. Here f is a scalar field and $\mathbf{v}$ is a vector field.

(18.8.6) $$\nabla \cdot (f\mathbf{v}) = (\nabla f) \cdot \mathbf{v} + f(\nabla \cdot \mathbf{v}). \qquad [\text{div}\,(f\mathbf{v}) = (\text{grad}\,f) \cdot \mathbf{v} + f(\text{div}\,\mathbf{v})]$$

(18.8.7) $$\nabla \times (f\mathbf{v}) = (\nabla f) \times \mathbf{v} + f(\nabla \times \mathbf{v}). \qquad [\text{curl}\,(f\mathbf{v}) = (\text{grad}\,f) \times \mathbf{v} + f(\text{curl}\,\mathbf{v})]$$

The verification of these identities is left to you in the exercises. $\square$

We know from Example 1 that $\nabla \cdot \mathbf{r} = 3$ and $\nabla \times \mathbf{r} = \mathbf{0}$ at all points of space. Now we can show that, if n is an integer, then, for all $\mathbf{r} \neq \mathbf{0}$,

(18.8.8)
$$\nabla \cdot (r^n \mathbf{r}) = (n + 3)r^n \quad \text{and} \quad \nabla \times (r^n \mathbf{r}) = \mathbf{0}. \qquad \dagger$$

Proof. Recall that $\nabla r^n = nr^{n-2}\mathbf{r}$. (16.1.5)   Using (18.8.6), we have

$$\nabla \cdot (r^n \mathbf{r}) = (\nabla r^n) \cdot \mathbf{r} + r^n(\nabla \cdot \mathbf{r})$$
$$= (nr^{n-2}\mathbf{r}) \cdot \mathbf{r} + r^n(3)$$
$$= nr^{n-2}(\mathbf{r} \cdot \mathbf{r}) + 3r^n = nr^n + 3r^n = (n + 3)r^n.$$

From (16.1.5) you can see that $r^n\mathbf{r}$ is a gradient. Its curl is therefore $\mathbf{0}$. (18.8.4) ☐

The Laplacean

From the operator ∇ we can construct other operators. The most important of these is the Laplacean $\nabla^2 = \nabla \cdot \nabla$. The Laplacean (named after the French mathematician Pierre-Simon Laplace) operates on scalar fields according to the following rule:

(18.8.9)
$$\nabla^2 f = \nabla \cdot (\nabla f) = \frac{\partial^2 f}{\partial x^2} + \frac{\partial^2 f}{\partial y^2} + \frac{\partial^2 f}{\partial z^2}. \qquad \dagger\dagger$$

Example 3. If $f(x, y, z) = x^2 + y^2 + z^2$, then

$$\nabla^2 f = \frac{\partial^2}{\partial x^2}(x^2 + y^2 + z^2) + \frac{\partial^2}{\partial y^2}(x^2 + y^2 + z^2) + \frac{\partial^2}{\partial z^2}(x^2 + y^2 + z^2)$$
$$= 2 + 2 + 2 = 6. \quad ☐$$

Example 4. If $f(x, y, z) = e^{xyz}$, then

$$\nabla^2 f = \frac{\partial^2}{\partial x^2}(e^{xyz}) + \frac{\partial^2}{\partial y^2}(e^{xyz}) + \frac{\partial^2}{\partial z^2}(e^{xyz})$$

$$= \frac{\partial}{\partial x}(yze^{xyz}) + \frac{\partial}{\partial y}(xze^{xyz}) + \frac{\partial}{\partial z}(xye^{xyz})$$

$$= y^2z^2e^{xyz} + x^2z^2e^{xyz} + x^2y^2e^{xyz}$$

$$= (y^2z^2 + x^2z^2 + x^2y^2)e^{xyz}. \quad ☐$$

$\dagger$ If n is positive and even, these formulas also hold at $\mathbf{r} = \mathbf{0}$.

$\dagger\dagger$ Sometimes you will see the Laplacean of f written Δf. Unfortunately this can be misread as the increment of f.

Example 5. To calculate $\nabla^2(\sin r) = \nabla^2(\sin \sqrt{x^2 + y^2 + z^2})$ we could write

$$\frac{\partial^2}{\partial x^2}(\sin \sqrt{x^2 + y^2 + z^2}) + \frac{\partial^2}{\partial y^2}(\sin \sqrt{x^2 + y^2 + z^2}) + \frac{\partial^2}{\partial z^2}(\sin \sqrt{x^2 + y^2 + z^2})$$

and proceed from there. The calculations are straightforward but lengthy. We will do it in a different way.

Recall that

$$\nabla^2 f = \nabla \cdot \nabla f \quad \text{(definition)} \qquad \nabla \cdot (f\mathbf{v}) = (\nabla f \cdot \mathbf{v}) + f(\nabla \cdot \mathbf{v}) \quad (18.8.6)$$
$$\nabla f(r) = f'(r) r^{-1} \mathbf{r} \quad (16.4.6) \qquad \nabla \cdot (r^n \mathbf{r}) = (n + 3) r^n \quad (18.8.8)$$

Using these relations, we have

$$\begin{aligned}
\nabla^2(\sin r) = \nabla \cdot (\nabla \sin r) &= \nabla \cdot [(\cos r) r^{-1} \, \mathbf{r}] \\
&= [(\nabla \cos r) \cdot r^{-1} \mathbf{r}] + \cos r \, (\nabla \cdot r^{-1} \mathbf{r}) \\
&= \{[(-\sin r) r^{-1} \mathbf{r}] \cdot r^{-1} \mathbf{r}\} + \cos r \, (2r^{-1}) \\
&= -\sin r + 2r^{-1} \cos r.
\end{aligned}$$

We leave it to you to justify each step. ☐

EXERCISES 18.8

Find $\nabla \cdot \mathbf{v}$ and $\nabla \times \mathbf{v}$.

1. $\mathbf{v}(x, y) = x\mathbf{i} + y\mathbf{j}$.

2. $\mathbf{v}(x, y) \, y\mathbf{i} + x\mathbf{j}$.

3. $\mathbf{v}(x, y) = \dfrac{x}{x^2 + y^2}\mathbf{i} + \dfrac{y}{x^2 + y^2}\mathbf{j}$.

4. $\mathbf{v}(x, y) = \dfrac{y}{x^2 + y^2}\mathbf{i} + \dfrac{x}{x^2 + y^2}\mathbf{j}$.

5. $\mathbf{v}(x, y, z) = x\mathbf{i} + 2y\mathbf{j} + 3z\mathbf{k}$.

6. $\mathbf{v}(x, y, z) = yz\mathbf{i} + xz\mathbf{j} + xy\mathbf{k}$.

7. $\mathbf{v}(x, y, z) = xyz\mathbf{i} + xz\mathbf{j} + z\mathbf{k}$.

8. $\mathbf{v}(x, y, z) = x^2 y\mathbf{i} + y^2 z\mathbf{j} + xy^2\mathbf{k}$.

9. $\mathbf{v}(\mathbf{r}) = r^{-2}\mathbf{r}$.

10. $\mathbf{v}(\mathbf{r}) = e^x \, \mathbf{r}$.

11. $\mathbf{v}(\mathbf{r}) = e^{r^2}(\mathbf{i} + \mathbf{j} + \mathbf{k})$.

12. $\mathbf{v}(\mathbf{r}) = e^{y^2}\mathbf{i} + e^{z^2}\mathbf{j} + e^{x^2}\mathbf{k}$.

13. Suppose that f is a differentiable function of one variable and $\mathbf{v}(x, y, z) = f(x)\mathbf{i}$. Determine $\nabla \cdot \mathbf{v}$ and $\nabla \times \mathbf{v}$.

14. Show that, if $\mathbf{v}$ is a differentiable field of the form $\mathbf{v}(\mathbf{r}) = f(x)\mathbf{i} + g(y)\mathbf{j} + h(z)\mathbf{k}$, then $\nabla \times \mathbf{v} = \mathbf{0}$.

15. Show that divergence and curl are *linear* operators:

$$\nabla \cdot (\alpha\mathbf{u} + \beta\mathbf{v}) = \alpha(\nabla \cdot \mathbf{u}) + \beta(\nabla \cdot \mathbf{v}) \quad \text{and} \quad \nabla \times (\alpha\mathbf{u} + \beta\mathbf{v}) = \alpha(\nabla \times \mathbf{u}) + \beta(\nabla \times \mathbf{v}).$$

16. (*Important*) Show that the gravitational field

$$\mathbf{F}(\mathbf{r}) = -\frac{GmM}{r^3} \mathbf{r}$$

has zero divergence and zero curl at each $\mathbf{r} \neq \mathbf{0}$.

Calculate the Laplacean $\nabla^2 f$.

17. $f(x, y, z) = x^4 + y^4 + z^4$.

18. $f(x, y, z) = xyz$.

19. $f(x, y, z) = x^2 y^3 z^4$.

20. $f(\mathbf{r}) = \cos r$.

21. $f(\mathbf{r}) = e^r$.

22. $f(\mathbf{r}) = \ln r$.

23. Given a vector field **u**, the operator $\mathbf{u} \cdot \nabla$ is defined by setting

$$(\mathbf{u} \cdot \nabla)f = \mathbf{u} \cdot \nabla f = u_1 \frac{\partial f}{\partial x} + u_2 \frac{\partial f}{\partial y} + u_3 \frac{\partial f}{\partial z}.$$

Calculate $(\mathbf{r} \cdot \nabla)f$ for (a) $f(\mathbf{r}) = r^2$. (b) $f(\mathbf{r}) = 1/r$.

24. (Based on Exercise 23) We can also apply $\mathbf{u} \cdot \nabla$ to a vector field **v** by applying it to each component. By definition

$$(\mathbf{u} \cdot \nabla)\mathbf{v} = (\mathbf{u} \cdot \nabla v_1)\mathbf{i} + (\mathbf{u} \cdot \nabla v_2)\mathbf{j} + (\mathbf{u} \cdot \nabla v_3)\mathbf{k}.$$

(a) Calculate $(\mathbf{u} \cdot \nabla)\mathbf{r}$ for an arbitrary vector field **u**.
(b) Calculate $(\mathbf{r} \cdot \nabla)\mathbf{u}$ given that $\mathbf{u} = yz\mathbf{i} + xz\mathbf{j} + xy\mathbf{k}$.

25. Show that, if $f(\mathbf{r}) = g(r)$ and g is twice differentiable, then

$$\nabla^2 f = g''(r) + 2r^{-1} g'(r).$$

26. Verify the following identities.

(a) $\nabla \cdot (f\mathbf{v}) = (\nabla f) \cdot \mathbf{v} + f(\nabla \cdot \mathbf{v})$.   (b) $\nabla \times (f\mathbf{v}) = (\nabla f) \times \mathbf{v} + f(\nabla \times \mathbf{v})$.

(c) $\nabla \times (\nabla \times \mathbf{v}) = \nabla(\nabla \cdot \mathbf{v}) - \nabla^2 \mathbf{v}$ where $\nabla^2 \mathbf{v} = (\nabla^2 v_1)\mathbf{i} + (\nabla^2 v_2)\mathbf{j} + (\nabla^2 v_3)\mathbf{k}$.

HINT: Begin (c) by writing out the ith-component of each side.

27. A scalar field $f = f(x, y, z)$ with continuous second partials is said to be *harmonic* on a solid T iff

$$\nabla^2 f = \frac{\partial^2 f}{\partial x^2} + \frac{\partial^2 f}{\partial y^2} + \frac{\partial^2 f}{\partial z^2} = 0 \qquad \text{on } T.$$

For what nonzero integers n is $f(\mathbf{r}) = r^n$ harmonic on every solid T that excludes the origin?

18.9 THE DIVERGENCE THEOREM

Green's theorem enables us to express a double integral over a Jordan region Ω with a piecewise-smooth boundary C as a line integral over C:

$$\iint_\Omega \left(\frac{\partial Q}{\partial x} - \frac{\partial P}{\partial y} \right) dx dy = \oint_C P \, dx + Q \, dy.$$

In vector terms this relation can be written

(18.9.1)
$$\iint_\Omega (\nabla \cdot \mathbf{v}) \, dx dy = \oint_C (\mathbf{v} \cdot \mathbf{n}) \, ds.$$

Here **n** is the outer unit normal and the integral on the right is taken with respect to arc length. (Section 18.4)

Proof. Set $\mathbf{v} = Q\mathbf{i} - P\mathbf{j}$. Then

$$\iint_\Omega (\nabla \cdot \mathbf{v}) \, dx dy = \iint_\Omega \left(\frac{\partial Q}{\partial x} - \frac{\partial P}{\partial y} \right) dx dy.$$

All we have to show then is that

$$\oint_C (\mathbf{v} \cdot \mathbf{n}) \, ds = \oint_C P \, dx + Q \, dy.$$

For C traversed counterclockwise, $\mathbf{n} = \mathbf{T} \times \mathbf{k}$ where $\mathbf{T}$ is the unit tangent vector. (Draw a figure.)   Thus

$$\mathbf{v} \cdot \mathbf{n} = \mathbf{v} \cdot (\mathbf{T} \times \mathbf{k}) = (-\mathbf{v}) \cdot (\mathbf{k} \times \mathbf{T}) = (-\mathbf{v} \times \mathbf{k}) \cdot \mathbf{T}.$$

$$\mathbf{a} \cdot (\mathbf{b} \times \mathbf{c}) = (\mathbf{a} \times \mathbf{b}) \cdot \mathbf{c}$$

Since $-\mathbf{v} \times \mathbf{k} = (P\mathbf{j} - Q\mathbf{i}) \times \mathbf{k} = P\mathbf{i} + Q\mathbf{j}$, we have $\mathbf{v} \cdot \mathbf{n} = (P\mathbf{i} + Q\mathbf{j}) \cdot \mathbf{T}$.   Therefore

$$\oint_C (\mathbf{v} \cdot \mathbf{n}) \, ds = \oint_C [(P\mathbf{i} + Q\mathbf{j}) \cdot \mathbf{T}] \, ds = \oint_C (P\mathbf{i} + Q\mathbf{j}) \cdot \mathbf{dr} = \oint_C P \, dx + Q \, dy. \quad \square$$

$$(18.4.6)$$

Green's theorem expressed as (18.9.1) has a higher dimensional analog that is known as the divergence theorem.†

THEOREM 18.9.2   THE DIVERGENCE THEOREM

Let T be a solid bounded by a closed surface S which, if not smooth, is piecewise smooth.   If the vector field $\mathbf{v} = \mathbf{v}(x, y, z)$ is continuously differentiable throughout T, then

$$\iiint_T (\nabla \cdot \mathbf{v}) \, dxdydz = \iint_S (\mathbf{v} \cdot \mathbf{n}) \, d\sigma$$

where $\mathbf{n}$ is the outer unit normal.

Proof.   We will carry out the proof under the assumption that S is smooth and that any line parallel to a coordinate axis intersects S at most twice.

Our first step is to express the outer unit normal $\mathbf{n}$ in terms of its direction cosines:

$$\mathbf{n} = \cos \alpha_1 \, \mathbf{i} + \cos \alpha_2 \, \mathbf{j} + \cos \alpha_3 \, \mathbf{k}.$$

Then

$$\mathbf{v} \cdot \mathbf{n} = v_1 \cos \alpha_1 + v_2 \cos \alpha_2 + v_3 \cos \alpha_3.$$

---

† Also called Gauss's theorem after the German mathematician Karl Friedrich Gauss (1777–1855).   Often referred to as "The Prince of Mathematicians," Gauss is regarded by many as one of the greatest geniuses of all time.

The idea of the proof is to show that

(1)
$$\iint_S v_1 \cos \alpha_1 \, d\sigma = \iiint_T \frac{\partial v_1}{\partial x} \, dx\,dy\,dz,$$

(2)
$$\iint_S v_2 \cos \alpha_2 \, d\sigma = \iiint_T \frac{\partial v_2}{\partial y} \, dx\,dy\,dz,$$

(3)
$$\iint_S v_3 \cos \alpha_3 \, d\sigma = \iiint_T \frac{\partial v_3}{\partial z} \, dx\,dy\,dz.$$

All three equations can be verified in much the same manner. We will carry out the details only for the third equation.

Figure 18.9.1

Let Ω_{xy} be the projection of T onto the xy-plane. (See Figure 18.9.1.) If $(x, y) \in \Omega_{xy}$, then, by assumption, the vertical line through (x, y) intersects S in at most two points, an upper point P^+ and a lower point P^-. (If the vertical line intersects S at only one point P, we set $P = P^+ = P^-$.) As (x, y) ranges over Ω_{xy}, the upper point P^+ describes a surface

$$S^+ : \quad z = f^+(x, y), \quad (x, y) \in \Omega_{xy} \qquad \text{(see the figure)}$$

and the lower point describes a surface

$$S^- : \quad z = f^-(x, y), \quad (x, y) \in \Omega_{xy}.$$

By our assumptions, f^+ and f^- are continuously differentiable, $S = S^+ \cup S^-$, and the solid T is the set of all points (x, y, z) with

$$f^-(x, y) \le z \le f^+(x, y), \qquad (x, y) \in \Omega_{xy}.$$

Now let γ be the angle between the positive z-axis and the upper unit normal. On S^+ the outer unit normal $\mathbf{n}$ is the upper unit normal. Thus on S^+

$$\gamma = \alpha_3 \qquad \text{and} \qquad \cos \alpha_3 \sec \gamma = 1.$$

On S^- the outer unit normal $\mathbf{n}$ is the lower unit normal. In this case

$$\gamma = \pi - \alpha_3 \qquad \text{and} \qquad \cos \alpha_3 \sec \gamma = -1.$$

Thus,

$$\iint_{S+} v_3 \cos \alpha_3 \, d\sigma = \iint_{\Omega_{xy}} v_3 \cos \alpha_3 \sec \gamma \, dxdy = \iint_{\Omega_{xy}} v_3 \, [x, y, f^+(x, y)] \, dxdy$$

and

$$\iint_{S-} v_3 \cos \alpha_3 \, d\sigma = \iint_{\Omega} v_3 \cos \alpha_3 \sec \gamma \, dxdy = -\iint_{\Omega} v_3 \, [x, y, f^-(x, y)] \, dxdy.$$

It follows that

$$\iint_{S} v_3 \cos \alpha_3 \, d\sigma = \iint_{S+} v_3 \cos \alpha_3 \, d\sigma + \iint_{S-} v_3 \cos \alpha_3 \, d\sigma$$

$$= \iint_{\Omega_{xy}} (v_3 \, [x, y, f^+(x, y)] - v_3 \, [x, y, f^-(x, y)]) \, dxdy$$

$$= \iint_{\Omega_{xy}} \left(\int_{f^-(x,y)}^{f^+(x,y)} \frac{\partial v_3}{\partial z} (x, y, z) \, dz \right) dxdy$$

$$= \iiint_{T} \frac{\partial v_3}{\partial z} (x, y, z) \, dxdydz.$$

This confirms equation (3). Equation (2) can be confirmed by projection onto the xz-plane; equation (1) can be confirmed by projection onto the yz-plane. ■

Divergence as Outward Flux per Unit Volume

Choose a point P and surround it by a closed ball N_ϵ of radius ϵ. According to the divergence theorem

$$\iiint_{N_\epsilon} (\nabla \cdot \mathbf{v}) \, dxdydz = \text{flux of } \mathbf{v} \text{ out of } N_\epsilon.$$

Thus

(average divergence of $\mathbf{v}$ on N_ϵ) × (volume of N_ϵ) = flux of $\mathbf{v}$ out of N_ϵ

and

$$\text{average divergence of } \mathbf{v} \text{ on } N_\epsilon = \frac{\text{flux of } \mathbf{v} \text{ out of } N_\epsilon}{\text{volume of } N_\epsilon}.$$

Taking the limit of both sides as ϵ shrinks to 0, we have

$$\text{divergence of } \mathbf{v} \text{ at } P = \lim_{\epsilon \to 0+} \frac{\text{flux of } \mathbf{v} \text{ out of } N_\epsilon}{\text{volume of } N_\epsilon}.$$

In this sense *divergence is outward flux per unit volume.*

Think of $\mathbf{v}$ as the velocity of a fluid. As suggested in Section 18.8, negative divergence at P signals an accumulation of fluid near P:

$$\nabla \cdot \mathbf{v} < 0 \text{ at } P \quad \Rightarrow \quad \text{flux out of } N_\epsilon < 0 \quad \Rightarrow \quad \text{net flow into } N_\epsilon.$$

Positive divergence at P signals a flow of liquid away from P:

$$\nabla \cdot \mathbf{v} > 0 \text{ at } P \quad \Rightarrow \quad \text{flux out of } N_\epsilon > 0 \quad \Rightarrow \quad \text{net flow out of } N_\epsilon.$$

Points at which the divergence is negative are called *sinks*; points at which the divergence is positive are called *sources*. If the divergence of $\mathbf{v}$ is 0 throughout, then the flow has no sinks and no sources and $\mathbf{v}$ is called *solenoidal*.

Solids Bounded by Two or More Closed Surfaces

The divergence theorem, stated for solids bounded by a single closed surface, can be extended to solids bounded by several closed surfaces. Suppose, for example, that we start with a solid bounded by a closed surface S_1 and extract from the interior of that solid a solid bounded by a closed surface S_2. The remaining solid, call it T and see Figure 18.9.2, has a boundary S that consists of two pieces: an outer piece S_1 and an inner piece S_2. The key here is to note that the *outer* normal for T points *out* of S_1 but *into* S_2. The divergence theorem can be proven for T by slicing T into two pieces T_1 and T_2 as in Figure 18.9.3 and applying the divergence theorem to each piece:

$$\iiint\limits_{T_1} (\nabla \cdot \mathbf{v}) \, dxdydz = \iint\limits_{\text{bdry of } T_1} (\mathbf{v} \cdot \mathbf{n}) \, d\sigma,$$

$$\iiint\limits_{T_2} (\nabla \cdot \mathbf{v}) \, dxdydz = \iint\limits_{\text{bdry of } T_2} (\mathbf{v} \cdot \mathbf{n}) \, d\sigma.$$

The triple integrals over T_1 and T_2 add up to the triple integral over T. When the surface integrals are added together, the integrals along the common cut cancel (because the normals are in opposite directions) and therefore only the integrals over S_1 and S_2 remain. Thus the surface integrals add up to the surface integral over $S = S_1 \cup S_2$ and the divergence theorem still holds:

$$\iiint\limits_{T} (\nabla \cdot \mathbf{v}) \, dxdydz = \iint\limits_{S} (\mathbf{v} \cdot \mathbf{n}) \, d\sigma.$$

Figure 18.9.2

Figure 18.9.3

An Application to Static Charges

Consider a point charge q somewhere in space. This charge creates around itself an *electric field* $\mathbf{E}$, which in turn exerts an electric force on every other nearby charge. If we center our coordinate system at q, then the electric field at $\mathbf{r}$ can be written

$$\mathbf{E}(\mathbf{r}) = q\,\frac{\mathbf{r}}{r^3}.$$

(This is found experimentally.) Note that this field has exactly the same form as the gravitational field: a constant multiple of $r^{-3}\mathbf{r}$. It follows from (18.8.8) that

$$\nabla \cdot \mathbf{E} = 0 \qquad \text{for all } \mathbf{r} \neq \mathbf{0}.$$

We are interested in the flux of $\mathbf{E}$ out of a closed surface S. We assume that S does not pass through q.

If the charge q is outside of S, then $\mathbf{E}$ is continuously differentiable on the region T bounded by S, and, by the divergence theorem,

$$\text{flux of } \mathbf{E} \text{ out of } S = \iiint_T (\nabla \cdot \mathbf{E})\,dxdydz = \iiint_T 0\,dxdydz = 0.$$

If q is inside of S, then the divergence theorem does not apply to T directly because $\mathbf{E}$ is not differentiable on all of T. We can circumvent this difficulty by surrounding q by a small sphere S_a of radius a and applying the divergence theorem to the region T' bounded on the outside by S and on the inside by S_a. (Figure 18.9.4)

Figure 18.9.4

Since $\mathbf{E}$ is continuously differentiable on T',

$$\iiint_{T'} (\nabla \cdot \mathbf{E})\,dxdydz = \text{flux of } \mathbf{E} \text{ out of } S + \text{flux of } \mathbf{E} \text{ into } S_a$$
$$= \text{flux of } \mathbf{E} \text{ out of } S - \text{flux of } \mathbf{E} \text{ out of } S_a.$$

Since $\nabla \cdot \mathbf{E} = 0$ on T', the triple integral on the left is zero and therefore

$$\text{flux of } \mathbf{E} \text{ out of } S = \text{flux of } \mathbf{E} \text{ out of } S_a.$$

The quantity on the right is easy to calculate: on S_a, $\mathbf{n} = \mathbf{r}/r$ and therefore

$$\mathbf{E} \cdot \mathbf{n} = q\,\frac{\mathbf{r}}{r^3} \cdot \frac{\mathbf{r}}{r} = \frac{q}{r^2} = \frac{q}{a^2}.$$

Thus

$$\text{flux of } \mathbf{E} \text{ out of } S_a = \int \int_{S_a} (\mathbf{E} \cdot \mathbf{n}) \, d\sigma = \int \int_{S_a} \frac{q}{a^2} \, d\sigma = \frac{q}{a^2} (\text{area of } S_a) = 4\pi q.$$

$$\underset{\text{area of } S_a = 4\pi a^2}{\uparrow}$$

It follows that the flux of $\mathbf{E}$ out of $S = 4\pi q$.

In summary, if $\mathbf{E}$ is the electric field of a point charge q and S is a closed surface that does not pass through q, then

$$\text{flux of } \mathbf{E} \text{ out of } S = \left\{ \begin{array}{ll} 0, & \text{if } q \text{ is outside of } S \\ 4\pi q, & \text{if } q \text{ is inside of } S \end{array} \right\}.$$

EXERCISES 18.9

Calculate the flux $\mathbf{v}$ out of the unit ball $x^2 + y^2 + z^2 \leq 1$ by applying the divergence theorem.

1. $\mathbf{v}(x, y, z) = x\mathbf{i} + y\mathbf{j} + z\mathbf{k}$. **2.** $\mathbf{v}(x, y, z) = (1 - x)\mathbf{i} + (2 - y)\mathbf{j} + (3 - z)\mathbf{k}$.

3. $\mathbf{v}(x, y, z) = x^2\mathbf{i} + y^2\mathbf{j} + z^2\mathbf{k}$. **4.** $\mathbf{v}(x, y, z) = (1 - x^2)\mathbf{i} - y^2\mathbf{j} + z\mathbf{k}$.

Verify the divergence theorem on the unit cube $0 \leq x \leq 1, 0 \leq y \leq 1, 0 \leq z \leq 1$ for the following vector fields.

5. $\mathbf{v}(x, y, z) = x\mathbf{i} + y\mathbf{j} + z\mathbf{k}$. **6.** $\mathbf{v}(x, y, z) = xy\mathbf{i} + yz\mathbf{j} + xz\mathbf{k}$.

7. $\mathbf{v}(x, y, z) = x^2\mathbf{i} - xz\mathbf{j} + z^2\mathbf{k}$. **8.** $\mathbf{v}(x, y, z) = x\mathbf{i} + xy\mathbf{j} + xyz\mathbf{k}$.

Use the divergence theorem to find the total flux out of the given solid.

9. $\mathbf{v}(x, y, z) = x\mathbf{i} + 2y^2\mathbf{j} + 3z^2\mathbf{k}; \quad 0 \leq x^2 + y^2 \leq 9, \quad 0 \leq z \leq 1$.

10. $\mathbf{v}(x, y, z) = xy\mathbf{i} + yz\mathbf{j} + xz\mathbf{k}; \quad 0 \leq x \leq 1, \quad 0 \leq y \leq 1 - x, \quad 0 \leq z \leq 1 - x - y$.

Calculate the total flux of $\mathbf{v}(x, y, z) = 2xy\mathbf{i} + y^2\mathbf{j} + 3yz\mathbf{k}$ out of the given solid.

11. The ball: $0 \leq x^2 + y^2 + z^2 \leq a^2$. **12.** The cube: $0 \leq x \leq a, 0 \leq y \leq a, 0 \leq z \leq a$.

13. What is the flux of $\mathbf{v}(x, y, z) = Ax\mathbf{i} + By\mathbf{j} + Cz\mathbf{k}$ out of a solid of volume V?

14. Let T be a basic solid with a piecewise-smooth boundary. Show that, if f is harmonic on T (defined in Exercise 27, Section 18.8), then the flux of ∇f out of T is zero.

15. Let S be a closed smooth surface with continuous unit normal $\mathbf{n} = \mathbf{n}(x, y, z)$. Show that

$$\int \int_S \mathbf{n} \, d\sigma = \left(\int \int_S n_1 \, d\sigma \right) \mathbf{i} + \left(\int \int_S n_2 \, d\sigma \right) \mathbf{j} + \left(\int \int_S n_3 \, d\sigma \right) \mathbf{k} = \mathbf{0}.$$

16. Let T be a solid with a piecewise-smooth boundary S and let $\mathbf{n}$ be the outer unit normal.
(a) Verify the identity $\nabla \cdot (f \nabla f) = \|\nabla f\|^2 + f(\nabla^2 f)$ and show that, if f is harmonic on T, then

$$\int \int_S (f f'_\mathbf{n}) \, d\sigma = \int \int \int_T \|\nabla f\|^2 \, dx\,dy\,dz$$

where $f'_\mathbf{n}$ is the directional derivative $\nabla f \cdot \mathbf{n}$.
(b) Show that, if g is continuously differentiable on T, then

$$\int \int_S (g f'_\mathbf{n}) \, d\sigma = \int \int \int_T [(\nabla g \cdot \nabla f) + g(\nabla^2 f)] \, dx\,dy\,dz.$$

17. Let T be a solid with a piecewise-smooth boundary. Show that, if f and g have continuous second partials, then the flux of $\nabla f \times \nabla g$ out of T is zero.

18. Let T be a solid with a piecewise-smooth boundary S. Express the volume of T as a surface integral over S.

19. Suppose that a solid T (boundary S, outer unit normal $\mathbf{n}$) is immersed in a fluid. The fluid exerts a pressure $p = p(x, y, z)$ at each point of S and therefore the solid T experiences a force. According to physics this force is given by the formula

$$\mathbf{F} = -\iint_S p\mathbf{n}\, d\sigma.$$

(The formula says that the force on the solid is the average pressure against S times the area of S.) Now choose a coordinate system with the z-axis vertical and assume that the fluid fills a region of space to the level $z = c$. The depth of a point (x, y, z) is then $c - z$ and we have $p(x, y, z) = \rho(c - z)$ where ρ is the weight density of the fluid (the weight per unit volume). Apply the divergence theorem to each component of $\mathbf{F}$ to show that $\mathbf{F} = W\mathbf{k}$ where W is the weight of the fluid displaced by the solid. We call this the *buoyant force* on the solid. (This shows that the object is not pushed from side to side by the pressure and verifies the *principle of Archimedes:* that the buoyant force on an object at rest in a fluid equals the weight of the fluid displaced.)

20. If $\mathbf{F}$ is a force applied at the tip of a radius vector $\mathbf{r}$, then the *torque,* or twisting strength, of $\mathbf{F}$ about the origin is given by the cross product $\boldsymbol{\tau} = \mathbf{r} \times \mathbf{F}$. From physics we learn that the total torque on the solid T of Exercise 19 is given by the formula

$$\boldsymbol{\tau}_{\text{Tot}} = -\iint_S [\mathbf{r} \times \rho(c - z)\mathbf{n}]\, d\sigma.$$

Use the divergence theorem to find the components of $\boldsymbol{\tau}_{\text{Tot}}$ and show that $\boldsymbol{\tau}_{\text{Tot}} = \bar{\mathbf{r}} \times \mathbf{F}$ where $\mathbf{F}$ is the buoyant force $W\mathbf{k}$ and $\bar{\mathbf{r}}$ is the centroid of T.

(This indicates that for calculating the twisting effect of the buoyant force we can view this force as being applied at the centroid. This is very important in ship design. Imagine, for example, a totally submerged submarine. While the buoyant force acts upward through the centroid of the submarine, gravity acts downward through the center of mass. Suppose the submarine should tilt a bit to one side as depicted in Figure 18.9.5. If the centroid lies above the center of mass, then the buoyant force acts to restore the submarine to an upright position. If, however, the centroid lies below the center of mass, then disaster. Once the submarine has tilted a bit, the buoyant force will make it tilt further. This kind of analysis also applies to surface ships, but in this case the buoyant force acts upward through *the centroid of the portion of the ship that is submerged.* This point is called the *center of flotation.* One must design and load a ship to keep the center of flotation above the center of mass. Putting a lot of heavy cargo on the deck, for instance, tends to raise the center of mass and destabilize the ship.)

Figure 18.9.5

18.10 STOKES'S THEOREM

We return to Green's theorem

$$\iint_{\Omega} \left(\frac{\partial Q}{\partial x} - \frac{\partial P}{\partial y} \right) dxdy = \oint_{C} P \, dx + Q \, dy.$$

Setting $\mathbf{v} = P\mathbf{i} + Q\mathbf{j} + R\mathbf{k}$, we have

$$(\nabla \times \mathbf{v}) \cdot \mathbf{k} = \begin{vmatrix} \mathbf{i} & \mathbf{j} & \mathbf{k} \\ \dfrac{\partial}{\partial x} & \dfrac{\partial}{\partial y} & \dfrac{\partial}{\partial z} \\ P & Q & R \end{vmatrix} \cdot \mathbf{k} = \frac{\partial Q}{\partial x} - \frac{\partial P}{\partial y}.$$

Thus in terms of $\mathbf{v}$, Green's theorem can be written

$$\iint_{\Omega} [(\nabla \times \mathbf{v}) \cdot \mathbf{k}] \, dxdy = \oint_{C} \mathbf{v}(\mathbf{r}) \cdot \mathbf{dr}.$$

Since any plane can be coordinatized as the xy-plane, this result can be phrased as follows: Let S be a flat surface in space bounded by a Jordan curve C. If $\mathbf{v}$ is continuously differentiable on S, then

$$\iint_{S} [(\nabla \times \mathbf{v}) \cdot \mathbf{n}] \, d\sigma = \oint_{C} \mathbf{v}(\mathbf{r}) \cdot \mathbf{dr}$$

where $\mathbf{n}$ is a unit normal for S and the line integral is taken in the *positive sense,* meaning, in the direction of the unit tangent $\mathbf{T}$ for which $\mathbf{T} \times \mathbf{n}$ points away from the surface. See Figure 18.10.1. (An observer marching along C in the attitude of $\mathbf{n}$ keeps the surface to his or her left.)

Figure 18.10.1

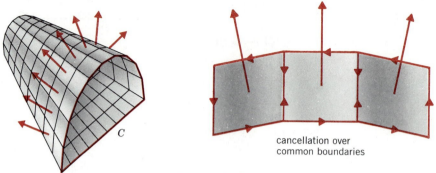

Figure 18.10.2

Figure 18.10.2 shows a *polyhedral surface S* bounded by a closed polygonal path C. The surface S consists of a finite number of flat faces $S_1, \ldots, S_N$ with polygonal boundaries $C_1, \ldots, C_N$ and unit normals $\mathbf{n}_1, \ldots, \mathbf{n}_N$. We choose these unit normals in a consistent manner; that is, they emanate from the same side of the surface. Now let $\mathbf{n} = \mathbf{n}(x, y, z)$ be a vector function of norm 1 which is $\mathbf{n}_1$ on S_1, $\mathbf{n}_2$ on S_2, $\mathbf{n}_3$ on S_3, etc. How $\mathbf{n}$ is defined on the line segments that join the different faces is immaterial. Suppose now that $\mathbf{v} = \mathbf{v}(x, y, z)$ is a vector function continuously differentiable on an open set that contains S. Then

$$\iint_S [(\nabla \times \mathbf{v}) \cdot \mathbf{n}] \, d\sigma = \sum_{i=1}^{N} \iint_{S_i} [(\nabla \times \mathbf{v}) \cdot \mathbf{n}_i] \, d\sigma = \sum_{i=1}^{N} \oint_{C_i} \mathbf{v}(\mathbf{r}) \cdot d\mathbf{r},$$

the integral over C_i being taken in the positive sense with respect to $\mathbf{n}_i$. Now when we add these line integrals, we find that all the line segments that make up the C_i but are not part of C are traversed twice and in opposite directions. (See the figure.) Thus these line segments contribute nothing to the sum of the line integrals and we are left with the integral around C. It follows that for a polyhedral surface S with boundary C

$$\iint_S [(\nabla \times \mathbf{v}) \cdot \mathbf{n}] \, d\sigma = \oint_C \mathbf{v}(\mathbf{r}) \cdot d\mathbf{r}.$$

This result can be extended to smooth surfaces with smooth bounding curves (see Figure 18.10.3) by approximating these configurations by polyhedral configurations of the type considered and using a limit process. In an admittedly informal way we have arrived at Stokes's theorem.

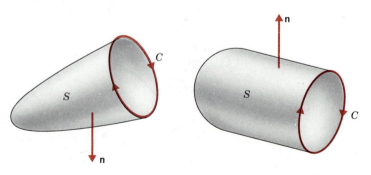

Figure 18.10.3

THEOREM 18.10.1   STOKES'S THEOREM†

Let S be a smooth surface with a smooth bounding curve C. If $\mathbf{v} = \mathbf{v}(x, y, z)$ is continuously differentiable on an open set that contains S, then

$$\iint_S [(\nabla \times \mathbf{v}) \cdot \mathbf{n}] \, d\sigma = \oint_C \mathbf{v(r)} \cdot \mathbf{dr}$$

where $\mathbf{n} = \mathbf{n}(x, y, z)$ is a unit normal that varies continuously on S and the line-integral is taken in the positive sense with respect to $\mathbf{n}$.

Problem 1.   Verify Stokes's theorem for

$$\mathbf{v} = -3y\mathbf{i} + 3x\mathbf{j} + z^4\mathbf{k}$$

taking S as the portion of the ellipsoid $2x^2 + 2y^2 + z^2 = 1$ that lies above the plane $z = 1/\sqrt{2}$.

Solution.   A little algebra shows that S is the graph of

$$f(x, y) = \sqrt{1 - 2(x^2 + y^2)}$$

with (x, y) restricted to the disc Ω: $x^2 + y^2 \leq \frac{1}{4}$. Now

$$\nabla \times \mathbf{v} = \begin{vmatrix} \mathbf{i} & \mathbf{j} & \mathbf{k} \\ \dfrac{\partial}{\partial x} & \dfrac{\partial}{\partial y} & \dfrac{\partial}{\partial z} \\ -3y & 3x & z^4 \end{vmatrix} = \left[\frac{\partial}{\partial x}(3x) + \frac{\partial}{\partial y}(3y) \right]\mathbf{k} = 6\mathbf{k}.$$

Taking $\mathbf{n}$ as the upper unit normal, we have

$$\iint_S [(\nabla \times \mathbf{v}) \cdot \mathbf{n}] \, d\sigma = \iint_S (6\mathbf{k} \cdot \mathbf{n}) \, d\sigma$$

$$(18.7.8) \longrightarrow \quad = \iint_\Omega (-(0)f'_x - (0)f'_y + 6) \, dx\,dy$$

$$= \iint_\Omega 6 \, dx\,dy = 6(\text{area of } \Omega) = \tfrac{3}{2}\pi.$$

The bounding curve for C is the set of all (x, y, z) with $x^2 + y^2 = \frac{1}{4}$ and $z = 1/\sqrt{2}$. We can parametrize C by setting

$$\mathbf{r}(u) = \tfrac{1}{2}\cos u \, \mathbf{i} + \tfrac{1}{2}\sin u \, \mathbf{j} + \frac{1}{\sqrt{2}}\mathbf{k}, \qquad u \in [0, 2\pi].$$

---

† The result was announced publicly for the first time by George Gabriel Stokes (1819–1903), an Irish mathematician and physicist who, like Green, was a Cambridge professor.

Since **n** is the upper unit normal, this parametrization gives C in the positive sense. Thus

$$\oint_C \mathbf{v}(\mathbf{r}) \cdot d\mathbf{r} = \int_0^{2\pi} (-\tfrac{3}{2} \sin u \, \mathbf{i} + \tfrac{3}{2} \cos u \, \mathbf{j} + \tfrac{1}{4}\mathbf{k}) \cdot (-\tfrac{1}{2} \sin u \, \mathbf{i} + \tfrac{1}{2} \cos u \, \mathbf{j}) \, du$$

$$= \int_0^{2\pi} (\tfrac{3}{4} \sin^2 u + \tfrac{3}{4} \cos^2 u) \, du = \int_0^{2\pi} \tfrac{3}{4} \, du = \tfrac{3}{2}\pi.$$

This is the value we obtained for the surface integral. ☐

Problem 2.   Verify Stokes's theorem for

$$\mathbf{v} = z^2\mathbf{i} - 2x\mathbf{j} + y^3\mathbf{k}$$

taking S as the upper half of the unit sphere $x^2 + y^2 + z^2 = 1$.

Solution.   We use the upper unit normal $\mathbf{n} = x\mathbf{i} + y\mathbf{j} + z\mathbf{k}$. Now

$$\nabla \times \mathbf{v} = \begin{vmatrix} \mathbf{i} & \mathbf{j} & \mathbf{k} \\ \dfrac{\partial}{\partial x} & \dfrac{\partial}{\partial y} & \dfrac{\partial}{\partial z} \\ z^2 & -2x & y^3 \end{vmatrix} = 3y^2\mathbf{i} + 2z\mathbf{j} - 2\mathbf{k}.$$

Therefore

$$\iint_S [(\nabla \times \mathbf{v}) \cdot \mathbf{n}] \, d\sigma = \iint_S [(3y^2\mathbf{i} + 2z\mathbf{j} - 2\mathbf{k}) \cdot (x\mathbf{i} + y\mathbf{j} + z\mathbf{k}) \, d\sigma$$

$$= \iint_S (3xy^2 + 2yz - 2z) \, d\sigma$$

$$= \iint_S 3xy^2 \, d\sigma + \iint_S 2yz \, d\sigma - \iint_S 2z \, d\sigma.$$

The first integral is zero because S is symmetric about the yz-plane and the integrand is odd with respect to x.   The second integral is zero because S is symmetric about the xz-plane and the integrand is odd with respect to y.   Thus

$$\iint_S [(\nabla \times \mathbf{v}) \cdot \mathbf{n}] \, d\sigma = - \iint_S 2z \, d\sigma = -2\bar{z} \,(\text{area of } S) = -2(\tfrac{1}{2})2\pi = -2\pi.$$

Exercise 11, Section 18.7 ⌐

This is also the value of the integral along the bounding base circle taken in the positive sense:

$$\oint_C \mathbf{v}(\mathbf{r}) \cdot d\mathbf{r} = \oint_C z^2 \, dx - 2x \, dy = -2 \oint_C x \, dy = -2 \int_0^{2\pi} \cos^2 u \, du = -2\pi. \quad ☐$$

Earlier you saw that the curl of a gradient is zero. Using Stokes's theorem we can prove a partial converse.

(18.10.2) | If a vector field $\mathbf{v} = \mathbf{v}(x, y, z)$ is continuously differentiable on an open convex set U and $\nabla \times \mathbf{v} = 0$ on all of U, then $\mathbf{v}$ is the gradient of some scalar field ϕ defined on U.

Proof. Choose a point $\mathbf{a}$ in U and for each point $\mathbf{x}$ in U define

$$\phi(\mathbf{x}) = \int_{\mathbf{a}}^{\mathbf{x}} \mathbf{v}(\mathbf{r}) \cdot d\mathbf{r}.$$

(This is the line integral from $\mathbf{a}$ to $\mathbf{x}$ taken along the line segment that joins these two points. We know that this line segment lies in U because U is convex.)

Since U is open, $\mathbf{x} + \mathbf{h}$ is in U for all $\mathbf{h}$ sufficiently small. Assume then that $\mathbf{h}$ is sufficiently small for $\mathbf{x} + \mathbf{h}$ to be in U. Since U is convex, the triangular region with vertices at $\mathbf{a}$, $\mathbf{x}$, $\mathbf{x} + \mathbf{h}$ lies in U. (See Figure 18.10.4.) Since $\nabla \times \mathbf{v} = 0$ on U, we can conclude from Stokes's theorem that

$$\int_{\mathbf{a}}^{\mathbf{x}} \mathbf{v}(\mathbf{r}) \cdot d\mathbf{r} + \int_{\mathbf{x}}^{\mathbf{x}+\mathbf{h}} \mathbf{v}(\mathbf{r}) \cdot d\mathbf{r} + \int_{\mathbf{x}+\mathbf{h}}^{\mathbf{a}} \mathbf{v}(\mathbf{r}) \cdot d\mathbf{r} = 0.$$

Figure 18.10.4

Therefore

$$\int_{\mathbf{x}}^{\mathbf{x}+\mathbf{h}} \mathbf{v}(\mathbf{r}) \cdot d\mathbf{r} = -\int_{\mathbf{x}+\mathbf{h}}^{\mathbf{a}} \mathbf{v}(\mathbf{r}) \cdot d\mathbf{r} - \int_{\mathbf{a}}^{\mathbf{x}} \mathbf{v}(\mathbf{r}) \cdot d\mathbf{r}$$

$$= \int_{\mathbf{a}}^{\mathbf{x}+\mathbf{h}} \mathbf{v}(\mathbf{r}) \cdot d\mathbf{r} - \int_{\mathbf{a}}^{\mathbf{x}} \mathbf{v}(\mathbf{r}) \cdot d\mathbf{r}.$$

By our definition of ϕ we have

$$\phi(\mathbf{x} + \mathbf{h}) - \phi(\mathbf{x}) = \int_{\mathbf{x}}^{\mathbf{x}+\mathbf{h}} \mathbf{v}(\mathbf{r}) \cdot d\mathbf{r}.$$

We can parametrize the line segment from $\mathbf{x}$ to $\mathbf{x} + \mathbf{h}$ by $\mathbf{r}(u) = \mathbf{x} + u\mathbf{h}$ with $u \in [0, 1]$. Therefore

$$\phi(\mathbf{x} + \mathbf{h}) - \phi(\mathbf{x}) = \int_0^1 [\mathbf{v}(\mathbf{r}(u)) \cdot \mathbf{r}'(u)]\, du$$

$$= \int_0^1 [\mathbf{v}(\mathbf{r}(u)) \cdot \mathbf{h}]\, du$$

Theorem 5.11.1 $\longrightarrow$ $= \mathbf{v}(\mathbf{r}(u_0)) \cdot \mathbf{h}$ for some u_0 in $[0, 1]$

$$= \mathbf{v}(\mathbf{x} + u_0\mathbf{h}) \cdot \mathbf{h} = \mathbf{v}(\mathbf{x}) \cdot \mathbf{h} + [\mathbf{v}(\mathbf{x} + u_0\mathbf{h}) - \mathbf{v}(\mathbf{x})] \cdot \mathbf{h}.$$

That $\mathbf{v} = \nabla\phi$ follows from observing that $[\mathbf{v}(\mathbf{x} + u_0\mathbf{h}) - \mathbf{v}(\mathbf{x})] \cdot \mathbf{h}$ is $o(\mathbf{h})$: as $\mathbf{h} \to \mathbf{0}$,

$$\frac{|[\mathbf{v}(\mathbf{x} + u_0\mathbf{h}) - \mathbf{v}(\mathbf{x})] \cdot \mathbf{h}|}{\|\mathbf{h}\|} \leq \frac{\|\mathbf{v}(\mathbf{x} + u_0\mathbf{h}) - \mathbf{v}(\mathbf{x})\|\,\|\mathbf{h}\|}{\|\mathbf{h}\|} = \|\mathbf{v}(\mathbf{x} + u_0\mathbf{h}) - \mathbf{v}(\mathbf{x})\| \to 0. \quad \square$$

The Normal Component of $\nabla \times \mathbf{v}$ as Circulation per Unit Area; Irrotational Flow

Interpret $\mathbf{v} = \mathbf{v}(x, y, z)$ as the velocity of a fluid flow. In Section 18.8 we stated that $\nabla \times \mathbf{v}$ measures the rotational tendency of the fluid. Now we can be more precise.

Take a point P within the flow and choose a unit vector $\mathbf{n}$. Let D_ϵ be the ϵ-disc that is centered at P and is perpendicular to $\mathbf{n}$. Let C_ϵ be the circular boundary of D_ϵ directed in the positive sense with respect to $\mathbf{n}$. (See Figure 18.10.5.) By Stokes's theorem

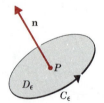

Figure 18.10.5

$$\iint\limits_{D_\epsilon} [(\nabla \times \mathbf{v}) \cdot \mathbf{n}]\, d\sigma = \oint_{C_\epsilon} \mathbf{v}(\mathbf{r}) \cdot d\mathbf{r}.$$

The line integral on the right is called *the circulation* of $\mathbf{v}$ around C_ϵ. Thus we can say that

$$\left(\begin{array}{c} \text{the average } \mathbf{n}\text{-component} \\ \text{of } \nabla \times \mathbf{v} \text{ on } D_\epsilon \end{array} \right) \times (\text{the area of } D_\epsilon) = \text{the circulation of } \mathbf{v} \text{ around } C_\epsilon.$$

It follows that

$$\text{the average } \mathbf{n}\text{-component of } \nabla \times \mathbf{n} \text{ on } D_\epsilon = \frac{\text{the circulation of } \mathbf{v} \text{ around } C_\epsilon}{\text{the area of } D_\epsilon}.$$

Taking the limit as ϵ shrinks to 0, you can see that

$$\text{the } \mathbf{n}\text{-component of } \nabla \times \mathbf{v} \text{ at } P = \lim_{\epsilon \to 0^+} \frac{\text{the circulation of } \mathbf{v} \text{ around } C_\epsilon}{\text{the area of } D_\epsilon}.$$

At each point P the component of $\nabla \times \mathbf{v}$ in any direction $\mathbf{n}$ is the circulation of $\mathbf{v}$ per unit area in the plane normal to $\mathbf{n}$. If $\nabla \times \mathbf{v} = \mathbf{0}$ identically, the fluid has no rotational tendency and the flow is called irrotational.

EXERCISES 18.10

Let S be the upper half of the unit sphere $x^2 + y^2 + z^2 = 1$ and take $\mathbf{n}$ as the upper unit normal. Find

$$\iint\limits_{S} [(\nabla \times \mathbf{v}) \cdot \mathbf{n}] \, d\sigma$$

(a) by direct calculation.　　(b) by Stokes's theorem.

1. $\mathbf{v}(x, y, z) = x\mathbf{i} + y\mathbf{j} + z\mathbf{k}$.　　　　**2.** $\mathbf{v}(x, y, z) = y\mathbf{i} - x\mathbf{j} + z\mathbf{k}$.

3. $\mathbf{v}(x, y, z) = z^2\mathbf{i} + 2x\mathbf{j} - y^3\mathbf{k}$.　　**4.** $\mathbf{v}(x, y, z) = 6xz\mathbf{i} - x^2\mathbf{j} - 3y^2\mathbf{k}$.

Let S be the triangular surface with vertices $(2, 0, 0), (0, 2, 0), (0, 0, 2)$ and take $\mathbf{n}$ as the upper unit normal. Find

$$\iint\limits_{S} [(\nabla \times \mathbf{v}) \cdot \mathbf{n}] \, d\sigma$$

(a) by direct calculation.　　(b) by Stokes's theorem.

5. $\mathbf{v}(x, y, z) = 2z\mathbf{i} - y\mathbf{j} + x\mathbf{k}$.　　**6.** $\mathbf{v}(x, y, z) = (x^2 + y^2)\mathbf{i} + y^2\mathbf{j} + (x^2 + z^2)\mathbf{k}$.

7. $\mathbf{v}(x, y, z) = x^4\mathbf{i} + xy\mathbf{j} + z^4\mathbf{k}$.

8. Show that, if $\mathbf{v} = \mathbf{v}(x, y, z)$ is continuously differentiable everywhere and its curl is identically zero, then

$$\int_{C} \mathbf{v}(\mathbf{r}) \cdot d\mathbf{r} = 0 \qquad \text{for every smooth closed curve } C.$$

9. Let $\mathbf{v} = y\mathbf{i} + z\mathbf{j} + x^2y^2\mathbf{k}$ and let S be the surface $z = x^2 + y^2$ from $z = 0$ to $z = 4$. Calculate the flux of $\nabla \times \mathbf{v}$ in the direction of the lower unit normal $\mathbf{n}$.

10. Let $\mathbf{v} = \frac{1}{2}y\mathbf{i} + 2xz\mathbf{j} - 3x\mathbf{k}$ and let S be the surface $y = 1 - (x^2 + z^2)$ from $y = -8$ to $y = 1$. Calculate the flux of $\nabla \times \mathbf{v}$ in the direction of the unit normal $\mathbf{n}$ with positive $\mathbf{j}$-component.

11. Let $\mathbf{v} = 2x\mathbf{i} + 2y\mathbf{j} + x^2y^2z^2\mathbf{k}$ and let S be the lower half of the ellipsoid

$$\frac{x^2}{4} + \frac{y^2}{9} + \frac{z^2}{27} = 1.$$

Calculate the flux of $\nabla \times \mathbf{v}$ in the direction of the upper unit normal $\mathbf{n}$.

12. Let S be a smooth closed surface and let $\mathbf{v} = \mathbf{v}(x, y, z)$ be a vector field with second partials continuous on an open convex set that contains S. Show that

$$\iint\limits_{S} [(\nabla \times \mathbf{v}) \cdot \mathbf{n}] \, d\sigma = 0$$

where $\mathbf{n} = \mathbf{n}(x, y, z)$ is the outer unit normal.

13. The upper half of the ellipsoid $\frac{1}{2}x^2 + \frac{1}{2}y^2 + z^2 = 1$ intersects the cylinder $x^2 + y^2 - y = 0$ in a curve C. Calculate the circulation of $\mathbf{v} = y^3\mathbf{i} + (xy + 3xy^2)\mathbf{j} + z^4\mathbf{k}$ around C by using Stokes's theorem.

14. The sphere $x^2 + y^2 + z^2 = a^2$ intersects the plane $x + 2y + z = 0$ in a curve C. Calculate the circulation of $\mathbf{v} = 2y\mathbf{i} - z\mathbf{j} + 2x\mathbf{k}$ about C by using Stokes's theorem.

15. The paraboloid $z = x^2 + y^2$ intersects the plane $z = y$ in a curve C. Calculate the circulation of $\mathbf{v} = 2z\mathbf{i} + x\mathbf{j} + y\mathbf{k}$ about C using Stokes's theorem.

16. The cylinder $x^2 + y^2 = b^2$ intersects the plane $y + z = a^2$ in a curve C. Assume $a^2 \geq b > 0$. Calculate the circulation of $\mathbf{v} = xy\mathbf{i} + yz\mathbf{j} + xz\mathbf{k}$ about C using Stokes's theorem.

17. Let S be a smooth surface with a smooth bounding curve C and let $\mathbf{a}$ be a fixed vector. Show that

$$\iint_S (2\mathbf{a} \cdot \mathbf{n})\, d\sigma = \oint_C (\mathbf{a} \times \mathbf{r}) \cdot d\mathbf{r}$$

where $\mathbf{n} = \mathbf{n}(x, y, z)$ is a unit vector that varies continuously over S and the line integral is taken in the positive sense with respect to $\mathbf{n}$.

18. Let S be a smooth surface with smooth bounding curve C. Show that, if ϕ and ψ are sufficiently differentiable scalar fields, then

$$\iint_S [(\nabla\phi \times \nabla\psi) \cdot \mathbf{n}]\, d\sigma = \oint_C (\phi\nabla\psi) \cdot d\mathbf{r}$$

where $\mathbf{n} = \mathbf{n}(x, y, z)$ is a unit normal that varies continuously on S and the line integral is taken in the positive sense with respect to $\mathbf{n}$.

19. Let S be a smooth surface with a smooth plane bounding curve C and let $\mathbf{v} = \mathbf{v}(x, y, z)$ be a vector field with second partials continuous on an open convex set that contains S. If S does not cross the plane of C, then Stokes's theorem for S follows readily from the divergence theorem and Green's theorem. Carry out the argument.

20. Our derivation of Stokes's theorem was admittedly nonrigorous. The following version of Stokes's theorem lends itself more readily to rigorous proof.

THEOREM 18.10.3

Let Γ be a Jordan region in the uv-plane with a piecewise-smooth boundary C_Γ given in a counterclockwise sense by a pair of functions $u = u(t)$, $v = v(t)$ with $t \in [a, b]$. Let $\mathbf{R}(u, v) = x(u, v)\mathbf{i} + y(u, v)\mathbf{j} + z(u, v)\mathbf{k}$ be a vector function with continuous second partials on Γ. Assume that $\mathbf{R}$ is one-to-one on Γ and that the fundamental vector product $\mathbf{N} = \mathbf{R}'_u \times \mathbf{R}'_v$ is never zero. The surface S: $\mathbf{R} = \mathbf{R}(u, v)$, $(u, v) \in \Gamma$ is a smooth surface bounded by the oriented space curve C: $\mathbf{r}(t) = \mathbf{R}(u(t), v(t))$, $t \in [a, b]$. If $\mathbf{v} = \mathbf{v}(x, y, z)$ is a vector field continuously differentiable on S, then

$$\iint_S [(\nabla \times \mathbf{v}) \cdot \mathbf{n}]\, d\sigma = \int_C \mathbf{v}(\mathbf{r}) \cdot d\mathbf{r}$$

where $\mathbf{n}$ is the unit normal in the direction of the fundamental vector product.

Give a detailed proof of the theorem.   HINT: Set $\mathbf{v} = v_1\mathbf{i} + v_2\mathbf{j} + v_3\mathbf{k}$. Then

$$\iint_S [(\nabla \times \mathbf{v}) \cdot \mathbf{n}]\, d\sigma = \iint_S [(\nabla \times v_1\mathbf{i}) \cdot \mathbf{n}]\, d\sigma + \iint_S [(\nabla \times v_2\mathbf{j}) \cdot \mathbf{n}]\, d\sigma + \iint_S [(\nabla \times v_3\mathbf{k}) \cdot \mathbf{n}]\, d\sigma$$

and

$$\int_C \mathbf{v}(\mathbf{r}) \cdot d\mathbf{r} = \int_C v_1\, dx + \int_C v_2\, dy + \int_C v_3\, dz.$$

Show that

$$\iint_S [(\nabla \times v_1\mathbf{i}) \cdot \mathbf{n}]\, d\sigma = \int_C v_1\, dx$$

by showing that both integrals can be written

$$\iint_\Gamma \left[\frac{\partial v_1}{\partial u} \frac{\partial x}{\partial v} - \frac{\partial v_1}{\partial v} \frac{\partial x}{\partial u} \right] du\, dv.$$

A similar argument (no need to carry it out) equates the integrals for v_1 and v_2 and proves the theorem.

CHAPTER HIGHLIGHTS

18.7 Surface Integrals

mass of a material surface (p. 1061)

average value on a surface, weighted average (p. 1063)

center of mass of a material surface (p. 1063)

smooth surface (p. 1066)

closed, piecewise-smooth surface (p. 1068)

surface integral (p. 1061)

centroid (p. 1063)

moment of inertia (p. 1064)

flux of a vector field (p. 1066)

one-sided surface, Möbius band (p. 1077)

18.8 The Vector Differential Operator ∇

the operator del: $\nabla = \dfrac{\partial}{\partial x}\mathbf{i} + \dfrac{\partial}{\partial y}\mathbf{j} + \dfrac{\partial}{\partial z}\mathbf{k}$

gradient of f: ∇f

divergence of $\mathbf{v}$: $\nabla \cdot \mathbf{v}$

curl of a gradient is zero (p. 1074)

$\nabla \cdot (f\mathbf{v}) = (\nabla f) \cdot \mathbf{v} + f(\nabla \cdot \mathbf{v})$

curl of $\mathbf{v}$: $\nabla \times \mathbf{v}$

divergence of a curl is zero (p. 1074)

$\nabla \times (f\mathbf{v}) = (\nabla f) \times \mathbf{v} + f(\nabla \times \mathbf{v})$

$\nabla \cdot (r^n \mathbf{r}) = (n + 3)r^n$,

Laplacean: $\nabla^2 f = \nabla \cdot \nabla f = \dfrac{\partial^2 f}{\partial x^2} + \dfrac{\partial^2 f}{\partial y^2} + \dfrac{\partial^2 f}{\partial z^2}$

$\nabla \times (r^n \mathbf{r}) = \mathbf{0}$

18.9 The Divergence Theorem

divergence theorem for a solid bounded by a single closed surface (p. 1078)

divergence as outward flux per unit volume (p. 1080)   sinks and sources (p. 1081)

divergence theorem for solids bounded by two or more closed surfaces (p. 1081)

an application to static charges (p. 1082)

buoyant force, principle of Archimedes (p. 1084)

18.10 Stokes's Theorem

positive sense along a curve bounding an open surface (p. 1085)

Stokes's theorem (p. 1087)

conditions under which $\nabla \times \mathbf{v} = \mathbf{0}$ implies $\mathbf{v} = \nabla f$

normal component of $\nabla \times \mathbf{v}$ as circulation per unit area   irrotational flow (p. 1090)

SOME ELEMENTARY TOPICS

A.1   SETS

A *set* is a collection of objects.   The objects in a set are called the *elements* (or *members*) of the set.

We might, for example, consider the set of capital letters appearing on this page, or the set of motorcycles licensed in Idaho, or the set of rational numbers.   Suppose, however, that we wanted to find the set of rational people.   Everybody might have a different collection.   Which would be the right one?   To avoid such problems we insist that sets be unambiguously defined.   Collections based on highly subjective judgments — such as "all good football players" or "all likeable children" — are not sets.

Notation

To indicate that an object x is in the set A, we write

$$x \in A.$$

To indicate that x is not in A, we write

$$x \notin A.$$

Thus

$$\sqrt{2} \in \text{the set of real numbers} \quad \text{but} \quad \sqrt{2} \notin \text{the set of rational numbers.}$$

Sets are often denoted by braces.   The set consisting of *a* alone is written {*a*}; that consisting of *a*, *b* is written {*a*, *b*}; that consisting of *a*, *b*, *c*, {*a*, *b*, *c*}; and so on.   Thus

$$0 \in \{0, 1, 2\}, \quad 1 \in \{0, 1, 2\}, \quad 2 \in \{0, 1, 2\}, \quad \text{but} \quad 3 \notin \{0, 1, 2\}.$$

We can also use braces for infinite sets:

$$\{1, 2, 3, \ldots\} \text{ is the set of positive integers,}$$
$$\{-1, -2, -3, \ldots\} \text{ is the set of negative integers,}$$
$$\{1, 2, 2^2, 2^3, \ldots\} \text{ is the set of powers of 2.}$$

Sets are often defined by a property.   We write {*x*: *P*} to indicate *the set of all x for which property P holds*.   Thus

{*x*: *x* > 2} is the set of all numbers greater than 2;

{*x*: *x*² > 9} is the set of all numbers whose squares are greater than 9;

{*p*/*q*: *p*, *q* integers, *q* ≠ 0} is the set of all rational numbers.

If *A* is a set, then {*x*: *x* ∈ *A*} is *A* itself.

Containment and Equality

If *A* and *B* are sets, then *A* is said to be *contained* in *B*, in symbols $A \subseteq B$, iff† every element of *A* is also an element of *B*.   For example,

the set of equilateral triangles ⊆ the set of all triangles,

the set of all college freshmen ⊆ the set of all college students,

the set of rational numbers ⊆ the set of real numbers.

If *A* is contained in *B*, then *A* is called a *subset* of *B*.   Thus

the set of equilateral triangles is a subset of the set of all triangles,

the set of college freshmen is a subset of the set of all college students,

the set of rational numbers is a subset of the set of real numbers.

Two sets are said to be *equal* iff they have exactly the same membership.   In symbols,

(A.1.1)
$$\boxed{A = B \quad \text{iff} \quad A \subseteq B \quad \text{and} \quad B \subseteq A.}$$

Examples

$$\{x: \ x^2 = 4\} = \{-2, 2\},$$
$$\{x: \ x^2 < 4\} = \{x: \ -2 < x < 2\},$$
$$\{x: \ x^2 > 4\} = \{x: \ x < -2 \ \text{or} \ x > 2\}. \quad \square$$

———

† By "iff" we mean "if and only if".   This expression is used so often in mathematics that it is convenient to have an abbreviation for it.

The Intersection of Two Sets

The set of elements common to two sets A and B is called the *intersection* of A and B and is denoted by $A \cap B$. The idea is illustrated in Figure A.1.1. In symbols,

(A.1.2)

$$x \in A \cap B \quad \text{iff} \quad x \in A \quad \text{and} \quad x \in B.$$

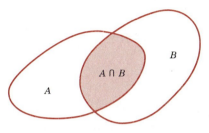

Figure A.1.1

Examples

1. If A is the set of all nonnegative numbers, and B is the set of all nonpositive numbers, then $A \cap B = \{0\}$.

2. If A is the set of all multiples of 3, and B is the set of all multiples of 4, then $A \cap B$ is the set of all multiples of 12.

3. If $A = \{a, b, c, d, e\}$, and $B = \{c, d, e, f\}$, then $A \cap B = \{c, d, e\}$.

4. If $A = \{x: \ x > 1\}$ and $B = \{x: \ x < 4\}$, then $A \cap B = \{x: \ 1 < x < 4\}$.   □

The Union of Two Sets

The *union* of two sets A and B, written $A \cup B$, is the set of elements that are either in A or in B. This does not exclude objects that are elements of both A and B. (See Figure A.1.2.) In symbols,

(A.1.3)

$$x \in A \cup B \quad \text{iff} \quad x \in A \quad \text{or} \quad x \in B.$$

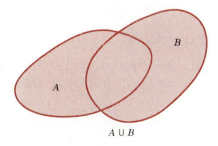

$A \cup B$

Figure A.1.2

Examples

1. If A is the set of all nonnegative numbers and B is the set of all nonpositive numbers, then $A \cup B$ is the set of all real numbers.

2. If $A = \{a, b, c, d, e\}$ and $B = \{c, d, e, f\}$, then $A \cup B = \{a, b, c, d, e, f\}$.

3. If $A = \{x: \ 0 < x < 1\}$ and $B = \{0, 1\}$, then $A \cup B = \{x: \ 0 \le x \le 1\}$.

4. If $A = \{x: \ x > 1\}$ and $B = \{x: \ x > 2\}$, then $A \cup B = \{x: \ x > 1\}$. ☐

The Empty Set

If the sets A and B have no elements in common, we say that A and B are *disjoint* and write $A \cap B = \varnothing$. We regard $\varnothing$ as a set with no elements and refer to it as *the empty set*.

Examples

1. If A is the set of all positive numbers, and B is the set of all negative numbers, then $A \cap B = \varnothing$.

2. If $A = \{0, 1, 2, 3\}$ and $B = \{4, 5, 6, 7, 8\}$, then $A \cap B = \varnothing$.

3. The set of all irrational rational numbers is empty; so is the set of all even odd integers; so is the set of real numbers with negative squares. ☐

The empty set $\varnothing$ plays a role in the theory of sets that is strikingly similar to the role played by 0 in the arithmetic of numbers. Without pursuing the matter very far, note that for numbers,

$$a + 0 = 0 + a = a, \qquad a \cdot 0 = 0 \cdot a = 0,$$

and for sets,

$$A \cup \varnothing = \varnothing \cup A = A, \qquad A \cap \varnothing = \varnothing \cap A = \varnothing.$$

Cartesian Products

If A and B are nonempty sets, then $A \times B$, the *Cartesian product* of A and B, is the set of all ordered pairs (a, b) with $a \in A$ and $b \in B$. In set notation

(A.1.4)
$$A \times B = \{(a, b): \ a \in A, b \in B\}.$$

Notice that $A \times B = B \times A$ iff $A = B$.

Examples

1. If $A = \{0, 1\}$ and $B = \{1, 2, 3\}$, then $A \times B = \{(0, 1), (0, 2), (0, 3), (1, 1), (1, 2), (1, 3)\}$ and $B \times A = \{(1, 0), (1, 1), (2, 0), (2, 1), (3, 0), (3, 1)\}$.

2. If A is the set of rational numbers and B is the set of irrational numbers, then $A \times B$ is the set of all pairs (a, b) with a rational and b irrational.   □

The Cartesian product $A \times B \times C$ consists of all ordered triples (a, b, c) with $a \in A$, $b \in B$, $c \in C$:

(A.1.5) $A \times B \times C = \{(a, b, c): a \in A, b \in B, c \in C\}.$

EXERCISES A.1   (The odd-numbered exercises have answers at the back of the book.)

For Exercises 1–20, take

$$A = \{0, 2\}, \quad B = \{-1, 0, 1\}, \quad C = \{1, 2, 3, 4\}, \quad D = \{2, 4, 6, 8, \ldots\},$$

and determine the following sets:

1. $A \cup B$.	**2.** $B \cup C$.	**3.** $A \cap B$.	**4.** $B \cap C$.
5. $B \cup D$.	**6.** $A \cup D$.	**7.** $B \cap D$.	**8.** $A \cap D$.
9. $C \cup D$.	**10.** $C \cap D$.	**11.** $A \times B$.	**12.** $B \times C$.
13. $B \times A$.	**14.** $C \times B$.	**15.** $A \times A \times B$.	**16.** $A \times B \times A$.
17. $A \cap (C \cap D)$.	**18.** $A \cap (B \cup C)$.	**19.** $A \cup (C \cap D)$.	**20.** $A \cup (B \cap C)$.

For Exercises 21–28, take

$$A = \{x: x > 2\}, \quad B = \{x: x \le 4\}, \quad C = \{x: x > 3\},$$

and determine the following sets:

21. $A \cup B$.	**22.** $B \cup C$.	**23.** $A \cap B$.	**24.** $B \cap C$.
25. $A \cup C$.	**26.** $A \cap C$.	**27.** $B \cup A$.	**28.** $C \cap B$.

29. Given that $A \subseteq B$, find (a) $A \cup B$. (b) $A \cap B$.
30. What can you conclude about A and B given that
    (a) $A \cup B = A$?   (b) $A \cap B = A$?   (c) $A \cup B = A$ and $A \cap B = A$?
31. List all the nonempty subsets of $\{0, 1, 2\}$.
32. Determine the number of nonempty subsets of a set A with n elements.

A.2   RADIAN MEASURE

Degree measure, traditionally used to measure the angles of a geometric figure, has one serious drawback. It is artificial. There is no intrinsic connection between a degree and the geometry of a circle. Why 360 degrees for one revolution? Why not 400? or 100?

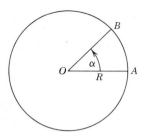

Figure A.2.1

There is another way of measuring angles that is more natural and lends itself better to the methods of calculus: measuring angles in *radians*. Figure A.2.1 shows a central angle in a circle of radius R. Suppose that you had never heard of degree measurement. How would you measure this angle? The most natural thing to do, it would seem, would be to measure the length of the arc from A to B and compare the length of this arc to the radius R. The *radian measure* of $\angle AOB$ is, by definition, *the length of $\overset{\frown}{AB}$ divided by R*:

(A.2.1) $\qquad \angle AOB$ measures α radians iff $\dfrac{\text{length of } \overset{\frown}{AB}}{R} = \alpha.$

Figure A.2.2 shows an angle serving as a central angle for two different circles. Using the smaller circle we find that the angle measures

$$\frac{\text{length of } \overset{\frown}{AB}}{R} \text{ radians.}$$

Using the larger circle we find that the angle measures

$$\frac{\text{length of } \overset{\frown}{A'B'}}{R'} \text{ radians.}$$

For our definition of radian measure to make sense, it must be true that

$$\frac{\text{length of } \overset{\frown}{AB}}{R} = \frac{\text{length of } \overset{\frown}{A'B'}}{R'}.$$

You can verify this relation by noting that sectors AOB and $A'OB'$ are similar figures.

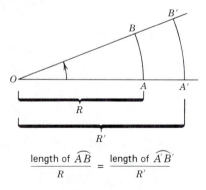

$$\frac{\text{length of } \overset{\frown}{AB}}{R} = \frac{\text{length of } \overset{\frown}{A'B'}}{R'}$$

Figure A.2.2

Since the circumference of a circle of radius R is $2\pi R$, a complete revolution comprises 2π radians; half a revolution (a straight angle) comprises π radians; a quarter revolution (a right angle) comprises $\frac{1}{2}\pi$ radians. See Figure A.2.3.

2π radians
a complete revolution

π radians
a straight angle

$\frac{1}{2}\pi$ radians
a right angle

Figure A.2.3

The conversion of radians to degrees and vice versa is made by noting that, if A is the measure of an angle in degrees and x is the measure of that same angle in radians, then

(A.2.1)
$$\frac{A}{360} = \frac{x}{2\pi}.$$

In particular

$$1 \text{ radian} = \frac{360}{2\pi} \text{ degrees} \cong 57.30 \text{ degrees}$$

and

$$1 \text{ degree} = \frac{2\pi}{360} \text{ radians} \cong 0.0175 \text{ radians}.$$

Problem 1. Convert $\frac{1}{10}\pi$ radians to degree measure.

Solution. We set

$$\frac{A}{360} = \frac{\frac{1}{10}\pi}{2\pi}$$

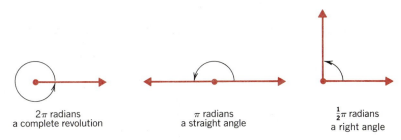

and find that $A = 360(\frac{1}{20}) = 18$. Therefore, $\frac{1}{10}\pi$ radians amounts to 18 degrees.   □

Table A.2.1 gives some common angles measured both in degrees and in radians.

					TABLE A.2.1				
degrees	0	30	45	60	90	120	135	150	180
radians	0	$\frac{1}{6}\pi$	$\frac{1}{4}\pi$	$\frac{1}{3}\pi$	$\frac{1}{2}\pi$	$\frac{2}{3}\pi$	$\frac{3}{4}\pi$	$\frac{5}{6}\pi$	π

The angles used most frequently in simple practice problems are those of a 30–60–90-degree triangle and those of a 45–45–90-degree triangle because at these angles we can evaluate the trigonometric functions without having to consult a table or a pocket calculator.

Problem 2. Find $\sec \frac{1}{6}\pi$. (That is, find the secant of an angle of $\frac{1}{6}\pi$ radians.)

Solution. Since

$$\tfrac{1}{6}\pi \text{ radians} = 30 \text{ degrees,}$$

we can use a 30–60–90-degree triangle:

$$\sec \tfrac{1}{6}\pi = \frac{\text{hypotenuse}}{\text{adjacent}} = \frac{2}{\sqrt{3}}. \quad \square$$

Problem 3. A 12-inch arc on a circle is subtended by a central angle of $40°$. What is the radius of the circle?

Solution. Let r be the radius of the circle measured in inches. The radian measure of the $40°$ angle is $12/r$. Thus

$$\frac{40}{360} = \frac{12/r}{2\pi}.$$

Solving this equation for r, we find that the radius measures $54/\pi$ inches. $\square$

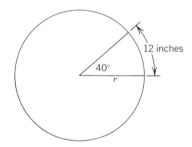

EXERCISES A.2

Convert to radian measure.

1. $30°$.
2. $135°$.
3. $270°$.
4. $120°$.
5. $10°$.
6. $210°$.
7. $225°$.
8. $300°$.

Convert to degree measure.

9. $\frac{1}{4}\pi$.
10. $\frac{7}{4}\pi$.
11. $\frac{7}{12}\pi$.
12. $\frac{2}{3}\pi$.
13. $\frac{1}{3}\pi$.
14. $\frac{11}{6}\pi$.
15. $\frac{5}{6}\pi$.
16. $\frac{1}{12}\pi$.

17. A wheel makes 15 complete revolutions in one minute.
 (a) Through how many degrees does the wheel turn in one second?
 (b) Through how many radians does the wheel turn in 20 seconds?

18. A wheel with spokes 10 centimeters long rolls along the ground, the spokes spinning at a rate of $120°$ per second.
 (a) Through how many radians does the wheel turn in one minute?
 (b) How far does the wheel roll in one minute?

19. How many radians are there in a central angle that subtends an arc of 30 centimeters on a circle of radius one meter?

20. How many radians are there in a central angle that subtends a one foot arc on a circle of radius 5 inches?

21. An arc of a circle of radius 10 inches is subtended by a central angle of $\frac{1}{7}\pi$ radians. What is the length of the arc in inches?

22. An arc of a circle of radius 8 centimeters is subtended by a central angle of $60°$. What is the length of the arc in centimeters?

23. A 10-inch arc on a circle is subtended by a central angle of $\frac{1}{3}\pi$ radians. What is the radius of the circle in inches?

Evaluate.

24. $\sin \frac{1}{4}\pi$.   **25.** $\cos \frac{1}{4}\pi$.   **26.** $\tan \frac{1}{6}\pi$.   **27.** $\csc \frac{1}{3}\pi$.

28. $\sec \frac{1}{6}\pi$.   **29.** $\sin \frac{1}{6}\pi$.   **30.** $\csc \frac{1}{4}\pi$.   **31.** $\tan \frac{1}{4}\pi$.

32. $\cot \frac{1}{4}\pi$.   **33.** $\sec \frac{1}{4}\pi$.   **34.** $\cos \frac{1}{3}\pi$.   **35.** $\cot \frac{1}{3}\pi$.

A.3   INDUCTION

Suppose that you were asked to show that a certain set S contains the set of positive integers. You could start by verifying that $1 \in S$, and $2 \in S$, and $3 \in S$, and so on, but even if each such step took you only one-hundredth of a second, you would still never finish.

To avoid such a bind, mathematicians use a special procedure called *induction*. That induction works is an *assumption* that we make.

AXIOM A.3.1   AXIOM OF INDUCTION

Let S be a set of integers. If

$$\text{(A) } 1 \in S \quad \text{and} \quad \text{(B) } k \in S \text{ implies } k + 1 \in S,$$

then all the positive integers are in S.

You can think of the axiom of induction as a kind of "domino theory." If the first domino falls (Figure A.3.1), and if each domino that falls causes the next one to fall, then, according to the axiom of induction, each domino will fall.

domino theory

Figure A.3.1

While we cannot prove that this axiom is valid (axioms are by their very nature assumptions and therefore not subject to proof), we can argue that it is *plausible*.

Let's assume that we have a set S that satisfies conditions (A) and (B). Now let's choose a positive integer m and "argue" that $m \in S$.

From (A) we know that $1 \in S$. Since $1 \in S$, we know from (B) that $1 + 1 \in S$, and thus that $(1 + 1) + 1 \in S$, and so on. Since m can be obtained from 1 by adding 1 successively $(m - 1)$ times, it *seems clear* that $m \in S$. ☐

As an example of this procedure, note that

$$5 = \{[(1 + 1) + 1] + 1\} + 1$$

and thus $5 \in S$.

Problem 1. Show that

$$\text{if} \quad 0 \le a < b \quad \text{then} \quad a^n < b^n \quad \text{for all positive integers } n.$$

Solution. Suppose that $0 \le a < b$, and let S be the set of positive integers n for which $a^n < b^n$.

Obviously, $1 \in S$. Let's assume now that $k \in S$. This assures us that $a^k < b^k$. It follows that

$$a^{k+1} = a \cdot a^k < a \cdot b^k < b \cdot b^k = b^{k+1}$$

and thus that $k + 1 \in S$.

We have shown that

$$1 \in S \quad \text{and that} \quad k \in S \quad \text{implies} \quad k + 1 \in S.$$

By the axiom of induction, we can conclude that all the positive integers are in S. ☐

Problem 2. Show that, if $x \ge -1$, then

$$(1 + x)^n \ge 1 + nx \quad \text{for all positive integers } n.$$

Solution. Take $x \ge -1$ and let S be the set of positive integers n for which

$$(1 + x)^n \ge 1 + nx.$$

Since

$$(1 + x)^1 \ge 1 + 1 \cdot x,$$

you can see that $1 \in S$.

Assume now that $k \in S$. By the definition of S,

$$(1 + x)^k \ge 1 + kx.$$

Since

$$(1 + x)^{k+1} = (1 + x)^k (1 + x) \ge (1 + kx)(1 + x) \qquad \text{(explain)}$$

and

$$(1 + kx)(1 + x) = 1 + (k + 1)x + kx^2 \geq 1 + (k + 1)x,$$

it follows that

$$(1 + x)^{k+1} \geq 1 + (k + 1)x$$

and thus that $k + 1 \in S$.

We have shown that

$$1 \in S \quad \text{and that} \quad k \in S \quad \text{implies} \quad k + 1 \in S.$$

By the axiom of induction, all the positive integers are in S. □

EXERCISES A.3

For Exercises 1–12, show that the statement holds for all positive integers n.

1. $2n \leq 2^n$.
2. $1 + 2n \leq 3^n$.
3. $n(n + 1)$ is divisible by 2.   [HINT: $(k + 1)(k + 2) = k(k + 1) + 2(k + 1)$.]
4. $n(n + 1)(n + 2)$ is divisible by 6.
5. $1 + 2 + 3 + \cdots + n = \frac{1}{2}n(n + 1)$.
6. $1 + 3 + 5 + \cdots + (2n - 1) = n^2$.
7. $1^2 + 2^2 + 3^2 + \cdots + n^2 = \frac{1}{6}n(n + 1)(2n + 1)$.
8. $1^3 + 2^3 + 3^3 + \cdots + n^3 = (1 + 2 + 3 + \cdots + n)^2$.   [HINT: Use Exercise 5.]
9. $1^3 + 2^3 + \cdots + (n - 1)^3 < \frac{1}{4}n^4 < 1^3 + 2^3 + \cdots + n^3$.
10. $1^2 + 2^2 + \cdots + (n - 1)^2 < \frac{1}{3}n^3 < 1^2 + 2^2 + \cdots + n^2$.

11. $\dfrac{1}{\sqrt{1}} + \dfrac{1}{\sqrt{2}} + \dfrac{1}{\sqrt{3}} + \cdots + \dfrac{1}{\sqrt{n}} > \sqrt{n}.$

12. $\dfrac{1}{1 \cdot 2} + \dfrac{1}{2 \cdot 3} + \dfrac{1}{3 \cdot 4} + \cdots + \dfrac{1}{n(n + 1)} = \dfrac{n}{n + 1}.$

13. For what integers n is $3^{2n+1} + 2^{n+2}$ divisible by 7?   Prove that your answer is correct.
14. For what integers n is $9^n - 8n - 1$ divisible by 64?   Prove that your answer is correct.
15. Find a simplifying expression for the product

$$\left(1 - \frac{1}{2}\right)\left(1 - \frac{1}{3}\right) \cdots \left(1 - \frac{1}{n}\right)$$

and verify its validity for all integers $n \geq 2$.
16. Find a simplifying expression for the product

$$\left(1 - \frac{1}{2^2}\right)\left(1 - \frac{1}{3^2}\right) \cdots \left(1 - \frac{1}{n^2}\right)$$

and verify its validity for all integers $n \geq 2$.
17. Prove that an N-sided convex polygon has $\frac{1}{2}N(N - 3)$ diagonals.
18. Prove that the sum of the angles in an N-sided convex polygon is $(N - 2)180°$.
19. Prove that all sets with n elements have 2^n subsets.   Count the empty set $\varnothing$ as a subset.

APPENDIX

SOME ADDITIONAL PROOFS

In this appendix we present some proofs that many would consider too advanced for the main body of the text. Some details are omitted. These are left to you.

The arguments presented in Sections B.1, B.2, and B.4 require some familiarity with the *least upper bound axiom*. This is discussed in Section 10.8. In addition, Section B.4 requires some understanding of *sequences*, for which we refer you to Sections 11.1 and 11.2.

B.1 THE INTERMEDIATE-VALUE THEOREM

> **LEMMA B.1.1**
>
> Let f be continuous on $[a, b]$. If $f(a) < 0 < f(b)$ or $f(b) < 0 < f(a)$, then there is a number c between a and b for which $f(c) = 0$.

Proof. Suppose that $f(a) < 0 < f(b)$. (The other case can be treated in a similar manner.) Since $f(a) < 0$, we know from the continuity of f that there exists a number ξ such that f is negative on $[a, \xi)$. Let

$$c = \text{lub } \{\xi: f \text{ is negative on } [a, \xi)\}.$$

Clearly, $c \leq b$. We cannot have $f(c) > 0$, for then f would be positive on some interval extending to the left of c, and we know that, to the left of c, f is negative. Incidentally this argument excludes the possibility $c = b$ and means that $c < b$. We cannot have $f(c) < 0$, for then there would be an interval $[a, t)$, with $t > c$, on which f is negative, and this would contradict the definition of c. It follows that $f(c) = 0$. ☐

THEOREM B.1.2   THE INTERMEDIATE-VALUE THEOREM

If f is continuous on $[a, b]$ and C is a number between $f(a)$ and $f(b)$, then there is at least one number c between a and b for which $f(c) = C$.

Proof.   Suppose for example that

$$f(a) < C < f(b).$$

(The other possibility can be handled in a similar manner.)   The function

$$g(x) = f(x) - C$$

is continuous on $[a, b]$.   Since

$$g(a) = f(a) - C < 0 \qquad \text{and} \qquad g(b) = f(b) - C > 0,$$

we know from the lemma that there is a number c between a and b for which $g(c) = 0$.   Obviously, then, $f(c) = C$. ☐

B.2   THE MAXIMUM-MINIMUM THEOREM

LEMMA B.2.1

If f is continuous on $[a, b]$, then f is bounded on $[a, b]$.

Proof.   Consider

$$\{x: \ x \in [a, b] \ \text{and} \ f \ \text{is bounded on} \ [a, x]\}.$$

It is easy to see that this set is nonempty and bounded above by b.   Thus we can set

$$c = \text{lub} \ \{x: \ f \ \text{is bounded on} \ [a, x]\}.$$

Now we argue that $c = b$.   To do so, we suppose that $c < b$.   From the continuity of f at c, it is easy to see that f is bounded on $[c - \epsilon, c + \epsilon]$ for some $\epsilon > 0$.   Being bounded on $[a, c - \epsilon]$ and on $[c - \epsilon, c + \epsilon]$, it is obviously bounded on $[a, c + \epsilon]$.   This contradicts our choice of c.   We can therefore conclude that $c = b$.   This tells us that f is bounded on $[a, x]$ for all $x < b$.   We are now almost through.   From the continuity of f, we know that f is bounded on some interval of the form $[b - \epsilon, b]$.   Since $b - \epsilon < b$, we know from what we have just proved that f is bounded on $[a, b - \epsilon]$.   Being bounded on $[a, b - \epsilon]$ and bounded on $[b - \epsilon, b]$, it is bounded on $[a, b]$. ☐

> ### THEOREM B.2.2   THE MAXIMUM-MINIMUM THEOREM
>
> If f is continuous on $[a, b]$, then f takes on both a maximum value M and a minimum value m on $[a, b]$.

Proof.   By the lemma, f is bounded on $[a, b]$.   Set

$$M = \text{lub} \{f(x): \ x \in [a, b]\}.$$

We must show that there exists c in $[a, b]$ such that $f(c) = M$.   To do this, we set

$$g(x) = \frac{1}{M - f(x)}.$$

If f does not take on the value M, then g is continuous on $[a, b]$ and thus, by the lemma, bounded on $[a, b]$.   A look at the definition of g makes it clear that g cannot be bounded on $[a, b]$.   The assumption that f does not take on the value M has led to a contradiction.   (That f takes on a minimum value m can be proved in a similar manner.)   □

B.3   INVERSES

> ### THEOREM B.3.1   CONTINUITY OF THE INVERSE
>
> Let f be a one-to-one function defined on an interval (a, b).   If f is continuous, then its inverse f^{-1} is also continuous.

Proof.   If f is continuous, then, being one-to-one, f either increases throughout (a, b) or it decreases throughout (a, b).   The proof of this assertion we leave to you.

Let's suppose now that f increases throughout (a, b).   Let's take c in the domain of f^{-1} and show that f^{-1} is continuous at c.

We first observe that $f^{-1}(c)$ lies in (a, b) and choose $\epsilon > 0$ sufficiently small so that $f^{-1}(c) - \epsilon$ and $f^{-1}(c) + \epsilon$ also lie in (a, b).   We seek $\delta > 0$ such that

$$\text{if} \quad c - \delta < x < c + \delta, \quad \text{then} \quad f^{-1}(c) - \epsilon < f^{-1}(x) < f^{-1}(c) + \epsilon.$$

This condition can be met by choosing δ to satisfy

$$f(f^{-1}(c) - \epsilon) < c - \delta \quad \text{and} \quad c + \delta < f(f^{-1}(c) + \epsilon)$$

for then, if $c - \delta < x < c + \delta$, then

$$f(f^{-1}(c) - \epsilon) < x < f(f^{-1}(c) + \epsilon)$$

and, since f^{-1} also increases,

$$f^{-1}(c) - \epsilon < f^{-1}(x) < f^{-1}(c) + \epsilon.$$

The case where f decreases throughout (a, b) can be handled in a similar manner.   □

THEOREM B.3.2 DIFFERENTIABILITY OF THE INVERSE

Let f be a one-to-one function defined on an interval (a, b). If f is differentiable and its derivative does not take on the value 0, then f^{-1} is differentiable and

$$(f^{-1})'(x) = \frac{1}{f'(f^{-1}(x))}.$$

Proof. (Here we use the characterization of derivative spelled out in Theorem 3.6.8.) Let x be in the domain of f^{-1}. We take $\epsilon > 0$ and show that there exists $\delta > 0$ such that

$$\text{if} \quad 0 < |t - x| < \delta, \quad \text{then} \quad \left| \frac{f^{-1}(t) - f^{-1}(x)}{t - x} - \frac{1}{f'(f^{-1}(x))} \right| < \epsilon.$$

Since f is differentiable at $f^{-1}(x)$ and $f'(f^{-1}(x)) \neq 0$, there exists $\delta_1 > 0$ such that

$$\text{if} \quad 0 < |y - f^{-1}(x)| < \delta_1, \quad \text{then} \quad \left| \frac{1}{\dfrac{f(y) - f(f^{-1}(x))}{y - f^{-1}(x)}} - \frac{1}{f'(f^{-1}(x))} \right| < \epsilon$$

and therefore

$$\left| \frac{y - f^{-1}(x)}{f(y) - f(f^{-1}(x))} - \frac{1}{f'(f^{-1}(x))} \right| < \epsilon.$$

By the previous theorem, f^{-1} is continuous at x and therefore there exists $\delta > 0$ such that

$$\text{if} \quad 0 < |t - x| < \delta, \quad \text{then} \quad 0 < |f^{-1}(t) - f^{-1}(x)| < \delta_1.$$

It follows from the special property of δ_1 that

$$\left| \frac{f^{-1}(t) - f^{-1}(x)}{t - x} - \frac{1}{f'(f^{-1}(x))} \right| < \epsilon. \quad \square$$

B.4 THE INTEGRABILITY OF CONTINUOUS FUNCTIONS

The aim here is to prove that, if f is continuous on $[a, b]$, then there is one and only one number I that satisfies the inequality

$$L_f(P) \leq I \leq U_f(P) \qquad \text{for all partitions } P \text{ of } [a, b].$$

DEFINITION B.4.1

A function f is said to be *uniformly continuous* on $[a, b]$ iff for each $\epsilon > 0$ there exists $\delta > 0$ such that

$$\text{if} \quad x, y \in [a, b] \quad \text{and} \quad |x - y| < \delta, \quad \text{then} \quad |f(x) - f(y)| < \epsilon.$$

For convenience, let's agree to say that *the interval* $[a, b]$ *has the property* P_ϵ iff there exist sequences $\{x_n\}$, $\{y_n\}$ satisfying

$$x_n, y_n \in [a, b], \qquad |x_n - y_n| < 1/n, \qquad |f(x_n) - f(y_n)| \geq \epsilon.$$

LEMMA B.4.2

If f is not uniformly continuous on $[a, b]$, then $[a, b]$ has the property P_ϵ for some $\epsilon > 0$.

Proof. If f is not uniformly continuous on $[a, b]$, then there is no $\delta > 0$ such that

$$\text{if} \quad x, y \in [a, b] \quad \text{and} \quad |x - y| < \delta, \qquad \text{then} \quad |f(x) - f(y)| < \epsilon.$$

The interval $[a, b]$ has the property P_ϵ for that choice of ϵ. The details of the argument are left to you. $\square$

LEMMA B.4.3

Let f be continuous on $[a, b]$. If $[a, b]$ has the property P_ϵ, then at least one of the subintervals $[a, \frac{1}{2}(a + b)]$, $[\frac{1}{2}(a + b), b]$ has the property P_ϵ.

Proof. Let's suppose that the lemma is false. For convenience, we let $c = \frac{1}{2}(a + b)$, so that the halves become $[a, c]$ and $[c, b]$. Since $[a, c]$ fails to have the property P_ϵ, there exists an integer p such that

$$\text{if} \quad x, y \in [a, c] \quad \text{and} \quad |x - y| < 1/p, \qquad \text{then} \quad |f(x) - f(y)| < \epsilon.$$

Since $[c, b]$ fails to have the property P_ϵ, there exists an integer q such that

$$\text{if} \quad x, y \in [c, b] \quad \text{and} \quad |x - y| < 1/q, \qquad \text{then} \quad |f(x) - f(y)| < \epsilon.$$

Since f is continuous at c, there exists an integer r such that, if $|x - c| < 1/r$, then $|f(x) - f(c)| < \frac{1}{2}\epsilon$. Set $s = \max\{p, q, r\}$ and suppose that

$$x, y \in [a, b], \qquad |x - y| < 1/s.$$

If x, y are both in $[a, c]$ or both in $[c, b]$, then

$$|f(x) - f(y)| < \epsilon.$$

The only other possibility is that $x \in [a, c]$ and $y \in [c, b]$. In this case we have

$$|x - c| < 1/r, \qquad |y - c| < 1/r,$$

and thus

$$|f(x) - f(c)| < \tfrac{1}{2}\epsilon, \qquad |f(y) - f(c)| < \tfrac{1}{2}\epsilon.$$

By the triangle inequality, we again have

$$|f(x) - f(y)| < \epsilon.$$

In summary, we have obtained the existence of an integer s with the property that

$$x, y \in [a, b], \quad |x - y| < 1/s \quad \text{implies} \quad |f(x) - f(y)| < \epsilon.$$

Hence $[a, b]$ does not have the property P_ϵ. This is a contradiction and proves the lemma. $\square$

THEOREM B.4.4

If f is continuous on $[a, b]$, then f is uniformly continuous on $[a, b]$.

Proof. We suppose that f is not uniformly continuous on $[a, b]$ and base our argument on a mathematical version of the classical maxim "Divide and Conquer."

By the first lemma of this section, we know that $[a, b]$ has the property P_ϵ for some $\epsilon > 0$. We bisect $[a, b]$ and note by the second lemma that one of the halves, say $[a_1, b_1]$, has the property P_ϵ. We then bisect $[a_1, b_1]$ and note that one of the halves, say $[a_2, b_2]$, has the property P_ϵ. Continuing in this manner, we obtain a sequence of intervals $[a_n, b_n]$, each with the property P_ϵ. Then for each n, we can choose $x_n, y_n \in [a_n, b_n]$ such that

$$|x_n - y_n| < 1/n \quad \text{and} \quad |f(x_n) - f(y_n)| \geq \epsilon.$$

Since

$$a \leq a_n \leq a_{n+1} < b_{n+1} \leq b_n \leq b,$$

we see that the sequences $\{a_n\}$ and $\{b_n\}$ are both bounded and monotonic. Thus they are convergent. Since $b_n - a_n \to 0$, we see that $\{a_n\}$ and $\{b_n\}$ both converge to the same limit, say l. From the inequality

$$a_n \leq x_n \leq y_n \leq b_n,$$

we conclude that

$$x_n \to l \quad \text{and} \quad y_n \to l.$$

This tells us that

$$|f(x_n) - f(y_n)| \to |f(l) - f(l)| = 0,$$

which contradicts the statement that $|f(x_n) - f(y_n)| \geq \epsilon$ for all n. $\square$

LEMMA B.4.5

If P and Q are partitions of $[a, b]$, then $L_f(P) \leq U_f(Q)$.

Proof. $P \cup Q$ is a partition of $[a, b]$ that contains both P and Q. It is obvious then that

$$L_f(P) \leq L_f(P \cup Q) \leq U_f(P \cup Q) \leq U_f(Q). \quad \square$$

From the last lemma it follows that the set of all lower sums is bounded above and has a least upper bound L. The number L satisfies the inequality

$$L_f(P) \leq L \leq U_f(P) \qquad \text{for all partitions } P$$

and is clearly the least of such numbers. Similarly, we find that the set of all upper sums is bounded below and has a greatest lower bound U. The number U satisfies the inequality

$$L_f(P) \leq U \leq U_f(P) \qquad \text{for all partitions } P$$

and is clearly the greatest of such numbers.

We are now ready to prove the basic theorem.

THEOREM B.4.6   THE INTEGRABILITY THEOREM

If f is continuous on $[a, b]$, then there exists one and only one number I that satisfies the inequality

$$L_f(P) \leq I \leq U_f(P) \qquad \text{for all partitions } P \text{ of } [a, b].$$

Proof. We know that

$$L_f(P) \leq L \leq U \leq U_f(P) \qquad \text{for all } P,$$

so that existence is no problem. We will have uniqueness if we can prove that

$$L = U.$$

To do this, we take $\epsilon > 0$ and note that f, being continuous on $[a, b]$, is uniformly continuous on $[a, b]$. Thus there exists $\delta > 0$ such that, if

$$x, y \in [a, b] \quad \text{and} \quad |x - y| < \delta, \qquad \text{then} \quad |f(x) - f(y)| < \frac{\epsilon}{b - a}.$$

We now choose a partition $P = \{x_0, x_1, \ldots, x_n\}$ for which max $\Delta x_i < \delta$. For this partition P, we have

$$U_f(P) - L_f(P) = \sum_{i=1}^{n} M_i \, \Delta x_i - \sum_{i=1}^{n} m_i \, \Delta x_i$$

$$= \sum_{i=1}^{n} (M_i - m_i) \, \Delta x_i$$

$$< \sum_{i=1}^{n} \frac{\epsilon}{b - a} \Delta x_i = \frac{\epsilon}{b - a} \sum_{i=1}^{n} \Delta x_i = \frac{\epsilon}{b - a} (b - a) = \epsilon.$$

Since

$$U_f(P) - L_f(P) < \epsilon \qquad \text{and} \qquad 0 \leq U - L \leq U_f(P) - L_f(P),$$

you can see that

$$0 \leq U - L < \epsilon.$$

Since ϵ was chosen arbitrarily, we must have $U - L = 0$ and $L = U$. □

B.5   THE INTEGRAL AS THE LIMIT OF RIEMANN SUMS

For the notation we refer to Section 5.13.

THEOREM B.5.1

If f is continuous on $[a, b]$, then

$$\int_a^b f(x)\, dx = \lim_{\|P\| \to 0} S^*(P).$$

Proof. Let $\epsilon > 0$. We must show that there exists $\delta > 0$ such that

$$\text{if} \quad \|P\| < \delta, \qquad \text{then} \quad \left| S^*(P) - \int_a^b f(x)\, dx \right| < \epsilon.$$

From the proof of Theorem B.4.6 we know that there exists $\delta > 0$ such that

$$\text{if} \quad \|P\| < \delta, \qquad \text{then} \quad U_f(P) - L_f(P) < \epsilon.$$

For such P we have

$$U_f(P) - \epsilon < L_f(P) \leq S^*(P) \leq U_f(P) < L_f(P) + \epsilon.$$

This gives

$$\int_a^b f(x)\, dx - \epsilon < S^*(P) < \int_a^b f(x)\, dx + \epsilon$$

and therefore

$$\left| S^*(P) - \int_a^b f(x)\, dx \right| < \epsilon. \quad \square$$

APPENDIX C

TABLES

TABLE 1.
Natural Logs

x	$\ln x$	x	$\ln x$	x	$\ln x$	x	$\ln x$
		3.0	1.099	6.0	1.792	9.0	2.197
0.1	-2.303	3.1	1.131	6.1	1.808	9.1	2.208
0.2	-1.609	3.2	1.163	6.2	1.825	9.2	2.219
0.3	-1.204	3.3	1.194	6.3	1.841	9.3	2.230
0.4	-0.916	3.4	1.224	6.4	1.856	9.4	2.241
0.5	-0.693	3.5	1.253	6.5	1.872	9.5	2.251
0.6	-0.511	3.6	1.281	6.6	1.887	9.6	2.262
0.7	-0.357	3.7	1.308	6.7	1.902	9.7	2.272
0.8	-0.223	3.8	1.335	6.8	1.917	9.8	2.282
0.9	-0.105	3.9	1.361	6.9	1.932	9.9	2.293
1.0	0.000	4.0	1.386	7.0	1.946	10	2.303
1.1	0.095	4.1	1.411	7.1	1.960	20	2.996
1.2	0.182	4.2	1.435	7.2	1.974	30	3.401
1.3	0.262	4.3	1.459	7.3	1.988	40	3.689
1.4	0.336	4.4	1.482	7.4	2.001	50	3.912
1.5	0.405	4.5	1.504	7.5	2.015	60	4.094
1.6	0.470	4.6	1.526	7.6	2.028	70	4.248
1.7	0.531	4.7	1.548	7.7	2.041	80	4.382
1.8	0.588	4.8	1.569	7.8	2.054	90	4.500
1.9	0.642	4.9	1.589	7.9	2.067	100	4.605
2.0	0.693	5.0	1.609	8.0	2.079		
2.1	0.742	5.1	1.629	8.1	2.092		
2.2	0.788	5.2	1.649	8.2	2.104		
2.3	0.833	5.3	1.668	8.3	2.116		
2.4	0.875	5.4	1.686	8.4	2.128		
2.5	0.916	5.5	1.705	8.5	2.140		
2.6	0.956	5.6	1.723	8.6	2.152		
2.7	0.993	5.7	1.740	8.7	2.163		
2.8	1.030	5.8	1.758	8.8	2.175		
2.9	1.065	5.9	1.775	8.9	2.186		

TABLE 2.
Exponentials (0.01 to 0.99)

x	e^x	e^{-x}	x	e^x	e^{-x}	x	e^x	e^{-x}
0.01	1.010	0.990	0.34	1.405	0.712	0.67	1.954	0.512
0.02	1.020	0.980	0.35	1.419	0.705	0.68	1.974	0.507
0.03	1.030	0.970	0.36	1.433	0.698	0.69	1.994	0.502
0.04	1.041	0.961	0.37	1.448	0.691	0.70	2.014	0.497
0.05	1.051	0.951	0.38	1.462	0.684	0.71	2.034	0.492
0.06	1.062	0.942	0.39	1.477	0.677	0.72	2.054	0.487
0.07	1.073	0.932	0.40	1.492	0.670	0.73	2.075	0.482
0.08	1.083	0.923	0.41	1.507	0.664	0.74	2.096	0.477
0.09	1.094	0.914	0.42	1.522	0.657	0.75	2.117	0.472
0.10	1.105	0.905	0.43	1.537	0.651	0.76	2.138	0.468
0.11	1.116	0.896	0.44	1.553	0.644	0.77	2.160	0.463
0.12	1.127	0.887	0.45	1.568	0.638	0.78	2.181	0.458
0.13	1.139	0.878	0.46	1.584	0.631	0.79	2.203	0.454
0.14	1.150	0.869	0.47	1.600	0.625	0.80	2.226	0.449
0.15	1.162	0.861	0.48	1.616	0.619	0.81	2.248	0.445
0.16	1.174	0.852	0.49	1.632	0.613	0.82	2.270	0.440
0.17	1.185	0.844	0.50	1.649	0.607	0.83	2.293	0.436
0.18	1.197	0.835	0.51	1.665	0.600	0.84	2.316	0.432
0.19	1.209	0.827	0.52	1.682	0.595	0.85	2.340	0.427
0.20	1.221	0.819	0.53	1.699	0.589	0.86	2.363	0.423
0.21	1.234	0.811	0.54	1.716	0.583	0.87	2.387	0.419
0.22	1.246	0.803	0.55	1.733	0.577	0.88	2.411	0.415
0.23	1.259	0.795	0.56	1.751	0.571	0.89	2.435	0.411
0.24	1.271	0.787	0.57	1.768	0.566	0.90	2.460	0.407
0.25	1.284	0.779	0.58	1.786	0.560	0.91	2.484	0.403
0.26	1.297	0.771	0.59	1.804	0.554	0.92	2.509	0.399
0.27	1.310	0.763	0.60	1.822	0.549	0.93	2.535	0.395
0.28	1.323	0.756	0.61	1.840	0.543	0.94	2.560	0.391
0.29	1.336	0.748	0.62	1.859	0.538	0.95	2.586	0.387
0.30	1.350	0.741	0.63	1.878	0.533	0.96	2.612	0.383
0.31	1.363	0.733	0.64	1.896	0.527	0.97	2.638	0.379
0.32	1.377	0.726	0.65	1.916	0.522	0.98	2.664	0.375
0.33	1.391	0.719	0.66	1.935	0.517	0.99	2.691	0.372

TABLE 3.
Exponentials (1.0 to 4.9)

x	e^x	e^{-x}	x	e^x	e^{-x}
1.0	2.718	0.368	3.0	20.086	0.050
1.1	3.004	0.333	3.1	22.198	0.045
1.2	3.320	0.301	3.2	24.533	0.041
1.3	3.669	0.273	3.3	27.113	0.037
1.4	4.055	0.247	3.4	29.964	0.033
1.5	4.482	0.223	3.5	33.115	0.030
1.6	4.953	0.202	3.6	36.598	0.027
1.7	5.474	0.183	3.7	40.447	0.025
1.8	6.050	0.165	3.8	44.701	0.024
1.9	6.686	0.150	3.9	49.402	0.020
2.0	7.389	0.135	4.0	54.598	0.018
2.1	8.166	0.122	4.1	60.340	0.017
2.2	9.025	0.111	4.2	66.686	0.015
2.3	9.974	0.100	4.3	73.700	0.014
2.4	11.023	0.091	4.4	81.451	0.012
2.5	12.182	0.082	4.5	90.017	0.011
2.6	13.464	0.074	4.6	99.484	0.010
2.7	14.880	0.067	4.7	109.947	0.009
2.8	16.445	0.061	4.8	121.510	0.008
2.9	18.174	0.055	4.9	134.290	0.007

TABLE 4.
Sines, Cosines, Tangents (Radian Measure)

x	$\sin x$	$\cos x$	$\tan x$	x	$\sin x$	$\cos x$	$\tan x$
0.00	0.000	1.000	0.000	0.42	0.408	0.913	0.447
0.01	0.010	1.000	0.010	0.43	0.417	0.909	0.459
0.02	0.020	1.000	0.020	0.44	0.426	0.905	0.471
0.03	0.030	1.000	0.030	0.45	0.435	0.900	0.483
0.04	0.040	0.999	0.040	0.46	0.444	0.896	0.495
0.05	0.050	0.999	0.050	0.47	0.453	0.892	0.508
0.06	0.060	0.998	0.060	0.48	0.462	0.887	0.521
0.07	0.070	0.998	0.070	0.49	0.471	0.882	0.533
0.08	0.080	0.997	0.080	0.50	0.479	0.878	0.546
0.09	0.090	0.996	0.090	0.51	0.488	0.873	0.559
0.10	0.100	0.995	0.100	0.52	0.497	0.868	0.573
0.11	0.110	0.994	0.110	0.53	0.506	0.863	0.586
0.12	0.120	0.993	0.121	0.54	0.514	0.858	0.599
0.13	0.130	0.992	0.131	0.55	0.523	0.853	0.613
0.14	0.140	0.990	0.141	0.56	0.531	0.847	0.627
0.15	0.149	0.989	0.151	0.57	0.540	0.842	0.641
0.16	0.159	0.987	0.161	0.58	0.548	0.836	0.655
0.17	0.169	0.986	0.172	0.59	0.556	0.831	0.670
0.18	0.179	0.984	0.182	0.60	0.565	0.825	0.684
0.19	0.189	0.982	0.192	0.61	0.573	0.820	0.699
0.20	0.199	0.980	0.203	0.62	0.581	0.814	0.714
0.21	0.208	0.978	0.213	0.63	0.589	0.808	0.729
0.22	0.218	0.976	0.224	0.64	0.597	0.802	0.745
0.23	0.228	0.974	0.234	0.65	0.605	0.796	0.760
0.24	0.238	0.971	0.245	0.66	0.613	0.790	0.776
0.25	0.247	0.969	0.255	0.67	0.621	0.784	0.792
0.26	0.257	0.966	0.266	0.68	0.629	0.778	0.809
0.27	0.267	0.964	0.277	0.69	0.637	0.771	0.825
0.28	0.276	0.961	0.288	0.70	0.644	0.765	0.842
0.29	0.286	0.958	0.298	0.71	0.652	0.758	0.860
0.30	0.296	0.955	0.309	0.72	0.659	0.752	0.877
0.31	0.305	0.952	0.320	0.73	0.667	0.745	0.895
0.32	0.315	0.949	0.331	0.74	0.674	0.738	0.913
0.33	0.324	0.946	0.343	0.75	0.682	0.732	0.932
0.34	0.333	0.943	0.354	0.76	0.689	0.725	0.950
0.35	0.343	0.939	0.365	0.77	0.696	0.718	0.970
0.36	0.352	0.936	0.376	0.78	0.703	0.711	0.989
0.37	0.362	0.932	0.388	0.79	0.710	0.704	1.009
0.38	0.371	0.929	0.399	0.80	0.717	0.697	1.030
0.39	0.380	0.925	0.411	0.81	0.724	0.689	1.050
0.40	0.389	0.921	0.423	0.82	0.731	0.682	1.072
0.41	0.399	0.917	0.435	0.83	0.738	0.675	1.093

TABLE 4   (continued)

x	$\sin x$	$\cos x$	$\tan x$	x	$\sin x$	$\cos x$	$\tan x$
0.84	0.745	0.667	1.116	1.22	0.939	0.344	2.733
0.85	0.751	0.660	1.138	1.23	0.942	0.334	2.820
0.86	0.758	0.652	1.162	1.24	0.946	0.325	2.912
0.87	0.764	0.645	1.185	1.25	0.949	0.315	3.010
0.88	0.771	0.637	1.210	1.26	0.952	0.306	3.113
0.89	0.777	0.629	1.235	1.27	0.955	0.296	3.224
0.90	0.783	0.622	1.260	1.28	0.958	0.287	3.341
0.91	0.790	0.614	1.286	1.29	0.961	0.277	3.467
0.92	0.796	0.606	1.313	1.30	0.964	0.267	3.602
0.93	0.802	0.598	1.341	1.31	0.966	0.258	3.747
0.94	0.808	0.590	1.369	1.32	0.969	0.248	3.903
0.95	0.813	0.582	1.398	1.33	0.971	0.238	4.072
0.96	0.819	0.574	1.428	1.34	0.973	0.229	4.256
0.97	0.825	0.565	1.459	1.35	0.976	0.219	4.455
0.98	0.830	0.557	1.491	1.36	0.978	0.209	4.673
0.99	0.836	0.549	1.524	1.37	0.980	0.199	4.913
1.00	0.841	0.540	1.557	1.38	0.982	0.190	5.177
1.01	0.847	0.532	1.592	1.39	0.984	0.180	5.471
1.02	0.852	0.523	1.628	1.40	0.985	0.170	5.798
1.03	0.857	0.515	1.665	1.41	0.987	0.160	6.165
1.04	0.862	0.506	1.704	1.42	0.989	0.150	6.581
1.05	0.867	0.498	1.743	1.43	0.990	0.140	7.055
1.06	0.872	0.489	1.784	1.44	0.991	0.130	7.602
1.07	0.877	0.480	1.827	1.45	0.993	0.121	8.238
1.08	0.882	0.471	1.871	1.46	0.994	0.111	8.989
1.09	0.887	0.462	1.917	1.47	0.995	0.101	9.887
1.10	0.891	0.454	1.965	1.48	0.996	0.091	10.983
1.11	0.896	0.445	2.014	1.49	0.997	0.081	12.350
1.12	0.900	0.436	2.066	1.50	0.997	0.071	14.101
1.13	0.904	0.427	2.120	1.51	0.998	0.061	16.428
1.14	0.909	0.418	2.176	1.52	0.999	0.051	19.670
1.15	0.913	0.408	2.234	1.53	0.999	0.041	24.498
1.16	0.917	0.399	2.296	1.54	1.000	0.031	32.461
1.17	0.921	0.390	2.360	1.55	1.000	0.021	48.078
1.18	0.925	0.381	2.427	1.56	1.000	0.011	92.620
1.19	0.928	0.372	2.498	1.57	1.000	0.001	1255.770
1.20	0.932	0.362	2.572	1.58	1.000	-0.009	-108.649
1.21	0.936	0.353	2.650				

TABLE 5.
Sines, Cosines, Tangents (Degree Measure)

x	$\sin x$	$\cos x$	$\tan x$	x	$\sin x$	$\cos x$	$\tan x$
0°	0.00	1.00	0.00	45°	0.71	0.71	1.00
1	0.02	1.00	0.02	46	0.72	0.69	1.04
2	0.03	1.00	0.03	47	0.73	0.68	1.07
3	0.05	1.00	0.05	48	0.74	0.67	1.11
4	0.07	1.00	0.07	49	0.75	0.66	1.15
5	0.09	1.00	0.09	50	0.77	0.64	1.19
6	0.10	0.99	0.11	51	0.78	0.63	1.23
7	0.12	0.99	0.12	52	0.79	0.62	1.28
8	0.14	0.99	0.14	53	0.80	0.60	1.33
9	0.16	0.99	0.16	54	0.81	0.59	1.38
10	0.17	0.98	0.18	55	0.82	0.57	1.43
11	0.19	0.98	0.19	56	0.83	0.56	1.48
12	0.21	0.98	0.21	57	0.84	0.54	1.54
13	0.22	0.97	0.23	58	0.85	0.53	1.60
14	0.24	0.97	0.25	59	0.86	0.52	1.66
15	0.26	0.97	0.27	60	0.87	0.50	1.73
16	0.28	0.96	0.29	61	0.87	0.48	1.80
17	0.29	0.96	0.31	62	0.88	0.47	1.88
18	0.31	0.95	0.32	63	0.89	0.45	1.96
19	0.33	0.95	0.34	64	0.90	0.44	2.05
20	0.34	0.94	0.36	65	0.91	0.42	2.14
21	0.36	0.93	0.38	66	0.91	0.41	2.25
22	0.37	0.93	0.40	67	0.92	0.39	2.36
23	0.39	0.92	0.42	68	0.93	0.37	2.48
24	0.41	0.91	0.45	69	0.93	0.36	2.61
25	0.42	0.91	0.47	70	0.94	0.34	2.75
26	0.44	0.90	0.49	71	0.95	0.33	2.90
27	0.45	0.89	0.51	72	0.95	0.31	3.08
28	0.47	0.88	0.53	73	0.96	0.29	3.27
29	0.48	0.87	0.55	74	0.96	0.28	3.49
30	0.50	0.87	0.58	75	0.97	0.26	3.73
31	0.52	0.86	0.60	76	0.97	0.24	4.01
32	0.53	0.85	0.62	77	0.97	0.22	4.33
33	0.54	0.84	0.65	78	0.98	0.21	4.70
34	0.56	0.83	0.67	79	0.98	0.19	5.14
35	0.57	0.82	0.70	80	0.98	0.17	5.67
36	0.59	0.81	0.73	81	0.99	0.16	6.31
37	0.60	0.80	0.75	82	0.99	0.14	7.12
38	0.62	0.79	0.78	83	0.99	0.12	8.14
39	0.63	0.78	0.81	84	0.99	0.10	9.51
40	0.64	0.77	0.84	85	1.00	0.09	11.43
41	0.66	0.75	0.87	86	1.00	0.07	14.30
42	0.67	0.74	0.90	87	1.00	0.05	19.08
43	0.68	0.73	0.93	88	1.00	0.03	28.64
44	0.69	0.72	0.97	89	1.00	0.02	57.29
45	0.71	0.71	1.00	90	1.00	0.00	—

ANSWERS TO ODD-NUMBERED EXERCISES

SECTION 1.3

1. $-\frac{2}{3}$ **3.** 0   **5.** -1 **7.** $-y_0/x_0$

9. slope 2, y-intercept -4

11. slope $\frac{1}{3}$, y-intercept 2

13. slope undefined, no y-intercept

15. slope $\frac{7}{3}$, y-intercept $\frac{4}{3}$

17. slope 0, y-intercept $\frac{5}{7}$ **19.** $y = 5x + 2$ **21.** $y = -5x + 2$

23. $y = 3$ **25.** $x = -3$ **27.** $y = 7$

29. $3y - 2x - 17 = 0$ **31.** $2y + 3x - 20 = 0$ **33.** $45°$

35. $90°$ **37.** approx. $143°$ **39.** $y = \frac{1}{3}\sqrt{3}\, x + 2$

41. $y = -\sqrt{3}\, x + 3$ **43.** $(\frac{1}{2}\sqrt{2}, \frac{1}{2}\sqrt{2})$, $(-\frac{1}{2}\sqrt{2}, -\frac{1}{2}\sqrt{2})$ **45.** $(3, 4)$

47. $(1, 1)$; approx. $39°$ **49.** $(-\frac{2}{23}, \frac{38}{23})$; approx. $17°$ **51.** $m = -\frac{5}{12}$

SECTION 1.4

1. $(-\infty, 1)$ **3.** $(-\infty, -3]$ **5.** $(-\infty, -\frac{1}{5})$

7. $(-1, 1)$ **9.** $(-\infty, 1) \cup (2, \infty)$ **11.** $(0, 1) \cup (2, \infty)$

13. $[0, \infty)$ **15.** $(-\infty, 2 - \sqrt{3}) \cup (2 + \sqrt{3}, \infty)$

17. $(-5 - 2\sqrt{6}, -5 + 2\sqrt{6})$ **19.** $(-\infty, 0) \cup (0, \infty)$ **21.** $(-1, 0) \cup (1, \infty)$

23. $(-\infty, 0] \cup (5, \infty)$ **25.** $(-\infty, -\frac{5}{3}) \cup (5, \infty)$ **27.** $(-3, -1) \cup (3, \infty)$

29. $(0, 2)$ **31.** $(-\infty, -6) \cup (2, \infty)$ **33.** $(-\infty, 0)$

35. $(-2, 0) \cup (2, \infty)$ **37.** $(-3, 1) \cup (3, \infty)$ **39.** $(1, 2) \cup (6, \infty)$

41. $(-\infty, 1) \cup (3, 5)$ **43.** $(-3, 0) \cup (0, 5)$ **45.** $(-\infty, 0) \cup (0, 1) \cup (3, \infty)$

47. $x < \sqrt{x} < 1 < \dfrac{1}{\sqrt{x}} < \dfrac{1}{x}$

49. $a^2 - 2ab + b^2 = (a - b)^2 \geq 0$; now add $2ab$ to both sides of the equation.

51. $b - a = (\sqrt{b} + \sqrt{a})(\sqrt{b} - \sqrt{a})$; since $\sqrt{b} + \sqrt{a} \geq 0$, $b - a$ and $\sqrt{b} - \sqrt{a}$ have the same sign.

53. With $0 \leq a \leq b$

$$a(1 + b) = a + ab \leq b + ab = b(1 + a).$$

Division by $(1 + b)(1 + a)$ gives the result.

SECTION 1.5

1. $(-2, 2)$

3. $(-\infty, -3) \cup (3, \infty)$

5. $(\frac{3}{2}, \frac{5}{2})$

7. $(-\frac{9}{4}, -\frac{7}{4})$

9. $(-1, 0) \cup (0, 1)$

11. $(\frac{3}{2}, 2) \cup (2, \frac{5}{2})$

13. $(-5, 3) \cup (3, 11)$

15. $(1, 2)$

17. $(-\frac{5}{8}, -\frac{3}{8})$

19. $(-\infty, -4) \cup (-1, \infty)$

21. $(-\infty, -\frac{8}{5}) \cup (2, \infty)$

23. $|x| < 3$

25. $|x - 2| < 5$

27. $|x + 2| < 5$

29. $0 < A \le \frac{3}{2}$

31. $A \ge 6$

33. $\big||a| - |b|\big|^2 = (|a| - |b|)^2 = |a|^2 - 2|a||b| + |b|^2 = a^2 - 2|a||b| + b^2 \le a^2 - 2ab + b^2 = (a - b)^2$

$\qquad\qquad (ab \le |ab|)$

Thus, $\big||a| - |b|\big| \le \sqrt{(a - b)^2} = |a - b|$.

SECTION 1.6

1. $x = 1, 3$

3. $x = -2$

5. $x = n\pi,$ n an integer

7. $x = n\pi,$ n an integer

9. dom $(f) = (-\infty, \infty)$; ran $(f) = [0, \infty)$

11. dom $(f) = (-\infty, \infty)$; ran $(f) = (-\infty, \infty)$

13. dom $(f) = (-\infty, 0) \cup (0, \infty)$; ran $(f) = (0, \infty)$

15. dom $(f) = (-\infty, 1]$; ran $(f) = [0, \infty)$

17. dom $(f) = (-\infty, 1]$; ran $(f) = [-1, \infty)$

19. dom $(f) = (-\infty, 1)$; ran $(f) = (0, \infty)$

21. dom $(f) = (-\infty, \infty)$; ran $(f) = [0, 1]$

23. dom $(f) = (-\infty, \infty)$; ran $(f) = [-2, 2]$

25. dom $(f) = \left\{x: x \ne \dfrac{2n + 1}{2}\, \pi, n \text{ an integer}\right\}$; ran $(f) = [1, \infty)$

27. horizontal line one unit above x-axis

29. line through the origin with slope 2

31. line through the origin with slope $\frac{1}{2}$

33. line through $(0, 2)$ with slope $\frac{1}{2}$

35. upper semicircle of radius 2 centered at the origin

37. Figure A.1.6.1

39. Figure A.1.6.2

41. Figure A.1.6.3

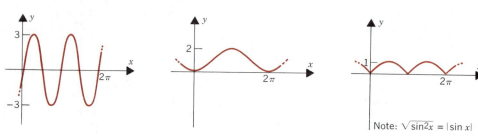

Figure A.1.6.1   Figure A.1.6.2   Figure A.1.6.3

Note: $\sqrt{\sin^2 x} = |\sin x|$

43. dom $(f) = (-\infty, 0) \cup (0, \infty)$; ran $(f) = \{-1, 1\}$; Figure A.1.6.4

45. dom $(f) = [0, \infty)$; ran $(f) = [1, \infty)$; Figure A.1.6.5

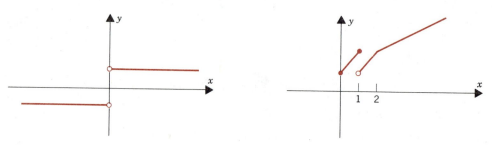

Figure A.1.6.4   Figure A.1.6.5

47. (a) $(6f + 3g)(x) = 9\sqrt{x}, \quad x > 0$ (b) $(fg)(x) = x - 2/x - 1, \quad x > 0$

(c) $(f/g)(x) = \dfrac{x+1}{x-2}, \quad 0 < x < 2 \text{ and } x > 2$

49. (a) $(f + g)(x) = \begin{cases} 1 - x, & x \le 1 \\ 2x - 1, & 1 < x < 2 \\ 2x - 2, & x \ge 2 \end{cases}$ (b) $(f - g)(x) = \begin{cases} 1 - x, & x \le 1 \\ 2x - 1, & 1 < x < 2 \\ 2x, & x \ge 2 \end{cases}$

(c) $(fg)(x) = \begin{cases} 0, & x < 2 \\ 1 - 2x, & x \ge 2 \end{cases}$

51. Figure A.1.6.6

53. Figure A.1.6.7

Figure A.1.6.6

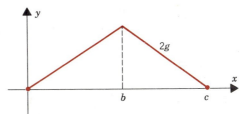

Figure A.1.6.7

55. Figure A.1.6.8

57. Figure A.1.6.9

Figure A.1.6.8

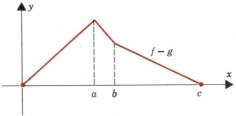

Figure A.1.6.9

59. odd   **61.** neither   **63.** even   **65.** odd

67. even   **69.** even; $(fg)(-x) = f(-x)g(-x) = f(x)g(x) = (fg)(x)$

71. (a) $f(x) = \begin{cases} 1, & x < -1 \\ -x, & -1 \le x < 0 \end{cases}$ (b) $f(x) = \begin{cases} -1, & x < -1 \\ x, & -1 \le x < 0 \end{cases}$

73. $g(-x) = f(-x) + f(-(-x)) = f(-x) + f(x) = g(x)$

SECTION 1.7

1. $(f \circ g)(x) = 2x^2 + 5$

3. $(f \circ g)(x) = \sqrt{x^2 + 5}$

5. $(f \circ g)(x) = x, \quad x \ne 0$

7. $(f \circ g)(x) = 1/x - 1$

9. $(f \circ g)(x) = \dfrac{1}{x^2 + 1}$

11. $(f \circ g \circ h)(x) = 4(x^2 - 1)$

13. $(f \circ g \circ h)(x) = (x^4 - 1)^2$

15. $(f \circ g \circ h)(x) = 2x^2 + 1$

17. $f(x) = \dfrac{1}{x}$

19. $f(x) = 2 \sin x$

21. $g(x) = \left(1 - \dfrac{1}{x^4}\right)^{2/3}$

23. $g(x) = 2x^3 - 1$; $g(x) = -(2x^3 - 1)$
25. $(f \circ g)(x) = |x|$; $(g \circ f)(x) = x$, $x \geq 0$
27. $(f \circ g)(x) = 2 \sin x$; $(g \circ f)(x) = \sin 2x$

29. $(f \circ g)(x) = \begin{cases} x^2, & x < 0 \\ 1 + x, & 0 \leq x < 1 \\ (1 + x)^2, & x \geq 1 \end{cases}$;

$(g \circ f)(x) = \begin{cases} 2 - x, & x \leq 0 \\ -x^2, & 0 < x < 1 \\ 1 + x^2, & x \geq 1 \end{cases}$

31. Figure A.1.7.1

	f_1	f_2	f_3	f_4	f_5	f_6
f_1	f_1	f_2	f_3	f_4	f_5	f_6
f_2	f_2	f_1	f_4	f_3	f_6	f_5
f_3	f_3	f_5	f_1	f_6	f_2	f_4
f_4	f_4	f_6	f_2	f_5	f_1	f_3
f_5	f_5	f_3	f_6	f_1	f_4	f_2
f_6	f_6	f_4	f_5	f_2	f_3	f_1

Figure A.1.7.1

SECTION 1.8

1. $f^{-1}(x) = \frac{1}{5}(x - 3)$ **3.** $f^{-1}(x) = \frac{1}{4}(x + 7)$ **5.** not one-to-one
7. $f^{-1}(x) = (x - 1)^{1/5}$ **9.** $f^{-1}(x) = [\frac{1}{3}(x - 1)]^{1/3}$ **11.** $f^{-1}(x) = 1 - x^{1/3}$
13. $f^{-1}(x) = (x - 2)^{1/3} - 1$ **15.** $f^{-1}(x) = x^{5/3}$ **17.** $f^{-1}(x) = \frac{1}{3}(2 - x^{1/3})$
19. $f^{-1}(x) = 1/x$ **21.** not one-to-one   **23.** $f^{-1}(x) = (1/x - 1)^{1/3}$
25. $f^{-1}(x) = (2 - x)/(x - 1)$ **27.** they are equal
29. Figure A.1.8.1   **31.** Figure A.1.8.2   **33.** Figure A.1.8.3

Figure A.1.8.1

Figure A.1.8.2

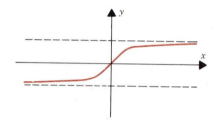

Figure A.1.8.3

SECTION 2.1

1. (a) 2   (b) -1 (c) does not exist   (d) -3
3. (a) does not exist   (b) -3 (c) does not exist   (d) -3
5. (a) does not exist   (b) does not exist   (c) does not exist   (d) 1
7. (a) 2   (b) 2   (c) 2   (d) -1
9. (a) -1 (b) -1 (c) -1 (d) undefined
11. (a) 0   (b) 0   (c) 0   (d) 0
13. $c = 0, 6$ **15.** -1 **17.** 4   **19.** 1
21. $\frac{3}{2}$ **23.** does not exist   **25.** 2   **27.** does not exist
29. 1   **31.** does not exist   **33.** 2   **35.** 2
37. 0   **39.** 1   **41.** 16   **43.** does not exist
45. does not exist   **47.** does not exist   **49.** $1/\sqrt{2}$ **51.** 1   **53.** $\frac{3}{2}$

SECTION 2.2

1. $\frac{1}{2}$ **3.** does not exist   **5.** $\frac{4}{5}\sqrt{5}$ **7.** 4
9. does not exist   **11.** -1 **13.** 1   **15.** 1

17. δ_1 and δ_2 **19.** $\frac{1}{2}\epsilon$ **21.** 2ϵ

23. Take $\delta = \frac{1}{2}\epsilon$. If $0 < |x - 4| < \frac{1}{2}\epsilon$, then $|(2x - 5) - 3| = 2|x - 4| < \epsilon$.

25. Take $\delta = \frac{1}{6}\epsilon$. If $0 < |x - 3| < \frac{1}{6}\epsilon$, then $|(6x - 7) - 11| = 6|x - 3| < \epsilon$.

27. Take $\delta = \frac{1}{3}\epsilon$. If $0 < |x - 2| < \frac{1}{3}\epsilon$, then $\big||1 - 3x| - 5\big| \le 3|x - 2| < \epsilon$.

29. Statements (b), (e), (g), and (i) are necessarily true.

31. (i) $\displaystyle\lim_{x\to3}\frac{1}{x-1} = \frac{1}{2}$ (ii) $\displaystyle\lim_{x\to3}\left(\frac{1}{x-1} - \frac{1}{2}\right) = 0$

    (iii) $\displaystyle\lim_{x\to3}\left|\frac{1}{x-1} - \frac{1}{2}\right| = 0$ (iv) $\displaystyle\lim_{h\to0}\frac{1}{(3+h)-1} = \frac{1}{2}$

33. (i) and (iii) of (2.2.5) with $l = 0$

35. Let $\epsilon > 0$. If $\displaystyle\lim_{x\to c} f(x) = l$, then there must exist $\delta > 0$ such that

(*)   if $0 < |x - c| < \delta$ then $|f(x) - l| < \epsilon$.

Suppose now that $0 < |h| < \delta$. Then $0 < |(c + h) - c| < \delta$, and thus by (*), $|f(c + h) - l| < \epsilon$. This proves that, if $\displaystyle\lim_{x\to c} f(x) = l$, then $\displaystyle\lim_{h\to0} f(c + h) = l$.

If, on the other hand, $\displaystyle\lim_{h\to0} f(c + h) = l$, then there must exist $\delta > 0$ such that

(**)   if $0 < |h| < \delta$ then $|f(c + h) - l| < \epsilon$.

Suppose now that $0 < |x - c| < \delta$. Then by (**), $|f(c + (x - c)) - l| < \epsilon$. More simply stated, $|f(x) - l| < \epsilon$. This proves that, if $\displaystyle\lim_{h\to0} f(c + h) = l$, then $\displaystyle\lim_{x\to c} f(x) = l$.

37. (a) Set $\delta = \epsilon\sqrt{c}$. By the hint

$$\text{if} \quad 0 < |x - c| < \epsilon\sqrt{c}, \quad \text{then} \quad |\sqrt{x} - \sqrt{c}| < \frac{1}{\sqrt{c}}|x - c| < \epsilon.$$

    (b) Set $\delta = \epsilon^2$. If $0 < x < \epsilon^2$, then $|\sqrt{x} - 0| = \sqrt{x} < \epsilon$.

39. Take $\delta = $ minimum of 1 and $\epsilon/7$. If $0 < |x - 1| < \delta$, then $0 < x < 2$ and $|x - 1| < \epsilon/7$. Therefore $|x^3 - 1| = |x^2 + x + 1||x - 1| < 7|x - 1| < 7(\epsilon/7) = \epsilon$.

41. Set $\delta = \epsilon^2$. If $3 - \epsilon^2 < x < 3$, then $-\epsilon^2 < x - 3$, $0 < 3 - x < \epsilon^2$ and therefore $|\sqrt{3 - x} - 0| < \epsilon$.

43. Suppose, on the contrary, that $\displaystyle\lim_{x\to c} f(x) = l$ for some particular c. Taking $\epsilon = \frac{1}{2}$, there must exist $\delta > 0$ such that,

$$\text{if} \quad 0 < |x - c| < \delta, \quad \text{then} \quad |f(x) - l| < \tfrac{1}{2}.$$

Let x_1 be a rational number satisfying $0 < |x_1 - c| < \delta$, and x_2 an irrational number satisfying $0 < |x_2 - c| < \delta$. (That such numbers exist follows from the fact that every interval contains both rational and irrational numbers.) Now $f(x_1) = 1$ and $f(x_2) = 0$. Thus we must have both $|1 - l| < \frac{1}{2}$ and $|0 - l| < \frac{1}{2}$. From the first inequality we conclude that $l > \frac{1}{2}$. From the second, we conclude that $l < \frac{1}{2}$. Clearly no such number l exists.

45. We begin by assuming that $\displaystyle\lim_{x\to c^+} f(x) = l$ and showing that $\displaystyle\lim_{h\to0} f(c + |h|) = l$.

Let $\epsilon > 0$. Since $\displaystyle\lim_{x\to c^+} f(x) = l$, there exists $\delta > 0$ such that

(*)   if $c < x < c + \delta$ then $|f(x) - l| < \epsilon$.

Suppose now that $0 < |h| < \delta$. Then $c < c + |h| < c + \delta$ and, by (*), $|f(c + |h|) - l| < \epsilon$. Thus $\displaystyle\lim_{h\to0} f(c + |h|) = l$.

Conversely let's assume that $\displaystyle\lim_{h\to0} f(c + |h|) = l$ and again take $\epsilon > 0$. Then there exists $\delta > 0$ such that

(**)   if $0 < |h| < \delta$ then $|f(c + |h|) - l| < \epsilon$.

Suppose now that $c < x < c + \delta$. Then $0 < x - c < \delta$ so that, by (**),

$$|f(x) - l| = |f(c + (x - c)) - l| < \epsilon.$$

Thus $\displaystyle\lim_{x\to c^+} f(x) = l$.

SECTION 2.3

1. (a) 3   (b) 4   (c) -2 (d) 0   (e) does not exist   (f) $\frac{1}{3}$

3. $\lim\limits_{x \to 4}\left[\left(\frac{1}{x} - \frac{1}{4}\right)\left(\frac{1}{x-4}\right)\right] = \lim\limits_{x \to 4}\left[\left(\frac{4-x}{4x}\right)\left(\frac{1}{x-4}\right)\right] = \lim\limits_{x \to 4}\frac{-1}{4x} = -\frac{1}{16};$

Theorem 2.3.2 does not apply since $\lim\limits_{x \to 4}\dfrac{1}{x-4}$ does not exist.

5. 3   **7.** -3 **9.** 5   **11.** 2

13. does not exist   **15.** $-\frac{23}{20}$ **17.** 0   **19.** -1

21. $\frac{1}{4}$ **23.** 0   **25.** 0   **27.** does not exist

29. does not exist   **31.** -1 **33.** 4   **35.** a/b

37. $\frac{5}{4}$ **39.** does not exist   **41.** 25   **43.** does not exist

45. 2   **47.** (a) 0   (b) $-\frac{1}{16}$ (c) 0   (d) does not exist

49. (a) 4   (b) -2 (c) 2   (d) does not exist   **51.** $f(x) = 1/x,\ g(x) = -1/x$

53. true   **55.** true   **57.** false   **59.** false

61. If $\lim\limits_{x \to c} f(x) = l$ and $\lim\limits_{x \to c} g(x) = l$, then

$$\lim\limits_{x \to c} h(x) = \lim\limits_{x \to c} \tfrac{1}{2}\{[f(x) + g(x)] - |f(x) - g(x)|\}$$
$$= \lim\limits_{x \to c} \tfrac{1}{2}[f(x) + g(x)] - \lim\limits_{x \to c}|f(x) - g(x)| = \tfrac{1}{2}(l + l) - \tfrac{1}{2}|l - l| = l.$$

A similar argument works for H.

63. Suppose on the contrary that the limit exists and is some number l: $\lim\limits_{x \to c} f(x) = l$. Since $\lim\limits_{x \to c} g(x) = 0$, we have

$$\lim\limits_{x \to c} [f(x)g(x)] = \left(\lim\limits_{x \to c} f(x)\right)\left(\lim\limits_{x \to c} g(x)\right) = l \cdot 0 = 0.$$

But since $f(x)g(x) = 1$ for all real x, we know that $\lim\limits_{x \to c} [f(x)g(x)] = 1$. This contradicts the uniqueness of limit.

SECTION 2.4

1. continuous   **3.** continuous

5. continuous   **7.** removable discontinuity

9. essential discontinuity   **11.** continuous

13. essential discontinuity   **15.** removable discontinuity at 2

17. no discontinuities   **19.** essential discontinuity at 1

21. no discontinuities   **23.** no discontinuities

25. essential discontinuities at 0 and 2

27. removable discontinuity at -2; essential discontinuity at 3

29. $f(1) = 2$ **31.** impossible

33. 4   **35.** $A - B = 3$ with $B \neq 3$

37. set for instance $f(x) = \begin{cases} 0, & x \leq \frac{1}{2} \\ 1, & x > \frac{1}{2} \end{cases}$ **39.** $f(5) = \frac{1}{6}$

41. $f(5) = \frac{1}{3}$ **43.** nowhere

45. $x = 0$, $x = 2$, and all nonintegral values of x

47. Refer to (2.2.5). Use the equivalence of (i) to (iv) setting $l = f(c)$.

49. Let $\epsilon > 0$ be arbitrary and let A be the billion points on which we have changed the value of f. In $A - \{c\}$ there is one point closest to c. Call it d and set $\delta_1 = |c - d|$. Note that if $0 < |x - c| < \delta_1$, then $f(x) = g(x)$.

Suppose now that g is continuous at c. Then there exists a positive number δ less than δ_1 such that

$$\text{if}\quad 0 < |x - c| < \delta, \qquad \text{then}\quad |g(x) - g(c)| < \epsilon.$$

But for such x, $f(x) = g(x)$. Therefore we see that

$$\text{if}\quad 0 < |x - c| < \delta, \qquad \text{then}\quad |f(x) - g(c)| < \epsilon.$$

This means that

$$\lim_{x \to c} f(x) = g(c)$$

and contradicts the assumption that f has an essential discontinuity at c.

SECTION 2.5

1. 3

3. $\frac{3}{5}$

5. 0

7. does not exist

9. $\frac{9}{5}$

11. $\frac{2}{3}$

13. 1

15. $\frac{1}{2}$

17. -4

19. 1

21. $\frac{3}{5}$

23. 0

25. $\dfrac{2}{\pi} \sqrt{2}$

27. -1

29. 0

31. We will show that

$$\lim_{x \to c} \cos x = \cos c \quad \text{by showing that} \quad \lim_{h \to 0} \cos (c + h) = \cos c.$$

Note that $\cos (c + h) = \cos c \cos h - \sin c \sin h$. We know that

$$\lim_{h \to 0} \cos h = 1 \quad \text{and} \quad \lim_{h \to 0} \sin h = 0.$$

Therefore

$$\lim_{h \to 0} \cos (c + h) = (\cos c)(\lim_{h \to 0} \cos h) - (\sin c)(\lim_{h \to 0} \sin h) = (\cos c)(1) - (\sin c)(0) = \cos c.$$

Here is a different proof. The addition formula for the sine gives $\cos x = \sin (\frac{1}{2}\pi + x)$. Being the composition of functions that are everywhere continuous, the cosine function is itself everywhere continuous. Therefore

$$\lim_{x \to c} \cos x = \cos c \quad \text{for all real } c.$$

33. $0 \le |xf(x)| \le M|x|$ for $x \ne 0$

35. $0 \le |f(x) - l| \le M|x - c|$ for $x \ne c$

37. $0 \le |x \sin (1/x)| \le |x|$ for $x \ne 0$

39. $0 \le |\sqrt{x} - \sqrt{c}| \le \dfrac{1}{\sqrt{c}}|x - c|$ for $x \ge 0$

SECTION 2.6

1.

3.

5.

7.

9.

11.

13. impossible by the intermediate-value theorem

15.

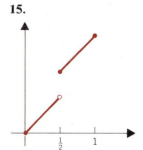

17. Set $g(x) = x - f(x)$. Since g is continuous on $[0, 1]$ and $g(0) \leq 0 \leq g(1)$, there exists c in $[0, 1]$ such that $g(c) = c - f(c) = 0$.

19. Set $h(x) = f(x) - g(x)$. Since h is continuous on $[0, 1]$ and $h(0) < 0 < h(1)$, there exists c between 0 and 1 such that $h(c) = f(c) - g(c) = 0$.

21. $m_{10} = 1.7314$ **23.** $m_{10} = 0.6181$ **25.** $m_{12} = 1.9997$ **27.** $m_{10} = 0.7392$

SECTION 2.7

1. does not exist   **3.** 0   **5.** 6   **7.** -1

9. 1   **11.** 0   **13.** 1   **15.** does not exist

17. does not exist   **19.** 0   **21.** $-\frac{1}{2}$ **23.** 1

25. $-\frac{1}{16}$ **27.** does not exist   **29.** 0   **31.** 0

33. $\frac{1}{6}$ **35.** $\frac{5}{2}$ **37.** $\frac{4}{9}$ **39.** -3

41. $-\frac{1}{2}$ **43.** $\frac{5}{2}$ **45.** $\frac{1}{3}$ **47.** does not exist

49. -3 **51.** 2   **53.** $\frac{1}{2}$ **55.** does not exist

57. $-\frac{3}{8}$ **59.** 6   **61.** 6   **63.** 1

65. (a) Figure A.2.7.1
    (b) (i) 1   (ii) 1   (iii) 1   (iv) 5
       (v) 5   (vi) 5   (vii) 9   (viii) 4
       (ix) does not exist
    (c) (i) at 1 and 2   (ii) at 1
       (iii) at 1   (iv) at 2

67. 0

69. Take $\delta = \frac{1}{5}\epsilon$. If $0 < |x - 2| < \frac{1}{5}\epsilon$, then $|(5x - 4) - 6| = 5|x - 2| < \epsilon$.

71. Take $\delta = \frac{1}{2}\epsilon$. If $0 < |x + 4| < \frac{1}{2}\epsilon$, then $||2x + 5| - 3| \leq 2|x + 4| < \epsilon$.

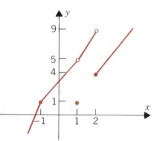

Figure A.2.7.1

73. Take $\delta = $ minimum of 4 and ϵ. If $0 < |x - 9| < \delta$, then $x - 5 > 0$ and $|x - 9| < \epsilon$. It follows that

$$|\sqrt{x - 5} - 2||\sqrt{x - 5} + 2| = |(x - 5) - 4| = |x - 9| < \epsilon$$

and, since $|\sqrt{x - 5} + 2| > 1$, that $|\sqrt{x - 5} - 2| < \epsilon$.

75. The polynomial $P(x) = x^5 - 4x + 1$ is continuous on $[0, 2]$. (Polynomials are everywhere continuous.) Since $P(0) < 7.21 < P(2)$, we know from the intermediate-value theorem that there is a number x_0 between 0 and 2 such that $P(x_0) = 7.21$.

SECTION 3.1

1. 0   **3.** -3 **5.** $5 - 2x$ **7.** $4x^3$

9. $\frac{1}{2}(x - 1)^{-1/2}$ **11.** $-2x^{-3}$ **13.** -6 **15.** $-\frac{1}{4}$ **17.** $\frac{3}{2}$

19. tangent $y - 4x + 4 = 0$; normal $4y + x - 18 = 0$ **21.** tangent $y + 3x - 16 = 0$; normal $3y - x - 8 = 0$

23. tangent $y + x = 0$; normal $y - x - 8 = 0$ **25.** tangent $y + x + 4 = 0$; normal $y - x - 2 = 0$

27. Figure A.3.1.1; $x = -1$ **29.** Figure A.3.1.2; $x = 0$ **31.** Figure A.3.1.3; $x = 1$

Figure A.3.1.1

Figure A.3.1.2

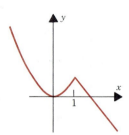

Figure A.3.1.3

33. 4

35. does not exist

37. Figure A.3.1.4

39. Figure A.3.1.5

41. Figure A.3.1.6

Figure A.3.1.4

Figure A.3.1.5

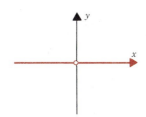

Figure A.3.1.6

43. Since $f(1) = 1$ and $\lim\limits_{x \to 1^+} f(x) = 2$, f is not continuous at 1 and thus, by (3.1.4), is not differentiable at 1.

45. (a) $f'(x) = \begin{cases} 2(x + 1), & x < 0 \\ 2(x - 1), & x > 0 \end{cases}$

(b) $\lim\limits_{h \to 0^-} \dfrac{f(0 + h) - f(0)}{h} = \lim\limits_{h \to 0^-} \dfrac{(h + 1)^2 - 1}{h} = \lim\limits_{h \to 0^-} (h + 2) = 2,$

$\lim\limits_{h \to 0^+} \dfrac{f(0 + h) - f(0)}{h} = \lim\limits_{h \to 0^+} \dfrac{(h - 1)^2 - 1}{h} = \lim\limits_{h \to 0^+} (h - 2) = -2$

In 47–53, there are many possible answers. Here are some.

47. $f(x) = c$, c any constant

49. $f(x) = |x + 1|$; $f(x) = \begin{cases} 0, & x \neq -1 \\ 1, & x = -1 \end{cases}$

51. $f(x) = 2x + 5$

53. $f(x) = \begin{cases} 0, & x \text{ rational} \\ 1, & x \text{ irrational} \end{cases}$

SECTION 3.2

1. -1

3. $55x^4 - 18x^2$

5. $2ax + b$

7. $\dfrac{2}{x^3}$

9. $3x^2 - 6x - 1$

11. $\dfrac{3x^2 - 2x^3}{(1 - x)^2}$

13. $\dfrac{2(x^2 + 3x + 1)}{(2x + 3)^2}$

15. $2x - 3$

17. $-\dfrac{2(3x^2 - x + 1)}{x^2(x - 2)^2}$

19. $-80x^9 + 81x^8 - 64x^7 + 63x^6$

21. $f'(0) = -\frac{1}{4}$, $f'(1) = -1$

23. $f'(0) = 0$, $f'(1) = -1$

25. $f'(0) = \dfrac{ad - bc}{d^2}$, $f'(1) = \dfrac{ad - bc}{(c + d)^2}$

27. $f'(0) = 3$

29. $f'(0) = \frac{20}{9}$

31. $2y - x - 8 = 0$

33. $y - 4x + 12 = 0$

35. $9y + 4x - 18 = 0$

37. $(-1, 27), (3, -5)$

39. $(-1, -\frac{5}{2}), (1, \frac{5}{2})$

41. $(2, 3)$

43. $(-2, -10)$

45. $(-1, -2), (\frac{5}{3}, \frac{50}{27})$

47. $\frac{425}{8}$

49. $A = -1, B = 0, C = 4$

51. Since

$$\left(\frac{f}{g}\right)(x) = \frac{f(x)}{g(x)} = f(x) \cdot \frac{1}{g(x)},$$

it follows from the product and reciprocal rules that

$$\left(\frac{f}{g}\right)'(x) = \left(f \cdot \frac{1}{g}\right)'(x) = f(x)\left(-\frac{g'(x)}{[g(x)]^2}\right) + f'(x) \cdot \frac{1}{g(x)} = \frac{g(x)f'(x) - f(x)g'(x)}{[g(x)]^2}.$$

53. $g(x) = [f(x)]^2 = f(x) \cdot f(x)$, $g'(x) = f(x)f'(x) + f(x)f'(x) = 2 f(x)f'(x)$

55. $A = -2, B = -8$

SECTION 3.3

1. $\dfrac{dy}{dx} = 12x^3 - 2x$

3. $\dfrac{dy}{dx} = 1 + \dfrac{1}{x^2}$

5. $\dfrac{dy}{dx} = \dfrac{1 - x^2}{(1 + x^2)^2}$

7. $\dfrac{dy}{dx} = \dfrac{2x - x^2}{(1 - x)^2}$

9. $\dfrac{dy}{dx} = \dfrac{-6x^2}{(x^3 - 1)^2}$

11. 2

13. $18x^2 + 30x + 5x^{-2}$

15. $\dfrac{-4t}{(t^2 - 1)^2}$

17. $\dfrac{2t^3(t^3 - 2)}{(2t^3 - 1)^2}$

19. $\dfrac{2}{(1 - 2u)^2}$

21. $-\left[\dfrac{1}{(u - 1)^2} + \dfrac{1}{(u + 1)^2}\right] = -2\left[\dfrac{u^2 + 1}{(u^2 - 1)^2}\right]$

23. $2x\left[\dfrac{1}{(1 - x^2)^2} + \dfrac{1}{x^4}\right]$

25. $\dfrac{-2}{(x - 1)^2}$

27. 47

29. $\frac{1}{4}$

31. $42x - 120x^3$

33. $-6x^{-3}$

35. $4 - 12x^{-4}$

37. 2

39. 0

41. $6 + 60x^{-6}$

43. $1 - 4x$

45. -24

47. -24

49. If $f(x) = g(x) = x$, then $(fg)(x) = x^2$, so that $(fg)''(x) = 2$, but $f(x)g''(x) + f''(x)g(x) = x \cdot 0 + 0 \cdot x = 0$.

51. (a) $x = 0$ (b) $x > 0$ (c) $x < 0$

53. (a) $x = -2, x = 1$ (b) $x < -2, x > 1$ (c) $-2 < x < 1$

55. The result is true for $n = 1$:

$$\frac{d^1 y}{dx^1} = \frac{dy}{dx} = -x^{-2} = (-1)^1 1! \, x^{-1-1}.$$

If the result is true for $n = k$:

$$\frac{d^k y}{dx^k} = (-1)^k k! \, x^{-k-1},$$

then the result is true for $n = k + 1$:

$$\frac{d^{k+1} y}{dx^{k+1}} = \frac{d}{dx}\left[\frac{d^k y}{dx^k}\right] = (-1)^k k! \, (-k - 1)x^{-k-2} = (-1)^{k+1}(k + 1)! \, x^{-(k+1)-1}.$$

SECTION 3.4

1. $\dfrac{dA}{dr} = 2\pi r, \quad 4\pi$

3. $\dfrac{dA}{dz} = z, \quad 4$

5. $-\frac{5}{36}$

7. $\dfrac{dV}{dr} = 4\pi r^2$

9. $x_0 = \frac{3}{4}$

11. (a) $\dfrac{3\sqrt{2}}{4} w^2$ (b) $\dfrac{\sqrt{3}}{3} z^2$

13. (a) $\frac{1}{2}r^2$ (b) $r\theta$ (c) $-4Ar^{-3} = -2\theta/r$

15. decreasing at the rate of 225 cm³/sec

17. $x = \frac{1}{2}$

SECTION 3.5

1. $x(5) = -6, v(5) = -7, a(5) = -2$, speed = 7

3. $x(2) = -4, v(2) = 6, a(2) = 12$, speed = 6

5. $x(1) = 6, v(1) = -2, a(1) = \frac{4}{3}$, speed = 2

7. $x(1) = 0, v(1) = 18, a(1) = 54$, speed = 18

9. never

11. at $-2 + \sqrt{5}$

13. at $2\sqrt{2}$

15. A

17. A

19. A and B

21. A

23. A and C

25. $(0, 2), (7, \infty)$

27. $(0, 3), (4, \infty)$

29. $(2, 5)$

31. $(0, 2 - \frac{2}{3}\sqrt{3}), (4, \infty)$

33. 576 ft

35. $v_0^2/2g$

37. clear from the energy equation $mgy + \frac{1}{2}mv^2 = C$ since y has the same value on both occasions

39. 9 ft/sec

41. (a) 2 sec   (b) 16 ft   (c) 48 ft/sec

43. (a) $\frac{1625}{16}$ ft   (b) $\frac{6475}{64}$ ft   (c) 100 ft

SECTION 3.6

1. $f(x) = x^4 + 2x^2 + 1$, $f'(x) = 4x^3 + 4x = 4x(x^2 + 1)$
$f(x) = (x^2 + 1)^2$, $f'(x) = 2(x^2 + 1)(2x) = 4x(x^2 + 1)$

3. $f(x) = 8x^3 + 12x^2 + 6x + 1$, $f'(x) = 24x^2 + 24x + 6 = 6(2x + 1)^2$
$f(x) = (2x + 1)^3$, $f'(x) = 3(2x + 1)^2(2) = 6(2x + 1)^2$

5. $f(x) = x^2 + 2 + x^{-2}$, $f'(x) = 2x - 2x^{-3} = 2x(1 - x^{-4})$
$f(x) = (x + x^{-1})^2$, $f'(x) = 2(x + x^{-1})(1 - x^{-2}) = 2x(1 + x^{-2})(1 - x^{-2}) = 2x(1 - x^{-4})$

7. $2(1 - 2x)^{-2}$

9. $20(x^5 - x^{10})^{19}(5x^4 - 10x^9)$

11. $4\left(x - \dfrac{1}{x}\right)^3\left(1 + \dfrac{1}{x^2}\right)$

13. $4(x - x^3 - x^5)^3(1 - 3x^2 - 5x^4)$

15. $200t(t^2 - 1)^{99}$

17. $-4(t^{-1} + t^{-2})^3(t^{-2} + 2t^{-3})$

19. $324x^3\left[\dfrac{1 - x^2}{(x^2 + 1)^5}\right]$

21. $2(x^4 + x^2 + x)(4x^3 + 2x + 1)$

23. $-\left(\dfrac{x^3}{3} + \dfrac{x^2}{2} + \dfrac{x}{1}\right)^{-2}(x^2 + x + 1)$

25. $3\left(\dfrac{1}{x + 2} - \dfrac{1}{x - 2}\right)^2\left[\dfrac{-1}{(x + 2)^2} + \dfrac{1}{(x - 2)^2}\right] = \dfrac{384x}{(x + 2)^4(x - 2)^4}$

27. -1

29. 0

31. $\dfrac{dy}{dt} = \dfrac{dy}{du}\dfrac{du}{dx}\dfrac{dx}{dt} = \dfrac{7(2t - 5)^4 + 12(2t - 5)^2 - 2}{[(2t - 5)^4 + 2(2t - 5)^2 + 2]^2}[4(2t - 5)]$

33. 16

35. 1

37. 1

39. 1

41. 2

43. 2

45. 0

47. $2xf'(x^2 + 1)$

49. $2f(x)f'(x)$

51. (a) $x = 0$ (b) $x < 0$ (c) $x > 0$

53. (a) $x = -1, x = 1$ (b) $-1 < x < 1$ (c) $x < -1, x > 1$

55. at 3

57. at 2 and $2\sqrt{3}$

59. If $p(x) = (x - a)^2 q(x)$, then $p'(x) = (x - a)[2q(x) + (x - a)q'(x)]$.

61. $2[(7x - x^{-1})^{-2} + 3x^2]^{-2}[(7x - x^{-1})^{-3}(7 + x^{-2}) - 3x]$

63. $3[(x + x^{-1})^2 - (x^2 + x^{-2})^{-1}]^2[2(x + x^{-1})(1 - x^{-2}) + (x^2 + x^{-2})^{-2}(2x - 2x^{-3})]$

65. by numerical work $f'(1) \cong -20$;   by the chain rule $f'(x) = 20x(x^2 - 2)^9$, $f'(1) = -20$

SECTION 3.7

1. $\dfrac{dy}{dx} = -3\sin x - 4\sec x \tan x$

3. $\dfrac{dy}{dx} = 3x^2 \csc x - x^3 \csc x \cot x$

5. $\dfrac{dy}{dt} = -2\cos t \sin t$

7. $\dfrac{dy}{du} = 2u^{-1/2}\sin^3 \sqrt{u} \cos \sqrt{u}$

9. $\dfrac{dy}{dx} = 2x \sec^2 x^2$

11. $\dfrac{dy}{dx} = 4(1 - \pi \csc^2 \pi x)(x + \cot \pi x)^3$

13. $\dfrac{d^2y}{dx^2} = -\sin x$

15. $\dfrac{d^2y}{dx^2} = \cos x(1 + \sin x)^{-2}$

17. $\dfrac{d^2y}{du^2} = 12\cos 2u(2\sin^2 2u - \cos^2 2u)$

19. $\dfrac{d^2y}{dt^2} = 8\sec^2 2t \tan 2t$

21. $\dfrac{d^2y}{dx^2} = (2 - 9x^2)\sin 3x + 12x\cos 3x$

23. $\dfrac{d^2y}{dx^2} = 0$

25. $\sin x$

27. $(27t^3 - 12t)\sin 3t - 45t^2 \cos 3t$

29. $3\cos 3x\, f'(\sin 3x)$

31. $y = x$

33. $y - \sqrt{3} = -4(x - \frac{1}{6}\pi)$

35. $y - \sqrt{2} = \sqrt{2}(x - \frac{1}{4}\pi)$

37. at π

39. at $\frac{1}{6}\pi, \frac{7}{6}\pi$

41. at $\frac{1}{2}\pi, \pi, \frac{3}{2}\pi$

43. at $\frac{1}{4}\pi, \frac{3}{4}\pi, \frac{5}{4}\pi, \frac{7}{4}\pi$

45. at $\frac{7}{6}\pi, \frac{11}{6}\pi$

47. at $\frac{1}{4}\pi, \frac{5}{4}\pi$

49. $(\frac{1}{2}\pi, \frac{2}{3}\pi), (\frac{7}{6}\pi, \frac{4}{3}\pi), (\frac{11}{6}\pi, 2\pi)$

51. $(0, \frac{1}{4}\pi), (\frac{7}{4}\pi, 2\pi)$

53. $(\frac{5}{6}\pi, \frac{3}{2}\pi)$

55. (a) $\dfrac{dy}{dt} = \dfrac{dy}{du}\dfrac{du}{dx}\dfrac{dx}{dt} = (2u)(\sec x \tan x)(\pi) = 2\pi \sec^2 \pi t \tan \pi t$

(b) $y = \sec^2 \pi t - 1$, $\quad \dfrac{dy}{dt} = 2 \sec \pi t (\sec \pi t \tan \pi t)\pi = 2\pi \sec^2 \pi t \tan \pi t$

57. (a) $\dfrac{dy}{dt} = \dfrac{dy}{du}\dfrac{du}{dx}\dfrac{dx}{dt} = 4[\tfrac{1}{2}(1-u)]^3(-\tfrac{1}{2}) \cdot (-\sin x) \cdot 2$

$\qquad\qquad = 4[\tfrac{1}{2}(1 - \cos 2t)]^3 \sin 2t = (4 \sin^6 t)(2 \sin t \cos t) = 8 \sin^7 t \cos t$

(b) $y = [\tfrac{1}{2}(1 - \cos 2t)]^4 = \sin^8 t$, $\quad \dfrac{dy}{dt} = 8 \sin^7 t \cos t$

59. $\dfrac{d^n}{dx^n} (\cos x) = \begin{cases} (-1)^{(n+1)/2} \sin x, & n \text{ odd} \\ (-1)^{n/2} \cos x, & n \text{ even} \end{cases}$

61. by numerical work $f'(0) \cong 0$;
by the chain rule $f'(x) = -2x \sin x^2$, $\quad f'(0) = 0$

SECTION 3.8

1. $\tfrac{3}{2}x^2(x^3 + 1)^{-1/2}$

3. $\dfrac{2x^2 + 1}{\sqrt{x^2 + 1}}$

5. $\dfrac{x}{(\sqrt[4]{2x^2 + 1})^3}$

7. $\dfrac{x(2x^2 - 5)}{\sqrt{2 - x^2}\,\sqrt{3 - x^2}}$

9. $\dfrac{1}{2}\left(\dfrac{1}{\sqrt{x}} - \dfrac{1}{x\sqrt{x}}\right)$

11. $\dfrac{1}{(\sqrt{x^2 + 1})^3}$

13. $\dfrac{1}{3}\left[\dfrac{1}{(\sqrt[3]{x})^2} - \dfrac{1}{(\sqrt[3]{x})^4}\right]$

15. (a) Figure A.3.8.1   (b) Figure A.3.8.2   (c) Figure A.3.8.3

Figure A.3.8.1

Figure A.3.8.2

Figure A.3.8.3

17. $-\dfrac{2b^2}{9(\sqrt[3]{a + bx})^5}$

19. $-a^2(a^2 - x^2)^{-3/2}$

21. $\dfrac{\sqrt{x} \cos \sqrt{x} - (x + 1) \sin \sqrt{x}}{4x\sqrt{x}}$

23. $\dfrac{1}{2\sqrt{x}}f'(\sqrt{x} + 1)$

25. $\dfrac{xf'(x^2 + 1)}{\sqrt{f(x^2 + 1)}}$

27. $\dfrac{1}{x}$

29. $\dfrac{1}{\sqrt{1 - x^2}}$

31. inelastic for $0 < P < 200$, unitary at $P = 200$, elastic for $200 < P < 300$

33. $\tfrac{1}{7}$ **35.** $\tfrac{1}{8}$ **37.** 63   **39.** $\tfrac{1}{63}$

41. from numerical work $f'(16) \cong 0.375$; from (3.8.4) $f'(x) = \tfrac{3}{4}x^{-1/4}$, $f'(16) = \tfrac{3}{8} = 0.375$

43. from numerical work $l \cong 0.125$; $\quad \lim\limits_{x \to 32} \dfrac{x^{2/5} - 4}{x^{4/5} - 16} = \lim\limits_{x \to 32} \dfrac{1}{x^{2/5} + 4} = \tfrac{1}{8} = 0.125$

SECTION 3.9

1. $\dfrac{dy}{dx} = \dfrac{2}{3}x^{-1/3}$

3. $\dfrac{dy}{dx} = -\dfrac{a}{x^2} + c$

5. $\dfrac{dy}{dx} = \dfrac{\sqrt{a}}{2}\left(\dfrac{1}{\sqrt{x}} - \dfrac{1}{x\sqrt{x}}\right)$

7. $\dfrac{dr}{d\theta} = -\dfrac{1}{\sqrt{1 - 2\theta}}$

9. $f'(x) = \dfrac{x}{(\sqrt{a^2 - x^2})^3}$

11. $\dfrac{dy}{dx} = -\dfrac{6b}{x^3}\left(a + \dfrac{b}{x^2}\right)^2$

13. $\dfrac{ds}{dt} = \dfrac{a^2 + 2t^2}{\sqrt{a^2 + t^2}}$

15. $\dfrac{dy}{dx} = \dfrac{4a^2x}{(a^2 - x^2)^2}$

17. $\dfrac{dy}{dx} = \dfrac{2x^2 - 8x - 1}{(1 + 2x^2)^2}$

19. $\dfrac{dy}{dx} = \dfrac{2a - bx}{2(\sqrt{a - bx})^3}$

21. $\dfrac{dy}{dx} = -\dfrac{c}{(1 + cx)^2}\sqrt{\dfrac{1 + cx}{1 - cx}}$

23. $\dfrac{dy}{dx} = \dfrac{2a^2x}{(a^2-x^2)^2}\sqrt{\dfrac{a^2-x^2}{a^2+x^2}}$

25. $\dfrac{dr}{d\theta} = -\dfrac{3a+2b\theta}{3\theta^2(\sqrt[3]{a+b\theta})^2}$

27. $\dfrac{dy}{dx} = -\dfrac{bx}{a\sqrt{a^2-x^2}}$

29. $\dfrac{dy}{dx} = -x^{-1/3}(a^{2/3}-x^{2/3})^{1/2}$

31. $\dfrac{dy}{dx} = 2x\sec x^2 \tan x^2$
33. $\dfrac{dy}{dx} = -6\csc^2 2x \cot^2 2x$
35. $\dfrac{dy}{dx} = -\sin x \cos(\cos x)$

37. 540 **39.** $\frac{1}{12}$ **41.** $\frac{5}{6}$ **43.** $\frac{3}{64}$
45. $-\frac{41}{90}$ **47.** 0 **49.** 20 **51.** 8
53. $-\frac{3}{2}\sqrt{3}$ **55.** $(1-\frac{1}{4}\pi)\sqrt{2}$

SECTION 3.10

1. $-\dfrac{x}{y}$
3. $-\dfrac{4x}{9y}$
5. $-\dfrac{x^2(x+3y)}{x^3+y^3}$

7. $\dfrac{2(x-y)}{2(x-y)+1}$
9. $\dfrac{y-\cos(x+y)}{\cos(x+y)-x}$
11. $\dfrac{16}{(x+y)^3}$

13. $\dfrac{90}{(2y+x)^3}$
15. $\dfrac{dy}{dx} = \dfrac{5}{8},\ \dfrac{d^2y}{dx^2} = -\dfrac{9}{128}$
17. $\dfrac{dy}{dx} = -\dfrac{1}{2},\ \dfrac{d^2y}{dx^2} = 0$

19. tangent $2x+3y-5=0$; normal $3x-2y+12=0$
21. tangent $x+2y+8=0$; normal $2x-y+1=0$
23. Let (x_0, y_0) be a point of the circle. If $x_0 = 0$, the normal line is the y-axis; if $y_0 = 0$, the normal line is the x-axis. For all other choices of (x_0, y_0) the normal line takes the form

$$y - y_0 = \frac{y_0}{x_0}(x - x_0) \qquad \text{which reduces to} \qquad y = \frac{y_0}{x_0}x.$$

In each case the normal line passes through the origin.
25. at right angles
27. The hyperbola and the ellipse intersect at the four pts $(\pm 3, \pm 2)$. For the hyperbola, $\dfrac{dy}{dx} = \dfrac{x}{y}$. For the ellipse, $\dfrac{dy}{dx} = -\dfrac{4x}{9y}$. The product of these slopes is therefore $-\dfrac{4x^2}{9y^2}$. At each of the points of intersection this product is -1.
29. $y-2x+12=0,\quad y-2x-12=0$

SECTION 3.11

1. (a) -2 units/sec (b) 4 units/sec
3. $-\frac{2}{27}$ m/min, $-\frac{8}{3}$ m²/min
5. $1/(50\pi)$ ft/min, $\frac{8}{5}$ ft²/min
7. 6 cm
9. decreasing 7 in.²/min
11. boat A
13. 10 ft³/hr
15. $r\omega$
17. $x(t) = r\cos(\omega t + \theta_0),\quad y(t) = r\sin(\omega t + \theta_0)$

19. for the sector: $A = \frac{1}{2}r^2\theta,\qquad \dfrac{dA}{dt} = \frac{1}{2}r^2\dfrac{d\theta}{dt} = \frac{1}{2}r^2\omega$ (constant)

for T: $A = \frac{1}{2}r^2\sin\theta,\qquad \dfrac{dA}{dt} = \frac{1}{2}r^2\cos\theta\dfrac{d\theta}{dt} = \frac{1}{2}r^2\omega\cos\theta$ (varies with θ)

for S: $A = \frac{1}{2}r^2(\theta - \sin\theta),\qquad \dfrac{dA}{dt} = \frac{1}{2}r^2(1-\cos\theta)\dfrac{d\theta}{dt} = \frac{1}{2}r^2\omega(1-\cos\theta)$ (varies with θ)

21. $\frac{1600}{3}$ ft/min
23. dropping $1/2\pi$ in./min
25. decreasing 0.04 rad/min
27. 5π mi/min
29. decreasing 0.12 rad/sec
31. increasing $\frac{4}{101}$ rad/min

SECTION 3.12

1. $\Delta V = (x + h)^3 - x^3$
    $= (x^3 + 3x^2h + 3xh^2 + h^3) - x^3$
    $= 3x^2h + 3xh^2 + h^3$
    $dV = 3x^2h$
    $\Delta V - dV = 3xh^2 + h^3$ (see figure)

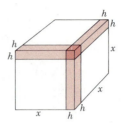

3. $10\frac{1}{30}$ taking $x = 1000$ 5. $1\frac{31}{32}$ taking $x = 16$ 7. 1.975 taking $x = 32$

9. 8.15 taking $x = 32$ 11. 0.719 13. 0.531

15. 1.6 17. $2\pi rht$ 19. error ≤ 0.01 ft

21. about 0.00153 sec   23. within $\frac{1}{2}\%$

25. (a) $x_{n+1} = \frac{1}{2}x_n + 12\left(\dfrac{1}{x_n}\right)$ (b) $x_4 \cong 4.89898$ 27. (a) $x_{n+1} = \frac{2}{3}x_n + \frac{25}{3}\left(\dfrac{1}{x_n}\right)^2$ (b) $x_4 \cong 2.92402$

29. (a) $x_{n+1} = \dfrac{x_n \sin x_n + \cos x_n}{\sin x_n + 1}$ (b) $x_4 \cong 0.73909$

31. (a) $x_{n+1} = \dfrac{2x_n \cos x_n - 2 \sin x_n}{2 \cos x_n - 1}$ (b) $x_4 \cong 1.89549$

33. (a) and (b)

35. $\displaystyle\lim_{h\to 0} \frac{g_1(h) + g_2(h)}{h} = \lim_{h\to 0} \frac{g_1(h)}{h} + \lim_{h\to 0} \frac{g_2(h)}{h} = 0 + 0 = 0$

$\displaystyle\lim_{h\to 0} \frac{g_1(h)g_2(h)}{h} = \lim_{h\to 0} h\,\frac{g_1(h)g_2(h)}{h^2} = \left(\lim_{h\to 0} h\right)\left(\lim_{h\to 0} \frac{g_1(h)}{h}\right)\left(\lim_{h\to 0} \frac{g_2(h)}{h}\right) = (0)(0)(0) = 0$

SECTION 3.13

1. (a) $x_0 = 2$ (b) $x_0 = \pm\sqrt{2}$ (c) $x_0 = \frac{1}{4}$ 3. (a) $(1, \frac{2}{3})$ (b) $(3, 2\sqrt{3})$ (c) $(\frac{1}{3}, \frac{2}{27}\sqrt{3})$

5. $y = \dfrac{1}{m}(x - b) + a$ 7. $b^n n!$ 9. decreasing 4 units/sec

11. $f'(2) = \displaystyle\lim_{h\to 0} \frac{(2 + h)^2 - 2^2}{h} = \lim_{h\to 0} \frac{4h + h^2}{h} = \lim_{h\to 0}(4 + h) = 4$

$f'(2) = \displaystyle\lim_{x\to 2} \frac{x^2 - 2^2}{x - 2} = \lim_{x\to 2} \frac{(x + 2)(x - 2)}{x - 2} = \lim_{x\to 2}(x + 2) = 4$

13. $\frac{1}{5}$ 15. -1 17. $y + x = 0, \quad y - 26x - 54 = 0$

19. decreasing 2.25 in./min   21. increasing 1.6 cm^2/min

23. increasing $10/\pi$ in.^{3}/min   25. $(0, \frac{1}{6}\pi), (\frac{1}{2}\pi, \frac{5}{6}\pi), (\frac{3}{2}\pi, 2\pi)$

27. $\dfrac{dS}{S} = \dfrac{12xh}{6x^2} = 2\dfrac{h}{x} = 0.002, \quad \dfrac{dV}{V} = \dfrac{3x^2h}{x^3} = 0.003$ 29. $\frac{1}{10}$ ft/min

31. $\frac{3}{2}a^2k$ units^2/sec   33. (a) about 0.0533 sec/in.   (b) 0.0106 sec

35. 984 ft   37. $C = \dfrac{1 - 2\sqrt{AB}}{B}; \left(\dfrac{\sqrt{AB}}{A}, \dfrac{1 - \sqrt{AB}}{B}\right)$ is the equilibrium point.

39. $A = 1, B = -3, C = 5$ 41. at $\frac{1}{3}\pi, \pi, \frac{5}{3}\pi$

43. increasing 15 mph or $15\sqrt{3}$ mph   45. (a) $(0, r), (0, -r)$ (b) $(r, 0), (-r, 0)$

SECTION 4.1

1. $c = \frac{3}{2}$ 3. $c = \frac{1}{3}\sqrt{39}$ 5. $c = \frac{1}{2}\sqrt{2}$ 7. $c = 0$

9. $f'(x) = \begin{cases} 2, & x \leq -1 \\ 3x^2 - 1, & x > -1 \end{cases}; \quad -3 < c \leq -1$ and $c = 1$

11. $\dfrac{f(1) - f(-1)}{(1) - (-1)} = 0$ and $f'(x)$ is never zero; f is not differentiable at 0

13. Set $P(x) = 6x^4 - 7x + 1$. If there existed three numbers $a < b < c$ at which $P(x) = 0$, then by Rolle's

theorem $P'(x)$ would have to be zero for some x in (a, b) and also for some x in (b, c). This is not the case: $P'(x) = 24x^3 - 7$ is zero only at $x = (\frac{7}{24})^{1/3}$.

15. Set $P(x) = x^3 + 9x^2 + 33x - 8$. Note that $P(0) < 0$ and $P(1) > 0$. Thus by the intermediate-value theorem there exists some number c between 0 and 1 at which $P(x) = 0$. If the equation $P(x) = 0$ had an additional real root, then by Rolle's theorem there would have to be some real number at which $P'(x) = 0$. This is not the case: $P'(x) = 3x^2 + 18x + 33$ is never 0 since the discriminant $b^2 - 4ac = (18)^2 - 12(33) < 0$.

17. Let c and d be two consecutive roots of the equation $P'(x) = 0$. The equation $P(x) = 0$ cannot have two or more roots between c and d, for then, by Rolle's theorem, $P'(x)$ would have to be zero somewhere between these two roots and thus between c and d. In this case, c and d would no longer be consecutive roots of $P'(x) = 0$.

19. If $x_1 = x_2$, then $|f(x_1) - f(x_2)|$ and $|x_1 - x_2|$ are both 0 and the inequality holds. If $x_1 \neq x_2$, then you know by the mean-value theorem that

$$\frac{f(x_1) - f(x_2)}{x_1 - x_2} = f'(c)$$

for some number c between x_1 and x_2. Since $|f'(c)| \leq 1$, you can conclude that

$$\left|\frac{f(x_1) - f(x_2)}{x_1 - x_2}\right| \leq 1 \quad \text{and thus that} \quad |f(x_1) - f(x_2)| \leq |x_1 - x_2|.$$

21. set, for instance, $f(x) = \begin{cases} 1, & a < x < b \\ 0, & x = a, b \end{cases}$

23. (a) Between any two times that the object is at the origin there is at least one instant when the velocity is zero. (b) On any time interval there is at least one instant when the instantaneous velocity equals the average velocity over the interval.

25. $f'(x_0) = \lim\limits_{y \to 0} \dfrac{f(x_0 + y) - f(x_0)}{y} = \lim\limits_{y \to 0} \dfrac{f'(x_0 + \theta y)y}{y} = \lim\limits_{y \to 0} f'(x_0 + \theta y) = \lim\limits_{x \to x_0} f'(x) = L$

　　　　　　　　　　　└─── by the hint　　　　　　　　　└─── by (2.2.5)

27. 2.99

SECTION 4.2

1. increases on $(-\infty, -1]$ and $[1, \infty)$, decreases on $[-1, 1]$
3. increases on $(-\infty, -1]$ and $[1, \infty)$, decreases on $[-1, 0)$ and $(0, 1]$
5. increases on $[-\frac{3}{4}, \infty)$, decreases on $(-\infty, -\frac{3}{4}]$　　　7. increases on $[-1, \infty)$, decreases on $(-\infty, -1]$
9. increases on $(-\infty, 2)$, decreases on $(2, \infty)$
11. increases on $(-\infty, -1)$ and $(-1, 0]$, decreases on $[0, 1)$ and $(1, \infty)$

13. increases on $[-\sqrt{5}, 0]$ and $[\sqrt{5}, \infty)$, decreases on $(-\infty, -\sqrt{5}]$ and $[0, \sqrt{5}]$

15. increases on $(-\infty, -1)$ and $(-1, \infty)$　　　17. decreases on $[0, \infty)$
19. increases on $[0, \infty)$, decreases on $(-\infty, 0]$
21. decreases on $(0, 3]$　　　23. increases on $[0, 2\pi]$
25. increases on $[\frac{2}{3}\pi, \pi]$, decreases on $[0, \frac{2}{3}\pi]$
27. increases on $[0, \frac{2}{3}\pi]$ and $[\frac{5}{6}\pi, \pi]$, decreases on $[\frac{2}{3}\pi, \frac{5}{6}\pi]$
29. $f(x) = \frac{1}{3}x^3 - x + \frac{8}{3}$　　　31. $f(x) = x^5 + x^4 + x^3 + x^2 + x + 5$
33. $f(x) = \frac{3}{4}x^{4/3} - \frac{2}{3}x^{3/2} + 1, \quad x \geq 0$　　　35. $f(x) = 2x - \cos x + 4$
37. increases on $(-\infty, -3)$ and $[-1, 1]$, decreases on $[-3, -1]$ and $[1, \infty)$
39. increases on $(-\infty, 0]$ and $[3, \infty)$, decreases on $[0, 1)$ and $[1, 3]$
41. (a) $M \leq L \leq N$　　　(b) none　　　(c) $M = L = N$
43. set, for instance, $f(x) = \begin{cases} 1, & x \text{ rational} \\ 0, & x \text{ irrational} \end{cases}$
45. Set $h(x) = f(x) - g(x)$. Since $h'(x) > 0$ for all x, h increases everywhere. Since $h(0) = 0$, $h(x) < 0$ on $(-\infty, 0)$ and $h(x) > 0$ on $(0, \infty)$. It follows that $f(x) < g(x)$ on $(-\infty, 0)$ and $f(x) > g(x)$ on $(0, \infty)$.

SECTION 4.3

1. no critical pts; no local extreme values
3. critical pts ± 1; local max $f(-1) = -2$, local min $f(1) = 2$
5. critical pts $0, \frac{2}{3}$; $f(0) = 0$ local min, $f(\frac{2}{3}) = \frac{4}{27}$ local max
7. no critical pts; no local extreme values   9. critical pt $-\frac{1}{2}$; local max $f(-\frac{1}{2}) = -8$
11. critical pts $0, \frac{3}{5}, 1$; local max $f(\frac{3}{5}) = 2^2 3^3/5^5$, local min $f(1) = 0$
13. critical pts $\frac{5}{8}, 1$; local max $f(\frac{5}{8}) = \frac{27}{2048}$
15. critical pts $-2, 0$; local max $f(-2) = 4$, local min $f(0) = 0$
17. critical pts $-2, -\frac{1}{2}, 1$; local max $f(-\frac{1}{2}) = \frac{9}{4}$, local min $f(-2) = f(1) = 0$
19. critical pts $-2, -\frac{12}{7}, 0$; local max $f(-\frac{12}{7}) = \frac{144}{49} (\frac{2}{7})^{1/3}$, local min $f(0) = 0$
21. critical pts $-\frac{1}{2}, 3$; local min $f(-\frac{1}{2}) = \frac{7}{8}$ 23. critical pt 1; local min $f(1) = 3$
25. critical pts $\frac{1}{4}\pi, \frac{5}{4}\pi$; local max $f(\frac{1}{4}\pi) = \sqrt{2}$, local min $f(\frac{5}{4}\pi) = -\sqrt{2}$
27. critical pts $\frac{1}{3}\pi, \frac{1}{2}\pi, \frac{2}{3}\pi$; local max $f(\frac{1}{2}\pi) = 1 - \sqrt{3}$, local min $f(\frac{1}{3}\pi) = f(\frac{2}{3}\pi) = -\frac{3}{4}$
29. critical pts $\frac{1}{3}\pi, \frac{5}{3}\pi$; local max $f(\frac{5}{3}\pi) = \frac{5}{4}\sqrt{3} + \frac{10}{3}\pi$, local min $f(\frac{1}{3}\pi) = -\frac{5}{4}\sqrt{3} + \frac{2}{3}\pi$
31. (i) f increases on $(c - \delta, c]$ and decreases on $[c, c + \delta)$.
    (ii) f decreases on $(c - \delta, c]$ and increases on $[c, c + \delta)$.
    (iii) If $f'(x) > 0$ on $(c - \delta, c) \cup (c, c + \delta)$, then, since f is continuous at c, f increases on $(c - \delta, c]$ and also on $[c, c + \delta)$. Therefore, in this case, f increases on $(c - \delta, c + \delta)$. A similar argument shows that, if $f'(x) < 0$ on $(c - \delta, c) \cup (c, c + \delta)$, then f decreases on $(c - \delta, c + \delta)$.
33. critical pts 1, 2, 3; local max $P(2) = -4$, local min $P(1) = P(3) = -5$
    Since $P'(x) < 0$ for $x < 0$, P decreases on $(-\infty, 0]$. Since $P(0) > 0$, P does not take on the value 0 on $(-\infty, 0]$.
    Since $P(0) > 0$ and $P(1) < 0$, P takes on the value 0 at least once on $(0, 1)$. Since $P'(x) < 0$ on $(0, 1)$, P decreases on $[0, 1]$. It follows that P takes on the value zero only once on $[0, 1]$.
    Since $P'(x) > 0$ on $(1, 2)$ and $P'(x) < 0$ on $(2, 3)$, P increases on $[1, 2]$ and decreases on $[2, 3]$. Since $P(1)$, $P(2)$, $P(3)$ are all negative, P cannot take on the value 0 between 1 and 3.
    Since $P(3) < 0$ and $P(100) > 0$, P takes on the value 0 at least once on $(3, 100)$. Since $P'(x) > 0$ on $(3, 100)$, P increases on $[3, 100]$. It follows that P takes on the value zero only once on $[3, 100]$.
    Since $P'(x) > 0$ on $(100, \infty)$, P increases on $[100, \infty)$. Since $P(100) > 0$, P does not take on the value 0 on $[100, \infty)$.

SECTION 4.4

1. critical pt -2; $f(-2) = 0$ endpt min and absolute min
3. critical pts 0, 2, 3; $f(0) = 1$ endpt max and absolute max, $f(2) = -3$ local min and absolute min, $f(3) = -2$ endpt max
5. critical pt $2^{-1/3}$; $f(2^{-1/3}) = 3 \cdot 2^{-2/3}$ local min
7. critical pts $\frac{1}{10}, 2^{-1/3}, 2$; $f(\frac{1}{10}) = 10 \frac{1}{100}$ endpt max and absolute max; $f(2^{-1/3}) = 3 \cdot 2^{-2/3}$ local min and absolute min, $f(2) = 4\frac{1}{2}$ endpt max
9. critical pts $0, \frac{3}{2}, 2$; $f(0) = 2$ endpt max and absolute max, $f(\frac{3}{2}) = -\frac{1}{4}$ local min and absolute min, $f(2) = 0$ endpt max
11. critical pts $-3, -2, 1$; $f(-3) = -\frac{3}{13}$ endpt max, $f(-2) = -\frac{1}{4}$ local min and absolute min, $f(1) = \frac{1}{5}$ endpt max and absolute max
13. critical pts $0, \frac{1}{4}, 1$; $f(0) = 0$ endpt min and absolute min, $f(\frac{1}{4}) = \frac{1}{16}$ local max, $f(1) = 0$ local min and absolute min
15. critical pts 2, 3; $f(2) = 2$ local max and absolute max, $f(3) = 0$ endpt min
17. critical pt 1; no extreme values
19. critical pts $0, \frac{5}{6}\pi, \pi$; $f(0) = -\sqrt{3}$ endpt min and absolute min, $f(\frac{5}{6}\pi) = \frac{7}{4}$ local max and absolute max, $f(\pi) = \sqrt{3}$ endpt min
21. critical pts $0, \pi$; $f(0) = 5$ endpt max and absolute max, $f(\pi) = -5$ endpt min and absolute min

23. critical pts $-\frac{1}{3}\pi$, 0; $f(-\frac{1}{3}\pi) = \frac{1}{3}\pi - \sqrt{3}$ endpt min and absolute min, no absolute max

25. critical pts 0, 1, 4, 7; $f(0) = 0$ endpt min, $f(1) = -2$ local min and absolute min, $f(4) = 1$ local max and absolute max, $f(7) = -2$ endpt min and absolute min

27. critical pts $-2, -1, 1, 3$; $f(-2) = 5$ endpt max, $f(-1) = 2$ local min and absolute min, $f(1) = 6$ local max and absolute max, $f(3) = 2$ local min and absolute min

29. critical pts $-3, -1, 0, 2$; $f(-3) = 2$ endpt max and absolute max, $f(-1) = 0$ local min, $f(0) = 2$ local max and absolute max, $f(2) = -2$ local min and absolute min

31. By contradiction. If f is continuous at c, then $f(c)$ is not a local maximum by the first-derivative test (4.3.3).

33. If f is not differentiable on (a, b), then f has a critical point at each point c in (a, b) where $f'(c)$ does not exist. If f is differentiable on (a, b), then there exists c in (a, b) where $f'(c) = (f(b) - f(a))/(b - a)$. (mean-value theorem) With $f(b) = f(a)$, we have $f'(c) = 0$ and thus c is a critical point of f.

35.
$$P(x) - M \geq a_0 x^n - (|a_1|x^{n-1} + \cdots + |a_{n-1}|x + |a_n| + M)$$

for $x > 0$ ⟶

$$\geq a_0 x^n - (|a_1| + \cdots + |a_{n-1}| + |a_n| + M) \geq 0 \quad \text{for } x \geq \left(\frac{|a_1| + \cdots + |a_{n-1}| + |a_n| + M}{a_0}\right)^{1/n}$$

for $x > 1$ ⟶

SECTION 4.5

1. 400 **3.** 20 by 10 ft **5.** 32

7. 100 by 150 ft with the divider 100 ft long

9. radius of semicircle $\dfrac{90}{12 + 5\pi} \cong 3.25$ ft; height of rectangle $\dfrac{90 + 30\pi}{12 + 5\pi} \cong 6.65$ ft

11. $x = 2$, $y = \frac{3}{2}$ **13.** $-\frac{5}{2}$ **15.** $\frac{10}{3}\sqrt{3}$ in. by $\frac{5}{3}\sqrt{3}$ in.

17. equilateral triangle with side 4 **19.** $(1, 1)$

21. height of rectangle $\frac{15}{11}(5 - \sqrt{3}) \cong 4.46$ in.; side of triangle $\frac{10}{11}(6 + \sqrt{3}) \cong 7.03$ in.

23. $\frac{5}{3}$ **25.** $(0, \sqrt{3})$ **27.** $5\sqrt{5}$ ft **29.** 54 by 72 in.

31. (a) use it all for the circle (b) use $28\pi/(4 + \pi) \cong 12.32$ in. for the circle

33. base radius $\frac{10}{3}$ and height $\frac{8}{3}$ **35.** 10 by 10 by 12.5 ft

37. equilateral triangle with side $2r\sqrt{3}$ **39.** base radius $\frac{1}{3}R\sqrt{6}$ and height $\frac{2}{3}R\sqrt{3}$

41. base radius $\frac{2}{3}R\sqrt{2}$ and height $\frac{4}{3}R$

43. \$160,000 **45.** $\tan\theta = m$ **47.** $6\sqrt{6}$ ft

SECTION 4.6

1. concave down on $(-\infty, 0)$, concave up on $(0, \infty)$

3. concave down on $(-\infty, 0)$, concave up on $(0, \infty)$; pt of inflection $(0, 2)$

5. concave up on $(-\infty, -\frac{1}{3}\sqrt{3})$, concave down on $(-\frac{1}{3}\sqrt{3}, \frac{1}{3}\sqrt{3})$, concave up on $(\frac{1}{3}\sqrt{3}, \infty)$; pts of inflection $(-\frac{1}{3}\sqrt{3}, -\frac{5}{36})$, $(\frac{1}{3}\sqrt{3}, -\frac{5}{36})$

7. concave down on $(-\infty, -1)$ and on $(0, 1)$, concave up on $(-1, 0)$ and on $(1, \infty)$; pt of inflection $(0, 0)$

9. concave up on $(-\infty, -\frac{1}{3}\sqrt{3})$, concave down on $(-\frac{1}{3}\sqrt{3}, \frac{1}{3}\sqrt{3})$, concave up on $(\frac{1}{3}\sqrt{3}, \infty)$; pts of inflection $(-\frac{1}{3}\sqrt{3}, \frac{4}{9})$, $(\frac{1}{3}\sqrt{3}, \frac{4}{9})$

11. concave up on $(0, \infty)$

13. concave down on $(-\infty, -2)$, concave up on $(-2, \infty)$; pt of inflection $(-2, 0)$

15. concave up on $(0, \frac{1}{4}\pi)$, concave down on $(\frac{1}{4}\pi, \frac{3}{4}\pi)$, concave up on $(\frac{3}{4}\pi, \pi)$; pts of inflection $(\frac{1}{4}\pi, \frac{1}{2})$ and $(\frac{3}{4}\pi, \frac{1}{2})$

17. concave up on $(0, \frac{1}{12}\pi)$, concave down on $(\frac{1}{12}\pi, \frac{5}{12}\pi)$, concave up on $(\frac{5}{12}\pi, \pi)$; pts of inflection $(\frac{1}{12}\pi, \frac{1}{2} + \frac{1}{144}\pi^2)$ and $(\frac{5}{12}\pi, \frac{1}{2} + \frac{25}{144}\pi^2)$

19. $d = \frac{1}{3}(a + b + c)$ **21.** $a = -\frac{1}{2}$, $b = \frac{1}{2}$ **23.** $A = 18$, $B = -4$

1. Figure A.4.7.1

3. Figure A.4.7.2

5. Figure A.4.7.3

Figure A.4.7.1

Figure A.4.7.2

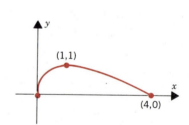

Figure A.4.7.3

7. Figure A.4.7.4

9. Figure A.4.7.5

11. Figure A.4.7.6

Figure A.4.7.4

Figure A.4.7.5

Figure A.4.7.6

13. Figure A.4.7.7

15. Figure A.4.7.8

17. Figure A.4.7.9

Figure A.4.7.7

Figure A.4.7.8

Figure A.4.7.9

19. Figure A.4.7.10   **21.** Figure A.4.7.11   **23.** Figure A.4.7.12

Figure A.4.7.10

Figure A.4.7.11

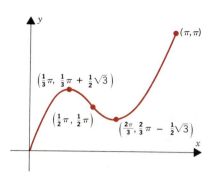

Figure A.4.7.12

25. Figure A.4.7.13   **27.** Figure A.4.7.14

Figure A.4.7.13

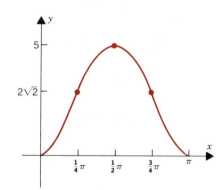

Figure A.4.7.14

SECTION 4.8

1. vertical: $x = \frac{1}{3}$; horizontal: $y = \frac{1}{3}$ **3.** vertical: $x = 2$; horizontal: none

5. vertical: $x = \pm 3$; horizontal: $y = 0$ **7.** vertical: $x = -\frac{4}{3}$; horizontal: $y = \frac{4}{9}$

9. vertical: $x = \frac{5}{2}$; horizontal: $y = 0$ **11.** vertical: none; horizontal: $y = \pm \frac{3}{2}$

13. vertical: $x = 1$; horizontal: $y = 0$ **15.** vertical: none; horizontal: $y = 0$

17. vertical: $x = (2n + \frac{1}{2})\pi$; horizontal: none

19. neither   **21.** cusp   **23.** tangent   **25.** neither

27. cusp   **29.** cusp   **31.** tangent   **33.** neither

35. Given a positive number ϵ, there exists a negative number K such that, if $x \le K$, then $|f(x) - l| < \epsilon$.

SECTION 4.9   [Rough sketches; not scale drawings]

1. Figure A.4.9.1

Figure A.4.9.1

3. Figure A.4.9.2

Figure A.4.9.2

5. Figure A.4.9.3

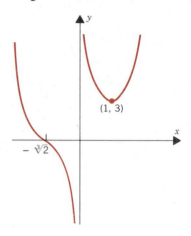

Figure A.4.9.3

7. Figure A.4.9.4

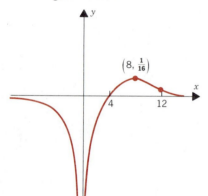

Figure A.4.9.4

9. Figure A.4.9.5

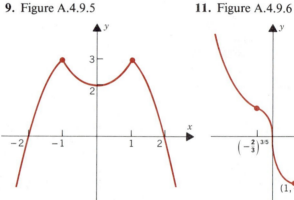

Figure A.4.9.5

11. Figure A.4.9.6

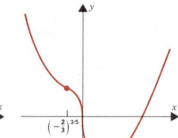

Figure A.4.9.6

13. Figure A.4.9.7

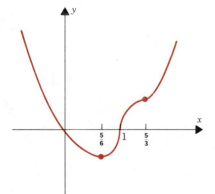

Figure A.4.9.7

15. Figure A.4.9.8

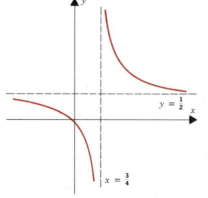

Figure A.4.9.8

17. Figure A.4.9.9

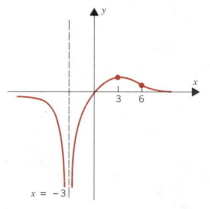

Figure A.4.9.9

19. Figure A.4.9.10

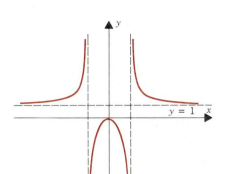

Figure A.4.9.10

21. Figure A.4.9.11

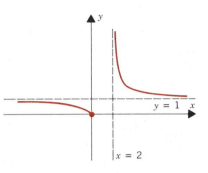

Figure A.4.9.11

23. Figure A.4.9.12

Figure A.4.9.12

25. Figure A.4.9.13

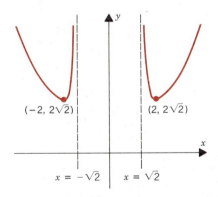

Figure A.4.9.13

27. Figure A.4.9.14

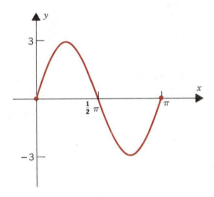

Figure A.4.9.14

29. Figure A.4.9.15

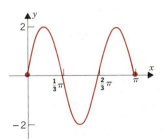

Figure A.4.9.15

31. Figure A.4.9.16

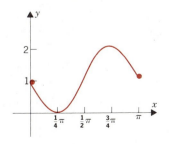

Figure A.4.9.16

33. Figure A.4.9.17

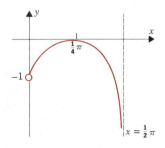

Figure A.4.9.17

SECTION 4.10

1. 1250 customers   **3.** $-b/a$ **5.** $x = y = 50$ **7.** radius $\frac{2}{3}R$, height $\frac{1}{3}H$

9. $\dfrac{\beta - b}{2(a + \alpha)}$ **11.** 29 lamps   **13.** 26 lamps

15. min for $f(Q_1) = Q_1 + 2\left(\dfrac{40 - 5Q_1}{10 - Q_1}\right)$ occurs at $Q_1 = 10 - \sqrt{20} \cong 5\frac{1}{2}$

17. $\frac{1}{2}(\beta - b)$ **19.** $2r^2$ **21.** 48   **23.** 220 by 360 ft

25. (a) $\frac{3}{2}\sqrt{6}, \frac{3}{2}\sqrt{6}, 2$ ft   (b) $3\sqrt{2}, 3\sqrt{2}, \frac{3}{2}$ ft   **27.** 2 by 4 by 11 in.

29. base radius $2\sqrt{6}$ in. and height $4\sqrt{3}$ in.   **31.** 1   **33.** 10 stories

35. (a) $(\frac{1}{2}[x_1 + y_1], \frac{1}{2}[x_1 + y_1])$ (b) $\frac{1}{2}\sqrt{2}|x_1 - y_1|$ **37.** $(\frac{1}{3}, \frac{1}{9}\sqrt{3})$

39. $\dfrac{a^{1/3}s}{a^{1/3} + b^{1/3}}$ **41.** $a(\sqrt{2} - 1)$ units to the left of B

43. cube with edge $\frac{2}{3}r\sqrt{3}$ **45.** $(\frac{1}{2}a\sqrt{2}, \frac{1}{2}b\sqrt{2})$

SECTION 5.2

1. $L_f(P) = \frac{5}{8}$, $U_f(P) = \frac{11}{8}$ **3.** $L_f(P) = \frac{9}{64}$, $U_f(P) = \frac{37}{64}$

5. $L_f(P) = \frac{17}{16}$, $U_f(P) = \frac{25}{16}$ **7.** $L_f(P) = \frac{7}{16}$, $U_f(P) = \frac{25}{16}$

9. $L_f(P) = \frac{3}{16}$, $U_f(P) = \frac{43}{32}$ **11.** $L_f(P) = \frac{1}{6}\pi$, $U_f(P) = \frac{11}{12}\pi$

13. (a) $L_f(P) \le U_f(P)$ but $3 \not\le 2$

(b) $L_f(P) \le \displaystyle\int_{-1}^{1} f(x)\,dx \le U_f(P)$ but $3 \not\le 2 \le 6$

(c) $L_f(P) \le \displaystyle\int_{-1}^{1} f(x)\,dx \le U_f(P)$ but $3 \le 10 \not\le 6$

15. (a) $L_f(P) = -3x_1(x_1 - x_0) - 3x_2(x_2 - x_1) - \cdots - 3x_n(x_n - x_{n-1})$
$U_f(P) = -3x_0(x_1 - x_0) - 3x_1(x_2 - x_1) - \cdots - 3x_{n-1}(x_n - x_{n-1})$
(b) $-\frac{3}{2}(b^2 - a^2)$

17. $L_f(P) = x_0^3(x_1 - x_0) + x_1^3(x_2 - x_1) + \cdots + x_{n-1}^3(x_n - x_{n-1})$
$U_f(P) = x_1^3(x_1 - x_0) + x_2^3(x_2 - x_1) + \cdots + x_n^3(x_n - x_{n-1})$
For each index i, $x_{i-1}^3 \le \frac{1}{4}(x_i^3 + x_i^2 x_{i-1} + x_i x_{i-1}^2 + x_{i-1}^3) \le x_i^3$ and thus by the hint
$x_{i-1}^3(x_i - x_{i-1}) \le \frac{1}{4}(x_i^4 - x_{i-1}^4) \le x_i^3(x_i - x_{i-1})$. Adding up these inequalities, we find that
$L_f(P) \le \frac{1}{4}(x_n^4 - x_0^4) \le U_f(P)$. Since $x_n = 1$ and $x_0 = 0$, the middle term is $\frac{1}{4}$. Thus the integral is $\frac{1}{4}$.

19. Let P be an arbitrary partition of $[0, 4]$. Since each $m_i = 2$ and each $M_i \ge 2$,

$$L_g(P) = 2\Delta x_1 + \cdots + 2\Delta x_n = 2(\Delta x_1 + \cdots + \Delta x_n) = 2 \cdot 4 = 8,$$
$$U_g(P) \ge 2\Delta x_1 + \cdots + 2\Delta x_n = 2(\Delta x_1 + \cdots + \Delta x_n) = 2 \cdot 4 = 8.$$

Thus $L_g(P) \le 8 \le U_g(P)$ for all partitions P of $[0, 4]$.

Uniqueness: Suppose that

$$(*) \qquad\qquad L_g(P) \le I \le U_g(P) \qquad \text{for all partitions } P \text{ of } [0, 4].$$

Since $L_g(P) = 8$ for all P, I is at least 8. Suppose now that $I > 8$ and choose a partition P of $[0, 4]$ with max $\Delta x_i < \frac{1}{5}(I - 8)$ and $0 = x_1 < \cdots < x_{i-1} < 3 < x_i < \cdots < x_n = 4$. Then

$$U_g(P) = 2\Delta x_1 + \cdots + 2\Delta x_{i-1} + 7\Delta x_i + 2\Delta x_{i+1} + \cdots + 2\Delta x_n$$
$$= 2(\Delta x_1 + \cdots + \Delta x_n) + 5\Delta x_i = 8 + 5\Delta x_i < 8 + \tfrac{5}{5}(I - 8) = I$$

and I does not satisfy $(*)$. This contradiction proves that I is not greater than 8 and therefore $I = 8$.

SECTION 5.3

1. (a) 5   (b) -2 (c) -1 (d) 0   (e) -4 (f) 1

3. With $P = \left\{1, \dfrac{3}{2}, 2\right\}$ and $f(x) = \dfrac{1}{x}$, we have $0.5 < \dfrac{7}{12} = L_f(P) \leq \displaystyle\int_1^2 \dfrac{dx}{x} \leq U_f(P) = \dfrac{5}{6} < 1$.

5. (a) $F(0) = 0$ (b) $F'(x) = x\sqrt{x+1}$ (c) $F'(2) = 2\sqrt{3}$

(d) $F(2) = \displaystyle\int_0^2 t\sqrt{t+1}\, dt$ (e) $-F(x) = \displaystyle\int_x^0 t\sqrt{t+1}\, dt$

7. (a) $\frac{1}{10}$ (b) $\frac{1}{9}$ (c) $\frac{4}{37}$ (d) $\dfrac{-2x}{(x^2+9)^2}$ **9.** (a) $\sqrt{2}$ (b) 0   (c) $-\frac{1}{4}\sqrt{5}$ (d) $-\dfrac{2x^2+1}{\sqrt{x^2+1}}$

11. (a) -1 (b) 1   (c) 0   (d) $-\pi \sin \pi x$

13. (a) Since $P_1 \subseteq P_2$, $U_f(P_2) \leq U_f(P_1)$ but $5 \nleq 4$.   (b) Since $P_1 \subseteq P_2$, $L_f(P_1) \leq L_f(P_2)$ but $5 \nleq 4$.

15. By the hint $F(b) - F(a) = F'(c)(b-a)$ for some c in (a, b). The desired result follows by observing that

$$F(b) = \int_a^b f(t)\, dt, \qquad F(a) = 0, \qquad \text{and} \qquad F'(c) = f(c).$$

17. Set $G(x) = \displaystyle\int_a^x f(t)\, dt$. Then $F(x) = \displaystyle\int_c^a f(t)\, dt + G(x)$. By (5.3.5) G, and thus F, is continuous on $[a, b]$, is differentiable on (a, b), and $F'(x) = G'(x) = f(x)$ for all x in (a, b).

SECTION 5.4

1. -2 **3.** 1   **5.** $\frac{14}{3}$ **7.** $\frac{32}{3}$

9. $\frac{2}{3}$ **11.** 0   **13.** $\frac{5}{72}$ **15.** $\frac{13}{2}$

17. $-\frac{4}{15}$ **19.** $\frac{1}{18}(2^{18} - 1)$ **21.** $\frac{1}{6}a^2$ **23.** $\frac{7}{4}$

25. $-\frac{1}{12}$ **27.** $\frac{21}{2}$ **29.** 1   **31.** 1

33. $2 - \sqrt{2}$ **35.** 0   **37.** (a) $\displaystyle\int_2^x \dfrac{dt}{t}$ (b) $-3 + \displaystyle\int_2^x \dfrac{dt}{t}$

39. $f(x)$ and $f(x) - f(a)$, respectively

SECTION 5.5

1. $\frac{9}{4}$ **3.** $\frac{38}{3}$ **5.** $\frac{47}{15}$ **7.** $\frac{5}{3}$ **9.** $\frac{1}{2}$

11. Figure A.5.5.1, area $= \frac{1}{3}$ **13.** Figure A.5.5.2, area $= \frac{9}{2}$ **15.** Figure A.5.5.3, area $= \frac{64}{3}$

Figure A.5.5.1

Figure A.5.5.2

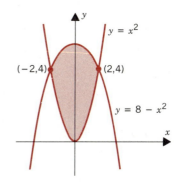

Figure A.5.5.3

17. Figure A.5.5.4, area = 10

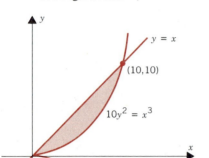

Figure A.5.5.4

19. Figure A.5.5.5, area = $\frac{32}{3}$

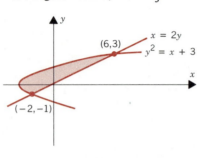

Figure A.5.5.5

21. Figure A.5.5.6, area = 4

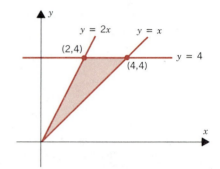

Figure A.5.5.6

23. Figure A.5.5.7, area = $2 + \frac{2}{3}\pi^3$

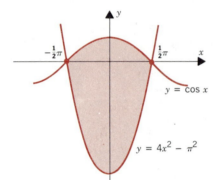

Figure A.5.5.7

25. Figure A.5.5.8, area = $\frac{1}{8}\pi^2 - 1$

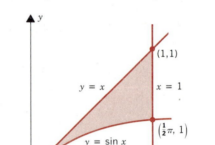

Figure A.5.5.8

SECTION 5.6

1. $-\dfrac{1}{3x^3} + C$

3. $\frac{1}{2}ax^2 + bx + C$

5. $2\sqrt{1+x} + C$

7. $\dfrac{1}{2}x^2 + \dfrac{1}{x} + C$

9. $\frac{1}{3}t^3 - \frac{1}{2}(a+b)t^2 + abt + C$

11. $\frac{2}{9}t^{9/2} - \frac{2}{5}(a+b)t^{5/2} + 2abt^{1/2} + C$

13. $\frac{1}{2}[g(x)]^2 + C$

15. $\frac{1}{2}\sec^2 x + C$

17. $-\dfrac{1}{4x+1} + C$

19. $x^2 - x - 2$

21. $\frac{1}{2}ax^2 + bx - 2a - 2b$

23. $3 - \cos x$

25. $x^3 - x^2 + x + 2$

27. $\frac{1}{12}(x^4 - 2x^3 + 2x + 23)$

29. $x - \cos x + 3$

31. $\frac{1}{3}x^3 - \frac{3}{2}x^2 - \frac{1}{3}x + 3$

33. $\dfrac{d}{dx}\left(\displaystyle\int f(x)\,dx\right) = f(x), \qquad \displaystyle\int \dfrac{d}{dx}[f(x)]\,dx = f(x) + C$

SECTION 5.7

1. (a) 34 units to the right of the origin   (b) 44 units

3. (a) $v(t) = 2(t+1)^{1/2} - 1$ (b) $x(t) = \frac{4}{3}(t+1)^{3/2} - t - \frac{4}{3}$

5. (a) 4.4 sec   (b) 193.6 ft

7. $[v(t)]^2 = (at + v_0)^2 = a^2t^2 + 2av_0t + v_0^2 = v_0^2 + 2a(\frac{1}{2}at^2 + v_0t) = v_0^2 + 2a[x(t) - x_0]$

$$x(t) = \frac{1}{2}at^2 + v_0t + x_0 \longrightarrow$$

9. 42 sec

11. $x(t) = x_0 + v_0 t + At^2 + Bt^3$

13. at $(\frac{160}{3}, 50)$

15. $A = -\frac{5}{2}, B = 2$

17. (a) at $t = \frac{11}{6}\pi$ sec   (b) at $t = \frac{13}{6}\pi$ sec

19. mean-value theorem

21. $v(t) = v_0(1 - 2tv_0)^{-1}$

SECTION 5.8

1. $\dfrac{1}{3(2 - 3x)} + C$

3. $\frac{1}{3}(2x + 1)^{3/2} + C$

5. $\dfrac{4}{7a}(ax + b)^{7/4} + C$

7. $-\dfrac{1}{8(4t^2 + 9)} + C$

9. $\frac{1}{75}(5t^3 + 9)^5 + C$

11. $\frac{4}{15}(1 + x^3)^{5/4} + C$

13. $-\dfrac{1}{4(1 + s^2)^2} + C$

15. $\sqrt{x^2 + 1} + C$

17. $-\frac{1}{5}(1 - x^3)^{5/3} + C$

19. $-\frac{5}{4}(x^2 + 1)^{-2} + C$

21. $-4(x^{1/4} + 1)^{-1} + C$

23. $-\dfrac{b^3}{2a^4}\sqrt{1 - a^4 x^4} + C$

25. $\frac{15}{8}$

27. $\frac{31}{2}$

29. 0

31. $\frac{1}{3}|a|^3$

33. $\frac{1}{6}a^3$

35. $\frac{2}{5}(x + 1)^{5/2} - \frac{2}{3}(x + 1)^{3/2} + C$

37. $\frac{1}{10}(2x - 1)^{5/2} + \frac{1}{6}(2x - 1)^{3/2} + C$

39. $\frac{1}{14}(y + 1)^{14} - \frac{1}{13}(y + 1)^{13} + C$

41. $\frac{1}{8}(t - 2)^8 + \frac{4}{7}(t - 2)^7 + \frac{2}{3}(t - 2)^6 + C$

43. $-\frac{1}{2}(t - 2)^{-2} - \frac{4}{3}(t - 2)^{-3} - (t - 2)^{-4} + C$

45. $\frac{5}{9}(x + 1)^{9/5} - \frac{5}{4}(x + 1)^{4/5} + C$

47. $\frac{16}{3}\sqrt{2} - \frac{14}{3}$

49. $-\frac{769}{112}$

SECTION 5.9

1. $\frac{1}{3}\sin(3x - 1) + C$

3. $-(\cot \pi x)/\pi + C$

5. $\frac{1}{2}\cos(3 - 2x) + C$

7. $-\frac{1}{5}\cos^5 x + C$

9. $-2\cos x^{1/2} + C$

11. $\frac{2}{3}(1 + \sin x)^{3/2} + C$

13. $\tan x + C$

15. $\frac{1}{8}\sin^4 x^2 + C$

17. $-\cot x - \frac{1}{3}\cot^3 x + C$

19. $\frac{1}{2}x - \frac{1}{12}\sin 6x + C$

21. $2(1 + \tan x)^{1/2} + C$

23. 0

25. $(\sqrt{3} - 1)/\pi$

27. $\frac{1}{4}$

29. $2 - \sqrt{2}$

31. π

33. $1/2\pi$

35. $(4\sqrt{3} - 6)/3\pi$

37. (a) $\frac{1}{2}\sec^2 x + C$ (b) $\frac{1}{2}\tan^2 x + C'$

(c) $\frac{1}{2}\sec^2 x + C = \frac{1}{2}(1 + \tan^2 x) + C = \frac{1}{2}\tan^2 x + (C + \frac{1}{2}) = \frac{1}{2}\tan^2 x + C'$; $C + \frac{1}{2}$ and C' each represent an arbitrary constant

39. πab

SECTION 5.10

1. yes; $\displaystyle\int_a^b [f(x) - g(x)] \, dx = \int_a^b f(x) \, dx - \int_a^b g(x) \, dx > 0$

3. yes; otherwise we would have $f(x) \le g(x)$ for all $x \in [a, b]$, and it would follow that

$$\int_a^b f(x) \, dx \le \int_a^b g(x) \, dx$$

5. no; take $f(x) = 0, g(x) = -1$ on $[0, 1]$

7. no; take, for example, any odd function on an interval of the form $[-c, c]$

9. no; $\displaystyle\int_{-1}^1 x \, dx = 0$ but $\displaystyle\int_{-1}^1 |x| \, dx \neq 0$

11. yes; $U_f(P) \ge \displaystyle\int_a^b f(x) \, dx = 0$

13. no; $L_f(P) \le \displaystyle\int_a^b f(x) \, dx = 0$

15. yes; $\displaystyle\int_a^b [f(x) + 1] \, dx = \int_a^b f(x) \, dx + \int_a^b 1 \, dx = 0 + b - a = b - a$

17. $\dfrac{2x}{\sqrt{2x^2 + 7}}$

19. $-f(x)$

21. $\dfrac{1}{x}$

23. $\dfrac{3}{1 + (2 + 3x)^{3/2}} - \dfrac{1}{3x^{2/3}(1 + x^{1/2})}$

25. (a) With P a partition of $[a, b]$

$$L_f(P) \le \int_a^b f(x)\, dx.$$

If f is nonnegative on $[a, b]$, then $L_f(P)$ is nonnegative and, consequently, so is the integral. If f is positive on $[a, b]$, then $L_f(P)$ is positive and, consequently, so is the integral.

(b) Take F as an antiderivative of f on $[a, b]$. Observe that

$$F'(x) = f(x) \quad \text{on } (a, b) \qquad \text{and} \qquad \int_a^b f(x)\, dx = F(b) - F(a).$$

If $f(x) \ge 0$ on $[a, b]$, then F is nondecreasing on $[a, b]$ and $F(b) - F(a) \ge 0$.
If $f(x) > 0$ on $[a, b]$, then F is increasing on $[a, b]$ and $F(b) - F(a) > 0$.

27. For all $x \in [a, b]$

$$-f(x) \le |f(x)| \qquad \text{and} \qquad f(x) \le |f(x)|.$$

It follows from II that

$$\int_a^b -f(x)\, dx \le \int_a^b |f(x)|\, dx \qquad \text{and} \qquad \int_a^b f(x)\, dx \le \int_a^b |f(x)|\, dx,$$

and, consequently, that

$$\left| \int_a^b f(x)\, dx \right| \le \int_a^b |f(x)|\, dx.$$

29. $H(x) = \displaystyle\int_{2x}^{x^3-4} \frac{x\, dt}{1+\sqrt{t}} = x \int_{2x}^{x^3-4} \frac{dt}{1+\sqrt{t}}$

$H'(x) = x \cdot \left[\dfrac{3x^2}{1+\sqrt{x^3-4}} - \dfrac{2}{1+\sqrt{2x}} \right] + 1 \cdot \displaystyle\int_{2x}^{x^3-4} \frac{dt}{1+\sqrt{t}}$

$H'(2) = 2\left[\dfrac{12}{3} - \dfrac{2}{3} \right] + \underbrace{\int_4^4 \frac{dt}{1+\sqrt{t}}}_{=0} = \dfrac{20}{3}$

SECTION 5.11

1. $A.V. = \frac{1}{2}mc + b, \quad x = \frac{1}{2}c$

3. $A.V. = 0, \quad x = 0$

5. $A.V. = 1, \quad x = \pm 1$

7. $A.V. = \frac{2}{3}, \quad x = 1 \pm \frac{1}{3}\sqrt{3}$

9. $A.V. = 2, \quad x = 4$

11. $A.V. = 0, \quad x = 0, \pi, 2\pi$

13. $A = $ average value of f on $[a, b] = \dfrac{1}{b-a} \displaystyle\int_a^b f(x)\, dx$

15. average of f' on $[a, b] = \dfrac{1}{b-a} \displaystyle\int_a^b f'(x)\, dx = \dfrac{1}{b-a}\left[f(x) \right]_a^b = \dfrac{f(b) - f(a)}{b-a}$

17. (a) 1 (b) $\frac{2}{3}\sqrt{3}$ (c) $\frac{7}{9}\sqrt{3}$

19. (a) The terminal velocity is twice the average velocity. (b) The average velocity during the first $\frac{1}{2}x$ seconds is one-third of the average velocity during the next $\frac{1}{2}x$ seconds.

21. (a) $M = \frac{2}{3}kL^{3/2}, \quad x_M = \frac{3}{5}L$ (b) $M = \frac{1}{3}kL^3, \quad x_M = \frac{1}{4}L$

23. $x_{M_2} = (2M - M_1)L/8M_2$ **25.** see answer to Exercise 15, Section 5.3

27. If f and g take on the same average value on every interval $[a, x]$, then

$$\frac{1}{x-a} \int_a^x f(t)\, dt = \frac{1}{x-a} \int_a^x g(t)\, dt.$$

Multiplication by $(x - a)$ gives

$$\int_a^x f(t)\, dt = \int_a^x g(t)\, dt.$$

Differentiation with respect to x gives $f(x) = g(x)$. This shows that, if the averages are the same on every interval, then the functions are everywhere the same.

SECTION 5.12

1. $\frac{2}{3}[(x - a)^{3/2} - (x - b)^{3/2}] + C$

3. $\frac{1}{2}(t^{2/3} - 1)^3 + C$

5. $x + \frac{8}{3}x^{3/2} + 2x^2 + C$

7. $\frac{5}{7}x^{7/5} - 2x + \frac{5}{3}x^{3/5} + C$

9. $\frac{2}{5}(2 - x)^{5/2} - \frac{4}{3}(2 - x)^{3/2} + C$

11. $\dfrac{2}{3b}(a + b\sqrt{y + 1})^3 + C$

13. $-\dfrac{1}{2[g(x)]^2} + C$

15. $\sqrt{1 + [g(x)]^2} + C$

17. $2\tan\theta - 2\sec\theta - \theta + C$

19. $\frac{1}{2}(\csc 2x - \cot 2x) + C$

21. $\dfrac{1}{4\pi}\sec^4 \pi x + C$

23. $-\dfrac{1}{2\pi}(1 + \sin^2 \pi x)^{-2} + C$

25. $\frac{13}{3}$

27. $\frac{39}{400}$

29. $y = \frac{1}{3}(x^2 + 1)^{3/2} + \frac{2}{3}$

31. (a) $\displaystyle\int_0^{x_0} [g(x) - y_0]\, dx$ (b) $\displaystyle\int_0^{x_0} [y_0 - f(x)]\, dx$ (c) $\displaystyle\int_0^{x_0} f(x)\, dx + \int_{x_0}^a g(x)\, dx$

(d) $\displaystyle\int_0^{x_0} \left[\left(\frac{c - b}{a}x + b\right) - g(x)\right] dx + \int_{x_0}^a \left[\left(\frac{c - b}{a}x + b\right) - f(x)\right] dx$ (e) $\displaystyle\int_{x_0}^a [f(x) - g(x)]\, dx$

33. Figure A.5.12.1, area = 8

35. Figure A.5.12.2, area = $\frac{125}{6}$

Figure A.5.12.1

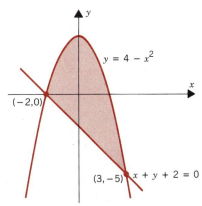

Figure A.5.12.2

37. Figure A.5.12.3, area = $\frac{3}{4}$

39. Figure A.5.12.4, area = $\frac{1}{2}a$

Figure A.5.12.3

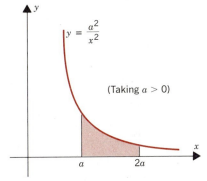

Figure A.5.12.4

41. necessarily holds: $L_g(P) \le \int_a^b g(x)\,dx < \int_a^b f(x)\,dx \le U_f(P)$

43. necessarily holds: $L_g(P) \le \int_a^b g(x)\,dx < \int_a^b f(x)\,dx$

45. necessarily holds: $U_f(P) \ge \int_a^b f(x)\,dx > \int_a^b g(x)\,dx$

47. $x(t) = A \sin(\omega t + \phi_0) + (v_0 - \omega A \cos \phi_0)t + x_0 - A \sin \phi_0$ **49.** $y = \dfrac{1}{1 + \sqrt{x}}$

51. $A \cong 0.615$:

$$L_f(P) = \tfrac{1}{100}\left(\tfrac{1}{1.1} + \tfrac{3}{1.2} + \tfrac{5}{1.3} + \tfrac{7}{1.4} + \tfrac{9}{1.5} + \tfrac{11}{1.6} + \tfrac{13}{1.7} + \tfrac{15}{1.8} + \tfrac{17}{1.9} + \tfrac{19}{2}\right) = \cong 0.595580$$
$$U_f(P) = \tfrac{1}{100}\left(\tfrac{1}{1} + \tfrac{3}{1.1} + \tfrac{5}{1.2} + \tfrac{7}{1.3} + \tfrac{9}{1.4} + \tfrac{11}{1.5} + \tfrac{13}{1.6} + \tfrac{15}{1.7} + \tfrac{17}{1.8} + \tfrac{19}{1.9}\right) \cong 0.634334$$
$$\tfrac{1}{2}[L_f(P) + U_f(P)] \cong \tfrac{1}{2}(0.595580 + 0.634334) \cong 0.615$$

53. $\dfrac{2x}{1 + x^4}$ **55.** $\sec x$ **57.** 1   **59.** $x_M = \tfrac{4}{9}a$ **61.** at $x = 1$

SECTION 5.13

1. Figure A.5.13.1

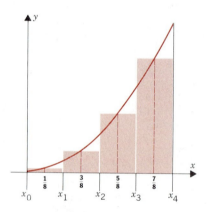

Figure A.5.13.1

3. (a) $\Delta x_1 = \Delta x_2 = \tfrac{1}{8}, \quad \Delta x_3 = \Delta x_4 = \Delta x_5 = \tfrac{1}{4}$
  (b) $\|P\| = \tfrac{1}{4}$
  (c) $m_1 = 0, m_2 = \tfrac{1}{4}, m_3 = \tfrac{1}{2}, m_4 = 1, m_5 = \tfrac{3}{2}$
  (d) $f(x_1^*) = \tfrac{1}{8}, f(x_2^*) = \tfrac{3}{8}, f(x_3^*) = \tfrac{3}{4}, f(x_4^*) = \tfrac{5}{4}, f(x_5^*) = \tfrac{3}{2}$
  (e) $M_1 = \tfrac{1}{4}, M_2 = \tfrac{1}{2}, M_3 = 1, M_4 = \tfrac{3}{2}, M_5 = 2$
  (f) $L_f(P) = \tfrac{25}{32}$ (g) $S^*(P) = \tfrac{15}{16}$
  (h) $U_f(P) = \tfrac{39}{32}$ (i) $\int_a^b f(x)\,dx = 1$

5. (a) $\dfrac{1}{n^2}(1 + 2 + \cdots + n) = \dfrac{1}{n^2}\left[\dfrac{n(n+1)}{2}\right] = \dfrac{1}{2} + \dfrac{1}{2n}$

  (b) $S_n^* = \dfrac{1}{2} + \dfrac{1}{2n}, \qquad \int_0^1 x\,dx = \left[\dfrac{1}{2}x^2\right]_0^1 = \dfrac{1}{2}$

  $\left| S_n^* - \int_0^1 x\,dx \right| = \dfrac{1}{2n} < \dfrac{1}{n} < \epsilon \quad \text{if} \quad n > \dfrac{1}{\epsilon}$

7. (a) $\dfrac{1}{n^4}(1^3 + 2^3 + \cdots + n^3) = \dfrac{1}{n^4}\left[\dfrac{n^2(n+1)^2}{4}\right] = \dfrac{1}{4} + \dfrac{1}{2n} + \dfrac{1}{4n^2}$

  (b) $S_n^* = \dfrac{1}{4} + \dfrac{1}{2n} + \dfrac{1}{4n^2}, \qquad \int_0^1 x^3\,dx = \left[\dfrac{1}{4}x^4\right]_0^1 = \dfrac{1}{4}$

  $\left| S_n^* - \int_0^1 x^3\,dx \right| = \dfrac{1}{2n} + \dfrac{1}{4n^2} < \dfrac{1}{n} < \epsilon \quad \text{if} \quad n > \dfrac{1}{\epsilon}$

9. $S^*(P) = \tfrac{1}{3}[\tfrac{1}{6}\cos(\tfrac{1}{6})^2 + \tfrac{3}{6}\cos(\tfrac{3}{6})^2 + \tfrac{5}{6}\cos(\tfrac{5}{6})^2 + \tfrac{7}{6}\cos(\tfrac{7}{6})^2 + \tfrac{9}{6}\cos(\tfrac{9}{6})^2 + \tfrac{11}{6}\cos(\tfrac{11}{6})^2] \cong -0.3991$
  $\int_0^2 x\cos x^2\,dx = \tfrac{1}{2}\sin 4 \cong -0.3784$

SECTION 6.1

1.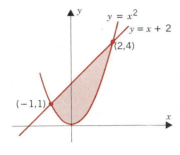

(a) $\int_{-1}^{2} [(x+2) - x^2]\, dx$

(b) $\int_{0}^{1} [\sqrt{y} - (-\sqrt{y})]\, dy + \int_{1}^{4} [\sqrt{y} - (y-2)]\, dy$

3.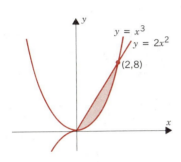

(a) $\int_{0}^{2} [2x^2 - x^3]\, dx$

(b) $\int_{0}^{8} \left[y^{1/3} - \left(\frac{1}{2} y\right)^{1/2} \right] dy$

5.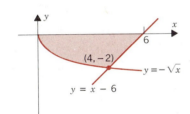

(a) $\int_{0}^{4} [0 - (-\sqrt{x})]\, dx + \int_{4}^{6} [0 - (x-6)]\, dx$

(b) $\int_{-2}^{0} [(y+6) - y^2]\, dy$

7.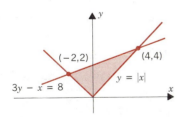

(a) $\int_{-2}^{0} \left[\frac{8+x}{3} - (-x) \right] dx + \int_{0}^{4} \left[\frac{8+x}{3} - x \right] dx$

(b) $\int_{0}^{2} [y - (-y)]\, dy + \int_{2}^{4} [y - (3y-8)]\, dy$

9.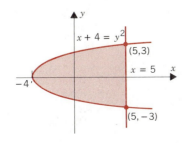

(a) $\int_{-4}^{5} [\sqrt{4+x} - (-\sqrt{4+x})]\, dx$

(b) $\int_{-3}^{3} [5 - (y^2 - 4)]\, dy$

11.

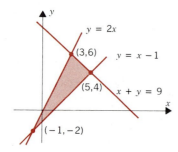

(a) $\displaystyle\int_{-1}^{3} [2x - (x-1)]\,dx + \int_{3}^{5} [(9-x) - (x-1)]\,dx$

(b) $\displaystyle\int_{-2}^{4} \left[(y+1) - \frac{1}{2}\,y \right] dy + \int_{4}^{6} \left[(9-y) - \frac{1}{2}\,y \right] dy$

13.

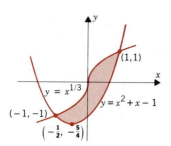

(a) $\displaystyle\int_{-1}^{1} [x^{1/3} - (x^2 + x - 1)]\,dx$

(b) $\displaystyle\int_{-5/4}^{-1} \left[\left(-\frac{1}{2} + \frac{1}{2}\sqrt{4y+5} \right) - \left(-\frac{1}{2} - \frac{1}{2}\sqrt{4y+5} \right) \right] dy$

$\displaystyle \qquad + \int_{-1}^{1} \left[\left(-\frac{1}{2} + \frac{1}{2}\sqrt{4y+5} \right) - y^3 \right] dy$

15. Figure A.6.1.1, area $= \frac{9}{8}$ **17.** Figure A.6.1.2, area $= 4$ **19.** Figure A.6.1.3, area $= \frac{5}{12}$

Figure A.6.1.1

Figure A.6.1.2

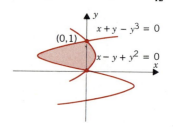

Figure A.6.1.3

SECTION 6.2

1. $\frac{1}{3}\pi$ **3.** $\frac{1944}{5}\pi$ **5.** $\frac{5}{14}\pi$ **7.** $\frac{3790}{21}\pi$

9. $\frac{72}{5}\pi$ **11.** $\frac{32}{3}\pi$ **13.** π **15.** $\dfrac{\pi^2}{24}(\pi^2 + 6\pi + 6)$

17. $\frac{16}{3}\pi$ **19.** $\frac{768}{7}\pi$ **21.** $\frac{2}{5}\pi$ **23.** $\frac{128}{3}\pi$

25. $\frac{16}{3}\pi$

27. (a) $\frac{16}{3}r^3$ (b) $\frac{4}{3}\sqrt{3}\,r^3$

29. (a) $\frac{512}{15}$ (b) $\frac{64}{15}\pi$ (c) $\frac{128}{15}\sqrt{3}$

31. (a) 32 (b) 4π (c) $8\sqrt{3}$

33. $\frac{4}{3}\pi ab^2$

35. $\frac{1}{3}\pi h(R^2 + rR + r^2)$

37. (a) $31\frac{1}{4}\%$ (b) $14\frac{22}{27}\%$

39. (a) 64π (b) $\frac{1024}{35}\pi$ (c) $\frac{704}{5}\pi$ (d) $\frac{512}{7}\pi$

SECTION 6.3

1. $\frac{2}{3}\pi$ **3.** $\frac{128}{5}\pi$ **5.** $\frac{2}{5}\pi$ **7.** 16π

9. $\frac{72}{5}\pi$ **11.** 36π **13.** 8π **15.** $\frac{1944}{5}\pi$

17. $\frac{5}{14}\pi$ **19.** $\frac{72}{5}\pi$ **21.** 64π **23.** $\frac{1}{3}\pi$

25. $\frac{4}{3}\pi ba^2$ **27.** $\frac{1}{4}\pi a^3\sqrt{3}$

29. (a) 64π (b) $\frac{1024}{35}\pi$ (c) $\frac{704}{5}\pi$ (d) $\frac{512}{7}\pi$

SECTION 6.4

1. $(\frac{12}{5}, \frac{3}{4})$, $V_x = 8\pi$, $V_y = \frac{128}{5}\pi$
3. $(\frac{3}{7}, \frac{12}{25})$, $V_x = \frac{2}{5}\pi$, $V_y = \frac{5}{14}\pi$
5. $(\frac{7}{3}, \frac{10}{3})$, $V_x = \frac{80}{3}\pi$, $V_y = \frac{56}{3}\pi$
7. $(\frac{3}{4}, \frac{22}{5})$, $V_x = \frac{704}{15}\pi$, $V_y = 8\pi$
9. $(\frac{2}{5}, \frac{2}{5})$, $V_x = \frac{4}{15}\pi$, $V_y = \frac{4}{15}\pi$
11. $(\frac{45}{28}, \frac{93}{70})$, $V_x = \frac{31}{5}\pi$, $V_y = \frac{15}{2}\pi$
13. $(3, \frac{3}{5})$, $V_x = \frac{40}{3}\pi$, $V_y = 24\pi$
15. $(\frac{5}{2}, 5)$
17. $(1, \frac{8}{5})$
19. $(\frac{10}{3}, \frac{40}{21})$
21. $(2, 4)$
23. $(-\frac{3}{5}, 0)$
25. (a) $(0, 0)$ (b) $(\frac{14}{5\pi}, \frac{14}{5\pi})$ (c) $(0, \frac{14}{5\pi})$
27. $V = \pi ab(2c + \sqrt{a^2 + b^2})$
29. (a) $(\frac{2}{3}a, \frac{1}{3}h)$ (b) $(\frac{2}{3}a + \frac{1}{3}b, \frac{1}{3}h)$ (c) $(\frac{1}{3}a + \frac{1}{3}b, \frac{1}{3}h)$
31. (a) $\frac{1}{3}\pi R^3 \sin^2 \theta\, (2 \sin \theta + \cos \theta)$ (b) $\dfrac{2R \sin \theta\, (2 \sin \theta + \cos \theta)}{3(\pi \sin \theta + 2 \cos \theta)}$

33. (a) The mass contributed by $[x_{i-1}, x_i]$ is approximately $\lambda(x_i^*)\Delta x_i$ where x_i^* is the midpoint of $[x_{i-1}, x_i]$. The sum of these contributions,

$$\lambda(x_1^*)\Delta x_1 + \cdots + \lambda(x_n^*)\Delta x_n,$$

is a Riemann sum, which as $\|P\| \to 0$, tends to the given integral.

(b) Take M_i as the mass contributed by $[x_{i-1}, x_i]$. Then $x_{M_i}M_i \cong x_i^* \lambda(x_i^*)\Delta x_i$ where x_i^* is the midpoint of $[x_{i-1}, x_i]$. Therefore

$$x_M M = x_{M_1}M_1 + \cdots + x_{M_n}M_n \cong x_1^*\lambda(x_1^*)\Delta x_1 + \cdots + x_n^*\lambda(x_n^*)\Delta x_n.$$

As $\|P\| \to 0$, the sum on the right converges to the given integral.

35. on the axis of the cone at distance $\frac{3}{4}h$ from the vertex
37. (a) $(\frac{2}{3}, 0)$ (b) $(0, \frac{5}{12})$
39. $(\frac{3}{8}a, 0)$

SECTION 6.5

1. (a) 25 ft-lb   (b) $\frac{225}{4}$ ft-lb
3. 1.95 ft
5. (a) $(6480\pi + 8640)$ ft-lb   (b) $(15,120\pi + 8640)$ ft-lb
7. (a) $\frac{11}{192}\pi r^2 h^2 \sigma$ ft-lb   (b) $(\frac{11}{192}\pi r^2 h^2 \sigma + \frac{7}{24}\pi r^2 hk\sigma)$ ft-lb
9. $GmM\left(\dfrac{1}{r_2} - \dfrac{1}{r_1}\right)$
11. 788 ft-lb
13. (a) $\frac{1}{2}\sigma l^2$ ft-lb   (b) $\frac{3}{2}l^2\sigma$ ft-lb
15. 20,800 ft-lb

SECTION *6.6

1. 2160 lb   3. $\frac{8000}{3}\sqrt{2}$ lb   5. 2560 lb   7. (a) 41,250 lb   (b) 41,250 lb
9. $F = \sigma \bar{x} A$ where A is the area of the submerged surface and $\bar{x}$ is the depth of its centroid

SECTION *6.7

1. $\dfrac{GmM}{h(L + h)}$
3. $\dfrac{GmMh}{(h^2 + R^2)^{3/2}}$
5. for h small compared to R, $\dfrac{GM}{(R + h)^2} \cong \dfrac{GM}{R^2} = g$

SECTION 7.2

1. $\ln 2 + \ln 10 \cong 2.99$
3. $2 \ln 4 - \ln 10 \cong 0.48$
5. $-\ln 10 \cong -2.30$
7. $\ln 8 + \ln 9 - \ln 10 \cong 1.98$
9. $\frac{1}{2}\ln 2 \cong 0.35$ 11. $5 \ln 2 \cong 3.45$
13. Figure A.7.2.1   15. 0.406
17. (a) 1.65   (b) 1.57   (c) 1.71
19. $x = e^2$ 21. $x = 1, e^2$
23. $x = 1$ 25. $l \cong 1$ 27. $l \cong 0$

Figure A.7.2.1

SECTION 7.3

1. domain $(0, \infty)$, $f'(x) = \dfrac{1}{x}$

3. domain $(-1, \infty)$, $f'(x) = \dfrac{3x^2}{x^3 + 1}$

5. domain $(-\infty, \infty)$, $f'(x) = \dfrac{x}{1 + x^2}$

7. domain all $x \neq \pm 1$, $f'(x) = \dfrac{4x^3}{x^4 - 1}$

9. domain $(0, \infty)$, $f'(x) = 1 + \ln x$

11. domain $(0, 1) \cup (1, \infty)$, $f'(x) = -\dfrac{1}{x(\ln x)^2}$

13. domain $(-1, \infty)$, $f'(x) = \dfrac{1 - \ln (x + 1)}{(x + 1)^2}$

15. $\ln |x + 1| + C$

17. $-\frac{1}{2} \ln |3 - x^2| + C$

19. $\dfrac{1}{2(3 - x^2)} + C$

21. $\ln \left| \dfrac{x + 2}{x - 2} \right| + C$

23. $\dfrac{-1}{\ln x} + C$

25. $\frac{2}{3} \ln |1 + x\sqrt{x}| + C$

27. 1   **29.** 1

31. $\frac{1}{2} \ln \frac{8}{5}$ **33.** $\frac{1}{2}(\ln 2)^2$ **35.** $g'(x) = (x^2 + 1)^2(x - 1)^5 x^3 \left(\dfrac{4x}{x^2 + 1} + \dfrac{5}{x - 1} + \dfrac{3}{x} \right)$

37. $g'(x) = \dfrac{x^4(x - 1)}{(x + 2)(x^2 + 1)} \left(\dfrac{4}{x} + \dfrac{1}{x - 1} - \dfrac{1}{x + 2} - \dfrac{2x}{x^2 + 1} \right)$

39. $g'(x) = \dfrac{1}{2} \sqrt{\dfrac{(x - 1)(x - 2)}{(x - 3)(x - 4)}} \left(\dfrac{1}{x - 1} + \dfrac{1}{x - 2} - \dfrac{1}{x - 3} - \dfrac{1}{x - 4} \right)$

41. $\frac{15}{8} - \ln 4$ **43.** $\ln 5$ ft   **45.** $(-1)^{n-1} \dfrac{(n - 1)!}{x^n}$ **47.** $(-1)^{n-1} \dfrac{(n - 1)!}{x^n}$

49. (a) for $t \in (1, x)$ (b) for $x > 1$

$$\frac{1}{t} < \frac{1}{\sqrt{t}}$$

$$\ln x = \int_1^x \frac{dt}{t} < \int_1^x \frac{dt}{\sqrt{t}} = \left[2\sqrt{t} \right]_1^x = 2(\sqrt{x} - 1)$$

(c) for $0 < x < 1$

$$0 < \ln \frac{1}{x} < 2 \left(\sqrt{\frac{1}{x}} - 1 \right) \quad \text{by (b)}$$

$$0 < -\ln x < 2 \left(\frac{1}{\sqrt{x}} - 1 \right)$$

$$2 \left(1 - \frac{1}{\sqrt{x}} \right) < \ln x < 0$$

$$2x \left(1 - \frac{1}{\sqrt{x}} \right) < x \ln x < 0$$

(d) Use (c) and the pinching theorem for one-sided limits.

51. (i) domain $(0, \infty)$ (ii) increases on $(0, \infty)$ (iii) no extreme values
(iv) concave down on $(0, \infty)$; no pts of inflection   (v) Figure A.7.3.1

53. (i) domain $(-\infty, 4)$ (ii) decreases throughout   (iii) no extreme values
(iv) concave down throughout; no pts of inflection   (v) Figure A.7.3.2

Figure A.7.3.1

Figure A.7.3.2

Figure A.7.3.3

Figure A.7.3.4

55. (i) domain $(0, \infty)$ (ii) decreases on $(0, 1/e]$, increases on $[1/e, \infty)$
(iii) $f(1/e) = -1/e$ local and absolute min
(iv) concave up throughout; no pts of inflection   (v) Figure A.7.3.3

57. (i) domain $(0, \infty)$ (ii) increases on $(0, 1]$, decreases on $[1, \infty)$
(iii) $f(1) = -\ln 2$ local and absolute max
(iv) concave down on $(0, \sqrt{2 + \sqrt{5}})$, concave up on $(\sqrt{2 + \sqrt{5}}, \infty)$; pt of inflection at $x = \sqrt{2 + \sqrt{5}}$
(v) Figure A.7.3.4

SECTION 7.4

1. $\dfrac{dy}{dx} = -2e^{-2x}$

3. $\dfrac{dy}{dx} = 2xe^{x^2 - 1}$

5. $\dfrac{dy}{dx} = e^x \left(\dfrac{1}{x} + \ln x \right)$

7. $\dfrac{dy}{dx} = -(x^{-1} + x^{-2})e^{-x}$

9. $\dfrac{dy}{dx} = \dfrac{1}{2}(e^x - e^{-x})$

11. $\dfrac{dy}{dx} = \dfrac{1}{2} e^{\sqrt{x}} \left(\dfrac{1}{x} + \dfrac{\ln \sqrt{x}}{\sqrt{x}} \right)$

13. $\dfrac{dy}{dx} = 2(e^{2x} - e^{-2x})$

15. $\dfrac{dy}{dx} = 4xe^{x^2}(e^{x^2} + 1)$

17. $\dfrac{dy}{dx} = x^2 e^x$

19. $\dfrac{dy}{dx} = \dfrac{2e^x}{(e^x + 1)^2}$

21. $\dfrac{dy}{dx} = 2(a - b)\dfrac{e^{(a+b)x}}{(e^{ax} + e^{bx})^2}$

23. $\dfrac{dy}{dx} = 4x^3$

25. $\frac{1}{2}e^{2x} + C$

27. $\dfrac{1}{k} e^{kx} + C$

29. $\frac{1}{2}e^{x^2} + C$

31. $-e^{1/x} + C$

33. $\frac{1}{2}e^{2x} - \frac{1}{2}e^{-2x} + 2x + C$

35. $\frac{1}{2}x^2 + C$

37. $-8e^{-x/2} + C$

39. $2\sqrt{e^x + 1} + C$

41. $\frac{1}{4}\ln(2e^{2x} + 3) + C$

43. $e - 1$

45. $\frac{1}{6}(1 - \pi^{-6})$

47. $2 - \dfrac{1}{e}$

49. $\ln \frac{3}{2}$

51. $\frac{185}{72} + \ln \frac{4}{9}$

53. $\frac{1}{2}e + \frac{1}{2}$

55. $e^{-0.4} = \dfrac{1}{e^{0.4}} \cong \dfrac{1}{1.49} \cong 0.67$

57. $e^{2.8} = (e^2)(e^{0.8}) \cong (7.39)(2.23) \cong 16.48$

59. 7.61

61. 23.10

63. $x''(t) = Ac^2 e^{ct} + Bc^2 e^{-ct} = c^2(Ae^{ct} + Be^{-ct}) = c^2 x(t)$

65. $\frac{1}{2}(3e^4 + 1)$

67. $e^2 - e - 2$

69. (i) domain $(-\infty, \infty)$ (ii) decreases on $(-\infty, 0]$, increases on $[0, \infty)$
(iii) $f(0) = 1$ local and absolute min   (iv) concave up everywhere   (v) Figure A.7.4.1

71. (i) domain $(-\infty, \infty)$ (ii) decreases on $(-\infty, -1]$, increases on $[-1, \infty)$
(iii) $f(-1) = -1/e$ local and absolute min
(iv) concave down on $(-\infty, -2)$, concave up on $(-2, \infty)$; pt of inflection $(-2, -2e^{-2})$
(v) Figure A.7.4.2

73. (i) domain $(-\infty, 0) \cup (0, \infty)$
(ii) increases on $(-\infty, 0)$, decreases on $(0, \infty)$ (iii) no extreme values
(iv) concave up on $(-\infty, 0)$ and on $(0, \infty)$ (v) Figure A.7.4.3

Figure A.7.4.1

horizontal aymptote $y = 0$

Figure A.7.4.2

Figure A.7.4.3

75. (a) $\left(\pm\dfrac{1}{a}, e\right)$ (b) $\dfrac{1}{a}(e-2)$ (c) $\dfrac{1+2a^2e}{a^3e}$

77. for $x > (n+1)!$

$$e^x > 1 + x + \cdots + \frac{x^{n+1}}{(n+1)!} > \frac{x^{n+1}}{(n+1)!} = x^n\left[\frac{x}{(n+1)!}\right] > x^n$$

79. Numerically, $8.15 \le l \le 8.16$. The limit is the derivative of $f(x) = e^{x^3}$ at $x = 1$; note that $f'(x) = 3x^2e^{x^3}$ and $f'(1) = 3e \cong 8.15485$.

SECTION 7.5

1. 6   **3.** $-\frac{1}{6}$ **5.** 0   **7.** 3   **9.** $\frac{3}{5}$ **11.** $-\frac{8}{5}$

13. $\log_p xy = \dfrac{\ln xy}{\ln p} = \dfrac{\ln x + \ln y}{\ln p} = \dfrac{\ln x}{\ln p} + \dfrac{\ln y}{\ln p} = \log_p x + \log_p y$

15. $\log_p x^y = \dfrac{\ln x^y}{\ln p} = y\dfrac{\ln x}{\ln p} = y\log_p x$ **17.** 0   **19.** 2

21. e^c, where $c = \dfrac{(\ln 2)^2}{\ln 2 - 1}$ **23.** $t_1 < \ln a < t_2$ **25.** $\dfrac{3^x}{\ln 3} + C$

27. $\frac{1}{4}x^4 - \dfrac{3^{-x}}{\ln 3} + C$ **29.** $\log_5 |x| + C$ **31.** $\dfrac{3}{\ln 4}(\ln x)^2 + C$

33. $\dfrac{10^{x^2}}{2\ln 10} + C$ **35.** $\dfrac{1}{e\ln 3}$ **37.** $\dfrac{1}{e}$

39. $f(x) = p^x$
$\ln f(x) = x \ln p$
$\dfrac{f'(x)}{f(x)} = \ln p$
$f'(x) = p^x \ln p$

41. $(x+1)^x\left[\dfrac{x}{x+1} + \ln(x+1)\right]$

43. $(\ln x)^{\ln x}\left[\dfrac{1 + \ln(\ln x)}{x}\right]$

45. $(x^2+2)^{\ln x}\left[\dfrac{2x\ln x}{x^2+2} + \dfrac{\ln(x^2+2)}{x}\right]$

47. Figure A.7.5.1   **49.** Figure A.7.5.2   **51.** Figure A.7.5.3   **53.** Figure A.7.5.4

Figure A.7.5.1   **Figure A.7.5.2**   **Figure A.7.5.3**   **Figure A.7.5.4**

55. domain $(-\infty, \infty)$; increasing on $(-\infty, 0]$, decreasing on $[0, \infty)$; $f(0) = 10$ local and absolute max

57. domain $[-1, 1]$; increasing on $[-1, 0]$, decreasing on $[0, 1]$; $f(0) = 10$ local and absolute max, $f(-1) = f(1) = 1$ endpt and absolute min

59. $\dfrac{1}{4\ln 2}$ **61.** 2   **63.** $\dfrac{45}{\ln 10}$ **65.** $(\ln 2)(\ln 10)$

67. $\dfrac{1}{3} + \dfrac{1}{\ln 2}$ **69.** $\log_{10} 7 = \dfrac{\ln 7}{\ln 10} \cong \dfrac{1.95}{2.30} \cong 0.85$

71. $\log_{10} 45 = \dfrac{\ln 45}{\ln 10} = \dfrac{\ln 9 + \ln 5}{\ln 10} \cong \dfrac{2.20 + 1.61}{2.30} \cong 1.66$

73. approx. 16.999999; $5^{(\ln 17)/(\ln 5)} = (e^{\ln 5})^{(\ln 17)/(\ln 5)} = e^{\ln 17} = 17$

75. approx. 54.59815; $16^{1/\ln 2} = (e^{\ln 16})^{1/\ln 2} = e^{(\ln 16)/(\ln 2)} = e^{4(\ln 2)/(\ln 2)} = e^4 \cong 54.59815$

SECTION 7.6

1. (a) \$411.06   (b) \$612.77   (c) \$859.14   **3.** (a) \$740.82   (b) \$670.32   (c) \$606.53
5. about $5\frac{1}{2}$%: $(\ln 3)/20 \cong 0.0549$ **7.** $20 + 5e^{-0.03t}$ lb

9. $200(\frac{4}{5})^{t^2/25}$ liters   **11.** a little more than $9\frac{1}{2}$ years: $\dfrac{4 \ln \frac{1}{2}}{\ln \frac{3}{4}} \cong 9.64$

13. $5(\frac{4}{5})^{5/2} \cong 2.86$ gms   **15.** $100[1 - (\frac{1}{2})^{1/n}]\%$

17. (a) $x_1(t) = 10^6 t,\quad x_2(t) = e^t - 1$

  (b) $\dfrac{d}{dt}[x_1(t) - x_2(t)] = \dfrac{d}{dt}[10^6 t - (e^t - 1)] = 10^6 - e^t.$

   This derivative is zero at $t = 6 \ln 10 \cong 13.8$.   After that the derivative is negative.
  (c) $x_2(15) < e^{15} = (e^3)^5 \cong 20^5 = 2^5(10^5) = 3.2(10^6) < 15(10^6) = x_1(15)$
    $x_2(18) = e^{18} - 1 = (e^3)^6 - 1 \cong 20^6 - 1 = 64(10^6) - 1 > 18(10^6) = x_1(18)$
    $x_2(18) - x_1(18) \cong 64(10^6) - 1 - 18(10^6) \cong 46(10^6)$
  (d) If by time t_1 EXP has passed LIN, then $t_1 > 6 \ln 10$.   For all $t \ge t_1$ the speed of EXP is greater than the
   speed of LIN: for $t \ge t_1 > 6 \ln 10$, $v_2(t) = e^t > 10^6 = v_1(t)$.

19. (a) $15(\frac{2}{3})^{1/2} \cong 12.25$ lb/in.^{2} (b) $15(\frac{2}{3})^{3/2} \cong 8.16$ lb/in.^{2}
21. (a) about \$6971   (b) about \$7627   (c) about \$8600   **23.** $176/\ln 2 \cong 254$ ft
25. about 284 million: $203(\frac{227}{203})^3 \cong 283.85$
27. about $16\frac{1}{2}$ years: $100 (\ln \frac{3}{2})^2 \cong 16.44$ **29.** 640 lb
31. Proceeding from the hint, we know from Theorem 7.6.1 that

$$f(t) + k_2/k_1 = Ae^{k_1(t + k_2/k_1)} = Ae^{k_1 t + k_2} = (Ae^{k_2})e^{k_1 t} = Ce^{k_1 t}.$$

$\qquad\qquad\qquad$ └── where A is an arbitrary constant   └── set $Ae^{k_2} = C$
  Therefore $f(t) = Ce^{k_1 t} - k_2/k_1$, C an arbitrary constant.
33. From Exercise 31 you can determine that

$$v(t) = \frac{32}{K}(1 - e^{-Kt}).$$

  At each time t, $1 - e^{-Kt} < 1$.   With $K > 0$,

$$v(t) = \frac{32}{K}(1 - e^{-Kt}) < \frac{32}{K}.$$

35. $\dfrac{2 \ln 3}{\ln \frac{4}{3}} - 2 \cong 5.64$ min

SECTION 7.7

1. $-xe^{-x} - e^{-x} + C$ **3.** $\frac{1}{3}x^3 \ln x - \frac{1}{9}x^3 + C$ **5.** $-\frac{1}{3}e^{-x^3} + C$
7. $-e^{-x}(x^2 + 2x + 2) + C$ **9.** $-2x^2(1 - x)^{1/2} - \frac{8}{3}x(1 - x)^{3/2} - \frac{16}{15}(1 - x)^{5/2} + C$
11. $\frac{1}{4}x^2 \ln x - \frac{1}{8}x^2 + C$ **13.** $2\sqrt{x+1} \ln(x+1) - 4\sqrt{x+1} + C$
15. $x(\ln x)^2 - 2x \ln x + 2x + C$ **17.** $3^x\left(\dfrac{x^3}{\ln 3} - \dfrac{3x^2}{(\ln 3)^2} + \dfrac{6x}{(\ln 3)^3} - \dfrac{6}{(\ln 3)^4}\right) + C$
19. $\frac{1}{15}x(x + 5)^{15} - \frac{1}{240}(x + 5)^{16} + C$ **21.** $x \sin x + \cos x + C$
23. $\frac{1}{10}x^2(x + 1)^{10} - \frac{1}{55}x(x + 1)^{11} + \frac{1}{660}(x + 1)^{12} + C$
25. $\frac{1}{2}e^x(\sin x - \cos x) + C$ **27.** $x \ln(1 + x^2) - 2x + 2 \tan^{-1} x + C$
29. $\dfrac{x^{n+1}}{n + 1} \ln x - \dfrac{x^{n+1}}{(n + 1)^2} + C$ **31.** $-\frac{1}{2}x^2 \cos x^2 + \frac{1}{2}\sin x^2 + C$
33. $e^x(x^4 - 4x^3 + 12x^2 - 24x + 24) + C$ **35.** $\bar{x} = 1/(e - 1),\quad \bar{y} = (e + 1)/4$
37. $\bar{x} = \frac{1}{2}\pi,\quad \bar{y} = \frac{1}{8}\pi$ **39.** (a) $M = (e^k - 1)/k$ (b) $x_M = [(k - 1)e^k + 1]/[k(e^k - 1)]$
41. $V = (2\pi/\alpha^2)(\alpha e^\alpha - e^\alpha + 1)$ **43.** $V = 4 - 8/\pi$
45. $V = 2\pi(e - 2)$ **47.** $\bar{x} = (e^2 + 1)/[2(e^2 - 1)]$

SECTION 7.8

1. $x(t) = \sin(8t + \frac{1}{2}\pi)$; $A = 1$, $f = 4/\pi$

3. $\pm 2\pi A/T$

5. $x(t) = (15/\pi) \sin \frac{1}{3}\pi t$

7. $x(t) = x_0 \sin(\sqrt{k/m}\, t + \frac{1}{2}\pi)$

9. at $x = \pm\frac{1}{2}\sqrt{3}\,x_0$

11. $\frac{1}{4}kx_0^2$

13. Set $y(t) = x(t) - 2$. Equation $x''(t) = 8 - 4x(t)$ can be written $y''(t) + 4y(t) = 0$. This is simple harmonic motion centered at $y = 0$, which is $x = 2$.

$$y(t) = A \sin(2t + \phi_0).$$

The condition $x(0) = 0$ gives $y(0) = -2$ and thus

(∗) $$A \sin \phi_0 = -2.$$

Since $y'(t) = x'(t)$ and $y'(t) = 2A \cos(2t + \phi_0)$, the condition $x'(0) = 0$ gives $y'(0) = 0$ and thus

(∗∗) $$2A \cos \phi_0 = 0.$$

Equations (∗) and (∗∗) are satisfied by $A = 2$, $\phi_0 = \frac{3}{2}\pi$. The equation of motion can therefore be written

$$y(t) = 2 \sin(2t + \frac{3}{2}\pi).$$

The amplitude is 2 and the period is π.

15. (a) $x''(t) + \omega^2 x(t) = 0$ with $\omega = r\sqrt{\pi\rho/m}$
(b) $x(t) = x_0 \sin(r\sqrt{\pi\rho/m}\, t + \frac{1}{2}\pi)$, taking downward as positive; $A = x_0$, $T = (2/r)\sqrt{m\pi/\rho}$

SECTION 7.9

1. $\frac{1}{3} \ln |\sec 3x| + C$

3. $(1/\pi) \ln |\csc \pi x - \cot \pi x| + C$

5. $\ln |\sin e^x| + C$

7. $\frac{1}{4} \ln |3 - 2 \cos 2x| + C$

9. $\frac{1}{3} e^{\tan 3x} + C$

11. $\frac{1}{2} \ln |\sec x^2 + \tan x^2| + C$

13. $\frac{1}{2}(\ln |\sin x|)^2 + C$

15. $x + 2 \ln |\sec x + \tan x| + \tan x + C$

17. $(1 + \cot x)^{-1} + C$

19. $\ln \frac{4}{3}$

21. $\frac{1}{2} \ln 2$

23. $\ln [(1 + \sqrt{2})/(\sec 1 + \tan 1)] \cong -0.345$

25. $\frac{1}{3}\pi - \frac{1}{2} \ln 3$

27. $\frac{1}{4}\pi - \frac{1}{2} \ln 2$

SECTION 7.10

1. 0

3. $\frac{1}{3}\pi$

5. $-\frac{1}{3}\pi$

7. 0

9. $-\frac{1}{6}\pi$

11. $\frac{1}{2}$

13. $-\frac{1}{4}\pi$

15. $\frac{1}{2}\sqrt{3}$

17. $\frac{1}{4}\pi$

19. $\frac{1}{2}\sqrt{3}$

21. 1.16

23. -0.46

25. 0.28

27. $1/\sqrt{1 + x^2}$

29. x

31. $\sqrt{1 + x^2}$

33. $\sqrt{1 - x^2}$

35. $\dfrac{1}{x}\sqrt{1 + x^2}$

37. $\dfrac{x}{\sqrt{1 - x^2}}$

39. $\dfrac{1}{x^2 + 2x + 2}$

41. $\dfrac{2x}{\sqrt{1 - x^4}}$

43. $\dfrac{2x}{\sqrt{1 - 4x^2}} + \sin^{-1} 2x$

45. $\dfrac{2 \sin^{-1} x}{\sqrt{1 - x^2}}$

47. $\dfrac{x - (1 + x^2) \tan^{-1} x}{x^2(1 + x^2)}$

49. $\dfrac{1}{(1 + 4x^2)\sqrt{\tan^{-1} 2x}}$

51. $\dfrac{1}{x[1 + (\ln x)^2]}$

53. $-\dfrac{r}{|r|\sqrt{1 - r^2}}$

55. $\dfrac{1}{1 + r^2}$

57. $\sqrt{\dfrac{c - x}{c + x}}$

59. $\dfrac{x^2}{(c^2 - x^2)^{3/2}}$

61. Set $au = x + b$, $a\, du = dx$.

$$\int \frac{dx}{\sqrt{a^2 - (x + b)^2}} = \int \frac{a\, du}{\sqrt{a^2 - a^2u^2}} = \int \frac{du}{\sqrt{1 - u^2}} = \sin^{-1} u + C = \sin^{-1}\left(\frac{x + b}{a}\right) + C$$

63. $\frac{1}{4}\pi$

65. $\frac{1}{4}\pi$

67. $\frac{1}{20}\pi$

69. $\frac{1}{24}\pi$

71. $\frac{1}{6}\pi$

73. $\tan^{-1} 2 - \frac{1}{4}\pi \cong 0.322$

75. set $u = \sin^{-1} x$, $dv = dx$ and integrate by parts

77. $\frac{1}{2} \sin^{-1} x^2 + C$

79. $\frac{1}{2} \tan^{-1} x^2 + C$

81. $\tan^{-1}(\frac{1}{3} \tan x) + C$

83. Set $\theta = \sin^{-1} x$. Then θ is the only number in $[-\frac{1}{2}\pi, \frac{1}{2}\pi]$ for which $\sin \theta = x$. Since $\cos(\frac{1}{2}\pi - \theta) = \sin \theta = x$ and $(\frac{1}{2}\pi - \theta) \in [0, \pi]$, $\cos^{-1} x = \frac{1}{2}\pi - \theta$. Therefore $\sin^{-1} x + \cos^{-1} x = \theta + (\frac{1}{2}\pi - \theta) = \frac{1}{2}\pi$.

85. By Exercise 83

$$\frac{d}{dx}(\cos^{-1} x) = \frac{d}{dx}(\tfrac{1}{2}\pi - \sin^{-1} x) = -\frac{d}{dx}(\sin^{-1} x) = -\frac{1}{\sqrt{1-x^2}}.$$

87.

$$y = \sec^{-1} x$$
$$\sec y = x$$
$$\sec y \tan y \frac{dy}{dx} = 1$$
$$\frac{dy}{dx} = \cos y \cot y = \left(\frac{1}{x}\right)\left(\frac{1}{x\sqrt{1-1/x^2}}\right) = \frac{1}{|x|\sqrt{x^2-1}}$$

89. estimate $\cong 0.523$, $\sin 0.523 \cong 0.499$

explanation: the integral $= \sin^{-1} 0.5$; therefore $\sin$ (integral) $= 0.5$

SECTION 7.11

1. $2x \cosh x^2$

3. $\dfrac{a \sinh ax}{2\sqrt{\cosh ax}}$

5. $\dfrac{1}{1 - \cosh x}$

7. $ab(\cosh bx - \sinh ax)$

9. $\dfrac{a \cosh ax}{\sinh ax}$

11. $\cosh^2 t - \sinh^2 t = \left(\dfrac{e^t + e^{-t}}{2}\right)^2 - \left(\dfrac{e^t - e^{-t}}{2}\right)^2 = \dfrac{e^{2t} + 2 + e^{-2t}}{4} - \dfrac{e^{2t} - 2 + e^{-2t}}{4} = 1$

13. $\cosh t \cosh s + \sinh t \sinh s = \left(\dfrac{e^t + e^{-t}}{2}\right)\left(\dfrac{e^s + e^{-s}}{2}\right) + \left(\dfrac{e^t - e^{-t}}{2}\right)\left(\dfrac{e^s - e^{-s}}{2}\right)$

$$= \tfrac{1}{4}(e^{t+s} + e^{s-t} + e^{t-s} + e^{-t-s} + e^{t+s} - e^{s-t} - e^{t-s} + e^{-t-s})$$
$$= \tfrac{1}{2}(e^{t+s} + e^{-(t+s)}) = \cosh(t+s)$$

15. $\cosh^2 t + \sinh^2 t = \left(\dfrac{e^t + e^{-t}}{2}\right)^2 + \left(\dfrac{e^t - e^{-t}}{2}\right)^2 = \tfrac{1}{4}(e^{2t} + 2 + e^{-2t} + e^{2t} - 2 - e^{-2t}) = \dfrac{e^{2t} + e^{-2t}}{2} = \cosh 2t$

17. absolute max -3

19. $[\cosh x + \sinh x]^n = \left[\dfrac{e^x + e^{-x}}{2} + \dfrac{e^x - e^{-x}}{2}\right]^n = [e^x]^n = e^{nx} = \dfrac{e^{nx} + e^{-nx}}{2} + \dfrac{e^{nx} - e^{-nx}}{2} = \cosh nx + \sinh nx$

21. $A = 2$, $B = \frac{1}{3}$, $c = 3$

23. $\dfrac{1}{a} \sinh ax + C$

25. $\dfrac{1}{3a} \sinh^3 ax + C$

27. $\dfrac{1}{a} \ln (\cosh ax) + C$

29. $-\dfrac{1}{a \cosh ax} + C$

31. $x \cosh x - \sinh x + C$

33. $\frac{1}{2}(\sinh x \cosh x + x) + C$

35. $A = \sinh 1 = \dfrac{e^2 - 1}{2e}$; $\quad \bar{x} = \dfrac{2}{e+1}$, $\quad \bar{y} = \dfrac{e^4 + 4e^2 - 1}{8e(e^2 - 1)}$

37. (a) $V_x = \pi$ (b) $V_y = \dfrac{2\pi}{e}(e - 2)$

SECTION *7.12

1. $2 \tanh x \, \text{sech}^2 x$

3. $\text{sech} \, x \, \text{csch} \, x$

5. $\dfrac{2e^{2x} \cosh(\tan^{-1} e^{2x})}{1 + e^{4x}}$

7. $\dfrac{-x \, \text{csch}^2(\sqrt{x^2 + 1})}{\sqrt{x^2 + 1}}$

9. $\dfrac{-\text{sech} \, x (\tanh x + 2 \sinh x)}{(1 + \cosh x)^2}$

11. $\dfrac{d}{dx}(\coth x) = \dfrac{d}{dx}\left(\dfrac{\cosh x}{\sinh x}\right) = \dfrac{\sinh^2 x - \cosh^2 x}{\sinh^2 x} = \dfrac{-1}{\sinh^2 x} = -\operatorname{csch}^2 x$

13. $\dfrac{d}{dx}(\operatorname{csch} x) = \dfrac{d}{dx}\left(\dfrac{1}{\sinh x}\right) = -\dfrac{\cosh x}{\sinh^2 x} = -\operatorname{csch} x \coth x$

15. (a) $\frac{3}{5}$ (b) $\frac{5}{3}$ (c) $\frac{4}{3}$ (d) $\frac{5}{4}$ (e) $\frac{3}{4}$

17. If $x \le 0$, the result is obvious. Suppose then that $x > 0$. Since $x^2 \ge 1$, we have $x \ge 1$. Consequently,

$$x - 1 = \sqrt{x-1}\,\sqrt{x-1} \le \sqrt{x-1}\,\sqrt{x+1} = \sqrt{x^2-1} \quad \text{and therefore} \quad x - \sqrt{x^2 - 1} \le 1.$$

19. $\dfrac{d}{dx}(\sinh^{-1} x) = \dfrac{d}{dx}[\ln(x + \sqrt{x^2+1})] = \dfrac{1 + \dfrac{x}{\sqrt{x^2+1}}}{x + \sqrt{x^2+1}} = \dfrac{1}{\sqrt{x^2+1}}$

21. $\dfrac{d}{dx}(\tanh^{-1} x) = \dfrac{1}{2}\dfrac{d}{dx}\left[\ln\left(\dfrac{1+x}{1-x}\right)\right] = \dfrac{1}{2} \cdot \dfrac{1}{\left(\dfrac{1+x}{1-x}\right)} \cdot \dfrac{2}{(1-x)^2} = \dfrac{1}{1-x^2}$

23. (a) $\tan \phi = \sinh x$
  $\phi = \tan^{-1}(\sinh x)$
  $\dfrac{d\phi}{dx} = \dfrac{\cosh x}{1 + \sinh^2 x} = \dfrac{\cosh x}{\cosh^2 x} = \dfrac{1}{\cosh x} = \operatorname{sech} x$

  (b) $\sinh x = \tan \phi$
  $x = \sinh^{-1}(\tan \phi)$
  $= \ln(\tan\phi + \sqrt{\tan^2\phi + 1})$
  $= \ln(\tan\phi + \sec\phi) = \ln(\sec\phi + \tan\phi)$

  (c) $x = \ln(\sec\phi + \tan\phi)$
  $\dfrac{dx}{d\phi} = \dfrac{\sec\phi\tan\phi + \sec^2\phi}{\tan\phi + \sec\phi} = \sec\phi$

SECTION 7.13

1. $2\sqrt{ab}$ **3.** $(\pm 1/\sqrt{2},\ 1/\sqrt{e})$ **5.** $x = 1/\sqrt{e}$ **7.** 1

9. $\frac{1}{2}\ln 2$ **11.** $\frac{1}{6}\pi$ **13.** $\frac{1}{4}\pi$ **15.** $-\frac{1}{2}$

17. average slope $= \dfrac{1}{b-a}\displaystyle\int_a^b \dfrac{d}{dx}(\ln x)\,dx = \dfrac{1}{b-a}(\ln b - \ln a) = \dfrac{1}{b-a}\ln\dfrac{b}{a}$

19. $a^2 \ln 2$ **21.** (a) $45°$ (b) between $63°$ and $64°$ (c) about $31°$

23. (a) $V_x = \frac{1}{3}\pi^2$ (b) $V_y = 2\pi$ **25.** $\bar{x} = 0,\quad \bar{y} = 1/\ln 2$

27. $\displaystyle\int_0^1 \dfrac{a}{1 + a^2 x^2}\,dx = \int_0^a \dfrac{du}{1 + u^2} = \int_0^a \dfrac{dx}{1 + x^2}$ (setting $u = ax,\ du = a\,dx$)

29. By the hint

$$\ln(m+1) - \ln m < \frac{1}{m} < \ln m - \ln(m-1)$$

$$\ln(m+2) - \ln(m+1) < \frac{1}{m+1} < \ln(m+1) - \ln m$$

$$\vdots \qquad \vdots$$

$$\ln(n+1) - \ln n < \frac{1}{n} < \ln n - \ln(n-1).$$

By addition

$$\underbrace{\ln(n+1) - \ln m}_{\ln\frac{n+1}{m}} < \frac{1}{m} + \frac{1}{m+1} + \cdots + \frac{1}{n} < \underbrace{\ln n - \ln(m-1)}_{\ln\frac{n}{m-1}}.$$

31. $\dfrac{d}{dx}[e^{-x}P(x)] = e^{-x}[p'(x) + \cdots + p^{(n)}(x)] - e^{-x}[p(x) + \cdots + p^{(n)}(x)] = e^{-x}p(x)$
  $\underset{\longrightarrow\ p^{(n+1)}(x)\,=\,0}{}$

33. $\bar{x} = \dfrac{\pi}{4(\pi - 2)}, \quad \bar{y} = \dfrac{\pi^2 - 8}{4(\pi - 2)}$

35. (a) $A = 1$ (b) $\bar{x} = \frac{1}{4}(e^2 + 1), \quad \bar{y} = \frac{1}{2}e - 1$ (c) $V_x = \pi(e - 2), \quad V_y = \frac{1}{2}\pi(e^2 + 1)$
     (d) $\bar{R} = \frac{1}{8}\sqrt{2}(e^2 - 2e + 5)$ (e) $V_l = 2\pi \bar{R} A = \frac{1}{4}\sqrt{2}\pi(e^2 - 2e + 5)$

37. (a) about \$3696   (b) about \$3423   **39.** (a) about \$2016   (b) about \$1918

41. (a) about \$1037   (b) about \$967   **43.** (a) $V_x = \dfrac{\pi}{8e^2}(e^4 - 4e^2 - 1)$ (b) $V_y = 2\pi/e$

45. (a) $n_1 \sin \theta_1 = n \sin \theta = n_2 \sin \theta_2$ (see Figure A.7.13.1)

     (b) Think of n and θ as functions of altitude y. Then

$$n \sin \theta = C.$$

Figure A.7.13.1

     Differentiation with respect to y gives

$$n \cos \theta \, \frac{d\theta}{dy} + \frac{dn}{dy} \sin \theta = 0, \qquad \cot \theta \, \frac{d\theta}{dy} + \frac{1}{n} \frac{dn}{dy} = 0$$

     and thus

$$\overbrace{\frac{1}{n} \frac{dn}{dy} - \cot \theta \, \frac{d\theta}{dy}}^{\alpha = \frac{1}{2}\pi - \theta} = \left(-\frac{dy}{dx}\right)\left(-\frac{d\alpha}{dy}\right) = \frac{d\alpha}{dy} \frac{dy}{dx} = \frac{d\alpha}{dx} = \overbrace{\frac{d^2 y/dx^2}{1 + (dy/dx)^2}}^{\tan \alpha = dy/dx, \quad \alpha = \tan^{-1}(dy/dx)}.$$

$$\underbrace{}_{n \sin \theta = C}$$

     (c) $1 + (dy/dx)^2 = 1 + \tan^2 \alpha = 1 + \cot^2 \theta = \csc^2 \theta = 1/\sin^2 \theta = (1/C)n^2$

     (d) $n(y) = k/|y + b|$ where b and k are constants, $k > 0$

SECTION 8.1

1. $-e^{2-x} + C$ **3.** $2/\pi$ **5.** $-\tan(1 - x) + C$

7. $\frac{1}{2}\ln 3$ **9.** $-\sqrt{1 - x^2} + C$ **11.** 0

13. $e - \sqrt{e}$ **15.** $\frac{1}{2}\ln(x^2 + 1) + C$ **17.** $\pi/4c$

19. $\frac{2}{3}\sqrt{3}\tan \theta + 1 + C$ **21.** $(1/a)\ln|ae^x - b| + C$ **23.** $-\frac{1}{2}(\cot 2x + \csc 2x) + C$

25. $\frac{1}{2}\ln[(x + 1)^2 + 4] - \frac{1}{2}\tan^{-1}(\frac{1}{2}[x + 1]) + C$

27. $\frac{1}{2}\sin^{-1} x^2 + C$ **29.** $\tan^{-1}(x + 3) + C$ **31.** $-\frac{1}{2}\cos x^2 + C$

33. $\tan x - x + C$ **35.** $-\frac{1}{2}x^2 e^{-x^2} - \frac{1}{2}e^{-x^2} + C$

SECTION 8.2

1. $\ln\left|\dfrac{x - 2}{x + 5}\right| + C$ **3.** $5 \ln|x - 1| - 3 \ln|x| + \dfrac{3}{x} + C$

5. $\dfrac{1}{4}x^4 + \dfrac{4}{3}x^3 + 6x^2 + 32x - \dfrac{32}{x - 2} + 80 \ln|x - 2| + C$

7. $5 \ln|x - 2| - 4 \ln|x - 1| + C$ **9.** $\dfrac{-1}{2(x - 1)^2} + C$

11. $\dfrac{3}{4}\ln|x - 1| - \dfrac{1}{2(x - 1)} + \dfrac{1}{4}\ln|x + 1| + C$

13. $\frac{1}{32} \ln \left| \frac{x-2}{x+2} \right| - \frac{1}{16} \tan^{-1} \frac{x}{2} + C$ **15.** $\frac{1}{2} \ln (x^2 + 1) + \frac{3}{2} \tan^{-1} x + \frac{5(1-x)}{2(x^2+1)} + C$

17. $\frac{1}{16} \ln \left[\frac{x^2 + 2x + 2}{x^2 - 2x + 2} \right] + \frac{1}{8} \tan^{-1}(x + 1) + \frac{1}{8} \tan^{-1}(x - 1) + C$

19. Note that

$$y = \frac{1}{x^2 - 1} = \frac{1}{2} \left[\frac{1}{x-1} - \frac{1}{x+1} \right] \quad \text{and thus} \quad \frac{d^0 y}{dx^0} = \left(\frac{1}{2} \right) (-1)^0 \, 0! \left[\frac{1}{(x-1)^{0+1}} - \frac{1}{(x+1)^{0+1}} \right].$$

The rest is a routine induction.

21. $\bar{x} = (2 \ln 2)/\pi, \qquad \bar{y} = (\pi + 2)/\pi$

23. (a) $C(t) = 2A_0 \left(\frac{t}{t+1} \right)$ (b) $C(t) = 4A_0 \frac{3^t - 2^t}{2(3^t) - 2^t}$ (c) $C(t) = A_0(m + n) \left(\frac{m^t - n^t}{m^{t+1} - n^{t+1}} \right)$

25. $F = \frac{2GmM}{L^2} \left[\frac{L}{h} + \ln \left(\frac{h}{L+h} \right) \right]$

SECTION 8.3

1. $\frac{1}{3} \cos^3 x - \cos x + C$ **3.** $\frac{1}{2}x - \frac{1}{12} \sin 6x + C$ **5.** $-\frac{1}{5} \cos^5 x + \frac{1}{7} \cos^7 x + C$

7. $\frac{1}{4} \sin^4 x - \frac{1}{6} \sin^6 x + C$ **9.** $\frac{1}{3} \sin^3 x - \frac{1}{5} \sin^5 x + C$

11. $\frac{3}{8}x - \frac{1}{4} \sin 2x + \frac{1}{32} \sin 4x + C$ **13.** $\frac{1}{2} \cos x - \frac{1}{10} \cos 5x + C$

15. $\frac{1}{2} \sin^4 x + C$ **17.** $\frac{3}{128} x - \frac{1}{128} \sin 4x + \frac{1}{1024} \sin 8x + C$

19. $\frac{5}{16}x - \frac{1}{4} \sin 2x + \frac{3}{64} \sin 4x + \frac{1}{48} \sin^3 2x + C$

21. $\sin x - \sin^3 x + \frac{3}{5} \sin^5 x - \frac{1}{7} \sin^7 x + C$ **23.** $\frac{1}{2} \sin x + \frac{1}{10} \sin 5x + C$

25. $\frac{3}{8} \tan^{-1} x + \frac{x}{2(x^2 + 1)} + \frac{x(1 - x^2)}{8(x^2 + 1)^2} + C$ **27.** $\frac{1}{2} \left[\tan^{-1} (x + 1) + \frac{x + 1}{x^2 + 2x + 2} \right] + C$

29. 1   **31.** 0   **33.** 0

SECTION 8.4

1. $\frac{1}{3} \tan 3x - x + C$ **3.** $(1/\pi) \tan \pi x + C$ **5.** $\frac{1}{2} \tan^2 x + \ln |\cos x| + C$

7. $\frac{1}{3} \tan^3 x + C$ **9.** $-\frac{1}{2} \csc x \cot x + \frac{1}{2} \ln |\csc x - \cot x| + C$

11. $-\frac{1}{3} \cot^3 x + \cot x + x + C$ **13.** $-\frac{1}{5} \csc^5 x + \frac{1}{3} \csc^3 x + C$

15. $-\frac{1}{6} \cot^3 2x - \frac{1}{2} \cot 2x + C$ **17.** $-\frac{1}{2} \cot x \csc x - \frac{1}{2} \ln |\csc x - \cot x| + C$

19. $\frac{1}{12} \tan^4 3x - \frac{1}{6} \tan^2 3x + \frac{1}{3} \ln |\sec 3x| + C$

21. $\frac{1}{4} \sec^3 x \tan x + \frac{3}{8} \sec x \tan x + \frac{3}{8} \ln |\sec x + \tan x| + C$ **23.** $\frac{1}{7} \tan^7 x + \frac{1}{5} \tan^5 x + C$

SECTION 8.5

1. $\sin^{-1} \left(\frac{x}{a} \right) + C$ **3.** $\frac{x}{5\sqrt{5 - x^2}} + C$

5. $\frac{1}{2}x \sqrt{x^2 - 1} - \frac{1}{2} \ln |x + \sqrt{x^2 - 1}| + C$ **7.** $2 \sin^{-1} \left(\frac{x}{2} \right) - \frac{1}{2}x \sqrt{4 - x^2} + C$

9. $\frac{1}{\sqrt{1 - x^2}} + C$ **11.** $\frac{x}{\sqrt{1 - x^2}} - \sin^{-1} x + C$

13. $-\frac{1}{3}(4 - x^2)^{3/2} + C$ **15.** $\ln (\sqrt{8 + x^2} + x) - \frac{x}{\sqrt{8 + x^2}} + C$

17. $\frac{1}{a} \ln \left| \frac{a - \sqrt{a^2 - x^2}}{x} \right| + C$ **19.** $\ln |x + \sqrt{x^2 - a^2}| + C$

21. $\frac{1}{2}e^x \sqrt{e^{2x} - 1} - \frac{1}{2} \ln (e^x + \sqrt{e^{2x} - 1}) + C$ **23.** $-\frac{1}{a^2 x} \sqrt{a^2 + x^2} + C$

25. $\frac{1}{a^2 x} \sqrt{x^2 - a^2} + C$ **27.** $\frac{1}{9} e^{-x} \sqrt{e^{2x} - 9} + C$ **29.** $-\frac{1}{2(x-2)^2} + C$

31. $-\frac{1}{3}(6x - x^2 - 8)^{3/2} + \frac{3}{2} \sin^{-1}(x-3) + \frac{3}{2}(x-3)\sqrt{6x - x^2 - 8} + C$

33. $\frac{x^2 + x}{8(x^2 + 2x + 5)} - \frac{1}{16} \tan^{-1}\left(\frac{x+1}{2}\right) + C$

35. $\sqrt{x^2 + 4x + 13} + \ln(x + 2 + \sqrt{x^2 + 4x + 13}) + C$

37. $\frac{1}{2}(x-3)\sqrt{6x - x^2 - 8} + \frac{1}{2} \sin^{-1}(x-3) + C$

39. $-\frac{1}{5}(8 - 2x - x^2)^{5/2} - \frac{243}{8} \sin^{-1}(\frac{1}{3}[x+1]) - \frac{1}{8}(x+1)(43 - 4x - 2x^2)(8 - 2x - x^2)^{1/2} + C$

41. $M = \ln(1 + \sqrt{2})$, $x_M = \frac{(\sqrt{2} - 1)a}{\ln(1 + \sqrt{2})}$

43. $A = \frac{1}{2}a^2[\sqrt{2} - \ln(\sqrt{2} + 1)]$; $\bar{x} = \frac{2a}{3[\sqrt{2} - \ln(\sqrt{2} + 1)]}$, $\bar{y} = \frac{(2 - \sqrt{2})a}{3[\sqrt{2} - \ln(\sqrt{2} + 1)]}$

45. $V_y = \frac{2}{3}\pi a^3$, $\bar{y} = \frac{3}{8}a$

SECTION 8.6

1. $-2(\sqrt{x} + \ln|1 - \sqrt{x}|) + C$ **3.** $2 \ln(\sqrt{1 + e^x} - 1) - x + 2\sqrt{1 + e^x} + C$

5. $\frac{2}{5}(1 + x)^{5/2} - \frac{2}{3}(1 + x)^{3/2} + C$ **7.** $\frac{2}{5}(x - 1)^{5/2} + 2(x - 1)^{3/2} + C$

9. $-\frac{1 + 2x^2}{4(1 + x^2)^2} + C$ **11.** $x + 2\sqrt{x} + 2 \ln|\sqrt{x} - 1| + C$

13. $x + 4\sqrt{x - 1} + 4 \ln|\sqrt{x - 1} - 1| + C$ **15.** $2 \ln(\sqrt{1 + e^x} - 1) - x + C$

17. $\frac{2}{3}(x - 8)\sqrt{x + 4} + C$ **19.** $\frac{1}{16}(4x + 1)^{1/2} + \frac{1}{8}(4x + 1)^{-1/2} - \frac{1}{48}(4x + 1)^{-3/2} + C$

21. $\frac{4b + 2ax}{a^2 \sqrt{ax + b}} + C$

SECTION 8.7

1. (a) 506   (b) 650   (c) 572   (d) 578   (e) 576

3. (a) 1.39   (b) 0.91   (c) 1.18   (d) 1.15   (e) 1.16

5. (a) $\frac{1}{4}\pi \cong 0.78$ (b) $\frac{1}{4}\pi \cong 0.79$

7. Such a curve passes through the three points (a_1, b_1), (a_2, b_2), (a_3, b_3) iff

$$b_1 = a_1^2 A + a_1 B + C, \qquad b_2 = a_2^2 A + a_2 B + C, \qquad b_3 = a_3^2 A + a_3 B + C,$$

which happens iff

$$A = \frac{b_1(a_2 - a_3) - b_2(a_1 - a_3) + b_3(a_1 - a_2)}{(a_1 - a_3)(a_1 - a_2)(a_2 - a_3)}, \qquad B = -\frac{b_1(a_2^2 - a_3^2) - b_2(a_1^2 - a_3^2) + b_3(a_1^2 - a_2^2)}{(a_1 - a_3)(a_1 - a_2)(a_2 - a_3)},$$

$$C = \frac{a_1^2(a_2 b_3 - a_3 b_2) - a_2^2(a_1 b_3 - a_3 b_1) + a_3^2(a_1 b_2 - a_2 b_1)}{(a_1 - a_3)(a_1 - a_2)(a_2 - a_3)}.$$

9. (a) $n \geq 8$ (b) $n \geq 4$ **11.** (a) $n \geq 238$ (b) $n \geq 19$

13. (a) $n \geq 51$ (b) $n \geq 7$ **15.** (a) $n \geq 37$ (b) $n \geq 5$

17. $f^{(4)}(x) = 0$ for all x; therefore by (8.7.2) the theoretical error is zero

SECTION 8.8

1. $\frac{10^{nx}}{n \ln 10} + C$ **3.** $\frac{1}{3}(2x + 1)^{3/2} + C$ **5.** $\ln|\sec e^x| + C$

7. $\frac{1}{ab} \tan^{-1}\left(\frac{ax}{b}\right) + C$ **9.** $x \ln(x\sqrt{x}) - \frac{3}{2}x + C$ **11.** $\frac{1}{3}(x^2 - 2)\sqrt{1 + x^2} + C$

13. $\frac{1}{2}x - \frac{n}{4\pi} \sin\left(\frac{2\pi}{n}x\right) + C$ **15.** $\frac{2\sqrt{x}}{a} - \frac{2b}{a^2} \ln|a\sqrt{x} + b| + C$

17. $-\frac{1}{3}\ln(2+\cos 3x)+C$ **19.** $\frac{1}{2}\tan^2 x+C$ **21.** $\tan^{-1}(2x-1)+C$

23. $a\sin^{-1}\left(\dfrac{x}{a}\right)-\sqrt{a^2-x^2}+C$ **25.** $-\left(\dfrac{1}{x}\right)\sqrt{a^2-x^2}-\sin^{-1}\left(\dfrac{x}{a}\right)+C$

27. $\ln|x+1|+\dfrac{1}{x+1}+C$ **29.** $\tan 2x-\sec 2x-x+C$

31. $\dfrac{x^2}{2}\ln(ax+b)-\dfrac{x^2}{4}+\dfrac{bx}{2a}-\dfrac{b^2}{2a^2}\ln(ax+b)+C$

33. $\frac{2}{3}(x+1)^{3/2}+\frac{2}{3}x^{3/2}+C$ **35.** $x-\tan^{-1}x+C$

37. $\frac{1}{2}x\sqrt{1-x^2}-\frac{1}{2}\sin^{-1}x+C$ **39.** $\frac{1}{4}\sinh 2x-\frac{1}{2}x+C$

41. $\frac{1}{2}x-\frac{1}{4}e^{-2x}+C$ **43.** $\frac{1}{2}(x^2-8)\tan^{-1}(x-3)-\frac{3}{2}\ln|x^2-6x+10|-\frac{1}{2}x+C$

45. $2\ln|x|-\ln(x^2+1)+C=\ln\left(\dfrac{x^2}{x^2+1}\right)+C$ **47.** $-\cot x-\frac{3}{2}x-\frac{1}{4}\sin 2x+C$

49. $2\ln(\sqrt{x^2+4}-2)-2\ln|x|+\sqrt{x^2+4}+C$ **51.** $\frac{1}{3}\ln(3-\sqrt{9-x^2})-\frac{1}{3}\ln|x|+C$

53. $(x-1)\ln(1-\sqrt{x})-\frac{1}{2}x-\sqrt{x}+C$ **55.** $\ln|1-\sqrt{1-e^{2x}}|-x+C$

57. $\sec x+\ln|\csc x-\cot x|+C$ **59.** $4\sin x-\frac{8}{3}\sin^3 x+C$

61. $-2\cos\frac{1}{2}x+\frac{4}{3}\cos^3\frac{1}{2}x-\frac{2}{5}\cos^5\frac{1}{2}x+C$ **63.** $-\csc x+C$

65. $\frac{1}{2}\ln|\sec x+\tan x|+C$ **67.** $\frac{1}{3}x^3\sin^{-1}x+\frac{1}{3}(1-x^2)^{1/2}-\frac{1}{9}(1-x^2)^{3/2}+C$

69. $\frac{1}{3}(x^2+2x-8)^{3/2}-\frac{1}{2}(x+1)\sqrt{x^2+2x-8}+\frac{9}{2}\ln|x+1+\sqrt{x^2+2x-8}|+C$

71. $\frac{7}{8}x+\frac{1}{32}\sin 4x-\frac{2}{3}\sin^3 x+C$ **73.** $\frac{3}{2}\sin^{-1}([8x+3]/\sqrt{41})+C$

SECTION 9.2

1. (a) $\frac{2}{13}$ (b) $\frac{29}{13}$

3. $(0, 1)$ is the closest pt; $(-1, 1)$ the farthest away   **5.** $\frac{17}{2}$

7. Adjust the sign of A and B so that the equation reads $Ax+By=|C|$.   Then we have

$$x\frac{A}{\sqrt{A^2+B^2}}+y\frac{B}{\sqrt{A^2+B^2}}=\frac{|C|}{\sqrt{A^2+B^2}}.$$

Now set

$$\frac{A}{\sqrt{A^2+B^2}}=\cos\alpha,\qquad \frac{B}{\sqrt{A^2+B^2}}=\sin\alpha,\qquad \frac{|C|}{\sqrt{A^2+B^2}}=p.$$

p is the length of $\overline{OQ}$, the distance between the line and the origin; α is the angle from the positive x-axis to the line segment $\overline{OQ}$.

9. (a) $4x+5y-3=0$ (b) $4x+5y-11=0$
(c) $4x+5y+17=0$ (d) $4x+5y+9=0$

11. (a) $(x-3)^2=(y-4)^3$ (b) $(x+3)^2=(y-4)^3$
(c) $(x-4)^2=(y+3)^3$ (d) $(x+4)^2=(y+3)^3$

13. decreasing $\frac{8}{5}b^2\pi$ units/min

SECTION 9.3

1. $y^2=8x$ **3.** $(x+1)^2=-12(y-3)$ **5.** $4y=(x-1)^2$ **7.** $(y-1)^2=-2(x-\frac{3}{2})$

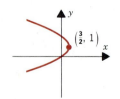

9. vertex $(0, 0)$
focus $(\frac{1}{2}, 0)$
axis $y = 0$
directrix $x = -\frac{1}{2}$

11. vertex $(0, -\frac{1}{2})$
focus $(0, -\frac{3}{8})$
axis $x = 0$
directrix $y = -\frac{5}{8}$

13. vertex $(-2, \frac{3}{2})$
focus $(-2, -\frac{1}{2})$
axis $x = -2$
directrix $y = \frac{7}{2}$

15. vertex $(\frac{3}{4}, -\frac{1}{2})$
focus $(1, -\frac{1}{2})$
axis $y = -\frac{1}{2}$
directrix $x = \frac{1}{2}$

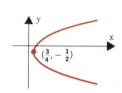

17. $(x - y)^2 = 6x + 10y - 9$

19. $(x + y)^2 = -12x + 20y + 28$

21. $P(x, y)$ is on the parabola with directrix l: $Ax + By + C = 0$ and focus $F(a, b)$ iff

$$d(P, l) = d(P, F) \quad \text{which happens iff} \quad \frac{|Ax + By + C|}{\sqrt{A^2 + B^2}} = \sqrt{(x - a)^2 + (y - b)^2}.$$

Square this last equation and simplify.

23. We can choose the coordinate system so that the parabola has an equation of the form $y = \alpha x^2$, $\alpha > 0$. One of the points of intersection is then the origin and the other is of the form $(c, \alpha c^2)$. We will assume that $c > 0$.

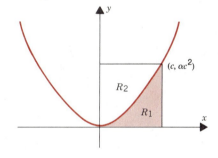

$$\text{area of } R_1 = \int_0^c \alpha x^2 \, dx = \tfrac{1}{3}\alpha c^3 = \tfrac{1}{3}A$$

area of $R_2 = A - \tfrac{1}{3}A = \tfrac{2}{3}A$.

25. $2y = x^2 - 4x + 7$, $\quad 18y = x^2 - 4x + 103$

27. $4c$

29. $A = \tfrac{8}{3}c^3$; $\quad \bar{x} = 0$, $\quad \bar{y} = \tfrac{3}{5}c$

31. $y = -\dfrac{16}{v_0^2}(\sec^2 \theta)\,x^2 + (\tan \theta)\,x$

33. $\tfrac{1}{16}v_0^2 \cos \theta \sin \theta$ ft

35. $\tfrac{1}{4}\pi$

37. $\dfrac{kx}{p(0)} = \tan \theta = \dfrac{dy}{dx}, \qquad y = \dfrac{k}{2p(0)}x^2 + C$

In our figure, $C = y(0) = 0$. Thus the equation of the cable is $y = kx^2/2p(0)$, the equation of a parabola.

39. Start with any two parabolas γ_1, γ_2. By moving them we can see to it that they have equations of the following form:

$$\gamma_1: \quad x^2 = 4c_1 y, \quad c_1 > 0; \qquad \gamma_2: \quad x^2 = 4c_2 y, \quad c_2 > 0.$$

Now we change the scale for γ_2 so that the equation for γ_2 will look exactly like the equation for γ_1. Set $X = (c_1/c_2)x$, $Y = (c_1/c_2)y$. Then

$$x^2 = 4c_2 y \implies (c_2/c_1)^2 X^2 = 4c_2(c_2/c_1)Y \implies X^2 = 4c_1 Y.$$

Now γ_2 has exactly the same equation as γ_1; only the scale, the units by which we measure distance, has changed.

SECTION 9.4

1. center $(0, 0)$
foci $(\pm\sqrt{5}, 0)$
length of major axis 6
length of minor axis 4

3. center $(0, 0)$
foci $(0, \pm\sqrt{2})$
length of major axis $2\sqrt{6}$
length of minor axis 4

5. center $(0, 1)$
 foci $(\pm\sqrt{5}, 1)$
 length of major axis 6
 length of minor axis 4

7. center $(1, 0)$
 foci $(1, \pm4\sqrt{3})$
 length of major axis 16
 length of minor axis 8

9. $\dfrac{x^2}{9} + \dfrac{y^2}{8} = 1$

11. $\dfrac{(x-1)^2}{16} + \dfrac{(y-6)^2}{25} = 1$

13. $\dfrac{(x-1)^2}{21} + \dfrac{(y-3)^2}{25} = 1$

15. $\dfrac{(x-3)^2}{25} + \dfrac{(y+1)^2}{9} = 1$

17. $d(F_1, F_2) + k = 2(c + a)$

19. $2\sqrt{\pi^2 a^4 - A^2}/\pi a$

21. $(5 \pm \frac{5}{21}\sqrt{5}, 0)$

23. $\frac{3}{5}$

25. $\frac{4}{5}$

27. E_1 is fatter than E_2, more like a circle

29. the ellipse tends to a line segment of length $2a$

31. $x^2/9 + y^2 = 1$

SECTION 9.5

1. $\dfrac{x^2}{9} - \dfrac{y^2}{16} = 1$

3. $\dfrac{y^2}{25} - \dfrac{x^2}{144} = 1$

5. $\dfrac{x^2}{9} - \dfrac{(y-1)^2}{16} = 1$

7. $16y^2 - \frac{16}{15}(x+1)^2 = 1$

9. center $(0, 0)$
 transverse axis 2
 vertices $(\pm 1, 0)$
 foci $(\pm\sqrt{2}, 0)$
 asymptotes $y = \pm x$

11. center $(0, 0)$
 transverse axis 6
 vertices $(\pm 3, 0)$
 foci $(\pm 5, 0)$
 asymptotes $y = \pm\frac{4}{3}x$

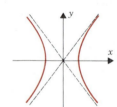

13. center $(0, 0)$
 transverse axis 8
 vertices $(0, \pm 4)$
 foci $(0, \pm 5)$
 asymptotes $y = \pm\frac{4}{3}x$

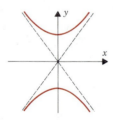

15. center $(1, 3)$
 transverse axis 6
 vertices $(4, 3)$ and $(-2, 3)$
 foci $(6, 3)$ and $(-4, 3)$
 asymptotes $y = \pm\frac{4}{3}(x - 1) + 3$

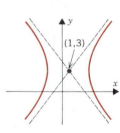

17. center $(1, 3)$
 transverse axis 4
 vertices $(1, 5)$ and $(1, 1)$
 foci $(1, 3 \pm \sqrt{5})$
 asymptotes $y = 2x + 1, \quad y = -2x + 5$

19. center $(0, 0)$, vertices $(1, 1)$ and $(-1, -1)$, foci $(\sqrt{2}, \sqrt{2})$ and $(-\sqrt{2}, -\sqrt{2})$, asymptotes $x = 0$ and $y = 0$, transverse axis $2\sqrt{2}$

21. $[2\sqrt{3} - \ln(2 + \sqrt{3})]ab$

23. $\frac{5}{3}$

25. $\sqrt{2}$

27. the branches of H_1 open up less quickly than the branches of H_2

29. the hyperbola tends to a pair of parallel lines separated by the transverse axis

31. about 0.25 miles west and 1.5 miles north of point A

SECTION *9.6

1. $\alpha = \frac{1}{4}\pi,$
$\frac{1}{2}(X^2 - Y^2) = 1$

3. $\alpha = \frac{1}{6}\pi,$
$4X^2 - Y^2 = 1$

5. $\alpha = \frac{1}{4}\pi,$
$2Y^2 + \sqrt{2}X = 0$

7. $\alpha = -\frac{1}{6}\pi,$
$Y^2 + X = 0$

9. $\alpha = \frac{1}{8}\pi,$ $\cos \alpha = \frac{1}{2}\sqrt{2 + \sqrt{2}},$ $\sin \alpha = \frac{1}{2}\sqrt{2 - \sqrt{2}}$

SECTION 9.7

1. parabola with vertex at $(0, -1)$ and focus at the origin
3. hyperbola with foci at $(5, \pm 2\sqrt{5})$ and transverse axis 4
5. ellipse with foci at $(-3 \pm \frac{1}{3}\sqrt{6}, 0)$ and major axis 2
7. parabola with vertex at $(\frac{3}{2}, -2)$ and focus at $(1, -2)$
9. ellipse with foci at $(\pm \frac{4}{15}, -2)$ and major axis $\frac{2}{3}$
11. hyperbola with foci at $(-1 \pm \frac{6}{7}\sqrt{21}, -4)$ and transverse axis $\frac{6}{7}\sqrt{35}$
13. the union of the parabola $x^2 = 4y$ and the ellipse $\frac{1}{9}x^2 + \frac{1}{4}y^2 = 1$

SECTION 10.1

1.–7. Figure A.10.1.1
11. $(1, 0)$
15. $(0, -3)$
17. $[1, \frac{1}{2}\pi + 2n\pi], [-1, \frac{3}{2}\pi + 2n\pi]$
19. $[3, \pi + 2n\pi], [-3, 2n\pi]$
21. $[2\sqrt{2}, \frac{7}{4}\pi + 2n\pi], [-2\sqrt{2}, \frac{3}{4}\pi + 2n\pi]$
23. $[8, \frac{1}{6}\pi + 2n\pi], [-8, \frac{7}{6}\pi + 2n\pi]$
25. $\sqrt{r_1^2 + r_2^2 - 2r_1 r_2 \cos(\theta_1 - \theta_2)}$
27. (a) $[\frac{1}{2}, \frac{11}{6}\pi]$ (b) $[\frac{1}{2}, \frac{5}{6}\pi]$ (c) $[\frac{1}{2}, \frac{7}{6}\pi]$
29. (a) $[2, \frac{2}{3}\pi]$ (b) $[2, \frac{5}{3}\pi]$ (c) $[2, \frac{1}{3}\pi]$
31. symmetry about the x-axis
33. no symmetry about the coordinate axes; no symmetry about the origin
35. symmetry about the origin
37. $r \cos \theta = 2$
39. $r^2 \sin 2\theta = 1$
41. $r = 4 \sin \theta$
43. $\theta = \pi/4$
45. $r = 1 - \cos \theta$
47. $r^2 = \sin 2\theta$
49. the horizontal line $y = 4$
51. the line $y = \sqrt{3}\, x$
53. the parabola $y^2 = 4(x + 1)$
55. the circle $x^2 + y^2 = 3x$
57. the line $y = 2x$
59. the vertical line $x = 0$

9. $(0, 3)$
13. $(-\frac{3}{2}, \frac{3}{2}\sqrt{3})$

Figure A.10.1.1

SECTION 10.2

1. Figure A.10.2.1

Figure A.10.2.1

3. Figure A.10.2.2

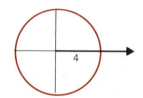

Figure A.10.2.2

5. Figure A.10.2.3

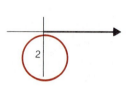

Figure A.10.2.3

7. Figure A.10.2.4

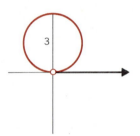

Figure A.10.2.4

9. Figure A.10.2.5

Figure A.10.2.5

11. Figure A.10.2.6

Figure A.10.2.6

13. Figure A.10.2.7

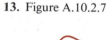

Figure A.10.2.7

15. Figure A.10.2.8

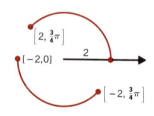

Figure A.10.2.8

17. Figure A.10.2.9

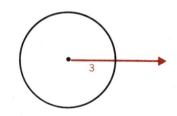

Figure A.10.2.9

19. Figure A.10.2.10

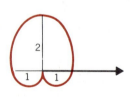

Figure A.10.2.10

21. Figure A.10.2.11

Figure A.10.2.11

23. Figure A.10.2.12

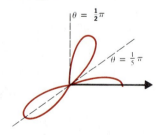

Figure A.10.2.12

25. Figure A.10.2.13   **27.** Figure A.10.2.14   **29.** Figure A.10.2.15

Figure A.10.2.13

Figure A.10.2.14

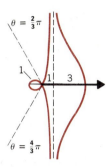

Figure A.10.2.15

31. Figure A.10.2.16

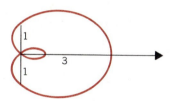

Figure A.10.2.16

SECTION *10.3

1. (a) $\frac{1}{2}$ unit to the right of the pole   (b) 2 units   (c) Figure A.*10.3.1
3. the parabola of Figure A.*10.3.1 rotated by π radians
5. (a) $e = \frac{2}{3}$ (b) 2 units to the left of the pole and $\frac{2}{5}$ units to the right of the pole   (c) $\frac{4}{5}$ units to the left of the pole   (d) $\frac{8}{5}$ units to the left of the pole   (e) $\frac{4}{5}\sqrt{5}$ units (about 1.79 units)   (f) $\frac{4}{3}$ units
(g) Figure A.*10.3.2

Figure A.*10.3.1

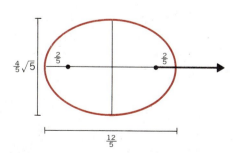

Figure A.*10.3.2

7. the ellipse of Figure A.*10.3.2 rotated by $\frac{1}{2}\pi$ radians

9. (a) $e = 2$ (b) 2 units to the right of the pole and 6 units to the right of the pole   (c) 4 units to the right
of the pole   (d) 8 units to the right of the pole   (e) 12 units   (f) Figure A.*10.3.3

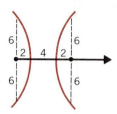

Figure A.*10.3.3

11. the hyperbola of Figure A.*10.3.3 rotated by $\frac{3}{2}\pi$ radians

13. ellipse: $x^2/48 + (y + 4)^2/64 = 1$ **15.** ellipse: $x^2/9 + (y - 4)^2/25 = 1$

SECTION 10.4

1. yes; $[1, \pi] = [-1, 0]$ and the pair $r = -1$, $\theta = 0$ satisfies the equation

3. yes; the pair $r = \frac{1}{2}$, $\theta = \frac{1}{2}\pi$ satisfies the equation

5. $[2, \pi] = [-2, 0]$.   The coordinates of $[-2, 0]$ satisfy the equation $r^2 = 4 \cos \theta$, and the coordinates of $[2, \pi]$
satisfy the equation $r = 3 + \cos \theta$.

7. $(0, 0)$, $(-\frac{1}{2}, \frac{1}{2})$ **9.** $(-1, 0)$, $(1, 0)$ **11.** $(0, 0)$, $(\frac{1}{4}, \pm\frac{1}{4}\sqrt{3})$

13. $(\frac{3}{2}, 2)$ **15.** $(0, 0)$, $(1, 0)$

SECTION 10.5

1. $\frac{1}{4}\pi a^2$ **3.** $\frac{1}{2}a^2$ **5.** $\frac{1}{2}\pi a^2$

7. $\frac{1}{4} - \frac{1}{16}\pi$ **9.** $\frac{3}{16}\pi + \frac{3}{8}$ **11.** $\frac{2}{3}a^2$

13. $\frac{1}{12}(3e^{2\pi} - 3 - 2\pi^3)$ **15.** $\frac{1}{4}(e^{2\pi} + 1 - 2e^{\pi})$

17. $\displaystyle\int_{\pi/6}^{5\pi/6} \frac{1}{2}([4 \sin \theta]^2 - [2]^2)\, d\theta$ **19.** $\displaystyle\int_{-\pi/3}^{\pi/3} \frac{1}{2}([4]^2 - [2 \sec \theta]^2)\, d\theta$

21. $2\left[\displaystyle\int_0^{\pi/3} \frac{1}{2}(2 \sec \theta)^2\, d\theta + \int_{\pi/3}^{\pi/2} \frac{1}{2}(4)^2\, d\theta\right]$ **23.** $\displaystyle\int_0^{\pi/3} \frac{1}{2}(2 \sin 3\theta)^2\, d\theta$

25. $2\left[\displaystyle\int_0^{\pi/6} \frac{1}{2}(\sin \theta)^2\, d\theta + \int_{\pi/6}^{\pi/2} \frac{1}{2}(1 - \sin \theta)^2\, d\theta\right]$

SECTION 10.6

1. $4x = (y - 1)^2$ **3.** $y = 4x^2 + 1$, $x \geq 0$ **5.** $9x^2 + 4y^2 = 36$

7. $1 + x^2 = y^2$ **9.** $y = 2 - x^2$, $-1 \leq x \leq 1$ **11.** $2y - 6 = x$, $-4 \leq x \leq 4$

13. $y = x - 1$ **15.** $xy = 1$ **17.** $y + 2x = 11$

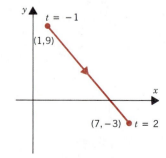

19. $x = \sin \frac{1}{2}\pi y$

21. $y^2 = x^2 + 1$

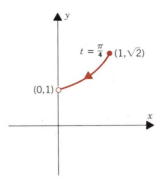

23. (a) $x(t) = -\sin 2\pi t, \quad y(t) = \cos 2\pi t$ (b) $x(t) = \sin 4\pi t, \quad y(t) = \cos 4\pi t$
     (c) $x(t) = \cos \frac{1}{2}\pi t, \quad y(t) = \sin \frac{1}{2}\pi t$ (d) $x(t) = \cos \frac{3}{2}\pi t, \quad y(t) = -\sin \frac{3}{2}\pi t$

25. $x(t) = \tan \frac{1}{2}\pi t, \quad y(t) = 2$ **27.** $x(t) = 3 + 5t, \quad y(t) = 7 - 2t$

29. $x(t) = \sin^2 \pi t, \quad y(t) = -\cos \pi t$ **31.** $x(t) = (2 - t)^2, \quad y(t) = (2 - t)^3$

33. $x(t) = t(b - a) + a, \quad y(t) = f(t(b - a) + a)$

35. $A = 2\pi r^2$ **37.** (a) $V_x = 3\pi^2 r^3$ (b) $V_y = 4\pi^3 r^3$

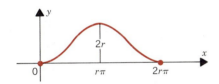

39. $x(t) = -a \cos t, \quad y(t) = b \sin t; \quad t \in [0, \pi]$

41. (a) paths intersect at $(6, 5)$ and $(8, 1)$ **43.** curve intersects itself at $(0, 0)$
     (b) particles collide at $(8, 1)$ **45.** curve intersects itself at $(0, 0)$ and $(0, \frac{3}{4})$

SECTION 10.7

1. $3x - y - 3 = 0$ **3.** $y = 1$ **5.** $3x + y - 3 = 0$

7. $2x + 2y - \sqrt{2} = 0$ **9.** $2x + y - 8 = 0$ **11.** $x - 5y + 4 = 0$

13. $x + 2y + 1 = 0$

15. Figure A.10.7.1; $x(t) = t, \quad y(t) = t^3;$ tangent line $y = 0$

17. Figure A.10.7.2; $x(t) = t^{5/3}, \quad y(t) = t;$ tangent line $x = 0$

19. Figure A.10.7.3;   (a) none;   (b) at $(2, 2)$ and $(-2, 0)$

Figure A.10.7.1

Figure A.10.7.2

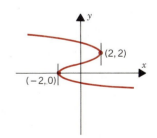

Figure A.10.7.3

21. Figure A.10.7.4;   (a) at $(3, 7)$ and $(3, 1)$;   (b) at $(-1, 4)$ and $(7, 4)$
23. Figure A.10.7.5;   (a) at $(-\frac{2}{3}, \pm\frac{2}{9}\sqrt{3})$;   (b) at $(-1, 0)$
25. Figure A.10.7.6;   (a) at $(\pm\frac{1}{2}\sqrt{2}, \pm 1)$;   (b) at $(\pm 1, 0)$

Figure A.10.7.4

Figure A.10.7.5

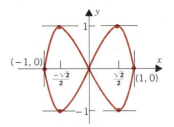

Figure A.10.7.6

27. $y = 0$, $(\pi - 2)y + 32x - 64 = 0$
29. The slope of $\overline{OP}$ is $\tan\theta_1$.   The curve $r = f(\theta)$ can be parametrized by setting

$$x(\theta) = f(\theta)\cos\theta, \qquad y(\theta) = f(\theta)\sin\theta.$$

Differentiation gives

$$x'(\theta) = -f(\theta)\sin\theta + f'(\theta)\cos\theta, \qquad y'(\theta) = f(\theta)\cos\theta + f'(\theta)\sin\theta.$$

If $f'(\theta_1) = 0$, then

$$x'(\theta_1) = -f(\theta_1)\sin\theta_1, \qquad y'(\theta_1) = f(\theta_1)\cos\theta_1.$$

Since $f(\theta_1) \neq 0$, we have

$$m = \frac{y'(\theta_1)}{x'(\theta_1)} = -\cot\theta_1 = -\frac{1}{\text{slope of } \overline{OP}}.$$

31. Figure A.10.7.7   **33.** Figure A.10.7.8   **35.** Figure A.10.7.9
37. -8 **39.** 2

Figure A.10.7.7

Figure A.10.7.8

Figure A.10.7.9

SECTION 10.8

1. lub $= 2$; glb $= 0$
3. no lub; glb $= 0$
5. lub $= 2$; glb $= -2$
7. no lub; glb $= 2$
9. lub $= 2\frac{1}{2}$; glb $= 2$
11. lub $= 1$; glb $= 0.9$
13. lub $= e$; glb $= 0$
15. lub $= \frac{1}{2}(-1 + \sqrt{5})$; glb $= \frac{1}{2}(-1 - \sqrt{5})$
17. no lub; no glb
19. no lub; no glb
21. glb $S = 0$, $0 \leq (\frac{1}{11})^3 < 0 + 0.001$
23. glb $S = 0$, $0 \leq (\frac{1}{10})^{2n-1} < 0 + (\frac{1}{10})^k$, $n > \frac{1}{2}(k + 1)$
25. Let $\epsilon > 0$.   The condition $m \leq s$ is satisfied by all numbers s in S.   All we have to show therefore is that there is some number s in S such that $s < m + \epsilon$.   Suppose on the contrary that there is no such number in S.   We then have $m + \epsilon \leq x$ for all $x \in S$, so that $m + \epsilon$ becomes a lower bound for S.   But this cannot happen, for it makes $m + \epsilon$ a lower bound that is *greater* than m, and by assumption, m is the *greatest* lower bound.

SECTION 10.9

1. $\sqrt{5}$ **3.** 7   **5.** $2\sqrt{3}$ **7.** $\frac{4}{3}$ **9.** $6 + \frac{1}{2}\ln 5$ **11.** $\frac{63}{8}$
13. $\ln(1 + \sqrt{2})$ **15.** $\frac{3}{2}$ **17.** $\frac{1}{3}\pi + \frac{1}{2}\sqrt{3}$
19. initial speed 2, terminal speed 4; $s = 2\sqrt{3} + \ln(2 + \sqrt{3})$
21. initial speed 0, terminal speed $\sqrt{13}$; $s = \frac{1}{27}(13\sqrt{13} - 8)$
23. initial speed $\sqrt{2}$, terminal speed $\sqrt{2}e^{\pi}$; $s = \sqrt{2}(e^{\pi} - 1)$
25. 2π **27.** $\sqrt{2}(e^{4\pi} - 1)$ **29.** $\frac{1}{2}\sqrt{5}(e^{4\pi} - 1)$
31. $4 - 2\sqrt{2}$ **33.** $\ln(1 + \sqrt{2})$ **35.** $c = 1$
37. $L = \displaystyle\int_a^b \sqrt{1 + \sinh^2 x}\, dx = \int_a^b \sqrt{\cosh^2 x}\, dx = \int_a^b \cosh x\, dx = A$
39. (a) Express GPE + KE as a function of t and verify that the derivative with respect to t is zero.
   (b) From (a) we know that throughout the motion

$$32my + \tfrac{1}{2}mv^2 = C.$$

At the time of firing $y = 0$ and $v = |v_0| = v_0$.   Therefore

$$32my + \tfrac{1}{2}mv^2 = \tfrac{1}{2}mv_0^2.$$

At impact $y = 0$, $\tfrac{1}{2}mv^2 = \tfrac{1}{2}mv_0^2$, and $v = v_0$.
41. $\sqrt{1 + [f'(x)]^2} = \sqrt{1 + \tan^2[\alpha(x)]} = |\sec[\alpha(x)]|$

SECTION 10.10

1. $L = 1$, $(\bar{x}, \bar{y}) = (\frac{1}{2}, 4)$, $A_x = 16\pi$ **3.** $L = 5$, $(\bar{x}, \bar{y}) = (\frac{3}{2}, 2)$, $A_x = 20\pi$
5. $L = 10$, $(\bar{x}, \bar{y}) = (3, 4)$, $A_x = 80\pi$
7. $L = \frac{1}{3}\pi$; $\bar{x} = 6/\pi$, $\bar{y} = 6(2 - \sqrt{3})/\pi$; $A_x = 4\pi(2 - \sqrt{3})$
9. $L = \frac{1}{3}\pi a$; $\bar{x} = 0$, $\bar{y} = 3a/\pi$; $A_x = 2\pi a^2$ **11.** $\frac{1}{9}\pi(17\sqrt{7} - 1)$
13. $\frac{61}{432}\pi$ **15.** $\pi[\sqrt{2} + \ln(1 + \sqrt{2})]$ **17.** $\frac{2}{5}\sqrt{2}\pi(2e^{\pi} + 1)$
19. See Figure A.10.10.1

$$\begin{aligned}
A &= \tfrac{1}{2}\theta s_2^2 - \tfrac{1}{2}\theta s_1^2 \\
&= \tfrac{1}{2}(\theta s_2 + \theta s_1)(s_2 - s_1) \\
&= \tfrac{1}{2}(2\pi R + 2\pi r)s = \pi(R + r)s
\end{aligned}$$

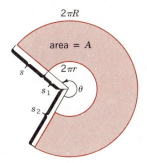

Figure A.10.10.1

21. (a) the 3, 4, 5 sides have centroids $(\frac{3}{2}, 0)$, $(4, 2)$, $(\frac{3}{2}, 2)$ (b) $\bar{x} = 2$, $\bar{y} = \frac{3}{2}$ (c) $\bar{x} = 2$, $\bar{y} = \frac{4}{3}$
   (d) $\bar{x} = \frac{13}{6}$, $\bar{y} = 2$ (e) $A = 20\pi$ (f) $A = 36\pi$
23. $4\pi^2 ab$
25. The band can be obtained by revolving about the x-axis the graph of a function

$$f(x) = \sqrt{r^2 - x^2}, \qquad x \in [a, b].$$

   A straightforward calculation shows that the surface area of the band is $2\pi r(b - a)$.
27. (a) $2\pi b^2 + \dfrac{2\pi ab}{e}\sin^{-1} e$ (b) $2\pi a^2 + \dfrac{\pi b^2}{e}\ln\left|\dfrac{1 + e}{1 - e}\right|$,
   where e is the eccentricity $c/a = \sqrt{a^2 - b^2}/a$
29. at the midpoint of the axis of the hemisphere
31. on the axis of the cone $\left(\dfrac{2R + r}{R + r}\right)\dfrac{h}{3}$ units from the base of radius r

SECTION 10.11

1. $x(\theta) = \overline{OB} - \overline{AB} = R\theta - R\sin\theta = R(\theta - \sin\theta)$, $\qquad y(\theta) = \overline{BQ} - \overline{QC} = R - R\cos\theta = R(1 - \cos\theta)$

3. (a) The slope at P is

$$m = \frac{y'(\theta)}{x'(\theta)} = \frac{\sin\theta}{1 - \cos\theta}.$$

The line tangent to the cycloid at P has equation

$$y - R(1 - \cos\theta) = \frac{\sin\theta}{1 - \cos\theta}[x - R(\theta - \sin\theta)].$$

The top of the circle is the point $(R\theta, 2R)$. Its coordinates satisfy the equation for the tangent:

$$2R - R(1 - \cos\theta) \overset{?}{=} \frac{\sin\theta}{1 - \cos\theta}[R\theta - R(\theta - \sin\theta)]$$

$$R(1 + \cos\theta) \overset{?}{=} \frac{R\sin^2\theta}{1 - \cos\theta}$$

$$1 - \cos^2\theta \overset{\checkmark}{=} \sin^2\theta.$$

(b) In view of the symmetry and repetitiveness of the curve we can assume that $\theta \in (0, \pi)$. Then

$$\tan\alpha = \frac{\sin\theta}{1 - \cos\theta} = \frac{\sin\theta}{2\sin^2\tfrac{1}{2}\theta} = \frac{\sin\tfrac{1}{2}\theta\cos\tfrac{1}{2}\theta}{\sin^2\tfrac{1}{2}\theta} = \cot\tfrac{1}{2}\theta$$

and $\alpha = \tfrac{1}{2}\pi - \tfrac{1}{2}\theta = \tfrac{1}{2}(\pi - \theta)$.

5. $8R$

7. $\bar{x} = \pi R$, $\quad \bar{y} = \tfrac{5}{6}R$

9. $V_y = 6\pi^3 R^3$

11. $A = \frac{64}{3}\pi R^2$

13. $\alpha = \tfrac{1}{2}\phi$ (radian measure)

15. (a) already shown more generally in Problem 7 of Section 10.9

(b) Combining $d^2s/dt^2 = -g\sin\alpha$ with $s = 4R\sin\alpha$, we have

$$\frac{d^2s}{dt^2} = -\frac{g}{4R}s.$$

This is simple harmonic motion with angular frequency $\omega = \tfrac{1}{2}\sqrt{g/R}$ and period $T = 2\pi/\omega = 4\pi\sqrt{R/g}$. (Section 7.8)

SECTION 10.12

1. Figure A.10.12.1, $A = 4$

3. Figure A.10.12.2, $A = \tfrac{9}{2}\pi$

5. $\tfrac{19}{27}$

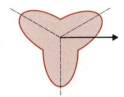

Figure A.10.12.1

Figure A.10.12.2

7. $\frac{335}{27}a$ $(a > 0)$

9. $\sqrt{2} + \ln(1 + \sqrt{2})$

11. $\tfrac{3}{2}\pi a$ $(a > 0)$

13. $\frac{52}{3}\pi p^2$

15. $\frac{47}{16}\pi a^2$

17. $\frac{2152}{15}\pi$

19. (a) $\tfrac{3}{2}a$ (b) $(a^2 + ab + b^2)/(a + b)$

21. $L = 6r$

23. $A = \tfrac{3}{8}\pi r^2$

25. $\bar{x} = -\tfrac{4}{5}a$, $\quad \bar{y} = 0$

*27. (a) $x(\theta) = R\theta - b\sin\theta$, $\quad y(\theta) = R - b\cos\theta$ (b) Figure A.10.12.3

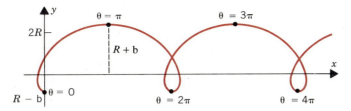

Figure A.10.12.3

SECTION 11.1

1. decreasing; bounded below by 0 and above by 2

3. increasing; bounded below by 1 but not bounded above

5. not monotonic; bounded below by 0 and above by $\frac{3}{2}$

7. decreasing; bounded below by 0 and above by 0.9

9. increasing; bounded below by $\frac{1}{2}$ but not bounded above

11. increasing; bounded below by $\frac{4}{5}\sqrt{5}$ and above by 2

13. increasing; bounded below by $\frac{2}{51}$ but not bounded above

15. decreasing; bounded below by 0 and above by $\frac{1}{2}(10^{10})$

17. increasing; bounded below by 0 and above by $\ln 2$

19. decreasing; bounded below by 1 and above by 4

21. increasing; bounded below by $\sqrt{3}$ and above by 2

23. decreasing; bounded above by -1 but not bounded below

25. increasing; bounded below by $\frac{1}{2}$ and above by 1

27. decreasing; bounded below by 0 and above by 1

29. decreasing; bounded below by 0 and above by $\frac{5}{6}$

31. decreasing; bounded below by 0 and above by $\frac{1}{2}$

33. decreasing; bounded below by 0 and above by $\frac{1}{3}\ln 3$

35. increasing; bounded below by $\frac{3}{4}$ but not bounded above

37. For $n \geq 5$

$$\frac{a_{n+1}}{a_n} = \frac{5^{n+1}}{(n+1)!} \cdot \frac{n!}{5^n} = \frac{5}{n+1} < 1 \qquad \text{and thus} \qquad a_{n+1} < a_n.$$

Sequence is not nonincreasing: $a_1 = 5 < \frac{25}{2} = a_2$.

39. boundedness: $0 < (c^n + d^n)^{1/n} < (2d^n)^{1/n} = 2^{1/n} d \leq 2d.$

monotonicity: $a_{n+1}^{n+1} = c^{n+1} + d^{n+1} = cc^n + dd^n < (c^n + d^n)^{1/n} c^n + (c^n + d^n)^{1/n} d^n$
$$= (c^n + d^n)^{1+(1/n)} = (c^n + d^n)^{(n+1)/n} = a_n^{n+1}.$$

Taking the $n + 1$th root of each side we have $a_{n+1} < a_n$. The sequence is monotonic decreasing.

41. $a_1 = 1,\ a_2 = \frac{1}{2},\ a_3 = \frac{1}{6},\ a_4 = \frac{1}{24},\ a_5 = \frac{1}{120},\ a_6 = \frac{1}{720};\quad a_n = 1/n!$

43. $a_1 = a_2 = a_3 = a_4 = a_5 = a_6 = 1;\quad a_n = 1$

45. $a_1 = 1,\ a_2 = 3,\ a_3 = 5,\ a_4 = 7,\ a_5 = 9,\ a_6 = 11;\quad a_n = 2n - 1$

47. $a_1 = 1,\ a_2 = 4,\ a_3 = 13,\ a_4 = 40,\ a_5 = 121,\ a_6 = 364;\quad a_n = \frac{1}{2}(3^n - 1)$

49. $a_1 = 1,\ a_2 = 4,\ a_3 = 9,\ a_4 = 16,\ a_5 = 25,\ a_6 = 36;\quad a_n = n^2$

51. $a_1 = 1,\ a_2 = 3,\ a_3 = 4,\ a_4 = 8,\ a_5 = 16,\ a_6 = 32;\quad a_n = 2^{n-1}\ (n \geq 3)$

53. $a_1 = 2,\ a_2 = 1,\ a_3 = 2,\ a_4 = 1,\ a_5 = 2,\ a_6 = 1;\quad a_n = \frac{1}{2}[3 - (-1)^n]$

55. $a_1 = 1,\ a_2 = 3,\ a_3 = 5,\ a_4 = 7,\ a_5 = 9,\ a_6 = 11;\quad a_n = 2n - 1$

57. First $a_1 = 2^1 - 1 = 1$. Next suppose $a_k = 2^k - 1$ for some $k \geq 1$. Then

$$a_{k+1} = 2a_k + 1 = 2(2^k - 1) + 1 = 2^{k+1} - 1.$$

59. First $a_1 = \dfrac{1}{2^0} = 1$. Next suppose $a_k = \dfrac{k}{2^{k-1}}$ for some $k \geq 1$. Then $a_{k+1} = \dfrac{k+1}{2k}\, a_k = \dfrac{k+1}{2k}\, \dfrac{k}{2^{k-1}} = \dfrac{k+1}{2^k}.$

SECTION 11.2

1. diverges
3. converges to 0
5. converges to 1
7. converges to 0
9. converges to 0
11. diverges
13. converges to 0
15. converges to 1
17. converges to $\frac{4}{9}$
19. converges to $\frac{1}{2}\sqrt{2}$
21. diverges
23. converges to 1
25. diverges
27. converges to 0
29. converges to $\frac{1}{2}$
31. converges to e^2
33. diverges
35. converges to 0
37. Use $|(a_n + b_n) - (l + m)| \le |a_n - l| + |b_n - m|$.
39. $\left(1 + \dfrac{1}{n}\right)^{n+1} = \left(1 + \dfrac{1}{n}\right)^n\left(1 + \dfrac{1}{n}\right)$. Note that $\left(1 + \dfrac{1}{n}\right)^n \to e$ and $\left(1 + \dfrac{1}{n}\right) \to 1$.
41. Imitate the proof given for the nondecreasing case in Theorem 11.2.6.
43. Let $\epsilon > 0$. Choose k so that, for $n \ge k$,
$$l - \epsilon < a_n < l + \epsilon, \quad l - \epsilon < c_n < l + \epsilon \quad \text{and} \quad a_n \le b_n \le c_n.$$
For such n,
$$l - \epsilon < b_n < l + \epsilon.$$
45. Use Theorem 11.2.12 with $f(x) = x^{1/p}$.
47. converges to 0
49. converges to 0
51. diverges
53. converges to 2
55. $l = 0, \ n = 32$
57. $l = 0, \ n = 4$
59. $l = 0, \ n = 7$
61. $l = 0, \ n = 65$

SECTION 11.3

1. converges to 1
3. converges to 0
5. converges to 0
7. converges to 0
9. converges to 1
11. converges to 0
13. converges to 1
15. converges to 1
17. converges to π
19. converges to 1
21. converges to 0
23. diverges
25. converges to 0
27. converges to e^{-1}
29. converges to 0
31. converges to 0
33. converges to e^x
35. converges to 0
37. $\sqrt{n+1} - \sqrt{n} = \dfrac{\sqrt{n+1} - \sqrt{n}}{\sqrt{n+1} + \sqrt{n}}(\sqrt{n+1} + \sqrt{n}) = \dfrac{1}{\sqrt{n+1} + \sqrt{n}} \to 0$.
39. $\displaystyle\lim_{n \to \infty} 2n \sin(\pi/n) = 2\pi$; the number $2n \sin(\pi/n)$ is the perimeter of a regular polygon of n sides inscribed in the unit circle; as n tends to ∞, the perimeter of the polygon tends to the circumference of the circle
41. $\frac{1}{2}$
43. $\frac{1}{8}$
45. (a) $m_{n+1} - m_n = \dfrac{1}{n+1}(a_1 + \cdots + a_n + a_{n+1}) - \dfrac{1}{n}(a_1 + \cdots + a_n)$

$$= \dfrac{1}{n(n+1)}\left[na_{n+1} - (a_1 + \cdots + a_n)\right] > 0 \ \text{ since } \{a_n\} \text{ is increasing.}$$

(b) We begin with the hint $m_n < \dfrac{|a_1 + \cdots + a_j|}{n} + \dfrac{\epsilon}{2}\left(\dfrac{n-j}{n}\right)$. Since j is fixed, $\dfrac{|a_1 + \cdots + a_j|}{n} \to 0$, and

therefore for n sufficiently large $\dfrac{|a_1 + \cdots + a_j|}{n} < \dfrac{\epsilon}{2}$. Since $\dfrac{\epsilon}{2}\left(\dfrac{n-j}{n}\right) < \dfrac{\epsilon}{2}$, we see that, for n

sufficiently large, $|m_n| < \epsilon$. This shows that $m_n \to 0$.

47. The numerical work suggests $l \cong 1$. Justification: Set $f(x) = \sin x - x^2$. Note that $f(0) = 0$ and for x close to 0, $f'(x) = \cos x - 2x > 0$. Therefore $\sin x - x^2 > 0$ for x close to 0 and $\sin(1/n) - 1/n^2 > 0$ for n large. Thus, for n large,

$$\frac{1}{n^2} < \sin\frac{1}{n} < \frac{1}{n}$$

$$|\sin x| \le |x| \ \text{ for all } x$$

$$\left(\frac{1}{n^2}\right)^{1/n} < \left(\sin\frac{1}{n}\right)^{1/n} < \left(\frac{1}{n}\right)^{1/n}$$

$$\left(\frac{1}{n^{1/n}}\right)^2 < \left(\sin\frac{1}{n}\right)^{1/n} < \frac{1}{n^{1/n}}.$$

As $n \to \infty$ both bounds tend to 1 and therefore the middle term also tends to 1.

SECTION 11.4

1. 0   **3.** 1   **5.** $\frac{1}{2}$ **7.** $\ln 2$

9. $\frac{1}{4}$ **11.** 2   **13.** $\dfrac{1+\pi}{1-\pi}$ **15.** e^2

17. π **19.** $-\frac{1}{2}$ **21.** -2 **23.** $\frac{1}{3}\sqrt{6}$

25. $-\frac{1}{8}$ **27.** $-\frac{1}{2}$ **29.** $\frac{1}{2}$ **31.** 1

33. 1   **35.** $\lim\limits_{x\to0}(2+x+\sin x)\neq0,\quad \lim\limits_{x\to0}(x^3+x-\cos x)\neq0$ **37.** $f(0)$

39. The numerical work suggests $l\cong-\frac{1}{6}$. Justification:

$$l = \lim_{x\to0^+}\frac{\sin x - x}{x^3} \overset{*}{=} \lim_{x\to0^+}\frac{\cos x - 1}{3x^2} \overset{*}{=} \lim_{x\to0^+}\frac{-\sin x}{6x} = -\tfrac{1}{6}.$$

SECTION 11.5

1. ∞ **3.** -1 **5.** ∞ **7.** $\frac{1}{5}$

9. 1   **11.** 0   **13.** ∞ **15.** $\frac{1}{3}$

17. e **19.** 1   **21.** $\frac{1}{2}$ **23.** $\frac{1}{2}$

25. 1   **27.** e^3 **29.** e **31.** 0

33. 1   **35.** 1   **37.** 0

39. Figure A.11.5.1; y-axis vertical asymptote   **41.** Figure A.11.5.2; x-axis horizontal asymptote

43. Figure A.11.5.3; x-axis horizontal asymptote

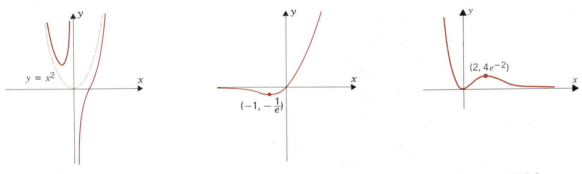

Figure A.11.5.1   **Figure A.11.5.2**   **Figure A.11.5.3**

45. $\dfrac{b}{a}\sqrt{x^2-a^2}-\dfrac{b}{a}x = \dfrac{\sqrt{x^2-a^2}+x}{\sqrt{x^2-a^2}+x}\left(\dfrac{b}{a}\right)(\sqrt{x^2-a^2}-x) = \dfrac{-ab}{\sqrt{x^2-a^2}+x}\to0$ as $x\to\infty$

47. for instance, $f(x)=x^3+\dfrac{1}{x}$ **49.** for instance, $f(x)=x+\dfrac{\sin x}{x}$ **51.** $\lim\limits_{x\to0^+}\cos x\neq0$

SECTION 11.6

1. 1   **3.** $\frac{1}{4}\pi$ **5.** diverges   **7.** 6

9. $\frac{1}{2}\pi$ **11.** 2   **13.** diverges   **15.** $-\frac{1}{4}$

17. π **19.** diverges   **21.** $\ln 2$ **23.** 4

25. diverges   **27.** diverges   **29.** diverges

31. (a) Figure A.11.6.1   (b) 1   (c) $\frac{1}{2}\pi$ (d) 2π (e) $\pi[\sqrt{2}+\ln(1+\sqrt{2})]$

33. (a) The interval $[0, 1]$ causes no problem. For $x\geq1$, $e^{-x^2}\leq e^{-x}$ and

$$\int_1^\infty e^{-x}\,dx \text{ is finite.} \quad\text{(b) } V_y=\int_0^\infty 2\pi x e^{-x^2}\,dx=\pi$$

Figure A.11.6.1

Figure A.11.6.2

35. (a) Figure A.11.6.2   (b) $\frac{4}{3}$ **37.** converges by comparison with $\int_0^\infty \dfrac{dx}{x^{3/2}}$

(c) 2π (d) $\frac{8}{7}\pi$

39. diverges since for x large the integrand is greater than $\dfrac{1}{x}$ and $\int_1^\infty \dfrac{1}{x}\,dx$ diverges

41. converges by comparison with $\int_1^\infty \dfrac{dx}{x^{3/2}}$ **43.** $L = (a\sqrt{1 + c^2}/c)e^{c\theta_1}$

SECTION 11.7

1. 1   **3.** 0   **5.** 1   **7.** 0   **9.** diverges

11. 0   **13.** 0   **15.** 0   **17.** 0

19. If $|a_n - l| < \epsilon$ for all $n \geq k$, then $|b_n - l| < \epsilon$ for all $n \geq k - 1$.

21. $l \cong 0.7390851$, $\cos(0.7390851) \cong 0.7390851$

23. $l \cong 0.7681691$, $\cos(\sin(0.7681691)) \cong 0.7681691$

25. Observe that $F(t) = \displaystyle\int_1^t f(x)\,dx$ is continuous and increasing, that $a_n = \displaystyle\int_1^n f(x)\,dx$ is increasing, and that

(∗) $$a_n \leq \int_1^t f(x)\,dx \leq a_{n+1} \qquad \text{for } t \in [n, n+1].$$

If $\displaystyle\int_1^\infty f(x)\,dx$ converges, then F, being continuous, is bounded and, by (∗), $\{a_n\}$ is bounded and therefore convergent. If $\{a_n\}$ converges, then $\{a_n\}$ is bounded and, by (∗), F is bounded. Being increasing, F is also convergent; i.e., $\displaystyle\int_1^\infty f(x)\,dx$ converges.

27. (a) If $\displaystyle\int_{-\infty}^\infty f(x)\,dx = L$, then both

$$\int_0^\infty f(x)\,dx = \lim_{c \to \infty} \int_0^c f(x)\,dx \qquad \text{and} \qquad \int_{-\infty}^0 f(x)\,dx = \lim_{c \to \infty} \int_{-c}^0 f(x)\,dx$$

exist and their sum is L. Thus

$$\lim_{c \to \infty} \int_{-c}^c f(x)\,dx = \lim_{c \to \infty}\left(\int_{-c}^0 f(x)\,dx + \int_0^c f(x)\,dx\right) = \lim_{c \to \infty}\int_{-c}^0 f(x)\,dx + \lim_{c \to \infty}\int_0^c f(x)\,dx = L.$$

(b) $\displaystyle\lim_{c \to \infty}\int_{-c}^c x\,dx = \lim_{c \to \infty}\left[\frac{x^2}{2}\right]_{-c}^c = \lim_{c \to \infty} 0 = 0$,   but $\displaystyle\int_{-\infty}^\infty x\,dx$ diverges since $\displaystyle\int_0^\infty x\,dx$ and $\displaystyle\int_{-\infty}^0 x\,dx$ both diverge.

29. \$2000

31. The numerical work suggests $l \cong -12.5$:

$$l = \lim_{x \to 0^+} \frac{\ln(1 + 5x) - 5x}{x^2} \overset{*}{=} \lim_{x \to 0^+} \frac{5/(1 + 5x) - 5}{2x} = \lim_{x \to 0^+} \frac{-25}{2(1 + 5x)} = -12.5.$$

33. $B = 2k_M I/a$

SECTION 12.1

1. 12   **3.** 15   **5.** -10 **7.** $\frac{15}{16}$ **9.** $-\frac{2}{15}$ **11.** $\frac{85}{64}$

13. $\displaystyle\sum_{k=1}^{n} m_k \Delta x_k$ **15.** $\displaystyle\sum_{k=1}^{n} f(x_k^*)\Delta x_k$ **17.** $\displaystyle\sum_{k=0}^{5} (-1)^k\, a^{5-k}\, b^k$

19. $\displaystyle\sum_{k=0}^{4} a_k x^{4-k}$ **21.** $\displaystyle\sum_{k=0}^{4} (-1)^k (k+1).x^k$ **23.** $\displaystyle\sum_{k=3}^{10} \frac{1}{2^k}, \quad \sum_{i=0}^{7} \frac{1}{2^{i+3}}$

25. $\displaystyle\sum_{k=3}^{10} (-1)^{k+1} \frac{k}{k+1}, \quad \sum_{i=0}^{7} (-1)^i \frac{i+3}{i+4}$

27. (a) $\displaystyle (1-x) \sum_{k=0}^{n} x^k = \sum_{k=0}^{n} (x^k - x^{k+1})$
$$= (1-x) + (x - x^2) + (x^2 - x^3) + \cdots + (x^n - x^{n+1}) = 1 - x^{n+1}.$$
(b) converges to $\frac{3}{2}$

29. $|a_n - l| < \epsilon$ for $n \geq k$ iff $|a_{n-p} - l| < \epsilon$ for $n \geq k + p$.

31. True for $n = 1$: $\displaystyle\sum_{k=1}^{1} k = 1 = \frac{1}{2}(1)(2)$. Suppose true for $n = p$. Then

$$\sum_{k=1}^{p+1} k = \sum_{k=1}^{p} k + (p+1) = \tfrac{1}{2}(p)(p+1) + (p+1) = \tfrac{1}{2}(p+1)(p+2) = \tfrac{1}{2}(p+1)[(p+1)+1]$$

and thus true for $n = p + 1$.

33. True for $n = 1$: $\displaystyle\sum_{k=1}^{1} k^2 = 1 = \frac{1}{6}(1)(2)(3)$. Suppose true for $n = p$. Then

$$\sum_{k=1}^{p+1} k^2 = \sum_{k=1}^{p} k^2 + (p+1)^2 = \tfrac{1}{6}(p)(p+1)(2p+1) + (p+1)^2$$
$$= \tfrac{1}{6}(p+1)(2p^2 + 7p + 6) = \tfrac{1}{6}(p+1)(p+2)(2p+3) = \tfrac{1}{6}(p+1)[(p+1)+1][2(p+1)+1]$$

and thus true for $n = p + 1$.

SECTION 12.2

1. $\frac{1}{4}$ **3.** $\frac{1}{2}$ **5.** $\frac{11}{18}$ **7.** $\frac{10}{3}$

9. $\frac{67000}{999}$ **11.** 4   **13.** $-\frac{3}{2}$ **15.** $\frac{1}{2}$

17. 24   **19.** $\displaystyle\sum_{k=1}^{\infty} \frac{7}{10^k} = \frac{7}{9}$ **21.** $\displaystyle\sum_{k=1}^{\infty} \frac{24}{100^k} = \frac{8}{33}$

23. $\displaystyle\sum_{k=1}^{\infty} \frac{112}{1000^k} = \frac{112}{999}$ **25.** $\displaystyle\frac{62}{100} + \frac{1}{100} \sum_{k=1}^{\infty} \frac{45}{100^k} = \frac{687}{1100}$

27. Let $x = .\overset{\frown}{a_1 a_2 \cdots a_n} \overset{\frown}{a_1 a_2 \cdots a_n} \cdots$. Then

$$x = \sum_{k=1}^{\infty} \frac{a_1 a_2 \cdots a_n}{(10^n)^k} = a_1 a_2 \cdots a_n \sum_{k=1}^{\infty} \left(\frac{1}{10^n}\right)^k = a_1 a_2 \cdots a_n \left[\frac{1}{1 - \frac{1}{10^n}} - 1\right] = \frac{a_1 a_2 \cdots a_n}{10^n - 1}.$$

29. $\displaystyle\frac{1}{1+x} = \frac{1}{1-(-x)} = \sum_{k=0}^{\infty} (-x)^k = \sum_{k=0}^{\infty} (-1)^k x^k$

31. $\displaystyle\sum_{k=0}^{\infty} x^{k+1}$ **33.** $\displaystyle\sum_{k=0}^{\infty} (-1)^k x^{2k+1}$ **35.** $\displaystyle\sum_{k=0}^{\infty} (-1)^k (2x)^{2k}$

37. $\displaystyle 4 + \frac{1}{3} \sum_{k=0}^{\infty} \left(\frac{1}{12}\right)^k = 4 + \frac{4}{11}$ o'clock   **39.** 12 ft

41. $\displaystyle\sum_{k=1}^{\infty} n_k \left(1 + \frac{r}{100}\right)^{-k}$ **43.** $\displaystyle s_n = \sum_{k=1}^{n} \ln\left(\frac{k+1}{k}\right) = \sum_{k=1}^{n} [\ln(k+1) - \ln k] = \ln(n+1) \to \infty$

45. (a) $s_n = \sum_{k=1}^{n} (d_k - d_{k+1}) = d_1 - d_{n+1} \to d_1$

(b) (i) $\sum_{k=1}^{\infty} \frac{\sqrt{k+1} - \sqrt{k}}{\sqrt{k(k+1)}} = \sum_{k=1}^{\infty} \left(\frac{1}{\sqrt{k}} - \frac{1}{\sqrt{k+1}} \right) = 1$ (ii) $\sum_{k=1}^{\infty} \frac{2k+1}{2k^2(k+1)^2} = \sum_{k=1}^{\infty} \frac{1}{2} \left(\frac{1}{k^2} - \frac{1}{(k+1)^2} \right) = \frac{1}{2}$

SECTION 12.3

1. converges; comparison $\Sigma\, 1/k^2$

3. converges; comparison $\Sigma\, 1/k^2$

5. diverges; comparison $\Sigma\, 1/(k+1)$

7. diverges; limit comparison $\Sigma\, 1/k$

9. converges; integral test

11. diverges; p-series with $p = \frac{2}{3} \le 1$

13. diverges; $a_k \not\to 0$

15. diverges; comparison $\Sigma\, 1/k$

17. diverges; $a_k \not\to 0$

19. converges; limit comparison $\Sigma\, 1/k^2$

21. diverges; integral test

23. converges; limit comparison $\Sigma\, 1/k^2$

25. diverges; limit comparison $\Sigma\, 1/k$

27. converges; limit comparison $\Sigma\, 1/k^{3/2}$

29. converges; integral test

31. (a) If $a_k/b_k \to 0$, then $a_k/b_k < 1$ for all $k \ge K$ for some K. But then $a_k < b_k$ for all $k \ge K$ and, since Σb_k converges, Σa_k converges. [The basic comparison test, theorem 12.3.5.]

(b) Similar to (a) except that this time we appeal to part (ii) of Theorem 12.3.5.

(c) $\Sigma a_k = \Sigma \dfrac{1}{k^2}$ converges, $\Sigma b_k = \Sigma \dfrac{1}{k^{3/2}}$ converges, $\dfrac{1/k^2}{1/k^{3/2}} = \dfrac{1}{\sqrt{k}} \to 0.$

$\Sigma a_k = \Sigma \dfrac{1}{k^2}$ converges, $\Sigma b_k = \Sigma \dfrac{1}{\sqrt{k}}$ diverges, $\dfrac{1/k^2}{1/\sqrt{k}} = \dfrac{1}{k^{3/2}} \to 0.$

(d) $\Sigma b_k = \Sigma \dfrac{1}{\sqrt{k}}$ diverges, $\Sigma a_k = \Sigma \dfrac{1}{k^2}$ converges, $\dfrac{1/k^2}{1/\sqrt{k}} = 1/k^{3/2} \to 0.$

$\Sigma b_k = \Sigma \dfrac{1}{\sqrt{k}}$ diverges, $\Sigma a_k = \Sigma \dfrac{1}{k}$ diverges, $\dfrac{1/k}{1/\sqrt{k}} = \dfrac{1}{\sqrt{k}} \to 0.$

33. (a) Set $f(x) = x^{1/4} - \ln x$. Then $f'(x) = \dfrac{1}{4} x^{-3/4} - \dfrac{1}{x} = \dfrac{1}{4x} (x^{1/4} - 4)$. Since $f(e^{12}) = e^3 - 12 > 0$ and

$f'(x) > 0$ for $x > e^{12}$, we know that $n^{1/4} > \ln n$ and therefore $\dfrac{1}{n^{5/4}} > \dfrac{\ln n}{n^{3/2}}$ for sufficiently large n.

Since $\Sigma \dfrac{1}{n^{5/4}}$ is a convergent p-series, $\Sigma \dfrac{\ln n}{n^{3/2}}$ converges by the basic comparison test.

(b) By L'Hospital's rule $\lim_{x \to \infty} \left[\left(\dfrac{\ln x}{x^{3/2}} \right) \Big/ \left(\dfrac{1}{x^{5/4}} \right) \right] = 0.$

SECTION 12.4

1. converges; ratio test

3. converges; root test

5. diverges; $a_k \not\to 0$

7. diverges; limit comparison $\Sigma\, 1/k$

9. converges; root test

11. diverges; limit comparison $\Sigma\, 1/\sqrt{k}$

13. diverges; ratio test

15. converges; comparison $\Sigma\, 1/k^{3/2}$

17. converges; comparison $\Sigma\, 1/k^2$

19. diverges; integral test

21. diverges; $a_k \to e^{-100} \ne 0$

23. diverges; limit comparison $\Sigma\, 1/k$

25. converges; ratio test

27. converges; comparison $\Sigma\, 1/k^{3/2}$

29. converges; ratio test   **31.** converges; ratio test: $a_{k+1}/a_k \to \frac{4}{27}$ **33.** $\frac{10}{81}$

35. Set $b_k = a_k r^k$. If $(a_k)^{1/k} \to \rho$ and $\rho < 1/r$, then

$$(b_k)^{1/k} = (a_k r^k)^{1/k} = (a_k)^{1/k}\, r \to \rho r < 1$$

and thus, by the root test, $\Sigma b_k = \Sigma a_k r^k$ converges.

SECTION 12.5

1. diverges; $a_k \not\to 0$

3. diverges; $a_k \not\to 0$

5. (a) does not converge absolutely; integral test
   (b) converges conditionally; Theorem 12.5.3

7. diverges; limit comparison $\Sigma\, 1/k$

9. (a) does not converge absolutely; limit comparison $\Sigma\, 1/k$
   (b) converges conditionally; Theorem 12.5.3

11. diverges; $a_k \not\to 0$

13. (a) does not converge absolutely; comparison $2\,\Sigma\, 1/\sqrt{k+1}$
    (b) converges conditionally; Theorem 12.5.3

15. converges absolutely (terms already positive); $\Sigma \sin\left(\dfrac{\pi}{4k^2}\right) \le \Sigma\, \dfrac{\pi}{4k^2} = \dfrac{\pi}{4}\, \Sigma\, \dfrac{1}{k^2}$ $(|\sin x| \le |x|)$

17. converges absolutely; ratio test

19. (a) does not converge absolutely; limit comparison $\Sigma\, 1/k$
    (b) converges conditionally; Theorem 12.5.3

21. diverges; $a_k \not\to 0$ 23. $\frac{10}{11}$

25. (a) 4   (b) 6

27. No. For instance, set $a_{2k} = 2/k$ and $a_{2k+1} = 1/k$.

29. See the proof of Theorem 12.8.2.

SECTION 12.6

1. $-1 + x + \frac{1}{2}x^2 - \frac{1}{24}x^4$

3. $-\frac{1}{2}x^2 - \frac{1}{12}x^4$

5. $1 - x + x^2 - x^3 + x^4 - x^5$

7. $x + \frac{1}{3}x^3 + \frac{2}{15}x^5$

9. $P_0(x) = 1, \quad P_1(x) = 1 - x, \quad P_2(x) = 1 - x + 3x^2, \quad P_3(x) = 1 - x + 3x^2 + 5x^3$

11. $\displaystyle\sum_{k=0}^{n} (-1)^k \frac{x^k}{k!}$

13. $\displaystyle\sum_{k=0}^{m} \frac{x^{2k}}{(2k)!}$ where $m = \dfrac{n}{2}$ and n is even

15. $79/48$ $(79/48 \cong 1.646)$

17. $5/6$ $(5/6 \cong 0.833)$

19. $13/24$ $(13/24 \cong 0.542)$

21. (a) 4   (b) 2   (c) 999

23. For $0 \le k \le n$, $P^{(k)}(0) = k!a_k$; for $k > n$, $P^{(k)}(0) = 0$. Thus $P(x) = \displaystyle\sum_{k=0}^{\infty} P^{(k)}(0)\,\frac{x^k}{k!}$.

25. $\displaystyle\sum_{k=0}^{\infty} \frac{a^k}{k!}\, x^k, \quad (-\infty, \infty)$

27. $\displaystyle\sum_{k=0}^{\infty} \frac{(-1)^k a^{2k}}{(2k)!}\, x^{2k}, \quad (-\infty, \infty)$

29. $\ln a + \displaystyle\sum_{k=1}^{\infty} \frac{(-1)^{k-1}}{ka^k}\, x^k, \quad (-a, a]$

31. $\ln 2 = \ln\left(\dfrac{1+\frac{1}{3}}{1-\frac{1}{3}}\right) \cong 2\left[\dfrac{1}{3} + \dfrac{1}{3}\left(\dfrac{1}{3}\right)^3 + \dfrac{1}{5}\left(\dfrac{1}{3}\right)^5\right] = \dfrac{842}{1215} \quad \left(\dfrac{842}{1215} \cong 0.693\right)$

33. routine; use $u = (x - t)^k$ and $dv = f^{(k+1)}(t)\, dt$

35. $f(0) = 0$ by definition

$$f'(0) = \lim_{x \to 0} \frac{f(x) - f(0)}{x - 0} = \lim_{x \to 0} \frac{e^{-1/x^2}}{x} = 0, \qquad f''(0) = \lim_{x \to 0} \frac{f'(x) - f'(0)}{x - 0} = \lim_{x \to 0} \frac{e^{-1/x^2}(2x^{-3})}{x} = 0$$

SECTION 12.7

1. $6 + 9(x - 1) + 7(x - 1)^2 + 3(x - 1)^3, \quad (-\infty, \infty)$

3. $-3 + 5(x + 1) - 19(x + 1)^2 + 20(x + 1)^3 - 10(x + 1)^4 + 2(x + 1)^5, \quad (-\infty, \infty)$

5. $\displaystyle\sum_{k=0}^{\infty} (-1)^k (\tfrac{1}{2})^{k+1}(x - 1)^k, \quad (-1, 3)$

7. $\dfrac{1}{5} \displaystyle\sum_{k=0}^{\infty} (\tfrac{2}{5})^k(x + 2)^k, \quad (-\tfrac{9}{2}, \tfrac{1}{2})$

9. $\displaystyle\sum_{k=0}^{\infty} \frac{(-1)^{k+1}}{(2k+1)!}(x - \pi)^{2k+1}, \quad (-\infty, \infty)$

11. $\displaystyle\sum_{k=0}^{\infty} \frac{(-1)^{k+1}}{(2k)!}(x - \pi)^{2k}, \quad (-\infty, \infty)$

13. $\displaystyle\sum_{k=0}^{\infty} \frac{(-1)^k}{(2k)!}\left(\frac{\pi}{2}\right)^{2k}(x - 1)^{2k}, \quad (-\infty, \infty)$

15. $\ln 3 + \displaystyle\sum_{k=1}^{\infty} \frac{(-1)^{k+1}}{k}(\tfrac{2}{3})^k(x - 1)^k, \quad (-\tfrac{1}{2}, \tfrac{5}{2}]$

17. $2 \ln 2 + (1 + \ln 2)(x - 2) + \sum\limits_{k=2}^{\infty} \dfrac{(-1)^k}{k(k-1)2^{k-1}} (x - 2)^k$

19. $\sum\limits_{k=0}^{\infty} \dfrac{(-1)^k}{(2k+1)!} x^{2k+2}$

21. $\sum\limits_{k=0}^{\infty} (k+2)(k+1) \dfrac{2^{k-1}}{5^{k+3}} (x + 2)^k$

23. $1 + \sum\limits_{k=1}^{\infty} \dfrac{(-1)^k 2^{2k-1}}{(2k)!} (x - \pi)^{2k}$

25. $\sum\limits_{k=0}^{\infty} \dfrac{n!}{(n-k)!\, k!} (x - 1)^k$

27. (a) $\dfrac{e^x}{e^a} = e^{x-a} = \sum\limits_{k=0}^{\infty} \dfrac{(x-a)^k}{k!}, \quad e^x = e^a \sum\limits_{k=0}^{\infty} \dfrac{(x-a)^k}{k!}$

　　(b) $e^{a+(x-a)} = e^x = e^a \sum\limits_{k=0}^{\infty} \dfrac{(x-a)^k}{k!}, \quad e^{x_1+x_2} = e^{x_1} \sum\limits_{k=0}^{\infty} \dfrac{x_2^k}{k!} = e^{x_1} e^{x_2}$

　　(c) $e^{-a} \sum\limits_{k=0}^{\infty} (-1)^k \dfrac{(x-a)^k}{k!}$

SECTION 12.8

1. $(-1, 1)$ **3.** $(-\infty, \infty)$ **5.** $\{0\}$ **7.** $[-2, 2)$

9. $\{0\}$ **11.** $[-\tfrac{1}{2}, \tfrac{1}{2})$ **13.** $(-1, 1)$ **15.** $(-10, 10)$

17. $(-\infty, \infty)$ **19.** $(-\infty, \infty)$ **21.** $(0, 4)$ **23.** $(-\tfrac{5}{2}, \tfrac{1}{2})$

25. Examine the convergence of $\Sigma |a_k x^k|$; for (a) use the root test and for (b) use the ratio test.

SECTION 12.9

1. $1 + 2x + 3x^2 + \cdots nx^{n-1} + \cdots$

3. $1 + kx + \dfrac{(k+1)k}{2!} x^2 + \cdots + \dfrac{(n+k-1)!}{n!(k-1)!} x^n + \cdots$

5. $\ln(1 - x^2) = -x^2 - \dfrac{1}{2} x^4 - \dfrac{1}{3} x^6 - \cdots - \dfrac{1}{n+1} x^{2n+2} - \cdots$

7. $1 + x^2 + \tfrac{2}{3}x^4 + \tfrac{17}{45}x^6 + \cdots$ **9.** -72

11. $\sum\limits_{k=0}^{\infty} \dfrac{(-1)^k}{(2k+1)!} x^{4k+2}$

13. $\sum\limits_{k=0}^{\infty} \dfrac{3^k}{k!} x^{3k}$

15. $2 \sum\limits_{k=0}^{\infty} x^{2k+1}$

17. $\sum\limits_{k=0}^{\infty} \dfrac{(k!+1)}{k!} x^k$

19. $\sum\limits_{k=1}^{\infty} \dfrac{(-1)^{k+1}}{k} x^{3k+1}$

21. $\sum\limits_{k=0}^{\infty} \dfrac{(-1)^k}{k!} x^{3k+3}$

23. $0.804 \le I \le 0.808$ **25.** $0.600 \le I \le 0.603$ **27.** $0.294 \le I \le 0.304$

29. e^{x^3} **31.** $3x^2 e^{x^3}$

33. Let $f(x)$ be the sum of these series; a_k and b_k are both $f^{(k)}(0)/k!$.

35. $0.0352 \le I \le 0.0359; \quad I = \tfrac{3}{6} - \tfrac{3}{8} \ln 1.5 \cong 0.0354505$

37. $0.2640 \le I \le 0.2643; \quad I = 1 - 2/e \cong 0.2642411$

SECTION 12.10

1. $1 + \tfrac{1}{2}x - \tfrac{1}{8}x^2 + \tfrac{1}{16}x^3 - \tfrac{5}{128}x^4$ **3.** $1 + \tfrac{1}{2}x^2 - \tfrac{1}{8}x^4$

5. $1 - \tfrac{1}{2}x + \tfrac{3}{8}x^2 - \tfrac{5}{16}x^3 + \tfrac{35}{128}x^4$ **7.** $1 - \tfrac{1}{4}x - \tfrac{3}{32}x^2 - \tfrac{7}{128}x^3 - \tfrac{77}{2048}x^4$

9. 9.8995 **11.** 2.0799 **13.** 0.4925

SECTION 12.11

1. $\tfrac{4}{3}$ **3.** $\tfrac{2}{3}$ **5.** 1

7. $\dfrac{x}{1 - x^5}, \quad |x| < 1$ **9.** $\dfrac{3x}{2(1 - x^2)}, \quad |x| < 1$ **11.** $2e$

13. diverges **15.** converges absolutely **17.** converges conditionally

19. converges absolutely

21. converges conditionally

23. $[\frac{9}{5}, \frac{11}{5})$

25. $(0, 2)$

27. $(-1, 1)$

29. $\{-1\}$

31. $(-9, 9)$

33. $(-3, 7)$

35. $\displaystyle\sum_{k=0}^{\infty} \frac{5^k}{k!} x^{2k+1}$

37. $\displaystyle\sum_{k=0}^{\infty} \frac{(-1)^k}{2k+1} x^{k+1}$

39. $\displaystyle\sum_{k=0}^{\infty} \frac{(-1)^k}{(2k+1)!} (x^{4k+3} + x^{4k+4})$

41. $1 + x + \frac{1}{2}x^2 - \frac{1}{8}x^4$

43. $1 + \frac{1}{2}x^2 + \frac{3}{8}x^4$

45. $0.493 \le I \le 0.500$

47. $4.081 \le \sqrt[3]{68} \le 4.084$

49. $2/e$

51. (a) diverges   (b) diverges   (c) 1

53. Set $a_n = x_n - \dfrac{\epsilon}{4^n}$, $b_n = x_n + \dfrac{\epsilon}{4^n}$. Then $\displaystyle\sum_{n=1}^{\infty} (b_n - a_n) = 2\epsilon \sum_{n=1}^{\infty} (\tfrac{1}{4})^n = \tfrac{2}{3}\epsilon < \epsilon$.

SECTION 13.1

1.

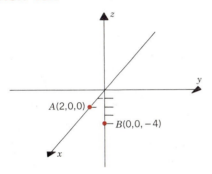

Figure A.13.1.1

length $\overline{AB}$: $2\sqrt{5}$
midpoint: $(1, 0, -2)$

3.

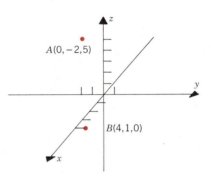

Figure A.13.1.2

length $\overline{AB}$: $5\sqrt{2}$
midpoint: $(2, -\frac{1}{2}, \frac{5}{2})$

5. $z = -2$

7. $y = 1$

9. $x = 3$

11. $x^2 + (y-2)^2 + (z+1)^2 = 9$

13. $(x-2)^2 + (y-4)^2 + (z+4)^2 = 36$

15. $(x-3)^2 + (y-2)^2 + (z-2)^2 = 13$

17. $(x-2)^2 + (y-3)^2 + (z+4)^2 = 25$

19. $(2, 3, -5)$

21. $(-2, 3, 5)$

23. $(-2, 3, -5)$

25. $(-2, -3, -5)$

27. $(2, -5, 5)$

29. $(-2, 1, -3)$

31. $(x-3)^2 + (y-3)^2 + (z-3)^2 = 9, \quad (x-7)^2 + (y-7)^2 + (z-7)^2 = 49$

33. (a) $x = a_1 + t(b_1 - a_1), \quad y = a_2 + t(b_2 - a_2), \quad z = a_3 + t(b_3 - a_3)$ (b) $t = \frac{1}{2}$

SECTION 13.3

1. $3\mathbf{i} - 4\mathbf{j} + 6\mathbf{k}$

3. $-3\mathbf{i} - \mathbf{j} + 8\mathbf{k}$

5. 5

7. 3

9. $\sqrt{6}$

11. (a) $\mathbf{a}, \mathbf{c}, \mathbf{d}$ (b) $\mathbf{a}, \mathbf{c}$ (c) $\mathbf{a}$ and $\mathbf{c}$ both have direction opposite to $\mathbf{d}$

13. (i) $\mathbf{a} + \mathbf{b}$ (ii) $-(\mathbf{a} + \mathbf{b})$ (iii) $\mathbf{a} - \mathbf{b}$ (iv) $\mathbf{b} - \mathbf{a}$

15. (a) $\mathbf{i} - 3\mathbf{j} + 10\mathbf{k}$ (b) $A = -2, B = \frac{3}{2}, C = -\frac{7}{2}$

17. $\alpha = \pm 3$

19. $\alpha = \pm \frac{1}{3}\sqrt{6}$

21. $\pm \frac{2}{13}\sqrt{13}(3\mathbf{j} + 2\mathbf{k})$

23. (a) the parallelogram is a rectangle
(b) simplify $\sqrt{(a_1 - b_1)^2 + (a_2 - b_2)^2 + (a_3 - b_3)^2} = \sqrt{(a_1 + b_1)^2 + (a_2 + b_2)^2 + (a_3 + b_3)^2}$

25. (a) $\|\mathbf{r} - \mathbf{a}\| = 3$ where $\mathbf{a} = a_1\mathbf{i} + a_2\mathbf{j} + a_3\mathbf{k}$
(b) $\|\mathbf{r}\| \le 2$ (c) $\|\mathbf{r} - \mathbf{a}\| \le 1$ where $\mathbf{a} = a_1\mathbf{i} + a_2\mathbf{j} + a_3\mathbf{k}$
(d) $\|\mathbf{r} - \mathbf{a}\| = \|\mathbf{r} - \mathbf{b}\|$ (e) $\|\mathbf{r} - \mathbf{a}\| + \|\mathbf{r} - \mathbf{b}\| = k$

SECTION 13.4

1. $\mathbf{a} \cdot \mathbf{b}$ **3.** $\mathbf{a} \cdot (\mathbf{b} + \mathbf{c})$

5. (a) $\mathbf{a} \cdot \mathbf{b} = 5,$ $\mathbf{a} \cdot \mathbf{c} = 8,$ $\mathbf{b} \cdot \mathbf{c} = 18$

(b) $\cos \sphericalangle (\mathbf{a}, \mathbf{b}) = \frac{1}{14} \sqrt{70},$ $\cos \sphericalangle (\mathbf{a}, \mathbf{c}) = \frac{8}{25} \sqrt{5},$ $\cos \sphericalangle (\mathbf{b}, \mathbf{c}) = \frac{9}{35} \sqrt{14}$

(c) $\text{comp}_b\, \mathbf{a} = \frac{5}{14} \sqrt{14},$ $\text{comp}_c\, \mathbf{a} = \frac{8}{5}$

(d) $\text{proj}_b\, \mathbf{a} = \frac{5}{14}(3\mathbf{i} - \mathbf{j} + 2\mathbf{k}),$ $\text{proj}_c\, \mathbf{a} = \frac{8}{25}(4\mathbf{i} + 3\mathbf{k})$

7. $\frac{1}{2}\mathbf{i} + \frac{1}{2}\sqrt{2}\mathbf{j} - \frac{1}{2}\mathbf{k}$ **9.** $\frac{1}{3}\pi$ **11.** $\frac{1}{3}\pi, \frac{2}{3}\pi, \frac{1}{4}\pi$

13. (a) $\text{proj}_b\, \alpha\mathbf{a} = (\alpha\mathbf{a} \cdot \mathbf{u_b})\mathbf{u_b} = \alpha(\mathbf{a} \cdot \mathbf{u_b})\mathbf{u_b} = \alpha\, \text{proj}_b\, \mathbf{a}$

(b) $\text{proj}_b(\mathbf{a} + \mathbf{c}) = [(\mathbf{a} + \mathbf{c}) \cdot \mathbf{u_b}]\mathbf{u_b}$
$$= (\mathbf{a} \cdot \mathbf{u_b} + \mathbf{c} \cdot \mathbf{u_b})\mathbf{u_b}$$
$$= (\mathbf{a} \cdot \mathbf{u_b})\mathbf{u_b} + (\mathbf{c} \cdot \mathbf{u_b})\mathbf{u_b} = \text{proj}_b\, \mathbf{a} + \text{proj}_b\, \mathbf{c}$$

15. (a) For $\mathbf{a} \neq 0$ the following statements are equivalent:

$$\mathbf{a} \cdot \mathbf{b} = \mathbf{a} \cdot \mathbf{c}, \qquad \mathbf{b} \cdot \mathbf{a} = \mathbf{c} \cdot \mathbf{a},$$

$$\mathbf{b} \cdot \frac{\mathbf{a}}{\|\mathbf{a}\|} = \mathbf{c} \cdot \frac{\mathbf{a}}{\|\mathbf{a}\|}, \qquad \mathbf{b} \cdot \mathbf{u_a} = \mathbf{c} \cdot \mathbf{u_a},$$

$$(\mathbf{b} \cdot \mathbf{u_a})\mathbf{u_a} = (\mathbf{c} \cdot \mathbf{u_a})\mathbf{u_a}, \qquad \text{proj}_a\, \mathbf{b} = \text{proj}_a\, \mathbf{c}. \qquad \mathbf{a} \cdot \mathbf{b} = \mathbf{a} \cdot \mathbf{c} \text{ but } \mathbf{b} \neq \mathbf{c}.$$

(b) $\mathbf{b} = (\mathbf{b} \cdot \mathbf{i})\mathbf{b} + (\mathbf{b} \cdot \mathbf{j})\mathbf{j} + (\mathbf{b} \cdot \mathbf{k})\mathbf{k} = (\mathbf{c} \cdot \mathbf{i})\mathbf{i} + (\mathbf{c} \cdot \mathbf{j})\mathbf{j} + (\mathbf{c} \cdot \mathbf{k})\mathbf{k} = \mathbf{c}.$

17. (a) Express the norms as dot products.

(b) The following statements are equivalent:

$$\mathbf{a} \perp \mathbf{b}, \qquad \mathbf{a} \cdot \mathbf{b} = 0, \qquad \|\mathbf{a} + \mathbf{b}\|^2 - \|\mathbf{a} - \mathbf{b}\|^2 = 0, \qquad \|\mathbf{a} + \mathbf{b}\| = \|\mathbf{a} - \mathbf{b}\|.$$

(c) By (b), the relation $\|\mathbf{a} + \mathbf{b}\| = \|\mathbf{a} - \mathbf{b}\|$ gives $\mathbf{a} \perp \mathbf{b}$. The relation $\mathbf{a} + \mathbf{b} \perp \mathbf{a} - \mathbf{b}$ gives

$$0 = (\mathbf{a} + \mathbf{b}) \cdot (\mathbf{a} - \mathbf{b}) = \|\mathbf{a}\|^2 - \|\mathbf{b}\|^2 \qquad \text{and thus} \qquad \|\mathbf{a}\| = \|\mathbf{b}\|.$$

The parallelogram is a square since it has two adjacent sides of equal length that meet at right angles.

19. $\pi - \alpha, \pi - \beta, \pi - \gamma$

21. If $\mathbf{a} \perp \mathbf{b}$ and $\mathbf{a} \perp \mathbf{c}$, then $\mathbf{a} \cdot \mathbf{b} = 0$ and $\mathbf{a} \cdot \mathbf{c} = 0$, so that

$$\mathbf{a} \cdot (\alpha\mathbf{b} + \beta\mathbf{c}) = \alpha(\mathbf{a} \cdot \mathbf{b}) + \beta(\mathbf{a} \cdot \mathbf{c}) = 0.$$

Thus $\mathbf{a} \perp \alpha\mathbf{b} + \beta\mathbf{c}$.

23. Existence of decomposition: $\mathbf{a} = (\mathbf{a} \cdot \mathbf{u_b})\mathbf{u_b} + [\mathbf{a} - (\mathbf{a} \cdot \mathbf{u_b})\mathbf{u_b}]$. Uniqueness of decomposition: suppose that $\mathbf{a} = \mathbf{a}_{\parallel} + \mathbf{a}_{\perp} = \mathbf{A}_{\parallel} + \mathbf{A}_{\perp}$. Then the vector $\mathbf{a}_{\parallel} - \mathbf{A}_{\parallel} = \mathbf{A}_{\perp} - \mathbf{a}_{\perp}$ is both parallel to $\mathbf{b}$ and perpendicular to $\mathbf{b}$. (Exercises 21 and 22.) Therefore it is zero. Consequently $\mathbf{A}_{\parallel} = \mathbf{a}_{\parallel}$ and $\mathbf{A}_{\perp} = \mathbf{a}_{\perp}$.

25. $x = 0, x = 4$ **27.** $\theta = \cos^{-1}\left(\frac{1}{3}\sqrt{3}\right) \cong 0.96$ radians

29. (a) The direction angles of a vector always satisfy $\cos^2 \alpha + \cos^2 \beta + \cos^2 \gamma = 1$, and, as you can check, $\cos^2 \frac{1}{4}\pi + \cos^2 \frac{1}{6}\pi + \cos^2 \frac{2}{3}\pi \neq 1$.

(b) The relation $\cos^2 \alpha + \cos^2 \frac{1}{4}\pi + \cos^2 \frac{1}{4}\pi = 1$ gives

$$\cos^2 \alpha + \frac{1}{2} + \frac{1}{2} = 1, \quad \cos \alpha = 0, \quad a_1 = \|\mathbf{a}\|\cos \alpha = 0.$$

31. $\mathbf{u} = \pm \frac{1}{165}\sqrt{165}\,(8\mathbf{i} + \mathbf{j} - 10\mathbf{k})$

33. Place center of sphere at the origin.

$$\overrightarrow{P_1 Q} \cdot \overrightarrow{P_2 Q} = (-\mathbf{a} + \mathbf{b}) \cdot (\mathbf{a} + \mathbf{b})$$
$$= -\|\mathbf{a}\|^2 + \|\mathbf{b}\|^2$$
$$= 0$$

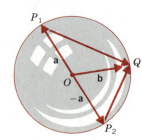

SECTION 13.5

1. -2 **3.** 0 **5.** 5 **7.** -6 **9.** no **11.** yes

13. yes **15.** $x = (ac - 1)/b$ **17.** $x = 0, \quad x = \frac{1}{2}(1 \pm \sqrt{5})$

19. $\begin{vmatrix} 4 & 5 & 6 \\ 1 & 2 & 3 \\ 7 & 8 & 9 \end{vmatrix} = - \begin{vmatrix} 1 & 2 & 3 \\ 4 & 5 & 6 \\ 7 & 8 & 9 \end{vmatrix}$ (interchanged two rows)

21. The second determinant is zero because two rows are the same.

23. $\begin{vmatrix} 1 & 2 & 3 \\ x & 2x & 3x \\ 4 & 5 & 6 \end{vmatrix} = x \begin{vmatrix} 1 & 2 & 3 \\ 1 & 2 & 3 \\ 4 & 5 & 6 \end{vmatrix} = 0$ ⌐—— two rows are the same

25. The expression on the left can be written

$$\lambda^2 \begin{vmatrix} 1 & 1 \\ a & b \end{vmatrix} - 2\lambda \begin{vmatrix} 1 & 1 \\ a & b \end{vmatrix} + 1 \begin{vmatrix} 1 & 1 \\ a & b \end{vmatrix} = (\lambda^2 - 2\lambda + 1) \begin{vmatrix} 1 & 1 \\ a & b \end{vmatrix} = (\lambda - 1)^2 \begin{vmatrix} 1 & 1 \\ a & b \end{vmatrix}.$$

27. $\begin{vmatrix} a_1 & a_2 & a_3 \\ b_1 & b_2 & b_3 \\ c_1 & c_2 & c_3 \end{vmatrix} = - \begin{vmatrix} a_1 & a_2 & a_3 \\ c_1 & c_2 & c_3 \\ b_1 & b_2 & b_3 \end{vmatrix} = \begin{vmatrix} c_1 & c_2 & c_3 \\ a_1 & a_2 & a_3 \\ b_1 & b_2 & b_3 \end{vmatrix} = c_1 \begin{vmatrix} a_2 & a_3 \\ b_2 & b_3 \end{vmatrix} - c_2 \begin{vmatrix} a_1 & a_3 \\ b_1 & b_3 \end{vmatrix} + c_3 \begin{vmatrix} a_1 & a_2 \\ b_1 & b_2 \end{vmatrix}$

29. $\begin{vmatrix} a & a^2 \\ b & b^2 \end{vmatrix} = a \begin{vmatrix} 1 & a \\ b & b^2 \end{vmatrix} = ab \begin{vmatrix} 1 & a \\ 1 & b \end{vmatrix}$

31. $\begin{vmatrix} a_1 + b_1 & a_2 + b_2 & a_3 + b_3 \\ b_1 & b_2 & b_3 \\ c_1 & c_2 & c_3 \end{vmatrix}$ ⌐—— by Exercise 28 $= \begin{vmatrix} a_1 & a_2 & a_3 \\ b_1 & b_2 & b_3 \\ c_1 & c_2 & c_3 \end{vmatrix} + \begin{vmatrix} b_1 & b_2 & b_3 \\ b_1 & b_2 & b_3 \\ c_1 & c_2 & c_3 \end{vmatrix}$

The second determinant is zero because two rows are the same.

SECTION 13.6

1. $-2\mathbf{k}$ **3.** $\mathbf{i} + \mathbf{j} + \mathbf{k}$ **5.** $-3\mathbf{i} - \mathbf{j} - 2\mathbf{k}$ **7.** -1

9. 0 **11.** 1 **13.** $\mathbf{i} + \mathbf{j} - 2\mathbf{k}$ **15.** -3

17. $5\mathbf{i} - 4\mathbf{j} - \mathbf{k}$ **19.** $\frac{3}{2}$ **21.** 1 **23.** $-2(\mathbf{a} \times \mathbf{b})$

25. $\mathbf{a} = \mathbf{0}$ **27.** $\begin{vmatrix} \alpha & \beta \\ \gamma & \delta \end{vmatrix} (\mathbf{a} \times \mathbf{b})$

29. $\mathbf{a} \cdot (\mathbf{b} \times \mathbf{c}) = (\mathbf{a} \times \mathbf{b}) \cdot \mathbf{c} = (\mathbf{c} \times \mathbf{a}) \cdot \mathbf{b} = (\mathbf{b} \times \mathbf{c}) \cdot \mathbf{a} = (\mathbf{a} \times -\mathbf{c}) \cdot \mathbf{b}$
$\mathbf{a} \cdot (\mathbf{c} \times \mathbf{b}) = \mathbf{c} \cdot (\mathbf{b} \times \mathbf{a}) = (-\mathbf{a} \times \mathbf{b}) \cdot \mathbf{c}$

31. $\mathbf{c} \times \mathbf{a} = \|\mathbf{a}\|^2 \mathbf{b}$

SECTION 13.7

1. P and Q

3. $\mathbf{r}(t) = (3\mathbf{i} + \mathbf{j}) + t\mathbf{k}$

5. $\mathbf{r}(t) = t(x_1\mathbf{i} + y_1\mathbf{j} + z_1\mathbf{k})$

7. $x(t) = 1 + t, \quad y(t) = -t, \quad z(t) = 3 + t$

9. $x(t) = 2, y(t) = t, z(t) = 3$

11. $\mathbf{r}(t) = (-\mathbf{i} + 2\mathbf{j} - 3\mathbf{k}) + t(2\mathbf{i} + \mathbf{j} + 4\mathbf{k})$

13. $P(1, 2, 0), \frac{1}{4}\pi$ rad

15. $P(1, 3, 1), \quad \cos^{-1}(\frac{1}{6}\sqrt{3}) \cong 1.28$ rad

17. $(x_0 - [d_1/d_3]z_0, y_0 - [d_2/d_3]z_0, 0)$

19. The lines are parallel.

21. $\mathbf{r}(t) = (2\mathbf{i} + 7\mathbf{j} - \mathbf{k}) + t(2\mathbf{i} - 5\mathbf{j} + 4\mathbf{k}), \quad 0 \le t \le 1$

23. $\mathbf{u} = -\frac{2}{3}\mathbf{i} + \frac{1}{3}\mathbf{j} + \frac{2}{3}\mathbf{k}, \quad 9 \le t \le 15$

25. triples of the form $X(u) = 3 + au, \ Y(u) = -1 + bu, \ Z(u) = 8 + cu$ with $2a - 4b + 6c = 0$

27. 1 **29.** $\sqrt{69/14} \cong 2.22$ **31.** $\sqrt{3} \cong 1.73$ **33.** (a) 1 (b) $\sqrt{3}$

35. $\mathbf{r}(t) = \frac{1}{11}(7\mathbf{i} + 4\mathbf{j} - \mathbf{k}) \pm t[\frac{1}{11}\sqrt{11}(\mathbf{i} - \mathbf{j} + 3\mathbf{k})]$

37. $0 < t < s$

SECTION 13.8

1. Q
3. $x - 4y + 3z - 2 = 0$
5. $3x - 2y + 5z - 9 = 0$
7. $\pm\frac{1}{62}\sqrt{62}(2\mathbf{i} - 3\mathbf{j} + 7\mathbf{k})$
9. $x/15 + y/12 + z/(-10) = 1$
11. $\frac{1}{2}\pi$
13. $\cos\theta = \frac{2}{21}\sqrt{42} \cong 0.617, \quad \theta \cong 0.91$ rad
15. coplanar
17. not coplanar
19. $\frac{2}{21}\sqrt{21} \cong 0.44$
21. $x + z = 2$
23. $(x - x_0)/d_1 = (y - y_0)/d_2, \quad (y - y_0)/d_2 = (z - z_0)/d_3$
25. $x(t) = t, \quad y(t) = t, \quad z(t) = -t$
27. $P(-\frac{19}{14}, \frac{15}{7}, \frac{17}{14})$
29. $10x - 7y + z = 0$
31. circle centered at P with radius $\|\mathbf{N}\|$
33. if $\alpha > 0$, then P_1 lies on the same side of the plane as the tip of $\mathbf{N}$;
    if $\alpha < 0$, then P_1 and the tip of $\mathbf{N}$ lie on opposite sides of the plane

SECTION 13.10

1. (a) $3\sqrt{10}$ (b) $(5, -\frac{3}{2}, \frac{3}{2})$ (c) $(11, -12, 9)$ or $(-1, 9, -6)$
   (d) $4x - 7y + 5z - 38 = 0$ (e) $x(t) = 6 + 4t, \quad y(t) = 6 - 7t, \quad z(t) = -4 + 5t$
   (f) $x(t) = 6 + 394t, \quad y(t) = 6 + 143t, \quad z(t) = -4 - 115t$
   (g) $x + 27y + 37z - 20 = 0$
3. $(1, 0, 1)$
5. 1
7. $2x + 3y - 4z - 19 = 0$
9. $7x - 3y - 5z + 18 = 0$
11. $x(t) = t - \frac{31}{7}, \quad y(t) = t + \frac{20}{7}, \quad z(t) = t$
13. $(-\frac{73}{5}, -\frac{37}{5}, -\frac{51}{5})$
15. $5x + 2y + 3z - 3 = 0$
17. 3

19.

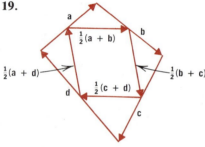

$\mathbf{a} + \mathbf{b} + \mathbf{c} + \mathbf{d} = \mathbf{0}$
gives
$\frac{1}{2}(\mathbf{a} + \mathbf{b}) = -\frac{1}{2}(\mathbf{c} + \mathbf{d})$
$\frac{1}{2}(\mathbf{b} + \mathbf{c}) = -\frac{1}{2}(\mathbf{a} + \mathbf{d})$

21. $\alpha = -2$
23. use the dot product
25. $11x + 4y + 8z - 3 = 0$

SECTION 14.1

1. $\mathbf{f}'(t) = 2\mathbf{i} - \mathbf{j} + 3\mathbf{k}$
3. $\mathbf{f}'(t) = -\dfrac{1}{2\sqrt{1-t}}\mathbf{i} + \dfrac{1}{2\sqrt{1+t}}\mathbf{j} + \dfrac{1}{(1-t)^2}\mathbf{k}$
5. $\mathbf{f}'(t) = \cos t\,\mathbf{i} - \sin t\,\mathbf{j} + \sec^2 t\,\mathbf{k}$
7. $\mathbf{i} + 3\mathbf{j}$
9. $(e - 1)\mathbf{i} + (1 - 1/e)\mathbf{k}$
11. $\frac{1}{2}\mathbf{i} + \mathbf{j}$
13. (a) $\mathbf{f}(t) = 3\cos t\,\mathbf{i} + 2\sin t\,\mathbf{j}$ (b) $\mathbf{f}(t) = 3\cos t\,\mathbf{i} - 2\sin t\,\mathbf{j}$
15. (a) $\mathbf{f}(t) = t\mathbf{i} + t^2\mathbf{j}$ (b) $\mathbf{f}(t) = -t\mathbf{i} + t^2\mathbf{j}$
17. $\mathbf{f}(t) = (1 + 2t)\mathbf{i} + (4 + 5t)\mathbf{j} + (-2 + 8t)\mathbf{j}, \quad 0 \le t \le 1$
19. $\mathbf{f}'(t_0) = \mathbf{i} + m\mathbf{j}, \quad \displaystyle\int_a^b \mathbf{f}(t)\,dt = \frac{1}{2}(b^2 - a^2)\mathbf{i} + A\mathbf{j}, \quad \int_a^b \mathbf{f}'(t)\,dt = (b - a)\mathbf{i} + (d - c)\mathbf{j}$
21. $\mathbf{f}(t) = t\mathbf{i} + (\frac{1}{3}t^3 + 1)\mathbf{j} - \mathbf{k}$
23. $\mathbf{f}(t) = e^{\alpha t}\mathbf{c}$
25. (a) if $\mathbf{f}'(t) = \mathbf{0}$ on an interval, then the derivative of each component is 0 on that interval, each component is constant on that interval, and therefore $\mathbf{f}$ itself is constant on that interval
    (b) set $\mathbf{h}(t) = \mathbf{f}(t) - \mathbf{g}(t)$ and apply part (a)
27. set $\mathbf{f}(t) = f_1(t)\mathbf{i} + f_2(t)\mathbf{j} + f_3(t)\mathbf{k}$, and apply (3.1.4) to f_1, f_2, f_3
29. no; as a counterexample set $\mathbf{f}(t) = \mathbf{i} = \mathbf{g}(t)$

SECTION 14.2

1. $\mathbf{f}'(t) = \mathbf{b}$, $\mathbf{f}''(t) = \mathbf{0}$
3. $\mathbf{f}'(t) = 2e^{2t}\mathbf{i} - \cos t\,\mathbf{j}$, $\mathbf{f}''(t) = 4e^{2t}\mathbf{i} + \sin t\,\mathbf{j}$
5. $\mathbf{f}'(t) = -2t\mathbf{i} + e^t(t+1)\mathbf{k}$, $\mathbf{f}''(t) = -2\mathbf{i} + e^t(t+2)\mathbf{k}$
7. $\mathbf{f}'(t) = \tfrac{1}{2}\sqrt{t}\,\mathbf{g}'(\sqrt{t}) + \mathbf{g}(\sqrt{t})$, $\mathbf{f}''(t) = \tfrac{1}{4}\mathbf{g}''(\sqrt{t}) + \tfrac{3}{4}(1/\sqrt{t})\mathbf{g}'(\sqrt{t})$
9. $-\sin t\, e^{\cos t}\mathbf{i} + \cos t\, e^{\sin t}\mathbf{j}$
11. $(\mathbf{a} \times \mathbf{d}) + (\mathbf{b} \times \mathbf{c}) + 2t(\mathbf{b} \times \mathbf{d})$
13. $(\mathbf{a} \cdot \mathbf{d}) + (\mathbf{b} \cdot \mathbf{c}) + 2t(\mathbf{b} \cdot \mathbf{d})$
15. $\mathbf{r}(t) = \mathbf{a} + t\mathbf{b}$
17. $\mathbf{r}(t) = \tfrac{1}{2}t^2\mathbf{a} + \tfrac{1}{6}t^3\mathbf{b} + t\mathbf{c} + \mathbf{d}$
19. $\mathbf{r}(t) \cdot \mathbf{r}'(t) = 0$, $\mathbf{r}(t) \times \mathbf{r}'(t) = \mathbf{k}$

21. $\dfrac{d}{dt}[\mathbf{f}(t) \times \mathbf{f}'(t)] = [\mathbf{f}(t) \times \mathbf{f}''(t)] + \underbrace{[\mathbf{f}'(t) \times \mathbf{f}'(t)]}_{\mathbf{0}} = \mathbf{f}(t) \times \mathbf{f}''(t)$

23. The following four statements are equivalent: $\|\mathbf{r}(t)\| = \sqrt{\mathbf{r}(t) \cdot \mathbf{r}(t)}$ is constant, $\mathbf{r}(t) \cdot \mathbf{r}(t)$ is constant,
$d/dt\,[\mathbf{r}(t) \cdot \mathbf{r}(t)] = 2[\mathbf{r}(t) \cdot \mathbf{r}'(t)] = 0$ identically, $\mathbf{r}(t) \cdot \mathbf{r}'(t) = 0$ identically.

25. $\dfrac{[\mathbf{f}(t + h) \times \mathbf{g}(t + h)] - [\mathbf{f}(t) \times \mathbf{g}(t)]}{h} = \left(\mathbf{f}(t + h) \times \left[\dfrac{\mathbf{g}(t + h) - \mathbf{g}(t)}{h}\right]\right) + \left(\left[\dfrac{\mathbf{f}(t + h) - \mathbf{f}(t)}{h}\right] \times \mathbf{g}(t)\right).$
Now take the limit as $h \to 0$.   (Appeal to Theorem 14.1.3.)

SECTION 14.3

1. $\pi\mathbf{j} + \mathbf{k}$, $\mathbf{R}(u) = (\mathbf{i} + 2\mathbf{k}) + u(\pi\mathbf{j} + \mathbf{k})$
3. $\mathbf{b} - 2\mathbf{c}$, $\mathbf{R}(u) = (\mathbf{a} - \mathbf{b} + \mathbf{c}) + u(\mathbf{b} - 2\mathbf{c})$
5. $4\mathbf{i} - \mathbf{j} + 4\mathbf{k}$, $\mathbf{R}(u) = (2\mathbf{i} + 5\mathbf{k}) + u(4\mathbf{i} - \mathbf{j} + 4\mathbf{k})$
7. The scalar components $x(t) = at$ and $y(t) = bt^2$ satisfy the equation $a^2 y(t) = a^2(bt^2) = b(at)^2 = b[x(t)]^2$ and generate the parabola $a^2 y = bx^2$.
9. (a) $P(0, 1)$ (b) $P(1, 2)$ (c) $P(-1, 2)$
11. The tangent line at $t = t_0$ has the form $\mathbf{R}(u) = \mathbf{r}(t_0) + u\mathbf{r}'(t_0)$. If $\mathbf{r}'(t_0) = \alpha\mathbf{r}(t_0)$, then

$$\mathbf{R}(u) = \mathbf{r}(t_0) + u\alpha\mathbf{r}(t_0) = (1 + u\alpha)\mathbf{r}(t_0).$$

The tangent line passes through the origin at $u = -1/\alpha$.
13. $P(1, 2, -2)$; $\cos^{-1}(\tfrac{1}{5}\sqrt{5}) \cong 1.11$ rad
15. (a) $\mathbf{r}(t) = a\cos t\,\mathbf{i} + b\sin t\,\mathbf{j}$ (b) $\mathbf{r}(t) = a\cos t\,\mathbf{i} - b\sin t\,\mathbf{j}$
   (c) $\mathbf{r}(t) = a\cos 2t\,\mathbf{i} + b\sin 2t\,\mathbf{j}$ (d) $\mathbf{r}(t) = a\cos 3t\,\mathbf{i} - b\sin 3t\,\mathbf{j}$
17. $\mathbf{r}(t) = (t^2 + 1)\mathbf{i} + t\mathbf{j}$, $t \geq 1$;   or, $\mathbf{r}(t) = \sec^2 t\,\mathbf{i} + \tan t\,\mathbf{j}$, $t \in [\tfrac{1}{4}\pi, \tfrac{1}{2}\pi)$
19. $\mathbf{r}(t) = \cos t\,\sin 3t\,\mathbf{i} + \sin t\,\sin 3t\,\mathbf{j}$, $t \in [0, \pi]$

21.

$y^3 = x^2$

23. $(2, 4, 8)$; $\cos^{-1}\left(\dfrac{24}{\sqrt{21}\sqrt{161}}\right) \cong 1.15$ rad
25. $\tfrac{1}{5}\sqrt{5}(-2\mathbf{i} + \mathbf{k})$, $-\mathbf{j}$, $x + 2z = \tfrac{1}{2}\pi$

There is no tangent vector at the origin.

27. $\tfrac{1}{14}\sqrt{14}(\mathbf{i} + 2\mathbf{j} + 3\mathbf{k})$, $\tfrac{1}{266}\sqrt{266}(-11\mathbf{i} - 8\mathbf{j} + 9\mathbf{k})$, $3x - 3y + z = 1$
29. $\mathbf{T}_1 = \dfrac{\mathbf{R}'(u)}{\|\mathbf{R}'(u)\|} = -\dfrac{\mathbf{r}'(a + b - u)}{\|\mathbf{r}'(a + b - u)\|} = -\mathbf{T}$. Then $\mathbf{T}_1'(u) = \mathbf{T}'(a + b - u)$ and thus $\mathbf{N}_1 = \mathbf{N}$.
31. Let $\mathbf{T}$ be the unit tangent at the tip of $\mathbf{R}(u) = \mathbf{r}(\phi(u))$ as calculated from the parametrization $\mathbf{r}$ and let $\mathbf{T}_1$ be the unit tangent at the same point as calculated from the parametrization $\mathbf{R}$. Then

$$\mathbf{T}_1 = \frac{\mathbf{R}'(u)}{\|\mathbf{R}'(u)\|} = \frac{\mathbf{r}'(\phi(u))\,\phi'(u)}{\|\mathbf{r}'(\phi(u))\,\phi'(u)\|} = \frac{\mathbf{r}'(\phi(u))}{\|\mathbf{r}'(\phi(u))\|} = \mathbf{T}.$$

$$\phi'(u) > 0 \underline{\qquad}\uparrow$$

This shows the invariance of the unit tangent. The invariance of the principal normal and the osculating plane follows directly from the invariance of the unit tangent.

SECTION 14.4

1. $2\pi\sqrt{a^2 + b^2}$ **3.** $\ln(1 + \sqrt{2})$ **5.** $\frac{1}{27}(13\sqrt{13} - 8)$

7. $\sqrt{2}(e^\pi - 1)$ **9.** $6 + \frac{1}{2}\sqrt{2}\ln(2\sqrt{2} + 3)$

11. differentiate $s(t) = \displaystyle\int_a^t \sqrt{[x'(u)]^2 + [y'(u)]^2 + [z'(u)]^2} \, du$

13. see Exercise 12, differentiate $s(x) = \displaystyle\int_a^x \sqrt{1 + [f'(t)]^2} \, dt$

15. Let L be the length as computed from r and L^* the length as computed from $\mathbf{R}$. Then

$$L^* = \int_c^d \|\mathbf{R}'(u)\| \, du = \int_c^d \|\mathbf{r}'(\phi(u))\| \, \phi'(u) \, du = \int_a^b \|\mathbf{r}'(t)\| \, dt = L.$$
$$t = \phi(u) \longrightarrow$$

17. $L \cong 0.503$

SECTION 14.5

1. v/r, v^2/r **3.** $\|\mathbf{r}''(t)\| = a^2|b\sin at| = a^2|y(t)|$

5. $y = \cos \pi x$, $0 \le x \le 2$ **7.** $x = \sqrt{1 + y^2}$, $y \ge -1$

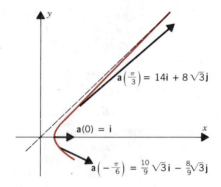

9. (a) (x_0, y_0, z_0) (b) $\alpha \cos \theta \, \mathbf{j} + \alpha \sin \theta \, \mathbf{k}$ (c) $|\alpha|$ (d) $-32\mathbf{k}$
    (e) arc from parabola $z = z_0 + (\tan \theta)(y - y_0) - 16(y - y_0)^2/(\alpha^2 \cos^2 \theta)$ in the plane $x = x_0$

11. (a) $\mathbf{r}'(0) = b\omega\mathbf{j}$ (b) $\mathbf{r}''(t) = \omega^2\mathbf{r}(t)$
    (c) The torque is $\mathbf{0}$ and the angular momentum is constant.

13. (a) $\mathbf{v}(t) = 2\mathbf{j} + (\alpha/m)t\mathbf{k}$ (b) $\mathbf{v}(t) = (1/m)\sqrt{4m^2 + \alpha^2 t^2}$ (c) $\mathbf{p}(t) = 2m\mathbf{j} + \alpha t\mathbf{k}$
    (d) $\mathbf{r}(t_1 = [2t + y_0]\mathbf{j} + [(\alpha/2m)t^2 + z_0]\mathbf{k}$, $t \ge 0$ $z = (\alpha/8m)(y - y_0)^2 + z_0$, $y \ge y_0$, $x = 0$

15. $F(t) = 2mk$

17. (a) $\pi b\mathbf{j} + \mathbf{k}$ (b) $\sqrt{\pi^2 b^2 + 1}$ (c) $-\pi^2 a\mathbf{i}$ (d) $m(\pi b\mathbf{j} + \mathbf{k})$
    (e) $m[b(1 - \pi)\mathbf{i} - 2a\mathbf{j} + 2\pi ab\mathbf{k}]$ (f) $-m\pi^2 a[\mathbf{j} - b\mathbf{k}]$

19. We have $m\mathbf{v} = m\mathbf{v}_1 + m\mathbf{v}_2$ and $\frac{1}{2}mv^2 = \frac{1}{2}mv_1^2 + \frac{1}{2}mv_2^2$. Therefore $\mathbf{v} = \mathbf{v}_1 + \mathbf{v}_2$ and $v^2 = v_1^2 + v_2^2$. Since

$$v^2 = \mathbf{v} \cdot \mathbf{v} = (\mathbf{v}_1 + \mathbf{v}_2) \cdot (\mathbf{v}_1 + \mathbf{v}_2) = v_1^2 + v_2^2 + 2(\mathbf{v}_1 \cdot \mathbf{v}_2),$$

we have $\mathbf{v}_1 \cdot \mathbf{v}_2 = \mathbf{0}$ and $\mathbf{v}_1 \perp \mathbf{v}_2$.

21. Here $\mathbf{r}''(t) = \mathbf{a}$, $\mathbf{r}'(t) = \mathbf{v}(0) + t\mathbf{a}$, $\mathbf{r}(t) = \mathbf{r}(0) + t\mathbf{v}(0) + \frac{1}{2}t^2\mathbf{a}$. If neither $\mathbf{v}(0)$ nor $\mathbf{a}$ is zero, the displacement $\mathbf{r}(t) - \mathbf{r}(0)$ is a linear combination of $\mathbf{v}(0)$ and $\mathbf{a}$ and thus remains on the plane determined by these vectors. The equation of this plane can be written

$$[\mathbf{a} \times \mathbf{v}(0)] \cdot [\mathbf{r} - \mathbf{r}(0)] = 0.$$

(If either $\mathbf{v}(0)$ or $\mathbf{a}$ is zero, the motion is restricted to a straight line; if both of these vectors are zero, the particle remains at its initial position $\mathbf{r}(0)$.)

23. $\mathbf{r}(t) = \mathbf{i} + t\mathbf{j} + (qE_0/2m)t^2\mathbf{k}$ **25.** $\mathbf{r}(t) = (1 + t^3/6m)\mathbf{i} + (t^4/12m)\mathbf{j} + \mathbf{k}$

27. $\dfrac{d}{dt}(\frac{1}{2}mv^2) = mv\,\dfrac{dv}{dt} = m\left(\mathbf{v}\cdot\dfrac{d\mathbf{v}}{dt}\right) = m\,\dfrac{d\mathbf{v}}{dt}\cdot\mathbf{v} = \mathbf{F}\cdot\dfrac{d\mathbf{r}}{dt} = 4r^2\left(\mathbf{r}\cdot\dfrac{d\mathbf{r}}{dt}\right) = 4r^2\left(r\,\dfrac{dr}{dt}\right) = 4r^3\,\dfrac{dr}{dt} = \dfrac{d}{dt}(r^4).$

Therefore $d/dt\,(\frac{1}{2}mv^2 - r^4) = 0$ and $\frac{1}{2}mv^2 - r^4$ is a constant E. Evaluating E from $t = 0$, we find that $E = 2m$. Thus $\frac{1}{2}mv^2 - r^4 = 2m$ and $v = \sqrt{4 + (2/m)r^4}$.

SECTION *14.6

1. about 61.5% of an earth year
3. set $x = r\cos\theta$, $y = r\sin\theta$
5. Substitute

$$r = \frac{a}{1 + e\cos\theta}, \qquad \left(\frac{dr}{d\theta}\right)^2 = \frac{(ae\sin\theta)^2}{(1 + e\cos\theta)^4}$$

into the right side of the equation and you will see that, with a and e^2 as given, the expression reduces to E.

SECTION 14.7

1. $\dfrac{e^{-x}}{(1 + e^{-2x})^{3/2}}$ **3.** $|\cos x|$ **5.** $\dfrac{|\sin x|}{(1 + \cos^2 x)^{3/2}}$

7. $\frac{5}{2}\sqrt{5}$ **9.** $5\sqrt{5}$ **11.** $(\frac{1}{2}\sqrt{2}, \frac{1}{2}\ln\frac{1}{2})$

13. $\dfrac{1}{(1 + t^2)^{3/2}}$ **15.** $\frac{1}{2}\sqrt{2}\,e^{-t}$ **17.** $\sqrt{2}$ **19.** $\dfrac{a^4b^4}{(b^4x^2 + a^4y^2)^{3/2}}$

21. follow the hint   **23.** $k = \frac{1}{3}\sqrt{2}\,e^{-t}$, $a_\mathbf{T} = \sqrt{3}\,e^t$, $a_\mathbf{N} = \sqrt{2}\,e^t$

25. $k = \dfrac{\sqrt{t^4 + 4t^2 + 1}}{(t^4 + t^2 + 1)^{3/2}}$, $a_\mathbf{T} = \dfrac{2t^3 + t}{\sqrt{t^4 + t^2 + 1}}$, $a_\mathbf{N} = \dfrac{\sqrt{t^4 + 4t^2 + 1}}{\sqrt{t^4 + t^2 + 1}}$

27. $\dfrac{e^{-a\theta}}{\sqrt{1 + a^2}}$ **29.** $\dfrac{3}{2\sqrt{2a^2(1 - \cos\theta)}} = \dfrac{3}{2\sqrt{2ar}}$ **33.** $9\rho^2 + s^2 = 16a^2$

31. (a) $s(\theta) = 4R\,|\cos\frac{1}{2}\theta|$ (b) $\rho(\theta) = 4R\sin\frac{1}{2}\theta$ (c) $\rho^2 + s^2 = 16R^2$

SECTION 15.1

1. dom (f) = the first and third quadrants, including the axes; ran $(f) = [0, \infty)$
3. dom (f) = the set of all points (x, y) not on the line $y = -x$; ran $(f) = (-\infty, 0)\cup(0, \infty)$
5. dom (f) = the entire plane; ran $(f) = (-1, 1)$
7. dom (f) = the first and third quadrants, excluding the axes; ran $(f) = (-\infty, \infty)$
9. dom (f) = the set of all points (x, y) with $x^2 < y$; in other words, the set of all points of the plane above the parabola $y = x^2$; ran $(f) = (0, \infty)$
11. dom (f) = the set of all points (x, y) with $-3 \le x \le 3$, $-2 \le y \le 2$ (a rectangle); ran $(f) = [-2, 3]$
13. dom (f) = the set of all points (x, y, z) not on the plane $x + y + z = 0$; ran $(f) = \{-1, 1\}$
15. dom (f) = the set of all points (x, y, z) with $|y| < |x|$; ran $(f) = (-\infty, 0]$
17. dom (f) = the set of all points (x, y) such that $x^2 + y^2 < 9$; in other words, the set of all points of the plane inside the circle $x^2 + y^2 = 9$; ran $(f) = [\frac{2}{3}, \infty)$
19. dom (f) = the set of all points (x, y, z) with $x + 2y + 3z > 0$; in other words, the set of all points in space that lie on the same side of the plane $x + 2y + 3z = 0$ as the point $(1, 1, 1)$; ran $(f) = (-\infty, \infty)$
21. dom (f) = all of space; ran $(f) = (0, \infty)$
23. dom (f) = the set of all points (x, y, z) with $-3 \le x \le 3$, $-2 \le y \le 2$, $-1 \le z \le 1$ (a rectangular solid); ran $(f) = [-2, 3]$
25. f assigns the number $\sqrt{x}$ to each real number $x \ge 0$; g assigns the number $\sqrt{x}$ to each point (x, y) of the plane with $x \ge 0$; h assigns the number $\sqrt{x}$ to each point (x, y, z) of space with $x \ge 0$

SECTION 15.2

1. a quadric cone   **3.** a parabolic cylinder
5. a hyperboloid of one sheet   **7.** sphere of radius 2 centered at the origin

9. an elliptic paraboloid **11.** a hyperbolic paraboloid
13. Figure A.15.2.1 **15.** Figure A.15.2.2 **17.** Figure A.15.2.3

Figure A.15.2.1

Figure A.15.2.2

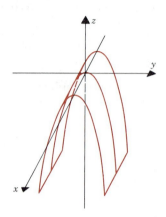

Figure A.15.2.3

19. elliptic paraboloid, xy-trace: the origin, xz-trace: the parabola $x^2 = 4z$, yz-trace: the parabola $y^2 = 9z$, surface has the form of Figure 15.2.5

21. a quadric cone, xy-trace: the origin, xz-trace: the lines $x = \pm 2z$, yz-trace: the lines $y = \pm 3z$, surface has the form of Figure 15.2.4

23. a hyperboloid of two sheets, xy-trace: none, xz-trace: the hyperbola $4z^2 - x^2 = 4$, yz-trace: the hyperbola $9z^2 - y^2 = 9$, surface has the form of Figure 15.2.3

25. (a) an elliptic paraboloid (vertex down if A and B are both positive, vertex up if A and B are both negative)
(b) a hyperbolic paraboloid
(c) the xy-plane if A and B are both zero; otherwise, a parabolic cylinder

27. $x^2 + y^2 - 4z = 0$ (paraboloid of revolution)
29. (a) a circle   (b) (i) $\sqrt{x^2 + y^2} = -3z$ (ii) $\sqrt{x^2 + z^2} = \frac{1}{3}y$
31. the line $5x + 7y = 30$ **33.** the circle $x^2 + y^2 = \frac{5}{4}$
35. the ellipse $x^2 + 2y^2 = 2$ **37.** the parabola $x^2 = -4(y - 1)$

SECTION 15.3

1. lines of slope 1: $y = x - c$; Figure A.15.3.1
3. parabolas, $y = x^2 - c$; Figure A.15.3.2
5. the y-axis and the lines $y = [(1 - c)/c]x$, the origin omitted throughout; Figure A.15.3.3

Figure A.15.3.1

Figure A.15.3.2

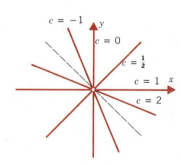

Figure A.15.3.3

7. the cubics $y = x^3 - c$; Figure A.15.3.4
9. the lines $y = \pm x$ and the hyperbolas $x^2 - y^2 = c$; Figure A.15.3.5
11. pairs of horizontal lines $y = \pm\sqrt{c}$ and the x-axis; Figure A.15.3.6

Figure A.15.3.4

Figure A.15.3.5

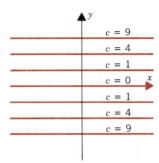

Figure A.15.3.6

13. the circles $x^2 + y^2 = e^c$ with c real; Figure A.15.3.7
15. the curves $y = e^{cx^2}$ with the point $(0, 1)$ omitted; Figure A.15.3.8
17. the coordinate axes and pairs of lines $y = \pm(\sqrt{1 - c}/\sqrt{c})x$, the origin omitted throughout; Figure A.15.3.9

Figure A.15.3.7

Figure A.15.3.8

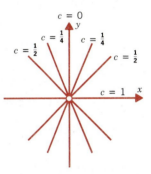

Figure A.15.3.9

19. $x + 2y + 3z = 0$, plane through the origin
21. $z = \sqrt{x^2 + y^2}$, the upper nappe of the circular cone $z^2 = x^2 + y^2$; Figure 15.2.4
23. the elliptic paraboloid $\dfrac{x^2}{(3\sqrt{2})^2} + \dfrac{y^2}{(2\sqrt{2})^2} = z$; Figure 15.2.5
25. (i) hyperboloid of two sheets; Figure 15.2.3 (ii) circular cone; Figure 15.2.4 (iii) hyperboloid of one sheet; Figure 15.2.2

SECTION 15.4

1. $\dfrac{\partial f}{\partial x} = 6x - y, \quad \dfrac{\partial f}{\partial y} = 1 - x$

3. $\dfrac{\partial \rho}{\partial \phi} = \cos\phi\cos\theta, \quad \dfrac{\partial \rho}{\partial \theta} = -\sin\phi\sin\theta$

5. $\dfrac{\partial f}{\partial x} = e^{x-y} + e^{y-x}, \quad \dfrac{\partial f}{\partial y} = -e^{x-y} - e^{y-x}$

7. $\dfrac{\partial g}{\partial x} = \dfrac{(AD - BC)y}{(Cx + Dy)^2}, \quad \dfrac{\partial g}{\partial y} = \dfrac{(BC - AD)x}{(Cx + Dy)^2}$

9. $\dfrac{\partial u}{\partial x} = y + z, \quad \dfrac{\partial u}{\partial y} = x + z, \quad \dfrac{\partial u}{\partial z} = x + y$

11. $\dfrac{\partial f}{\partial x} = z\cos(x - y), \quad \dfrac{\partial f}{\partial y} = -z\cos(x - y), \quad \dfrac{\partial f}{\partial z} = \sin(x - y)$

13. $\dfrac{\partial \rho}{\partial \theta} = e^{\theta+\phi}\,[\cos{(\theta - \phi)} - \sin{(\theta - \phi)}]$, $\dfrac{\partial \rho}{\partial \phi} = e^{\theta+\phi}\,[\cos{(\theta - \phi)} + \sin{(\theta - \phi)}]$

15. $\dfrac{\partial h}{\partial x} = 2f(x)\,f'(x)\,g(y)$, $\dfrac{\partial h}{\partial y} = [f(x)]^2\,g'(y)$

17. $\dfrac{\partial f}{\partial x} = (y^2 \ln z)z^{xy^2}$, $\dfrac{\partial f}{\partial y} = (2xy \ln z)z^{xy^2}$, $\dfrac{\partial f}{\partial z} = xy^2 z^{xy^2-1}$

19. $f_x(0, e) = 1$, $f_y(0, e) = e^{-1}$ **21.** $f_x(1, 2) = \frac{2}{9}$, $f_y(1, 2) = -\frac{1}{9}$

23. $f_x(x, y) = 2xy$, $f_y(x, y) = x^2$ **25.** $f_x(x, y) = \dfrac{2}{x}$, $f_y(x, y) = \dfrac{1}{y}$

27. $f_x(x, y) = -\dfrac{1}{(x - y)^2}$, $f_y(x, y) = \dfrac{1}{(x - y)^2}$

29. $f_x(x, y, z) = y^2 z$, $f_y(x, y, z) = 2xyz$, $f_z(x, y, z) = xy^2$

31. $x\dfrac{\partial u}{\partial x} + y\dfrac{\partial u}{\partial y} = x(4Ax^3 + 4Bxy^2) + y(4Bx^2 y + 4Cy^3) = 4(Ax^4 + 2Bx^2 y^2 + Cy^4) = 4u$

33. r **35.** follows from $\dfrac{\partial P}{\partial V} = -\dfrac{kT}{V^2}$ and $\dfrac{\partial P}{\partial T} = \dfrac{k}{V}$

37. (a) $50\sqrt{3}$ in.^{2} (b) $5\sqrt{3}$ (c) 50 (d) $\frac{5}{18}\pi$ in.^{2} (e) -2

39. (a) y_0-section: $\mathbf{r}(x) = x\mathbf{i} + y_0\mathbf{j} + f(x, y_0)\mathbf{k}$

  tangent line: $\mathbf{R}(t) = [x_0\mathbf{i} + y_0\mathbf{j} + f(x_0, y_0)\mathbf{k}] + t\left[\mathbf{i} + \dfrac{\partial f}{\partial x}(x_0, y_0)\mathbf{k}\right]$

(b) x_0-section: $\mathbf{r}(y) = x_0\mathbf{i} + y\mathbf{j} + f(x_0, y)\mathbf{k}$

  tangent line: $\mathbf{R}(t) = [x_0\mathbf{i} + y_0\mathbf{j} + f(x_0, y_0)\mathbf{k}] + t\left[\mathbf{j} + \dfrac{\partial f}{\partial y}(x_0, y_0)\mathbf{k}\right]$

(c) For (x, y, z) in the plane

$$[(x - x_0)\mathbf{i} + (y - y_0)\mathbf{j} + (z - f(x_0, y_0))\mathbf{k}] \cdot \left[\left(\mathbf{i} + \dfrac{\partial f}{\partial x}(x_0, y_0)\mathbf{k}\right) \times \left(\mathbf{j} + \dfrac{\partial f}{\partial y}(x_0, y_0)\mathbf{k}\right)\right] = 0.$$

From this it follows that

$$z - f(x_0, y_0) = (x - x_0)\dfrac{\partial f}{\partial x}(x_0, y_0) + (y - y_0)\dfrac{\partial f}{\partial y}(x_0, y_0).$$

41. (a) Set $u = ax + by$. Then $\dfrac{\partial w}{\partial x} = ag'(u)$ and $\dfrac{\partial w}{\partial y} = bg'(u)$.

(b) Set $u = x^m y^n$. Then $\dfrac{\partial w}{\partial x} = mx^{m-1}y^n\,g'(u)$ and $\dfrac{\partial w}{\partial y} = nx^m y^{n-1}\,g'(u)$.

SECTION 15.5

1. interior $= \{(x, y): 2 < x < 4, 1 < y < 3\}$ (the inside of the rectangle)
boundary $=$ the union of the four line segments that bound the rectangle
set is closed; Figure A.15.5.1

3. interior $=$ the entire set (region between two concentric circles)
boundary $= \{(x, y): x^2 + y^2 = 1$ or $x^2 + y^2 = 4\}$ (the two circles)
set is open; Figure A.15.5.2

5. interior $= \{(x, y): 1 < x^2 < 4\} = \{(x, y): -2 < x < 1\} \cup \{(x, y): 1 < x < 2\}$
            (two vertical strips without the boundary lines)
boundary $= \{(x, y): x = -2, x = -1, x = 1,$ or $x = 2\}$ (four vertical lines)
set is neither open nor closed; Figure A.15.5.3

7. interior $= \{(x, y): y < x^2\}$ (region below the parabola)
boundary $= \{(x, y): y = x^2\}$ (the parabola)
set is closed; Figure A.15.5.4

Figure A.15.5.1

Figure A.15.5.2

Figure A.15.5.3

Figure A.15.5.4

9. interior = $\{(x, y, z): x^2 + y^2 < 1, \ 0 < z < 4\}$
(the inside of a cylinder)
boundary = the total surface of the cylinder
(the curved part, the top, the bottom)
set is closed; Figure A.15.5.5

Figure A.15.5.5

11. (a) ϕ (b) S (c) closed
13. interior = $\{x: 1 < x < 3\}$, boundary = $\{1, 3\}$; set is closed
15. interior = the entire set, boundary = $\{1\}$; set is open
17. interior = $\{x: |x| > 1\}$, boundary = $\{1, -1\}$; set is neither open nor closed
19. interior = ϕ, boundary = {the entire set} $\cup \{0\}$; the set is neither open nor closed

SECTION 15.6

1. $\dfrac{\partial^2 f}{\partial x^2} = 2A, \quad \dfrac{\partial^2 f}{\partial y^2} = 2C, \quad \dfrac{\partial^2 f}{\partial y \partial x} = \dfrac{\partial^2 f}{\partial x \partial y} = 2B$

3. $\dfrac{\partial^2 f}{\partial x^2} = Cy^2 e^{xy}, \quad \dfrac{\partial^2 f}{\partial y^2} = Cx^2 e^{xy}, \quad \dfrac{\partial^2 f}{\partial y \partial x} = \dfrac{\partial^2 f}{\partial x \partial y} = Ce^{xy}(xy + 1)$

5. $\dfrac{\partial^2 f}{\partial x^2} = 2, \quad \dfrac{\partial^2 f}{\partial y^2} = 4(x + 3y^2 + z^3), \quad \dfrac{\partial^2 f}{\partial z^2} = 6z(2x + 2y^2 + 5z^3)$

$\dfrac{\partial^2 f}{\partial x \partial y} = \dfrac{\partial^2 f}{\partial y \partial x} = 4y, \quad \dfrac{\partial^2 f}{\partial z \partial x} = \dfrac{\partial^2 f}{\partial x \partial z} = 6z^2, \quad \dfrac{\partial^2 f}{\partial z \partial y} = \dfrac{\partial^2 f}{\partial y \partial z} = 12yz^2$

7. $\dfrac{\partial^2 f}{\partial x^2} = \dfrac{1}{(x + y)^2} - \dfrac{1}{x^2}, \quad \dfrac{\partial^2 f}{\partial y^2} = \dfrac{1}{(x + y)^2}, \quad \dfrac{\partial^2 f}{\partial y \partial x} = \dfrac{\partial^2 f}{\partial x \partial y} = \dfrac{1}{(x + y)^2}$

9. $\dfrac{\partial^2 f}{\partial x^2} = 2(y + z), \quad \dfrac{\partial^2 f}{\partial y^2} = 2(x + z), \quad \dfrac{\partial^2 f}{\partial z^2} = 2(x + y)$; the second mixed partials are all $2(x + y + z)$

11. $\dfrac{\partial^2 f}{\partial x^2} = y(y - 1)x^{y-2}, \quad \dfrac{\partial^2 f}{\partial y^2} = (\ln x)^2 \, x^y, \quad \dfrac{\partial^2 f}{\partial y \partial x} = \dfrac{\partial^2 f}{\partial x \partial y} = x^{y-1}(1 + y \ln x)$

13. $x^2 \dfrac{\partial^2 u}{\partial x^2} + 2xy \dfrac{\partial^2 u}{\partial x \partial y} + y^2 \dfrac{\partial^2 u}{\partial y^2} = x^2 \left(\dfrac{-2y^2}{(x + y)^3} \right) + 2xy \left(\dfrac{2xy}{(x + y)^3} \right) + y^2 \left(\dfrac{-2x^2}{(x + y)^3} \right) = 0$

15. (a) $\dfrac{\partial^2 f}{\partial t^2} - c^2 \dfrac{\partial^2 f}{\partial x^2} = 0 - c^2(0) = 0$ (b) $\dfrac{\partial^2 f}{\partial t^2} - c^2 \dfrac{\partial^2 f}{\partial x^2} = c^2 k^2 f(x, t) - c^2 [k^2 f(x, t)] = 0$

(c) use Exercise 40, Section 15.4

17. (a) mixed partials are 0   (b) mixed partials are $g'(x)\,h'(y)$
    (c) by the hint mixed partials for each term $x^m y^n$ are $mnx^{m-1}y^{n-1}$

19. (a) no, since $\dfrac{\partial^2 f}{\partial x \partial y} \neq \dfrac{\partial^2 f}{\partial y \partial x}$ (b) no, since $\dfrac{\partial^2 f}{\partial x \partial y} \neq \dfrac{\partial^2 f}{\partial y \partial x}$ for $x \neq y$

21.
$$\frac{\partial^3 f}{\partial x^2 \partial y} = \frac{\partial}{\partial x}\left(\frac{\partial^2 f}{\partial x \partial y}\right)$$
by definition

$$= \frac{\partial}{\partial x}\left(\frac{\partial^2 f}{\partial y \partial x}\right) = \frac{\partial^2}{\partial x \partial y}\left(\frac{\partial f}{\partial x}\right) = \frac{\partial^2}{\partial y \partial x}\left(\frac{\partial f}{\partial x}\right) = \frac{\partial}{\partial y}\left(\frac{\partial^2 f}{\partial x^2}\right) = \frac{\partial^3 f}{\partial y \partial x^2}$$

(15.6.5)   by definition   (15.6.5)   by definition

23. (a) 0   (b) 0   (c) $\dfrac{m}{1+m^2}$ (d) 0   (e) $\dfrac{f'(0)}{1+[f'(0)]^2}$
    (f) $\frac{1}{4}\sqrt{3}$ (g) does not exist

25. (a) $\dfrac{\partial g}{\partial x}(0,0) = \lim_{h \to 0} \dfrac{g(h,0) - g(0,0)}{h} = \lim_{h \to 0} 0 = 0, \qquad \dfrac{\partial g}{\partial y}(0,0) = \lim_{h \to 0} \dfrac{g(0,h) - g(0,0)}{h} = \lim_{h \to 0} 0 = 0$

    (b) as (x,y) tends to $(0,0)$ along the x-axis, $g(x,y) = g(x,0) = 0$ tends to 0;
       as (x,y) tends to $(0,0)$ along the line $y = x$, $g(x,y) = g(x,x) = \frac{1}{2}$ tends to $\frac{1}{2}$

27. For $y \neq 0$, $\dfrac{\partial f}{\partial x}(0,y) = \lim_{h \to 0} \dfrac{f(h,y) - f(0,y)}{h} = \lim_{h \to 0} \dfrac{y(y^2 - h^2)}{h^2 + y^2} = y$.

Since $\dfrac{\partial f}{\partial x}(0,0) = \lim_{h \to 0} \dfrac{f(h,0) - f(0,0)}{h} = \lim_{h \to 0} 0 = 0$, we have $\dfrac{\partial f}{\partial x}(0,y) = y$ for all y.

For $x \neq 0$, $\dfrac{\partial f}{\partial y}(x,0) = \lim_{h \to 0} \dfrac{f(x,h) - f(x,0)}{h} = \lim_{h \to 0} \dfrac{x(h^2 - x^2)}{x^2 + h^2} = -x$.

Since $\dfrac{\partial f}{\partial y}(0,0) = \lim_{h \to 0} \dfrac{f(0,h) - f(0,0)}{h} = \lim_{h \to 0} 0 = 0$, we have $\dfrac{\partial f}{\partial y}(x,0) = -x$ for all x.

Therefore $\dfrac{\partial^2 f}{\partial y \partial x}(0,y) = 1$ for all y and $\dfrac{\partial^2 f}{\partial x \partial y}(x,0) = -1$ for all x.

In particular, $\dfrac{\partial^2 f}{\partial y \partial x}(0,0) = 1$ while $\dfrac{\partial^2 f}{\partial x \partial y}(0,0) = -1$.

SECTION 16.1

1. $\nabla f = e^{xy}[(xy + 1)\mathbf{i} + x^2\mathbf{j}]$ **3.** $\nabla f = (6x - y)\mathbf{i} + (1 - x)\mathbf{j}$
5. $\nabla f = 2xy^{-2}\,\mathbf{i} - 2x^2y^{-3}\mathbf{j}$
7. $\nabla f = z\cos(x - y)\mathbf{i} - z\cos(x - y)\mathbf{j} + \sin(x - y)\mathbf{k}$
9. $\nabla f = (y + z)\mathbf{i} + (x + z)\mathbf{j} + (x + y)\mathbf{k}$ **11.** $\nabla f = e^{x-y}[(1 + x + y)\mathbf{i} + (1 - x - y)\mathbf{j}]$
13. $\nabla f = e^x[\ln y\,\mathbf{i} + y^{-1}\,\mathbf{j}]$ **15.** $\nabla f = \dfrac{AD - BC}{(Cx + Dy)^2}[y\mathbf{i} - x\mathbf{j}]$
17. $\nabla f = -\mathbf{i} + 18\mathbf{j}$ **19.** $\nabla f = -\frac{1}{2}\sqrt{2}(\mathbf{i} + 2\mathbf{j} + \mathbf{k})$
21. (a) $(1/r^2)\mathbf{r}$ (b) $[(\cos r)/r]\mathbf{r}$ (c) $(e^r/r)\mathbf{r}$
23. (a) Let $\mathbf{c} = c_1\mathbf{i} + c_2\mathbf{j} + c_3\mathbf{k}$. First, we take $\mathbf{h} = h\mathbf{i}$. Since $\mathbf{c} \cdot \mathbf{h}$ is $o(h)$,
$$0 = \lim_{\mathbf{h} \to 0} \frac{\mathbf{c} \cdot \mathbf{h}}{\|\mathbf{h}\|} = \lim_{h \to 0} \frac{c_1 h}{h} = c_1.$$
    Similarly, $c_2 = 0$ and $c_3 = 0$.
    (b) $(\mathbf{y} - \mathbf{z}) \cdot \mathbf{h} = [f(\mathbf{x} + \mathbf{h}) - f(\mathbf{x}) - \mathbf{z} \cdot \mathbf{h}] + [\mathbf{y} \cdot \mathbf{h} - f(\mathbf{x} + \mathbf{h}) + f(\mathbf{x})] = o(\mathbf{h}) + o(\mathbf{h}) = o(\mathbf{h})$,
    so that, by part (a), $\mathbf{y} - \mathbf{z} = \mathbf{0}$.

25. (a) In Section 15.6 we showed that f was not continuous at $(0, 0)$. It is therefore not differentiable at $(0, 0)$.

(b) For $(x, y) \neq (0, 0)$, $\dfrac{\partial f}{\partial x} = \dfrac{2y(y^2 - x^2)}{(x^2 + y^2)^2}$. As (x, y) tends to $(0, 0)$ along the y-axis, $\partial f/\partial x = 2/y$ tends to ∞.

SECTION 16.2

1. $-2\sqrt{2}$

3. $\frac{1}{5}(7 - 4e)$

5. $\frac{1}{4}\sqrt{2}(a - b)$

7. $\frac{2}{3}\sqrt{6}$

9. $-3\sqrt{2}$

11. $-(x^2 + y^2)^{-1/2}$

13. (a) $\sqrt{2}[a(B - A) + b(C - B)]$ (b) $\sqrt{2}[a(A - B) + b(B - C)]$

15. $-\frac{7}{5}\sqrt{5}$

17. $\nabla f = f'(x_0)\mathbf{i}$. If $f'(x_0) \neq 0$, the gradient points in the direction in which f increases: to the right if $f'(x_0) > 0$, to the left if $f'(x_0) < 0$.

19. (a) $\displaystyle\lim_{h \to 0} \frac{f(h, 0) - f(0, 0)}{h} = \lim_{h \to 0} \frac{\sqrt{h^2}}{h} = \lim_{h \to 0} \frac{|h|}{h}$ does not exist

(b) no; by Theorem 16.2.5 f cannot be differentiable at $(0, 0)$

21. (a) $-\frac{2}{3}\sqrt{97}$ (b) $-\frac{8}{3}$ (c) $-\frac{26}{3}\sqrt{2}$

23. (a) its projection onto the xy-plane is the curve $y = x^3$ from $(1, 1)$ to $(0, 0)$

(b) its projection onto the xy-plane is the curve $y = -2x^3$ from $(1, -2)$ to $(0, 0)$

25. its projection onto the xy-plane is the curve $(b^2)^{a^2} x^{b^2} = (a^2)^{b^2} y^{a^2}$ from (a^2, b^2) to $(0, 0)$

27. the curve $y = \ln |\sqrt{2} \sin x|$ in the direction of decreasing x

29. (a) 16 (b) 4 (c) $\frac{16}{17}\sqrt{17}$

(d) The limits computed in (a) and (b) are not directional derivatives. In (a) and (b) we have, in essence, computed $\nabla f(2, 4) \cdot \mathbf{r}_0$ taking $\mathbf{r}_0 = \mathbf{i} + 4\mathbf{j}$ in (a) and $\mathbf{r}_0 = \frac{1}{4}\mathbf{i} + \mathbf{j}$ in (b). In neither case is $\mathbf{r}_0$ a unit vector.

SECTION 16.3

1. (a) $f(x, y, z) = a_1 x + a_2 y + a_3 z + C$

(b) $f(x, y, z) = g(x, y, z) + a_1 x + a_2 y + a_3 z + C$

3. (a) U is not connected

(b) (i) $g(\mathbf{x}) = f(\mathbf{x}) - 1$ (ii) $g(\mathbf{x}) = -f(\mathbf{x})$ (c) U is not connected

5. Since f is continuous at $\mathbf{a}$ and $f(\mathbf{a}) = A$, there exists $\delta > 0$ such that

$$\text{if } \|\mathbf{x} - \mathbf{a}\| < \delta \quad \text{and} \quad \mathbf{x} \in \Omega, \quad \text{then} \quad |f(\mathbf{x}) - A| < \epsilon.$$

Whether $\mathbf{a}$ is on the boundary of Ω or in the interior of Ω, there exists $\mathbf{x}_1$ in the interior of Ω within δ of $\mathbf{a}$. That implies that $|f(\mathbf{x}_1) - A| < \epsilon$ and thus that $f(\mathbf{x}_1) < A + \epsilon$. A similar argument shows the existence of $\mathbf{x}_2$ with the desired property.

SECTION 16.4

1. e^t

3. $3t^2 - 5t^4$

5. $2\omega(b^2 - a^2) \sin \omega t \cos \omega t + b\omega$

7. $\sin 2t - 3 \cos 2t$

9. $e^{t/2}(\frac{1}{2} \sin 2t + 2 \cos 2t) + e^{2t}(2 \sin \frac{1}{2}t + \frac{1}{2} \cos \frac{1}{2}t)$

11. $1 - 4t + 6t^2 - 4t^3$

13. increasing $\frac{1288}{3}\pi$ in.^{3}/sec

15. $\dfrac{\partial u}{\partial s} = 2s \cos^2 t - t \sin s \cos t - st \cos s \cos t$

$\dfrac{\partial u}{\partial t} = -2s^2 \sin t \cos t + st \sin s \sin t - s \sin s \cos t$

17. $\dfrac{\partial u}{\partial s} = 2s \cos^2 t - \sin(t - s) \cos t + s \cos t \cos(t - s) + 2t^2 \sin s \cos s$

$\dfrac{\partial u}{\partial t} = -2s^2 \sin t \cos t + s \sin(t - s) \sin t - s \cos t \cos(t - s) + 2t \sin^2 s$

19. $\dfrac{d}{dt}[f(\mathbf{r}(t))] = \left[\nabla f(\mathbf{r}(t)) \cdot \dfrac{\mathbf{r}'(t)}{\|\mathbf{r}'(t)\|}\right] \|\mathbf{r}'(t)\| = f'_{\mathbf{u}(t)}(\mathbf{r}(t)) \|\mathbf{r}'(t)\|$ where $\mathbf{u}(t) = \dfrac{\mathbf{r}'(t)}{\|\mathbf{r}'(t)\|}$

21. (a) $(\cos r)\dfrac{\mathbf{r}}{r}$ (b) $(r\cos r + \sin r)\dfrac{\mathbf{r}}{r}$ **23.** (a) $(r\cos r - \sin r)\dfrac{\mathbf{r}}{r^3}$ (b) $\left(\dfrac{\sin r - r\cos r}{\sin^2 r}\right)\dfrac{\mathbf{r}}{r}$

25. (a) $\dfrac{\partial u}{\partial r} = \dfrac{\partial u}{\partial x}\dfrac{\partial x}{\partial r} + \dfrac{\partial u}{\partial y}\dfrac{\partial y}{\partial r} = \dfrac{\partial u}{\partial x}\cos\theta + \dfrac{\partial u}{\partial y}\sin\theta,$ $\dfrac{\partial u}{\partial\theta} = \dfrac{\partial u}{\partial x}\dfrac{\partial x}{\partial\theta} + \dfrac{\partial u}{\partial y}\dfrac{\partial y}{\partial\theta} = \dfrac{\partial u}{\partial x}(-r\sin\theta) + \dfrac{\partial u}{\partial y}(r\cos\theta)$

(b) $\left(\dfrac{\partial u}{\partial r}\right)^2 = \left(\dfrac{\partial u}{\partial x}\right)^2\cos^2\theta + 2\dfrac{\partial u}{\partial x}\dfrac{\partial u}{\partial y}\cos\theta\sin\theta + \left(\dfrac{\partial u}{\partial y}\right)^2\sin^2\theta,$

$\dfrac{1}{r^2}\left(\dfrac{\partial u}{\partial\theta}\right)^2 = \left(\dfrac{\partial u}{\partial x}\right)^2\sin^2\theta - 2\dfrac{\partial u}{\partial x}\dfrac{\partial u}{\partial y}\cos\theta\sin\theta + \left(\dfrac{\partial u}{\partial y}\right)^2\cos^2\theta,$

$\left(\dfrac{\partial u}{\partial r}\right)^2 + \dfrac{1}{r^2}\left(\dfrac{\partial u}{\partial\theta}\right)^2 = \left(\dfrac{\partial u}{\partial x}\right)^2(\cos^2\theta + \sin^2\theta) + \left(\dfrac{\partial u}{\partial y}\right)^2(\sin^2\theta + \cos^2\theta) = \left(\dfrac{\partial u}{\partial x}\right)^2 + \left(\dfrac{\partial u}{\partial y}\right)^2$

27. $\dfrac{\partial z}{\partial x} = -\dfrac{2xz^3 + y}{z^2(4z + 3x^2)},$ $\dfrac{\partial z}{\partial y} = -\dfrac{x + 2y}{z^2(4z + 3x^2)}$

29. (a) Figure A.16.4.1

(b) $\dfrac{\partial u}{\partial r} = \dfrac{\partial u}{\partial x}\left(\dfrac{\partial x}{\partial w}\dfrac{\partial w}{\partial r} + \dfrac{\partial x}{\partial t}\dfrac{\partial t}{\partial r}\right) + \dfrac{\partial u}{\partial y}\left(\dfrac{\partial y}{\partial w}\dfrac{\partial w}{\partial r} + \dfrac{\partial y}{\partial t}\dfrac{\partial t}{\partial r}\right) + \dfrac{\partial u}{\partial z}\left(\dfrac{\partial z}{\partial w}\dfrac{\partial w}{\partial r} + \dfrac{\partial z}{\partial t}\dfrac{\partial t}{\partial r}\right)$

To obtain $\dfrac{\partial u}{\partial s}$, replace each r by s.

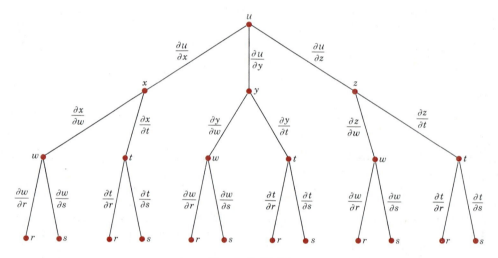

Figure A.16.4.1

31. (a) We write $u = u(x, y);\ x = t,\ y = y(t).$

By (16.4.3), $\dfrac{du}{dt} = \dfrac{\partial u}{\partial x}\dfrac{dx}{dt} + \dfrac{\partial u}{\partial y}\dfrac{dy}{dt}.$ Since $u(t, y(t)) = 0$ for all t, $\dfrac{du}{dt} = 0.$ Since $x = t$, $\dfrac{dx}{dt} = 1$ and

$\dfrac{dy}{dt} = \dfrac{dy}{dx}.$ Thus we have $\dfrac{\partial u}{\partial x} + \dfrac{\partial u}{\partial y}\dfrac{dy}{dx} = 0.$ Since $\partial u/\partial y \neq 0$, $\dfrac{dy}{dx} = -\dfrac{\partial u/\partial x}{\partial u/\partial y}.$

(b) $\dfrac{dy}{dx} = -\dfrac{2xy^4 + \cos(x + y)}{4x^2y^3 + \cos(x + y)}$

33. $\dfrac{\partial \mathbf{u}}{\partial s} = \dfrac{\partial \mathbf{u}}{\partial x}\dfrac{\partial x}{\partial s} + \dfrac{\partial \mathbf{u}}{\partial y}\dfrac{\partial y}{\partial s},$ $\dfrac{\partial \mathbf{u}}{\partial t} = \dfrac{\partial \mathbf{u}}{\partial x}\dfrac{\partial x}{\partial t} + \dfrac{\partial \mathbf{u}}{\partial y}\dfrac{\partial y}{\partial t}$

SECTION 16.5

1. normal vector $\mathbf{i} + \mathbf{j}$; tangent vector $\mathbf{i} - \mathbf{j}$
tangent line $x + y + 2 = 0$; normal line $x - y = 0$
3. normal vector $\sqrt{2}\,\mathbf{i} - 5\mathbf{j}$; tangent vector $5\mathbf{i} + \sqrt{2}\,\mathbf{j}$
tangent line $\sqrt{2}\,x - 5y + 3 = 0$; normal line $5x + \sqrt{2}\,y - 6\sqrt{2} = 0$

5. normal vector $7\mathbf{i} - 17\mathbf{j}$; tangent vector $17\mathbf{i} + 7\mathbf{j}$
   tangent line $7x - 17y + 6 = 0$; normal line $17x + 7y - 82 = 0$
7. normal vector $\mathbf{i} - \mathbf{j}$; tangent vector $\mathbf{i} + \mathbf{j}$
   tangent line $x - y - 3 = 0$; normal line $x + y + 1 = 0$

9. 0   11. $4x - 5y + 4z = 0$ 13. $x + ay - z - 1 = 0$

15. $2x + 2y - z = 0$ 17. $b^2c^2x_0x - a^2c^2y_0y - a^2b^2z_0z - a^2b^2c^2 = 0$

19. $(a^2/b, b^2/a, 3ab)$ 21. $(0, 0, 0)$ 23. $(\frac{1}{3}, \frac{11}{6}, -\frac{1}{12})$

25. $\dfrac{x - x_0}{\partial f/\partial x \,(x_0, y_0, z_0)} = \dfrac{y - y_0}{\partial f/\partial y \,(x_0, y_0, z_0)} = \dfrac{z - z_0}{\partial f/\partial z \,(x_0, y_0, z_0)}$

27. the tangent planes meet at right angles and therefore the normals ∇F and ∇G must meet at right angles:

$$\frac{\partial F}{\partial x}\frac{\partial G}{\partial x} + \frac{\partial F}{\partial y}\frac{\partial G}{\partial y} + \frac{\partial F}{\partial z}\frac{\partial G}{\partial z} = 0$$

29. $\frac{9}{2}a^3$ $(V = \frac{1}{3}Bh)$ 31. approx 0.528 rad

33. $3x + 4y + 6z = 22$, $6x + y - z = 11$

35. $(1, 1, 2)$ lies on both surfaces and the normals at this point are perpendicular

37. (a) $3x + 4y + 6 = 0$ (b) $\mathbf{r}(t) = (4t - 2)\mathbf{i} - 3t\mathbf{j} + (43t^2 - 16t + 6)\mathbf{k}$
    (c) $\mathbf{R}(s) = (2\mathbf{i} - 3\mathbf{j} + 33\mathbf{k}) + s(4\mathbf{i} - 3\mathbf{j} + 70\mathbf{k})$
    (d) $4x - 18y - z = 29$ (e) $\mathbf{r}(t) = t\mathbf{i} - (\frac{3}{4}t + \frac{3}{2})\mathbf{j} + (\frac{35}{2}t - 2)\mathbf{k}; l = l'$

SECTION 16.6

1. $(1, 0)$ gives a local max of 1   3. $(1, -1)$ is a saddle point   5. $(-2, 1)$ gives a local min of -2

7. no stationary points and no local extreme values

9. $(1, 1)$ gives a local min of -3 11. $(-2, 2)$ is a saddle point   13. $(0, 0)$ is a saddle point

15. $(3, 4)$ gives absolute min of $\frac{1}{5}$; $(1, 1)$ gives absolute max of $\frac{1}{2}\sqrt{2}$

17. absolute max of 4 occurs at $(-1, 0)$ and $(1, 0)$; there is no absolute min

19. $(1, 1)$ gives absolute min of 0; $(-\sqrt{2}, -\sqrt{2})$ gives absolute max of $6 + 4\sqrt{2}$

21. absolute min of 0 occurs at each point (x, x) with $0 \le x \le 4$; absolute max of 144 occurs at $(0, 12)$

23. $(1, 1)$ gives absolute min of 10; $(2, 8)$ gives absolute max of 68

25. (a) $(0, 0)$ (b) no local extremes as $(0, 0)$ is a saddle point
    (c) $(1, 0)$ and $(-1, 0)$ give absolute max of $\frac{1}{4}$; $(0, 1)$ and $(0, -1)$ give absolute min of $-\frac{1}{9}$

27. $\left(\dfrac{x_1 + x_2 + x_3}{3}, \dfrac{y_1 + y_2 + y_3}{3} \right)$ 29. $1 - \frac{1}{2}\sqrt{3}$ 31. \$1 per dozen blades, \$2 per razor

SECTION 16.7

1. $(4, -2)$ gives a local min of -10 3. $(\frac{10}{3}, \frac{8}{3})$ gives a local max of $\frac{28}{3}$

5. $(0, 0)$ is a saddle point; $(2, 2)$ gives a local min of -8

7. $(1, \frac{3}{2})$ is a saddle point; $(5, \frac{27}{2})$ gives a local min of $-\frac{117}{4}$

9. no stationary points and no local extreme values

11. $(0, n\pi)$ for integral n are saddle points; no local extreme values

13. $(1, -1)$ and $(-1, 1)$ are saddle points; no local extreme values

15. $(\frac{1}{2}, 4)$ gives a local min of 6   17. $(1, 1)$ gives a local min of 3   19. $\frac{1}{27}$

21. (a) saddle point, $f(x, y) = 0$ along the plane
    curve $y = x^{2/3}$ (see figure)
    (b) $(0, 0)$ gives a local max of 3

23. $\theta = \frac{1}{6}\pi$, $x = (2 - \sqrt{3})P$, $y = \frac{1}{6}(3 - \sqrt{3})P$

25. $\frac{2}{3}\sqrt{6}$ 27. $4 \times 4 \times 6$ m

29. (a) $y = x - \frac{2}{3}$ (b) $y = \frac{14}{13}x^2 - \frac{19}{13}$

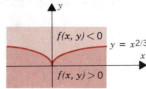

SECTION 16.8

1. 2

3. $-\frac{1}{2}ab$

5. $\frac{2}{9}\sqrt{3}ab^2$

7. 1

9. $\frac{1}{9}\sqrt{3}abc$

11. $19\sqrt{2}$

13. $\frac{1}{27}abc$

15. 1

17. $|D|(A^2 + B^2 + C^2)^{-1/2}$

19. $4A^2(a^2 + b^2 + c^2)^{-1}$, where A is the area of the triangle and a, b, c, are the sides

21. $(2^{-1/3}, -2^{-1/3})$

23. hint is given

25. $Q_1 = 10{,}000$, $Q_2 = 20{,}000$, $Q_3 = 30{,}000$

SECTION 16.9

1. $df = (3x^2y - 2xy^2)\,\Delta x + (x^3 - 2x^2y)\,\Delta y$

3. $df = (\cos y + y \sin x)\,\Delta x - (x \sin y + \cos x)\,\Delta y$

5. $df = \Delta x - (\tan z)\,\Delta y - (y \sec^2 z)\,\Delta z$

7. $\Delta u = -7.15$, $du = -7.5$

9. $22\frac{249}{352}$ taking $u = x^{1/2}y^{1/4}$, $x = 121$, $y = 16$, $\Delta x = 4$, $\Delta y = 1$

11. $\frac{1}{14}\sqrt{2}\pi$ taking $u = \sin x \cos y$, $x = \pi$, $y = \frac{1}{4}\pi$, $\Delta x = -\frac{1}{7}\pi$, $\Delta y = -\frac{1}{20}\pi$

13. $dz = -\frac{1}{90}$, $\Delta z = -\frac{1}{93}$

15. decreases about 13.6π in.^{2}

17. (a) $\Delta h = -\dfrac{(2r + \Delta r)h}{(r + \Delta r)^2}\,\Delta r$, $\Delta h \cong -\left(\dfrac{2h}{r}\right)\Delta r$

(b) $\Delta h = -\dfrac{(2r + h + \Delta r)}{r + \Delta r}\,\Delta r$, $\Delta h \cong -\left(\dfrac{2r + h}{r}\right)\Delta r$

19. $2.23 \le s \pm |\Delta s| \le 2.27$

SECTION 16.10

1. $f(x, y) = \frac{1}{2}x^2y^2 + C$

3. $f(x, y) = xy + C$

5. not a gradient

7. $f(x, y) = \sin x + y \cos x + C$

9. $f(x, y) = e^x \cos y^2 + C$

11. $f(x, y) = xye^x + e^{-y} + C$

13. $f(x, y) = \sqrt{x^2 + y^2} + C$

15. $f(x, y) = \frac{1}{3}x^3 \sin^{-1} y + y - y \ln y + C$

17. $f(x, y) = Ce^{x+y}$

19. (a), (b), (c) routine; (d) $f(x, y, z) = x^2 + yz + C$

21. $f(x, y, z) = x^2 + y^2 - z^2 + xy + yz + C$

23. $\mathbf{F(r)} = \nabla\left(G\dfrac{mM}{r}\mathbf{r}\right)$

SECTION 17.1

1. $a_2 - a_1$

3. $(a_2 - a_1)(b_2 - b_1)$

5. $a_2^2 - a_1^2$

7. $(a_2^2 - a_1^2)(b_2 - b_1)$

9. $2n(a_2 - a_1) - 3m(b_2 - b_1)$

11. $(a_2 - a_1)(b_2 - b_1)(c_2 - c_1)$

13. $a_{111} + a_{222} + \cdots + a_{nnn} = \displaystyle\sum_{p=1}^{n} a_{ppp}$

SECTION 17.2

1. $L_f(P) = 2\frac{1}{4}$, $U_f(P) = 5\frac{3}{4}$

3. (a) $L_f(P) = \displaystyle\sum_{i=1}^{m}\sum_{j=1}^{n}(x_{i-1} + 2y_{j-1})\Delta x_i\Delta y_j$, $U_f(P) = \displaystyle\sum_{i=1}^{m}\sum_{j=1}^{n}(x_i + 2y_j)\Delta x_i\Delta y_j$

(b) $I = 4$; the volume of the prism bounded above by the plane $z = x + 2y$ and below by R

5. $L_f(P) = -\frac{7}{24}$, $U_f(P) = \frac{7}{24}$

7. (a) $L_f(P) = \displaystyle\sum_{i=1}^{m}\sum_{j=1}^{n}4x_{i-1}y_{j-1}\Delta x_i\Delta y_j$, $U_f(P) = \displaystyle\sum_{i=1}^{m}\sum_{j=1}^{n}4x_iy_j\Delta x_i\Delta y_j$ (b) $I = b^2d^2$

9. (a) $L_f(P) = \displaystyle\sum_{i=1}^{m}\sum_{j=1}^{n}3(x_{i-1}^2 - y_j^2)\Delta x_i\Delta y_j$, $U_f(P) = \displaystyle\sum_{i=1}^{m}\sum_{j=1}^{n}3(x_i^2 - y_{j-1}^2)\Delta x_i\Delta y_j$ (b) $I = bd(b^2 - d^2)$

SECTION 17.3

1. $\displaystyle\iint\limits_{\Omega}\int dx\,dy = \int_a^b \phi(x)\,dx$

SECTION 17.4

1. 1 **3.** $\frac{9}{2}$ **5.** $\frac{1}{24}$ **7.** -2 **9.** $\frac{1}{4}\pi^2 + \frac{1}{64}\pi^4$

11. $\frac{2}{27}$ **13.** $\frac{512}{15}$ **15.** 0 **17.** $\frac{1}{4}(e^4 - 1)$

19. Figure A.17.4.1; $\displaystyle\int_0^1 \int_{y^{1/2}}^{y^{1/4}} f(x, y)\,dx\,dy$

21. Figure A.17.4.2; $\displaystyle\int_{-1}^0 \int_{-x}^1 f(x, y)\,dy\,dx + \int_0^1 \int_x^1 f(x, y)\,dy\,dx$

23. Figure A.17.4.3; $\displaystyle\int_1^2 \int_1^y f(x, y)\,dx\,dy + \int_2^4 \int_{y/2}^y f(x, y)\,dx\,dy + \int_4^8 \int_{y/2}^4 f(x, y)\,dx\,dy$

Figure A.17.4.1

Figure A.17.4.2

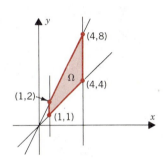

Figure A.17.4.3

25. 9 **27.** $\frac{1}{160}$

29. Figure A.17.4.4; $\frac{2}{3}(\cos\frac{1}{2} - \cos 1)$ **31.** Figure A.17.4.5; $1 - \ln 2$

Figure A.17.4.4

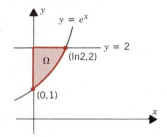

Figure A.17.4.5

33. $\ln 4 - \frac{1}{2}$ **35.** 4 **37.** $\frac{2}{15}$ **39.** 3π

41. $\frac{1}{2}(e - 1)$ **43.** $\frac{1}{12}(e - 1)$

45. $\displaystyle\iint\limits_{R}\int f(x)g(y)\,dx\,dy = \int_c^d \int_a^b f(x)g(y)\,dx\,dy = \int_c^d \left(\int_a^b f(x)g(y)\,dx\right)dy = \int_c^d g(y)\left(\int_a^b f(x)\,dx\right)dy$

$$= \left(\int_a^b f(x)\,dx\right)\left(\int_c^d g(y)\,dy\right)$$

47. We have $R: -a \le x \le a, \ c \le y \le d$. Set $f(x, y) = g_y(x)$. For each fixed $y \in [c, d]$, g_y is an odd function. Thus

$$\int_{-a}^a g_y(x)\,dx = 0. \qquad\qquad (5.8.2)$$

Therefore

$$\iint_R f(x,\,y)\,dxdy = \int_c^d \int_{-a}^a f(x,\,y)\,dx\,dy = \int_c^d \int_{-a}^a g_y(x)\,dx\,dy = \int_c^d 0\,dy = 0.$$

49. Note that $\Omega = \{(x,\,y):\ 0 \le x \le y,\ \ 0 \le y \le 1\}$. Set $\Omega' = \{(x,\,y):\ 0 \le y \le x,\ \ 0 \le x \le 1\}$.

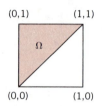

(0,1)   (1,1)

Ω

(0,0)   (1,0)

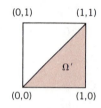

(0,1)   (1,1)

Ω'

(0,0)   (1,0)

$$\iint_\Omega f(x)f(y)\,dxdy = \int_0^1 \int_0^y f(x)f(y)\,dx\,dy$$

$$= \int_0^1 \int_0^x f(y)f(x)\,dy\,dx$$

x and y are dummy variables

$$= \iint_{\Omega'} f(x)f(y)\,dxdy.$$

Note that Ω and Ω' don't overlap and their union is the unit square R: $\{(x,\,y):\ 0 \le x \le 1,\ 0 \le y \le 1\}$.

If $\displaystyle\int_0^1 f(x)\,dx = 0$, then

$$0 = \left(\int_0^1 f(x)\,dx\right)\left(\int_0^1 f(y)\,dy\right) = \iint_R f(x)f(y)\,dxdy$$

by Exercise 45

$$= \iint_\Omega f(x)f(y)\,dxdy + \iint_{\Omega'} f(x)f(y)\,dxdy$$

$$= 2\iint_\Omega f(x)f(y)\,dxdy$$

and therefore $\displaystyle\iint_\Omega f(x)f(y)\,dxdy = 0.$

51. Let M be the maximum value of $|f(x,\,y)|$ on Ω.

$$\int_{\phi_1(x+h)}^{\phi_2(x+h)} = \int_{\phi_1(x+h)}^{\phi_1(x)} + \int_{\phi_1(x)}^{\phi_2(x)} + \int_{\phi_2(x)}^{\phi_2(x+h)}$$

$$|F(x+h) - F(x)| = \left| \int_{\phi_1(x+h)}^{\phi_2(x+h)} f(x,\,y)\,dy - \int_{\phi_1(x)}^{\phi_2(x)} f(x,\,y)\,dy \right|$$

$$= \left| \int_{\phi_1(x+h)}^{\phi_1(x)} f(x,\,y)\,dy + \int_{\phi_2(x)}^{\phi_2(x+h)} f(x,\,y)\,dy \right| \le \left| \int_{\phi_1(x+h)}^{\phi_1(x)} f(x,\,y)\,dy \right| + \left| \int_{\phi_2(x)}^{\phi_2(x+h)} f(x,\,y)\,dy \right|$$

$$\le |\phi_1(x) - \phi_1(x+h)|M + |\phi_2(x+h) - \phi_2(x)|M.$$

The expression on the right tends to 0 as h tends to 0 since ϕ_1 and ϕ_2 are continuous.

SECTION 17.5

1. (a) $\pi \sin 1 \cong 0.84\pi$ (b) $\pi(\sin 2 - \sin 1) \cong 0.07\pi$ **3.** (a) $\frac{2}{3}$ (b) $\frac{14}{3}$

5. $\frac{1}{3}\pi$ **7.** $\frac{1}{15}\pi$ **9.** $b^3\pi$ **11.** $\frac{16}{3}\sqrt{3}\,\pi$

13. $\frac{2}{3}(8 - 3\sqrt{3})\pi$ **15.** 2π **17.** $\frac{1}{3}\pi a^2 b$

SECTION 17.6

1. $I_x = \frac{1}{12}MW^2,\ \ I_y = \frac{1}{12}ML^2,\ \ I_z = \frac{1}{12}M(L^2 + W^2);$ $K_x = \frac{1}{6}\sqrt{3}\,W,\ \ K_y = \frac{1}{6}\sqrt{3}\,L,\ \ K_z = \frac{1}{6}\sqrt{3}\sqrt{L^2 + W^2}$

3. $x_M = \frac{1}{6}L, \quad y_M = 0$

5. $I_x = I_y = \frac{1}{4}MR^2, \quad I_z = \frac{1}{2}MR^2; \quad K_x = K_y = \frac{1}{2}R, \quad K_z = \frac{1}{2}\sqrt{2}R$

7. center the disc at a distance $\sqrt{I_0 - \frac{1}{2}MR^2}/\sqrt{M}$ from the origin

9. $I_x = \frac{1}{4}Mb^2, \quad I_y = \frac{1}{4}Ma^2, \quad I_z = \frac{1}{4}M(a^2 + b^2)$

11. (a) $\frac{1}{4}M(r_2^2 + r_1^2)$ (b) $\frac{1}{4}M(r_2^2 + 5r_1^2)$ (c) $\frac{1}{4}M(5r_2^2 + r_1^2)$

13. $\frac{1}{2}M(r_2^2 + r_1^2)$ **15.** $x_M = 0, \quad y_M = R/\pi$

17. on the diameter through P at a distance $\frac{6}{5}R$ from P

19. Suppose Ω, a basic region of area A, is broken up into n basic regions $\Omega_1, \ldots, \Omega_n$ with areas $A_1, \ldots, A_n$. Then

$$\bar{x}A = \iint_\Omega x \, dxdy = \sum_{i=1}^{n} \left(\iint_{\Omega_i} x \, dxdy \right) = \sum_{i=1}^{n} \bar{x}_i A_i = \bar{x}_1 A_1 + \cdots + \bar{x}_n A_n.$$

The second formula follows in the same manner.

SECTION 17.7

1. they are equal   **3.** $\frac{1}{8}a^2b^2c$

5. $\bar{x} = \dfrac{A^2BC - a^2bc}{ABC - abc}, \quad \bar{y} = \dfrac{AB^2C - ab^2c}{ABC - abc}, \quad \bar{z} = \dfrac{ABC^2 - abc^2}{ABC - abc}$

7. $M = \frac{1}{2}ka^4$ where K is the constant of proportionality for the density function

9. $I_z = \frac{2}{3}Ma^2$

SECTION 17.8

1. abc **3.** $\frac{2}{3}$

5.
$$\iiint_\Pi f(x)g(y)h(z) \, dxdydz = \int_{c_1}^{c_2} \left[\int_{b_1}^{b_2} \left(\int_{a_1}^{a_2} f(x)g(y)h(z) \, dx \right) dy \right] dz$$
$$= \int_{c_1}^{c_2} \left[\int_{b_1}^{b_2} g(y)h(z) \left(\int_{a_1}^{a_2} f(x) \, dx \right) dy \right] dz$$
$$= \int_{c_1}^{c_2} \left[h(z) \left(\int_{a_1}^{a_2} f(x) \, dx \right) \left(\int_{b_1}^{b_2} g(y) \, dy \right) dz \right]$$
$$= \left(\int_{a_1}^{a_2} f(x) \, dx \right) \left(\int_{b_1}^{b_2} g(y) \, dy \right) \left(\int_{c_1}^{c_2} h(z) \, dz \right)$$

7. 8   **9.** $(\frac{2}{3}a, \frac{2}{3}b, \frac{2}{3}c)$ **11.** Figure A.17.8.1   **13.** $(\frac{1}{2}, \frac{1}{3}, \frac{1}{3})$

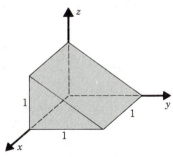

Figure A.17.8.1

15. $\displaystyle\int_{-r}^{r} \int_{-\sqrt{r^2-x^2}}^{\sqrt{r^2-x^2}} \int_{-\sqrt{r^2-(x^2+y^2)}}^{\sqrt{r^2-(x^2+y^2)}} k(r - \sqrt{x^2 + y^2 + z^2}) \, dz \, dy \, dx$

17. $\displaystyle\int_{0}^{1} \int_{-\sqrt{x-x^2}}^{\sqrt{x-x^2}} \int_{-2x-3y-10}^{1-y^2} dz \, dy \, dx$

19. $\displaystyle\int_{-1}^{1}\int_{-2\sqrt{2-2x^2}}^{2\sqrt{2-2x^2}}\int_{3x^2+y^2/4}^{4-x^2-y^2/4} k(z-3x^2-\tfrac{1}{4}y^2)\,dz\,dy\,dx$

21. $V=\tfrac{8}{3}$, $(\tfrac{11}{10},\tfrac{9}{4},\tfrac{11}{20})$ **23.** $V=\tfrac{27}{2}$, $(\tfrac{1}{2},\tfrac{3}{2},\tfrac{12}{5})$ **25.** $V=\tfrac{1}{6}abc$, $(\tfrac{1}{4}a,\tfrac{1}{4}b,\tfrac{1}{4}c)$

27. (a) $\tfrac{1}{3}M(a^2+b^2)$ (b) $\tfrac{1}{12}M(a^2+b^2)$ (c) $\tfrac{1}{3}Ma^2+\tfrac{1}{12}Mb^2$

29. $M=\tfrac{1}{2}k$, $(\tfrac{7}{12},\tfrac{34}{45},\tfrac{37}{90})$ **31.** (a) 0 by symmetry   (b) $\tfrac{4}{3}\pi a_4$

33. $M=\tfrac{128}{15}k$ **35.** $M=\tfrac{135}{4}k$, $(\tfrac{1}{2},\tfrac{9}{5},\tfrac{12}{5})$

37. (a) $V=\displaystyle\int_{0}^{6}\int_{z/2}^{3}\int_{x}^{6-x} dy\,dx\,dz$

(b) $V=\displaystyle\int_{0}^{3}\int_{0}^{2x}\int_{x}^{6-x} dy\,dz\,dx$

(c) $V=\displaystyle\int_{0}^{6}\int_{z/2}^{3}\int_{z/2}^{y} dx\,dy\,dz+\int_{0}^{6}\int_{3}^{(12-z)/2}\int_{z/2}^{6-y} dx\,dy\,dz$

39. (a) $V=\displaystyle\iint_{\Omega_{yz}} 2y\,dydz$ (b) $V=\displaystyle\iint_{\Omega_{yz}}\left(\int_{-y}^{y} dx\right) dydz$

(c) $V=\displaystyle\int_{0}^{4}\int_{-\sqrt{4-y}}^{\sqrt{4-y}}\int_{-y}^{y} dx\,dz\,dy$ (d) $V=\displaystyle\int_{-2}^{2}\int_{0}^{4-z^2}\int_{-y}^{y} dx\,dy\,dz$

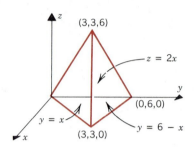

SECTION 17.9

1. $M=\tfrac{1}{2}k\pi R^2 h^2$ **3.** $\tfrac{1}{2}MR^2$

5. Inverting the cone and placing the vertex at the origin, we have

$$V=\int_{0}^{h}\int_{0}^{2\pi}\int_{0}^{(R/h)z} r\,dr\,d\theta\,dz=\tfrac{1}{3}\pi R^2 h.$$

7. $\tfrac{3}{10}MR^2$ **9.** $\tfrac{1}{2}\pi$ **11.** $\tfrac{1}{4}k\pi$ **13.** $\tfrac{1}{6}(8-3\sqrt{3})\pi$

15. $\tfrac{32}{9}a^3$ **17.** $\tfrac{1}{36}a^2(9\pi-16)$ **19.** $\tfrac{1}{32}\pi$ **21.** $\tfrac{1}{3}\pi(2-\sqrt{3})$

SECTION 17.10

1. $(\sqrt{3},\tfrac{1}{4}\pi,\cos^{-1}[\tfrac{1}{3}\sqrt{3}])$ **3.** $(\tfrac{3}{4},\tfrac{3}{4}\sqrt{3},\tfrac{3}{2}\sqrt{3})$

5. the circular cylinder $x^2+y^2=1$; the radius of the cylinder is 1 and the axis is the z-axis

7. the lower nappe of the cone $z^2=x^2+y^2$

9. horizontal plane one unit above the xy-plane   **11.** $V=\tfrac{4}{3}\pi R^3$ **13.** $V=\tfrac{2}{3}\alpha R^3$

15. $M=\tfrac{1}{6}k\pi h[(r^2+h^2)^{3/2}-h^3]$ **17.** (a) $\tfrac{2}{5}MR^2$ (b) $\tfrac{7}{5}MR^2$

19. (a) $\tfrac{2}{5}M\left(\dfrac{R_2^5-R_1^5}{R_2^3-R_1^3}\right)$ (b) $\tfrac{2}{3}MR^2$ (c) $\tfrac{5}{3}MR^2$ **21.** $V=\tfrac{2}{3}\pi(1-\cos\alpha)a^3$

23. (a) $\rho=2R\cos\phi$ (b) $0\le\theta\le 2\pi$, $0\le\phi\le\tfrac{1}{4}\pi$, $R\sec\phi\le\rho\le 2R\cos\phi$ **25.** $V=\tfrac{1}{3}(16-6\sqrt{2})\pi$

27. Encase T in a spherical wedge W. W has spherical coordinates in a box Π that contains S. Define f to be zero outside of T. Then $F(\rho,\theta,\phi)=f(\rho\sin\phi\cos\theta,\rho\sin\phi\sin\theta,\rho\cos\phi)$ is zero outside of S and

$$\iiint_{T} f(x,y,z)\,dxdydz=\iiint_{W} f(x,y,z)\,dxdydz$$

$$=\iiint_{\Pi} F(\rho,\theta,\phi)\,\rho^2\sin\phi\,d\rho d\theta d\phi=\iiint_{S} F(\rho,\theta,\phi)\,\rho^2\sin\phi\,d\rho d\theta d\phi.$$

29. $\mathbf{F}=\dfrac{GmM}{R^2}(\sqrt{2}-1)\mathbf{k}$

SECTION 17.11

1. $ad - bc$ **3.** $2(v^2 - u^2)$ **5.** $u^2v^2 - 4uv$ **7.** abc

9. r **11.** $w(1 + w\cos v)$ **13.** $\frac{1}{2}$ **15.** 0

17. $\frac{2}{3}$ **19.** (a) $A = 3\ln 2$ (b) $\bar{x} = \dfrac{7}{9\ln 2}$, $\bar{y} = \dfrac{14}{9\ln 2}$

21. $I_x = \frac{4}{75}M,\ \ I_y = \frac{14}{75}M,\ \ I_z = \frac{18}{75}M$ **23.** $A = \frac{32}{15}$

25. $A = \pi/\sqrt{65}$ **27.** $J = ab\alpha r \cos^{\alpha-1}\theta \sin^{\alpha-1}\theta$ **29.** $V = \frac{4}{3}\pi abc$

31. $I_x = \frac{1}{5}M(b^2 + c^2),\ \ I_y = \frac{1}{5}M(a^2 + c^2),\ \ I_z = \frac{1}{5}M(a^2 + b^2)$ **33.** $\frac{1}{2}\pi^{3/2}$

SECTION 18.1

1. (a) 1 (b) -2 **3.** 0 **5.** (a) $-\frac{17}{6}$ (b) $\frac{17}{6}$

7. -8 **9.** $-\pi$ **11.** (a) 1 (b) $\frac{23}{21}$ **13.** $\frac{1}{3}$

15. $\displaystyle\int_C \mathbf{q} \cdot d\mathbf{r} = \int_a^b [\mathbf{q} \cdot \mathbf{r}'(u)]\, du = \int_a^b \frac{d}{du}[\mathbf{q} \cdot \mathbf{r}(u)]\, du = [\mathbf{q} \cdot \mathbf{r}(b)] - [\mathbf{q} \cdot \mathbf{r}(a)] = \mathbf{q} \cdot [\mathbf{r}(b) - \mathbf{r}(a)]$

$\displaystyle\int_C \mathbf{r} \cdot d\mathbf{r} = \int_a^b [\mathbf{r}(u) \cdot \mathbf{r}'(u)]\, du = \frac{1}{2}\int_a^b \frac{d}{du}[\mathbf{r}(u) \cdot \mathbf{r}(u)]\, du = \frac{1}{2}\int_a^b \frac{d}{du}(\|\mathbf{r}(u)\|^2)\, du = \frac{1}{2}(\|\mathbf{r}(b)\|^2 - \|\mathbf{r}(a)\|^2)$

17. $\displaystyle\int_C \mathbf{f}(\mathbf{r}) \cdot d\mathbf{r} = \int_a^b [\mathbf{f}(\mathbf{r}(u)) \cdot \mathbf{r}'(u)]\, du = \int_a^b [f(u)\mathbf{i} \cdot \mathbf{i}]\, du = \int_a^b f(u)\, du$

19. $|W| = $ area of ellipse

21. force at time $t = m\mathbf{r}''(t) = m(2\beta\mathbf{j} + 6\gamma t\mathbf{k});\ \ \ W = (2\beta^2 + \frac{9}{2}\gamma^2)m$ **23.** 0

SECTION 18.2

1. 0 **3.** -1 **5.** 0 **7.** 0 **9.** 0 **11.** 4

13. Set $f(x, y, z) = g(x)$ and C: $\mathbf{r}(u) = u\mathbf{i},\ \ u \in [a, b]$. In this case, $\nabla f(\mathbf{r}(u)) = g'(x(u))\mathbf{i} = g'(u)\mathbf{i}$ and $\mathbf{r}'(u) = \mathbf{i}$, so that

$$\int_C \nabla f(\mathbf{r}) \cdot d\mathbf{r} = \int_a^b [\nabla f(\mathbf{r}(u)) \cdot \mathbf{r}'(u)]\, du = \int_a^b g'(u)\, du$$

and

$$f(\mathbf{r}(b)) - f(\mathbf{r}(a)) = g(b) - g(a).$$

The statement $\displaystyle\int_C \nabla f(\mathbf{r}) \cdot d\mathbf{r} = f(\mathbf{r}(b)) - f(\mathbf{r}(a))$ reduces to $\displaystyle\int_a^b g'(u)\, du = g(b) - g(a)$.

15. $W = mG\left(\dfrac{1}{r_2} - \dfrac{1}{r_1}\right)$

SECTION *18.3

1. If f is continuous, then $-f$ is continuous and has antiderivatives u. The scalar fields $U(x, y, z) = u(x)$ are potential functions for $\mathbf{F}$:

$$\nabla U = \frac{\partial U}{\partial x}\mathbf{i} + \frac{\partial U}{\partial y}\mathbf{j} + \frac{\partial U}{\partial z}\mathbf{k} = \frac{du}{dx}\mathbf{i} = -f\mathbf{i} = -\mathbf{F}.$$

3. The scalar field $U(x, y, z) = cz + d$ is a potential energy function for $\mathbf{F}$. We know that the total mechanical energy remains constant. Thus, for any times t_1 and t_2,

$$\tfrac{1}{2}m[v(t_1)]^2 + U(\mathbf{r}(t_1)) = \tfrac{1}{2}m[v(t_2)]^2 + U(\mathbf{r}(t_2)).$$

This gives

$$\tfrac{1}{2}m[v(t_1)]^2 + cz(t_1) + d = \tfrac{1}{2}m[v(t_2)]^2 + cz(t_2) + d.$$

Solve this equation for $v(t_2)$ and you have the desired formula.

5. (a) We know that $-\nabla U$ points in the direction of maximum decrease of U. Thus $\mathbf{F} = -\nabla U$ attempts to drive objects toward a region where U has lower values.
  (b) At a point where U has a minimum, $\nabla U = \mathbf{0}$ and therefore $\mathbf{F} = \mathbf{0}$.
7. (a) By conservation of energy $\frac{1}{2}mv^2 + U = E$. Since E is constant and U is constant, v is constant.
  (b) ∇U is perpendicular to any surface where U is constant. Obviously so is $\mathbf{F} = -\nabla U$.

SECTION 18.4

1. $\frac{1}{2}$ **3.** $\frac{9}{2}$ **5.** $\frac{11}{6}$ **7.** 2   **9.** 2   **11.** 3
13. (a) $M = 2ka^2$, $x_M = y_M = \frac{1}{8}a(\pi + 2)$ (b) $I_x = ka^4 = \frac{1}{2}Ma^2$
15. (a) $I_z = 2ka^4 = Ma^2$ (b) $I = \frac{1}{3}ka^4 = \frac{1}{6}Ma^2$
17. (a) $L = 2\pi\sqrt{a^2 + b^2}$ (b) $x_M = y_M = 0$, $z_M = \pi b$ (c) $I_x = I_y = \frac{1}{6}M(3a^2\pi + 8b^2\pi^2)$, $I_z = Ma^2$
19. $M = \frac{2}{3}\pi k \sqrt{a^2 + b^2}(3a^2 + 4\pi^2 b^2)$

SECTION 18.5

1. 2π **3.** $a^2 b$ **5.** 7π **7.** $5a\pi r^2$
9. 0   **11.** πab **13.** $(c - a)A$ **15.** $3\pi r^2$
17. Taking Ω to be of type II, we have

$$\int\int_\Omega \frac{\partial Q}{\partial x}(x, y)\, dxdy = \int_c^d \int_{\phi_3(y)}^{\phi_4(y)} \frac{\partial Q}{\partial x}(x, y)\, dx\, dy = \int_c^d \{Q[\phi_4(y), y] - Q[\phi_3(y), y]\}\, dy$$

$$* = \int_c^d Q[\phi_4(y), y]\, dy - \int_c^d Q[\phi_3(y), y]\, dy.$$

Set $C_3: \mathbf{r}_3(u) = \phi_3(u)\mathbf{i} + u\mathbf{j}$, $u \in [c, d]$ and $C_4: \mathbf{r}_4(u) = \phi_4(u)\mathbf{i} + u\mathbf{j}$, $u \in [c, d]$. Then

$$\oint_C Q(x, y)\, dy = \int_{C_4} Q(x, y)\, dy - \int_{C_3} Q(x, y)\, dy = \int_c^d Q[\phi_4(u), u]\, du - \int_c^d Q[\phi_3(u), u]\, du.$$

Comparison with $*$ proves the result.

19. $\oint_{C_1} = \oint_{C_2} + \oint_{C_3}$ **21.** (a) 0   (b) 0
23. If Ω is the region bounded by C, then

$$\oint_C \mathbf{v} \cdot d\mathbf{r} = \oint \frac{\partial\phi}{\partial x}\, dx + \frac{\partial\phi}{\partial y}\, dy = \int\int_\Omega \left\{\frac{\partial}{\partial x}\left(\frac{\partial\phi}{\partial y}\right) - \frac{\partial}{\partial y}\left(\frac{\partial\phi}{\partial x}\right)\right\}\, dxdy$$

is zero by equality of mixed partials.

SECTION 18.6

1. $4[(u^2 - v^2)\mathbf{i} - (u^2 + v^2)\mathbf{j} + 2uv\mathbf{k}]$ **3.** $2(\mathbf{j} - \mathbf{i})$
5. $\mathbf{r}(x, z) = x\mathbf{i} + g(x, z)\mathbf{j} + z\mathbf{k}$, $(x, z) \in \Omega$ **7.** $x^2/a^2 + y^2/b^2 + z^2/c^2 = 1$; ellipsoid
9. $x^2/a^2 - y^2/b^2 = z$; hyperbolic paraboloid
11. $\mathbf{r}(u, v) = v\cos u\mathbf{i} + v\sin u\mathbf{j} + f(v)\mathbf{k}$; $0 \le u \le 2\pi$, $a \le v \le b$
13. area of $\Gamma = A_\Omega \sec\gamma$ **15.** $\frac{1}{2}\sqrt{a^2 b^2 + a^2 c^2 + b^2 c^2}$ **17.** $\frac{1}{6}\pi(17\sqrt{17} - 1)$
19. $\frac{1}{6}\pi[(4a^2 + 1)^{3/2} - (a^2 + 1)^{3/2}]$ **21.** $\frac{1}{15}(36\sqrt{6} - 50\sqrt{5} + 32)$ **23.** 4π

25. (a) $\int\int_\Omega \sqrt{\left[\frac{\partial g}{\partial y}(y, z)\right]^2 + \left[\frac{\partial g}{\partial z}(y, z)\right]^2 + 1}\, dydz = \int\int_\Omega \sec[\alpha(y, z)]\, dydz$

where α is the angle between the unit normal with positive $\mathbf{i}$ component and the x-axis

(b) $\int\int_\Omega \sqrt{\left[\frac{\partial h}{\partial x}(x, z)\right]^2 + \left[\frac{\partial h}{\partial z}(x, z)\right]^2 + 1}\, dxdz = \int\int_\Omega \sec[\beta(x, z)]\, dxdz$

where β is the angle between the unit normal with positive $\mathbf{j}$ component and the y-axis

27. (a) $\mathbf{N}(u, v) = v \cos u \sin \alpha \cos \alpha \mathbf{i} + v \sin u \sin \alpha \cos \alpha \mathbf{j} - v \sin^2 \alpha \mathbf{k}$ (b) $A = \pi s^2 \sin \alpha$

29. $A = \sqrt{A_1^2 + A_2^2 + A_3^2}$

31. (a) $\frac{1}{4}\sqrt{2}\pi[\sqrt{6} + \ln(\sqrt{2} + \sqrt{3})]$ (b) $\frac{1}{2}a^2[\sqrt{2e^{4\pi} + 1} - \sqrt{3} + 2\pi + \ln(1 + \sqrt{3}) - \ln(1 + \sqrt{2e^{4\pi} + 1})]$

SECTION 18.7

1. $\frac{1}{2}[\sqrt{2} + \ln(1 + \sqrt{2})]$ **3.** $2\sqrt{2} - 1$ **5.** $\frac{1}{3}[2\sqrt{2} - 1]$

7. $\frac{1}{2}\sqrt{3}\, a^2 k$ **9.** $\frac{1}{12}\sqrt{3}\, a^4 k$ **11.** $(0, 0, \frac{1}{2}a)$

13. 2 **15.** $\frac{4}{3}\pi a^3$ **17.** 0

19. $\frac{1}{2}a^3$ **21.** 0 **23.** $-\frac{3}{2}$

25. $2\pi l a^2$ **27.** $\frac{8}{35}$ **29.** $-\frac{4}{63}$

31. $\bar{x} = \bar{y} = 0, \quad \bar{z} = \frac{2}{3}s \cos \alpha$ **33.** $x_M = y_M = 0, \quad z_M = \frac{3}{4}$

35. no answer required   **37.** $x_M = \frac{11}{9}$

39. Total flux out of the solid is 0. It is clear from a diagram that the outer unit normal to the cylindrical side of the solid is given by $\mathbf{n} = x\mathbf{i} + y\mathbf{j}$, in which case $\mathbf{v} \cdot \mathbf{n} = 0$. The outer unit normals to the top and bottom of the solid are $\mathbf{k}$ and $-\mathbf{k}$, respectively. So, here as well, $\mathbf{v} \cdot \mathbf{n} = 0$ and the total flux is 0.

41. $(4\sqrt{2} - \frac{7}{2})\pi$

SECTION 18.8

1. $\nabla \cdot \mathbf{v} = 2, \quad \nabla \times \mathbf{v} = \mathbf{0}$ **3.** $\nabla \cdot \mathbf{v} = 0, \quad \nabla \times \mathbf{v} = \mathbf{0}$

5. $\nabla \cdot \mathbf{v} = 6, \quad \nabla \times \mathbf{v} = \mathbf{0}$ **7.** $\nabla \cdot \mathbf{v} = yz + 1, \quad \nabla \times \mathbf{v} = -x\mathbf{i} + xy\mathbf{i} + (1 - x)z\mathbf{k}$

9. $\nabla \cdot \mathbf{v} = 1/r^2, \quad \nabla \times \mathbf{v} = \mathbf{0}$

11. $\nabla \cdot \mathbf{v} = 2(x + y + z)e^{r^2}, \quad \nabla \times \mathbf{v} = 2e^{r^2}[(y - z)\mathbf{i} - (x - z)\mathbf{j} + (x - y)\mathbf{k}]$

13. $\nabla \cdot \mathbf{v} = f'(x), \quad \nabla \times \mathbf{v} = \mathbf{0}$ **15.** use components

17. $\nabla^2 f = 12(x^2 + y^2 + z^2)$ **19.** $\nabla^2 f = 2y^3 z^4 + 6x^2 yz^4 + 12x^2 y^3 z^2$

21. $\nabla^2 f = e^r(1 + 2r^{-1})$ **23.** (a) $2r^2$ (b) $-1/r$

25. $\nabla^2 f = \nabla^2 g(r) = \nabla \cdot (\nabla g(r)) = \nabla \cdot (g'(r)r^{-1}\mathbf{r})$

$\qquad = [(\nabla g'(r)) \cdot r^{-1}\mathbf{r}] + g'(r)(\nabla \cdot r^{-1}\mathbf{r})$

$\qquad = \{[g''(r)r^{-1}\mathbf{r}] \cdot r^{-1}\mathbf{r}\} + g'(r)(2r^{-1}) = g''(r) + 2r^{-1}g'(r)$

27. $n = -1$

SECTION 18.9

1. $\displaystyle \iint_S (\mathbf{v} \cdot \mathbf{n})\, d\sigma = \iiint_T (\nabla \cdot \mathbf{v})\, dxdydz = \iiint_T 3\, dxdydz = 3V = 4\pi$

3. $\displaystyle \iint_S (\mathbf{v} \cdot \mathbf{n})\, d\sigma = \iiint_T (\nabla \cdot \mathbf{v})\, dxdydz = \iiint_T 2(x + y + z)\, dxdydz.$

The flux is zero since the function $f(x, y, z) = 2(x + y + z)$ satisfies the relation $f(-x, -y, -z) = -f(x, y, z)$ and T is symmetric about the origin.

5.

face	$\mathbf{n}$	$\mathbf{v} \cdot \mathbf{n}$	flux	
$x = 0$	$-\mathbf{i}$	0	0	
$x = 1$	$\mathbf{i}$	1	1	
$y = 0$	$-\mathbf{j}$	0	0	
$y = 1$	$\mathbf{j}$	1	1	total flux = 3
$z = 0$	$-\mathbf{k}$	0	0	
$z = 1$	$\mathbf{k}$	1	1	

$\displaystyle \iiint_T (\nabla \cdot \mathbf{v})\, dxdydz = \iiint_T 3\, dxdydz = 3V = 3$

7.

face	n	$v \cdot n$	flux
$x = 0$	$-i$	0	0
$x = 1$	i	1	1
$y = 0$	$-j$	xz	fluxes add up to 0 total flux = 2
$y = 1$	j	$-xz$	
$z = 0$	$-k$	0	0
$z = 1$	k	1	1

$$\iiint_T (\nabla \cdot v)\, dxdydz = \iiint_T 2(x + z)\, dxdydz = 2(\bar{x} + \bar{z})\, V = 2(\tfrac{1}{2} + \tfrac{1}{2})1 = 2$$

9. flux $= \iiint_T (1 + 4y + 6z)\, dxdydz = (1 + 4\bar{y} + 6\bar{z})\, V = (1 + 0 + 3)9\pi = 36\pi$

11. 0   **13.** $(A + B + C)V$

15. Let T be the solid enclosed by S and set $n = n_1 i + n_2 j + n_3 k$.

$$\iint_S n_1\, d\sigma = \iint_S (i \cdot n)\, d\sigma = \iiint_T (\nabla \cdot i)\, dxdydz = \iiint_T 0\, dxdydz = 0.$$

Similarly, $\iint_S n_2\, d\sigma = 0$ and $\iint_S n_3\, d\sigma = 0.$

17. A routine computation shows that $\nabla \cdot (\nabla f \times \nabla g) = 0$.   Therefore

$$\iint_S [(\nabla f \times \nabla g) \cdot n]\, d\sigma = \iiint_T [\nabla \cdot (\nabla f \times \nabla g)]\, dxdydz = 0.$$

19. Set $F = F_1 i + F_2 j + F_3 k$.

$$F_1 = \iint_S [\rho(z - c)i \cdot n]\, d\sigma = \iiint_T [\nabla \cdot \rho(z - c)i]\, dxdydz = \iiint_T \underbrace{\frac{\partial}{\partial x}[\rho(z - c)]}_{0}\, dxdydz = 0.$$

Similarly $F_2 = 0$.

$$F_3 = \iint_S [\rho(z - c)k \cdot n]\, d\sigma = \iiint_T [\nabla \cdot \rho(z - c)k]\, dxdydz$$

$$= \iiint_T \frac{\partial}{\partial z}[(\rho(z - c))]\, dxdydz = \iiint_T \rho\, dxdydz = W.$$

SECTION 18.10

For Exercises 1 and 3: $n = xi + yj + zk$ and $C: r(u) = \cos ui + \sin uj,\ u \in [0, 2\pi]$.

1. (a) $\iint_S [(\nabla \times v) \cdot n]\, d\sigma = \iint_S (0 \cdot n)\, d\sigma = 0$

   (b) S is bounded by the unit circle $C: r(u) = \cos ui + \sin uj,\ u \in [0, 2\pi]$.

   $\oint_C v(r) \cdot dr = 0$ since v is a gradient.

3. (a) $\iint_S [(\nabla \times v) \cdot n]\, d\sigma = \iint_S [(-3y^2 i + 2zj + 2k) \cdot (xi + yj + zk)]\, d\sigma$

$$= \iint_S (-3xy^2 + 2yz + 2z)\, d\sigma$$

$$= \underbrace{\iint_S (-3xy^2)\, d\sigma}_{\substack{0 \text{ by symmetry}}} + \underbrace{\iint_S 2yz\, d\sigma}_{\substack{0 \text{ by symmetry}}} + \underbrace{\iint_S 2z\, d\sigma = 2\pi}_{\substack{2\bar{z}A = 2\pi \\ \text{by Ex. 11,} \\ \text{Section 18.7}}}$$

(b) $\displaystyle \oint_C \mathbf{v}(\mathbf{r}) \cdot d\mathbf{r} = \oint_C z^2\, dx + 2x\, dy = \oint_C 2x\, dy = \int_0^{2\pi} 2\cos^2 u\, du = 2\pi$

For Exercises 5 and 7 take S: $z = 2 - x - y$ with $0 \le x \le 2$, $0 \le y \le 2 - x$ and C as the triangle $(2, 0, 0)$, $(0, 2, 0)$, $(0, 0, 2)$. Then $C = C_1 \cup C_2 \cup C_3$ with

$$C_1:\ \mathbf{r}_1(u) = 2(1 - u)\mathbf{i} + 2u\mathbf{j}, \quad u \in [0, 1],$$
$$C_2:\ \mathbf{r}_2(u) = 2(1 - u)\mathbf{j} + 2u\mathbf{k}, \quad u \in [0, 1],$$
$$C_3:\ \mathbf{r}_3(u) = 2(1 - u)\mathbf{k} + 2u\mathbf{i}, \quad u \in [0, 1].$$

$\mathbf{n} = \frac{1}{3}\sqrt{3}\,(\mathbf{i} + \mathbf{j} + \mathbf{k})$.

5. (a) $\displaystyle \iint_S [(\nabla \times \mathbf{v}) \cdot \mathbf{n}]\, d\sigma = \iint_S \frac{1}{3}\sqrt{3}\, d\sigma = \frac{1}{3}\sqrt{3}A = 2$

(b) $\displaystyle \oint_C \mathbf{v}(\mathbf{r}) \cdot d\mathbf{r} = \left(\int_{C_1} + \int_{C_2} + \int_{C_3} \right) \mathbf{v}(\mathbf{r}) \cdot d\mathbf{r} = -2 + 2 + 2 = 2$

7. (a) $\displaystyle \iint_S [(\nabla \times \mathbf{v}) \cdot \mathbf{n}]\, d\sigma = \iint_S [y\mathbf{k} \cdot \frac{1}{3}\sqrt{3}(\mathbf{i} + \mathbf{j} + \mathbf{k})]\, d\sigma = \frac{1}{3}\sqrt{3} \iint_S y\, d\sigma = \frac{1}{3}\sqrt{3}\,\bar{y}A = \frac{4}{3}$

(b) $\displaystyle \oint_C \mathbf{v}(\mathbf{r}) \cdot d\mathbf{r} = \left(\int_{C_1} + \int_{C_2} + \int_{C_3} \right) \mathbf{v}(\mathbf{r}) \cdot d\mathbf{r} = (\frac{4}{3} - \frac{32}{5}) + \frac{32}{5} + 0 = \frac{4}{3}$

9. 4π 11. 0 13. $\pm\frac{1}{8}\pi$ 15. $\pm\frac{1}{4}\pi$

17. Straightforward calculation shows that

$$\nabla \times (\mathbf{a} \times \mathbf{r}) = \nabla \times [(a_2 z - a_3 y)\mathbf{i} + (a_3 x - a_1 z)\mathbf{j} + (a_1 y - a_2 x)\mathbf{k}] = 2\mathbf{a}.$$

19. In the plane of C, the curve C bounds some Jordan region that we call Ω. The surface $S \cup \Omega$ is a piecewise-smooth surface that bounds a solid T. Note that $\nabla \times \mathbf{v}$ is continuously differentiable on T. Thus, by the divergence theorem,

$$\iiint_T [\nabla \cdot (\nabla \times \mathbf{v})]\, dx\,dy\,dz = \iint_{S \cup \Omega} [(\nabla \times \mathbf{v}) \cdot \mathbf{n}]\, d\sigma$$

where $\mathbf{n}$ is the outer unit normal. Since the divergence of a curl is identically zero, we have

$$\iint_{S \cup \Omega} [(\nabla \times \mathbf{v}) \cdot \mathbf{n}]\, d\sigma = 0.$$

Now $\mathbf{n}$ is $\mathbf{n}_1$ on S and $\mathbf{n}_2$ on Ω. Thus

$$\iint_S [(\nabla \times \mathbf{v}) \cdot \mathbf{n}_1]\, d\sigma + \iint_\Omega [(\nabla \times \mathbf{v}) \cdot \mathbf{n}_2]\, d\sigma = 0.$$

This gives

$$\iint\limits_{S} [(\nabla \times v) \cdot n_1] \, d\sigma = \iint\limits_{\Omega} [(\nabla \times v) \cdot (-n_2)] \, d\sigma = \oint_{C} v(r) \cdot dr$$

where C is traversed in a positive sense with respect to $-n_2$ and therefore in a positive sense with respect to n_1.   ($-n_2$ points toward S)

SECTION A.1

1. $\{-1, 0, 1, 2\}$

3. $\{0\}$

5. $\{-1, 0, 1, 2, 4, 6, 8, \ldots\}$

7. $\varnothing$

9. $\{1, 2, 3, 4, 6, 8, \ldots\}$

11. $\{(0, -1), (0, 0), (0, 1), (2, -1), (2, 0), (2, 1)\}$

13. $\{(-1, 0), (0, 0), (1, 0), (-1, 2), (0, 2), (1, 2)\}$

15. $\left\{ \begin{array}{l} (0, 0, -1), (0, 0, 0), (0, 0, 1), (0, 2, -1), (0, 2, 0), (0, 2, 1) \\ (2, 0, -1), (2, 0, 0), (2, 0, 1), (2, 2, -1), (2, 2, 0), (2, 2, 1) \end{array} \right\}$

17. $\{2\}$

19. $\{0, 2, 4\}$

21. the set of real numbers

23. $\{x: 2 < x \le 4\}$

25. A

27. the set of real numbers

29. (a) B (b) A

31. $\{0\}, \{1\}, \{2\}, \{0, 1\}, \{0, 2\}, \{1, 2\}, \{0, 1, 2\}$

SECTION A.2

1. $\frac{1}{6}\pi$ radians

3. $\frac{3}{2}\pi$ radians

5. $\frac{1}{18}\pi$ radians

7. $\frac{5}{4}\pi$ radians

9. $45°$

11. $105°$

13. $60°$

15. $150°$

17. (a) $90°$ (b) 10π radians

19. 0.3 radians

21. $\frac{10}{7}\pi$ in.

23. $30/\pi$ in.

25. $\frac{1}{2}\sqrt{2}$

27. $\frac{2}{3}\sqrt{3}$

29. $\frac{1}{2}$

31. 1

33. $\sqrt{2}$

35. $\frac{1}{3}\sqrt{3}$

SECTION A.3

1. Let S be the set of integers for which the statement is true. Since $2(1) \le 2^1$, S contains 1.
   Assume now that $k \in S$. This tells us that $2k \le 2^k$, and thus
   $$2(k + 1) = 2k + 2 \le 2^k + 2 \le 2^k + 2^k = 2(2^k) = 2^{k+1}.$$
   $$\underset{\underline{\qquad} (k \ge 1)}{\uparrow}$$
   This places $k + 1$ in S.
   We have shown that
   $$1 \in S \quad \text{and that} \quad k \in S \text{ implies } k + 1 \in S.$$
   It follows that S contains all the positive integers.

3. Let S be the set of integers for which the statement is true. Since $(1)(2) = 2$ is divisible by 2, $1 \in S$.
   Assume now that $k \in S$. This tells us that $k(k + 1)$ is divisible by 2 and therefore
   $$(k + 1)(k + 2) = k(k + 1) + 2(k + 1)$$
   is also divisible by 2. This places $k + 1 \in S$.
   We have shown that
   $$1 \in S \quad \text{and that} \quad k \in S \text{ implies } k + 1 \in S.$$
   It follows that S contains all the positive integers.

5. Use $1 + 2 + \cdots + k + (k + 1) = (1 + 2 + \cdots + k) + (k + 1)$
   $$= \tfrac{1}{2}k(k + 1) + (k + 1)$$
   $$= (k + 1)(\tfrac{1}{2}k + 1) = \tfrac{1}{2}(k + 1)(k + 2) = \tfrac{1}{2}(k + 1)[(k + 1) + 1].$$

7. Use $1^2 + 2^2 + \cdots + k^2 + (k + 1)^2 = \tfrac{1}{6}k(k + 1)(2k + 1) + (k + 1)^2$
   $$= \tfrac{1}{6}(k + 1)[k(2k + 1) + 6(k + 1)]$$
   $$= \tfrac{1}{6}(k + 1)(2k^2 + 7k + 6)$$
   $$= \tfrac{1}{6}(k + 1)(k + 2)(2k + 3) = \tfrac{1}{6}(k + 1)[(k + 1) + 1][2(k + 1) + 1].$$

9. By Exercise 8 and Exercise 5

$$1^3 + 2^3 + \cdots + (n-1)^3 = [\tfrac{1}{2}(n-1)n]^2 = \tfrac{1}{4}(n-1)^2 n^2 < \tfrac{1}{4}n^4$$

and

$$1^3 + 2^3 + \cdots + n^3 = [\tfrac{1}{2}n(n+1)]^2 = \tfrac{1}{4}n^2(n+1)^2 > \tfrac{1}{4}n^4.$$

11. Use $\dfrac{1}{\sqrt{1}} + \dfrac{1}{\sqrt{2}} + \dfrac{1}{\sqrt{3}} + \cdots + \dfrac{1}{\sqrt{n}} + \dfrac{1}{\sqrt{n+1}} > \sqrt{n} + \dfrac{1}{\sqrt{n+1}} > \sqrt{n} + \dfrac{1}{\sqrt{n+1}+\sqrt{n}}\left(\dfrac{\sqrt{n+1}-\sqrt{n}}{\sqrt{n+1}-\sqrt{n}}\right) = \sqrt{n+1}.$

13. Let S be the set of integers for which the statement is true. Since

$$3^{2(1)+1} + 2^{1+2} = 27 + 8 = 35$$

is divisible by 7, $1 \in S$.
  Assume now that $k \in S$. This tells us that

$$3^{2k+1} + 2^{k+2} \text{ is divisible by 7.}$$

It follows that

$$3^{2(k+1)+1} + 2^{(k+1)+2} = 3^2 \cdot 3^{2k+1} + 2 \cdot 2^{k+2} = 9 \cdot 3^{2k+1} + 2 \cdot 2^{k+2} = 7 \cdot 3^{2k+1} + 2(3^{2k+1} + 2^{k+2})$$

is also divisible by 7. This places $k + 1 \in S$.
  We have shown that

$$1 \in S \quad \text{and that} \quad k \in S \quad \text{implies} \quad k + 1 \in S.$$

It follows that S contains all the positive integers.

15. For all positive integers $n \geq 2$,

$$\left(1 - \frac{1}{2}\right)\left(1 - \frac{1}{3}\right) \cdots \left(1 - \frac{1}{n}\right) = \frac{1}{n}.$$

To see this let S be the set of integers n for which the formula holds. Since $1 - \tfrac{1}{2} = \tfrac{1}{2}$, $2 \in S$. Suppose now that $k \in S$. This tells us that

$$\left(1 - \frac{1}{2}\right)\left(1 - \frac{1}{3}\right) \cdots \left(1 - \frac{1}{k}\right) = \frac{1}{k}$$

and therefore that

$$\left(1 - \frac{1}{2}\right)\left(1 - \frac{1}{3}\right) \cdots \left(1 - \frac{1}{k}\right)\left(1 - \frac{1}{k+1}\right) = \frac{1}{k}\left(1 - \frac{1}{k+1}\right) = \frac{1}{k}\left(\frac{k}{k+1}\right) = \frac{1}{k+1}.$$

This places $k + 1$ in S and verifies the formula for $n \geq 2$.

17. From the figure, observe that adding a vertex V_{N+1} to an N-sided polygon increases the number of diagonals by $(N-2) + 1 = N - 1$.
  Then use the identity

$$\tfrac{1}{2}N(N-3) + (N-1) = \tfrac{1}{2}(N+1)(N+1-3).$$

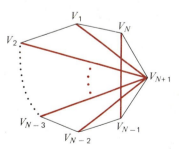

19. To go from k to $k + 1$, take $A = \{a_1, \ldots, a_{k+1}\}$ and $B = \{a_1, \ldots, a_k\}$. Assume that B has 2^k subsets: $B_1, B_2, \ldots, B_{2^k}$. The subsets of A are then $B_1, B_2, \ldots, B_{2^k}$ together with

$$B_1 \cup \{a_{k+1}\}, \; B_2 \cup \{a_{k+1}\}, \; \ldots, \; B_{2^k} \cup \{a_{k+1}\}.$$

This gives $2(2^k) = 2^{k+1}$ subsets for A.

INDEX

inverse trigonometric functions

47. $\int \sin^{-1} x \, dx = x \sin^{-1} x + \sqrt{1 - x^2} + C$

48. $\int \cos^{-1} x \, dx = x \cos^{-1} x - \sqrt{1 - x^2} + C$

49. $\int \tan^{-1} x \, dx = x \tan^{-1} x - \frac{1}{2} \ln (1 + x^2) + C$

50. $\int \cot^{-1} x \, dx = x \cot^{-1} x + \frac{1}{2} \ln (1 + x^2) + C$

51. $\int \sec^{-1} x \, dx = x \sec^{-1} x - \ln |x + \sqrt{x^2 - 1}| + C$

52. $\int \csc^{-1} x \, dx = x \csc^{-1} x + \ln |x + \sqrt{x^2 - 1}| + C$

53. $\int x \sin^{-1} x \, dx = \frac{1}{4}(2x^2 - 1) \sin^{-1} x + x\sqrt{1 - x^2} + C$

54. $\int x \cos^{-1} x \, dx = \frac{1}{4}(2x^2 - 1) \cos^{-1} x - x\sqrt{1 - x^2} + C$

55. $\int x \tan^{-1} x \, dx = \frac{1}{2}(x^2 + 1) \tan^{-1} x - \frac{1}{2}x + C$

56. $\int x \cot^{-1} x \, dx = \frac{1}{2}(x^2 + 1) \cot^{-1} x + \frac{1}{2}x + C$

57. $\int x \sec^{-1} x \, dx = \frac{1}{2}x^2 \sec^{-1} x - \frac{1}{2}\sqrt{x^2 - 1} + C$

58. $\int x \csc^{-1} x \, dx = \frac{1}{2}x^2 \csc^{-1} x + \frac{1}{2}\sqrt{x^2 - 1} + C$

hyperbolic functions

59. $\int \sinh x \, dx = \cosh x + C$

60. $\int \cosh x \, dx = \sinh x + C$

61. $\int \tanh x \, dx = \ln (\cosh x) + C$

62. $\int \coth x \, dx = \ln |\sinh x| + C$

63. $\int \operatorname{sech} x \, dx = \tan^{-1} (\sinh x) + C$

64. $\int \operatorname{csch} x \, dx = \ln |\tanh \frac{1}{2}x| + C$

65. $\int \operatorname{sech}^2 x \, dx = \tanh x + C$

66. $\int \operatorname{csch}^2 x \, dx = -\coth x + C$

67. $\int \operatorname{sech} x \tanh x \, dx = -\operatorname{sech} x + C$

68. $\int \operatorname{csch} x \coth x \, dx = -\operatorname{csch} x + C$

69. $\int \sinh^2 x \, dx = \frac{1}{4} \sinh 2x - \frac{1}{2}x + C$

70. $\int \cosh^2 x \, dx = \frac{1}{4} \sinh 2x + \frac{1}{2}x + C$

71. $\int \tanh^2 x \, dx = x - \tanh x + C$

72. $\int \coth^2 x \, dx = x - \coth x + C$

73. $\int x \sinh x \, dx = x \cosh x - \sinh x + C$

74. $\int x \cosh x \, dx = x \sinh x - \cosh x + C$

inverse trigonometric functions

47. $\int \sin^{-1} x \, dx = x \sin^{-1} x + \sqrt{1 - x^2} + C$

48. $\int \cos^{-1} x \, dx = x \cos^{-1} x - \sqrt{1 - x^2} + C$

49. $\int \tan^{-1} x \, dx = x \tan^{-1} x - \frac{1}{2} \ln (1 + x^2) + C$

50. $\int \cot^{-1} x \, dx = x \cot^{-1} x + \frac{1}{2} \ln (1 + x^2) + C$

51. $\int \sec^{-1} x \, dx = x \sec^{-1} x - \ln |x + \sqrt{x^2 - 1}| + C$

52. $\int \csc^{-1} x \, dx = x \csc^{-1} x + \ln |x + \sqrt{x^2 - 1}| + C$

53. $\int x \sin^{-1} x \, dx = \frac{1}{4}(2x^2 - 1) \sin^{-1} x + x\sqrt{1 - x^2} + C$

54. $\int x \cos^{-1} x \, dx = \frac{1}{4}(2x^2 - 1) \cos^{-1} x - x\sqrt{1 - x^2} + C$

55. $\int x \tan^{-1} x \, dx = \frac{1}{2}(x^2 + 1) \tan^{-1} x - \frac{1}{2}x + C$

56. $\int x \cot^{-1} x \, dx = \frac{1}{2}(x^2 + 1) \cot^{-1} x + \frac{1}{2}x + C$

57. $\int x \sec^{-1} x \, dx = \frac{1}{2}x^2 \sec^{-1} x - \frac{1}{2}\sqrt{x^2 - 1} + C$

58. $\int x \csc^{-1} x \, dx = \frac{1}{2}x^2 \csc^{-1} x + \frac{1}{2}\sqrt{x^2 - 1} + C$

hyperbolic functions

59. $\int \sinh x \, dx = \cosh x + C$

60. $\int \cosh x \, dx = \sinh x + C$

61. $\int \tanh x \, dx = \ln (\cosh x) + C$

62. $\int \coth x \, dx = \ln |\sinh x| + C$

63. $\int \text{sech} \, x \, dx = \tan^{-1} (\sinh x) + C$

64. $\int \text{csch} \, x \, dx = \ln |\tanh \frac{1}{2} x| + C$

65. $\int \text{sech}^2 x \, dx = \tanh x + C$

66. $\int \text{csch}^2 x \, dx = -\coth x + C$

67. $\int \text{sech} \, x \tanh x \, dx = -\text{sech} \, x + C$

68. $\int \text{csch} \, x \coth x \, dx = -\text{csch} \, x + C$

69. $\int \sinh^2 x \, dx = \frac{1}{4} \sinh 2x - \frac{1}{2}x + C$

70. $\int \cosh^2 x \, dx = \frac{1}{4} \sinh 2x + \frac{1}{2}x + C$

71. $\int \tanh^2 x \, dx = x - \tanh x + C$

72. $\int \coth^2 x \, dx = x - \coth x + C$

73. $\int x \sinh x \, dx = x \cosh x - \sinh x + C$

74. $\int x \cosh x \, dx = x \sinh x - \cosh x + C$